Control
of Proliferation
in Animal Cells

Control
of Proliferation
in Animal Cells

edited by

Bayard Clarkson, M.D.
Sloan-Kettering Institute for Cancer Research

Renato Baserga, M.D.
Temple University School of Medicine

COLD SPRING HARBOR CONFERENCES ON CELL PROLIFERATION
VOLUME 1

Cold Spring Harbor Laboratory
1974

Control of Proliferation in Animal Cells

© 1974 by Cold Spring Harbor Laboratory
All rights reserved

International Standard Book Number 0-87969-111-5
Library of Congress Catalog Card Number 73-88195

Printed in the United States of America

Cover and book design by Emily Harste

Contents

IV. BIOCHEMISTRY OF THE CELL CYCLE

Cyclic Nucleotides

Activation of Chromosomal Proteins

V. PROLIFERATION KINETICS, DIFFERENTIATION, AND EXTERNAL INFLUENCES

VI. SUMMARY

Preface

When the idea of holding a meeting at Cold Spring Harbor Laboratory on control of proliferation of animal cells was first suggested to Jim Watson about 5 years ago, it was temporarily put aside because of the doubt that anything of much value would come out of a meeting on a subject so diverse and still having only a rather flimsy scientific foundation. However during the next few years as the Laboratory became more deeply engaged in cancer research, it became increasingly evident that many new lines of investigation were converging around problems concerned with regulation of cell growth, clearly a central part of the cancer problem. When Jim decided last year that the time was ripe and asked us to organize the meeting, we began to appreciate the difficulties in trying to deal with such a multifaceted subject; in trying to encompass all the relevant aspects we were afraid the meeting would suffer from diffuseness and not meet the traditionally high standards of other Laboratory meetings on more sharply focused topics. Despite these concerns it was decided to try to cover as many basic aspects of control of proliferation as possible in the hope that a comprehensive overview would be most instructive. Rather than providing a general review, it was further decided to emphasize recent findings of importance to bring everyone up to date on where the frontiers of knowledge stand in each area and to encourage exchange of ideas between investigators in adjacent fields.

The meeting was organized around five broad topics, but a great deal of overlap occurred and some investigations did not fit nicely into any of the major categories. The five major topics chosen for consideration were: (1) growth factors, including nutritional and environmental influences (except those involving the hematopoietic system); (2) changes in proliferative behavior occurring during viral transformation and behavior of revertants and mutants; (3) the cell surface, including structure and biochemistry of the cell membrane, enzymatic modifications leading to mitogenesis, chemical changes accompanying cell transformation, and a special session on lymphocytes and the immune system; (4) biochemical events during the cell cycle and especially during transition from the resting to the proliferative state; and (5) proliferation kinetics, differentiation and external influences, including growth and microenvironmental factors influencing hematopoiesis and effects of cytotoxic agents.

Leading investigators in each of these general areas were invited to give papers. Whereas it was originally intended to limit the meeting to about 40 speakers, the list quickly grew to double that number as we considered all of the new work we felt should be presented; but even with this absolute limit we are fully aware and regret that there are many investigators who have made valuable contributions who could not be included. Although we tried to have adequate representation for each of the major topics, there was no attempt to present a balanced program, and it will be noted that there is much greater emphasis on some areas than others. This imbalance resulted in part because there were more new advances in some fields (e.g., cell surface, cyclic nucleotides), in part because progress in some areas had already been brought up to date in recent well-publicized conferences (e.g., oncogenic viruses, cancer therapy), and in part because of the inability of leading investigators in some fields to attend or because they chose to speak on some topic other than that which we had anticipated (e.g., effects of specific drugs during the cell cycle, chalones, etc.).

In general, the response was very gratifying since the great majority of invitees accepted, and it was particularly fortunate that so many of the most prominent investigators in their fields were able to attend. The value of the latter cannot be overemphasized, since not only did they provide continuity and place new research in proper perspective, but they served an important tutorial function in repeatedly reminding some of the younger investigators of the pitfalls of drawing conclusions after examining only one facet of a problem without giving sufficient attention to other variables that might influence their results. Throughout the conference they gave repeated warnings that some of the supposedly isolated systems designed to test just one variable were in reality extremely complicated and dependent on the interplay of multiple factors which were not always well controlled.

Although the central theme of the meeting was cancer and much attention was given to characterizing the proliferative abnormalities of neoplastic cells, fully half of the papers were mainly concerned with factors controlling growth of normal cells. This is, of course, entirely reasonable since better understanding of normal growth control is essential to understanding the defects underlying cancerous growth, and conversely, the study of abnormal cell behavior in turn may often lead to better appreciation of normal control mechanisms. No particular cell types or systems were chosen for special emphasis, but it naturally evolved that most investigators were working with fibroblasts, lymphocytes, or hematopoietic cells. It was recognized that it is often difficult to interpret studies of cell proliferation in intact animals because of the enormous complexity of the system and that understanding of basic mechanisms controlling cell division will most likely emerge from studies in well-defined in vitro systems. Nevertheless it was felt very important to emphasize that cells may behave quite differently in vivo and that too great a concentration on studies in simplified in vitro systems may sometimes give rise to misleading or erroneous conclusions.

Judging from the feedback we have received from those who attended all or most of the sessions, there were two main reactions: (1) that the meeting was both timely and instructive in that it fulfilled its main purposes of stimulating exchange of ideas, focusing attention on specific problems, and encouraging new approaches to solving them; and (2) that, at least for many, it was a humbling experience in that not only did they come to appreciate more fully the many different levels of growth regulation and the extreme complexity of the control systems at each level, but they also better realized the awesome task of trying to unravel the intimate interrelationships which exist between the multiple factors operating in the whole animal. However instead of being overwhelmed, most investigators professed to be enlightened by the meeting, and they came away more determined than ever to solve the many remaining problems; some even told us that they were planning to redirect their research programs on the basis of what they had learned.

Thus the meeting was apparently as successful as we had hoped, and we are grateful to all of the participants for making it so. We wish to thank Helen Parker for her indispensable help in arranging the meeting and especially for remaining cheerful and unflappable despite the many last minute changes we introduced, Judy Gordon for editing the manuscripts and speeding publication of this book, and Bob Pollack for much help and advice in organizing the meeting. We also wish to thank Jim for allowing us to organize the meeting and for his invaluable advice and help in assuring its success. Finally we want to express our immense gratitude to Michael Stoker for agreeing to take on the admittedly nearly impossible task of trying to summarize the conglomerate we had put together.

Bayard Clarkson
Renato Baserga

Some Effects of Environmental pH on Cellular Metabolism and Function

Harry Eagle, M.D.

Department of Cell Biology, Albert Einstein College of Medicine
Bronx, New York 10461

Although the pH of cell cultures in bicarbonate-buffered media may vary markedly in relation to the loss of CO_2, the cell density, the frequency of feeding, and the amount of acid produced by the specific cell strain (Fig. 1), the possible effects of that variation have been largely ignored. In recent years it has, however, become clear that the metabolism of cultured cells is in fact profoundly modified by environmental pH (Mackenzie, Mackenzie and Beck 1961; Ceccarini and Eagle 1971a, b; Rubin 1971). The addition to the medium of nonvolatile buffers (Good et al. 1966; Eagle 1971) (Table 1) effects partial pH stabilization and has permitted the evaluation of the effect of environmental pH on various aspects of cellular growth and function. As will be here summarized, the rate of growth, the maximum population density, the synthesis of collagen, S-100 protein and globulin, virus production, cellular fusion and hybridization, and the rescue of SV40 virus have all proved pH-sensitive in varying degree; and one may anticipate that the list will be extended as other parameters of cellular growth and function are examined.

Cellular Growth and "Contact" Inhibition

The optimum pH for the growth of cultured cells varies markedly according to cell type (Table 2) (Fig. 2). In general, virus-transformed or cancer human cells tended to have a significantly more acid pH optimum than did normal fibroblasts; and monkey and rat strains generally grew well over a considerably broader pH range than did many of the human or mouse cells.

At the optimum pH cells not only grow more rapidly, but ultimately attain a higher population density (Fig. 3). Although normal diploid cells are subject to a population-dependent inhibition of growth, often termed "contact" inhibition, the absolute population density at which growth is so inhibited is clearly pH-dependent. Stabilized cultures will resume growth if the pH is adjusted to its optimal level (Fig. 4); and almost complete growth inhibition can be induced by shifting the pH to a suboptimal level (Fig. 5). Further, since the maximum population

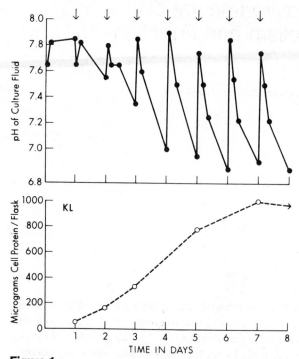

Figure 1
pH variation in a human fibroblast (KL) culture in medium containing 24 mM NaHCO$_3$ and changed daily (arrows, change of medium: 3 ml in a 15-ml stoppered container). Reproduced with permission from Eagle 1971 (copyright by the American Association for the Advancement of Science).

density at the optimal pH is a large multiple of that necessary to form a so-called "complete" monolayer of cells, intercellular contact as such is clearly not the sole determinant of growth arrest.

Multiple factors may thus determine the cessation of growth in crowded cell cultures. Three in particular, (a) the nutritional state of the cells, (b) the pH of the medium, and (c) the concentration of serum apparently operate independently to determine growth arrest. When the population has stabilized in cultures which have not been refed, or in which the pH of the medium has not been controlled, or in which the serum concentration has been kept at suboptimal levels, then refeeding, pH adjustment, or increasing the serum concentration respectively results in a burst of growth and eventual stabilization of population density at a higher plateau (Fig. 6). Appropriate adjustment of a second variable results in a second burst of growth; and the third variable effects yet another increase. The order in which these changes are made does not affect the final result.

The concentration of serum necessary for the initiation of cellular growth varies according to cell type (Temin 1966). For any one cell strain, however, the effective concentration of serum was independent of the environmental pH (Fig. 7). Further the pH optimum for cell growth was independent of the serum concentration (Eagle 1973) (Fig. 8).

Table 1
Buffers for Cell Culture

	pKa	6.5	6.8	7.1	7.4	7.7	8.0	8.3	For general use[b]
BIS TRIS	6.46	10[a]							
PIPES	6.8	10	10	10					
BES[c]	7.15		15	10	10				10
TES	7.5				10	15	10		
HEPES	7.55			10	15	15	15		15
HEPPS (EPPS)	8.0						10	10	10
TRICINE	8.15							15	
BICINE	8.35							10	
Na_2HPO_4	6.7	10	10	2					
$NaHCO_3$[d]	6.3	1	2	5	10	20	40	60	15

From Eagle 1971. Copyright by the American Association for the Advancement of Science.

Buffer concentrations (mM) recommended for media at indicated pH. Concentrations of organic buffers should be reduced if toxicity is noted for specific cell line, and reduced also for primary cultures, direct from tissues. Buffers are in addition to the $NaHCO_3$ of the medium (15mM in closed containers; concentrations as indicated for open containers in CO_2 incubator).

[a] Conveniently added as 1% of 1 M stock solution, except for PIPES (500 mM stock). Some of these buffers are strongly acidic or basic and medium must be adjusted to desired pH with NaOH or HCl.

[b] Although this buffer combination is moderately effective over the pH range 7.0–8.0, the pH fluctuations will be somewhat greater than with the combinations suggested for a specific pH range.

[c] May be substituted by MOPS, with a pKa of 7.2.

[d] For open containers in CO_2 incubator with 2–5% CO_2 atmosphere.

Synthesis of Nonstructural Proteins

The pH optimum for collagen synthesis by mouse and human fibroblasts coincided with the pH optimum for cellular growth (pH 7.2 and 7.7, respectively) (Nigra, Martin and Eagle 1973). The conversion of pro-collagen to collagen was independent of pH in the range 6.8–8.0. The cross-linking of collagen was, however, pH-sensitive. These cross links are formed from lysine- and hydroxylysine-derived aldehydes, the formation of which in the cell is catalyzed by lysyl-oxidase; and it is perhaps more than a coincidence that the optimum pH for the action of the isolated enzyme (7.8) is approximately the same as that for cross-linking in cell culture, measured by the extractibility of the collagen from cell layers grown at varying pH.

Preliminary experiments indicate that the optimum pH for globulin synthesis and secretion by mouse myeloma cells (Eagle and Scharff unpublished) also corresponded essentially to that optimal for the growth of the specific cell (pH 7.35). The synthesis of S-100 protein by rat astrocytes was, however, optimal at approximately pH 6.4–6.8, significantly more acid than that optimal for the growth of the specific cell (pH 7.15) (Pfeiffer and Eagle unpublished).

Table 2

pH Optima for Growth of Mammalian Cells (as determined by cell protein in 6–11 days)

Species	Cell type	Specific strain		Optimum pH for growth[a]
Human	Normal			
	embryonic lung fibroblast	WI 38		7.65; 7.7; 7.7
	embryonic skin fibroblast	KL		7.5–7.7; 7.5; 7.5–7.8
	skin fibroblasts (normal)	MS2		7.7; 7.6
	(homocystinuria)	Penny		7.7
	(homocystinuria)	Renee		7.6–7.9; 7.45–7.95
	Cancer	HeLa		6.9–7.4; 7.0–7.9; 6.9–7.6
		KB		7.0; 7.0
	Virus (SV40)-transformed	WI 18VA		7.3
	fibroblasts	WI 26VA		7.3–7.6; 7.3–7.5
Hamster	Normal			
	whole embryo fibroblast	NIL 2[b]		7.2; 7.1–7.6
	baby kidney	BHK[b]		7.4; 6.8–7.6
Monkey	Normal			
	green monkey kidney	Primary culture		6.65–7.5
		CV 1[b]		6.5–7.5; 6.7–7.7; 6.6–7.65
	Virus (adeno)-transformed	AGMK-adeno		6.6–7.4; 6.8–7.1
Mouse	Normal			
	total embryo	Primary culture		7.15–7.5
	skin fibroblast	929[b]		7.1–7.3; 6.9–7.9 (!)
	skin fibroblast	3T3[b]		7.4–7.7; 7.4–7.65; 7.6
	skin fibroblast	Cl 1-D		7.35; 7.35
	Cancer			
	renal adenocarcinoma	RAG		6.4–7.4; 7.1–7.3; 7.15; 7.35
	myeloma	425		7.35; 6.9–7.5
	Virus (SV40)-transformed	SV 3T3		7.1–7.3; 6.8–7.5
Rabbit	Normal	Lens epithelium[b]		6.85
Rat	Normal			
	lung fibroblast	BL		7.35
	liver epithelium	E3		7.7; 7.8
		G1		7.55–7.85
		B1		7.5–7.9; 7.45–7.9
	Cancer			
	hepatoma	HTC		7.5
	glial tumor	C6		7.1; 7.15; 7.15–7.85
Hybrids	mouse-human	Cl 1D × 18VA	Clone 1	7.0–7.65; 7.3
			Clone 2	7.05–7–8; 7.3
	mouse-human	RAG × WI 38		7.2

From Eagle (1973) with permission of *J. Cell. Physiol.*

[a] Results in individual experiments. The ranges (e.g., pH 6.7–7.7) indicate experiments in which there was a broad optimal range, rather than a well-defined peak (cf. Fig. 1, 2).

[b] Isolated from normal tissue but subsequently underwent "spontaneous" transformation in culture.

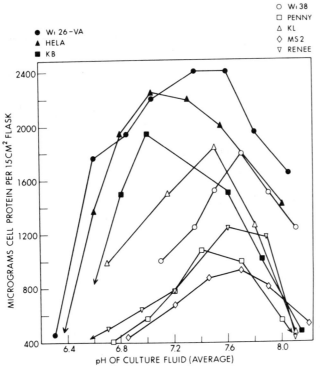

Figure 2
pH optima for the growth of human cells. From Eagle (1973) with permission
of *J. Cell. Physiology*.

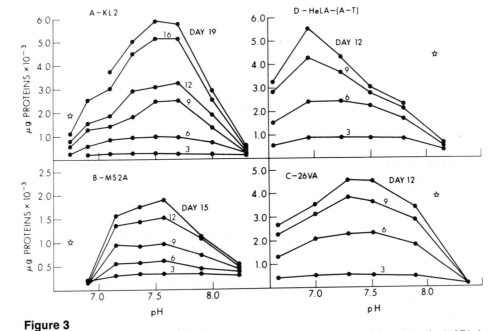

Figure 3
Cellular growth as a function of the environment pH. From Ceccarini and Eagle (1971a)
with permission of *Proc. Nat. Acad. Sci.*

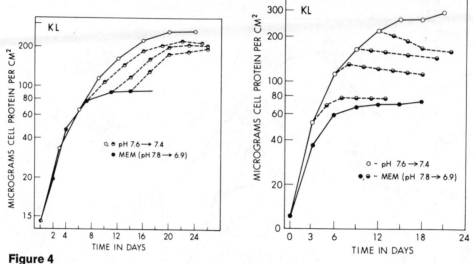

Figure 4
Reversal of "contact" inhibition in a human fibroblast culture by buffers which maintained the pH at optimal range (7.6→7.4). From Ceccarini and Eagle (1971b) with permission of *Nature New Biology*.

Figure 5
Induction of "contact" inhibition in a human fibroblast culture by shifting from buffered (pH 7.6) to unbuffered medium. From Ceccarini and Eagle (1971b) with permission of *Nature New Biology*.

Synthesis of Reovirus

The optimum pH for the synthesis of reovirus also did not correspond to that optimal for cellular growth. The pH optima for the growth of the rat liver and mouse fibroblast strains used in these experiments were approximately 7.8 and 7.2, respectively. With both strains, however, the pH optimum for virus release was in the range 6.8–7.2 (cf. Fig. 9) (Fields and Eagle 1973). The possibility may be considered that the pH optimum for reovirus synthesis reflects the activity of virus-specific enzymes.

Cellular Fusion and Hybridization

With the cells used in these experiments, the pH of the medium had a modest effect on the efficiency of cellular fusion as mediated by either Sendai virus or lysolecithin, with approximately a 3- to 5-fold increase at pH 7.8–8.0 as compared with pH 6.8–7.2 (Croce, Koprowski and Eagle 1972). There was, however, a pronounced effect on the efficiency of cellular hybridization following fusion, i.e., the formation of viable hybrids from those fused cells. When cells fused at either 7.2 or 8.0 were then incubated at varying pH, the eventual yield of hybrid colonies increased progressively with pH to reach a maximum at pH 7.8–8.0, which was as much as two logs higher than the yield observed at pH 7.1–7.2 (Fig. 10). At yet higher pH, the yield of hybrids fell off, and all the cells died when the initial pH was in excess of pH 8.4 (Croce, Koprowski and Eagle 1972).

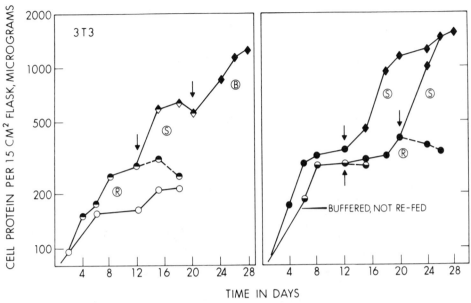

Figure 6
The sequential reversal of "contact" inhibition in cultures of a mouse fibroblast by re-feeding (R), increasing the concentration of serum (S), or buffering the medium (B).

Whatever the nature of the pH-sensitive step in cellular hybridization, it is expressed in the first 4 to 8 days after cellular fusion. If cells were kept at pH 8.0 for varying periods before shifting back to pH 7.2, the percentage of viable hybrids increased continuously with the duration of the initial incubation at 8.0 to reach a maximum after 4 to 8 days. Conversely, if cells were shifted up to the optimal pH after varying times at pH 7.2, the ultimate yield of hybrid colonies fell off progressively, depending upon the duration of the original incubation at 7.2 (Fig. 11). The nature of the pH-sensitive step(s) remains to be established.

Virus Rescue

One of the most striking effects of environmental pH on cultured cells has been on the rescue of SV40 virus by fusion of transformed mouse (mKS BU 100) or hamster (BTH) cells with a permissive monkey cell (CV-1). As in the case of hybridization, the degree of virus rescue increased progressively with increasing pH to reach a maximum at a pH (8.2–8.4) which was ultimately lethal to both cell types. In some experiments, there was as much as a four-log difference in the titer of virus rescued at, e.g., pH 6.8 and at 8.4 (Calothy et al. 1973) (Fig. 12).

Conceivably the same mechanism may underlie the favorable effects of alkaline pH on hybridization and virus rescue. The differences in optimal pH for the two phenomena may simply reflect the fact that the pH which is optimal for virus rescue in a short-term experiment does not regularly permit the sustained cellular multiplication obviously necessary for the formation of the hybrid clone.

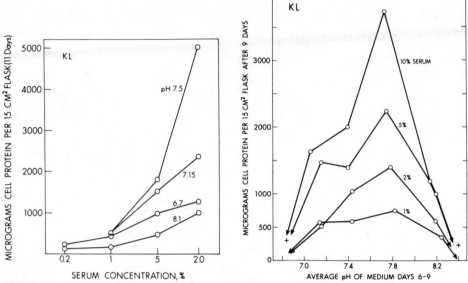

Figure 7

Failure of medium pH to affect the serum requirement for the growth of a human fibroblast. From Eagle (1973) with permission of *J. Cell. Physiology*.

Figure 8

Failure of serum concentration in the medium to affect the pH optimum for the growth of a human fibroblast. From Eagle (1973) with permission of *J. Cell. Physiology*.

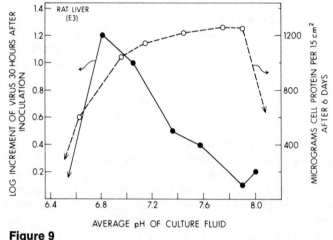

Figure 9

The contrasting effects of environmental pH on the growth of a rat liver cell (o– – –o) and on the yield of reovirus (•————•). From Fields and Eagle (1973) with permission of *Virology*.

8

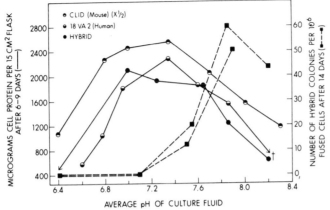

Figure 10

The contrasting effects of environmental pH on cellular growth (◑, ◐, ○) and on cellular hybridization by Sendai virus (■). From Croce, Koprowski and Eagle (1972) with permission of *Proc. Nat. Acad. Sci.*

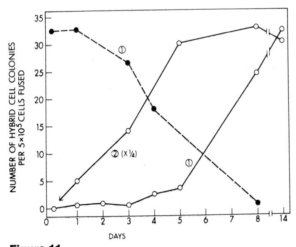

Figure 11

The effect of changing environmental pH after cell fusion on the development of hybrid cell colonies. (○———○) Stepdown to pH 7.2 after varying periods at pH 8.0; (●— — —●) stepup to pH 8.0 after varying periods at pH 7.2. From Croce, Koprowski and Eagle (1972) with permission of *Proc. Nat. Acad. Sci.*

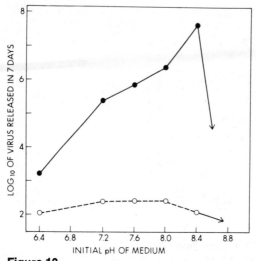

Figure 12
The effect of environmental pH on virus rescue after the fusion of SV40-transformed mouse cells (mKS Bu 100) with monkey cells (CV-1) (●——●). Open circles indicate virus rescue when the same cells were cocultivated rather than fused. From Calothy et al. (1973) with permission of *Proc. Nat. Acad. Sci.*

DISCUSSION

The preceding catalog of pH-sensitive aspects of cellular metabolism and function is surprising with respect to the regularity and magnitude of the effect and the frequent sharpness of the pH optimum. The list will undoubtedly expand as other parameters of cellular growth and function are examined. Self-evidently when, for example, the degree of cellular multiplication, or the amount of specialized protein synthesized, or the titer of virus produced is important to the outcome of an experiment, the environmental pH must be controlled if the results are to be interpretable; and even the imperfect control provided by the nonvolatile buffers listed in Table 1 is helpful in this connection.

It seems clear that cellular contact as such is not the primary determinant of growth arrest in crowded cell cultures. The maximum population density achieved by a given cell strain varies markedly with the pH of the medium, the optimum pH varying from strain to strain in the range pH 6.8–7.8. The maximum population density also varies with the concentration of serum in the medium, but these two factors act independently. The environmental pH does not affect the concentration of serum necessary to initiate growth; and conversely, the concentration of serum has no effect on the pH optimum. By appropriate adjustment of these two variables, the maximum population density achieved by a given strain ("contact" inhibition) can be modified at will; and at the optimum serum concentration and optimum pH some normal cells achieve population densities comparable with those achieved by transformed or malignant cells.

It is not surprising to find that the optimum pH for the synthesis of some specialized proteins (globulin, collagen), but not all (S-100 protein), corresponds to that for the growth of the specific cell strain. For the rest of the parameters here examined, the pH optimum varied widely and unpredictably. The relatively sharp optimum for the production of reovirus, and the similarity of that optimum in two strains which varied widely with respect to the optimum for cellular growth, suggests the involvement of virus-specific enzymes. There is no present explanation for the fact that in the systems here studied the pH optimal for cellular fusion, cellular hybridization, and virus rescue was at an alkaline range (pH 8–8.4) which was considerably higher than that optimal for cellular growth.

One would, of course, like to know the basis for these pH effects. Studies are continuing on this and related aspects of the pH control of cellular metabolism and function.

REFERENCES

Calthoy, G., C. M. Croce, V. Defendi, H. Koprowski and H. Eagle. 1973. Effect of environmental pH on rescue of simian virus 40. *Proc. Nat. Acad. Sci.* **70:**366.

Ceccarini, C. and H. Eagle. 1971a. pH as a determinant of cellular growth and contact inhibition. *Proc. Nat. Acad. Sci.* **68:**229.

————. 1971b. Induction and reversal of contact inhibition of growth by pH modification. *Nature New Biol.* **233:**271.

Croce, C. M., H. Koprowski and H. Eagle. 1972. Effect of environmental pH on the efficiency of cellular hybridization. *Proc. Nat. Acad. Sci.* **69:**1953.

Eagle, H. 1971. Buffer combinations for mammalian cell culture. *Science* **174:**500.

————. 1973. The effect of environmental pH on the growth of normal and malignant cells. *J. Cell. Physiol.* **82:**1.

Fields, B. N. and H. Eagle. 1973. The pH dependence of reovirus synthesis. *Virology* **52:**581.

Good, N. E., G. D. Winget, W. Winter, T. N. Connolly, S. Izawa and R. M. M. Singh. 1966. Hydrogen ion buffers for biological research. *Biochemistry* **5:**467.

Mackenzie, C. G., J. B. Mackenzie and P. Beck. 1961. The effect of pH on growth, protein synthesis, and lipid-rich particles of cultured mammalian cells. *J. Biophys. Biochem. Cytol.* **9:**141.

Nigra, T. P., G. R. Martin and H. Eagle. 1973. The effect of environmental pH on collagen synthesis by cultured cells. *Biochem. Biophys. Res. Commun.* **53:**272.

Rubin, H. 1971. Growth regulation in cultures of chick embryo fibroblasts. *Ciba Fndn. Symp., growth control in cell cultures* (ed. G. E. W. Wolstenholme and J. Knight) pp. 127–149.

Temin, H. M. 1966. Studies on carcinogenesis by avian sarcoma viruses. III. The differential effect of serum and polyanions on multiplication of uninfected and converted cells. *J. Nat. Cancer Inst.* **37:**167.

Serum Factors and Growth Control

Robert W. Holley

The Armand Hammer Center for Cancer Biology
The Salk Institute, San Diego, California 92112

Studies of the control of growth of mammalian cell lines in culture have led me to the following conclusions and hypotheses:

1. The phenomenon that is often referred to as "contact inhibition of growth" is not due to contacts and is not an inhibition and is better described as "density-dependent regulation of growth."
2. Growth control of mammalian cells involves an interaction among many growth controlling factors. In special situations a single factor can control growth, but an alteration of the situation will often lead to control by a different factor. Probably most growth situations involve several interacting factors.
3. Changes in the cell membrane probably are responsible for malignant growth. Changes in the cell membrane, which result from the effects of chemical or physical carcinogens or tumorigenic viruses on the genetic material, can alter the response of the cell to all of the many different fatcors that are observed to control growth in culture.

The evidence that supports these conclusions and hypotheses is summarized briefly below.

Density-dependent Regulation of Growth

We have studied this phenomenon extensively with 3T3 cells. The growth of 3T3 cells is normally controlled by serum factors (Todaro, Lazar and Green 1965; Holley and Kiernan 1968). The density to which 3T3 cells grow, under conditions in which the medium is depleted, is determined by the initial concentration of serum in the medium (Fig. 1A) (Holley and Kiernan 1971). Higher cell densities are attained under conditions of frequent medium changes (Fig. 1B), which avoids depletion of serum factors, but the final cell density is again dependent on the concentration of serum (Holley and Kiernan 1971). The evidence that contacts between cells are not involved directly in growth control is as follows: (a) In low concentrations of serum (Fig. 1A), 3T3 cells stop growing as sparse cells. (b) The

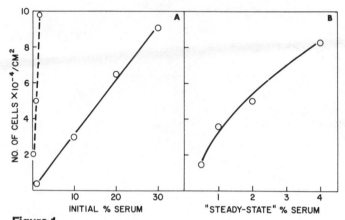

Figure 1

Dependence of cell density on serum concentration. **A,** the cell density is that attained after growth had largely ceased (4 days) in media with the serum concentration shown. (——) 3T3 cells; (– – –) SV3T3 cells. **B,** the cell density is that attained by 3T3 cells on a 12-mm coverslip after growth had ceased (5 days) in 10 ml medium, fluid being changed daily, with the serum concentration shown. From Holley and Kiernan (1971) with permission of the Ciba Foundation, London.

relationship between serum concentration and cell density (Fig. 1A) shows no discontinuity at the density at which cells become confluent. (c) Cells grown to high density in high serum under "steady state" conditions (Fig. 1B) decrease to a lower cell density if the culture medium is changed to a lower serum concentration, indicating that there is an equilibrium between serum concentration and cell density. (d) Breaking the contacts between quiescent 3T3 cells is insufficient to initiate growth (Lipton et al. 1971). Also it appears that inhibitors do not play a significant role in controlling growth of 3T3 cells, since addition of fresh serum to used medium or mixing used and fresh medium gives no evidence that growth is controlled by an inhibitor. Thus there is much evidence that density-dependent regulation of growth of 3T3 cells is related to the high serum requirement of these cells. It appears that density-dependent regulation of growth of 3T3 cells is very similar to regulation of population density by available food supply, a phenomenon that is very common in biology.

Growth Control: Interaction of Many Factors

Fractionation of serum gives a highly purified factor, probably a low molecular weight protein, that controls the final cell density of 3T3 cells (Holley and Kiernan 1971). Addition of this purified factor to medium with 2% calf serum, for example, gives growth of 3T3 cells to a final density equivalent to that obtained with medium that contains 10% serum. However the purified factor cannot replace serum completely, it does not initiate growth in quiescent 3T3 cells, and it does not increase the rate of growth of cells that are growing slowly because of a low serum concentration (Holley and Kiernan unpublished). Therefore it is clear that the growth of 3T3 cells is controlled by more than one serum factor. By manipulating the levels of factors, control of continued growth can be distinguished from initiation of growth and from growth rate. Fractionation of serum has shown that the initiation of DNA synthesis in quiescent 3T3 cells involves at least four other serum factors, two of them heat labile and two heat stable (Holley and Kiernan unpublished). Serum also contains a factor required for "survival" of 3T3 cells (Paul, Lipton and Klinger 1971), a factor that stimulates migration of the cells on the substratum (Lipton et al. 1971), and a factor that stimulates the uptake of phosphate ion (Cunningham and Pardee 1969).

In addition to all the serum factors that affect 3T3 cell growth, many other serum factors and related materials affect the growth of other cells (Lieberman and Ove 1958; Jones 1966; Holmes 1967; Levi-Montalcini and Angeletti 1968; Stanley, Robinson and Ada 1968; Van Wyk et al. 1971; Pierson and Temin 1972; Frank 1972; Adamson, Herington and Bornstein 1972; Rozengurt and Pardee 1972; Bürk 1973; Dulak and Temin 1973; Houck and Cheng 1973).

It has also become clear that the growth of 3T3 cells can be controlled by low molecular weight compounds. Some of these low molecular weight compounds are furnished by the serum as free compounds, some are bound to serum protein carriers, and others are normally present in excess in the growth medium. Following the observations of Ley and Tobey (1970), we have found that limitation of any one of nine amino acids (arginine, glutamine, histidine, isoleucine, lysine, phenylalanine, tryptophan, tyrosine, and valine) in the growth medium preferentially arrests 3T3 cell growth with the G_1 content of DNA. Arrest of growth of sparse 3T3 cells in high serum in G_1 (G_0) takes place without loss of viability if the entire complement of amino acids is allowed to fall to approximately 0.1% of the concentrations normally present in Dulbecco and Vogt's modified Eagle's medium. Subsequent addition of amino acids leads to a synchronous initiation of DNA synthesis approximately 15 hours later. A similar arrest and reinitiation of growth of sparse 3T3 cells in high serum takes place if the phosphate ion concentration is restricted and then raised. For maximum synchrony of reinitiation of growth, a serum factor that stimulates the uptake of phosphate must be added as well as phosphate. In addition to amino acids and phosphate ion, the initiation of DNA synthesis in sparse quiescent 3T3 cells by serum requires the presence of approximately 0.3 mM potassium ion, 0.1 mM calcium ion and 0.04 mM magnesium ion.

Table 1 summarizes the many types of factors that are known to control the growth of mammalian cells in culture. Any one factor can initiate growth in the laboratory, if the situation is such that this one factor is limiting growth. Whether all of these factors are involved in regulating growth in vivo is uncertain, but the potential exists.

Changes in the Cell Membrane and Malignant Growth

All of the growth regulatory factors listed in Table 1 can be involved, one way or another, with the cell membrane. Therefore changes in the cell membrane could account for the altered growth control seen in malignant cells. Certain types of evidence support this conclusion.

Studies of growth control in temperature-sensitive transformed cells suggest that membrane changes are involved in the changes in growth control that accompany temperature shifts. For example, a temperature-sensitive SV3T3 cell (Renger and Basilico 1972) that ordinarily shows restricted growth at 39°C has been found to

Table 1
Factors Controlling Growth of Mammalian Cells

Low molecular weight nutrients
Serum factors serving as carriers of nutrients
Serum factors stimulating uptake of nutrients
Serum factors affecting cyclic AMP levels
Serum factors affecting spreading and movement
Serum factors with unknown functions

continue to grow at 39°C (Fig. 2) if the medium is supplemented with a serum fraction that initiates DNA synthesis in quiescent 3T3 cells. In contrast, a temperature-sensitive (*ts-3*) PyBHK cell (Dulbecco and Eckhart 1970) that ordinarily shows restricted growth at 39°C continues to grow at 39°C if the medium is supplemented with biotin and iron (Fig. 2). (Iron and biotin are normally supplied to cells by protein carriers in serum [Messmer 1973; Vallotton et al. 1965].) In both instances it is known that the temperature shifts are accompanied by membrane changes. In both instances the temperature shifts reinstate growth controls that exist in the original untransformed cell and that can involve the cell membrane.

There appears to be evidence to exclude the alternative hypothesis that the internal control mechanisms for arrest of cells in G_1 (G_0) are lost in malignant cells. Glinos and Werrlein (1972) have shown that L929 cells become quiescent when they are grown to high cell densities with daily medium changes. We have confirmed their results with a tumor cell line, a mouse myeloma, XS63·5. In this instance, preliminary results suggest that arrest in G_1 (G_0) takes place when a low molecular weight serum factor becomes limiting in the medium. Addition of fresh serum to the quiescent tumor cells leads to the initiation of DNA synthesis after approximately 18 hours. Thus it is clear that regulatory mechanisms for arrest of growth exist in at least some tumor cells.

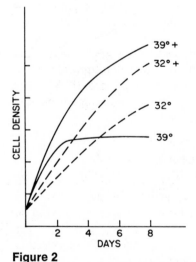

Figure 2
Growth curves of temperature-sensitive transformed cells with (+) and without addition of supplements to the medium as indicated in the text.

SUMMARY

The simplest hypothesis for control of growth that is consistent with all the above evidence is that the factors listed in Table 1 control growth by affecting the availability of nutrients inside the cell, and that it is the concentrations of certain critical nutrients inside the cell that actually control growth (Holley 1972). It is clear now that the control of normal mammalian cell growth by low molecular weight nutrients is much more important than has been appreciated. How important nutrients are in malignant cell growth remains to be determined. If certain of the growth controlling factors listed in Table 1 act by mechanisms other than by affecting the availability of nutrients, these mechanisms will, of course, also be altered by changes in the cell membrane and they may also contribute to malignant growth.

Whatever may be one's view of the conclusions and hypotheses mentioned in this paper, it is clear that serum contains many factors that play important roles in the control of mammalian cell growth. Much work is needed to isolate these factors and to study their mechanisms of action.

Acknowledgment

This research was supported in part by the American Cancer Society (BC-30), The National Cancer Institute, NIH (CA11176)(72-3207), and the National Science Foundation (GB 17912). The author is an American Cancer Society Professor of Molecular Biology.

REFERENCES

Adamson, L. F., A. C. Herington and J. Bornstein. 1972. Evidence for the selection by the membrane transport system of intracellular or extracellular amino acids for protein synthesis. *Biochim. Biophys. Acta* **282:**352.

Bürk, R. R. 1973. A factor from a transformed cell line that affects migration. *Proc. Nat. Acad. Sci.* **70:**369.

Cunningham, D. D. and A. B. Pardee. 1969. Transport changes rapidly initiated by serum addition to "contact inhibited" 3T3 cells. *Proc. Nat. Acad. Sci.* **64:**1049.

Dulak, N. C. and H. M. Temin. 1973. Multiplication-stimulating activity for chick embryo fibroblasts from rat liver cell conditioned medium: A family of small polypeptides. *J. Cell. Physiol.* **81:**161.

Dulbecco, R. and W. Eckhart. 1970. Temperature-dependent properties of cells transformed by a thermosensitive mutant of polyoma virus. *Proc. Nat. Acad. Sci.* **67:**1775.

Frank, W. 1972. Cyclic 3':5' AMP and cell proliferation in cultures of embryonic rat cells. *Exp. Cell Res.* **71:**238.

Glinos, A. D. and R. J. Werrlein. 1972. Density dependent regulation of growth in suspension cultures of L-929 cells. *J. Cell. Physiol.* **79:**79.

Holley, R. W. 1972. A unifying hypothesis concerning the nature of malignant growth. *Proc. Nat. Acad. Sci.* **69:**2840.

Holley, R. W. and J. A. Kiernan. 1968. "Contact inhibition" of cell division in 3T3 cells. *Proc. Nat. Acad. Sci.* **60:**300.

―――. 1971. Studies of serum factors required by 3T3 and SV3T3 cells. *Ciba Fndn. Symp., growth control in cell cultures* (ed. G. E. W. Wolstenholme and J. Knight) p. 3. Churchill Livingstone, London.

Holmes, R. 1967. Preparation from human serum of an alpha-one protein which induces the immediate growth of unadapted cells in vitro. *J. Cell Biol.* **32:**297.

Houck, J. C. and R. F. Cheng. 1973. Isolation, purification, and chemical characterization of the serum mitogen for diploid human fibroblasts. *J. Cell. Physiol.* **81**:257.

Jones, R. O. 1966. The *in vitro* effect of epithelial growth factor on rat organ cultures. *Exp. Cell Res.* **43**:645.

Levi-Montalcini, R. and P. U. Angeletti. 1968. Nerve growth factor. *Phys. ol. Rev.* **48**:534.

Ley, K. D. and R. A. Tobey. 1970. Regulation of initiation of DNA synthesis in Chinese hamster cells. II. Induction of DNA synthesis and cell division by isoleucine and glutamine in G_1 arrested cells in suspension culture. *J. Cell Biol.* **47**:453.

Lieberman, I. and P. Ove. 1958. A protein growth factor for mammalian cells in culture. *J. Biol. Chem.* **233**:637.

Lipton, A., I. Klinger, D. Paul and R. W. Holley. 1971. Migration of mouse 3T3 fibroblasts in response to a serum factor. *Proc. Nat. Acad. Sci.* **68**:2799.

Messmer, T. O. 1973. Nature of the iron requirement for Chinese hamster V79 cells in tissue culture medium. *Exp. Cell Res.* **77**:404.

Paul, D., A. Lipton and I. Klinger. 1971. Serum factor requirements of normal and simian virus 40-transformed 3T3 mouse fibroblasts. *Proc. Nat. Acad. Sci.* **68**:645.

Pierson, R. W., Jr. and H. M. Temin. 1972. The partial purification from calf serum of a fraction with multiplication-stimulating activity for chicken fibroblasts in cell culture and with nonsuppressible insulin-like activity. *J. Cell. Physiol.* **79**:319.

Renger, H. C. and C. Basilico. 1972. Mutation causing temperature-sensitive expression of cell transformation by a tumor virus. *Proc. Nat. Acad. Sci.* **69**:109.

Rozengurt, E. and A. B. Pardee. 1972. Opposite effects of dibutyryl adenosine 3':5' cyclic monophosphate and serum on growth of Chinese hamster cells. *J. Cell. Physiol.* **80**:273.

Stanley, E. R., W. A. Robinson and G. L. Ada. 1968. Properties of the colony-stimulating factor in leukemic and normal mouse serum. *Aust. J. Exp. Biol. Med. Sci.* **46**:715.

Todaro, G. J., G. K. Lazar and H. Green. 1965. The initiation of cell division in a contact-inhibited mammalian cell line. *J. Cell. Comp. Physiol.* **66**:325.

Vallotton, M., U. Hess-Sander and F. Leuthardt. 1965. Fixation spontanée de la biotine à une proteine dans le serum-humain. *Helv. Chim. Acta* **48**:126.

Van Wyk, J. J., K. Hall, J. L. Van den Brande and R. P. Weaver. 1971. Further purification and characterization of sulfation factor and thymidine factor from acromegalic plasma. *J. Clin. Endocrinol.* **32**:389.

Control of Multiplication of Normal and Rous Sarcoma Virus-transformed Chicken Embryo Fibroblasts by Purified Multiplication-stimulating Activity with Nonsuppressible Insulin-like and Sulfation Factor Activities

Howard M. Temin and Gary L. Smith

McArdle Laboratory, University of Wisconsin, Madison, Wisconsin 53706

Norman C. Dulak

Department of Biochemistry, University of Kansas Medical Center
Kansas City, Kansas 66103

Under standard conditions of cell culture, which include excess glucose and amino acids and maintenance of a standard pH and temperature, the multiplication of chicken embryo fibroblasts is completely dependent upon the amount of serum in the medium (Temin 1966). Chicken fibroblasts stop multiplying when the ability of the serum in the medium to stimulate multiplication is depleted. (Under certain conditions glucose or amino acids in the medium are depleted or toxic factors accumulate. These conditions also lead to cessation of cell multiplication.) The factor in serum which is depleted by cell multiplication and which stimulates cell multiplication when it is added to cells is called multiplication-stimulating activity (MSA).

MSA acts on IGl phase (interphase Gl) cells stimulating them into S phase and then to mitosis (Temin 1971). This cycle is described in Fig. 1.

The multiplication of chicken embryo fibroblasts infected and transformed by Rous sarcoma virus is also controlled by the level of MSA in the medium. However under conditions in which the amount of MSA is limiting for cell multiplication, chicken cells transformed by RSV multiply more than uninfected chicken cells (Temin 1966, 1967b). The increased multiplication of RSV-transformed chicken cells in the presence of limited amounts of MSA appears related to an increased efficiency of utilization of MSA (Temin 1969). One mark of this increased efficiency of utilization of MSA is the difference in response of RSV-transformed chicken cells and uninfected chicken cells to the withdrawal of MSA from the medium. Under these circumstances, as described in Fig. 1, normal chicken cells complete a replicative cell cycle and return to the IG1 phase of the cell cycle. Upon withdrawal of MSA from the medium, RSV-transformed chicken cells appear unable to enter the IG1 phase of the cell cycle, but they continue through the replicative cell cycle, that is, they enter S phase without fresh stimulation with MSA (see Fig. 2). However in the continued absence of MSA, RSV-transformed chicken cells die (Temin 1968, 1969). (It is not known in which stage of the replicative cell cycle the RSV-transformed cells die.)

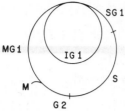

Figure 1
Life cycle of normal chicken fibroblasts. S, G2, and M have their usual mean-
ings. MG1 is mitotic G1 phase; IG1 is interphase G1; and SG1 is stimulated
G1 phase. (Modified from Temin 1971.) The outer circle represents the replicative cell cycle. IG1 has also been called G0 (see Baserga this volume). Cells
that no longer can enter the replicative cell cycle are called nondividing cells.
They are not in IG1 or G0.

Purification of MSA

Because of our interest in knowing more about the chemistry of MSA and in being
able to replace serum with a defined molecule, we started several years ago to
purify MSA from calf serum. MSA was assayed by its ability to stimulate the incorporation of labeled thymidine into stationary chicken cells 10 to 13 hours after
the addition of MSA (see Fig. 3). Our early attempts to purify MSA from calf
serum met with no success. Then we realized that since insulin had been shown to
possess MSA (Temin 1967a, 1968), MSA might be fractionated by the techniques
used for the purification of nonsuppressible insulin-like activity. With the use of
such techniques, MSA was purified approximately 5000-fold from serum (Pierson
and Temin 1972).

The observation that a line of buffalo rat liver cells, cloned by Hayden Coon and
called by him 3A (Coon 1968), could multiply in the absence of serum suggested
that these cells produced MSA. On assay, serum-free medium conditioned by the
growth of Coon's buffalo rat liver cells (CRL) was found to contain as much or
more MSA than calf serum (Dulak and Temin 1973a). Table 1 shows the amounts
of MSA in this conditioned medium and the steps in its purification. The Dowex-50
chromatography is primarily for concentration of MSA and removal of the peptides
from tryptose phosphate broth. In acetic acid-urea polyacrylamide gel electro-

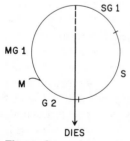

Figure 2
Life cycle of RSV-transformed chicken fibroblasts. Abbreviations are as in
Fig. 1. In the absence of MSA, these cells continue in the replicative cycle and
then die.

Table 1

Purification of CRL MSA

Stage of purification	Protein (mg)	MSA (units)	Specific activity
Medium	210	150,000	700
Dowex-50	15	85,000	6000
Acetic acid-urea PAGE	2–3	80,000	30,000
SDS PAGE	0.5–0.7	30,000	35,000

One liter of serum-free medium from cultures of Coon's buffalo rat liver cells was treated as described in Dulak and Temin (1973a). MSA was assayed as described in the legend to Fig. 3. One unit is that amount of MSA required to cause stimulation of thymidine incorporation equal to that produced by 1 mg of serum proteins. (Some data is from Dulak and Temin 1973a.) PAGE is polyacrylamide gel electrophoresis. SDS is sodium dodecyl sulfate.

phoresis, MSA is the fastest migrating component. This preparation can be separated into at least four rapidly migrating polypeptides by polyacrylamide gel electrophoresis in sodium dodecyl sulfate (Dulak and Temin 1973b).

Properties of Purified CRL MSA

Purified CRL MSA is a small polypeptide with a molecular weight of about 10,000 in sodium dodecyl sulfate polyacrylamide gel electrophoresis and Sephadex chromatography. It is inactivated by β-mercaptoethanol, dithiothreitol, or trypsin. It is resistant to incubation at pH 2 or in sodium dodecyl sulfate. At neutral pH's in the absence of sodium dodecyl sulfate, it adsorbs to glass, Sephadex, etc. Therefore most manipulations have been carried out in dissociating conditions. CRL MSA has no trypsin or anti-trypsin activity.

Figure 3 presents a dose-response curve for purified CRL MSA and for calf serum. The CRL MSA is much more active per mg protein than calf serum in stimulating DNA synthesis. (Note the difference in scale on the abscissa.) However the CRL MSA can apparently stimulate only half as many cells as the MSA in native calf serum. The same effect is seen with CRL conditioned medium. The reasons for this difference between CRL MSA and calf serum are not known. Recent experiments indicate that this apparent difference between CRL MSA and calf serum depends on the source of the chicken cells, since cells from some chicken embryos did not exhibit this difference.

Table 2 presents the data from Fig. 3 and from measurements of two other types of activity of CRL MSA and serum. CRL MSA has both nonsuppressible insulin-like activity, that is, insulin-like activity not neutralized by antibody to insulin, and sulfation factor activity. The multiplication-stimulating activity and the sulfation factor activity are purified to about the same extent with respect to serum. Whether all of the separate polypeptides of CRL MSA have insulin-like and sulfation factor activities has not yet been determined.

CRL MSA resembles somatomedin purified from human serum (Hall and Van Wyk 1973) in having nonsuppressible insulin-like and sulfation factor activities. It also resembles human plasma somatomedin in competing with labeled insulin for

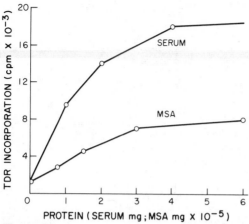

Figure 3

Dose-response curves of calf serum and CRL MSA. The indicated amounts of calf serum or CRL MSA in 3 ml of ET medium were added to duplicate cultures of stationary chicken embryo fibroblasts. The amount of incorporation of labeled thymidine in a one-hour pulse was determined 12 hours later. (The scales for calf serum and for CRL MSA differ by a factor of 10^5.)

binding to receptors of rat liver cell membranes (Van Wyk, personal communication). In all cases, the ratios of the different activities were the same for both molecules.

Purified CRL MSA had one other property of whole serum (Temin et al. 1972). It stimulated the migration of chicken embryo fibroblasts into a "wound" made in a cell monolayer.

Stimulation of Cell Multiplication

Purified CRL MSA supported the multiplication of chicken and duck embryo fibroblasts. (More CRL MSA was required to stimulate duck cells than was required for chicken cells.) It also supported limited multiplication of RSV-infected chicken embryo fibroblasts. However there was considerable death of RSV-infected chicken embryo fibroblasts in serum-free medium containing CRL MSA. Therefore in order to compare multiplication of uninfected and RSV-infected

Table 2

Activities of CRL MSA and Calf Serum

	MSA (units/mg)	NSILA (munits/mg)	Sulfation factor activity (units/mg)
Serum	1		0.02
CRL MSA	35,000	200	5000

A preparation of calf serum and of CRL MSA purified as described in Table 1 were assayed for MSA on chicken embryo fibroblasts, for nonsuppressible insulin-like activity (NSILA) on rat epididymal fat pad cells, and for sulfation factor activity on hypophysectomized rat rib cartilages. (Data taken from Dulak and Temin 1973a.)

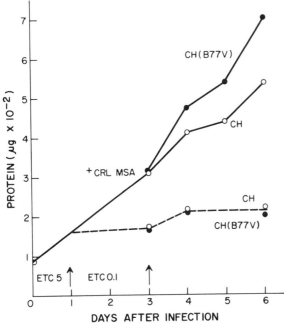

Figure 4
Multiplication of uninfected and RSV-infected chicken cells in CRL MSA-containing medium. Cultures of chicken embryo fibroblasts were inoculated with control medium or B77 virus and were overlaid with 3 ml of ET medium containing 5% calf serum. After incubation overnight, the medium was replaced with 5 ml of ET medium containing 0.1% calf serum and no or 0.8 μg/ml CRL MSA. The medium was replaced again on day 3. At the indicated times, duplicate cultures were analyzed for protein.

chicken cells, it was necessary to have 0.1% calf serum present in the medium. This concentration of calf serum did not stimulate noticeable cell multiplication (see Fig. 4). In addition, because of the requirement for normal cell cycle activation of RSV transformation (Humphries and Temin 1972), newly infected cultures were exposed to 5% calf serum for one day. The medium was then replaced by medium with 0.1% calf serum and no or 0.8 μg/ml of CRL MSA. There was significant multiplication of both types of cells, with greater multiplication of the RSV-infected cells (Fig. 4). This difference was reproducible.

Detailed study was then made of the requirements for stimulation of DNA synthesis in addition to MSA (Table 3). It was found that both amino acids and glucose were required. This result is consistent with the sensitivity of serum stimulation of DNA synthesis to inhibitors of protein synthesis and of glycolysis (Temin 1971).

CRL MSA appears to have no effect on increasing the rate of transport of amino acids. However it did increase the rate of transport of glucose soon after addition.

The kinetics of stimulation of DNA synthesis by CRL MSA and by calf serum were also studied (Fig. 5). It was found that there was a similar lag before the start of DNA synthesis in both cases. The peak of synthesis was also at about the same time in both cases.

Table 3

Requirements for Stimulation of
DNA Synthesis by CRL MSA

Additions	dT incorp. cpm/hr ($\times 10^{-3}$)
None	0.2
AA	1.2
Glucose	0.6
AA + Glucose	3.0
MEM + TP	2.9

CRL MSA (0.75 μg) was added to duplicate cultures of stationary chicken embryo fibroblasts in 3 ml of Earle's balanced salt solutions without glucose; MEM + TP, Eagle's minimal essential acids of Eagle's MEM; glucose, 3 mg; AA and glucose; MEM + TP, Eagle's minimal essential medium plus 20% tryptose phosphate broth. The amount of incorporation in a one-hour pulse was determined 12 hours later.

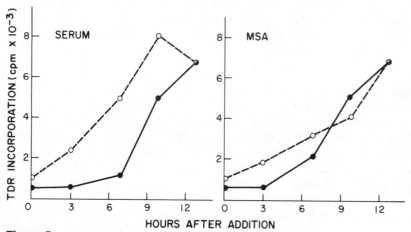

Figure 5

Kinetics of stimulation of DNA synthesis by calf serum and CRL MSA. Stationary cultures of chicken embryo fibroblasts were exposed to 3 ml of ET medium containing 4% calf serum or 1 μg/ml CRL MSA. At the times indicated (●———●) the rate of incorporation of labeled thymidine in one hour was determined in duplicate cultures. At the same times, medium was removed from duplicate parallel cultures and replaced with 3 ml of ET medium. The rate of incorporation of labeled thymidine in one hour was determined in these cultures at 13 hours. It is plotted at the time of the medium change (o– – –o).

24

591.8761   C768n

C.1

However, the effects of withdrawal of MSA and serum were different. As is described in Fig. 1, after the withdrawal of serum, some cells (those in the SG1 phase of the cell cycle) are committed to DNA synthesis even though they have not yet started DNA synthesis (Fig. 5, left). By contrast, after withdrawal of CRL MSA, no more cells started DNA synthesis (Fig. 5, right). Therefore there is apparently no SG1 phase after stimulation by CRL MSA. The length of time from IG1 to S, however, appears to increase by an amount of time equal to that of the SG1 phase induced by serum MSA. Therefore the time from IG1 to S appears to be the same for both serum and MSA. This difference does not appear to be related to the MSA concentration since in the experiment of Fig. 5 there was the same increase in rate of DNA synthesis in both sets of cultures. The difference might represent one between crude and purified MSA, between calf and rat MSA, or between serum and liver MSA.

Removal of MSA after several hours prevents the stimulation of DNA synthesis. Therefore the mechanism of action of MSA is not to trigger an irreversible event, but to initiate and maintain a reversible process which culminates in initiation of DNA synthesis.

SUMMARY

An electrophoretically purified polypeptide fraction from serum-free medium conditioned by growth of a line of rat liver cells has multiplication-stimulating activity. The fraction, CRL MSA, also has nonsuppressible insulin-like activity and sulfation factor activity. CRL MSA, therefore, resembles somatomedin from human plasma.

CRL MSA can support the multiplication of chicken embryo fibroblasts. RSV-infected chicken embryo fibroblasts multiplied more than uninfected chicken embryo fibroblasts in a medium with 0.1% calf serum and 0.8 μg/ml CRL MSA.

The stimulation of DNA synthesis by CRL MSA requires amino acids and glucose in the medium. The kinetics of stimulation are somewhat different from those of calf serum.

Acknowledgments

The research reported here was supported by Public Health Service Research Grant CA-07175 from the National Cancer Institute and Training Grant CA-5002 from the National Cancer Institute. HMT holds Research Career Development Award CA-8182 from the National Cancer Institute.

REFERENCES

Coon, H. G. 1968. Clonal culture of differentiated rat liver cells. *J. Cell Biol.* **39**:29a (abstr.).

Dulak, N. C. and H. M. Temin. 1973a. A partially purified polypeptide fraction from rat liver cell conditioned medium with multiplication-stimulating activity for embryo fibroblasts. *J. Cell. Physiol.* **81**:153.

———. 1973b. Multiplication-stimulating activity for chicken embryo fibroblasts from rat liver cell conditioned medium: A family of small polypeptides. *J. Cell. Physiol.* **81**:161.

Hall, K. and J. J. Van Wyk. 1973. Somatomedin. *Current topics in experimental endocrinology,* ed. Martini and James. Academic Press, New York. (in press)

Humphries, E. H. and H. M. Temin. 1972. Cell cycle-dependent activation of Rous sarcoma virus-infected stationary chicken cells: Avian leukosis virus group-specific antigens and ribonucleic acid. *J. Virol.* **10:**82.

Pierson, R. W., Jr. and H. M. Temin. 1972. The partial purification from calf serum of a fraction with multiplication-stimulating activity for chicken fibroblasts in cell culture and with non-suppressible insulin-like activity. *J. Cell. Physiol.* **79:**319.

Temin, H. M. 1966. Studies on carcinogenesis by avian sarcoma viruses. III. The differential effect of serum and polyanions on multiplication of uninfected and converted cells. *J. Nat. Cancer Inst.* **37:**167.

———. 1967a. Studies on carcinogenesis by avian sarcoma viruses. VI. Differential multiplication of uninfected and of converted cells in response to insulin. *J. Cell. Physiol.* **69:**377.

———. 1967b. Control by factors in serum of multiplication of uninfected cells and cells infected and converted by avian sarcoma viruses. *Growth-regulating substances for animal cells in culture* (ed. V. Defendi and M. Stoker) *Wistar Inst. Monogr.* **7:**103.

———. 1968. Carcinogenesis by avian sarcoma viruses. X. The decreased requirement for insulin-replaceable activity in serum for cell multiplication. *Int. J. Cancer* **3:**771.

———. 1969. Control of cell multiplication in uninfected chicken cells and chicken cells converted by avian sarcoma viruses. *J. Cell. Physiol.* **74:**9.

———. 1971. Stimulation by serum of multiplication of stationary chicken cells. *J. Cell. Physiol.* **78:**161.

Temin, H. M., R. W. Pierson, Jr. and N. C. Dulak. 1972. The role of serum in the control of multiplication of avian and mammalian cells in culture. *Growth, nutrition, and metabolism of cells in culture* (ed. V. I. Cristafalo and G. Rothblat) pp. 50–81. Academic Press, New York.

Some Properties of a Migration Factor from a Transformed Cell Line

R. R. Bürk

Imperial Cancer Research Fund Laboratories
Lincoln's Inn Fields, London WC2A 3PX

One of the earliest reported differences between normal and tumor cells was a loss of contact inhibition of movement by the tumor cells. This, it has been thought, may be connected with the ability of tumor cells to invade the surrounding tissue. As this invasiveness accounts for the malignancy of tumors and would seem to be a prerequisite of metastasis, considerable effort has been directed towards an understanding of the normal phenomenon of contact inhibition. It was found that there was apparently a phenomenon of contact inhibition of growth in normal cells, but not in neoplastically transformed cells, which was thought to parallel contact inhibition of movement. A possible explanation of contact inhibition of growth was depletion of the nutrients in the intercellular spaces of dense cultures. Experiments in which monolayer cultures were wounded suggested that the effects were very localized and that a migrating cell escaped from the inhibition in the monolayer. Migration seemed to be a part of the physiology of replication. A substance which stimulates growth would therefore stimulate migration. Most of the characteristics of transformed fibroblasts would be explained if it could be shown that the cells produce their own factor for the initiation of proliferation, since such a factor would overcome inhibition of growth and of movement.

We have chosen to use 3T3 mouse fibroblasts as our normal cell line because the differences consequent on transformation by SV40 are probably the best documented of all cell lines. It has been shown that transformation of 3T3 cells results in growth to high density, in an elevated mitotic index in a confluent layer, in an elevated thymidine labeling index in a confluent layer, in increased mobility and migration, in increased deoxyglucose transport, in increased uridine transport, in lowered cAMP concentrations associated with the cells and in increased phosphate transport. Each of these differences is also associated with growing, compared with nongrowing, normal 3T3 cells. If there is one factor in serum which initiates the growth of 3T3 cells and produces these changes, then it may be that transformed 3T3 cells produce something similar themselves and so do not require the factor from serum. Transformed 3T3 cells have a lower, actually different requirement for serum for growth (Paul et al. 1971) which would be consistent with their

producing their own factor. The continuous production of such a proliferation initiation factor and its continuous action would lead to the continuous presence of
these characteristics of growth and transformation. It is convenient to call such a
factor a transformation factor.

A necessary property of a transformation factor is that it is produced by the
transformed cell but not by its normal counterpart. To facilitate this comparison
we have used BHK fibroblasts, which though transformable have a high saturation
density, and their SV40, polyoma, and hamster sarcoma virus-transformed derivatives as sources of factors which we have tested on normal 3T3. SV28 is a subline
of BHK which has been transformed by SV40 virus. It is of particular interest
because SV28 injected into hamsters forms tumors which metastasize to give in-
vasive secondaries. Balb/c-3T3 and A31 were the origin of the 3T3 cells we use.
We have found an overgrowth factor that stimulates growth of 3T3 to high density
and is present in the medium of both BHK and SV40-transformed BHK. Rubin
(1970) earlier found an overgrowth stimulating factor present in normal and Rous
sarcoma virus-transformed chick cells that stimulates the growth of chick embryo
fibroblasts to higher density. Our overgrowth factor and Rubin's overgrowth
stimulating factor are therefore not transformation factors.

A migration factor that may stimulate the overgrowth of 3T3 cells has been
described from SV28 cells (SV40-transformed BHK), which could not be
obtained from BHK (Bürk 1973). This could be a transformation factor. In this
paper we describe the partial purification of this SV28 migration factor, some of its
chemical properties, the activities of the most purified fraction, and its sources.

Purification

Assays

Migration and overgrowth assays were performed as described previously (Bürk
1973).

The method was to seed 2×10^7 SV28 cells in 220 ml medium in 80-oz. Win-
chester bottles. The bottles were rolled at 1 rpm at 37°C. The medium was replaced on the 4th and 5th days by fresh medium. On the 6th day it was replaced by
200 ml of medium without serum but with tryptose phosphate broth. On the 7th day
the medium was harvested and stored at −20°C, and the cells were trypsinized
from the bottles and counted. Three to five liters of medium were pooled for the
starting material. The medium was adjusted to pH 6.0 with 1.0 N HCl and then to
pH 8.9 with 1 N NaOH. The precipitate formed was centrifuged and discarded.
The supernatant (fraction I) was applied to a column of Dowex AG50WX2 (25 ×
6.4 cm). The column was washed with 3 liters of 0.15 M NaCl 0.1 M Na_2HPO_4 pH
9.0 and then with one liter 0.15 M NaCl 0.1 M Na_3PO_4 pH 12.2. The active fractions were pooled (fraction II) and dialyzed against three changes of 25 volumes
of 10 mM formic acid, lyophilized and the residue extracted three times with 0.1 M
formic acid to give about 2.5 ml fraction III. Fraction III was applied to a column
of G75 Sephadex (bed vol. 165 ml, 1.7 cm diam.) and eluted with 0.02 M HCl,
0.14 M NaCl pH 1.7, collecting 1.8-ml fractions. The fractions with migration factor
activity were pooled (fraction IV), dialyzed against 10 mM formic acid, lyophilized
and dissolved in 2.5 ml 0.1 M formic acid. This was applied to another column of
G75 Sephadex (bed vol. 175 ml, 1.2 cm diam.) and eluted with 0.02 M HCl, 0.14
M NaCl pH 1.7, collecting 1.8 ml-fractions. The fractions were assayed for migration activity and pooled to give fraction V.

Table 1

Purification of Migration Factor

	Volume (ml)	Protein (mg/ml)	Activity (u/ml)	Sp. Act. (u/mg)	+ PO$_4$* Activity (u/ml)	Sp. Act. (u/mg)	Total u
Serum		86	236	3	261	3	
Hg medium	3200	0.196	65	330	72	370	230,000
II	250	0.475	566	1190	540	1140	135,000
IV	20	0.230	733	3190	2740	11,900	54,800
Hq medium	3200	0.122	82	672	81	664	259,000
V	16	0.045	1042	23,200	2830	62,900	45,300

Purification of migration factor from two batches (Hg, Hq) of SV28 medium. u = unit which stimulates migration of one cell per 1.7 mm of wound edge.

* Phosphate was added to increase concentration from 0.9 to 2.9 mM.

Table 1 records the progress of the purification of two out of many batches of SV28 medium. The specific activity of serum is included for comparison. It can be seen that the starting medium had about a hundredfold higher specific activity than calf serum and fraction V had about 20,000-fold higher specific activity than calf serum. The starting medium seems to contain more than one migration factor. One of them attaches to the column (60–70% of the activity) and one is found in the effluent. Dialysis of fraction II reduced the activity considerably but the addition of 2 mM phosphate to the medium restored the activity. The migration activity in the starting medium is little affected by added phosphate, but as the purification proceeds the dependence on added phosphate increases. Figure 1 shows the effect of adding different amounts of fraction V to the assay system. It also shows the linearity of the assay. One μg of protein was effective in producing detectable migration in the presence of phosphate. If one assumes absolute purity for migration factor, which is not true, and a molecular weight of 40,000, then factor stimulates migration at 5×10^{-9} M, which corresponds to about 1×10^7 molecules per cell. For comparison this is at least one order of magnitude less active than insulin in systems which are insulin-sensitive and rather more than one order of magnitude more effective than beef insulin in this system.

Chemical Properties

Isoelectric Point

Fraction IV prepared as described previously (Bürk 1973) from G75 Sephadex was added to a pH 3–10 ampholyte mixture and a voltage applied for 24 hours to focus the material in an LKB column according to the manufacturers instructions. Figure 2 shows clearly one peak with an isoelectric point of about 10.4.

Molecular Weight

Bovine serum albumin (MW 68,000), egg albumin (MW 43,000) and cytochrome c (MW 14,000) were run in the 1.2-cm diameter G75 Sephadex column as markers (Fig. 3) in 0.02 M HCl, 0.14 M NaCl, pH 1.7. Migration factor elutes

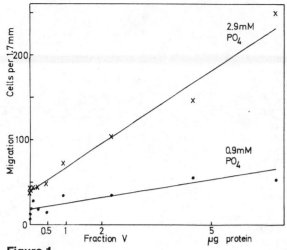

Figure 1
Effect of adding different amounts of migration factor to the assay.

relative to these markers in a position which would be expected for a globular protein of about 40,000 daltons.

Stability

Lyophilized fraction II was weighed into 10 mg aliquots and some dissolved immediately in 0.1 M formic acid and frozen at −20°C and some put into a screw-capped vial, sealed with adhesive tape, left on the bench for 14 days and then dissolved in 0.1 M formic acid. Both samples were assayed together and had the same migration factor activity. Lyophilized fraction IV dissolved in 0.1 M sodium acetate pH 6.9 was stable to 56°C for 30 minutes and to boiling for 3 minutes. In the purification procedures migration factor survived pH 1–pH 12 at 4°C for several hours without detectable loss in activity. It may, however, have been rather insoluble around neutral pH at low ionic strength, e.g. 10^{-3}M sodium acetate.

Pepsin Sensitivity

The sensitivity to protease was conveniently tested with pepsin since fraction V was at pH 1.7 and pepsin has a pH optimum at pH 2 which also meant it could

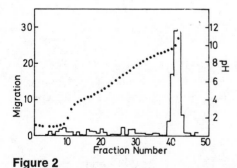

Figure 2
Isoelectric focusing of migration factor. Migration, cells per 1.7 mm of wound edge shown as histogram.

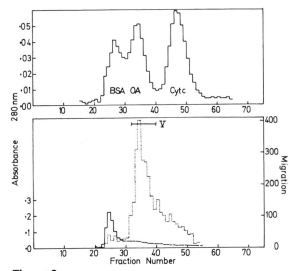

Figure 3
Chromatography of migration factor on G75 Sephadex. Histogram, absorbance at 280 nm. Broken histogram, migration factor activity (cells per 1.7 mm of wound edge).

be added to the cells at pH 7.2 and have very little effect. Table 2 shows that migration factor was stable at pH 2 for 2 hours at 37°C and that pepsin at 10 μg/ml completely inactivated it.

Activities of Fraction V

Since the absorbance at 280 nm of the fractions from the second G75 Sephadex column did not correspond to the migration activity of these fractions (Fig. 4), the migration factor was not pure. However when the ability of these fractions to promote overgrowth of 3T3 was assayed, it was found that there is overgrowth activity which corresponds well with the migration activity (Fig. 4). Also when the number of mitoses per 0.87 mm² in the monolayer of the plates used for the migration assay was counted, the mitotic activity was most in the fractions with the most migration activity (Fig. 4). The cells in the cultures treated with the fractions

Table 2
Effect of Pepsin on Migration Factor Activity

Treatment	Migration
Untreated fraction V	145
Fraction V incubated in buffer	139
Pepsin incubated in buffer	7
Fraction V incubated with pepsin	14
Fraction V and pepsin incubated separately and both added to assay	102

Fraction V and pepsin (10 μg/ml) were incubated separately or together at pH 2 for 2 hours at 37°C. Migration is cells per 1.7 mm of wound edge.

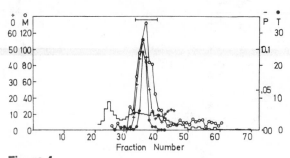

Figure 4
Activities associated with migration factor fractionated on G75 Sephadex. Bed
volume 175 ml, diameter 1.2 cm, buffer 0.02 M HCl, 0.14 M NaCl pH 1.7.
Fraction volume 1.8 ml. Histogram, absorbance at 280 nm. (○) Migration, cells
per 1.7 mm of wound edge. (●) Dividing cells. (+) Overgrowth, 10^4 cells per
5-cm petri plate.

with most migration factor activity had an altered morphology, looking rather like
SV3T3.

Fraction V stimulated 3T3 cells to incorporate tritiated thymidine not only in
those which migrate, but also in those in the layer (Table 3). There was, however,
a marked edge effect on the proportion of thymidine-labeled nuclei. A wounded
monolayer of 3T3 cells treated with 4% serum had uniformly 3% of cells with
labeled nuclei at the edge and in the layer. With low amounts of fraction V (0.06
μg/ml, which is less than 1.5×10^{-9} M), 86% of the cells were labeled at the
edge, whereas 5 mm from the edge 6% were labeled. With 0.6 μg/ml, uniformly
85% were labeled at the edge and in the layer (Bürk unpublished). Drs. Rozengurt and Jimenez de Asua have shown that the fractions which stimulated migration also stimulated uridine transport into 3T3 cells 20–25 minutes after the

Table 3
Effect of Migration Factor
on Thymidine Incorporation

		Layer	
	Wound	Edge	5 mm
Blank	4	< 0.3	0.3
+ PO_4	27	4	3
HgV	17	1	1
+ PO_4	35	19	7
HgV + Serum	83	60	58
+ PO_4	77	63	58
Serum	71	36	14
+ PO_4	75	54	28

Table shows percent labeled nuclei (18–24 hours after addition) among cells which had moved into the wound, among
those in the layer on the edge of the wound, and among
those in the layer 5 mm from the edge of the wound.

addition. Compared with calf serum fraction V was 5000-fold more active since 1.2 μg of fraction V produced the same effect as 6 mg of calf serum protein. Unlike calf serum, migration factor produced only a small increase (50%) in phosphate transport. Perhaps as Rozengurt and Jimenez de Asua have shown for calf serum, migration factor is in some way partly dependent on phosphate transport for its activity, which might explain the effect of raised phosphate in the medium on migration factor activity.

Sources of Migration Factor

Fraction III has been prepared as described previously (Bürk 1973) from the medium of a number of cell lines and then applied to G75 Sephadex. Migration factor activity was never found in medium from the untransformed parental BHK cells, nor was it found in the medium of polyoma virus-transformed or hamster sarcoma virus-transformed BHK cells. It was not found in the medium of SV3T3 either. It has been found in one out of three batches of medium from 28/13T. 28/13T is an early passage of SV28 which differs most notably from SV28 in that the tumors of SV28 reliably metastasize, whereas those of 28/13T in some experiments do and in others do not metastasize.

There are other migration factors in the medium of SV28 cells. One probably comes from calf serum and one seems to be present in the medium of normal BHK cells. These factors are consistent with the migration activity of whole serum and the observation that normal BHK cells migrate without added factor. The migration factor discussed in this paper was not obtained from SV3T3, showing it is not necessarily associated with SV40 transformation. It was not found in the medium of polyoma or hamster sarcoma virus-transformed BHK, showing it is not necessarily associated with transformation of BHK. The migration factor does come from a transformed cell and can promote many of the characteristics of transformation in normal 3T3 cells and so falls within the definition of a transformation factor.

DISCUSSION

Migration factor is an operational term derived from the assay. Time lapse cinematography shows that the cells do, in fact, migrate over the line without hindrance into the wound area. They do not, for instance, detach from the monolayer and float across the line to reattach, nor is the movement due to cytokinesis at mitosis. Mitosis is reached in 24 hours with added phosphate. The simplest model for all the activities of fraction V is that there is one factor which stimulates the normal sequence of events of cell replication. Thus uridine transport is stimulated after about 15 minutes (Rozengurt and Jimenez de Asua), then movement and changes in culture morphology (which are admittedly unquantitated), thymidine incorporation, mitosis and proliferation in monolayers to high density (overgrowth). This is summarized in Fig. 5. In this scheme membrane is assumed to be the first thing affected. We have not detected any proteolytic activity on azocoll. Caffeine and high levels of dibutyryl cyclic AMP were found to inhibit the expression of migration factor activity, suggesting a role of cAMP in cell migration as has been previously shown by Johnson et al. (1972). Thus it would be worth looking for an

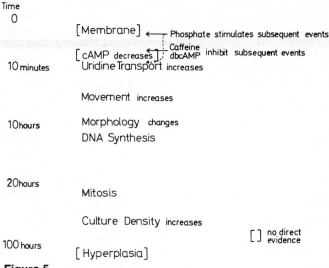

Time

0

[Membrane] ←—— Phosphate stimulates subsequent events

[cAMP decreases] Caffeine dbcAMP inhibit subsequent events

10 minutes   Uridine Transport increases

Movement increases

10 hours   Morphology changes
DNA Synthesis

20 hours
Mitosis

Culture Density increases
[ ] no direct evidence

100 hours
[Hyperplasia]

Figure 5

Sequence of events after addition of migration factor.

effect of migration factor on the concentration of cAMP in the cells. Migration factor should be very useful in studying the role of cAMP in the initiation of the replication sequence. If the cAMP concentration is unaffected while the sequence is initiated, then cAMP is merely a side effect of serum and proteolytic enzymes. If it is affected, then the in vitro effect of migration factor on adenyl cyclase and the phosphodiesterase can be explored.

The SV28 migration factor is different by Sephadex chromatography from the migration factor we extracted from calf serum and from BHK medium. This factor seems to be smaller than the migration factor reported by Lipton et al. (1971). It is larger than insulin and the insulin-like factors reported by Pierson and Temin (1972) to stimulate overgrowth of chick fibroblasts. It also differs in heat stability from the overgrowth stimulating factor Rubin found released by Rous sarcoma-infected chick fibroblasts (Rubin 1970).

The association of a topological effect on thymidine labeling with migration is reminiscent of the association between density inhibition of growth and contact inhibition of movement. It seems a migration factor might explain the correlated loss of density inhibition of growth and contact inhibition of movement observed after transformation. The continuous production of migration factor by SV28 cells could explain the transformed characteristics of SV28 cells. Even their invasiveness and metastasis might be due to an increased ability to migrate. It may be possible to decide whether migration factor does explain the transformed characteristics of SV28 cells if specific antibodies can be produced against purified migration factor and if these will inactivate factor in SV28 cells.

Acknowledgment

I thank Drs. E. Rozengurt and L. Jimenez de Asua for much helpful discussion and Miss M. Grimaldi for her able technical assistance.

REFERENCES

Bürk, R. R. 1973. A factor from a transformed cell line that affects cell migration. *Proc. Nat. Acad. Sci.* **70**:369.

Johnson, G. S., W. D. Morgan and I. Pastan. 1972. Regulation of cell motility by cyclic AMP. *Nature* **235**:54.

Lipton, A., I. Klinger, D. Paul and R. W. Holley. 1971. Migration of mouse 3T3 fibroblasts in response to a serum factor. *Proc. Nat. Acad. Sci.* **68**:2799.

Paul, D., A. Lipton and I. Klinger. 1971. Serum factor requirements of normal and simian virus 40-transformed 3T3 mouse fibroblasts. *Proc. Nat. Acad. Sci.* **68**:645.

Pierson, R. W. and H. M. Temin. 1972. The partial purification from calf serum of a fraction with multiplication-stimulating activity for chicken fibroblasts in cell culture and with nonsuppressible insulin-like activity. *J. Cell. Physiol.* **79**:319.

Rubin, H. 1970. Overgrowth stimulating factor released from Rous sarcoma cells. *Science* **167**:1271.

Amino Acids and the Control
of Nuclear DNA Replication in Liver

John Short, Ned B. Armstrong, Michael A. Kolitsky,
Robert A. Mitchell, Reuben Zemel, and Irving Lieberman

Department of Anatomy and Cell Biology
University of Pittsburgh School of Medicine
and Veterans Administration Hospital, Pittsburgh, Pennsylvania 15261

The cells of the mammalian liver respond to partial hepatectomy by replicating their nuclear DNA and forming new cells. In the rat, DNA synthesis in hepatocytes begins at about 12 hours after the operation. The 12-hr period appears to be devoted to carrying out changes in RNA and protein metabolism that are requisite to the later formation of DNA (Fujioka, Koga and Lieberman 1963; Lieberman 1969).

The stimuli that are evoked by partial hepatectomy and that induce the pre-replicative events leading to DNA synthesis are not yet known. It is clear, however, that the information is carried to the residual liver cells by the blood (Moolten and Bucher 1967).

To learn about the chemistry and mechanisms of action of the stimuli, means have been sought for stimulating liver cells without removing a part of the liver. It was hoped that the nature of the nonsurgical manipulation would shed some light on the regulatory process. Leduc (1949) described a wave of mitotic activity in liver within a day after she shifted mice from a low- to a high-protein diet. We have found that DNA replication and cell division take place in hepatocytes when unoperated rats are shifted from a protein-free diet to a diet containing protein or amino acids (Short et al. 1973).

The purpose of this report is to show the evidence that some amino acids play an important role in controlling DNA synthesis and cell multiplication in liver and to consider the preliminary probes that have been made to discover the connection between amino acids and DNA formation.

NUTRITIONAL SHIFT

DNA synthesis was induced in the liver upon shifting unoperated rats from a protein-free diet to a diet containing casein, and the size of the response was related to the concentration of protein (Fig. 1). Conversely, the stimulatory effect

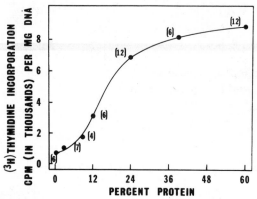

Figure 1

Stimulation of liver DNA synthesis in intact rats after a shift from a protein-free to a protein-containing diet (Short et al. 1973). Female albino rats (Fisher 344, about 150 g) that had been freely fed pellets of Purina Laboratory Chow (24% protein) were kept on a protein-free mash (glucose, corn starch, vitamins, salts, and corn oil) for 3 days. At the end of this time (8 AM) food was removed and at 4 PM the animals were given mashes containing the indicated concentrations of casein. The following morning (8 AM) each rat received 5 μCi of [³H]thymidine in the tail vein and liver samples were taken 1 hr later. The specific activity of nuclear DNA was estimated as previously described (Short et al. 1972). Each point represents the average of the results with 4 to 12 rats as shown. The average specific activity of liver nuclear DNA of animals that were maintained on the Purina Chow was 570 cpm/mg DNA.

of the high-protein meal could be dampened by adding increasing levels of casein to the preparatory diet (Fig. 2).

Radioautographic analysis of histological sections of [³H]thymidine-labeled liver from nutritionally shifted animals established the following points (Table 1):

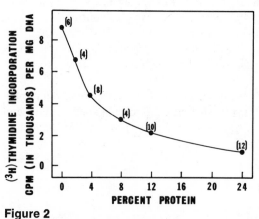

Figure 2

Effect of adding protein to the preparatory mash on the stimulation of liver DNA synthesis by a high-protein diet (Short et al. 1973). The conditions were the same as for Fig. 1 except that the preparatory diet contained varying concentrations of casein, as shown. The high-protein mash was 50% casein.

1. All the radioactivity was in hepatocyte nuclei.
2. Just as after 70% hepatectomy, the label incorporated by the intact livers was proportional to the number of hepatocyte nuclei that were forming DNA.
3. The rate of DNA synthesis by an individual nucleus was the same as in regenerating liver after 70% hepatectomy, as judged by the relationships between the radioactivity incorporated and the percent of labeled nuclei as well as by the number of grains over a labeled nucleus.
4. The number of cells in mitosis was increased and about 60% of the mitotic cells were labeled with [³H]thymidine. The labeling of the dividing cells was taken to mean that the DNA formed after the dietary shift was for the replication of the entire nuclear genome.

Table 1 also shows the results with protein-deprived rats that underwent 70% hepatectomy and were not fed protein after the operation. The increases in DNA synthesis and mitotic cells were similar to those in the well-fed, partially hepatectomized animals.

LEVELS AND UPTAKE OF AMINO ACIDS

In the rat, more than 80% of the blood that feeds the liver comes directly from the viscera in the hepatic portal vein. The amino acid levels in the portal blood

Table 1

Radioautographic Analysis and Mitotic Counts

Treatment	*DNA synthesis*				*Mitoses*	
	cpm (thousands/ mg DNA)	*% labeled nuclei*	*cpm (thousands/ mg DNA/ % labeled nuclei)*	*grains/ labeled nucleus*	*%*	*% labeled*
None	5.3	0.3	17.7	20	0	
Protein-free diet[a]	4.9	0.3	16.3	17	0	
40% protein diet[b]	73	3.9	18.7	18	0.3	57
Protein-free diet, 70% hepatectomy[c]	161	7.5	21.5	22	0.4	67
70% hepatectomy	185	10.8	17.1	20	0.45	54

DNA synthesis was measured by the incorporation of [³H]thymidine (100μCi) given at 18 hr after the nutritional shift or 70% hepatectomy. No mitotic trapping agent was used, liver samples were taken 3 hr after the injection of label, and nuclear labeling and mitoses were scored in parenchymal cells only. The values shown are the averages of the results with two animals.

[a] The rats were fed a protein-free mash for 3 days.

[b] After 3 days on a protein-free diet, the rats were fed a mash containing 40% casein.

[c] The rats were fed a protein-free diet before (3 days) and after the operation.

reflect the state of intestinal absorption as well as the picture in the peripheral circulation. Amino acids absorbed from the gut come not only from ingested protein, but a considerable quantity is derived from enzymes secreted into the gastrointestinal tract and from proteins released from dead or dying mucosal cells (Nasset 1957).

The levels of the amino acids in hepatic portal serum from rats deprived of protein for 3 days were reduced but most of the reductions were small (Table 2). The concentrations of many of the amino acids were markedly elevated, however, as early as 2.5 hours after the deficient animals were given a meal with 40% casein. Indeed, the increases were much greater than in rats that had been shifted from a 24% to a 40% protein diet. Finally, portal sera from sham- and partially hepatectomized rats had similar amino acid patterns.

Tews and Woodcock (1969) studied the effect of diet on the entry into liver cells in vitro of α-aminoisobutyrate (AIB), a nonmetabolizable amino acid that is not incorporated into protein (Noall et al. 1957) and that is transported against a concentration gradient (Crawhill and Segal 1968). They described a slightly reduced rate of uptake by liver slices from weanling rats fed a protein-free diet for one day and an increased rate after feeding a high-protein meal. We have obtained similar results with liver in vivo. Thus the distribution ratios of [^{14}C]AIB between

Table 2

Levels of Amino Acids in Hepatic Portal Serum

	No treat-ment	Protein-free	0% → 40% protein	24% → 40% protein	Sham hepa-tectomy	70% hepa-tectomy
			(μ moles/ml)			
Ala	1.4	1.2	2.3	2.4	1.1	1.1
Asp	0.34	0.14	0.80	0.39	0.27	0.31
Arg	0.02	0.01	<0.01	<0.01	0.01	0.02
Cys	<0.01	<0.01	0.04	<0.01	0.02	0.12
Gly	0.61	0.52	0.70	0.84	0.52	0.45
Glu	0.16	0.15	0.25	0.12	0.11	0.20
His	0.07	0.05	0.07	0.07	0.07	0.07
Ile	0.22	0.11	0.36	0.28	0.13	0.12
Leu	0.31	0.20	0.56	0.38	0.24	0.22
Lys	0.12	0.05	0.07	0.07	0.07	0.07
Met	0.11	0.09	0.20	0.12	0.06	0.09
Phe	0.11	0.09	0.15	0.16	0.09	0.12
Pro	0.30	0.21	0.61	0.46	0.63	0.70
Ser	0.54	0.50	0.90	0.55	0.51	0.40
Thr	0.48	0.45	1.5	0.57	0.33	0.54
Tyr	0.12	0.10	0.22	0.13	0.10	0.12
Val	0.32	0.21	0.71	0.44	0.24	0.23

Deproteinization was with 7% sulfosalicylic acid. No treatment refers to serum from rats kept on a 24% protein diet; protein-free to serum taken after 3 days on the deficient diet; 0% → 40% protein and 24% → 40% protein to samples taken 2.5 hr after shifts from a protein-free (3 days) or a 24% protein diet to a 40% casein mash, respectively; and sham hepatectomy and partial hepatectomy to samples collected at 5 hr after the operations. Each value is the average of the results with three rats.

liver and portal serum were 0.7, 0.5, and 1.0, respectively, for samples from un-treated and protein-deficient (3 days) rats and from protein-deficient animals that had been given a 40% casein mash 5 hours previously.

AIB was used to compare the rates of transport of an amino acid by liver cells in vivo after sham- and partial hepatectomy. For these experiments, the results with the total liver were corrected for the radioactivity contributed by the fluid of the hepatic extracellular space (assumed to be identical with portal serum). The size of the extracellular space of blotted liver, measured as the chloride space, has been found to be 25% for intact and partially hepatectomized rats (Harkness 1952; Ove et al. 1967). In fair agreement with the chloride measurements, spaces of 21 and 24% have now been estimated with [14C]inulin for liver at 5 hours after sham- and 70% hepatectomy, respectively.

The cells of the liver remnant at 5 hours after partial hepatectomy transported AIB much more actively than did the cells of the control organ (Table 3). On the other hand, the uptake of [14C]-O-methyl-D-glucose, a nonmetabolizable sugar (Campbell and Young 1952) that is not concentrated by the liver (Csaky and Glenn 1958), was the same with intact and regenerating liver cells. Measurements with several 3H-labeled natural amino acids were consistent with the results obtained with AIB. In every case, the acid-soluble radioactivity in the liver cells was greater after partial hepatectomy than after the sham operation. Identical labeling

Table 3

Transport of Amino Acids In Vivo by Intact and Regenerating Liver

Labeled compound	Labeling period (min)	Sham hepatectomy			70% hepatectomy		
		portal serum	total liver	liver cell	portal serum	total liver	liver cell
		(cpm in thousands/ml serum or gr, wet liver)					
AIB	2.5	62	33	20	64	166	151
	5	58	47	35	56	236	223
	10	27	48	42	35	309	301
Methylglucose	2.5	36	27	19	35	25	17
	5	32	23	16	34	23	15
Gly	2.5	83	88	71	93	191	169
His		108	396	373	155	734	697
Met		146	152	121	179	220	177
Phe		76	58	42	95	78	55
Thr		75	144	128	91	268	246
Trp		561	218	100	385	452	360
Val		318	134	67	218	188	136

At 5 hr after the operations, each rat was given in the tail vein 1 μCi of [14C]AIB or [14C]3-O-methyl-D-glucose or 10 μCi of a 3H-labeled natural amino acid. After 2.5 to 10 min blood flow to a portion of the liver (about 0.5 gr of the left lateral lobe) was stopped with a hemostat and blood was taken from the hepatic portal vein. The liver sample was then cut off, blotted to remove excess blood, and homogenized in 7% sulfosalicylic acid. The supernatant liver fractions and the portal sera were counted in INSTA-GEL (Packard). The values for "liver cell" were calculated from "total liver" by subtracting 21 (sham hepatectomy) or 24% (partial hepatectomy) of the radioactivity of the corresponding portal serum. Each value is the average of the results with three rats.

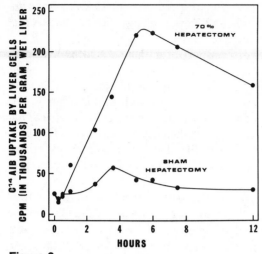

Figure 3
The rate of uptake of [^{14}C] AIB by liver cells in vivo as a function of time after
70% and sham hepatectomy. The conditions were as for Table 3. Labeling was
for 5 min and the values shown are the averages of the results with three animals
and were corrected for the contribution of the extracellular fluid.

of muscle was found in sham- and partially hepatectomized rats with AIB and
the natural amino acids.

Following partial hepatectomy, the rate of uptake of AIB by the residual liver
cells rose linearly after a lag period of 30 minutes and reached a peak of almost
10 times the normal at about 6 hours after the operation, whereupon it slowly
declined (Fig. 3). An increase in amino acid transport also occurred after sham
hepatectomy but it was smaller and it lasted for a short time only. The figure does
not show the results on AIB uptake by liver cells as a function of time after the
removal of 70% of the liver of protein-deficient rats. The same increase in transport occurred as with rats that had been kept on a 24% protein diet.

AMINO ACID REQUIREMENTS

DNA formation could be induced in the liver of the protein-deficient rat by feeding
10% of a mixture of 16 amino acids, but the effect was smaller and more variable
than after feeding 40% protein. Diets with higher levels of amino acids were not
eaten well. Since thyroid hormones stimulate DNA synthesis in the intact liver
(Short et al. 1972), the effect of 3, 3′, 5-triiodo-L-thyronine (T3) given in conjunction with the amino acid diet was studied (Table 4). Better and more consistent responses were obtained than with the amino acids alone.

The [^{3}H]thymidine incorporated after the combined treatment with T3 and
amino acids was entirely for the synthesis of nuclear DNA in parenchymal liver
cells. Cell division was greatly increased and, in some of the liver samples, more
than 90% of the mitotic cells were labeled with [^{3}H]thymidine.

Liver DNA synthesis in protein-deficient animals given T3 and amino acids
began at the same time as after a shift to a 50% casein meal (Short et al. 1973)

Table 4

Effects of T3 and Amino Acids on Hepatic
DNA Synthesis in Protein-deficient Rats

Treatment and additions to diet[a]	DNA synthesis[b] (cpm/mg DNA)
None	550 (390–820)
T3	1380 (450–2910)
Amino acids	2240 (840–4650)
T3 + amino acids	7460 (5420–9930)

[a] The rats were fed a protein-free diet for 3 days. At the end of this time (8 AM) food was removed and at 7 PM the animals were injected subcutaneously with 100 μg of T3 and were offered the protein-free mash supplemented with 10% of a mixture of 16 amino acids (glycine and L-amino acids in equimolar amounts except for tryptophan (one-fifth) and cystine (one-half); aspartate, glutamate, asparagine, and glutamine were omitted).

[b] Hepatic DNA synthesis was measured by labeling the rats for 1 hr with 5 μCi of [3H]thymidine beginning at 14 hr after the dietary shift. The values shown are the averages of the results with five rats and the ranges of the individual determinations are given in parentheses.

and several hours sooner than after 70% hepatectomy (Fig. 4). The brevity of the prereplicative period after the dietary shifts has not yet been explained. The possibility that the deficient rats had already carried out some of the prereplicative changes essential for DNA replication did not seem to be correct since the kinetics of liver DNA formation in protein-deficient rats after 70% hepatectomy was indistinguishable from that in rats kept on a 24% protein diet.

The specific amino acid requirements for hepatic DNA synthesis in protein-deficient rats were determined with the help of T3 (Table 5). The amino acids fell into three groups: (1) the nonessential amino acids as well as arginine, histidine, leucine, and phenylalanine could be omitted from the inductive mash with little effect on DNA formation; (2) omission of lysine, methionine, or tryptophan resulted in about a 70% reduction in the response of the liver; and (3) isoleucine, threonine, and valine were essential for the increase in DNA synthesis. Based on these results, a mixture of the six required amino acids was tested and, as the table shows, it was as effective as the 16 compounds.

To learn whether the amino acids themselves acted in the induction of hepatic DNA synthesis, catabolic intermediates were substituted for some of the required amino acids in the test diets (Table 6). α-Aminobutyrate and isobutyrate were unable to substitute for threonine and valine, respectively, and kynurenine enhanced DNA formation only slightly in tryptophan-deficient rats. No attempt was made to determine how much of the catabolites reached the liver cells.

DISCUSSION

There is no trivial basis for the ability of amino acids to stimulate hepatic DNA synthesis and mitosis in protein-deficient animals. No loss of liver DNA or cells

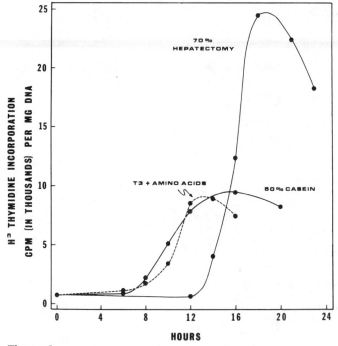

Figure 4

Liver DNA synthesis as a function of time after giving protein-deficient rats
T3 and a mash containing 10% of a mixture of 16 amino acids. The conditions
were as for Table 4. At the time indicated, each rat was injected with 5 μCi of
[³H]thymidine and liver samples were taken 1 hr later. Each point is the average
of the measurements with four rats. For comparison, also shown are the results
after 70% hepatectomy of rats kept on a 24% protein diet and those previously
obtained with protein-deficient rats given 50% casein (Short et al. 1973).

occurs even in rats kept on a protein-free diet for as long as 28 days (Kosterlitz
1947; Munro 1968). Liver protein is decreased by about one-fourth during the
first two days on the deficient diet but little further reduction occurs thereafter
(Addis, Poo and Lew 1936). The breakdown of muscle protein provides an
adequate supply of amino acids so that the ability of the liver to make protein is
not depressed (Garrow 1959; Wannemacher 1961; Drysdale, Olafsdottir and
Munro 1968). Finally, the synthesis of liver DNA in the deficient animals after
70% hepatectomy, and with the same kinetics as in the well-fed rat, is strong
evidence that the liver is not damaged.

Hepatic proteins do not appear to be lost in a random fashion during a period of
amino acid deprivation. The activities and the number of molecules of several of
the enzymes of amino acid catabolism are greatly reduced (Knox and Greengard
1964; Schimke and Doyle 1970), whereas enzymes not directly concerned with
amino acid breakdown may show little or no change in activity (Ely and Ross
1951; Fisher 1954). It is possible that decreases in some of the activities that
degrade amino acids are critical events that help to prepare the liver of the protein-
deprived rat to form DNA in response to amino acids.

One of the responsibilities of the liver is to regulate the circulating levels of

Table 5

Amino Acid Requirements for Hepatic
DNA Synthesis in Protein-deficient Rats

Omissions	DNA synthesis (cpm/mg DNA)	Δ
Amino acids	1590 (300–4490) (24)	
None	8270 (3010–14,100) (24)	6680
Arg*	6580 (5070–12,500) (6)	4990
His	6600 (1430–15,900) (8)	5010
Leu	6160 (3960–11,300) (6)	4570
Phe	6800 (900–13,900) (8)	5210
Nonessentials	7330 (3520–15,200) (12)	5740
Lys	4000 (860–9200) (8)	2410
Met	3780 (960–7390) (8)	2190
Trp	3600 (2420–5380) (18)	2010
Ile	2460 (480–5150) (6)	870
Thr	1870 (250–3410) (12)	280
Val	1650 (330–3070) (12)	60
Nonessentials, Arg, His, Leu, Phe	8910 (3050–15,800) (16)	7320

The conditions were as described for Table 4 except that all the rats received 100 μg of T3. The values shown are the averages of the results obtained, and the ranges of the individual determinations and the number of animals tested are given in parentheses.

* The ten amino acids named in the table are considered to be essential for the rat (Rose 1938).

amino acids, except perhaps for the branched chain compounds (McMenamy et al. 1962; Miller 1962). After a high protein meal or partial hepatectomy, the liver is called upon to dispose of a large quantity of amino acids. Not only do both conditions increase the number of molecules that each cell must handle, but they raise the rate of amino acid transport into the liver cells, most strikingly in the regenerating organ.

Stimulation of the transport of model amino acids into normal liver in response to a variety of hormones has been studied in vivo (Noall et al. 1957; Sanders and Riggs 1967), with the perfused organ (Chambers, Georg and Bass 1968), and with liver slices (Tews, Woodcock and Harper 1970). Glucagon, insulin, and corticosteroids act directly on the liver to increase amino acid uptake, and it seems safe to assume that one or more of these hormones is involved in raising the rate of transport of amino acids after a high-protein meal and partial hepatectomy. As regards glucagon and insulin, it may be worth pointing out that all the blood from the pancreas empties into the hepatic portal vein, so that the residual cells after partial hepatectomy may serve as a target for three times more of the hormones than in the intact animal even if endocrine secretion is unaltered.

Peraino, Blake and Pitot (1965) studied the amino acids required to induce the hepatic catabolizing enzymes, ornithine transaminase and threonine dehydrase, whose levels are regulated by the protein content of the diet. Using protein-deficient rats, they found that a mixture of the ten essential amino acids gave the greatest elevations in the enzyme activities but that the omission of any one of the

Table 6

Inability of Intermediates of Threonine, Valine, and
Tryptophan Breakdown to Substitute for the Amino Acids

Omission	Addition	DNA synthesis (cpm/mg DNA)	Δ
Amino acids		1480 (490–3130)(16)	
None		7230 (3230–12,400)(18)	5750
Thr		1540 (660–2520)(6)	60
Thr	α-aminobutyrate	1660 (230–2270)(6)	180
None	α-aminobutyrate	7870 (4610–11,770)(4)	6390
Val		1490 (95–2960)(8)	10
Val	isobutyrate	2010 (330–4130)(8)	530
None	isobutyrate	6840 (2590–10,630)(4)	5360
Trp		3480 (2150–6670)(12)	2000
Trp	kynurenine	4500 (2210–10,300)(8)	3020

The conditions were as for Table 5. In some cases, as shown, the mashes were supplemented with 1.7% of the sodium salts of D, L-α-aminobutyrate or isobutyrate or with 0.25% of L-kynurenine. The values shown are the averages of the results obtained, and the ranges of the individual determinations and the number of animals tested are given in parentheses.

compounds, with the exception of tryptophan, reduced only little or not at all the stimulatory effect of the mixture.

In contrast to the results with the enzymes, isoleucine, threonine, and valine, rather than tryptophan, were most stringently required for hepatic DNA synthesis in protein-deficient animals. The possibility was considered that the essentiality of isoleucine and valine stemmed only from a block in the intestinal or hepatic uptake of the limiting endogenous amino acid by the other branched chain compounds. This, however, does not seem to be the case. The levels of valine in portal serum and liver from rats fed a mixture lacking the amino acid were 60 and 45%, respectively, of those in samples from animals given a mixture containing all the branched chain compounds. In addition, the rate of incorporation of [³H]histidine into total liver protein was not depressed in animals that had been fed 10% mixtures of amino acids that lacked isoleucine or valine (or threonine).

The studies with the protein-deficient rats, as well as those with the partially hepatectomized animals, suggest that a close relationship exists between amino acid metabolism and the regulation of nuclear DNA replication in liver. At a superficial level, it can be hypothesized that one of the requirements for the induction of hepatic DNA formation is to exceed the capacity of the liver to metabolize amino acids. Such a condition does seem to prevail when a high-protein meal is fed to a protein-deficient animal and when 70% of the liver is removed. In both cases, disturbances take place in the hepatic pools of some free amino acids and other ninhydrin-positive compounds (Short et al. 1973) and up to fifty times more radioactivity than in normal liver is found in a variety of anionic intermediates after a brief labeling period with some tritiated amino acids (for example, tryptophan and methionine).

There is yet no clue on how an excessive amino acid burden is translated into a stimulation of DNA synthesis in liver. The possibility that specific amino acid catabolites accumulate and trigger the critical prereplicative changes is weakened by the inability of breakdown products to substitute for the amino acids themselves.

By whatever means amino acids act to cause nuclear DNA synthesis in liver, at some point they must have a direct or an indirect effect on the nucleus. One of the bridges between amino acids and the nucleus that was considered is a protein that is made so rapidly by stimulated mitochondria that it spills over into the cytosol. This possibility was attractive for the following reasons: thyroid hormone not only causes changes in the labeling pattern of mitochondrial proteins (Volfin, Kaplay and Sanadi 1969), but it also induces some DNA formation in intact liver; the rate of incorporation of [³H]leucine in vivo into the particulate fraction of liver mitochondria is almost doubled during the first 12 hours after 70% hepatectomy (T. Umeda, J. Short and I. Lieberman, unpublished work); and a majority of the amino acid breakdown reactions in liver take place in the mitochondrion.

Chloramphenicol was used to test for the hypothetical mitochondrial protein. By giving the antibiotic at 4-hr intervals from the time of the operation, mitochondrial protein synthesis in liver in vivo could be suppressed by 90% during the entire prereplicative period. Despite this, the ability of the residual hepatocytes to enter the S period was only little affected (T. Umeda, J. Short and I. Lieberman, unpublished work). It seems most unlikely, therefore, that mitochondrial protein synthesis is involved in the regulation of nuclear DNA synthesis.

Acknowledgments

These studies were supported by grants from the National Cancer Institute and the American Cancer Society. We thank Mrs. Kay Susser for carrying out the amino acid analyses.

REFERENCES

Addis, T., L. J. Poo and W. Lew. 1936. The rate of protein formation in the organs and tissues of the body. *J. Biol. Chem.* **116:**343.

Campbell, P. N. and F. G. Young. 1952. Metabolic studies with 3-methyl glucose. *Biochem. J.* **52:**439.

Chambers, J. W., R. H. Georg and A. D. Bass. 1968. Effects of catecholamines and glucagon on amino acid transport in the liver. *Endocrinology* **83:**1185.

Crawhill, J. C. and S. Segal. 1968. Transport of some amino acids and sugars in rat-liver slices. *Biochim. Biophys. Acta* **163:**163.

Csaky, T. Z. and J. E. Glenn. 1958. Distribution of 3-methylglucose in organs of rat. *Proc. Soc. Exp. Biol. Med.* **98:**400.

Drysdale, J. W., E. Olafsdottir and H. N. Munro. 1968. Effect of ribonucleic acid depletion on ferritin induction in rat liver. *J. Biol. Chem.* **243:**552.

Ely, J. O. and M. H. Ross. 1951. Effects of a protein-free diet on the alkaline and acid phosphatase activity of the liver of the rat. *Nature* **168:**323.

Fisher, R. B. 1954. Protein metabolism. *Methuen Monogr.,* Methuen and Company, London.

Fujioka, M., M. Koga and I. Lieberman. 1963. Metabolism of ribonucleic acid after partial hepatectomy. *J. Biol. Chem.* **238:**3401.

Garrow, J. S. 1959. The effect of protein depletion on the distribution of protein synthesis in the dog. *J. Clin. Invest.* **38:**1241.

Harkness, R. D. 1952. Changes in the liver of the rat after partial hepatectomy. *J. Physiol.* (London) **117**:267.

Knox, W. E. and O. Greengard. 1964. The regulation of some enzymes of nitrogen metabolism—an introduction to enzyme physiology. *Advanc. Enzyme Regul.* **3**:247.

Kosterlitz, H. W. 1947. The effects of changes in dietary protein on the composition and structure of the liver cell. *J. Physiol.* (London) **106**:194.

Leduc, E. H. 1949. Mitotic activity in the liver of the mouse during inanition followed by refeeding with different levels of protein. *Amer. J. Anat.* **84**:397.

Lieberman, I. 1969. In *Biochemistry of cell division* (ed. R. Baserga) p. 119. Charles C. Thomas, Springfield, Ill.

McMenamy, R. H., W. C. Shoemaker, J. E. Richmond and D. Elwyn. 1962. Uptake and metabolism of amino acids by the dog liver perfused in situ. *Amer. J. Physiol.* **202**:407.

Miller, L. L. 1962. In *Amino acid pools* (ed. J. T. Holden) p. 708. Elsevier, New York.

Moolten, F. L. and N. L. R. Bucher. 1967. Regeneration of rat liver: Transfer of humoral agent by cross circulation. *Science* **158**:272.

Munro, H. N. 1968. Role of amino acid supply in regulating ribosome function. *Fed. Proc.* **27**:1231.

Nasset, E. S. 1957. Role of the digestive tract in the utilization of protein and amino acids. *J. Amer. Med. Ass.* **164**:172.

Noall, M. W., T. R. Riggs, L. M. Walker and H. N. Christensen. 1957. Endocrine control of amino acid transfer. *Science* **26**:1002.

Ove, P., S. Takai, T. Umeda and I. Lieberman. 1967. Adenosine triphosphate in liver after partial hepatectomy and acute stress. *J. Biol. Chem.* **242**:4963.

Peraino, C., R. L. Blake and H. C. Pitot. 1965. Studies on the induction and repression of enzymes in rat liver. *J. Biol. Chem.* **240**:3039.

Rose, W. C. 1938. The nutritive significance of the amino acids. *Physiol. Rev.* **18**:109.

Sanders, R. B. and T. R. Riggs. 1967. Modification by insulin of the distribution of two model amino acids in the rat. *Endocrinology* **80**:29.

Schimke, R. T. and D. Doyle. 1970. Control of enzyme levels in animal tissues. *Annu. Rev. Biochem.* **39**:929.

Short, J., R. F. Brown, A. Husakova, J. R. Gilbertson, R. Zemel and I. Lieberman. 1972. Induction of deoxyribonucleic acid synthesis in the liver of the intact animal. *J. Biol. Chem.* **247**:1757.

Short, J., N. B. Armstrong, R. Zemel and I. Lieberman. 1973. A role for amino acids in the induction of deoxyribonucleic acid synthesis in liver. *Biochem. Biophys. Res. Commun.* **50**:430.

Tews, J. K. and N. A. Woodcock. 1969. Factors affecting the transport of amino acids by rat liver slices. *Fed. Proc.* **28**:301.

Tews, J. K., N. A. Woodcock and A. E. Harper. 1970. Stimulation of amino acid transport in rat liver slices by epinephrine, glucagon, and adenosine 3′,5′-monophosphate. *J. Biol. Chem.* **245**:3026.

Volfin, P., S. S. Kaplay and D. R. Sanadi. 1969. Early effect of thyroxine *in vivo* on rapidly labeled mitochondrial fractions and respiratory control. *J. Biol. Chem.* **244**:5631.

Wannemacher, R. W., Jr. 1961. Incorporation of ^{35}S from L-methionine in tumor-bearing and protein-depleted rats. *Proc. Soc. Exp. Biol. Med.* **107**:277.

Stimulators, Enzyme Induction and the Control of Liver Growth

Thomas S. Argyris

Department of Pathology, Upstate Medical Center, Syracuse, New York 13210

The mechanism by which growth is controlled in adult mammalian tissue and organs is still not known, although it has been under investigation for some time. Many theories have been suggested to explain the mechanism of the control of growth. The theories which have been proposed may be arbitrarily divided into two groups, and they tend to represent the bias of the milieu in which the investigators who proposed them worked. The first group of theories has as its basis that special growth controlling substances (either stimulators or inhibitors), whose primary function is to control growth, exist in the body. The second group of theories has as its basis that functional demand controls growth. Implicit in this approach is that it is not necessary to postulate special growth controlling substances. Changes in the levels of ordinary molecules which are involved in cell homeostasis is sufficient to control growth. (For reviews see Abercrombie 1957; Adolph 1972; Argyris 1964, 1968a, 1969a, 1972; Argyris et al. 1969; Bullough 1962, 1965; Goss 1964; Weiss 1955.) It is interesting to note that those individuals who have held that the control of growth is due to special growth controlling substances have been embryologists, or have been strongly influenced by the embryologist's notion of "inducers" which supposedly control growth and differentiation in the embryo. Those who have held that functional demand controls growth have been primarily investigators studying how various functional changes can stimulate organ growth, such as occurs after dietary changes, the removal of one of a pair of organs, or the removal of a portion of an unpaired organ.

Our laboratory has long been interested in the problem of the control of growth in adult mammalian tissues and organs. Because we have felt that the problem of the control of growth has not sufficiently matured to produce a single investigative model whose study would give us the answer to the question of how growth is controlled, we have used a comparative approach studying a variety of model systems of induced growth, such as damage-induced growth, functional hypertrophy, compensatory hypertrophy, and tumor-induced growth, with the hope that comparisons would act catalytically to single out key processes in growth control (Argyris 1964, 1968a, 1969a, 1969b, Argyris et al. 1969).

49

In this paper I wish to focus on our work attempting to relate the induction of the drug-metabolizing enzymes in rat liver to the triggering of liver growth. We hope to show that this model system may well be the best approach for arriving, at least, at one molecular explanation for how functional demand may exercise control over organ size and therefore growth.

Nature of the Induction of the Drug-metabolizing Enzymes by Drugs

The injection of phenobarbital (PB), 3-methylcholanthrene (MC) or a wide variety of other drugs results in an increase in the drug-metabolizing enzyme activity in the livers of rats and other mammals. (For review see Conney 1967.) Associated with the increase in the drug-metabolizing enzyme activity is a significant increase in liver weight, ranging from 20–50% depending on the drug and regimen used (Argyris 1971; Argyris and Magnus 1968; Barka and Popper 1967; Conney 1967; Kunz et al. 1966a). Therefore the exciting possibility is raised that increasing the drug-metabolizing enzyme activity turns on liver growth. If so, this would be an example of how functional demand controls organ size and therefore organ growth. This is essentially the problem I will be discussing. Before I do so it is useful to present some background information about the nature of the drug-metabolizing enzymes, the character of their increase after drug induction, and the nature of the liver growth induced.

The drug-metabolizing enzymes are actually a microsomal enzyme system which is composed essentially of three parts: an NADPH-cytochrome c reductase, a lipid component identified as phosphatidylcholine, and a CO-binding hemoprotein usually referred to as cytochrome-P-450 (Conney 1967; Lu et al. 1969, 1971, 1973; Remmer 1970). The primary physiological substrates for the drug-metabolizing enzymes within the body are the steroid hormones (Conney 1967; Kuntzman 1969). The drug-metabolizing enzyme system in the intestine and liver also metabolize a wide variety of substances which are in the "normal" diets animals eat and a variety of metabolic products of the naturally occurring bacterial flora in the intestinal lumen (Wattenberg 1972). To the extent one wishes to consider these normally occurring substances, these compounds also can be considered as naturally occurring substrates for the drug-metabolizing enzyme system.

In the liver the microsomal drug-metabolizing enzyme system is located in the hepatocytes (Gielen and Nebert 1971; Henderson and Dewaide 1969; Holtzman 1972). There is no data to indicate whether drug-metabolizing enzymes are present in bile duct cells. Recently there is suggestive evidence that some drug-metabolizing enzyme activity may be present in littoral cells or other stromal cells in the liver (Cantrell and Bresnick 1972; Gielen and Nebert 1971). More work is needed to completely identify what cell types in the liver show drug-metabolizing enzyme activity and to what extent.

Administration of a wide variety of drugs results in increases in drug-metabolizing enzyme activity (Conney 1967). Figure 1 shows what happens if 50 mg/kg b wt of phenobarbital is injected daily into immature male rats. By the first day after treatment is begun, there is an increase in demethylase activity in the liver. We use demethylase activity as a representative activity of the entire drug-metabolizing enzyme system, much as one uses acid phosphatase as an indicator of lysosomal enzyme activity. (There are obvious limitations to the use of a single enzyme activity as an indicator of what an entire system or family of enzymes is doing, and this

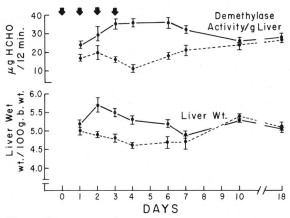

Figure 1

The effect of multiple intraperitoneal injections (indicated by arrows) of 50 mg/kg b wt of phenobarbital on aminopyrine demethylase activity and liver wet weight in immature male rats. Controls were injected with saline (– – –). Sucrose homogenates were prepared and the enzyme in the 9000 × g supernatant was assayed. (Adapted from Argyris 1969a; Argyris and Magnus 1968.)

should be kept in mind.) Similar results are obtained if one measures enzyme activity per mg of protein, of the homogenate, the 9000 × g supernatant fraction, or of the microsomal fraction (Argyris and Magnus 1968; Argyris 1969a; Conney 1967). Once phenobarbital treatment is stopped, the demethylase activity returns to normal. Associated with the increase in demethylase activity is an increase in liver weight, which parallels the increase in enzyme activity (Fig. 1). Moreover just as demethylase activity returns to normal after cessation of phenobarbital treatment, so does liver weight. Thus the intriguing possibility is raised that the increase in the drug-metabolizing enzyme activity is coupled to the triggering of liver growth in a way which might be physiologically significant. Before addressing ourselves to this question, let us first ask (1) what is the increase in demethylase activity due to, and is it specific? (2) Does the increase in liver weight represent real liver growth, and if so, is it brought about by cell proliferation and/or cell enlargement?

An answer to the second question requires that we define liver growth. We recognize the complexity of this problem because liver growth may involve a multitude of factors, such as changes in cell number and cell size, not only of the hepatocytes, bile duct cells, and littoral cells, but also of the various supportive tissue cells, especially fibroblasts, as well as the synthesis of many intercellular substances. The best overall measurement of liver growth, no matter what cell types are involved or what combination of cell proliferation, cell enlargement, or the synthesis of intercellular substances is used to bring about the liver growth, is the increase in wet weight and total liver protein. Thus we will use parallel increases in both wet weight and total liver protein as our indicator of liver growth.

Requirement for Enzyme Synthesis

The increase in drug-metabolizing enzyme activity is due largely to enzyme synthesis. Increases in drug-metabolizing enzyme activity are prevented by inhibitors of RNA and protein synthesis (Conney 1967; Conney and Gilman 1963; Ernster and Orrenius 1965; Orrenius et al. 1965). Increases in drug-metabolizing enzyme activity are associated with increases in amino acid incorporation (Arias et al. 1969; Conney 1967; Kuriyama et al. 1969; Schimke et al. 1968; Staubli et al. 1969). However some of the increase in drug-metabolizing enzyme activity, especially if more than one injection of phenobarbital is given, may be due to enzyme stabilization (Jick and Shuster 1966; Kuriyama et al. 1969; Shuster and Jick 1966). Both the NADPH-cytochrome c reductase and cytochrome P-450 increase in amount. It is not clear whether or not the phospholipid portion is increased (Ernster and Orrenius 1965; Glaumann 1970).

The induction of the drug-metabolizing enzyme system by drugs is not a part of a nonspecific increase in all enzymes in the liver cell following drug administration (Kunz et al. 1966b; Tepperman and Tepperman 1967). Indeed even within the microsomal fraction not all enzymes are increased. For example, there is either a decrease or no increase in glucose-6-phosphatase, IDPase, ATPase, NADH-cytochrome c reductase, and cytochrome b_5 (Barka and Popper 1967; Ernster and Orrenius 1965; Glaumann 1970; Glazer and Sartorelli 1971, 1972; Koransky et al. 1969; Kuriyama et al. 1969). Thus we may conclude that the increase in drug-metabolizing enzyme activity requires enzyme synthesis and that it shows considerable specificity. The exact degree of specificity remains to be determined.

Nature of the Liver Growth Induced by Drugs

The increase in liver weight associated with the induction of the drug-metabolizing enzyme system represents growth. Figure 2 shows that there is an increase in the total protein of the liver which roughly follows the increase in wet weight. Many others before us have shown similar increases in total protein (Conney 1967; Kunz et al. 1966a). We have also shown that there is an increase in dry liver weight (Argyris and Magnus 1968). The homogenate protein/g of liver does not change; however the microsomal protein/g of liver increases and this will be further discussed below (Conney 1967). The increase in total protein is accompanied by an increase in total RNA, DNA, and nuclear count (Fig. 2). If one calculates the data as percent of control, the percent increase in total DNA and nuclear count is approximately the same as the percent increase in total protein, suggesting that the growth of the liver can be accounted for by an increase in cell number (Argyris 1969a; Argyris and Magnus 1968). Figure 2 alerts us also to the problem that immature rats are growing and that this growth affects the parameters we study. This is especially true when one looks at the changes in nuclear count (Fig. 2). It is obvious that the saline-treated immature male rats show a decrease in the number of nuclei as a consequence of normal growth (Argyris and Magnus 1968). This raises the question if liver growth after phenobarbital injection should not be studied in adult male rats, where normal liver growth has essentially ceased. As we shall see below however, the cellular mechanisms bringing about the induced liver growth vary with the age of the rat.

Initially we stated that a wide variety of drugs and other susbtances induce the drug-metabolizing enzyme system and growth of rat liver. The inducers of the drug-metabolizing enzyme system are often arbitrarily divided into two major groups: those substances which induce a wide variety of detoxifying reactions, called general inducers, and those substances which induce a very limited number of reactions,

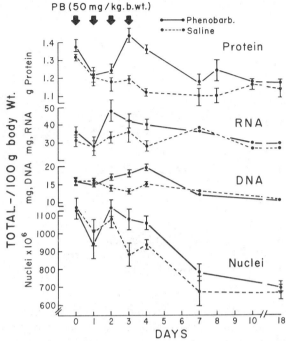

Figure 2

The effect of multiple intraperitoneal injections (indicated by arrows) of phenobarbital, in saline, on liver growth. All the parameters are given as the amount
per liver per 100 g b wt of rat in order to eliminate variations due to body
weight. The protein was measured with the Biuret reaction using albumin
(Sigma) as a standard. RNA and DNA were extracted using a Schmidt-Thann-
hauser extraction procedure. RNA was determined with the Orcinol reaction
with purified yeast RNA (Sigma) as a standard. DNA was measured with the
diphenylamine reaction using purified calf thymus DNA (Sigma) as a standard.
The number of nuclei was determined by taking aliquots of the liver homogenate
and counting the number of nuclei in a hemocytometer. (Adapted from Argyris
1968b, 1969a; Argyris and Magnus 1968.)

called specific inducers (Alveres et al. 1968; Conney 1967). Phenobarbital is an
example of a general inducer. The question arises if specific inducers also trigger
liver growth. The answer to this question is that indeed a number of specific inducers, such as 3-methylcholanthrene, have been shown to induce liver growth
(Argyris 1969a; Argyris and Layman 1969; Conney et al. 1956). In Fig. 3 we see
the results of a series of experiments in which male immature rats were injected
once with 1 mg of 3-methylcholanthrene. The administration of 3-methylcholan-
threne results in liver growth as evidenced by the increases in liver weight and total
liver protein. Again as after phenobarbital treatment, the liver growth appears to
be produced by cellular proliferation, as evidenced by the increases in the total
nuclear count. Not shown is a comparable increase in total liver DNA (Argyris
and Layman 1969). Thus both general and specific inducers of the drug-metabolizing enzyme system stimulate liver growth.

The association of liver growth with the induction of enzymes is not limited to

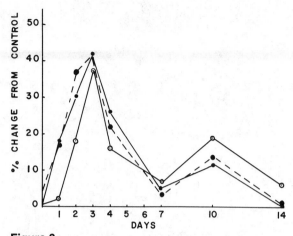

Figure 3

The effect of one intraperitoneal injection of 3-methylcholanthrene on (●———●) rat liver wet weight, (●– – –●) total protein, and (o———o) total number of nuclei. See Fig. 2 for methods. (Adapted from Argyris and Layman 1969.)

the induction of the drug-metabolizing enzyme system. Schimke (1962) has clearly demonstrated that feeding rats a high protein diet results in the induction of the urea cycle enzymes and an increase in liver weight and total protein. Moreover he has shown that the induction of the urea cycle enzymes has considerable specificity. We have confirmed the basic findings of Schimke (1962) and have shown that in immature male rat liver there is an increase in the total cell number (Argyris 1969a, 1971).

The question next arises as to whether the inductions of the drug-metabolizing enzyme system by drugs, and the urea cycle enzymes by the feeding of a high-protein diet, are "nonspecific" inductions which are due to the liver growth. That this is not the case is shown in Fig. 4. Feeding a high protein diet results in an increase in the urea cycle enzymes, as evidenced by increases in ornithine transcarbamylase activity and liver growth. However the drug-metabolizing enzyme system is not induced, as evidenced by the lack of increase in the demethylase activity. On the other hand, daily injections of phenobarbital induce demethylase activity and liver growth, but not ornithine transcarbamylase activity. When both phenobarbital and high-protein diet are given, there is an additive effect on liver growth and both enzymes are increased.

Our evidence so far strongly suggests that the induction of some enzyme systems is coupled to the triggering of liver growth. Before we review the evidence of how tightly coupled these two events are, let us discuss in greater detail the nature of the liver growth produced by drug administration. After multiple injections of 50 mg/kg b wt of phenobarbital, the liver growth is accountable by the increase in cell number, as evidenced by the changes in the total nuclear count (Fig. 5). Increasing the dosage of phenobarbital to 100 mg/kg b wt results in a doubling of liver wet weight and total protein (the latter not shown) and total DNA. However, the total nuclear count increases only about 25%. This suggests that now cell enlargement is significantly contributing to the induced liver growth in addition to cell proliferation. On the other hand, if adult male rats are injected with 100 mg/kg b wt of pheno-

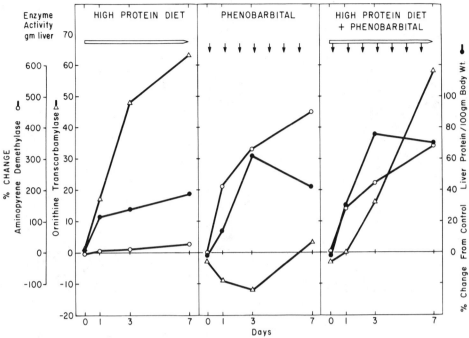

Figure 4

The effects of intraperitoneal injections of 100 mg/kg b wt of phenobarbital, feeding of a 64% protein diet, or both on ornithine transcarbamylase and aminopyrine demethylase activities, and total liver protein. Control rats were fed a 15% protein diet and injected with saline. (Adapted from Argyris 1971.)

barbital, the increase in liver weight is about 25% (Fig. 6). Similar increases occur in total protein (Argyris 1971). All of it is accountable by cell enlargement, in contrast to immature rats where a 25% increase in liver weight is brought about entirely by cell proliferation (Fig. 5). The increase in total DNA in adult rats probably largely represents increases in the number of polyploid cells (Paulini et al. 1970, 1971; Staubli et al. 1969; Schulte-Hermann et al. 1968). Similarly after a single injection of 3-methylcholanthrene in immature rats, the principal contributor to the increased liver mass is cell proliferation, whereas in adult rats it is cell enlargement (Argyris and Layman 1969; Hopkinson and Argyris 1972). Thus it is clear that the cellular mechanisms brought into play to effect the induced liver growth are drastically influenced by the dosage of the drug administered and by the age of the animal.

Multiple injections of 100 mg/kg b wt of phenobarbital induce an intense proliferation of the smooth endoplasmic reticulum beginning about the second day after injection, and smooth endoplasmic reticulum packs the hepatocytes by 5 days (Burger and Herdson 1966; Orrenius et al. 1965; Orrenius and Ericsson 1966). The increased proliferation of the smooth endoplasmic reticulum is accompanied by an increase in microsomal protein, RNA, and phospholipid per g of liver (Glaumann 1970; Orrenius et al. 1965). Staubli et al. (1969) have suggested that increase in cell size produced by appropriate doses of phenobarbital can be accounted for by the increase in the volume of the smooth endoplasmic reticulum. Proliferation of endoplasmic reticulum is not a necessary concomitant of cell en-

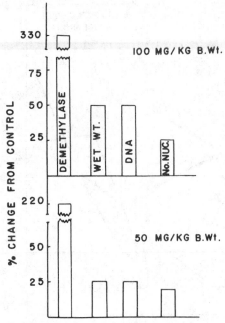

Figure 5

Comparison of the effects of different dosages of phenobarbital on aminopyrine demethylase activity, liver weight, total DNA and number of nuclei.
For methods see legend of Fig. 1 and 2. Phenobarbital was injected once daily
for 4 days and the immature male rats killed 24 hr after the last injection.
(Adapted from Argyris 1969a; Argyris and Magnus 1968; Augenlicht 1971.)

largement after drug treatment, since cell enlargement after 3-methylcholanthrene
treatment is not accompanied by a significant increased proliferation of the endoplasmic reticulum, nor increases in microsomal protein or RNA/g of liver (Conney and Gilman 1963; Fouts and Rogers 1965; Glaumann 1970). It is not clear if
low doses of phenobarbital, where liver growth can be accounted for by cell pro-

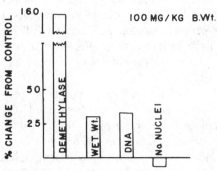

Figure 6

Effects of multiple intraperitoneal injections of phenobarbital on aminopyrine
demethylase activity, liver weight, total RNA, and the number of nuclei in adult
male rat livers. Phenobarbital was injected once daily for 4 days and the rats
killed 24 hr after the last injection. (Adapted from Argyris 1968b.)

liferation alone, are associated with significant increases in the endoplasmic reticulum (Argyris and Magnus 1968).

Evidence for Coupling of the Induction of the Drug-metabolizing Enzymes and Liver Growth

We are now ready to ask how closely is the induction of the drug-metabolizing enzyme system coupled to the triggering of liver growth.

As we have already discussed (Fig. 1 and 2), multiple injections of phenobarbital induce the drug-metabolizing enzymes and trigger liver growth. Similar results have been found by many others for a variety of drugs and regimens (Augenlicht 1971; Barka and Popper 1967; Conney 1967; Golberg 1966; Kunz et al. 1966b; Schlicht et al. 1968; Schulte-Hermann 1968; Staubli et al. 1969). Single injections of phenobarbital also induce the drug-metabolizing enzymes and liver growth (Miller 1969). The kinetics of increase in liver weight and total protein follow closely the increase in drug-metabolizing enzyme activity, whether one gives single or multiple injections of phenobarbital. In current experiments in which we have studied the changes in the drug-metabolizing enzyme activity and liver weight and protein during the first few hours after phenobarbital treatment, we have found that the induction of the drug-metabolizing enzyme activity occurs prior to significant increases in liver weight or total protein. Within 5 days after a single injection of phenobarbital or 5–10 days after multiple injections of phenobarbital, the drug-metabolizing enzyme activity returns to normal and so does liver weight and total protein (Argyris 1969a; Argyris and Magnus 1968; Koransky et al. 1969; Kuriyama et al. 1969; Miller 1969; Orrenius and Ericsson 1966; Wilson and Fouts 1966). Single injections of 3-methylcholanthrene in immature and adult rats induce the drug-metabolizing enzymes and increases in liver weight and total protein, the latter following the former closely (Argyris and Layman 1969; Hopkinson and Argyris 1972). Moreover in immature rats the increase in the drug-metabolizing activity after a single injection of 3-methylcholanthrene occurs much faster than it occurs after phenobarbital treatment, often reaching a peak within 24 hours (Argyris and Layman 1969; Conney et al. 1956). Accordingly the increase in liver weight and total protein also occurs earlier and follows the increase in drug-metabolizing activity faithfully (Argyris and Layman 1969). As after cessation of phenobarbital administration, the drug-metabolizing enzyme activity, liver weight and total protein return to normal within 10–14 days after injection of 3-methylcholanthrene. To the limited extent studied, there is a dosage relationship between the increase of the drug-metabolizing enzyme activity and the increase in liver weight and total protein after phenobarbital treatment (Argyris and Magnus 1968; Augenlicht 1971; Kunz et al. 1966b). Finally, the injection of 50 mg/kg b wt of phenobarbital results in an increase in the drug-metabolizing enzyme activity, liver weight and total protein, followed by a return to normal of these parameters within 5 days after phenobarbital treatment. If one then gives a second injection of phenobarbital there will be a reinduction of the drug-metabolizing enzyme activity and a second stimulation of liver growth as evidenced by increased liver weight and total protein. Again these parameters will together return to normal within 5 or 6 days (Miller 1969). Thus we conclude that there is considerable circumstantial evidence to closely link the induction of the drug-metabolizing enzymes to the triggering of liver growth. There is no critical evidence that indicates that the induction of the drug-metabolizing enzymes is causally related to the triggering of liver growth. However as we shall argue below, this is not necessary in order to consider the coupling of enzyme induction to liver growth to be physiologically significant.

Additive Effects

As we have already pointed out, the drug-metabolizing enzyme system is composed of three major parts, NADPH-cytochrome c reductase, a phospholipid, and cytochrome P-450 (Lu et al. 1969). The question arises as to whether there is any evidence to suggest that it is the change in only one of the three components which is important in the coupling of the induction of the drug-metabolizing enzymes to the triggering of liver growth. This is a question for which no clear answer can be given, but there is suggestive information that it is a change in the cytochrome P-450 which may be preferentially important in the triggering of liver growth. Sladek and Mannering (1969a) have recently shown that if rats are pretreated with a general inducer such as phenobarbital and a specific inducer such as 3-methylcholanthrene, there is an additive effect on the drug-metabolizing enzyme activity. There is no additive effect on the drug-metabolizing enzyme activity if two specific inducers are used, such as 3-methylcholanthrene and benzpyrene. Others have shown that combined treatment with two general inducers also does not result in additive effects on the drug-metabolizing enzyme activity (Conney 1967; Gielen and Nebert 1971). Furthermore Sladek and Mannering (1969b) have shown that 3-methylcholanthrene induces a P-450 which is different from that induced by phenobarbital. It has been denoted as P_1-450 or P-448. Additive effects on the induction of the drug-metabolizing enzyme activity also have recently been obtained with liver cells in culture when phenobarbital and benzpyrene are given together (Gielen and Nebert 1971).

Recently in an extensive series of investigations Lu et al. (1969, 1971, 1972, 1973) have isolated the three components of the drug-metabolizing enzyme system and have been able to reconstitute the drug-metabolizing enzyme system in vitro and get full drug-metabolizing enzyme activity. In addition they have shown through mixing experiments, in which the components were derived from phenobarbital- or 3-methylcholanthrene-pretreated rats, that it is the cytochrome P-450 which determines the nature of the drug-metabolizing enzyme activity of the reconstituted enzyme system. If the P-450 comes from phenobarbital pretreated rats, then the reconstituted system behaves like a phenobarbital derived system, even though the NADPH-cytochrome c reductase and lipid portions come from 3-methylcholanthrene-treated rats. On the other hand, if the cytochrome P-450 comes from 3-methylcholanthrene-pretreated rats, the reconstituted drug-metabolizing enzyme system behaves as if it comes from 3-methylcholanthrene-treated rats even though the NADPH-cytochrome c reductase and lipid portions are from phenobarbital-pretreated rats. Thus the evidence is suggestive that it may be the cytochrome P-450 which controls the specificity of the drug-metabolizing enzyme system. Since in the experiments of Sladek and Mannering (1969a) additive effects on drug-metabolizing enzyme activity were obtained only when phenobarbital and 3-methylcholanthrene were used, and the only known difference between these two systems is the nature of the cytochrome P-450, it suggests that it is the difference in the two types of cytochrome P-450's or the pathways leading to their synthesis which is important in resulting in additive effects on drug-metabolizing enzyme activity.

If our hypothesis that the induction of the drug-metabolizing enzymes is coupled to triggering of liver growth is correct, then we should be able to produce additive effects on liver growth by the combined treatment of rats with phenobarbital and 3-methylcholanthrene. Moreover if this occurs, it would be suggestive evidence that it is change in the P-450 and/or changes in the pathway leading to its induction which in some way results in the triggering of liver growth. Recently Augenlicht (1971) in our laboratory pretreated rats with maximal doses of phenobarbital until the growth response of the liver was saturated. He then injected 3-methylcholanthrene and produced an additive effect on liver weight and total protein. Moreover in control rats which continued to be injected with phenobarbital or phenobarbital and corn oil, the latter the vehicle for 3-methylcholanthrene, there was no increase in liver weight and total liver protein. If changes in the cytochrome P-450 are critical in the triggering of liver growth in rats treated with drugs, then specifically inhibiting the cytochrome P-450 should inhibit drug-induced liver growth. Such experiments are extremely difficult to do for obvious reasons, such as difficulty in finding an inhibitor that selectively inhibits cytochrome P-450 and has no independent effect on liver growth. However Schulte-Hermann et al. (1972) have presented suggestive evidence that SKF-525A and CFT-1201, both inhibitors of cytochrome P-450, inhibit drug-induced liver cell proliferation. In one series of experiments, rats were pretreated with BHT, an inducer of the drug-metabolizing enzyme system and liver cell proliferation, and CFT-1201 was given at appropriate intervals (Table 1). CFT-1201 treatment abolished the BHT increases in [³H]thymidine incorporation into DNA, the number of [³H]thymidine-labeled nuclei, and mitotic activity. Moreover CFT-1201 had no inhibitory effect on any of these parameters in regenerating liver after partial hepatectomy (Table 2), suggesting that CFT-1201 does not have an independent action on inhibiting cell proliferation. These interesting experiments are important, but they must be extended before we can accept them as having demonstrated that specific inhibition of cytochrome P-450 prevents drug-induced cell proliferation.

The Coupling of Function to Control of Growth

We may summarize our discussion by saying that we believe that the evidence is strong that the induction of the drug-metabolizing enzyme system is coupled to the triggering of liver growth (Argyris 1969a, 1971; Argyris and Magnus 1968;

Table 1

Effect of CFT-1201 on BHT-induced Proliferation of Rat Hepatocytes

				Per 1000 hepatocytes	
No. rats	Inducer	CFT-1201 (mg/kg)	dpm/μg DNA	Labeled hepatocytes	Hepatocyte mitoses
2	—	—	21	1.5	0
2	BHT	—	324	115.0	1.5
2	—	80 + 40	25	1.5	0
5	—	120 + 40	22 ± 5	6 ± 5	0.5 ± 0.5

BHT (3,5,di tert butyl-4-hydroxytoluene) 500 mg/kg b wt was injected 48 and 24 hr prior to sacrifice. Diethylaminoethylphenyldiallylacetate (CFT-1201B) was injected 18 hr prior to sacrifice. Data are averages or average ± standard deviation. Adapted from Schulte-Hermann et al. 1972.

Table 2

Effect of CFT-1201 on Hepatocyte Proliferation Induced by Partial Hepatectomy

			Per 1000 hepatocytes	
No. rats	CFT-1201 (mg/kg)	dpm/μg DNA	Labeled hepatocytes	Hepatocyte mitoses
5	—	440 ± 144	297 ± 91	30 ± 5
5	120	421 ± 240	243 ± 144	24 ± 20

Rats were partially hepatectomized and sacrificed 26 hr later. Diethylaminoethylphenyldiallylacetate (CFT-1201) was injected 19 hr prior to sacrifice. Tritiated thymidine was injected 24 hr after partial hepatectomy. Average ± standard deviation. Adapted from Schulte-Hermann et al. 1972.

Schulte-Hermann et al. 1972). The evidence is also suggestive that the induction of the urea cycle enzymes is coupled to the triggering of liver growth. Does this mean that the induction of all enzymes in the liver is coupled to triggering of liver growth? We feel this is unlikely. Indeed, it is our working hypothesis that there is a "peck order" of enzymes whose induction is coupled to the triggering of liver growth. Some will be strongly coupled, others weakly, and still others not at all. The determination of this "peck order" becomes an important problem if we are to understand how enzyme induction is related to liver growth.

The evidence also indicates that liver growth is not always associated with the induction of the drug-metabolizing enzyme system. Partial hepatectomy results in liver growth. Yet at the time of intense cell proliferation and sharp increases in total protein, the drug-metabolizing enzyme activity decreases (Conney 1967; Fouts et al. 1961; Glazer and Sartorelli 1971; Henderson and Kersten 1970). Likewise many hepatomas show a marked decrease in drug-metabolizing enzyme activity, yet they are proliferating and increasing their total protein (Adamson and Fouts 1961; Conney 1967; Rogers et al. 1967). However it is interesting to note that under appropriate conditions the drug-metabolizing enzyme activity can be increased in regenerating liver (Becker and Lane 1968; Hilton and Sartorelli 1970; Henderson and Kersten 1970), as well as liver weight (Gram et al. 1968) and mitotic activity (Japundzic et al. 1967). Induction of the drug-metabolizing enzyme system has also been reported in some tumors (Watanabe et al. 1970; Conney 1967).

Enzyme induction is not the only functional change which can be coupled to liver growth. Our working hypothesis is similar to that proposed by Adolph (1972), that a variety of functional activities, when changed sufficiently, will place demands on an organ and result in its growth. For example, decreases in plasma albumin following plasmapheresis result in an intense proliferative response by the liver (Sudweeks and Hill 1967). It is assumed that the decrease in plasma albumin somehow triggers liver growth. An increase in free amino acids (Ferris and Clark 1972; Fausto 1972; Short et al. 1972, 1973), increased synthesis of UMP (Fausto 1972), and a drop in the plasma proteins (Glinos 1958) are metabolic changes that have been suggested as being able to stimulate liver growth in the regenerating liver after partial hepatectomy. Short et al. (1972) have in fact induced DNA synthesis in livers of rats given blood from partially hepatectomized rats. They have also shown that a diet of appropriate mixture of amino acids fed to rats maintained on a protein-free diet can also stimulate liver DNA synthesis (Short et al. 1973). Finally we may ask, if metabolic changes can stimulate liver growth, is it not possible that prolonged functional changes resulting in chronic functional demands may result in neoplastic growth? Recently Becker (1971) has advanced this notion for the induction of some hepatomas and plasma cell tumors and reminds us of the well known fact that prolonged hormonal stimulation can lead to neoplasia in a number of endocrine organs.

Thus sufficient evidence is accumulating to suggest that a wide variety of functional changes can trigger liver growth. The question which next arises is how can the pathway that links the functional change to the triggering of growth be unraveled. This will depend on the functional change and the nature of the growth induced. In the case of the coupling of the induction of the drug-metabolizing enzyme system and liver growth, we offer the following suggestion.

On the molecular level growth is largely, but not entirely, reducible to protein synthesis. This is true whether one is concerned with the synthesis of proteins for cell replication, enlargement, or the synthesis of intercellular substances. In the case of drug-induced liver growth, we have already reviewed the evidence that the incearse in drug-metabolizing enzyme activity is due largely to protein synthesis. Therefore to determine the link between the induction of the drug-metabolizing enzyme system and the triggering of liver growth, we must know the steps in the activation of the protein synthetic machinery necessary for the induction of the drug-metabolizing enzyme system. This must include the steps leading to the turning on of the genome and to the synthesis and putting together of the polyribosomes involved in the synthesis of the drug-metabolizing enzyme system. We must then determine how turning on this sequence of events activates the protein synthetic machinery for the general protein synthesis necessary for growth. We have taken our first steps on this road with the study of the changes in the free and bound polyribosomes and the synthesis of the ribosomal RNA during the induction of the drug-metabolizing enzymes and liver growth under conditions in which these processes are temporally separable (Argyris and Heinemann 1973; Hopkinson and Argyris 1972).

The study of the induction of the drug-metabolizing enzymes and its coupled liver growth contributes to our general theories of growth control by presenting one example of the induction of growth in which special growth-controlling substances, be they stimulators or inhibitors, need not be postulated to explain the data. Changes in the levels of chemical substances involved in the maintenance of homeostasis is probably sufficient to explain our results. Triggering of liver growth by feeding of a high-protein diet shows us how changing levels of amino acids in the liver can control liver growth. The induction of liver growth by the administration of drugs suggests how changing levels of drugs, and by inference, changing levels of the naturally occurring substrates for the drug-metabolizing enzyme system, such as steroid hormones and toxic products of digestion and of the bacterial flora in the intestine, can control liver growth. Of course the fact that special growth-controlling substances or inducers are not necessary to explain our results does not guarantee that they do not normally operate in the control of this form of induced growth. The simplest explanation is in all probability the correct one, if and only if, the background of knowledge upon which our new information is tested is adequate and correct. Our background about the relationship of the metabolic changes to the control of growth is so small that we must be careful not to assume with confidence that the simplest explanation is probably the correct one. Moreover even if we are correct in the assumption that no special growth-controlling substances are involved in the induction of growth by the functional changes that we have studied, it does not mean that special growth-controlling substances are not involved in other examples of functional induced growth, such as in erythropoiesis where erythropoietin appears to act as a specific growth stimulator (Adolph 1972; Goss 1964), or in other forms of induced growth such as damage- or tumor-induced growth (Argyris 1968a).

Acknowledgments

The author's research was supported by grants from the National Science Foundation (GB12554), Syracuse University Research Institute, and NIH General Research Support Grant to Upstate Medical Center RR5402.

REFERENCES

Abercrombie, M. 1957. Localized formation of new tissue in an adult mammal. *The biological action of growth substances, Symp. Soc. Exp. Biol.* **11:**235. Academic Press, New York.

Adamson, R. H. and J. R. Fouts. 1961. The metabolism of drugs by hepatic tumors. *Cancer Res.* **21:**667.

Adolph, E. F. 1972. Physiological adaptations, hypertrophies and superfunctions. *Amer. Scientist* **60:**608.

Alvares, A. P., G. R. Schilling and R. Kuntzman. 1968. Differences in the kinetics of benzpyrene hydroxylation by hepatic drug-metabolizing enzymes from phenobarbital and 3-methylcholanthrene-treated rats. *Biochim. Biophys. Res. Commun.* **30:**588.

Argyris, T. S. 1964. Wound healing and the control of growth of the skin. *Advanc. Biol. Skin* **5:**231.

———. 1968a. Growth induced by damage. *Advanc. Morphog.* **7:**1.

———. 1968b. Liver growth associated with the induction of aminopyrine demethylase activity after phenobarbital treatment in adult male rats. *J. Pharmacol. Exp. Therap.* **164:**405.

———. 1969a. Enzyme induction and the control of growth. *Repair and regeneration* (ed. J. E. Dunphy and W. Van Winkle) p. 201. McGraw-Hill, New York.

———. 1969b. Hair growth induced by damage. *Advanc. Biol. Skin* **9:**339.

———. 1971. Additive effects of phenobarbital and high protein diet on liver growth in immature male rats. *Develop. Biol.* **25:**293.

———. 1972. Chalones and the control of normal, regenerative, and neoplastic growth of the skin. *Amer. Zoologist* **12:**137.

Argyris, T. S. and R. E. Heinemann. 1973. Free and bound ribosomes in methylcholanthrene induced liver growth. *Proc. Amer. Ass. Cancer Res.* **14:**6.

Argyris, T. S. and D. L. Layman. 1969. Liver growth associated with the induction of demethylase activity after injection of 3-methylcholanthrene in immature male rats. *Cancer Res.* **29:**549.

Argyris, T. S. and D. R. Magnus. 1968. The stimulation of liver growth and demethylase activity following phenobarbital treatment. *Develop. Biol.* **17:**187.

Argyris, T. S., M. E. Trimble and R. Janicki. 1969. Control of induced kidney growth. *Compensatory renal hypertrophy* (ed. W. Nowinski and R. Goss) p. 45. Academic Press, New York.

Arias, I., D. Doyle and R. T. Schimke. 1969. Studies on the synthesis and degradation of proteins of endoplasmic reticulum of rat liver. *J. Biol. Chem.* **244:**3303.

Augenlicht, L. H. 1971. The effect of phenobarbital and 3-methylcholanthrene on the stimulation of liver growth and the induction of liver 3-methyl-4-monomethylamino-azobenzene demethylase activity in the immature male rat. Thesis, Syracuse University, Syracuse, New York.

Barka, T. and H. Popper. 1967. Liver enlargement and drug toxicity. *Medicine* **46:**103.

Becker, F. 1971. Cell function: Its importance in chemical carcinogenesis. *Fed. Proc.* **30:**1736.

Becker, F. and B. Lane. 1968. Regeneration of the mammalian liver. VI. Retention of phenobarbital-induced cytoplasmic alterations in dividing hepatocytes. *Amer. J. Pathol.* **52:**211.

Bullough, W. S. 1962. The control of mitotic activity in adult mammalian tissues. *Biol. Rev.* (Cambridge) **37:**307.

———. 1965. Mitotic and functional homeostasis: A speculative review. *Cancer Res.* **25:**1683.

Burger, P. C. and P. B. Herdson. 1966. Phenobarbital-induced fine structural changes in rat liver. *Amer. J. Pathol.* **48:**793.

Cantrell, E. and E. Bresnick. 1972. Benzpyrene hydroxylase activity in isolated parenchymal and nonparenchymal cells of rat liver. *J. Cell Biol.* **52:**316.

Conney, A. H. 1967. Pharmacological implications of microsomal enzyme induction. *Pharmacol. Rev.* **19:**317.

Conney, A. H. and A. G. Gilman. 1963. Puromycin inhibition of enzyme induction by 3-methylcholanthrene and phenobarbital. *J. Biol. Chem.* **238:**3682.

Conney, A. H., E. C. Miller and J. A. Miller. 1956. The metabolism of methylated aminoazo dyes. V. Evidence for induction of enzyme synthesis in the rat by 3-methylcholanthrene. *Cancer Res.* **16:**450.

Ernster, L. and S. Orrenius. 1965. Substrate-induced synthesis of the hydroxylating enzyme system of liver microsomes. *Fed. Proc.* **24:**1190.

Fausto, N. 1972. The conversion of orotic acid into uridine-5-monophosphate by isolated perfused normal and regenerating rat livers. *Biochem. J.* **129:**811.

Ferris, G. M. and J. B. Clark. 1972. Early changes in plasma and hepatic free amino acids in partially hepatectomized rats. *Biochem. Biophys. Acta* **273:**73.

Fouts, J. R. and L. A. Rogers. 1965. Morphological changes in the liver accompanying stimulation of microsomal drug metabolizing enzyme activity by phenobarbital, chlordane, benzpyrene or methylcholanthrene in rats. *J. Pharmacol. Exp. Therap.* **147:**112.

Fouts, J. R., R. L. Dixon and R. W. Schultice. 1961. The metabolism of drugs by regenerating liver. *Biochem. Pharmacol.* **7:**265.

Gielen, J. E. and D. W. Nebert. 1971. Aryl hydrocarbon hydroxylase induction in mammalian liver cell culture. I. Stimulation of enzyme activity in nonhepatic cells and in hepatic cells by phenobarbital, polycyclic hydrocarbons, and 2,2-bis(*p*-chlorophenyl)1,1,1-trichloroethane. *J. Biol. Chem.* **246:**5189.

Glazer, R. I. and A. C. Sartorelli. 1971. Induction by phenobarbital of reduced nicotinamide-adenine dinucleotide phosphate (NADPH) cytochrome C reductase in regenerating rat liver. *Biochem. Pharmacol.* **20:**3521.

———. 1972. The effect of phenobarbital on the synthesis of nascent protein on free and membrane-bound polyribosomes of normal and regenerating liver. *Mol. Pharmacol.* **8:**701.

Glaumann, H. 1970. Chemical and enzymatic composition of microsomal subfractions from rat liver after treatment with phenobarbital and 3-methylcholanthrene. *Chem. Biol. Interactions* **2:**369.

Glinos, A. D. 1958. The mechanism of liver growth and regeneration. *The chemical basis of development* (ed. W. D. McElroy and B. Glass) p. 813. The Johns Hopkins Press, Baltimore.

Golberg, L. 1966. Liver enlargement produced by drugs: Its significance. *Proc. Eur. Soc. for the Study of Drug Toxicity* (ed. S. J. Alcock et al.) vol. 7, p. 171. Excerpta Medica Foundation, New York.

Goss, R. 1964. *Adaptive growth.* Academic Press, New York.

Gram, T., A. M. Guarino, F. E. Greene, P. L. Gigon and J. R. Gillette. 1968. Effect of partial hepatectomy on the responsiveness of microsomal enzymes and cytochrome P-450 to phenobarbital or 3-methylcholanthrene. *Biochem. Pharmacol.* **17:**1769.

Henderson, P. T. and J. H. Dewaide. 1969. Metabolism of drugs in isolated rat hepatocytes. *Biochem. Pharmacol.* **18:**2087.

Henderson, P. T. and K. J. Kersten. 1970. Metabolism of drugs during rat liver regeneration. *Biochem. Pharmacol.* **19:**2343.

Hilton, J. and A. C. Sartorelli. 1970. Induction of microsomal drug-metabolizing enzymes in regenerating liver. *Advanc. Enzyme Regul.* **8:**153.

Holtzman, J. 1972. Metabolism of drugs by isolated hepatocytes. *Biochem. Pharmacol.* **21:**581.

Hopkinson, J. and T. Argyris. 1972. Changes in ribonucleic acid metabolism associated with methylcholanthrene-induced adult rat liver growth. *Fed. Proc.* **31:**611.

Japundzic, M., B. Knezevic, V. Djordjevic-Camba and I. Japundzic. 1967. The influence of phenobarbital-Na on the mitotic activity of parenchymal liver cells during rat liver regeneration. *Exp. Cell Res.* **48:**163.

Jick, H. and L. Shuster. 1966. The turnover of microsomal reduced nicotinamide adenine dinucleotide phosphate-cytochrome C reductase in the livers of mice treated with phenobarbital. *J. Biol. Chem.* **241:**5366.

Koransky, W., S. Magour, G. Noack and R. Schulte-Hermann. 1969. Uber den Einfluss induzierender substanzen auf Fremdstoff-oxydasen und andere redoxenzyme der leber. *Naunyn-Schmiedebergs Arch. Pharmakol. Exp. Pathol.* **263:**281.

Kuntzman, R. 1969. Drugs and enzyme induction. *Annu. Rev. Pharmacol.* **9:**21.

Kunz, W., G. Schaude, W. Schmid and M. Seiss. 1966a. Stimulation of liver growth by drugs. I. Morphological analysis. *Proc. Eur. Soc. for the Study of Drug Toxicity* (ed. S. Alcock et al.) vol. 1, p. 113. Excerpta Medica Foundation, New York.

Kunz, W., G. Schaude, H. Schmissek, W. Schmid and M. Siess. 1966b. Stimulation of liver growth by drugs. II. Biochemical analysis. *Proc. Eur. Soc. for the Study of Drug Toxicity,* vol. 7, p. 138. Excerpta Medica Foundation, New York.

Kuriyama, Y., T. Omura, P. Siekevitz and G. E. Palade. 1969. Effects of phenobarbital on the synthesis and degradation of the protein components of rat liver microsomal membranes. *J. Biol. Chem.* **244:**2017.

Lu, A. Y., K. W. Junk and M. J. Coon. 1969. Resolution of the cytochrome P-450 containing ω-hydroxylation system of liver microsomes into three components. *J. Biol. Chem.* **244:**3714.

Lu, A., R. Kuntzman, S. West and A. H. Conney. 1971. Reconstituted liver microsomal enzyme system that hydroxylates drugs, other foreign compounds and endogenous substrates. I. Determination of substrate specificity by the cytochrome P-450 and P-448 fractions. *Biochem. Biophys. Res. Commun.* **42:**1200.

Lu, A. Y., M. Jacobson, W. Levin, S. B. West and R. Kuntzman. 1972. Reconstituted liver microsomal enzyme system that hydroxylates drugs, other foreign compounds, and endogenous substrates. IV. Hydroxylation of aniline. *Arch. Biochem. Biophys.* **153:**294.

Lu, A. Y., W. Levin, S. B. West, M. Jacobson, D. Ryan, R. Kuntzman and A. H. Conney. 1973. Reconstituted liver microsomal enzyme system that hydroxylates drugs, other foreign compounds, and endogenous substrates. VI. Different substrate specificities of the cytochrome P-450 fractions from control and phenobarbital-treated rats. *J. Biol. Chem.* **248:**456.

Miller, A. 1969. Liver growth associated with the induction and reinduction of liver aminopyrine demethylase activity after phenobarbital treatment of immature male rats. Thesis, Syracuse University, Syracuse, New York.

Orrenius, S. and J. L. E. Ericsson. 1966. Enzyme-membrane relationship in phenobarbital induction of synthesis of drug-metabolizing enzyme system and proliferation of endoplasmic membranes. *J. Cell Biol.* **28:**181.

Orrenius, S., J. L. E. Ericsson and L. Ernster. 1965. Phenobarbital-induced synthesis of the microsomal drug-metabolizing enzyme system and its relationship to the proliferation of endoplasmic membranes. *J. Cell Biol.* **25:**627.

Paulini, K., G. Beneke and R. Kulka. 1970. Lebervergrösserung durch phenobarbital in abhangigkeit vom lebensalter. *Beitr. Path.* **141:**327.

Paulini, K., K. Grimmel and G. Beneke. 1971. Die anwendbarkeit der "critical mass" theorie fur die phenobarbitalinduzierte lebervergröBerung. *Beitr. Path.* **142:**129.

Remmer, H. 1970. The role of the liver in drug metabolism. *Amer. J. Med.* **49:**617.

Rogers, L. A., H. P. Morris and J. R. Fouts. 1967. The effect of phenobarbital on drug metabolic enzyme activity, ultrastructure and growth of a "minimal deviation" hepatoma (Morris 7800). *J. Pharmacol. Exp. Therap.* **157:**227.

Schimke, R. T. 1962. Adaptive characteristics of urea cycle enzymes in the rat. *J. Biol. Chem.* **237:**459.

Schimke, R. T., R. Ganschow, D. Doyle and I. M. Arias. 1968. Regulation of protein turnover in mammalian tissues. *Fed. Proc.* **27:**1223.

Schlicht, I., W. Koransky, S. Magour and R. Schulte-Hermann. 1968. Grösse und DNA-Synthese der Leber unter dem einfluss Körperfremder stoffe. *Naunyn-Schmiedeberg's Arch. Pharmakol. Exp. Pathol.* **261:**26.

Schulte-Hermann, R., R. Thom, I. Schlicht and W. Koransky. 1968. Zahl und ploidiegrad der zellkerne der leber unter dem einfluss Körperfremder stoffe. *Naunyn-Schmiedeberg's Arch. Pharmakol. Exp. Pathol.* **261:**42.

Schulte-Hermann, R., I. Schlicht, W. Koransky, C. Lebrel, C. Eulenstedt and M. Zimek. 1972. Selective inhibition of liver-cell proliferation by CFT 1201 and SKF 525A. *Naunyn-Schmiedebergs Arch. Pharmakol. Exp. Pathol.* **273:**109.

Short, J., R. F. Brown, A. Kusakova, J. R. Gilbertson, R. Zemel and I. Lieberman. 1972. Induction of deoxyribonucleic acid synthesis in the liver of the intact animal. *J. Biol. Chem.* **247:**1757.

Short, J., N. B. Armstrong, R. Zemel and I. Lieberman. 1973. A role for amino acids in the induction of desoxyribonucleic acid synthesis in liver. *Biochem. Biophys. Res. Commun.* **50:**430.

Shuster, L. and H. Jick. 1966. The turnover of microsomal protein in the livers of phenobarbital-treated mice. *J. Biol. Chem.* **241:**5361.

Sladek, N. E. and G. J. Mannering. 1969a. Induction of drug metabolism. Differences in the mechanisms by which polycyclic hydrocarbons and phenobarbital produce their inductive effects on microsomal N-demethylating systems. *Mol. Pharmacol.* **5:**174.

———. 1969b. Induction of drug metabolism. II. Qualitative differences in the microsomal N-dimethylating systems stimulated by polycyclic hydrocarbons and by phenobarbital. *Mol. Pharmacol.* **5:**186.

Staubli, W., R. Hess and E. R. Weibel. 1969. Correlated morphometric and biochemical studies on the liver cell. II. Effects of phenobarbital on rat hepatocytes. *J. Cell Biol.* **42:**92.

Sudweeks, A. D. and R. B. Hill, Jr. 1967. Control of liver cell replication by albumin need. *J. Cell Biol.* **34:**404.

Tepperman, H. M. and J. Tepperman. 1967. Comparison of drug-metabolizing and olefin-forming systems in rat liver. *Amer. J. Physiol.* **213:**400.

Watanabe, M., V. R. Potter and H. P. Morris. 1970. Benzpyrene hydroxylase activity and its induction by methylcholanthrene in Morris hepatomas, host livers, in adult livers, and in rat liver during development. *Cancer Res.* **30:**263.

Wattenberg, L. W. 1972. Dietary modification of intestinal and pulmonary aryl hydrocarbon hydroxylase activity. *Toxicol. Appl. Pharmacol.* **23:**741.

Weiss, P. 1955. Specificity in growth control. *Biological specificity and growth* (ed. E. G. Butler) p. 195. Princeton University Press, Princeton, New Jersey.

Wilson, J. T. and J. R. Fouts. 1966. The effect of actinomycin D on post-phenobarbital activity of hepatic microsomal drug-metabolizing enzymes in the rat. *J. Biol. Chem.* **241:**4810.

Early Stages in Estrogen Control of Cell Proliferation

Francesco Bresciani, Giovanni A. Puca, Ernesto Nola, and Vincenzo Sica

Instituzioni di Patologia generale, 1ª Facolta di Medicina e Chirurgia
Università di Napoli, Napoli 80138

One of the main aspects of estrogen action is the regulation of cell proliferation and growth in responsive tissues. In the uterus (endometrium) or mammary gland, to cite two of the most studied targets for estrogen, cell death rate will prevail over cell birth rate after removal of the ovary, with the number of cells in the organ steadily decreasing to a minimum. Upon administration of estrogen to an ovariectomized animal, the contrary will occur, with the increase of cell birth rate brought about by both an increase in the fraction of cells engaged in proliferation (growth fraction) and a speeding up of the process of cell replication (cell cycle) (Bresciani 1964, 1971; Epifanova 1965).

The increase in the number of cells engaged in cell replication upon administration of estrogen to an ovariectomized adult animal (or a prepuberal one) is initially a synchronous phenomenon, with a large number of noncycling cells (G_0 cells) being started and passing through the DNA replication process (S phase) and mitosis at about the same time. This situation lends itself to the study of the mechanism of induction of DNA synthesis and mitosis, and much work (reviewed by Tata 1968; Hamilton 1971; Glasser, Chytil and Spelsberg 1972) has been carried out to establish the sequence of events which takes place in an estrogen-stimulated organ. As shown in Table 1, the basic macromolecular events in the uterus are similar to those occurring in other systems of stimulated DNA synthesis and mitosis (reviewed by Stein and Baserga 1972).

Over the years, scientists interested in the control of cell division have progressively shifted the focus of their interest from mitosis to DNA synthesis, to synthesis of enzymes required for DNA synthesis and so forth, moving backwards as soon as it was realized that the phenomenon under investigation was the result of some previous cellular change. The critical changes occurring with a latency time of a few minutes to one hour after administration of the hormone have been actively investigated, but to date a definite understanding of the mechanism and interrelationship of these early effects has not been reached. We know that an increased synthesis of RNA, including specific mRNA, is among the earliest events registered,

67

Table 1

Sequence of Basic Macromolecular Events in Estrogen-stimulated Uterus

Latency time (hr)[a]		Events
	1.	Estrogen
10^{-2}–10^{-1}	2.	Actinomycin insensitive steps
10^{-1}–1	3.[b]	Increase in chromatin template capacity, DNA-dependent RNA synthesis, RNA polymerase activity, specific protein synthesis, concentration of nuclear acidic proteins
1–10	4.	Increase of protein synthesis in free ribosomes
10–10^2	5.	Synthesis of templates for enzymes related to DNA synthesis
	6.	Marked increase in ribosomal RNA synthesis
	7.	Enzymes for DNA synthesis
	8.	DNA synthesis, histone synthesis
	9.	Mitosis

[a] Order of magnitude

[b] The interrelationship and sequence of events in step 3 are not yet understood.

but the mechanism by which this increased transcription is brought about, and its relationship to other early events, is still elusive.

There is hope of throwing light on the nature of the initial act(s) starting the chain of events leading to cell division by studying the interaction of the estrogen molecule with the cell. By physically following the regulatory molecule into the cell, one may indeed select the cellular molecules involved in the interaction and from their study understand the relevant physiological changes.

Interaction of 17β-Estradiol with Cellular Proteins

Since the discovery that 17β-estradiol is concentrated and retained by the uterus and vagina (Jensen and Jacobson 1962), many laboratories have helped expand and progress in the understanding of the physical basis of this phenomenon. Work in this field has recently been reviewed by Jensen and DeSombre (1972). There appears to be general agreement that the incoming estrogen molecule is bound by cytoplasmic receptor proteins and that the complex, after some modification, enters the nucleus and binds to chromatin.

Although the general lines of 17β-estradiol interaction with the target cell are agreed upon, many questions remain open. They concern the existence of different molecular forms of receptor and their interrelationship, the mechanism of translocation of the protein-bound hormone from cytoplasm to the nucleus, the nature of the chromatin component with which the estrogen-receptor complex interacts and the consequences of this interaction.

Our recent work (Puca et al. 1970; 1971a, b; 1972) has been directed especially toward purification and characterization of cellular estrogen-binding proteins (EB-proteins) with high affinity for 17β-estradiol (K_{ass} in excess of 10^9 liters/mole at $+4°C$). These proteins are also referred to as estrogen receptors. It is hoped that such studies will help find the answer to some of the above questions. The results of our work using calf uterus, a conveniently large source of estrogen-binding proteins, are summarized in Table 2. Starting from fresh uterine cytosol, either not warmed or warmed for 30 minutes at 36°C in the presence of 17β-estradiol, purification procedures always yield not more than three distinct molecular forms.

When Ca^{++} is lacking in cytosol (EDTA added), the purified EB-protein sediments at 8.6S and is 67 Å in Stokes radius when analyzed at salt concentrations equal to or lower than 0.1 M KCl (or NaCl); in contrast at 0.2 M KCl or higher salt concentrations, the sedimentation rate is 5.3S and the Stokes radius 54 Å. Calculation of the molecular weight gives about 240,000 daltons for the 8.6S form and 120,000 for the 5.3S, and it was thus concluded that in low salt two 5.3S molecules associate to give the 8.6S state. From frictional ratio considerations, it derives that the association occurs along the longer molecular axis. Because by changing the salt concentration of the medium one can reversibly produce one or the other of the two forms, we refer to this system as the 5.3–8.6S EB-protein system. However, considering that the 8.6S is stable only at salt concentrations lower than that found in utero (Cole 1950), the in vivo existence of this state is doubtful; it may be the result of dilution of uterine cytosol with low salt solutions.

When Ca^{++} (optimally 4 mM) is also present in high-salt cytosol, the result of purification is a molecule which sediments at 4.5S and is 33 Å in Stokes radius, both in low-salt and high-salt mediums. The molecular weight is about 60,000 daltons and the frictional ratio (1.25) is smaller than those of the 5.3S and 8.6S forms. The electrofocusing pattern of this 4.5S EB-protein consistently shows a double-spiked peak (pI 6.6 and 6.8), suggesting that there may be two proteins of similar molecular weight but slightly different in electrical properties. In crude cytosol formation of the 4.5S protein coincides with disappearance of the 5.3–8.6S forms; experiments with purified proteins have shown beyond doubt that the 4.5S

Table 2

Physical Characteristics of Partially Purified
Estrogen-binding Proteins from Calf Uterus

| | Native | | |
	low salt (< 0.1 M)*	high salt (> 0.2 M)*	Derived
$S_{20, w}$	8.6 ± 0.05	5.3 ± 0.1	4.5 ± 0.01
Stokes radius (Å)	67	54	33
Molecular weight	238,000	118,000	61,000
f/f_o	1.65	1.67	1.25
I.P.	6.2		6.6, 6.8, (7.0)

Data from Puca et al. (1970; 1971a, b; 1972).

Sedimentation coefficients are average ± SE of a number of independent measures by sucrose gradient centrifugation. Stokes radii were obtained by reverse exclusion chromatography on calibrated columns. Molecular weight and frictional ratio were calculated assuming a partial specific volume $\bar{v} = 0.725$ cm³/g. Isoelectric points were measured by electrofocusing. I.P. of native EB-protein in high-salt could not be assessed due to incompatibility of high ionic strength with electrofusing.

* Between 0.1 and 0.2 M salt (KCl, NaCl) both the 8.6S and 5.3S states are present. The ionic strengths refer to 2–4° temperatures; at physiological temperature the ionic strength required for the 5.3S state to exist is expectedly lower.

derives from the latter forms. It is also interesting that the 4.5S protein has much less tendency to aggregate than the larger 5.3–8.6S.

As stated above, when uterus, either in vivo or in vitro, is exposed to 17β-estradiol, the hormone rapidly concentrates in the nucleus. By adding 0.4 M KCl in Tris-HCl (0.1 M) pH 8.5 to an equal volume of uterine nuclear fraction and gently stirring at +4°C for 30–60 minutes, one succeeds in solubilizing protein-bound estradiol from the nuclei. The protein in the complex has the same high affinity for 17β-estradiol as the cytoplasmic receptor proteins. This nuclear estradiol-protein complex sediments at 5–6S when centrifuged as crude extract on a sucrose gradient and is commonly referred to as nuclear "5S" (Jensen et al. 1967; Puca and Bresciani 1968). However after passing through purification procedures, the sedimentation rate of the complex becomes 4.5S, and also by all other molecular properties, including the double-spiked electrofocusing pattern, this nuclear receptor complex is indistinguishable from the cytoplasmic 4.5S EB-protein (Puca, Nola and Bresciani 1970).

The Receptor Transforming Factor (RTF)

The Ca^{++}-induced transformation of 5.3–8.6S EB-protein into 4.5S is mediated by a separate macromolecular factor of cytoplasm, the Receptor Transforming Factor or RTF (Puca et al. 1971a; 1972). This factor is separated from EB-protein by fractional precipitation with $(NH_4)_2SO_4$; its transforming activity is Ca^{++}-dependent and requires salt concentration at which the 5.3–8.6S system is all or mostly in the 5.3S state.

Taking advantage of the difference in isoelectric point between the 4.5S product and the precursor 5.3S EB-protein (see Table 1), we have now devised a fast and reliable method for separating these two proteins. It is based on the finding that DEAE-cellulose (DE-52) at pH 8.3 quantitatively retains both the precursor and the derived EB-proteins, but it releases only the 4.5S product when washed with 0.12 M KCl. By using this method to measure receptor-transforming activity, further studies of molecular and functional properties of RTF were carried out. An up-to-date summary of known properties of RTF is given in Table 3. The factor appears to be a hydrolase acting on peptide bonds and ester bonds involving the carboxyl group of L-arginine, as based on the following results:

1. Several hundredfold purified RTF preparations attack casein, producing acid-soluble, tyrosin-containing material.
2. The proteolytic activity on α-casein and the receptor transforming activity coincide on fractionation by exclusion chromatography, exchange chromatography on DEAE-cellulose, and on electrofocusing.
3. Both activities are Ca^{++}-activated and are destroyed upon incubation with Ca^{++} in the absence of substrate.
4. Both show pH optimum at 8.7 and are not inhibited by phenyl-methyl-sulfonyl-fluoride and di-isopropyl-fluorophosphate.

Trypsin, too, is able to attack the 5.3–8.6S EB-protein, producing EB-fragments sedimenting at about 4.5S (Erdos 1968). However RTF is definitely different from trypsin, not only on the basis of molecular parameters, as shown in Table 3, but also because the product of mild tryptic hydrolysis, although similar in sedimentation rate and molecular weight, shows an electrofocusing pattern distinctly different

Table 3
Receptor Transforming Factor (RTF)

Source calf uterus   *Location* cytosol

Molecular parameters Stokes r. 46 Å $S_{20,w}$ 6.4 MW 115,000 pI 4.9

Specificity and kinetic constants

substrate	K_m	conditions
native EB-protein (5.3S)	1.25×10^{-8}	pH 7.5, 4°C
α-casein*	1.25×10^{-5}	pH 7.5, 22°C
N-benzoylarginine methyl ester	$\sim 2.25 \times 10^{-2}$	pH 7.5, 4°C
L-leucylglycylglycine	not a substrate	pH 7.5, 4°C
N-benzoyltyrosinamide	not a substrate	pH 7.5, 4°C

Hydrolyses peptide bonds and esters at bonds involving the carboxyl group of L-arginine. Hemoglobin, ovoalbumin, plasma albumin are not attacked.

pH optimum 8.7

Ions Ca^{++} is required for activity. Sr^{++} and Mn^{++} may in part substitute for Ca^{++}. Mg^{++} is ineffective. RTF is labile in presence of Ca^{++}.

Inhibitors none known (phenyl-methyl-sulfonylfluoride and di-isopropyl-fluorophosphate have no effect).

Data from Puca et al. (1971a; 1972) and unpublished. Molecular parameters were assessed on partially purified preparation of RTF by the methods listed in legend to Table 2.

* Assuming a MW of 60,000 daltons for α-casein.

from that of the RTF product. Finally, trypsin is active in producing fragments also under low-salt conditions and in the absence of Ca^{++} and, contrary to formation of 4.5S by RTF, formation of tryptic fragments is inhibited by the trypsin inhibitors phenyl-methyl-sulfonylfluoride and di-isopropyl-fluorophosphate.

Even upon lengthy incubation, the final product of RTF action on the 5.3S EB-protein always shows the same molecular characteristics and the same electrofocusing pattern. One may therefore suggest that formation of 4.5S is a case of limited proteolysis by a hydrolase with high affinity for the native 5.3S EB-protein, as shown by the high K_m at $+4°C$. The 8.7 pH optimum of RTF activity excludes a lysosomal origin for this enzyme. The high specificity for its receptor substrate, its nonlysosomal origin, and its limited proteolytic activity on the receptor substrate all speak against the RTF having a generic catabolic role.

A summary of findings regarding molecular transformation and interrelationship among estrogen-receptor complexes in cytosol and nuclear extract is presented in Table 4.

Nuclear Translocation of Cytoplasmic Estrogen Receptor Proteins

It has been proven beyond a reasonable doubt that estrogen receptor proteins such as found in the cytoplasm are unable to penetrate the nucleus if not complexing an estrogen ligand. As previously stated, however, one does not succeed in obtaining both the native 5.3–8.6S and the derived 4.5S EB-proteins from nuclei.

Table 4

A Summary of Findings Regarding Molecular Transformation and
Interrelationship among Estrogen-Receptor Complexes in Cytosol
and Nuclear Extract of Calf Uterus

Cytosol

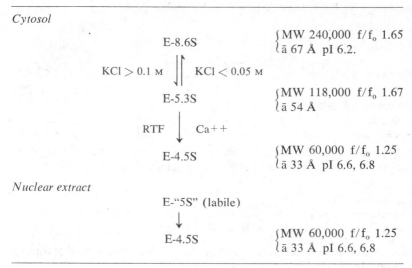

E-8.6S $\begin{cases} \text{MW } 240,000 \ \ f/f_0 \ 1.65 \\ \bar{a} \ 67 \ \text{Å} \ \ \text{pI } 6.2. \end{cases}$

KCl > 0.1 M $\quad$ KCl < 0.05 M

E-5.3S $\begin{cases} \text{MW } 118,000 \ \ f/f_0 \ 1.67 \\ \bar{a} \ 54 \ \text{Å} \end{cases}$

RTF $\quad$ Ca++

E-4.5S $\begin{cases} \text{MW } 60,000 \ \ f/f_0 \ 1.25 \\ \bar{a} \ 33 \ \text{Å} \ \ \text{pI } 6.6, 6.8 \end{cases}$

Nuclear extract

E-"5S" (labile)

E-4.5S $\begin{cases} \text{MW } 60,000 \ \ f/f_0 \ 1.25 \\ \bar{a} \ 33 \ \text{Å} \ \ \text{pI } 6.6, 6.8 \end{cases}$

E: estrogen ligand. RTF: receptor transforming factor.

One obtains a labile "5S" complex which, after partial purification, results in being
the same as cytoplasmic 4.5S (Puca, Nola and Bresciani 1970). Furthermore,
accumulation of nuclear estrogen-receptor complex is a temperature-dependent
process accompanied by a decrease of the higher molecular weight precursor
(Brecher et al. 1967; Gorski et al. 1968).

The above results and the finding of a Receptor Transforming factor may be
included in a comprehensive hypothesis. One may envisage the following sequence
of events: (1) the estrogen enters the cytoplasm and forms a complex with the
native 5.3S EB-protein, which is confined in the cytoplasm possibly because of
being anchored to a structure; (2) the Ca++-activated RT factor acts on the
5.3S-estradiol complex splitting off a fragment, the 4.5S-estradiol complex; (3)
the derivative, mobile 4.5S-estrogen complex penetrates the nucleus; (4) the
4.5S-estradiol complex interacts with chromatin. This hypothesis is part of the
conclusive scheme in Fig. 7. It does not include the 8.6S state as present in the
cell because ionic strength in vivo is higher than that required in vitro for formation
of such a state, as well as of larger aggregates. The hypothesis could include interaction of the 4.5S with other cellular components, to form a faster sedimenting
"5S" either in the cytoplasm before nuclear penetration (Brecher et al. 1970) or
after the 4.5S complex has entered the nucleus. However the labile "5S" complex,
which is unable to stand even mild purification procedures, may well be the result
of unspecific interaction of the 4.5S complex with other molecules that happen to
be present in crude cytosol or nuclear extract. Such an interpretation is strengthened by results presented in the remainder of this paper showing that the 4.5S-
estrogen complex is able to interact specifically with nuclear components without
the necessity of further transformation. For these reasons the "5S" complex is of
doubtful significance and is excluded from our hypothesis.

Interaction of Estrogen-Receptor Complex with Nuclear Components

Identification of the nuclear acceptor site for the estrogen-receptor complex from cytoplasm would furnish important information concerning the mechanism by which nuclear activity relevant to estrogen action is initiated. To date, several reports have been published suggesting that the estrogen-receptor complex interacts with DNA (Toft 1972; Yamamoto and Alberts 1972; Clemens and Kleinsmith 1972; King and Gordon 1972) and one suggesting that the complex interacts with an 80S ribonucleoprotein in the nucleus (Liao, Tehming and Tymoczko 1973). Furthermore there are studies with progesterone receptors pointing to interaction of this receptor-hormone complex with acidic proteins of chromatin (O'Malley et al. 1972).

When the number of acceptor sites on DNA was computed, the result was about 500 such sites per nucleus (King and Gordon 1972). This is only a small fraction of the estrogen receptor molecules in cytoplasm which have been estimated at up to 100,000 per uterine cell (Jensen and DeSombre 1972). Such a limited number of nuclear acceptor sites is at discrepancy with the ability of the nucleus to accept nearly all estrogen-receptor complex from cytoplasm after a large dose of estradiol in vivo (Jensen and DeSombre 1972) or in vitro (Giannopoulos and Gorski 1971). Indeed, experiments with isolated nuclei in vitro show that a nucleus is able to bind at least twice the amount of estrogen-receptor complex in its cytoplasm (Jensen and DeSombre 1972).

We wish to report on some experiments that we have carried out on the problem of nuclear acceptor sites for the estradiol-receptor complex. These studies have been carried out with calf uterus fractions, using the rationale of maintaining in vitro about the same cytoplasm/nuclear ratio as in vivo. Difficulties arising from insolubility of most nuclear components at salt concentrations that do not interfere with stability of estrogen-receptor complex interaction with nuclear structure (0.2–0.4 M KCl) have been circumvented by coupling such nuclear components with CNBr-activated Sepharose.

Fresh calf uterus nuclei were purified according to Spelsberg, Steggles and O'Malley (1971) and fractionated as follows. Twenty volumes of 2 M NaCl in 50 mM phosphate buffer were added to one volume of nuclei, and after mixing by means of an Ultraturrax at low speed, the mixture was stirred for 20 minutes at $+4°C$. Centrifugation for 1 hour at $105,000 \times g$ followed. The sediment was resuspended in Tris-HCl pH 7.5 (10 mM) containing KCl (0.1 M) and EDTA (1 mM), by means of an Ultraturrax at high speed, and labeled *fraction A*. The supernatant was decanted and dialyzed against a large volume of 0.025 M phosphate buffer pH 7.1 until its conductivity became equivalent to 0.1 M NaCl (about 20 hours at $+4°C$) and thereafter centrifuged at $46,000 \times g$ for 20 minutes. The supernatant, containing material still soluble at 0.1 M NaCl, was labeled *fraction B*. The sediment was solubilized by adding 2 M NaCl in 0.025 M phosphate buffer pH 7.1 and labeled *fraction C*. Fractions B and C were coupled to CNBr-activated Sepharose-4B, according to Cuatrecasas (1970). The substituted Sepharoses were repeatedly washed with 2 M NaCl, water and finally incubated in 2 M glycine-NaOH buffer pH 9.4 at room temperature for 2 hours to deactivate nonsubstituted Sepharose. More than 90% of the proteins in fractions remained bound to Sepharose. The experiment was so carried out as to make one gram of packed Sepharose bind about the equivalent of proteins present in nuclei from one gram of fresh uterus.

Denatured calf thymus DNA-Sepharose (1.5 mg/g of packed Sepharose) was prepared according to Poonian, Schlabach and Weissbach (1971).

When fraction A was incubated for one hour at $+4°C$ with crude (whole cytosol) or partially purified (4.5S) preparations of [³H]estradiol-receptor complex and the suspended material was thereafter sedimented by centrifugation, no significant

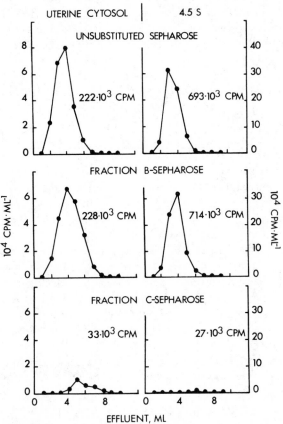

Figure 1

Binding of estrogen-receptor complex by Sepharose. Fraction B and fraction C were prepared, as described in text, from the nuclear fraction of 15 g of fresh calf uterus and coupled to separate batches of CNBr-activated Sepharose-4B by incubation at 4°C for 24 hr. More than 90% of the protein in the fraction was coupled to the Sepharose matrix. After washing (2 M NaCl) and deactivating unsubstituted Sepharose, 3-ml columns were prepared and exhaustively washed with 10^{-2} M Tris-HCl buffer pH 7.5 containing KCl (0.12 M) and EDTA (10^{-3} M). ³H-E₂-labeled cytosol corresponding to 0.5 g of fresh uterus, or labeled 4.5S purified as described elsewhere (Puca et al. 1972) and corresponding to 1.5 g of fresh calf uterus, was filtered through the columns. Unsubstituted, deactivated Sepharose columns were used as control. The same pH 7.5 buffer as above was used as solvent for the samples applied to the columns and for subsequent elution. All operations were carried at $+4°C$ and at the constant flow speed of 15 ml/hr. One-ml fractions were collected and their radioactivity measured. Only fraction C-Sepharose retains estrogen-receptor complexes.

binding of estrogen-receptor complex by the particulate matter was detected; incubation in low (0.05 M) or high molarity (0.4 M) KCl, in the presence or absence of Ca^{++} or Mn^{++} or Mg^{++} always gave the same negative result. Also, no binding was detected by columns of either unsubstituted or fraction B-substituted Sepharose; on the contrary, fraction C Sepharose retained virtually all the estrogen-receptor complex applied, whether it was the crude cytosol preparation or purified 4.5S. These results are presented in Fig. 1. In another experiment, calf thymus DNA-Sepharose was found lacking a significant ability to bind estrogen-receptor complex under the conditions of the present study. This experiment is shown in Fig. 2.

A study of estrogen-receptor complex bound as a function of increasing concentrations of fractions A, B and C is presented according to Schatchard in Fig. 3. Fraction C clearly contains high affinity acceptor sites; from the steeper part of

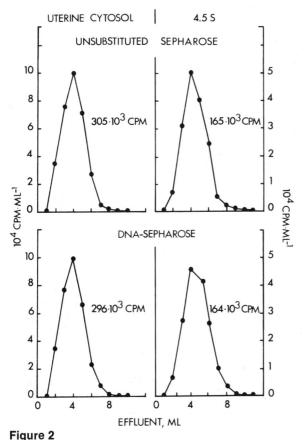

Figure 2

Binding of estrogen-receptor complex by DNA-Sepharose. Calf thymus DNA (Sigma) was heat-denatured and coupled to Sepharose according to Poonian et al. (1971). One and one-half mg of DNA was coupled per ml of packed Sepharose. Unsubstituted, deactivated Sepharose columns were used as control. Except for some differences in the amount of estrogen-receptor complex applied, all other operations were carried out as described in legend to Fig. 1. DNA-Sepharose does not retain significant amounts of estrogen-receptor complexes.

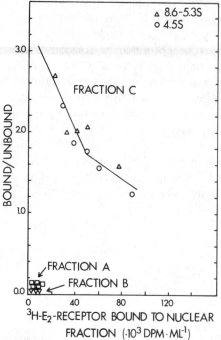

Figure 3

Schatchard plots of purified 4.5S estrogen-receptor complex bound by either calf uterus nuclear fraction A, or B, or C as a function of ligand concentration. For fraction A ($\square$) bound complex was separated from unbound by centrifugation as described in text. Fractions B ($\triangledown$) and C ($\circ$, $\triangle$) were coupled to Sepharose and experimental conditions were as described in legend to Fig. 1, except that the binding was tested batchwise. Different symbols for fraction C represent two separate experiments. Only fraction C contains high-affinity acceptor sites for estrogen-receptor complex.

the curve one obtains a K_{ass} in excess of 10^9 liters/mole of binding sites. Further, from the intercept on the abscissa of the steeper part of the curve, one can compute the concentration of high-affinity acceptor sites per gram of fresh calf uterus to be $4–11 \times 10^{13}$ (range of three experiments).

The following further studies were carried out to improve characterization of the estrogen-receptor binding sites in fraction C. Fraction C-Sepharose was incubated with nucleases or proteolytic enzymes. As shown in Table 5, pronase and trypsin destroy the ability of fraction C to bind estrogen-receptor complex, while DNase and RNase are ineffective. It was also found that hydroxyapatite columns at pH 7.0 do not retain the acceptor sites in fraction C (Fig. 4). On the contrary, carboxymethylcellulose at pH 4.0 (or 6.0) does retain the acceptor sites in fraction C; acceptor activity is not eluted from the column by 0.4 M NaCl but it is eluted by 0.03 N HCl (Fig. 5). These results thus indicate that the acceptor sites in fraction C are protein in nature and that this protein is basic.

Since the estrogen-receptor complex is acidic and the identified acceptor site basic, we have compared fraction C-Sepharose (fraction C was partially purified by carboxymethylcellulose chromatography) with Sepharoses substituted with basic proteins (protamine or an histone fraction) known to interact with estrogen-

Table 5

Ability to Bind $[^3H]E_2$-4.5S Estrogen-Receptor Complex by Sepharose-coupled Fraction C after Incubation with Different Enzymes

Enzyme	cpm bound	% of control
None (control)	103,000	
DNase	90,000	90
RNase	92,000	91
Pronase	4,000	4
Trypsin	10,400	10

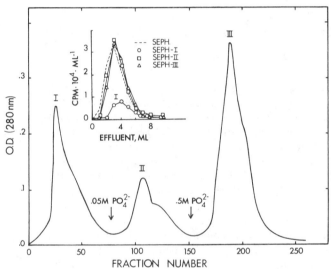

Figure 4

Hydroxyapatite chromatography of fraction C from calf uterus nuclei. A column (2.5 × 45 cm) containing 60 gm of HAP (Biorad) was equilibrated at room temperature with phosphate buffer (10^{-3} M) pH 7.0, containing NaCl (2 M) and urea (5 M). Fraction C from 6 g of uterine nuclear fraction was dissolved in the same buffer as above and applied to the column. Elution was carried out with the same phosphate buffer. At arrows the phosphate concentration of the eluting solution was changed as indicated. OD was continuously monitored. Effluent was collected in 7.4-ml fractions and flow rate was 30 ml/hour at all times. Fraction I (breakthrough peak) contains protein with an aa-acidic/aa-basic ratio of 0.6; in fraction II this ratio is 1.4. Minute amounts of RNA are present in fractions I and II. Fraction III contains nucleic acids. The three fractions were separately collected and, to eliminate urea, extensively dialyzed against 20 mM phosphate buffer pH 7.1 containing NaCl (1 M). The three fractions, HAP-CI, HAP-CII and HAP-CIII, were coupled to CNBr-Sepharose as described in legends to Fig. 1 and 2. Ability of the substituted Sepharoses to bind estrogen-receptor complex was tested and the results are presented in the insert. Only fraction HAP-CI contains acceptor sites for the estrogen-receptor complex.

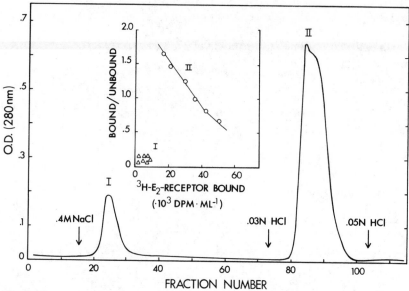

Figure 5

Caboxymethylcellulose chromatography of fraction C from calf uterus nuclei. A column (2.5 × 45 cm) containing 75 g of carboxymethylcellulose (CM-52) was equilibrated at room temperature with 0.1 M acetic acid containing 0.03 N NaOH (pH 4.0). Fraction C from 5 g of calf uterus nuclear fraction was dissolved in 0.2 N HCl, instead of 2 M NaCl. After centrifugation for 30 min at 40,000 × g, the 0.2 N HCl-dissolved material was dialyzed for 24 hr at +4°C against the same acetic acid/NaOH solution as above and applied to the column. Initial elution was carried out with the same solution. At arrows the eluting solution was changed as indicated. OD was continuously monitored. Effluent was collected in 10-ml fractions and flow rate was 50 ml/hr at all times. Fractions CM-CI and CM-CII were collected separately and dialyzed extensively against 20 mM phosphate buffer pH 7.2, containing NaCl (2 M). Equal amounts of protein from the two fractions were coupled with Sepharose as described in text and the ability of the substituted Sepharoses to retain 4.5S estrogen-receptor complex was tested at various ligand concentrations as described in legend to Fig. 3. The results of these last tests are plotted according to Schatchard in the insert. Only fraction CM-CII contains acceptor sites with high affinity for the estrogen-receptor complex.

receptor complex (King, Gordon and Steggles 1970). Furthermore α-casein, an acidic protein, was also found to bind and was tested. The results of two independent batchwise experiments, carried out at 0.1 and 0.4 M KCl and with purified 4.5S estrogen-receptor complex, are plotted according to Schatchard in Fig. 6. From these experiments the following conclusions can be drawn: (1) it is confirmed that the acceptor sites in fraction C have a high affinity for purified 4.5S estrogen-receptor complex; (2) 0.4 M KCl, a salt concentration which is known to solubilize estrogen-receptor complex bound to nuclei, abolishes this high affinity; (3) the fraction of calf thymus histones, salmon sperm protamine and α-casein all bind estrogen-receptor complex, but with far lower affinity than fraction C, the association constant being in their case 10^2 liters per mole at most.

A summary of presently known properties of the nuclear basic acceptor site is

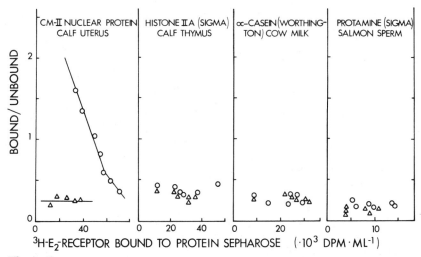

Figure 6

Comparison of binding affinity of proteins known to interact with estrogen-receptor complex. Proteins were coupled to Sepharose. The experiments were carried out batchwise, as described in legend to Fig. 3, and using purified 4.5S complex. The following amounts of Sepharose-coupled proteins were used in each test: carboxymethylcellulose-purified fraction C from calf uterus nuclei, 20 μg; histone IIa, 50 μg; α-casein, 150 μg; protamine, 12 μg. Parallel tests were carried out at 0.1 M ($\circ$) and 0.4 M ($\triangle$) KCl. Although all proteins interact with estrogen-receptor complex, only fraction C shows acceptor sites with high affinity; at 0.4 M KCl this high affinity is lost. K_{ass} of CM-II fraction C at 0.12 M KCl is at least 10^9 liters/mole at $+4°C$. K_{ass} of other proteins is at most 10^2 liters/mole at $+4°C$.

presented in Table 6. Latest data (not shown) indicate that the acceptor does not bind estrogen-free 4.5S receptor.

A peculiarity of the basic acceptor site, described in this paper, is that its capacity is in excess of receptor proteins in the cytoplasm. Free estrogen-binding sites in calf uterus cytosol were computed by us to vary from 5 to 12×10^{12} sites per gram of fresh tissue, a concentration which is about one-tenth that of nuclear basic acceptor sites in the same amount of tissue. This situation is contrary to that of binding by DNA, where the receptor molecules are from 20- to 200-fold in excess of DNA capacity.

Of course such a large concentration of nuclear acceptor sites is difficult to reconcile with specific regulation of the activity of one or a few genes by the estrogen-receptor complex. It rather suggests a broader nuclear involvement. Modification of a large number of nuclear sites by the estrogen-receptor complex from cytoplasm could justify the estrogen-induced increase of template activity of uterine chromatin in ovariectomized rats, which starts within 15 minutes of administration of the hormone, is more than $+30\%$ at 30 minutes and about $+40\%$ at one hour (Glasser, Chytil and Spelsberg 1972). These are large increases, suggesting that a sizable portion of DNA is made available for transcription. Of course such a large burst of synthetic activity could well include, as it seems to do (Glasser et al. 1972), production of specific RNA sequences.

Table 6
Known Properties of the Nuclear Basic Acceptor
Site for Estrogen-Receptor Complex

Source calf uterus nuclear fraction

Chemical properties
   Destroyed by proteases but not nucleases
   Eluted from nuclei by 2 M NaCl in 0.05 M phosphate buffer pH 7
   Present in the precipitate after dilution to 0.1 M NaCl
   Not retained by hydroxyapatite at pH 7.0
   Retained by carboxymethylcellulose at pH 4.0
      Eluted by 0.03 N HCl

Binding activity
   Affinity* for E-4.5S complex: K_{ass} 1–3 $\times$ 10^9 liters/mole at 4° C
   No significant binding of estrogen-free 4.5S
   Binding sites per gm of fresh tissue: 4–11 $\times$ 10^{13}

---

* When acceptor is coupled to CNBr-activated Sepharose

SUMMARY

The early stages of 17β-estradiol interaction with the uterus are (1) association of the incoming hormone with a specific receptor in the cytoplasm, (2) transformation and translocation of the estrogen-receptor complex to the nucleus, and (3) interaction of the complex with chromatin. Within this general framework, we report on progress in the study of recently identified receptor-transforming factor (RTF) of cytoplasm that acts on the native estrogen-receptor complex (5.3S) to produce a smaller complex (4.5S) which is found in the nucleus. The RTF has now been partially purified. It is a protein which sediments at 6.4S, has a Stokes radius of 46 Å, a molecular weight of 115,000 and an isoelectric point of 4.9. It has a high affinity for the native estrogen-receptor complex and shows a Ca^{++}-dependent proteolytic activity on α-casein but not on native or denatured bovine plasma albumin, ovalbumin or hemoglobin. The pH optimum is 8.7. The RTF-induced size decrease of the native estrogen-receptor complex appears to be a case of limited hydrolysis and may be instrumental in the cytoplasm to nucleus translocation of the protein-bound hormone. Furthermore we report on studies on interaction of estrogen-receptor complexes with uterine nuclear components. Difficulty arising from insolubility of most nuclear components at low ionic strength was circumvented by coupling with CNBr-activated Sepharose. Nuclear binding sites have been identified which show an affinity in excess of 10^9 liters per mole (+4°C) for estrogen-receptor complex, but not free receptor, at 0.1 M KCl. All affinity is virtually lost at 0.4 M KCl. These sites are protein in nature, are soluble at 2 M NaCl pH 7.1, but precipitate when KCl concentration is decreased to 0.1 M. They are not retained by hydroxyapatite at pH 7.0 but are retained by CM-cellulose at pH 4.0 or 6.0. Hydrochloric acid 0.03 N is required for their elution from CM-cellulose. It is concluded that these estrogen-receptor-complex acceptor sites consist of basic protein. Protamine, histone IIa (calf thymus), and α-casein coupled to Sepharose also associate with estrogen-receptor complexes but with affinities on the order of 10^2 liters/mole at +4°C at most. All findings to date are included in a

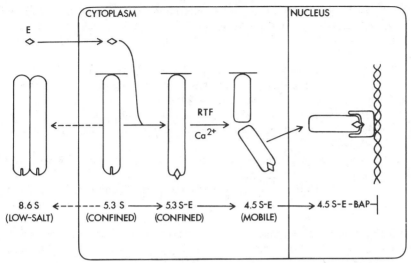

Figure 7

Schematic presentation of hypothesis concerning the early stages of estrogen interaction with a uterine cell. Estrogen (E) enters the cell and interacts with the 5.3S receptor, which is confined to cytoplasm. Confinement to cytoplasm is assumed to depend on association of 5.3S with some cytoplasmic structure. Activation of Receptor Transforming Factor (RTF) by Ca^{++} results in production of a mobile 4.5S-estrogen fragment, which enters the nucleus and specifically interacts with a Basic Acceptor Protein (BAP). The larger 8.6S state, present only in cytosol prepared using low-salt buffers, is excluded from the physiological sequence.

comprehensive hypothesis (Fig. 7) on the early stages of estradiol interaction with a target cell.

Acknowledgment

This paper was written while one of the authors (F.B.) was Visiting Professor at the Department of Pathology, Temple University School of Medicine, Philadelphia, Pennsylvania. F. B. wishes to thank the Chairman of the Department, Dr. Renato Baserga, for his kind hospitality. The research was supported by the Consiglio Nazionale delle Ricerche, Rome.

REFERENCES

Brecher, P. I., R. Vigersky, H. S. Wotiz and H. H. Wotiz. 1967. An *in vitro* system for the binding of estradiol to rat uterine nuclei. *Steroids* **10:**635.

Brecher, P. I., M. Numata, E. R. DeSombre and E. V. Jensen. 1970. Conversion of uterine 4S estradiol-receptor complex to 5S complex in a soluble system. *Fed. Proc.* **29:**249.

Bresciani, F. 1964. DNA synthesis in alveolar cells of the mammary gland: Acceleration by ovarian hormones. *Science* **146:**653.

————— 1971. Ovarian steroid control of cell proliferation in the mammary gland and cancer. *Basic action of sex steroids on target organs*, p. 130. S. Karger, New York.

Clemens, L. E. and L. J. Kleinsmith. 1972. Specific binding of the oestradiol-receptor complex to DNA. *Nature New Biol.* **237:**204.

Cole, D. F. 1950. The effects of oestradiol on the rat uterus. *J. Endocrinol.* **7:**12.

Cuatrecasas, P. 1970. Protein purification by affinity chromatography. Derivations of agarose and polyacrylamide beads. *J. Biol. Chem.* **245:**3059.

Epifanova, O. I. 1965. Hormones and the reproduction of cells. (Israel Program for Scientific Translations, Ltd. and the National Science Foundation, Washington, D.C.) Translated from Russian.

Erdos, T. 1968. Properties of a uterine oestradiol receptor. *Biochem. Biophys. Res. Commun.* **32:**338.

Giannopoulos, G. and J. Gorski. 1971. Estrogen receptors. Quantitative studies on transfer of estradiol from cytoplasmic to nuclear binding sites. *J. Biol. Chem.* **246:** 2524.

Glasser, S. R., F. Chytil and T. C. Spelsberg. 1972. Early effects of estradiol-17β on the chromatin and activity of DNA-dependent RNA polymerase (I and II) of rat uterus. *Biochem. J.* **130:**947.

Gorski, J., D. Toft, G. Shyamala, D. Smith and A. Notides. 1968. Hormone receptors: Studies on the interaction of estrogen with the uterus. *Recent Progr. Hormone Res.* **24:**45.

Hamilton, T. H. 1971. Control by estrogen of genetic transcription and translation. *Basic action of sex steroids on target organs*, p. 56. S. Karger, New York.

Jensen, E. W. and E. R. DeSombre. 1972. Mechanism of action of female sex hormones. *Annu. Rev. Biochem.* **41:**789.

Jensen, E. V. and H. I. Jacobson. 1962. Basic guides to the mechanism of estrogen action. *Recent Progr. Hormone Res.* **18:**387.

Jensen, E. V., D. J. Hurst, E. R. DeSombre and P. W. Jungblut. 1967. Sulfhydryl groups and estradiol-receptor interaction. *Science* **158:**385.

King, R. J. B. and J. Gordon. 1972. Involvement of DNA in the acceptor mechanism for uterine oestradiol receptor. *Nature New Biol.* **240:**185.

King, R. J. B., J. Gordon and A. W. Steggles. 1970. Receptor-polycation interaction in relation to oestradiol binding in uterus. *Exc. Med. Int. Congr. Series* **219:**394.

Liao, S., L. Tehming and J. L. Tymoczko. 1973. Ribonucleoprotein binding of steroid-"receptor" complexes. *Nature New Biol.* **241:**211.

O'Malley, B. W., T. C. Spelsberg, W. T. Schrader, F. Chytil and A. W. Steggles. 1972. Mechanism of interaction of hormone-receptor complex with the genome of a eukaryotic target cell. *Nature* **235:**141.

Poonian, S. M., A. J. Schlabach and A. Weissbach. 1971. Covalent attachment of nucleic acid to agarose. *Biochemistry* **10:**4241.

Puca, G. A. and F. Bresciani. 1968. A receptor molecule for estrogens from rat uterus. *Nature* **218:**967.

Puca, G. A., E. Nola and F. Bresciani. 1970. Partial purification and preliminary characterization of an estrogen binding protein (estrogen receptor) from calf uterus nuclear fraction. *Research on steroids,* vol. 4, p. 319. Pergamon Press, Oxford.

Puca, G. A., E. Nola, V. Sica, and F. Bresciani. 1971a. Studies on isolation and characterization of estrogen binding proteins of calf uterus. *Advances in the biosciences,* vol. 7, p. 97. Pergamon Press, Oxford.

—————. 1971b. Estrogen binding proteins of calf uterus. Partial purification and preliminary characterization of two cytoplasmic proteins. *Biochemistry* **10:**3769.

————— 1972. Estrogen binding proteins of calf uterus. Interrelationship between various forms and identification of a receptor-transforming factor. *Biochemistry* **11:**4157.

Spelsberg, T. C., A. W. Steggles and B. W. O'Malley. 1971. Progesterone-binding components of chick oviduct. III. Chromatin acceptor sites. *J. Biol. Chem.* **246:**4188.

Stein, G. and R. Baserga. 1972. Nuclear proteins and the cell cycle. *Advanc. Cancer Res.* **15:**287.

Tata, J. R. 1968. Hormone regulation of growth and protein synthesis. *Nature* **219:**331.

Toft, D. 1972. The interaction of uterine estrogen receptors with DNA. *J. Steroid Biochem.* **3:**515.

Yamamoto, K. R. and B. M. Alberts. 1972. *In vitro* conversion of estradiol-receptor protein to its nuclear form: Dependence on hormore and DNA. *Proc. Nat. Acad. Sci.* **69:**2105.

Molecular Mechanisms of Steroid-mediated Alterations in Target Cell Growth and Differentiation: Induction of Specific Messenger RNAs

Bert W. O'Malley, Lawrence Chan, Stephen E. Harris,
John P. Comstock, Jeffrey M. Rosen, and Anthony R. Means

Department of Cell Biology, Baylor College of Medicine
Texas Medical Center, Houston, Texas 77025

The manner by which steroid hormones regulate growth and differentiation of target tissues has been the topic of numerous investigations. One model system which has been particularly useful in this regard is the oviduct of the immature chick (O'Malley et al. 1969; Means and O'Malley 1972). Administration of estrogen to these animals results in a rapid and pronounced increase in mitosis, DNA synthesis and consequent cytodifferentiation (Kohler, Grimley and O'Malley 1969; Oka and Schimke 1969; Socher and O'Malley 1973). Moreover one of the new cell types that appears will synthesize large quantities of the specific protein ovalbumin. This protein is easily quantitated by chemical and immunochemical techniques and serves as an excellent biochemical marker for hormone-induced differentiation of oviduct cells. Thus estrogenic steroid hormones regulate both biochemical and morphological differentiation of the oviduct in a highly ordered fashion. In this same tissue a second steroid hormone (progesterone) acts as a modulator of oviduct function by causing induction of another specific protein (avidin) in the absence of readily demonstrable influences on cell division or hypertrophy.

Our laboratory has previously generated a series of indirect experiments consistent with the suggestion that both estrogenic and progestational steroids act in the oviduct to alter gene transcription in a manner which leads to the production of mRNA for the specific proteins ovalbumin and avidin, respectively. Actinomycin D blocks these steroid hormone-mediated responses. Furthermore administration of these hormones has resulted in quantitative changes in rapidly labeled nuclear RNA and RNA polymerase (McGuire and O'Malley 1968) and qualitative changes in the populations of newly synthesized nuclear RNA as measured by nearest-neighbor dinucleotide analysis and DNA-RNA hybridization (O'Malley et al. 1969; O'Malley and McGuire 1968; Liarakos, Rosen and O'Malley 1973). Our present methods now allow a direct assessment of tissue mRNA levels. The present study will demonstrate that hormone-induced increases in differentiation-specific oviduct proteins are temporally preceded by increased intracellular concentrations of the specific mRNAs coding for these proteins.

85

MATERIALS AND METHODS

Animals

Seven-day-old female Rhode Island Red Chicks (unstimulated/undifferentiated) received daily subcutaneous injections of 5 mg diethylstilbestrol (DES) in sesame oil. To produce differentiated oviducts for restimulation experiments, chicks received daily injections of 5 mg DES for two weeks; they were then withdrawn from all hormones for two weeks before again receiving a single subcutaneous injection of 1 mg DES. In experiments using progesterone, the steroid was administered as a single injection to chicks pretreated with DES (5 mg/day) for 12 days. Chicks were killed by cervical dislocation at various times after this injection as described in the text and figure legends.

Assay of Ovalbumin and Avidin

Methods for determining tissue levels of ovalbumin (Chan, Means and O'Malley 1973) and avidin (O'Malley 1967) have been previously described.

In Vitro Labeling Conditions

The magnum portions of the oviducts were removed and cut into 5–10 mg pieces. The tissue was then transferred to a 25-ml Erlenmyer flask and incubated in 2 ml of Medium 199 without amino acids, but containing 1.2 mg per ml of $NaHCO_3$ and 5 μCi of L-[^{3}H]lysine (sp. act. 19 Ci/mmole) (Amersham-Searle Corporation). Incubation was at 37°C in the presence of 95% O_2–5% CO_2 and flasks were shaken continuously. After 2 hours of incubation, the tissues were removed from the medium, weighed, and a 5% (w/v) homogenate was prepared in 10 mM sodium phosphate, 15 mM sodium chloride (pH 7.5). The homogenate was then centrifuged at 105,000 x g for 60 minutes at 4°C. The supernatant fluid was used for determination of radioactive ovalbumin and trichloracetic acid (TCA) precipitable radioactivity (Chan et al. 1973).

Isolation of Partially Purified Oviduct mRNA

Total oviduct nucleic acid was prepared from chicks at various times after hormone administration by methods reported previously (Means et al. 1972; Rosenfeld et al. 1972a, b). The nucleic acid preparations were treated with DNase (electrophoretically pure, Worthington Biochem.) at a concentration of 2 μg/ml in 2 mM Mg^{++}-acetate, 10 mM Tris acetate pH 7.0 for 60 minutes at 4°C. The mixture was then again extracted with phenol and precipitated from ethanol. The DNase-treated RNA was then filtered on nitrocellulose membrane filters (Millipore) and eluted off the filters as previously described (Rosenfeld et al. 1972b; Brawerman, Mendecki and Lee 1972). The oviducts from about 30 chicks were grouped together and used for each RNA preparation.

Specific Protein Synthesis in the Rabbit Reticulocyte Lysate System

The assay contained in a final volume of 0.5 ml: rabbit reticulocyte lysate (1:2, v/v), 0.2 ml; ATP (pH 7.0), 1.0 mM; GTP (pH 7.0), 0.2 mM; phosphoenol pyruvate (pH 7.0), 7.5 mM; pyruvate kinase, 0.3 IU; $MgCl_2$, 2.0 mM; Tris-HCl (pH 7.4 at 23°C), 20 mM; KCl, 100 mM; 20 μM each of 19 amino acids (minus valine); L-[^{14}C]valine, 10 μM (572 dpm/pmole) and 5–10 μg of the filtered nucleic acid. Incubation was for 120 minutes at 25°C. Aliquots (25 μl) were removed from each tube, precipitated in trichloracetic acid and counted by the standard procedures (Means et al. 1972); 400 μl aliquots were used to measure the radioactive ovalbumin or avidin synthesized by immunoprecipitation methods as previously described (Means et al. 1972; O'Malley et al. 1972).

RESULTS

Translation of Ovalbumin mRNA in a Rabbit Reticulocyte Lysate System

The only definitive way to prove the existence of ovalbumin mRNA was to demonstrate that it would support the unambiguous synthesis of ovalbumin on heterologous ribosomes under cell-free conditions. In order to accomplish this goal we chose to use a protein synthesis system derived from rabbit reticulocyte lysate. Indeed it was possible to demonstrate the translation of ovalbumin mRNA in this system (Means et al. 1972). Proof that the reaction product was authentic ovalbumin has been gained by several procedures: (1) interaction with a specific antiserum against purified ovalbumin; (2) solubilization of the immunoprecipitate and analysis on SDS acrylamide gels; (3) ion-exchange chromatography on carboxymethylcellulose followed by reprecipitation with antiovalbumin; and (4) peptide maps (Means et al. 1972; Rhoads, McKnight and Schimke 1971).

Table 1 shows that the ovalbumin mRNA activity is specific for RNA isolated from oviduct of estrogen-stimulated chicks. Messenger RNA activity is primarily found in the 8–18S fraction of polysomal RNA and can also be detected in RNA extracts of whole oviduct homogenates. The amount of synthesis is increased by addition of protein synthesis initiation factors. Moreover inhibitors of chain initiation, such as edeine or aurintricarboxylic acid, or of general protein synthesis, such as puromycin or cycloheximide, completely block ovalbumin synthesis directed by the oviduct mRNA fraction. Ribonuclease destroys the messenger activity, whereas deoxyribonuclease has no effect. Neither steroid hormones in absence or presence of specific receptors, nor cyclic AMP have any direct effect on the in vitro translation of ovalbumin mRNA.

Estrogen Induction of Oviduct Growth and Synthesis of Ovalbumin

Continuous administration of estrogenic hormone to unstimulated chicks results in a 500- to 1000-fold increase in oviduct mass during a 3- to 4-week period. The magnum of the oviduct is unique in its responses to estrogen and upon administration of this steroid undergoes marked cytodifferentiation. Eventually three distinct types of epithelial cells differentiate from the primitive cells of the immature oviduct mucosa (Kohler, Grimley and O'Malley 1969; Oka and Schimke 1969). Two of these cell types synthesize cell-specific proteins, which can be readily measured by biochemical and immunochemical techniques and therefore can be used as markers for the differentiation process. Thus tubular gland cells synthesize the major egg-white proteins such as ovalbumin, conalbumin, ovomucoid and lysozyme (O'Malley et al. 1969; Palmiter 1972), whereas the goblet cells synthesize avidin in response to administration of another steroid hormone, progesterone (O'Malley et al. 1969; Kohler et al. 1969; Oka and Schimke 1969).

Prior to hormone administration, no ovalbumin can be detected in oviduct cells. Following an injection of estrogen, parasynchronous waves of mitosis occur

Table 1

Ovalbumin Synthesis in Reticulocyte Lysate: Specificity
of mRNA and Effects of Various Compounds

Addition to lysate	Ovalbumin synthesized (cpm)
None	0
Total brain RNA	0
Total liver RNA	0
0–4S oviduct RNA	0
19–28S oviduct RNA	165
8–18S oviduct RNA	5410
Total oviduct RNA	6952
Total oviduct RNA + edeine	327
Total oviduct RNA + aurintricarboxylic acid	388
Total oviduct RNA + puromycin	233
Total oviduct RNA + cycloheximide	361
Total oviduct RNA + RNase	390
Total oviduct RNA + DNase	6765
Total oviduct RNA + 17β-estradiol	6481
Total oviduct RNA + estrogen receptor	6663
Total oviduct RNA + progesterone	6365
Total oviduct RNA + progesterone receptor	6870
Total oviduct RNA + cyclic AMP	7170
Total oviduct RNA − initiation factors	3386

Total nucleic acid preparations were extracted from hen brain, liver and
oviduct and partially purified on Millipore filters as previously described
(Rosenfeld et al. 1972b; Means et al. 1972). Oviduct polyribosomes were
also prepared and the RNA fractions obtained by previously published
methods (Means et al. 1972). Each RNA sample was assayed for ovalbumin mRNA content in a reticulocyte lysate translation system (Means et
al. 1972). Incubation was for 2 hours at 25°C. In most assays the lysate
was supplemented by addition of a rabbit reticulocyte initiation factor
preparation (Means et al. 1972). When fractions of polysomal RNA were
utilized, 50 μg samples were added to the translation system. All total RNA
assays contained 10 μg of the Millipore filter-treated RNA preparation.
Other compounds were added and were present at the following final concentrations: edeine, 7 μM; aurintricarboxylic acid, 10^{-4} M; puromycin,
10^{-3} M; cycloheximide, 10^{-3} M; RNase, 0.1 μg per ml; DNase, 5 μg/ml;
17β-estradiol, 10^{-9} M; estrogen receptor, 10 μl of a 35% $(NH_4)_2SO_4$ cut
of oviduct cytosol prelabeled with 10^{-9} M 17β-estradiol; progesterone,
10^{-9} M; progesterone receptor, 10 μl of a 35% $(NH_4)_2SO_4$ cut of oviduct
cytosol prelabeled with 10^{-9} M progesterone; and cyclic AMP, 10^{-6} M.
Background radioactivity in the immunoprecipitate obtained in the absence
of added mRNA ranged from 200–300 cpm and has been subtracted from
the values shown.

every 18 hours (Socher and O'Malley 1973). At 24–48 hours of estrogen stimulation, ovalbumin can be detected and intracellular concentration of this protein
increases progressively over the next 12 days. In addition when ovalbumin and
ovalbumin mRNA are measured in the same tissue samples during continuous
estrogen stimulation (Fig. 1), a striking correlation over a period of 0 to 17 days
can be demonstrated between ovalbumin accumulation in the estrogen-stimulated
oviduct and its mRNA activity (Comstock et al. 1972). This relationship requires

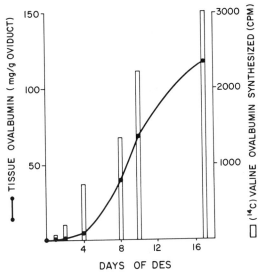

Figure 1
Effects of estrogen on ovalbumin mRNA activity and tissue concentrations of ovalbumin. Immature chicks received daily injections of estrogen. Tissue levels of ovalbumin were determined immunochemically. Messenger RNA was partially purified on Millipore filters and assayed for its ability to promote the synthesis of ovalbumin in a heterologous protein synthesis system. From Comstock et al. (1972) with permission of *Proc. Nat. Acad. Sci.*

clarification since during this same time of differentiation and growth there are other dramatic changes occurring, particularly in the cellular content of total ribonucleic acids. In order to better understand the hormonal regulation of ovalbumin synthesis, we next used oviduct slices from chicks which had been withdrawn from estrogen for 14 days and then killed at various times following readministration of a single dose of this steroid.

Rate of Ovalbumin Synthesis in Oviduct Tissue

At 2 weeks after estrogen withdrawal, a basal level of synthesis of ovalbumin has been shown to exist but at a very low rate (Chan et al. 1973). However, when similar experiments were repeated with liver slices, there were no counts in the immunoprecipitate, indicating that the small basal rate of synthesis of ovalbumin in chicks withdrawn from estrogen for 2 weeks is real and not an artifact of the assay system. Nevertheless within 3 hours following a single 1 mg injection of DES, there is a definite increase in the rate of ovalbumin synthesis (Fig. 2). The rate reaches a peak at 18–20 hours and returns to baseline by 36 hours. The rate of synthesis of total oviduct protein in the same tissue reaches a maximum slightly earlier, at 12 hours. Moreover there is a preferential synthesis of ovalbumin over other proteins since the rate of the former increases about 10-fold, whereas that for total protein only increases about 4-fold. The increased rate of incorporation of L-[^{3}H]lysine after DES administration has been shown to result from an absolute increase in the rate of synthesis of ovalbumin rather than from changes in the L-lysine pool size within the protein synthesizing compartment (Chan et al. 1973).

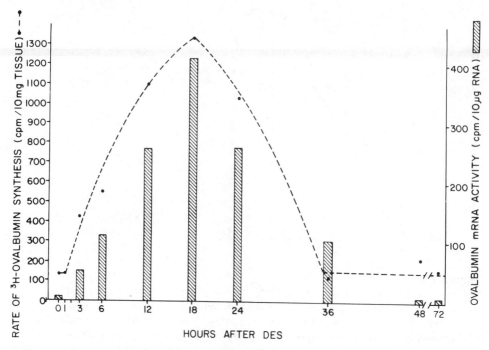

Figure 2

Effect of estrogen on levels of ovalbumin mRNA activity as compared to the rate of ovalbumin synthesis in oviduct. Withdrawn chicks were given a single subcutaneous injection of 1 mg DES in sesame oil at zero time. For each time point after the injection 30 chicks were sacrificed, and total RNA was extracted and filtered as previously described (Rosenfeld et al. 1972a,b; O'Malley et al. 1972); mRNA activity was assayed in the reticulocyte lysate system (Means et al. 1972). Each reaction tube contained 10 μg exogenous mRNA and the components described in Methods, using [^{14}C]valine as a label. mRNA activity is represented by the bars and rate of protein synthesis (L-[^{3}H]lysine incorporation/10 mg tissue/2 hr) is superimposed (dashed line) for comparison. The background radioactivity in the mRNA assays ranged from 200–250 cmp and has been subtracted from the values shown. From Chan et al. (1973) with permission of *Proc. Nat. Acad. Sci.*

Estrogen Effects on Ovalbumin mRNA Levels

As shown above (Fig. 1) a good correlation exists between ovalbumin mRNA activity and ovalbumin synthesis and accumulation after repeated injections of DES (Comstock et al. 1972). In the experiment shown in Fig. 2, the early kinetics of the induction of ovalbumin mRNA after a single injection of DES was examined by assaying total extractable oviduct mRNA in the reticulocyte assay system. Significant amounts of ovalbumin mRNA activity were not detected at 0 hour. Messenger RNA activity was first demonstrable at 3 hours after a single injection of 1 mg DES. This activity increases with time after the injection and peaks at approximately 18 hours. Thereafter it again declines to undetectable levels by 48 hours (Chan et al. 1973).

An excellent correlation exists between extractable mRNA activity as directly quantified by the reticulocyte cell-free system and intracellular mRNA activity as reflected by the rate of incorporation of L-[³H]lysine into ovalbumin (Fig. 2). There is some variation in the response curve which is apparently dependent on the dose of DES. Thus larger doses of estrogenic compounds produce a more extended curve since tissue levels of hormone remain elevated longer. An estimate of the half-life for ovalbumin mRNA can be made from the slope of mRNA activity fall-off shown in Fig. 1 and 2. One-half of the activity has disappeared by 8–10 hours following the peak activity at 18 hours, suggesting an initial 8–10 hr $T_{1/2}$ for ovalbumin mRNA. This apparent $T_{1/2}$, however, probably represents an overestimation since a small amount of continued synthesis of ovalbumin mRNA may well be occurring on the descending limb of the activity curve. More recent experiments have suggested that these specific mRNAs may have complex decay curves, as a subpopulation of mRNA appears to have an extended $T_{1/2}$.

Synthesis of a DNA Sequence Complementary to Ovalbumin mRNA

Ovalbumin messenger RNA was partially purified from total oviduct nucleic acid isolated from estrogen-stimulated hens by sequential adsorption of poly(A)-containing mRNA to Millipore filters, fractionation on sucrose gradients, and readsorption of an 18S fraction to Millipore filters. This material was tested in a heterologous cell-free protein synthesizing system and shown to contain a high concentration of ovalbumin mRNA activity. The final RNA fraction was enriched approximately 100-fold for ovalbumin mRNA activity and poly(A) content, compared to the total nucleic acid extract. The mRNA was incubated in an in vitro reaction system with RNA-directed DNA polymerase isolated from avian myeloblastosis virus (AMV) to produce a complementary DNA (cDNA) copy. The reaction system containing mRNA, ³H-labeled deoxyribonucleotides, and actinomycin D was dependent on the addition of oligo-dT primer and completely inhibited by the addition of RNase. The [³H]cDNA produced, analyzed on alkaline sucrose gradients, ranged in size from 60 to 1600 nucleotides. The average size of the product was 220 nucleotides. When the [³H]cDNA was reacted with excess ovalbumin mRNA, 90% of the [³H]cDNA formed a stable hybrid with the ovalbumin mRNA, indicating that the [³H]cDNA was indeed a complementary copy of the mRNA. A fraction of this [³H]cDNA, isolated from the alkaline sucrose gradients containing approximately 400 nucleotides, was then used in a DNA excess reannealing experiment. Whole chick oviduct DNA, sheared to 400 nucleotides, was mixed with a small amount of the [³H]cDNA (unlabeled DNA:[³H]cDNA ratio of 4×10^6:1) and reannealing experiments were then carried out at 60°C in 0.14 M phosphate buffer, pH 6.8. At a DNA C_0t of approximately 5000, greater than 70% of the [³H]cDNA formed stable duplexes with the excess sheared, denatured oviduct DNA. Complementary [³H]DNA reannealed with chick DNA with an apparent $C_0t_{1/2}$ of 480 (Fig. 3). Under these conditions single copy (unique sequence) DNA reassociated with a $C_0t_{1/2}$ of approximately 460. Thus the kinetics of reassociation of the complementary [³H]DNA suggests that the ovalbumin gene is not amplified but rather is present only once in the chick genome. These results have important regulatory implications in that under conditions of estrogen stimulation, approximately 65% of the total protein synthesized by oviduct is ovalbumin. These data then suggest that estrogen may act at the level of transcription to stimulate production of numerous mRNA copies from a single gene. This ultimately results in a high intracellular concentration of both ovalbumin mRNA and subsequently ovalbumin itself.

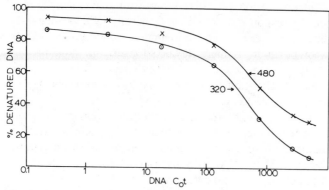

Figure 3

Reannealing of [³H]cDNA with unlabeled chick DNA. A complementary [³H]DNA copy of the partially purified ovalbumin mRNA was synthesized using AMV RNA-directed DNA polymerase. Reannealing of the [³H]cDNA with a vast excess of unlabeled sheared chick oviduct DNA (4×10^6/liter) was performed in 0.14 M phosphate buffer, pH 6.8 at 60°C in sealed 1.0-ml ampules. The reaction was terminated by diluting the samples 1:4 into deionized H_2O to yield a phosphate concentration of 0.03 M and samples were fractionated on hydroxylapatite columns into single- and double-stranded DNA. The A_{260} (o) and radioactivity (X) of each fraction were determined.

Progesterone Effects on Avidin mRNA Levels

Following estrogen-mediated growth and differentiation, the oviduct responds to progesterone in a more limited and specific manner, i.e., the induction of synthesis of the specific protein, avidin. Figure 4 shows the results of experiments in which estrogen-pretreated chicks were given a single injection of progesterone at zero time and levels of avidin mRNA and oviduct content of avidin protein were then quantified at various time points over the next 240 hours. A derivative plot of avidin accumulation is shown in Fig. 4 to reflect the endogenous rate of avidin synthesis. Neither avidin mRNA nor avidin itself can be detected in oviduct tissue prior to administration of progesterone. Following injection of the steroid, messenger RNA was first detected at 6 hours, while accumulation of avidin protein was noted at 6–12 hours. Avidin mRNA levels increase sharply prior to the peak rate of avidin synthesis. Intracellular accumulation of avidin continues until 36–48 hours, at which time avidin mRNA is less than 50% of the peak level at 18–24 hours. Disappearance of mRNA for avidin is complex and probably reflects at least two populations of intracellular mRNA. Avidin was again unmeasurable at 120 and 240 hours and correspondingly, avidin mRNA was also undetectable.

DISCUSSION AND CONCLUSION

Inhibitors of RNA synthesis at appropriate concentrations can block the induction of specific oviduct proteins by steroid hormones (O'Malley 1967). It has always been assumed from these experiments that RNA synthesis is required for the process. We have now reported that in the chick oviduct estrogen induces the appearance of ovalbumin mRNA activity which parallels in vivo ovalbumin ac-

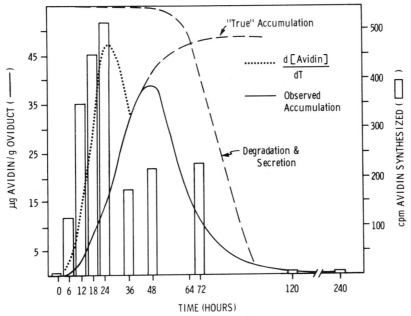

Figure 4

Effect of progesterone on levels of avidin mRNA activity and accumulation of avidin in oviduct. Estrogen-pretreated chicks (12 days) were given a single subcutaneous injection of 1 mg progesterone in sesame oil at zero time. For each time point 10 chicks were sacrificed and all but 2 oviducts were used for the extraction and partial purification of avidin mRNA (O'Malley et al. 1972); the mRNA activity for avidin was assayed in the reticulocyte lysate system (Means et al. 1972; Rosenfeld et al. 1972a,b). Avidin was measured in a 105,000 g supernatant fluid prepared from the remaining 2 oviducts by the [^{14}C]biotin assay (O'Malley 1967; Korenman and O'Malley 1968). mRNA activity is shown by the bars and tissue accumulation of avidin is represented by the solid and dotted lines. Data represent average values from three separate experiments.

cumulation (Fig. 1), and withdrawal of estrogen treatment results in disappearance of ovalbumin mRNA activity. The present communication also documents the appearance of ovalbumin mRNA activity within 3 hours of a single estrogen injection (Fig. 2). The changes in mRNA activity correlate well with the changes in the rate of ovalbumin synthesis. These data indicate that the intracellular accumulation of mRNA for ovalbumin is the rate-limiting factor in the estrogen induction of this protein in the chick oviduct. Furthermore the experiments employing a [^{3}H]DNA probe complementary to ovalbumin mRNA indicate that estrogen acts to cause repeated transcription of a (single) gene to produce the large intracellular concentrations of ovalbumin (65% of total protein) found in oviduct cells.

Similar conclusions were drawn from experiments performed to determine the mode of action of another steroid hormone, progesterone. We have previously reported the de novo induction of synthesis of the specific oviduct protein avidin in response to progesterone administration (O'Malley et al. 1969; O'Malley 1967). This response is of considerable interest because progesterone results in very little change in net cellular RNA or protein synthesis (O'Malley 1967; Korenman and

O'Malley 1968). This report reveals that following administration of the steroid hormone and coincident with detectable avidin synthesis, the specific mRNA for avidin can be shown to appear in the tissue.

The female sex steroids can exert regulatory effects on cell division and growth in target tissues. They may do so by regulating the synthesis, activity, and possibly even the degradation of tissue enzymes and structural proteins. Each response, nevertheless, appears to be dependent on the synthesis of nuclear RNA. In many instances, the steroid actually promotes a qualitative change in the base composition and sequence of the RNA synthesized by the target cell, implying a specific effect on gene transcription. Most important, we now have direct quantitative evidence that sex steroids cause a net increase in the intracellular levels of specific messenger RNA molecules in target tissues.

It thus appears that we see an evolving pattern of steroid hormone action (Fig. 5) consisting of

1. uptake of the hormone by the target cell and binding to a specific cytoplasmic receptor protein;
2. transport of the steroid-receptor complex to the nucleus;
3. binding of this "active" complex to specific "acceptor" sites on the genome (chromatin DNA and nonhistone acidic protein);
4. activation of the transcriptional apparatus resulting in the appearance of new RNA species, including specific mRNAs;
5. transport of the hormone-induced RNA to the cytoplasm, resulting in synthesis of new proteins on cytoplasmic ribosomes;
6. the occurrence of the specific steroid-mediated "functional response" characteristic of that particular target tissue.

A final objective in elucidation of the mechanism of steroid hormone action must involve a careful delineation of the biochemistry of the transfer of information held

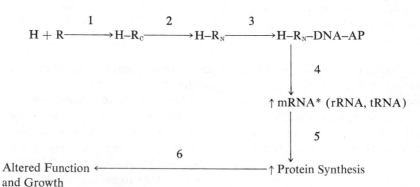

Figure 5

Sequence of biochemical events leading to steroid hormone-induced alterations in target tissue function and growth. See text for description.

by the steroid hormone-receptor complex to the nuclear transcription apparatus. We should ultimately learn whether this hormone–receptor complex exerts a specific regulatory effect on nuclear RNA metabolism (1) by direct effects on chromatin template, leading to increased gene transcription and thus RNA synthesis; (2) by activation of the polymerase complex itself; (3) by inhibition of RNA breakdown; or (4) by regulation of intranuclear processing of large precursor molecules to biologically active sequences and transport of RNA from the nucleus to the cytoplasmic sites of cellular protein synthesis. However considering our sum total of presently available evidence such as the magnitude of the inductive response, the rate of accumulation of mRNA in the tissue, the estimated intracellular turnover ($T_{1/2}$) of the mRNA and the baseline level of mRNA prior to stimulation, it appears likely that the steroid hormone acts at least in part to increase the rate of synthesis of specific mRNAs. The overall results are consistent with the hypothesis that generation of mRNA is a rate-limiting factor in steroid hormone-mediated induction of differentiation-specific proteins.

REFERENCES

Brawerman, G., J. Mendecki and S. Y. Lee. 1972. A procedure for the isolation of mammalian messenger ribonucleic acid. *Biochemistry* **11**:637.

Chan, L., A. R. Means and B. W. O'Malley. 1973. Rates of induction of specific translatable messenger RNAs for ovalbumin and avidin by steroid hormones. *Proc. Nat. Acad. Sci.* **70**:1870.

Comstock, J. P., G. C. Rosenfeld, B. W. O'Malley and A. R. Means. 1972. Estrogen-induced changes in translation, and specific messenger RNA levels during oviduct differentiation. *Proc. Nat. Acad. Sci.* **69**:2377.

Kohler, P. O., P. M. Grimley and B. W. O'Malley. 1969. Estrogen-induced cytodifferentiation of the ovalbumin secreting glands of the chick oviduct. *J. Cell Biol.* **40**:8.

Korenman, S. G. and B. W. O'Malley. 1968. Progesterone action: Regulation of avidin biosynthesis by hen oviduct *in vivo* and *in vitro*. *Endocrinology* **83**:11.

Liarakos, C. D., J. M. Rosen and B. W. O'Malley. 1973. Effect of estrogen on gene expression in the chick oviduct. II. Transcription of chick ^{3}H-unique DNA as measured by hybridization in RNA excess. *Biochemistry* **12**:2809.

McGuire, W. L. and B. W. O'Malley. 1968. Ribonucleic acid polymerase activity of the chick oviduct during steroid-induced synthesis of a specific protein. *Biochim. Biophys. Acta* **157**:187.

Means, A. R. and B. W. O'Malley. 1972. Mechanism of estrogen action: Early transcriptional and translational events. *Metabolism* **21**:357.

Means, A. R., J. P. Comstock, G. C. Rosenfeld and B. W. O'Malley. 1972. Ovalbumin messenger RNA of chick oviduct: Partial characterization, estrogen dependence, and translation *in vitro*. *Proc. Nat. Acad. Sci.* **69**:1146.

Oka, T. and R. T. Schimke. 1969. Interaction of estrogen and progesterone in chick oviduct development. I. Antagonistic effect of progesterone on estrogen-induced proliferation and differentation of tubular gland cells. *J. Cell Biol.* **41**:816.

O'Malley, B. W. 1967. *In vitro* hormonal induction of a specific protein (avidin) in chick oviduct. *Biochemistry* **6**:2546.

O'Malley, B. W. and W. L. McGuire. 1968. Studies on the mechanism of estrogen-mediated tissue differentiation: Regulation of nuclear transcription and induction of new RNA species. *Proc. Nat. Acad. Sci.* **60**:1527.

O'Malley, B. W., W. L. McGuire, P. O. Kohler and S. G. Korenman. 1969. Studies on the mechanism of steroid hormone regulation of synthesis of specific proteins. *Rec. Progr. Hormone Res.* **25**:105.

O'Malley, B. W., G. C. Rosenfeld, J. P. Comstock and A. R. Means. 1972. Steroid hormone induction of a specific translatable messenger RNA. *Nature New Biol.* **240**:45.

Palmiter, R. D. 1972. Regulation of protein synthesis in chick oviduct. I. Independent regulation of ovalbumin, conalbumin, ovomucoid, and lysozyme induction. *J. Biol. Chem.* **247**:6450.

Rhoads, R. E., G. S. McKnight and R. T. Schimke. 1971. Synthesis of ovalbumin in a rabbit reticulocyte cell-free system programmed with hen oviduct ribonucleic acid. *J. Biol. Chem.* **246**:7407.

Rosen, J. M., C. D. Liarakos and B. W. O'Malley. 1973. Effect of estrogen on gene expression in the chick oviduct. I. DNA–DNA renaturation studies. *Biochemistry* **12**:2803.

Rosenfeld, G. C., J. P. Comstock, A. R. Means and B. W. O'Malley. 1972a. Estrogen-induced synthesis of ovalbumin messenger RNA and its translation in a cell-free system. *Biochem. Biophys. Res. Commun.* **46**:1695.

———. 1972b. A rapid method for the isolation and partial purification of specific eucaryotic messenger RNA's. *Biochem. Biophys. Res. Commun.* **47**:387.

Socher, S. H. and B. W. O'Malley. 1973. Estrogen-mediated cell proliferation during chick oviduct development and its modulation by progesterone. *Develop. Biol.* **30**:411.

Control of Mammalian Cell Growth in Culture: The Action of Protein and Steroid Hormones as Effector Substances

Hugo A. Armelin,* K. Nishikawa,† and Gordon H. Sato

Biology Department, University of California at San Diego
La Jolla, California 92037

The basic mechanisms controlling mammalian cell proliferation are essentially still unknown in spite of the many studies developed in recent years. Several lines of evidence indicate that tropic hormones (proteins like gonadotropins, or steroids like estrogens) are involved in the control of mammalian cell proliferation in vivo. These facts suggest that the target cells for tropic hormones would make ideal systems for studies of cell growth control in general. However many practical difficulties inherent to in vivo experiments have limited the understanding of tropic hormone-dependent tissue growth. Most difficulties of in vivo experiments would be circumvented by isolation of the target cells in culture. This is one of the ideas behind the present work, whose main aim is to study physiological growth regulatory mechanisms utilizing methodological advantages offered by cell culture. Cell cultures here refers to permanent cell lines, which can be readily cloned and are amenable to available genetic, cytological or biochemical approaches.

The first steps in this work were directed towards the two following objectives: (a) identification of growth regulatory substances, which are effective in culture at concentrations compatible with physiological levels of in vivo regulators, and (b) development of cell systems that retain in culture a growth response to effector substances.

Cell lines, established and cultured for many years with serum as a limiting factor, have been selected for a low or no requirement for factors and hormones for growth. Consequently long-cultured cells probably lack many of the responses to growth control effectors operative in vivo, rendering them poor model systems in the study of physiologically significant growth regulatory mechanisms. Therefore if cell culture is going to be used as an approach, cells retaining growth response mechanisms must be chosen. Unfortunately general procedures for the development of growth-responsive cell lines are not available. An exception, perhaps, is the establishment of fibroblast lines (Todaro and Green 1963). "Contact-inhibited" fibroblasts are cells that exhibit in culture growth control mechanisms

---

* On leave from Instituto de Quimica, Universidade de São Paulo, São Paulo, Brazil.
† On leave from Institute for Protein Research, Osaka University, Osaka, Japan.

97

probably operative in vivo for this kind of cell. Hormone-dependent tumors from experimental animals contain cells that still retain a physiologically significant growth response. This kind of tumor has been used in our laboratory as sources of nonfibroblastic responsive cells. With mammary and ovarian tumors as starting material, we have been able to establish useful growth responsive cell lines (see Clark et al. 1972; Armelin and Sato 1973). Requirements for growth in culture of fibroblasts and established lines of growth responsive epithelial cells have been studied by us with the aim of defining the basic components of growth regulating reactions in mammalian cells. Our present results suggest that: (a) the pituitary gland might be a source of protein growth factors, which are distinct from the classical pituitary hormones, (b) glucocorticoids seem to be involved in growth regulation of several cell types, and (c) cGMP might be an intracellular mediator of the growth response. A summary of the present experimental evidence which supports these hypotheses is presented below.

EXPERIMENTAL RESULTS AND DISCUSSION

Minimal Medium

In order to assay for growth-promoting activity of hormones or growth factors, it was essential to define a minimal medium. A problem arises from the fact that serum is a general requirement for mammalian cells in culture and at the same time serum is a natural source of hormones and growth factors. Consequently we needed a battery of sera depleted in specific hormones or factors. However no procedures were available for specific extraction of hormones from serum without affecting other activities required for cell growth. The situation was even worse with respect to serum growth factors, whose existence was still essentially a postulate. Considering these facts we have chosen an empirical approach to these problems. At present two serum treatments are used with good results, charcoal extraction and passage through a carboxymethylcellulose column. The charcoal extraction procedure (serum with 10 mg/ml of Norit A, Sigma Company, and 1 mg/ml of Dextran T40 for 30 minutes at 55°C) efficiently extracts steroids (monitored with labeled steroid), extracts thyroxin poorly (30% of the labeled hormone remains after extraction), does not extract more than 2% of total proteins and also extracts fatty acids (inferred from results of Chen 1967). The carboxymethylcellulose passage (10 × diluted serum, pH 6.5, room temperature) eliminates 1–2% of total fetal calf serum protein, which adheres to the column and is eluted with 1 M NaCl. Charcoal extraction and CMC chromatography render the serum unable to promote growth of some cell types. The growth activity of treated serum can be reconstituted partially or completely by addition of certain pituitary protein fractions, glucocorticoids, or both. An overall view of these observations is presented in Table 1 and a closer examination of the results with different cell types will be presented in the next sections. Clark et al. (1972) suggested that affinity chromatography (anti-LH covalently bound to Sepharose) specifically eliminated ovarian cell growth factor from fetal calf serum. Our present results indicate that the anti-LH Sepharose column probably adsorbed nonspecifically "ovarian cell growth factor," which happens to be a basic or neutral protein that can be efficiently adsorbed to a CMC column.

Table 1
Growth in Minimal Media: Pituitary Protein Factor and Glucocorticoid Effects

Cell system		10% ChCS					10% CMC-FCS			
		No addition	+ Pituitary protein factor	+ Pituitary factor + hydrocortisone	+ Hydrocortisone	10% CS	No addition	+ Pituitary protein factor	+ Hydrocortisone	10% FCS
Ovary (epithelial)	Uncloned population	NGD	NGD	G++	G	G++	NGD	G++	NGD	G++
	Clone MF2B	G+	G+	G+	G+	G++	G+	G++		G++
Mammary (epithelial)	Clone 102S-1A	NGD	NGD	G++	G++	G+	G		G++	G+
Fibroblast	3T3	NGR	G+	G++	NGR	G++	G			
	SV3T3	G++	G++	G++	G++	G++				G++

NGR: nonpopulation growth, cells resting, minimal DNA synthesis, high viability.

NGD: nonpopulation growth, cell death, high DNA synthesis, viability constantly decreasing.

G, G+ and G++: poor, intermediate and full growth; this represents an arbitrary classification of growth extent only for the purpose of description in this table.

These observations came from a large number of experiments performed over a period of more than a year. Hydrocortisone concentration ranged from 0.01–0.1 μg/ml. Pituitary protein factor refers to NIH-LH-B8 ranging from 0.1–2 μg/ml. Culture regimen: hormones added only at the plating day; growth followed for 4–12 days; initial inoculum 5–50 $\times$ 10^3 cells/plate. Clone MF2B is a variant of ovarian cells obtained after a FdU treatment (Armelin and Sato 1973).

Growth Responsive Cell Systems

Ovarian Cells (Epithelial)

Clark et al. (1972) established in culture a line of ovarian cells responsive for growth to gonadotropins and glucocorticoids. We have studied these cells extensively (see Armelin and Sato 1973) and, at present, the following points seem to be clearly established:

1. The cell population does not increase in 10% CMC-FCS medium, but population growth is observed with addition of an impure preparation of LH (NIH-LH-B8).
2. Pure LH is ineffective, and the active species is probably a basic or neutral protein of pituitary origin.
3. Population growth is proportional to the amount of added NIH-LH-B8, and this is the basis of a quantitative assay, as indicated by the results in Fig. 1a.
4. Cells are not quiescent in minimal media. [³H]thymidine uptake into DNA/cell is the same in CMC-FCS or CMC-FCS+LH medium. The essential result of factor deficiency in CMC-FCS medium is cell death as shown by the colony development assay in Fig. 1b.
5. Fluorodeoxyuridine (FdU) can change the growth requirement of these ovarian cells by a still unknown mechanism (Armelin and Sato 1973). FdU-modified clones are able to grow in CMC-FCS medium (Fig. 1c) but may show some increase in growth rate with added NIH-LH-B8.
6. Ovarian cells are also unable to grow in ChCS media. In this case addition of

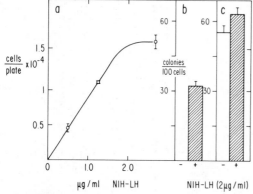

Figure 1

Effect of NIH-LH-B8 on growth of ovarian cells.

   (a) Ovarian cells (uncloned population) were plated (5 × 10³ cells/plate) in 10% CMC-FCS. NIH-LH-B8 was added at the time of plating at different concentrations. Two days later cells were trypsinized and counted (Coulter Counter). Cells/plate in plates with NIH-LH-B8 minus cells/plate in plates with no addition was plotted against NIH-LH-B8 concentration. Plates without hormone had 5 ± 0.2 × 10³ cells/plate.

   (b) Ovarian cells (uncloned population) and (c) ovarian cells of MF2B clone (modified by FdU, see Armelin and Sato 1973) were plated (150 cells/plate) in 10% CMC-FCS in the presence or absence of NIH-LH-B8 (2 µg/ml). Seven days later colonies were fixed with formalin, stained with crystal violet and counted.

NIH-LH-B8 is not sufficient to stimulate population growth. However the combination of LH plus hydrocortisone added to ChCS medium stimulates population growth. Maximum glucocorticoid activity is obtained with concentrations of 2×10^{-8} M. Progesterone can substitute for the glucocorticoid activity at higher concentrations and sex steroids seem to be ineffective.

In conclusion we can say that this line of ovarian cells shows an absolute requirement for an unknown protein pituitary factor (Gospodarowicz, Jones and Sato have recently purified to homogeneity an active molecular species from pituitary tissue) and glucocorticoids for growth in culture. The effective concentration range is a physiological one.

Mammary Cells (Epithelial)

Clones from a hormone-dependent rat mammary adenocarcinoma (tumor developed by R. Iglesias 1970) have been isolated in culture (Armelin unpublished). These cells are epithelial, have been in culture for over 500 generations, and show a remarkable growth response to glucocorticoids. One clone of these cells has been studied in detail. It grows very slowly in ChCS medium (doubling time 60 hours or more), and addition of hydrocortisone at 10^{-9} M decreases the doubling time to 18 hours. The maximum response (doubling time 13 hours) is achieved with concentrations of 6×10^{-8} M. Progesterone is 60 times less effective than hydrocortisone, and estradiol and testosterone are without effect. In the absence of steroids, these cells also do not become quiescent.

Fibroblasts

We have recently observed that pituitary extracts stimulate the growth of 3T3 cells (mouse fibroblast). This growth promoting activity is (a) not dialysable, (b) thermolabile, (c) protease sensitive, and (d) enhanced by hydrocortisone (Armelin 1973). NIH-LH-B8 has a very high growth promoting specific activity according to the quantitative assays, described in Fig. 2. Highly purified LH, however, is not active. The active species seems to be a basic protein that contaminates this hormone preparation. Hydrocortisone can increase the activity of NIH-LH-B8 by twofold and is effective at 10^{-8} M, but the glucocorticoid seems to be ineffective by itself. 3T3 cells in 1% CS were treated with NIH-LH-B8. LH was added 2 days after plating at 0.4 μg/ml 10 or 180 minutes after addition, the cells were washed thoroughly with 1% CS and then incubated with fresh 1% CS. Both treatments are equally effective in triggering a wave of DNA synthesis, which is indicated by a sharp increase in [^{3}H]thymidine uptake between 12 and 18 hours after addition of the hormone preparation. These results suggest that a pituitary protein factor is able to trigger cell division in a resting fibroblast. Attempts to purify the active fraction are under way.

The Intracellular Growth Mediator

Several laboratories are involved in searching for an intracellular mediator of the growth response. Among the candidates considered are the nucleotides cAMP (Hsie and Puck 1971; Johnson, Friedman and Pastan 1971; Sheppard 1971), ppGpp (Mamont et al. 1972) and cGMP (Hadden et al. 1972). The main approaches used so far have been to analyze the effects of addition of high levels of

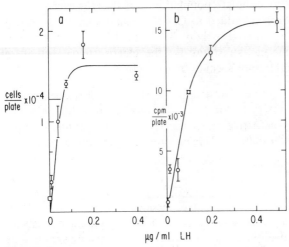

Figure 2
Effect of NIH-LH-B8 on growth of 3T3 cells.

(a) Cells were plated (2×10^3 cells/plate) in 10% CS; at 24 hr the medium was changed to 1% CS; at 48 hr hormones were added (0.1 μg/ml hydrocortisone; NIH-LH-B8 at different concentrations) and at 96 hr the cells were counted in a Coulter Counter. (o) Cells/plate in plates with NIH-LH-B8 + hydrocortisone minus cells/plate in plates with no hormone was plotted against NIH-LH-B8 concentration; (□) cells/plate in plates with hydrocortisone alone minus cells/plate in plates with no hormone. Plates with no hormone had $4.6 \pm 0.1 \times 10^3$ cells/plate at 96 hr.

(b) Cells were plated (6×10^4 cells/plate) and cultured as described for (a) except that 12 hr after hormone addition [^{3}H]thymidine was added; uptake into DNA was measured after 6 hr of incorporation. Cpm/plate in plates with NIH-LH-B8 + hydrocortisone minus cpm/plate in plates with no hormone was plotted against NIH-LH-B8 concentration. Plates with no hormone had 2800 ± 300 cpm/plate.

nucleotides to culture medium and to measure the intracellular levels of the possible mediator in populations growing at different rates. We wish to present some of our preliminary results which are pertinent to this problem.

The first kind of result comes from experiments in which addition of dibutyryl cAMP (10^{-3} M) or theophylline (10^{-3} M) or both to 3T3 and SV3T3 cultures causes growth inhibition. In the case of the transformed fibroblast, extensive cell death is caused by this treatment whereas the "normal" fibroblast undergoes growth inhibition without appreciable loss of viability. With mammary cells in 10% ChCS plus 10^{-7} M hydrocortisone (doubling time 14 hours), the results can be summarized as follows: (a) Dibutyryl cAMP (2×10^{-3} M) inhibited growth by 20%; (b) theophylline (10^{-4} M) inhibited growth by 50%; (c) theophylline (10^{-4} M) plus dibutyryl cAMP (10^{-3} M) inhibited growth by 100%; (d) theophylline (10^{-3} M) killed the great majority of cells in 3 to 4 days; (e) in all cases growth inhibition involved extensive cell death. Ovarian cells seem to be less sensitive to these treatments. In a rich medium (12.5% horse serum plus 2.5% fetal calf serum; doubling time 15 hours) the growth of these cells was not affected by theophylline at 10^{-4} M, and at 10^{-3} M 50–60% inhibition of growth was observed. This kind of experiment indicated to us that dibutyryl cAMP and theophylline can inhibit growth of virus-transformed fibroblasts or epithelial tumor cells. The cells, however, are not blocked in a resting state and extensive cell killing is always involved in the process.

Cyclic AMP and cGMP intracellular levels were measured using a radioimmunoassay (according to Steiner, Parker and Kipnis 1972; reagents obtained from Collaborative Research Inc.). The nucleotides were extracted with cold 10% TCA immediately after aspirating medium from the plates and the assays were performed without prior separation of the nucleotides. Sparse cultures of slow growing or resting 3T3 were used. Cells were plated in 10% CS medium. Ten hours later they were shifted to 1% CS medium and two days later assayed for cyclic nucleotides. Under these conditions 3T3 cells contain 29.0 pmoles of cAMP/mg protein and 30 times less cGMP (0.98 pmoles/mg protein). Addition of NIH-LH-B8 (1 μg/ml) for 30 minutes caused a decrease of 10% in cAMP and 2.2-fold increase in cGMP. Addition of NIH-LH-B8 (1 μg/ml) plus hydrocortisone (10^{-7} M) for 30 minutes induced a decrease of 30% in cAMP and a remarkable 20-fold increase in cGMP. After 15 hours with LH plus hydrocortisone, cAMP was at the same level observed at 30 minutes (although there was an increase of 50% in protein and 15% in RNA/plate), and cGMP was still sixfold the level observed in plates with no additions. These nucleotides were also assayed in the mammary cells, whose growth is stimulated by glucocorticoids. In presence of 10^{-7} M hydrocortisone these cells contain 13.8 pmoles cAMP/mg protein and 2–10 times this value in the absence of the hormone. Hydrocortisone, however, had no effect on cGMP levels. The cellular content of this nucleotide is 0.31 $\pm$ 0.01 and 0.37 $\pm$ 0.12 pmoles/mg protein in the presence and absence of hormones, respectively. In these experiments the nucleotides were extracted 24 hours after plating the cells in 10% ChCS medium with and without hormones. Hence though hydrocortisone seems to be an active growth regulator for 3T3 cells and this mammary cell line, the cGMP level apparently is only affected in the case of the fibroblast. These observations suggest the interesting possibility that cGMP might be involved in the mechanism of transition from a "resting" to a "proliferative" cellular state (perhaps G_0 to G_1). Since the mammary cell is a "nonresting" cell, it would bypass this mechanism and cGMP would not be involved. Hadden et al. (1972) have already suggested clearly that cGMP might be a key intracellular mediator of the lymphocyte mitogenic response, and it is of interest that this is also a "resting" to "proliferative" state transition. Of course further experimentation is required to critically test this hypothesis.

CONCLUSION

We are convinced that growth control in cell cultures involves hormones and unknown serum factors as extracellular regulatory substances. In order to establish these cultures as generally useful model systems, it will be necessary to show that these same regulatory substances are effective in vivo. Initial information about in vivo growth regulation could come from experiments in which these cells are grown in animals. Experiments are in progress testing the ability of these ovarian and mammary cell lines to develop tumors in animals which are hormonally conditioned.

Acknowledgments

This work was supported by the U.S. Public Health Service (GM 17019) and the National Science Foundation (15788 GB). Protein hormones for these studies were kindly supplied by the Hormone Distribution Program NIAMD of the National Institutes of Health. H.A.A. was supported by the Fundacão de Amparo A Pesquisa Do Estado de São Paulo (FAPESP) and a grant from the Alfred P. Sloan Foundation.

REFERENCES

Armelin, H. A. 1973. Pituitary extracts and steroid hormones in the control of 3T3 cell growth. *Proc. Nat. Acad. Sci.* (in press).

Armelin, H. A. and G. Sato. 1973. Cell cultures as model systems for the study of growth control. *World Symp., Model studies in chemical carcinogenesis,* Baltimore 1972. Proceedings in *The biochemistry of disease,* ed. E. Faber (in press)

Chen, R. F. 1967. Removal of fatty acids from serum albumin by charcoal treatment. *J. Biol. Chem.* **242**:173.

Clark, J. L., K. L. Jones, D. Gospodarowicz and G. H. Sato. 1972. Growth response to hormones by a new rat ovary cell line. *Nature New Biol.* **236**:180.

Hadden, J. W., E. M. Hadden, M. K. Haddox and N. D. Goldberg. 1972. 3,5-cGMP: A possible intracellular mediator of mitogenic influences in lymphocytes. *Proc. Nat. Acad. Sci.* **69**:3024.

Hsie, A. and T. T. Puck. 1971. Morphological transformation of Chinese hamster cells by dibutyril cyclic AMP and testosterone. *Proc. Nat. Acad. Sci.* **68**:358.

Iglesias, R. 1970. Transplantable endocrine tumors. *Oncology* (ed. R. L. Clark et al.) Proc. 10th Int. Cancer Congr. vol. 1, p. 300.

Johnson, G. S., R. M. Friedman and I. Pastan. 1971. Restoration of several morphological characteristics of normal fibroblasts in sarcoma cells treated with cAMP. *Proc. Nat. Acad. Sci.* **68**:425.

Mamont, P., A. Hershko, R. Kram, L. Schacter, J. Lust and G. M. Tomkins. 1972. The pleiotypic response in mammalian cells: Search for an intracellular mediator. *Biochem. Biophys. Res. Commun.* **48**:1378.

Sheppard, J. R. 1971. Restoration of contact-inhibited growth to transformed cells by dibutyril cAMP. *Proc. Nat. Acad. Sci.* **68**:1316.

Steiner, A. L., C. W. Parker and D. M. Kipnis. 1972. Radioimmunoassay for cyclic nucleotides. *J. Biol. Chem.* **247**:1106.

Todaro, G. J. and H. Green. 1963. Quantitative studies of the growth of mouse embryo cells in culture and their development into established lines. *J. Cell Biol.* **17**:299.

Initiation of Division of Density-inhibited Fibroblasts by Glucocorticoids

Dennis D. Cunningham, Cornelia R. Thrash, and Russell D. Glynn

Department of Medical Microbiology, College of Medicine
University of California, Irvine, California 92664

Our studies on the effects of cortisol on fibroblast division were prompted by several lines of information suggesting that it might modify division of fibroblasts by decreasing a cell surface proteolytic activity. First, there are reports that cortisol inhibits DNA synthesis and division of malignant L929 mouse fibroblasts (Ruhmann and Berliner 1965; Pratt and Aronow 1966). In addition cortisol can inhibit release of lysosomal enzymes, presumably by stabilizing lysosomal membranes (Weissman and Thomas 1964). In view of reports that proteases can bring about a loss of density-dependent growth control (Burger 1970; Sefton and Rubin 1970), we thought that cortisol might inhibit division of L929 cells possibly by inhibiting release of proteases from lysosomes to the cell exterior.

We first examined the effects of cortisol on growth of untransformed, viral-transformed, and spontaneously transformed fibroblasts (Thrash and Cunningham 1973). Unexpectedly cortisol stimulated DNA synthesis and division of density-inhibited fibroblasts. In contrast, it had very little effect on DNA synthesis or division of spontaneously transformed or viral-transformed fibroblasts. This apparent specificity was not just a consequence of the very slow growth of the density-inhibited fibroblasts: transformed fibroblasts remained unresponsive to cortisol when their growth was greatly slowed by placing them in medium containing low levels of serum. In addition initiation of density-inhibited fibroblasts was specific in terms of the steroid added: only steroids with glucocorticoid activity brought about the response. This selective initiation by a well defined molecule should facilitate studies at the biochemical level on events leading to cell proliferation.

MATERIALS AND METHODS

Cells were grown as previously described (Cunningham 1972) in Dulbecco-Vogt-modified Eagle's medium containing 10% calf serum (3T3, 3T6, Py3T3 and SV3T3 cells), 10% fetal calf serum (L cells), or 3% calf serum (early passage human diploid foreskin fibroblasts). We found no evidence of mycoplasma con-

105

tamination of any cells following radioautography with [^{3}H]thymidine (Nardone et al. 1965). Density-inhibited cultures of 3T3 cells were routinely prepared by plating 1.8×10^4 cells per cm^2 and allowing them to grow to their saturation density (3.0 to 3.7×10^4 cells per cm^2) over a three-day period. Cell number was monitored by suspending cells with 0.05% trypsin, diluting with serum and counting in a hemacytometer. DNA synthesis was measured by incubating cells for 15 minutes with [^{3}H]thymidine (2.5 μCi/ml; 0.4 Ci/mmole). The cells were then washed once with cold phosphate-buffered saline, five times with 10% trichloroacetic acid, and dissolved in 0.5 M KOH. Radioactivity in neutralized aliquots was determined by liquid scintillation. Protein was measured as previously described (Lowry et al. 1951).

RESULTS

Effects of Cortisol on Division of Density-inhibited and Transformed Fibroblasts

Initiation of division of density-inhibited 3T3 cells by cortisol is shown in the right panel of Fig. 1. 3T3 cells were grown to their saturation density in medium containing 10% calf serum, and cortisol was added to final concentrations of 0.07 μM and 14 μM at the time indicated by the arrow. This treatment brought about a large increase in cell number: 14 μM cortisol produced an 85% increase in four days. Cortisol brought about a similar initiation of division of early passage human diploid foreskin fibroblasts, demonstrating that the effect was not unique to the established 3T3 line. Treatment of density-inhibited 3T3 cells with cortisol also brought about a marked increase in DNA synthesis, which was detectable at 12 hours and peaked at about 24 hours. This stimulation was observed over a broad cortisol concentration range, plateauing at maximal levels between 0.5 and 28 μM (Thrash and Cunningham 1973). A large stimulation of DNA synthesis by cortisol also occurred in slowly growing 3T3 cells limited by 0.3% serum in the medium.

We also examined the effects of cortisol on division of transformed fibroblasts (Fig. 1, left). Addition of 14 μM cortisol (indicated by the arrow) to growing confluent L or Py3T3 cells had very little effect on cell division. To exclude the possibility that an effect of cortisol might be obscured by rapid growth, we limited the growth of transformed cells by low serum, added cortisol, and monitored cell number. These experiments are shown in Fig. 2. Growth of 3T6 cells was limited by changing to medium containing 0.1% serum at day zero. As shown in the upper panel, growth continued for one day after the medium change and then stopped. Cortisol addition at day one had no effect on cell number during the following three days. The middle panel shows a similar experiment with Py3T3 cells. In this case, serum-free medium was placed on the cells at day zero. These cells continued to grow very slowly during the next four days. Cortisol addition at day one slightly inhibited cell division relative to controls during the following three days. The effect of cortisol on slowly growing SV3T3 cells is shown in the bottom panel of Fig. 2. Here we changed to medium containing 0.03% serum at day zero. The number of cells attached to the plate increased slowly after this medium change. However a significant number of cells detached from the plate into the medium. Accordingly we plotted attached cells, cells in medium, and total cells

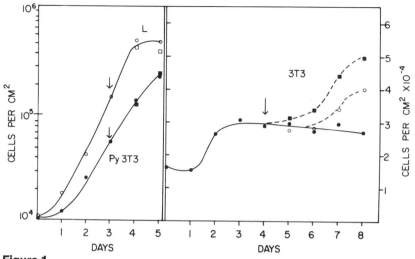

Figure 1

Effect of cortisol on confluent 3T3, Py3T3 and L cells. Cortisol was added at times indicated by the arrows. *Left:* (o) L cells; (□) L cells + 14 μM cortisol; (•) Py3T3 cells; (■) Py3T3 cells + 14 μM cortisol. *Right:* (•) 3T3 cells; (o– – –o) 3T3 cells + 0.07 μM cortisol; (■– – –■) 3T3 cells + 14 μM cortisol. From Thrash and Cunningham (1973) with permission of *Nature*.

per plate for the SV3T3 cells. The bottom panel shows that in accord with data on the other transformed lines, cortisol-treated SV3T3 cultures had slightly fewer total cells per plate compared to untreated controls after three days. However cortisol slightly increased the number of cells that remained attached to the growth surface. This latter result is consistent with the finding that glucocorticoids increase adhesiveness of hepatoma tissue culture cells (Ballard and Tomkins 1969). Although we have not yet checked the viability of the cells in the medium, we conclude that cortisol had little effect on the division of slowly growing SV3T3 cells.

Taken together these results demonstrate that cortisol initiates division of nongrowing density-inhibited fibroblasts but that it has very little effect on division of slowly growing transformed fibroblasts limited by low serum.

Effects of Other Steroids on DNA Synthesis

We checked the effect of a number of different steroids on DNA synthesis in density-inhibited 3T3 cells to determine if the response was specific for glucocorticoids. We measured DNA synthesis rather than cell number, since even barely detectable increases in cell number are preceded by easily measurable increases in the rate of DNA synthesis in this system. Rates of thymidine incorporation into an acid-insoluble product were measured 24 hours after adding each steroid. This time corresponded to the peak of DNA synthesis after adding cortisol.

The specificity of the initiation by cortisol was demonstrated by our finding that cholesterol, progesterone, estradiol, testosterone, and tetrahydrocortisol, an inactive metabolite of cortisol, had no significant effect on DNA synthesis at 24 hours over broad steroid concentration ranges. To rule out the possibility that a peak of DNA synthesis appeared earlier or later than the 24-hour cortisol peak, we measured

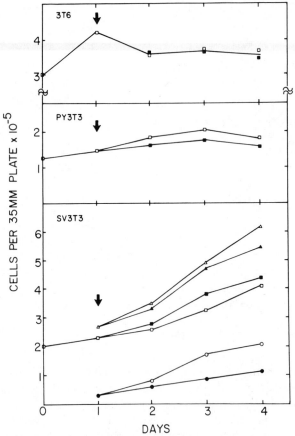

Figure 2

Effect of cortisol on slowly growing transformed fibroblasts. Cells were allowed to attach to plates in medium containing 10% serum. After 5 to 16 hr the cells were rinsed twice with serum-free medium. We then added medium containing serum levels indicated under Results. Cortisol was added to a final concentration of 2.8μM at the time indicated by arrows. ($\square$) Attached cells in control cultures; ($\blacksquare$) attached cells in cortisol-treated cultures; ($\circ$) SV3T3 cells in medium in control cultures; ($\bullet$) SV3T3 cells in medium in cortisol-treated cultures; ($\triangle$) total SV3T3 cells (attached plus medium) in control cultures; ($\blacktriangle$) total SV3T3 cells (attached plus medium) in cortisol-treated cultures.

rates of [³H]thymidine incorporation at 8-hour intervals from 0 to 48 hours after adding cholesterol or progesterone. Neither steroid produced measurable increases at these times. We are currently checking the other inactive steroids at these additional times.

Four different steroids with glucocorticoid activity initiated DNA synthesis in density-inhibited 3T3 cells (Table 1). The concentration of each steroid that produced a maximal increase in DNA synthesis was determined by dose-response experiments. These maximal increases are presented in Table 1. The ability of these steroids to stimulate density-inhibited 3T3 cells correlated with the glucocorticoid potency of these compounds in vivo.

Table 1
Stimulation of DNA Synthesis in
Confluent 3T3 Cells by Glucocorticoids

	Optimal conc. (μM)	*[³H]thymidine incorp. (cpm)*
No addition		313
Triamcinolone acetonide	0.23	2480
Dexamethasone	0.26	1968
Cortisol	2.8	2099
Corticosterone	2.9	1389

Duration of Cortisol Treatment Required to Initiate DNA Synthesis

These experiments were carried out to determine if only brief treatment with cortisol was sufficient to initiate DNA synthesis, or whether a relatively prolonged interaction between cortisol and density-inhibited 3T3 cells was required for the proliferative response. The results are shown in Fig. 3. The abscissa shows the duration of cortisol treatment. Cortisol was added at time zero, and at the indicated times we removed this medium, rinsed the cells, and added cortisol-free medium which had supported growth of parallel 3T3 cultures to confluency. The rate of DNA synthesis in all cultures was measured at 26 hours. This time corresponded to the peak of DNA synthesis even in cultures treated for only 6.5 hours with cortisol. There was only a slight "triggering" effect after brief treatment with cortisol. It had to be present for 8.5 hours to bring about a half-maximal response.

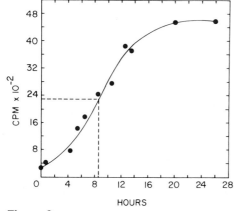

Figure 3
Duration of cortisol treatment and DNA synthesis in density-inhibited 3T3 cells. Confluent 3T3 cells were treated with 2.8 μM cortisol for varying times as described under Results. DNA synthesis was measured by incorporation of [³H]thymidine at 26 hr as described in Materials and Methods.

An analogous experiment was carried out using radioautography with [³H]thymidine to measure the number of cells making DNA at 26 hours. This experiment supported the above conclusion. An 8-hour treatment with cortisol was required to bring about a half-maximal response. The number of cells making DNA at 26 hours after a 4-hour cortisol treatment was only 12% of the number after a 26-hour cortisol treatment.

Comparison of Serum and Cortisol Effects on Cell Protein and Cell Number

Density-inhibited 3T3 cells treated with levels of cortisol sufficient to initiate DNA synthesis and division appeared to become smaller. Thus cortisol seemed to increase cell number without bringing about an equivalent increase in total cell mass. We checked this possibility by measuring cell number and protein after initiating density-inhibited 3T3 cells with cortisol, fresh serum, and cortisol plus fresh serum.

Addition of 0.1 ml fresh serum to the 2.0 ml of medium on density-inhibited 3T3 cells brought about a 60% increase in cell number and a 90% increase in cell protein after three days (Fig. 4). This treatment brought about an 18% increase in protein per cell. In contrast, addition of cortisol to a final concentration of 2.8 μM caused a 90% increase in cell number but only a 25% increase in cell protein. Cortisol treatment brought about a 36% decrease in protein per cell compared to controls.

The data in Fig. 4 also show that the stimulatory effects of fresh serum and cortisol on density-inhibited 3T3 cells were additive. This was true for both the increase in cell number (upper panel) and cell protein (lower panel).

Concanavalin A Agglutinability and Initiation of Cell Division

Treatment of density-inhibited 3T3 cells with levels of cortisol which initiated division brought about an early increase in concanavalin A (conA)-specific agglutinability (Thrash and Cunningham 1973). Changes in the cell surface measured by increased agglutinability with conA and certain other plant lectins have been linked to escape from density-dependent growth control. This relationship was strengthened by the findings that brief protease treatment at the surface of density-inhibited 3T3 cells brings about a temporary loss of density-dependent growth control and an increased agglutinability by these lectins (for review see Burger 1973). The increased conA agglutinability we observed after cortisol treatment, however, was not always reproducible. We therefore treated density-inhibited 3T3 cells for 10 minutes with broad concentration ranges of pronase to further test the relationship of a temporary loss of density-dependent growth control to increased conA agglutinability. These pronase treatments brought about a graded increase in conA-specific agglutinability ranging from control values of 25% of cells agglutinated to plateau values of about 80%. However none of these pronase concentrations brought about an increase in cell number within 84 hours, even though pronase-treated cells remained responsive to the stimulatory action of fresh serum and cortisol. Therefore the surface change measured by increased conA-specific agglutinability is not an event sufficient by itself to bring about cell division (Glynn, Thrash and Cunningham, 1973).

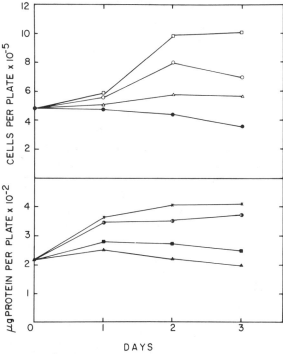

Figure 4

Effects of cortisol and serum on protein and cell number of density-inhibited 3T3 cells. 3T3 fibroblasts were grown to confluency as described in Materials and Methods. Fresh serum (final concentration 10%) and/or cortisol (final concentration 2.8 μM) were then added at day zero. *Upper panel,* cell number: (●) control, (△) plus serum, (○) plus cortisol, (□) plus serum and cortisol. *Lower panel,* total protein: (▲) control, (■) plus cortisol, (●) plus serum, (×) plus serum and cortisol.

DISCUSSION

These results show that physiological levels of glucocorticoids can initiate DNA synthesis and division in density-inhibited 3T3 and early passage human diploid foreskin fibroblasts. This initiation is specific in terms of the steroid added, since cholesterol, progesterone, estradiol, testosterone and tetrahydrocortisol do not bring about the response. In addition the ability of steroids to initiate DNA synthesis in density-inhibited 3T3 cells is correlated with the glucocorticoid potency of these compounds in vivo. The response of confluent 3T3 cultures to this stimulation by cortisol is characterized by a large increase in cell number, but only a small increase in total cell protein. After initiation by cortisol, protein per cell decreases about 35%. Glucocorticoids therefore bring about only a small increase in total cell mass of density-inhibited cultures. Our results further show that a relatively long exposure to cortisol is required to initiate DNA synthesis in density-inhibited 3T3 cells. Cortisol treatment for about 8 hours is required for half-maximal increases in both the rate of DNA synthesis and the number of cells making DNA at 26 hours. Cultures treated for 4 hours with cortisol have only 12% as many cells making DNA at 26 hours as do cultures treated with cortisol over the entire 26-hour period.

We previously reported that cortisol treatment of density-inhibited 3T3 cells brings about an increase in conA-specific agglutinability that can be detected as early as one hour (Thrash and Cunningham 1973). This increase, however, is not very reproducible. In addition experiments briefly summarized here show that the surface change measured by increased conA agglutinability is not an event that is sufficient by itself to initiate cell division.

We are considering several possibilities that might account for the lack of a stimulatory effect of cortisol on transformed 3T3 cells which are growth limited by low serum. (1) Stimulation by cortisol might require higher levels of serum than were employed in our experiments on transformed cells. For example, stimulation might depend on prior binding of cortisol to a serum protein (perhaps transcortin). Alternatively, cortisol might initiate the proliferative response, but continuation of it might require the action of factors in serum. However it should be noted that cortisol initiates DNA synthesis in 3T3 cells which are growth limited by 0.3% serum. (2) If cortisol initiates DNA synthesis and division only in cells in the G_0 state, serum-limited transformed cells might be unresponsive to cortisol because they are randomly distributed throughout the cell cycle or do not stop in an equivalent G_0 state. (3) Transformed cells might be unresponsive because of alterations in glucocorticoid-specific cell components. Along with other alternatives, we are examining the possibility that transformed cells have an altered cytosol receptor. Alterations of this kind have already been demonstrated for glucocorticoid-resistant cells selected from sensitive cell populations (Pratt and Ishii 1972; Rosenau et al. 1972).

Acknowledgments

This work was supported by a grant from the National Cancer Institute of the USPHS (CA12306). C.R.T. was supported by a USPHS training grant (GM-02063-02) to the Dept. of Molecular Biology and Biochemistry. We thank Mr. Tom Ho for technical assistance.

REFERENCES

Ballard, P. L. and G. M. Tomkins. 1969. Hormone induced modification of the cell surface. *Nature* **224**:344.

Burger, M. M. 1970. Proteolytic enzymes initiating cell division and escape from contact inhibition of growth. *Nature* **227**:170.

———. 1973. Surface changes in transformed cells detected by lectins. *Fed. Proc.* **32**:91.

Cunningham, D. D. 1972. Changes in phospholipid turnover following growth of 3T3 mouse cells to confluency. *J. Biol. Chem.* **247**:2464.

Glynn, R. D., C. R. Thrash and D. D. Cunningham. 1973. Maximal concanavalin A-specific agglutinability without loss of density-dependent growth control. *Proc. Nat. Acad. Sci.* (in press)

Lowry, O. H., N. J. Rosebrough, A. L. Farr and R. J. Randall. 1951. Protein measurement with the folin phenol reagent. *J. Biol. Chem.* **193**:265.

Nardone, R. M., J. Todd, P. Gonzales and E. V. Gaffney. 1965. Nucleoside incorporation into strain L cells: Inhibition by pleuropneumonia-like organisms. *Science* **149**:1100.

Pratt, W. B. and L. Aronow. 1966. The effect of glucocorticoids on protein and nucleic acid synthesis in mouse fibroblasts growing *in vitro. J. Biol. Chem.* **241:**5244.

Pratt, W. B. and D. N. Ishii. 1972. Specific binding of glucocorticoids *in vitro* in the soluble fraction of mouse fibroblasts. *Biochemistry* **11:**1401.

Rosenau, W., J. D. Baxter, G. G. Rousseau and G. M. Tomkins. 1972. Mechanism of resistance to steroids: Glucocorticoid receptor defect in lymphoma cells. *Nature New Biol.* **237:**20.

Ruhmann, A. G. and D. L. Berliner. 1965. Effect of steroids on growth of mouse fibroblasts *in vitro. Endocrinology* **76:**916.

Sefton, B. M. and H. Rubin. 1970. Release from density dependent growth inhibition by proteolytic enzymes. *Nature* **227:**843.

Thrash, C. R. and D. D. Cunningham. 1973. Stimulation of division of density inhibited fibroblasts by glucocorticoids. *Nature* **242:**399.

Weissman, G. and L. Thomas. 1964. The effects of corticosteroids upon connective tissue and lysosomes. *Rec. Progr. Hormone Res.* **20:**215.

Hormonal Control of Cellular Growth

Carol Sibley, Ulrich Gehring,
Henry Bourne,* and Gordon M. Tomkins

Department of Biochemistry and Biophysics and *Departments of Medicine
and Pharmacology, University of California, San Francisco, California 94143

A major effort in our laboratory has been the attempt to understand the mechanisms of glucocorticoid action. For this work we have established a line of cultured rat hepatoma (HTC) cells in which this group of steroids induces the synthesis of a specific enzyme, tyrosine aminotransferase (EC 2.6.1.5.). The steps thus far elucidated in this process include the penetration of the steroid into the cell, its association with specific cytoplasmic receptor proteins, and conformational changes in the receptor–steroid complex, followed by association of the complex with specific DNA-containing nuclear sites (reviewed in Baxter et al. 1973). This latter interaction results in the accumulation of polyribosomes active in the synthesis of the inducible enzyme (Scott, Shields and Tomkins 1972). We have also identified a gene product responsible for the inactivation of tyrosine aminotransferase messenger RNA (Tomkins et al. 1972).

It would obviously be very useful to explore cell–hormone interactions by a genetic approach in addition to the biochemical techniques used thus far. To do this, however, one has to be able to select mutants in which various steps in this process are altered. Since tyrosine aminotransferase induction does not, under ordinary circumstances, result in changes in cellular growth, we have recently begun a series of investigations using cell lines in which both the glucocorticoids and cyclic AMP not only inhibit growth but cause cell death (Rosenau et al. 1972; Gehring, Mohit and Tomkins 1972). Cells resistant to the killing action of these effectors might well be defective in the various aspects of hormone action and should therefore be of value for studies on the mechanisms of hormone action.

MATERIALS AND METHODS

Cell Lines

The mouse lymphoma line S49.1T.B4 (derived from a Balb/c mouse) was obtained from Ms. Ruth Epstein at the Salk Institute. These cells contain θ antigen and are TL-positive, indicating their thymic origin. The EL4 lymphoma line (derived from a C57BL mouse) and the immunoglobulin-secreting myeloma line CL4 (from a Balb/c mouse) were obtained from Dr. B. Mohit. The cells were cultivated in Dulbecco's modified minimal essential medium supplemented with 10% heat-inactivated fetal calf serum in an atmosphere of 10% CO_2. Cloning of S49 cells was carried out using a feeder layer of mouse embryo fibroblasts (Pluznik and Sachs 1965; Coffino et al. 1972). Cell-free binding of steroid-specific receptors was measured in crude cytosol preparations using either activated charcoal (Rousseau, Baxter and Tomkins 1972) or a DEAE filter paper method (Santi et al. 1973).

RESULTS AND DISCUSSION

S49 lymphoma cells plated in agar on mouse fibroblast feeder layers clone with an efficiency close to 100%. If, however, the medium contains the synthetic glucocorticoid dexamethasone, the cloning efficiency is greatly diminished (Fig. 1). Death resulting from treatment with glucocorticoids can also be quantitated by counting cells with a Coulter Counter or by recording their ability to exclude trypan blue (Rosenau et al. 1972; Gehring, Mohit and Tomkins 1972). Populations of steroid-resistant S49 cells have been derived by growing them in the presence of increasing concentrations of dexamethasone (Rosenau et al. 1972). Experiments were carried out to obtain a quantitative estimate of the frequency of steroid-resistant variants

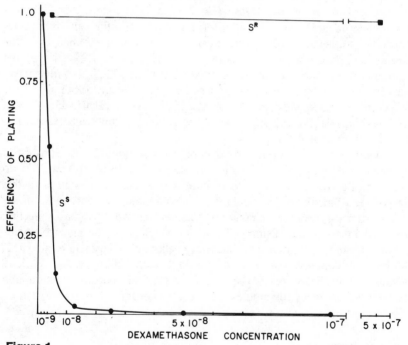

Figure 1
Growth response of steroid-sensitive and resistant clones to dexamethasone. Appropriate dilutions of sensitive (•) and resistant (■) clones were plated in the presence of the indicated concentrations of dexamethasone. The number of clones per plate was counted at the end of 10 days. The efficiency of plating shown is the average of five identical plates.

in the "wild-type" sensitive population and to determine whether the transition from sensitivity to resistance is a random event (Luria and Delbrück 1943). The data (Sibley and Tomkins 1973) show that the transition indeed occurs at random, at a frequency of approximately once every 3×10^5 cell divisions, and is independent of the presence of the selective agent. Steroid-resistant cells, once selected, give rise to stably resistant progeny capable of growing in the presence of high concentrations of dexamethasone (Fig. 1).

These results suggest that steroid resistance arises as a result of a mutation. This view is strengthened by experiments with several mutagens that caused a significant increase in the frequency of appearance of steroid-resistant clones (Sibley and Tomkins 1973).

Previous studies on mixed populations of steroid-resistant S49 lymphoma cells had indicated that such cells had lost the steroid binding activity of the specific cytoplasmic receptors (Rosenau et al. 1972). To determine whether this is also the case when the cells are cloned in the presence of the steroid, about 60 resistant clones were picked from a recently cloned population and the total specific cellular binding and the intracellular distribution of the specifically bound steroid were determined. The data in Fig. 2 show that by this method two general classes of steroid-resistant cells are selected: those in which overall cellular uptake is diminished, and those in which it is normal.

The crude cell fractionation allows a further subdivision of the latter class of resistant cells into those with a normal distribution of bound steroid (about 40% particulate and 60% soluble under these conditions) and those in which there is a decrease in the fraction of bound steroid localized in the crude nuclear fraction. Cell-free studies on the formation of complexes between dexamethasone and the specific cytoplasmic glucocorticoid receptors performed on representative clones have shown that cells with a decreased total uptake are deficient in receptor activity (Fig. 3). Those clones of resistant cells that retained the labeled hormone in whole cell experiments, however, possess receptors with normal steroid binding properties (Fig. 3). This is true whether or not nuclear localization of the receptor-steroid complex takes place in a given clone.

To facilitate discussion, we refer to steroid-resistant cells (S^R) which are defective in receptor activity as r^-. Those which contain receptor but have diminished nuclear localization of the receptor-steroid complex are designated r^+n^-. Those clones in which receptor-steroid complex is formed and accumulated in the nucleus, but which are nonetheless resistant to glucocorticoid killing, are called "deathless" and designated d^-. In the present studies about half of the steroid-resistant clones contain the cytoplasmic receptor and half are r^-. Of the r^+ resistant clones, a certain proportion is defective in nuclear uptake, r^+n^-, while the remainder is, by these criteria, $r^+n^+d^-$. Steroid-resistant $r^+n^+d^-$ lymphoid cells have also been described in a previous study (Gehring, Mohit and Tomkins 1973).

The present findings show that mutants in hormonal regulation can be isolated and analyzed. Although thus far the results have not shed new light on the mechanisms of hormone action, they at least confirm, by an entirely different set of observations, previous biochemical studies. Clearly these investigations would be facilitated by an understanding of the killing mechanism itself. Our work on this process is as yet quite preliminary, but nevertheless it complements the more detailed biochemical analysis of the lympholytic reaction carried out by other investigators (e.g., Makman, Dvorkin and White 1971; Hallahan, Young and Munck 1973). It seemed

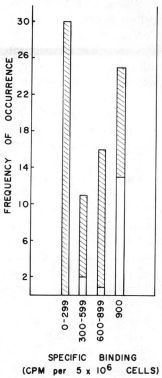

SPECIFIC   BINDING
(CPM per 5 x 10⁶   CELLS)

Figure 2

Steroid binding in vivo by steroid-sensitive and resistant (shaded portion)
lymphoma clones. Duplicate samples of 5×10^6 cells of each clone were exposed
at 37°C for 40 min to 3 or 4×10^{-9} M [³H]dexamethasone (35 Ci/mmole) in
the presence and absence of a 10^4-fold excess of nonradioactive dexamethasone.
The total cellular radioactivity was assessed in aliquots following centrifugation
($800 \times g$ for 2 min) and washing with isotonic saline (pH 7.6) at 0°C. The
remaining cells were disrupted and the distribution of label between the crude
nuclear and soluble fractions determined. The amount of specifically bound
steroid is given by the difference between the radioactive contents of the comparable samples with and without competitive unlabeled steroid (Rousseau,
Baxter and Tomkins 1972). The data are arbitrarily put into four different
groups.

advisable to determine first whether the primary action of the steroids is to deprive
the cells of essential components or whether the hormones induce products which
are themselves lethal (for example, by interfering with the uptake of required substances). In our laboratory this question has been approached using cell hybridization techniques. In this study cells of a steroid-sensitive, S^s ($r^+n^+d^+$), myeloma
line CL4 were fused with cells of a receptor-containing mouse lymphoma line,
EL4, which despite nuclear localization of the receptor steroid complex is not killed
by dexamethasone, i.e., S^R ($r^+n^+d^-$). The hybrid clones studied were susceptible
to the killing action of dexamethasone (Gehring, Mohit and Tomkins 1972), suggesting that the d^+ killing function of the CL4 parent can overcome the "deathless" condition of the EL4 cell line.

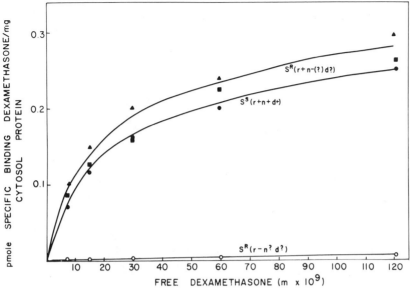

Figure 3

Specific steroid binding by extracts of steroid-sensitive and resistant lymphoma clones. Fresh cell pellets were homogenized at 2°C in hypotonic buffer (20 mM Tricine, 2 mM CaCl$_2$, 1 mM MgCl$_2$ pH 7.8) and spun at 100,000 × g for one hour. These particle-free supernatants were incubated at 2°C for 90 min with the indicated concentration of [^{3}H]dexamethasone (35 Ci/mmole) with and without 5 × 10^{-5} M unlabeled dexamethasone as competitor. The difference between these is plotted as specific binding. Values are the average of 3 to 4 determinations. Protein concentrations were determined by the method of Lowry et al. (1951). The resistant clones were either r$^-$ (○) or r$^+$. One of the latter (▲) was shown to be r$^+$ n$^-$ by the technique described in the legend to Fig. 2.

A further suggestion as to the nature of the killing substance was provided by different types of experiments. Glucocorticoid treatment of thymus cells inhibits the cellular uptake of glucose, uridine, and amino acids. These effects are similar to those of serum deprivation on cultured fibroblasts (Hershko et al. 1971; Cunningham and Pardee 1969); and this "pleiotypic" response (Hershko et al. 1971) has recently (Kram, Mamont and Tomkins 1973; Kram and Tomkins 1973) been related to the accumulation of cyclic AMP in serum-deprived fibroblasts. The apparent similarity of the responses of fibroblasts to serum starvation and of lymphoid cells to glucocorticoids raised the possibility that these steroids might also increase the concentration of the cyclic nucleotide, which could then be the mediator of cell death. Earlier work (Granner et al. 1968) had shown that the glucocorticoids do not stimulate adenylate cyclase directly and that cyclic AMP is not a direct intermediate in the steroid induction of tyrosine aminotransferase. However it seemed possible that steroids might augment intracellular cyclic AMP levels indirectly (Manganiello and Vaughan 1972). We therefore inquired whether dexamethasone increases cyclic AMP levels. The experiments shown in Table 1 indicate that within 8 hours of exposure of S49 cells to the steroid there was approximately a 2-fold increase in cyclic nucleotide concentration. Since these results do not

Table 1

Effect of Dexamethasone on Cyclic AMP Levels in S49 Lymphoma Cells

	Time (hr)	cyclic AMP (pmoles/10⁶ live cells)	Viability of cells (%)
10^{-4} M theophylline	0	1.3	92
	2	1.4	—
	5	1.7	—
	8	2.3	86
	24	2.1	47
No theophylline	0	1.2	92
	8	2.9	87
	24	1.6	55

100 ml cultures of cells (10^6 cells/ml) were incubated in the presence of 10^{-6} M dexamethasone in the presence or absence of theophylline. At the times indicated cells were collected by centifugation and the cyclic AMP content was determined by the method of Gilman (1970). Cell viability was determined by the trypan blue exclusion test.

indicate whether this increase is sufficient to produce cell death, we decided to test the effects of exposing S49 cells directly to the cyclic nucleotide. Figure 4 shows that the addition of the dibutyryl derivative of cyclic AMP to cultures causes cell death, consistent with the possibility that steroids and cyclic nucleotides might somehow interact in the killing reaction. Several types of relationships between these effectors might be imagined. For instance, since the steroid increases cyclic

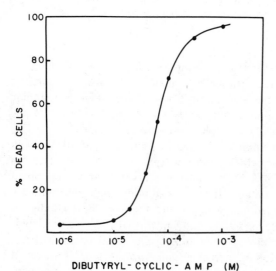

DIBUTYRYL - CYCLIC - A M P (M)

Figure 4

Effect of dibutyryl cyclic AMP on S49 lymphoma cells. Cells were seeded at a density of 3×10^5/ml in 5 ml medium containing 5×10^{-4} M theophylline to which dibutyryl cyclic AMP had been added at various concentrations. After 48 hr cell viability was determined by trypan blue exclusion in triplicate cultures.

AMP levels, the cyclic nucleotide itself might cause the death of the cells. Alternatively the steroid might induce the synthesis of macromolecules that kill cells independent of the cyclic nucleotide while cyclic AMP would augment the lethal activity. Finally, the steroid- and cyclic AMP-mediated cell killing might be completely independent. To test these possibilities a population of cells resistant to cyclic AMP was selected (Daniel, Litwack and Tomkins 1973) by growing S49 cells in increasing concentrations of the dibutyryl derivative of the cyclic nucleotide.

To show that resistance to the exogenous cyclic nucleotide was not due simply to its exclusion from resistant cells, experiments were performed to test the effects on cellular growth of intracellularly generated cyclic AMP. Isoproterenol (Daniel, Bourne and Tomkins 1973) and cholera enterotoxin (H. Bourne, unpublished) increase intracellular cyclic AMP levels in both sensitive and resistant cell populations. These agents slow the growth and decrease the viability of sensitive, but not resistant, cells (Table 2). On the basis of these experiments it seemed that the intracellular actions of cyclic AMP must be blocked in the resistant cells. This supposition was confirmed (Daniel, Litwack and Tomkins 1973) by the finding (Fig. 5) that cytoplasmic extracts from resistant cells bind added radioactive cyclic

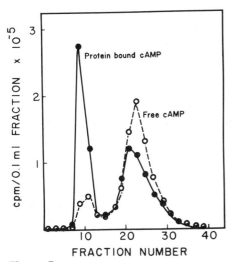

Figure 5

Gel filtration of [³H]cAMP-bound cytosols from dibutyryl cAMP sensitive and resistant cells. Centrifuged packed cells (1–2 ml) were homogenized with an equal volume of 0.25 M sucrose in 0.05 M Tris HCl buffer (pH 7.4) containing 3 mM MgCl₂ and 4 mM 2-mercaptoethanol. Particulate material was removed by a 90-min centrifugation in a Spinco preparative ultracentrifuge at 40,000 rpm (105,000 × g). The same amounts (25 mg of protein) of the resulting cytosols from both dibutyryl cAMP sensitive (●) and resistant (○) cells were incubated for 60 min at 0°C with 0.5 μM [³H]cAMP in the presence of 5 mM theophylline. The macromolecular-bound [³H]cAMP was then separated from free [³H]cAMP by filtration through a 1.2 × 20 cm Sephadex G-25 column equilibrated with 0.05 M Tris HCl buffer (pH 7.5). One-ml fractions were collected and samples of 0.1 ml were counted for radioactivity in a scintillation counter with a mixture of Triton X100-water-Omnifluor-toluene (25:4:0.4:70). From Daniel, Litwack and Tomkins (1973) with permission of *Proc. Nat. Acad. Sci.*

nucleotide much less than do comparable extracts from sensitive cells. Additional experiments revealed that this defect in resistant cells was accompanied by diminished activity of the cyclic AMP-mediated protein kinase. These results, together with the absence of the effects of both exogenous and endogenous cyclic nucleotide on the cellular growth of resistant cells, suggest that the resistant populations do not respond to cyclic AMP because of the deficiency in the cyclic AMP-binding protein and its associated kinase.

Although the molecular basis of this deficiency is not as yet known, several characteristics of the resistant population are of particular interest with respect to cellular growth control. The resistant population has a significantly faster doubling time (16 hours) than the wild-type sensitive population (19–20 hours) (see Table 2), and resistant populations grow to higher densities than do sensitive populations. These results are consistent with the idea presented in other reports in this volume and with our own studies (Kram, Mamont and Tomkins 1973; Kram and Tomkins 1973) that cyclic AMP inhibits cellular growth by controlling macromolecular synthesis and precursor uptake in fibroblasts. Furthermore cyclic AMP-resistant lymphoma cells stick much less to the substratum than the sensitive cells do if they are deprived of serum.

More relevant to our present concerns, however, is the question of the steroid sensitivity of the cyclic AMP-resistant populations. We find that cyclic AMP-resistant S49 and CL4 cells are indeed killed by dexamethasone, which is inconsistent with the possibility that the steroid-mediated increase in intracellular cyclic AMP leads to cell death. The effects of cyclic AMP on the growth of both "deathless" and "receptorless" steroid-resistant S49 and "deathless" EL4 cells was tested. The cyclic nucleotide either inhibits or kills these cells, suggesting that the steroid and cyclic AMP promote the death of lymphoid cells by independent mechanisms.

CONCLUSIONS

The results in the paper show that the growth inhibitory properties of the glucocorticoids, their so-called "catabolic effects," can be studied in cell culture. These actions

Table 2

Effect of Adenylate Cyclase Stimulators
on S49 Lymphoma Cells

	Cells ($\times 10^{-5}$/ml) after 48 hr incubation			
	db-cAMP-sensitive		db-cAMP-resistant	
	alive	dead	alive	dead
No drug	9.2	0.6	18.6	3.0
Isoproterenol				
(1×10^{-6} M)	3.3	3.3	13.5	2.0
Cholera enterotoxin				
(30 ng/ml)	3.8	3.0	16.2	2.2

Cells at a density of 2.5×10^5/ml were incubated under conditions previously described (Daniel, Bourne and Tomkins 1973) in the presence of theophylline (10^{-3} M). Cell viability was determined by the trypan blue exclusion test.

are mediated by glucocorticoid receptor molecules similar to those involved in enzyme induction in hepatoma cells, suggesting that the killing is the result of the enhanced synthesis of new macromolecules. The results also show that certain immunocytes behave like fibroblasts in that their growth is inhibited by cyclic AMP. Although the glucocorticoids produce a small and very late increase in the cyclic AMP concentration in cultured lymphoma cells, it does not appear that these increases are the basis of the steroid killing. Cells resistant to cyclic AMP display other "nonselected" changes in their phenotype, such as enhanced growth rate and ability to grow to higher density, consistent with the actions of cyclic AMP as a negative regulator of cell growth. Finally and perhaps most significant, mutants in various hormone-mediated cellular functions can be derived and analyzed by a combination of genetic and biochemical techniques.

Acknowledgments

This study was supported by grants no. GM 17239 and GM 19527 of the National Institute of General Medical Sciences of the NIH and contract no. CP 33332 within the Virus Cancer Program of the National Cancer Institute of NIH.

REFERENCES

Baxter, J. D., G. G. Rousseau, S. J. Higgins and G. M. Tomkins. 1973. Mechanism of glucocorticoid hormone action and of regulation of gene expression in cultured mammalian cells. *The biochemistry of gene expression in higher organisms* (ed. J. K. Pollak and J. W. Lee) pp. 206–224. Australia and New Zealand Book Co., Sydney, Australia.

Coffino, P., R. Baumal, R. Laskov and M. Scharff. 1972. Cloning of mouse myeloma cells and detection of rare variants. *J. Cell. Physiol.* **79**:429.

Cunningham, D. D. and A. B. Pardee. 1969. Transport changes rapidly initiated by serum addition to "contact inhibited" 3T3 cells. *Proc. Nat. Acad. Sci.* **64**:1049.

Daniel, V., H. R. Bourne and G. M. Tomkins. 1973. Altered metabolism and effects of endogenous cyclic AMP in cultured cells deficient in cyclic AMP binding proteins. *Nature New Biol.* **244**:167.

Daniel, V., G. Litwack and G. M. Tomkins. 1973. Induction of cytolysis of cultured lymphoma cells by adenosine 3':5'-cyclic monophosphate and the isolation of resistant variants. *Proc. Nat. Acad. Sci.* **70**:76.

Gehring, U., B. Mohit and G. M. Tomkins. 1972. Glucocorticoid action on hybrid clones derived from cultured myeloma and lymphoma cell lines. *Proc. Nat. Acad. Sci.* **69**:3124.

———. 1973. Interactions of glucocorticoids with cultured lymphoid cells. *Excerpta Medica* (in press).

Gilman, A. G. 1970. A binding protein assay for adenosine 3':5'-cyclic monophosphate. *Proc. Nat. Acad. Sci.* **67**:305.

Granner, D., L. R. Chase, G. D. Aurbach and G. M. Tomkins. 1968. Tyrosine aminotransferase: Enzyme induction independent of adenosine 3',5'-monophosphate. *Science* **162**:1018.

Hallahan, C., D. A. Young and A. Munck. 1973. Time course of early events in the action of glucocorticoids in rat thymus cells *in vitro*. *J. Biol. Chem.* **248**:2922.

Hershko, A., P. Mamont, R. Shields and G. M. Tomkins. 1971. The pleiotypic response. *Nature New Biol.* **232**:206.

Kram, R. and G. M. Tomkins. 1973. Pleiotypic control by cyclic AMP: Interaction with cyclic GMP and possible role of microtubules. *Proc. Nat. Acad. Sci.* **70:**1659.

Kram, R., P. Mamont and G. M. Tomkins. 1973. Pleiotypic control by adenosine 3':5'-monophosphate: A model for growth control in animal cells. *Proc. Nat. Acad. Sci.* **70:**1432.

Lowry, O. H., N. J. Rosebrough, A. L. Farr and R. J. Randall. 1951. Protein measurement with the folin phenol reagent. *J. Biol. Chem.* **246:**710.

Luria, S. and M. Delbrück. 1943. Mutations of bacteria from virus sensitivity to virus resistance. *Genetics* **28:**491.

Makman, M., B. Dvorkin and A. White. 1971. Evidence for induction by cortisol *in vitro* of a protein inhibitor of transport and phosphorylation processes in rat thymocytes. *Proc. Nat. Acad. Sci.* **68:**1269.

Manganiello, V. and M. Vaughan. 1972. An effect of dexamethasone on adenosine 3',5'-monophosphate content and adenosine 3',5'-monophosphate phosphodiesterase activity of cultured hepatoma cells. *J. Clin. Invest.* **51:**2763.

Pluznik, D. and L. Sachs. 1965. The cloning of normal "mast" cells in tissue culture. *J. Cell. Comp. Physiol.* **66:**319.

Rosenau, W., J. D. Baxter, G. G. Rousseau and G. M. Tomkins. 1972. Mechanism of resistance to steroids: Glucocorticoid receptor defect in lymphoma cells. *Nature New Biol.* **237:**20.

Rousseau, G. G., J. D. Baxter and G. M. Tomkins. 1972. Glucocorticoid receptors: Relations between steroid binding and biological effects. *J. Mol. Biol.* **67:**99.

Santi, D. V., C. H. Sibley, E. Perriard, G. M. Tomkins and J. D. Baxter. 1973. A filter assay for steroid hormone receptors. *Biochemistry* **12:**2412.

Scott, W. A., R. Shields and G. M. Tomkins. 1972. Mechanism of hormonal induction of tyrosine aminotransferase studied by measurement of concentration of growing enzyme molecules. *Proc. Nat. Acad. Sci.* **69:**2937.

Sibley, C. H. and G. M. Tomkins. 1973. *Proceedings of 13th international congress of genetics, abstract* p. 253.

Tomkins, G. M., B. B. Levinson, J. D. Baxter and L. Dethlefsen. 1972. Further evidence for posttranscriptional control of inducible tyrosine aminotransferase synthesis in cultured hepatoma cells. *Nature New Biol.* **239:**9.

Two Classes of Revertants Isolated from SV40-transformed 3T3 Mouse Cells

Arthur Vogel, Jan Oey,* and Robert Pollack

Cold Spring Harbor Laboratory, Cold Spring Harbor, New York 11724

A major unresolved problem is to determine which, if any, of the selective and non-selective differences between 3T3 and SV40-transformed 3T3 cells are the consequence of the continued expression of integrated viral genes (Eckhart, Dulbecco and Burger 1971). Our approach to this problem has been to isolate and characterize sublines that have reverted to resemble 3T3 in some or all of these differences, and to then compare SV40 viral expression in revertants to that in transformants.

Properties of Transformed Cells

Transformants isolated for their ability to grow to high cell densities on top of 3T3 monolayers (Todaro and Green 1963; Todaro, Green and Goldberg 1964) also grow to high saturation densities when cloned and propagated in 10% calf serum on plastic. We term clones with these two properties density transformants.

Normal 3T3 cells grow poorly in 1% calf serum (Holley and Kiernan 1968; Dulbecco 1970), gamma-globulin-free serum (Jainchill and Todaro 1970) or gamma-globulin-free serum depleted by incubation with confluent Balb-3T3 cells (Smith, Scher and Todaro 1971). Following infection with SV40, colonies of cells arise with a diminished serum requirement for growth. These colonies can grow in either 1% calf serum (Risser and Pollack this volume) or agamma-depleted calf serum (Smith et al. 1971). We term clones having this reduced serum requirement for growth serum transformants. Serum transformants may also be density transformed, but a significant fraction maintain a low saturation density in 10% calf serum (Scher and Nelson-Rees 1971; Risser and Pollack this volume).

SV40 infection of 3T3 also yields a small minority of cells with the ability to grow in methylcellulose or soft agar (Black 1966). These colonies can be recovered with a micropipette and cells from these colonies are capable of establishing new colonies when suspended in Methocel.

---

* Present address: University of Konstanz, Konstanz, Germany

SV40-transformed cells contain an intranuclear SV40-specific tumor antigen (Black et al. 1963). Transformed cells have a lower intracellular cyclic AMP concentration than normal cells (Otten, Johnson and Pastan 1971; Sheppard 1972; Seifert and Paul 1972). While neither the presence of viral products nor the changes in cyclic AMP concentration have yet been used to select transformants directly, both are found to have occurred in most transformants derived by the above selective assays.

Selection of Revertants

Variant sublines selected specifically for reversion in one of the transformed properties have been described (Pollack, Green and Todaro 1968; Culp, Grimes and Black 1971; Culp and Black 1972; Wyke 1971; Ozanne and Sambrook 1971; Ozanne 1973; Vogel and Pollack 1973; Vogel, Risser and Pollack 1973). Such revertant lines are isolated by negative selection from populations of transformed cells exhibiting all three transformed growth properties. In this technique culture conditions are such that cells with a single transformed property are killed, but cells resembling 3T3 in that property survive and can be recovered after removal of the toxic agent.

Density revertants grow to low saturation densities in excess amounts of serum (Pollack et al. 1968; Culp et al. 1971; Vogel et al. 1973), anchorage revertants are unable to form colonies in Methocel (Wyke 1971), and serum revertants cannot grow in 1% calf serum (Vogel and Pollack 1973).

We have asked whether reversion in one transformed property necessarily leads to reversion in the other transformed properties. Also we have attempted to characterize the amount and degree of expression of the SV40 genome in the revertant lines in order to ask whether the revertant properties are accompanied by a defect in SV40 expression. All the revertant lines discussed below were derived from a single clone of SV40-transformed 3T3 cells, SV101.

RESULTS
Growth Properties of Serum Revertants

Normal and transformed cells show markedly different responses to 1% calf serum (Table 1). Normal 3T3 cells grow poorly in 1% calf serum with a doubling time of 90 hours, whereas SV101 grows relatively well with a doubling time of 30 hours.

Serum revertants are isolated by plating SV101 in either 1% calf serum or 10% agamma-depleted calf serum (Smith et al. 1971) and adding BrdU to kill dividing cells (Vogel and Pollack 1973). Revertants isolated in 1% calf serum are designated as LsSV, and revertants isolated in 10% agamma-depleted serum are designated as AγSV. Neither type of serum revertant can grow at all in 1% calf serum (Fig. 1).

Although these lines were selected only for serum reversion, they have also reverted to a low saturation density (Table 1). This second reversion is not expected. Its consistent occurrence suggests that a high serum requirement for growth is not compatible with growth to a high cell density.

The two classes of serum revertants differ in their anchorage requirement for growth. Revertants isolated in 1% calf serum (LsSV1 and LsSV2) do not form

Table 1
Growth Properties of Mouse Cell Lines in 10% and 1% Calf Serum

	Line	Anchorage (growth in Methocel*)	Saturation density in 10% calf serum (cells/ cm² × 10⁴)	Doubling time in calf serum (hr)	
				1%	*10%*
Parent lines	3T3	0.001	5	90	21
	SV101	20	>45 (peels)	30	16
Serum revertants	AγSV4	2	15	>120	22
	AγSV5	11	10	>120	21
	LsSV1	0.001	9	>120	22
	LsSV2	0.04	12	>120	22
Density revertants	FlSV101	0.01	9	36	22
	BuSV2	0.02	13	50	25
	BuSV3	0.05	15	45	24

Cells were seeded in 4 ml of Methocel medium containing 10% calf serum and incubated for 3 weeks, with 4 ml of fresh medium added every week. Colonies larger than 0.3 mm in diameter were counted (Vogel et al. 1973).

Cells were plated at $0.1–0.2 \times 10^4$ cells/cm² in medium containing the appropriate serum concentration. Counts were done daily by trypsinizing a plate and counting on a Coulter Counter. The medium was changed every 3 days.

* Visible spherical colonies (greater than 300 micron diameter) per 100 cells cultured in Methocel for 21 days.

colonies in Methocel, while revertant lines isolated in agamma-depleted serum form colonies in Methocel with a high efficiency (Table 1). Thus the Aγ revertants represent a unique class of cells which are transformed in only the anchorage property (Table 2).

Growth Properties of Density Revertants

Density revertants were isolated by plating transformed cells in 10% calf serum and killing the cells capable of growing in dense culture with FdU (Pollack et al. 1968) or BrdU (Vogel and Pollack 1973). FlSV101 is a density revertant isolated with FdU, and BuSV2 and BuSV3 are revertants isolated with BrdU. These lines have reduced saturation densities, comparable to 3T3 (Table 1).

Reversion in saturation density can be accomplished without affecting the transformed serum requirement, since density revertants grow in 1% calf serum with a doubling time similar to that of SV101 (Fig. 1). Density revertants show that the properties of serum transformation and density transformation are closely, but not inextricably, linked, since in these lines density inhibition of growth is compatible with a low serum requirement.

Relation of Saturation Density to Serum Requirement

For all lines examined saturation density is dependent on serum concentration (Fig. 2). The data in Fig. 2 are in agreement with the hypothesis that alterations in

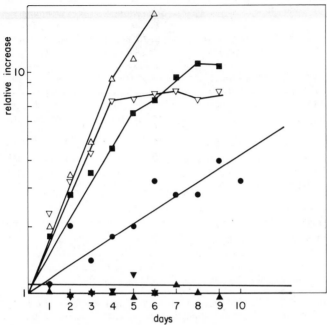

Figure 1
Growth of serum revertants in 1% calf serum. Cells were plated at 0.1–0.2 ×
10^4 cells/cm^2 in 35-mm dishes. Cells per plate were determined daily on a
Coulter Counter. The medium was changed every 3 days. (•) 3T3; (△) SV101;
(▽) FlSV101; (■) BuSV3; (▲) LsSV2; (▼) AγSV5.

saturation density are the result of changes in serum requirement (Holley and
Kiernan 1968). However, while the slopes of the graphs for SV101, 3T3 and the
density revertants BuSV3 and FlSV101 are similar (Fig. 2a), the serum revertants
LsSV2 and AγSV5 have steeper slopes than all the other lines (Fig. 2b).

This implies that serum revertants utilize serum in a manner different from either
the parent lines or the density revertants. Specifically these data distinguish two
classes of cells, which have similar saturation densities in 10% calf serum but
respond quite differently to 1% calf serum. The serum revertants LsSV2 and
AγSV5 fail to grow in 1% calf serum and grow to low saturation density in 10%
calf serum, whereas the density revertants FlSV101 and BuSV3 have a low saturation density in 10% calf serum despite their ability to grow well in 1% calf serum.

Normal 3T3 cells cease DNA synthesis and mitosis upon forming a confluent
monolayer, but SV101 cells continue cell division after confluence is reached (Fig.
3; Table 3). We have examined the regulation of cell division in 10% calf serum
of one density revertant, FlSV101 (Pollack and Vogel 1973). At confluence
FlSV101 cultures show a marked decrease in the fraction of mitotic cells (Fig. 3;
Table 3). While 15% of FlSV101 cells synthesize DNA at confluence, the rate of
DNA synthesis at confluence is only a few percent of the rate of synthesis in
logarithmically growing cells (Table 3). We conclude that FlSV101 regulates its
saturation density in a manner similar to 3T3.

Table 2

Properties of Revertants Isolated from SV101

Line	Saturation density[a]	Serum requirement[b]	Anchorage requirement[c]
3T3	Normal	Normal	Normal
SV101	Transformed	Transformed	Transformed
Density revertants selected with:			
FdU	Normal	Transformed	Normal
BrdU	Normal	Transformed	Normal
Serum revertants selected in:			
1% calf serum	Normal	Normal	Normal
Aγ-depleted calf serum	Normal	Normal	Transformed

[a] Normal cells have saturation densities less than 15×10^4 cells/cm² in 10% calf serum.

[b] Assayed by growth in 1% calf serum. Normal cells have doubling times greater than 90 hr; transformed cells double in 35 hr or less.

[c] Assayed by ability to form a colony in Methocel. Normal cells do not form colonies in Methocel; transformed cells do.

We do not know how the serum revertants maintain a low saturation density. They may show a density-dependent decrease in DNA synthesis at confluence, or they may continue to proliferate and shed at confluence (Scher and Nelson-Rees 1971).

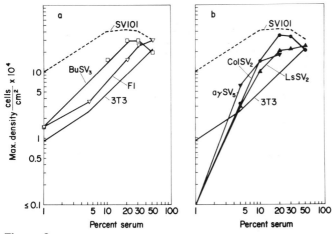

Figure 2

Saturation density as a function of serum concentration. Cell growth was determined as described above in varying concentrations of serum. ColSV2 is a revertant isolated with colchicine (Vogel et al. 1973). From Vogel and Pollack (1973) with permission of *J. Cell. Physiol.*

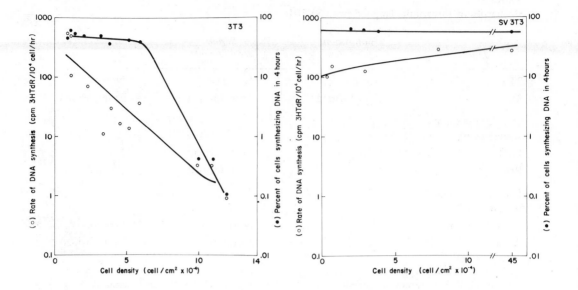

Figure 3

Fraction of cells synthesizing DNA and rate of thymidine incorporation at different cell densities. Cells were inoculated into 9.6-cm² dishes containing sterile 12-mm circular glass coverslips (area = 1.13 cm²). After 4 hr exposure to 2.5 μCi/ml [³H]thymidine in the presence of 10^{-5} M cold thymidine, coverslips were washed twice in PBS, fixed in ethanol-acetic acid (2:1), air-dried, and mounted cell-side up on microscope slides. The slides were dipped in melted (40°C) Kodak NTB-2 emulsion, air-dried for 2 hr, stored at 4°C for 48 hr, developed in Dektol, and permitted to dry and harden overnight. The developed slides were then stained in hematoxylin (Harris), blued in LiCl₂, cleared

through alcohol to xylene, mounted in Permount, and examined at 500×, phase contrast (Zeiss). Silver grains were seen only overlying nuclei, which had been stained blue by the hematoxylin. Cytoplasmic localization of [³H]thymidine, a sign of contamination by mycoplasma, was never observed (see Fig. 2).

After exposure to 2.5 μCi/ml [³H]thymidine in the presence of 10^{-5} M cold thymidine for periods from 1 to 6 hr, cells were solubilized with 1 ml 1 N NaOH and the DNA was precipitated with an equal volume of cold 20% TCA. The precipitates were trapped on glass-fiber filters and counted. The rate of DNA synthesis at a given cell density was obtained from a plot of [³H]thymidine incorporated versus incorporation time.

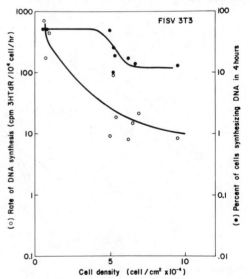

Table 3

Mitotic Fraction of Cell Lines at Different Cell Densities

Cell density	Percent of cells in mitosis/hr[a]		
	3T3	SV101	FlSV101
Sparse	3.7	4.2	5.4
Partly confluent	0.08		0.48
Fully confluent	0.04	7.6	0.18
Residual activity[b]	1.1	180	3.3

[a] After exposure to 0.01 μg/ml Velban (Gibco) for 2 or 3 hr, attached cells were trypsinized, pooled with cells floating in the medium, centrifuged, swelled in 0.38% KCl, fixed in methanol-acetic acid (3:1) overnight, recentrifuged and spread onto clean, wet, cold slides. The slides were stained with Giemsa, dehydrated, mounted and examined at 400×. At least 1000 nuclei were counted for each determination. The counts were normalized to fraction in mitosis/hr.

[b] Mitotic index at confluence as percent of rate in sparse culture.

Alterations in DNA Content per Cell

All revertants isolated from SV101 have more chromosomes and more DNA per cell than the transformed parent line (Fig. 5). Using the Los Alamos Flow Microfluorometer, we have found that SV101 and 3T3 have, as expected for established cell lines, more DNA per cell than normal somatic mouse cells (Fig. 4). If primary mouse embryo fibroblasts are assumed to have twice the haploid amount of DNA (2c), then 3T3 and SV101 each have three times the haploid amount (3c) (Fig. 4). When serum and density revertants derived from SV101 were examined, all were found to contain more DNA than either 3T3 or SV101 (Fig. 4, 5). For example, FlSV101 density revertant cells had a DNA content equivalent to 6.5c (Fig. 4).

The average chromosome number per cell also increased in all revertants (Fig. 5). However the variation in chromosome number per cell was too great in these aneuploid lines to permit us to estimate whether the increase in chromosome number was equivalent to the increase in DNA for each line.

These increases in DNA and chromosomes are not merely an inadvertant consequence of the selection procedure per se. The lines aγ61, aγ12d, aγ256 survived negative selection with BrdU in agamma-depleted serum, but still grew in 1% calf serum and grew dense in 10% calf serum. These lines did not acquire any DNA or chromosome increase (Fig. 5). Also dense sublines derived from FlSV101 have regained the lower chromosome number characteristic of SV101 (Pollack, Wolman and Vogel 1970). Apparently the increases we observe are not nonspecific alterations but are specifically associated with the revertant phenotype.

Intracellular Concentration of Cyclic AMP

Cyclic AMP has been implicated as a possible mediator of cell growth (Otten et al. 1971; Sheppard 1972). To investigate the role that cAMP may play in the regulation of cell growth, we have measured intracellular cyclic AMP levels in

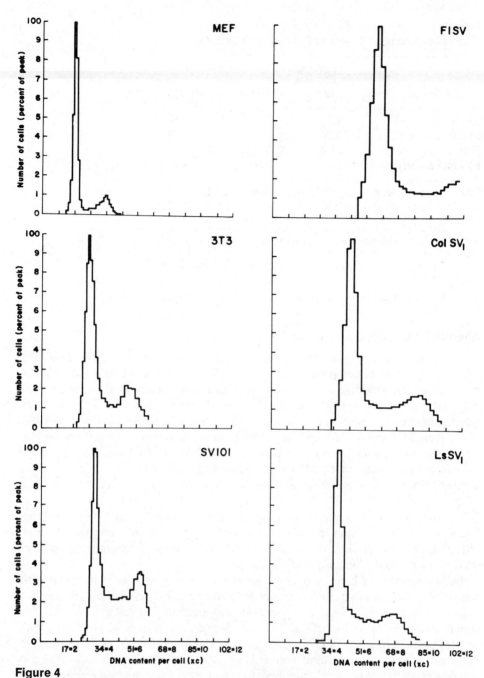

Figure 4

The distributions of DNA content (in arbitrary units) from Flow microfluorometry of growing cell lines. On the abscissa, the arbitrary scale is converted to multiples of the haploid amount (h) of mammalian cell DNA. For each line, the left-most peak represents cells in G_1. *MEF*, normal primary mouse embryo fibroblasts; *3T3*, normal established mouse cell line; *SV101*, SV40-transformed clone derived from 3T3; *FlSV*, density revertant derived by FdU selection from SV101; *ColSV*$_1$, density revertant derived by colchicine selection from SV101 (Vogel, Risser and Pollack 1973); *LsSV*$_1$, serum revertant derived by BrdU selection from SV101.

132

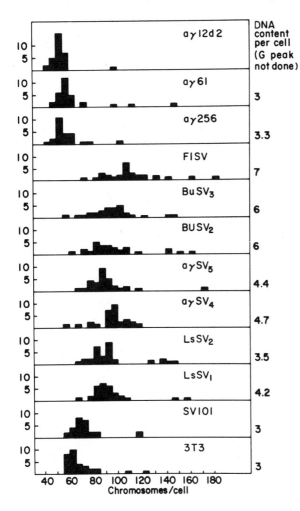

Figure 5

Chromosomes and DNA content per cell. aγ12d2, aγ61, and aγ256 are lines which survived the serum selection but can still grow in 1% calf serum. DNA content per cell expressed as a multiple of the haploid DNA content of primary mouse fibroblasts (see Fig. 4). From Vogel and Pollack (1973) with permission of *J. Cell. Physiol.*

three growth conditions: sparse culture in excess serum, confluence in excess serum, and sparse culture in low concentrations of serum. The first condition compares the cyclic AMP levels among cultures that are all growing at similar rates. The latter two conditions are restrictive for some lines but not others. Therefore they compare cells able to grow versus cells unable to grow.

Sparse Density in 10% Calf Serum

3T3, SV101 and all the revertants grow with approximately the same doubling time in 10% calf serum. Although the doubling times are similar, the intracellular cyclic AMP levels in these cells are not (Table 4). Normal 3T3 cells have a cAMP level twice that of SV101. This finding agrees with other published reports (Otten et al. 1971; Sheppard 1972). With the exception of the Aγ serum revertants, reversion in either the serum or density property is accompanied by reversion to a high cAMP concentration (Table 4).

Confluent Densities in 10% Calf Serum

3T3 and the density revertants cease to increase in cell number at confluent cell densities, but SV101 cells continue to grow. The cAMP concentration in confluent cultures of 3T3 has been reported to increase with increasing cell density (Otten et

Table 4
Cyclic AMP Concentrations in Sparse Cultures of
Mouse Cell Lines in Excess or Restrictive Serum

		cAMP level (pmoles/mg protein)[a]			
		10% calf serum		1% calf serum	
	Line	range[b]	avg	range[b]	avg
Parent lines	3T3	20–28	25	75–100	83
	SV101	8–14	12	25–40	31
Serum revertants	LsSV1	17–22	20	52–70	68
	LsSV2	21–30	24	75–80	78
	aγSV4	10–15	13	60–75	67
	aγSV5	10–16	15	60–85	72
Density revertants	FlSV101	21–25	23	31–43	37
	BuSV2	17–25	21	30–46	34

[a] Cyclic AMP was determined by the Gilman assay.

[b] Four experiments, 4 determinations each, measured at 2 days after plating. Cell densities were 5×10^4–2×10^5 cells/plate, except SV101, which was 1×10^5–5×10^5.

al. 1971). When we determined cyclic AMP concentrations at different cell densities in 10% calf serum, we found not an increase, but a decrease in cyclic AMP at confluence (Table 5). This decrease occurred in all lines tested, even those that ceased to increase in cell number at confluence. Even the Aγ serum revertants, which might have been expected to show an increase in cAMP at confluence, maintained their low level.

Table 5
Cyclic AMP Concentration vs. Cell Density

		cAMP concentration	
Line	Cell density (cells/28-cm² dish)	pmoles/mg protein	pmoles/10⁶ cells
3T3	2.5×10^5	21	12
	1.5×10^6	17	9
SV101	7.5×10^5	11	7
	1.5×10^7	8	1.5
LsSV2	2.5×10^5	21	13
	3.0×10^6	10	4
aγSV4	2.5×10^5	10	6
	3.0×10^6	6	2

Cells were seeded at 10^5 cells per 60-mm dish. cAMP was determined 2 days after plating. Upon reaching confluence the medium was changed and the cAMP level was determined 3 days later.

Sparse Density in 1% Calf Serum

This condition restricts the growth of the serum revertants and 3T3, but does not markedly affect the growth of SV101 or the density revertants. Cell lines unable to grow well in 1% calf serum showed a marked increase in cAMP concentration (Table 4). The concentration of cAMP in restricted lines exceeded 70 pmoles/mg protein, which was more than twice the concentration found in lines (such as SV101 and the density revertants) that grew in 1% calf serum (Table 4).

These data permit three conclusions. First, growing cultures maintain different levels of cAMP, and these levels are higher in lines that have the ability to regulate their growth at high cell density. Second, the process of density inhibition of cell growth does not lead to a further rise in cAMP in such cultures at confluence. Third, cells that are unable to grow in sparse culture in 1% calf serum because of an increased requirement for serum increase their cAMP concentration markedly; this increase correlates with the serum dependence, and does not correlate with the density inhibition, of these lines.

SV40 in Revertant Cell Lines

All of the revertants contain SV40 T antigen detectable by immunofluorescence (Table 6) and SV40-specific RNA (Table 7). The number of SV40 genomes per haploid amount of cell DNA was found to be the same in the density revertant FlSV101 as the parent SV101 line (Ozanne, Sharp and Sambrook 1973; Ozanne et al. 1973).

Most revertants yielded little or no infectious SV40 after fusion with permissive BSC-1 cells (Watkins and Dulbecco 1967). SV40 virus recovered after fusion was wild type with respect to its ability to transform 3T3 cells (Table 6). Mutant SV40, capable of growth on BSC-1 cells but defective in its ability to transform 3T3 cells, has not been rescued from any of the revertants.

Reinfection of the revertants with SV40 does not restore the high saturation density, nor does it confer the ability to grow in 1% calf serum (Table 6). However the revertants are retransformable by murine sarcoma virus, so they are capable of responding to an oncogenic virus (Table 6). This result extends the previous report of Renger (Renger 1972).

SV101 contains SV40 RNA sequences specific for 80% of the "early" strand of SV40 DNA and less than 10% of the "late" strand (Sambrook, Sharp and Keller 1972). The amount and strand specificity of SV40 RNA in the revertants are similar to those of the transformed parent (Table 7).

With one exception all variation in transcription seen among the revertants is within the range of variations seen among different SV40-transformed lines (Sambrook et al. 1972). The exception is in FlSV101, which contains less SV40-specific RNA than any of the other T antigen positive cell lines (Table 7). Despite this single exception, it is clear that reversion does not require a failure in transcription of SV40-specific DNA sequences.

DISCUSSION

The two classes of revertants, serum and density, demonstrate that it is possible to affect saturation density without altering the serum property, but that to alter the serum property is to simultaneously alter saturation density. A virus-transformed

Table 6

Rescue of SV40 from Revertant Lines

	Cell line	T antigen	PFU/10⁶ mouse cells after 6 days	PFU/TFU	Retransformed by	
					SV40	MSV
Parent lines	3T3	−	not done		+	+
	SV101	+	2000	1600	NT	NT
Serum revertants	LsSV1	+	0		−	NT
	LsSV2	+	0		NT	+
	aγSV4	+	0		−	+
	aγSV5	+	20	2000	−	NT
Density revertants	FlSV101	+	0		−	+
	BuSV2	+	0		NT	NT
	BuSV3	+	20	2000	−	NT

SV40 T antigen was assayed as immunofluorescence. Cells were fixed in cold methanol: acetone (1:2) for 10 min. The fixed cells were then stained with hamster anti-T antibody and counterstained with fluorescein-conjugated goat anti-hamster antibody.

One million transformed or revertant cells were fused with 10⁶ BSC-1 cells, plated and incubated for 6 days. The plates were then freeze-thawed twice, sonicated and plaqued on BSC-1 cells.

SV40 transformation was done according to Todaro et al. (1964) and MSV transformation was done according to Renger (1972).

Nt = not tested.

Table 7

Transcription of SV40 DNA
in Mouse Cell Lines

	% DNA sequences present in RNA[a]	
	E strand	L strand
3T3	0	0
SV101	81	4
FlSV101	40[b]	5[b]
BuSV2	59	1
BuSV3	64	4
LsSV1	42	1
LsSV2	50	1
aγSV3	70	10
aγSV4	78	12
aγSV5	60	10

[a] Taken from Ozanne et al. (1973). Hybridization was done as described in Sambrook et al. (1972).

[b] With this line it was impossible to saturate the SV40 DNA with RNA.

line with a high saturation density and normal serum requirement has not yet been isolated.

Two conditions that restrict cell growth have been described: confluent cell density in 10% serum and sparse density in 1% serum. Cyclic AMP levels are clearly dependent on the serum concentration and reflect the ability of a line to grow in a given serum concentration (Table 4). Normal 3T3 cells and the serum revertants cannot grow in 1% calf serum and have very high cAMP levels in 1% calf serum, whereas SV101 and the density revertants can grow in 1% calf serum and display similar low cAMP levels in 1% calf serum.

Normal 3T3 cells maintain a low saturation density by stopping DNA synthesis and mitosis upon forming a confluent monolayer. The density revertant FlSV101 shows a density-dependent decrease in rate of DNA synthesis, in the fraction of cells synthesizing DNA, and in the fraction of cells in mitosis.

Although FlSV101 does respond to contact, the shut off of DNA synthesis is not as "tight" as in confluent 3T3 cells. We do not yet know how the serum revertants maintain a low saturation density in 10% calf serum.

Revertants contain SV40 T antigen, and the amount and strand specificity of the SV40-specific RNA in the revertants resemble those of the transformed parent. The revertant phenotypes therefore are not the result of the inability to transcribe the SV40 genome. The fact that nondefective SV40 can be rescued from the revertants implies that the reversion in transformed growth properties does not require all copies of any single SV40 gene in the cell to be defective.

Acknowledgments

This work was supported by funds from the National Institutes of Health Cancer Center Grant #1-P01-CA13106-01. We thank Drs. S. Cram, H. Crissman, D. Petersen and M. van Dilla for their help with the Los Alamos Flow Microfluorometer apparatus, Rex Risser for his discussions, and Sue Arelt and Bonnie Mitchell for excellent technical assistance.

A.V. is supported by NIH training grant 5 T05 GM01668.

REFERENCES

Black, P. 1966. Transformation of mouse cell line 3T3 by SV40: Dose response relationship and correlation with SV40 tumor antigen production. *Virology* **28:**760.

Black, P., W. Rowe, H. Turner and R. Huebner. 1963. A specific complement-fixing antigen present in SV40 tumor and transformed cells. *Proc. Nat. Acad. Sci.* **50:**1148.

Culp, L. and P. Black. 1972. Contact inhibited revertant cell lines isolated from simian virus 40-transformed cells. III. Concanavelin A-selected revertant cells. *J. Virol.* **9:**611.

Culp, L., W. Grimes and P. Black. 1971. Contact inhibited revertant cell lines. I. Biologic, virologic and chemical properties. *J. Cell Biol.* **50:**682.

Dulbecco, R. 1970. Topoinhibition and serum requirement of transformed and untransformed cells. *Nature* **227:**802.

Eckhart, W., R. Dulbecco and M. Burger. 1971. Temperature-dependent surface changes in cells infected or transformed by a thermosensitive mutant of polyoma virus. *Proc. Nat. Acad. Sci.* **68:**283.

Holley, R. and J. Kiernan. 1968. Contact inhibition of cell division in 3T3 cells. *Proc. Nat. Acad. Sci.* **60:**300.

Jainchill, J. and G. Todaro. 1970. Stimulation of cell growth *in vitro* by serum with and without growth factor; relation to contact inhibition and viral transformation. *Exp. Cell Res.* **59**:137.

Otten, J., G. Johnson, and I. Pastan. 1971. Cyclic AMP levels in fibroblasts: Relationship to growth rate and contact inhibition of growth. *Biochem. Biophys. Res. Commun.* **44**:1192.

Ozanne, B. 1973. Variants of 3T3 cells transformed by SV40 that are resistant to concanavalin A. *J. Virol.* **12**:79.

Ozanne, B. and J. Sambrook. 1971. Isolation of lines of cells resistant to agglutination by concanavalin A from 3T3 cells transformed by SV40. *Lepetit Colloq. Biol. Med.* **2**:247.

Ozanne, B., P. Sharp and J. Sambrook. 1973. Transcription of simian virus 40. II. Hybridization of RNA extracted from different lines of transformed cells to the separated strands of SV40 DNA. *J. Virol.* **12**:90.

Ozanne, B., A. Vogel, P. Sharp, W. Keller and J. Sambrook. 1973. Transcription of SV40 DNA sequences in different transformed cell lines. *Lepetit Colloq. Biol. Med.* **4**. (in press).

Pollack, R. and A. Vogel. 1973. Isolation and characterization of revertant cell lines. II. Contact inhibition in a polyploid density revertant line derived from SV40 transformed 3T3 mouse cells. *J. Cell. Physiol.* **82**:93.

Pollack, R., H. Green and G. Todaro. 1968. Growth control in cultured cells: Selection of sublines with increased sensitivity to contact inhibition and decreased tumor-producing capacity. *Proc. Nat. Acad. Sci.* **60**:126.

Pollack, R., S. Wolman and A. Vogel. 1970. Reversion of virus transformed cell lines: Hyperploidy accompanies retention of viral genes. *Nature* **228**:967.

Renger, H. 1972. Temperature sensitive SV40 transformants retransformed by murine sarcoma virus at non-permissive temperature. *Nature New Biol.* **240**:19.

Sambrook, J., P. Sharp and W. Keller. 1972. Transcription of simian virus 40. I. Separation of the strands of SV40 DNA and hybridization of the separated strands to RNA extracted from lytically infected and transformed cells. *J. Mol. Biol.* **70**:57.

Scher, C. and W. Nelson-Rees. 1971. Direct isolation and characterization of "flat" SV40 transformed cells. *Nature New Biol.* **233**:263.

Seifert, W. and D. Paul. 1972. Levels of cyclic AMP in sparse and dense cultures of growing and quiescent 3T3 cells. *Nature New Biol.* **240**:281.

Sheppard, J. 1972. Differences in the cAMP levels in normal and transformed cells. *Nature* **236**:14.

Smith, H., C. Scher and G. Todaro. 1971. Induction of cell division in medium lacking serum growth factor by SV40. *Virology* **44**:359.

Stoker, M. 1968. Abortive transformation by polyoma virus. *Nature* **218**:234.

Todaro, G. and H. Green. 1963. Quantitative studies on the growth of mouse embryo cells in culture and their development into established lines. *J. Cell Biol.* **17**:299.

Todaro, G., H. Green and B. Goldberg. 1964. Transformation of properties of an established cell line by SV40 and polyoma virus. *Proc. Nat. Acad. Sci.* **51**:66.

Vogel, A. and R. Pollack. 1973. Isolation and characterization of revertant cell lines. IV. Direct selection of serum-revertant sublines of SV40-transformed 3T3 mouse cells. *J. Cell. Physiol.* (in press).

Vogel, A., R. Risser and R. Pollack. 1973. Isolation and characterization of revertant cell lines. III. Isolation of density-revertants of SV40-transformed 3T3 cells using colchicine. *J. Cell. Physiol.* (in press).

Watkins, J. and R. Dulbecco. 1967. Production of SV40 virus in heterokaryons of transformed and susceptible cells. *Proc. Nat. Acad. Sci.* **57**:1396.

Wyke, J. 1971. A method of isolating cells incapable of multiplication in suspension culture. *Exp. Cell. Res.* **66**:203.

Biological Analysis of Clones
of SV40-infected Mouse 3T3 Cells

Rex Risser and Robert E. Pollack

Cold Spring Harbor Laboratory, Cold Spring Harbor, New York 11724

A variety of selective assays have been used to monitor the transformation of normal cells into tumor cells. These assays take advantage of the ability of transformed cells to grow to high saturation densities (Todaro, Green and Goldberg 1964), to grow in the absence of anchorage to glass or plastic (Macpherson and Montagnier 1964; Stoker et al. 1968), to form colonies or foci on monolayers of normal cells (Temin and Rubin 1958), to grow in reduced concentrations of serum components (Holley and Kiernan 1968; Jainchill and Todaro 1970; Smith, Scher and Todaro 1971), and to grow in disoriented patterns (Stoker and Abel 1962).

The transformation of mouse 3T3 cells by SV40 is usually monitored by the first assay and is a low-frequency event. At very high virus doses (10^6 infectious units/cell) only 50% of the cells are transformed in this assay (Todaro and Green 1966; Black 1966). When such transformants are tested in other assays, they are found to have reduced serum requirements, to form colonies on normal monolayers, and to form colonies in agar or methylcellulose suspension. Cells which show all of these transformed growth properties we term standard transformants.

Recently, however, Smith et. al (1971) and Scher and Nelson-Rees (1971) have isolated SV40-transformed cells which differ from standard transformants. These cells are primarily altered in their serum requirements, though they have saturation densities comparable to 3T3. The existence of such cells demonstrated that SV40 infection could yield cells with only a partially transformed character. This raised the possibility that selective transformation assays were not detecting the full range of cellular alterations that were induced by SV40.

To isolate all possible types of SV40 transformants, a nonselective scan of clones of 3T3 cells arising after SV40 infection was carried out. Forty clones of SV40-infected 3T3 cells were picked without regard to morphology and analyzed in each transformation assay.

A density transformation assay carried out simultaneously on these infected cells showed a standard transformation frequency of 10%. However the nonselective scan demonstrated that fully 90% of the clones differed from 3T3 cells in their ability to utilize low concentrations of serum.

In addition to this change 25% of the clones showed intermediate growth properties in other assays and another 25% had growth properties comparable to those of standard transformants. A more thorough description of all clones will appear elsewhere.

Scheme of the Experiment

A recently cloned 3T3 line was infected at subconfluence (2×10^4 cells/cm²) with SV40 or mock-infected with Dulbecco's modified Eagles medium (DME). The virus was prepared from a low-multiplicity passage on BSC-1 monkey cells of a triply plaque-purified clone of SV40 (Strain 776 originally NIH). The virus was purified by standard procedures and banded twice to equilibrium in CsC1 (Black, Crawford and Crawford 1964). After collection it was dialyzed against serum-free medium (Dulbecco) and used immediately for infection. A separate aliquot of banded virus was used to determine the number of physical particles by optical density (Koch et al. 1967). The multiplicity of infection in this experiment was 2×10^3 pfu/cell, which corresponded to a physical multiplicity of 5×10^5–1×10^6 particles/cell.

The cells were allowed to recover from infection in medium with 10% calf serum. The next day they were trypsinized and replated at different dilutions in a standard density transformation assay (Sambrook and Pollack 1973). Medium (10% Colorado calf serum in DME) was changed every third day for two weeks. Some plates were then stained and others used for cloning. Dilution plates having between 10 and 20 colonies were scanned and well-isolated colonies were chosen without regard to colony morphology.

Each colony was photographed and then picked using trypsin within a steel cloning cyclinder. Clones were grown up and passaged weekly at a density of about 5×10^2 cells/cm² for one month. Forty isolated colonies were picked for experimental clones. Four isolated colonies from mock-infected plates and the 3T3 parental line served as control normal clones. Four dense foci picked from infected low-dilution plates and the standard transformant SV101 served as control standard transformants.

From stained plates the transformation frequency per colony was 11%. Of the clones picked 15% were judged to be transformed by the morphological criterion of high cell density. The plating efficiencies of virus-infected or mock-infected cells were 83% and 91%, respectively.

Properties of Clones of SV40-infected 3T3 Cells

Viral Antigens

Twenty-five of the 40 clones were tested for the presence of viral capsid antigen by indirect immunofluorescence. All clones were found to be negative, whereas BSC-1 cells infected three days earlier with SV40 fluoresced brightly in the nucleus.

SV40-specific T antigen was assayed by indirect immunofluorescence. Cells were plated onto 9-mm coverslips at a density of 2–4 $\times$ 10⁴ cells/cm². Twelve hours before staining, coverslips were placed in DME containing only 1% calf serum. To stain, the coverslips were rinsed in phosphate-buffered saline (PBS), fixed in cold acetone and incubated one hour with hamster antibody to SV40 T antigen (Flow

Laboratories). Coverslips were then rinsed with PBS and incubated with goat antibody to hamster gammaglobulin, which had been conjugated to fluorescein (Antibodies, Inc.). After rinsing in PBS the coverslips were examined at 400× magnification under dark-field ultraviolet illumination (Zeiss).

Three patterns of staining were seen: negative, positive and intermediate. Half of the 40 clones were negative by immunofluorescence, as were 3T3 cells and all mock-infected clones. In SV101 and standard transformants more than 95% of the cells stained uniformly positive; ten experimental clones stained identically to these transformants. In the remaining ten experimental clones, the staining pattern was heterogeneous. In such clones 10–30% of the cells stained as brightly as standard transformants, 50–80% of the cells stained weakly, and 10–20% of the cells did not show any nuclear fluorescence.

The validity of the staining pattern was tested by staining mixtures of SV101 cells and 3T3 cells. In such mixtures only positive or negative cells were seen; furthermore the percentage of positive cells depended on the proportion of SV101 cells in the original mixture. When SV101 and 3T3 cells were mixed in a ratio of 1 to 5, 215/1000 cells were positive; when SV101 and 3T3 cells were mixed in a ratio of 1 to 100, 11/1000 cells were positive. The cloning procedure was also checked by plating a mixture of SV101 and 3T3 (ratio 1 to 4) very sparsely and allowing colonies to form on coverslips. Seven of the 25 colonies scored were positive and 18 were negative. No intermediate or mixed colonies were seen.

T antigen levels have also been checked by complement fixation on equal numbers of cells of various clones. When two uniformly positive clones and SV101 were compared to eight T antigen intermediate clones, the complement-fixing titer of the three positive clones was tenfold greater than that of any of the intermediate clones. Two T antigen negative clones were also tested; both had background complement-fixing titers comparable to that of 3T3. Thus the indirect immunofluorescence and complement fixation assays both revealed similar levels of T antigen in the clones. In the intermediate clones a wide variation from cell to cell in intensity of staining was revealed by immunofluorescence.

Growth Properties in 10% and 1% Calf Serum

The saturation density and doubling times of each clone were obtained from growth curves derived from cultures in which cells were seeded sparsely on plastic dishes. Typical growth curves are given in Fig. 1. In 10% calf serum 3T3 cells grew to a density of 7.5×10^4 cells/cm^2 with a doubling time of 23 hours. SV101 cells grew to a density of 50×10^4 cells/cm^2 with a doubling time of 16 hours. Cells from one experimental clone, SVR-Cl 8–2, grew to a density of 15×10^4 cells/cm^2 with a doubling time of 20 hours.

In 1% calf serum these lines grew to densities of 0.9, 10 and 3×10^4 cells/cm^2 with doubling times of 76, 26 and 29 hours, respectively. Vogel and Pollack (1973) have found that the doubling times of cell lines in 1% calf serum can be used to distinguish a normal from transformed response to low serum. By this criterion SVR-Clone 8–2 is like the standard SV40-transformant SV101 in its utilization of serum to grow; it is, however, T antigen negative.

A compilation of data from each clone on saturation densities in 10% calf serum and doubling times in 1% calf serum is shown in Fig. 2 and 3. The saturation densities of control normal clones or standard transformed clones show little spread in

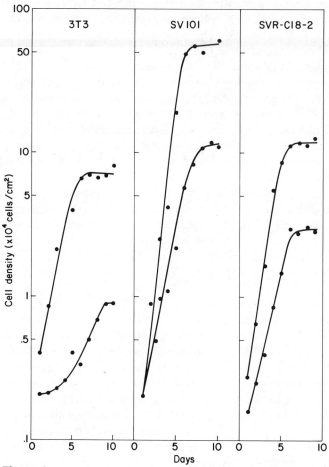

Figure 1

Growth curves of SV101, 3T3 and experimental clone SVR-Cl 8-2. *Upper curve,* growth in 10% calf serum-DME; *lower curve,* growth in 1% calf serum-DME. Cells were seeded at a density of 1–2 × 10³ cells/cm² in 2 ml of medium onto a 35-mm Falcon petri dish. Medium was changed every third day and cell counts were taken daily using a Coulter Counter.

values (Fig. 2); however, the 40 experimental clones show a spectrum of values ranging from that of 3T3 to that of standard transformants. In most cases the saturation density of a clone correlates with its T antigen level. Positive clones have high densities; intermediate clones have intermediate densities, and negative clones have densities closer to that of 3T3.

In contrast, the doubling time in 1% calf serum bears little or no correlation to the T antigen level of a given clone (Fig. 3). Only five of the experimental clones have doubling times comparable to 3T3 or mock controls; the remaining 35 double as efficiently in 1% calf serum as do transformants. The four experimental clones with doubling times of greater than 60 hours correspond to the four clones with saturation densities in 10% calf serum of < 8 × 10⁴ cells/cm².

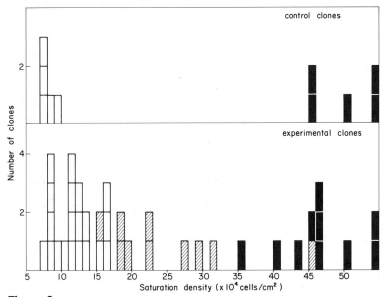

Figure 2

Histogram of saturation densities of control and experimental clones in 10% calf serum-DME. Black rectangles, T antigen positive clones; hatched rectangles, T antigen intermediate clones; white rectangles, T antigen negative clones.

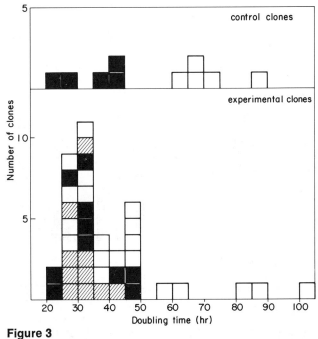

Figure 3

Histogram of doubling times of control and experimental clones during logarithmic growth in 1% calf serum-DME. Black rectangles, T antigen positive clones; white rectangles, T antigen intermediate clones; hatched rectangles, T antigen negative clones.

Anchorage Properties

Anchorage assays on each clone were carried out according to the method of Stoker et al. (1968) using methylcellulose. The colony-forming efficiencies of each clone in methylcellulose are presented in Fig. 4. As with saturation densities, a considerable range of colony-forming efficiencies can be seen. T antigen positive clones plated with an efficiency of 10–50%, whereas intermediate clones plated with an efficiency of 0.5–10%. One intermediate line (SVR-Cl 6–3), however, formed colonies with a very low efficiency.

Additional information on anchorage dependence of various clones was obtained from measurements on individual cells or colonies. Figure 5 presents the results of such measurements as histograms of the number of colonies vs. colony size. From these data we conclude that most cells from clones with intermediate plating efficiencies (and intermediate T antigen) are capable of a number of divisions in methylcellulose, though they do not form a visible colony in the 3-week assay time. In methylcellulose the T antigen intermediate clone least capable of division, SVR-Cl 6–3, corresponds to that with the lowest plating efficiency (0.002%). 3T3

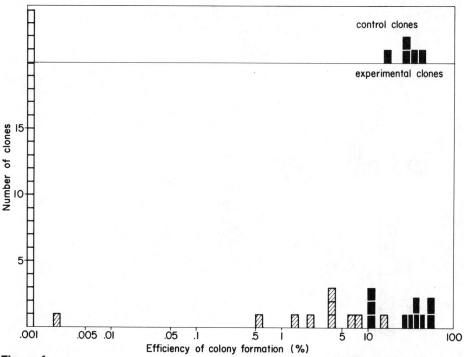

Figure 4
Histogram of efficiency of colony formation of control and experimental clones in 1.17% Methocel, 10% CS-DME. Medium and agar plates were prepared according to Stoker et al. (1968). Cells (10^5, 10^4, 10^3 or 10^2 in 4 ml Methocel medium) were plated in quadruplicate onto plates containing 3 ml solid agar. An additional 4 ml Methocel medium was added at 1 and 2 weeks. Colonies of 100–200 cells or greater were scored at three weeks with dissecting microscope. Plates having 30–200 colonies were used for scoring. Black rectangles, T antigen positive clones; hatched rectangles, T antigen intermediate clones; white rectangles, T antigen negative clones.

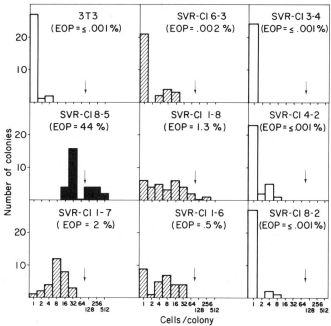

Figure 5

Histograms of colony sizes of various clones grown in Methocel. At 3 weeks 30 colonies were measured using a reticle eyepiece at random on plates inoculated with 10^4 cells. The eyepiece was calibrated using pollen grains of 13–90 μ in diameter. From measurements of 3T3 and SV101 the day after plating the mean cell diameter was 19.8 μ. The volume increase was calculated from the colony diameter and from this the colony size was inferred. Black rectangles, T antigen positive clones; hatched rectangles, T antigen intermediate clones; white rectangles, T antigen negative clones.

and other T antigen negative clones undergo few, if any, divisions in methylcellulose.

Colony Formation on Monolayers of Normal Cells

The ability of cells from each clone to form visible colonies on monolayers of 3T3 cells or on plastic dishes was tested. The plating efficiency of all clones on plastic dishes ranged from 10–60% with an average of about 40%. Standard transformants and most clones which are uniformly T antigen positive plated on plastic or on 3T3 monolayers with about equal efficiencies (Fig. 6). Intermediate clones, however, showed a wide range in their plating efficiency on 3T3 monolayers as compared to their plating efficiency on plastic. In general, the higher the saturation density of the intermediate clone, the higher its plating efficiency on 3T3 monolayers. A similar correlation of monolayer plating with saturation density has been seen by Pollack, Green and Todaro (1968).

T antigen negative clones did not form visible colonies on normal monolayers. It must be pointed out, however, that the density of these clones is only about twice that of 3T3. Colonies of that density might grow on a 3T3 monolayer but be indistinguishable from the monolayer itself.

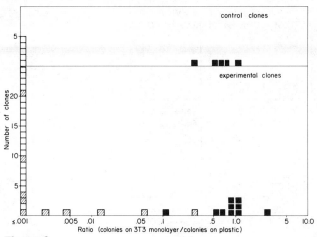

Figure 6

Histograms of efficiencies of colony formation on 3T3 monolayers for control and experimental clones. Cells 10^2 or 10^3 were plated onto confluent 3T3 monolayers or plastic dishes (60-mm) in 10% calf serum-DME in triplicate. Medium was changed every third day and plates fixed and stained with Harris-hemotoxylin on the tenth day. The plating efficiencies on plastic vary from 20–60%, averaging 40%. The data is normalized to the plating efficiency on plastic. Black rectangles, T antigen positive clones; hatched rectangles, T antigen intermediate clones; white rectangles, T antigen negative clones.

Stability of Growth Properties of SV40 Clones

One month after the initial cloning, six clones were tested in each assay. The clones were passaged for 4 months and then recloned. Both parental and recloned lines were retested in each assay one month after the recloning. The data in Table 1 demonstrate that the growth properties measures in this series of assays are quite stable.

Specifically the T antigen negative clones that initially grew well in 1% calf serum (e.g., SVR-Cl 4–2, 5–7 and 8–2) continued to do so even after recloning. The T antigen intermediate clone, SVR-Cl 6–3, was tested for saturation density in 10% calf serum, growth in 1% calf serum, plating in methylcellulose, and plating on 3T3 monolayers. By all of these tests both the reclone and the initial clone after 5 months of continuous passage were very similar to the initial clone after 1 month of continuous passage.

A more thorough investigation has been made of T antigen intermediate clones. Five clones showing heterogeneous T antigen by immunofluorescence and lower levels of T antigen by complement fixation were recloned. Eighteen subclones were picked and their T antigen staining characterized by indirect immunofluorescence. Approximately half of the clones showed a wide variation in staining pattern, much like the parental cells. The other subclones showed a uniform pattern of staining; the intensity of this staining was considerably less than that of standard transformants. Other growth properties of these subclones will be described elsewhere (Risser and Pollack in prep.). In no case was a T antigen negative subclone generated.

Table 1

Stability of Growth Characteristics of SV40-transformed 3T3 Clones

Line	T antigen			Saturation density (cells/cm² × 10⁴)			Doubling time in 1% CS (hr)			EOP in Methocel (colonies/100 cells)			Ratio EOP on monolayers to EOP on plastic		
	1 mo.	5 mo.	5 mo. re-clone	1 mo.	5 mo.	5 mo. re-clone	1 mo.	5 mo.	5 mo. re-clone	1 mo.	5 mo.	5 mo. re-clone	1 mo.	5 mo.	5 mo. re-clone
4-2	–	–	–	8	12.5	10.0	24	37	47	.001	NT†	.001	.01	NT	.01
5-7	–	–	–	15	16	15	40	31	31	.001	NT	.001	.01	NT	.01
6-3	I*	I	I	9	15	15	33	29	29	.01	.001	.001	.01	NT	.01
8-1	+	+	+	60	50	70	20	24	22	11	NT	20	1.0	NT	.7
8-2	–	–	+	10.5	12	13.5	40	29	29	.001	NT	.001	.01	NT	.01
8-4	+	+	+	50	45	60	32	27	30	11	NT	28.5	2.6	NT	2.0

* I = intermediate. † NT = not tested.

Table 2

Classes of SV40 Clones

Class	% of Clones	T antigen	Saturation density (cells/cm² × 10⁴)	Doubling time in 1% CS (hr)	Anchorage (colonies in Methocel per 100 cells)	Ratio EOP on monolayers to EOP on plastic
I Normal	12	–	8.5 (7.6–9.5)	78 (55–100)	≤ .001	≤ .001
II Transformed	25	+	47 (35–60)	34 (24–45)	32 (11–58)	.94 (.1–2.6)
III Serum-transformed	38	–	15 (9.5–16.5)	34 (26–49)	≤ .001	≤ .001
IV Intermediate-transformed	25	±	25 (15–45)	32 (29–41)	4 (.5–14.6)	.03 (.001–.25)
Control						
Normal		–	7.8 (7–9)	70 (60–86)	≤ .001	≤ .001
Transformed		+	53 (47–60)	37 (26–43)	25 (16–30)	.7 (.25–1.1)

DISCUSSION

The nonselective nature of this assay has allowed us to isolate many transformants which have not been previously described. In fact we have found that most clones arising after SV40 infection are altered at least with respect to serum requirement for growth.

The variety of clones obtained in this manner is presented in Table 2. Clones of the first class, which are indistinguishable from 3T3 in these assays, may be un-affected by the virus or may be similar to the cryptic transformants of Smith et al. (1971) if they carry SV40-specific sequences in unexpressed form (Smith, Gelb and Martin 1972). They are presently being tested for SV40-specific DNA sequences.

Clones of the second class correspond to the standard transformants that have been described by many authors. The high proportion of these clones may be due in part to delayed transformation (Stoker 1963; Todaro and Green 1966). In fact four of the clones looked normal by morphological criteria when they were picked. Subsequent to their cloning they converted to transformants, perhaps be-cause of the more rapid doubling time of transformed cells in 10% calf serum (16 hr vs. 23 hr). Reconstruction experiments on 3T3 and SV101 are currently in progress to test this hypothesis.

The third class of clones consists of cells which are altered in their ability to utilize low concentration of serum. The saturation density of these clones is about twice that of 3T3. The mechanism by which this saturation density is maintained has not yet been investigated. We do not yet know if class III clones contain viral-specific DNA sequences.

Serum transformants are expected from the work of others (Smith, Scher and Todaro 1971; Scher and Nelson-Rees 1971). What is surprising is the high pro-portion of clones which they represent and the lack of viral T antigen in these clones. Among a less extensive catalog of cells, Smith et al. (1971) reported one T antigen negative, serum-transformed clone and one T antigen positive, serum-transformed clone. It must be pointed out, however, that the assay used by Smith et al. (1971) differs from the assay used here in that it requires cells to form visible colonies in 10% agamma-depleted calf serum when plated sparsely.

The last class of clones does not represent a uniform category of cells, but rather cells with a spectrum of growth properties intermediate between 3T3 cells and SV101 cells. The pattern of staining and level of complement-fixing T antigen is significantly lowered from that of standard transformants. The saturation densities of such clones show a spectrum of values, much as seen by Scher and Nelson-Rees (1971). Intermediate saturation densities seen in this class of clones may well be the result of an equilibrium between cell growth and cell death and detachment as Scher and Nelson-Rees have demonstrated for their lines. The intermediate growth of these clones in methylcellulose medium demonstrates that such clones are not comprised of a mixture of normal and standard transformed cells, but rather are comprised of cells each with a limited anchorage independence, similar to the abortive transformants of Stoker (1968). Such cells differ from abortive trans-formants in that they continue to grow slowly throughout the three-week assay. It will be quite important to know if the few dense colonies which such lines form on 3T3 monolayers in fact represent segregation of standard transformants.

We have shown in this work that the effect of SV40 on 3T3 cells is more varied and of a greater extent than previously thought to be the case. Whether the variety of cell growth patterns seen in this study is a result of an inherent instability of 3T3 cells due to their aneuploid chromosome complement is not yet clear. This range could result from a mechanism of SV40 transformation that effected each cell somewhat differently, leading to the simultaneous appearance of every different type of transformant. Alternatively, the range may be the result of secondary alterations that become possible only after a primary virus-mediated alteration, such as serum utilization, has occurred.

In any event the results reported here demonstrate that SV40 is capable of bringing about a variety of stable changes in growth properties in the majority of infected cells. Which of these changes, if any, is related to the uncontrolled proliferation of a cell in an animal host is not yet clear.

Acknowledgments

We thank Mary Weber for performing the complement fixation assays and Nancy Hopkins and Art Vogel for their encouragement and discussions. R.R. was supported by NIH predoctoral fellowship 4F01-GM-49503. This work was supported by a center grant from the National Cancer Institute 5-P01-CA13106.

REFERENCES

Black, P. H. 1966. Transformation of mouse cell line 3T3 by SV40: Dose response relationship and correlation with SV40 tumor antigen production. *Virology* **28:**760.

Black, P. H., E. M. Crawford and L. V. Crawford. 1964. The purification of simian virus 40. *Virology* **24:**381.

Holley, R. W. and J. A. Kiernan. 1968. "Contact inhibition" of cell division in 3T3 cells. *Proc. Nat. Acad. Sci.* **60:**300.

Jainchill, J. and G. J. Todaro. 1970. Stimulation of cell growth *in vitro* by serum with and without growth factor. *Exp. Cell Res.* **59:**137.

Koch, M. A., H. J. Eggers, F. A. Anderer, H. D. Schlumberger and H. Frank. 1967. Structure of simian virus 40: I. Purification and physical characterization of the virus particle. *Virology* **32:**503.

Macpherson, I. and L. Montagnier. 1964. Agar suspension culture for the selective assay of cells transformed by polyoma virus. *Virology* **23:**291.

Pollack, R. E., H. Green and G. J. Todaro. 1968. Growth control in cultured cells: Selection of sublines with increased sensitivity to contact inhibition and decreased tumor-producing ability. *Proc. Nat. Acad. Sci.* **60:**126.

Sambrook, J. and R. E. Pollack. 1973. Cell transformation. *Methods in Enzymology* (in press)

Scher, C. D. and W. A. Nelson-Rees. 1971. Direct isolation and characterization of "flat" SV40-transformed cells. *Nature New Biol.* **233:**263.

Smith, H. S., L. D. Gelb and M. A. Martin. 1972. Detection and quantitation of simian virus 40 genetic material in abortively transformed Balb/3T3 clones. *Proc. Nat. Acad. Sci.* **69:**152.

Smith, H. S., C. D. Scher and G. J. Todaro. 1971. Induction of cell division in medium lacking serum growth factor by SV40. *Virology* **44:**359.

Stoker, M. 1963. Delayed transformation by polyoma virus. *Virology* **20:**366.

————. 1968. Abortive transformation by polyoma virus. *Nature* **218:**234.

Stoker, M. and P. Abel. 1962. Conditions affecting transformation by polyoma virus. *Cold Spring Harbor Symp. Quant. Biol.* **27:**375.

Stoker, M., C. O'Neill, S. Berryman and V. Waxman. 1968. Anchorage and growth regulation in normal and virus-transformed cells. *Int. J. Cancer* **3:**683.

Temin, H. M. and H. Rubin. 1958. Characteristics of an assay for Rous sarcoma virus and Rous sarcoma cells in tissue culture. *Virology* **6:**669.

Todaro, G. J. and H. Green. 1966. High frequency of SV40 transformation of mouse cell line 3T3. *Virology* **28:**756.

Todaro, G. J., H. Green and B. D. Goldberg. 1964. Transformation of properties of an established cell line by SV40 and polyoma virus. *Proc. Nat. Acad. Sci.* **51:**66.

Vogel, A. and R. E. Pollack. 1973. Isolation and characterization of revertant cell lines. IV: Direct selection of serum-revertant sublines of SV40 transformed 3T3 mouse cells. *J. Cell. Physiol.* (in press).

Changes of Regulation of Host DNA Synthesis and Viral DNA Integration in SV40-infected Cells

K. Hirai, G. Campbell, and V. Defendi

The Wistar Institute of Anatomy and Biology
Philadelphia, Pennsylvania 19104

The expression of oncogenic capacity of DNA tumor viruses is the result of a series of unique interactions between the virus and the host cell. Three such processes can be singled out: (1) incomplete expression of the viral genome resulting in abortive infection, (2) induction of cellular DNA synthesis, and (3) integration of the viral DNA into host cell DNA. The first process is dependent upon the host cell's activity and the other two are dependent upon the physical and functional properties of the viral genome. The net results of these processes is that the growth regulation of normal cells is altered to a different level of intra- and intercellular controls; it is the expression of these alterations which we recognize as the transformed phenotype.

This paper is primarily concerned with the virus-host cell interactions which are virus dependent. The induction of cell DNA synthesis by small DNA tumor viruses has been demonstrated in cells that had withdrawn from the replicating cycle either reversibly or irreversibly (Lehman and Defendi 1970). The actual mechanism by which such initiation of DNA synthesis is triggered is unknown although several possibilities have been suggested (Winocour 1969). Two aspects are worth considering: First, the process of induction of DNA synthesis by virus infection and the events which precede entrance into S are not different from those observed when physiological stimuli are applied to a resting population, even including early stimulation of nuclear acidic protein synthesis (Rovera, Baserga and Defendi 1972). Second, the result of the stimulation, at least in cells undergoing abortive infection, is compatible with cell survival and replication. It is then difficult to visualize what possibly unique role the induction of cell DNA synthesis may have in the transformation process. It is clear, however, that there is a functional dependence; i.e., when there is no induction of cell DNA synthesis, there is no transformation. A possibly significant alteration in the control of DNA synthesis became apparent when we found that within one cell generation after infection by SV40 of secondary cultures of Chinese hamster embryonic cells (Lehman and Defendi 1970) or mouse peritoneal macrophages (Lehman, Mauel and Defendi 1971) a fraction of the cell population undergoes two periods of DNA synthesis

151

with no intervening mitosis. These events result in the production of a population of viable cells that have twice the normal DNA content and are tetraploid. The fate and biological significance of this polyploid population is not yet understood.

In this paper we outline experiments designed to examine the biological significance of these polyploid cells by monitoring two specific relationships at various times during the virus-host cell interaction: (1) the fate of the viral genome and its relationship to the host DNA, (2) the degree of polyploidy and the appearance of the transformed state. Both of these relationships have a direct bearing on the hypothesis illustrated in Fig. 1. In this scheme the phenomena that occur after SV40 infection are arranged in chronological and possibly causal sequence. Without excluding the role of virus-specified products, the hypothesis states that viral DNA integration is the primary event leading to the production of altered (transformed) cells through an initial and temporary disruption of control of cell DNA synthesis.

METHODS

Unless stated otherwise, secondary cultures of Chinese hamster (CHE) embryo cells were used; the cells were infected at a multiplicity of infection (MOI) of 30–50 plague-forming units (PFU). The proportion of infected cells was determined by immunofluorescence staining for SV40 T antigen (Lehman and Defendi 1970). DNA synthesis was estimated by incorporation of [^{3}H]thymidine into TCA-insoluble material and/or by radioautography. Integration of incoming SV40 DNA (we do not have any evidence for viral replication in this host system) was determined by hybridization of SV40 [^{3}H]cRNA with cellular DNA of the Hirt's sediment (Hirt 1966) or with the fast-sedimenting DNA fraction of nuclear DNA separated by velocity sedimentation in alkaline sucrose or glycerol gradient (Hirai, Lehman and Defendi 1971a). The extent of reinitiation of DNA synthesis was measured by the appearance of heavy-heavy (HH) DNA in cultures treated with 5-bromodeoxyuridine (BrdU 5 μg/ml), 5-fluorodeoxyuridine (FdU, 15 μg/ml)

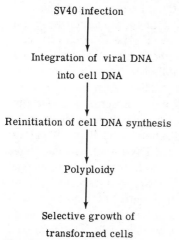

Figure 1
Proposed process of cell transformation by SV40.

and colcemide (0.05 μg/ml) to block mitosis (Hirai et al. 1971b). As we have already demonstrated the HH DNA represents that fraction of the infected cell population which twice replicates its complete DNA content (Hirai et al. 1971b).

RESULTS

Relation between Induction of Cell DNA Synthesis and Viral DNA Integration

In the SV40-infected Chinese hamster cells, integrated viral DNA becomes detectable about 10–15 hours post infection, at the time when the DNA synthesis is stimulated. The amount of integrated SV40 DNA under our experimental conditions reaches a plateau at about 30 hours (Fig. 2). Actually, as shown later, the amount decreases in successive cell passage according to culture conditions. What is important is that the time of integration is clearly defined, with no further increase even though nonintegrated viral DNA may survive in the cytoplasm for several days. We have already shown that integration depends upon the integrity of the viral genome to the same extent as it depends upon its capacity to induce cell DNA synthesis, and that the level of integration is dependent upon the multiplicity of infection (Hirai et al. 1971a). In a parallel way different physiological conditions influence the induction of DNA synthesis and the level of integration. For example, the stimulation of DNA synthesis in both infected and control cultures is dependent upon the concentration of fetal calf serum in the medium, the enhancement factor in the infected culture being higher at a low serum concentration (Fig. 3). In addition the proportion of integrated viral DNA increases threefold when the serum concentration is raised from 0 to 5% (Table 1). All of these experiments indicate that there is a good concurrence between the two phenomena in terms of their order of appearance, their dependence upon the functioning viral genome, and upon the culture conditions.

It is possible, however, to discriminate between the two phenomena by the selective inhibition of one. We have previously shown that when the cell DNA synthesis is inhibited by D-arabinosyl cytosine, integration of viral DNA still occurs at the same extent as in untreated cultures (Hirai et al. 1971a). In a further analysis we have used an SV40 *ts* mutant, TS 101, which at nonpermissive temperature in monkey cells fails to induce T and V antigen, viral DNA and cell DNA synthesis (Robb and Martin 1972). In the CV-1 cells at the nonpermissive temperature (40 and 42°C) no viral functions could be detected, and in particular no DNA synthesis induction or integration (Table 2). In the CHE cells not all experiments could be conducted at the optimal nonpermissive conditions because Chinese hamster cells did not survive at the temperature of 42°C, but at 40°C when cell DNA synthesis was not induced, integration still occurred. It is then possible to demonstrate that viral DNA is integrated even when the induction of cell DNA synthesis is abolished either by drugs or by the use of conditional lethal virus mutant. This indicates that the integration of viral DNA does not depend upon the replication of the cell DNA and that at least functionally it precedes the induction of cell DNA synthesis.

Relation between Induction of HH DNA and Viral DNA Integration

As indicated above a unique characteristic of cell DNA synthesis induced by SV40 is that DNA synthesis is reinitiated in a fraction of the cell population; this was

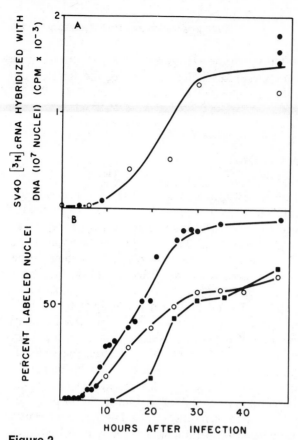

Figure 2

Chronological sequences of SV40 DNA integration into the cell DNA and induction of synthesis of cellular DNA and T antigen. (Partially adapted from Hirai et al. 1971a).

A, Confluent monolayers of primary Chinese hamster cells were infected with SV40 at an MOI of 30, split at a ratio of 1:2 and maintained in 8% fetal calf serum. At various times thereafter nuclei were isolated. To separate the cell DNA from SV40 two methods were employed: (1) Nuclei were layered on alkaline sucrose gradient (10–30%) and then centrifuged 5 hr at 83,000 × g. Fractions V and VI of the alkaline gradient (○) corresponding to the cell DNA were isolated as described (Hirai et al. 1971a); (2) cell DNA isolated by the Hirt's method (●) was hybridized with SV40 [³H]cRNA. 80–100 μg of the cell DNA immobilized on the membrane filter was incubated for 20 hr at 65°C in a vial containing, in 1 ml of 6 × SSC, tritiated SV40 cRNA (10^5 cpm) and 1 mg SDS. After hybridizations the vial was kept at room temperature for 3 hr, the filter was washed in 2 × SSC and incubated with 10 μg/ml pancreatic ribonuclease in 2 ml 2 × SSC for 1 hr at room temperature; the filter was washed in 2 × SSC and incubated 15 min at 65°C in 3 ml 2 × SSC. The fluid was then removed by suction from the vial kept at 65°C. The filter was then dried and the radioactivity counted in a toluene-based scintillation mixture by a Beckman liquid scintillation spectrometer. After counting the filter was washed with toluene and then dried. The amount of the DNA on the filter was determined by the diphenylamine methods. The background radioactivity (100 μg of Chinese hamster cell DNA) was 179 cpm and was subtracted from the observed values.

B, Assay of DNA synthesis by radioautography. (●———●) SV40 infection, 30 PFU/cell; (○———○) mock infection; (■———■) T antigen measured by immunofluorescence.

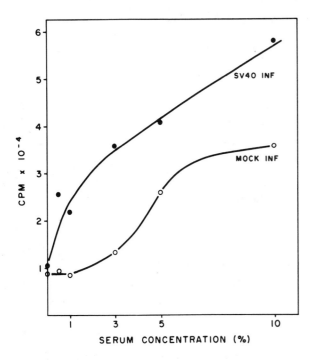

Figure 3

Effect of serum concentration on induction of DNA synthesis by SV40. Confluent monolayers of secondary Chinese hamster cells were infected with 50 PFU SV40 per cell and then labeled with [³H]BrdU (0.1 μCi/5 μg/ml) in the presence of FdU (15 μg/ml). At 42 hr p.i. the cells (5 $\times$ 10⁶) were harvested and the radioactivity of the TCA precipitate was counted.

demonstrated by the appearance of tetraploid metaphases at the time of the first mitosis after infection and by the finding of HH DNA when normal cycling through mitosis was inhibited. The kinetics of HH DNA synthesis is illustrated in Fig. 4, where it is shown that the first HH DNA appears at approximately 27 hours after infection, a time which, because of kinetic consideration of the cell system, is compatible with the assumption that a new period of S is reinitiated. The total amount and the percentage of HH DNA (of the newly synthesized cell DNA) synthesized during the first 48 hours post infection is directly proportional to the amount of the integrated viral DNA (Table 1). Thus not only is there a correlation between viral integration and induction of cell DNA synthesis, but more importantly, between viral integration and the alteration of control of DNA synthesis as expressed by the appearance of the HH DNA.

If the relation is more than coincidental, one would then expect that the integrated DNA should be found in the HH DNA. This was tested in the following experiment. DNA was extracted from Hirt's sediment of the infected culture, the different classes of DNA were separated by a CsCl density gradient equilibrium centrifugation, and the various fractions hybridized with the SV40 [³H]cRNA. No DNA hybridizable with SV40 cRNA was found in the HH DNA class of the Hirt's supernatant, indicating that the HH DNA does not represent replicated free SV40

Table 1

Effect of Serum Concentration on the Synthesis of Cell DNA
and Integration of SV40 DNA into DNA of Infected CHE Cells

Serum conc. (%)	Infection	Total ^3H cpm in gradient	^3H-label cpm in the position of			Hybridized SV40 [^3H]cRNA (cpm/100 μg DNA)
			HL DNA	HH DNA	(% of total)	
0	SV40	4394	4002	197	(4.5)	384
1	SV40	12217	11637	965	(7.9)	525
5	SV40	74916	63223	9215	(13.3)	1260
5	Mock	35962	33821	82	(0.2)	

Cell DNA synthesis spans the Total ^3H cpm and ^3H-label columns; *Integration†* heads the Hybridized SV40 column.

* Confluent monolayers of secondary Chinese hamster cells were infected with 50 PFU SV40 per cell. After 2 hr of absorption they were exposed to fresh medium containing a mixture of unlabeled BrdU and [^3H]BrdU (1 μCi/μg/ml) in the presence of 5-FdU (15 μg/ml) at various concentrations of serum. At 16 hr p.i. colcemide (0.05 μg/ml) was added. At 48 hr p.i. the total DNA was extracted from 5×10^6 cells and analyzed by a CsCl density gradient equilibrium centrifugation (212,000 × g in a Beckman Spinco SW 50.1 rotor at 25°C for 48 hr). The nucleic acids were precipitated with 5% TCA in the presence of 200 μg bovine serum albumin as a carrier and collected on the membrane filter. Radioactivity was measured in a Beckman liquid scintillation spectrophotometer.

† At 48 hr p.i. nuclei were isolated from parallel cultures. A total of 4–6 × 10⁶ nuclei were layered on an alkaline sucrose gradient and centrifuged as described in the legend of Fig. 2. 100 μg of the cell DNA on the filter was hybridized with SV40 [^3H]cRNA (1.7 × 10⁵ cpm). The radioactivity bound to normal Chinese hamster cell DNA (102 cpm) was subtracted from the observed value.

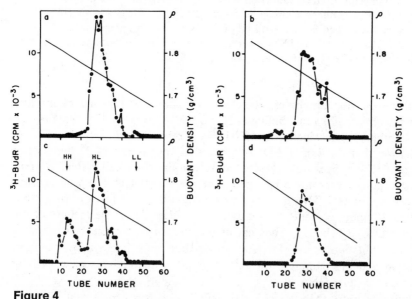

Figure 4

CsCl density gradient equilibrium centrifugations of [^3H]BrdU-labeled cell DNA extracted from Chinese hamster cells at various times after infection with SV40 (adapted from Hirai et al. 1971b). Confluent monolayers of secondary cells were infected with 50 PFU SV40 per cell and then labeled with [^3H]BrdU (1 μCi/5 μg/ml) in the presence of FdU (15 μg/ml). At **(a)** 16, **(b)** 27, **(c)** 46 hr p.i. the monolayers were harvested. The SDS-phenol-extracted DNA was analyzed by CsCl density gradient equilibrium centrifugation. **(d)** Mock-infected cells 46 hr p.i.

Table 2

Results of Infection of Permissive and Nonpermissive
Cells by Wild-Type and *ts* Mutant (TS 101) SV40

Temp (°C)	Chinese hamster (nonpermissive cell)					
	T antigen		*Cell DNA synthesis*		*Integration*	
	wt	*ts*	*wt*	*ts*	*wt*	*ts*
33	++	++	++	++	+	+
40	++	++	++	−	+	+
42	++	+	NS*	NS	NS	NS

Temp (°C)	CV-1 (permissive cell)							
	T antigen		*SV40 replication*		*Cell DNA synthesis*		*Integration*	
	wt	*ts*	*wt*	*ts*	*wt*	*ts*	*wt*	*ts*
33	++	++	++	++	+	++	+	+
40	++	±	++	−	++	−	ND†	ND
42	++	−	++	−	+	−	+	−

Confluent monolayers of secondary Chinese hamster and CV-1 cells were infected at an MOI of 30 PFU and 1 PFU, respectively, of SV40 wild-type or SV40 TS 101 with an adsorption period of 2 hr at 37°C. The Chinese hamster cell cultures were then maintained at the permissive or at the nonpermissive temperature. T antigen was determined by immunofluorescence; viral DNA and cellular DNA synthesis was measured by the radioactivity of the TCA precipitate from the supernatant and sedimented fraction of the Hirt's extract from cells labeled with [³H]thymidine 3 hr from 48–72 hr p.i. In Chinese hamster cells DNA synthesis was determined by radioautography after 3-hr labeling with [³H]thymidine at various times after infection. For the integration experiments the cell DNA was separated in an alkaline glycerol gradient (10–30% in 0.3 N NaOH, 0.01 M ethylenediamine tetra-acetate, 0.5 M NaCl). The hybridization procedure was similar to that described in Fig. 2. For the integration experiments the CV-1 cells were treated with D-arabinosylcytosine (15 μg/ml) to inhibit viral replication.

* Cells did not survive   † Not done

DNA. It is quite evident (Fig. 5) that the viral DNA is found in the HH DNA to a larger extent than in any of the other fractions; except for the lighter of the heavy-light (HL) fraction, the enrichment is severalfold. The finding that the viral DNA is also integrated in the HL DNA poses some restriction to the hypothesis that viral DNA integration always results in reinitiation of DNA synthesis. However, the following should be considered: Neither the virus infection nor the cell population is synchronous; thus some of the HL DNA at 48 hours could well become HH later on. In addition we have evidence that the level of integrated viral DNA decreases after cell passage. Therefore we may presume that there are different sites of integration, some of which could be of no consequence and others which, activated, could be responsible for the abnormal reduplication of the ce DNA.

Relation between Polyploidy and the Transformed Phenotype

The next point to determine was the biological significance of the synthesis of HH DNA or of the tetraploid cells that are the direct result of this phenomenon. We have looked at this problem in several ways. First, if the induction of tetraploid cells is related to transformation, one should expect that the transformed colonies would be tetraploid or contain a large proportion of tetraploid cells.

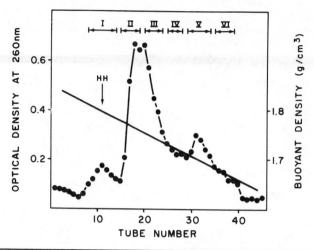

Fraction in gradient	Average buoyant density (g/cm³)	The amount of recovered DNA immobilized on filter	Hybridized ³H SV40 cRNA	
			Total radio-activity (cpm)	cpm per 100 μg of DNA
1st centrifugation				
I (HH)	1.79	63	652	1039
II (HL)	1.76	189	545	288
III (HL)	1.75	188	975	518
IV (HL)	1.72	85	807	949
V (LL)	1.70	111	237	213
VI (LL)	1.68	43	77	167

Figure 5

Distribution of SV40 [³H]cRNA hybridizable DNA in classes of DNA of different densities. Confluent monolayers of secondary Chinese hamster cells were infected with 30 PFU SV40 per cell and exposed to fresh medium containing a mixture of BrdU (5 μg/ml) and FdU (15 μg/ml). At 16 p.i. colcemide (0.05 μg/ml) was added. At 48 hr p.i. nuclei were isolated and subjected to the selective extraction method of Hirt to isolate the cell DNA. The SDS-phenol-extracted DNA in CsCl solution (p = 1.75 g/cm³) was centrifuged at 147,000 × g in a Beckman Spinco SW 50-1 rotor for 68 hr at 25°C. Two-drop fractions were collected and pooled as indicated by the roman numerals. Half volume of each fraction was used for DNA-RNA hybridization as described in the legend of Fig. 2.

A clone of Chinese hamster kidney cells (CHK Cl 6) which maintains high susceptibility to SV40 infection was developed; this clone has diploid chromosome content and when tested at various times, the percentage of tetraploid cells averaged 2.5% with no instances found higher than 3.5% (Fig. 6). Sparse cell monolayers were infected with SV40 at an MOI of 30 PFU; 24 hours later the cells were

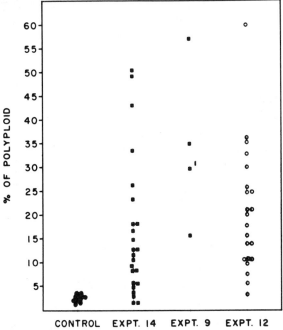

Figure 6

Percentage of polyploidy in Chinese hamster kidney clones noninfected (•) and SV40-transformed. Transformed colonies selected by the agar method (▪); transformed colonies selected by the morphological phenotype in monolayer (○). For each point 75 to 200 metaphases were scored.

harvested and suspended in soft agar medium for the selection of transformed colonies (Macpherson and Montagnier 1964). Approximately 2 weeks later colonies were removed with a micropipet, vigorously resuspended in medium and plated on coverslips in petri dishes (Exp. 14 and 9). As soon as sufficient cells grew (2–5 days) the cultures were treated for chromosome analysis. It is estimated that the cells were examined between 8–15 generations after the initial transforming events. It is important to note that in the scoring of mitosis as many cells as possible were assigned to either diploid or tetraploid classes with minimal selection for well-spread metaphases; thus even mitosis up to near triploid would have been scored as diploid. As illustrated in Fig. 6, in the great majority of colonies analyzed the percentage of tetraploid metaphases was much higher than in the parental noninfected population, with great spread of distribution. It is also clear, however, that not all metaphases from the transformed colonies were tetraploid at the time when the first analysis was made. Whether this was due largely to chromosome segregation that had occurred during intercurrent generations or to the fact that some transformed colonies are truly still diploid can only be determined by detailed karyological analysis. The efficiency of SV40 transformation with these cells by the agar selection method is of the order of 70–100/10^5 cells at the stated multiplicity of infection; thus the possibility could not be excluded that the tetraploid cells already present in the parental population could be the preferred target for transformation. To reduce this possibility secondary or tertiary cultures of Chinese hamster embryo cells—a population that also has a 2–3% proportion of "natural" polyploid cells and that, after SV40 infection at MOI of 30 PFU, yields transformed colonies that can be scored on the basis of their morphological characteristics on a plastic monolayer with a frequency of 8% (Smith, Defendi and Wigglesworth 1973)—were infected and seeded at low dilutions in large petri dishes. Ten days later colonies that appeared phenotypically transformed were isolated and transferred to coverslips as indicated above (Fig. 6, Exp. 12). The distribution of tetraploid frequency in this group of colonies is very similar to that of the previous experiments, indicating that transformed colonies do not originate exclusively from infected, preexisting tetraploid cells.

Another way of looking at the problem is through a converse experiment determining whether tetraploid cells preferentially give origin to transformed colonies. Fox and Levine (1971) have determined enrichment of progenitors of transformed colonies in 3T3 cells induced into DNA synthesis after SV40 infection. In selecting the tetraploid cells at the first mitosis after infection, however, considerable problems arise due to the evolution of the infected cell population (Fig. 7). As the infected cultures are subdivided the following events are observed: The proportion of T antigen-positive cells declines quite rapidly, so that by the second or third passage it falls to 5–12%; only after a lag period does the proportion of T positive cells increase again to reach the 100% level by the 25–30th day. This phenomenon has been observed in several similar systems (Diamond 1967; Oxman and Black 1966). Since there is no obvious cell death, this suggests that either the gene coding for the T antigen is segregated among the daughter cells, or more probably that cells which were T antigen positive lose the capacity to express it to a detectable level. From initial experiments it appears that the reduction is higher if the cells are allowed to replicate rapidly. Since the subsequent increase is not due to reinfection of cells, it must represent the progressive emergence of a "transformed" cell population. Concurrently the amount of detectable integrated viral DNA per 100 μg of cell DNA decreases to an almost imperceptible amount; its level then slightly increases to stabilize itself at that amount found in more permanently transformed cell clones. The decrease of integrated viral DNA indicates that either redundant integrated genomes are excised or that some cells in which viral DNA was integrated lose it completely. The fact that the capability of synthesizing T antigen is lost as well and that lines originated from "absortively transformed" cells are negative for viral genome (Dieckman et al. pers. comm.) would favor the latter interpretation. During this period the progression of tetraploid metaphases either declines slightly or remains at a low level; only later does it reach a higher level.

Because of these reasons fractionation of the cell population was done at the time in the life of the culture when a considerable proportion of the cells were stably T antigen positive. An infected population at the time when 30% of the cells were T antigen positive and 20% of the metaphases were polyploid was fractionated in an albumen "sta put" gradient, which separates cells on the basis of volume (Miller and Phillips 1969). In a preliminary experiment it was found that the tetraploid population sediments more rapidly than a diploid one and that a good separation of cells with high DNA content may be obtained (Fig. 8). Fractions were pooled and tested for the proportion of T antigen positive, for the frequency of tetraploid metaphases, and for the transforming capacity. It is evident that in the first two pooled fractions, which actually contain only a minority of the total cell population, the polyploid cells are concentrated as well as cells capable of producing transformed colonies (Fig. 9).

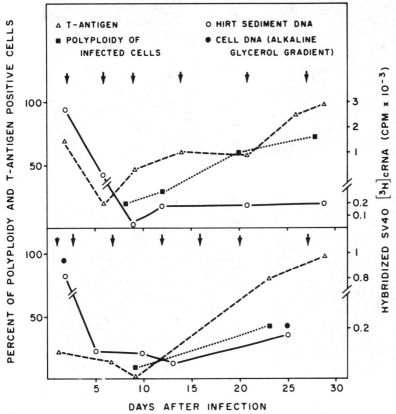

Figure 7

Detection of polyploidy, T antigen and SV40 genomes in SV40-infected Chinese hamster cells at different passages. Tertiary cultures of Chinese hamster embryo cells were infected with SV40 at a multiplicity of 30 PFU per cell. At different times after infection, whenever the monolayer become confluent, the cells were split at different ratios. T antigen ($\triangle$) was detected by immunofluorescence. Polyploidy ($\blacksquare$) was determined by treating cells with colchicine for 5 hr and counting the proportion of polyploid metaphase in at least 200 cells. SV40 genome which was associated with cell DNA was detected by DNA-RNA hybridization; the cell DNA was isolated by the Hirt method ($\circ$) and alkaline glycerol gradient ($\bullet$).

Even from these preliminary experiments, it is quite clear that the cell populations containing more tetraploid cells originate more transformed colonies than cell populations containing few tetraploid cells.

DISCUSSION AND CONCLUSION

In these experiments we have demonstrated that, at least in the host system we have investigated, there is an alteration in the control of DNA synthesis after SV40 infection which is expressed in the appearance of polyploidy. We have also obtained good suggestive evidence that reinitiation of DNA synthesis may be the result of the viral DNA integration and that the polyploid cells could be the source

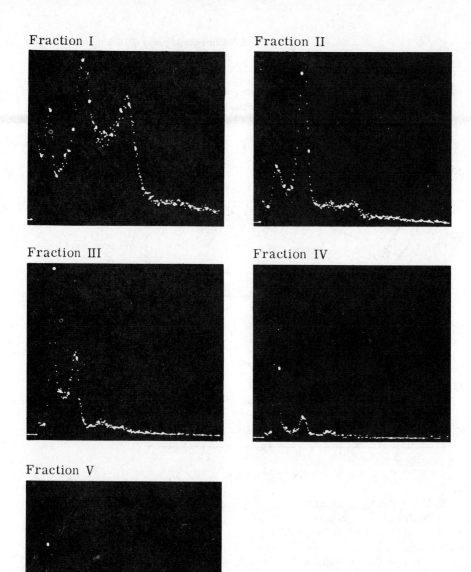

Figure 8

DNA content distribution in SV40-infected Chinese hamster cells. This series of graphs represents the result of an experiment designed to analyze the DNA content of combined fractions derived from BSA gradient using Flow micro-fluorometric techniques (Kramer et al. 1972). These fractions correspond to those described in Fig. 9.

Abscissa: Channel No. corresponding to the amount of fluorescence/single cell.

Ordinate: Relative frequency of cells displaying specific fluorescence values. Fraction I shows three peaks of DNA content corresponding to (*left*) 2C, (*center*) 4C, and (*right*) 8C values. Successive fractions show the disappearance first of 8C values, and then 4C values, and the predominance in fraction V of 2C values. These analyses were performed by Dr. Marvin Van Dilla and his colleagues at Biomedical Division, Lawrence Livermore Laboratories, California.

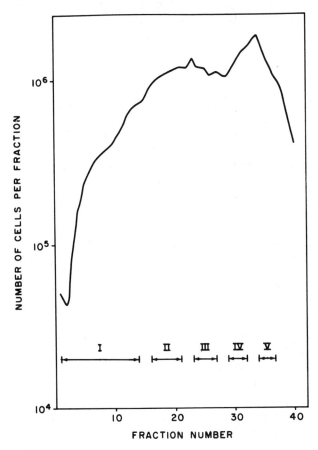

	T antigen	Polyploid	% Transf. colonies		
			S	M	L
Unfractionated	30%	20%			
I	95%	83.5%	0.20	0.03	0.01
II	95%(clon.)	73%	0.05	0.01	0.004
III	30%(clon.)	27%	0.014	0.002	0
IV	0.8%		0.0009	0	0
V	0%		–	–	–

Figure 9

Confluent monolayers of secondary Chinese hamster cells were infected with SV40 (MOI 30). At about 3 weeks after infection (see text) cultures were trypsinized and then separated on a gradient of 1–5% bovine serum albumin dissolved in Spinner Medium (total volume 1600 ml) for 3 hr in the cold. The gradient was fractionated into 40-ml aliquots. Each aliquot was counted using Model B Coulter Counter (amplification = 8, aperture current = 1). Fractions 1–14, 16–21, 23–27, 29–32, and 34–37 were combined to give a total of five fractions. Cells from the various fractions were tested for their properties by methods described previously. Colonies were scored in three classes: small (S), medium (M), and large (L).

of the transformed population. On these bases we feel that the hypothesis illustrated in Fig. 1 is valid. With the present knowledge it cannot be stated that every DNA virus-induced transformation has to proceed through the same sequence of events. However we have similar, though more fragmentary, findings in mouse peritoneal macrophages infected by SV40 (Lehman, Mauel and Defendi 1971) and perhaps in hamster kidney cells infected with SV40 virus (May, May and Weil 1971).

The biological effect of integration of an exogenous DNA on the regulation of eukaryotes is unknown, but in bacteria it has been shown that integration of the phage *Mu* DNA might produce mutation if it is located in a position which destroys the continuity of a functional gene (Taylor 1963; Boram and Abelson 1971). Furthermore even transcription of host DNA is altered after phage integration with "reading through" from the prophage DNA into the bacterial DNA (see Yarmolinski 1971). A similar evidence has been found in monkey cells infected by SV40, from which giant SV40-specific RNA molecules have been isolated, containing both cellular and host sequences (Jaenish 1972). This most probably results from integration of the viral DNA that also occurs in permissive cells (Hirai and Defendi 1972). More directly pertinent to the problem concerned here are the findings that, in a *ts* mutant for initiation of DNA replication of *E. coli,* integration of an episome *F* (Nishimura et al. 1971) or of bacteriophage *P2* (Lindahl, Hirota and Jacob 1971) can restore initiation of DNA synthesis at the restricted temperature. Thus precedents exist in the bacterial system for integration resulting in alteration of transcription of host genes or of taking over the host replicons by the viral DNA. When a conditional lethal mutant for initiation of DNA synthesis of suitable mammalian cells is isolated, this hypothesis can be more rigorously tested.

The evidence for a direct lineage from tetraploid to transformed cells is good but not as yet conclusive; obtaining a pure population of tetraploid cells or detecting stable transformed cells prior to the occurrence of extensive segregation of chromosomes in the successive generation after infection is still difficult. As already indicated (Lehman and Defendi 1970) the results of the karyological analysis of stably SV40 transformed Chinese hamster or Syrian hamster cells or of primary tumors, which are heteroploid or tetraploid, are not incompatible with a direct relationship between polyploidization and transformation.

The central elements of interpretation for the mechanism of transformation in the present hypothesis are the phenomenon of the viral DNA integration and the chromosomal imbalance which results from the initial polyploidization. It is improbable that polyploidization and the resulting chromosomal segregation is sufficient per se to produce cell transformation; for example, in preliminary experiments tetraploid BHK21 cells, obtained by one cycle of treatment with cytochalasin B (Defendi and Stoker 1973), did not acquire the capacity to grow in soft agar (Defendi and Stoker unpubl.). On the other hand, several recent findings (Harris et al. 1969; Hitsumachi, Rabinowitz and Sachs 1971; Pollack, Wolman and Vogel 1970) have indicated that chromosomal balance regulates the expression of malignancy and the phenotype of tumor cells in which viral genes are integrated and functioning.

Acknowledgments

We thank Paula Kardos, Jay Miller and Richard Walsh for excellent technical assistance.

This investigation was supported in part by Public Health Service Research grants CA-10815 from the National Cancer Institute and RR 05540 from the Division of Research Resources; a research grant #VC-73N from the American Cancer Society Inc.; and the support of the Commonwealth of Pennsylvania. Dr. Defendi is the recipient of an American Cancer Society Research Professorship (PRP #47).

REFERENCES

Boram, W. and J. Abelson. 1971. Bacteriophage Mu integration: On the mechanism of Mu-induced mutations. *J. Mol. Biol.* **62:**171.

Defendi, V. and M. G. P. Stoker. 1973. General polyploidy produced by cytochalasin B. *Nature New Biol.* **242:**24.

Diamond, L. 1967. Transformation of simian virus 40-resistant hamster cells with an adenovirus 7-simian virus 40 hybrid. *J. Virol.* **1:**1109.

Fox, T. O. and A. J. Levine. 1971. Relationship between virus-induced cellular deoxyribonucleic acid synthesis and transformation by simian virus 40. *J. Virol.* **7:**473.

Harris, H., O. J. Miller, G. Klein, P. Worst and T. Tachibana. 1969. Suppression of malignancy by cell fusion. *Nature* **223:**363.

Hitotsumchi, S., Z. Rabinowitz and L. Sachs. 1971. Chromosomal control of reversion in transformed cells. *Nature* **231:**511.

Hirai, K. and V. Defendi. 1972. Integration of simian virus 40 deoxyribonucleic acid into the deoxyribonucleic acid of permissive monkey kidney cells. *J. Virol.* **9:**705.

Hirai, K., J. Lehman and V. Defendi. 1971a. Integration of simian virus 40 deoxyribonucleic acid into the deoxyribonucleic acid of primary infected Chinese hamster cells. *J. Virol.* **8:**708.

————. 1971b. Reinitiation within one cell cycle of the deoxyribonucleic acid synthesis induced by simian virus 40. *J. Virol.* **8:**828.

Hirt, B. 1966. Evidence for semiconservative replication of circular polyoma DNA. *Proc. Nat. Acad. Sci.* **55:**997.

Jaenish, R. 1972. Evidence for SV40 specific RNA containing virus and host specific sequences. *Nature New Biol.* **235:**46.

Kraemer, P. M., L. L. Deaven, H. H. Crissman and M. A. Van Dilla. 1972. DNA constancy despite variability in chromosome number. *Advance Cell Mol. Biol.* **2:**47.

Lehman, J. M. and V. Defendi. 1970. Changes in deoxyribonucleic acid synthesis regulation in Chinese hamster cells infected with simian virus 40. *J. Virol.* **6:**738.

Lehman, J. M., J. Mauel and V. Defendi. 1971. Regulation of DNA synthesis in macrophages infected with simian virus 40. *Exp. Cell Res.* **67:**230.

Lindahl, G., Y. Hirota and F. Jacob. 1971. On the process of cellular division in *Escherichia coli:* Replication of the bacterial chromosome under control of prophage P2. *Proc. Nat. Acad. Sci.* **68:**2407.

Macpherson, I. and L. Montagnier. 1964. Agar suspension culture for the selective assay of cells transformed by polyoma virus. *Virology* **23:**291.

May, E., P. May and R. Weil. 1971. Analysis of the events leading to SV40-induced chromosome replication and mitosis in primary mouse kidney cell cultures. *Proc. Nat. Acad. Sci.* **68:**1208.

Miller, R. G. and R. A. Phillips. 1969. Separation of cells by velocity sedimentation. *J. Cell. Physiol.* **73:**191.

Nishimura, Y., L. Caro, C. M. Berg and Y. Hirota. 1971. Chromosome replication in *Escherichia coli*. IV. Control of chromosome replication and cell division by an integrated episome. *J. Mol. Biol.* **55:**441.

Oxman, M. N. and P. H. Black. 1966. Inhibition of SV40 T antigen formation by interferon. *Proc. Nat. Acad. Sci.* **55:**1133.

Pollack, R. S. Wolman and A. Vogel. 1970. Reversion of virus-transformed cell lines hyperploidy accompanies retention of viral genes. *Nature* **228:**938, 967.

Robb, J. A. and R. C. Martin. 1972. Genetic analysis of simian virus 40. III. Characterization of a temperature sensitive mutant blocked at an early stage of productive infection in monkey cells. *J. Virol.* **9:**956.

Rovera, G., R. Baserga and V. Defendi. 1972. Early increase in nuclear acidic protein synthesis after SV40 infection. *Nature New Biol.* **237:**240.

Smith, B. J., V. Defendi and N. M. Wigglesworth. 1973. The effect of dibutyryl cyclic AMP on transformation by oncogenic viruses. *Virology* **51:**230.

Taylor, A. L. 1963. Bacteriophage induced mutation in *Escherichia coli*. *Proc. Nat. Acad. Sci.* **50:**1043.

Winocour, E. 1969. Some aspects of the interaction between polyoma virus and cell DNA. *Advanc. Virus Res.* **14:**153.

Yarmolinsky, M. B. 1971. Alternative modes of prophage insertion and excision. *Advanc. Biosci.* **8:**31. Pergamon Press.

Host Cell Control of Viral Transformation

**Claudio Basilico, Hartmut C. Renger,
Stuart J. Burstin, and Daniela Toniolo**

Department of Pathology, New York University School of Medicine
New York, New York 10016

It is important to define and clarify the regulation of the various properties of virally transformed cells, namely, how viral and cellular functions control the expression of transformation.

This problem has been approached in a variety of ways. Through the use of viral mutants it has been possible to demonstrate that most of the characteristics of Rous sarcoma virus-transformed cells are under the control of viral genes (Martin 1970; Kawai and Hanafusa 1971). Some investigators have directed their attention to the study of the properties of transformed cells, trying to determine how some alterations in basic requirements, such as serum, do in turn affect other parameters of cell growth (Holley and Kiernan 1968). Other investigators have studied "revertants" of transformed cells in an attempt to understand the underlying mechanism which leads to expression of the malignant phenotype (Pollack, Green and Todaro 1968).

We have taken the approach of isolating cells that are conditional for the expression of transformed cell characteristics and attempting to understand how this conditional state is determined.

We devised a method of selecting for SV40-transformed cells whose transformed phenotype was temperature sensitive. The details of the isolation procedure have been published (Renger and Basilico 1972). Originally such procedure was carried out with 3T3 mouse cells infected by SV40 that had been mutagenized with nitrosoguanidine. If the properties of transformed cells upon which our selection procedure was based were exclusively under the control of viral genes, the procedure should have led to the isolation of cells transformed by viral mutants specifically affected in "regulatory" genes. Alternatively it could have led to the isolation of cells transformed by wild-type SV40 genomes, but which nevertheless expressed the transformed phenotype in a temperature-sensitive manner, presumably because of a mutation in cellular genes.

This paper will deal with the possible mechanisms underlying the behavior of the ts SV40-transformed 3T3 cells we have isolated. These cells, by most in vitro criteria, express the transformed phenotype when growing at 32°C but not at 39°C.

Since the lines isolated so far are apparently transformed by wild-type virus, they represent mutations in a host gene(s) which controls the expression of malignant transformation.

Properties of the *ts* Transformants

We will summarize here the main properties of the *ts* transformants (Table 1), since most of them have already been described in detail (Renger and Basilico 1972; Noonan et al. 1973; Renger and Basilico 1973).

Saturation Density

At 39°C the *ts* SV3T3 reach saturation densities similar to that of normal 3T3 cells, but at 32°C they attain the high density of transformed cells.

Colony-forming Ability on 3T3 Monolayers

SV40-transformed 3T3 cells from colonies when plated on top of a monolayer of normal 3T3 cells, whereas normal cells will not form colonies under those conditions.

When the colony-forming ability on top of 3T3 monolayers of the *ts* transformants was tested, it was found to be 30- to 100-fold higher at 32°C than at 39°C.

It is worth mentioning here that the normal efficiency of plating of these cells is not affected by temperature, and their generation time is shorter at 39°C than at 32°C. This shows that these cells are not temperature sensitive for growth.

Density Inhibition of DNA Synthesis

In normal 3T3 cells density inhibition of growth is accompanied by inhibition of DNA synthesis. SV40-transformed 3T3 cells do not respond to this type of growth control. As a result DNA synthesis and division continue even in dense cultures.

We examined the effect of cell density on the DNA synthesizing activity of the *ts* transformants at 32 and 39°C. The results of those experiments showed that at 39°C both the rate of DNA synthesis and the frequency of DNA synthesizing cells

Table 1
Properties of the *ts* SV3T3 Cells

	Saturation* density		Colony-forming† ability on 3T3		Agglutination‡ by ConA		T antigen	
	39°C	32°C	39°C	32°C	39°C	32°C	39°C	32°C
3T3	0.8	1.0	0	0	>1500	>1500	—	—
SV3T3 wt	9.0	10.0	53.0	51.0	50	50	+	+
ts H6	1.25	10.5	0.47	25.5			+	+
ts H6-15	1.0	8.5	0.75	78.0	>1000	50	+	+
ts H1	1.5	8.5	0.2	12.5	>1000	50	+	+

* Cells $\times$ 10^{-5}/cm² of surface area.

† Percent of cells plated yielding colonies.

‡ Concentration (μg/ml) necessary for half-maximal agglutination.

greatly decreased at confluence, approaching values similar to those of confluent 3T3 cells. At 32°C the rate of DNA synthesis remains high even after confluence.

Surface Properties

SV40-transformed cells, due to surface alterations, show enhanced agglutination by plant lectins, such as concanavalin A (ConA) or wheat germ agglutinin (WGA) (Burger 1969). We tested three of our lines with respect to agglutinability and binding by ConA and WGA. It was found that these cell lines were agglutinable when growing at 32°C, but at 39°C their agglutinability was as low as that of normal cells (Table 1). Binding of radioactive ConA was also 4- to 5-fold higher at 32 than at 39°C, a difference similar to that found between normal and standard SV40-transformed 3T3 cells at any temperature (Noonan et al. 1973). This finding shows that the surface configuration of the *ts* transformants is also temperature dependent.

Properties of the Transforming Virus

All of the *ts* transformants tested were found to be free of infectious virus. They synthesize SV40-specific T antigen at any temperature (Table 1). To test whether the temperature-sensitive expression of the transformed phenotype was due to the fact that those cells had been transformed by a temperature-sensitive SV40 mutant, virus was rescued from four of the cell lines by means of Sendai-assisted fusion with permissive monkey cells.

Invariably the rescued viruses behaved exactly like wild-type SV40, as they were capable of multiplication and transformation both at 39 and 32°C with approximately the same efficiency (Renger and Basilico 1972). Furthermore virus rescue by fusion took place both at 39 and at 32°C.

It would seem, therefore, that the *ts* transformants owe their behavior to a cellular, rather than a viral, mutation. A corollary of this conclusion was that we should have been able to isolate cells with similar properties also from 3T3 populations that had been infected with wild-type SV40, and not with mutagenized virus, as it had been done in the original selection experiments. This was indeed the case (Renger and Basilico 1973).

Determinants of the *ts* Expression of Transformation in the *ts* SV3T3 Cells

The data presented so far show that by a number of parameters the *ts* transformants behave in vitro like transformed cells when growing at 32°C, but at 39°C they approximate the behavior of normal cells. The reason for this seems to be a host cell mutation, which interferes with the expression of transformation at 39°C. It can be asked, What are the main determinants of these differences in behavior? To elucidate this question we have performed a number of experiments.

Cyclic AMP

It has been suggested that variations in cyclic AMP concentration could be a critical factor determining the phenotype of the cells. Virally transformed cells have generally a lower cyclic AMP content than their normal counterpart, and dibutyryl cyclic AMP added to growing cells has been reported to alter cell growth (Johnson, Friedman and Pastan 1971; Sheppard 1971, 1972). We tested the cyclic AMP content of the *ts* SV3T3 cells growing at 32° and at 39°C, and we compared their values with those of normal or standard transformed 3T3 cells (Table 2). It can be seen that the cyclic AMP levels of the *ts* SV3T3 cells do not vary with the temperature of growth, although the values are somewhat intermediate between those of normal and those of transformed cells. It can probably be concluded that variations in total intracellular cyclic AMP do not play a major role in determining the different behavior of the *ts* SV3T3 cells at the two temperatures.

Phenomena Accompanying Transition from the Transformed to the "Normal" State

When nonconfluent cultures of *ts* SV3T3 cells were shifted from 32 to 39°C, the cells grew to confluence and growth stopped. However when the cells were transferred to 39°C after they had already grown beyond confluence, not only cell growth stopped but the cultures shed a large number of cells into the medium. This phenomenon started about 24 hours after shift and seemed to subside only after the number of cells attached had been reduced to that characteristic of the saturation density at 39°C, i.e., $\sim 7 \times 10^4$ cells/cm^2 (Fig. 1).

The *ts* cells that detach upon shift-up are not viable. On the other hand, the majority of the cells that do not detach remain viable, since if these cells are reshifted to 32°C growth resumes and the cultures attain again a high density (Renger and Basilico 1973).

It can be asked whether normal 3T3 cells would display a similar behavior. Therefore we plated 3T3 cells at a concentration (5×10^6/50-mm plate) about 5 times higher than their normal saturation density. The number of cells attached never reached the number of cells plated, and in a few days it was down to the characteristic saturation density of this line (Renger and Basilico 1973).

The phenomenon we observed by shifting dense cultures of *ts* SV3T3 to 39°C could be explained on the basis of surface alterations. The surface of *ts* SV3T3 cells changes upon shift from 32 to 39°C. It could be thought that the surface configuration at 39°C is such that these cells cannot maintain contact with one another in a multilayer state but need a solid substrate upon which to attach. The sudden change from one state to another in absence of such a substrate could cause loss of viability.

Table 2
Cyclic AMP Content of the *ts* SV3T3 Cells

Cells	Condition	39°C	32°C
3T3	Sparse	23.0	
	Confluent	16.5	23.3
SV3T3	Sparse	7.0	
(R-17)	Confluent	6.0	5.7
ts H1	Sparse	12.0	10.5
	Confluent	10.0	9.0
ts H6-15	Sparse	13.5	14.3
	Confluent	10.8	13.9

Picomoles/mg protein; the assay was performed using the "binding protein" method of Gilman (1970).

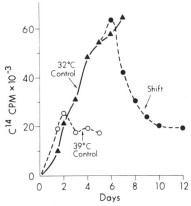

Figure 1

Effect of shift-up on the cell density of the *ts* transformants at 39°C. *ts* H6-15 cells were plated at approximately 3×10^4 cells/cm² in medium containing [^{14}C]thymidine (0.1 μCi, 10 μg/ml) and incubated at 39°C (—○—) and at 32°C (—▲—). At the times indicated the amount of label incorporated into cells attached to the plate was determined. Radioactivity is directly proportional to cell number. At day 5 several plates growing at 32°C were shifted to 39°C (—●—). Medium containing the same amount of thymidine with the same specific activity was changed every 3 days.

If this hypothesis were true, one would expect that the inability of 3T3 and of *ts* SV3T3 cells at 39°C to form colonies when plated on a confluent monolayer of 3T3 cells could result from their inability to attach to such a substrate. However we found that attachment was not inhibited when a *small* (10^3–10^4) number of cells were plated on a confluent 3T3 monolayer (Renger and Basilico 1973).

It is known that SV40-transformed 3T3 cells have lower serum requirements than untransformed 3T3 (Jainchill and Todaro 1970). Therefore a second interpretation of the shedding of dead cells upon shift of dense *ts* SV3T3 to 39°C would ascribe it to changes in the ability of the cells to utilize serum factors. We found that when 3T3 cells are grown to high densities in medium containing 50% serum, this density ($\sim 3 \times 10^6$ cells/50-mm plate) cannot be maintained if the medium is changed to 10% serum. Under these conditions the number of cells per culture decreases within 3 or 4 days to that characteristic of the saturation density reached in medium containing 10% serum ($\sim 1 \times 10^6$) (Renger and Basilico 1973). Similarly a drastic change in the serum requirements of the *ts* SV3T3 after shift to 39°C could make them unable to maintain a high density as that would require very high serum concentrations.

The serum would contribute survival factors that could regulate the ability of the cells to utilize nutrients. In that case the cell number would decrease to the point at which the concentration of serum factors is sufficient to maintain the viability of the cells present. This interpretation would be in agreement with the fact that we have been able to prevent substantially the cell detachment upon shift-up by supplementing the medium with 50% serum, and would imply that the serum properties of our cells are different at 39 and at 32°C. This was substantiated by experiments where the growth of *ts* SV3T3 was tested in various concentrations of serum at the two temperatures (Fig. 2). Whereas SV3T3 cells grow well at 39

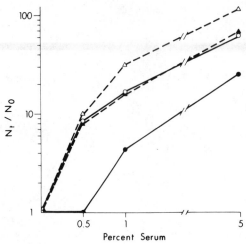

Figure 2
Growth of *ts* SV3T3 cells in medium containing different concentrations of serum at 32°C and 39°C. *ts* H6-15 and SV3T3 101 cells were seeded at approximately 3×10^4 cells/60-mm plate and incubated for 16 hr in medium containing 0.5% serum at 32°C and 39°C. Then the cell number per plate was determined (N_0), and the medium was changed to one containing 0.5%, 1%, or 5% serum, respectively. Four days later at 39°C and 7 days later at 32°C the cell number was determined (N_t).

(○)	ts H6-15	32°C
(▲)	SV101	39°C
(•)	ts H6-15	39°C
(△)	SV101	32°C

and 32°C in 5% serum, and somewhat less but with no temperature differences at 1% and 0.5%, *ts* SV3T3 cells grow like SV3T3 at 32°C in the different serum concentrations, but at 39°C growth at 0.5% and 1% is extremely reduced. Growth of normal 3T3 cells in these serum concentrations is practically zero at both temperatures. From this experiment it can be concluded that serum requirements of the *ts* SV3T3 cells at 39°C are higher than at 32°C. These higher serum requirements are probably due to a requirement for normal cell factor(s) as, in an experiment performed by Dr. R. Holley, fraction A from serum (Kaplan and Bartholomew 1972) was able to stimulate growth of the *ts* SV3T3 at 39°C in the presence of 0.4% serum, whereas at 32°C growth took place with or without addition of the serum factor(s).

These experiments suggest that the *ts* SV3T3 cells are also closer to normal with respect to serum requirements when growing at 39 than at 32°C, although this difference is somewhat less pronounced than that obtained with other parameters of transformation.

It could be thought that serum requirements are monitored by the same cellular control systems that determine other growth parameters. If this were the case a leaky mutation, whether cellular or viral, could strongly modify properties of transformation such as density inhibition, surface changes, etc., but could affect to a much lesser extent the serum requirements. Alternatively one could think that the serum properties are altered first in a cell upon transformation, and that even a

Table 3

Properties of Temperature-dependent Serum Variants of SV40-Transformed 3T3

	Efficiency of plating*				Growth (N_4/N_0)†				T antigen	
	10% serum		1% serum		10% serum		1% serum			
Cells	39°C	32°C	39°C	32°C	39°C	32°C	39°C	32°C	39°C	32°C
SV3T3 wt	60	50	10	10	40	30	10.3	9.6	+	+
3T3	20	25	<.01	<.01	18.2	15.0	2.0	5.0	−	−
SV F1D	10	20	<.01	10	17.3	15.5	1.93	8.0	+	+
SV 23A	10	20	<.01	15	10.6	11.5	2.30	10.6	+	+

* Percent of cells plated which formed colonies in medium containing 10% or 1% calf serum.

† Fold increase in cell number after 4 days of incubation.

small change in this central "control" would bring about a considerable alteration of the phenotype with respect to other properties (Holley 1972).

In order to test this hypothesis we are trying to isolate SV40-transformed cells which are temperature sensitive with respect to serum requirements. The selection method is based on the killing of transformed cells capable of growing in low serum at 39°C. We succeeded in isolating two transformed lines, which grow in low serum at 32 but not at 39°C (Table 3). However preliminary experiments suggest that the high serum requirements of these cells at 39°C are not directed towards the serum factors necessary for the growth of normal cells, but reflect a higher requirement for the same factors that are necessary for the growth of transformed cells.

Clearly the answer to the question of the relationship between serum requirements and other properties of transformed cells will necessitate the isolation of the appropriate mutants. As of now, this matter is in our opinion quite unresolved.

Behavior of the ts *Transformants upon Superinfection*
with Oncogenic Viruses

Attempts to retransform *ts* SV3T3 cells with SV40 at 39°C using high multiplicities of infection were not successful. This is consistent with the interpretation that the cells represent host mutants that do not allow the expression of the SV40-transformed phenotype at 39°C.

However when we challenged the *ts* SV3T3 cells at 39°C with murine sarcoma virus, an RNA tumor virus, it was found that they were fully susceptible to retransformation, and that the MSV-transformed *ts* SV3T3 were now completely temperature independent with respect to the expression of transformed growth characteristics (Renger 1972). Leukemia virus alone did not cause any morphological transformation.

This finding is important in that it shows that the *ts* SV3T3 are not incapable of expressing transformation at 39°C under any circumstance. Rather it suggests that the mutation they carry is in a regulatory pathway and that MSV overcomes in some way this block. It was then considered of interest to test the effect of polyoma infection on the phenotype of these cells. Polyoma virus multiplies in mouse cells and

Table 4

Growth of Polyoma Virus in *ts* H6-15 Cells

Mode of infection	Virus yield (PFU/culture)	
	39°C*	32°C†
Virus	6×10^7	2×10^7
Viral DNA	2.4×10^7	3.8×10^6

* Determined at 48 hr p. i.

† Determined at 72 hr p. i.

transformation is a very rare event. However since viral transformation functions are also expressed during a lytic infection, such a study was nevertheless considered useful. Growing *ts* SV3T3 cells can support the multiplication of polyoma virus both at 39 and at 32°C (Table 4). We tested whether polyoma infection of confluent cultures of *ts* SV3T3 cells at 39°C induced cellular DNA synthesis and whether virus production would follow. It was found that infection with polyoma at high m.o.i. was capable of inducing cellular DNA synthesis in a large proportion of the cells at 39°C (Table 5) and virus production also followed. It appears, therefore, that polyoma infection can overcome density inhibition of DNA synthesis of these cells at 39°C.

We also tested the effect of high m.o.i. of SV40 on the cellular DNA synthesis of *ts* SV3T3 cells at 39°C. It was found that also infection with SV40 was capable of stimulating cellular DNA synthesis in these cultures (Table 5). The extent of induction (frequency of induced cells) was somewhat less than in 3T3 cells, but nevertheless at least half of the cells appeared to be induced to synthesize DNA.

This finding was somewhat surprising, since these cells cannot be retransformed by SV40 at 39°C, and in addition they already carry integrated SV40 genomes.

Table 5

Induction of Cellular DNA Synthesis by Polyoma and SV40 in Confluent *ts* SV3T3 Cells at 39°C

Virus*	Frequency of DNA synthesizing cells (%)†	Frequency (%) of cells synthesizing		
		Py T antigen‡	Py V antigen‡	SV40 T antigen
0	11	0	0	100
Py	73	66	31	100
SV40	50	0	0	100

* m.o.i. = 200 PFU/cell.

† Determined by labeling the cells with [³H]thymidine from 36 to 48 hr after infection.

‡ Determined by immunofluorescence at 24 and 45 hr after infection, respectively.

It is possible that this apparent contradiction could be due to a dosage effect. The block of *ts* SV3T3 cells to the expression of the SV40-induced effect could be stringent if the number of SV40 genomes in the cells is limited, but could be overcome by the large multiplicity of infection ($\sim$ 20,000 particles/cell) used in the induction experiments. This possibility is undergoing investigation.

Control of the Expression of Viral Transformation

It is clear from the results described above that the host cell can exert control over the expression of SV40-induced transformation, and that this control is probably at the level of a regulatory mechanism. In addition the fact that it is possible to isolate cells expressing transformation in a temperature-sensitive manner suggests the involvement of a host cell protein. What do these findings suggest with regard to the general mechanism of cell transformation by viruses?

The expression of the transformed phenotype is presumably under the control of a viral protein, which is coded by the integrated viral genome (Martin 1970; Kawai and Hanafusa 1971; Dulbecco and Eckhart 1970). The mechanism of action of this protein is unknown. A complex type of mechanism is suggested by the evidence collected from our *ts* SV40-transformed cells. In principle they could be cells that are deficient in some components necessary for the phenotypic expression of transformation. This does not seem to be the case, as the *ts* SV3T3 can be retransformed by MSV and undergo some temporary change characteristic of transformation upon polyoma infection. On the other hand, resistance to retransformation expression could be due to a change in some regulatory substance which makes it unable to interact with the "transforming factor." Such cells should be transformable by other agents than the one to which they are found to be originally resistant. Our *ts* SV3T3 cells, as already mentioned, can be retransformed by MSV at 39°C, but not by SV40. Recently Stephenson, Reynolds and Aaronson (1973) have isolated some phenotypically normal MSV-transformed 3T3 cells which, however, contain a transforming MSV. Interestingly these cells can be retransformed by SV40 but not by MSV. A simplified model accounting for these facts would be the following: If the normal growth regulation in a cell was maintained by one or more regulatory proteins, with a mechanism of action similar to that of a bacterial "repressor," one could depict the normal cell as being in a repressed state. Viral transformation could be caused by the synthesis of a viral transforming protein, which acts as an "inducer" and would maintain the cells in a permanently derepressed state. Expression of transformation in a cell containing an intact viral genome could then be blocked at many different stages. Synthesis of an inactive transforming protein by the virus would, of course, result in a normal phenotype. At later stages the interaction of the viral inducer with the regulatory protein could be critical. In the *ts* SV40 transformants the regulatory protein could be temperature sensitive in its ability to bind the viral protein. The cell would then behave at 39°C like a normal cell. Superinfection by different viruses could easily cause retransformation because their transforming protein, although similar in function, has a different amino acid sequence and therefore could still bind the cellular protein.

A number of questions must be answered before any plausible model of the mechanism of viral transformation can be constructed. The data presented in this paper suggest a complex interaction of viral and cellular functions, which eventually determine the malignant phenotype.

Acknowledgments

This investigation was supported by grant CA11893 from the National Cancer Institute and by grant VC-99A from the American Cancer Society. C. B. is a Scholar of the Leukemia Society.

REFERENCES

Burger, M. M. 1969. A difference in the architecture of the surface membrane of normal and virally transformed cells. *Proc. Nat. Acad. Sci.* **62:**994.

Dulbecco, R. and W. Eckhart. 1970. Temperature dependent properties of cells transformed by a thermo-sensitive mutant of polyoma virus. *Proc. Nat. Acad. Sci.* **67:**1775.

Gilman, A. G. 1970. A protein binding assay for cyclic AMP. *Proc. Nat. Acad. Sci.* **67:**305.

Holley, R. W. 1972. A unifying hypothesis concerning the nature of malignant growth. *Proc. Nat. Acad. Sci.* **69:**2840.

Holley, R. W. and J. A. Kiernan. 1968. "Contact inhibition" of cell division in 3T3 cells. *Proc. Nat. Acad. Sci.* **60:**300.

Jainchill, J. L. and G. J. Todaro. 1970. Stimulation of cell growth in vitro by serum with and without growth factor. *Exp. Cell Res.* **59:**137.

Johnson, G. S., R. L. Friedman and I. Pastan. 1971. Restoration of several morphological characteristics of normal fibroblasts in sarcoma cells treated with adenosine 3′,5′-cyclic monophosphate and its derivates. *Proc. Nat. Acad. Sci.* **68:**425.

Kaplan, A. E. and J. C. Bartholomew. 1972. Study of the growth response of normal and SV40-transformed 3T3 mouse fibroblasts with serum fractions obtained by use of organic solvents. *Exp. Cell Res.* **73:**262.

Kawai, S. and H. Hanafusa. 1971. The effects of reciprocal changes in temperature on the transformed state of cells infected with a Rous sarcoma virus mutant. *Virology* **46:**470.

Martin, A. S. 1970. Rous sarcoma virus: A function required for the maintenance of the transformed state. *Nature* **227:**1021.

Noonan, K. D., H. C. Renger, C. Basilico and M. M. Burger. 1973. Surface changes in temperature-sensitive SV40 transformed cells. *Proc. Nat. Acad. Sci.* **70:**347.

Pollack, R. E., H. Green and G. J. Todaro. 1968. Growth control in cultured cells: Selection of sublines with increased sensitivity to contact inhibition and decreased tumor-producing ability. *Proc. Nat. Acad. Sci.* **60:**126.

Renger, H. C. 1972. Retransformation of temperature-sensitive SV40 transformants by murine sarcoma virus at non-permissive temperature. *Nature New Biol.* **240:**19.

Renger, H. C. and C. Basilico. 1972. Mutation causing temperature-sensitive expression of cell transformation by a tumor virus. *Proc. Nat. Acad. Sci.* **69:**109.

———. 1973. Temperature sensitive SV40 transformed cells: Phenomena accompanying transition from the transformed to the "normal" state. *J. Virol.* **11:**702.

Sheppard, J. 1971. Restoration of contact inhibited growth to transformed cells by dibutyryl-cyclic AMP. *Proc. Nat. Acad. Sci.* **68:**1316.

———. 1972. Difference in cyclic AMP levels in normal and transformed cells. *Nature New Biol.* **236:**14.

Stephenson, J. R., R. K. Reynolds and S. A. Aaronson. 1973. Characterization of morphologic revertants of murine and avian sarcoma virus-transformed cells. *J. Virol.* **11:**218.

Isolation and Characterization of Two Classes of ConA-resistant SV3T3 Cells

Brad Ozanne and Mirka Lurye

Cold Spring Harbor Laboratory, Cold Spring Harbor, New York 11724

Transformation of 3T3 cells by Simian Virus 40 results in the cells displaying a new set of heritable traits which are characteristic of the transformed state. In contrast to their ancestral 3T3 cells, SV40 transformed 3T3 cells (SV3T3) no longer exhibit density-dependent inhibition of growth, but instead grow into multilayers of cells reaching saturation densities many-fold greater than those attained by 3T3 cells (Todaro and Green 1964). SV3T3 cells also have a decreased serum requirement (Todaro et al. 1967; Holley and Kiernan 1968), lower levels of intracellular cAMP (Otten, Johnson and Pastan 1971; Sheppard 1972), and increased efficiency of colony formation when grown in Methocel-containing medium (Stoker 1968). The cell surface of SV3T3 cells has also been altered. New surface antigens can be detected (see review Sambrook 1972) and the SV3T3 cells differ from 3T3 cells in their interactions with the plant lectins concanavalin A (ConA) (Inbar and Sachs 1969) and wheat germ agglutinin (WGA) (Burger and Goldberg 1967; Burger 1969). These lectins interact with sugar moieties on the cell surface and at low concentrations cause the SV3T3 cells to agglutinate. 3T3 cells, on the other hand, remain dispersed even in the presence of high lectin concentrations. Furthermore SV3T3 cells are more readily killed than are 3T3 cells when ConA is added to the culture medium (Shoham, Inbar and Sachs 1970; Ozanne and Sambrook 1971a; Culp and Black 1972; Ozanne 1973).

Since the SV3T3 cells were transformed by a virus, it is felt that the changes in the phenotype of the cells following transformation result from an interaction of the virus and the cell. In the case of SV40 transformation it is known that the virus genome has become stably integrated into the cellular DNA (Westphal and Dulbecco 1968; Sambrook et al. 1968; Gelb, Kohne and Martin 1971; Ozanne, Sharp and Sambrook 1973) and is expressed since both viral-specific RNA (Sambrook 1972; Khoury et al. 1973; Ozanne, Sharp and Sambrook 1973) and antigens (Sambrook 1972) can be detected in the transformed cells. Though it has yet to be shown for SV40 transformation, presumably some continued virus function is required for the expression of the transformed state. Just what this function could be and how it interacts with the cell to result in transformation is totally obscure.

In attempts to sort out the virus-cell interaction causing transformation, a variety of revertants of SV3T3 cells have been isolated by employing selective procedures which prey upon specific differences between SV3T3 and 3T3 cells. Under conditions where 3T3 cells cannot grow, drugs, such as fluoro- or bromodeoxyuridine, and UV light have been used to select against the cells that can synthesize DNA (Pollack, Green and Todaro 1968; Culp and Black 1972; Vogel and Pollack 1973). Such revertants usually have simultaneously lost many other characteristics of transformed cells besides the one selected against. Most of the revertants isolated have regained growth control to varying degrees in that they have saturation densities similar to those of 3T3 cells. These variants have also reverted in their interaction with lectins in that they have become nonagglutinable by ConA or WGA (Pollack and Burger 1969; Vogel, Hough and Pollack unpubl.).

Pollack and Burger (1969), using a suite of normal, transformed, and revertant mouse cells, demonstrated an inverse correlation between the saturation density of a cell line and its susceptibility to lectin-mediated agglutination. When the lectin ConA has been used as a selective agent, ConA-resistant SV3T3 cells have been isolated, which also behave as revertants (Ozanne and Sambrook 1971; Culp and Black 1972; Wollman and Sachs 1973; Ozanne 1973). This paper extends the characterization of several previously reported ConA-resistant SV3T3 cells (Ozanne 1973) and reports the isolation of a new class of ConA-resistant SV3T3 cells. This new class grows to high saturation densities like SV3T3 cells, but interacts with ConA as do 3T3 cells.

MATERIALS AND METHODS

Cells

Swiss 3T3 cells (R. Pollack, Cold Spring Harbor Laboratory) and SV40-transformed 3T3 cells (SV3T3) (W. Eckhart, Salk Institute) were grown in Dulbecco's modification of Eagle's medium supplemented with 10% calf serum (Gibco, New York). Monkey cells BSC-1 (T. Benjamin, Harvard Medical School) and MA-134 (J. Pagano, University of North Carolina) were grown in Dulbecco's modification of Eagle's medium supplemented with 10% fetal bovine serum. All cells were carried on plastic tissue culture plates (Nunc) at 37°C in 5% CO_2 atmosphere.

Virus

SV40 strain 77 (Gerber 1962) was grown on BSC-1 cells. The Rauscher strain of murine sarcoma/leukemia virus was kindly provided by S. Aaronson (National Institutes of Health). UV-inactivated Sendai virus was a gift from H. Coon (National Institutes of Health).

Selection of ConA-resistant SV3T3 Cells

The ConA-resistant SV3T3 cells behaving as revertants were isolated as previously described (Ozanne and Sambrook 1971a; Ozanne 1973). Briefly, to 2×10^6 cells 24 hours after plating, ConA was added to give a final concentration of 300 μg/ml. Twenty-four hours later the cells that had rounded up were removed by repeated washings with phosphate-buffered saline, fresh medium was added and the remaining cells were allowed to grow into colonies. Two weeks after the ConA treatment, flat-looking clones were isolated and their properties tested.

Dense-growing ConA-resistant SV3T3 cells were isolated as follows. Subconfluent SV3T3 cells were removed from plates by trypsinization and suspended at a concentration of 2×10^5 cells/ml in complete medium containing 300 μg/ml ConA. Ten ml of this medium was added to 90-mm plastic tissue culture dishes and incubated for 24 hr at 37°C. After this incubation period the floating cells were removed by repeated washings with PBS. Fresh medium was added to the dishes and changed every 3 days for 2 weeks, at which time dense-looking clones were isolated, grown up and tested for their resistance to ConA.

Saturation Densities

Saturation densities were determined from growth curves as described previously (Ozanne 1973). The saturation density for a cell line was determined by calculating the number of cells per cm² 2 days after the cells had ceased growing after reaching confluency. The generation times for the various cell lines were determined during log phase growth. Growth curves in 1% serum were performed as described in Ozanne (1973).

Growth in Methocel

The cells to be tested were suspended in Methocel medium containing 10% calf serum and plated onto dishes with a thin coat of hard agar covering the bottom and sides of the dish (Stoker 1968). Between 2.5×10^2 and 2.5×10^5 cells were added to each plate. Fresh medium containing Methocel was added once a week for 3 weeks, at which time visible colonies were counted.

cAMP Assay

The cyclic AMP content of the cell lines was determined as performed by Sheppard (1972). Sparse cultures were used 2 days after seeding and the value obtained for each cell line was the average of three experiments each done in triplicate.

T Antigen and Virus Rescue

T antigen was determined by immunofluorescence using fluorescent antisera from Flow Laboratories, Rockville, Md.

Virus was rescued by fusing the SV40-containing mouse cells with BSC-1 cells using UV-inactivated Sendai virus as described by Watkins and Dulbecco (1967).

Infectivity and Transformation Assays

SV40 was plaqued on BSC-1 cells as described by Kimura and Dulbecco (1972). The focus-forming ability of MSV stocks was determined on 3T3 cells as described by Aaronson, Janchill and Todaro (1970).

Transformation assays with SV40 were carried out as described by Aaronson and Todaro (1968).

Agglutination Assays

Agglutination assays for ConA and WGA were carried out as described by Ozanne and Sambrook (1971a). The cells from 48-hour sparse cultures were removed from the plate by repeated washings with 5×10^{-4} M EDTA in PBS. The cells were washed and resuspended at 2×10^6 cells/ml in PBS and used in the agglutination assay. Agglutination was scored on a serological scale. Half-maximal agglutination equals 50–65% of the cells in clumps.

Binding of Radioactive ConA to Cells

[^{125}I]ConA was prepared using lactoperoxidase as described by Ozanne and Sambrook (1971b). The binding assays were performed on 2×10^6 cells at 20°C for 15–20 minutes in the presence and absence of the hapten for ConA, α-methylglucose, as described in Ozanne and Sambrook (1971b). The amount of ConA specifically bound to the cells was determined from difference of counts bound to the cells in the presence and absence of hapten. The [^{125}I]ConA used had a specific activity of 1500 cpm/μg of protein.

DNA Determination and Karyotype Analysis

The rate of DNA synthesis was measured as described by Ozanne (1973) in both sparse and confluent cultures by pulse-labeling the cells for 4 hours with [^{3}H]thymidine.

Karyotype analysis was carried out by following the procedure of Pollack, Vogel and Wolman (1970). At least 30 metaphases were counted for each cell line.

DNA content per cell was determined by Arthur Vogel using the method described by Van Dilla et al. (1969).

Assay for RNA-dependent DNA Polymerase

RNA-dependent DNA polymerase was assayed according to Kelloff, Hatanka and Gilden (1972). Ten days after 3T3 cells or ConA revertants were infected with MSV/MLV with a titer of 3×10^4 ffu/ml on 3T3 cells, the media was collected and clarified by centrifugation for 10 minutes at 10,000 rpm. The virus was then pelleted at 100,000 g for 1 hour. The pellets were resuspended in 0.05 ml Tris buffer and assayed for RNA-dependent DNA polymerase.

RESULTS

Isolation of ConA-resistent SV3T3 Cells

We previously reported isolation of ConA-resistant SV3T3 cells having characteristic revertant behavior (Ozanne and Sambrook 1971a; Ozanne 1973). Such variants arise at a frequency of about one in 10^5 cells and have a flat morphology. Of 63 flat-looking clones picked several weeks after being exposed to 300 μg/ml ConA for 24 hours, only nine proved to be resistant to ConA upon subsequent testing and maintained a flat morphology. None of the 54 clones isolated, which grew to high saturation densities, proved to be resistant to the toxic effects of ConA. However by modifying the selection procedure as mentioned in Materials and Methods, dense-growing variants resistant to the toxic effects of ConA were isolated. Of 20 dense clones picked, three were resistant to ConA; these variants appear to be more polygonal in shape than their parent SV3T3 cells and to grow in a more ordered fashion.

The properties of both classes of ConA-resistant SV3T3 cells are stable in that after prolonged periods in culture their characteristics, such as saturation density and resistance to ConA, appear to be unchanged. All of the variants have been subcloned several times in the absence of ConA. The subclones' characteristics do not differ significantly from those of the parent cell lines.

Saturation Densities

Like 3T3 cells, the ConA-selected revertants cease to grow after they reach a mono-layer. The parent SV3T3 cells, however, continue to grow into multilayers of cells and reach a saturation density ten to twenty times greater than that attained by 3T3 cells. The ConA-selected revertants have saturation densities about 1.5 times that of 3T3 cells, and one revertant (CA^r30) has a saturation density twice that of 3T3 cells (Table 1). These revertants appear to recognize a monolayer for the following reasons. If the media is changed after the cells have reached confluency, the revertants, like 3T3 cells, show only a slight increase in cell number over the next 24–48 hours (Ozanne 1973). Also if the cells are seeded at such a density so that monolayers are formed after one round of division, the ConA revertants behave like 3T3 cells in that upon achieving their saturation densities, the number of cells in the monolayer remains constant (Ozanne 1973).

The ConA-resistant dense-growing clone CA^rD4A, however, acts like SV3T3 cells in its ability to overgrow the monolayer and reaches a saturation density essentially equal to that of SV3T3 cells (Table 1).

From the growth characteristics of these variants it appears that two classes of ConA-resistant SV3T3 cells can be obtained, those that refuse to form multilayers of cells and have low saturation densities and those which overgrow the monolayer and grow to high saturation densities.

Serum Requirement

After transformation by SV40, 3T3 cells require less serum to sustain growth (Holley and Kiernan 1968). SV3T3 cells are capable of growing in 1% serum

Table 1
Growth Characteristics of ConA-Resistant SV3T3 Cells

Cell line	Sat. density[a] (10^4 cells/cm^2)	Generation time[b] (hr) in 10%	Generation time[b] (hr) 1% serum	Colony formation in Methocel	% Log phase rate[c] of DNA synthesis
3T3	8.5	23	> 80	< 0.001%	1
SV3T3	76	20	26	40	65.8
CA^r30	16	22	27	0.01	13
CA^r32	10	21	24	0.005	20.5
CA^r41	10.7	22	31	0.003	27.5
CA^rD4A	69	20	26	1	

[a] The saturation densities were calculated from growth curves performed as described in the Materials and Methods section of this paper.

[b] The generation times of the cell lines listed were calculated from the logarithmic phase of growth of the cells determined by growth curves as described in Materials and Methods.

[c] The procedures for measuring the rate of DNA synthesis of the cell lines was determined by the uptake of [³H]thymidine into the cells. The rate of thymidine incorporation for the sparse and confluent cells was described in Materials and Methods. These rates were used to calculate the % log phase rate of DNA synthesis for confluent cultures as a measure of the degree of density-dependent inhibition of DNA synthesis of a particular cell line displayed.

with almost the same generation time as in 10% serum (Table 1). 3T3 cells, on the other hand, either fail to grow at all, or do so at a much reduced rate—a generation time of greater than 80 hours in 1% serum as compared to 20–23 hours in 10% serum. The ConA-resistant SV3T3 cells, both the ConA revertants, and clone CA^{r}D4A retain the transformed cell serum requirement since they grow in 1% serum with only a minor increase in generation time. So even though the ConA revertants act like 3T3 cells in their inability to overgrow a monolayer, they have the serum requirement of SV3T3 cells. In this regard they are similar to other revertants (Smith, Scher and Todaro 1971; Vogel and Pollack 1973), which also have low saturation densities and low serum requirements.

DNA Synthesis at Confluency

3T3 cells in resting monolayers synthesize DNA at a much reduced rate when compared to subconfluent growing 3T3 cells (Colby and Romano 1973; Pollack and Vogel 1973; Ozanne 1973). SV3T3 cells at confluency, however, continue to synthesize DNA at essentially the same rate as logarithmically growing SV3T3 cells do. As shown in Table 1, the ConA revertants continue synthesizing DNA after reaching their saturation densities, albeit at reduced rates, ranging from 13 to 27.5% the rates measured during log phase growth. It appears then that the ConA revertants fall into a class of revertants (Smith et al. 1971; Pollack and Vogel 1973) in which the recognition of a monolayer is not equivalent to density-dependent inhibition of DNA synthesis. We do not know what proportion of the ConA revertants in the monolayer are synthesizing DNA or whether they continue through the cell cycle to mitosis. Presumably this occurs since cells seem to be sloughed into the media.

Growth in Methocel

Stoker (1968) demonstrated that transformed cells could grow into clones while suspended in Methocel-containing media, but normal cells could not, presumably because they needed to attach to a substrate in order to grow. The ConA revertants, like 3T3 cells and other SV3T3 revertants (Vogel and Pollack 1973), failed to form colonies when grown in Methocel under conditions where 40% of the SV3T3 cells formed colonies after 3 weeks of growth (Table 1). The dense clone CA^{r}D4A was intermediate between 3T3 and SV3T3 cells in its ability to form clones in Methocel (Table 1). The significance of this decrease in the efficiency of CA^{r}D4A cells to form clones in Methocel is unclear, but may relate to the cell's more polygonal morphology and increased orientation in culture compared to the parent SV3T3 cells.

Cyclic AMP

Several groups have shown that SV40-transformed 3T3 cells have lower intracellular levels of cAMP than do 3T3 cells (Otten et al. 1971; Sheppard 1972). The ConA revertants have the same cAMP levels as do 3T3 cells (Table 2). Other revertants (Vogel, Oey and Pollack this volume) also show increased levels of cAMP. Since these revertants do not seem to shut off DNA synthesis at confluency, it seems that 3T3-like levels of cAMP alone are insufficient to render the cells

Table 2
Cyclic AMP Level

Cell line	cAMP (pmole/mg protein)
SV3T3	9
3T3	27
CA^{r}30	26
CA^{r}32	25
CA^{r}41	25

sensitive to density-dependent inhibition of growth, but may be important in maintaining morphology and determining interaction with lectins (Sheppard 1972).

Agglutinability

When the ConA revertants were tested for their ability to be agglutinated by either ConA or WGA, CA^{r}32 and CA^{r}41 failed to agglutinate at high concentrations of these lectins, as did 3T3 cells (Table 3). CA^{r}30 cells agglutinated more readily, but still required several times the concentration of either lectin necessary to agglutinate SV3T3 cells. The dense-growing ConA-resistant SV3T3 cells CA^{r}D4A were not agglutinable by high concentrations of ConA or WGA (Table 3).

Trypsin has been shown to render 3T3 cells as agglutinable by ConA (Inbar and Sachs 1969) or WGA (Burger 1969) as transformed 3T3 cells. After exposure to 0.001% trypsin for 2 minutes at 37°C, both the ConA revertants and CA^{r}D4A cells became agglutinable by either ConA or WGA.

Resistance to Toxic Effects of ConA

The ConA-resistant SV3T3 variants were tested to determine their resistance to the toxic effects of ConA. After being incubated with 300 μg/ml ConA for 24 hours, the dead cells were washed from the plate and the remaining cells were serially diluted and allowed to grow into colonies. Three weeks after the treatment the clones were stained and counted to determine the rate of survival of the treated cells (Ozanne 1973). About 1% of the cells of 3T3, CA^{r}32, and CA^{r}41 and 0.01% of the cells of CA^{r}30 grew into colonies; only 0.001% of the SV3T3 cells survived. The CA^{r}D4A cells had the highest survival rate, with 10% of the treated cells able to form colonies after treatment (Table 3).

From these results it seems to hold that 3T3 cells and revertants of SV3T3 cells with low saturation densities are more resistant to the toxic effects of ConA than are wild-type SV3T3 cells, but cells with high saturation densities may also be resistant to the toxic effects of ConA. Whether the two classes of ConA-resistant cells achieved their resistance by the same mechanism, however, is unclear.

Binding of Radioactive ConA

The number of ConA binding sites for ^{125}I-labeled ConA on 3T3 cells, SV3T3 cells, and the ConA-revertant cells was determined as described in Materials and

Table 3

Interaction of ConA-resistant Cells with ConA and WGA

Cell line	ConA[a]		WGA[a]		% Survival[b]	[125I]ConA molecules/cell[c]
	Not treated	trypsin	Not treated	trypsin		
3T3	> 750	10	> 500	5	1	1.2×10^7
SV3T3	10	10	10	5	.001	1.7×10^7
CA[r]30	500	10	250	5	.01	1.5×10^7
CA[r]32	> 750	10	> 500	5	1	1.4×10^7
CA[r]41	> 750	10–20	> 500	5	1	1.5×10^7
CA[r]D4A	> 750	20			10	

[a] Values given are μg/ml lectin required to give half-maximal agglutination. See Materials and Methods for details of assay.

[b] The % survival of cells after 24 hr exposure to 300 μg/ml ConA.

[c] The number of [125I]ConA molecules per cell was determined as described by Ozanne and Sambrook (1971b). A brief description of the procedure is in the Materials and Methods section of the text. These numbers were calculated from binding curves from Ozanne (1973).

Methods. There is little variation in the number of ConA molecules bound by the three types of cells—about 1.2–1.7 $\times$ 10[7] molecules per cell (Table 3). This figure agrees well with other determinations for the number of ConA molecules bound to 3T3 and SV3T3 cells (Arndt-Jovin and Berg 1971; Ozanne and Sambrook 1971b), and the density of ConA binding sites on these cells is about the same as found on lymphocytes and fat cells (Cuatrecasas 1973).

Karotype Analysis and DNA Content

Pollack, Vogel and Wolman (1970) reported that revertants isolated from SV3T3 cells had an increased number of chromosomes per cell. Such an increase in chromosome number seems to be a common trait for revertants isolated from SV3T3 cells (Ozanne and Sambrook 1971a; Culp and Black 1972; Ozanne, Sharp and Sambrook 1973; Vogel and Pollack 1973). The ConA revertants were also found to have a slightly increased number of chromosomes when compared to their parent SV3T3 cells (Table 4).

Since the ConA revertants have an increased number of chromosomes per cell, they should show an increased amount of DNA per cell. When the DNA content of the ConA revertants was determined, they were found to have more DNA per cell than their parent SV3T3 cells (Table 4). It is not known why the revertant cells show almost a doubling in their DNA content while displaying such a slight increase in their chromosome number.

T Antigen and Rescue of SV40 Virus

To see if the ConA-selected revertants still contain a functional SV40 genome, the cells were tested for the presence of SV40-specific T antigen by immunofluorescence. All of the cells of the three revertant lines were as positive as the parent SV3T3 cells. Since SV40 virus cannot grow in 3T3 cells and cannot be induced from SV3T3

Table 4
Copies of SV40 Genome per Cell

Cell line	Modal chromosome number	DNA/cell[a]	Copies SV40 DNA/diploid amount cell DNA[b]	Copies SV40 DNA per cell[b]	% Virus genome transcribed[b]	
					E strand	L strand
3T3	67	1.5	0	0	0	0
SV3T3	70	1.6	8–9	13–14	70	15
CA^{r}30	72	2.9	8–9	23–26	70	10
CA^{r}32	75	2.9	8–9	23–26	75	12
CA^{r}41	76	2.7	8–9	22–24	78	10

[a] Compared with mouse embryo fibroblasts.

[b] From Ozanne, Sharp and Sambrook (1973).

cells, it is unknown whether the virus in the SV3T3 cells is wild-type. However the virus can be rescued from SV3T3 cells by fusing them with monkey cells in which the virus can grow. To determine whether the alteration in the phenotype of the ConA-selected revertants was due to a change in the integrated virus genome, fusions were performed between the variants and BSC-1 cells. Infectious SV40 was rescued from the parent SV3T3 cells and from each of the ConA-selected revertants as shown in Table 5. The titer of virus rescued from the variants was lower than that rescued from the parent cells. Rather than attribute this to an alteration in the virus, it seems more likely to reflect a variation in the efficiency of heterokaryon formation between the different revertants and BSC-1 cells.

The rescued virus was grown to high titer by several passages on MA-134 cells and tested for its ability to transform 3T3 cells. The virus rescued from the revertants was as an efficient transforming agent as the virus rescued from SV3T3 cells (Table 5). From this data it appears that the ConA revertants contain some wild-type SV40 genomes.

Table 5
Rescue of SV40 from Cells

Cell line	T antigen	Virus rescued[a] (pfu/10^6 cells)	% 3T3 cells transformed[b] by rescued virus
SV3T3	+	9.5×10^3	3
3T3	−		
CA^{r}30	+	40	2.1
CA^{r}32	+	5	1.1
CA^{r}41	+	2.7×10^3	3.2

[a] The titer of the rescued virus was determined by plaque formation on BSC-1 cells as described in Materials and Methods.

[b] The rescued virus was grown to titers of around 10^6 pfu/ml by two passages on MA-134 cells prior to being tested for its transforming ability on 3T3 cells.

Retransformation

Revertants of SV3T3 cells seem to be resistant to retransformation by SV40 virus (Vogel and Pollack 1973). When the ConA revertants are infected with enough SV40 to transform 3% of the control 3T3 cells, no transformants were observed (Table 6). The reason for this is unknown and there are no assurances that the virus did, in fact, infect the cells.

Whatever the mechanism(s) is that renders SV3T3 revertants resistant to retransformation by SV40, it does not necessarily prevent their transformation by murine sarcoma virus. Recently Renger (1972) has shown that temperature-sensitive revertants of SV3T3 cells and flat revertants isolated by Pollack (Pollack, Green and Todaro 1968) form foci when infected with MSV, as do the revertants isolated by Vogel and Pollack (pers. commun.). However when the ConA revertants were infected with titers of MSV sufficient to transform 3T3 cells with a high efficiency, only a few foci developed (Table 6).

In the case of MSV it does not appear that the cells are resistant to infection by the virus. Under conditions where the number of MSV foci observed on the transformants is 1000-fold less than seen on 3T3 cells, the titer of MSV produced by the revertants is 1% that produced by the infected 3T3 cells. This difference in yield of virus does not seem to reflect an inability of the cells to produce virus since the amount of viral-specific RNA-dependent DNA polymerase found in the supernatants of the infected revertant cells is about 20% of that found in the MSV-transformed 3T3 cells. Thus the ConA revertants of SV3T3 cells appear to resist transformation by both homologous and heterologous viruses and in this respect are unique among revertants.

Table 6

Retransformation of ConA-Resistant Cells

Cell line	SV40[a]	MSV[b]	ffu/ml media[c]	cpm [3H]TTP incorporated per ml media[d]
3T3	3.1%	3×10^4	5×10^5	10,839
CA^{r}30	0	4	2.1×10^3	2356
CA^{r}32	0	7	5×10^3	3189
CA^{r}41	0	3	3×10^3	2015

[a] Percent transformants of 3T3 and revertants after infection with SV40 with a titer of 5×10^6 pfu/ml.

[b] Focus forming units/ml of MSV titered on 3T3 cells and the revertants.

[c] Titer of MSV on 3T3 cells. The virus stocks were grown on 3T3 cells or the revertants and taken 10 days after infection with 3×10^4 ffu/ml of MSV.

[d] Virus was prepared as described in the Materials and Methods section of the text and assayed for RNA-dependent DNA polymerase as described by Kelloff, Hatanka and Gilden (1972). Virus suspension 0.05 ml was incubated with an equal volume of reaction mixture containing 0.1 M Tris buffer pH 8.3, 6 mM Mg acetate, 0.12 M NaCl, 0.04 M DTT, poly (rA): oligo (dT) 25 μg/ml (final concentration) and 6.0 mM [3H]TTP. The reaction mixture was incubated for 45 min at 37°C and the acid-insoluble material was collected on Millipore filters and counted. Each value is the average of 3 determinations.

DISCUSSION

The ConA resistant SV3T3 cells described in this paper fall into two classes. In the first class the cells behave as revertants in that they have a flat morphology and low saturation densities. The cells in the second class act more like the parent SV3T3 cells since they retain the ability to grow to high saturation densities. It is unknown whether the alteration in the two sorts of cells, which rendered them resistant to the toxic effects of ConA, occurred by the same mechanism. However both classes of variants interact with ConA as do 3T3 cells. Neither cell type is agglutinated by high concentrations of ConA, but become susceptible to ConA-mediated agglutination following trypsinization. The cells of both classes are resistant to the toxic effects of ConA and in the case of the revertants, the cells bind the same number of ConA molecules per cell as do 3T3 or SV3T3 cells. From such results it seems that the alteration which rendered the cells resistant to the toxic effects of ConA does not involve a general loss of ConA receptor sites. Since the ConA-revertant cells interact with WGA as do 3T3 cells, and it has been shown that ConA and WGA do not compete for the same receptor sites on the cell surface (Ozanne and Sambrook 1971b; Jansons, Sakamoto and Burger 1973), a more plausible explanation may be that the cells have undergone a change which restricts the ConA receptor sites from aggregating into clusters, as has been observed on the surface of SV3T3 cells, but holds them in the dispersed configuration seen on 3T3 cells (Nicolson 1973). Other explanations are also possible (Culp and Black 1972).

Regardless of the change leading to ConA resistance, the variants also have altered growth characteristics compared to the parent SV3T3 cells. The dense-growing variant CA^{r}D4A, though retaining an SV3T3-like saturation density, grows in a more ordered pattern on the plate. The revertant clones (CA^{r}30, CA^{r}32 and CA^{r}41) behave akin to 3T3 cells in having low saturation densities. However in contrast to 3T3 cells, the variants continue to synthesize DNA after reaching their saturation density.

From the growth characteristics of clone CA^{r}D4A, it appears that an SV3T3-like interaction with the lectin ConA is not required in order for a cell line to grow to a high saturation density. However 3T3-like behavior of a cell with regard to lectins may still be necessary, though not sufficient, for a cell line to maintain a low saturation density.

Since SV40-specific T antigen can be detected and wild-type SV40 can be rescued from the ConA-selected revertants, it would appear that the alteration observed in the variants' phenotype resulted from changes in the cell and not the virus. However this conclusion cannot be reached from such results since we know that the parent SV3T3 cell (Table 4) contains multiple copies of the SV40 genome (Ozanne, Sharp and Sambrook 1973), and there exist no assurances that the virus rescued was responsible for maintaining the transformed state. Thus it is possible that the ConA revertants could have arisen from a deficiency in a virus function that maintains the cell in a transformed state. However the failure of the ConA revertants to be transformed by the heterologous virus MSV strongly suggests that an altered cellular function is responsible for the revertant phenotype. It can be shown that the failure of ConA revertants to become transformed by MSV is not due to an inability of the virus to enter the cell. Infectious MSV can be recovered from the infected revertant cells and the amount of RNA-dependent DNA polymerase activity present in the media of the infected revertant cells is almost equivalent to that found in the media from 3T3 cells transformed by the same virus stock. The results imply that, whatever the alteration was that resulted in the revertant phenotype of the cells, it also prevents the cells from being retransformed by MSV. The nature of this putative cellular change is totally obscure.

From the properties of the ConA revertants and the properties of revertants of SV3T3 isolated by other procedures (Pollack et al. 1968; Vogel and Pollack 1973), it appears that several traits characteristic of the transformed state—saturation density, agglutinability by lectins, cAMP levels and anchorage requirement—revert simultaneously with a frequency of about one cell in 10^5. However two traits of SV3T3 cells—the low serum requirement and lack of density-dependent inhibition of DNA synthesis—revert, if at all, with a much lower frequency. Even using selective conditions directed against DNA synthesizing cells, such as FdU or BrdU and UV light, it has not yet proved possible to isolate variants that have reverted for these two properties. The reason for the difference in the ease of reversions of these two sets of characteristics of the transformed state remains unclear. A possible explanation for the existence of different classes of characteristics of SV3T3 cells, as defined by revertants, may be that a function of SV40 is directly responsible for the release of density-dependent inhibition of DNA synthesis, which also alters the cell's serum requirement. However this function may only indirectly enhance the cell's chance to express the second set of characteristics of the transformed state that enables the cell to be recognized as a full-blown transformant. After all, one of the first occurrences after infection of a 3T3 cell by SV40 is the induction of cellular DNA synthesis, whereas the other characteristics of transformation are expressed only if the cells are allowed to undergo several divisions.

Acknowledgments

We wish to thank Rex Risser for help with the Methocel growth, Art Vogel for determining the DNA content of the cells and Joe Sambrook for his discussion. This work was supported by Grant #CA 11432 from the National Cancer Institute, Public Health Service.

REFERENCES

Aaronson, S. A. and G. J. Todaro. 1968. Basis for the acquisition of malignant potential by mouse cells cultivated *in vitro*. *Science* **162**:1024.

Aaronson, S. A., J. L. Jainchill and G. J. Todaro. 1970. Murine sarcoma virus transformation of Balb/3T3 cells: Lack of dependence on murine leukemia virus. *Proc. Nat. Acad. Sci.* **66**:1236.

Arndt-Jovin, D. J. and P. Berg. 1971. Quantitative binding of ^{125}I-concanavalin A to normal and transformed cells. *J. Virol.* **8**:716.

Burger, M. M. 1969. A difference in the architecture of the surface membrane of normal and virally transformed cells. *Proc. Nat. Acad. Sci.* **62**:994.

Burger, M. M. and A. R. Goldberg. 1967. Identification of a tumor-specific determinant on neoplastic cell surfaces. *Proc. Nat. Acad. Sci.* **57**:359.

Cuatrecasas, P. 1973. Interaction of wheat germ agglutinin and concanavalin A with isolated fat cells. *Biochemistry* **12**:1312.

Culp, L. A. and P. H. Black. 1972. Contact inhibited revertant cell lines isolated from simian virus 40 transformed cells. *J. Virol.* **9**:611.

Gelb, L. D., D. E. Kohne and M. A. Martin. 1971. Quantitation of simian virus 40 sequences in African green monkey, mouse and virus-transformed cell genomes. *J. Mol. Biol.* **57**:129.

Gerber, P. 1962. An infectious deoxyribonucleic acid derived from vacuolating virus (SV40). *Virology* **16**:96.

Holley, R. W. and J. A. Kiernan. 1968. "Contact inhibition" of cell division in 3T3 cells. *Proc. Nat. Acad. Sci.* **60**:300.

Inbar, M. and L. Sachs. 1969. Interaction of the carbohydrate-binding protein concanavalin A with normal and transformed cells. *Proc. Nat. Acad. Sci.* **63**:148.

Jansons, V. K., C. K. Sakamoto and M. M. Burger. 1973. Isolation and characterization of agglutinin receptor sites. II. Studies on the interaction with other lectins. BBA **291**: 136–143.

Kelloff, G. J., M. Hatanka and R. V. Gilden. 1972. Assay of C-type virus infectivity by measurement of RNA-dependent DNA polymerase activity. *Virology* **48**:266.

Khoury, G., J. C. Byrne, K. Takemoto and M. Martin. 1973. Patterns of simian virus 40 deoxyribonucleic acid transcription. II. In transformed cells. *J. Virol.* **11**:54.

Kimura, G. and R. Dulbecco. 1972. Isolation and characterization of temperature-sensitive mutants of simian virus 40. *Virology* **49**:394.

Nicolson, G. 1973. Temperature-dependent mobility of concanavalin A sites on tumor cell surfaces. *Nature New Biol.* **243**:218.

Otten, J., G. Johnson and I. Pastan. 1971. Cyclic AMP levels in fibroblasts: Relationship to growth rate and contact inhibition of growth. *Biochem. Biophys. Res. Commun.* **44**:1192.

Ozanne, B. 1973. Variants of 3T3 cells transformed by SV40 that are resistant to concanavalin A. *J. Virol.* **12**:79.

Ozanne, B. and J. Sambrook. 1971a. Isolation of lines of cells resistant to agglutination by concanavalin A from 3T3 cells transformed by SV40. *Lepetit Colloq. Biol. Med.* **2**:248.

————. 1971b. Binding of radioactivity labeled concanavalin A and wheat germ agglutinin to normal and virus-transformed cells. *Nature New Biol.* **232**:156.

Ozanne, B., P. A. Sharp and J. Sambrook. 1973. Transcription of simian virus 40. II. Hybridization of RNA extracted from different lines of transformed cells to the separated strands of SV40 DNA. *J. Virol.* **12**:90.

Pollack, R. E. and M. M. Burger. 1969. Surface specific characteristics of a contact inhibited cell line containing the SV40 viral genome. *Proc. Nat. Acad. Sci.* **62**:1074.

Pollack, R. E. and A. Vogel. 1973. Isolation and characterization of revertant cell lines. II. Growth control of a polyploid revertant line derived from SV40-transformed 3T3 mouse cells. *J. Cell. Phys.* **82**:93.

Pollack, R. E., H. Green and G. Todaro. 1968. Growth control in cultured cells: Selection of sublines with increased sensitivity to contact inhibition and decreased tumor producing ability. *Proc. Nat. Acad. Sci.* **60**:126.

Pollack, R. E., A. Vogel and S. Wolman. 1970. Reversion of virus-transformed cell lines: Hyperploidy accompanies retention of viral genes. *Nature* **228**:938, 967.

Renger, H. C. 1972. Retransformation of temperature sensitive SV40 transformants by murine sarcoma virus at non-permissive temperature. *Nature New Biol.* **240**:19.

Sambrook, J. 1972. Transformation by polyoma virus and simian virus 40. *Advanc. Cancer Res.* **16**:141.

Sambrook, J., H. Westphal, P. R. Srinivasan and R. Dulbecco. 1968. The integrated state of SV40 DNA in transformed cells. *Proc. Nat. Acad. Sci.* **60**:1288.

Sheppard, J. 1972. Difference in the cAMP levels in normal and transformed cells. *Nature* **236**:14.

Shoham, J., M. Inbar and L. Sachs. 1970. Differential toxicity on normal and transformed cells *in vitro* and inhibition of tumor development *in vivo* by concanavalin A. *Nature* **227**:1244.

Smith, H. S., C. P. Scher and G. J. Todaro. 1971. Induction of cell division in medium lacking serum growth factor by SV40. *Virology* **44**:359.

Stoker, M. 1968. Abortive transformation by polyoma virus. *Nature* **218**:234.

Todaro, G. J. and H. Green. 1964. An assay for cellular transformation by SV40. *Virology* **23:**117.

Todaro, G. J., Y. Matsuya, S. Bloom, A. Robbins and H. Green. 1967. Stimulation of RNA synthesis and cell division in resting cells by a factor present in serum. *Wistar Inst. Symp. Monogr.* **7:**87.

Van Dilla, M. A., T. T. Trujillo, P. F. Mullaney and J. R. Coulter. 1969. A method of rapid fluorescence measurement. *Science* **163:**1213.

Vogel, A. and R. Pollack. 1973. Isolation and characterization of revertant cell lines. IV. Direct selection of serum-revertant sublines of SV40-transformed 3T3 mouse cells. *J. Cell. Phys.* (in press).

Watkins, J. F. and R. Dulbecco. 1967. Production of SV40 virus in heterokaryons of transformed and susceptible cells. *Proc. Nat. Acad. Sci.* **58:**1396.

Westphal, H. and R. Dulbecco. 1968. Viral DNA in polyoma and SV40-transformed cell lines. *Proc. Nat. Acad. Sci.* **59:**1158.

Wollman, J. and L. Sachs. 1972. Mapping of sites on the surface of mammalian cells. II. Relationship of sites for concanavalin A and an ornithine-leucine copolymer. *J. Membrane Biol.* **10:**1.

Studies on a Mammalian Cell Mutant with a Temperature-Sensitive Leucyl-tRNA Synthetase

C. P. Stanners and L. H. Thompson*

The Ontario Cancer Institute, 500 Sherbourne Street, Toronto, Canada

The synthesis and function of the protein-synthesizing machinery of animal cells has been correlated with the state of growth in a variety of different systems (Kay, Levinthal and Cooper 1969; Lee, Vaughan and Abrams 1970; Stanners and Becker 1971). Whether these changes actually determine critical events such as the initiation of DNA synthesis remains to be seen, but there is little doubt that they are at least intimately involved in such processes. Animal cells with conditional mutations in genes that determine the protein-synthesizing machinery are therefore of great interest, and our discovery of one such mutant and some of its cellular and molecular properties is the subject of this article.

Genetic Properties

As described in detail elsewhere (Thompson, Harkins and Stanners 1973) temperature-sensitive mutants of the Chinese hamster ovary (CHO) line were selected from mutagenized wild-type cultures by killing cells capable of growth at 38.5°C with [³H]thymidine, followed by clonal outgrowth of survivors at 34°C. One of the survivors, *ts*Hl, was extremely temperature sensitive for growth and colony formation: at 34°C doubling times and plating efficiencies were almost normal; whereas at 38.5°C growth ceased almost immediately followed by cellular disintegration, and plating efficiencies of recently cloned cultures were approximately 10^{-7}. Revertants could be isolated at frequencies of 10^{-5} to 10^{-7}, and the reversion frequency could be increased tenfold by treatment with the mutagen ethylmethane-sulfonate. Extensive proliferation at 34°C had no effect on the temperature sensitivity, and every one of 20 subclones of *ts*Hl showed similar temperature sensitivity. Thus *ts*Hl is a stable clone, with low leakiness and reversion and satisfies most of the criteria for true genetic mutants (Thompson and Baker 1973).

* Present address: Biomedical Division, Lawrence Livermore Laboratory, P. O. Box 808, Livermore, California 94550.

Molecular Phenotype

We first measured the rates of incorporation of radioactive precursors of DNA, RNA and protein as a function of time after shifting tsH1 cells to the nonpermissive temperature. The rate of protein synthesis ([^{14}C]valine incorporation) was rapidly and profoundly depressed; at 39.5°C it declined with a "half-life" of about 5 minutes to a value of 5% of the rate at 34°C. The decline was not due to failure of the precursor to enter the intracellular amino acid pool (Thompson et al. 1973). The decline was also reversible with similar kinetics when the cells were cooled again to 34°C after 30 minutes at 39.5°C, and both decline and recovery could occur in the absence of appreciable RNA synthesis (i.e., in the presence of actinomycin D at 10 μg/ml). DNA synthesis ([^{14}C]thymidine incorporation) showed an immediate but less rapid and less profound decline, and RNA synthesis ([^{14}C]uridine incorporation) was little affected until several hours after the shift to 39.5°C. These results implied to us that the temperature-sensitive lesion in tsH1 was probably in some component of the protein-synthesizing machinery. The decline in DNA synthesis could be explained as a secondary consequence of a primary decline in protein synthesis. We interpreted the kinetics of the decline and reversal in protein synthesis to reflect the rates of denaturation and renaturation of a temperature-sensitive component of the protein-synthesizing machinery.

We next studied polysome function in tsH1 cells incubated at various temperatures. The results (described below) led to the hypothesis that the protein-synthesizing machinery of tsH1 cells was being effectively starved for amino acids, and that the severity of starvation increased with increasing temperature to a point much further than would be expected from failure to take up amino acids from the medium. Using bacterial mutants as a precedent, we suspected that the lesion in tsH1 was either in one of the enzymes that add amino acids to tRNA (synthetases) or in tRNA molecules themselves. Aware of the difficulties often encountered in microbial systems in demonstrating temperature sensitivity of enzymes in cell-free systems (Campbell, Söll and Richardson 1972), we devised an assay for the pools of aminoacyl-tRNA complexes using whole cells. The assay involved incubating tsH1 cells at 34 and 40.5°C in complete growth medium containing [^{14}C]-labeled amino acids one at a time, and including cycloheximide and actinomycin D to suppress entry of the label into protein and RNA. After phenol extraction nucleic acids were precipitated with acid and the incorporated radioactivity determined. To show that the label was in aminoacyl-tRNA complexes, portions of the extracts were given mild alkali or RNase treatment before precipitation. For all but one amino acid (cysteine), more than 95% of the radioactivity was rendered acid soluble. The results are shown in Table 1. There was a striking reduction (about 300-fold) in the pool of leucyl-tRNA at 40.5°C, while approximately equal pools of aminoacyl-tRNA were seen for all other amino acids at the two temperatures. Wild-type cells and a revertant of tsH1 showed equal levels of leucyl-tRNA at both temperatures. Control experiments demonstrated similar uptake of leucine into the acid-soluble intracellular pools at both temperatures.

These results indicated that the lesion in tsH1 was specifically related to a failure to form leucyl-tRNA complexes at the nonpermissive temperature, but did not distinguish between a faulty leucyl-tRNA synthetase or faulty tRNA^{leu} molecules. To distinguish between these two possibilities aminoacyl-tRNA synthetase activities were measured in a cell-free system using excess rat liver tRNA as substrate. With crude cytoplasmic extracts of tsH1 cells, we were able to demonstrate that the leucyl

Table 1

Amino Acid Incorporation into Amino Acyl-tRNA In Vivo

Cell line	L-[^{14}C] amino acid	cpm Incorporated at	
		34°C	40.5°C
tsH1	alanine	26	33
	arginine	199	148
	asparagine	94	100
	aspartic acid	17	19
	cysteine	28	24
	glutamine	130	129
	glycine	68	61
	histidine	101	87
	isoleucine	85	75
	leucine	**150**	**0.2**
	lysine	208	178
	methionine	165	128
	phenylalanine	103	84
	proline	52	41
	serine	169	150
	threonine	115	100
	tyrosine	100	83
	tryptophan	99	81
	valine	178	193
WT	**leucine**	**98**	**93**
tsH1 revertant	**leucine**	**88**	**87**

Cultures at 1.5 × 10⁷ cells/ml in growth medium were incubated at 34 or 40.5°C for 30 min. Cycloheximide at 200 μg/ml and actinomycin at 10 μg/ml were then added. After 15 min incubation L-[^{14}C]amino acid was added to 5 μCi/ml (specific activities may be obtained from the published concentrations of amino acids in the medium [Stanners et al. 1971] except for asparagine [50 μg/ml] and glutamine [2.8 μg/ml]) and the incubation continued for a further 60 min. Incubation was terminated by rapid chilling to 0°C and the cells were centrifuged and resuspended in cold phosphate-buffered saline. The cells were disrupted by sonication and nucleic acid extracted by vortex mixing with phenol followed by chloroform containing 1% isoamyl alcohol. The aqueous phase was divided into two equal fractions and one was treated with 0.1 N NaOH. After 10 min both fractions were treated with 10% cold TCA and the precipitates collected and counted as described in the legend of Fig. 1. The values in the table represent the difference between the acid-precipitable cpm with and without alkali treatment averaged for duplicate cultures.

synthetase activity was lower than that of wild-type cells even at 34°C, and that the residual leucyl synthetase activity was markedly temperature sensitive (Thompson et al. 1973). Low activity even at the permissive temperature implies a "shaky" enzyme, as has been observed for bacterial ts synthetase mutants (Low et al. 1971). The leucyl synthetase activity from wild-type cells and valyl and phenylalanyl synthetase activities from tsH1 cells showed no temperature sensitivity between 34 and 45°C. Although these results were not obtained using purified enzyme, they strongly support the contention that the molecular phenotype of *ts*Hl can be equated to a temperature-sensitive leucyl-tRNA synthetase.

Phenotypic Modification

The rate of decline and final level of protein synthesis in *ts*Hl cells was shown to depend on the actual value of the nonpermissive temperature (Thompson et al. 1973). Thus at 38.5, 39.5 and 40.5°C the final rates of [^{14}C]valine incorporation were found to be 15%, 5% and 2%, respectively, of the rate at 34°C. We interpret this to be due to a progressively smaller pool of leucyl-tRNA due, in turn, to progressively lower leucyl-tRNA synthetase activity with increasing temperature.

The above results were obtained using complete growth medium containing leucine at 52 μg/ml. A second phenotypic modifier was found to be the extracellular concentration of leucine. The ratio of the number of intracellular leucine molecules to tRNA^{leu} molecules in *ts*Hl cells in normal growth medium is, by calculation and measurement, approximately 500:1. Yet increasing the extracellular leucine concentration has the effect of shifting the temperature sensitivity for growth and colony formation to higher temperatures, and decreasing the extracellular leucine concentration has the opposite effect (S.J. Molnar pers. comm.). Thus though there appears to be a great excess of leucine in *ts*Hl cells under normal conditions, further increases in this level must lead to increases in leucyl-tRNA formed by residual synthetase activity.

The above results on phenotypic modification are of interest in relation to a number of new, independently isolated mutant lines which also appear to be defective in the leucyl-tRNA synthetase. Prior to our knowledge of the lesion in *ts*Hl, a selection procedure was developed involving suicide with [^{3}H]leucine in leucine-free medium at 38.5°C which, in reconstruction experiments, could recover *ts*Hl cells mixed at low frequency (e.g., 10^{-5}) in cultures of wild-type cells. When this method of selection was applied to mutagenized wild-type cultures, additional *ts* mutants were obtained in two independent experiments. Several of these isolates have been tested and each appears to be defective in the leucyl-tRNA synthetase. One could speculate that this repeated recovery of synthetase mutants for leucine might be due, at least partially, to accentuation of the phenotypes of such mutants by leucine starvation. Other factors that might explain the results are the following: (1) a very limited degree of hemizygosity in CHO cells, (2) the existence of a genetic region having a high degree of mutability, and (3) a leucyl-tRNA synthetase in wild-type cells which is already slightly temperature sensitive.

Attachment of Ribosomes to mRNA

When cultured animal cells enter the stationary phase of growth, there is an increase in the proportion of free ribosomes and a compensating decrease in the proportion of ribosomes on polysomes (Levine et al. 1965). We have shown that, for normal hamster embryo fibroblasts, the observation is most likely explained by a reduction in the probability of attachment of ribosomes to mRNA (Stanners and Becker 1971). Similar conclusions regarding the generation of free ribosomes have been reached by Fan and Penman (1970) for mitotic HeLa cells and by Vaughan, Pawlowski and Forchhammer (1971) for HeLa cells starved for various amino acids. It was, therefore, of interest to study the effect of the *ts*Hl aminoacyl syn-

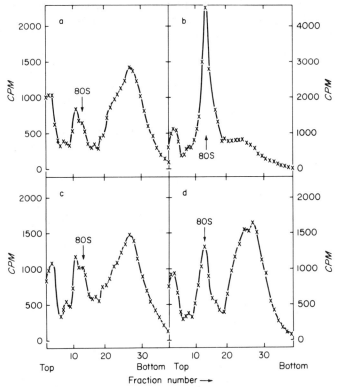

Figure 1

Effect of temperature on polysomes of *ts*Hl cells. A suspension culture of *ts*Hl cells was grown for several days at 34°C in growth medium (α-MEM (Stanners et al. 1971) plus 10% fetal calf serum) containing [14C]uridine at 0.02 μCi/ml, 10 μg/ml, followed by growth overnight at 34°C in medium containing un-labeled uridine. While in the exponential phase of growth, portions of the culture were **(a)** left at 34°C, **(b)** incubated for 20 min at 38.5°C, **(c)** treated with cycloheximide at 100 μg/ml for 2 min at 34°C, then incubated for 20 min at 38.5°C, and **(d)** incubated for 20 min at 38.5°C, then treated with cycloheximide at 0.5 μg/ml for 30 min. Rapid temperature shifts were achieved by mixing with an appropriate volume of conditioned medium at a higher temperature. Cyto-plasmic extracts containing about 90% of the [14C]uridine incorporated by the cells were prepared essentially as described previously (Stanners and Becker 1971). These were layered over 11.5 ml 5% to 50% w/w linear sucrose gradients in buffer (0.01 M NaCl, 0.001 M MgCl$_2$, 0.01 M Tris pH 7.4) and centrifuged for 105 min at 32,000 rpm and 4°C in the SW41 rotor of the Beckman L2-50 ultracentrifuge. Fractions were collected, precipitated with cold trichloracetic acid (TCA) and the precipitates collected on nitrocellulose filters; these were counted using a low background gas-flow counter.

thetase lesion on ribosomes and polysomes. To label the ribosomes *ts*Hl cells were incubated at 34°C with [14C]uridine for several generations. Labeled cultures in the exponential phase of growth at 34°C were heated rapidly to 38.5 or to 39.5°C and, as a function of time thereafter, samples were taken and cytoplasmic extracts were prepared and subjected to analysis by velocity sedimentation on sucrose density gradients. Typical radioactivity profiles are shown in Fig. 1. After 20 minutes at 38.5°C there was an increase in the proportion of free ribosomes at the expense of the polysomes. No change in the ribosome distribution was observed with wild-type cells heated to 38.5 from 34°C (Fig. 2). The shift towards free ribosomes in *ts*H1 cells did not occur if ribosome "run off" was prevented by the addition of a high concentration of cycloheximide prior to heating the culture to 38.5°C (Fig. 1); and cycloheximide at a low concentration (0.5 μg/ml), sufficient merely to slow the rate of ribosome motion across polysomes (Stanners and Becker 1971), could reverse the shift toward free ribosomes when added after 20 minutes at 38.5°C (Fig. 1). These results are diagnostic of a reduced attachment probability of the ribosomes to mRNA at the nonpermissive temperature (Stanners and Becker 1971; Fan and Penman 1970). Another possibility, that ribosomes detach prematurely from the polysomes at leucine codons thereby releasing incomplete polypeptides, is unlikely since the size distribution of polypeptides released by the polysomes at 34 and 38.5°C as determined by SDS-gel electrophoresis was found to be very similar.

A priori, one would expect a leucyl synthetase defect to lead to a slow-down of ribosome movement across polysomes, since the ribosomes would have to wait

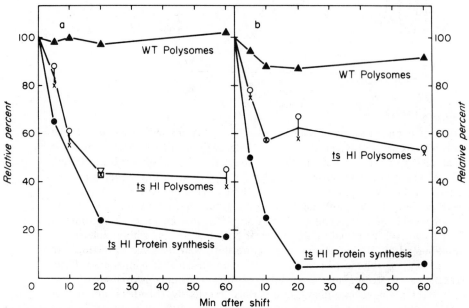

Figure 2
Quantitative decline in polysomes and protein synthesis with time after temperature shift. At time zero, exponential suspension cultures of *ts*H1 and wild-type (WT) cells were shifted from **(a)** 34 to 38.5°C and **(b)** 34 to 39.5°C and at various times samples were taken to determine the distribution of polysomes and free ribosomes as described in the legend of Fig. 1. The proportion of ribosomes on polysomes was calculated from the sucrose density gradient profiles and was expressed as a percentage of the proportion at time zero (34°C). To give an indication of experimental variability, two typical independent experiments are shown for the *ts*H1 polysomes. Also shown is the relative rate of protein synthesis of *ts*H1 cells as a function of time after the temperature shift, determined by the TCA-precipitable cpm after incubation of the cells in complete medium containing [14C]valine for sufficient time to equilibrate the intracellular valine pool with the medium.

longer than normal for leucyl-tRNA at leucine codons randomly distributed along mRNA molecules. This effect would tend to shift the ribosome-polysome distribution towards polysomes, if anything, the opposite of what was observed. One could propose a model for the observation involving initiation factors, but at this stage it is probably better just to note that the failure to charge tRNA with a given amino acid can lead to the generation of free ribosomes, as has been observed for cells starved for amino acids, mitotic cells and for cells in stationary phase.

Ribosome Function on Polysomes

When the proportion of ribosomes in polysomes at various times after shifting tsHl cells from 34 to 38.5°C was estimated quantitatively, it was immediately apparent that the reduction in ribosome attachment could not account for the entire reduction in the rate of protein synthesis. These results, shown in Fig. 2, indicate a discrepancy of more than a factor of 2 at 38.5°C and a factor of about 10 at 39.5°C. Thus the polysomes that remained at nonpermissive temperatures could not have been functioning normally. Two measures of ribosome function were applied: the transit time of ribosomes across mRNA, and the size distribution of primary polypeptides produced by the polysomes. The transit time was measured by comparison of the radioactivity in nascent polypeptides on polysomes with that in completed polypeptides, as a function of time after addition of [^{14}C]valine, using methods described previously (Stanners 1968; Fan and Penman 1970).[1] The average transit time at 34°C for tsHl was found to be about 2 minutes, at 38.5°C it was about 3 minutes, at 38.8°C, 5 minutes and at 39.5°C, 10 minutes. These values indicate that the rate of ribosome motion across the polysomes slows down with increasing temperature, as might be expected assuming that tRNA charged with leucine becomes increasingly scarce with increasing temperature. They do not, however, account quantitatively for the reduction in protein synthesis, especially at 39.5°C. At this temperature the reduction in bound ribosomes was about twofold and in ribosome motion fivefold, giving a net effect of tenfold, whereas protein synthesis was reduced by 20-fold. This discrepancy was resolved by the finding that at 39.5°C the nascent protein per ribosome was about one-half of the value at 34°C. Analysis by SDS-polyacrylamide gel electrophoresis of the labeled polypeptides released from the polysomes at 34 and 39.5°C revealed a dramatic shift towards smaller polypeptides at 39.5°C (Fig. 3). A consistent explanation of these observations is that incomplete polypeptide chains are released occasionally at leucine codons, but that the ribosomes remain bound to the mRNA.

Using tsHl cells under extremely nonpermissive conditions (39.5°C in medium lacking leucine), we also have preliminary evidence for amino acid substitutions at leucine codons and for the accumulation of very large polypeptides. The latter could conceivably be analogous to the large polypeptide precursors observed in cells infected with certain RNA viruses (Jacobson, Asso and Baltimore 1970). If such precursors existed in normal uninfected cells, amino acid substitutions could produce an aberrant secondary and tertiary structure leading to faulty processing. These studies require direct confirmation, however, perhaps by infection of tsHl

[1] In all of these experiments complete conditioned growth medium was used; the relative rates of protein synthesis, etc., were obtained from rates of [^{14}C]valine incorporation after the specific activity of the intracellular valine pool had reached the specific activity of the medium.

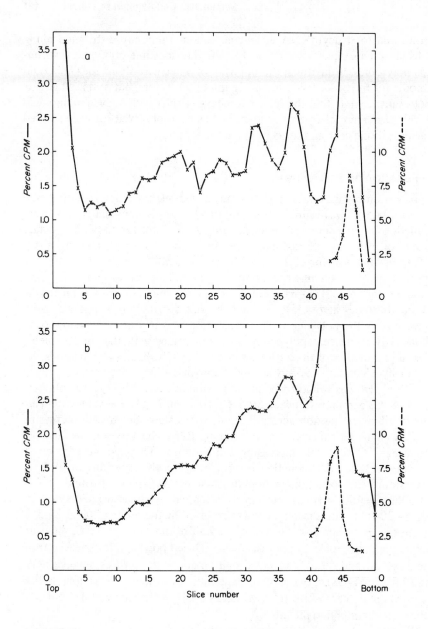

Figure 3
Electrophoretic profiles of polypeptides synthesized by *ts*Hl cells at **(a)** 34°C
and at **(b)** 39.5°C. A suspension culture of *ts*Hl cells was incubated for 15 min
at 34°C in growth medium lacking valine, then divided in two and incubated **(a)**
at 34°C and **(b)** at 39.5°C for 20 min. The cultures were then incubated for 30
min with [³H]valine 19 Ci/mɴ at 10 μCi/ml (34°C) or at 100 μCi/ml
(39.5°C). After rapid chilling to 0°C, cytoplasmic extracts were prepared,
treated with SDS and mercaptoethanol and heated to 100°C, then subjected to
SDS-polyacrylamide (7%) gel electrophoresis, slicing, counting and analysis
as described previously (Becker and Stanners 1972). Similar results were obtained when organelles, particulates and ribosomes and polysomes were removed
from the extracts before SDS treatment by high speed centrifugation.

cells with an RNA virus, where the effect of the lesion on the synthesis of a limited number of well-characterized viral polypeptides could be studied.

Protein Turnover

*ts*Hl cells kept at 39.5°C lose their colony-forming ability (assayed at 34°C) fairly rapidly after a lag of several hours and eventually disintegrate. Death and disintegration could possibly be the consequence of a net loss of cellular protein at 39.5°C due to a greater rate of degradation than synthesis. We therefore investigated protein turnover in *ts*Hl cells under nonpermissive conditions. To do this cells were labeled with [^{14}C]valine for 1 hour at 34°C, then the extracellular label was removed by washing the cells in conditioned growth medium; the intracellular label was removed by exchange at 34°C with unlabeled valine in growth medium followed by further washing. The labeled cells were then incubated at 34 or 39.5°C in conditioned growth medium and, at various times, aliquots were removed and the total acid-soluble cpm in both cells and medium was measured. The results are shown in Fig. 4. The rate of turnover was about 4% of the acid-insoluble cpm in the first hour at 34°C and about 8% at 39.5°C. Since the synthesis of proteins is only about 0.02% per hour at 39.5°C, the rate of degradation greatly exceeds the rate of synthesis, and this could account for cell death.

The increase in the rate of turnover from 34 to 39.5°C is also of interest. It is probably not a consequence of the *ts* lesion in *ts*Hl, as wild-type cells showed a similar increase with temperature (Fig. 4). This is unlike the situation in bacteria, where inactivation of aminoacyl-tRNA synthetases can have a large effect on protein turnover (Goldberg 1971).

Regarding the turnover of proteins synthesized at nonpermissive temperatures in *ts*Hl, interpretation of the results is complicated by the fact that cellular proteins have a spectrum of degradation rates and that the more unstable ones are labeled first. A hint of this can be seen in Fig. 4: the curves show continually decreasing slopes with time, as the more unstable proteins decay first. Thus proteins labeled for 1 hour at 39.5°C in *ts*Hl cells might degrade faster because they could be equivalent to a 3-minute label at 34°C. When allowance was made for this, it was found that the degradation of polypeptides synthesized at 39.5° showed two major components: one which degraded extremely rapidly, being essentially entirely acid soluble by 30 minutes, and a second which showed normal degradation rates. Preliminary experiments indicate that it is the fragments (see Fig. 3) generated at nonpermissive temperatures that degrade so rapidly.

Preliminary Studies on Coupling of Macromolecular Synthesis

It is well known that in stringent strains of bacteria, starvation for amino acids elicits a number of different and seemingly unrelated molecular responses, including a rapid reduction in RNA synthesis and the generation of guanosine tetraphosphate. It was conceivable that such strong "coupling" could have been missed in animal cells (Vaughan 1972) because, due to the continual supply of amino acids from intracellular protein turnover, starvation for amino acids reduces protein synthesis by a factor of only 3 to 5.

We first looked for guanosine tetraphosphate using a protocol similar to that which has been successful for its detection in stringent bacteria (Fiil, von Meyen-

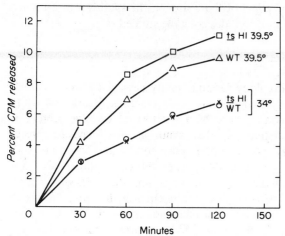

Figure 4

Release of acid-soluble radioactivity from prelabeled proteins at 34 and 39.5°C by wild-type and *ts*Hl cells. Suspension cultures of wild-type (WT) and *ts*Hl cells were labeled with [^{14}C]valine at 2 μCi/ml, 46 μg/ml, in growth medium for 1 hr at 34°C. Unincorporated label in the medium and in the intracellular pool was removed as described in the text. The cultures were then divided and incubated in suspension at 34 and at 39.5°C in complete growth medium and at various times samples were taken and added to an equal volume of 20% TCA. The precipitate was removed by centrifugation and filtration and the radioactivity in precipitate and filtrate was determined by liquid scintillation counting. The cpm in the filtrate (acid soluble) was expressed as a percentage of the cpm in the precipitate. The percentage at time zero was 1–2% and was subtracted from all values.

burg and Friesen 1972). The experiment was done in collaboration with Dr. J. D. Friesen at York University, Toronto. No guanosine tetraphosphate or other unusual nucleotides could be detected in *ts*Hl incubated at 34 or 39.5°C. [^{14}C]Uridine incorporation is also unaffected for several hours after *ts*Hl cells have been put at 39.5°C (Thompson et al. 1973). These negative results do not, however, exclude the existence of guanosine tetraphosphate in animal cells. CHO cells are an established transformed line and may have lost some of the controls over macromolecular synthesis that exist in normal cells.

We next studied the effect of the synthetase lesion on the rate of synthesis of DNA. Many workers have shown that both the entry of cells into the DNA synthesis phase of the cell cycle and the continuation of DNA synthesis in this phase require protein synthesis. Weintraub and Holtzer (1972) have noted that inhibition of protein synthesis in chick erythroblasts with cycloheximide produces a 50% inhibition of DNA synthesis, and Weintraub (1972) has shown an interesting correlation between the inhibition of histone and DNA synthesis. He has suggested that when histone synthesis is completely inhibited, DNA synthesis can continue at only 50% of the normal rate since preexisting histone can provide only 50% of the requirement for new histone at the growing point. Preliminary experiments have been carried out on the rate of incorporation of labeled thymidine into *ts*Hl cells as a function of time after shifting the cells from 34 to 39.5°C. Typical results are shown in Fig. 5. It can be seen that there was a greater than twofold increase in the rate of [³H]thymidine incorporation during the first 10 minutes after shifting to 39.5°C, followed by a decline to a value of 50% of the rate at 34°C. The initial increase in rate could be due to an increased rate of DNA synthesis at the higher temperature before the decline in protein synthesis had time to exert an effect or due to an increase in the efficiency of utilization of exogenous thymidine. To overcome this difficulty in interpretation, *ts*Hl cells were incubated in medium containing hypoxanthine (15 μg/ml), amethopterin (0.25 μg/ml), and thymidine (10μg/ml) which forces them to use only exogenous thymidine. A small amount of [³H]thymidine was added to the medium and the cells were incubated at 34°C until the intracellular thymidine pool had equilibrated with the medium (about 10 minutes), after which incorporation into acid-insoluble material was linear. At 20 minutes the culture was shifted to 39.5°C and the incorporation was measured for a further 40 minutes. The rate of incorporation increased immediately after the shift, then decreased to a plateau, confirming the rate measurements shown in Fig. 5. In two independent experiments the final rates at 39.5°C were 42% and 47% of the rate

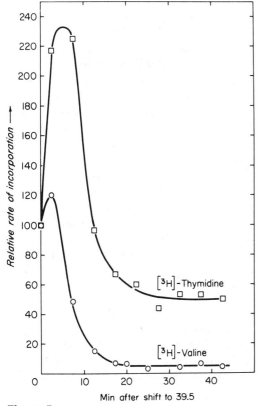

Figure 5

Incorporation of [¹⁴C]thymidine into *ts*Hl cells with time after temperature shift. At time zero, exponentially growing suspension cultures of *ts*Hl cells were shifted from 34 to 39.5°C and at various times samples were taken and the cold TCA-insoluble incorporation of [³H]thymidine at 17 μCi/ml, 1 μg/ml and [³H]valine, 17 μCi/ml, 46 μg/ml was measured for a 4-min interval.

at 34°C. In a parallel experiment using wild-type cells, the rate of [^{3}H]thymidine incorporation at 39.5°C was 12% higher than the rate at 34°C, which is exactly the same as the increase in the growth rate at 39.5°C. If the results for *ts*Hl cells are corrected for this temperature effect, the effective rate at 39.5°C would be about 40% the rate at 34°C, which is quite close to the 50% predicted by Weintraub's model. These results indicate a rather close coupling between the expression of the synthetase lesion and DNA synthesis in *ts*Hl cells.

CONCLUSIONS

We have demonstrated that a non-leaky, low-reverting mutant of CHO cells, which is temperature sensitive for protein synthesis, has a specific temperature-sensitive defect in the charging of tRNA with leucine. In a cell-free system using crude cytoplasmic extracts the activity of the leucyl-tRNA synthetase was found to be low and temperature sensitive. The extent of the depression in protein synthesis could be modified by temperature and by the external concentration of leucine, being greater for higher temperatures and lower leucine concentrations.

At 38.5°C there is a marked increase in the proportion of free ribosomes and a decrease in the proportion of ribosomes bound to polysomes. This effect is similar to that seen in mitotic cells, cells starved for amino acids, and for cells in stationary phase and appears to be due to a reduced attachment probability of ribosomes to mRNA. The transit time of the ribosomes across mRNA in polysomes increases with increasing temperature. At 39.5°C, many polypeptide chains are terminated prematurely, resulting in the generation of polypeptide fragments.

The turnover of preexisting proteins is normal at nonpermissive temperatures. Since the rate of degradation of proteins can exceed the rate of synthesis, this might account for the cell death and disintegration observed under these conditions. Preliminary experiments indicate that polypeptide fragments generated at 39.5°C are degraded extremely rapidly and that normal length polypeptides synthesized under nonpermissive conditions show more or less normal turnover rates.

A preliminary survey of possible couplings between the synthesis of various macromolecules in mutant cells indicates a tight coupling between expression of the lesion and thymidine incorporation and no coupling with uridine incorporation. No guanosine tetraphosphate or other unusual nucleotides could be detected under permissive or nonpermissive conditions.

Acknowledgments

We thank Mr. J. L. Harkins, Mr. W. A. Mehring and Miss Jennifer Pell for excellent technical assistance. The work was supported by grants MT-1877 and MT-4734 from the Medical Research Council of Canada, by contract number 72-2051(c) from the National Cancer Institute, NIH, USPHS, and by grants from the National Cancer Institute of Canada.

REFERENCES

Becker, H. and C. P. Stanners. 1972. Control of macromolecular synthesis in proliferating and resting Syrian hamster cells in monolayer culture. III. Electrophoretic patterns of newly synthesized proteins in synchronized proliferating cells and resting cells. *J. Cell. Physiol.* **80:**51.

Campbell, J. L., L. Söll and C. C. Richardson. 1972. Isolation and partial characterization of a mutant of *Eschericia coli* deficient in DNA polymerase II. *Proc. Nat. Acad. Sci.* **69:**2090.

Fan, H. and S. Penman. 1970. Regulation of protein synthesis in mammalian cells. II. Inhibition of protein synthesis at the level of initiation during mitosis. *J. Mol. Biol.* **50:**655.

Fiil, N. P., K. von Meyenburg and J. D. Friesen. 1972. Accumulation and turnover of guanosine tetraphosphate in *Eschericia coli. J. Mol. Biol.* **71:**769.

Goldberg, A. L. 1971. A role of aminoacyl-tRNA in the regulation of protein breakdown in *Eschericia coli. Proc. Nat. Acad. Sci.* **68:**362.

Jacobson, M. F., J. Asso and D. Baltimore. 1970. Further evidence on the formation of poliovirus proteins. *J. Mol. Biol.* **49:**657.

Kay, J. E., B. G. Levinthal and H. L. Cooper. 1969. Effects of inhibition of ribosomal RNA synthesis on the stimulation of lymphocytes by PHA. *Exp. Cell Res.* **54:**94.

Lee, M. J., M. H. Vaughan and R. Abrams. 1970. Nature of the ribonucleic acid formed prior to deoxyribonucleic acid synthesis in kidney cortex cells cultured directly from the rabbit. *J. Biol. Chem.* **245:**4525.

Levine, E. M., Y. Becker, C. W. Boone and H. Eagle. 1965. Contact inhibition, macromolecular synthesis, and polyribosomes in cultured human diploid fibroblasts. *Proc. Nat. Acad. Sci.* **53:**350.

Low, B., F. Gates, T. Goldstein and D. Söll. 1971. Isolation and partial characterization of temperature-sensitive *Eschericia coli* mutants with altered leucyl- and seryl-transfer ribonucleic acid synthetases. *J. Bacteriol.* **108:**742.

Stanners, C. P. 1968. Polyribosomes of hamster cells. Transit time measurements. *Biophys. J.* **8:**231.

Stanners, C. P. and H. Becker. 1971. Control of macromolecular synthesis in proliferating and resting Syrian hamster cells in monolayer culture. I. Ribosome function. *J. Cell. Physiol.* **77:**31.

Stanners, C. P., G. L. Eliceiri and H. Green. 1971. Two types of ribosomes in mouse-hamster hybrid cells. *Nature New Biol.* **230:**52.

Thompson, L. H. and R. M. Baker. 1973. Isolation of mutants of cultured mammalian cells. *Methods in cell physiology* (ed. D. M. Prescott), vol. 6. Academic Press, New York (in press).

Thompson, L. H., J. L. Harkins and C. P. Stanners. 1973. A mammalian cell mutant with a temperature sensitive leucyl-tRNA synthetase. *Proc. Nat. Acad. Sci.* (in press).

Vaughan, M. H., Jr. 1972. Comparison of regulation of synthesis and utilization of 45S ribosomal precursor RNA in diploid and heteroploid human cells in response to valine deprivation. *Exp. Cell Res.* **75:**23.

Vaughan, M. H., Jr., P. J. Pawlowski and J. Forchhammer. 1971. Regulation of protein synthesis initiation in HeLa cells deprived of single essential amino acids. *Proc. Nat. Acad. Sci.* **68:**2057.

Weintraub, H. 1972. A possible role of histone in the synthesis of DNA. *Nature* **240:**449.

Weintraub, H. and H. Holtzer. 1972. Fine control of DNA synthesis in developing chick red blood cells. *J. Mol. Biol.* **66:**13.

Bromodeoxyuridine Dependence—
Characterization of the Mutant Cells

Richard L. Davidson* and Michael D. Bick[†]

*Department of Medicine, Children's Hospital Medical Center
and Department of Microbiology and Molecular Genetics, Harvard Medical School
†Department of Biological Chemistry, Harvard Medical School
Boston, Massachusetts 02115

We recently reported the isolation of mutant mammalian cell lines requiring high concentrations of the drug 5-bromodeoxyuridine (BrdU) for growth (Davidson and Bick 1973). These cells seem to represent the only known case of this mutation, which we have called BrdU dependence. Since BrdU is generally toxic to all cells except those which do not incorporate it, the requirement of BrdU for growth suggested that the dependent cells had undergone an alteration in some basic process involved in cell growth. The properties of the BrdU-dependent cells will be described in this paper.

The BrdU-dependent cells were isolated in experiments that had been designed to establish new lines of BrdU-resistant cells. The parental cell line, which was BrdU sensitive, was the Syrian hamster melanoma line RPMI 3460. These cells had been maintained in culture for several years, during which time they had continued to produce pigment. A pigmented clone called W1 was isolated from 3460 and the W1 cells were used for the experiments. During the course of the experiments, it was observed that the W1 cells were even more sensitive to BrdU than the uncloned 3460 population. In order to inhibit almost completely the growth of 3460 cells, it was necessary to use concentrations of BrdU greater than 10^{-5} M. In contrast the growth of W1 cells was almost totally inhibited at 10^{-6} M BrdU. The reason for the increased sensitivity of W1 cells to BrdU, and whether there is any relationship between this increased sensitivity and the BrdU-dependence of the cells ultimately isolated, remain to be determined.

The basic growth medium for all of the cells was Dulbecco's modified Eagle's medium supplemented with 10% fetal calf serum. The W1 cells were first exposed to medium containing 10^{-5} M BrdU. The frequency with which cells survived in this medium was approximately 1×10^{-4}. The cells from approximately 40 colonies developing in this medium were pooled, and the cells were passed (at 2 to 3 week intervals) two more times in medium with 10^{-5} M BrdU and then two times in medium containing 10^{-4} M BrdU. After the second passage at the higher concentration, large colonies of rapidly growing cells were observed. Several colonies were picked and subcultured individually in medium with 10^{-4} M BrdU, and the cells were maintained thereafter in medium with that concentration of BrdU. The lines were called B1, B2, B3, B4, etc. Because of the repeated passages in BrdU before the isolation of the colonies, it is not known whether the colonies are of independent origin.

All of the cell lines isolated in BrdU were unpigmented, consistent with previous observations that BrdU prevents the differentiation of pigment cells (Coleman et al. 1968). The various lines were maintained in culture for at least 6 weeks after their isolation, and one of the lines, called B4, was maintained in culture for more than 6 months, undergoing more than 75 cell divisions. The B4 cells were found to have high levels of thymidine kinase activity, unlike most cells able to survive in BrdU. The B4 cells were also found to have approximately twice as many chromosomes as the parental melanoma cells. Most of the experiments described below were performed with the B4 line and its derivatives.

Requirement for BrdU for Growth

When the cells of the different lines isolated in BrdU were shifted from medium with to medium without BrdU, it quickly became apparent that the cells were not simply BrdU resistant. In the absence of BrdU cell growth was markedly inhibited. This inhibition was observed with seven different lines that were transferred into medium lacking BrdU. The data for three lines are presented in Table 1. Although the cells grew at varying rates in the presence of BrdU (generation times of 48–72 hours), all of the cells grew much more slowly in its absence (generation times of 112–168 hours). The cells thus seemed to require BrdU for maximal growth.

Experiments were performed to determine whether the small increase in cell number in cultures grown in medium lacking BrdU was due to the slow proliferation of all of the cells or to the more rapid proliferation of a minority of the cells which did not require BrdU. Cells of the B4 line were cloned in medium with and without BrdU. The cloning efficiency in the two media was about the same (ap-

Table 1

Requirement for BrdU for Maximal Growth

Cell line[a]	Cell Number[b]	
	+ BrdU	− BrdU
B4	150×10^4	4×10^4
B10	72×10^4	8×10^4
B13	28×10^4	2×10^4
B4-A	330×10^4	12×10^4
B4-D	580×10^4	10×10^4

[a] B4, B10, and B13 are independent clones of cells isolated in medium containing 10^{-4} M BrdU. B4-A and B4-D are subclones of the B4 line isolated in the same medium.

[b] 60-mm Falcon plastic tissue culture dishes were inoculated with 10^4 cells in medium with 10^{-4} M BrdU or without BrdU. The cells were counted after 14 days.

proximately 10%). However in the presence of BrdU, the colonies attained an average size of about 200 cells in 14 days, whereas the colonies attained an average size of only about 10 cells in the absence of the drug. B4 subclones developing in medium with BrdU were subcultured and tested for BrdU dependence. Like the B4 line, all of the cells exhibited the requirement for BrdU and, similar to the parental line, they all underwent a small increase in cell number in the absence of BrdU. The data for two subclones are presented in Table 1. The results suggested that all of the cells in the BrdU-dependent populations were able to multiply slowly, at least for a few generations, after the removal of BrdU.

Experiments were undertaken with the B4 cells to study the quantitative and qualitative aspects of the requirement for BrdU. It was found that the rate of growth of the B4 cells was related to the concentration of BrdU in the medium (see Table 2). At 10^{-7} M BrdU the cells grew as if there were no drug present at all. As the concentration of BrdU was increased up to 10^{-4} M, the cells grew progressively faster. At 3×10^{-4} M BrdU cell growth was inhibited. The effect of BrdU on cell morphology was also studied in this experiment. B4 cells growing in 10^{-4} M BrdU resembled in morphology the parental (BrdU sensitive) cells in the absence of BrdU. The B4 cells in the absence of BrdU were much more flattened and spread out. At intermediate concentrations of BrdU, as the concentration was increased, the cells became progressively less and less flattened and spread out. This effect of the drug on the cell morphology of BrdU-dependent cells is the opposite of that observed when BrdU-sensitive cells are exposed to BrdU. When BrdU-sensitive cells are exposed to BrdU, they become more flattened and spread out (Silagi and Bruce 1970).

In order to study the specificity of the requirement for BrdU, the BrdU-dependent cells were cultured in medium containing thymidine or related compounds. In medium containing thymidine as an additive, B4 cells grew no better than in medium with no additive at all. This showed that the dependent cells were not using BrdU to compensate for an inability to synthesize thymidine. It was also found that, even though thymidine could not satisfy the requirement for BrdU, thymidine would compete with BrdU and inhibit cell growth. The inhibition could be reversed by increasing the concentration of BrdU.

The BrdU-dependent cells were grown in medium containing as additive either

Table 2

Effect of BrdU Concentration
on Growth Rate

	Cell Number*
	12×10^4
1×10^{-7} M	12×10^4
1×10^{-6} M	75×10^4
1×10^{-5} M	180×10^4
1×10^{-4} M	445×10^4
3×10^{-4} M	2×10^4

* 60-mm dishes were inoculated with 10^4 cells of the B4 line. The cells were counted after 14 days.

deoxyuridine, iododeoxyuridine, uridine, bromouracil, or bromouridine. In no case was there an increase in the cell number greater than that in medium without additives. This suggested that the requirement for BrdU was specific for both the bromine atom and the deoxy sugar.

BrdU Incorporation into DNA of Dependent Cells

The specificity of the requirement for BrdU suggested that it was being incorporated into the DNA of the dependent cells. In order to test this the buoyant density of the DNA was determined in neutral CsCl density gradients and compared to the DNA of the parental cells. The dependent cells were labeled with either [^{3}H]thymidine, [^{3}H]BrdU, or ^{32}P, and the DNA was isolated (by phenol extraction and RNase treatment) and run on neutral CsCl gradients. The DNA of the Syrian hamster melanoma cells was found to have a mean buoyant density of 1.693 g/ml. In contrast the DNA from the dependent cells (regardless of the labeled precursor used) had a mean buoyant density of 1.753 g/ml. This large increase (0.06 g/ml) in buoyant density indicated that there was a significant replacement of thymine by bromouracil (BrUra) in the DNA of the dependent cells.

The DNA of the dependent cells was also run on alkaline CsCl gradients, and it was seen that the single-stranded DNA banded with a single peak. The facts that the buoyant density of the DNA on neutral gradients was the same regardless of the precursor used for labeling and that the two strands of DNA did not separate on alkaline gradients suggested that the BrUra was evenly distributed throughout both strands of the DNA. This conclusion was strengthened by the results of irradiating the DNA of the dependent cells with ultraviolet light, a treatment known to be much more damaging to BrUra-substituted DNA than to unsubstituted DNA (Cleaver 1968). The DNA of the dependent cells and of the normal parental cells was irradiated with a germicidal lamp and run on alkaline sucrose gradients to detect single-strand breaks. Whereas the UV treatment had relatively little effect on the molecular weight distribution of the DNA of the normal cells, the DNA of the dependent cells was reduced totally to small fragments (molecular weight less than 2×10^5). These results suggested that there were very few, if any, long stretches of DNA in the dependent cells that did not contain some BrUra.

The amount of BrUra in the DNA of the dependent cells was calculated on the basis of the change in buoyant density. The density of the normal (unsubstituted) DNA indicated that the DNA contained 30% thymine. It had been shown that the density of poly(dA-dBrUra) is 0.2 g/ml greater than the density of poly(dA-dT) (Wake and Baldwin 1962). Since 100% substitution of thymine by BrUra in the synthetic DNA, which was 50% thymine, resulted in an increase in density of 0.20 g/ml, it was calculated that an increase in density of 0.06 g/ml for hamster DNA that was 30% thymine should correspond to the replacement of 50% of the thymine residues by BrUra in the DNA of the dependent cells. As indicated above, the results of the labeling and irradiation experiments suggested that the BrUra was evenly distributed throughout the DNA.

A "revertant" subline, able to grow rapidly in the absence of BrdU, was isolated from the B4 line. This was accomplished by passaging the cells several times in the absence of BrdU. The density of the DNA of the revertant cells, called B4-E, had dropped from 1.753 to 1.693 g/ml, identical to that of the unsubstituted DNA of the original BrdU-sensitive cells. The B4-E cells were exposed once again to BrdU (10^{-4} M), and approximately 10% of the cells were able to grow rapidly in the presence of BrdU. A new subline of cells able to grow in BrdU, called B4-E-B, was isolated, and the cells were maintained in BrdU. The B4-E-B cells were compared to the B4 cells in terms of BrdU incorporation and dependence. The density of the DNA of the B4-E-B cells was similar to that of the B4 cells, indicating that the two cell lines incorporated similar amounts of BrUra into their DNA. However, unlike the B4 cells, the B4-E-B cells grew as rapidly in the absence as in the presence of BrdU. Thus the B4-E-B cells resembled the B4 cells in their ability to grow in medium with BrdU and to incorporate it into the DNA, but were unlike the B4 cells in that they were no longer BrdU dependent. The significance of these observations is not clear at present.

Since half of the thymine in the dependent cells was replaced by BrUra, the question was raised as to whether the cells could grow with thymine totally replaced by BrUra. To test this, dependent cells (of the B4 line) were exposed to medium containing aminopterin (4×10^{-7} M), which inhibits the de novo pathway of thymidine biosynthesis, plus BrdU (10^{-5} M). In the presence of aminopterin plus BrdU in medium with no thymidine, the cells should be forced to replicate their DNA with BrUra completely replacing thymine, and the cells would grow only if they were viable with totally substituted DNA. (Since aminopterin blocks purine as well as pyrimidine biosynthesis, the medium also contained 10^{-4} M hypoxanthine.) When the cells were exposed to medium containing aminopterin plus BrdU, the growth of the cells was almost completely inhibited. However by repeated passaging and cloning of the cells in medium with aminopterin plus BrdU, it was possible to isolate a line of cells which did grow well in this medium. This line of cells, called HAB (to indicate that it was isolated in hypoxanthine-aminopterin-BrdU-containing medium), has been maintained in continuous multiplication in medium with aminopterin plus BrdU for more than 100 generations and has shown no decrease in the ability to grow.

In order to determine the degree of replacement of thymine by BrUra in the HAB cells, the buoyant density of the DNA was determined in neutral CsCl density gradients, as described previously for the B4 cells. The density of the DNA was found to be 1.799 g/ml (see Table 3). Using the same type of calculations as described above, based on the observed density difference between poly(dA-dT) and poly(dA-dBrUra), the density of the DNA in the HAB cells suggested that approximately 90% of the thymine residues in the DNA of these cells had been replaced by BrUra.

Since the density of the DNA of the HAB cells suggested that there was still some thymine in their DNA, attempts were made to eliminate any remaining sources of thymidine and thereby increase the BrUra substitution to 100%. The possibility was considered that there could be trace amounts of thymidine in the fetal calf serum in the medium, which could serve as the source for a small amount of thymine in the DNA. To decrease any possible uptake of thymidine from the serum, the concentration of BrdU in the medium was increased tenfold (to 10^{-4} M) while the aminopterin was kept constant, and the cells were also grown in medium supplemented with dialyzed fetal calf serum plus the usual concentrations of aminopterin and BrdU. Neither of these modifications, either separately or together, had a significant effect on the density of the DNA. Another possibility was that the aminopterin block was not completely inhibiting the de novo pathway of thymidine

Table 3
BrdU Incorporation in HAB Cells

Cell line	Medium additive[a]			DNA density[b]
	Aminopterin	BrdU	FdU	
3460				1.693
B4		1×10^{-4}		1.753
HAB	4×10^{-7}	1×10^{-5}		1.799
HAB	4×10^{-7}	1×10^{-4}		1.802
HAB	4×10^{-7}	1×10^{-5}	1×10^{-5}	1.800
HAB	1.6×10^{-6}	4×10^{-5}		1.800
HAB-B		1×10^{-4}		1.765

[a] Molar concentrations

[b] The density of the DNA (g/ml) was determined by centrifugation in neutral CsCl gradients.

biosynthesis in the cells. The cells were therefore grown in medium with the concentration of aminopterin increased fourfold (to 1.6×10^{-6} M) or in medium containing aminopterin and also FdU (10^{-5} M), another powerful inhibitor of thymidine biosynthesis. Again it was observed that neither modification resulted in a change in the density of the DNA (see Table 3).

The above observations on the density of the DNA of the HAB cells can be interpreted in various ways. It is possible that the HAB cells produced or obtained enough thymidine in some way so that there was still some thymine in the DNA, approximately 10% of the normal amount, while the rest (90%) was replaced by BrUra. Another possibility is that the density of 1.800 g/ml is the density of Syrian hamster DNA when it is totally substituted. The calculation of 90% substitution of thymine by BrUra was based on the density shift observed for the total substitution of a synthetic copolymer, which contained 50% thymine, and involved the assumption that the total substitution of thymine by BrUra in a native DNA preparation with a different thymine content would cause a corresponding increase in DNA density directly related to the thymine content. Since total substitution in a synthetic molecule of 50% thymine resulted in an increase in density of 0.02 g/ml, it was assumed that total substitution in a native molecule of 30% thymine should result in an increase in density of 0.12 g/ml. This assumption may be incorrect. Experiments are presently in progress to determine directly (by base analysis) whether there is any thymine in the DNA of the HAB cells, or whether there has been total replacement of thymine by BrUra.

Although the exact degree of BrUra substitution has not as yet been determined for the HAB cells, the results suggested that there was at least 90% substitution of thymine by BrUra in HAB cells, in contrast to the 50% substitution in B4 cells. Experiments were therefore performed to determine whether the HAB cells required such a high level of substitution for survival or whether the cells were only tolerant to such a level of substitution. The HAB cells were grown for several generations in medium with BrdU but without aminopterin, so that the cells could begin to synthesize thymidine again. When the DNA of these cells (called HAB-B) was analyzed, the density was found to have dropped to 1.765 g/ml, corresponding to approximately 60% substitution (Table 3). The degree of substitution in the HAB-B cells was close to that in the B4 cells. Since the degree of substitution in the HAB cells dropped when they were provided with a source of thymidine, it was concluded that, even though the HAB cells incorporated much more BrUra into their DNA than did the B4 cells, they did not require this increased level of substitution for survival. The HAB-B cells were also tested for BrdU dependence. When BrdU was removed from the medium, cell growth was inhibited. Thus the HAB cells were still BrdU dependent.

DISCUSSION

This paper has summarized the characteristics of a new type of mutation in mammalian cells—BrdU dependence. It was shown that the requirement for BrdU was specific for both the bromine atom and the deoxy sugar. In the dependent cells first isolated, 50% of the thymine residues in the DNA were replaced by BrUra. From one of the clones of dependent cells, a subline was isolated which grew in the presence of aminopterin plus BrdU. In these cells at least 90%, and possibly all, of the thymine was replaced by BrUra. The cells with at least 90% BrUra substitution have survived in continuous cultivation for over a hundred generations.

The existence and properties of the BrdU-dependent cells raise a number of questions. The two main questions are: (1) How can the cells tolerate such high levels of substitution of thymine by BrUra? and (2) Why are the cells dependent upon BrdU?

The question of BrdU tolerance is raised by the many disruptive effects that BrdU has been observed to have on cells. BrdU has been shown to act as a mutagen (Freese 1959), to photosensitize DNA, to suppress differentiated functions, to induce latent viruses (Lowy et al. 1971), and to affect the strength of the binding of regulator molecules (Lin and Riggs 1972). Mammalian cell lines able to survive in BrdU have been isolated. However the survival of such cells is generally associated with a modification which prevents BrdU incorporation, either through the loss of the enzyme thymidine kinase (Kit et al. 1963) or through a deficiency in the thymidine uptake system (Breslow and Goldsby 1969). The BrdU-dependent cells described in this paper have both thymidine kinase activity and an active thymidine uptake system, and they incorporate BrdU into the DNA. BrdU-tolerant cells that replace some of their thymine by BrUra have been isolated in the past (Toliver and Simon 1967). However when such cells were exposed to FdU plus BrdU in attempts to force them to grow with thymine totally replaced by BrUra, the cells did not survive. This treatment is similar to the aminopterin plus BrdU selection used to isolate the HAB cells described above, which cells survived with at least 90% replacement, and possibly total replacement, of thymine by BrUra. The ability of the cells to continue growing with such a high level of substitution raises the possibility that they may have undergone some modification which permits them to escape the mutagenic effects of BrUra in DNA. It is possible that in these cells BrUra is always replicated and transcribed correctly as thymine.

It is not presently known whether there is a relationship between BrdU dependence and BrdU tolerance in the mutant cells described in this paper. However there is some suggestive evidence, from the study of the "revertant" cells able to grow in the absence of BrdU, that at least in some cases BrdU dependence and BrdU tolerance can be separated. As described above a significant fraction of the revertant population had maintained, during a long period of growth in the absence of BrdU, the ability to grow rapidly in high concentrations of BrdU and to incorporate BrUra into the DNA. Nevertheless these cells, which had retained the ability to survive with highly substituted DNA (approximately 50%), did not exhibit any dependence on BrdU for growth.

The study of the BrdU-dependent cells has not as yet provided evidence to explain why the cells require BrdU. A number of possible explanations could be proposed, based on effects of BrdU that have been observed in other systems. Thus BrdU dependence could involve the action of BrdU as a mutagen, as a suppressor of differentiated functions, or as an inducer of latent viruses. It is also possible, however, that BrdU dependence is not related to the action of BrdU itself or to the effects of its incorporation into DNA. Instead the modification resulting in the requirement for BrdU could have occurred in the systems which carry out DNA replication and transcription. For example, there could be a mutation in a DNA polymerase so that the enzyme would function only with BrdU, not with thymidine. The elucidation of the mechanism of BrdU dependence could therefore result in the positive identification of an enzyme which functions to synthesize DNA in vivo.

Acknowledgments

This work was supported by NIH Grants HD 04807, HD 06276, AI 08186-03, and NSF Grant 31118X. M.D.B. is a Damon Runyon fellow.

REFERENCES

Breslow, R. and R. Goldsby. 1969. Isolation and characterization of thymidine transport mutants of Chinese hamster cells. *Exp. Cell Res.* **55:**339.

Cleaver, J. 1968. Repair, replication and degradation of bromouracil-substituted DNA in mammalian cells after irradiation with ultraviolet light. *Biophys. J.* **8:**775.

Coleman, A., D. Kankel, I. Werner and J. Coleman. 1968. Cellular differentiation *in vitro:* Perturbation by halogenated deoxyribonucleosides. *J. Cell Biol.* **39:**27A.

Davidson, R. L. and M. Bick. 1973. Bromodeoxyuridine dependence—a new mutation in mammalian cells. *Proc. Nat. Acad. Sci.* **70:**138.

Freese, E. 1959. The specific mutagenic effect of base analogues on phage T4. *J. Mol. Biol.* **1:**87.

Kit, S., D. Dubbs, L. Piekarski and T. Hsu. 1963. Deletion of thymidine kinase activity from L cells resistant to bromodeoxyuridine. *Exp. Cell Res.* **31:**297.

Lin, S. and A. Riggs. 1972. *Lac* operator analogues: Bromodeoxyuridine substitution in the *lac* operator affects the rate of dissociation of the *lac* repressor. *Proc. Nat. Acad. Sci.* **69:**2574.

Lowy, D., W. Rowe, N. Teich and J. Hartley. 1971. Murine leukemia virus: High frequency activation *in vitro* by 5-iododeoxyuridine and 5-bromodeoxyuridine. *Science* **174:**155.

Silagi, S. and S. Bruce. 1970. Suppression of malignancy and differentiation in melanotic melanoma cells. *Proc. Nat. Acad. Sci.* **66:**72.

Toliver, A. and E. Simon. 1967. DNA synthesis in 5-bromouracil tolerant HeLa cells. *Exp. Cell Res.* **45:**603.

Wake, R. and R. Baldwin. 1962. Physical studies on the replication of DNA *in vitro.* *J. Mol. Biol.* **5:**201.

Restriction of Antigen Mobility in the Plasma Membranes of Some Cultured Fibroblasts

Michael Edidin and Arthur Weiss

Biology Department, The Johns Hopkins University
Baltimore, Maryland 21218

Evidence has accumulated during the past 5 years indicating that the surface membrane of animal cells is sufficiently fluid to allow rotational and translational movements of membrane lipids and proteins. Lipid movements have been estimated in terms of diffusion constants (Scandella, Devaux and McConnell 1972) or microviscosities (Rudy and Gittler 1972). More or less quantitative data on protein movement in membranes has been obtained by following the intermixing of the surface antigens of newly formed heterokaryons (Frye and Edidin 1970), by timing the spread of labeled myotube membrane antigens (Edidin and Fambrough 1973), by timing the rotational relaxation of bleached rhodopsin (Cone 1972), and recently by determining the lateral movement of this protein (Poo and Cone 1973). Additionally a number of experiments have dealt qualitatively with rearrangement of surface antigens or membrane particles induced by changes in pH (Pinto da Silva 1972) or the binding of multivalent ligands, lectins and antibodies to a cell (Taylor et al. 1971; Singer 1973; Yahara and Edelman 1972). The latter leads to accumulation of antigen at one pole of the cell and is termed "capping" (Taylor et al. 1971); it appears to involve both free diffusion of membrane molecules in the plane of the membrane and cell metabolic activity.

We have recently suggested that this metabolic activity causes the flow of membrane from the leading ruffled edge of migrating cells toward their center (Edidin and Weiss 1972); such a flow has been inferred by Abercrombie and coworkers from their observations on the movement of particles stuck to the cell surface (Abercrombie, Heaysman and Pergrum 1970; Harris 1972). Our analysis suggested that the absence of capping behavior in a given cell type would not necessarily mean that the cell's surface proteins were immobile. However we did expect that cells with prominent ruffled membranes would show vigorous capping behavior. The fibroblast line that we had successfully capped with anti-H-2 antibodies has relatively poorly developed ruffled membranes (Domnina et al. 1972), and it might be thought that cells with greater ruffles would cap more readily than those used previously. In this paper we present evidence that, contrary to expectations, several well-ruffled cell types failed to cap, despite extensive treatment with anti-H-2 anti-

213

bodies and antiglobulin. Furthermore the cells that fail to cap are slow to intermix their surface antigens when included in heterokaryons of several types.

MATERIALS AND METHODS

3T3, 3T12, SV40-transformed 3T3 (SV3T3), and flat revertant SV3T3 (FRSV-3T3), all of Balb/c origin (Aaronson and Todaro 1968), were obtained from Dr. Stephen Roth of this department. They were cultured in Dulbecco's modified Eagle's medium with 10% calf serum. Cultures of all types were passed before reaching confluence, usually every 3 days for transformed cells and every 7 days for 3T3.

Human kidney fibroblasts were obtained from Mr. Bill Smith and Dr. L. Parks, Department of Surgery, The Johns Hopkins University. The cells were maintained in Waymouth's medium plus 20% fetal calf serum and used as primary or secondary cultures. Clone 1d of mouse L cells and the SV40-transformed line WI-18VA-2 have been used by us before and their maintenance is described in previous papers, as is our preparation of antisera to mouse H-2 antigens and of fluorescent anti-globulins (Frye and Edidin 1970; Edidin and Weiss 1972). Our xeno-antibody to human surface antigens has been found to be directed against protein antigens (Edidin, unpublished). All reagents used to stain heterokaryons were absorbed on Sendai virus before use and were specific for either mouse or human surface antigens.

Cap formation was induced by coating cells, either in suspension or attached to glass or plastic, with anti-H-2 or anti-species antibodies at $0°C$. After washing in Hepes-buffered Hank's solution, the cells were reacted with fluorescent antiglobulin, again at $0°C$. Excess fluorescent reagent was washed away and the cells were warmed to $37°C$ for 45–180 minutes to induce cap formation. They were then chilled again and either examined live (suspended cells) or briefly fixed in 70% ethanol (attached cells) before scoring cells for localization of fluorescence. Suspended cells with all fluorescent stain concentrated in one area of the membrane, or attached cells staining in the cell body but not at the tips of processes, were counted as capped.

Our procedures for cell fusion and for staining for heterokaryons have been described previously (Frye and Edidin 1970).

RESULTS

Contrary to expectations 3T3 fibroblasts failed to form caps after coating with anti-H-2^d antibody and antiglobulin. Since antibody concentration and time of incubation at $37°C$ have been shown to affect the extent of capping of a cell population (Taylor et al. 1971), 3T3 cells were tested with three different dilutions of antiserum and were incubated for twice the usual time used for cap induction (Table 1). Despite this extended treatment the cells neither speckled nor capped. Similar results were seen with WI-38 human fibroblasts and with a normal mouse fibroblast strain, MKA, derived from adult strain A mouse kidney. In contrast SV3T3 cells capped readily in suspension and on substrate, though their capping was sensitive to antiserum dilution (Table 1). The prominent ruffles of 3T3 and

Table 1

Capping of Cells after Reaction with Anti-H-2 Antibody and Antiglobulin

Anti-H-2^d serum dilution	% 3T3 capped*		% SV3T3 capped†	
	in suspension	on substrate	in suspension	on substrate
undiluted	0	0	15	30
1/5	0	0	45	39
1/15	0		> 50	39

* Two cycles of incubation with antisera followed by 90 minutes at 37°C.

† One cycle of incubation with antisera followed by 45 minutes at 37°C.

WI-38, compared to SV3T3 and Cl 1d, led us to expect them to cap readily. Since failure to cap might be due to restriction of antigen mobility in the plane of the membrane rather than to an inability to drive aggregates of mobile antigens from the cell periphery, antigen migration was tested in heterokaryons. The mobilities of H-2 and human species antigens were measured by fusing mouse and human cells and counting the number of heterokaryons with completely intermixed surface antigens and the number bearing segregated antigens at various times after fusion.

The rate of antigen intermixing, in terms of loss of segregated heterokaryons in the cross 3T3 × WI-38, is shown in Fig. 1. For comparison a curve for loss of segregates in Cl 1d × VA-2 heterokaryons is also shown. The rate of intermixing of 3T3 × WI-38 heterokaryon surface antigens is about 2.5 times slower than that for Cl 1d × VA-2 heterokaryons. Also the 3T3 × WI-38 intermixing rate is not linear over the entire time of the experiment. The curve seems to be leveling at around 3 hours after initiating fusion, suggesting a gradually increasing restriction of antigen mobility in the remaining segregated heterokaryons.

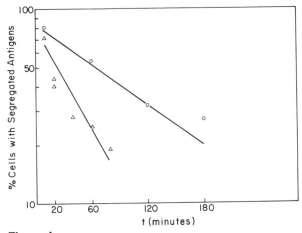

Figure 1

Loss from the population of all heterokaryons of cells with segregated mouse and human cell surface antigens. The cells fused: (△) Cl 1d × VA-2; (○) 3T3 × WI-38.

Further fusion experiments were done to measure the loss of segregated hetero-karyons in populations formed from both contact-inhibited and transformed cells. Data from these experiments are plotted in Fig. 2 and 3, together with reference lines for fast and slow rates of mixing shown in Fig. 1. It appears that if one parent in a fusion does not cap, then regardless of the second cell type, heterokaryons will show some restriction of mobility of surface antigens. On the other hand, fusion of 3T3 with primary or secondary adult human fibroblasts (Fig. 3) yields heterokaryons with an extremely low rate of antigen mixing. In three experiments most of the heterokaryons remained segregated at times when even the population of 3T3 × WI-38 heterokaryons contained a considerable proportion of cells with completely intermixed surface antigens.

Finally, though not tested for capping, morphological revertants of SV3T3, FRSV3T3, also appear to form heterokaryons whose surface proteins are of low mobility (Fig. 3).

DISCUSSION

Previous work on the phenomenon of "capping" (ligand-induced accumulation of cell surface proteins in one portion of the membrane) led to the inference that, be-sides a multivalent ligand, cap formation requires both mobility of membrane mole-cules and ability of a cell to drive small aggregates of surface molecules and ligands in a constant direction, away from the cell's leading, ruffled edge (Edidin and Weiss 1972; dePetris and Raff 1972). From this analysis we expected that cells with ruffles more prominent than those borne by cells known to cap would be capable of rapid and complete capping. In fact the experiments reported indicate that cells may not cap, even if they display large membrane ruffles. This implies either that the model proposed for cap formation is not complete, or that the mobility of membrane proteins is limited in cells that fail to cap. While we may still have not

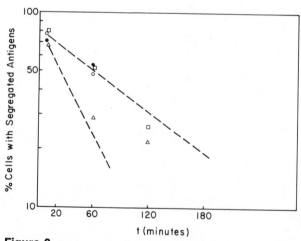

Figure 2

Loss of segregated heterokaryons in populations formed from (○) SV3T3 × WI-38; (△) SV3T3 × VA-2; (□) 3T3 × VA-2; (●) 3T12 × WI-38. The lines of Fig. 1 are included for reference.

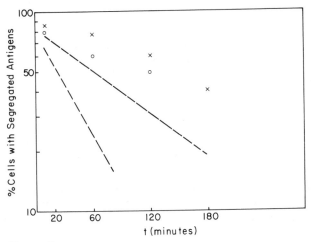

Figure 3
Loss of segregated heterokaryons in populations formed from (×) 3T3 ×
secondary human fibroblasts; (o) FRSV3T3 × WI-38. As before the lines of
Fig. 1 are included for reference.

correctly described capping mechanisms in fibroblasts, it does appear that the
glycoprotein H-2 transplantation antigens (Nathanson 1970) and human species
antigens, apparently protein, are of restricted mobility in fibroblasts that fail to cap.

The restriction is demonstrated when cell surface antigens are examined in virus-
induced heterokaryons. This system clearly indicates the possibility of surface
protein movements in the plane of the membrane. The movement appears to involve
only diffusional processes, being dependent upon temperature but not upon continued generation of ATP in the cell or upon protein or amino sugar synthesis
(Frye and Edidin 1970). Hence restrictions on antigen mixing in heterokaryon
membranes ought to reflect only the degree of interference with protein diffusion
in that membrane.

Our data raise two issues: the relationship between social behavior and antigen
mobility and the mechanism that may be involved in restricting antigen mobility.
It appears that cells showing density-dependent inhibition of growth and movement
(Aaronson and Todaro 1968) have restricted mobilities of plasma membrane
proteins. Such inhibition of movement in "normal" as compared to "transformed"
cells is concordant with some other observations on the arrangements of plasma
membrane molecules in 3T3 and transformed 3T3 cells. Thus the clustering of
concanavalin A binding sites seen in SV3T3 after binding lectin at room or higher
temperatures, though originally thought to reflect the arrangement of sites in the
untreated cell (Nicolson 1971), now appears to be due to the aggregation of
mobile sites by the lectin. Cells reacted with lectin after adequate fixation, or
entirely at 4°C, fail to show clustered sites (Rosenblith et al. 1973). Untransformed 3T3 cells appear to bear random lectin sites, which are not aggregated by
lectin whatever the temperature of its application (Nicolson 1971; Rosenblith et al.
1973).

The concordance is marred by a report that the concanavalin A binding sites
of even nontransformed 3T3 are aggregated by lectin and remain randomly dispersed only if reacted with ferritin ConA after fixation (Raff and dePetris this

volume). Perhaps the difference in the two results is a reflection of the precise way in which 3T3 cells were prepared and labeled. In our hands at the resolution level of the light microscope, H-2 antigens of 3T3 cells appear to be immobile by one test, capping, while they are slowly mobilized in heterokaryons.

Another comparison of transformed and density-sensitive 3T3 cells was made by Roth and coworkers (Roth and White 1972). They found that transformed cells were able to self-glycosylate their surfaces when given appropriate nucleotide sugars. In contrast to this cis-glycosylation by 3T12 and SV3T3, 3T3 cells were able only to glycosylate adjacent cells and incorporated labeled nucleotide sugars poorly when incubated in a stirred suspension. This observation was interpreted as indicating proximity of glycosyltransferases and macromolecular acceptor molecules on transformed cells and the separation of these molecules on contact inhibited cells. The difference between the two states could be due to differential restrictions on enzyme and receptor mobility rather than to different fixed arrangements of the two (Roth 1973).

Despite these concordances we feel that extrapolation from our data and that of others to general statements about relative mobility of surface molecules in tumor cells should be limited. Measurements must be made of a great many other cell types before we have an inkling of the relationship between mobility of membrane components and cell social behavior and tumorigenicity.

Another point of speculation, for which we have even less information than the first, concerns mechanisms of restriction of antigen mobility. Unlike mobility measurements using labeled stearic acid derivatives (Scandella et al. 1972) or lipid-soluble fluorescent dyes (Rudy and Gittler 1972), our measurements do not directly probe membrane lipids, but rather are made on proteins embedded in the lipid continuum. Thus while protein mobility, if not due to cell metabolism, must be based on diffusion in a fluid lipid phase, restriction of mobility does not imply an alteration in membrane fluidity. Alteration of fluidity by substitution of more highly saturated fatty acids for unsaturated chains in phospholipids is one of several possibilities for inhibiting protein motion in the membrane. It seems as likely that other mechanisms, involving anchoring from above or beneath the membrane or aggregation of proteins within the membrane, might also come into play in restricting translational diffusion (Edidin 1972). Here again our speculation is barely based in fact and must await experimental investigation of the degree of restrictions imposed upon membrane proteins in a variety of cells, as well as some knowledge of drug effects on this restriction.

Acknowledgments

This work was supported by grant AM 11202 from the National Institutes of Health. This is publication number 731 from the Department of Biology.

REFERENCES

Aaronson, S. A. and G. J. Todaro. 1968. Development of 3T3-like lines from Balb/c mouse embryo cultures: Transformation susceptibility to SV40. *J. Cell. Physiol.* **72**:141.

Abercrombie, M., J. E. M. Heaysman and S. Pergrum. 1970. The locomotion of fibroblasts in culture. II. "Ruffling." *Exp. Cell Res.* **60**:437.

Cone, R. A. 1972. Rotational diffusion of rhodopsin in the visual receptor membrane. *Nature New Biol.* **236**:39.

dePetris, S. and M. Raff. 1972. Distribution of immunoglobulin on the surface of mouse lymphoid cells as determined by immunoferritin electron microscopy. Antibody-induced, temperature-dependent redistribution and implications for membrane structure. *Eur. J. Immunol.* **2**:523.

Domnina, L. V., V. Y. Ivanova, L. B. Margolis, L. V. Olshevskaja, Y. A. Rovensky, J. M. Vasiliev and I. M. Gelfand. 1972. Defective formation of lamellar cytoplasm by neoplastic fibroblasts. *Proc. Nat. Acad. Sci.* **69**:248.

Edidin, M. 1972. Aspects of plasma membrane fluidity. *Membrane research* (ed. C. F. Fox) p. 15. Academic Press, New York.

Edidin, M. and D. Fambrough. 1973. Fluidity of the surface of cultured muscle cell fibers. Rapid lateral diffusion of marked surface antigens. *J. Cell Biol.* **57**:27.

Edidin, M. and A. Weiss. 1972. Antigen cap formation in cultured fibroblasts: A reflection of membrane fluidity and of cell motility. *Proc. Nat. Acad. Sci.* **69**:2456.

Frye, L. D. and M. Edidin. 1970. The rapid inter-mixing of cell surface antigens after formation of mouse-human heterokaryons. *J. Cell Sci.* **7**:319.

Harris, A. 1972. Surface movement in fibroblast locomotion. *Acta Protozoologica* **11**:145.

Nathanson, S. G. 1970. Biochemical properties of histocompatibility antigens. *Ann. Rev. Genet.* **4**:69.

Nicolson, G. 1971. Difference in the topology of normal and tumor cell membranes as shown by different distributions of ferritin-conjugated concanavalin A on their surfaces. *Nature New Biol.* **233**:244.

Pinto da Silva, P. 1972. Translational mobility of the membrane intercalated particles of human erythrocyte ghosts. pH-dependent, reversible aggregation. *J. Cell Biol.* **53**:777.

Poo, M.-M. and R. A. Cone. 1973. Lateral diffusion of rhodopsin in the visual receptor membrane. *J. Supramol. Structure* (in press).

Rosenblith, J. Z., T. E. Ukena, H. H. Yin, R. D. Berlin and M. J. Karnovsky. 1973. A comparative evaluation of the distribution of concanavalin A binding sites on the surfaces of normal, virally-transformed and protease-treated fibroblasts. *Proc. Nat. Acad. Sci.* **70**:1625.

Roth, S. 1973. A molecular model for cell interactions. *Quart. Rev. Biol.* (in press)

Roth, S. and D. White. 1972. Intercellular contact and cell-surface galactosyltransferase activity. *Proc. Nat. Acad. Sci.* **69**:485.

Ruby, B. and C. Gittler. 1972. Microviscosity of the cell membrane. *Biochim. Biophys. Acta* **288**:225.

Scandella, C. J., P. Devaux and H. M. McConnell. 1972. Rapid lateral diffusion of phospholipids in rabbit sarcoplasmic reticulum. *Proc. Nat. Acad. Sci.* **69**:2056.

Singer, S. J. 1973. Are cell membranes fluid? *Science* **180**:983.

Taylor, R., P. Duffus, M. Raff and S. dePetris. 1971. Redistribution and pinocytosis of lymphocyte surface immunoglobulin molecules induced by anti-immunoglobulin antibody. *Nature* **233**:1225.

Yahara, I. and G. M. Edelman. 1972. Restriction of the mobility of lymphocyte immunoglobulin receptors by concanavalin A. *Proc. Nat. Acad. Sci.* **69**:608.

Effect of Ferritin and Other Visual Markers on Hybrid Antibody-induced Topographical Displacement of Cell Surface Components

Christopher W. Stackpole, Ulrich Hämmerling, Michael E. Lamm,*
Michael P. Lardis, and Joseph Lumley-Frank

Memorial Sloan-Kettering Cancer Center and *Department of Pathology
New York University School of Medicine, New York, N.Y. 10021

Ferritin-conjugated antibodies have been used widely to locate cell surface antigens by electron microscopy and to analyze the organization of these constituents within the cell surface membrane (Davis and Silverman 1968; Aoki et al. 1969; Stackpole 1971; Kourilsky et al. 1971; Davis et al. 1971; Karnovsky, Unanue and Leventhal 1972). However a variety of labeling patterns have been obtained for the same antigens on similar cell types, depending upon the method used to label with ferritin. For example, when H-2 alloantigens on mouse lymphoid cells are labeled indirectly with unconjugated anti-H-2 alloantiserum followed by rabbit anti-mouse immunoglobulin (Ig) chemically coupled to ferritin, large "patches" of ferritin in densely packed multiple layers, separated by large expanses of completely unlabeled cell surface, are observed (Aoki et al. 1969; Karnovsky et al. 1972). Ferritin applied by the indirect hybrid antibody method (Hämmerling et al. 1968) similarly labels H-2 alloantigens in widely separated patches, but in this case the patches typically consist of a monolayer of markers (Aoki et al. 1969; Stackpole et al. 1971). Direct conjugation of ferritin to anti-H-2 alloantibody results in the appearance of small clusters of ferritin distributed uniformly (or randomly) over the entire cell surface (Davis et al. 1971). A similar distribution of H-2 alloantigens on erythrocyte ghosts was obtained by indirect labeling of alloantibody with ferritin-conjugated rabbit anti-mouse Ig antibody (Nicolson, Hyman and Singer 1971).

These discrepancies in labeling patterns may largely be explained by redistribution effects occurring at the cell surface (Taylor et al. 1971; Davis 1972). Redistribution apparently can occur because of a high degree of fluidity within the cell surface membrane at physiological temperatures, allowing extensive lateral movements of membrane constituents (Frye and Edidin 1970; Singer and Nicolson 1972). The appearance of large patches of label on the cell surface results from attachment of bivalent antibody to cell surface components and precedes the development of polar "caps" of label (Taylor et al. 1971; Loor, Forni and Pernis 1972; Unanue, Perkins and Karnovsky 1972). Similarly Davis (1972) demonstrated that uniformly distributed ferritin-conjugated anti-H-2 alloantibody on

221

mouse lymphoid cells can be sequestered into distinct patches by the addition of a secondary bivalent antibody.

Since monovalent antibody fragments cannot induce formation of patches and caps, molecular cross-linking and aggregation of surface components within a fluid membrane matrix has been proposed to explain induced redistribution (de-Petris and Raff 1972). However hybrid rabbit anti-mouse Ig/rabbit anti-ferritin antibody, which is virtually free of bivalent anti-mouse Ig/anti-mouse Ig antibody contaminants (formed by homologous recombination during preparation of hybrid antibody) and is therefore monovalent for an individual specificity, can displace cell surface components in the same manner as bivalent anti-mouse Ig antibody (Stackpole et al. 1973). This suggested that factors in addition to molecular cross-linking are involved in topographical displacement of cell surface constituents.

The effect of hybrid antibodies on redistribution of cell surface components can be analyzed by immunofluorescence microscopy either with or without the addition of a visual marker, a technique which is not possible by electron microscopy. Specific attachment of ferritin or another visual marker, southern bean mosaic virus (SBMV) (Hämmerling et al. 1969), to cell-bound hybrid antibodies was found to enhance the redistributing effect of these antibodies (Stackpole et al. 1973).

We present here further studies on the effects of hybrid antibodies and visual markers on topographical displacement of cell surface components.

EXPERIMENTAL PROCEDURES

Mouse alloantiserum to H-2^b alloantigen specificities was prepared by immunizing (C57BL/6/H-2^{k*} × AKR)F$_1$ mice with the C57BL ascites leukemia EL4. Anti-H-2^k alloantiserum was obtained by immunizing AKR/H-2^{b*} mice with normal lymphoid cells from AKR mice. These alloantisera do not recognize any H-2 antigen specificities in common and were therefore used in double-labeling experiments. Alloantiserum to TL alloantigens (TL.1,2,3) was raised by immunizing (A/TL^{-*} × C57BL/6)F$_1$ mice with the A strain spontaneous leukemia, ASL1.

Concanavalin A (ConA) was obtained from Miles Laboratories, Kankakee, Ill. (3 × crystallized) and was further purified by elution from a Sephadex G-50 column with D-glucose (Agrawal and Goldstein 1967). Ferritin-conjugated ConA (ConA/F) was prepared according to Nicolson and Singer (1971). Anti-H-2^b alloantibody was coupled to ferritin (anti-H-2/F) by the basic method of Singer and Schick (1961), as adapted to mouse antibody by Davis and Silverman (1968). Procedures for the conjugation of anti-H-2^b alloantibody, ConA, and rabbit antibodies to fluorescein isothiocyanate are presented elsewhere (Stackpole et al. 1973).

Hybrid rabbit anti-mouse Ig/anti-ferritin antibody was prepared as described previously (Hämmerling et al. 1968). Mouse alloantisera to H-2^b and H-2^k alloantigens were coupled to benzylpenicilloyl (BzPen) or hydroxyiodonitrophenyl-acetyl (NIP) hapten groups, and hybrid anti-BzPen/anti-ferritin and anti-NIP/anti-SBMV antibodies prepared according to Lamm et al. (1972).

---

* C57BL/6/H-2^k is a congenic stock on a C57BL/6 (H-2^b) genetic background, but with an H-2^k allele from the AKR strain substituted for the H-2^b allele. AKR/H-2^b is a congenic stock on an AKR (H-2^k) background, with an H-2^b allele from the C57BL/6 strain substituted for the H-2^k allelle. A/TL$^-$ is a congenic stock on an A (TL+) background.

Viable cell suspensions (5×10^6 cells) were labeled by incubating in 0.5 ml volumes of immune reagent diluted in Earle's balanced salt solution (EBSS), followed by two washes with EBSS. All incubations were for 30 minutes unless indicated otherwise. Double-labeling and preparation of labeled cells for electron microscopy have been described previously (Lamm et al. 1972).

RESULTS AND DISCUSSION

Double-labeling of H-2 Alloantigens

Lamm et al. (1972) devised a method for specifically labeling two different surface antigens on the same cell, using hapten-conjugated mouse alloantibodies and hybrid antibodies directed against the hapten groups. This enables hybrid antibodies to distinguish between alloantibodies that would be indistinguishable with conventional hybrid antibodies (Hämmerling et al. 1968). With this method H-2^b and H-2^k alloantigens on lymph node cells from H-2^b/H-2^k heterozygous mice were labeled in separate sectors (Lamm et al. 1972).

To determine whether this observed segregation of ferritin and SBMV is due to topographical segregation of H-2^b and H-2^k alloantigens on the cell surface, the following control was performed. Lymph node cells from homozygous H-2^b mice (C57BL/6) were incubated simultaneously with two samples of the same anti-H-2^b alloantiserum, one sample coupled to BzPen hapten, the other coupled to NIP hapten. The cells were then incubated with hybrid anti-BzPen/anti-ferritin and anti-NIP/anti-SBMV antibodies simultaneously and finally incubated with ferritin and SBMV. All incubations were at 4°C. Although the same alloantigens were being labeled, the two markers were again separated into sectors (Fig. 1a). Two possibilities that might explain these results are (a) steric hindrance or interference between the haptenized alloantibodies, hybrid antibodies, or markers. Haptenization of the alloantibodies may couple ten or more hapten groups to each alloantibody molecule (Hämmerling, Stackpole and Koo 1973) and could result in electrostatic imbalances and consequently repulsion between the BzPen/alloantibody and NIP/alloantibody molecules, or in physical configurations conducive to steric hindrance of attachment to alloantigen determinants. Hybrid antibodies and visual markers should behave as in conventional labeling. (b) Alloantigens are independently displaced laterally within the cell surface membrane and aggregated into patches by the separate sets of immune reagents. Unconjugated alloantibodies should not be able to displace alloantigens alone, at the low temperature used for labeling (Stackpole et al. 1973), although haptenized alloantibodies may behave differently. On the other hand, hybrid antibodies can displace alloantigen-alloantibody complexes, and markers attached specifically to hybrid antibodies enhance this displacement.

One way to test these possibilities was to react the same cell type with anti-H-2^b/BzPen and anti-H-2^b/NIP alloantibodies as above, then to replace the anti-BzPen/anti-ferritin and anti-NIP/anti-SBMV hybrid antibodies with anti-mouse Ig/anti-ferritin and anti-mouse Ig/anti-SBMV hybrids, and finally incubate with ferritin and SBMV. A complete intermixing of visual markers was obtained (Fig. 1b), indicating that steric hindrance was not involved in segregation of markers.

The most likely explanation for the observed segregation of labels is therefore lateral displacement and aggregation of membrane-bound alloantigens induced by

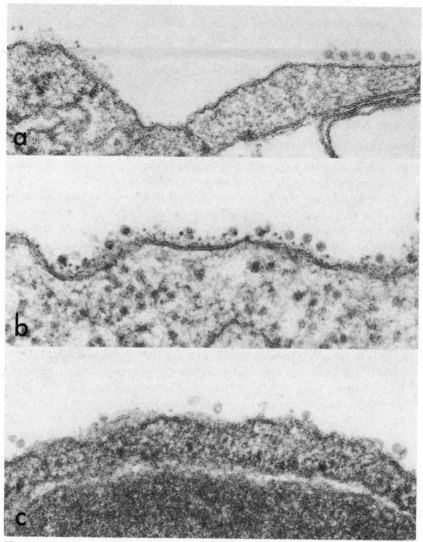

Figure 1
(a) C57BL/6 lymph node cell double-labeled for H-2^b alloantigens with anti-H-2^b/BzPen and anti-H-2^b/NIP haptenized alloantibodies, hybrid antibodies directed against the haptens, and ferritin and SBMV. The smaller marker is ferritin, the larger SBMV. (b) Double-labeling as above, but with hybrid antibodies directed against mouse Ig substituted for anti-hapten hybrids. (c) Double-labeling for H-2^b alloantigens on a cell reacted with both haptenized alloantibodies, fixed with glutaraldehyde, then reacted with anti-hapten hybrid antibodies and ferritin and SBMV. 120,000 ×

attachment of the immune reagents and/or visual markers. The following evidence suggests that alloantigens are not displaced significantly by haptenized alloantibodies alone. If cells are reacted with haptenized alloantibody, then fixed with glutaraldehyde, hybrid antibody and visual markers attach in expected amounts, indicating that the haptens are resistant to aldehyde fixation. To determine the distribution of H-2^b alloantigens after attachment of anti-H-2^b/BzPen and anti-H-2^b/ NIP, cells were fixed with 1% glutaraldehyde in 0.1 M phosphate buffer for 10 minutes following attachment of the haptenized alloantibodies. To avoid subsequent nonspecific attachment of hybrid antibodies or visual markers to the cell surface, excess glutaraldehyde sites were saturated by incubating cells in phosphate buffer containing 5% fetal bovine serum. Fixed cells were then incubated with anti-BzPen/anti-ferritin and anti-NIP/anti-SBMV hybrid antibodies, and then with ferritin and SBMV. Label was much more uniformly distributed over the cell surface than is typical of labeling with hybrid antibodies, and ferritin and SBMV were intermixed (Fig. 1c).

Displacement of surface alloantigens at low temperatures is therefore not due to alloantibody, but rather to hybrid antibody and/or visual markers. Hybrid antibody must be largely responsible for the observed displacement since in the control experiment, illustrated in Fig. 1b, ferritin and SBMV did not induce segregation of intermixed hybrid antibodies.

Patch Formation at Low Temperature

Alloantigens which are represented weakly on mouse lymphoid cells, such as H-2 and TL alloantigens on thymocytes, appear to be located on the cell surface as widely separated dense patches when labeled with visual markers applied indirectly with hybrid antibodies (Aoki et al. 1969; Stackpole et al. 1971). The same labeling pattern is obtained even at very low temperatures (Stackpole et al. 1973), suggesting that a high degree of fluidity within the cell surface membrane is not necessary for significant lateral displacement of membrane constituents to occur.

H-2^b alloantigens on thymocytes labeled with fluorescent anti-H-2^b alloantiserum at −5°C in the presence of high concentrations (100 mM) of sodium azide (to prevent endocytosis of label) are uniformly distributed over the entire cell surface (Stackpole et al. 1973). When hybrid anti-mouse Ig/anti-ferritin antibody is added at −5°C, the labeling pattern remains unchanged, but subsequent incubation with ferritin results in the appearance of conspicuous patches of fluorescence label although the remainder of the cell surface is uniformly labeled (Fig. 2, inset).

A sample of these cells was fixed with 1% glutaraldehyde in 0.1 M cacodylate buffer and postfixed with 1% cacodylate-buffered OsO$_4$ at −5°C, then prepared for electron microscopy in the usual manner. The cell surface was labeled in widely separated dense patches of ferritin, with the remainder of the surface completely free of ferritin (Fig. 2). Endocytosis of ferritin was not observed. The indication from these results is that, at least at very low temperatures, ferritin specifically attached to cell-bound hybrid antibody causes significant displacement of H-2^b alloantigens. This displacement does not involve all of the alloantigen-alloantibody complexes, either because of the viscosity of the cell surface membrane which would restrict lateral movement of membrane constituents, or because of inefficient binding of hybrid antibody or ferritin at such low temperatures. Since less of the cell surface is labeled with ferritin at −5°C than at 4°C, inefficient binding of label is the probable explanation.

The double-labeling experiments described above indicate that hybrid antibodies and markers can bind to alloantigen-alloantibody complexes fixed in a state in which large-scale aggregation cannot occur. Nevertheless a large multivalent marker

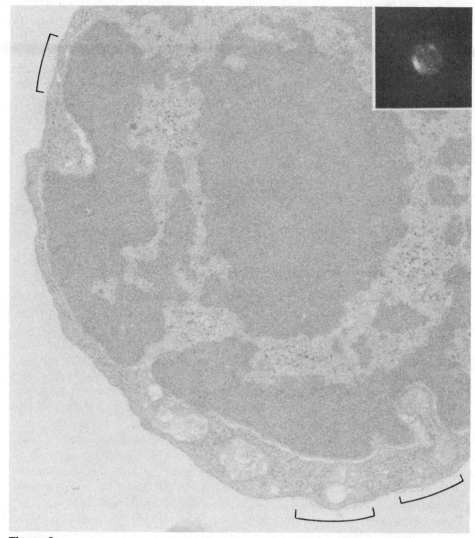

Figure 2
C57BL/6 thymus cell reacted with fluorescent anti-H-2[b] alloantibody, hybrid antibody and ferritin and fixed, all at −5°C. Patches of ferritin are bracketed. 36,000 × *Inset:* Cell processed similarly but not fixed, photographed at −5°C, showing fluorescent labeling pattern after reaction with ferritin. 1080 ×

such as ferritin could readily cross-link and aggregate surface components by acquiring multiple antibody attachments if available.

Effect of ConA on Labeling of Alloantigens with Hybrid Antibody and Ferritin

Yahara and Edelman (1971) demonstrated that redistribution of surface Ig on mouse lymphocytes into patches and caps, induced by bivalent rabbit anti-mouse Ig antibody, could be reversibly inhibited by incubation of the cells with ConA prior to labeling with anti-Ig antibody. While the mechanism of inhibition is unclear, it appeared that surface Ig molecules were being "frozen" in near-native configuration either by physical restriction of lateral movement or by some sort of reversible alteration in the nature of the cell surface due to attachment of ConA. Incubation of mouse lymphoid cells with ConA at 37°C, either before or after incubation with alloantiserum, similarly inhibits displacement of H-2^b, TL.1,2,3 and Thy-1.2 allo-antigens into patches and caps by either bivalent or hybrid anti-mouse Ig antibody (Stackpole et al. unpubl.). In all cases uniform fluorescent labeling of surface components was obtained unless the cell-bound ConA was removed by incubating labeled cells with α-methyl-D-mannoside at 37°C, after which patchy or capped labeling patterns were observed on all cells.

We were interested in determining whether labeling of a weak antigen system with hybrid antibody and ferritin could be obtained if the antigen was "frozen" into a uniform distribution on the cell surface. Under these conditions little multiple binding, cross-linking, or aggregation would be expected to occur.

Thymocytes from the congenic mouse stock C57BL/6/TL$^+$ were incubated with 100 μg/ml ConA at 37°C, then incubated sequentially with anti-TL.1,2,3 alloantiserum, hybrid anti-mouse Ig/anti-ferritin antibody, and ferritin. Ferritin on these cells is very sparsely distributed over the entire cell surface, usually in clusters of two to five molecules (Fig. 3a). The ferritin is attached specifically to TL alloantigens, as indicated by the complete absence of label on C57BL/6

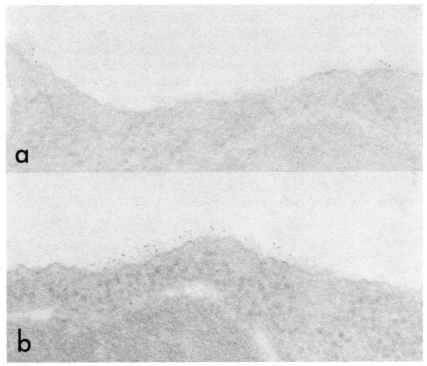

Figure 3
(a) Thymus cell reacted with anti-TL alloantiserum, then with ConA, and finally with hybrid antibody and ferritin. (b) Similar reaction, but following addition of ferritin the cells were incubated with mannoside. 100,000 $\times$

(TL^-) thymus cells treated in exactly the same manner. When TL^+ cells are incubated with ConA, then labeled for TL alloantigens with alloantiserum, hybrid antibody, and ferritin, and finally incubated with 40 mM mannoside at 37°C, ferritin labels the cell surface in a typically patchy pattern (Fig. 3b), indicating the reversibility of the inhibition of displacement of these alloantigens. Approximately the same percentage of the cell surface (about 1%) is labeled following removal of the ConA as when ConA is not used at all; therefore hybrid antibody and ferritin are able to attach to TL alloantigen-alloantibody complexes to a similar extent when displacement cannot occur as when it can occur. The labeling pattern obtained when ConA is used may represent ferritin attached to multiple hybrid antibodies, and hence TL alloantigen-alloantibody complexes, locally aggregated. However it is more likely that, as in double-labeling of fixed cells, hybrid antibodies and markers can label complexes that are not aggregated to a significant extent, if aggregation is prevented from occurring.

Cap Formation Induced by Ferritin-conjugated Primary Reagents

Since ferritin or SBMV attached specifically to cell-bound hybrid antibody enhances capping of surface Ig and alloantigens on mouse lymphoid cells (Stackpole et al. 1973), it was of interest to determine whether ferritin, directly conjugated to a primary reagent (that reagent which attaches directly onto the cell surface component), which alone does not induce rapid cap formation, would have any effect on the capacity of that reagent to induce capping. Negligible capping of $H-2^b$ alloantigens on lymph node cells occurs when the alloantigens are labeled indirectly with unconjugated anti-$H-2^b$ alloantibody at 37°C followed by fluorescent rabbit anti-Fc fragment of mouse Ig (to discern alloantibody from surface Ig, which does not react with anti-Fc antibody) at 4°C, or directly with fluorescent anti-$H-2^b$ alloantibody at 37°C (Table 1). Ferritin-conjugated anti-$H-2^b$ alloantibody, however, induced a significant amount of capping of $H-2^b$ alloantigens when applied at 37°C followed by fluorescent rabbit anti-ferritin at 4°C. Similar results were obtained with ConA. Capping of ferritin-conjugated primary reagents apparently does not represent a novel type of displacement of cell surface constituents, since longer incuba-

Table 1

Effect of Ferritin-conjugated Alloantiserum and Concanavalin A on Cap Formation

Cell surface component	Cell	Primary reagent	% Cells capped
$H-2^b$ alloantigen	C57BL/6 lymph node	anti-$H-2^b$	4
		anti-$H-2^b$/Fl	1
		anti-$H-2^b$/F	12
ConA receptors	Balb/c spleen	ConA	2
		ConA/Fl	2
		ConA/F	14

/Fl designates a fluorescent reagent, /F a ferritin-conjugated reagent. Incubations were for 30 min at 37°C; unconjugated reagents were detected by subsequent incubation with fluorescent rabbit anti-mouse Fc or anti-ConA; ferritin-conjugated reagents were detected with fluorescent rabbit anti-ferritin.

tions of cells with unconjugated or fluorescent anti-H-2^b antibody or ConA (greater than 90 minutes) at 37°C results in comparable levels of capping. Ferritin conjugation therefore presumably accentuates the rate of cap formation by a primary reagent by increasing the amount of cross-linking of cell surface molecules, since the chemical conjugation procedure probably results in coupling of more than one anti-H-2^b alloantibody or ConA molecule to each ferritin molecule (Hämmerling et al. 1973).

Immuno-Electronmicroscopic Labeling with Ferritin-conjugated Alloantibody

Davis et al. (1971) labeled H-2 alloantigens on mouse lymphoid cells with anti-H-2 alloantibody directly conjugated to ferritin and obtained uniformly distributed clusters of label (3–10 ferritin molecules per cluster) over the entire cell surface. We have similarly labeled H-2^b alloantigens with ferritin-conjugated anti-H-2^b alloantibody, and even at 0°C, large clusters or patches of label randomly distributed over the entire cell surface were observed (Fig. 4). The clusters of ferritin are considerably larger than those observed by Davis et al. (1971), and the size of the spaces between the clusters is also greater, suggesting that in these two cases the ferritin-conjugated alloantibodies are pulling alloantigens together into aggregates to different extents. The degree of induced aggregation may depend upon the number of alloantibody molecules attached to each individual ferritin molecule (a greater number of antibody molecules per ferritin might result in greater aggregating efficiency), as well as on the ratio of ferritin-conjugated to unconjugated alloantibody molecules (a large amount of unconjugated antibody might interfere with cross-linking and aggregation of alloantigens by ferritin-conjugated antibody). It

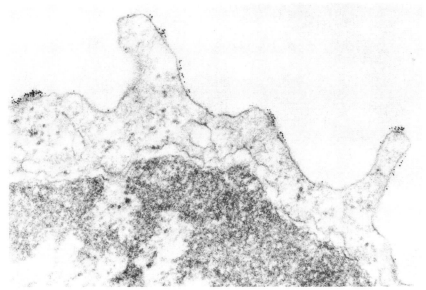

Figure 4
H-2^b alloantigens on a lymph node cell labeled at 0°C with ferritin-conjugated alloantibody. 21,000 ×

is possible, therefore, that the more efficient the conjugation of alloantibody to ferritin, the more displacement and aggregation of alloantigen will occur.

Mechanism of Displacement of Surface Components

Figure 5 depicts schematically a possible sequence of events resulting from labeling cell surface alloantigens with hybrid antibody and ferritin, based on the experiments described here and previous studies (Stackpole et al. 1973). Attachment of alloantibody (Fig. 5A) does not result in significant cross-linking of randomly distributed alloantigens at low temperatures. At elevated temperatures lateral diffusion of membrane constituents enables individual alloantigens to gradually come in contact and become cross-linked and aggregated by the bivalent alloantibody molecules, leading to patch and cap formation. This process occurs more rapidly if alloantibody molecules are multiply bound to ferritin.

The second step in the labeling sequence is the attachment of hybrid antibody to the alloantigen-alloantibody complexes (Fig. 5B). Hybrid antibody induces aggregation of these complexes except at very low temperatures, patch formation at low to moderate temperatures, and capping at higher temperatures. Displacement and aggregation may not be due to molecular cross-linking but may occur because of electrostatic alterations of the labeled components, which normally should repel each other (Singer and Nicolson 1972), conformational changes, or other factors.

The final step in the labeling sequence is the attachment of ferritin (Fig. 5C). Ferritin accentuates the displacement of hybrid antibody-bound complexes, even at very low temperatures, probably because of the tendency for such a large mole-

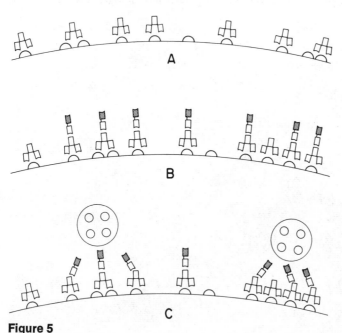

Figure 5
Schematic diagram of alloantibody-hybrid antibody-ferritin labeling sequence. See text for details.

cule to cross-link by acquiring multiple antibody attachments. Factors in addition to cross-linking might also contribute to the effect of ferritin on displacement of surface components, such as alteration of the normal electrostatic charge on these molecules or induction of conformational changes.

The implication from these studies is that whenever ferritin or other large visual markers are used for topographic analysis of cell surface components by electron microscopy, even when coupled directly to a primary reagent, the results must be interpreted with extreme caution because of the possibility of displacement effects.

REFERENCES

Agrawal, B. B. L. and I. J. Goldstein. 1967. Protein-carbohydrate interaction. VI. Isolation of concanavalin A by specific adsorption on cross-linked dextran gels. *Biochim. Biophys. Acta* **147**:262.

Aoki, T., U. Hämmerling, E. de Harven, E. A. Boyse and L. J. Old. 1969. Antigenic structure of cell surfaces: An immunoferritin study of the occurrence and topology of H-2, θ, and TL alloantigens on mouse cells. *J. Exp. Med.* **130**:979.

Davis, W. C. 1972. H-2 antigen on cell membranes: An explanation for the alteration of distribution by indirect labeling techniques. *Science* **175**:1006.

Davis, W. C. and L. Silverman. 1968. Localization of mouse H-2 histocompatibility antigen with ferritin-labeled antibody. *Transplantation* **6**:535.

Davis, W. C., M. A. Alspaugh, J. H. Stimpfling and R. L. Walford. 1971. Cellular surface distribution of transplantation antigens: Discrepancy between direct and indirect labeling techniques. *Tissue Antigens* **1**:89.

dePetris, S. and M. C. Raff. 1972. Distribution of immunoglobulin on the surface of mouse lymphoid cells as determined by immunoferritin electron microscopy. Antibody-induced, temperature-dependent redistribution and its implications for membrane structure. *Eur. J. Immunol.* **2**:523.

Frye, L. D. and M. Edidin. 1970. The rapid intermixing of cell surface antigens after formation of mouse-human heterokaryons. *J. Cell Sci.* **7**:319.

Hämmerling, U., C. W. Stackpole and G. C. Koo. 1973. Hybrid antibodies for labeling cell surface antigens. *Methods in cancer research,* ed. H. Busch. Academic Press, New York. (in press).

Hämmerling, U., T. Aoki, E. de Harven, E. A. Boyse and L. J. Old. 1968. Use of hybrid antibody with anti-γG and anti-ferritin specificities in locating cell surface antigens by electron microscopy. *J. Exp. Med.* **128**:1461.

Hämmerling, U., T. Aoki, H. A. Wood, L. J. Old, E. A. Boyse and E. de Harven. 1969. New visual markers of antibody for electron microscopy. *Nature* **223**:1158.

Karnovsky, M. J., E. R. Unanue and M. Leventhal. 1972. Ligand-induced movement of lymphocyte membrane macromolecules. II. Mapping of surface moieties. *J. Exp. Med.* **136**:907.

Kourilsky, F. M., D. Silvestre, C. Neauport-Sautes, J. Dausset and J. P. Levy. 1971. Ultrastructural localization of HL-A antigens at cell surface. *Transpl. Proc.* **3**:1203.

Lamm, M. E., G. C. Koo, C. W. Stackpole and U. Hämmerling. 1972. Hapten-conjugated antibodies and visual markers used to label cell-surface antigens for electron microscopy: An approach to double labeling. *Proc. Nat. Acad. Sci.* **69**:3732.

Loor, F., L. Forni and B. Pernis. 1972. The dynamic state of the lymphocyte membrane. Factors affecting the distribution and turnover of surface immunoglobulins. *Eur. J. Immunol.* **2**:203.

Nicolson, G. L. and S. J. Singer. 1971. Ferritin-conjugated plant agglutinins as specific saccharide stains for electron microscopy: Application to saccharides bound to cell membranes. *Proc. Nat. Acad. Sci.* **68:**942.

Nicolson, G. L., R. Hyman and S. J. Singer. 1971. The two-dimensional topographical distribution of H2-histocompatibility alloantigens on mouse red blood cell membranes. *J. Cell Biol.* **50:**905.

Singer, S. J. and G. L. Nicolson. 1972. The fluid mosaic model of the structure of cell membranes. *Science* **175:**720.

Singer, S. J. and A. F. Schick. 1961. The properties of specific stains for electron microscopy prepared by the conjugation of antibody molecules with ferritin. *J. Biophys. Biochem. Cytol.* **9:**519.

Stackpole, C. W. 1971. Topography of cell surface antigens. *Transpl. Proc.* **3:**1199.

Stackpole, C. W., T. Aoki, E. A. Boyse, L. J. Old, J. Lumley-Frank and E. de Harven. 1971. Cell surface antigens: Serial sectioning of single cells as an approach to topographical analysis. *Science* **172:**472.

Stackpole, C. W., L. T. DeMilio, U. Hämmerling, J. B. Jacobson and M. P. Lardis. 1973. Hybrid antibody-induced topographical redistribution of surface immunoglobulins, alloantigens and concanavalin A receptors on mouse lymphoid cells. *Proc. Nat. Acad. Sci.* (in press)

Taylor, R. B., P. H. Duffus, M. C. Raff and S. dePetris. 1971. Redistribution and pinocytosis of lymphocyte surface immunoglobulin molecules induced by anti-immunoglobulin antibody. *Nature New Biol.* **233:**225.

Unanue, E. R., W. D. Perkins and M. J. Karnovsky. 1972. Ligand-induced movement of lymphocyte membrane macromolecules. I. Analysis by immunofluorescence and ultrastructural radioautography. *J. Exp. Med.* **136:**885.

Yahara, I. and G. M. Edelman. 1971. Restriction of the mobility of lymphocyte immunoglobulin receptors by concanavalin A. *Proc. Nat. Acad. Sci.* **69:**608.

Design and Function of Site-specific Particle Arrays in the Cell Membrane

Peter Satir and Birgit Satir

Department of Physiology-Anatomy, University of California
Berkeley, California 94720

In 1966 Branton first proposed that the freeze-fracture technique cleaves biological membranes along natural internal planes to expose two complementary membrane fracture faces. This hypothesis, supported by a considerable body of decisive recent evidence (Pinto da Silva and Branton 1970; Tillack and Marchesi 1970), has proven extremely productive. It is now clear that freeze-fracture yields unique information about the organization of the interior of the membrane in a form that is readily visualized.

In model membranes composed of lipid bilayers (Deamer and Branton 1967) and in peripheral nerve myelin, the fracture faces produced by cleavage are smooth and entirely aparticulate. Evidently such membranes preferentially fracture along the middle of the lipid bilayer, where hydrophobic bonds alone hold the opposed monolayers together. The smooth background of the fracture faces of most cell membranes in all likelihood represents regions, similar to the models, where a lipid bilayer predominates. Since such regions are extensive, the overall structure of most membranes studied by freeze-fracture appears to be in accord with the Danielli-Davson (1935) or Robertson (1959) models of unit membrane construction.

Embedded in the smooth background in natural membranes other than myelin are particles of various sizes. Since there are many particles in physiologically active membranes, and since the particles disappear in the fracture face as the membrane is digested with protease (Engstrom 1970), the particles are thought to be protein intercalations into the hydrophobic bilayer, some of which may represent important membrane enzymes.

An unexpected dividend of the freeze-fracture technique has been that the organization of such particles in the fracture face is in many cases not only highly local (site-specific), but also often characteristically related to functional membrane differentiations, such as cell junctions, cilia or secretion granules (Kreutziger 1968; Gilula, Branton and Satir 1970; Gilula and Satir 1972; B. Satir, Schooley and Satir 1972b, 1973).

In our laboratory we have been studying some of the more prominent particle arrays related to the differentiations just mentioned, in terms that may illuminate

233

aspects of membrane structure, growth, fluidity and sometimes function. The present article summarizes our recent efforts.

MATERIAL

We describe particularly, but not exclusively, three different systems that may be taken as models of cell or epithelial systems in general: (1) the lateral cells of the freshwater mussel, *Elliptio complanatus,* gill epithelium, (2) the ciliate protozoan *Tetrahymena pyriformis* and (3) mid- and hindgut cells of the termite, *Zootermopsis nevadensis.* Details of dissection and techniques of preparation of this material for freeze-fracture and electron microscopy are given in our earlier papers (Gilula et al. 1970; Gilula and Satir 1971, 1972; B. Satir et al. 1973; Satir and Fong 1973).

RESULTS

Size, Arrangement, Packing and the Particle Partition Coefficient of Selected Arrays

After freeze-fracture, individual particles may appear on either of the two complementary fracture faces (Fig. 1). Usually more particles adhere to the cytoplasmic (or, by convention, A) face than to the extracellular (or B) face. Most particles are about 100 Å in diameter and appear to be randomly spaced on the membrane. However there are certain regions where particles are absent (exclusion zones) or where they are packed regularly with regard to their neighbors (arrays). Some known arrays are listed in Table 1 and illustrated by electron micrographs in Fig. 2–5. Arrays may occupy a minute portion ($<<1\%$) or be a significant fraction (5% or more) of the cell membrane (Table 2). In an array, particles may separate or be fused into long or short chains. Where the array particles are larger than the background, they usually appear to be comprised of subunits. Single particles may be close-packed, as in the gap junction where particle density is over 100 per 10^4 nm^2, or clearly separated and spaced, as in the septate junction, where there are only 35 or 36 particles in the same area (Table 1). Particles in an array are often, but not always, surrounded by an exclusion zone giving a halo effect, which makes the array more visible against the background. Arrays appear in such forms as rosettes, rings or annuli, rectangular patches, linear rows and belts.

For every particle on one fracture face, in theory there exists a matching depression on the complementary fracture face of the same membrane. This is true where both halves of a single fracture have been recovered, replicated and examined (Chalcroft and Bullivant 1970; Wehrli, Muhlethaler and Moor 1970). For arrays, complementarity is readily confirmed without double replica, since the characteristic spacing of depressions is easily detected (Gilula et al. 1970; B. Satir et al. 1973).

The particle partition coefficient (K_p) may be defined as the ratio of the number of particles adhering to face A to the number adhering to face B in a single array. For localized arrays, composed of a small number of particles, the entire array is counted for the computation of partition coefficient; for large, extensive belt-like arrays with regular packing patterns, a standard area on both fracture faces is

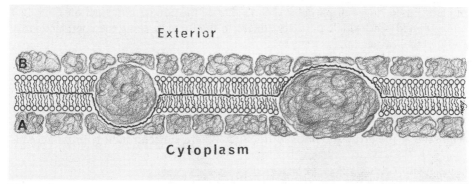

Exterior

B

A

Cytoplasm

Figure 1

Diagram of proposed typical unit membrane structure in a "fluid-mosaic membrane," ~ 75 Å thick. Because of this thickness the hydrophobic center must be coated on either side with a reasonable number of proteins. Mucoprotein or cell coat material that may be intimately associated with some membrane surfaces is not shown. Jagged line indicates freeze-fracture plane through center of lipid bilayer that cleaves membrane into complementary A and B sides. The line deviates around protein intercalations penetrating into the hydrophobic zone that then appear as particles on either fracture face. Regularly packed intercalations form particle arrays. The particle partition coefficient (K_p) compares frequency of deviation of fracture plane to either side of a single particle in an array. From Satir and Gilula (1973) with permission of *Annual Review of Entomology*.

Table 1

Partition Coefficients of Selected Particle Arrays

Array	Organism	Ref.	C_A	C_B	$A + B$	K_p	n
Fusion rosette	*Tetrahymena* (W)	(1)					
Central particle			0.5[a]	0.4	1	1.25	1
Outer ring			8.2[a]	1.1	9	7.5	6
Annulus	*Tetrahymena*	(1)	90[b]	6	96	15.0	12
Ciliary patches	*Tetrahymena* (EO 12)	(2)	13.2[c]	2.1	15	6.3	—
	(H 9-6)		12.6[c]	1.3	14	9.7	—
	(CA 10)		12.3[c]	2.7	15	4.5	—
	Paramecium		11.7[c]	2.7	14	4.3	—
Ciliary necklace	*Tetrahymena* (W)	(1)	6.2[d]	4.5	11	1.25	1
	Elliptio	(3)	4.7[d]	2.6	7	1.8	1
Gap junction (A)	*Elliptio*	(4)	123[e]	3.8	127	32	18
Septate junction (A)	*Elliptio*	(5)	33[e]	2	35	16.5	9
	Zootermopsis	(6)	30.8[e]	4.2	35	7.1	5
Gap junction (B)	*Zootermopsis*	(6)	10[e]	72	82	0.14	10[g]
Continuous junction	*Zootermopsis*	(6)	64[f]	24	88	2.7	2

References: (1) B. Satir et al. 1973. (2) B. Satir and Schooley, unpublished. (3) Gilula and Satir 1972. (4) Gilula and Satir 1971. (5) Gilula et al. 1970. (6) Satir and Fong 1973.

[a] particles per rosette   [b] per annulus   [c] per patch   [d] per strand per doublet   [e] per

10^4 nm² junctional area (6 × 6 array)   [f] per μm ridge length   [g] B side association counted. Fused ridges of particles such as that of the continuous junction (Fig. 2a) can be treated, since single particles adhere to the furrows along the complementary fracture face. In general,

$$K_p = C_A/C_B$$

where C_A is the concentration of particles on face A, and C_B is the concentration on face B. Values of K_p are given in Table 1 for selected arrays.

In Table 1 the column $A + B$ computes the total number of particles within the membrane at the array site prior to fracture. Where concentration is in dimensions

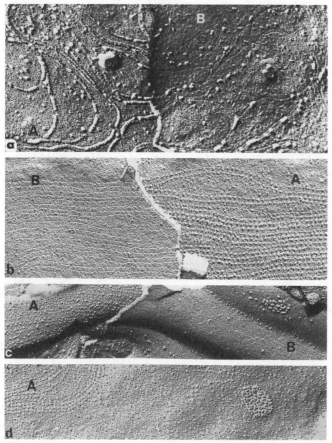

Figure 2

Selected particle arrays: cell junctions.

(a) Continuous junction of termite midgut cells. Furrows with occasional single particles on fracture face B fit together with broken ridges on face A in the complete membrane. 105,000 ×

(b) Septate junction of mussel gill cells. Particles are found spaced in regular rows on fracture face A. The few particles found in line with the matching depressions on face B correspond to occasional particles missing along the rows. 72,000 ×

(c) Septate and gap junctions of termite hindgut cells. Gap junction particles adhere mainly to fracture face B ($K_p < 1$). 75,000 ×

(d) Septate and gap junctions of mussel gill cells. Gap junction particles adhere mainly to fracture face A ($K_p > 1$). 60,000 ×

of particles per array, this is given by adding $C_A + C_B$ and rounding off to the nearest whole integer.

Array Types on One Cell

There may be many different types of arrays found on a single cell. A conspicuous example of comingling of arrays is found on termite hindgut cells where gap junctions are set in the interstices of a conspicuous septate junction (Fig. 2c). The K_p values of these arrays can be markedly different so that the gap junction particles that adhere mainly to fracture face B ($K_p < 1$) are easily discernable, whereas septate junction particles that adhere mainly to face A ($K_p > 1$) are rarely seen (Fig. 2c). Large numbers of the smaller, circumscribed arrays, such as the gap junction and the fusion rosette, are typically present on a single surface. The placement of these is never entirely random and may be highly ordered. In *Tetrahymena* ciliary necklace rows together with fusion rosettes mark 1° meridians, whereas rosette rows alone mark 2° meridians running along the cell cortex. Cortical pattern is apparently anticipated by internal membrane differentiation in such cells.

Exomembrane Relationships

Particle arrays represent those portions found within the hydrophobic layer of the membrane of much larger differentiations that extend into hydrophilic zones far outside of the membrane proper. We have reconstructed the larger pathway in several cases.

The ciliary necklace, for example, forms one part of a "microtubule-membrane complex" (Gilula and Satir 1972) that attaches a ciliary basal body or centriole to the cell membrane. In the cytoplasm underlying the membrane we demonstrate the presence of cuplike extensions that connect the necklace region to the midwall of each microtubule doublet. In the best studied case the number, position, and spacing of these cups per microtubule doublet is identical to that of the necklace strands. We conclude that the ciliary necklace itself is the expression of these cuplike structures within the membrane.

In cell junctions part of the larger pathway is usually extracellular (Satir and Gilula 1973). With the exception of the tight junction, each junctional type is characterized by a differentiation of the intercellular gap in the junctional region that facilitates point by point correspondence of the cell membranes at apposing sides of the gap. These extracellular differentiations range from 20–40 Å high, 90 Å diameter cylinders in gap junctions to the elaborate multimicrometer long, 150 Å high pleated sheets of the septate junction. The particles are that part of a continuous path from the cytoplasm of one cell to an adjacent cell that bridges the hydrophobic interior of each membrane. Although the K_p of many junctions is > 1, there is still exact direct correspondence between particle position and the local structure of the extracellular space. We conclude that there are loci at the surfaces of membranes that correspond to the particle arrays and that may serve as nucleation sites for exomembrane differentiations.

Relation to Cell Histotype and Phylogeny

Since arrays are associated with particular functions or organelle subdivisions of the cell membrane, in a given tissue cells of identical histotype usually have identical arrangements of arrays, but neighboring cells of unlike histotype may also share certain arrays, including, obviously, the cell junctions between them.

The K_p values of similar arrays on cells of widely different species and even phyla are remarkably similar in the cases we have so far examined: for the ciliary necklace in *Tetrahymena* vs. *Elliptio,* for the patches in *Paramecium* vs. *Tetrahymena,* and for the septate junction (type A) in *Elliptio* vs. *Zootermopsis* (Table 1). Especially, the array particle concentrations on the A fracture face (C_A), for which the most data are available, are similar within each pair. The total concentrations of array particles within the membranes ($A + B$) are also similar.

Where arrays are associated with homologous organelles, such similarities may perhaps be anticipated. For example, fusion arrays in both trichocyst and cell membranes of *Paramecium* (Janish 1972; B. Satir, Schooley and Kung 1972) clearly match in type and position those of the *Tetrahymena* mucocyst, the homologous organelle.

However for cell junctions the situation may not be quite so simple in that there may be two distinct classes of fracture for what we identify, by other criteria, as one class of junction. The example of gap junctions A, characteristic of vertebrates and molluscs on one hand (Fig. 2d), vs. B, characteristic of arthropods (Fig. 2c), is best known but not unique. Here the K_p's differ qualitatively (Table 1).

Junctional type appears closely related to phylogeny, and it may be the phylogenetic considerations are paramount for these arrays. For instance, the tight junction is absent in invertebrates so far studied, and septate and continuous junctions are absent in vertebrates (Satir and Gilula 1973).

In general the appearance of even similar arrays of the same K_p in corresponding cells of phylogenetically related organisms is more similar than in those of phylogenetically distant organisms. For example, within limits the number of strands of the ciliary necklace is taxonomically diagnostic for organisms so far studied; two strands for ciliates, three or four for lamellibranch molluscs, and six or more for vertebrates (Gilula and Satir 1972).

Tetrahymena provides us with a good example of an array that appears on different strains of a single cell (B. Satir and Schooley, in prep.). This is found above the ciliary necklace as a rectangular patch of 100 Å diameter particles (Fig. 3). Each patch overlies one microtubular doublet of the cilium. Such patches are absent

---

Figure 3
Selected particle arrays: the ciliary necklace and patches.

(a) *Elliptio* lateral cilium, fracture face A. The necklace is three-stranded with evident scallops. 100,000 ×

(b) *Tetrahymena* (strain W), fracture face A, oral cilia. The cilia are patchless and the necklace is composed of two strands. 108,000 ×

(c) *Paramecium,* fracture face A, somatic cilia. The cilia bear patches as multiple rows of three particles corresponding to one doublet. Note that the number of rows per patch is variable, suggesting that rows are added one by one as the patch develops. Some rows are missing particles that have adhered to the B fracture face. The ciliary necklace is two-stranded. 102,000 ×

(d) *Tetrahymena* (strain E012), somatic cilia. The cilia bear patches and a two-stranded necklace. Both A and B fracture face images of the patches are seen. Although the B fracture face contains virtually no patch particles (at p), many necklace particles are present (arrows). This is a consequence of the difference in particle partition coefficients of the two arrays. 102,000 ×

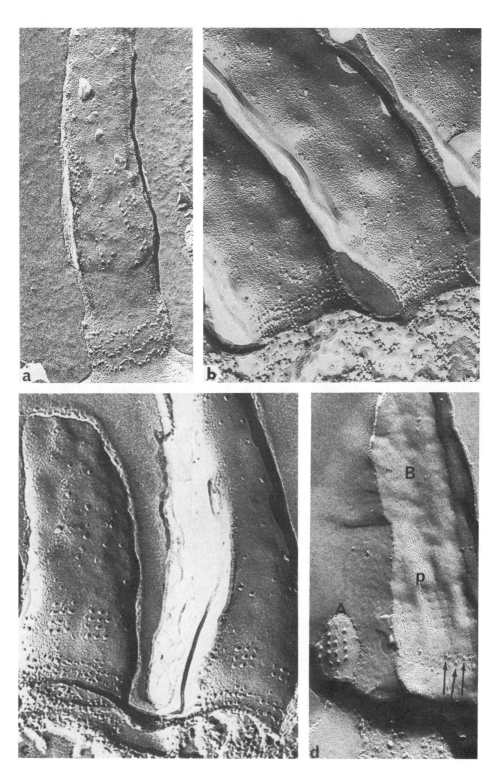

on cilia of strain W (B. Satir et al. 1972a) (Fig. 3b) and strain E (Sattler and Stae-helin 1972), but present in other strains (Fig. 3d and Table 1) and in *Paramecium* as well (Fig. 3c). In contrast the ciliary necklace is found in all strains thus far examined. The K_p of the patches varies over only a twofold range for all strains studied, while C_A and the total numbers of particles per patch essentially remain constant. In *Paramecium* these parameters closely match values for *Tetrahymena* strain CA10. The presence or absence of patches potentially may be a manipu-latable membrane marker that provides a tool for examining genetic control of local membrane structure.

Strain differences as regards the ability to form gap junctions are also present between fibroblast lines and may be responsible for the presence or absence of "metabolic cooperation" between cell lines (Gilula, Reeves and Steinbach 1972).

Array Morphopoiesis

Arrays are not stable throughout the life of the cell and may appear and disappear with membrane changes, which may be physiological or morphogenetic. Three cases of the morphopoiesis of arrays have been studied in our laboratory.

1. In logarithmic growing *Tetrahymena,* the number of somatic cilia increases throughout the cell cycle in linear fashion (Nanney 1971). The appearance of a new cilium is preceded by the differentiation of the ciliary necklace at the site of outgrowth. The necklace particles do not all appear at the same time (B. Satir et al. 1972a, 1973); instead, small rows of particles first appear, and one strand of the necklace assembles before the other; outgrowth of the axoneme follows.

In developing cilia in the mouse oviduct, some strands of the necklace appear before outgrowth of the cilium, but other strands are added later (Davidson, Dirksen and Satir, in prep.). The insertion of particles into a developing array thus seems to be independent of membrane growth. It seems probable that the appear-ance of the necklace coincides with the establishment of the "microtubule-mem-brane" complex.

2. Similarly in cilia of patchy strains of *Tetrahymena* and *Paramecium,* the patches are not always of uniform appearance. In particular the number of rows in the rectangular patch is variable, from 0–6 (rarely more than 6), although the number of columns is constant, always three. Arrays consisting of 3×2, 3×3, 3×4, and 3×5 particles are seen in Fig. 3c and d. The arrays are not always uniform around a single axoneme. We interpret such variation to mean that rows consisting of three particles are added stepwise, one at a time to each patch. Morphopoiesis of the patch continues after ciliary axonemal outgrowth and is in-dependent of morphopoiesis of the ciliary necklace (B. Satir and Schooley, in prep.).

3. In the developing sea urchin blastula, Gilula (1971) has shown that the septate junction arises by stepwise assembly. This array has a $K_p < 1$ and small rows of particles are first seen on the B fracture face. As development proceeds in-dividual rows elongate and simultaneously several rows appear under one another at the same site. Thin sections now reveal variable numbers of septa between ad-jacent cells. Finally, the rows encircle the cell in the traditional septate junction girdle. It is interesting to note that in junctional differentiation, membranes of two adjacent cells must differentiate synchronously on a point to point basis.

Membrane Fluidity, Fusion and Growth

B. Satir et al. (1973) present a detailed case for the importance of particle arrays in a secretory process in *Tetrahymena*. In this case as the membranes fuse, the particles comprising the fusion rosette (Fig. 4), the site of attachment of the secretory vesicle, separate and finally disappear at the fusion lip, while the annulus of particles that accompanies the vesicle spreads out into the plasma membrane (Fig. 5). During the fusion process the distances between array particles are altered in the rosette and later in the annulus. The particles apparently float apart as new membrane pushes up between them. The separation of array particles is an indication of membrane fluidity. The mucocyst undergoes a shape change at the instant of fusion from an elongate sac (Fig. 5a,b) to a sphere (Fig. 5c), where a 60–70% increase in surface area occurs (Table 2). This may be due to membrane stretch at explosion or to nonspecific fusion of small cytoplasmic vesicles with expanding organelles.

The morphology suggests that at fusion the secretory vesicle membrane is incorporated in bulk into the plasma membrane (Fig. 5c). The amount of membrane surface added by one fusion event is given in Table 2. Stereo high voltage electron microscopy of vesicles fixed in different stages of discharge range downwards from this value (B. Satir and Hama 1973). The number of fusion rosettes per somatic cilium approximates ten, six along the 2° meridian, four along the 1° meridian. If all these rosettes represent possible fusion events during one cell cycle, and if the vesicle membrane is not recycled by reverse pinocytosis, then the amount of bulk incorporation from mucocyst fusion would be sufficient to account for approximately one surface doubling per generation. The number of particles incorporated

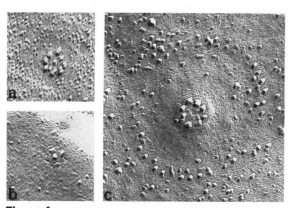

Figure 4

Selected particle arrays: fusion rosettes.

(a) *Tetrahymena*, fracture face A of the plasma membrane. The central particle and 8 peripheral rosette particles are seen. 104,000 ×

(b) *Tetrahymena*, fracture face B of the plasma membrane, showing a second rosette. This time the central particle, together with a single peripheral particle, has adhered to the B fracture face. 104,000 ×

(c) *Paramecium*, fracture face A of the plasma membrane. The rosette corresponds to the location of the trichocyst tip. Note the exclusion zone around the rosette and the incomplete rings of particles circumscribing this zone. 110,000 ×

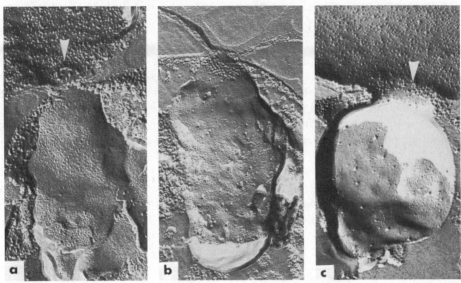

Figure 5

Membrane fusion in *Tetrahymena*.

(a) Relative position of mucocyst and rosette (arrow) prior to fusion. As seen here, fracture face B of the mucocyst membrane bears few annulus particles. 96,000 ×. From B. Satir et al. (1973) with permission of *J. Cell Biology*.

(b) Fracture face A of the mucocyst membrane, showing the annulus prior to fusion. Note the elongate sac shape of the organelle. 72,000 ×. From B. Satir et al. (1972b) with permission of *Nature*.

(c) Fracture face A of the mucocyst and plasma membranes after fusion. The membranes are continuous at the fusion lip which bears the annulus (arrow), now somewhat expanded. The mucocyst is now spherical in shape. 72,000 ×

into the plasma membrane during one fusion event is insufficient to account for the non-array particle background of the plasma membrane.

DISCUSSION

Particle Arrays in Membrane Function

Particle arrays are genetically controlled, site-specific differentiations within the hydrophobic portion of the membrane of animal cells. Present evidence suggests that the particles are protein and that, in an array, particles correspond in a one to one fashion to sites on at least one and probably both membrane surfaces. The arrays are diagnostic for certain organelles or membrane-related functions. In cell junctions they are postulated to provide regular hydrophilic channels for ion and possibly metabolic flow between cells. Such channels require the piercing of the hydrophobic zone and a point by point match of two adjacent membranes across the intercellular gap.

One important membrane differentiation to which specific arrays correspond is the cilium. In ciliated cells the ciliary membrane often comprises over 50% of the entire membrane surface (Table 2). The ciliary necklace (Satir and Gilula 1970;

Table 2

Estimates of Total Cell Membrane Area and
Related Parameters for Two Animal Cells

	Elliptio gill lateral cell	Tetrahymena
length (a)[a]	27	50
width or diameter (b)	3	25
height (h)	12	
basic membrane area (A_b)[b]	882	3332
cilia: number (h_c)	200	450 + 141[c]
length (h_c)	14	8
radius (r_c)	0.125	0.125
total ciliary membrane (A_c)[d]	2198	3711
microvilli: number (n_m)	250	
length (h_m)	1	
radius (r_m)	0.05	
total microvillar membrane (A_m)[e]	78.5	
total cell membrane area ($A = A_b + A_c + A_m$)	3159	7043
mucocyst: number (n_s)		4500[f]
surface area: unexploded (S_u)		1.0
exploded (S_e)		1.6
total mucocyst membrane (A_s)[g]		7200
septate junction height (h_j)	3	
total junction area (A_j)[h]	180	
rosette: surface area (S_r)		0.0028
total rosette surface area (A_r)[i]		12.6

[a] Linear measurements in μm; areas in μm^2.

[b] Lateral cell $A_b = 2$ (ab + ah + bh); *Tetrahymena* $A_b = \dfrac{\pi b^2}{2} + \dfrac{\pi}{2}$ absin^{-1}e/e
where e = 0.86 and sin $^{-1}$ e = 1.03 (oral area simplified for calculation).

[c] Somatic cilia (initial) = 450 (Nanney 1971); oral cilia = 141 (Nilsson and Williams 1966).

[d] $A_c = 2\pi r_c h_c n_c$ [e] $A_m = 2\pi r_m h_m n_m$

[f] Based on B. Satir et al. (1973) Fig. 10 showing 6 fusion rosettes per 1° meridian and 4 per 2° meridian per somatic cilium.

[g] $A_s = S_e n_s = 4500$ μm^2 = .64 A   [h] $A_j = 2h_j$ (a + b)   [i] $A_r = S_r n_s$

Gilula and Satir 1972) is as constant a feature of somatic cilia as the 9 + 2 pattern of microtubules. The number of strands in the necklace is species characteristic. Development precedes outgrowth of the ciliary axoneme but is presumably concomitant with the formation or attachment of a basal body to the cell surface. Such attachment may be important in the regulation of growth vs. differentiation of nonciliated cells by removing one centriole as an initiator of spindle formation, producing instead single short 9 + 0 axonemes. Formation of the necklace appears to be a membrane feature accompanying and perhaps preceding or determining cortical patterning in protozoa. The "microtubule-membrane complex," of which the necklace is a critical portion, has been postulated to play a role in the feedback control of ciliary beat, since microtubule sliding accompanying beat (Satir, 1968, 1972) would distort the necklace pattern, but evidence for such distortion is not yet conclusive.

Unlike the necklace, the particle patches on the ciliary membrane are characteristic of only a few protozoan cilia. Even among different strains of the same protozoan, the patches are not always present. Their function is presently unknown. The patches apparently arise after the formation of the necklace and after outgrowth of the axoneme. Patches and necklace are two different arrays on the same organelle; this is reflected by differences in the K_p value for the respective array particles. Similar differences between these arrays were noted by Speth and Wunderlich (1972).

The most convincing demonstration of the importance of particle arrays in membrane function comes from the discovery of fusion rosettes in *Tetrahymena* (B. Satir et al. 1971, 1973). The rosette is a specific array that marks the initial point of membrane contact and reorganization during mucocyst secretion. The particles of the rosette remain together until fusion occurs, and then they move apart and eventually disappear from the fracture face. The fusion event probably requires competence of both partner membranes since the rosette is matched by an annulus of particles that develop as the mucocyst membrane matures. The annulus may assist in restricting the reorganizational events of membrane fusion to a specific part of the membrane at the fusion lip (Fig. 5c). The mucocyst membrane flows into the plasma membrane at this point. These changes in rosette and annulus arrays during fusion illustrate that membrane fluidity is a crucial aspect of membrane function in the secretory event.

Membrane fluidity has been demonstrated microscopically at the membrane surface by Frye and Edidin (1970), Nicolson (1972) and others and in the fracture face by Pinto da Silva (1972) and by us (B. Satir et al. 1972; Ojakian and Satir, 1973). Because the membrane is fluid, not only individual particles (Pinto da Silva 1972) but also particle arrays in stable associations can float intact in it. In this way the extending ciliary membrane may appear to flow around the formed necklace, mucocyst membrane may flow around the annulus as incorporation proceeds, or the position of a cell junction (Pitelka et al. 1973) can change as the cell changes shape.

Particle Arrays in Membrane Growth

Cell growth is characterized by increase in cell volume. Volume increase could originally be accommodated without corresponding surface increase of the cell membrane by a change in cell shape. At some point, however, a true increase in cell membrane surface area must accompany continued growth. How much of such an increase could be accounted for by stretching of the existing membrane is not known. Gross stretching would seem to require changes in unit membrane dimension and probably does not occur extensively.

True membrane growth requires addition of new lipid and protein moieties to existing structure. For the cell membrane this may be accomplished by two mechanisms: either (1) by addition of single molecules of lipid or protein or small lipoprotein micelles, or (2) by bulk incorporation of preformed vesicles. Of course, such vesicles themselves would be formed by micellar incorporation in, for example, the Golgi region of the cell (Chaplowski and Band 1971; Nozawa and Thompson 1971b; Levine, Higgins and Barnett 1972).

In log phase *Tetrahymena,* as in all dividing cells in nutritionally adequate media, membrane growth evidently accompanies the cell cycle. Nozawa and Thompson (1971a, b) have studied the incorporation of labeled lipids into defined membrane fractions in *Tetrahymena.* They find that it takes perhaps one or two generations of growth for label that originally is incorporated into membranes of the microsomal fraction to find its way into the ciliary membrane. They propose a scheme of micellar incorporation whereby access to the ciliary membrane is very restricted. However an alternative explanation consistent with their data would be that most label is added to the cell membrane by bulk incorporation during mucocyst explosion and that, in the course of about a generation of growth, cell membrane becomes ciliary membrane as new cilia appear. This should be susceptible of experimental resolution in the near future.

In growing multinucleate amebae, Satir and Zeuthen (1961) found that cellular growth rate, measured as increase in reduced weight (RW) with time, depended upon the initial cell surface present at the birth of the ameba. One interpretation of this finding is that, at birth, a given area of cell membrane bears a genetically fixed number of particle arrays, including sites for membrane fusion, which determine overall membrane surface increase and hence growth rate.

It seems likely that bulk incorporation will prove to be one significant mechanism of cell membrane area increase in several animal cells (Chaplowski and Band 1971; Levine et al. 1972) including *Tetrahymena.* It is not likely to be the exclusive mechanism in any case. The number of particles incorporated into the plasma membrane during one fusion event does not seem to be sufficient to account for the non-array particle background of the plasma membrane. Particles within an array can come and go; particularly they are added in micellar fashion. Where do all these particles come from? Three separate possibilities or their combination may be considered: (1) de novo synthesis and micellar insertion of particles into the membrane background at varying points in the cell cycle, (2) conformational changes of surface proteins that are a continuous part of the membrane background, or (3) association and disassociation of existing particles, particularly transformation between background particles and arrays. For example, particle fusion could account for the appearance of larger particles and fused rows, and disassociation could account for disappearance of the rosette particles at the fusion lip.

Significance of Partition Coefficient

The K_p provides a quantitative measure of particle association in an array. For some particles, such as the central particle of the fusion rosette, K_p is approximately unity. We interpret this to mean that the chance of the fracture plane passing to either side of the particle is essentially the same; e.g., there is no special association of the particle with one side of the membrane. However in many arrays K_p differs significantly from unity. There are several possible interpretations of this finding:

1. The single particles in such arrays are placed in physically different positions within the membrane but not specially bonded to one or the other surface. Speth

and Wunderlich (1972) suggest this possibility for the qualitative differences they observe between necklace and patch particles in *Tetrahymena* cilia.

2. Glutaraldehyde causes artifactual linkage of array particles or their surface extensions. Dempsey, Bullivant and Watkins (1973) suggest that such changes occur with fixation in particle association with A and B faces in capillary epithelium.

3. The different K_p's represent different natural strengths of association of the particles, or more probably, their surface extensions to one side at the membrane.

One interesting finding is that for each cell examined here, all K_p's are simple multiples of the one closest to unity. We define the factor relating the K_p's as the association number (n) of the array (Table 1). For any single type of array, as far as has been determined, this factor is constant at least under our conditions which normally include brief glutaraldehyde fixation. We think it likely that the array particles are held together in a large stable complex in the unfixed membrane. Particles seem to be added to an array in a typical assembly pattern. Since the particles are probably protein, there is a good probability that such associations are stabilized by cooperative weak bonds, responsible for quaternary structure in other protein associations. Assuming that the particles are equivalent or quasi-equivalent (Casper 1966) in the array lattice, the association number would depend only upon the number of particles effectively bonded together, as determined by the number and type of weak bonds between single units. Covalent bonds are not preferentially broken by the fracturing process.

Since the particles in an array often appear to be quite separate (center to center spacing ~ 200 Å) rather than close-packed, the associations are not found with the hydrophobic portion of the membrane, but rather in the hydrophilic zones into which the particles project or even in the exomembrane pathway.

This study confirms and extends important aspects of presently accepted models of membrane structure, discussed by Branton and Deamer (1972), Singer and Nicolson (1972), Satir and Gilula (1973) and others, that conveniently can be labeled the "fluid mosaic model" of the cell membrane (Singer and Nicolson 1972). The particle arrays provide important and incontrovertible evidence for mosaicism of the membrane. As we have shown here, this mosaicism has implications in terms of physiological function as well as in the control of organelle design and region differentiation.

SUMMARY

In freeze-fracture studies of animal cells, cell membranes cleave along an internal hydrophobic layer to reveal localized, site-specific particle arrays on complementary fracture faces, where individual particles adhere to either the cytoplasmic (A) or the extracellular (B) membrane face. Together with particle size and packing parameters, the ratio of particles on faces A/B, designated the particle partition coefficient (K_p), is diagnostic for each array. Corresponding cells of similar histotype have diagnostically similar arrays; the K_p values for one type of array in different species are usually similar. The arrays represent those portions found within the membrane of much larger differentiations, such as cell junctions, cilia and secretion granules, and they may function in intercellular communication, sensory transduction or membrane fusion by providing part of a structural pathway through the membrane.

Several different arrays with K_p varying from ~ 0.2–30 may be found within a few square microns on a single membrane. In protozoa such placement corresponds to the cortical pattern; for example, in *Tetrahymena* ciliary necklace rows mark 1° meridians and fusion rosette rows mark 2° meridians. Arrays assemble within the membrane in a stepwise fashion by accumulation of particles, indicating point by point membrane differentiation. Changes in particle associations within arrays reflect dynamic processes within the membrane, including membrane fluidity and growth.

Acknowledgments

We expressly thank Caroline Schooley of the Electron Microscope Laboratory who collaborated with us in a significant part of the work presented here. We are grateful to Drs. C. Kung and E. Orias for stock cultures of *Paramecium* and *Tetrahymena*. We thank Ivy Hsieh and Don Pardoe for technical and photographic assistance. This work was supported by USPHS grant HL 13849.

A portion of this work was completed at the Zoological Institute and the Department of Fine Morphology, University of Tokyo, Japan and was supported by the NSF US-Japan Cooperative Program and fellowships from the John Simon Guggenheim Foundation and the USPHS (GM 54452).

REFERENCES

Branton, D. 1966. Fracture faces of frozen membranes. *Proc. Nat. Acad. Sci.* **55**:1048.

Branton, D. and D. Deamer. 1972. Membrane structure. *Protoplasmatologia Wien.* Springer-Verlag, New York.

Casper, D. H. D. 1966. Design principles in organized biological structures. *Principles of biomolecular organization* (ed. G. E. W. Wolstenholme) pp. 7–34. Little Brown, Boston.

Chalcroft, J. P. and S. Bullivant. 1970. An intepretation of liver cell membrane and junction structure based on observation of freeze-fracture replicas of both sides of the membrane. *J. Cell Biol.* **47**:49.

Chalplowski, F. J. and R. N. Band. 1971. Assembly of lipids into membranes of *Acanthameba palestinensis*. II. The origin and fate of glycerol ^{3}H-labeled phospholipids of cellular membranes. *J. Cell Biol.* **50**:634.

Danielli, J. R. and H. A. Davson. A contribution to the theory of permeability of thin films. *J. Cell. Physiol.* **5**:495.

Deamer, D. and D. Branton. 1967. Fracture planes in an ice-layer model membrane system. *Science* **158**:655.

Dempsey, G. P., S. Bullivant and W. B. Watkins. 1973. Endothelial cell membranes: Polarity of particles as seen by freeze-fracturing. *Science* **179**:190.

Engstrom, L. H. 1970. Structure in the erythrocyte membrane. Ph.D. thesis, Univ. of Calif., Berkeley.

Frye, L. D. and M. Edidin. 1970. The rapid intermixing of cell surface antigens after formation of mouse-human heterokaryons. *J. Cell Sci.* **7**:319.

Gilula, N. B. 1971. Studies on the septate junction. Ph.D. thesis, Univ. of Calif., Berkeley.

Gilula, N. B. and P. Satir. 1971. Septate and gap junctions in molluscan gill epithelium. *J. Cell Biol.* **53**:869.

————. 1972. The ciliary necklace: A ciliary membrane specialization. *J. Cell Biol.* **53**:494.

Gilula, N. B., D. Branton and P. Satir. 1970. The septate junction: A structural basis for intercellular coupling. *Proc. Nat. Acad. Sci.* **67**:213.

Gilula, N. B., O. R. Reeves and A. Steinbach. 1972. Intercellular communication: Metabolic coupling, ionic coupling and cell contacts. *Nature* **235**:262.

Janish, R. 1972. Pellicle of *Paramecium caudatum* as revealed by freeze-etching. *J. Protozool.* **19**:470.

Kreutziger, G. O. 1968. Freeze-etching of intercellular junctions of mouse liver. *Proc. Electron Microscope Soc. Amer.*, 26th Meeting, p. 234.

Levine, A. M., J. A. Higgins and R. J. Barnett. 1972. Biogenesis of plasma membranes in salt glands of salt-stressed domestic ducklings: Localization of acyltransferase activity. *J. Cell Sci.* **11**:855.

Nanney, D. L. 1971. The pattern of replication of cortical units in *Tetrahymena*. *Develop. Biol.* **26**:296.

Nicolson, G. L. 1972. Topography of membrane concanavalin A sites modified by proteolysis. *Nature New Biol.* **239**:193.

Nilsson, J. R. and N. E. Williams. 1966. An electron microscope study of the oral apparatus of *Tetrahymena pyriformis*. *Compt. Rend. Trav. Lab. Carlsberg* **35**:119.

Nozawa, Y. and G. A. Thompson. 1971a. Studies of membrane formation in *Tetrahymena pyriformis*. II. Isolation and lipid analysis of cell fractions. *J. Cell Biol.* **49**:712.

————. 1971b. Studies of membrane formation in *Tetrahymena pyriformis*. III. Lipid incorporation into various cellular membranes of logarithmic phase cells. *J. Cell Biol.* **49**:722.

Ojakian, G. K. and P. Satir. 1973. Particle movements during stacking of *Chlamydomonas* chloroplast membranes. *J. Cell Biol.* (in press)

Pinto da Silva, P. 1972. Translational mobility of the membrane intercalated particles of human erythrocyte ghosts: pH-dependent, reversible aggregation. *J. Cell Biol.* **53**:777.

Pinto da Silva, P. and D. Branton. 1970. Membrane splitting in freeze-etching. *J. Cell Biol.* **45**:598.

Pitelka, D. R., S. T. Hamamoto, J. G. Duafala and M. K. Nemanic. 1973. Cell contacts in the mouse mammary gland. I. Normal gland in postnatal development and the secretory cycle. *J. Cell Biol.* **56**:797.

Robertson, J. D. 1959. The ultrastructure of cell membranes and their derivatives. *Biochem. Soc. Symp.* **16**:3.

Satir, B. and K. Hama. 1973. High voltage electron microscopy of *Tetrahymena* membranes (Abstract). Proc. 3rd Int. Congress High Voltage Electron Microscopy, Oxford.

Satir, B., C. Schooley and C. Kung. 1972. Internal membrane specializations in *Paramecium aurelia*. *J. Cell Biol.* **55**:227a.

Satir, B., C. Schooley and P. Satir. 1971. Membrane specializations and reorganization during secretion in *Tetrahymena*. Abstracts, 11th Ann. Meet. Amer. Soc. Cell Biol., New Orleans. p. 260.

————. 1972a. The ciliary necklace in *Tetrahymena*. *Acta Protozool.* **11**:291.

————. 1972b. Membrane reorganization during secretion in *Tetrahymena*. *Nature* **235**:53.

————. 1973. Membrane fusion in a model system: Mucocyst secretion in *Tetrahymena*. *J. Cell Biol.* **56**:153.

Satir, P. 1968. Studies on cilia. III. Further studies on the cilium tip and a "sliding filament" model of ciliary motility. *J. Cell Biol.* **39**:77.

————. 1972. The sliding microtubule hypothesis of ciliary motion. *Acta Protozool.* **11**:279.

Satir, P. and I. Fong. 1973. Cell junctions in insects. 25th Symp. Jap. Soc. Cell Biol. (in press).

Satir, P. and N. B. Gilula. 1970. Freeze etch of cilia. *J. Cell Biol.* **47**:149a.

————. 1973. The fine structure of membrane and intercellular communication in insects. *Annu. Rev. Entomol.* **18**:143.

Satir, P. and E. Zeuthen. 1961. Cell cycle and the relationship of growth cycle to reduced weight (RW) in the giant ameba *Chaos chaos* L. *Compt. Rend. Lab. Carlsberg* **32**:241.

Sattler, C. A. and L. A. Staehelin. 1972. Ciliary and plasma membrane differentiation of *Tetrahymena. J. Cell Biol.* **55**:228a.

Singer, S. J. and G. L. Nicolson. 1972. The fluid mosaic model of the structure of cell membranes. *Science* **175**:720.

Speth, V. and F. Wunderlich. 1972. Evidence for different dispositions of particles associated with freeze-etched membranes. *Protoplasma* **75**:341.

Tillack, T. W. and V. T. Marchesi. 1970. Demonstration of the outer surface of freeze etched red blood cell membranes. *J. Cell Biol.* **45**:649.

Wehrli, E., K. Muhlethaler and H. Moor. 1970. Membrane structure as seen with a double replica method for freeze fracturing. *Exp. Cell Res.* **59**:336.

Factors Influencing the Dynamic Display of Lectin Binding Sites on Normal and Transformed Cell Surfaces

Garth L. Nicolson

The Salk Institute for Biological Studies
San Diego, California 92112

Normal and transformed cells are characterized by their growth properties in culture; the transformed cells show much less density-dependent inhibition. This property appears to be mediated through the cell plasma membrane, and it has been proposed that this is the site of modification after transformation that accounts for tumorigenicity as well as altered cell growth behavior (Pardee 1971; Holley 1972).

In their studies on the structural differences between the surface membranes of normal and transformed cells, Aub et al. (1963, 1965) found that a wheat germ (*Triticum vulgaris*) preparation agglutinated certain transformed cells but not their normal counterparts. Burger and Goldberg (1967) purified the agglutinating activity and characterized the responsible molecule as a plant lectin, wheat germ agglutinin, that bound to cell surface N-acetyl-D-glucosamine-like residues. To explain the difference in lectin-mediated agglutination between normal and transformed cells, Burger (1969) advanced the possibility that the agglutinability of transformed cells was due to an exposure of "cryptic" lectin binding sites during transformation. The sites were reasoned to be cryptic or masked on normal cells because brief proteolysis rendered normal cells as agglutinable as transformed cells but did not enhance the agglutinability of transformed cells (Burger 1969).

Since that time several alternate theories have been proposed to explain differential agglutinability by lectins and to account for the subsequent finding that normal, protease-treated, and transformed cells have equivalent numbers of lectin sites per cell* (Cline and Livingston 1971; Ozanne and Sambrook 1971; Inbar, Ben-Bassat and Sachs 1972; Arndt-Jovin and Berg 1971; Sela et al. 1971; Nicolson 1973c). Inbar et al. (1972) proposed that cell agglutination was due to special metabolically linked lectin agglutination sites (distinct from lectin binding sites) which are altered on the transformed cell surface. Simultaneously we proposed that

---

* One recent exception is the finding that ^{3}H-labeled ConA reveals a greater number of ConA binding sites per unit surface area on polyoma-transformed 3T3 compared to 3T3 cells (Noonan et al. 1973). On the basis of cell surface area, SV40-transformed 3T3 cells (which are smaller than 3T3 cells) have more ConA binding sites than 3T3 cells, although the number of ConA binding sites per cell is approximately equivalent (Ozanne and Sambrook 1971; Arndt-Jovin and Berg 1971).

the difference in agglutinability might be due to a clustering in the topographic distribution of lectin binding sites that allows multiple lectin cross-bridging between adjacent cells (Nicolson 1971; Singer and Nicolson 1972).

Cell agglutination is probably determined by a complex relationship between the agglutinating molecules' structure, their number and surface distribution, and several cell surface properties such as charge repulsive forces, etc. (Nicolson 1972). I will discuss these parameters and others (cell surface rigidity, cellular processes, mobility of lectin receptors, etc.) and their relationships to lectin-mediated cell agglutination here. Two plant lectins purified by affinity chromatography were used for the experimental portion of this report: concanavalin A (ConA), specifically inhibited by α-D-mannopyranosyl-like saccharides (Agrawal and Goldstein 1967; So and Goldstein 1967), and *Ricinus communis* agglutinin of 120,000 mol. wt. (RCA), specifically inhibited by β-D-galactopyranosyl-like saccharides (Drysdale et al. 1968; Nicolson and Blaustein 1972).

Lectin-mediated Cell Agglutinability

A wide variety of plant lectins have been isolated and characterized as having "tumor-specific" agglutinating properties (reviews: Burger 1970; Sharon and Lis 1972); however cell agglutinability by lectins is not restricted to transformed cells. But it does appear that several types of transformed cells are more readily agglutinated than their normal counterparts at equivalent lectin concentrations. These lectins include ConA (*Canavalia enisformis*) (Inbar and Sachs 1969a,b), soy bean (*Glycine max*) agglutinin (Sela et al. 1970), pea (*Pisum sativum*) agglutinin (Vesely et al. 1972), castor bean (*Ricinus communis*) agglutinin (Nicolson and Blaustein 1972) and others (Tomita et al. 1970; Borek et al. 1973). Incidentally not all types of transformed cells show these differential agglutination properties (Aub et al. 1965; Liske and Franks 1968; Gantt et al. 1969). Several types of normal cells are readily agglutinated: erythrocytes (review: Mäkela 1957), lymphocytes (Aub et al. 1965; Liske and Franks 1968), sperm (Nicolson and Yanagimachi 1972; Uhlenbruck and Herrmann 1972), embryonic cells (Moscona 1971; Sivak 1971), and nononcogenic virus-infected cells (Becht et al. 1972; Poste and Reeve 1972; Tevethia et al. 1972).

Normal cells can be rendered more agglutinable after brief treatment with low concentrations of proteolytic enzymes. Burger (1969) and others found that trypsin and other proteolytic enzymes render normal cells as agglutinable as transformed cells. Since the increased lectin agglutinability of trypsinized and transformed cells is not related to the total number of lectin binding sites (Cline and Livingston 1971; Ozanne and Sambrook 1971; Arndt-Jovin and Berg 1971; Sela et al. 1971; Inbar et al. 1972; Nicolson 1973c), it is doubtful that cryptic sites are unmasked by transformation or protease treatment.

Lectin-mediated agglutinability appears to change during cell growth in culture. Goto et al. (1972) noticed that murine 3T6 cells, which do not possess the property of density-dependent inhibition, are less agglutinable with ConA when cells are grown to confluency compared to sparsely growing cells. This also appears to be true for SV3T3 cells and to a lesser extent with normal 3T3 cells (Table 1). ConA- or RCA-mediated cell agglutinability decreases at confluency or at the point where the cells make contact in culture (Tables 1 and 2).

Table 1
Cell Density-dependent Agglutination of Normal and SV40-transformed Fibroblasts by Concanavalin A

	Saccharide inhibitor*	Concentration of concanavalin A ($\mu g/ml$)[†]								
		1200	600	300	150	75	30	15	8	4
Sparse 3T3		++	+	±	0	0	0	0	0	0
Sparse 3T3	α-MM	0	0	0	0	0	0	0	0	0
Confluent 3T3		±	0	0	0	0	0	0	0	0
Confluent 3T3	α-MM	0	0	0	0	0	0	0	0	0
Sparse SV3T3		++++	++++	++++	++++	+++	+++	++	++	+
Sparse SV3T3	α-MM	0	0	0	0	0	0	0	0	0
Confluent SV3T3		++++	++++	++++	+++	++	++	+	0	0
Confluent SV3T3	α-MM	0	0	0	0	0	0	0	0	0

* 0.05 M α-methyl-D-mannoside

† Agglutination scored after 20 minutes at room temperature by the method of Sela et al. 1970. For methods see Nicolson and Lacorbiere 1973.

Table 2
Cell Density-dependent Agglutination of Normal and SV40-transformed Fibroblasts by *Ricinus communis* Agglutinin

	Saccharide inhibitor*	Concentration of Ricinus communis agglutinin ($\mu g/ml$)[†]								
		15	7.5	3.7	1.8	0.9	0.45	0.22	0.11	0.05
Sparse 3T3		++	++	+	±	0	0	0	0	0
Sparse 3T3	β-lac	0	0	0	0	0	0	0	0	0
Confluent 3T3		++	+	0	0	0	0	0	0	0
Confluent 3T3	β-lac	0	0	0	0	0	0	0	0	0
Sparse SV3T3		++++	++++	++++	++++	+++	++	++	+	+
Sparse SV3T3	β-lac	0	0	0	0	0	0	0	0	0
Confluent SV3T3		++++	+++	++	++	+	0	0	0	0
Confluent SV3T3	β-lac	0	0	0	0	0	0	0	0	0

Data from Nicolson and Lacorbiere 1973

* 0.1 M β-lactose

† Agglutination scored after 20 minutes at room temperature by the method of Sela et al. 1970. For methods see Nicolson and Lacorbiere 1973.

Cell Density-dependent Changes in Lectin Binding Sites

Cell-to-cell contact of normal cells in culture results in distinct biochemical alterations in the cell surfaces that apparently do not occur after transformation. Certain glycolipids, such as galactosylgalactosylglucosyl-, N-acetylneuraminosylgalactosylglycosyl- and N-acetylneuraminosyl (N-acetylneuraminosyl) galactosylglucosylceramide, increase on cell-to-cell contact in normal hamster BHK cells, but not on transformed BHK cells (Hakomori 1970; Robbins and Macpherson 1971). Recently Forssman and hematoside glycolipids were found to increase at the early stages of cell contact in hamster NIL cells (Kijimoto and Hakomori 1972). These findings originally led Hakomori (1970) to propose that on normal cells, but not their transformed counterparts, certain glycolipids are "extended" or have certain specific terminal saccharides added at cell contact. Roth and White (1972) found that 3T3 fibroblasts modify their surfaces at contact by increasing the rate of transfer of D-galactose from externally added UDP-D-galactose to cell surface galactose acceptors, possibly via galactosyltransferases acting on acceptors present on adjacent cell membranes. When spontaneously transformed 3T12 cells were examined, they were readily able to transfer D-galactose independent of cell contact, presumably to acceptors present on the same membrane. They also found that sparse 3T3 cells, removed from substratum and suspended in fresh media for 3.5 hours, transfer approximately one-fourth as much galactose to cell surface acceptors compared to the same cells present in a loose pellet. This was interpreted as evidence for a difference in accessibility between the galactose acceptors on normal versus transformed cells, the normal cell galactose acceptors becoming accessible when cells reached confluency or made contact. However it remains to be proven that adequate supplies of UDP-D-galactose are excreted by cells in situ for cell-to-cell glycosylation to occur. Also in 3.5 hours at 37°C a large portion of the plasma membrane can be replaced.

In order to relate the number of lectin binding saccharide sites to the presence or absence of cell contact, we examined the quantities of ^{125}I-labeled lectins that bound to normal or transformed cells grown to various cell culture densities during a 10 minute incubation at 4°C. Both transformed cell lines examined (SV3T3 and 3T12) bound amounts of [^{125}I]ConA or [^{125}I]RCA that were independent of culture density or cell-to-cell contact (Table 3). When 3T3 cells were examined, there was a slight increase in ConA binding sites and a 2.5- to 3-fold increase in RCA binding sites at cell-to-cell contact. After cell contact 3T3 and SV3T3 cells possessed equal numbers of RCA sites, whereas 3T12 cells had twice the RCA sites compared to 3T3 cells (Table 3).

The increase in RCA binding sites under all culture conditions was associated with cell-to-cell contact. When intercellar contacts began to appear, or when the cell cultures were confluent, the number of RCA binding sites increased. Because various serum concentrations or plating conditions did not change the results, this increase in RCA binding sites was not due to serum factor exhaustion (Nicolson and Lacorbiere 1973). These contact-dependent changes in RCA binding sites on 3T3 cells are consistent with the observations of D-galactose transfer via surface galactosyltransferases (Roth and White 1972), but alternate routes of arriving at this result, such as plasma membrane turnover, are also possible.

Table 3
Cell Density-dependent Binding of ^{125}I-labeled Agglutinins to Normal and Transformed Fibroblasts

Cell line	Cell density ($\times 10^{-5} \; cm^{-2}$)	Cell contact	Lectin	Specific binding per 10^6 cells (cpm)*	Ratio cpm confluent to sparse cells†	P§
3T3	0.09	−	RCA	18,900 ± 2400		
	1.20	+	RCA	50,100 ± 4100	2.65	<0.001
		−	ConA	26,300 ± 2600		
		+	ConA	28,100 ± 3100	1.07	NS¶
SV3T3	0.10	−	RCA	58,700 ± 3300		
	2.50	+	RCA	64,300 ± 4100	1.09	NS¶
		−	ConA	27,200 ± 3400		
		+	ConA	28,700 ± 3100	1.05	NS¶
3T12	0.03	−	RCA	108,600 ± 6900		
	2.10	+	RCA	118,600 ± 5800	1.09	NS¶

Portions of the data from Nicolson and Lacorbiere 1973. For methods see Nicolson 1973.

* ± Standard deviation. Control samples incubated in the presence of saccharide inhibitor averaged 8–12% (RCA) or 10–20% (ConA) of raw data. All incubations were for 10 minutes at 4°C. The samples were then washed twice and the cell pellets counted in a gamma scintillation counter.

† Ratio of mean cpm ^{125}I-lectin bound to 10^6 touching cells compared to 10^6 sparse cells.

§ Significance (t) test for comparison of two means.

¶ Not significant.

Clustering of Lectin Binding Sites

The fact that (a) confluent 3T3 cells are less agglutinable than their SV40-transformed derivatives while there are only slight differences in their total numbers of lectin binding sites and (b) the agglutinability of 3T3 cells decreases slightly at confluency, although certain lectin binding sites (RCA) increase in number, suggested that the surface distribution of these sites might be an important factor in determining cell agglutination (Nicolson 1971). Experimental evidence for a topographic change in the distribution of ConA binding sites was first obtained utilizing ferritin-conjugated ConA (Nicolson and Singer 1971) labeling to isolated plasma membrane preparations of transformed and normal fibroblasts at room temperature (Nicolson 1971). The ConA binding sites on mounted SV3T3 cell plasma membranes were found to be more clustered when compared to the same sites on 3T3 membrane surfaces. Subsequently Martinez-Palomo et al. (1972) and Bretten et al. (1972) found that ConA labeling of cells in situ, followed by peroxidase binding to surface-bound ConA, resulted in a more patchy or clustered distribution of the ConA peroxidase product on transformed cells compared to normal cells.

Evidence for the involvement of the clustered ConA binding sites in cell agglutination was demonstrated with trypsinized normal cells, which are as agglutinable as transformed cells. Trypsinized 3T3 cells were agglutinated with ferritin-ConA; as the cells agglutinated into small cell aggregates, they were carefully removed by micromanipulation, washed and fixed with glutaraldehyde. After preparation for conventional embedding and thin sectioning through the cell aggregates, the ferritin-ConA was found in clusters at the sites of contact between the agglutinated cells (Nicolson 1972). Similarly ferritin-ConA agglutination of SV3T3 cells (Fig. 1) resulted in ferritin-ConA clusters at the sites of cell contacts (Fig. 2).

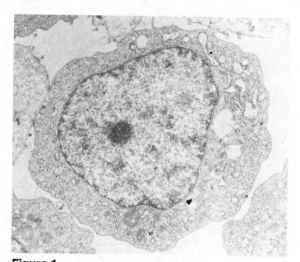

Figure 1

Agglutination of SV40-transformed 3T3 cells with ferritin-conjugated concanavalin A. The cell aggregates were removed by micromanipulation, washed and fixed in 1.5% buffered glutaraldehyde followed by post-fixation in 1% buffered osmium tetroxide. The fixed cell aggregates were dehydrated and embedded. Section through an aggregate stained with uranyl acetate. × 8000.

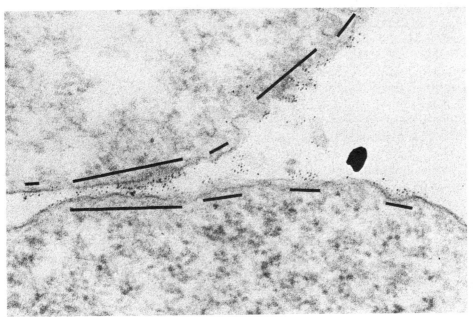

Figure 2
Sites of cell contact in an aggregate of SV3T3 cells agglutinated with ferritin-conjugated concanavalin A. The concentration of ferritin between the agglutinated cells is higher than on the surrounding membrane regions. × 82,000.

Addition of the inhibitor α-methyl-D-mannoside to the cell aggregates a short time after they had formed easily dispersed them, indicating specific agglutination by the ferritin-ConA reagent. That the ferritin-ConA was trapped nonspecifically between the cells during agglutination was ruled out by agglutinating the SV3T3 cells with a mixture of RCA, ferritin-ConA and its inhibitor α-methyl-D-mannoside. No effect on RCA-mediated agglutination occurred; but the ferritin-ConA was blocked from binding to cell surfaces by its specific inhibitor and was not trapped between the agglutinated cells. This result supports the hypothesis that lectin-mediated agglutination occurs directly through the formation of multiple lectin cross-bridges and not via some indirect mechanism that does not directly involve lectin molecules (Nicolson 1972).

It has been shown that cell surface antigens are capable of lateral diffusion in the membrane (Frye and Edidin 1970; Taylor et al. 1971; Davis 1972; Edidin and Weiss 1972; de Petris and Raff 1972; Kourilsky et al. 1972; Karnovsky et al. 1972; Unanue et al. 1972). Using fluorescent lectin and lectin-anti-lectin indirect techniques, Comoglio and Guglielmone (1972) reported that ConA binding sites appear to be uniformly distributed on cell surfaces at low temperatures, but quickly migrate into patches or clusters upon warming the cell. Three groups have recently examined the distribution of ConA binding sites at low temperature or on aldehyde-fixed normal and transformed cells and have found that the ConA binding sites are inherently dispersed on these cells, but are capable of rearranging into patches or clusters when the unfixed ConA-labeled cells are warmed (Inbar et al. 1973; Nicolson 1973d; Rosenblith et al. 1973). These observations (using ConA reagents) seem to be true for other lectin binding sites as well. Aldehyde fixation of

SV3T3 cells reduces RCA-mediated agglutinability without reducing the total number of RCA binding sites (G.L. Nicolson unpublished) and prevents rearrangement of RCA binding sites into clusters (Fig. 3). SV3T3 cells incubated with 1–2 μg/ml RCA at 0°C, then fixed with formaldehyde and indirectly labeled with fluorescent anti-RCA show uniform "ring" fluorescence (Fig. 3a), indicating that a dispersed distribution of RCA binding sites probably exists on these cells before RCA labeling. If the RCA-labeled cells are warmed to 37°C for 10–15 minutes before aldehyde fixation and labeling with fluorescent anti-RCA, the cells are discontinuously fluorescent with patches or clusters visible (Fig. 3b). Labeling was specific since substitution of ConA for RCA and then labeling with fluorescent anti-RCA produced little cell fluorescence (Fig. 3c). 3T3 cells treated in a similar fashion do not show patchy distributions of RCA binding sites unless the aldehyde fixation step is omitted and an indirect fluorescent anti-RCA label allowed to cluster the membrane-bound RCA (Table 4). Thus the lectin binding sites on normal 3T3 cells are capable of being clustered under appropriate conditions, but are not clustered by the simple attachment of RCA to the cell surface. These findings and those of others (Inbar et al. 1973; Nicolson 1973d; Rosenblith et al. 1973) indicate that lectin binding sites are apparently more mobile on transformed

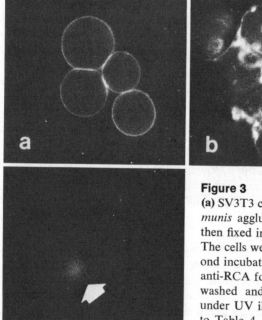
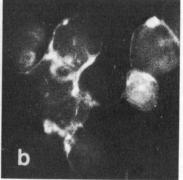

Figure 3
(a) SV3T3 cells were treated with 1 μg/ml *R. communis* agglutinin at 0°C for 15 min, washed and then fixed in 2% formaldehyde (first incubation). The cells were incubated for 10 min at 37°C (second incubation) and then labeled with fluorescent anti-RCA for 60 min at 0° (third incubation) and washed and examined with a light microscope under UV illumination. For details see the legend to Table 4. The cells show uniform or ring fluorescence. (b) Same as in (a) except that the cells were not fixed in 2% formaldehyde until after the second incubation. The cells show patchy or discontinuous fluorescence. (c) Same as in (a) except that concanavalin A (20 μg/ml) was substituted for RCA. Arrow indicates some cell autofluorescence. × 2600.

Table 4

Distribution of *Ricinus communis* Binding Sites on Normal and SV40-transformed Fibroblast Surfaces by Indirect Immunofluorescence

Cell line	First incubation (R. communis agglutinin)				Second incubation				Third incubation (Fl-anti-R. communis agglutinin)		Distribution of fluorescence
	prior fixation	temp (°C)	time (min)	concen (μg/ml)	prior fixation	temp (°C)	time (min)	subsequent fixation	temp (°C)	time (min)	
3T3	–	0	15–30	1–2	–	0	10	+	0	60	Uniform
	+	0	15–30	1–2	–	37	10	+	0	60	Uniform
	–	0	15–30	1–2	–	37	10	+	0	60	Uniform
	–	0	15–30	1–2	–	0	10	–	20	15	Patchy
	–	0	15–30	50	–	0	10	+	0	60	Uniform
	–	0	15–30	50	–	37	10	+	0	60	Uniform
SV3T3	–	0	15–30	1–2	–	0	10	+	0	60	Uniform
	+	0	15–30	1–2	–	37	10	+	0	60	Uniform
	–	0	15–30	1–2	–	37	10	+	0	60	Patchy
	–	0	15–30	1–2	+	37	10	–	0	60	Uniform
	–	0	15–30	50	–	0	10	+	0	60	Uniform
	–	0	15–30	50	–	37	10	+	0	60	Uniform/Patchy

Formaldehyde-fixed (2%, 15 min at 20°C) or unfixed cells were treated with *R. communis* agglutinin for 15–30 min at 0°C (first incubation). Some of the unfixed cells were fixed in 3% buffered formaldehyde for 15 min. The cells were washed once and incubated at 0 or 37°C (second incubation). Most of the cells were fixed in 2% buffered formaldehyde for 15 min. Cells were treated with fluorescent-labeled anti-*R. communis* agglutinin γG (100 μg/ml) for 15–60 min (third incubation). The cells were washed and examined in a Leitz Ortholux microscope using UV illumination with UG-1 excitation and K-430 barrier filters.

cell surfaces than on the plasma membranes of normal cells from which they were derived (see also Edidin and Weiss this volume).

At high RCA labeling concentrations (50 μg/ml), both 3T3 and SV3T3 cells are agglutinated and both types of cells show uniform fluorescence when treated with fluorescent anti-RCA (Table 4). This finding may be similar to those of Yahara and Edelman (1972) where high concentrations of ConA prevented clustering of Ig determinants on lymphoid cells induced by anti-Ig. At high RCA concentrations (50 μg/ml), approximately 70–80% of the RCA binding sites are occupied by [125I]RCA compared to less than 5–10% at low RCA concentrations (1–2 μg/ml). This is interpreted to indicate that high surface densities of bound RCA exist when cells are treated with RCA concentrations above that required for agglutination, but below complete saturation. When low enough concentrations of RCA are used so that only a few of the total RCA binding sites are occupied, lateral rearrangements may be necessary to increase local densities of bound RCA so that multiple cross-bridges can form between adjacent cells. Thus lectin-induced RCA binding site clustering, per se, is not responsible for cell agglutination; rather it probably enhances cell agglutination at low lectin concentrations by presenting proper local densities of membrane-bound lectin molecules to adjacent cells. At high lectin concentrations where most of RCA binding sites are occupied, the average density of surface-bound RCA is probably adequate for agglutination without the necessity of RCA binding site clustering (Fig. 4).

Lectin Binding Site Mobility in a Fluid Membrane Environment

The surface membranes of mammalian cells have been presented as fluid, dynamic structures (Fluid Mosaic Membrane Model [Singer and Nicolson 1972]). In this model the fluidity of the membrane allows lateral diffusion of certain membrane components. For example, phospholipids in the membrane are thought to be in rapid lateral diffusion (e.g., $D \cong 10^{-8}$ cm sec^{-1} for phospholipid lateral diffusion under physiological conditions; Kornberg and McConnell 1971a; Scandella et al. 1972). Other components, such as membrane intercalated particles revealed by freeze-cleavage (Pinto da Silva 1972; Tillack et al. 1972) or membrane glycoproteins carrying specific antigenic determinants, may also be rapidly diffusing in the membrane plane (e.g., $D \cong 10^{-9}$ cm sec^{-1} for certain muscle fiber antigens; Edidin and Fambrough 1973). There is at least one clear example of different membrane antigenic components moving laterally at different rates after fusion of cell heterokaryons (Frye and Edidin 1970).

One might ask, then, why are the lectin sites on normal cells incapable of rapid lateral diffusion compared to their transformed derivatives? There may be several reasons for this observation: (a) the intrinsic fluidity of the lipid bilayer is altered after transformation to a more fluid state; (b) the lectin binding determinant is structurally modified by transformation to a component which diffuses more rapidly in the membrane; or (c) the mobility of specific membrane components expressing lectin binding sites is controlled, in part, by attached peripheral membrane structures. There is some evidence that renders the first two proposals less likely than the third. Extraction of total membrane lipids from transformed cells does not suggest consistent, significant (or major) alterations from the bulk normal cell membrane lipids (Weiss 1967). Since there is no evidence suggesting that the lipid phase transition is altered or that the lipid fluidity is decreased, this proposal does

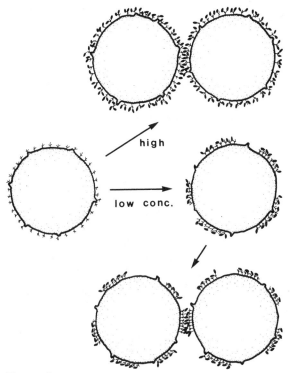

Figure 4

Model to explain the differences in *R. communis*-mediated cell agglutinability
and lectin binding site distributions at high and low lectin concentrations.
Lectin binding sites are initially expressed randomly across the cell surface. At
low lectin concentrations and high (room or physiological) temperatures, subsaturating numbers of polyvalent lectin molecules bind and induce surface
topographical changes ("clustering") in lectin binding sites due to cross-linking.
Some of the clustered sites (with bound lectin molecules) are eventually involved in the formation of multiple cross-bridges between different cells that
overcome cell repulsive forces. At high lectin concentrations and low or high
temperatures saturating numbers of polyvalent lectin molecules bind without
inducing a topographical change in lectin binding sites and directly agglutinate
cells.

not seem likely, but it remains to be eliminated. There is scant evidence, however,
suggesting that lectin receptors are modified after cell transformation. Janson and
Burger (1973) isolated and partly purified a lectin binding receptor for wheat
germ agglutinin from L1210 mouse leukemia cells and found that antisera made
against the wheat germ agglutinin receptor reacted with transformed L1210, PyBHK
and Py3T3 cells, but not with normal lymphocytes. Since normal 3T3 and BHK
cells were not examined by Janson and his colleagues, this observation may not be
true for the PyBHK/BHK or Py3T3/3T3 fibroblast pairs. They noted that normal
lymphocytes were nonagglutinable with wheat germ agglutinin, even after trypsinization; this is not true for the other cell pairs. Ozanne and Sambrook (1971) claim
that there is no difference between the total number of wheat germ agglutinin binding sites on similar fibroblast normal and transformed cell pairs.

Some data exists for the third proposal on cytoplasmic interference in the mobility of specific surface components. Yin et al. (1972) found that microtubule-disrupting drugs, such as colchicine and vinblastine, affect ConA-mediated agglutin-ability of fibroblasts and also of polymorphonuclear leukocytes (Berlin and Ukena 1972). At the concentrations used (10^{-6} M), these drugs should not have drastic effects directly on the lipid bilayer (Seeman et al. 1973), but this point has to be carefully ruled out. These drugs also affect the endocytosis of transport receptors, which are normally distinct from the endocytosed (internalized) membrane regions of the plasma membrane (Ukena and Berlin 1972).

The role of cytoplasmic peripheral membrane components in maintaining mem-brane shape and topography is not well defined, but available morphological and biochemical data support the hypothesis that a cell filament system is involved. McNutt et al. (1971) found that growing, sparse 3T3 cells do not have as an extensive microfilament/microtubule system associated with the plasma membrane as confluent 3T3 cells. SV3T3 cells had less membrane-associated filament net-work than either sparse or confluent 3T3, but a flat revertant of SV3T3 was mor-phologically similar to 3T3 cells. In this system (3T3/SV3T3) microfilament/ microtubule proteins may restrain the mobility of surface receptors that are linked across the membrane (trans-membrane control [Singer and Nicolson 1972; Nicol-son 1973b]). This may explain the decrease in lectin-mediated agglutinability of 3T3 cells at confluency.

Evidence for microfilament involvement in trans-membrane control of surface topography comes from experiments with cytochalasin B. Cytochalasin B is re-ported to break up cell microfilament systems (Schroeder 1968; Wessels et al. 1971) and to modify the lectin-mediated agglutination of some types of cells (Kaneko et al. 1973). However utilization of cytochalasin B to disrupt cell micro-filament systems and affect surface display of membrane components will have to be verified by alternate means due to recent findings that indicate cytochalasin B also modifies certain membrane transport systems (Cohn et al. 1972; Kletzien and Perdue 1973). Goldman (1972) found that there is not a simple relationship be-tween the action of cytochalasin B and microfilament disruption. Hamster BHK21 cells treated with cytochalasin B did not uniformly and specifically respond to the drug, and after treatment many cells extruded their nuclei, while some cells showed no demonstrable morphological alteration. As mentioned above, microtubular drugs have also been reported to modify lectin-mediated cell agglutination of certain cells (Berlin and Ukena 1972; Yin et al. 1972); they also seem to affect the reversibility of agglutination by specific saccharide inhibitors, suggesting that their effects may be, in part, nonspecific (Nicolson 1973e). Microtubular drugs do not appear to affect the lectin-mediated agglutination of ascites tumors (Kaneko et al. 1973) or other lymphoid tumors (Nicolson unpubl.). Thus experiments using microfilament and microtubule drugs to demonstrate cytoplasmic trans-membrane control of cell surface components by cell contractile proteins must be cautiously interpreted.

Direct evidence for cytoplasmic trans-membrane control of cell surface topog-raphy comes from "model" cell membrane systems such as the human erythrocyte ghost. The human erythrocyte expresses only a few (glyco) proteins at the extracel-lular surface (Phillips and Morrison 1971; Steck 1972; Bretscher 1971). One of these glycoproteins, the sialoglycoprotein or glycophorin, apparently traverses the membrane and is expressed at both membrane surfaces (Bretscher 1971; Steck 1972; Segrest et al. 1973). It is structurally a single polypeptide chain with various oligosaccharides attached to the N-terminal region, which is comprised of predominantly hydrophilic amino acid residues and is expressed extracellularly. An internal sequence appears to be predominantly hydrophobic, and at the C-terminal end of the molecule the primary sequence is again hydrophilic. Both the N-terminal and C-terminal ends of the molecule are exposed to [^{125}I]iodine labeling by lactoperoxidase in leaky ghosts that allow lactoperoxidase entry, but the C-terminal region cannot be labeled in impermeable ghosts or intact cells, indicating that this region is expressed intracellularly (Segrest et al. 1973). Glycophorin may be directly or indirectly associated with spectrin, a fibrous peripheral protein (Marchesi et al. 1969; Mazia and Ruby 1968; Clarke 1971) attached to the erythrocyte membrane inner surface (Nicolson et al. 1971). This association comes from cross-linking data (Ji 1973) and from electron microscopic data that indicate spectrin can control the topographic distribution of glycophorin (Nicolson and Painter 1973; Nicolson 1973a,b). Spectrin trans-membrane control of glycophorin's sialic acid topography was shown by sequestering affinity purified anti-spectrin inside resealed erythrocyte ghosts. The purified antibodies bound exclusively to spectrin at the inner membrane surface and caused aggregation of sialic acid residues (overwhelming on glycophorin) at the outer membrane surface. The trans-membrane action of anti-spectrin required intact γG antibodies (Fab would not substitute) and was time- and concentration-dependent (Nicolson and Painter 1973). In an analogous fashion microfilament/microtubule systems or other peripheral membrane proteins may be attached to integral membrane glycoproteins or glycoprotein-complexes that are expressed at both membrane surfaces. This would allow a cell to control its surface topography by controlling the contraction and attachment of a microfilament/microtubule system (Fig. 5).

Factors Affecting Cell Agglutination

There appear to be several interrelated phenomena that determine whether cell agglutination will occur in any given system. These include the biochemical nature of the agglutinating molecules, the number of molecules involved in agglutination, the mobility of the agglutination sites in the membrane, cell surface structures, cell repulsive forces, peripheral membrane component interference, cell rigidity, etc. The agglutinating molecules must have the proper specificity and a high enough cell binding constant for successful agglutination to occur. If the binding constant is not high, as in the case of most lectins, multiple interaction is required, leading to the probable formation of multiple cross-bridges between cells. Most lectins are polyvalent which should aid in cell agglutination. The number of cell surface agglutination sites and the number of surface-bound agglutination molecules are important to the agglutination process (Hoyer and Trabold 1970), probably by determining whether a sufficient number of multiple cross-bridges can form. In experiments with RCA-mediated agglutination of SV3T3 cells the distribution of RCA binding sites using high RCA concentrations (50 μg/ml) with immunofluorescent techniques was uniform compared with the patchy distributions obtained when low concentrations (1 μg/ml) were used (Table 4). Increasing local surface concentrations of agglutinating molecules would also be expected to aid in agglutination, so the mobility of the agglutination sites and their ability to be clustered into higher densities should enhance agglutination (Nicolson 1971). However under the appropriate conditions mobility and clustering are probably not important, if the

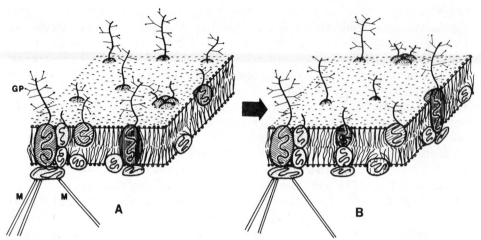

Figure 5

Fluid Mosaic Membrane Model (Singer and Nicolson 1972) with hypothetical cytoplasmic restraints. Certain membrane components (glycoproteins, glycolipids and lipids) are rapidly diffusing in the membrane plane under physiological conditions (A → B), while others such as the glycoprotein complex (*GP*), illustrated at the left, have their mobility impeded by membrane-associated peripheral protein structures (*M*) in the cell cytoplasm.

proper density of agglutination molecules can be attained by simply increasing the total number of agglutination sites (this also appears to be an important factor in the RCA-mediated agglutination of neuraminidase-treated cells where the number of RCA binding sites increases substantially after enzyme treatment [Nicolson 1973c]), or by increasing the number of bound agglutination molecules by simply increasing their concentration in solution (Fig. 4). Cell surface structures such as microvilli may enter into cell agglutination by literally trapping other cells more effectively or presenting specialized surfaces to other cells that are different from the remaining membrane surface in density or topography of receptors, etc. Similarly cell rigidity or deformability may affect agglutinability by determining the amount of opposed surface area that can be brought into play by a reduction in the local membrane radius of curvature during the agglutination process (Weiss 1965). For example, strong centrifugation of cells causes their deformation and enhances agglutination, probably by increasing the amount of surface in contact. Cell charge repulsion forces are extremely important in reducing spontaneous cell aggregation (Weiss 1969). Cells are net negatively charged, and reducing their charge density (zeta potential) at the hydrodynamic slip surface favors cell association (Pollack et al. 1965). Finally as mentioned above, peripheral membrane restraints may determine, in part, the mobility and topographic distribution of agglutination sites.

The various factors involved in cell agglutination may or may not oppose one another, but agglutination will occur when the sum of aggregation forces outweighs the repulsive forces. Thus one or several of the factors involved in agglutination could actually oppose agglutination, but not prevent it, if other forces favoring agglutination overcome cell repulsion. Cells can be fixed with aldehyde reagents that literally freeze the mobility of certain lectin binding sites without destroying them

(Rosenblith et al. 1973; Inbar et al. 1973; Nicolson 1973d; Table 4). The fixation also increases cell rigidity which would be expected to decrease cell deformation and perhaps reduce cell agglutinability. Nevertheless aldehydre-fixed cells are still agglutinable, but a much greater number of ^{125}I-labeled lectin molecules must be bound (using higher lectin concentrations) in order to overcome cell repulsive forces. In this case the density of bound lectin molecules probably reaches a critical point, above which cell agglutination occurs.

SUMMARY

The dynamic behavior of lectin sites on normal and transformed cells was investigated using ^{125}I-labeled concanavalin A and *R. communis* agglutinin and fluorescent-labeled antibodies against these lectins. Concanavalin A- or *R. communis*-mediated agglutinability of murine 3T3 cells and their SV40-transformed derivatives (SV3T3) decreased at cell confluency. The number of concanavalin A binding sites was independent of cell culture density; the number of *R. communis* binding sites on 3T3 cells increased 2.5–3 times at cell-to-cell contact, but remained constant on SV3T3 cells and also on spontaneously transformed 3T12 cells. The distribution and mobility of *R. communis* sites on 3T3 and SV3T3 cells were investigated by labeling unfixed and aldehyde-fixed cells at 0°C, incubating the *R. communis* agglutinin-labeled cells at various temperatures, and then labeling with fluorescent anti-*R. communis* agglutinin. Examination of the cells by fluorescence microscopy revealed that unfixed or fixed cells labeled at 0°C and incubated at 0°C show uniform fluorescence, indicating that an inherently dispersed distribution of lectin binding sites exists on the cell surface. Treating unfixed cells with low concentrations of *R. communis* agglutinin (1–2 μg/ml, 0°C) and then incubating them at 37°C for 10 minutes prior to fluorescent labeling revealed a change from uniform to patchy fluorescence on SV3T3, but not 3T3 cells. This indicates greater mobility of *R. communis* sites on the transformed cell surface. Using fixed cells or high concentrations of lectin (50 μg/ml, 0°C) prevented the lectin-induced redistribution of fluorescence from uniform to patchy on SV3T3 cells. Since both 3T3 and SV3T3 cells are agglutinable when treated with 50 μg/ml *R. communis* agglutinin and both types of cells show inherently uniform *R. communis* binding site distributions, lectin-induced redistribution or clustering of lectin binding sites, per se, is not essential for cell agglutination. However at low lectin concentrations where only a few lectin binding sites are occupied, lateral rearrangements may be necessary to increase local densities of cell surface-bound lectin molecules so that multiple cross-bridges can form between adjacent cells. The involvement of clustered lectin sites in cell agglutination at low lectin concentrations was demonstrated by agglutinating SV3T3 cells with ferritin-conjugated concanavalin A. The ferritin was found in clusters at the sites of contact between the agglutinated cells. The difference in lectin-mediated agglutinability of SV3T3 compared to 3T3 cells was discussed in terms of several factors which control lectin-mediated cell agglutination: nature of the lectin molecule, total number of lectin binding sites and the number of sites occupied, mobility of lectin binding sites and ability of the sites to be clustered, interference of cell surface structures, cell charge repulsive forces, cell membrane rigidity and peripheral trans-membrane restraints on the molecules bearing lectin binding sites.

Acknowledgments

The original work reported in this paper was supported by a contract from the Tumor Immunology Program of the National Cancer Institute, USPHS and grants from the National Science Foundation, the New York Cancer Research Institute and the Armand Hammer Fund for Cancer Research. I thank M. Lacorbiere for excellent technical assistance and A. Brodginski for help in preparing the manuscript.

REFERENCES

Agrawal, B. B. L. and I. J. Goldstein. 1967. Protein-carbohydrate interaction. VI. Isolation of concanavalin A by specific adsorption on cross-linked dextran gels. *Biochim. Biophys. Acta* **147**:262.

Arndt-Jovin, D. J. and P. Berg. 1971. Quantitative binding of ^{125}I-concanavalin A to normal and transformed cells. *J. Virol.* **8**:716.

Aub, J. C., B. H. Sanford and M. N. Cote. 1965. Studies on the reactivity of tumor and normal cells to a wheat germ agglutinin. *Proc. Nat. Acad. Sci.* **54**:396.

Aub, J. C., C. Tieslau and A. Lankester. 1963. Reaction of normal and tumor cell surfaces to enzymes. I. Wheat-germ lipase and associated mucopolysaccharides. *Proc. Nat. Acad. Sci.* **50**:613.

Becht, H., R. Rott and H. D. Klenk. 1972. Effect of concanavalin A on cells infected with enveloped RNA viruses. *J. Gen. Virol.* **14**:1.

Berlin, R. D. and T. E. Ukena. 1972. Effect of colchicine and vinblastine on the agglutination of polymorphonuclear leucocytes by concanavalin A. *Nature New Biol.* **238**:120.

Borek, C., M. Grob and M. M. Burger. 1973. Surface alterations in transformed epithelial and fibroblastic cells in culture: A disturbance of membrane degradation versus biosynthesis? *Exp. Cell Res.* **77**:207.

Bretscher, M. S. 1971. Major protein which spans the human erythrocyte membrane. *J. Mol. Biol.* **59**:351.

————. 1972. Asymmetrical lipid bilayer structure for biological membranes. *Nature New Biol.* **236**:11.

Bretton, R., R. Wicker and W. Bernhard. 1972. Ultrastructural localization of concanavalin A receptors in normal and SV40-transformed hamster and rat cells. *Int. J. Cancer* **10**:397.

Burger, M. M. 1969. A difference in the architecture of the surface membrane of normal and virally transformed cells. *Proc. Nat. Acad. Sci.* **62**:994.

————. 1970. Changes in the chemical architecture of transformed cell surfaces. *Permeability and function in biological membranes* (ed. L. Bolis et al.) p. 107. North-Holland, Amsterdam.

Burger, M. M. and A. R. Goldberg. 1967. Identification of a tumor-specific determinant on neoplastic cell surfaces. *Proc. Nat. Acad. Sci.* **56**:359.

Clarke, M. 1971. Isolation and characterization of a water-soluble protein from bovine erythrocyte membranes. *Biochem. Biophys. Res. Commun.* **45**:1063.

Cline, M. J. and D. C. Livingston. 1971. Binding of ^{3}H-concanavalin A by normal and transformed cells. *Nature New Biol.* **232**:155.

Cohn, R. H., S. D. Banerjee, E. R. Shelton and M. R. Bernfield. 1972. Cytochalasin B: Lack of effect on mucopolysaccharide synthesis and selective alterations in precursor uptake. *Proc. Nat. Acad. Sci.* **69**:2865.

Comoglio, P. M. and R. Guglielmone. 1972. Two dimensional distribution of concanavalin A receptors molecules on fibroblast and lymphocyte plasma membranes.
FEBS Letters **27**:256.

Davis, W. D. 1972. H-2 antigen on cell membranes: An explanation for the alternation
of distribution by indirect labeling techniques. *Science* **175**:1006.

dePetris, S. and M. Raff. 1972. Distribution of immunoglobulin on the surface of mouse
lymphoid cells as determined by immunoferritin electron microscopy. Antibody-
induced, temperature-dependent redistribution and implications for membrane structure. *Eur. J. Immunol.* **2**:523.

Drysdale, R. G., P. R. Herrick and D. Franks. 1968. The specificity of the haemagglutinin of the castor bean, *Ricinus communis*. *Vox Sanguinis* **15**:194.

Edidin, M. and D. Fambrough. 1973. Fluidity of the surface of cultured cell muscle
fibers. Rapid lateral diffusion of marked surface antigens. *J. Cell Biol.* **57**:27.

Edidin, M. and A. Weiss. 1972. Antigen cap formation in cultured fibroblasts: A reflection of membrane fluidity and of cell motility. *Proc. Nat. Acad. Sci.* **69**:2456.

Frye, L. D. and M. Edidin. 1970. The rapid inter-mixing of cell surface antigens after
formation of mouse-human heterokaryons. *J. Cell Sci.* **7**:319.

Gantt, R. R., J. R. Martin and V. J. Evans. 1969. Agglutination of *in vitro* cultured
neoplastic and non-neoplastic cell lines by a wheat germ agglutinin. *J. Nat. Cancer
Inst.* **42**:369.

Goldman, R. D. 1972. The effects of cytochalasin B on the microfilaments of baby
hamster kidney (BHK-21) cells. *J. Cell Biol.* **52**:246.

Goto, M., Y. Kataoka, K. Goto, T. Yodoyama and H. Sato. 1972. Decrease in agglutinability of cultured tumor cells to concanavalin A at the plateau of cell growth.
Gann **63**:505.

Hakomori, S. 1970. Cell-density dependent changes in glycolipids of fibroblasts and loss
of this response in transformed cells. *Proc. Nat. Acad. Sci.* **67**:1741.

Holley, R. W. 1972. A unifying hypothesis concerning the nature of malignant growth.
Proc. Nat. Acad. Sci. **69**:2840.

Hoyer, L. W. and N. C. Trabold. 1970. The significance of erythrocyte antigen site
density. I. Hemagglutination. *J. Clin. Inves.* **49**:87.

Inbar, M. and L. Sachs. 1969a. Structural difference in sites on the surface membrane of
normal and transformed cells. *Nature* **223**:710.

———. 1969b. Interaction of the carbohydrate-binding protein concanavalin A with
normal and transformed cells. *Proc. Nat. Acad. Sci.* **63**:1418.

Inbar, M., H. Ben-Bassat and L. Sachs. 1972. A specific metabolic activity on the surface membrane in malignant cell-transformation. *Proc. Nat. Acad. Sci.* **69**:2748.

Inbar, M., C. Huet, A. R. Oseroff, H. Ben-Bassat and L. Sachs. 1973. Inhibition of
lectin agglutinability by fixation of the cell surface membrane. *Biochim. Biophys.
Acta* **311**:594.

Janson, V. K. and M. M. Burger. 1973. Isolation and characterization of agglutinin
receptor sites. II. Isolation and partial purification of a surface membrane receptor
for wheat germ agglutinin. *Biochim. Biophys. Acta* **291**:1127.

Ji, T. H. 1973. Cross-linking sialoglycoproteins of human erythrocyte membranes.
Biochem. Biophys. Res. Commun. **53**:508.

Kaneko, I., H. Satoh and T. Ukita. 1973. Effect of metabolic inhibitors on the agglutination of tumor cells by concanavalin A and *Ricinus communis* agglutinin. *Biochem.
Biophys. Res. Commun.* **50**:1087.

Karnovsky, M. J., E. R. Unanue and M. Leventhal. 1972. Ligand-induced movement
of lymphocyte membrane macromolecules. II. Mapping of surface moieties. *J. Exp.
Med.* **136**:907.

Kijimoto, S. and S. I. Hakomori. 1972. Contact-dependent enhancement of net synthesis of Forssman glycolipid antigen and hematoside in NIL cells at the early stage of cell-to-cell contact. *FEBS Letters* **25**:38.

Kletzien, R. F. and J. F. Perdue. 1973. Inhibition of sugar transport in chick embryo fibroblasts by cytochalasin B. Evidence for a membrane-specific effect. *J. Biol. Chem.* **248**:711.

Kornberg, R. D. and H. M. McConnell. 1971a. Lateral diffusion of phospholipids in a vesicle membrane. *Proc. Nat. Acad. Sci.* **68**:2564.

————. 1971b. Inside-outside transitions of phospholipids in vesicle membranes. *Biochemistry* **10**:111.

Kourilsky, F. M., C. Silvestre, C. Neauport-Sautes, Y. Loosfelt and J. Dausset. 1972. Antibody-induced redistribution of HL-A antigens at the cell surface. *Eur. J. Immunol.* **2**:249.

Liske, R. and D. Franks. 1968. Specificity of the agglutinin in extracts of wheat germ. *Nature* **217**:860.

Mäkela, O. 1957. *Studies in hemagglutinins of leguminosae seeds,* p.1. Weilin and Goos, Helsinki.

Marchesi, V. T., E. Steers, Jr., T. W. Tillack and S. L. Marchesi. 1969. Some properties of spectrin. A fibrous protein isolated from red cell membrane. *The red cell membrane, structure and function* (ed. G. A. Jamieson and T. J. Greenwald) p. 117. Lippincott, Philadelphia.

Marchesi, V. T., T. W. Tillack, R. L. Jackson, J. P. Segrest and R. E. Scott. 1972. Chemical characterization and surface orientation of the major glycoprotein of the human erythrocyte membrane. *Proc. Nat. Acad. Sci.* **69**:1445.

Martinez-Palomo, A., R. Wicker and W. Bernhard. 1972. Ultrastructural detection of concanavalin surface receptors in normal and in polyoma-transformed cells. *Int. J. Cancer* **9**:676.

Mazia, D. and A. Ruby. 1968. Dissolution of erythrocyte membranes in water and comparison of the membrane protein with other structural proteins. *Proc. Nat. Acad. Sci.* **61**:1005.

McNutt, N. S., L. A. Culp and P. H. Black. 1971. Contact-inhibited revertant cell lines isolated from SV40-transformed cells. II. Ultrastructural study. *J. Cell Biol.* **50**:691.

Moscona, A. A. 1971. Embryonic and neoplastic cell surfaces: Availability of receptors for concanavalin A and wheat germ agglutinin. *Science* **171**:905.

Nicolson, G. L. 1971. Difference in the topology of normal and tumor cell membranes as shown by different distributions of ferritin-conjugated concanavalin A on their surfaces. *Nature New Biol.* **233**:244.

————. 1972. Topography of cell membrane concanavalin A-sites modified by proteolysis. *Nature New Biol.* **239**:193.

————. 1973a. Anionic sites of human erythrocyte membranes: Effects of trypsin, phospholipase C and pH on the topography of positively charged colloidal particles. *J. Cell Biol.* **57**:373.

————. 1973b. Cis and trans-membrane control of cell surface topography. *J. Supramol. Struct.* (in press).

————. 1973c. Neuraminidase unmasking and the failure of trypsin to unmask β-D-galactose-like sites on erythrocyte, lymphoma and normal and SV40-transformed 3T3 fibroblast cell membranes. *J. Nat. Cancer Inst.* **50**:1443.

————. 1973d. Temperature-dependent mobility of concanavalin A sites on tumour cell surfaces. *Nature New Biol.* **243**:218.

————. 1973e. The relationship of a fluid membrane structure to cell agglutination and surface topography. *Series Haematologica* **6**:275.

Nicolson, G. L. and J. Blaustein. 1972. The interaction of *Ricinus communis* agglutinin with normal and tumor cell surfaces. *Biochim. Biophys. Acta* **266**:543.

Nicolson, G. L. and M. Lacorbiere. 1973. Cell contact-dependent increase in membrane D-galactopyranosyl-like residues on normal, but not virus- or spontaneously-transformed murine fibroblasts. *Proc. Nat. Acad. Sci.* **70**:1672.

Nicolson, G. L. and R. G. Painter. 1973. Anionic sites of human erythrocyte membranes. II. Anti-spectrin induced trans-membrane aggregation of the binding sites for positively charged colloidal particles. *J. Cell Biol.* (in press)

Nicolson, G. L. and S. J. Singer. 1971. Ferritin-conjugated plant agglutinins as specific saccharide stains for electron microscopy: Application to saccharides bound to cell membranes. *Proc. Nat. Acad. Sci.* **68**:942.

Nicolson, G. L. and R. Yanagimachi. 1972. Terminal saccharides on sperm plasma membranes: Identification by specific agglutinins. *Science* **177**:276.

Nicolson, G. L., V. T. Marchesi and S. J. Singer. 1971. The localization of spectrin on the inner surface of human red blood cell membranes by ferritin-conjugated antibodies. *J. Cell Biol.* **51**:265.

Noonan, K. D., H. C. Renger, C. Basilico and M. M. Burger. 1973. Surface changes in temperature-sensitive simian virus 40-transformed cells. *Proc. Nat. Acad. Sci.* **70**:347.

Ozanne, B. and J. Sambrook. 1971. Binding of radioactively labelled concanavalin A and wheat germ agglutinin to normal and virus-transformed cells. *Nature New Biol.* **232**:156.

Pardee, A. 1971. The surface membrane as regulator of animal cell division. *In Vitro* **7**:95.

Phillips, D. R. and M. Morrison. 1971. Exposed protein on the intact human erythrocyte. *Biochemistry* **10**:1766.

Pinto da Silva, P. 1972. Translational mobility of the membrane intercalated particles of human erythrocyte ghosts. pH-dependent, reversible aggregation. *J. Cell Biol.* **53**:777.

Pollack, W., H. J. Hager, R. Reckel, D. A. Toren and H. O. Singher. 1965. A study of the forces involved in the second stage of hemagglutination. *Transfusion* **5**:158.

Poste, B. and P. Reeve. 1972. Agglutination of normal cells by plant lectins following infection with non-oncogenic viruses. *Nature New Biol.* **237**:113.

Robbins, P. W. and I. A. Macpherson. 1971. Glycolipid synthesis in normal and transformed animal cells. *Proc. Roy. Soc. London* B **177**:49.

Rosenblith, J. A., T. E. Ukena, H. H. Yin, R. D. Berlin and M. J. Karnovsky. 1973. A comparative evaluation of the distribution of concanavalin A-binding sites on the surfaces of normal, virally-transformed and protease-treated fibroblasts. *Proc. Nat. Acad. Sci.* **70**:1625.

Roth, S. and D. White. 1972. Intercellular contact and cell-surface galactosyltransferase activity. *Proc. Nat. Acad. Sci.* **69**:485.

Scandella, C. J., P. Devaux and H. M. McConnell. 1972. Rapid lateral diffusion of phospholipids in rabbit sarcoplasmic reticulum. *Proc. Nat. Acad. Sci.* **69**:2056.

Schroeder, T. E. 1968. Cytokinesis: Filaments in the cleavage furrow. *Exp. Cell Res.* **53**:272.

Segrest, J. P., I. Kahne, R. L. Jackson and V. T. Marchesi. 1973. Major glycoprotein of the human erythrocyte membrane: Evidence for an amphipathic molecular structure. *Arch. Biochem. Biophys.* **155**:167.

Seeman, P., M. Chau-wong and S. Moyyen. 1973. Membrane expansion by vinblastine and strychnine. *Nature New Biol.* **241**:22.

Sela, B., H. Lis, N. Sharon and L. Sachs. 1970. Different locations of carbohydrate-containing sites in the surface membrane of normal and transformed mammalian cells. *J. Membrane Biol.* **3**:267.

———. 1971. Quantitation of *N*-acetyl-D-galactosamine-like sites on the surface membrane of normal and transformed mammalian cells. *Biochim. Biophys. Acta* **249**:564.

Sharon, N. and H. Lis. 1972. Lectins: Cell-agglutinating and sugar-specific proteins. *Science* **177**:949.

Singer, S. J. and G. L. Nicolson. 1972. The fluid mosaic model of the structure of cell membranes. *Science* **175**:720.

Sivak, A. 1971. Agglutinin interaction with embryonic and adult cell surfaces. *Science* **173**:264.

Steck, T. L. 1972. The organization of proteins in human erythrocyte membranes. *Membrane research* (ed. C. F. Fox) p. 71. Academic Press, New York.

So, L. L. and I. J. Goldstein. 1967. Protein-carbohydrate interaction. IX. Application of the quantitative hapten inhibition technique to polysaccharide-concanavalin A interaction. Some comments on the forces involved in concanavalin A-polysaccharide interaction. *J. Immunol.* **99**:158.

Taylor, R., P. Duffus, M. Raff and S. de Petris. 1971. Redistribution and pinocytosis of lymphocyte surface immunoglobulin molecules induced by anti-immunoglobulin antibody. *Nature* **233**:1225.

Tevethia, S. S., S. Lowry, W. E. Rawls, J. L. Melnick and V. McMillan. 1972. Detection of early cell surface changes in herpes simplex virus infected cells by agglutination with concanavalin A. *J. Gen. Virol.* **15**:93.

Tillack, T. W., R. E. Scott and V. T. Marchesi. 1972. The structure of erythrocyte membranes studied by freeze-etching. II. Localization of receptors for phytohemagglutinin and influenza virus to the intramembranous particles. *J. Exp. Med.* **135**:1209.

Tomita, M., T. Osawa, Y. Sakura and T. Ukita. 1970. On the surface structure of murine ascites tumors. I. Interactions with various phytoagglutinins. *Int. J. Cancer* **6**:283.

Uhlenbruck, G. and W. P. Herrmann. 1972. Agglutination of normal, coated and enzyme-treated human spermatozoa with heterophile agglutinins. *Vox Sanguinis* **23**:444.

Unanue, E. R., W. D. Perkins and M. J. Karnovsky. 1972. Ligand-induced movement of lymphocyte membrane macromolecules. I. Analysis by immunofluorescence and ultrastructural radioautography. *J. Exp. Med.* **136**:885.

Vesely, P., G. Entlicher and J. Kocourek. 1972. Pea phytohemagglutinin selective agglutination of tumour cells. *Experientia* **28**:1085.

Weiss, L. 1965. Studies on cell deformability. I. Effect of surface charge. *J. Cell Biol.* **26**:735.

———. 1967. The cell periphery, metastasis and other contact phenomena. *Frontiers of biology*, p. 264. North-Holland, Amsterdam.

———. 1969. The cell periphery. *Int. Rev. Cytol.* **26**:63.

Wessells, N. K., B. S. Spooner, J. F. Ash, M. O. Bradly, M. A. Luduena, E. L. Taylor, J. T. Wrenn and K. M. Yamada. 1971. Microfilaments in cellular and developmental processes. *Science* **171**:135.

Yahara, I. and G. M. Edelman. 1972. Restriction of the mobility of lymphocyte immunoglobulin receptors by concanavalin A. *Proc. Nat. Acad. Sci.* **69**:608.

Yin, H. H., T. E. Ukena and R. D. Berlin. 1972. Effect of colchicine, colcemid and vinblastine on the agglutination by concanavalin A of transformed cells. *Science* **178**:867.

Distribution and Mobility of Membrane Macromolecules: Ligand-induced Redistribution of Concanavalin A Receptors and Its Relationship to Cell Agglutination

Martin C. Raff and Stefanello dePetris*

Medical Research Council Neuroimmunology Project
Zoology Department, University College London, London WC1E 6BT

Livio Mallucci

Microbiology Department, Guy's Hospital Medical School, London SE1

It has been two years since the discovery that multivalent antibodies can induce a striking rearrangement of the specific membrane antigens to which they bind on the surface of living cells (Taylor et al. 1971). Immunofluorescence and immuno-ferritin electron microscopic studies of the distribution and antibody-induced redistribution of immunoglobulin receptors on lymphocytes led us to propose a general model of plasma membrane structure for isolated cells (dePetris and Raff 1972; Raff and dePetris 1972). It suggested that membranes are fluid and that surface macromolecular units, composed of one or a few identical or unidentical protein subunits, are randomly dispersed in the plane of the unperturbed membrane, and that they can be passively aggregated into clusters (and sometimes actively carried to one pole of the cell forming a "cap"), to an extent determined by the degree of cross-linking affected by the binding of multivalent antibodies or ligands. Subsequent studies of alloantigens on lymphocytes (Davis 1972; dePetris and Raff 1973) and lectin receptors on lymphocytes (dePetris and Raff 1973), liver cells (dePetris, Smith and Raff unpublished), myelin fragments and synaptosomes (Matus, dePetris and Raff 1973) have supported the model and suggest that all biological membranes are fluid at physiological temperatures.

DISTRIBUTION OF ALLOANTIGENS AND CONCANAVALIN A RECEPTORS ON LYMPHOCYTES

When living thymus lymphocytes are labeled directly with fluorescein-conjugated anti-θ antibody, the fluorescence is diffusely distributed as a uniform ring around the cell (Ashman and Raff 1973); adding a second layer of rabbit anti-mouse immunoglobulin antibody (R-anti-MIg) causes the fluorescent label to cluster together into patches (Raff and dePetris 1973). Similarly when the distribution of the θ alloantigen on mouse thymus lymphocytes is studied by immunoferritin

---

* Present address: Basel Institute of Immunology, Basel, Switzerland

electron microscopy and the anti-θ antibody is visualized with monovalent Fab fragments of R-anti-MIg conjugated to ferritin (FT), the ferritin marker is randomly dispersed over the cell surface (Fig. 1). If divalent R-anti-MIg-FT is used as the second antibody layer instead of the Fab R-anti-Mig-FT, the ferritin is distributed in patches (Fig. 2) or in polar "caps," depending on whether the labeling is done at 0–4°C or at 20–37°C, respectively. A dispersed distribution is also seen if anti-θ coated cells are fixed with glutaraldehyde and then labeled with divalent R-anti-MIg-FT (Fig. 1). Thus we conclude that θ is diffusely distributed on the thymocyte surface and can be induced to redistribute into clusters or caps by two layers of divalent antibody. The fact that divalent anti-θ antibody itself does not induce patch formation suggests that the θ determinant is represented only once, or possibly twice, on any independently mobile macromolecular unit, precluding lattice formation with anti-θ antibody alone.

Unlike anti-θ, anti-TL 1, 2, 3 antiserum, which is directed against three different thymus-leukemia (TL) alloantigen specificities probably present on the same membrane macromolecule (Boyse and Old 1969), is able to form patches on TL-bearing mouse thymus lymphocytes, as shown using Fab R-anti-MIg-FT to visualize anti-TL antibodies. However if the cells are prefixed with formaldehyde before exposing them to anti-TL and R-anti-MIg-FT, the ferritin molecules are found dispersed over the cell surface. Antibodies directed against H-2K alloantigens, probably reacting with several different specificities, show an intermediate pattern of distribution with some dispersed molecules and some patches of variable size when labeled with Fab R-anti-MIg-FT, but form large patches and caps when further cross-linked by divalent anti-MIg. Similar observations on H-2 alloantigens have been reported by Davis (1972).

These observations leave little doubt that the patchy distributions of alloantigens previously reported (Aoki et al. 1969; Stackpole et al. 1971), using the indirect hybrid-antibody immunoferritin technique, were artifactually induced by the reagents used for labeling—either the divalent alloantibodies themselves or the multivalent ferritin or virus markers. This is further supported by our studies involving the labeling of mouse spleen lymphocytes with concanavalin A conjugated to ferritin (ConA-FT), which is expected to bind to the great majority of lymphocyte membrane glycoproteins (Allan, Auger and Crumpton 1972). When ConA-FT is used at a subsaturating concentration (less than 1 mg FT/ml) to minimize cross-linking, or if the cells are prefixed with glutaraldehyde, the ferritin molecules are

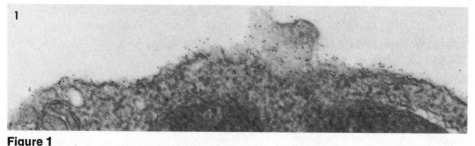

Figure 1
Diffuse distribution of θ antigen on mouse thymus cells. Thymus cells were incubated with anti-θ antiserum at 22°C, washed, fixed with 3% glutaraldehyde and then labeled with ferritin-conjugated rabbit anti-mouse Ig antibody. $\times$ 64,000.

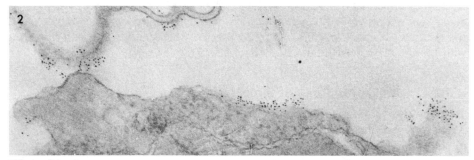

Figure 2
Patched distribution of θ antigen on mouse thymus cell induced by divalent anti-mouse Ig antibody conjugate on unfixed cells. Thymus cells were incubated with anti-θ antiserum at 22°C, washed, labeled with ferritin-conjugated rabbit anti-mouse Ig antibody at 6°C, washed and then fixed with glutaraldehyde. × 64,000.

randomly dispersed. On the other hand if the dispersed ConA-FT molecules are further cross-linked on unfixed cells by anti-FT antibody, patches and caps are formed.

In more recent studies done in collaboration with Bruce Smith and Andrew Matus on nonlymphoid cells and tissues (cultured liver cells, synaptosomes, and myelin fragments), we have similarly found ConA-FT-binding macromolecules randomly dispersed in membranes that had been prefixed with glutaraldehyde, but clustered in unfixed membranes where an additional layer of anti-FT antibody was used to further cross-link the ConA-FT. Thus most, if not all, macromolecules on the free surface of many biological membranes appear to be randomly dispersed in the unperturbed membrane and able to redistribute in the plane of the membrane when suitably cross-linked, as had been predicted (dePetris and Raff 1972; Raff and dePetris 1972).

CONCANAVALIN RECEPTORS ON FIBROBLAST-DERIVED CELLS

Recent experiments using ConA-FT to label isolated plasma membranes from fibroblast-derived cells (3T3) were interpreted as indicating that ConA-binding glycoproteins were spontaneously aggregated in the membranes of SV40-transformed (Nicolson 1971) or trypsinized (Nicolson 1972) 3T3 cells, but were dispersed in membranes of normal 3T3 cells. In view of the studies reviewed above, it seemed more likely to us that the observed clustering of ConA receptors was induced by the binding of the multivalent ConA-FT and that the differences observed between the normal cells and the transformed or trypsinized cells reflected differences in the redistribution ability of the membrane glycoproteins, rather than in the arrangement of these molecules in the unperturbed membrane. Since it was proposed that the differences in distribution of membrane glycoproteins in these cells might account for the generally observed differences in their agglutinability by a variety of lectins (Nicolson 1971, 1972), which in turn has been postulated to be related to their growth characteristics (Burger 1973), the question seemed an important one to resolve.

In order to investigate this point we have studied the topography of ConA receptors on normal, trypsinized (0.0001% for 3 minutes at 37°C) or polyoma virus-transformed (PV) 3T3 cells, PV-TT3 cells (Mallucci 1971) and on secondary and tertiary mouse embryo fibroblasts using ConA-FT. Cells were detached from culture dishes with 0.02% EDTA, labeled in suspension, washed and fixed with 1% OsO_4 (dePetris, Raff and Mallucci 1973).

Normal Distribution of ConA Receptors

To study the "normal" arrangement of ConA receptors, cells were prefixed with 3% glutaraldehyde for 2–3 hours at 20°C prior to labeling with ConA-FT in order to immobilize the surface proteins without inactivating the carbohydrate ConA receptors. The amount of labeling was not significantly different on the various cell types, and on all cells the ferritin molecules were uniformly dispersed and distributed over all parts of the membrane, including microvilli (Fig. 3). Thus as predicted by the general model of membrane structure, there was no spontaneous clustering of ConA receptors on trypsinized or transformed cells.

ConA-FT-induced Redistribution of ConA Receptors

When unfixed cells were labeled with ConA-FT at 20°C, the distribution of ferritin was very different from that seen on prefixed cells. The ferritin molecules were clustered in patches of various sizes, separated by areas with little or no labeling (Fig. 4). Although the degree of patching varied to some extent from experiment to experiment, in all experiments the amount of labeling and the extent of patching, although difficult to quantitate with precision, were not appreciably different in normal, trypsinized and transformed cells. This was also true when patching was purposely increased by adding an additional layer of anti-ferritin antibody to cross-link the ConA-FT molecules, or when patching was minimized (but still appreciable) by labeling the cells at 0–4°C.

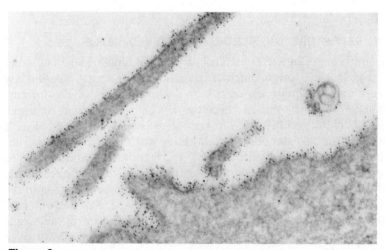

Figure 3
Diffuse distribution of ConA-ferritin conjugate on the surface of a trypsinized 3T3 fibroblast. The cells were prefixed with 3% glutaraldehyde at 22°C for 2 hours, washed and incubated with the conjugate at 22°C. In many points the membrane is cut tangentially. × 53,000.

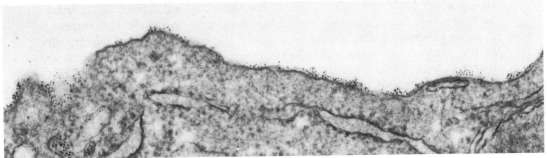

Figure 4
Patched distribution of ConA-ferritin conjugate on an unfixed normal 3T3 fibroblast. The cell was incubated with the conjugate at 22°C, washed and fixed with glutaraldehyde. In the upper part of the figure there is considerable pinocytosis of the labeled membrane. × 64,000.

These experiments confirmed the prediction that clustering of ConA receptors is induced by the binding of multivalent ConA, in agreement with the recent observations of others using peroxidase (Rowlett, Wicker and Bernhard 1973), hemocyanin (Smith and Revel 1972; Rosenblith et al. 1973) or fluorescein (Nicolson 1973; Inbar and Sachs 1973) labeling of ConA. However to our surprise we failed to demonstrate appreciable differences in the redistribution ability of these receptors on normal, transformed and trypsinized cells, despite the fact that the trypsinized 3T3 cells and polyoma virus-transformed 3T3 and TT3 cells were much more agglutinable by ConA than normal 3T3 cells and embryo fibroblasts were intermediate in agglutinability. Our findings thus raised two questions: (1) Why were we unable to detect the differences in ConA receptor distribution (or redistribution) on these different cells that others have reported, and, more importantly, (2) can differences in lectin receptor redistribution ability account for differences in lectin-induced agglutinability? Before considering these questions, it is worth discussing the various factors that can affect the extent of lectin-induced redistribution of membrane glycoproteins.

Factors Determining the Extent of Lectin-induced Redistribution of Membrane Glycoproteins

As discussed above there is increasing evidence that ligand-induced aggregation of membrane macromolecules is determined by the cross-linking of the potentially mobile macromolecules by the multivalent binding of the ligand, in a manner analogous to a precipitation reaction. Our results support the idea that this also applies to redistribution of ConA receptors, both in unperturbed and in mildly trypsinized membranes. Therefore the occurrence and extent of redistribution would be related to the valence of the lectin and the binding characteristics of the various ConA-binding surface molecules, such as the number of accessible receptor sites per molecule or mobile unit. For example, in principle if the number of accessible sites per membrane glycoprotein were one, only small clusters consisting of a number of membrane molecules equal to or smaller than the valence of the lectin could be formed, whereas chains or two-dimensional lattices could be made from molecules bearing two or more sites. On the other hand the presence of repeated and properly arranged receptor sites on the same individual membrane glycoprotein, which could bind to most or all the sugar binding sites of an individual ConA molecule, could decrease the efficiency of cross-linking. In this respect it should be emphasized that the measurement of the number of lectin binding sites per unit area of cell using radiolabeled lectins gives no information about the number of lectin binding sites per membrane glycoprotein molecule or their intra-molecular arrangement.

Another important factor which could influence the final redistribution pattern is the extent of receptor mobility in the plane of the membrane. On general grounds and by analogy with phenomena observed in lymphocytes, clustering of surface molecules could be the result of either passive diffusion or active movements requiring metabolic energy (dePetris and Raff 1972). The first type of movement would be determined mainly by the intrinsic "viscosity" of the membrane, which in turn would depend on the viscosity of the lipid milieu and possibly to some extent on interactions of membrane glycoproteins with other membrane or cytoplasmic proteins (e.g., microfilaments). The active transport of membrane macromolecules must entail the modification of these putative interactions and/or the establishment of new ones, and would be expected in general to cause a more rapid and larger displacement of the macromolecules. An example of this second type of movement is probably the centripetal movement ("capping") of ConA-hemocyanin complexes observed at 37°C by Smith and Revel (1972) in granulocytes, or of ConA and ConA-FT in lymphocytes (Unanue, Perkins and Karnovsky 1972); an example of the first is the clustering of ConA-FT observed by Nicolson in isolated membranes of trypsinized (1972) and transformed (1971) 3T3 cells.

Discrepency in Results Concerning Redistribution of ConA Receptors

Several investigators have found a patchy distribution of ConA receptors on transformed (Nicolson 1971; Martinez-Palomo, Wicker and Bernhard 1972; Inbar and Sachs 1973; Rosenblith et al. 1973) and trypsinized cells (Nicolson 1972), while finding the receptors uniformly distributed on normal cells of the same type, whereas no such differences were observed in our experiments and those of Smith and Revel (1972). The explanation for this discrepancy could involve any of the factors discussed above that influence redistribution and could be related to differences in the ConA conjugates, labeling procedures, and/or cell lines used. It could be, for example, that ConA-FT is a more effective cross-linking complex than native ConA or ConA-fluorescein; however in preliminary studies we have found a similar degree of patching on normal, polyoma virus-transformed, and trypsinized 3T3 cells labeled at 20 or 37°C with ConA or wheat germ agglutination (WGA) conjugated to fluorescein. The reason(s) why clustering of ConA receptors was not observed by Nicolson (1971, 1972) on normal 3T3 cell membranes using ConA-FT may be related to the differences in ConA-FT conjugates, the shorter incubation period (1–3 minutes) and/or the fact that isolated membranes were studied. It is unlikely that the patching we observed on normal cells was secondarily induced by agglutination of the cells during washing and pelleting, since in experiments where cells were diluted and fixed before centrifugation, or washed and fixed simultaneously through a glutaraldehyde gradient, the patchy distributions seen were not different from those seen in conventionally washed cells.

Perhaps a likely explanation for the discrepancies is the variation between different cell lines from laboratory to laboratory. This is suggested by the fact that of two groups using apparently the same technique (i.e., conA-hemocyanin) to visualize the ultrastructural distribution of ConA receptors, one observed a dispersed distribution (Rosenblith et al. 1973) while the other found a patchy distribution (Smith and Revel 1972) of these receptors on normal 3T3 cells. In addition in studies using ConA and peroxidase to label intact cells on glass, different distributions of peroxidase labeling were observed between a proportion of transformed cells and normal cell lines from hamster (Martinez-Palomo et al. 1972) but not from rat (Bretten, Wicker and Bernhard 1972). In the rat cell lines a proportion of both normal and transformed cells showed a patchy distribution of peroxidase. It thus seems likely that transformation or trypsinization may sometimes, but not invariably, be associated with a readily detectable increase in the redistribution ability of lectin receptors, making it unlikely that this property is an essential feature of the transformed phenotype. When differences in redistribution ability are found, it would seem unwise to attribute them exclusively to differences in membrane glycoprotein mobility until valency considerations and other factors have been excluded.

Do Differences in Lectin Receptor Redistribution Ability Account for Differences in Lectin-induced Agglutinability?

In studies involving a large number of different cell lines and a variety of different plant lectins it has been generally (although not invariably) found that normal cells require much higher concentrations of lectins in order to be agglutinated than do transformed cells of the same line (Aub, Sanford and Cote 1963; Burger and Goldberg 1967, Inbar and Sachs 1969). Increased agglutination by lectins can also be induced in normal cells by low concentrations of trypsin (Burger 1969; Inbar and Sachs 1969). The reasons for these differences in agglutinability are still unknown. Although there may be instances where enhanced agglutinability is associated with an increased density of lectin receptors on the cell surface, resulting from new synthesis or "unmasking" of "buried" receptors (Inbar, Ben-Bassat and Sachs 1972), there are an increasing number of studies indicating that this is usually not the case. Thus when measured under conditions similar to those used for agglutination assays (e.g., 20°C for 20–30 minutes), normal, trypsinized and transformed cells bind similar amounts of radiolabeled lectin (Sela et al. 1971; Arndt-Jovin and Berg 1971; Cline and Livingston 1971; Ozanne and Sambrook 1971). In our experiments, for example, as judged from the density of labeling in thin sections, the amount of ConA-FT binding per unit area was not appreciably different in the various cell types studied. In view of the recent demonstrations of differences in the arrangements of bound ConA on trypsinized and transformed cells compared to normal cells (Nicolson, 1971, 1972; Inbar and Sachs 1973; Rosenblith et al. 1973), it is not surprising that they have been proposed as the explanation for differences in agglutination. The newly modified version of this hypothesis would be that differential agglutinability reflects differential redistribution ability of membrane lectin receptors (Nicolson 1973; Inbar and Sachs 1973). If, as seems likely, agglutination is due (in part, at least) to the cross-linking of cells by the multivalent ConA, then redistribution of ConA receptors into patches, before or after cell-to-cell contact, would be expected to increase the efficiency of bridging and thus enhance agglutination. The decreased agglutination observed with ConA at low temperatures (Inbar, Ben-Bassat and Sachs 1971) or when glutaraldehyde-fixed cells are studied (Inbar et al. 1973; dePetris, Raff and Mallucci 1973) is consistent with this hypothesis.

However we have not been able to find differences in redistribution ability with ConA-FT despite marked differences in agglutinability with native ConA. During labeling for electron microscopy with ConA-FT at 20°C at high cell concentration (10^7/ml), there was usually grossly visible agglutination of transformed and trypsinized 3T3 cells, whereas this was not the case with normal 3T3 cells. In two experiments where agglutination was monitored microscopically during labeling with ConA-FT at 4°C, modest but significant agglutination occurred with trypsinized and transformed cells after 25 minutes, but none was seen with normal cells; yet all three cell preparations showed similar degrees of patching, which was appreciable although considerably less than that seen at 20°C. These results suggest that clustering of receptors, although perhaps important and even necessary for agglutination of these cells, may not be the only factor determining differential agglutination, at least when induced by ConA-FT. This is supported by our finding that in many instances the average density of ConA-FT on the surface of normal (unagglutinated) cells was not appreciably different from that observed in areas of contact between agglutinated transformed or trypsinized cells.

What other factors may determine differential agglutinability? The valency considerations of lectin and surface glycoproteins discussed previously would be expected to play an important role, although lectin-binding characteristics of membrane glycoproteins that would favor receptor redistribution may not necessarily favor agglutination. It is unlikely that metabolically dependent processes (Inbar et al. 1971) play a role in determining the agglutinability of the cells we have studied since heat-killing them did not change their agglutination properties (dePetris, Raff and Mallucci 1973). In addition sodium azide and cytochalasin B did not inhibit the agglutination of trypsinized or transformed cells (dePetris, Raff and Mallucci 1973). On the other had the various fibroblast lines we studied showed obvious differences in the morphology of their surfaces, which could have influenced their agglutinability. Thus embryo fibroblasts and 3T3 cells (normal and trypsinized) had a number of relatively short microvilli (probably accentuated by the EDTA used to detach the cells from the culture dishes [Huet and Herzberg 1973]), which were frequently glued against the membrane or to each other by the conjugate, resulting in the "burial" of much of the ConA-FT. The transformed cells had fewer and more slender microvilli and these were less often agglutinated to each other, leaving most of the label exposed on the cell surface. Whereas these morphological differences could account for the differential agglutination of normal versus transformed cells, they could not explain agglutinability differences between normal and trypsinized cells that had similar surface morphology. Surface charge would be expected to influence agglutination, and the finding of diminished sialic acid in the membranes of PV-3T3 compared to normal 3T3 cells (Grimes 1970) is consistent with the notion that differences in zeta potential on these cells may play a role in determining their agglutinability.

Lack of Correlation between Receptor Redistribution and Agglutination

In preliminary studies with L1210 mouse lymphoma cells (carried in ascites form), which are not agglutinated by ConA (up to 500 μg/ml) but are readily agglutinated by WGA (as low as 10 μg/ml) (Burger 1973), we have not found a correlation between clustering of lectin receptors visualized by immunofluorescence and lectin-induced agglutination. When L1210 cells were exposed to 500 μg/ml of ConA-fluorescein (ConA-Fl) (kindly supplied by M. F. Greaves) or WGA-Fl (kindly supplied by M. Crumpton) for 30 minutes at 4 or 20°C, only the WGA-treated cells agglutinated. At this point the cells were fixed with 2% glutaraldehyde for 15 minutes at 20°C, washed twice, and the distribution of the fluorescein-labeled lectin was determined. The fluorescence appeared diffuse at 4°C (Fig. 5) and somewhat patchy at 20°C (Fig. 6) in both WGA-Fl- and ConA-Fl-treated cells, despite the fact that the former were markedly agglutinated at both temperatures and the latter did not agglutinate at either temperature. In addition fixing the cells with 3% glutaraldehyde for 2 hours at 20°C did not prevent the striking agglutination induced by WGA, although fixation did delay the agglutination. In this system at least, it appears that agglutination can occur without receptor redistribution, and receptor clustering can occur without agglutination ensuing.

CONCLUSIONS

Surface glycoproteins are diffusely and randomly distributed in the unperturbed fluid membrane of dissociated cells, including normal, trypsinized and transformed fibroblast-derived cells. Multivalent ligands can cross-link membrane macromolecules and thus induce them to redistribute in the plane of the membrane. ConA-FT induced readily detectable clustering of ConA-binding macromolecules on the surface of normal, SV-transformed and trypsinized 3T3 cells, SV-TT3 cells and mouse embryo fibroblasts. Despite marked differences in ConA-induced agglutinability of these cells, no appreciable differences were detected in the density of ConA-FT binding or in the redistribution induced by the ConA-FT on the various cell types. Similarly we could not demonstrate a correlation between receptor redistribution and agglutination of L1210 lymphoma cells by ConA-fluorescein and WGA-fluorescein.

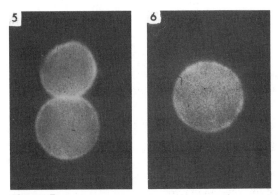

Figure 5
Diffuse and smooth distribution of fluorescein-conjugated wheat germ agglutinin on the surface of two agglutinated L1210 lymphoma cells at 4°C. Cells fixed with glutaraldehyde after labeling.

Figure 6
Slightly patched distribution of fluorescein-conjugated concanavalin A on isolated cells of L1210 lymphoma at 20°C. Cells fixed with glutaraldehyde after labeling.

Since lectin-induced agglutination probably can be influenced by a variety of factors, including density, valency and redistribution ability of membrane receptors, cell morphology, and surface charge, it may be unwise to look for a single factor determining all differences in agglutinability. More importantly it would seem premature to suggest that any of these factors are important in the control of cell growth in vitro or in vivo.

Acknowledgment

L.M. was supported by the Cancer Research Campaign.

REFERENCES

Allan, D., J. Auger and M. J. Crumpton. 1972. Glycoprotein receptors for concanavalin A isolated from pig lymphocyte plasma membrane by affinity chromatography in sodium deoxycholate. *Nature New Biol.* **236**:23.

Aoki, T., U. Hämmerling, E. deHarven, E. A. Boyse and L. J. Old. 1969. Antigenic structure of cell surfaces. An immunoferritin study of the occurrence and topography of H-2, θ and TL alloantigens on mouse cells. *J. Exp. Med.* **130**:979.

Arndt-Jovin, D. J. and P. Berg. 1971. Quantitative binding of ^{125}I-concanavalin A to normal and transformed cells. *J. Virol.* **8**:716.

Ashman, R. F. and M. C. Raff. 1973. Direct demonstration of theta-positive antigen-binding cells with antigen-induced movement of thymus-dependent cell receptors. *J. Exp. Med.* **137**:69.

Aub, J. C., B. H. Sanford and M. H. Cote. 1963. Reactions of normal and tumor cell surfaces to enzymes. I. Wheat-germ lipase and associated mucopolysaccharides. *Proc. Nat. Acad. Sci.* **50**:613.

Boyse, E. A. and L. J. Old. 1969. Some aspects of normal and abnormal surface genetics. *Ann. Rev. Genet.* **3**:269.

Bretten, R., R. Wicker and W. Bernhard. 1972. Ultrastructural localization of concana-valin A receptors on normal and SV40-transformed hamster and rat cells. *Int. J. Cancer* **10**:397.

Burger, M. M. 1969. A difference in the architecture of the surface membrane of normal and virally transformed cells *Proc. Nat. Acad. Sci.* **62**:994.

———. 1973. Surface changes in transformed cells detected by lectins. *Fed. Proc.* **32**:91.

Burger, M. M. and A. R. Goldberg. 1967. Identification of a tumor-specific determina-tion on neoplastic cell surfaces. *Proc. Nat. Acad. Sci.* **57**:359.

Cline, M. J. and D. C. Livingston. 1971. Binding of ^{3}H-Concanavalin A by normal and transformed cells. *Nature New Biol.* **232**:155.

Davis, W. C. 1972. H-2 antigen on cell membranes: An explanation for the alteration of distribution by indirect labeling techniques. *Science* **175**:1006.

dePetris, S. and M. C. Raff. 1972. Distribution of immunoglobulin on the surface of mouse lymphoid cells as determined by immunoferritin electron microscopy. Anti-body-induced, temperature-dependent redistribution and its implications for mem-brane structure. *Eur. J. Immunol.* **2**:523.

———. 1973. Ultrastructural distribution and redistribution of alloantigens and con-canavalin A receptors on the surface of mouse lymphocytes. *Eur. J. Immunol.* (in press).

dePetris, S., M. C. Raff and L. Mallucci. 1973. Ligand-induced redistribution of con-canavalin A receptors on normal trypsinized and transformed fibroblasts. *Nature New Biol.* **244**:275.

Grimes, W. J. 1970. Sialic acid transferases and sialic acid levels in normal and trans-formed cells. *Biochemistry* **9**:5083.

Huet, C. and M. Herzberg. 1973. Effects of enzymes and EDTA on ruthenium red and concanavalin A labelling of the cell surface. *J. Ultrastruct. Res.* **42**:186.

Inbar, M. and L. Sachs. 1969. Interaction of the carbohydrate-binding protein concanavalin A with normal and transformed cells. *Proc. Nat. Acad. Sci.* **63**:1418.

Inbar, M. and L. Sachs. 1973. Mobility of carbohydrate containing sites on the surface membrane in relation to the control of cell growth. *FEBS Letters* **32**:124.

Inbar, M., H. Ben-Bassat and L. Sachs. 1971. A specific metabolic activity on the surface membrane in malignant cell transformation. *Proc. Nat. Acad. Sci.* **68**:2748.

————. 1972. Membrane changes associated with malignancy. *Nature New Biol.* **236**:3.

Inbar, M., C. Huet, A. R. Oseroff, H. Ben-Bassat and L. Sachs. 1973. Inhibition of lectin agglutinability by fixation of the cell surface membrane. *Biochim. Biophys. Acta* **311**:594.

Mallucci, L. 1971. Binding of concanavalin A to normal and transformed cells as detected by immunofluorescence. *Nature New Biol.* **233**:241.

Martinez-Palomo, A., R. Wicker and W. Bernhard. 1972. Ultrastructural detection of concanavalin surface receptors in normal and polyoma-transformed cells. *Int. J. Cancer* **9**:676.

Matus, A., S. dePetris and M. C. Raff. 1973. Mobility of concanavalin A receptors in myelin and synaptic membranes. *Nature New Biol.* **244**:278.

Nicolson, G. L. 1971. Difference in topology of normal and tumor cell membranes shown by different surface distributions of ferritin-conjugated concanavalin A. *Nature New Biol.* **233**: 244.

————. 1972. Topography of membrane concanavalin A sites modified by proteolysis. *Nature New Biol.* **239**:193.

————. 1973. Temperature-dependent mobility of concanavalin A sites on tumor cell surfaces. *Nature* **243**:218.

Ozanne, B. and J. Sambrook. 1971. Binding of radioactively labelled concanavalin A and wheat germ agglutinin to normal and virus-transformed cells. *Nature New Biol.* **232**:156.

Raff, M. C. and S. dePetris. 1972. Antibody-antigen reactions at the lymphocyte surface: Implications for membrane structure, lymphocyte activation and tolerance induction. *Cell interactions* (ed. L. G. Silvestri) p. 237. North-Holland, Amsterdam.

————. 1973. Movement of lymphocyte surface antigens and receptors: The fluid nature of the lymphocyte plasma membrane and its immunological significance. *Fed. Proc.* **32**:48.

Rosenblith, J. Z., H. H. Yin, T. E. Ukena, R. D. Berlin and M. J. Karnovsky. 1973. Topographical alteration of surface binding sites by concanavalin A in virally-transformed cells. *Fed. Proc.* **32**:881 (abstract).

Rowlett, C., R. Wicker and W. Bernhard. 1973. Ultrastructural distribution of concanavalin A receptors on hamster embryo and adenovirus tumor cell cultures. *Int. J. Cancer* **11**:314.

Sela, B., H. Lis, N. Sharon and L. Sachs. 1971. Quantitation of *N*-acetyl-D-galactosamine-like sites on the surface of normal and transformed mammalian cells. *Biochim. Biophys. Acta* **249**:546.

Smith, S. B. and J.-P. Revel. 1972. Mapping of concanavalin A binding sites on the surface of several cell types. *Develop. Biol.* **27**:434.

Stackpole, C. W., T. Aoki, E. A. Boyse, L. J. Old, J. Lumley-Frank and E. deHarven. 1971. Cell surface antigens: Serial sectioning of single cells as an approach to topographical analysis. *Science* **172**:472.

Taylor, R. B., W. P. H. Duffus, M. C. Raff and S. dePetris. 1971. Redistribution and pinocytosis of lymphocyte surface immunoglobulin molecules induced by anti-immunoglobulin antibody. *Nature New Biol.* **233**:225.

Unanue, E. R., W. D. Perkins and M. J. Karnovsky. 1972. Ligand-induced movement of lymphocyte membrane macromolecules. I. Analysis by immunofluorescence and ultrastructural radioautography. *J. Exp. Med.* **136**:885.

Mobility of Lectin Sites on the Surface Membrane and the Control of Cell Growth and Differentiation

Leo Sachs, Michael Inbar and Meir Shinitzky

Departments of Genetics and Biophysics, Weizmann Institute of Science
Rehovot, Israel

Molecules that bind specifically to carbohydrate-containing sites on the surface membrane can be used to elucidate changes in the surface membrane associated with changes in the regulation of cell growth (Inbar and Sachs 1969a; Burger 1969; Sela et al. 1970). Using as a probe the carbohydrate-binding protein concanavalin A (ConA) (Sumner and Howell 1936; Edelman et al. 1972), differences between normal and malignant transformed cells have been shown in ConA-induced cell agglutinability (Inbar and Sachs 1969a; Inbar, Ben-Bassat and Sachs 1971a, 1972a, 1973a; Ben-Bassat, Inbar and Sachs 1970), the number and distribution of ConA binding sites (Inbar and Sachs 1969b, 1973; Ben-Bassat et al. 1971; Shoham and Sachs 1972; Inbar et al. 1972a; Nicolson 1972), the location of amino acid and carbohydrate transport sites (Inbar et al. 1971b), ConA-induced cell toxicity (Shoham, Inbar and Sachs 1970; Inbar et al. 1972b; Wollman and Sachs 1972), and membrane stability and the level of cellular ATP (Vlodavsky, Inbar and Sachs 1973). Changes in the distribution of ConA binding sites (Ben-Bassat et al. 1971; Inbar and Sachs 1973) and the movement of antigens on the cell surface (Taylor et al. 1971; Loor, Forni and Pernis 1972; Edidin and Weiss 1972) have indicated that receptors can be mobile in a fluid surface membrane (Singer and Nicolson 1972). In the present paper we summarize our studies (Inbar and Sachs 1973; Inbar et al. 1973b,c,d,e; Shinitzky et al. 1973) to determine the mobility of ConA binding sites on the surface membrane as a probe for membrane fluidity of specific sites in relation to the regulation of cell growth in (A) normal and transformed fibroblasts, as examples of cells that form a solid tissue, (B) normal lymphocytes and lymphoma cells, as examples of cells that are in suspension in vivo and (C) in relation to the normal differentiation of myeloid leukemic cells to macrophages and granulocytes. We will also present data on the mobility on normal lymphocytes of the sites for the lectins from wheat germ and soybean.

METHODS

Cells and Cell Cultures

The transformed fibroblasts used were a line derived from a simian virus 40-induced golden hamster tumor. The normal fibroblasts were from tertiary cultures of golden hamster embryos. Normal and transformed fibroblasts were cultured in Eagle's medium with a fourfold concentration of amino acid and vitamins and 10% fetal calf serum. For the experiments normal and transformed fibroblasts at 3–4 days after seeding were dissociated with a 0.02% EDTA solution (Inbar and Sachs 1969a) by incubation for 15–30 minutes at 37°C, and the dissociated cells were then washed three times with phosphate-buffered saline (pH 7.2) (PBS). Normal lymphocytes were obtained from lymph nodes of 6- to 8-week-old male CR/RAR rats. Lymphocytes were collected by teasing the tissue apart and allowing the pieces to sediment. The lymphoma cells were from an ascites form of a Moloney virus-induced lymphoma grown in strain A mice. Adult mice were inoculated intraperitoneally with 10^5 cells, and the cells were used 10 days after inoculation. The normal lymphocytes and the lymphoma cells were collected from animals in PBS and used in the experiments after washing three times with PBS. The experiments with the myeloid leukemic cells were carried out with a tissue culture line that consists of two types of clones. One type of clone contains cells (D^+) that can be induced to undergo normal differentiation to mature macrophages and granulocytes. The other type of clone contains cells (D^-) that could not be induced to differentiate (Fibach, Hayashi and Sachs 1973). The myeloid leukemic cells were cultured in Eagle's medium with 10% inactivated horse serum (56°C for 30 minutes). For the experiments cells at 3–4 days after seeding were washed three times with PBS.

Assay for Agglutination

Concanavalin A (ConA) (Miles-Yeda) at a concentration of 30 mg/ml was kept as a solution in PBS containing 1 M NaCl at $-20°C$. To test for agglutination, 0.5 ml ConA, diluted at different concentrations in PBS, was mixed with 0.5 ml cell suspension in a 35-mm petri dish. The density and size of aggregates was scored in a scale from $-$ to $++++$ after 30 minutes of incubation. The agglutination was specific since it was completely inhibited when ConA was preincubated with 0.1 M α-methyl-D-mannopyranoside (α-MM) as a hapten inhibitor.

Assay for Binding of Radioactive ConA

ConA was labeled with [³H]acetic anhydride by the method of Miller and Great (1972). The labeled ConA was purified by an affinity chromatography on a Sephadex G-100 column and kept as a solution in PBS containing 1 M NaCl at $-20°C$. For binding of [³H]ConA to cells, 0.5 ml of [³H]ConA, diluted at different concentrations in either PBS or PBS containing 0.1 M α-MM, was mixed with 0.5 ml cell suspension in a centrifuge tube and incubated for 30 minutes. The cells were then washed three times with 5 ml PBS, the pellet dissolved in 0.1 N NaOH and the radioactivity counted in Triton scintillation fluid. To calculate the amount of [³H]ConA bound specifically, the amount bound in the presence of α-MM was subtracted from the amount bound in the absence of α-MM. The results on ConA binding are given as specific binding. Total cell protein was measured by the method of Lowry et al. (1951).

Assay for Binding of Fluorescent ConA

Fluorescein isothiocyanate-conjugated ConA (F-ConA) (Miles-Yeda) at the ratio of 1.86 fluorescein to protein was kept as a solution in PBS containing 1 M NaCl at $-20°C$. For the experiments cells were incubated with F-ConA for 30 minutes, the cells washed with PBS, and the fluorescence determined with a Leitz Ortholux microscope with transmitted UV light. With all cell types tested, 95–100% of the cells were stained at the concentration of 100 μg F-ConA/ml. The binding of F-ConA to the membrane was specific, since it was completely inhibited when F-ConA was preincubated with 0.1 M α-MM as a hapten inhibitor. Rotational diffusion analysis of fluorescent ConA and the other two fluorescent lectins was carried out as described (Shinitzky et al. 1973).

RESULTS

Specific Binding Sites for ConA on the Surface Membrane

In order to obtain direct evidence for the assumption that specific binding sites for ConA are located on the cell surface membrane, we have examined the interaction of Sepharose-conjugated ConA beads (Miles-Yeda) with malignant lymphoma cells. Cells were incubated with Sepharose-conjugated ConA beads in the presence or absence of α-MM as a specific hapten inhibitor for ConA. The results indicate that in the absence of α-MM most of the cells were bound to the beads (Fig. 1A). The interaction of the cells with the beads was specific since it was completely inhibited by preincubation of the beads with 0.1 M α-MM (Fig. 1B). Ferritin-conjugated ConA molecules were used to obtain further evidence for the presence

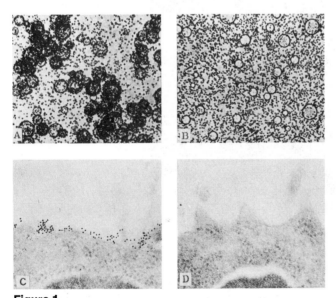

Figure 1
Binding of lymphoma cells to Sepharose-conjugated ConA beads **(A)** in the absence of α-MM, **(B)** in the presence of α-MM. Binding of ferritin-conjugated ConA molecules to the surface membrane of lymphoma cells **(C)** in the absence of α-MM, **(D)** in the presence of α-MM.

of specific binding sites for ConA on the surface membrane. Lymphoma cells were incubated with ferritin-conjugated ConA in the presence and absence of α-MM. The results indicate that the binding sites for ConA are distributed on the surface membrane (Fig. 1C) and that the binding of ferritin-conjugated ConA was specific since it was inhibited by pretreatment with α-MM.

Final Distribution of ConA Binding Sites after Binding of ConA

In order to determine the final distribution of ConA binding sites on the surface membrane, the interaction of F-ConA with normal and malignant transformed fibroblasts was examined. The experiments have indicated that in 99% of the transformed fibroblasts, the surface binding of F-ConA was in clusters of fluorescence over the cell surface (Fig. 2C). Preincubation of the transformed cells with NaN$_3$ or DNP did not inhibit the cluster formation. However most of the normal fibroblasts gave a diffuse semirandom fluorescence covering the surface membrane (Fig. 2B). Binding of F-ConA after fixation of the fluid surface membrane of normal and transformed fibroblasts with aldehyde or LaCl$_3$ resulted in an apparently complete random distribution of ConA binding sites (Fig. 2A).

Experiments on binding of F-ConA to the surface membrane of normal lymphocytes and lymphoma cells have indicated that the surface binding of F-ConA in 99% of the lymphoma cells was in small or large clusters of fluorescence that formed an incompleted ring on the cell periphery (Fig. 2C) as in the transformed

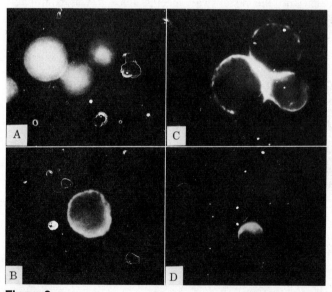

Figure 2
Distribution of F-ConA on the surface membrane of normal and transformed cells. **A,** Lymphoma cells after fixation with 2.5% glutaraldehyde. Similar results were obtained with normal lymphocytes, normal fibroblasts and transformed fibroblasts. The binding of F-ConA shows an apparently complete random distribution. **B,** Distribution of F-ConA of the type seen with normal fibroblasts, the formation of a semirandom distribution. **C,** Distribution of F-ConA of the type seen with transformed fibroblasts and lymphoma cells, the formation of clusters. **D,** Cap formation in normal lymphocytes.

fibroblasts. However about 30% of the normal lymphocytes gave a polar fluorescence cap, covering about half of the cell surface area (Fig. 2D). Preincubation of cells with NaN₃ or DNP inhibited cap formation in normal lymphocytes but did not inhibit cluster formation in the lymphoma cells (Table 1). Addition of NaN₃ or DNP to cells after binding of F-ConA resulted in dissociation of the caps, but not the clusters. Cap but not cluster formation, therefore, requires energy. Formation of caps was also inhibited by low temperature (Table 1). Binding of F-ConA after fixation of the surface membrane of normal lymphocytes and lymphoma cells with aldehyde or $LaCl_3$ resulted in an apparently complete random distribution of ConA binding sites (Fig. 2A).

Fluorescent wheat germ and soybean agglutinins did not produce caps in either normal lymphocytes or malignant lymphoma cells. In normal lymphocytes that showed caps with ConA, the remainder of the cell still stained with fluorescent wheat germ agglutinin.

Redistribution of ConA Binding Sites Induced by ConA

Differences in the final distribution of F-ConA on the surface membrane could be due to movement of ConA molecules, movement of membrane sites in the absence of ConA, or movement of ConA binding sites to form a new distribution only after interaction with ConA molecules. The experiments were carried out at saturation conditions where all the membrane sites were occupied by ConA, excluding the possibility that the differences were due to movement of ConA molecules. The second possibility was excluded by the following experiment. Incubation of normal lymphocytes with F-ConA at 0°C or 37°C resulted in binding of F-ConA, but the cells only formed caps at 37°C. To determine whether the formation of caps at 37°C is a result of movement of membrane sites without ConA, normal lymphocytes were incubated at 0°C and 37°C for 30 minutes, followed by aldehyde fixation at the two temperatures; F-ConA was added after fixation. The results showed that cap formation was completely abolished when ConA was added after fixation, indicating that caps were not formed in the absence of ConA.

The binding of F-ConA after fixation of the cell membrane (Fig. 2A) suggests that the binding sites for ConA are floating in a fluid membrane in a random distribution in normal and transformed cells. This random distribution can be changed by interaction with ConA molecules, and the final distribution of sites

Table 1

Inhibition of Movement of ConA Binding Sites

Treatment	Inhibition of cluster formation	Inhibition of cap formation
Formaldehyde 10%	+	+
Glutaraldehyde 2.5%	+	+
$LaCl_3$ 10^{-2}M	+	+
Low temperature 0°C	−	+
NaN₃ 10^{-2}M	−	+
DNP 10^{-3}M	−	+

+ = Inhibition; − = not inhibited.

Table 2

Induction of Cap Formation in Myeloid Leukemic Cells (D⁻)

	Treatment	Cells with cap (%)
Nontrypsinized cells	F-ConA	5 ± 0.5
	F-ConA + anti-ConA	22 ± 2
	F-ConA + glycogen	25 ± 2
Trypsinized cells	F-ConA	45 ± 2
	F-ConA + anti-ConA	73 ± 5
	F-ConA + glycogen	75 ± 5

Cells were treated with 1 μg purified trypsin for 15 minutes at 37°C. Anti-ConA antibodies and glycogen (100 μg) were added to cells after binding of F-ConA.

was different in the different cell types studied (Fig. 2). Cells from a mouse myeloid leukemia cell line (D⁻), in which about 5% of the cells formed caps with F-ConA, were used to obtain further evidence on the redistribution of ConA membrane sites by ConA molecules. The results show that when a higher cross-linking was induced by adding anti-ConA antibodies or glycogen after binding of ConA, the percent of cells with caps increased from about 5% to 25%. Trypsinization of the cells increased cap formation to about 45%, and addition of anti-ConA antibodies or glycogen increased the percentage of trypsinized cells with caps to about 75% (Table 2). These data support the conclusion that the redistribution of membrane sites is induced by ConA and also show that trypsinization of the cells increased the mobility of these sites.

Mobility of ConA Binding Sites

Differential Agglutination of Normal and Transformed Cells

The experiments on binding of F-ConA to normal and transformed cells indicate that the degree of site mobility increased from no or almost no change in the random distribution in normal fibroblasts (Fig. 2B) to the formation of clusters in the transformed fibroblasts and lymphoma cells (Fig. 2C) to the formation of caps in normal lymphocytes (Fig. 2D). Agglutination experiments with ConA indicate that only the cells with the intermediate degree of site mobility that form clusters are highly agglutinated cells. Normal fibroblasts with a low degree of site mobility and normal lymphocytes with a high degree of mobility have a low agglutinability (Table 3). In each cell system a similar number of radioactively labeled ConA

Table 3

ConA Binding Sites and Cell Agglutination

Cell type	Final distribution	Agglutination by ConA
Normal fibroblasts	Semi-random	±
Transformed fibroblasts and lymphoma cells	Clusters	++++
Normal lymphocytes	Caps	±

Table 4

Specific Binding of [³H]ConA to the Surface Membrane
of Normal Lymphocytes and Lymphoma Cells

Conc. [³H]ConA (μg/ml)	Input (cpm/10⁷ cells)	Specific binding (cym/1000 μg cell protein)	
		Normal lymphocytes	Lymphoma cells
1	1215	1235	985
2.5	2782	2650	2180
5	5800	4200	4020
10	11717	5170	5010
50	56385	6580	7030
100	116615	7150	7570

Similar results were obtained with normal and malignant transformed fibroblasts.

molecules were bound to normal and transformed cells per unit protein (Table 4).
The results indicate that the formation of a threshold amount of cluster formation
of ConA binding sites is required for agglutination. Further evidence for this assumption was obtained by producing the three degrees of site mobility in the same
cell type. Normal lymphocytes were treated with $LaCl_3$ or NaN_3 and the untreated
and treated cells tested for ConA agglutinability. The untreated normal lymphocytes
with caps and the $LaCl_3$-treated lymphocytes with a random distribution showed a
low ($\pm$) degree of agglutination by ConA. However the NaN_3-treated lymphocytes,
which had a clustered distribution, also had a high degree of agglutinability (Fig. 3).

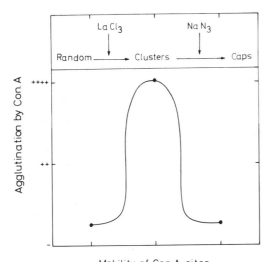

Figure 3

Mobility of ConA binding sites and the agglutinability of normal lymphocytes
by ConA. The untreated normal lymphocytes with caps and the $LaCl_3$-treated
lymphocytes with random distribution showed a low degree of agglutination ($\pm$).
However the NaN_3-treated lymphocytes with the clustered distribution showed
a high degree of agglutination ($++++$).

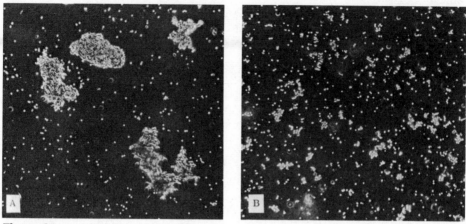

Figure 4

Agglutinability of transformed cells after fixation of the fluid surface membrane. **A,** Unfixed lymphoma cells. **B,** Lymphoma cells after fixation with 2.5% glutaraldehyde. Agglutination was also inhibited by cell fixation with 10% formaldehyde or 10^{-2}M LaCl$_3$. Similar results were also obtained with transformed fibroblasts.

Formation of clusters of ConA binding sites on the surface membrane of transformed cells was inhibited by fixation of the fluid state of the membrane. Fixation also inhibited cell agglutination by ConA (Fig. 4), although both the fixed and unfixed cells bound a similar number of radioactively labeled ConA molecules (Fig. 5).

Normal Differentiation of Myeloid Leukemic Cells to Macrophages and Granulocytes

The experiments with mouse leukemia have been carried out with a tissue culture line of myeloblastic leukemia cells that consists of two types of clones. One type of clone contains cells (D$^+$) that can be induced to undergo normal differentiation to mature macrophages and granulocytes. The other type of clone contains cells (D$^-$) that could not be induced to differentiate. In soft agar the D$^-$ cells form compact colonies (Fig. 6A), whereas the D$^+$ cells form diffuse colonies due to migration of the differentiated cells (Fig. 6B). In order to determine possible differences in the location and mobility of ConA binding sites on the surface membrane, we examined the interaction of F-ConA with D$^-$ and D$^+$ cells at 3–4 days after seeding. The results indicate that in D$^-$ clones about 95% of the cells showed the surface binding of F-ConA in the form of a ring of clusters on the cell periphery (Fig. 2C). However about 50% of the cells in D$^+$ clones and about 5% of the cells in D$^-$ clones showed a polar fluorescence cap (Fig. 2D). Similar results were obtained with four D$^-$ and four D$^+$ clones (Fig. 7). Measurements of the cell protein and surface area have indicated that D$^-$ and D$^+$ cells have a similar content of cellular protein and surface area per cell. The binding of radioactively labeled ConA molecules has indicated that D$^-$ and D$^+$ cells bind a similar number of ConA molecules per unit cell protein and appear to have a similar affinity for ConA (Fig. 8). The results show a difference in the mobility of ConA binding sites in these two types of cells and suggest that a gain of the ability of myeloid

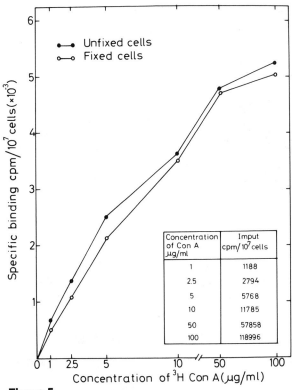

Concentration of Con A μg/ml	Imput cpm/10⁷ cells
1	1188
2.5	2794
5	5768
10	11785
50	57858
100	118996

Figure 5

Specific binding of [³H]ConA molecules to the surface membrane after fixation with 10% formaldehyde. (●) Unfixed cells, (○) fixed cells.

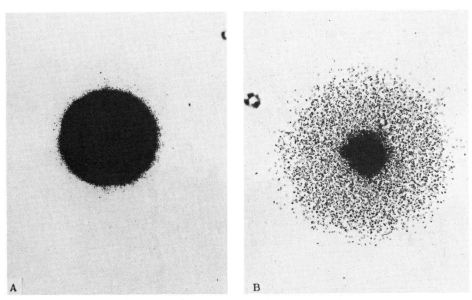

Figure 6

D⁻ and D⁺ colonies grown in soft agar. **A,** Compact colony of D⁻ cells. **B,** Dispersed colony of D⁺ cells.

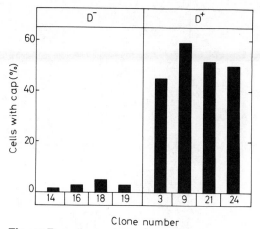

Figure 7
Cap formation after binding of F-ConA in four D$^-$ and four D$^+$ clones.

leukemic cells to undergo normal differentiation is associated with an increase in the mobility of ConA binding sites on the surface membrane.

Rotational Diffusion of Lectins Bound to the Cell Surface Membrane

Thermal motion of a specific site on a biomembrane can in principle be analyzed by determining the fluorescence polarization characteristics of a fluorophore which is specifically attached to it. This method is especially useful for studies on the mobility of receptor sites with the aid of specific fluorescent hormones, antigens or lectins.

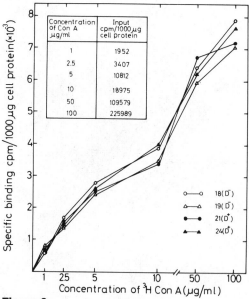

Figure 8
Specific binding of [^{3}H]ConA to the surface membrane of D$^-$ and D$^+$ cells.

The relation between thermal motion and the degree of fluorescence polarization is given by the Perrin equation

$$\frac{r_o}{r} = 1 + \frac{3\tau}{\rho} \tag{1}$$

r is the fluorescence anisotropy term which is defined by

$$r = \frac{I_{\parallel} - I_{\perp}}{I_{\parallel} + 2I_{\perp}} \tag{2}$$

where $I_{\parallel}$ and $I_{\perp}$ are the fluorescence intensities polarized parallel and perpendicular to the polarized excitation beam. r_o is the upper limit value of r where no rotations take place during the lifetime of excited state, τ. ρ is the rotational relaxation time, which is defined as the average time required for a displacement by a mean angle of arc cos $1/e$ (68° 25′). For rigid structures ρ is directly proportional to the viscosity, η, and is inversely proportional to the absolute temperature, T, in which cases $1/r$ should increase linearly with $T \times \tau/\eta$. In the case of fluorescence-labeled proteins two distinct rotations contribute to the depolarization of the fluorescence: the internal rotations of the fluorescence probe and the rotations of the protein molecule. To each of these rotations a value of a rigid structure can be assigned.

A convenient way of determinging ρ is by measuring r and τ in a series of solutions of increasing viscosity at a constant temperature (Shinitzky et al. 1973; Inbar et al. 1973c). The results are typical for Perrin plots of fluorescence-labeled proteins. At $\eta_0\tau/\eta < 1$, the fluorescence depolarization is caused almost exclusively by internal rotations of the fluorescein side chain since the rotation of the protein molecules during the excited state lifetime remain frozen. On the other hand, at $\eta_0\tau/\eta > 2$, where $1/r$ increases linearly, it can be assumed that the statistical weight of each orientation of the excited fluorescein moieties remains constant so that the changes in $1/r$ reflect only the rotations of the protein molecule. The straight line of this region extrapolates to a value of $r'_o = 0.209$, which corresponds to the case when the rotations during the lifetime of excited state of F-ConA are frozen but the fluorescein moiety rotates freely.

The presence of 0.1 M α-methyl mannoside was found to increase the r value of F-ConA in PBS by less than 1%, which indicates that the fluorescence properties of the fluorescein side chain are unaffected by the bound saccharide. Thus the value of $r'_o = 0.209$ is the upper limit of the fluorescence anisotropy of F-ConA bound to a cell surface. This value and the value of $\tau = 3.9$ nsec permit the evaluation of the rotational relaxation time of F-ConA bound to a membrane merely by determining its fluorescence anisotropy r (see Eq.1). As a criterion for site mobility one can use the ratio ρ_0/ρ, the "degree of mobility," where ρ_0 and ρ are the rotational relaxation times of F-ConA when free in solution and when bound to a cell surface.

In order to determine directly the degree of mobility of ConA binding sites on the surface membrane, we have defined a quantitative term for the degree of mobility that extends from 0, the immobilized state, to 1, the fully mobile state. The data obtained with this method (Inbar et al. 1973c) support the conclusion that the binding sites for ConA on the surface membrane are mobile and that different cell types have different degrees of mobility (Table 5).

The mobilities of F-ConA, fluorescent wheat germ and fluorescent soybean agglutinins bound to normal lymphocytes were 0.82, 0.37 and 0.43, respectively

Table 5

Mobility of ConA Binding Sites Determined by
Fluorescence Polarization Analysis

Cells	Degree of mobility*
Normal fibroblasts	0.48
Transformed fibroblasts	0.79
Normal lymphocytes	0.82
Lymphoma cells	0.36

* 0 = the immobilized state; 1 = fully mobile state

(Table 6). The lower degree of mobility of the wheat germ and soybean agglutinins
was associated with the lack of cap formation by these two lectins and their lack
of activation of DNA synthesis in normal lymphocytes (Inbar et al. 1973a;
Shinitzky et al. 1973).

CONCLUSIONS

Our results (Inbar and Sachs 1973; Inbar et al. 1973b,c,d,e; Shinitzky et al. 1973)
indicate that carbohydrate-containing structures, which are associated with binding
sites for ConA, can be mobile on the cell surface and that in the cells studied there
was a difference in the mobility of these structures in normal and malignant transformed cells. Using the mobility of ConA binding sites as a probe for fluidity, the
results show that in cells which form a solid tissue in vivo, the transformation of
normal into malignant cells is associated with an increase in membrane fluidity of
the carbohydrate-containing structures where the ConA sites are located. However in cells that are in suspension in vivo, the malignant transformation is associated with a decreased fluidity of these structures on the surface membrane. This
increased fluidity of specific sites on the membrane in the transformed fibroblasts
can explain their lack of contact inhibition, their ability to grow in soft agar, and
their malignancy in a solid tissue. The studies with normal lymphocytes have

Table 6

Mobility of Lectins on the Lymphocyte
Surface Membrane Determined by
Fluorescence Polarization Analysis

Lectin	Degree of mobility
F-ConA	0.83
F-WGA*	0.37
F-SBA**	0.43

* Fluorescein isothiocyanate-conjugated wheat
germ agglutinin (Miles-Yeda).

** Fluorescein isothiocyanate-conjugated soybean agglutinin (Miles-Yeda).

shown that in these cells there is a higher degree of mobility of the ConA sites than the sites for the lectins from wheat germ on soybean. These two lectins, in contrast to ConA, also do not activate normal lymphocytes to undergo DNA synthesis (Inbar et al. 1973a).

The results also show that there is a difference in the fluid state of the ConA binding structures on the surface membrane in myeloid leukemic cells that can be induced to differentiate normally (D$^+$ cells) and myeloid leukemic cells that could not be induced to differentiate (D$^-$ cells). A higher degree of fluidity was associated with a gain of the ability of myeloid leukemic cells to undergo normal differentiation to mature macrophages and granulocytes. The increased fluidity in D$^+$ cells was associated with an increased migration of the differentiated cells in soft agar. Differences in membrane fluidity of specific sites may explain differences in the ability of cells to respond to other differentiation-inducing stimuli, in the cellular response to hormones, and differences in cell migration in embryonic development and carcinogenesis.

Acknowledgment

This study was supported by Contract No. NCI-69-2014 with the Virus Cancer Program of the National Cancer Institute, National Institutes of Health.

REFERENCES

Ben-Bassat, H., M. Inbar and L. Sachs. 1970. Requirement for cell replication after SV40 infection for a structural change of the cell surface membrane. *Virology* **40**:854.

———. 1971. Changes in the structural organization of the surface membrane in malignant cell transformation. *J. Membrane Biol.* **6**:183.

Burger, M. M. 1969. A difference in the architecture of the surface membrane of normal and virally transformed cells. *Proc. Nat. Acad. Sci.* **62**:994.

Edelman, G. M., B. A. Cunningham, G. N. Reeke, J. W. Becker, M. J. Waxdal and J. L. Wang. 1972. The covalent and three-dimensional structure of concanavalin A. *Proc. Nat. Acad. Sci.* **69**:2580.

Edidin, M. and A. Weiss. 1972. Antigen cap formation in cultured fibroblasts: A reflection of membrane fluidity and cell motility. *Proc. Nat. Acad. Sci.* **69**:2456.

Fibach, E., M. Hayashi and L. Sachs. 1973. Control of normal differentiation of myeloid leukemic cells to macrophages and granulocytes. *Proc. Nat. Acad. Sci.* **70**:343.

Inbar, M. and L. Sachs. 1969a. Interaction of the carbohydrate binding protein concanavalin A with normal and transformed cells. *Proc. Nat. Acad. Sci.* **63**:1418.

———. 1969b. Structural differences in sites on the surface membrane of normal and transformed cells. *Nature* **223**:710.

———. 1973. Mobility of carbohydrate containing sites on the surface membrane in relation to the control of cell growth. *FEBS Letters* **32**:124.

Inbar, M., H. Ben-Bassat and L. Sachs. 1971a. A specific metabolic activity on the surface membrane in malignant cell transformation. *Proc. Nat. Acad. Sci.* **68**:2748.

———. 1971b. Location of amino acid and carbohydrate transport sites in the surface membrane of normal and transformed mammalian cells. *J. Membrane Biol.* **6**:195.

———. 1972a. Membrane changes associated with malignancy. *Nature New Biol.* **236**:3.

———. 1972b. Inhibition of ascites tumor development by concanavalin A. *Int. J. Cancer* **9**:143.

————. 1973a. Temperature sensitive activity on the surface membrane in the activation of lymphocytes by lectins. *Exp. Cell Res.* **76**:143.

————. 1973b. Difference in the mobility of lectin sites on the surface membrane of normal lymphocytes and malignant lymphoma cells. *Int. J. Cancer* **12**:93.

Inbar, M., M. Shinitzky and L. Sachs. 1973c. Rotational relation time of concanavalian A bound to the surface membrane of normal and malignant transformed cells. *J. Mol. Biol.* (in press)

Inbar, M., H. Ben-Bassat, E. Fibach and L. Sachs. 1973d. Mobility of carbohydrate-containing structures on the surface membrane and the normal differentiation of myeloid leukemic cells to macrophages and granulocytes. *Proc. Nat. Acad. Sci.* (in press)

Inbar, M., H. Ben-Bassat, C. Huet, A. R. Oseroff and L. Sachs. 1973e. Inhibition of lectin agglutinability by fixation of the cell surface membrane. *Biochim. Biophys. Acta* **311**:594.

Loor, F., L. Forni and B. Pernis. 1972. The dynamic state of the lymphocyte membrane. Factors affecting the distribution and turnover of surface immunoglobulins. *Eur. J. Immunol.* **2**:203.

Lowry, O. H., N. J. Rosenbrough, A. L. Farr and R. L. Randall. 1951. Protein measurement by the folin phenol reagent. *J. Biol. Chem.* **193**:265.

Miller, I. R. and H. Great. 1972. Protein labelling by acetylation. *Biopolymers* **11**:2533.

Nicolson, G. L. 1972. Topography of membrane concanavalin A sites modified by proteolysis. *Nature New Biol.* **239**:193.

Sela, B., H. Lis, N. Sharon and L. Sachs. 1970. Different location of carbohydrate containing sites in the surface membrane of normal and transformed mammalian cells. *J. Membrane Biol.* **3**:267.

Shinitzky, M., M. Inbar and L. Sachs. 1973. Rotational diffusion of lectins bound to the surface membrane of normal lymphocytes. *FEBS Letters* **34**:247.

Shoham, J. and L. Sachs. 1972. Differences in the binding of fluorescent concanavalin A to the surface membrane of normal and transformed cells. *Proc. Nat. Acad. Sci.* **69**:2479.

Shoham, J., M. Inbar and L. Sachs. 1970. Differential toxicity on normal and transformed cells *in vitro* and inhibition of tumor development *in vivo* by concanavalin A. *Nature* **227**:1244.

Singer, S. H. and G. L. Nicolson. 1972. The fluid mosaic model of the structure of cell membranes. *Science* **175**:720.

Sumner, J. B. and S. F. Howell. 1936. The identification of the hemagglutinin of the jack bean with concanavalin A. *J. Bacteriol.* **32**:227.

Taylor, R. B., W. P. H. Duffus, M. C. Raff and S. dePetris. 1971. Redistribution and pinocytosis of lymphocyte surface immunoglobulin molecules induced by anti-immunoglobulin antibody. *Nature New Biol.* **233**:225.

Vlodavsky, I., M. Inbar and L. Sachs. 1973. Membrane changes and adenosine triphosphate content in normal and malignant transformed cells. *Proc. Nat. Acad. Sci.* **70**:1780.

Wollman, Y. and L. Sachs. 1972. Mapping of sites on the surface membrane of mammalian cells. II. Relationship of sites for concanavalin A and an ornithine, leucine copolymer. *J. Membrane Biol.* **10**:1.

Differences in Lectin Agglutinability of Normal and Transformed Cells in Interphase and Mitosis

Jacob Shoham* and Leo Sachs

Department of Genetics, Weizmann Institute of Science, Rehovot, Israel

Differences in cell surface structure between normal and transformed cells can be demonstrated by agglutination with several plant lectins, including concanavalin A (Inbar and Sachs 1969a), wheat germ agglutinin (Burger 1969) and soybean agglutinin (Sela et al. 1970). As more than 90% of the cells in regular cultures are in interphase, these results can be taken as representative of this phase in the cell cycle. With the aid of fluorescein-conjugated concanavalin A (F-ConA), used in nonsaturation conditions, it was shown that normal and transformed cells undergo different cyclic changes in their surface membranes (Shoham and Sachs 1972). Transformed cells in interphase and normal cells in mitosis acquire surface fluorescence with F-ConA at high rate, whereas normal cells in interphase and transformed cells in mitosis acquire surface fluorescence at a lower rate.

In the present study, we have correlated the observed differences in surface fluorescence with the agglutinability of these cells and with the number of ConA molecules bound to the cell surface in these nonsaturation conditions. Cells were cultured under conditions in which a similar number of radioactively labeled ConA molecules were bound to normal and transformed cells at saturation (Ben-Bassat, Inbar and Sachs 1971).

METHODS

Cells and Cell Cultures

The cells used in these experiments were normal cells from secondary cultures of golden hamster embryos, lines of golden hamster cells transformed in vitro by Rous sarcoma virus or after treatment with the chemical carcinogen dimethylnitrosamine (Huberman, Salzberg and Sachs 1968), hamster cells transformed by polyoma virus and their revertant subline 3-1 (Rabinowitz and Sachs 1968, 1972), transformed cells from the mouse cell line 3T3 transformed by SV40 or doubly

---

* Present address: Department of Cytogenetics, Chaim Sheba Medical Center, Tel Hashomer, Israel.

transformed by polyoma and SV40. Cells were cultured in plastic petri dishes in Eagle's medium with 4-fold concentration of amino acids and vitamins and 10% fetal calf serum. The surface area of the cells was measured microscopically on samples of 100 cells.

Lectins

Purified ConA (Miles-Yeda) was labeled with fluorescein isothiocyanate (Shoham and Sachs 1972) or with tritiated acetic anhydride (Inbar, Ben-Bassat and Sachs 1973) by the method of Miller and Great (1972). Fluorescent and radioactive ConA (F-ConA and [^{3}H]ConA, respectively) retained the specific agglutination activity to the same degree as unlabeled ConA. By competition experiments it was shown that their binding capacity is also similar. Wheat germ agglutinin was used as crude extract from wheat germ lipase (Calbiochem). Purified soybean agglutinin was prepared as described (Lis et al. 1966).

Assays for Agglutination and Binding

The assays were performed on suspended cells. Interphase cell samples were prepared by dissociation of cultures with 0.02% EDTA solution (Inbar and Sachs 1969a). Mitotic cells were collected by manual agitation of unsynchronized subconfluent cultures cultivated in calcium-free medium with 7% fetal calf serum (modified from Robbins and Marcus 1964). The percent of transformed cells that were in mitosis was 75–95%, whereas in normal hamster cell samples, there were 60–65% in mitosis. The interphase or mitotic cell samples were suspended to the proper concentrations and used for agglutination assay (Inbar and Sachs 1969a) and for binding of [^{3}H]ConA and F-ConA (Inbar and Sachs 1973; Inbar et al. 1973; Shoham and Sachs 1972). Some of the experiments on the binding of F-ConA to mitotic cells were carried out on cover slides (Shoham and Sachs 1972).

RESULTS

Lectin Agglutinability of Mitotic Cells

ConA and wheat germ agglutinin agglutinated normal hamster cells in mitosis but not in interphase (Table 1). Transformed hamster and 3T3 cells of the four lines tested were agglutinated by these lectins in interphase but not in mitosis. Soybean agglutinin, which agglutinated transformed 3T3 cells in interphase, did not agglutinate these cells in mitosis. In no case were hamster cells agglutinated by soybean agglutinin (Sela et al. 1970). The appropriate sugars, α-methyl-D-mannopyranoside for ConA, N-acetyl-D-glucosamine for wheat germ agglutinin, and N-acetyl-D-galactosamine for soybean agglutinin, inhibited the agglutination. All the cells were cultured and tested under the same conditions. Treatment of the mitotic cells with EDTA solution did not change the results. Transformed 3T3 cells in mitosis had a tendency to spontaneous clump formation, but this began later than the lectin-induced agglutination.

Preparations with normal cells in mitosis were agglutinated by ConA at a slower rate than transformed cells in interphase (Fig. 1). This difference can be attributed to the presence of about 40% normal interphase cells in these mitotic preparations. Normal mitotic cells were not agglutinated by ConA at 4°C, but when returned to

Table 1

Agglutination of Normal and Transformed Cells
in Interphase and Mitosis by Three Lectins

			Degree of agglutination		
Cell type	Phase in cell cycle	% Cells in this phase	ConA	wheat germ agglutinin	soybean agglutinin
H-normal	Interphase	95	−	−	−
	Mitosis	62	++	+++	−
H-DMNA	Interphase	90	+++	+++	−
	Mitosis	78	−	−	−
	Mitosis-trypsin treated	78	++++	++++	−
H-Rous	Interphase	92	+++	+++	−
	Mitosis	84	−	−	−
3T3-SVPV	Interphase	91	++	+++	+++
	Mitosis	86	−	−	−
3T3-SV	Interphase	90	+++	+++	+++
	Mitosis	81	−	−	−

Normal, Rous- and dimethylnitrosamine (DMNA)-transformed hamster (H) cells, and 3T3 cells transformed by SV40 or by SV40 and polyoma virus were cultured in flasks. Mitotic cells were prepared by manual agitation and interphase cells by EDTA dissociation of the monolayer in the flasks.

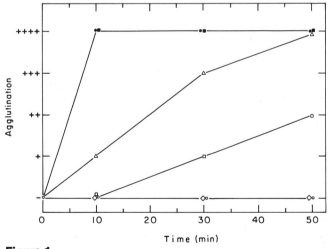

Figure 1

Agglutination of cells with 500 μg/ml ConA. Normal cells in interphase ($\diamond$) and mitosis ($\square$). Transformed cells in interphase ($\triangle$) and mitosis (o). Normal interphase ($\blacksquare$) and transformed mitotic ($\bullet$) cells, treated with 10 μg/ml trypsin for 10 minutes.

24°C, they exhibited the same agglutination as cells that had not been kept at 4°C (Inbar et al. 1971). Transformed mitotic cells treated with purified trypsin (10 μg/ml for 10 minutes) were agglutinated by ConA, as is the case with normal interphase cells (Fig. 1).

Binding of Labeled ConA to the Cell Surface

The binding of F-ConA and [³H]ConA was compared in nonsaturation conditions. The binding of F-ConA was evaluated visually by scoring cells with surface fluorescence (Shoham and Sachs 1972). The number of [³H]ConA molecules specifically bound to the cell surface was calculated from the counts of samples incubated in the absence of the sugar inhibitor minus those in its presence (Inbar and Sachs 1969b).

Cells were incubated with 0.5–10 μg/ml F-ConA or [³H]ConA for 30 minutes (Fig. 2). With 1 and 2.5 μg F-ConA, the highest percentage of fluorescent cells was found among transformed interphase and normal mitotic cells, which were very similar to each other in this respect, the next highest with transformed mitotic cells and the least with normal interphase cells. The largest differences were observed with 1 μg. With [³H]ConA no differences were found at all concentrations tested. The comparison of [³H]ConA and F-ConA binding data (Fig. 2) indicated that an average of about 1000 molecules per μ² cell surface area are needed for the appearance of surface fluorescence in transformed interphase and normal mitotic cells, 2000 molecules for transformed mitotic cells, and 4000–5000 molecules for normal interphase cells. The number of bound ConA molecules is linear with the concentrations 0.5 to 10 μg (it tends to plateau with 50 μg or more). The acquisition of surface fluorescence is nonlinear and gives a sigmoid curve.

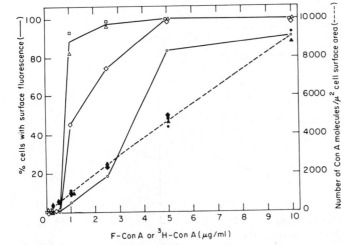

Figure 2
Binding of fluorescent and [³H]ConA to the cell surface as a function of concentration after 30 minutes incubation. Open symbols, percent cells with surface fluorescence; closed symbols, number of ConA molecules bound per μ² cell surface area. (o, •) Normal interphase cells; (□, ■) normal mitotic cells; (△, ▲) transformed interphase cells; (◇, ◆) transformed mitotic cells. The cells were used four days after seeding and were tested at 24°C.

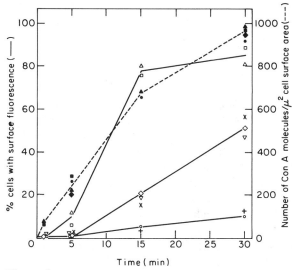

Figure 3
Binding of 1 μg/ml fluorescent and [³H]ConA to the cell surface as a function of time. Open symbols, percent cells with surface fluorescence; closed symbols, number of ConA molecules bound per μ^2 cell surface area. (o, •) Normal interphase cells; (□, ■) normal mitotic cells; (△, ▲) transformed interphase cells; (◇, ♦) transformed mitotic cells; (×) revertant interphase cells. These cells were used 4 days after seeding and tested at 24°C. (▽) Transformed interphase cells one day after seeding. (+) Transformed interphase cells 4 days after seeding tested at 4°C.

By using short incubation times, the rate of binding of 1 μg F-ConA and [³H]ConA was also compared (Fig. 3). The cells again differed in the rate of appearance of surface fluorescence but not in the rate of binding of ConA molecules. It is interesting to note that only the high rate of fluorescence appearance, which occurred in transformed interphase and normal mitotic cells, could be correlated with agglutinability of these cells (Table 2). This correlation was also found in other situations. Trypsinized normal cells in interphase become agglutinable and acquire

Table 2
Time of Acquisition of Surface Fluorescence and Cell Agglutinability

Cell type	Time* (minutes)	Agglutination
Trypsin-treated normal and transformed interphase	5	++++
Transformed interphase	10	+++
Normal mitotic	10	++
Transformed mitotic	30	—
Revertant interphase	30	—
Normal interphase	> 60	—

* Required in 50% of the cells with 1 μg/ml F-ConA.

cell surface fluorescence at a high rate. Transformed interphase cells are non-agglutinable during the first day after seeding (Ben-Basset et al. 1971) and the cells in the present experiments were always used at 4 days after seeding. A revertant subline of polyoma-transformed cells (Rabinowitz and Sachs 1968, 1972) is also nonagglutinable. In both cases the rate of acquisition of surface fluorescence is similar to that of transformed cells in mitosis, and all three are in intermediate positions between normal and transformed interphase cells (Table 2, Fig. 3).

With 50 μg [^{3}H]ConA for 30 minutes, the amount of ConA molecules bound specifically to the cell surface reached a plateau, again without difference between normal and transformed cells in either interphase or mitosis (Table 3). At these saturation conditions about 12,000 molecules were bound per μ^2 cell surface area of both types of mitotic and interphase cells. No differences among these cells could be demonstrated with 25 μg or more of F-ConA.

Using fluorescein-conjugated wheat germ agglutinin, Fox, Sheppard and Burger (1971) have shown that normal cells in mitosis exhibit surface fluorescence similar to that of transformed interphase cells. Changes in glycoprotein and/or glycolip content (Glick, Gerner and Warren 1971) were demonstrated in KB cells in mitosis.

CONCLUSIONS

Our results indicate cyclic changes in the cell surface which are different for normal and transformed cells. Normal cells are agglutinated by ConA and the lectin from wheat germ during mitosis but not in interphase. After incubation with F-ConA at nonsaturation conditions the normal mitotic cells acquire surface fluorescence at the same rate as transformed interphase cells. Transformed cells that agglutinate by these lectins in interphase do not agglutinate in mitosis. The rate of surface fluorescence acquisition by these transformed cells in mitosis was also lower than the rate with transformed interphase cells. No differences were found among these cells in the amount of bound [^{3}H]ConA molecules in saturation and nonsaturation conditions. These experiments indicate that the number of ConA binding sites on the cell surface, as well as their affinity to ConA, does not remarkably differ among these cells.

The demonstrated differences in the percentage of cells with surface fluorescence can be attributed to the clustering of bound fluorescent ConA molecules. With low concentrations of F-ConA there is a greater chance for the fluorescent molecules to be located in close proximity to each other, when the sites are clustered, and this represents a more concentrated light source, easily perceived by the eye. The threshold conditions for the appearance of cell surface fluorescence may reflect the relative degree of cluster formation, which was high with transformed interphase and normal mitotic cells and lower with transformed mitotic and normal interphase cells. Only the high degree of cluster formation permits ConA agglutination. It has been shown (Inbar and Sachs 1973) that the formation of clusters is induced by ConA binding of ConA, and there is an increased mobility of the ConA sites on the surface of transformed interphase compared to normal interphase fibroblasts.

Acknowledgment

This work was supported by Contract No. 69-2014 with the Virus Cancer Program of the National Cancer Institute, United States Public Health Service.

Table 3

Binding of $[^3H]$ConA to Normal and Transformed Cells in Interphase and Mitosis

Cell type	Phase in cell cycle	cpm per 2×10^6 cells $[^3H]$ConA $(\mu g/ml)$				Cell surface area (μ^2)	No. molecules per μ^2 cell surface at saturation $(\times 10^{-3})$
		1	5	50	250		
H-normal	Interphase	640	3272	9791	11654	970	12
	Mitosis	773	3500	11153	13027	1130	11.6
H-DMNA	Interphase	593	3347	10215	11245	830	13.5
	Mitosis	662	3912	11284	12702	1020	12.5

The cells were prepared as in Table 1 and incubated with the listed concentrations of $[^3H]$ConA for 30 minutes at 24°C. H = hamster; DMNA = dimethylnitrosamine-transformed.

REFERENCES

Ben-Bassat, H., M. Inbar and L. Sachs. 1971. Changes in the structural organization of the surface membrane in malignant cell transformation. *J. Membrane Biol.* **6**:183.

Burger, M. M. 1969. A difference in the architecture of the surface membrane of normal and virally transformed cells. *Proc. Nat. Acad. Sci.* **62**:994.

Fox, T. O., J. R. Sheppard and M. M. Burger. 1971. Cyclic membrane changes in animal cells: Transformed cells permanently display a surface architecture detected in normal cells only during mitosis. *Proc. Nat. Acad. Sci.* **68**:244.

Glick, M. C., E. W. Gerner and L. Warren. 1971. Changes in the carbohydrate content of the KB cell during the growth cycle. *J. Cell. Physiol.* **77**:1.

Huberman, E., S. Salzberg and L. Sachs. 1968. The *in vitro* induction of an increase in cell multiplication and cellular life span by the water-soluble carcinogen dimethyl-nitrosamine. *Proc. Nat. Acad. Sci.* **59**:77.

Inbar, M. and L. Sachs. 1969a. Interaction of the carbohydrate-binding protein concanavalin A with normal and transformed cells. *Proc. Nat. Acad. Sci.* **63**:1418.

————. 1969b. Structural difference in sites on the surface membrane of normal and transformed cells. *Nature* **223**:710.

————. 1973. Mobility of carbohydrate containing sites on the surface membrane in relation to the control of cell growth. *FEBS Letters* **32**:124.

Inbar, M., H. Ben-Bassat and L. Sachs. 1971. A specific metabolic activity on the surface membrane in malignant cell transformation. *Proc. Nat. Acad. Sci.* **68**:2748.

————. 1973. Temperature sensitive activity on the surface membrane in the activation of lymphocytes by lectins. *Exp. Cell Res.* **76**:143.

Lis, H., N. Sharon and E. Katchalski. 1966. Soybean hemagglutinin, a plant glycoprotein. I. Isolation of a glycopeptide. *J. Biol. Chem.* **241**:684.

Miller, I. R. and H. Great. 1972. Protein labeling by acetylation. *Biopolymers* **11**:2533.

Rabinowitz, Z. and L. Sachs. 1968. Reversion of properties in cells transformed by polyoma virus. *Nature* **220**:1203.

————. 1972. The formation of variants with a reversion of properties of transformed cells. VI. Stability of the reverted state. *Int. J. Cancer* **9**:334.

Robbins, E. and P. I. Marcus. 1964. Mitotically synchronized mammalian cells. *Science* **144**:1152.

Sela, B. A., H Lis, N. Sharon and L. Sachs. 1970. Different locations of carbohydrate containing sites in the surface membrane of normal and transformed mammalian cells. *J. Membrane Biol.* **3**:267.

Shoham, J. and L. Sachs. 1972. Differences in the binding of fluorescent concanavalin A to the surface membrane of normal and transformed cells. *Proc. Nat. Acad. Sci.* **69**:2479.

Cell Surface and Growth Control of Chick Embryo Fibroblasts in Culture

Antti Vaheri, Erkki Ruoslahti, and Tapani Hovi

Departments of Virology and of Serology and Bacteriology
University of Helsinki, SF-00290 Helsinki 29, Finland

The evolution of multicellular organisms from monocellular units demands mechanisms for the control of cell growth and movement. It is predictable that the cell surface is centrally involved in this regulation and several lines of experimental evidence support this view. Cell surface proteins and carbohydrates seem to be crucial for recognition both of distant and local signals affecting growth control. Receptors for several mitogens, such as hormones and lectins, are localized on the cell membrane (Cuatrecasas 1972; Chase and Miller 1973) and interaction with the surface can be sufficient to trigger the sequence of events leading to cell proliferation (Greaves and Bauminger 1972). It is apparent that these cellular recognition mechanisms have an important function both in differentiation and control of cell proliferation and that in malignantly transformed cells they are largely inoperative.

The aim of our recent work has been to define some of the molecular mechanisms operative in growth and recognition phenomena. We use density-inhibited cultures of chick embryo fibroblasts as the experimental model. This system responds to minute amounts of certain substances, presumed to act on the cell surface, by proliferation. Early changes in membrane function precede proliferation in activated cells. Our data on the role of cyclic purine nucleotides in growth control of these cells suggest that variation in surface-mediated regulatory mechanisms may exist between different types of cells.

PROCEDURES

Density-inhibited chick embryo fibroblasts (Sefton and Rubin 1970) were prepared by seeding 10^6 cells from trypsinized primary cultures in 5 ml of medium 199 containing 2% tryptose phosphate broth and 2% chicken serum in 20 cm² plastic petri dishes. After 48 hours at 38.5°C the cultures were confluent (with about 1.5×10^6 cells per culture) and had a low rate of DNA synthesis (Vaheri, Ruoslahti and Nordling 1972).

The uptake of labeled substances by cells was assayed according to Hatanaka and Hanafusa (1970) with some modification (Vaheri et al. 1973a). The cultures were washed twice with warm Hanks balanced salt solution without glucose and then incubated for 10 minutes at 38.5°C with 2 ml of the same solution containing the labeled compound diluted to 6×10^{-6} M. The cells were washed rapidly four times with cold Hanks solution and their radioactivity and protein content was determined.

Incorporation of [^{3}H]thymidine, [^{3}H]uridine and a [^{3}H]amino acid mixture into DNA, RNA and protein, respectively, was performed using 60 minute pulses and 1 μCi/ml.

Insulin (26–27 IU/mg, Novo, Copenhagen) and cyclic GMP were determined by radioimmunoassays (Pharmacia, Uppsala; Collaborative Research, Inc., Mass.). Cyclic AMP levels in cells were assayed from trichloroacetic acid extracts by competitive protein binding (Gilman 1970) using rabbit skeletal muscle as the source of cyclic AMP binding protein (Reimann, Walsh and Krebs 1971).

RESULTS

Release from Growth Control by Insulin

Density-inhibited chick fibroblasts responded with proliferation to the addition of substances such as serum, trypsin or papain in accordance with what is known from similar studies by other workers (Burger 1970; Sefton and Rubin 1970). We have recently reported (Vaheri et al. 1972, 1973a) that insulin, a well-established growth factor (Temin 1967), also releases chick fibroblasts from density-dependent inhibition of growth (DDI). Microgram amounts of insulin released the cells from the G_1 phase in an almost synchronous wave of proliferation (Table 1). The early stages of reinitiation of proliferation could be monitored by assaying of cell volume and sugar transport. Within 2 to 3 hours the uptake of D-[^{3}H]2-deoxyglucose or D-[^{3}H]glucosamine increased two- to threefold but not that of [^{3}H]leucine or [^{3}H]thymidine. The increase in sugar uptake is one of several surface properties that rapidly growing cells, or cells released from DDI, transiently share with transformed cells, as discussed previously (Vaheri et al. 1973b).

As the insulin-fibroblast model seemed well suited for analysis of reinitiation of

Table 1

Sequence of Events after Activation of Density-inhibited
Chick Embryo Fibroblasts by Insulin

Change	Hours after insulin addition
Increase in sugar uptake	1–2
Increase in cell volume	1–2
Stimulation of protein and RNA synthesis	1–3
Increase in [^{3}H]dT incorporation	
start	4–5
maximum	8–9
Peak of mitosis	12–13

growth, we wanted to establish that interaction with the cell surface is sufficient to activate the resting cell. That this may be the case was indicated by the finding that under the conditions giving maximal cell proliferation, less than 0.002–0.2% of insulin was taken up by the cells as measured by radioimmunoassay or using immunoreactive ^{125}I-labeled insulin (Vaheri et al. 1973a). Attempts to substantiate this using insolubilized insulin were not successful. Insulin coupled to agarose (Axén, Porath and Ernback 1967) and repeatedly washed at low and high pH 8 M urea and phosphate buffer stimulated cells but only insofar as insulin was released in soluble form from the particles. Radioimmunoassay showed that, whereas only 0.1% of the coupled insulin was released in 20 hours in buffered saline, up to 10% became soluble in the presence of cells and serum. Similar results were obtained when the release was measured as detachment of radioactivity from Sepharose particles to which ^{125}I-labeled insulin had been coupled. We have recently had similar experiences with insolubilized papain (Enzite-EMA, Miles-Yeda). It seems that any experiments involving such insolubilized preparations should be interpreted with caution.

Stimulation of Fibroblasts by Mitogens

Our recent data (Vaheri et al., unpublished) indicate that three different mitogens, lipopolysaccharide (LPS), the pokeweed lectin (PWM) and tuberculin (PPD), reinitiate proliferation of stationary fibroblasts (Table 2). The results show that the stimulatory effect of these substances, generally implied as lymphocyte mitogens, is not restricted to lymphoid cells. The doses needed for optimal effect are 10,000 times (LPS) or 10 times (PWM, PPD) lower than those giving full activation of lymphocytes (Andersson et al. 1973; Nilsson, Sultzer and Bullock 1973).

The three mitogens active in the fibroblast system are known to activate thymus-independent (B) lymphocytes, whereas the two selective T cell mitogens concanavalin A (ConA) and the *Phaseolus vulgaris* lectin preparations were inactive in our system. ConA, which activates isolated fat cells simulating insulin in its mode of action (Cuatrecases and Tell 1973), did not even increase sugar uptake in

Table 2

Activation of Density-inhibited Cultures of Chick
Embryo Fibroblasts by Lymphocyte Mitogens

Substance	*Min. dose giving full effect (ng/ml)*	*Rate [³H]dT incorp. at 12 hr (ratio to control)*	*Cell number at 30 hr ($\times 10^{-6}$)*
Lipopolysaccharide Ra	0.1	11.2	3.13
Crude pokeweed mitogen	100*	8.3	2.97
Tuberculin	3000	7.4	2.81
Neuraminidase	300	9.0	2.63
Insulin	1000	20.9	3.66
Trypsin	3000	21.7	3.75
Control		1.0	1.71

* The value indicates the protein concentration in the crude preparation.

our system. Chick embryo fibroblasts are agglutinated by high concentrations of ConA (Burger and Martin 1972), indicating that it does bind to those cells. It seems possible that the divalent ConA lectin molecules immobilize the cell membrane and prevent its activation.

As little as 100 pg/ml of LPS was sufficient to stimulate density-inhibited chick embryo fibroblasts. One of the LPS preparations, LPS-Ra from a *Salmonella typhimurium* strain (Vaheri et al., unpublished), had similar effects at 100 pg/ml and 100 μg/ml. LPS preparations with quite different side chains gave practically identical results, indicating that the architecture of the sugar chain is not critical. This is in accordance with the findings by Andersson et al. (1973) that in B lymphocyte activation the lipid A moiety common to LPS's from different gram-negative bacteria is responsible for the total mitogenic effect of intact LPS. Brailow-sky et al. (1973) showed that LPS preparations in high doses, 1–10 μg/ml, in-hibited the proliferation of transformed but not of normal rat embryo fibroblasts. This reaffirms that detailed analysis of the LPS–cell interaction could contribute to the assessment of the role the cell surface has in growth control.

Stimulators with Reversible or Irreversible Effect

Short exposure to a mitogen may be sufficient to initiate the complete sequence of events leading to proliferation, or a continuous cell-stimulator interaction may be needed for full response. We have tested the different chick embryo fibroblast mitogens for reversibility of their effect. The results were similar whether assayed by [³H]dT incorporation (Table 3), cell counting or sugar uptake. The effect of the potent stimulators, insulin and trypsin, is reversed by removal of the mitogen-containing culture medium. Rapidly dividing cultures (12 hours after exposure to insulin or trypsin) can be brought to rest in this way. Such cultures may be switched on again by either one of these substances. On the other hand the effect of some other stimulators (neuraminidase and the lymphocyte mitogens) is largely

Table 3

Reversible and Irreversible Stimulators of Cell Proliferation in Density-inhibited Cultures of Chick Embryo Fibroblasts

		Rate of [³H]dT incorporation at 16 hr (cpm/culture)	
Substance	*Concentration*	*No change of medium*	*Change of medium*
Control		1440	1630
Calf serum	5%	28540	1530
Insulin	5 μg/ml	27850	1210
Trypsin	5 μg/ml	30650	1170
Neuraminidase	1 μg/ml	11320	10790
Lipopolysaccharide	0.001 μg/ml	13960	13650

* Three hours after addition of the stimulatory substance to the medium the cell monolayers were washed twice with medium 199 and the medium was replaced with conditioned medium from unstimulated density-inhibited cultures.

irreversible; however, our experiments so far do not differentiate between a true triggering effect or simple uptake of the stimulator.

The chick embryo fibroblast and murine 3T3 fibroblast models differ in the reversibility of the effect of trypsin and other proteolytic enzymes. Change of the culture medium after addition of trypsin removed the mitogenic stimulus from the chick embryo fibroblast system, whereas, according to Burger (1970), a very brief proteolytic treatment leads to mitosis in 3T3 cells. The variety of proteins apparently cleaved when trypsin is used to stimulate cells makes it difficult to evaluate the significance of such differences.

Enhancing Effect of Cyclic Purine Nucleotides on Chick Embryo Fibroblasts—Possible Role of Cyclic GMP in Stimulation

Several recent reports suggest a crucial role for cyclic AMP in regulation of fibroblast proliferation: Addition of high levels (1–10 mM) of cyclic AMP or its butyryl derivatives to culture medium slows down the rate of proliferation of several mammalian cell lines (Johnson and Pastan 1972). Evidence from certain clones of the murine 3T3 cell line suggests that in a culture approaching confluency, cell-to-cell contacts induce an increased intracellular concentration of cyclic AMP, which is assumed to mediate the observed decline of proliferation (Burger et al. 1972). Our data (Hovi and Vaheri 1973) indicate that, while the cyclic purine nucleotides are associated with proliferation control in chick embryo fibroblasts, differences from 3T3 cells are obvious. This conclusion is derived from the following ·experimental observations.

1. Under slightly restrictive conditions (2% serum, pH 7.4) the proliferation of secondary chick embryo fibroblasts was significantly enhanced (Table 4) by 10^{-4} M concentrations of cyclic AMP, cyclic GMP, and certain other purine nucleotides. Several criteria indicated that these enhancing effects were on proliferation rather than on cell attachment.
2. Under less restrictive or nutritionally "optimal" conditions (2% serum and pH 7.1 or 5% serum and 10% tryptose phosphate) the stimulatory effects were

Table 4

Enhancing Effect of Cyclic Purine Nucleotides on Growth of Chick Embryo Fibroblasts

Substance	*Concentration*	*Cell number at 72 hr* $(\times 10^5)$
Control		9.60
Cyclic AMP	10^{-3} M	14.21
	10^{-4} M	14.94
Dibutyryl cyclic AMP	10^{-4} M	12.96
Cyclic GMP	10^{-6} M	12.85
Adenosine	10^{-4} M	13.25
Inosine-5′-monophosphate	10^{-4} M	15.63

5×10^5 cells were initially seeded per dish. The medium contained 2% calf serum and the cultures were maintained at 37° C and pH 7.4.

minimal, but no inhibitory effect on cell proliferation was seen under any conditions with doses as high as 1 mM of cyclic AMP or its mono- or dibutyryl derivatives. Addition of these components to culture medium caused a rapid increase in the cellular cyclic AMP concentration. These levels remained high for at least one cell cycle.

3. When chick embryo fibroblasts were released from DDI by either insulin, serum, trypsin or neuraminidase, we detected no significant alteration of intracellular cyclic AMP levels, assayed at frequent intervals from 0.5 minutes to 24 hours after activation.

4. Our recent experiments (Hovi, Keski-Oja and Vaheri, in prep.) suggest that a rapid increase of intracellular cyclic GMP accompanies the release of chick embryo fibroblasts from density-dependent inhibition of growth. This is in accordance with observations on lectin-induced blast transformation of human lymphocytes (Hadden et al. 1972), suggesting that cyclic GMP may in some cells have functions related to those of cyclic AMP in 3T3 cells.

CONCLUDING REMARKS

We have found that exceedingly small amounts of LPS and certain other lymphocyte mitogens stimulate density-inhibited chick embryo fibroblasts. This suggests that cells possess "receptors" through which a cycle of events leading to proliferation can be triggered in a highly specific way. The integration of LPS with lipid bilayers seems to be well documented (Rothfield and Romeo 1971). It appears unlikely that integration of LPS molecules in any site of the membrane is sufficient to trigger the cell to proliferation. The fluid state of the plasma membrane (Singer and Nicholson 1972) favors rapid interaction of integrated LPS with other membrane components. In a recent abstract Springer, Adye and Jirgensons (1973) report isolation from human red cells of a physiochemically homogeneous lipoglycoprotein, which interacted strongly with the LPS's of all the gram-negative bacteria tested. This also suggests that high affinity LPS receptors may occur in cells.

The chick embryo fibroblast and murine 3T3 cell models respond differently to added cyclic AMP. The growth of chick embryo fibroblasts is enhanced by this and several other purine nucleotides, whereas added cyclic AMP slows down the proliferation of 3T3 cells (Johnson and Pastan 1972). A rapid decrease of intracellular cyclic AMP levels seems to accompany the release of 3T3 cells from DDI (Sheppard 1972, Burger et al. 1972). We detected no significant change in chick embryo fibroblasts in the level of cyclic AMP after addition of any mitogen. Instead an increase in intracellular cyclic GMP seems to accompany initiation of proliferation of chick embryo fibroblasts. It is entirely possible that the recognition sites for mitogenic signals may be coupled in different cell types to different effector systems such as the adenyl cyclase in the plasma membrane. It appears that receptors for the different mitogens, such as lectins and hormones, may exist in most or all cells, but in a target cell the number of high affinity receptors is high and a change in their conformation is able to affect the effector system.

Although exogenous mitogens have served as useful probes for the characterization of mechanisms involved in the control of cell proliferation, little is known about the molecules involved in these phenomena in vivo. As suggested by Goldschneider and Moscona (1972) tissue-specific cell surface antigens make good candidates for molecules with such functions. In order to learn about the nature of the cell surface components involved in recognition and regulatory processes, we have used brief proteolytic digestion to solubilize proteins from the surface of chick or human embryo fibroblasts and studied the released material immunochemically. This has led to the identification (Ruoslahti et al. 1973) of a major fibroblast-specific cell surface protein that is also present in the homologous serum. We are currently evaluating the possible role of such proteins in the regulation of cell proliferation.

Acknowledgments

We thank Miss Leena Aura, Miss Pirkko Korpela and Miss Anja Nieminen for their competent technical assistance and Miss Leena Salminen for helping with the manuscript. This research was supported by grants from the Finnish Medical Research Council and the Finnish Cancer Foundation.

REFERENCES

Andersson, J., F. Melchers, C. Galanos and O. Lüderitz. 1973. The mitogenic effect of lipopolysaccharide on bone marrow-derived mouse lymphocytes. Lipid A as the mitogenic part of the molecule. *J. Exp. Med.* **137**:943.

Axén, R., J. Porath and S. Ernback. 1967. Chemical coupling of peptides and proteins to polysaccharides by means of cyanogen halides. *Nature* **214**:1302.

Brailowsky, C., M. Trudel, R. Lallier and V. N. Nigram. 1973. Growth of normal and transformed rat embryo fibroblasts. Effects of glycolipids from *Salmonella minnesota* R mutants. *J. Cell Biol.* **57**:124.

Burger, M. M. 1970. Proteolytic enzymes initiating cell division and escape from contact inhibition of growth. *Nature* **227**:170.

Burger, M. M. and G. S. Martin. 1972. Agglutination of cells transformed by Rous sarcoma virus by wheat germ agglutinin and concanavalin A. *Nature New Biol.* **237**:9.

Burger, M. M., B. M. Bombik, B. M. Breckenbridge and J. R. Sheppard. 1972. Growth control and cyclic alterations of cyclic AMP in the cell cycle. *Nature New Biol.* **239**:161.

Chase, P. S. and F. Miller. 1973. Preliminary evidence for the structure of concanavalin-A binding site on human lymphocytes that induces mitogenesis. *Cell. Immunol.* **6**:132.

Cuatrecasas, P. 1972. Properties of the insulin receptor isolated from liver and fat cell membranes. *J. Biol. Chem.* **247**:1980.

Cuatrecasas, P. and G. P. E. Tell. 1973. Insulin-like activity of concanavalin A and wheat germ agglutinin: Direct interactions with insulin receptors. *Proc. Nat. Acad. Sci.* **70**:485.

Gilman, A. G. 1970. A protein binding assay for adenosine 3′:5′-cyclic monophosphate. *Proc. Nat. Acad. Sci.* **67**:305.

Goldschneider, I. and A. A. Moscona. 1972. Tissue-specific cell-surface antigens in embryonic cells. *J. Cell Biol.* **53**:435.

Greaves, M. F. and S. Bauminger. 1972. Activation of T and B lymphocytes by insoluble phytogens. *Nature New Biol.* **235**:67.

Hadden, J. W., E. M. Hadden, M. K. Haddox and N. D. Goldberg. 1972. Guanosine 3′:5′-cyclic monophosphate: A possible mediator of mitogenic influences in lymphocytes. *Proc. Nat. Acad. Sci.* **69**:3024.

Hatanaka, M. and H. Hanafusa. 1970. Analysis of a functional change in membrane in the process of cell transformation by Rous sarcoma virus; alteration in the characteristics of sugar transport. *Virology* **41**:647.

Hovi, T. and A. Vaheri. 1973. Cyclic AMP and cyclic GMP enhance growth of chick embryo fibroblasts. *Nature* (in press).

Johnson, G. S. and I. Pastan. 1972. Role of 3':5'-adenosine monophosphate in regulation of morphology and growth of transformed and normal fibroblasts. *J. Nat. Cancer Inst.* **48**:1377.

Nilsson, B. S., B. M. Sultzer and W. W. Bullock. 1973. Purified protein derivative of tuberculin induces immunoglobulin production in normal mouse spleen cells. *J. Exp. Med.* **137**:127.

Reimann, E. M., D. A. Walsh and E. G. Krebs. 1971. Purification and properties of rabbit skeletal muscle adenosine 3':5'-monophosphate-dependent protein kinases. *J. Biol. Chem.* **246**:1986.

Rothfield, L. and D. Romeo. 1971. Role of lipids in the biosynthesis of the bacterial cell envelope. *Bact. Rev.* **35**:14.

Ruoslahti, E., A. Vaheri, P. Kuusela and E. Linder. 1973. Fibroblast surface antigen: A new serum protein. *Biochem. Biophys. Acta* (in press).

Sefton, B. M. and H. Rubin. 1970. Release from density dependent growth inhibition by proteolytic enzymes. *Nature* **227**:843.

Sheppard, J. R. 1972. Difference in cyclic adenosine 3':5'-monophosphate levels in normal and transformed cells. *Nature New Biol.* **236**:14.

Singer, S. J. and G. L. Nicolson. 1972. The fluid mosaic model of the structure of cell membranes. *Science* **175**:720.

Springer, G. F., J. C. Adye and B. Jirgensons. 1973. Endotoxin receptor from human red cells. *Fed. Proc.* **32**:673.

Temin, H. M. 1967. Studies on carcinogenesis by avian sarcoma viruses. VI. Differential multiplication of uninfected and of converted cells in response to insulin. *J. Cell. Physiol.* **69**:377.

Vaheri, A., E. Ruoslahti and S. Nordling. 1972. Neuraminidase stimulates division and sugar uptake in density-inhibited cell cultures. *Nature New Biol.* **85**:211.

Vaheri, A., E. Ruoslahti, T. Hovi and S. Nordling. 1973a. Stimulation of density-inhibited cell cultures by insulin. *J. Cell. Physiol.* **81**:355.

———. 1973b. Cell surface and initiation of proliferation. *The biology of fibroblasts* (ed. E. Kulonen and J. Pikkarainen) Academic Press, New York.

The Transformed Cell Surface: An Analysis of the Increased Lectin Agglutinability and the Concept of Growth Control by Surface Proteases

K. W. Talmadge, K. D. Noonan, and M. M. Burger

Biocenter, University of Basel, Klingelbergstrasse 70
CH 4056 Basel, Switzerland

Within the last ten years, a number of investigators have employed plant agglutinins (lectins) as probes for studying the cell surface of normal and transformed cells (Aub, Tieslau and Lankester 1963; Burger and Goldberg 1967; Inbar and Sachs 1969; Sharon and Lis 1972). This work has been given impetus by the finding that, whereas cells transformed by DNA or RNA tumor viruses agglutinate at very low concentrations of agglutinin, the normal parent cell line agglutinates only at much higher concentrations of the same agglutinin (Burger 1969; Burger and Martin 1972). It has also been found that normal cells treated with very low concentrations of proteases agglutinate at the same low concentration of agglutinin as do the transformed cells (Burger 1969; Inbar and Sachs 1969). This work has strongly suggested that a structural alteration exists in the cell surface of the transformed cell which can be imitated in normal cells by mild protease treatment.

Three hypotheses were originally considered as possible explanations for the lectin-mediated agglutination of transformed cells: an increase in the number of receptor sites present in the surface membrane, de novo synthesis of agglutinin receptor sites, or a conformational rearrangement of the cell surface such as to move the agglutinin receptor sites into a configuration permitting agglutinin binding and subsequent agglutination (Burger 1970a). According to the third model the agglutinin receptor sites on the transformed cell surface would be in a conformation that allowed agglutinin binding and subsequent agglutination. In normal cells, however, less binding would occur and trypsinization would expose the agglutinin receptor sites, allow agglutinin binding and permit agglutination. This model came to be known as the "cryptic-site hypothesis" (Burger 1970a).

Several agglutinin receptors have been isolated and partially purified (Janson and Burger 1973; Wray and Walborg 1971). However it must be stressed that these receptors have not yet been sufficiently characterized, and therefore it has not been ruled out that there are qualitative differences between the receptors from transformed and untransformed cells. Preliminary results indicate that both the transformed and the untransformed cell types seem to contain a similar number of agglutinin receptor sites. This data, plus the fact that protease-treated normal

313

cells agglutinate at the same lectin concentrations as transformed cells, made the first two explanations unlikely.

One of the best tests of the validity of the "cryptic-site hypothesis" would have been a demonstration that transformed cells bind significantly more agglutinin molecules than do the normal cells. However a variety of laboratories (Cline and Livingston 1971; Ozanne and Sambrook 1971; Arndt-Jovin and Berg 1971; Ben-Bassat, Inbar and Sachs 1971) have suggested that there is little or no difference in agglutinin binding between normal and transformed cells. We have recently demonstrated that, if care is taken to avoid endocytosis and nonspecific binding of the agglutinin molecule to the cell surface, transformed cells bind more than three times more concanavalin A than do the normal cells (Noonan and Burger 1973a). Table 1 demonstrates that at 0°C polyoma virus-transformed 3T3 mouse fibroblasts bind at least twice as much concanavalin A at saturation if compared to the 3T3 cells and if the number of sites is calculated on a per cell basis. Since transformed cells often differ in their size from their parental normal cells and since agglutination depends on the density of sites on the cell surface and not the total number of sites per cell, comparisons between site densities or sites per protein are more meaningful. If the number of binding sites are compared, therefore, per μg total cell protein, the transformed Py3T3 cells have three times more concanavalin A receptors; or if the comparisons are made per μ^2 surface area, then the Py3T3 cells have almost six times more concanavalin A binding sites. At room temperature where pinocytosis and other phenomena increase the "binding," this difference disappears.

Increased lectin binding has been demonstrated in three other instances so far using the binding assay at low temperature.

A temperature-sensitive simian virus 40-transformed 3T3 fibroblast line was isolated by Renger and Basilico (1972). The defect that prevents expression of the transformed state at 38°C seems not to be in the oncogenic virus present in the cell but rather in the function of a temperature-sensitive product coded by the host cell. At 32°C the cell grows and behaves like a transformed cell and agglutinates well with concanavalin A and wheat germ agglutinin. At 38°C however the same cell does not agglutinate as well as the untransformed 3T3 fibroblast and

Table 1

Molecules ^{3}H-ConA Bound to 3T3 and Py3T3 Cells

	0° C		22°C	
	Py3T3	*3T3*	*Py3T3*	*3T3*
Molecules bound/cell	1.3×10^6	6.6×10^5	2.7×10^7	4.1×10^7
Molecules bound/μg protein	4.8×10^8	1.8×10^8	9.9×10^9	1.1×10^{10}
Molecules bound/μ^2 surface area	1.7×10^3	3.0×10^2	3.5×10^4	1.8×10^4

The number of molecules ConA bound at saturation was determined from the known specific activity of 1.7×10^7 cpm per mg ConA. 3T3 mouse fibroblasts contained 37.2 μg protein per 1×10^4 cells; polyoma virus-transformed 3T3 fibroblasts contained 27.2 μg protein per 1×10^4 cells.

The surface area of 2200 μ^2 per 3T3 and 750 μ^2 per Py3T3 cell was determined according to procedures given in Noonan and Burger (1973a).

binds four times less labeled concanavalin A than at 32°C (Noonan et al. 1973).

If untransformed 3T3 cells are briefly treated with a low dose of any of a series of proteases, they will become as agglutinable as the virally, chemically or X-ray transformed cells. An increase in concanavalin A binding of twofold per μg cell protein is found following these protease treatments (Noonan and Burger 1973a).

3T3 fibroblasts were shown to bind more fluorescein-labeled wheat germ agglutinin during mitosis and perhaps also shortly after mitosis (Fox, Sheppard and Burger 1971). This observation was corroborated by Shoham and Sachs (1972) with fluorescent concanavalin A, and we have recently also shown that isolated mitotic 3T3 cells required about five to eight times less wheat germ agglutinin or concanavalin A for half-maximal agglutination. It can be calculated from Fig. 1 that mitotic cells bind about three times more [³H]concanavalin A. Using ferritin-labeled concanavalin A and electron microscopy, Nicolson and Singer (1971) have similarly seen a small increase in agglutinin binding between transformed and normal cells. However they attributed the better agglutinability to a topographical difference, i.e., to the presence of receptor site clusters in transformed cell surfaces. The limited but significant increase in available receptor sites may be sufficient to explain the increased agglutinability of some transformed cells; for others

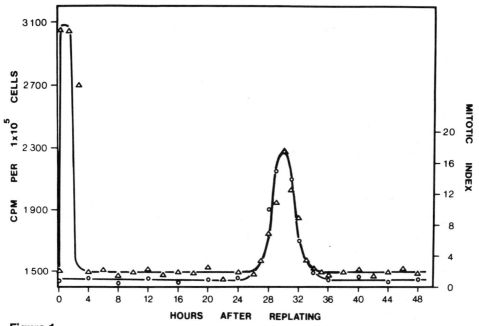

Figure 1

³H-ConA binding to 3T3 cells in mitosis. In this particular experiment the 3T3 cells were synchronized by replating the cells to a lower cell density from confluency (time 0).

Metaphase figures were counted with a phase microscope every hour and the percent of total cells expressed as mitotic index. Note that the ordinate does not begin at 0 and that the cells bind after replating, i.e., a protease treatment, temporarily more lectin. At the peak of mitotic activity 18% of the cells bind 810 cpm more per 10^5 cells, which amounts to more than three times higher binding capacity of mitotic as compared to interphase cells. From Noonan, Levine and Burger (1973) with permission of *J. Cell Biology*.

it may be just contributing and for some it may not explain the increased agglutinability at all.

Two other pieces of evidence have recently appeared that further complicate the interpretation of the mechanism which leads to agglutination of transformed cells. Inbar, Ben-Bassat and Sachs (1971) have first reported and we have confirmed (Noonan and Burger 1973b) that transformed cells and trypsin-treated normal cells do not agglutinate with concanavalin A at 0°C. Furthermore Nicolson and Singer (1971) have demonstrated that the pattern of ferritin-labeled concanavalin A binding to the surface of normal cells is random and homogenous, whereas the pattern of ferritin-labeled concanavalin A on the transformed cell surface is patchy and nonrandomly distributed (Nicolson 1971). This has led Nicolson and Singer (1971) to suggest that prior to the addition of concanavalin A, the agglutinin receptor sites on the normal cells are in a different arrangement as compared to the transformed cell surface and that it is this difference in spatial arrangement which accounts for the preferential agglutination of transformed cells (Nicolson 1972). Nicolson (1972) has suggested that such a clustering of the agglutinin receptor sites would allow for a better cross-linking between cells and thereby overcome the electrostatic forces normally keeping two cells apart. Two other laboratories have similar data suggesting that clumping of the agglutinin receptor sites may be responsible for agglutination (Ben-Bassat, Inbar and Sachs 1971; Bretton, Wicker and Bernhard 1972); but other investigators using hemocyanin as a visual marker do not detect significant differences in the topographical distribution of concanavalin A in transformed and normal cell surfaces (Smith and Revel 1972). A thorough investigation into the possible contributions of ferritin or the fixation and preparation procedures to the patchy site distribution will still be required before the clustering of lectins with their receptor sites can be accepted as a general phenomenon.

We have recently considered in a review that secondary effects, which occur after the binding of the lectin to the carbohydrate site on the cell surface, may also influence the outcome of the agglutination (Burger 1973a). One possibility among those was the mobility, or the distribution due to the mobility, of the receptor sites (distributability). Several experiments begin to indicate that this so-called mobility may be relevant for the agglutination process. Proteins seem to be free to move in the lateral plane of the membrane (Frye and Edidin 1970; Taylor et al. 1971; Edidin and Weiss 1972; Yahara and Edelman 1972; Loor, Forni and Pernis 1972). Lectins may therefore cross-link freely moving receptor sites within the plane of the cell surface, thus creating the clusters, rather than making the clusters visible, that were assumed to preexist in the electron microscope after being labeled with ferritin concanavalin A.

We have recently found that treatment with the bivalent fixative glutaraldehyde prevented the agglutinability of transformed cells although these conditions did not interfere with the binding of the lectin (Table 2). This experiment was interpretated to indicate that the cross-linking and denaturing of surface proteins and thereby presumably the immobilization of the proteins abolished agglutinability. As mentioned before Inbar, Ben Bassat and Sachs (1971) have reported that agglutination with concanavalin A did not proceed at 0°C and they suggested that a temperature-sensitive component in the membrane, like an enzyme, may be necessary for the agglutination. We have carried out some detailed temperature-dependence studies for agglutination by concanavalin A over the range from 0–37°C

Table 2

Effect of Mild Glutaraldehyde Fixation on Agglutination

	Cells agglutinated (50 μg ConA/ml after 15 min)	Molecules bound (per μg protein)
Py3T3 cells[a]	10%	4.6×10^8
3T3 cells[b]	10%	3.8×10^8
3T3 + 0.001% trypsin[c]	15%	3.8×10^8

In part from Noonan and Burger (1973b).
Conditions basically the same as in Table 1.

[a] Preincubated with 0.05% glutaraldehyde for 15 min at 37°C and washed three times with phosphate-buffered saline.

[b] Preincubated with 0.05% glutaraldehyde for 15 min at 37°C, washed three times with phosphate-buffered saline, treated with 0.001% trypsin for 5 min and 0.005% soybean inhibitor added.

[c] For 5 min, 0.05% soybean inhibitor added and then incubated with 0.5% glutaraldehyde for 15 min.

and found a rather sharp transition point between 12 and 18°C (Noonan and Burger 1973b). This observation suggests that below the freezing point for lipids agglutination does not occur, indicating again that concanavalin A agglutination may require some freedom of mobility.

None of these observations, however, indicate that the increase in agglutinability in transformed cells would be caused by an increased mobility of macromolecules within the membrane. In the case of wheat germ agglutinin, mobility seems hardly to matter since most cells tested agglutinate well at 4°C.

In Table 3 are listed some of the possible surface alterations that could influence agglutination. At the present time we have to be aware that most of these possibilities may also explain the increased agglutinability of transformed cells. The two most conspicious candidates are small differences in the number of available receptor sites and differences in the degree of molecular mobility within the surface membrane.

If the membrane change as detected by agglutination is important for growth control, then it would be predicted that a mild protease treatment that leads to a transient agglutinability of untransformed cells could also transiently release untransformed cells from density-dependent inhibition of growth. This prediction has been borne out and demonstrated in the case of mouse 3T3 fibroblasts (Burger 1970b) and chick embryo fibroblasts (Sefton and Rubin 1970). A mild protease treatment caused a growth stimulation and the cells went through one round of cell division. Thereafter they reached their usual resting state and their surfaces returned to the nonagglutinable state after repair of the protease effect, preventing further rounds of cell division. Not all parameters seem to have been worked out so far since the growth response in 3T3 cells is not yet fully reproducible.

These results suggest the existence of a proteolytic activity associated with the surface membranes of transformed cells that might be important in the release from contact inhibition of growth (growth control). Two different approaches have been used to measure this proteolytic activity and to assess its relevance for

Table 3

Survey of Possible Surface Alterations Influencing Agglutinability

Primary alterations present prior to and independent of agglutinin-surface interaction

Qualitative changes in chemical nature of receptor site

Quantitative changes in amount of receptor present in surface membrane

Topographical changes in position of receptor

    Perpendicular to surface plane: transformed cell receptors available from outside; normal cell receptors cryptic

    Within the surface plane: transformed cell receptors inhomogenously distributed (clustered)

Secondary alterations expressed only during binding and agglutination process

Distribution of receptor sites

    Mobility of receptor sites

    Repulsion or attraction between receptor sites

Cell-cell interaction after agglutinin binding

    Availability of bound agglutinin to neighboring cell

        Not available for sterical reasons

        Not available since lectin valencies absorbed by multivalent receptor molecules

    Zetapotential

    Lipophilic interactions

    Flexibility of surface membrane

In part from Burger (1973a).

growth control. Schnebli (1972) examined intact monolayers of Py3T3 and normal 3T3 cells for their ability to produce TCA-soluble material from ^{14}C Chlorella protein. The transformed cells were able to degrade the protein three to ten times faster than the normal cells, although the final level of degradation of the protein was the same for both cell types. Also Unkeless et al. (1973) and Ossowski et al. (1973) have reported that a number of different fibroblast cultures, transformed by either DNA or RNA viruses, show increased fibrinolytic activity as compared to the normal counterpart cultures.

The other approach has been to look at the effect of various protease inhibitors on the growth characteristics of transformed cells. A number of protease inhibitors with different modes of action were found to inhibit the growth of several different transformed mouse cell lines. Normal cells required considerably higher doses for inhibition of growth (Schnebli and Burger 1972). However the question remains in this study as to where in the cell these inhibitors are acting. The growth inhibition could be a consequence of uptake of these molecules and action within the cells, or they could be interacting with a component on the cell surface which is important in a critical event in growth control. In order to answer this question we have directed our attention to the interaction of Py3T3 cells with the protease inhibitor ovomucoid, which was covalently bound to polymer particles.

The proteins used in these studies were attached to the insoluble support of polyacrylamide beads, Bio-Gel P-10, 400 mesh. These spherical beads have a diameter of less than 37 μ and are of comparable size as the Py3T3 cells that are attached to the culture dish. Ovomucoid has a MW of 28,000 and is excluded from the matrix of Bio-Gel P-10 and thus should be present only on the surface of the

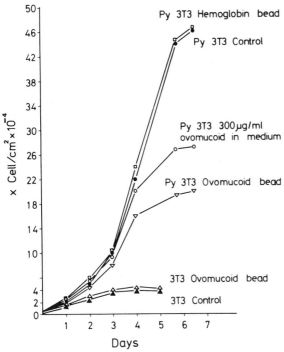

Figure 2

Growth curves for Py3T3 and 3T3 cells in the presence and absence of ovomucoid beads. The cells were plated in Dulbecco's modified Eagle's medium containing 10% calf serum in 3.5-cm petri dishes (Falcon) at a cell density of 2–6 $\times$ $10^3/cm^2$. After 24 hr the medium was replaced with 2 ml medium containing 5% calf serum and the beads were added in 100–200 μl calcium- and magnesium-free phosphate-buffered saline (CMF-PBS). Every 2 or 3 days thereafter the medium was changed and beads were replaced in each plate. Cell counts were made every 24 hr by removing the beads with several washings of CMF-PBS and the cells released from the plates with trypsin-EDTA. The cells were counted in a hemocytometer where the beads could easily be distinguished from cells. The number of beads added was generally between 5 $\times$ 10^5–10 $\times$ 10^5 per plate, which covers the plate with a monolayer of beads. It is to be noted that while ovomucoid beads did not stick to the cells, hemoglobin beads did show some unspecific attachment to the cells. The beads were sterilized by washing with 70% ethanol and then washing extensively under sterile conditions with CMF-PBS.

beads. The proteins were coupled to the *p*-aminobenzamidoethyl derivative of the polyacrylamide beads via the diazonium salt intermediate according to the procedures of Inman and Dintzis (1969). This derivative allows the protein to extend from the bead matrix, which is of importance here, as the ovomucoid may be interacting with a component on the cell surface.

Four preparations of the polyacrylamide bead-coupled ovomucoid have been tested for their ability to inhibit the growth of Py3T3 cells. Figure 2 gives an example of the growth curves for Py3T3 and 3T3 cells in the presence and absence of ovomucoid beads. Table 4 summarizes the results from a number of experiments.

Table 4

Details of Growth Inhibition Experiments with Substituted Polyacrylamide Beads

Addition	No. of Expts.	Inhibition at saturation (%)	No. of bead batches	Protein/10^6 beads (μg)	Protein/ml medium (μg)
Ovomucoid beads	10	57 ± 6	4	310, 100, 420, 130	155, 50, 210, 65
Ovomucoid (300 μg/ml)	6	49 ± 14			
Uncoated beads	3	2 ± 3			
Hemoglobin beads	4	1 ± 3	1	410	205
Ovalbumin beads	1	10	1	280	140
Bovine serum albumin beads	1	7	1	300	150
Fetuin beads	1	2	1	80	40

The growth experiments were performed as described in Fig. 2. Protein attached to the beads was estimated in either of two ways. In one case the difference in protein content of the supernatant before and after reacting the protein solution with the beads was determined. The other and more accurate way was to label the ovomucoid with [³H]acetic anhydride and then to determine the radioactivity released from a known number of the beads following acid hydrolysis. The results from the two procedures were comparable.

These results demonstrate that "insolubilized" ovomucoid has retained the ability to interact with the cells in such a way that a substantial but not complete inhibition of growth of the transformed cells occurs. There is little or no effect on the normal 3T3 cells. Uncoated or hemoglobin-coated beads have no effect on the growth properties of the Py3T3 cells. The hemoglobin beads used (Fig. 2) contain approximately 420 μg protein per 10^6 beads and the ovomucoid beads 310 μg per 10^6 beads. A 57% inhibition of growth of the Py3T3 cells has been obtained with ovomucoid beads containing as little as 100 μg protein per 10^6 beads, which corresponds to a maximum "concentration" of 50 μg ovomucoid per ml of medium. This is to be compared with a concentration of 300 μg/ml soluble ovomucoid, which still gave less growth inhibition than the beads. The inhibition becomes apparent at the earliest after 3 or 4 days (Fig. 2) and would seem to require at least four to five generations before the effect can be observed. If the ovomucoid beads are washed off the plates after the 4th or 5th day, the cells recover again their normal growth properties; but it takes about 2 or 3 days to regain the density of untreated Py3T3 cells. Studies with beads to which four other proteins and glycoproteins were attached indicate that the ovomucoid beads were the only ones to give a significant growth-inhibitory response (Table 4).

Incubation of beads coupled to labeled ovomucoid with or without cells shows very little release of radioactivity during the incubation period (Table 5). At most 4% of the total radioactivity is recovered in the soluble fraction and less yet in the cellular pellet. This is several orders of magnitude less than what was necessary to obtain a similar effect with soluble ovomucoid. Similar results have been obtained with several different preparations of labeled ovomucoid beads.

To determine if the observed growth-inhibitory action of the ovomucoid beads was due to the binding and inactivation of growth factors in the serum or other possible effects upon the medium, ovomucoid beads and hemoglobin beads were incubated with the medium for 36 hours at 37°C. The supernatants obtained after this incubation were then added to Py3T3 cells and growth monitored. The supernatants from the preincubations of ovomucoid beads had an impaired capability to sustain growth. A loss in growth-sustaining capability was, however, also seen with hemoglobin beads and with control medium incubated for 36 hours

Table 5
Ovomucoid Leaking from Beads

			Ovomucoid	
	Treatment	*Fraction measured*	*cpm $\times$ 10^{-2}*	*μg/ml*
Ovomucoid beads	1500 $\times$ g	supernatant	25	1.1
	1500 $\times$ g	pellet	1110	46.2
Ovomucoid beads + Py3T3 cells	wash	cells	10	0.4
	1500 $\times$ g	supernatant	40	1.6
	1500 $\times$ g	pellet	1030	42.9

9×10^5 ovomucoid coated beads (100 μg/10^6 beads) were incubated for 48 hr in 2 ml medium containing 5% calf serum with and without 1 $\times$ 10^6 Py3T3 cells. Released radioactivity was measured after treatments indicated. The radioactivity in the pellets was determined after acid hydrolysis.

except that it was lower. The specific ovomucoid inhibition, however, accounts for only about 13% of the effect that is obtained using the ovomucoid beads (Table 6).

The results clearly suggest that growth inhibition does not involve uptake of the ovomucoid and that it may be related to a critical event in the plasma membrane. The most attractive hypothesis is that the ovomucoid is inhibiting a proteolytic enzyme on the cell surface. An alternative explanation could be that the ovomucoid beads are interacting with a component that has been excreted from the cells.

Since Py3T3 cells are quite a bit smaller than 3T3 cells, they reach a monolayer at about 4–6 $\times$ 10^4 cells per cm^2 and do not pile up before 1–2 $\times$ 10^5 cells per cm^2. In the presence of the ovomucoid beads the Py3T3 cells stop growing at a density of around 2 $\times$ 10^5 cells per cm^2, i.e., just before the point where they would start piling up. Since the beads are spherical and there is approximately one bead per cell only, and particularly since the beads cannot reach the cell surface where it is attached to the culture dish, the actual surface area in contact with beads is probably much less than one-half of the total. Nevertheless the magnitude of the growth inhibition by the ovomucoid beads was considerably greater than that obtained with the soluble ovomucoid, which should reach also the lower face of the cell surface where it is only attached via some pseudopodes to the culture dish (see Table 4). This fact is very suggestive and can actually be used as additional evidence that the ovomucoid beads are interacting at the level of the plasma membrane and not via ovomucoid that was released from the beads and entered the cell. The beads would in effect present the ovomucoid in concentrated form to the cell surface. Other explanations cannot be excluded at the present time, as for instance, the possibility that multiple ovomucoid sites can attach to multiple sites on the cell surface simultaneously and increase the efficiency by acting in a cooperative or integrated manner.

Table 6

Growth-sustaining Capacity of Medium Pretreated with Ovomucoid Beads

	Growth inhibition at saturation (%)	Specific ovomucoid bead inhibition as compared to hemoglobin beads
Medium + ovomucoid beads	20 ± 5	13 ± 5
Medium + hemoglobin beads	11 ± 5	0
Medium	6 ± 5	0
Ovomucoid beads	57 ± 6	100

The same number of hemoglobin and ovomucoid beads were added separately to medium containing 5% calf serum and the suspensions incubated for 36 hr at 37°C. In 4 experiments the concentration of beads used in the preincubation varied from 1 $\times$ 10^5 to 10 $\times$ 10^5 beads/ml medium. After preincubation the suspensions were centrifuged at 1500 $\times$ g and the supernatants were then added to Py3T3 cells and growth studies performed as described in Fig. 2.

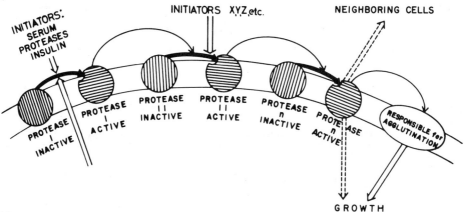

Figure 3

Cascade of proteolytic factors acting as a chain of events between growth-stimulatory agents on the membrane and subsequent parts of the same growth stimulation mechanism inside the cell. Serum, proteases, insulin or perhaps also mitosis (arrow from inside the cell) may activate the same first inactive protease into an active form. As an alternative, these or any of a series of initiators may also act at different places before the cascade or within the cascade. Each activated protease in turn activates another one until the nth one. The last one may directly initiate intracellular processes or indirectly, via surface membrane, affect alterations like the increased agglutinability. The increased agglutinability could have been triggered by earlier steps in the cascade, and any of them may be capable of acting on proteins or cells in the immediate environment of this particular surface membrane. From Burger (1973b) with permission of MIT Press.

We have earlier mentioned (Burger 1970b, 1971) and recently elaborated (Burger 1973b) that the common denominator of some of the multitude of treatments that can trigger growth in the cell cultures may be a cascading system of proteases that are membrane attached (see Fig. 3). We have also pointed out that such cascading systems are not purely hypothetical but are precedented in nature in at least the mechanisms leading to complement lysis and blood clotting (Burger 1973b). It is therefore quite interesting that fibrinolytic activity was recently found to be increased in transformed cells in a two-component system requiring both a cell factor and a serum factor (Unkeless et al. 1973; Ossowski et al. 1973). Such a multistep model has heuristic value: On one hand it would explain how many different treatments of cell cultures might start the chain reaction that eventually leads to the same phenomenon, namely, growth promotion. On the other hand such a multistep process would permit multiple and subtle control, which would certainly be an important requirement for as vital a phenomenon as cell growth known to be subject to a multitude of conditions and controlling factors.

Acknowledgments

This work was supported by a grant and a contract SVCP with the NCI of the National Institutes of Health of the U.S.A. In addition KWT was supported by Public Health Service Research Grant CA 51981 from the National Cancer Institute.

REFERENCES

Arndt-Jovin, P. J. and P. Berg. 1971. Quantitative binding of I^{125} concanavalin A to normal and transformed cells. *J. Virol.* **8**:716.

Aub, J. C., C. Tieslau and A. Lankester. 1963. Reaction of normal and tumor cell surfaces to enzymes. I. Wheat-germ lipase and associated mucopolysaccharides. *Proc. Nat. Acad. Sci.* **50**:613.

Ben-Bassat, H., M. Inbar and L. Sachs. 1971. Changes in the structural organization of the surface membrane in malignant cell transformation. *J. Membrane Biol.* **6**:183.

Bretton, R., R. Wicker and W. Bernhard. 1972. Ultrastructural localization of concanavalin A receptors in normal and SV40-transformed hamster and rat cells. *Int. J. Cancer* **10**:397.

Burger, M. M. 1969. A difference in the architecture of the surface membrane of normal and virally transformed cells. *Proc. Nat. Acad. Sci.* **62**:994.

―――. 1970a. Changes in the chemical architecture of transformed cell surfaces. *Permeability and function of biological membranes* (ed. L. Bolis et al.) p. 107. North-Holland, Amsterdam.

―――. 1970b. Proteolytic enzymes initiating cell division and escape from contact inhibition of growth. *Nature* **227**:170.

―――. 1971. The significance of the surface structure changes for growth control under crowded conditions. *Ciba Fndn. Symp., Growth control in cell culture* (ed. G. E. W. Wolstenholme and J. Knight) p. 45. Churchill Livingstone, London.

―――. 1973a. Surface changes in transformed cells detected by lectins. *Fed. Proc.* **32**:91.

―――. 1973b. The surface membrane and cell-cell interactions. *The neurosciences: Intensive study program,* MIT Press (in press).

Burger, M. M. and A. R. Goldberg. 1967. Identification of a tumor-specific determinant on neoplastic cell surfaces. *Proc. Nat. Acad. Sci.* **57**:359.

Burger, M. M. and G. L. Martin. 1972. Agglutination of cells transformed by Rous sarcoma virus by wheat germ agglutinin and concanavalin A. *Nature New Biol.* **237**:9.

Cline, M. J. and D. C. Livingston. 1971. Binding of ^{3}H-Concanavalin A by normal and transformed cells. *Nature New Biol.* **232**:155.

Edidin, M. and A. Weiss. 1972. Antigen cap formation in cultured fibroblasts: A reflection of membrane fluidity and of cell motility. *Proc. Nat. Acad. Sci.* **69**:2456.

Fox, T. O., J. R. Sheppard and M. M. Burger. 1971. Cyclic membrane changes in animal cells: Transformed cells permanently display a surface architecture detected in normal cells only during mitosis. *Proc. Nat. Acad. Sci.* **68**:244.

Frye, L. D. and M. Edidin. 1970. The rapid intermixing of the cell surface antigens after formation of mouse-human heterokaryons. *J. Cell Sci.* **7**:319.

Inbar, M. and L. Sachs. 1969. Interaction of the carbohydrate-binding protein concanavalin A with normal and transformed cells. *Proc. Nat. Acad. Sci.* **63**:1418.

Inbar, M., H. Ben-Bassat and L. Sachs. 1971. A specific metabolic activity on the surface of membranes in malignant cell transformation. *Proc. Nat. Acad. Sci.* **68**:2748.

Inman, J. K. and H. M. Dintzis. 1969. The derivatization of cross-linked polyacrylamide beads. Controlled interaction of functional groups for the preparation of special-purpose biochemical adsorbents. *Biochemistry* **8**:4074.

Janson, V. K. and M. M. Burger. 1973. Isolation and characterization of agglutinin receptor sites. II. Isolation and partial purification of surface membrane receptors for wheat germ agglutinin. *Biochim. Biophys. Acta* **291**:127.

Loor, M., J. Forni and M. Pernis. 1972. The dynamic state of the lymphocyte membrane. Factors affecting the distribution and turnover of surface immunoglobulin. *Eur. J. Immunol.* **2**:203.

Nicolson, G. L. 1971. Difference in topology of normal and tumour cell membranes shown by different surface distribution of ferritin-conjugated concanavalin A. *Nature New Biol.* **223**:244.

———. 1972. Topography of membrane concanavalin A sites modified by proteolysis. *Nature New Biol.* **239**:193.

Nicolson, G. L. and S. J. Singer. 1971. Ferritin-conjugated plant agglutinins as specific saccharide stains for electron microscopy: Application to saccharide bound to cell membranes. *Proc. Nat. Acad. Sci.* **68**:942.

Noonan, K. D. and M. M. Burger. 1973a. Binding of [^{3}H]concanavalin A to normal and transformed cells. *J. Biol. Chem.* **248**:4286.

———. 1973b. The relationship of concanavalin A binding to lectin initiated cell agglutination. *J. Cell Biol.* (in press)

Noonan, K. D., A. J. Levine and M. M. Burger. 1973. Cell cycle dependent changes in the surface membrane as detected with ^{3}H-concanavalin A. *J. Cell Biol.* **58**:491.

Noonan, K. D., H. C. Renger, C. Basilico and M. M. Burger. 1973. Surface changes in temperature-sensitive simian virus 40-transformed cells. *Proc. Nat. Acad. Sci.* **70**:347.

Ossowski, L., J. C. Unkeless, A. Tobia, J. P. Quigley, D. B. Rifkin and E. Reich. 1973. An enzymatic function associated with transformation of fibroblasts by oncogenic viruses. Mammalian fibroblast cultures transformed by DNA and RNA tumor viruses. *J. Exp. Med.* **137**:112.

Ozanne, B. and J. Sambrook. 1971. Binding of radioactively labelled concanavalin A and wheat germ agglutinin to normal and virus-transformed cells. *Nature New Biol.* **232**:156.

Renger, H. C. and C. Basilico. 1972. Mutation causing temperature sensitive expression of cell transformation by a tumor virus. *Proc. Nat. Acad. Sci.* **69**:109.

Schnebli, H. P. 1972. A protease-like activity associated with malignant cells. *Schweiz. Med. Wschr.* **102**:1194.

Schnebli, H. P. and M. M. Burger. 1972. Selective inhibition of growth of transformed cells by protease inhibitors. *Proc. Nat. Acad. Sci.* **69**:3825.

Sefton, B. M. and H. Rubin. 1970. Release from density dependent growth inhibition by proteolytic enzymes. *Nature* **277**:843.

Sharon, N. and H. Lis. 1972. Lectins, cell agglutinating and sugar-specific proteins. *Science* **177**:949.

Shoham, J. and L. Sachs. 1972. Differences in the binding of fluorescent concanavalin A to the surface membrane of normal and transformed cells. *Proc. Nat. Acad. Sci.* **69**:2479.

Smith, S. B. and J. P. Revel. 1972. Mapping of concanavalin A binding site on the surface of several cell types. *Develop. Biol.* **27**:434.

Taylor, R., P. Duffus, M. C. Raff and S. dePetris. 1971. Redistribution and pinocytosis of lymphocyte surface immunoglobin molecules induced by anti-Ig antibody. *Nature New Biol.* **233**:225.

Unkeless, J. C., A. Tobia, L. Ossowski, J. P. Quigley, D. B. Rifkin and E. Reich. 1973. An enzymatic function associated with transformation of fibroblasts by oncogenic viruses. I. Chick embryo fibroblast cultures transformed by avian RNA tumor viruses. *J. Exp. Med.* **137**:85.

Wray, V. P. and E. F. Walborg. 1971. Isolation of tumor cell surface binding sites for concanavalin A and wheat germ agglutinin. *Cancer Res.* **31**:2072.

Yahara, I. and G. Edelman. 1972. Restriction of the mobility of lymphocyte immunoglobulin receptors by concanavalin A. *Proc. Nat. Acad. Sci.* **69**:608.

Growth Inhibition of Tumor Cells by Protease Inhibitors: Consideration of the Mechanisms Involved

Hans Peter Schnebli

Friedrich Miescher-Institut, Basel, Switzerland

Mild treatment with proteases has been shown to temporarily release normal cultured fibroblasts from contact inhibition (Burger 1970; Sefton and Rubin 1970). Increased (Bosmann 1972; Schnebli 1972) or altered (Kazakova et al. 1972) hydrolase activities have been found in malignant cells. Furthermore Unkeless et al. (1973) and Ossowski et al. (1973) have found that transformed but not normal cells induce a fibrinolytic activity in the serum in which they are cultured. All of this led to the suggestion that a protease-like activity might be responsible or at least required for the loss of contact inhibition in transformed cells. If this hypothesis were correct, it follows that protease inhibitors should affect the growth and saturation density of transformed cells in culture.

Indeed, it was found that a number of protease inhibitors selectively inhibit the growth of transformed cells (Schnebli and Burger 1972). As an example, the effect of N-α-tosyl-L-lysyl-chloromethane (TLCK) on the growth of SV3T3 and 3T3 cells is shown in Fig. 1. A strong inhibition of growth of the transformed cells is produced by 50 μg/ml of TLCK and the saturation density of this culture is about the same as that of 3T3 cells. Py3T3 and 3T12 cells were similarly sensitive to TLCK (not shown), but the growth of 3T3 cells was not affected by TLCK concentrations of up to 50 μg/ml.

In addition to TLCK (an active site titrant for trypsin), a number of other protease inhibitors with different specificities and modes of action also affect the growth of transformed 3T3 cells selectively. These include N-α-tosyl-L-arginyl-methyl-ester (TAME), a substrate and competitive inhibitor for trypsin; N-α-tosyl-L-phenylalanyl-chloromethane (TPCK), an active site titrant for chymotrypsin; and ovomucoid and Trasylol (a pancreatic protease inhibitor), which form strong macromolecular complexes with trypsin.

Protease inhibitors not only inhibit the growth of transformed mouse cells (Schnebli and Burger 1972; Prival 1972), but also affect the growth and morphology of hamster tumor cells (Goetz, Weinstein and Roberts 1972). M. Weber (pers. commun.) has shown that TLCK reversibly reduces the hexose transport of RSV-transformed chick fibroblasts to levels closer to those of un-

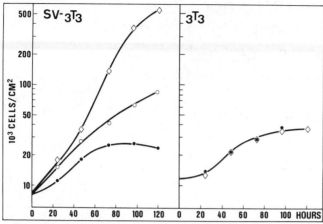

Figure 1

Growth curves of SV3T3 and 3T3 cells in the presence and absence of TLCK. The cells were plated in 3.5-cm petri dishes (Falcon) 48 hours before the addition of the inhibitors. At the beginning (0 hours) and every 24 hours thereafter the medium with or without inhibitor (1.5 ml containing 10% fetal calf serum, Microbiol. Ass.) was changed in each plate. Each point represents the average of five counts on two plates; duplicates generally agreed to within less than 10%.

($\diamond$) Control; (o) 25 μg/ml TLCK; ($\bullet$) 50 μg/ml TLCK. From Schnebli and Burger (1972) with permission of *Proc. Nat. Acad. Sci.*

transformed cells. In addition it has been shown that protease inhibitors suppress dimethylbenzanthracene-induced and phorbol ester-promoted tumorigenesis in mouse skin (Troll, Klassen and Janoff 1970; Hozumi et al. 1972).

All of this supports the postulate that an inhibitor-sensitive protease-like activity is required by malignant transformed cells for unrestrained growth, i.e., loss of contact inhibition. In this report we attempt to define the target of the protease inhibitors and probe their ability to restore contact inhibition in transformed cells.

MATERIALS AND METHODS

Cells

3T3 mouse fibroblasts and their polyoma- and SV40-transformed (Py3T3, SV3T3) derivative lines were kindly supplied by Dr. H. Green, M.I.T., Cambridge, Mass. The cells were grown in Dulbecco's modified Eagle's medium containing 10% calf serum (Gräub, Bern, Switzerland). Where noted, fetal calf serum (Microbiological Associates, Bethesda, Md. or Colorado Serum Company) replaced the calf serum. Stocks were passaged two to three times weekly, and care was taken to keep the normal cells at low densities. Cells were checked for PPLO regularly and were free of contamination throughout the study.

Growth Curves and Saturation Densities

Cells were grown in 3.5-cm petri dishes (Falcon). The cells were trypsinized in 1 ml of trypsin-EDTA (Gibco No. 530) and counted directly in a Neubauer hemocytometer or by use of a Coulter Counter after a 10- to 20-fold dilution in saline.

The following protease inhibitors were used: TAME (N-α-tosyl-L-arginine methylester); TLCK (N-α-tosyl-L-chloromethane, often called tosyl-lysyl-chloromethylketone); TPCK (N-α-tosyl-L-phenylalanyl-chloromethane, often called tosylphenylalanyl-chloromethylketone) all from Calbiochem, and ovomucoid II O from Sigma.

TAME-Hydrolase Assay

The cells used for hydrolase assays were removed from the culture vessels with EDTA and washed three times with either PBS or Hepes buffered medium. Although we found that similar results were obtained using cells previously subcultured with EDTA alone or with the usual trypsin-EDTA method, we routinely used cells that had been last passaged in the absence of trypsin.

Cells were broken by freeze-thawing five times in 5 volumes of distilled water. The soluble "supernatant" fraction was obtained by centrifugation at $40,000 \times g$ for 20 minutes. The pellet of this centrifugation was washed three times in about 10 volumes PBS and centrifuged as above to give the "particulate" fraction.

TAME hydrolysis was measured by a modification of the method of Roffmann, Sanocka and Troll (1970). [³H]TAME, labeled in the methyl group, was used as substrate. Since the ester is virtually insoluble in toluene, the radioactive methanol produced in the reaction could be extracted directly into a toluene-based scintillant solution and counted.

The reaction was performed in Dulbecco's modified Eagle's medium in which the bicarbonate was replaced by 45 mM Hepes buffer, pH 7.4. The reaction mixture (0.1 ml) contained 1 mM [³H]TAME (152,000 cpm/μmole) and one of the following sources of enzyme: serum, 5 μl; whole cells, 1–8×10^5; or cell fragments, 50–400 μg protein. After incubation at 37°C in a closed scintillation vial for 0–60 minutes, the reaction was stopped by freezing in liquid nitrogen. A twentyfold excess of cold TAME and 15 ml of a toluene scintillant solution were then added. The water phase was allowed to thaw and the [³H]methanol was counted after extraction into the toluene phase by shaking.

Thymidine incorporation was measured in 3.5-cm petri dish cultures. 0.5 μCi [³H]thymidine (5 Ci/mmole, Amersham) was added to each culture. After 60 minutes the cultures were rinsed twice with PBS and extracted for 60 minutes with cold 10% TCA, washed three times with 10% TCA (10 minutes each) and once with absolute ethanol. The cells were solubilized with 1 ml 1% SDS and transferred to scintillation vials. To count, 0.5 ml Soluene™ (Packard) and 15 ml Bray's solution were added. Cell numbers were determined in parallel cultures in order to calculate the incorporation on a per cell basis.

RESULTS AND DISCUSSION

Target of the Protease Inhibitors

It appeared likely that TAME would act through inhibition of a hydrolase, presumably by competing with its natural substrate. We therefore looked for a TAME hydrolyzing activity as a potential target for the inhibitors.

Table 1

TAME Hydrolyzing Activity of Sera and Cellular Growth

	TAME hydrolyzed (μmoles/ml/hr)	Saturation density (cells/cm^2)	
		3T3	SV3T3
Calf serum	7.66	42,000	300,000
Calf serum, 56°C, 60 min	0.46	37,000	
Calf serum, TLCK-treated	0.05	38,000	320,000
Fetal calf serum	2.05	43,000	196,000
Fetal calf serum, TLCK-treated	0.04	45,000	205,000

For the determination of saturation densities, the cells were plated (10,000 cells/cm^2) in Dulbecco's modified Eagle's medium containing 10% of the sera indicated and allowed to grow without change of medium until no further increase in cell number occurred (5–8 days). All sera were dialyzed against PBS. One sample was heat inactivated at 56°C for 1 hr; TLCK treatment involved incubation of serum with 1 mg per ml TLCK for 2 hr at 37°C and subsequent dialysis against PBS. [^{3}H]TAME hydrolysis was measured as described in Methods.

Both the cells and the sera used in media are able to split TAME at a measurable rate. However it became clear that the TAME hydrolase* activity and the growth supporting capacity of sera are not related (Table 1): the saturation densities of 3T3 or SV3T3 cells were not affected when the TAME hydrolase in serum was eliminated by either heat inactivation or TLCK treatment. It was interesting to note that calf sera always had a three to six times higher hydrolase activity than did fetal sera. Conditioning of sera on either 3T3 or SV3T3 cells did not increase their hydrolase activity, indicating that the cells neither excrete TAME hydrolases nor activate serum enzymes.

That the inhibitors exert their effect on the cells directly and not by interacting with serum components in the medium could also be shown in a different way (Table 2): TLCK treatment of serum did not affect the growth of SV3T3 cells, but cells treated with TLCK for one hour, washed and grown in inhibitor-free medium remained inhibited for at least 24 hours.

TAME Hydrolases in Cells

The hydrolysis of [^{3}H]TAME by intact SV3T3 cells proceeds in a biphasic manner: a rapid phase, lasting less than 20 seconds, and a second phase at least 200 times slower. Figure 2 (left) shows that a second addition of substrate only affects the rate of the second phase and does not cause another burst of product release. The amount of substrate cleaved during the rapid phase is proportional to the number of cells in the assay (Fig. 3) and is equal to approximately 3×10^9 molecules per cell. This plateau is not the result of substrate exhaustion or sequestering, since only a small percentage of the substrate is destroyed during the rapid phase and exogenous trypsin cleaves the remainder at the same rate as in the absence of cells

---

* It is not intended to introduce "TAME hydrolase" as a term, but before the nature of these enzymes is known, it may serve as an operational definition.

Table 2

Effects of TLCK on Serum and Cells

	Cells/cm² (after 72 hrs)
Control (dialyzed serum)	247,000
TLCK, 100 μg/ml	18,400
TLCK-treated, dialyzed serum	258,000
Cells, TLCK-treated but grown on control medium	33,200

SV3T3 cells were plated (2000 cells/cm²) 48 hr before the experiment; media were changed at the beginning and every 24 hr during the experiment. Cells were counted 72 hr after the first medium change. All media contained 10% fetal calf serum (Microbiol. Ass.); the serum was dialyzed extensively against phosphate-buffered saline (pH 7.4) or treated with 1 mg/ml of TLCK for 1 hr at 37°C before exhaustive dialysis. In the last experiment the cells were exposed to TLCK (100 μg/ml) for only 1 hr at the beginning of each day of the experiment; after this exposure, the cells were washed twice and grown in control medium. From Schnebli and Burger (1972) with permission of *Proc. Nat. Acad. Sci.*

(Fig. 2, left). The results are consistent with, but do not prove the presence of at least two TAME hydrolyzing activities in these cells and suggest that the rapid phase enzyme is inactivated in the course of the reaction.

Cells, broken by freezing and thawing, were separated into a supernatant and a particulate fraction. The time course of TAME hydrolysis catalyzed by the particulate fraction was found to be very similar to that of whole cells: again a "rapid phase" is observed and the reaction appears to account quantitatively for the measured hydrolysis in whole cells (Fig. 2, right). The hydrolysis during the rapid phase is proportional to the concentration of particulate fraction (up to about 6 mg protein/ml). The rapid phase activity could not be removed from the particulate by repeated washing with PBS.

In addition to the particulate enzyme, there appears a "soluble" hydrolase with different properties: the reaction is roughly linear with time and proportional to the concentration of soluble fraction (Fig. 2, right). Its K_M for TAME is 1.3 mM.

Since in whole cells the soluble enzyme appears not to be accessible to TAME, and since the kinetics of the reaction catalyzed by the particulate suggested a TAME-dependent self-inactivation, the particulate enzyme appeared as the more likely target of the protease inhibitors and was further investigated.

3T3, SV3T3 and Py3T3 cells did not differ significantly when compared with respect to the rapid phase in intact cells or particulate fractions, nor with respect to the slow reaction of the particulate or the soluble enzyme. This does not necessarily mean that there are no differences in the cells in culture, since enzymes could be unmasked or activated during the process of removing cells from the surface on which they grow and during the washing and handling in serum-free solutions. Still it was disappointing not to find a difference between the activities in normal and transformed cells.

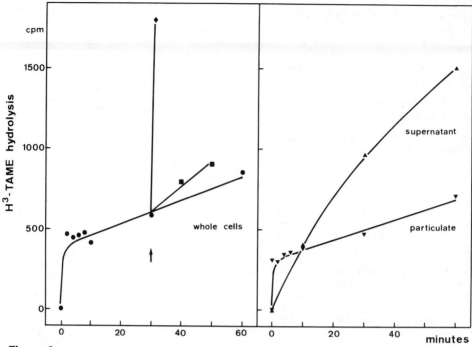

Figure 2

TAME hydrolysis by SV3T3 cells. *Left,* the time course of the reaction with whole cells
(8 × 10⁶/ml) (●). At the arrow either 5 μg/ml trypsin was added (♦) or the concentration of [³H]TAME was increased from 1 to 2 mM (■). *Right,* the time course of
the reactions catalyzed by the particulate (▼) and soluble fractions (▲) corresponding
to 8 × 10⁶ cells/ml. All values are corrected for a small nonenzymatic hydrolysis. For
details of the assay procedure, see Methods.

The effect of protease inhibitors on the rapid phase activity in the particulate
from SV3T3 was then investigated. TLCK readily inhibits this activity at concentrations that are also effective in inhibiting growth of transformed cells (i.e., below
100 μg/ml). TPCK was also found to be inhibitory but at concentrations that are
slightly higher than those needed in growth experiments. Ovomucoid did not inhibit
the activity. The rapid phase activity may thus be the target for TAME and TLCK,
possibly also for TPCK, but apparently not for ovomucoid. There are reasons to
believe that ovomucoid acts as a growth inhibitor by an altogether different mechanism. ¹²⁵I-labeled ovomucoid was found to bind to both glass and plastic surfaces
much more strongly than to cells. This suggests the possibility that ovomucoid coats
and alters the surface of the culture vessel and thus influences the growth of transformed cells (Maciero-Coelho and Avrameas 1972). To test this possibility we
"precoated" petri dishes by incubation for one hour with ovomucoid in complete
medium, after which the dishes were washed three times. We found that the
growth of cells plated in these dishes was inhibited. However to obtain an equal
inhibition as in cultures where ovomucoid was continually present, two- to threefold
higher concentrations were needed for the coating.

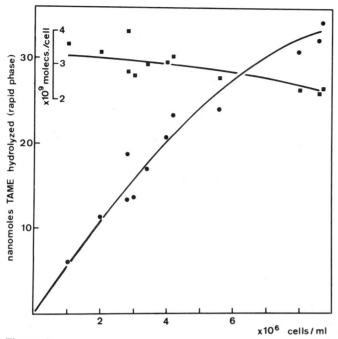

Figure 3

TAME hydrolysis by intact SV3T3 cells. Correlation of the amount of [³H]TAME hydrolyzed during the rapid phase with the concentration of cells in assay (●). Insert shows the number of molecules of [³H]TAME split per cell during the rapid phase as a function of cell concentration (■).

Do Protease Inhibitors Restore Contact Inhibition?

It has already been shown that protease inhibitor-treated cultures of transformed cells have lowered saturation densities (Fig. 1), a point which has been carefully worked out by Prival (1972). Treatment with protease inhibitors causes transformed cells to attain a flatter, more spread-out morphology (Fig. 4). Similar morphological alterations have also been reported by Goetz et al. (1972) for hamster cells and by Weber (pers. commun.) for transformed chick cells.

TLCK-treated SV3T3 cells become less agglutinable with wheat germ agglutinin than untreated SV3T3 cells (Table 3), but it must be noted that they still agglutinate more readily than do normal 3T3 cells. Virtually the same result has been obtained by Prival (1972) using TPCK.

Since it is known that contact-inhibited cells rest in the G_1 phase (Nilausen and Green 1965), we investigated the capacity of protease inhibitor-treated cells to incorporate thymidine. Thymidine incorporation is less sensitive than proliferation of SV3T3 cells; i.e., even in TLCK-treated cultures where little or no net increase in cell number occurs, there is still considerable thymidine incorporation (Table 4). By comparison the incorporation into 3T3 cells is drastically reduced after they become confluent (Table 4). This demonstrates that TLCK does not stop the cells from entering the S phase. Preliminary data on the DNA content of individual

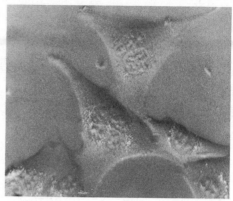
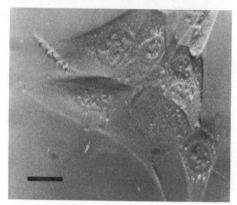

Figure 4
Untreated and TLCK-treated SV3T3 cells.
SV3T3 (about 10,000 cells/cm²) were
treated for 24 hours with TLCK. *Left,* untreated controls; *center,* 30 μg/ml TLCK;
right, 60 μg/ml TLCK. Photographs were
taken of live cells in culture on a Wild M
20 light microscope using a new relief contrast technique (Albrecht-Bühler 1973).
Insert bar corresponds to 25 μ. Photographs courtesy Dr. G. Albrecht-Bühler.

Table 3
Agglutination of Protease Inhibitor-
Treated Cells with WGA

	Amt. required for half-max. agg. (μg/ml)
3T3	500
SV3T3	45
SV3T3 TLCK-treated	120

Cells were grown in the absence or presence
of 50 μg/ml TLCK for 24 hr prior to the
agglutination experiment. The agglutination
test was performed and scored as described
by Burger and Goldberg (1967).

Table 4

TLCK Effect on Proliferation and Thymidine Incorporation

Cell line	TLCK in medium ($\mu g/ml$)	Proliferation (divisions/24 hr)	Thymidine incorp. ($cpm/10^3$ cells)
SV3T3	0 (control)	1.18	125 ± 24
	30	0.60	115 ± 32
	60	0.09	60 ± 26
	90	-0.15	36 ± 23
3T3			
log phase	—	0.90	108 ± 27
contact-inhibited	—	0.02	0.9 ± 2

SV3T3 cells were plated (2000 cells/cm²) and allowed to grow for 48 hr. The medium was then changed to medium containing TLCK at the concentrations indicated. After 24 hr [³H]thymidine incorporation was measured (1 hr). In parallel cultures cell numbers were determined at 8–14 hr intervals (5 points within 48 hr), from which the number of cell divisions per 24 hr could be calculated. For comparison 3T3 cells were plated at 10,000 cells/cm²; [³H]thymidine incorporation was measured in the log phase (27 hr after plating; 21,000 cells/cm²) and 24 hr after reaching confluency (5 days, 42,000 cells/cm²).

TLCK-treated SV3T3 cells determined by cytophotometry also show that treated cells, even if strongly inhibited in their growth, are not arrested in G_1. With high concentrations of TLCK there appears to be an increase in hyperploid cells.

We conclude that protease inhibitors fail to induce contact inhibition in the sense that they do not arrest the cells in the G_1 phase of the cell cycle. It follows that the inhibitor-sensitive activity in transformed cells is not necessarily required for the escape from contact inhibition. This point is further supported by the finding that both the insulin-stimulated overgrowth in 3T3 cells and the log-phase growth of 3T3 cells are less sensitive to TLCK than is the growth of SV3T3 (Table 5).

Table 5

Effect of TLCK on 3T3, SV3T3 and Insulin-stimulated 3T3

	Inhibition of growth (%)		
	30 $\mu g/ml$	60 $\mu g/ml$	90 $\mu g/ml$
3T3	0	2	18
SV3T3	64	88	92
3T3 insulin-stimulated overgrowth	0	14	24

The first two experiments were done as described in the legend of Fig. 1; the results are given as % reduction of cell numbers after 72 hr. In the third experiment 3T3 cells were plated at 10,000 cells/cm²; after three days, when they reach a stationary density of about 40,000 cells/cm², insulin was added to a final concentration of 20 $\mu g/ml$. Within 48 hr the cells "overgrow," i.e., reach 40–60% higher densities than parallel untreated controls. The above results are given as % reduction of this "overgrowth."

SUMMARY

A number of protease inhibitors have previously been shown to selectively inhibit the growth of transformed cells in tissue culture. As a possible cellular target of these inhibitors a TAME-hydrolyzing activity was found in SV3T3 and Py3T3 cells, but also in 3T3 cells. This activity appears to be inactivated by the substrate TAME and is sensitive to TLCK and TPCK but not to ovomucoid.

The protease inhibitors restore several "criteria" for contact inhibition in transformed cells; i.e., they reduce the saturation density and the agglutinability by lectins and induce "flat" morphology, but they fail to arrest the cells in the G_1 phase. It is thus doubtful that contact inhibition is truly restored by the inhibitors. It also raises doubts as to whether the inhibitor-sensitive activity plays a key role in the escape from contact inhibition, as has been suggested previously.

Acknowledgment

I wish to thank Dr. Gisela Haemmerli (Zurich) for performing the cytophotometric DNA assays.

REFERENCES

Albrecht-Bühler, G. 1973. One-sided oblique illumination in light microscopy: The effect of increasingly incoherent illumination. *Trans. Amer. Microsc. Soc.* (in press).

Bosmann, H. B. 1972. Elevated glycosidases and proteolytic enzymes in cells transformed by RNA tumor virus. *Biochim. Biophys. Acta* **264**:339.

Burger, M. M. 1970. Proteolytic enzymes initiating cell division and escape from contact inhibition of growth. *Nature* **227**:170.

Burger, M. M. and A. R. Goldberg. 1967. Identification of a tumor-specific determinant on neoplastic cell surfaces. *Proc. Nat. Acad. Sci.* **57**:359.

Goetz, I. E., C. Weinstein and E. Roberts. 1972. Effects of protease inhibitors on growth of hamster tumor cells. *Cancer Res.* **32**:2469.

Hozumi, M., M. Ogawa, T. Sugimura, T. Takeuchi and H. Umezawa. 1972. Inhibition of tumorigenesis in mouse skin by leupeptin, a protease inhibitor from actinomycetes. *Cancer Res.* **32**:1725.

Kazakova, O. V., V. N. Orekhovich, L. Purchot and J. M. Schuck. 1972. Effect of cathepsins D from normal and malignant tissues on synthetic peptides. *J. Biol. Chem.* **247**:4224.

Maciero-Coelho, A. and S. Avrameas. 1972. Modulation of cell behaviour *in vitro* by the substratum in fibroblastic and leukemic mouse cell lines. *Proc. Nat. Acad. Sci.* **69**:2469.

Nilausen, K. and H. Green. 1965. Reversible arrest of growth in G_1 of an established fibroblast line (3T3). *Exp. Cell Res.* **40**:166.

Ossowski, L., J. C. Unkeless, A. Tobia, J. P. Quigley, D. B. Rifkin and E. Reich. 1973. An enzymatic function associated with transformation of fibroblasts by oncogenic viruses. Mammalian fibroblast cultures transformed by DNA and RNA tumor viruses. *J. Exp. Med.* **137**:112.

Prival, J. T. 1972. Surface membrane proteins of normal and transformed mouse fibroblasts. Thesis, M.I.T. p. 120–135.

Roffman, S., U. Sanocka and W. Troll. 1970. Sensitive proteolytic enzyme assay using differential solubilities of radioactive substrates and products in biphasic systems. *Anal. Biochem.* **36**:11.

Schnebli, H. P. 1972. A protease-like activity associated with malignant cells. *Schweiz. Med. Wschr.* **102**:1194.

Schnebli, H. P. and M. M. Burger. 1972. Selective inhibition of growth of transformed cells by protease inhibitors. *Proc. Nat. Acad. Sci.* **69**:3825.

Sefton, B. M. and H. Rubin. 1970. Release from density dependent growth inhibition by proteolytic enzymes. *Nature* **227**:843.

Troll, W., A. Klassen and A. Janoff. 1970. Tumorigenesis in mouse skin: Inhibition by synthetic inhibitors of proteases. *Science* **169**:1211.

Unkeless, J. C., A. Tobia, L. Ossowski, J. P. Quigley, D. B. Rifkin and E. Reich. 1973. An enzymatic function associated with transformation of fibroblasts by oncogenic viruses. I. Chick embryo fibroblast cultures transformed by avian RNA tumor viruses. *J. Exp. Med.* **137**:85.

Effects of Protease Inhibitors
on Growth of 3T3 and SV3T3 Cells

Iih-Nan Chou, Paul H. Black, and Richard Roblin

Departments of Medicine and Microbiology and Molecular Genetics
Harvard Medical School and Infectious Disease Unit
Massachusetts General Hospital, Boston, Massachusetts 02114

Untransformed fibroblasts in tissue culture exhibit a characteristic growth pattern evidenced by the cessation of cell division upon formation of a confluent monolayer. This phenomenon has been termed contact inhibition or density-dependent inhibition of growth (Todaro et al. 1965; Stoker and Rubin 1967). In contrast cells transformed by an oncogenic virus, such as simian virus 40 (SV40), are capable of continuing cell division beyond confluency and multilayered cell growth.

Several proteolytic enzymes have been shown to alter the growth characteristics and the surface properties of untransformed cells. A mild proteolytic enzyme treatment of confluent untransformed fibroblasts in culture can result in initiation of cell division and release from contact inhibition of growth (Burger 1970; Sefton and Rubin 1970). A similar proteolytic enzyme treatment renders the untransformed cells more agglutinable by plant lectins (Burger 1969; Inbar and Sachs 1969). In addition Rubin (1970a,b) has suggested that the overgrowth stimulating factor (OSF) produced by Rous sarcoma virus-transformed cells might be a proteolytic enzyme because of the similarity between the activity of OSF and that of trypsin. These results suggest that increased proteolytic activity in transformed cells might be responsible for the lack of contact-inhibited growth and some of the cell surface alterations associated with malignant transformation of cells. Indeed, Schnebli (1972) reported such an increase in protease-like activity in DNA virus-transformed cells.

If this protease hypothesis is correct, then inhibition of this activity by protease inhibitors might be able to reverse the unrestrained growth of transformed cells. Prival (1971) found that the saturation density, but not the growth rate, of transformed cells was reduced by a chymotrypsin inhibitor, N-tosyl-L-phenylalanylchloromethylketone (TPCK), and that the TPCK-treated transformed cells became less agglutinable than the untreated transformed cells. Schnebli and Burger (1972) reported that five protease inhibitors, including TPCK and a trypsin inhibitor N-tosyl-L-lysyl-chloromethylketone (TLCK), caused selective inhibition of the growth of transformed, but not untransformed, cells. They concluded that those in-

339

hibitors blocked a protease-like activity that is required for the unrestrained growth of transformed cells.

We have carried out similar experiments to study the effects of TPCK and TLCK on the growth of SV40-transformed Swiss and Balb/c 3T3 cells. We find that TPCK causes a nonselective, dose-dependent and rapidly reversible inhibition of the growth of our SV3T3 cells. In addition we have shown that TPCK causes a dose-dependent and reversible inhibition of both protein and DNA synthesis. Our results strongly suggest that the observed growth inhibition in the presence of TPCK is a consequence of the overall inhibition of protein synthesis, rather than the selective inhibition of a protease-like activity.

MATERIALS AND METHODS

Cells and Media

Four cloned cell lines, untransformed Swiss and Balb/c 3T3 cells and their SV40-transformed counterparts, were employed in this study. Swiss 3T3 cells were originally obtained from Dr. G. Todaro and from this line a cloned SV40-transformed line was derived by Dr. P. Black (Black 1966). The Swiss 3T3 cells were used between their 15th and 20th passages and the transformed cells between their 10th and 25th passages in our laboratory. Both Balb/c 3T3 (clone A31) and SV3T3 (clone SVT2) cells (Aaronson and Todaro 1968) were received from Dr. S. Aaronson, the former at 40 generations and the latter at 150 generations. Both lines were used between their 12th and 18th passages in our laboratory. Cells were usually passaged two to three times weekly and special care was taken to prevent overgrowth of the untransformed cells. All cells were grown in 8- or 32-oz. glass bottles in Eagle's minimal essential medium containing a fourfold concentration of vitamins and amino acids (MEM $\times$ 4) and supplemented with 10% (v/v) fetal bovine serum (FBS) and 250 U/ml penicillin plus 250 μg/ml streptomycin before use.

TPCK and TLCK Inhibition Studies

TPCK solution was freshly prepared at 5 mg/ml in phosphate-buffered saline (PBS) containing 0.1 N NaOH according to Prival (1971) and added with stirring to MEM $\times$ 4 (FBS and antibiotics free) to give the desired concentrations. After immediate adjustment of the pH to that of MEM $\times$ 4 (7.8), the medium was quickly filtered and then supplemented with 10% FBS and antibiotics. TPCK solutions prepared in this way retain a minimum of 85% of their potential to inactivate chymotrypsin, compared with equal concentrations of TPCK dissolved in methanol (Shaw 1970). TLCK was dissolved in MEM $\times$ 4 at 10 mg/ml and added to growth medium at the desired concentrations without pH adjustment since addition of TLCK solution did not affect the pH of the medium. All control media were prepared in exactly the same way except they contained no TPCK or TLCK.

Trypsinized SV3T3 cells were usually plated at 4×10^5 cells/60-mm Falcon plastic tissue culture dish containing 5 ml MEM $\times$ 4 plus 10% FBS and antibiotics. Untransformed 3T3 cells were plated at $1-2 \times 10^5$ cells/dish in the same way as for transformed cells. Twenty to 24 hours later, the medium was removed and immediately replaced with control and TPCK- or TLCK-containing media. The media with or without inhibitor were changed every day and cell growth was followed by daily determination of cell counts. Cells were trypsinized and counted in an automatic laser beam cell counter, CYTOGRAF Model 6302 (Bio/physics Systems, Inc., Mahopac, New York). For viability tests cell suspensions were mixed with 0.4% Trypan Blue (1:1), stained for 5–10 minutes and counted in a hemocytometer. The number of cells floating in the medium was also determined by counting either the pooled medium directly in the CYTOGRAF or the suspension of pelleted cells using a hemocytometer.

Incorporation of Tritiated Amino Acids

^{3}H-labeled L-amino acid mixture in PBS (500 μCi/ml) was added to the plates at a final concentration of 2.5 μCi/ml. After one hour of incubation, the medium was decanted and the cells were washed three times with PBS, extracted three times each with 5 ml cold 5% TCA and finally the entire cell monolayer was dissolved in 1 ml 0.5 N NaOH. Aliquots of the sample were neutralized with 4 N HC1, precipitated with 5% TCA and heated at 90°C in a water bath for 30 minutes. The hot TCA-precipitable materials were collected on glass fiber filter discs (Whatman, GF/C) and the dried discs were counted in 10 ml toluene scintillation fluid in a Packard Tri Carb Liquid Scintillation Spectrometer. The radioactivity of TCA-soluble fractions was determined by counting aliquots of the TCA extracts in 10 ml Bray's scintillation fluid (Bray 1960).

Incorporation of Tritiated Thymidine

For quantitative incorporation measurements [^{3}H]thymidine in PBS (100 μCi/ml) was added to the plates at a final concentration of 5 μCi/ml for one hour of labeling. At the end of labeling the medium was decanted and the cells were washed twice each with 5 ml cold Dulbecco's Tris buffer pH 7.4 (Dulbecco and Vogt 1962), extracted three times each with 5 ml cold 5% TCA, washed five times with cold 95% ethanol and finally dissolved in 1 ml 0.5 N NaOH. Aliquots of NaOH-dissolved samples were precipitated with cold 5% TCA and the TCA-insoluble materials were collected, dried and counted as described above. For radioautography the cells were labeled in the presence of 5 μCi/ml of [^{3}H]thymidine for 4 hours and processed in the same way as for quantitative incorporation measurement through the ethanol washing step. The plates were air-dried overnight, covered with Kodak AR10 stripping film and developed according to a previously described procedure (Culp and Black 1972). The cells used throughout this study were found to be free from mycoplasma contamination by periodic assay of the cells using the same radioautographic procedure (Culp and Black 1972).

Uptake of L-Phenylalanine

The procedure described by Hare (1967) was modified as follows. At 4 and 24 hours after addition of TPCK at 200 μg/ml, the plates were fluid-changed to 5 ml TPCK (200 μg/ml) and L-[^{3}H]phenylalanine (1 μCi/ml)-containing medium that had previously been adjusted to pH 7.4. The plates were then incubated at room temperature without CO_2 gas for 1, 2, 3, 4, 5 and 10 minutes respectively. At the end of each incubation period, the medium was quickly decanted and the cells were washed four times with cold PBS, extracted three times each with 2 ml cold 5% TCA for one minute and the unextracted residue dissolved in 1 ml 0.5 N NaOH. Aliquots of TCA extracts were counted directly in 10 ml Bray's scintillation fluid.

The total radioactivity in all three TCA extracts was used as a measure of uptake (transport) of L-phenylalanine by the cells.

Chemicals

TPCK, TLCK, chymotrypsin and benzoyltyrosine ethylester were obtained from Sigma Chemical Company. L-[³H]phenylalanine (0.5 mCi/ml, sp. act. 7 Ci/mmole) was purchased from Schwartz-Mann. L-[³H]amino acid mixture (1 mCi/ml) and [³H]thymidine (1 mCi/ml, sp. act. 20 Ci/mmole) were obtained from New England Nuclear Corporation. Cycloheximide was obtained from Nutritional Biochemical Corporation.

RESULTS

Growth Inhibition by TPCK

The effects of TPCK on growth of Swiss SV3T3 cells are markedly concentration dependent. Growth of Swiss SV3T3 cells was unaffected by up to 50 μg TPCK/ml medium, and it was only slightly inhibited by TPCK at 100 μg/ml (0.285 mM). TPCK at 150 μg/ml causes a marked decrease in growth rate (Fig. 1), but the cells continue growing and eventually reach the same cell density as the untreated control cultures. When TPCK is added at 200 μg/ml, the growth inhibition becomes so severe that, after about 24 hours, the cells stop increasing in total cell number. This results in the appearance of a growth "plateau" (Fig. 1) at a cell density which approximates the saturation density of "contact-inhibited" untransformed 3T3 cells. This "plateau" may be maintained for 5–7 days and the percentage of viable cells (90–96% of the total cells from the monolayer) remains essentially constant throughout the plateau period. Furthermore the TPCK-mediated growth inhibition is rapidly reversible by removing the TPCK-containing medium from the cultures and replacing it with fresh medium (Fig. 1). Upon removal of TPCK, the cell number approximately doubles in the next 24 hours, showing that the cells retain their viability even after several days in the presence of TPCK. TPCK concentrations of 250 μg/ml or higher lead to rapid cell death.

These effects on Swiss SV3T3 cells prompted us to study the growth of untransformed cells in the presence of TPCK. The results showed that the TPCK inhibitory effect is not selective against transformed cells since the growth of untransformed 3T3 cells is also inhibited by TPCK. At 50 μg TPCK/ml medium, only the growth rate, but not the final cell density, of untransformed 3T3 cells is affected. A growth "plateau" at 2×10^5 cells/dish can be produced by treatment with TPCK at 100 μg/ml, but this growth "plateau" is considerably below the normal saturation density of this 3T3 cell line. We observed that TPCK at 150 and 200 μg/ml was toxic to our Swiss 3T3 cells.

We have also studied the inhibitory effect of TPCK on growth of Balb SV3T3 cells and found that the growth rate of Balb SV3T3 cells was also decreased by TPCK at 150 and 200 μg/ml, although to a lesser extent than that of Swiss SV3T3 cells. Furthermore TPCK at 200 μg/ml failed to produce a growth "plateau" with Balb SV3T3 cells as it did with Swiss SV3T3 cells. Thus this Balb SV3T3 cell line is somewhat less sensitive to TPCK inhibition than the Swiss SV3T3 line. Because of the limited solubility of TPCK in aqueous solutions, the effect of TPCK concentrations higher than 300 μg/ml on cell growth could not be determined.

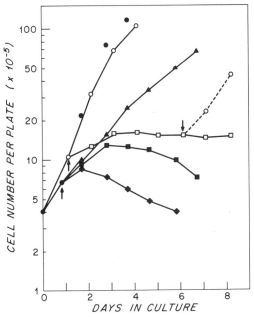

Figure 1

Effects of TPCK on growth of Swiss SV3T3 cells; composite curve showing results of two experiments. Cells were plated at 4×10^5 cells/dish on day 0 and TPCK treatment was initiated at time indicated by the upward arrow. In the first experiment (open symbols) cells were treated with the following TPCK concentrations: (o) 0 μg/ml, (□) 200 μg/ml. At the time indicated by the downward arrow TPCK medium was replaced with control medium without TPCK; the subsequent cell growth is indicated by a dotted line and open circles. In the second experiment (closed symbols) cells were treated with: (●) 0 μg/ml TPCK, (▲) 150 μg/ml TPCK, (■) 200 μg/ml TPCK, and (◆) 250 μg/ml TPCK.

In the presence of TPCK at 150 or 200 μg/ml, many cells appear to lose or contract their long, fiber-like cytoplasmic projections and look cucumber-shaped with knoblike structures on the cell surface and periphery. The TPCK-treated cells are more heavily stained by Giemsa and show a markedly increased number of binucleate cells in addition to the more frequent formation of multinucleated giant cells. Upon removal of TPCK, some of the fiber-like processes may reappear within as early as 6–8 hours and most cells reacquire their normal appearance within 24 hours after removal of TPCK.

TPCK Inhibition of Macromolecular Synthesis

Since TPCK at 200 μg/ml causes severe growth inhibition and a growth "plateau" phenomenon with Swiss SV3T3 cells, we investigated whether protein and DNA synthesis were also affected by TPCK. The results (Table 1) show that TPCK at 100 and 200 μg/ml inhibits the incorporation of amino acids into TCA-precipitable materials by 60 and 70%, respectively. When TPCK is added at 250 μg/ml, the inhibition of protein synthesis reaches 75% and cell death ensues.

Table 1

TPCK Inhibition of ^{3}H-Amino Acid and [^{3}H]Thymidine
Incorporation by Swiss SV3T3 Cells

TPCK ($\mu g/ml$)	Amino acid incorporation[a]		Thymidine incorporation[a]	
	$(cpm/10^6\ cells)$[b]	(% of control)	$(cpm/10^6\ cells)$[b]	(% of control)
0	2706	100	353750	100
100	1134	42		
150	990	36		
200	846	31	174480	49[c]
250	647	24	52120	15[c]
300	434	16		

[a] Experiments done at 22 hours after initiation of TPCK treatment.

[b] Average cpm from two determinations on samples from duplicate plates.

[c] By radioautography the percentage of labeled nuclei was approximately 30–35% and 24–28% for TPCK at 200 and 250 $\mu g/ml$, respectively.

An inhibition of thymidine incorporation in TPCK-treated cells is also shown in Table 1. TPCK at 200 and 250 $\mu g/ml$ causes 50 and 85% inhibition of DNA synthesis, respectively, 22 hours after treatment. By radioautography 30–35% and 24–28% of the cells were able to incorporate thymidine in the presence of TPCK at 200 and 250 $\mu g/ml$, respectively. When the cells were treated with TPCK at 200 $\mu g/ml$ for 3 days, cell death started to occur and at this point the thymidine incorporation was reduced to about 24–28% of the untreated control, while radioautography showed that about 20–24% of cells were incorporating thymidine.

As mentioned previously the TPCK-mediated growth inhibition and the "plateau" phenomenon are rapidly reversed following removal of the inhibitor from growth medium. We have found that the inhibition of amino acid incorporation is also readily reversible; within 2 hours after removal of TPCK, the amino acid incorporation increased to about the level of untreated cells.

Although the incorporation of amino acids into TCA-insoluble material is strongly inhibited by TPCK, the incorporation into acid-soluble fractions ("pools") remains essentially unaffected by TPCK (data not shown). Since TPCK is a structural analog of phenylalanine, we thought that TPCK might inhibit protein synthesis by selectively interfering with the phenylalanine transport system, in effect starving the cells for phenylalanine and thus inhibiting protein synthesis. However comparison of the kinetics of L-[^{3}H]phenylalanine uptake by untreated and TPCK-treated (200 $\mu g/ml$, 24 hr) Swiss SV3T3 cells did not reveal any decrease in the rate of L-[^{3}H]phenylalanine uptake (Fig. 2). The small decrease in rate of L-[^{3}H]phenyl-alanine uptake after 4 hours of TPCK treatment (Fig. 2) does not appear to be of sufficient magnitude to account for the decrease in protein synthesis. These results help to rule out the possibility that the observed inhibition of amino acid incorporation into proteins is due to the impaired transport and thus the unavailability of L-phenylalanine.

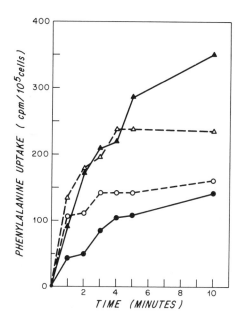

Figure 2
Phenylalanine uptake by TPCK-treated SV3T3 cells. Swiss SV3T3 cells were plated at 4×10^5 cells/dish and grown for 20 hours before fluid-changing to either fresh control medium or fresh medium containing 200 μg/ml TPCK. L-[³H]phenylalanine uptake was determined after 4 hours (o, ●) and 24 hours (△, ▲) of TPCK treatment. Open symbols, controls; closed symbols, TPCK-treated.

Effect of Cycloheximide on Cell Growth

The finding that TPCK inhibits protein synthesis suggested that the observed growth inhibition might be linked to the overall inhibition of amino acid incorporation by TPCK. If this were so, other inhibitors of protein synthesis might affect cell growth in much the same way that TPCK does. Figure 3 presents a family of growth curves which show that low concentrations of cycloheximide cause dose-dependent inhibitory effects on the growth of Swiss SV3T3 cells. If the appropriate concentration of cycloheximide is chosen, a "plateau" phenomenon, which mimics that observed with TPCK at 200 μg/ml, is observed. This cycloheximide-mediated growth "plateau" can also be rapidly reversed, even up to 6 days after cycloheximide treatment, simply by removing the inhibitor from growth medium. We also have determined that during the growth "plateau" caused by 0.5–0.6 μg cycloheximide/ml medium, the ³H-amino acid and [³H]thymidine incorporation are inhibited by about 95% and 85%, respectively. These results indicate that in the case of cycloheximide, the inhibition of protein synthesis can result in reversible growth "plateau."

Partial Release by Serum of TPCK-mediated
Growth Inhibition of SV3T3 Cells

TPCK might inhibit the growth of transformed cells by inactivating serum protein factors required for cell growth. Because of the limited solubility of TPCK, we have been unable to treat serum with sufficient concentrations of TPCK to determine whether TPCK acts on the cells directly or interacts with serum components of the medium to cause inhibition of cell growth. However we have observed that upon increasing the serum content of the growth medium from 10% to 15 or 20%, the growth inhibition caused by 200 μg/ml TPCK can be partially overcome. As shown in Fig. 4, in the presence of 10% serum, TPCK at 150 μg/ml causes an inhibition level that is equivalent to that caused by 200 μg TPCK/ml medium but

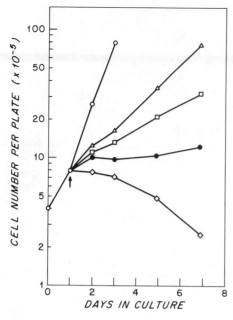

Figure 3

Effect of cycloheximide on growth of Swiss SV3T3 cells. Cells were plated at 4×10^5 cells/dish on day 0 and, at the time indicated by the arrow, medium containing cycloheximide was added at the following concentrations: (o) 0 μg/ml, ($\triangle$) 0.1 μ/ml, ($\square$) 0.2 μg/ml, ($\bullet$) 0.5 μg/ml and ($\diamond$) 2.0 μg/ml.

only in the presence of 15% or more serum. However this release of TPCK-mediated growth inhibition by serum is effective only to a limited extent, in that 20% serum gives no increased effect over 15% serum when TPCK is present at 200 μg/ml (Fig. 4).

Lack of Growth Inhibition by TLCK

Growth curves of Balb/c 3T3 and SV3T3 cells in the presence of various concentrations of TLCK are shown in Fig. 5. The growth of Balb SV3T3 cells is essentially unaffected by TLCK up to 100 μg/ml (0.27 mM). Although there is a slight decrease in the rate of Balb SV3T3 cell multiplication at 100 μg TLCK/ml medium, the cells do not stop growing and eventually reach approximately the same cell density as the control cultures. Figure 5 also shows that untransformed Balb/c 3T3 cells are as insensitive to TLCK treatment as transformed SV3T3 cells; all cultures reach the same saturation density (about 2×10^6/60-mm plate) in the presence of TLCK up to 100 μg/ml. TLCK at 150 μg/ml is slightly inhibitory to growth of Balb/c 3T3 cells. We have also observed that Swiss SV3T3 cells are as insensitive to TLCK treatment as Balb SV3T3 cells. In the presence of 60–100 μg TLCK/ml medium, although the growth rate is slower, Swiss SV3T3 cells can eventually reach a cell density of $12–14 \times 10^6$ cells/60-mm plate before the cell sheets peel off the dish. At 200 μg/ml TLCK becomes toxic to both transformed and untransformed 3T3 cells.

However our failure to observe growth inhibition with TLCK should be interpreted with caution, since Shaw et al. (1965) have shown that TLCK has a high rate of spontaneous hydrolysis at pH values greater than 6.0. From data in their paper, we calculate that as much as 50% of the TLCK may be spontaneously hydrolyzed in 25 minutes at pH 7.5. Thus although we added freshly prepared TLCK-containing medium each day to the SV3T3 cell cultures (Fig. 5), there may still have been insufficient TLCK persisting in the culture medium to halt the growth of the transformed cells.

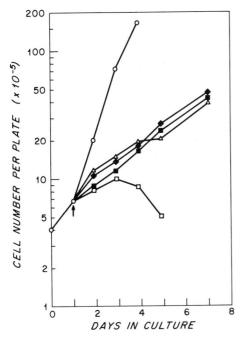

Figure 4

Effect of increased serum concentrations on TPCK inhibition of SV3T3 cell growth. Swiss SV3T3 cells were plated at 4×10^5 cells/dish and, at the time indicated by the arrow, fluid-changed to fresh medium containing varying amounts of fetal calf serum and TPCK as follows: (o) 0 µg/ml TPCK and 10, 15 or 20% serum, (□) 200 µg/ml TPCK and 10% serum, (△) 150 µg/ml TPCK and 10% serum, (■) 200 µg/ml TPCK and 15% serum, and (◆) 200 µg/ml TPCK and 20% serum.

DISCUSSION

Our results demonstrate that TPCK at concentrations below 100 µg/ml is without inhibitory effect on SV3T3 cell growth; at concentrations higher than 100 µg/ml TPCK causes marked growth inhibition of SV3T3 cells. This TPCK-mediated growth inhibition is concentration dependent, reversible, and can be partially overcome by serum, but it is apparently not selective against transformed cells since

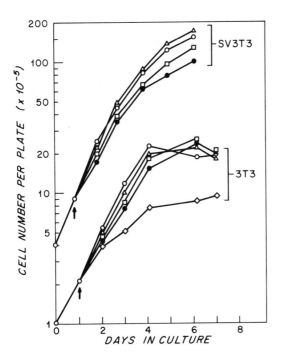

Figure 5

Effect of TLCK on growth of Balb 3T3 and Balb SV3T3 cells. 3T3 and SV3T3 cells were plated at 1×10^5 and 4×10^5 cells/dish respectively on day 0 and, at the time indicated by the arrow, fluid-changed to medium containing TLCK at the following concentrations: (o) 0 µg/ml, (△) 25 µg/ml, (□) 50 µg/ml, (•) 100 µg/ml and (◇) 150 µg/ml.

growth of untransformed cells is similarly affected. We have also shown that doses of TPCK that are required to stop cell growth also inhibit both protein and DNA synthesis. Our data suggest that the TPCK-mediated inhibition of protein synthesis is directly responsible for the observed growth inhibition of the cells. The demonstration that cycloheximide, a known protein synthesis inhibitor, can cause growth inhibition and a "plateau" phenomenon which mimics those caused by TPCK is consistent with this interpretation. However TPCK is known to have other effects on cells. For example, TPCK (but not TLCK) has also recently been shown to inhibit post-translational cleavage of high molecular weight viral precursor polypeptides in Sindbis virus-infected chick embryo fibroblasts (Pfefferkorn and Boyle 1972) and in polio virus-infected HeLa cells (Summers et al. 1972). If such post-translational cleavage were also required for the functioning of cellular proteins involved in cell growth, inhibition of this activity by TPCK could lead to growth inhibition.

The mechanism by which TPCK inhibits protein synthesis in mammalian cells is not known. From kinetic studies on the rate of L-phenylalanine uptake, we conclude that TPCK-mediated inhibition of protein synthesis is probably not due to decreased transport and consequent unavailability of L-phenylalanine for protein synthesis. TPCK (but not TLCK) is known to inhibit polyphenylalanine synthesis by cell-free bacterial extracts, apparently by irreversible inactivation of the ability of one of the elongation factors to bind phenylalanyl-tRNA (Jonak et al. 1973; Richman and Bodley 1973). It may be that TPCK inhibits protein synthesis in mammalian cells by a similar mechanism.

The TPCK-induced growth "plateau" phenomenon observed with SV3T3 cells (Fig. 1) does not correspond to reacquisition of the characteristic "contact inhibition" of growth of untransformed 3T3 cells for several reasons. First, SV3T3 cells at an inhibited cell density of 1.5×10^6 cells/60-mm plate do not show extensive cell-to-cell contact and have not yet formed a confluent monolayer. Second, during the growth "plateau" phase, about 30% of the TPCK-treated SV3T3 cells continue to synthesize DNA. Finally, we find an approximately constant number of predominantly dead cells floating in medium every day during the "plateau" period. Thus we propose that the TPCK-induced growth "plateau" is the result of a dynamic equilibrium between limited cell multiplication and cell death.

A growth "plateau" with SV3T3 cells can also be produced at a lower cell density (5×10^5 cells/plate) simply by plating SV3T3 cells at 1×10^5 cells/60-mm plate and treating the cells with an appropriate concentration of TPCK 24 hours later. Under these conditions TPCK at 200 μg/ml can produce a growth "plateau" that lasts for 3 days before cell death becomes apparent. These results suggest that the ratio of TPCK concentration to cell density at the time of TPCK treatment plays an important role in determining the final cell density and whether or not a growth "plateau" can be achieved. If an appropriate ratio of TPCK concentration to cell density is achieved at the time of TPCK treatment, the cells may undergo one more cell division cycle and then stop increasing in total cell number, resulting in a growth "plateau."

TLCK concentrations up to 100 μg/ml show little, if any, inhibitory effects on growth of our Balb SV3T3 cells. At 200 μg/ml TLCK becomes toxic to the cells. We have also found that our Swiss SV3T3 cells are as insensitive to TLCK treatment as Balb SV3T3 or 3T3 cells. We therefore conclude that the growth inhibition caused by TLCK is not selective for either of the SV3T3/3T3 cell pairs that we examined.

The effects of TPCK and TLCK on transformed cell growth which we have observed differ from those previously reported by Schnebli and Burger (1972). In their experiments growth of their SV3T3 cells was arrested, and a growth "plateau" achieved at TLCK concentrations of 50 μg/ml medium. This concentration of TLCK had no inhibitory effects on the growth of their untransformed 3T3 cells. Similarly they reported selective inhibition of the growth of SV3T3 cells by TPCK at 10 μg/ml medium. From these and other results they concluded that TPCK and TLCK act by blocking a protease-like activity that is required for the unrestrained growth of transformed cells. In contrast we observed no selective inhibition of the growth of our SV3T3 cell lines by either TPCK or TLCK. Differences in the way medium containing TPCK was prepared do not explain differences between these previous results and ours, since TPCK dissolved in dimethyl sulfoxide and added to culture medium at concentrations up to 25 μg/ml also had no inhibitory effect on the growth of our Swiss SV3T3 cells. Under our experimental conditions doses of TPCK that were required to halt a net increase in cell number also caused a substantial inhibition of protein synthesis. We suggest that this inhibition of protein synthesis leads to death of a fraction of the cell population and that the growth "plateaus" we observed are the result of roughly equal rates of cell multiplication and cell death.

Our results show that selective growth inhibition of transformed cells by TPCK and TLCK is not a general property of all SV40-transformed 3T3 cells and thus not an invariable consequence of cell transformation by SV40 virus. At a minimum we can conclude that a TPCK- or TLCK-inhibitable proteolytic activity cannot be exclusively responsible for loss of "contact inhibition" in SV40-transformed fibroblasts. However it may be that other types of proteolytic enzymes play a role in loss of "contact inhibition" in virus-transformed cells. In particular a fibrinolytic activity has recently been described (Unkeless et al. 1973; Ossowski et al. 1973) that is considerably enhanced in fibroblasts transformed by either oncogneic RNA or DNA viruses. This fibrinolytic activity is generated by the interaction of a cellular factor and a serum factor (probably plasminogen), and, although it has a specificity resembling that of trypsin, it is not inhibited by high concentrations of either TLCK or TPCK. While the role of this fibrinolytic activity in loss of cellular growth control in virus-transformed fibroblasts remains to be evaluated, its properties illustrate the exquisite specificity of proteolytic interactions and suggest caution in the interpretation of studies with a few selected protease inhibitors.

Acknowledgments

This research was supported by Grants VC-31 from the American Cancer Society and CA 10126-06 from the National Institutes of Health. I.N.C. holds an NIH Postdoctoral Fellowship (1-FO2-CA54314-01) and R.R. is a Faculty Research Associate (PRA-75) of the American Cancer Society.

REFERENCES

Aaronson, S. and G. Todaro. 1968. Basis for acquisition of malignant potential by mouse cells cultivated in vitro. *Science* **164**:1024.

Black, P. H. 1966. Transformation of mouse cell line 3T3 by SV40: Dose response relationship. *Virology* **28**:760.

Bray, G. A. 1960. A simple efficient liquid scintillator for counting aqueous solutions in a liquid scintillation counter. *Anal. Biochem.* **1**:279.

Burger, M. M. 1969. A difference in the architecture of the surface membrane of normal and virally transformed cells. *Proc. Nat. Acad. Sci.* **62**:994.

Burger, M. M. 1970. Proteolytic enzymes initiating cell division and escape from contact inhibition of growth. *Nature* **227**:170.

Culp, L. A. and P. H. Black. 1972. Contact-inhibited revertant cell lines isolated from simian virus 40-transformed cells. III. Concanavalin A-selected revertant cells. *J. Virol.* **9**:611.

Dulbecco, R. and M. Vogt. 1962. Studies on cells rendered neoplastic by polyoma virus: The problem of the presence of virus-related materials. *Virology* **16**:41.

Hare, J. D. 1967. Location and characteristics of the phenylalanine transport mechanism in normal and polyoma-transformed hamster cells. *Cancer Res.* **27**:2357.

Inbar, M. and L. Sachs. 1969. Structural difference in sites on the surface membrane of normal and transformed cells. *Nature* **223**:710.

Jonak, J., J. Sedeáček and I. Rychlik. 1973. Mode of action of *N*-tosyl-L-phenylalanyl-chloromethane on the elongation protein-synthesizing S_3 factor from *Bacillus stearothermophilus. Biochem. Biophys. Acta* **294**:322.

Ossowski, L., J. C. Unkeless, A. Tobia, J. P. Quigley, D. B. Rifkin and E. Reich. 1973. An enzymatic function associated with transformation of fibroblasts by oncogenic viruses. II. Mammalian fibroblast cultures transformed by DNA and RNA tumor viruses. *J. Exp. Med.* **137**:112.

Pfefferkorn, E. R. and M. K. Boyle. 1972. Selective inhibition of the synthesis of Sindbis virion proteins by an inhibitor of chymotrypsin. *J. Virol.* **9**:187.

Prival, J. T. 1971. Surface membrane proteins of normal and transformed mouse fibroblasts. Ph.D. thesis, Massachusetts Institute of Technology.

Richman, N. and J. W. Bodley. 1973. Irreversible inhibition of the interaction between elongation factor Tu and phenylalanyl transfer ribonucleic acid by L-1-tosylamido-2-phenylethyl chloromethyl ketone. *J. Biol. Chem.* **248**:381.

Rubin, H. 1970a. Overgrowth stimulating factor released from Rous sarcoma cells. *Science* **167**:1271.

————. 1970b. Overgrowth-stimulating activity of disrupted chick embryo cells and cells infected with Rous sarcoma virus. *Proc. Nat. Acad. Sci.* **67**:1256.

Schnebli, H. P. 1972. A protease-like activity associated with malignant cells. *Schweiz. Med. Wechenschr.* **102**:1194.

Schnebli, H. P. and M. M. Burger. 1972. Selective inhibition of growth of transformed cells by protease inhibitors. *Proc. Nat. Acad. Sci.* **69**:3825.

Sefton, B. M. and H. Rubin. 1970. Release from density dependent growth inhibition by proteolytic enzymes. *Nature* **227**:843.

Shaw, E. 1970. Selective chemical modification of proteins. *Physiol. Rev.* **50**:224.

Shaw, E., M. Mares-Guia and W. Cohen. 1965. Evidence for an active-center histidine in trypsin through use of a specific reagent, 1-chloro-3-tosylamido-7-amino-2-heptanone, the chloromethyl ketone derived from *N*-α-tosyl-L-lysine. *Biochemistry* **4**:2219.

Stoker, M. P. G. and H. Rubin. 1967. Density dependent inhibition of cell growth in culture. *Nature* **215**:171.

Summers, D. F., E. N. Shaw, M. L. Stewart and J. V. Maizel. 1972. Inhibition of cleavage of large poliovirus specific precursor proteins in infected HeLa cells by inhibition of proteolytic enzymes. *J. Virol.* **10**:880.

Todaro, G. J., G. K. Lazar and H. Green. 1965. Inhibition of cell division in a contact inhibited mammalian cell line. *J. Cell. Comp. Physiol.* **66**:325.

Unkeless, J. C., A. Tobia, L. Ossowski, J. P. Quigley, D. B. Rifkin and E. Reich. 1973. An enzymatic activity associated with transformation of fibroblasts by oncogenic viruses. I. Chick embryo fibroblast cultures transformed by avian RNA tumor viruses. *J. Exp. Med.* **137**:85.

Tumor-associated Fibrinolysis

E. Reich

The Rockefeller University, New York, N.Y. 10021

All primary cultures and nearly all known vertebrate cell lines require macromolecular components of serum for growth; and, although their serum requirement is lower than that of normal cells, cells transformed by oncogenic agents do not grow in the absence of serum (Temin, Pierson and Dulak 1972; Holley this volume). Since the protein concentration of the extravascular and extracellular fluid compartments may be low in comparison with that of the plasma, it is reasonable to conjecture that the initiation of growth by either normal or transformed cells would require that increased levels of plasma proteins gain local access to the extracellular milieu. Such increased penetration of protein into tissue spaces might be brought about by direct physical damage to blood vessels, as in wounds, by changes in vascular permeability, such as those induced by inflammation (e.g., carcinogenic promoters), or by humoral agents released from tissue cells or formed in the blood. The brain tumors present a persuasive example of modified vascular permeability associated with growth. The normal brain, where cell growth is at a minimum, is immersed in cerebrospinal fluid, in which the protein concentration is 0.2% of that in plasma. In patients with brain tumors the concentration of protein in cerebrospinal fluid may rise by nearly two orders of magnitude, and this increase is clearly the result of changes in permeability since the proteins in question are entirely derived from plasma and the cerebrospinal fluid usually remains free of blood cells.

While this reasoning is admittedly speculative, it appeared to deserve some attention owing to the neglected work of A. Fischer (1925). Fischer compared the growth of normal and neoplastic explants in culture and found that malignant tissue lysed the substrate—a plasma clot—whereas the normal tissue did not. Since the lysis of plasma clots is a process catalyzed by enzymes, Fischer's observation implies an enzymatic difference between normal and malignant cells, a possibility that merits reinvestigation and analysis.

Correlation of Transformation and Fibrinolysis and Generality of the Phenomenon

To test whether Fischer's phenomenon could be reproduced under conditions of modern cell culture, normal and transformed fibroblasts were grown on thin films of fibrin (Unkeless et al. 1973a; Ossowski et al. 1973b). The disappearance of fibrin can be observed visually or by monitoring the degradation of [^{125}I]fibrin, and both experimental approaches confirm Fischer's findings: virus-transformed fibroblasts show rapid fibrinolysis, but normal controls do not. The induction of fibrinolysis is closely correlated with transformation as shown by the following facts:

1. Fibrinolysis is stimulated only when fibroblasts are infected by transforming viruses. Infection with any of a wide range of cytolytic viruses is not followed by induction of fibrinolysis; this is also the case both for nontransforming avian leukosis viruses and for the "temperate" paramyxovirus SV5 (Unkeless et al. 1973a).
2. When chick embryo fibroblasts are infected with mutants of Rous sarcoma virus that are temperature sensitive with respect to transformation, there is no fibrinolysis at the high, nonpermissive temperature. Following transfer of the cultures to the low, permissive temperature, the onset of fibrinolysis occurs within less than one hour of incubation, and it is therefore a very early event in the expression of transformation (Unkeless et al. 1973a; Rifkin and Bader unpubl.).
3. Fibrinolysis is associated with transformation of avian and mammalian cells in culture by either DNA or RNA viruses, it is observed in primary cultures of chemically induced mammary carcinomas, hepatomas, and skin cancers, and it is found in a variety of human and animal tumor cell lines. The association of fibrinolysis and neoplasia is therefore quite general (Unkeless et al. 1973a; Ossowski et al. 1973b and unpubl.; Rifkin et al. unpubl.), and it is independent of the nature of the transforming stimulus.

Determination of Serum Specificity in Fibrinolysis

Although enhanced fibrinolysis by malignant cells is demonstrable under a wide range of experimental conditions, optimal expression of this property depends on the nature of the serum supplement, and a characteristic spectrum of activating and nonactivating sera is obtained for different transformed cultures. The differences in activity of various sera are due both to complex interactions involving the fibrinolytic factors themselves and to the effects of serum inhibitors, and no adequate quantitative description of these interactions has been obtained to date. Nevertheless the pattern of activating sera is a useful empirical parameter for comparing the serum specificity of cells from different species, as well as cells from a single species transformed by different agents. Tests of this kind show that serum specificity is determined by the cell type, and not by the transforming agent (Ossowski et al. 1973b). Thus the serum specificity of mouse cells transformed by SV40 virus differs from that of hamster cells transformed by SV40, but it is identical with that of mouse cells transformed by mouse sarcoma virus.

Mechanism of Induction of Fibrinolysis

The mechanism by which neoplastic cells initiate fibrinolysis has been analyzed and shown to depend on two protein factors, one of which is present in all vertebrate sera, the other being released by cells following transformation (Unkeless et al., 1973; Ossowski et al. 1973b). The serum factor has been identified as the known zymogen, plasminogen, a protein with molecular weight of approximately 90,000 daltons (Quigley et al. 1973). The cell factor is itself an arginine-specific protease of molecular weight 38,000; it is a serine protease that is inhibited by DFP, and it hydrolyzes a single peptide bond in plasminogen, thereby generating the active fibrinolytic protease, plasmin (Unkeless et al. 1973b). The cell factor is therefore a plasminogen activator. Differences in the efficiency of activation of a series of pure plasminogens by a single cell factor account for some of the differences in activating properties of various sera. For example, chicken serum does not promote fibrinolysis by transformed hamster cells (Ossowski et al. 1973b) and chicken plasminogen is not activated by hamster cell factor (Quigley et al. 1973).

The release of cell factor, which accumulates in the culture fluids, is the basic cause of the fibrinolytic activity of transformed cultures. We do not know whether the cell factor is newly synthesized or merely activated following transformation; however the cell factor is not detectable in homogenates of normal fibroblasts, whereas a substantial concentration is present in transformed cells, where it is tightly bound to one of the sedimentable, membranous cellular particulates in the post-nuclear fraction.

Relationship between Fibrinolysis and Parameters of Transformation

Since enhanced fibrinolysis appears to be the first reliable and general enzymatic change associated with transformation, it is of interest to correlate the activity of the fibrinolytic system with the expression of different parameters of the transformed phenotype. Tests of this kind can be performed by using inhibitors of the fibrinolytic enzyme, by comparing the effects of activating and nonactivating sera, and by depleting sera of plasminogen (Ossowski et al. 1973a,b and unpubl.). From the results obtained to date it appears likely that characteristic changes in cell morphology, colony formation in semi-solid media, and enhanced cell motility all require the presence of an intact fibrinolytic system. Recent observations (Ossowski et al. 1973a) also show that transformed cells produce receptors for plasmin that are not detectable in normal cultures, although their chemical nature has not yet been identified. Finally when normal cells are cocultivated with transformed cells under conditions that permit activation of the fibrinolytic system, the normal cells develop morphological changes characteristic of transformation (Ossowski et al. 1973c). All of these findings indicate that some of the parameters ordinarily associated with transformation are determined, at least in part, by the activation of the fibrinolytic system, and it appears likely that some of the serum effects on these processes will be attributable to differences in the levels of activity of these enzymes. For example, the serum factor recently reported to be toxic for transformed cells (Paul 1973) is probably plasminogen. A fact deserving emphasis is that plasminogen activation occurs either at the external surface of the cell or in the adjacent extracellular milieu. In this context it is tempting to ask whether the formation of such a freely diffusible active protease, which is capable of attacking cell surface proteins, could account for the broad spectrum of antigenic specificities that is characteristic of neoplastic tissues. The relationship, if any, between fibrinolysis and saturation density of cultures is under study.

Benign Neoplasms

Only a single class of benign tumors, the rat mammary fibroadenoma, has been tested for production of cell factor. Like normal lactating and nonlactating mammary tissue, the fibroadenoma appears not to produce the cell factor. In contrast, the mammary carcinoma induced by dimethylbenzanthracene in the same strain of rats uniformly produces high levels of plasminogen activator.

There is ample clinical evidence, in the form of disorders in bleeding, clotting and fibrinogen metabolism, to indicate that the fibrinolytic system is operating at pathological levels in human malignant disease. Although there is no basis for assuming a link between fibrinolysis and tumor growth, invasiveness or metastasis, these are obvious subjects for future study. Other promising clinical applications are in the area of diagnosis and, perhaps, therapy aimed at the activity of the cell factor. The development of these applications, as well as thorough exploration of the tumor cell biology, all depend on large-scale preparation and isolation of cell factor. Although this activator is produced in very low amounts (1–10 μg/liter), the methodology for production and isolation on a large scale has been developed and is being applied.

Although the active fibrinolysin, plasmin, has particularly high affinity for fibrinogen and fibrin, it is also a highly potent trypsin-like enzyme that effectively degrades a wide variety of soluble and insoluble peptides. The activation of plasminogen in normal and pathological states is therefore possibly of significance, especially in relation to tissue destruction in chronic inflammation and to other processes such as thrombosis, hematostasis, lymphocyte activation and hemopoiesis. With this in mind we have examined the behavior of selected cell types and have found that activated macrophages and blood monocytes, which may have a role in some or all of these processes, produce large amounts of a cell factor that is indistinguishable from that released by transformed cells (Gordon et al. unpubl.). The implications of this observation may be of some interest.

Acknowledgments

The studies summarized in this abstract have been performed in collaboration with K. Danø, S. Gordon, G. M. Kellerman, D. Loskutoff, L. Ossowski, A. Piperno, J. P. Quigley, D. B. Rifkin, A. Tobia, and J. C. Unkeless and have been supported in part by grants from the American Cancer Society and the National Institutes of Health.

REFERENCES

Fischer, A. 1925. Beitrag zur Biologie der Gewebezellen. Eine vergleichendbiologische Studie der normalen und malignen Gewebezellen, *in vitro. Arch. Entwicklungsmech. Org.* (Wilhelm Roux) **104**:210.

Ossowski, L., J. P. Quigley, G. M. Kellerman and E. Reich. 1973a. Fibrinolysis associated with oncogenic transformation: Requirement of plasminogen for correlated changes in cellular morphology, colony formation in agar and cell migration. *J. Exp. Med.* (in press)

Ossowski, L., J. C. Unkeless, A. Tobia, J. P. Quigley, D. B. Rifkin and E. Reich. 1973b. An enzymatic function associated with transformation of fibroblasts by oncogenic viruses. II. Mammalian fibroblast cultures transformed by DNA and RNA tumor viruses. *J. Exp. Med.* **137:**112.

Ossowski, L., J. P. Quigley and E. Reich. 1973c. Fibrinolysis associated with oncogenic transformation: Morphological correlates. *J. Biol. Chem.* (in press)

Paul, D. 1973. Factor in acidified serum toxic for malignant cells in culture. *Nature New Biol.* **242:**186.

Quigley, J. P., L. Ossowski and E. Reich. 1973. Plasminogen: The serum proenzyme activated by factors from malignant cells. *J. Biol. Chem.* (in press)

Temin, H. M., R. W. Pierson, Jr. and N. C. Dulak. 1972. The role of serum in the control of multiplication of avian and mammalian cells in culture. *Growth, nutrition, and metabolism of cells in culture,* vol. 5, p. 50–81. Academic Press, New York.

Unkeless, J. C., A. Tobia, L. Ossowski, J. P. Quigley, D. B. Rifkin and E. Reich. 1973a. An enzymatic function associated with transformation of fibroblasts by oncogenic viruses. I. Chick embryo fibroblast cultures transformed by avian RNA tumor viruses. *J. Exp. Med.* **137:**85.

Unkeless, J. C., K. Danø, G. M. Kellerman and E. Reich. 1973b. Fibrinolysis associated with oncogenic transformation: Partial purification and characterization of the cell factor—a plasminogen activator. *J. Biol. Chem.* (in press)

Surface Alterations and Mitogenesis in Lymphocytes

Gerald M. Edelman

The Rockefeller University, New York, New York 10021

Although lymphoid cells have several singular properties related to clonal selection by antigens during the immune response, they also share fundamental mechanisms of growth control with other specialized tissues. Lymphoid cells are particularly suitable for studying certain features of mitogenesis and growth control because they are readily available in a dissociated state, have gene products of known structure as well as several known surface markers, and can be stimulated by specific antigens and a variety of mitogenic agents. Furthermore the binding by lymphoid cells of antigens, antibodies, and lectins induces a variety of surface changes that may be of significance for the study of cell–cell interactions.

Recent studies (Taylor et al. 1971; Yahara and Edelman 1972; dePetris and Raff 1972) suggest that the distribution and mobility of lymphocyte surface receptors may be under control of specific structures within the cell. The purpose of this paper is to review some of the evidence for the modulation of cell surface receptor mobility by mitogenic lectins such as concanavalin A. I shall then consider some hypotheses to explain this modulation as well as the possibility that it may be related to the initial events of mitogenic stimulation.

Clonal Selection and Mitogenesis in Lymphocytes

It is now generally accepted that antigens elicit antibody production by means of clonal selection (Cold Spring Harbor Symp. Quant. Biol. 1967). Although each lymphoid cell possesses antibodies of a single specificity on its surface, the entire population of different antigen binding cells expresses a large diversity of antibodies with different binding sites (Edelman 1971; Edelman and Gall 1969). Binding of a particular antigen to some of the antigen binding cells results in triggering of a subpopulation of cells having antibodies of sufficiently high affinity; these cells then mature, divide, and produce more antibodies of the same kind. As many as 1 in 10^2 cells of a spleen cell population may bind a particular antigen but only 1 in 10^5 cells is antigen-reactive, and for this reason it is difficult to study directly the mitotic process in response to an antigen.

Fortunately this process may be studied by other means, for the capacity to be triggered is to some extent specificity-independent. A large number of different agents in addition to antigens are mitogenic for lymphocytes. These agents include mitogenic lectins (plant proteins with specificity for various carbohydrates) (Sharon and Lis 1972), lipopolysaccharides (Andersson et al. 1972), and chemical reagents such as periodate (Novogrodsky and Katchalski 1971). Although their detailed structure is unknown, the surface receptors for these various mitogenic agents are in general different from each other, at least in their carbohydrate portions. Because the events of lymphocyte stimulation with each mitogen are similar, however, it is probable that mitogenesis is mediated by a final common pathway. The availability of a variety of mitogens provides an opportunity to study the stimulation of lymphocytes without extensively purifying them according to the specificity of their Ig receptors.

There are two requirements for mitogenic stimulation: (1) The antigen or mitogenic agent must bind to the appropriate receptors in the proper conformation or arrangement. (2) After this binding, coupling to the chain of metabolic alterations preceding transformation and mitosis must take place. These alterations include immediate changes in lipid turnover (Fisher and Mueller 1971), in cyclic GMP levels (Hadden et al. 1972), and in transport of ions and small molecules (Quastel and Kaplan 1970). Somewhat later there are changes in RNA and DNA metabolism (Cooper 1972). It should be noted that binding events involve changes in seconds or minutes, whereas metabolic events occur in minutes to hours. Moreover irreversible stimulation does not in general occur until 12 hours after the initial binding; i.e., if the lectin is specifically removed before that time, transformation will not occur (Novogrodsky and Katchalski 1971). Thus stimulation involves a complex series of events, the detailed kinetics of which remain to be worked out.

Even before the kinetics are known, it is of definite value to ask whether the lectin must act first at the cell surface, and if so, to consider the effects it can induce at that surface. It is known that only some of the glycoprotein receptors on the cell surface are responsive to stimulation, for several lectins can bind to glycoproteins but are not mitogenic. Of a given population of receptors responsive to a mitogenic lectin such as ConA, as few as 6% have to be bound to induce transformation (Inbar, Ben-Bassat and Sachs 1973). There is, in addition, some evidence to suggest that lectins act directly at the cell surface, for covalent coupling of ConA and PHA (Andersson, Möller and Sjöberg 1972a,b; Andersson et al. 1972; Greaves and Bauminger 1972) at solid surfaces does not abolish their ability to stimulate cells.

In determining how surface interactions might be transmitted to the interior of the cell, the grouping, mobility, and attachment of the lectin receptors must be considered. Recent findings indicate that the lipid portion of the cell membrane is fluid (Hubell and McConnell 1969), that cell receptors are mobile (Frye and Edidin 1970), and that immunoglobulins and other receptors on the lymphocyte may be cross-linked by specific divalent antibodies to form patches and ultimately caps at one pole of the cell (Taylor et al. 1971; dePetris and Raff 1972; Yahara and Edelman 1972). dePetris and Raff (1972) have proposed that the patches are formed by diffusion but that formation of caps from patches depends upon cell movement and metabolism. Receptors of an individual type can be "patched" and "capped" independently of other receptors and thus, with one exception to be discussed in detail here, they appear to be independent of each other; i.e., they are not physically associated in terms of their mobility. Although there is no direct structural knowledge about the detailed mode of anchorage of any of the receptors in the membrane, certain studies indicate that membrane proteins may go through the membrane (Bretscher 1971) or be connected to structures within the membrane. Despite the absence of details about the connection of surface receptors to structures in the cytoplasm, valuable inferences can be drawn by using lectins to agglutinate cells or otherwise alter their function. We have recently found a quite striking effect of ConA on the mobility of receptors that sheds light on the nature of receptor anchorage and receptor interactions.

Immobilization of Cell Surface Receptors by ConA

Native concanavalin A strongly affects the ability of cell receptors to form patches and caps (Yahara and Edelman 1972, 1973). If ConA is added in doses greater than 5 μg/ml to lymphocytes at 21 or 37°C prior to treatment with anti-Ig, then both patch formation and cap formation are inhibited (Fig. 1). This effect, which is dose dependent, may be reversed by addition of α-methyl-D-mannoside, a competitive inhibitor of ConA. In contrast, if ConA is added to the cells at 4°C and excess ConA is washed away, and the cells are then brought to 37°C, ConA forms patches and caps with its own receptors. Native tetravalent ConA therefore has two antagonistic actions that depend both on the temperature and on the concentration of the lecin. The same effect is seen for capping of antigens on T cells by anti-θ antibodies (Yahara and Edelman 1973).

A number of questions must be answered in order to explain these effects:

1. For what dose range is the inhibition observed, i.e., how many molecules of ConA must be bound before an appreciable effect is seen?

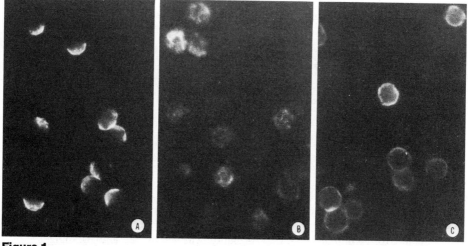

Figure 1

Labeling patterns of cells with fl-anti-Ig and with fl-ConA. **(A)** Cells incubated with fl-anti-Ig (80 μg/ml) at 21°C for 30 min showing caps. **(B)** Prior addition of NaN$_3$ (10 mM) showing patches. **(C)** Prior addition of ConA (100 μg/ml) showing diffuse patterns.

2. Is the inhibition mainly of cap formation or of patch formation and does it depend upon metabolism?
3. Is the inhibition reversible after removing bound ConA and is it mediated at the cell surface?
4. Does the multivalent ConA molecule interact with more than one kind of receptor at the same time?
5. What structural features of the molecule are responsible for the activity?
6. What is the nature of the cellular structures that modulate receptor mobility?

Dose Range and Temperature Dependence

As shown in Fig. 2, the effect of ConA is seen at doses as low as 2 μg/ml at 21°C, and it is virtually complete at 20 μg/ml. ConA also inhibits the capping of antibody receptors induced by forming a sandwich of antigen and antibody (Table 1). Similar effects are seen in the prevention of patching and capping by antibodies to the θ antigen of thymocytes (Table 1). ConA also inhibits patch and cap formation by its own receptors at 21 and 37°C.

In contrast to these findings, if ConA is bound at 4°C, the excess ConA is washed away, and the cells with bound ConA are brought to a temperature of 37°C, patching and capping of the ConA receptors occurs (Table 2). The results of the binding–washing procedure at 4°C show that capping of ConA receptors is optimal at a binding level of 3 $\times$ 10^5 molecules per cell and the curve resembles a precipitation curve (Fig. 3). At a given concentration of ConA, fewer molecules are bound to splenic lymphocytes at 4°C than at 37°C. Nevertheless, except for a small amount of capping at a concentration of 2 μg/ml, the ConA inhibition at 37°C is seen at levels of binding ranging from 10^5 to 10^6 molecules per cell (Fig. 3).

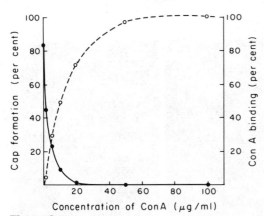

Figure 2

Effect of ConA on cap formation ($\bullet$——$\bullet$). 2 $\times$ 10^7 cells /ml were incubated with various concentrations of ConA in PBS-BSA at 21°C for 10 min. Fl-anti-Ig was added to the mixture to give a concentration of 80 μg/ml, and the mixture was incubated for 30 min at 21°C. Cap formation was determined after washing cells.

Binding of ^{125}I-ConA to cells (o– – –o). 2 $\times$ 10^7 cells/ml were incubated with various concentrations of ^{125}I-ConA (2.4 $\times$ 10^4 cpm/μg) at 21°C for 30 min in PBS-BSA. Cells were washed with PBS-BSA and the radioactivity was determined.

Table 1

Effect of ConA on Patch and Cap Formation

Cell	Inducer	% Stained cells	conc. ConA ($\mu g/ml$)	% Caps[d]	% Inhibition[e]
Spleen[a]	fl-anti-Ig	45	0	92	0
			5	25	73
			50	0	100
Thymocytes[b]	anti-θ + fl-anti-Ig	98	0	88	0
			5	10	89
			50	0	100
Spleen[c]	DNP-BSA + fl-anti-DNP-BGG	7	0	90	0
			5	45	50
			50	0	100

[a] Mouse spleen cells (2×10^7/ml) were incubated with 80 μg/ml fl-anti-Ig in the presence or absence of ConA at 37°C for 10 min in PBS-BSA.

[b] Mouse thymocytes (2×10^7/ml) were incubated with 1:20 dil. anti-θ C3H serum at 21°C for 15 min, washed and incubated with fl-anti-Ig at 37°C for 10 min with or without ConA.

[c] Spleen cells (2×10^7/ml) were incubated with 20 μg/ml DNP-BSA at 4°C for 30 min in PBS-BSA-NaN$_3$, washed and incubated with 100 μg/ml fl-anti-DNP-BGG at 37°C for 15 min in PBS-BSA with or without ConA.

[d] Caps are expressed as the percentage of total stained cells showing caps.

[e] Calculated for cap forming cells, taking control values as 100%.

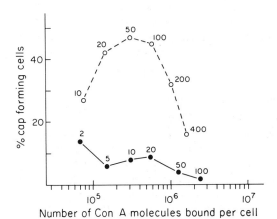

Figure 3

Relation of cap formation to the number of ConA molecules bound to lymphoid cells at 4°C (o) and at 37°C (•). Numbers above each point refer to the concentration of ConA used in μg/ml.

Table 2

Cap Formation Induced by ConA
under Various Conditions

conc. ConA ($\mu g/ml$)	I^a (%)	II^b (%)	III^c (%)
1	5–12	11	
2	3–9	14	
5	0.2–6	6	
10	0.3–3	8	27
20	0.2	9	42
50	0.2	4	47
100	0.2	2	45
200		0	32
400		0	16

At concentrations of 1–2 $\mu g/ml$ cap forming cells were determined radioautographically (Yahara and Edelman 1973); at concentrations of 5–10 $\mu g/ml$ cap forming cells were determined either radioautographically or by direct labeling with fl-ConA. At all other concentrations cells were determined by direct fluorescent labeling.

[a] Mouse spleen cells ($1 \times 10^7/ml$) were incubated with various concentrations of ConA at 37°C for 6 hr in HBSS-FBS.

[b] Cells were incubated with ConA at 37°C for 30 min, washed and incubated at 37°C for another 30 min in HBSS-FBS.

[c] Cells were incubated with ConA at 4°C for 30 min, washed in the cold and incubated at 37°C for 30 min in HBSS-FBS.

These experiments show that the antagonistic actions of ConA depend on the temperature, the concentration, and state of the lymphoid cell. The results cannot, however, be explained by the fact that only about one-third as many molecules of ConA are bound per cell at 4°C as compared to 37°C, for even when equal numbers of molecules are bound at the two different temperatures, the inhibition at 37°C is still seen (Fig. 3).

Stage of Inhibition by ConA

A number of questions arise concerning the stage in cap formation at which ConA acts and the dependence of the ConA effect upon cellular metabolism. The evidence indicates that after addition of ConA patches are not formed and, therefore, that cap formation is inhibited because prior patch formation is inhibited. Experiments on cap formation after formation of patches (Yahara and Edelman 1973) show, however, that added ConA also decreases the *rate* of cap formation. Thus although the main effect is to prevent patch formation, ConA also appears in some degree to inhibit the process leading from patches to caps. It seems unlikely that the inhibition of patch formation (which precedes cap formation) depends directly upon cellular metabolism, for preincubation of cells with ConA in the presence of NaN_3 at 37°C, followed by removal of NaN_3 still did not result either in patch or cap formation (Yahara and Edelman 1973).

Reversibility and Surface Effects

It appears that the *initial* effect of ConA depends upon events at the surface, for the inhibition is reversible (Fig. 4). When bound ConA is removed from its receptors by addition of alpha-methyl-mannoside (a competitive inhibitor of binding), patch and cap formation reappear. The system responsible for inhibition of receptor mobility therefore can recover from the effects of ConA binding, and the lymphoid cell is not killed by the interaction under the conditions of these experiments.

Receptors Other Than ConA Receptors

The inhibition phenomena might be attributable to the interaction of ConA with two or more completely different kinds of receptors, e.g., its own receptor and IgM receptor antibody. Although we have found that ConA interacts in vitro to precipitate mouse IgM, the experiments with cells suggest that this interaction does not occur at the cell surface. ConA receptors and IgM receptors can undergo patch

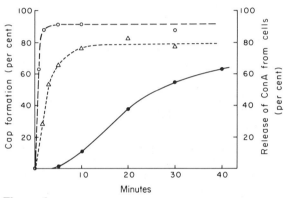

Figure 4

Cap formation after release of ConA from cells ($\bullet$——$\bullet$). 2×10^7 cells/ml were incubated with ConA (100 μg/ml) at 0°C for 10 min in PBS-BSA and fl-anti-Ig (80 μg/ml) was added to the mixture. The mixture was then incubated at 0°C for 30 min and centrifuged. The cells were washed with PBS-BSA and resuspended in the same volume of PBS-BSA at 21°C, to which α-methyl-D-mannoside was added to give a concentration of 40 mM. Aliquots were pipetted at various times after the addition of α-methyl-D-mannoside, the cells were washed, and cap formation was determined.

Rate of release of ^{125}I-ConA from cells after the addition of α-methyl-D-mannoside (o——o). 2×10^7 cells/ml were incubated with 100 μg/ml of ^{125}I-ConA (2.4×10^4 cpm/μg) at 21°C for 10 min in PBS-BSA. Unlabeled anti-Ig was added to this mixture to give a final concentration of 80 μg/ml. After incubation at 21°C for 30 min, the mixture was centrifuged and the cells were resuspended in PBS-BSA. Aliquots of the mixture were transferred to small tubes containing α-methyl-D-mannoside in PBS-BSA (final concentration 40 mM) and were incubated at 21°C for various times; the radioactivity was then determined.

Cap formation after removal of NaN$_3$ ($\triangle$– – –$\triangle$). 2×10^7 cells/ml were incubated with fl-anti-Ig (80 μg/ml) for 30 min at 21°C in PBS-BSA containing NaN$_3$ (3 mM). Cells were collected by centrifugation and resuspended in PBS-BSA at 21°C with a decrease in NaN$_3$ concentration to 30 μM. Cells were washed and at various times cap formation was determined.

and cap formation independently of each other (Yahara and Edelman 1973; Karnovsky and Unanue 1973). Moreover at concentrations where capping and patching are completely inhibited by bound ConA, anti-Ig antibodies bind to cellular IgM receptors almost as well as they bind in the absence of ConA (Yahara and Edelman 1972).

These experiments provide more or less clear-cut answers to the first four questions posed at the beginning of this section. It is the last two questions that are of deep significance: What molecular features of ConA account for its activity and what cellular structures modulate mobility of surface receptors? In the remainder of this paper, I shall consider some experiments bearing upon these two questions.

Structural Properties and Activity of ConA and Its Derivatives

Knowledge of the molecular structure of a lectin is important in determining how it binds to the cell surface and how it induces the molecular transformation required for stimulation. With a knowledge of the three-dimensional structure of a lectin, for example, various amino acid side chains at the molecular surface may be deliberately modified by group reagents. The activities of the modified lectin derivatives may then be observed in various assays of their effects on cell surfaces and cell function.

Studies (Edelman et al. 1972) on the subunit and three-dimensional structure of ConA indicate that at pH 7 and above it is a tetramer with four saccharide binding sites (Fig. 5). This suggests that receptor cross-linkage by the tetravalent ConA molecule may play an important role in its action on cells. Furthermore because it is multivalent, ConA may function to cross-link receptors between two different cells and agglutinate them. For these reasons it is of particular interest to determine the effects of a change in valence on the inhibition of the mobility of cell surface receptors.

Recent experiments (Gunther et al. 1973) indicate that the valence of ConA may be changed by suitable chemical treatment. Succinylation of ConA by treatment with succinic anhydride in two steps of derivatization leads to the attachment of an average of 10 succinyl groups per subunit. The molecular weight of succinyl-ConA at pH 7.4 is 56,000 daltons, half the value obtained for the native protein under identical conditions. Equilibrium dialysis indicates that there are four binding sites for α-methylglucose on the tetrameric native protein and two binding sites on the dimeric succinyl derivative; moreover the affinity constants of ConA and succinyl-ConA for this sugar were both approximately $2 \times 10^3/1$ mole. The capacity of the succinyl-ConA to bind to Sephadex and to react with antibodies to ConA was also similar to that of the native lectin, suggesting that no gross changes in the tertiary structure of the carbohydrate binding site occurred during derivatization. Titration of the number of receptors for ConA and succinyl-ConA on sheep erythrocytes showed that both ConA and its succinyl derivative were bound in approximately equal numbers; approximately the same number of molecules of ConA and succinyl-ConA were also bound by splenic lymphocytes. All of these findings (Table 3) indicate that the major effect of succinylation is to alter the aggregation properties of the ConA subunit so that it forms dimers rather than tetramers.

Unlike the native molecule, dimeric succinyl-ConA lacks both the capacity to induce patches and caps on splenic lymphocytes and the capacity to inhibit patch and cap formation by its own and other receptors. More pertinent to the question

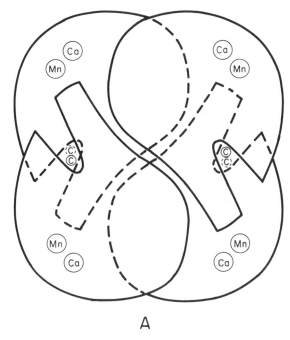

A

Figure 5
Structure of concanavalin A. **(A)** Schematic representation of the tetrameric structure of ConA viewed down the crystallographic *c* axis. The proposed binding sites for manganese, calcium and saccharides are indicated by *Mn, Ca,* and *C,* respectively. **(B)** Alpha-carbon backbone of the ConA tetramer oriented as in *A*. **(C)** Side view of the ConA tetramer shown in *B*.

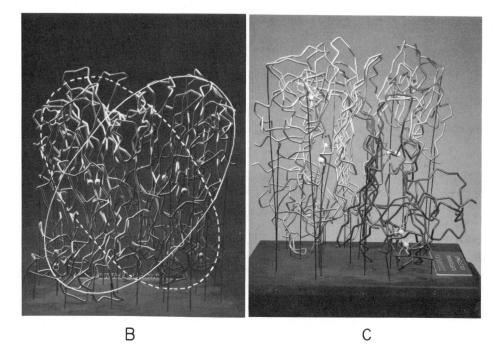

B   C

concerning the mechanism of inhibition of patch formation is the finding that treatment with anti-ConA can restore the ability of bound succinyl-ConA to restrict the movement of immunoglobulin receptors. Spleen cells were incubated with 50 μg/ml succinyl-ConA at room temperature. After 15 minutes the cells were washed and anti-ConA and fl-anti-Ig were added successively so that the final concentration

Table 3

Comparison of the Biological Activities of ConA and Succinyl-ConA

Property	ConA	Succinyl-ConA
Number of binding sites per cell		
sheep erythrocytes	1.1×10^6	2.8×10^6
mouse spleen cells	1.4×10^6	4.4×10^6
Agglutination		
sheep erythrocytes	$1 \mu g/ml$	$> 500 \mu g/ml$
sheep erythrocytes +		
succinyl-ConA ($330 \mu g/ml$)	$8 \mu g/ml$	
mouse spleen cells	$4.5 \mu g/ml$	$40 \mu g/ml$
Percentage of cells forming		
lectin receptor caps		
lectin ($5 \mu g/ml$, $37°C$)	$0.2–2\%$	0%
lectin ($100 \mu g/ml$, $37°C$)	$\leqslant 0.2\%$	0%
lectin ($170 \mu g/ml$)		
preincubated in ice bath, washed,		
then brought to $37°C$	62%	
lectin ($20 \mu g/ml$) + anti-ConA		
($100 \mu g/ml$)	18%	82%
Percent inhibition of anti-Ig capping		
lectin ($100 \mu g/ml$)	100%	0%
lectin ($50 \mu g/ml$) + anti-ConA		
($100 \mu g/ml$)	100%	40%
Mitogenesis		
lectin ($5 \mu g/ml$)	positive	positive
lectin ($50 \mu g/ml$)	negative	positive

was 100 $\mu g/ml$ for each of these reagents. After this treatment, inhibition of anti-Ig cap formation increased from less than 1% to about 40% (Table 3). Treatment with anti-ConA also increased the number of cap-forming cells seen with native ConA. For example, when cells preincubated with 20 $\mu g/ml$ ConA were then incubated with 100 $\mu g/ml$ fl-anti-ConA at $37°C$, cap formation occurred immediately in 18% of the cells. Without anti-ConA, less than 1% of the cells formed caps under these conditions. Anti-ConA itself has no effect on cap formation induced by anti-Ig, and conversely, the presence of anti-Ig neither enhances nor inhibits cap formation by succinyl-ConA plus fl-anti-ConA. These results are summarized in Fig. 6.

What is the major factor in the alteration of ConA activity by succinylation? Apparently the dimer cannot achieve the degree of cross-linking of cellular receptors necessary for reactions such as agglutination, induction of cap formation by ConA receptors, and inhibition of immunoglobulin cap formation. The failure to inhibit receptor mobility may be the result of a change in valence or a change in the surface charge of the molecule. At the concentration necessary for agglutination, however, ConA covers only about 1% of the cell surface. This is probably not

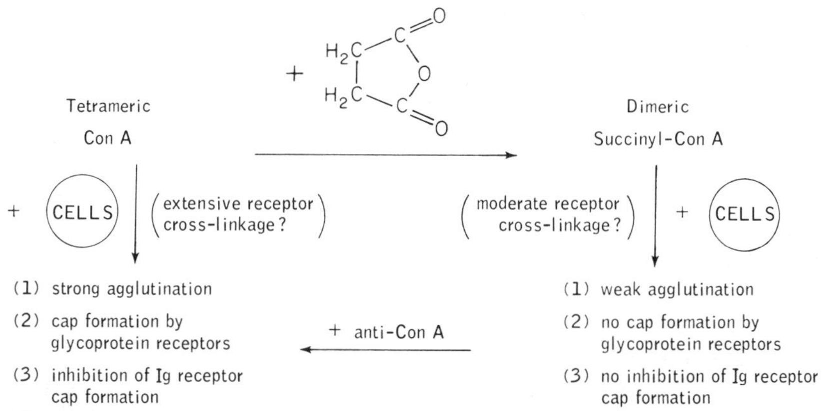

Figure 6
Schematic comparison of the biological properties of native ConA and suc-cinyl-ConA. The activities listed include those immediate cellular reactions mediated by the lectin.

enough for the altered charge to affect the agglutination process. Most convincing, however, is the observation that the addition of anti-ConA to cells *after* succinyl-ConA has been bound can bring about all three phenomena shown in Fig. 6. Fab fragments of antibodies to ConA do not restore these phenomena and anti-ConA alone has no effect. It seems likely, therefore, that treatment with anti-ConA is equivalent to an increase in the effective valence of the bound succinyl-ConA mole-cules and that the main effects do not result from changes in net charge.

All of these experiments suggest that cross-linkage of lectin receptors plays a large role in the inhibition of the mobility of cell surface receptors as well as in other activities of mitogenic lectins. They leave unanswered, however, the question concerning the nature of the cellular structures that modulate receptor mobility. Recent observations (Edelman, Yahara and Wang 1973) on certain drugs that reverse the inhibition of mobility may provide a clue to the nature of these structures.

Reversal of the Mobility Inhibition Effect by Colchicine and Related Drugs

From the observations that I have so far described, we can conclude that binding of tetravalent ConA leads to restriction of the mobility of several receptors includ-ing its own, that this restriction is reversible, and that it depends in part upon the valence of the molecule. There are three mechanisms that might explain these phenomena: (1) the binding of a relatively small number of ConA molecules leads to a phase transition in the lipids of the membrane, increasing their transition tem-perature and decreasing its fluidity; (2) the binding of ConA to its receptors leads to the formation of membrane structures that trap other receptors or prevent their movement; (3) the binding of ConA alters cellular structures attached to receptors in such a way as to restrict receptor movement.

The first possibility does not seem likely, for 50% of the cells show inhibition of patching when an average of no more than 1.5×10^5 molecules are bound per cell, occupying less than 1% of the cell surface. Furthermore it seems probable that lipids of the membrane undergo phase separation, rather than a single phase transition, and that the membrane never exists as a sheet of solid lipid bilayer. In any case this possibility can be tested by measuring the relaxation time of spin labels (Hubell and McConnell 1969) in the presence and absence of ConA. The second possibility also appears somewhat unlikely. Although it has been noted that binding of ConA leads to a redistribution of lectin receptors (Unanue, Perkins and Karnovsky 1972), given the size of the clusters formed and their number at the effective doses, it does not seem probable that they could alter the movement of the various other cell receptors. Although such clusters might "trap" other receptors, there is no evidence that they could do so efficiently or, in fact, that they do so at all. I have already mentioned evidence, for example, to indicate that ConA and IgM receptors on the same cell do not interact.

It seems more likely that ConA binding modifies a common cytoplasmic structure to which some of the cell surface receptors and possibly the membrane itself are anchored or attached. In accord with this suggestion, it has been found (Edelman, Yahara and Wang 1973) that colchicine, colcemid, vinblastine and vincristine will partially reverse the effect of ConA on receptor mobility and thus permit the formation again of both ConA caps and anti-Ig caps (Table 4). Colchicine does not bind

Table 4

Effects of Various Drugs on the Inhibition by
ConA of Cap Formation in Splenic Lymphocytes

Treatment	*fl-anti-Ig* (*100 μg/ml*)[a]	*fl-anti-Ig* (*100 μg/ml*) + *ConA* (*100 μg/ml*)[b]	*fl-ConA* (*100 μg/ml*)[c]
		% Cap forming cells with	
Control	85	2	2
Colchicine[d] (10⁻⁴ M)	87	22	31
Colcemid (10⁻⁴ M)	88	25	24
Vinblastine[d] (10⁻⁴ M)	91	55	42
Vincristine (10⁻⁴ M)	83	15	19
Low temperature (4°C)[e]	88	30	45
Cytochalasin B	62		1

[a] In order to test for cap formation by immunoglobulin receptors, the percent of cap forming cells obtained with fluorescein-labeled anti-immunoglobulin (fl-anti-Ig) was measured.

[b] In order to test for the inhibition by ConA of immunoglobulin receptor cap formation, the percent of cap forming cells obtained with fl-anti-Ig was measured in the presence of ConA.

[c] In order to test for cap formation by ConA receptors, the percent of cap forming cells obtained with fluorescein-labeled ConA (fl-ConA) was measured.

[d] Colchicine (10⁻⁴ M) and vinblastine (10⁻⁴ M) did not affect the amount of ^{125}I-ConA bound to splenic lymphocytes.

[e] The amount of ^{125}I-ConA bound to splenic lymphocytes at 4°C was 30% of the value obtained at 37°C.

to ConA nor does it cause disaggregation of its subunits. Furthermore colchicine does not inhibit either ConA-saccharide interactions or the cell binding activity of ConA. Although it remains to be seen whether colchicine alters the fluidity of the lipid portion of the membrane, these observations suggest an alternative interpretation: ConA binding, and cross-linking of its receptors, may affect the association-dissociation equilibrium of cytoplasmic structures interacting with the cell surface receptors, which in turn may affect the mobility of the receptors. Microtubular proteins (Weisenberg, Borisy and Taylor 1968) appear to be good candidates for this role because they have been shown to be sensitive to all four of the drugs listed in Table 4, because they have singular association-dissociation properties, and because they are ubiquitous in the cell.

Although it is intriguing to consider this notion that ConA binding causes alterations in a common protein anchorage to which some of the cell surface receptors are attached, it should be noted that no extensive microtubular structures have been observed directly under the plasma membrane, except perhaps in the case of lens epithelial cells in culture (Piatigorsky, Rothschild and Wollberg 1973). We have found that cytochalasin B, which affects cap formation but not patch formation (Taylor et al. 1971; dePetris and Raff 1972) by altering microfilaments, has no effect on inhibition of receptor mobility by ConA. We can tentatively conclude, therefore, that although it is possible that a filament structure other than microtubules may mediate the inhibition of receptor mobility, this structure is not likely to be the same as the structure mediating cap formation.

An Hypothesis on Receptor Anchorage and the Modulation of Receptor Mobility

Any model to account for the restriction of receptor mobility in terms of variations in anchorage of these receptors to structures on, in, or under the surface membrane must account minimally for several key observations:

1. the inhibition of the mobility of several receptors by doses of ConA far lower than those required to saturate all the ConA binding sites;
2. the failure of divalent ConA to inhibit receptor mobility and the restoration of the inhibition by antibodies to ConA;
3. the effect of temperature on the capping and patching behavior of tetravalent ConA;
4. the release from mobility-inhibition by colchicine and Vinca alkaloids.

As I have already indicated, one reasonable working hypothesis to explain these observations requires that some of the cell surface receptors are anchored on a common assembly which is on, in, or under the lymphocyte plasma membrane. Although the essential features of a model based on this hypothesis would hold regardless of the exact location of this assembly, the available evidence suggests that it is cytoplasmic in origin. The observations implicate a protein that can be altered by colchicine and Vinca alkaloids; although its identity is unknown, this colchicine-binding protein (CBP) may be related to certain of the actin-like proteins (Berl, Puszkin and Nicklas 1973) or microtubules (Weisenberg, Borisy and Taylor 1968).

The detailed hypothesis (Edelman et al. 1973) incorporates the following assumptions: (1) Certain surface receptors interact reversibly with colchicine-binding proteins, possibly the microtubular assemblies of the cytoplasm. If A is the anchored state of the receptors (attached to the CBP), and F is the free state of the receptors (not attached to the CBP), these two states are assumed to exist in an equilibrium $A \rightleftharpoons F$. Through this anchorage the distribution of the receptors on the cell surface is affected by the state of the CBP. A similar suggestion has been made by Berlin and Ukena (1972), who found that colchicine and vinblastine inhibited the agglutination of fibroblasts and polymorphonuclear leukocytes by ConA. (2) Not only is the distribution of the cell receptors affected by the state of the CBP but, conversely, the mobility and state of this assembly are affected by cross-linking interactions and aggregations of particular receptors. This provides a means by which receptor states can be communicated to the interior of the cell. The valence of external ligands can therefore be a critical factor in cell surface-cytoplasmic interactions. (3) The mobility of the membrane or its receptors is affected by the state of the CBP, and therefore, alteration of the CBP by one set of cell surface receptors may affect the movement of the other receptors. (4) Finally, the equilibrium between the two states (A and F) of the receptors is affected by colchicine and related agents. Alteration of the equilibrium may occur either because structures such as microtubules are dissociated by these agents (Weisenberg et al. 1968) or because receptors are released from attachment to the CBP assembly or both.

This hypothesis (Fig. 7) can account for the inhibitory action of tetravalent ConA at relatively low doses, for the lack of inhibition by divalent ConA, and for the restoration of the inhibition after treatment with anti-ConA, because it assumes

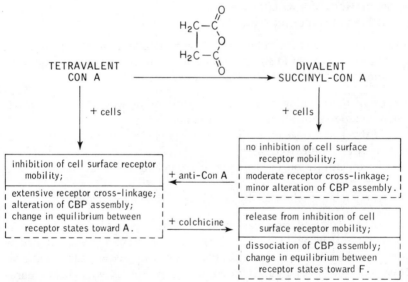

Figure 7
Summary of the effects of binding ConA or succinyl-ConA on mouse splenic lymphocytes. The experimentally observed effect on restriction of receptor mobility is boxed by solid lines. The hypothesized changes in the colchicine-binding protein assembly (CBP) and the shift in equilibrium of receptor states between A (anchored to the CBP) and F (free from the CBP) are boxed by dotted lines. The addition of anti-ConA to succinyl-ConA bound on cells mimics the effects of both ConA and succinyl-ConA plus anti-ConA.

that cross-linkage of relatively small numbers of particular surface receptors perturbs the CBP and alters the mobility of the other receptors. It obviously explains the effects of colchicine and related drugs. Finally it can also explain the apparently paradoxical observations on the effects of ConA at high and low temperatures. According to the hypothesis, binding of tetravalent ConA at 37°C followed by cross-linking of receptors in the A state is assumed to modify the mobility of the membrane. The results at low temperature may be related to the fact that microtubules can dissociate at low temperatures (Behnke and Forer 1967). Dissociation of the CBP with production of receptors in the F state could therefore account for the finding that ConA can cap its own receptors after binding at 4°C (Yahara and Edelman 1973; Karnovsky and Unanue 1973; Unanue et al. 1972), for at that temperature, it may react mainly with receptors in the F state. If reassociation of the CBP assembly occurred more slowly than cap formation after returning to 37°C, cap formation would be expected even for ConA receptors. In fact it has been found that cap formation occurs in minutes (Yahara and Edelman 1972; Taylor et al. 1971; dePetris and Raff 1972), whereas there is some evidence that microtubular reassembly can take as long as hours (Borisy, Olmsted and Klugman 1972).

The proposed hypothesis assumes a means of coupling between certain receptors (such as ConA receptors) and an ordered state of the CBP assembly and therefore it is important to consider how the cell surface receptors might be linked to this common network. Several experimental observations suggest how such a linkage might occur. First, ConA receptors bound to this lectin also appear in clusters (Smith and Revel 1972). Second, protein receptors or cell surface molecules may extend through the membrane (Bretscher 1971). Finally, colchicine-binding proteins have been found in association with membranes (Stadler and Franke 1972). Together with our assumptions, these observations suggest two possible means of coupling—direct and indirect (Fig. 8). Direct coupling can occur either through intramembranous particles or independently of them by extension of the lectin receptors through the membrane to interact noncovalently with proteins such as those of the CBP. Indirect coupling would act strictly via the intramembranous particles, implying that the system depends upon interaction of those particles to which receptors are attached with those particles to which the CBP is attached. At present there is no direct evidence to support either of these modes of coupling but they are sufficiently specific to warrant an experimental search. An additional possibility is that lectin binding and receptor aggregation leads to a change in membrane transport or of enzymatic activity, which in turn may lead to alterations in the mobility of the CBP or its interaction with receptors.

Surface Alterations and Mitogenesis

It is obviously premature to construct a detailed hypothesis on the relationship between surface events and mitogenesis. Mitogenesis is a complex series of events with a long delay between the initial signal from the mitogen and the final metabolic responses of the cell leading to replication and cytokinesis. Moreover the effects of mitogens are complex, showing a very strong dose dependence as well as possible requirements for factors from other cells. The role of transport alterations, endocytosis of lectins and comitogenic factors remains to be assessed in detail. Nevertheless there is some justification for suggesting that the surface events that I have described may be connected with early events of mitogenesis.

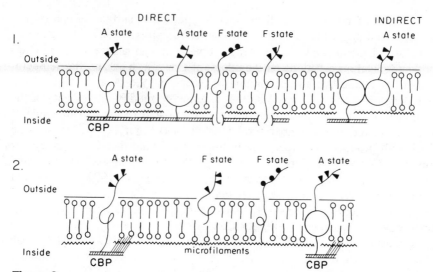

Figure 8

Diagram illustrating some hypothesized interactions between cell surface receptors and colchicine-binding proteins (CBP).

Receptors are assumed to interact with the CBP in an equilibrium consisting of two states, anchored (A) and free (F). As indicated in (1), receptors may interact directly or indirectly via the intramembranous particles. The arrangement shown implies that most receptors are able to interact in this fashion. Alternatively as shown in (2), only certain glycoprotein receptors may interact with the CBP and the CBP may interact with membrane-associated microfilamentous arrays.

An examination of the dose-response curve of ConA-induced lymphocyte stimulation shows that the immobilization phenomenon occurs largely in the region of ConA-induced *inhibition* of mitogenesis. At 2–3 μg/ml ConA, the optimal dose range for mitogenesis, about 80% of the cells show receptor movement (Fig. 9). At higher doses of ConA, however, both mitogenic response and cell surface receptor mobility are inhibited. Another set of observations also indicates that valence and receptor cross-linkage may be important in altering the mitogenic signal. Divalent succinyl-ConA has been found to be as potent a mitogen for mouse spleen lymphocytes as the native lectin (Fig. 10). Under the conditions of the assay, both ConA and succinyl-ConA showed maximum stimulation at concentrations as low as 3–6 μg/ml as measured by [^{3}H]thymidine incorporation, and the increase in [^{3}H]thymidine incorporation as a function of the concentration of succinyl-ConA was the same as that for the native protein. A striking difference was found at higher concentrations, however: the dose response curve for ConA fell off rapidly with increasing doses, but the response to succinyl-ConA did not decrease until the concentration exceeded 100 μg/ml. These observations suggest that the dimer is sufficient for stimulation and that the response to succinyl-ConA is a saturation phenomenon in which, beyond a certain dose, increased amounts of the mitogen have no further effect. Thus, divalent succinyl-ConA is mitogenic, does not inhibit receptor mobility, and unlike the tetravalent molecule, does not lead to inhibition of mitogenesis in the low dose range. Parenthetically, because binding of succinyl-

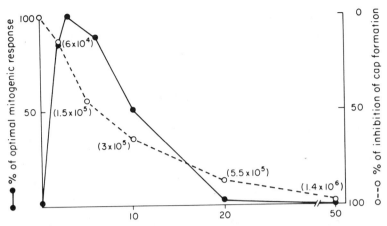

Figure 9

Comparison of the dose-response curve of the inhibition by ConA of anti-immunoglobulin cap formation with the dose-response curve of mitogenic stimulation by ConA of mouse spleen cells. Mitogenesis was measured by the incorporation of [^{3}H]thymidine. Cap formation was measured using fluorescein-labeled anti-immunoglobulin (100 μg/ml). Succinyl-ConA does not inhibit cap formation even at a lectin concentration of 100 μg/ml (Gunther et al. 1973). The number in parentheses denotes the number of ConA subunits bound per cell at 37°C. (●——●) Mitogenic stimulation; (o– – –o) inhibition of cap formation.

ConA does not result in capping even after long times of incubation, it appears likely that capping is not required for mitogenesis.

In addition to the correlation of valence effects with the properties of the mitogenesis dose-response curve, there also appears to be a correlation of the mitogenic activity of some lectins with their capacity to immobilize the cell surface receptors. Preliminary experiments show that several mitogenic lectins (ConA, PHA, lentil, and pea lectins, and extracts from fava and black turtle beans) inhibited cap

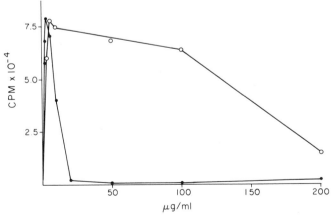

Figure 10

Dose-response curve of the incorporation of [^{3}H]thymidine in the stimulation of mouse spleen cells by ConA (●——●) and succinyl-ConA (o– – –o).

formation. In contrast, nonmitogenic lectins (wheat germ agglutinin, extracts of Idaho red, small California white and pink beans) did not inhibit cap formation. This suggests the possibility that nonmitogenic lectins may attach to receptors that are not directly connected with the CBP system.

Finally preliminary experiments have shown that the mitogenic activity of ConA is inhibited by drugs such as colchicine, vinblastine, and vincristine (Table 5) at concentrations as low as 10^{-6}M. This effect is not attributable to inhibition of DNA synthesis, for we have found that thymidine incorporation can take place in the presence of the drugs.

A simple interpretation consistent with these observations is that mitogenic stimulation involves the formation of micropatches containing relatively mobile receptors in reversible equilibrium with the CBP assembly. The formation of these micropatches may lead to alterations in the CBP and also initiate the various metabolic events in stimulation. At higher concentrations ConA may inhibit the mitogenic response by extensive alteration of the CBP, instantaneously "freezing" the cell surface receptors and preventing the formation of micropatches; eventually this state would lead to cell death. A divalent lectin such as succinyl-ConA cannot immobilize the receptors and is therefore mitogenic at high concentrations without being toxic (Fig. 2). Colchicine and Vinca alkaloids may act to inhibit mitogenesis by affecting the postulated CBP assembly in such a way that, even if micropatches are formed, they cannot alter cytoplasmic function via this assembly.

A good deal remains to be done to rule out the possibility that colchicine and the Vinca alkaloids may act at different sites than are postulated here, for these drugs can act in a variety of ways. For example, they have been reported to reduce incorporation of uridine into RNA and of certain amino acids into proteins (Creasey and Markin 1965). Synthesis of RNA has been found to be inhibited by

Table 5
Effects of Various Drugs on the Mitogenic Activity of ConA

	% Optimal mitogenic response[a]
Control	100
Colchicine[b] (10^{-4}M)	14
Vinblastine[b] (10^{-4}M)	2
Vincristine (10^{-4}M)	15

[a] Optimal mitogenic response was obtained by culturing mouse spleen cells with 3 μg/ml ConA as described by Gunther et al. (1973).

[b] Colchicine (10^{-4} M) and vinblastine (10^{-4} M) did not affect the amount of ^{125}I-ConA bound to splenic lymphocytes.

Vinca alkaloids (Wagner and Roizman 1968) and RNase and DPNase activities have been found to be increased after colchicine treatment (Erbe et al. 1966). In some cases vinblastine has exhibited a lethal action on cultured mammalian cells (Madoc-Jones and Mauro 1968). Cilia induction (Stubblefield and Brinkley 1966) and fragmentation and redistribution of the Golgi apparatus have been observed after the use of these spindle inhibitors (Robbins and Gonatos 1964).

Although these different effects on various cells must be carefully studied in the lymphoid cell itself, it should be noted that the effects of colchicine and Vinca alkaloids on the mobility of lymphocyte receptors are immediate (i.e. occur in seconds or minutes) and are in some cases reversible. Moreover no cell death has been observed in the doses used. In the case of the effects of these drugs on mitogenesis, however, the multiple actions of these drugs must be much more extensively explored and isolated from each other.

CONCLUSION

The striking observation of the inhibition of lymphocyte surface receptor mobility after binding of certain lectins suggests the possibility that there are structures that modulate the mobility and distribution of receptors in the membrane. Clearly the most direct test of the anchorage hypothesis proposed here is to demonstrate the existence of a colchicine-binding assembly that is connected directly or indirectly to some of the cell surface receptors or to the inner lamella of the cell membrane. Such structures may be visualized by electron microscopic radioautography using antibodies to actin-like proteins and tubulins. Even if such structures are seen, however, it is important to demonstrate their interaction with membrane receptors either in situ or by isolating the appropriate components and showing their interactions in vitro. Demonstrating the possible connection of such structures with the events of mitogenesis may prove to be more elusive and may have to await the design of more refined assays.

If the hypothesis concerning the anchorage of cell receptors were to receive support from such experiments, it would have a number of implications for cell–ligand and cell–cell interactions. It may, for example, provide a reasonable basis for understanding how certain receptors might remain relatively fixed on the membrane while others move. It certainly would help to explain avidity effects in which clusters of receptors participate in specific interactions, whereas the dispersed receptors may not. Such effects may account for the stability of certain cell–cell interactions. Finally because of the possible relationship of microtubules to morphogenesis and of microfilaments to cell movement, the present hypothesis may help to provide a connection between specific surface signals and alterations of cellular form. Obviously this depends upon validation or disproof of the working hypotheses proposed here as well as upon the demonstration of the inhibition of receptor mobility in cell lines other than lymphocytes.

Acknowledgment

This work was supported by USPHS Grants AI 09273, AI 11378 and AM 04256 from the National Institutes of Health.

REFERENCES

Andersson, J., G. Möller and O. Sjöberg. 1972a. Selective induction of DNA synthesis in T and B lymphocytes. *Cell. Immunol.* **4**:381.

———. 1972b. Induction of immunoglobulin and antibody synthesis *in vitro* by lipopolysaccharides. *Eur. J. Immunol.* **2**:349.

Andersson, J., G. M. Edelman, G. Möller and O. Sjöberg. 1972. Activation of B lymphocytes by locally concentrated concanavalin A. *Eur. J. Immunol.* **2**:233.

Behnke, O. and A. Forer. 1967. Evidence for four classes of microtubules in individual cells. *J. Cell Sci.* **2**:169.

Berl, S., S. Puszkin and W. J. Nicklas. 1973. Actomyosin-like protein in brain. *Science* **179**:441.

Berlin, R. D. and T. E. Ukena. 1972. Effects of colchicine and vinblastine on the agglutination of polymorphonuclear leucocytes by concanavalin A. *Nature New Biol.* **238**:120.

Borisy, G. G., J. B. Olmsted and R. A. Klugman. 1972. *In vitro* aggregation of cytoplasmic microtubule subunits. *Proc. Nat. Acad. Sci.* **69**:2890.

Bretscher, M. 1971. Major human erythrocyte glycoprotein spans the cell membrane. *Nature New Biol.* **231**:229.

Cold Spring Harbor Symp. Quant. Biol. 1967. *Antibodies.* (vol. 32).

Cooper, H. L. 1971. Biochemical alterations in resting cells. *The cell cycle and cancer* (ed. R. Baserga) p. 191. Marcel Dekker, New York.

———. 1972. Studies on RNA metabolism during lymphocyte activation. *Transpl. Rev.* **11**:1.

Creasey, W. A. and M. E. Markin. 1965. Biochemical effects of the Vinca alkaloids. III. The synthesis of ribonucleic acid and the incorporation of amino acids in Ehrlich ascites cells *in vitro*. *Biochim. Biophys. Acta* **103**:635.

dePetris, S. and M. C. Raff. 1972. Distribution of immunoglobulins on the surface of mouse lymphoid cells as determined by immunoferritin electron microscopy. Antibody-induced, temperature-dependent redistribution and its implications for membrane structure. *Eur. J. Immunol.* **2**:523.

Edelman, G. M. 1971. Antibody structure and molecular immunology. *Ann. N.Y. Acad. Sci.* **190**:5.

Edelman, G. M. and W. E. Gall. 1969. The antibody problem. *Ann. Rev. Biochem.* **38**:415.

Edelman, G. M., I. Yahara and J. L. Wang. 1973. Receptor mobility and receptor-cytoplasmic interactions in lymphocytes. *Proc. Nat. Acad. Sci.* **70**:1442.

Edelman, G. M., B. A. Cunningham, G. N. Reeke Jr., J. W. Becker, M. J. Waxdal and J. L. Wang. 1972. The covalent and three-dimensional structure of concanavalin A. *Proc. Nat. Acad. Sci.* **69**:2580.

Erbe, W., J. Preiss, R. Seifert and H. Helz. 1966. Increase in RNase and DPNase activities in ascites tumor cells induced by various cytostatic agents. *Biochim. Biophys. Res. Commun.* **23**:392.

Fisher, D. B. and G. C. Mueller. 1971. Studies on the mechanism by which phytohemagglutinin rapidly stimulates phospholipid metabolism of human lymphocytes. *Biochim. Biophys. Acta* **248**:434.

Frye, C. D. and M. Edidin. 1970. The rapid intermixing of cell surface antigens after formation of mouse-human heterokaryons. *J. Cell Sci.* **7**:319.

Greaves, M. F. and S. Bauminger. 1972. Activation of T and B lymphocytes by insoluable phytomitogens. *Nature New Biol.* **235**:67.

Gunther, G. R., J. L. Wang, I. Yahara, B. A. Cunningham and G. M. Edelman. 1973. Concanavalin A derivatives with altered biological activities. *Proc. Nat. Acad. Sci.* **70**:1012.

Hadden, J. W., E. M. Hadden, M. K. Haddox and N. O. Goldberg. 1972. Guanosine 3':5'-cyclic monophosphate: A possible intracellular mediator of mitogenic influences in lymphocytes. *Proc. Nat. Acad. Sci.* **69:**3024.

Hubell, W. L. and H. M. McConnell. 1969. Motion of steroid spin labels in membranes. *Proc. Nat. Acad. Sci.* **63:**16.

Inbar, M., H. Ben-Bassat and L. Sachs. 1973. Temperature-sensitive activity on the surface membrane in the activation of lymphocytes by lectins. *Exp. Cell Res.* **76:**143.

Karnovsky, M. J. and E. R. Unanue. 1973. Mapping and migration of lymphocyte surface macromolecules. *Fed. Proc.* **32:**55.

Madoc-Jones, H. and F. Mauro. 1968. Interphase action of vinblastine and vincristine: Differences in their lethal action through the mitotic cycle of cultured mammalian cells. *J. Cell Physiol.* **72:**185.

Novogrodsky, A. and E. Katchalski. 1971. Induction of lymphocyte transformation by periodate. *Biochim. Biophys. Acta* **228:**579.

Piatigorsky, J., S. S. Rothschild and M. Wollberg. 1973. Stimulation by insulin of cell elongation and microtubule assembly in embryonic chick-lens epithelia. *Proc. Nat. Acad. Sci.* **70:**1195.

Quastel, M. R. and J. G. Kaplan. 1970. Early stimulation of potassium uptake in lymphocytes treated with PHA. *Exp. Cell Res.* **63:**230.

Robbins, E. and N. K. Gonatos. 1964. Histochemical and ultrastructural studies on HeLa cell cultures exposed to spindle inhibitors with special reference to the interphase cell. *J. Histochem. Cytochem.* **12:**704.

Sharon, N. and H. Lis. 1972. Lectins: Cell-agglutinating and sugar-specific proteins. *Science* **177:**949.

Smith, S. B. and J. B. Revel. 1972. Mapping of concanavalin A binding sites on the surface of several cell types. *Develop. Biol.* **27:**434.

Stadler, J. and W. W. Franke. 1972. Colchicine-binding proteins in chromatin and membranes. *Nature New Biol.* **237:**237.

Stubblefield, E. and B. R. Brinkley. 1966. Cilia formation in Chinese hamster fibroblasts *in vitro* as a response to colcemid treatment. *J. Cell Biol.* **30:**645.

Taylor, R. B., W. P. H. Duffus, M. C. Raff and S. dePetris. 1971. Redistribution and pinocytosis of lymphocyte surface immunoglobulin molecules induced by anti-immunoglobulin. *Nature New Biol.* **233:**225.

Unanue, E. R., W. D. Perkins and M. J. Karnovsky. 1972. Ligand-induced movement of lymphocyte membrane macromolecules. I. Analysis by immunofluorescence and ultrastructural radioautography. *J. Exp. Med.* **136:**885.

Wagner, E. K. and B. Roizman. 1968. Effect of the Vinca alkaloids on RNA synthesis in human cells *in vitro*. *Science* **162:**569.

Weisenberg, R. C., G. G. Borisy and E. W. Taylor. 1968. The colchicine-binding protein of mammalian brain and its relation to microtubules. *Biochemistry* **7:**4466.

Yahara, I. and G. M. Edelman. 1972. Restrictions of the mobility of lymphocyte immunoglobulin receptors by concanavalin A. *Proc. Nat. Acad. Sci.* **69:**608.

————. 1973. The effects of concanavalin A on the mobility of lymphocyte surface receptors. *Exp. Cell Res.* (in press)

Control of Proliferation in the Immune System

John L. Fahey

Department of Microbiology and Immunology, University of California
Los Angeles, California 90024

Immune systems of adult organisms have the unique characteristic of existing essentially in an embryonic state. The analogy is made because the immune system contains stem cells which give rise to immune competent cells (ICC), i.e., cells similar to embryonic cells in being partially differentiated but not having realized their full potential in terms of cell number or their full level of function, which is making quantities of antibody to carry out the effector arms of the immune response. If the immune system existed in a fully developed state of responsiveness for all potential antigens, the mass of cells would be immense, a situation incompatible with the normal function of other tissues and organs.

By existing as a system with great potential but with realization of only a part of that potential at any one time, the immune system is able by economical means to maintain the necessary diversity and capacity to respond to a great variety of demands. If a few cells are to meet the needs of urgent challenge from the external environment, such as an aggressive infectious organism, the antibody-producing cells must increase rapidly. Clearly facilitation of proliferation (and differentiation) will be an important part of an effective immune response. As a corollary, flexible but ultimately very effective limitations on cell numbers must operate to control such a situation.

Understanding of the workings of the immune system has come along many routes. Many of these involve studies in whole animals, albeit often inbred animals subjected to replacement of immune cells from selected organs. Quantitative and increasingly sophisticated techniques are used to regulate quantities of antigen and reacting cell and to assess immune response. Even though immune function is usually assessed in complex situations, carefully done studies have contributed a great deal of insight into the components of the immune response.

COMPONENTS OF THE IMMUNE RESPONSE

The immune response is determined by the interaction of antigen (immunogen) and responding clones of immune competent cells (ICC). The role of these components may be facilitated by helper systems and by processes of recruitment. Each of these areas can be considered as promoting or facilitating the immune response. Quantitative factors in each area may have an appreciable effect on the dimensions of the immune response.

The relationships between various components of the antibody response are indicated in a very tentative way in Fig. 1. These areas are briefly introduced below and are described in more detail in many textbooks and reviews.

Antigen/Immunogen

In context of this presentation, reference will be almost interchangeable between immunogen, a substance capable of eliciting an immune response, and antigen, the substance reacting with the product (antibody or immune active cell) of the immune response. Immunogen should be recognized as containing unique chemical configurations (antigenic determinants) which determine specificity of immune response, as well as additional components which determine distribution and catabolism of the substance, interaction with cells, and the type of immune response.

The amount of antigen will determine the number of ICC clones initially stimulated. The duration of stimulation is important in terms of the amount of proliferation that will go on and the number of cells eventually able to make antibody. Continued presence of immunogen is needed for continuing proliferation of the antibody-forming cells. If antigen input stops, then proliferation ceases and antibody formation falls off.

Responding Clones

Number of Primary Immune Component Clones

Most immunogens that have been investigated stimulate many clones of immune competent cells. This is evident in the diversity of specific antibody molecules found in the circulation, molecules differing in electrophoretic mobility, class and subclass (e.g., IgG1, IgG2, etc.) and in binding affinity for the antigenic determinant.

Immunogens are presumed to stimulate a diversity of antibodies that fit in various ways and with different exactness to the contours of the antigenic determinant. The

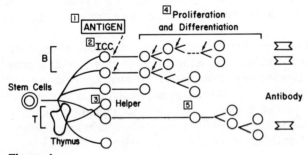

Figure 1

Components and characteristics of the immune response. Each is indicated as numbered: (1) antigen/immunogen, (2) immune competent cells (ICC), responding clones, (3) helper systems, (4) proliferation and differentiation, and (5) recruitment.

antigen receptors of the immune competent cell appear to be reflected in the anti-body molecules they eventually produce. Since each clone seems to make but one molecular form of antibody molecule, the large variety of antibodies developing in response to most antigens reflects extensive heterogeneity in the responding population, e.g., a large number of ICC clones. It has not been possible to be precise about the number of clones responding to any one antigen. Under unusual circumstances this number may be one (Askonas, Williamson and Wright 1971) indicated by a homogeneous antibody response.

There is some evidence that a primary immune response requires one or more intermediary steps, and probably one or more cell divisions, before appreciable immune response develops. Thus proliferation and maturation may be linked at some stages of the immune reaction.

Memory Cell Number

A primary immune response usually gives rise to antibody-forming cells and another set of cells that make little or no antibody after initial exposure to antigen. These are long-lived memory cells. This last population is ready to respond rapidly to a new exposure to antigen. They enter quickly into a rapidly dividing pool of antibody-producing cells. Indeed for many years the quicker and higher antibody response to second antigen administration has been distinguishable from the initial or primary response and has been termed the secondary response.

Helper Systems

Qualitative characteristics of the antigenic substance will determine its effectiveness in stimulating proliferation. Those features which determine distribution in the body, persistence, resistance to degradation, etc., are important. Factors such as the carrier effectiveness of nonantigenic parts of the immunogen will determine the participation of helper cells and cooperating cells in developing the full immune response. Some immunogens apparently are able to stimulate immune competent B lymphoid cells directly, without assistance of another cell system.

Many immunogens, however, require the cooperation of the T lymphoid system—so-called Helper Cells—in order to stimulate specific B-cell proliferation and anti-body production. Indeed there is evidence that three separate cell systems—B cells, T cells and macrophages—are required for antibody production in response to certain antigens (Tan and Gordon 1971; Gisler and Dukor 1972). T cells and macrophage may function to provide an optimal presentation of antigenic deter-minants to the immune competent B cell. The mechanisms of action, furthermore, are not known with certainty.

Proliferation

Expansion of the immune response to a specific antigen occurs via proliferation of the responding clones. The number of divisions and the rate of occurrence vary tremendously with local conditions. As many as ten divisions were regularly ob-served in response to a single large antigenic challenge in an already primed animal. Division occurred as frequently as 8–9 hours in both primary and secondary responses (Makinodan et al. 1969; Perkins, Sado and Makinodan 1969). These figures should be regarded as representative of a special experimental situation. In other circumstances the division time may be slower and the number of divisions lower, or even much greater. With repeated and continued exposure to antigen, many divisions would be expected.

The already primed B lymphocyte, the memory cell, seems to be waiting in a G_1 (G_0) phase of the cell cycle. The stimulus of antigen exposure can produce a wave of DNA synthesis prior to the appearance of mitosis (Saunders 1969). While it is not certain that all antigen-responding cells are in G_1 phase, many and perhaps all of those involved in a secondary response arise from cells in the G_1 portion of the cell cycle.

The variety of cell cycle times observed in responding cells indicates that there is no obligatory proliferation rate. Most of the variability in division time is attributable to the time spent in the G_1 phase of the cell cycle.

Antigen initiates lymphoid cell division by interacting with a specific receptor. How this cell surface reaction provides an unambiguous signal for cell division has been the subject of much speculation, but the mechanisms are still not known. Many studies have used the broader stimulus of phytomitogens such as phytohemagglutin (PHA), concanavalin and other substances. The mitogens instigate complex changes in the membrane, in protein and RNA metabolism of the cytoplasm, in the level of cAMP and eventually in DNA synthesis, blastogenesis and cell division. As yet a specific mechanism of action for these mitogens and for antigen has not been identified.

Recruitment

The immune response may involve additional cells other than those dividing and producing antibody in the initial response to antigen. This phenomenon is called recruitment. Significant data was obtained by Mackinodan and coworkers (1969), who measured the number of antibody-forming cells at short time intervals following an intense immune stimulation. In this situation many of the steps of increase in antibody-forming cells indicate that the number of active cells more than doubled. Such step increases, which were as great as four, indicate that recruitment of cells into the immune response from outside the original clones probably occurred.

Several general mechanisms of recruitment may operate. One of these involves bringing into the immune response lymphoid cells which have the capacity to respond to antigen, but which need some additional supportive factor or change in the environment that is provided by the initial divisions of other cells responding to antigen. Thus in this very special situation, intense immune response of an already primed population may produce factors which recruit cells that are less capable of responding and make them fully able to join into the immune response and adding to the sum of immune active cells.

Alternatively some of the cells detected later may have to go through several divisions before the specific immune response can be detected. This postulates an antigen response and cell divisions that would not be detected initially because antibody production only developed at a later period in time. Such cells would appear then to be new antibody-forming cells, but in fact would have been stimulated in the original response and are detected only later. Another possibility to explain recruitment would be the involvement of by-stander cells not programmed to respond to the specific antigen. Agents such as an RNA fraction (Fishman and Adler 1963; Askonas and Rhodes 1965; Bell and Dray 1971) or transfer factor (Lawrence 1969) have been described and postulated to act as instructive agents, which would pass from directly stimulated cells into other uninvolved cells and thus increase the number of cells responding to antigen stimulation.

PROCESSES LIMITING THE IMMUNE RESPONSE

The immune response, once initiated and maintained by antigen, could lead to excessive and undesirable amounts of plasma cells and antibody if not restricted or modulated. For example, a single immune competent cell that went through 50 divisions would give rise to about 10^{15} daughter cells, approximately one ton of a single clone. Manifestly such increases cannot be tolerated.

A large number of clonal divisions (50 or more) is not at all hard to imagine. A B lymphoid cell clone division sequence maintained in mice by repeated antigen administration was calculated to have involved 60 to 90 division steps (Williamson and Askonas 1972). Many chronic or recurrent parasitic infections, such as malaria, would present continuous antigenic stimulation in man under conditions that could lead to as many divisions in several clones if proliferation was not limited.

Clearly regulatory mechanisms must exist which ultimately limit the proliferation of immune active cells. Several characteristics of the postulated regulatory mechanisms can be inferred from a consideration of the roles of the immune system.

1. Limitation systems will become operational only after the immune response has developed. Such a constraint is necessary so that needed immune reactions can develop rapidly and fully enough to meet sudden demands such as those occurring with infections by pathogenic organisms.
2. Regulatory mechanisms will become increasingly active as the dimension of an immune response grows. Controlling reactions have to limit individual responses with increasing effectiveness in order to contain large-scale responses within acceptable bounds.
3. At least some of the regulatory mechanisms will have to be specific for individual clones. The host would be ill-served if there were indiscriminant, severe limitation on all immune responses. Alternatively the restriction could be directed at clones that had proliferated and increased past some critical mass.
4. Multiple regulatory systems can be expected to limit immune responses. The eventual limitation of immune response would seem to be too important to entrust to a single controlling apparatus.

A number of processes are known that serve to limit the immune response. These include feedback inhibition of antigen stimulation, termination of proliferation with maturation, diversion of a portion of each clone into memory cells, clonal dominance, and senescence or aging of activated clones. There is also evidence for an intracellular control mechanism operating at the translational level in immunoglobulin synthesis. Two additional means of limiting immune response may involve T cells through restriction in helper cell numbers or function (i.e., the helper T cells which facilitate response to immunogenic material) or through the development of suppressor T cells. These cells develop in response to antigenic stimulation and interfere with the optimal development of antibody-forming cells in the B lymphoid cell system. An additional process of limitation would be that controlling the development of immunocompetent cells from stem cell pools. These processes limiting the immune response are discussed briefly below. Their approximate level of operation in the immune response is illustrated in Fig. 2.

Feedback Control

Since immune response depends on the presence of immunogen, any means of limiting the antigen or antigen access to immune competent cells serves to limit the dimension to the immune response. One means of reducing the antigens is to have them bound and covered by specific antibody. Thus the product of the immune response, specific antibody, can reduce the stimulus to continued immune response (Uhr and Möller 1968; Schwartz 1971). This is represented in Fig. 2, where antibody is shown bound to antigenic determinants. Of course antibody could act anywhere in the afferent arc to prevent antigenic material from interacting with immune responsive cells. This kind of mechanism has been tested by passively infusing antibody to a nonimmune animal coincidental with or preceding the administration of antigen. The immune response of the recipient is less than that found in the absence of antibody. Furthermore removal of antibody formed during an immune response will allow the immune response to be prolonged. Thus antibody appears to act as the principal agent in a mechanism of feedback control of the immune response.

Another means by which antibodies, together with antigen, can inhibit the immune response has been described in systems where antigen-antibody complexes have been identified (Diener and Feldman 1970; Sjögren et al. 1971). These antigen-antibody complexes appear to react with lymphocytes to produce a kind of central immune suppression, which may not allow the cells to express their immune ability.

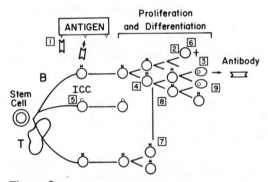

Figure 2
Processes limiting the immune response. Approximate site of action is indicated as numbered: (1) feedback control by antibody binding of antigen, (2) exhaustion of clonal proliferation capacity, (3) maturation, (4) diversion, memory cells, (5) clonal dominance, (6) Gompertzian kinetics of proliferation, (7) suppressor T cells, (8) suppressor factors, and (9) intracellular control of immunoglobulin production.

Exhaustion of Clonal Proliferation Capacity

In vitro studies of mammalian cells in culture by Hayflick (1965) and Martin and associates (1970) indicated that normal mammalian cells had a limited proliferation capacity, i.e., on the order of 50 to 90 generations. Once this had been achieved, cells no longer proliferated and the clone would die. This was frequently observed with embryonic fibroblast cultures in vitro. It was not observed, however, with human lymphoid cell lines established in vitro, which appeared to be derived from normal lymphoid cells (Fahey and Finegold 1967; Moore 1970).

A recent study of antigen dependent clones that proliferate in vivo under the influence of antigen has indicated that the proliferation capacity of mammalian (mouse) B cells is similarly limited (Williamson and Askonas 1972), i.e., not more than 90 divisions. Cells of the immune system thus may have a built-in limitation on their proliferation potential. The mechanism of exhaustion or senescence of proliferating ability, however, is not known.

Maturation

A very important feature of the immune response is development of mature antibody-forming cells. This is seen in the change from the small lymphocyte (the immune competent cell) to the plasma cell (the antibody factory). The lymphocyte has a small amount of cytoplasm and dense cromatin in the nucleus. Following antigenic stimulation there is enlargement of the cell, division occurs, and with subsequent divisions cells become clearly identified as plasma cells with plentiful endoplasmic reticulum. Plasma cells have many of the morphologic features common to all cells that synthesize protein for secretion outside the cell (Feldman 1972).

Functionally the change from the small lymphocyte to an antibody-(immunoglobulin) secreting plasma cell is accompanied by change in the amount of immunoglobulin formed. The rate of synthesis of immunoglobulin in the lymphocyte is relatively small in relation to the amount of immunoglobulin that can be synthesized in the mature plasma cell. Measurements of immunoglobulin synthesis in plasma cell tumors and in clones of immunoglobulin-producing cells in vitro (Fahey and Finegold 1967; Hiramoto, McGhee and Hamlin 1972) indicate that about 1000 molecules per cell per second are secreted.

The mechanisms by which the cell changes from an incompletely developed state to one with a lively synthesis of specific protein are not defined. A number of hypotheses have been proposed (Bretscher and Cohn 1968) that are worth considering because this system indeed offers possibilities as a model system of differentiation.

Maturation is accompanied under some circumstances by cessation of proliferation. One mechanism for limiting proliferation would be the loss of membrane receptor for antigen. Thus the cells would no longer be accessible to the necessary drive of antigen: receptor interaction. This loss of receptor has been described in certain murine plasma cell tumors (Cowan and Milstein 1972) but not in the E9 antigen-dependent clone (Askonas, Williamson and Wright 1971). Other mechanisms promoting immunoglobulin synthesis while limiting cell proliferation can be imagined. Specific understanding will certainly develop with further investigations.

Diversion

Antigen stimulation not only can induce development of antibody synthesis but can also cause cells of a lymphoid clone to be diverted into an unactive memory cell population. Such a diversion reduces the number of antibody-forming cells that result from any single exposure to antigen, but does increase the reservoir of immune cells for subsequent response to antigen.

The factors controlling diversion of potential antibody-forming cells into a memory cell population have been difficult to assess. In an antigen-dependent clone system, Williamson and Askonas (1972) have estimated that approximately 50% of the daughter cells stimulated by antigen end up in the memory cell population. Whether diversion to memory cell population results from a characteristic of the specific ICC, or to lack of specific assistance in the response system, such as from a helper cell, or to something about the physical state of the antigen or to some other cause remains unknown. Diversion, however, is an important consideration in the quantitative aspects of immune response to a specific antigen.

Clonal Dominance

Clonal dominance is a phenomenon in which the response to antigen of one clone (or a few clones) of immune cells excludes other clones that might be able to produce an effective immune response. It has been best defined in hapten or other systems with well-defined antigens. One manifestation of this has been termed "original antigenic sin" (Fazekas de St. Groth 1967). Here antigenic stimulation with one antigen so effects the immune system that subsequent exposure to related (but not identical) immunogen results primarily in antibodies directed at the first antigen.

Clonal dominance has been best defined in hapten or other systems with well-defined antigens. In one system Askonas and Williamson (1972) found that the quantity and (presumably) high affinity and continued presence of the antibody were important in maintaining clonal dominance. Apparently immature precursor cells of other clones were unable to proliferate or differentiate in the presence of relatively large amounts of high affinity antibody.

Gompertzian Kinetics of Proliferation

The growth of individual organs and tissues has been observed to follow a pattern of growth known as Gompertzian kinetics (Laird 1965). This pattern is based on a relatively rapid rate of growth when the cell numbers are small and a progressively slower rate of growth as the number of cells increases. Cell numbers eventually reach a plateau.

A similar growth pattern is observed for antibody-producing cells of the immune response. The plateau, however, is short-lived as factors operating to reduce the individual immune response become effective.

This general biologic phenomenon is relevant to the possible limitations of growth in a specific immune response. Although the mechanisms of this growth limitation (Gompertzian kinetics) have not been defined, it seems probable that they operate in lymphoid cell systems. Indeed lymphoid neoplasms show this type of growth kinetics (Clarkson and Fried 1971). Thus the limitations on growth of an immune response may operate at three levels: general biologic control, lymphoid tissue, and specific immune response.

Suppressor Cells

Great emphasis in past studies has been on conditions that help in the development of immune responses to antigens. Recent investigations, however, have revealed that antigen may also activate suppressor cells (Gershon and Kondo 1971; Baker et al. 1970). These are cells which reduce and limit immune response to antigen.

Suppressor cells are T lymphoid cells which are stimulated to function by antigen. They act to reduce the dimensions of antibody response to antigen, but the means by which the suppressor T cells interact with B cells is not yet defined.

Suppressor Factors

The immune response may cause production of factors that inhibit immune cell proliferation. In several systems inhibitory activity has been identified in fluids bathing stimulated lymphocytes, and the inhibitory agent shown not to be antibody.

Human lymphocytes cultured with PHA release a soluble substance capable of inhibiting DNA synthesis in a variety of cells (Green et al. 1970). Also lymphocytes stimulated in delayed hypersensitivity reactions (Dwyer and Kantor 1973) appear to produce an inhibitory substance. Soluble substances also have been identified from immune cells stimulated in vitro by antigen (Ambrose 1969). It is not clear whether or not these factors are responsible for any of the suppressor effects of T cells noted above.

Regulation at Translational Level

Feedback control of intracellular immunoglobulin synthesis has been demonstrated in recent experiments by Stevens and Williamson (1937a). They used the murine plasmacytoma 5563, which synthesizes an IgG2a myeloma protein and assigns about 10–15% of the protein synthesis to this specific product. Immunoglobulin (myeloma protein) was allowed to accumulate in the cells when secretion of myeloma protein from the cell was stopped by cooling cells to $25°C$. Synthesis continued but at reduced rate. When cells were warmed to $37°C$, the synthesis of immunoglobulin H chain decreased relative to L chain and total protein. The opposite effect was observed when the intracellular immunoglobulin (myeloma protein) level was depleted. The inverse correlation between intracellular product (H_2L_2 myeloma protein) level and synthesis of immunoglobulin H chain suggests a feedback control, probably at the translational level.

The mechanism for the feedback control was further defined by Stevens and Williamson (1973b), who found that the completed H_2L_2 myeloma protein formed a complex with the mRNA for immunoglobulin H chains. Specific H chain synthesis was quantitatively assessed after injection of mRNA into oocytes together with varying amounts of H_2L_2 protein. The mRNA complexed to H_2L_2 was repressed for H chain synthesis.

This system relates primarily to a differentiated function, specific immunoglobulin formation. This example, however, may have broader significance as a possible control mechanism for cell proliferation. Perhaps other specific intracellular components (necessary for cell division) also bind to their mRNA under some circumstances and inhibit translation and limit cell proliferation.

STUDIES WITH LYMPHOID CELLS IN CONTINUOUS CULTURE

Lymphoid cell behavior has been investigated by taking advantage of the opportunities presented by the availability of human lymphoid cells in tissue culture. These cell lines can be initiated from normal human subjects (as well as those with disease) and from primates, but have not yet been initiated from experimental animals such as the rodents. These cells have many of the characteristics of normal lymphoblastoid cells (Moore 1970; Buell, Sox and Fahey 1971).

The lymphoid cells in culture appear to be representative of fairly early cells in the lymphocyte-plasma cell series. They might be equivalent to cells only a small part of the way down the sequence represented in Fig. 1. Morphologically they resemble lymphoblastoid cells. Many are already committed to immunoglobulin synthesis, but assign only a part of their protein synthesis machinery to immunoglobulin synthesis.

In tissue culture the cells will grow at logarithmic rates, doubling about every 24 hours if the concentration is maintained in the range of $0.3-1.0 \times 10^6$ cells/ml. If the culture is allowed to grow to full density and nutrition is maintained, the cell number will plateau at $2-3 \times 10^6$/ml. Thus the lymphoid cell lines can be studied in stationary as well as log phase of growth.

The functional capacities of lymphoid cell lines under various growth conditions were evaluated by measuring protein and nucleic acid synthesis. Both total protein and specific immunoglobulin were assessed (Fahey, Buell and Sox 1971). The cells in stationary phase were shown to be in the G_1 (G_0) phase of the cell cycle and to have a reduced rate of protein and nucleic acid synthesis.

Immunoglobulin and total protein synthesis were measured in logarithmic and stationary phases of growth (Sox and Quinn, to be published). Cells in the stationary phase synthesized protein at about one-third of the rate found for cells in exponential growth. On the other hand, the proportion of total protein synthesis represented as immunoglobulin remained about the same (Table 1). Thus immunoglobulin synthesis was decreased in the stationary phase, but in proportion to all protein synthesis in the IM-9 cell line.

Detailed investigations of immunoglobulin synthesis in the human lymphoid cells revealed that immunoglobulin synthesis in the human lymphoid cells was concen-

Table 1

Immunoglobulin Synthesis in Lymphoid Cells in Culture

Growth phase	Secreted	Intracellular	Total
Log	60	3.0	4.7
Stationary	54	4.1	5.5

Protein synthesis was measured as [^{14}C]leucine incorporated per 10^6 cells. Values are given as percent IgG of all protein and are the averages of two experiments.

trated in a portion of the cell cycle, late G_1 and early S, as illustrated in Fig. 3A (Buell and Fahey 1969; Takahashi et al. 1969).

This was confirmed in several cell lines—IM-1, IM-9 and Wil-2—in our laboratories and, independently, with other cell lines in other laboratories and shown to apply to IgG and IgM. A similar restriction for myeloma protein synthesis to late G_1 and early S phase was described for a mouse plasma cell tumor line by Byars and Kidson (1970).

These original studies relating immunoglobulin synthesis to the cell cycle were done with chemical synchronization techniques. The cell lines were treated with double blockage with excess thymidine or with thymidine and colchimid in sequence to achieve metabolic uniformity. Because of the possibility that these procedures might have disordered cellular processes and caused an artificial localization of immunoglobulin synthesis, other means of cell synchronizing were sought.

Human lymphoid cells were shown to be synchronized by allowing them to achieve stationary phase growth and then diluting the cell concentration about fivefold (Lerner and Hodge 1971; Buell, Sox and Fahey 1971). In stationary phase cultures the lymphoid cells are arrested in a G_1 phase. Fresh medium and culture conditions allow them to pass through the cell cycle in a synchronous manner. Under these conditions and with the cell lines studied, immunoglobulin synthesis was found to increase in a wave during late G_1 and early S phases of the cell cycle.

A different pattern of immunoglobulin synthesis has been observed recently with other cell lines. Cowan and Milstein (1972) found that one murine plasma cell tumor in culture appeared to synthesize myeloma protein throughout the cell cycle. In a human cell line synthesizing an IgM-λ product, Watanabe, Yagi and Pressman (1973) found two peaks of immunoglobulin synthetic activity during the cell cycle, one in the late G_1 phase and the other in the G_2 phase (Fig. 3B). Minimal activity was found in M and early G_1. The secretion activity did not show as much change.

These findings emphasize that immunoglobulin synthesis does not have a uniform relationship to the division cycle in all cells. Immunoglobulin synthesis does not seem to be tightly linked to other metabolic events in the cycle of cell division, at least not in some cells in tissue culture. It might be wondered if this finding is in any way associated with malignant or other transformation of the lines that continually synthesize immunoglobulin. Whether these changes reflect differences in levels of maturation of the lymphoid-plasma cell series or malignant transformation or some other event will have to be determined by future studies.

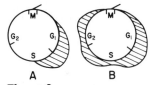

Figure 3
Immunoglobulin synthesis during the cell cycle. Illustration of two patterns of synthesis observed in lymphoid or plasma cells established in tissue culture. Time of immunoglobulin synthesis indicated by shaded zones. **(A)** Synthesis concentrated in late G_1 and early S phases of the cell cycle; **(B)** more or less continuous synthesis with waves of increase in late G_1 and in G_2 and a reduction at mitosis.

Lymphoid cells established in vitro tissue culture lines present many of the features of antigen-sensitive cells in vivo. The antigen-sensitive cells and the lymphoid cultures have the capacity to make specific immunoglobulin products. Tannenberg (1967) and Saunders (1969) have demonstrated a wave of antibody (immunoglobulin) synthesis prior to cell division. Cultured lymphoid cells diluted in concentration from stationary phase also show a wave of immunoglobulin production before DNA synthesis or mitosis as noted above. Similarly antigen-stimulated cells and diluted stationary lymphoid cultures show several synchronous doublings. This synchrony and the waves of immunoglobulin and DNA synthesis before the first mitosis indicate that the responsive cells in both systems are all in approximately the same stage of the cell cycle, i.e., in G_1 (G_0). Finally in both systems response becomes limited after a number of divisions.

These similarities between the behavior of human lymphoblastoid cells in tissue culture and immune responses in vivo indicate the possibility of common pathways of immune cell stimulation and of growth limitation. The availability of functional lymphoid and plasma cells in tissue culture has already proved useful in developing new insights into immune function and can be expected to continue to do so.

Acknowledgment

This work was supported by National Institutes of Health Research Grant CA 12800.

REFERENCES

Ambrose, C. T. 1969. Regulation of the secondary antibody response in vitro. Enhancement by actinomycin D and inhibition by a macromolecular product of stimulated lymph node cultures. *J. Exp. Med.* **130**:1003.

Askonas, B. A. and J. M. Rhodes. 1965. Immunogenicity of antigen-containing ribonucleic acid preparations from macrophages. *Nature* **205**:470.

Askonas, B. A. and A. R. Williamson. 1972. Dominance of a cell clone forming antibody to DNP. *Nature* **238**:339.

Askonas, B. A., A. R. Williamson and B. E. G. Wright. 1971. Selection of a single antibody-forming cell clone and its propagation in syngeneic mice. *Proc. Natl. Acad. Sci.* **67**:1398.

Baker, P. J., P. W. Strashab, D. F. Amsbaugh, B. Prescott and R. F. Barth. 1970. Evidence for the existence of two functionally distinct types of cells which regulate the antibody response to type III pneumococcal polysaccharide. *J. Immunol.* **105**:1581.

Bell, C. and S. Dray. 1971. Conversion of non-immune rabbit spleen cells by ribonucleic acid of lymphoid cells from an immunized rabbit to produce IgM and IgB antibody of foreign heavy-chain allotype. *J. Immunol.* **107**:83.

Buell, D. N. and J. L. Fahey. 1969. Limited periods of gene expression in immunoglobulin-synthesizing cells. *Science* **164**:1524.

Buell, D. N., H. C. Sox and J. L. Fahey. 1971. Immunoglobulin production by proliferating lymphoid cells. *Developmental aspects of the cell cycle* (ed. I. L. Cameron) pp. 279–296. Academic Press, N.Y.

Bretscher, P. A. and M. Cohn. 1968. Minimal model for the mechanism of antibody induction and paralysis by antigen. *Nature* **220**:444.

Byars, N. and C. Kidson. 1970. Programmed synthesis and export of immunoglobulin by synchronized myeloma cells. *Nature* **226**:648.

Clarkson, B. D. and J. Fried. 1971. Changing concepts of treatment in acute leukemia. *Med. Clinics North Amer.* **55**:561.

Cowan, N. J. and C. Milstein. 1972. Automatic monitoring of biochemical parameters in tissue culture. *Biochem. J.* **128**:445.

Diener, E. and M. Feldman. 1970. Antibody-mediated suppression of the immune response in vitro. *J. Exp. Med.* **132**:31.

Dwyer, J. M. and F. S. Kantor. 1973. Regulation of delayed hypersensitivity. *J. Exp. Med.* **137**:32.

Fahey, J. L. and I. Finegold. 1967. Synthesis of immunoglobulins in human lymphoid cell lines. *Cold Spring Harbor Symp. Quant. Biol.* **27**:283.

Fazekas de St. Groth, S. 1967. Cross recognition and cross reactivity. *Cold Spring Harbor Symp. Quant. Biol.* **32**:525.

Feldman, J. D. 1972. Morphology of lymphocytes and plasma cells. *Hematology* (ed. Williams, Beutler, Ersler, and Rundles) p. 752. McGraw-Hill, New York.

Fishman, M. and F. L. Adler. 1963. Antibody formation initiated in vitro. II. Antibody synthesis in X-irradiated recipients of diffusion chamber containing nucleic acid derived from macrophages incubated with antigen. *J. Exp. Med.* **117**:595.

Gershon, R. K. and K. Kondo. 1971. Infectious immunological tolerance. *Immunology* **21**:903.

Gisler, R. H. and P. Dukor. 1972. A three-cell mosaic culture: in vitro immune response by a combination of pure B and T cells with peritoneal macrophages. *Cell. Immunol.* **4**:341.

Green, J. A., S. R. Cooperband, J. A. Rutstein and S. Kibrick. 1970. Inhibition of target cell proliferation by supernatants from cultures of human peripheral lymphocytes. *J. Immunol.* **105**:48.

Hayflick, L. 1965. The limited in vitro lifetime of human diploid cell strains. *Exp. Cell Res.* **37**:614.

Hiramoto, R. N., J. R. McGhee and N. M. Hamlin. 1972. Measurement of antibody release from single cells. *J. Immunol.* **109**:961.

Laird, A. K. 1965. Dynamics of relative growth. *Growth* **29**:249.

Lawrence, H. S. 1969. Transfer factor. *Advanc. Immunol.* **11**:196.

Lerner, R. A. and L. D. Hodge. 1971. Gene expression in synchronized lymphocytes: Studies on the control of synthesis of immunoglobulin polypeptides. *J. Cell. Physiol.* **77**:265.

Makinodan, T., T. Sado, D. L. Groves and G. Price. 1969. Growth patterns of antibody-forming cell populations. *Current Topics Microbiol. Immunol.* **49**:80.

Martin, G. M., C. A. Sprague and E. J. Epstein. 1970. Replicative lifespan of cultivated human cells. *Lab. Invest.* **23**:86.

Moore, G. E. 1970. The culture of human lymphocytoid cell lines. *Methods Cancer Res.* **5**:423.

Perkins, E. H., T. Sado and T. Makinodan. 1969. Recruitment and proliferation of immunocompetent cells during the log phase of the primary antibody response. *J. Immunol.* **103**:668.

Saunders, G. C. 1969. Maturation of hemolysin-producing cell clones. The kinetics of the induction period of an in vitro hemolysin response to erythrocyte antigen. *J. Exp. Med.* **130**:543.

Schwartz, R. S. 1971. Immunoregulation by antibody. *Progr. Immunol.* **1**:1081.

Sjögren, H. O., I. Hellström, S. C. Bansai and K. E. Hellström. 1971. Suggestive evidence that "blocking antibodies" of tumor-bearing individuals may be antigen-antibody complexes. *Proc. Nat. Acad. Sci.* **68**:1372.

Stevens, R. H. and A. R. Williamson. 1973a. Translational control of immunoglobulin synthesis. I. Repression of heavy chain synthesis. *J. Mol. Biol.* **78**:505.

————. 1973b. Translational control of immunoglobulin synthesis. II. Cell-free interaction of myeloma immunoglobulin with RNA. *J. Mol. Biol.* **78:**517.

Takahashi, M., Y. Yagi, G. E. Moore and D. Pressman. 1969. Immunoglobulin production in synchronized cultures of human hematopoietic cell lines. I. Variation of cellular immunoglobulin level with generation cycle. *J. Immunol.* **103:**834.

Tan, T. and J. Gordon. 1971. Participation of three cell types in the anti-SRBC response in vitro. *J. Exp. Med.* **133:**520.

Tannenberg, W. J. K. 1967. Induction of 19S antibody synthesis without stimulation of cellular proliferation. *Nature* **214:**293.

Uhr, J. W. and G. Möller. 1968. Regulatory effect of antibody on the immune response. *Advanc. Immunol.* **9:**81.

Watanabe, S., Y. Yagi and D. Pressman. 1973. Immunoglobulin production in synchronized cultures of human hematopoietic cell lines. II. Variation of synthetic and secretion activities during the cell cycle. *J. Immunol.* **111:**797.

Williamson, A. R. and B. A. Askonas. 1972. Senescence of an antibody-forming cell clone. *Nature* **238:**337.

Immunoglobulin M Synthesis in Resting (G$_0$) and in Mitogen-activated B Lymphocytes

Fritz Melchers, Louis Lafleur, and Jan Andersson

Basel Institute for Immunology, Basel, Switzerland

In the absence of stimulation by antigens or mitogens, lymphocytes are quiescent cells. Morphologically they are small cells with little cytoplasm in which only the surface membrane is visible as membranous structure in electron microscopic pictures. These small cells synthesize very little, if any, DNA and do not divide. Such quiescent cells are thought to rest in a distinct period of the cell cycle called the G$_0$ period (Baserga 1969). Lymphocytes, primarily those from the bone marrow (B lymphocytes), synthesize immunoglobulins (Ig) (Miller and Mitchell 1969). In quiescent, G$_0$ lymphocytes the synthesized Ig is inserted into the surface (plasma) membrane where it presumably serves as receptor molecule for antigen recognition (Greaves and Hogg 1971).

After exposure to either antigens or mitogens, previously quiescent lymphocytes undergo numerous divisions, which lead to the formation of clones of cells. In these clones plasma cells develop with an enlarged cytoplasm containing membranous structures such as the rough and the smooth endoplasmic reticulum and Golgi apparatus, visible in electron micrographs (Rifkind et al. 1962; dePetris, Karlsbad and Pernis 1963). Plasma cells synthesize increased quantities of Ig molecules and actively secrete them. Stimulation of B lymphocytes therefore results in an expansion of the synthesis of the phenotype Ig. One may see this specialization of lymphoid cells as a reprogramming (Tsanev 1973) to active secretion of Ig.

We measure synthesis, surface deposition and molecular conformation of Ig molecules in resting G$_0$ lymphocytes. We monitor changes in these biochemical parameters with increasing time of mitogenic stimulation of the B cells. By inhibition of DNA synthesis we can distinguish between activation to proliferation and activation to active secretion of Ig.

IgM IN RESTING G$_0$ LYMPHOCYTES

Source and Purification of Small B Lymphocytes

Small mouse B lymphocytes can be obtained from the spleen. We use as a source of B cells spleens of mice with congenital thymic aplasia ("nude" mice) (Pantelouris 1968). We thus exclude the contamination by thymus-derived (T) lymphocytes, which occurs in spleens of normal mice. Most of the spleen cells are mitogen- and antigen-reactive. Small lymphocytes are, however, contaminated by a small number of large antibody secreting cells, detected as "background" plaque-forming cells. The large cells have a 50 to 100 times higher rate of synthesis of IgM (Andersson, Lafleur and Melchers, unpublished) and therefore obscure analyses of the rate of synthesis and turnover and of other molecular parameters of IgM in the population of small cells. Thus small cells must be purified of large cells. This is done by velocity sedimentation fractionation (Miller and Phillips 1969).

Immunoglobulin Class Produced by Small Resting B Lymphocytes

Small B cells from spleens of "nude" mice synthesize IgM, but not IgG or IgA. Thus only $H\mu$ chain-specific antisera, but not $H\gamma$ or $H\alpha$ chain-specific antisera, precipitate radioactive material from cells labeled for 4 hours with radioactive leucine. Reduced and alkylated radioactive polypeptide chains contained in the $H\mu$ chain-specific serological precipitates show mobilities in SDS-urea-polyacrylamide gels coinciding with those of reference $H\mu$ and L chains (Melchers and Andersson 1973).

Synthesis of IgM

Synthesis of IgM in small B lymphocytes and its release from the cells is followed by the incorporation of radioactive leucine during 4 hours into material which is precipitable by IgM-specific antisera in cellular lysates and extracellular fluids (Fig. 1a). Synthesis of radioactive IgM continuously increases during the 4-hour period of labeling. IgM does not reach a constant specific activity inside the cells. This indicates that the intracellular pool of IgM molecules has not been equilibrated with radioactive molecules during these 4 hours.

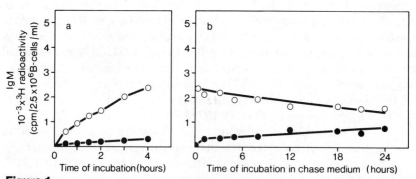

Figure 1

(a) Time-course of incorporation of [³H]leucine (60 μCi/ml, 38 Ci/m$\textsc{m}$) into intracellular (o———o) and secreted (•———•) IgM produced by small B lymphocytes. Serological detection of radioactive IgM in intracellular lysates and extracellular fluids of B cells is described in detail elsewhere (Melchers and Andersson 1973).

(b) Rate of disappearance (o———o) of leucine-labeled IgM from the 4-hr labeled small B cells and rate of appearance (•———•) of the labeled IgM in the extracellular fluid during a 24-hr chase period in nonradioactive medium.

Small G_0 lymphocytes, up to 4 hours of labeling, keep over 90% of their radioactive IgM bound to the cells. Active secretion of IgM is absent. Biosynthetic incorporation of radioactive leucine into B cells labels many proteins. IgM represents only 1–3% of the total incorporated radioactivity.

We can estimate how many IgM molecules have been made in 4 hours within one quiescent B cell. For this estimation it is assumed that all cells have the same synthetic activity and the specific radioactivity of each of the estimated 120 leucine residues in 7–8S IgM (see below) synthesized during the 4 hours equals the specific radioactivity of [³H]leucine in the incubation medium (38 Ci/mM) (Melchers 1971a; Andersson and Melchers 1973). From the amount of radioactivity incorporated into IgM within 4 hours (Fig. 1a) we can calculate that one small B cell has synthesized between 1000 and 2000 7–8S IgM molecules in that time. It will therefore take a small B cell 40 to 200 hours to synthesize half of the estimated $3–15 \times 10^4$ receptor IgM molecules (Rabellino et al. 1971) found on the surface of that cell. The estimated 1000–2000 IgM molecules represent the lower limit for the number of molecules which one small B cell may synthesize within 4 hours, since the specific activity of the cellular leucine pool is unknown.

Turnover of IgM

For turnover measurements small B lymphocytes are labeled for 4 hours with radioactive leucine, then transferred into nonradioactive chase medium. Turnover of IgM is then monitored by the disappearance of leucine-labeled IgM from the cells and by the appearance of this radioactive IgM in the supernatant nonradioactive chase medium (Fig. 1b). The median time of disappearance of radioactive IgM from the small cells and of appearance in the supernatant medium varies between 20 and 80 hours in different experiments. Turnover of IgM molecules in resting G_0 lymphocytes thus constitutes shedding from the cells into the supernatant medium. IgM molecules are not taken back into the cells and are intact in their polypeptide structure, i.e., not degraded by proteases, when they appear in the supernatant medium (see below for the size of the molecules). Wilson, Nossal and Lewis (1972), Marchalonis, Cone and Atwell (1972), and Vitetta and Uhr (1972) have obtained median disappearance times for surface-bound Ig of spleen lymphocytes between 4 and 8 hours, i.e., much shorter times than we find.

The median time of disappearance of IgM from small B cells, 20 to 80 hours, agrees reasonably well with the time it takes to synthesize half of the $3–15 \times 10^4$ surface receptor Ig molecules on a small cell (see above).

We can estimate the number of messenger RNA (mRNA) molecules in a small B lymphocyte that is sufficient to provide the sites for the production of these IgM molecules. For this estimation we assume that one ribosome can make 13 peptide bonds per second (Kepes and Beguin 1966; Maaløe and Kjeldgaard 1966). One $H\mu$ chain may have about 700 amino acids; therefore it takes nearly one minute to synthesize one $H\mu$ chain. Since the L chain mRNA has been estimated to contain 6 ribosomes and the $H\gamma$ chain mRNA 12 to 14 ribosomes (Williamson and Askonas 1966; Shapiro et al. 1966), the $H\mu$ chain mRNA may contain 18 to 20 ribosomes. One $H\mu$ chain mRNA could therefore make 20 $H\mu$ chains at the same time. Thus one small B cell containing *one* $H\mu$ chain mRNA can make 20 $H\mu$ chains per minute, or 4.8×10^4 $H\mu$ chains in 40 hours. Therefore one small B cell may not need much more than one mRNA actively engaged in $H\mu$ chain synthesis and one engaged in L chain synthesis at any given time in the cell to provide, with the assumed rate of peptide bond formation, the amount of IgM molecules which it turns over and sheds from the surface membrane. If $H\mu$ and L mRNA translation is less efficient than assumed above, the number of mRNA molecules is higher. These mRNA molecules may be stable in resting G_0 lymphocytes or may continuously be synthesized and degraded, i.e., be unstable.

Stability of IgM Synthesis in the Presence of Actinomycin D

Actinomycin D selectively inhibits DNA-dependent RNA synthesis. Actinomycin D is expected not to influence the capacity of small cells to synthesize IgM if all of the factors involved in synthesis of IgM, including the mRNAs for $H\mu$ chain and L chain, are stable. Actinomycin D is expected to decrease the capacity to synthesize IgM if any one of the factors involved in IgM synthesis is unstable in the presence of the inhibitor.

From the results of such an experiment (Fig. 2a) it is evident that synthesis of IgM in small B cells is unstable in the presence of actinomycin D. In fact IgM synthesis decays even more rapidly than the sum of syntheses of all proteins in the cell, evident in the drop of the values for ratios of synthetic rates of IgM over those of total protein (Fig. 2a, lower part). These experiments therefore do not permit conclusions about the life span of mRNA for $H\mu$ chain in small cells. We conclude that continued DNA-dependent RNA synthesis is requisite in small, resting B lymphocytes for the maintenance of IgM synthesis. Lerner et al. (1972) have reached the opposite conclusion for the stability of the mRNAs for $H\mu$ and L chains in unstimulated lymphocytes.

Size of IgM

The size of IgM molecules is measured in two ways, which distinguish between those IgM molecules in which the polypeptide chains are associated to a given size but are not linked together by disulfide bridges (analyzed by method 1), and other molecules in which the associated polypeptide chains are also linked to each other by disulfide bridges (analyzed by methods 1 and 2). Method 1 constitutes sedimentation analysis of radioactive material contained in cellular lysates and extracellular fluids of labeled B cells on sucrose gradients (Melchers 1972). With this method radioactive IgM is first sedimented according to its size, then precipitated by specific antiserum for detection in the different fractions of the gradients. For method 2 radioactive IgM is purified from cell lysates and extracellular fluids by precipitation with specific antisera. The isolated serological precipitates are then dissociated in 8 M urea, 0.1% sodium dodecyl sulfate and as such analyzed for size of the precipitated radioactive material on 2.5% cross-linked polyacrylamide-agarose composite gels (Peacock and Dingman 1968).

Resting G_0 lymphocytes contain over 90% of their radioactive IgM, labeled within 4 hours, as 7–8S subunits ($H\mu_2$-L_2) or as $H\mu$-L precursor material. No difference in size is detected between method 1 and 2 of analysis. Therefore these polypeptide chains are all associated *and* linked by disulfide bridges.

Radioactive IgM shedded into the supernatant medium from cells labeled for 4 hours, then chased in nonradioactive medium for 6 hours, is up to 80% 7–8S subunits and 10% $H\mu$-L precursor material. Again the polypeptide chains are not

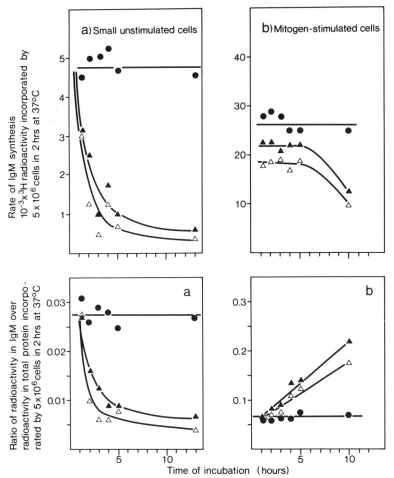

Figure 2

The effect of actinomycin D (5 μg/ml = $\triangle$; 0.5 μg/ml = $\blacktriangle$; no inhibitor = $\bullet$) on the rate of IgM synthesis (upper part) **(a)** in small, resting G_0 B lymphocytes, **(b)** in 70-hr mitogen-stimulated B cells. [^{3}H]leucine was incorporated for 2 hr into cells pretreated for various periods of time with actinomycin D. The values of IgM synthesis are plotted at the median time of leucine labeling, i.e., 1 hr after the initiation of the labeling.

In the lower part of the figure ratios are given of the rates of IgM synthesis over the rates of synthesis of all cellular proteins at different time periods of incubation in the presence of actinomycin D.

only associated, but also linked by disulfide bridges (Melchers and Andersson 1973 and unpublished). These size analyses are in agreement with results obtained by Marchalonis, Cone and Atwell (1972) and by Vitetta and Uhr (1972).

Carbohydrate Composition of IgM

Radiochemical analysis of IgM molecules labeled by radioactive precursors of carbohydrate moieties such as those found in the glycoprotein IgM (Shimizu et al.

1971) permits the qualitative analysis of the presence or absence of certain sugar residues in IgM (Melchers 1970, 1971a; Parkhouse and Melchers 1971; Andersson and Melchers 1973).

We find that radioactive mannose labels both intracellular and shed IgM of resting B lymphocytes. Radioactive fucose and galactose, however, label neither the intracellular nor the shedded IgM molecules during a 4-hour labeling period (Andersson, Lafleur and Melchers, unpublished). From such results we conclude in analogy to earlier reports with other types of B cells (quoted above) that IgM inside of and shed from small B lymphocytes contains the core sugars mannose and glucosamine, but not the penultimate galactose and the terminal fucose residues common to branched carbohydrate moieties of the 19S form of IgM actively secreted by plasma cell-like B cells (see below). The $H\mu$ chains, but not the L chains, of subunit IgM carry the carbohydrate moieties that are labeled by radioactive mannose.

Quantitation of Surface-bound IgM Molecules

Two methods are used to quantitate surface-bound immunoglobulin. Both employ the binding of antibodies with specificities for mouse Ig to viable B lymphocytes.

In method 1 ^{125}I-labeled F_{ab} fragments with specificities for mouse Ig are bound in saturating concentrations to B cells. With the known specific radioactivity of the $[^{125}I]F_{ab}$ preparation we can quantitate that between 5 and 15 $\times$ 10^5 molecules of F_{ab} are bound to one B cell.

In method 2 (Melchers and Andersson unpublished), surface-bound Ig is complexed on intact cells with whole rabbit antibodies having specificities for mouse Ig. Prior to the binding of the anti-mouse Ig antibodies, the distribution of Ig molecules on the surface membrane of resting B lymphocytes is diffuse (Taylor et al. 1971; Loor, Forni and Pernis 1972). Immunofluorescent studies have shown that interaction with polyvalent anti-Ig antibodies induces aggregation of these Ig molecules first into small aggregates ("spots"), then into larger aggregates ("caps") found over the uropod of the lymphocyte. Aggregation, in analogy to the prozone effect in immunoprecipitation, is inhibited at increasing concentrations of anti-Ig antibodies. This suggests that lattice formation, just as in solution, takes place in the surface membrane between the Ig molecules and the polyvalent anti-Ig antibodies. Diffusion of Ig molecules in the surface membrane is compatible with the view that the surface membrane of animal cells is quasi-fluid (Singer and Nicholson 1972; Gitler 1972).

We use this precipitation reaction in the surface membrane between mouse Ig molecules and anti-mouse Ig antibodies to distinguish surface IgM from labeled intracellular IgM in B lymphocytes. For this we label B cells through biosynthetic incorporation of radioactive leucine. Anti-Ig antibodies will bind to such labeled cells and precipitate only those labeled IgM molecules which face the outside of intact cells within the surface membrane. They will not be able to reach IgM molecules located in intracellular compartments.

On small B cells labeled for 4 hours with radioactive leucine 45% of all cellular radioactive IgM can be precipitated by the anti-mouse Ig antibodies (Fig. 3). During the first 6 hours after the pulse of radioactive leucine in nonradioactive chase medium most of the labeled IgM disappears from the intracellular compartments. Cells after the 6-hour chase period have mostly surface-bound labeled IgM.

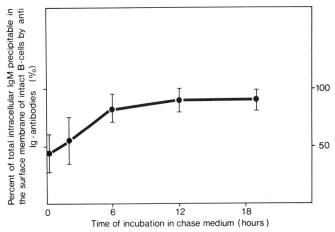

Figure 3

The amount of surface-bound, leucine-labeled IgM molecules expressed as percent of the total cellular pool of labeled IgM molecules during a 24-hr chase period in nonlabeled medium. Surface-bound IgM was determined by the precipitation reaction with polyvalent anti-Ig antibodies on intact cells (see text). If, as at 0 hr of chase time, 45% of the radioactive IgM molecules are removed by precipitation on the intact cells with anti-Ig antibodies, we conclude that the rest (55%) must be in intracellular compartments not exposed to the outside of the cells.

From the surface membrane labeled IgM molecules disappeared by shedding into the extracellular fluid (turnover) with a median disappearance time between 20 and 80 hours (see above). The median appearance time in the surface membrane of the 1000 to 2000 radioactive IgM molecules synthesized within the 4-hour labeling period is therefore short compared to the median disappearance time of these IgM molecules from the cells. The surface membrane, with 3 to 15×10^4 Ig molecules (Rabellino et al. 1971), therefore appears to be the main cellular pool of IgM molecules in small, resting B lymphocytes.

The biochemical properties of IgM in resting G_0 B lymphocytes are summarized in Table 1.

IgM IN MITOGEN-STIMULATED LYMPHOCYTES

B Cell Mitogens

Mitogens stimulate a large portion (between 20% and 70%) of all lymphocytes. Locally concentrated concanavalin A (Andersson, Sjöberg and Möller 1972a) and bacterial lipopolysaccharides (LPS; Andersson et al. 1972) stimulate B lymphocytes, but not T (thymus-derived) lymphocytes, to DNA synthesis, proliferation and immunoglobulin secretion. The mitogenic part of LPS resides in the lipid A portion (Lüderitz et al. 1972) of the LPS molecule (Andersson et al. 1973). The initial step in mitogenesis may therefore be an insertion of the lipid A portion of LPS into the lipid bilayer of the surface membrane of B lymphocytes. Knowledge of the molecular structure of LPS and the high reproducibility of activation of small B lymphocytes by LPS are our two main reasons for using this mitogen.

Table 1
Biochemical Parameters of IgM in B Lymphocytes

	Unstimulated, resting G_0 B cells	*Mitogen-stimulated, activated B cells*
Synthetic rate (IgM mol/4 hr)	1000–2000	150,000–200,000
Ratio of synthetic rates IgM/total protein (radioactive leucine)	0.01–0.03	0.06–0.12 in cells 0.4 –0.7 secreted
Size of IgM		
inside cells	7–8S + precursors	7–8S + precursors
outside cells	7–8S	19S
on surface	7–8S	7–8S
Carbohydrate content of Hμ		
inside cells	core*	core
outside cells	core	core + branch*
on surface	core	core
Turnover of IgM		
median disappearance time from cells	20–80 hr	4 hr
appearance in extracellular fluid	shedded	actively secreted
Sensitivity of IgM synthesis to actinomycin D	sensitive	resistant
Cell content of IgM molecules	$3–15 \times 10^4$ molecules	$2–5 \times 10^6$ molecules
main intracellular pool	over 90% in the surface membrane	70% in intracellular membranous compartment (rough endoplasmic reticulum)

* Core sugars: mannose and glucosamine; branch sugars: galactose and fucose.

Activation of DNA Synthesis

DNA synthesis, as measured by the uptake of radioactive thymidine by B cells, commences after a lag period of 12 to 24 hours after initiation of mitogenic stimulation. The rate of DNA synthesis thereafter continues to increase up to 72 hours after initiation of stimulation (Melchers and Andersson 1973).

Activation of Protein and IgM Synthesis and Secretion

Concomitant with the increase in DNA synthesis, the number of immunoglobulin secreting cells, detected in the Jerne plaque assay (Jerne, Nordin and Henry 1963), starts to increase at 24 hours after stimulation.

Synthesis of IgM and of other proteins and their release from the cells is followed, as in small, resting B lymphocytes, by the incorporation of radioactive leucine during a period of usually 4 hours into acid-precipitable or serologically precipitable material in cellular lysates and extracellular fluids. In the absence of LPS synthesis and release of IgM and of other proteins remains unchanged during a 72-hour culture period. In the continued presence of LPS rates of intracellular protein synthesis and simultaneously of IgM synthesis increase 10 to 15 hours after stimulation. Active secretion (more clearly defined below in comparison to shedding) of IgM and of other proteins is increased between 24 and 30 hours after stimulation (Andersson and Melchers 1973). The results are expressed in Fig. 4a as ratios of the rates of synthesis or of secretion of IgM over those of all proteins made. These ratios increase with increasing time of stimulation. Proteins other than IgM are made and secreted by B cells after activation, but IgM synthesis and secretion increases selectively over synthesis and secretion of other proteins in the cells. Since the number of plaque-forming, immunoglobulin-secreting cells increases with time of stimulation, we interpret the increase in the ratios of synthesis and secretion rates to mean that more cells within the population change from nonsecretors to secretors, rather than that a given cell in the population synthesizes and secretes more IgM and less other protein with increasing length of stimulation (Melchers and Andersson 1973).

After 72 hours of mitogenic stimulation, the B cell cultures produce and secrete as much IgM as does an IgM-producing and secreting mouse plasma cell tumor (Parkhouse and Askonas 1969; Parkhouse and Melchers 1971). This makes it likely that a large proportion of all cells in the culture actively secrete IgM after 72 hours of stimulation. The average secretory capacity of a B cell stimulated for 72 hours can be calculated as the number of molecules of IgM secreted per unit time, with the same assumptions made for the calculations of the synthetic capacity of small B cells. An average B cell stimulated for 72 hours contains between 2 and 5×10^6 molecules 7–8S IgM (see below for size) and secretes 3 to 5×10^4 molecules 19S IgM per hour. This synthetic capacity is comparable to that of IgG-producing plasma cells (Melchers 1970). Jerne (1967) has calculated that between 5000 and 10,000 mRNA molecules each for H and for L chains must exist in such a cell.

Activation of Synthesis of "Branch" Carbohydrate Portions of IgM Molecules

Radioactive mannose, galactose and fucose are used as precursors to label biosynthetically specific "core" or "branch" sugar residues of immunoglobulins (Melchers 1970, 1971; Parkhouse and Melchers 1971; Andersson and Melchers 1973).

Radioactive mannose labels, as in resting lymphocytes, intracellular and extracellular IgM. In addition, however, radioactive galactose and fucose also label IgM molecules in their respective galactose and fucose residues of carbohydrate moieties, which appear on secreted, but not on intracellular IgM after mitogenic stimulation (Andersson and Melchers 1973). These secreted IgM molecules therefore contain the "branch" sugars galactose and fucose. Again, ratios of synthesis and secretion of fucose- and of galactose-labeled IgM and of those of all labeled carbohydrate-containing macromolecules secreted from cells increase with time of stimulation (Fig. 4b, c).

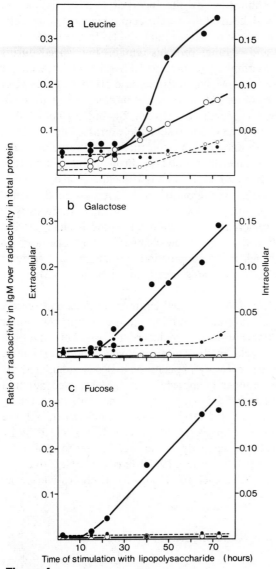

Figure 4

Ratios of the rates of synthesis (O,o) and secretion (●,•) of IgM over those of total protein synthesis and secretion at various times after mitogenic stimulation (20 μg bacterial lipopolysaccharide/ml/5 $\times$ 10⁶ B cells). Control data (– – –) are from cells labeled in the absence of the mitogen.

    (a) [³H]leucine-labeled cells;
    (b) [³H]galactose-labeled cells;
    (c) [³H]fucose-labeled cells.

The increases in the ratios of synthetic and secreting rates for polypeptide and carbohydrate portions of IgM can be taken as parameters to monitor the stage of activation of a B cell population. The appearance of secreted fucose- or galactose-labeled IgM is a further sign of B cell activation to Ig secretion, since IgM molecules shed from resting B lymphocytes are not labeled by these two sugars.

Turnover and Induction of Active Secretion

Changes in turnover of IgM molecules in B cells are observed after mitogenic stimulation. For such turnover measurements B cells are pulse-labeled, usually for 4 hours, with radioactive leucine and thereafter transferred back into nonradioactive chase medium containing the mitogen. Disappearance of labeled IgM from the labeled cells and appearance of the IgM in the extracellular fluid is measured at different time intervals during a 22-hour chase period.

In Fig. 5 (lines b,c,d) disappearance of IgM is shown from B cells stimulated for 30, 50 and 70 hours. Disappearance of IgM in these stimulated cells is compared to that in resting lymphocytes (line a; see also Fig. 1). The rates of disappearance from the cells (and, not shown, of appearance in the extracellular medium) are biphasic in stimulated B cells. The slow rate of disappearance of IgM is similar to that observed in small resting lymphocytes. The rapid rate of disappearance of IgM appears only after stimulation and thus represents another sign for B cell activation. With increasing time of stimulation, more IgM molecules with the rapid rate of turnover are synthesized. While IgM molecules with the rapid rate of turnover increase at least by a factor of 20 in 72-hour stimulated cells, IgM synthesis with the slow rate of turnover increases only 3 to 5 times during that time. Thus increased IgM synthesis after stimulation produces primarily rapidly turning-over IgM. The median disappearance time for IgM with the rapid rate of turnover is estimated to be around 4 hours.

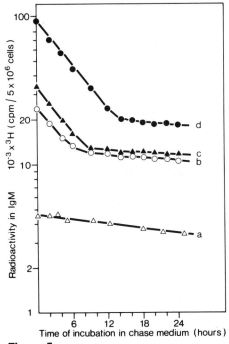

Figure 5

Disappearance of 4-hr leucine-labeled IgM from (a) unstimulated resting small B cells and (b, c, d) from mitogen-stimulated B cells (20 μg LPS/ml/5 × 10⁶ cells). Stimulation was for (b) 30 hr, (c) 50 hr, (d) 70 hr. The data are not corrected for possible increases in synthetic rates and cell contents of radioactive IgM due to cell division.

IgM secreted during the rapid phase of turnover is 19S pentameric IgM. Thus pentamerization of IgM appearing in extracellular fluids represents another sign of B cell activation.

The rapid disappearance of IgM from our activated B cells is reminiscent of rates measured by other laboratories (Wilson, Nossal and Lewis 1972; Marchalonis, Cone and Atwell 1972; Vitetta and Uhr 1972) in unstimulated B cells. Since their IgM appears outside the cell as 7–8S subunits, whereas ours is polymerized to 19S pentamers, comparisons of the various methods of turnover and size measurement will have to be made to clarify these discrepancies.

Stability of IgM Synthesis in the Presence of Actinomycin D

In contrast to small B cells, mitogen-stimulated B cells show IgM synthetic rates that are stable up to 5 hours in the presence of actinomycin D (Fig. 2b, upper part). In fact IgM synthesis is more stable than syntheses of other cellular proteins, evident from the increase in the ratios of rates of synthesis of IgM over those of all proteins in the cells (Fig. 2b, lower part). The change from actinomycin D-sensitive to actinomycin D-resistant IgM synthesis can therefore be taken as another sign that small B cells have undergone stimulation to increased synthesis and to active secretion. The observations of Lerner et al. (1972) that IgM synthesis is stable in actinomycin D-treated, unstimulated lymphocytes may be expected if stimulated, large cells contaminate the unstimulated cell populations.

Changes in the Number of Surface-bound IgM Molecules

Quantitations of surface-bound IgM and estimations of the distribution of cellular IgM between intracellular compartment and the surface membrane are followed with increasing length of mitogenic stimulation by the two methods described with small B cells.

Within 20 minutes after the addition of mitogen, 7–8S IgM molecules on the surface membrane aggregate to complexes much larger than 19S pentameric IgM. These complexes may contain other cellular material, and it is not known at present whether they contain also the mitogen, bacterial lipopolysaccharide. Binding of ^{125}I-labeled anti-mouse Ig F_{ab} to such mitogen-treated cells decreases to 10% of the original binding capacity found before addition of the mitogen. Biosynthetically labeled, surface-bound IgM can no longer be precipitated in the surface membrane of intact cells by anti-Ig antibodies.

Between 6 and 12 hours after the initiation of mitogenic stimulation, binding capacities for ^{125}I-labeled anti-mouse Ig F_{ab} reappear on the stimulated B cells. Biosynthetically labeled IgM can be precipitated again on the surface of the intact cells by anti-Ig antibodies after 10 to 20 hours, i.e., at the time when an increase in intracellular de novo synthesis of IgM is observed after mitogenic stimulation. Between 10 and 20 hours after the initiation of stimulation, the same amount of IgM appears on the surface as was originally present on the unstimulated small cells.

Beyond 20 hours of stimulation more and more IgM molecules appear on the surface membrane. This is concluded from the finding that (a) higher concentrations of anti-Ig antibodies are needed to precipitate an optimal amount of IgM molecules on the surface of the cells, and (b) that more ^{125}I-labeled anti-mouse Ig F_{ab} molecules are bound per cell. Figure 6a shows how much more anti-Ig

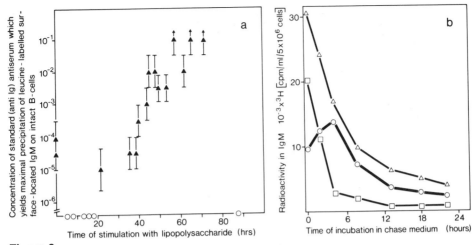

Figure 6

Increase in the number of surface-located IgM molecules during mitogenic stimulation of B cells. **(a)** The concentrations of our standard anti-Ig antiserum which yield maximal precipitation of labeled IgM molecules on the surface of intact B cells stimulated for various periods of time with mitogen. For the precipitation reaction B cells, exposed to mitogen for various periods of time, were labeled for 4 hr with radioactive leucine, then incubated in chase medium in the presence of mitogen for another 22 hr. The pulse-chased cells were then incubated with different dilutions of the anti-Ig antiserum, excess antiserum washed away and then reacted at room temperature for 30 min. Radioactive IgM in cellular lysates without anti-Ig antiserum (total) and radioactive IgM remaining in nonsedimentable lysates of cells exposed to the anti-Ig antiserum (intracellular Ig) were determined by serologic precipitation with Hμ-specific antisera (Melchers and Andersson 1973). The difference between "total" and "intracellular" Ig is taken as the amount of radioactive Ig exposed to anti-Ig antibodies on the surface of the intact cells, forming insoluble, sedimentable precipitates with them during the 30-min incubation period at room temperature. Brackets denote the range of antiserum dilutions which gave maximal precipitation of surface-located IgM.

(b) The amount of surface-located, leucine-labeled IgM molecules on 44-hr stimulated B cells is expressed as percent of the total cellular pool of labeled IgM molecules during a 24-hr chase period in nonlabeled medium (see also Fig. 3). From "total" cellular IgM (△) intracellular (□) and surface-located (○) labeled IgM are distinguished by the precipitation reaction with anti-Ig antibodies on intact cells.

antibodies are needed for optimal precipitation on the intact cells at various times after mitogenic stimulation. Exact estimations as to how much more IgM molecules appear on the surface are difficult to make because the quantitative relationship between the number of antibody molecules that bind IgM molecules arranged in certain patterns on the surface membrane is unknown.

The increased number of IgM molecules on the surface membrane of stimulated B cells shows a rapid rate of turnover. In B cells stimulated for 44 hours with mitogen, 30% of all IgM molecules labeled within 4 hours by radioactive leucine are on the surface, while 70% are in intracellular compartments (Fig. 6b). During the first 4 hours in nonradioactive chase medium, labeled IgM molecules disappear from intracellular compartments with a median disappearance time of 2 hours. They appear in the surface membrane, since the amount of radioactive IgM there increases during the first 4 hours. The surface membrane therefore is an intermediate station of IgM molecules leaving the cell with a rapid rate. From the disappearance of labeled IgM molecules from the surface membrane in the time between 4 and 13 hours in chase medium, this rapid rate is estimated to have a median disappearance time around 5 hours. We estimate from the amount of radioactive IgM that has left the cells after 13 hours in chase medium that at least 60 to 70% of all surface-bound IgM in stimulated B cells turns over rapidly. The observed increase in the number of surface-bound IgM molecules after stimulation is therefore largely due to active secretion of IgM molecules via the surface membrane after mitogenic stimulation.

The biochemical properties of mitogen-stimulated, activated B lymphocytes are summarized in Table 1.

MITOGENIC STIMULATION OF B CELLS TO ACTIVE SECRETION OF IgM IN ABSENCE OF DNA SYNTHESIS

Hydroxyurea inhibits the biosynthesis of DNA in mammalian cells without primarily affecting RNA or protein synthesis. The inhibition is probably due to interference with the enzyme ribonucleoside diphosphate reductase, which reduces ribonucleotides to deoxyribonucleotides (Krakoff, Brown and Reichard 1968); it is complete within 10 minutes and can be reversed to original levels of DNA synthesis within 3 hours. During these 3 hours protein synthesis and secretion is not inhibited. Incubation of B cells with hydroxyurea does not affect cellular viability during a 48-hour time period in the presence or absence of mitogen.

Small, resting G_0 lymphocytes, in which DNA synthesis has been inhibited to 99.5% by 10 mM hydroxyurea, do respond to mitogenic stimulation by the formation of antibody-secreting, plaque-forming cells (Andersson, Lafleur and Melchers unpublished). The activation to active secretion can also be monitored by the biochemical parameters which we take as signs of B cell activation. Thus protein and IgM synthesis increases (Fig. 7a). Active secretion of 19S pentameric IgM is induced. The secreted, polymerized IgM contains the branch carbohydrate residues, galactoses and fucoses. Ratios of the rate of synthesis of IgM over rates of syntheses of the other cellular proteins increase (Fig. 7b).

The activation to increased IgM synthesis and to active secretion is accomplished in the absence of DNA synthesis within the first 24 hours of stimulation and occurs therefore *earlier* than in cells which are stimulated by the same dose of mitogen, but in which DNA synthesis is *not* inhibited.

Thus it appears that "mitogens," such as bacterial lipopolysaccharides, provide B cells with two signals of induction, one for the induction of DNA synthesis and cell proliferation, and another for the initiation of increased synthesis and active secretion of IgM. We can separate the signal for active secretion from the signal for proliferation by the inhibition of DNA synthesis. We should also be able to induce B lymphocytes to DNA synthesis and proliferation only, if induction to increased synthesis and active secretion can be inhibited.

Stimulation of B lymphocytes therefore balances between activation to proliferation and activation to increased synthesis and active secretion of IgM. The magnitude of an immune response appears governed by factors which influence this balance.

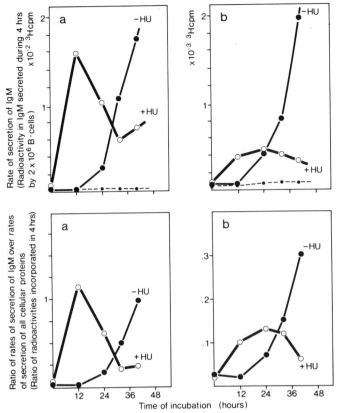

Figure 7

Mitogenic stimulation of resting G_0 lymphocytes to active secretion in the absence (o) and in the presence (•) of DNA synthesis. DNA synthesis was inhibited by 10 mM hydroxyurea (HU) from the beginning of mitogenic stimulation. *Upper part:* IgM is measured as secreted [³H]fucose + [³H]galactose-labeled protein **(a)** or as secreted [³H]leucine-labeled protein **(b)** by serological precipitation with Hμ-specific antisera as described elsewhere (Melchers and Andersson 1973). Ratios of radioactivities in IgM over those in all cellular proteins secreted from the cells after different periods of stimulation in the absence and presence of hydroxyurea are given in the lower parts of the Figure as measures for the relative rate of secretion of labeled IgM compared to other labeled cellular proteins.

CONCLUDING REMARKS

A given antigen will stimulate only a very small portion (less than 0.02%) of all B lymphocytes. This heterogeneity of B lymphocytes in antigenic stimulation has thus far prevented biochemical studies on the activation of B lymphocytes. The discovery of B cell-specific mitogens (Andersson, Sjöberg and Möller 1972b) has now opened the possibility of following biochemical changes that occur in B lymphocytes after stimulation.

Specialization of B cells after activation is primarily to synthesize more Ig molecules and to actively secret them. Proliferation into clones of cells helps to enlarge the response. Changes in synthesis, turnover, surface deposition and molecular conformation of IgM molecules can be regarded as directly connected to changes in the function of B cells in their response to a stimulus, such as an antigen. Our detailed knowledge of the functions of lymphocytes and of the structure of Ig made by these lymphocytes, and the available highly specified serological analysis methods for Ig molecules enable us to correlate functional changes in B lymphocytes with defined molecular changes. Our analyses may therefore help to clarify the mechanisms by which cells achieve selective increases of phenotypic expressions of some genes over others. It is reasonable to expect that synthetic rates for other proteins, including surface membrane proteins, will increase after mitogenic stimulation. Such proteins may be topologically and functionally closely connected with IgM in B cells. The search for such proteins with increased rates of synthesis after mitogenic stimulation may therefore help the search for functions which influence IgM synthesis, surface deposition and secretion.

IgM synthesis, surface deposition and secretion in small B cells can be regarded as biochemical processes that are part of the overall process of surface membrane biogenesis. IgM synthesis in small cells is in balance with the synthesis of other proteins destined to be incorporated into the surface membrane of the cell (Melchers and Andersson 1973). After activation IgM is synthesized more than other cellular proteins. Active secretion may result from the displacement of other membrane proteins in lipid bilayers by more and more IgM molecules. For this specialization for active secretion of IgM a cellular membranous apparatus is developed in B cells which strikingly resembles that of other secretory cells. It appears possible that activation of a variety of exocrine and endocrine gland cells to active secretion involves a common cellular and molecular principle by which different specific mRNAs are more efficiently translated into specific proteins, characteristic for the type of activated cell secreting them. Our experiments suggest that B lymphocytes can activate, through the action of mitogen, this principle of active secretion in the absence of DNA synthesis.

Acknowledgments

The able technical assistance of Miss Monica Gidlund, Miss Dorothee Jablonski and Miss Ingrid Möllegard is gratefully acknowledged.

REFERENCES

Andersson, J. and F. Melchers. 1973. Induction of immunoglobulin M synthesis and secretion in bone marrow-derived lymphocytes by locally concentrated concanavalin A. *Proc. Nat. Acad. Sci.* **70**:4016.

Andersson, J., O. Sjöberg and G. Möller. 1972a. Induction of immunoglobulin and antibody synthesis in vitro by lipopolysaccharides. *Eur. J. Immunol.* **2**:349.

———. 1972b. Mitogens as probes for immunocyte activation and cellular cooperation. *Transplant. Rev.* **11**:131.

Andersson, J., G. M. Edelman, G. Möller and O. Sjöberg. 1972. Activation of B-lymphocytes by locally concentrated concanavalin A. *Eur. J. Immunol.* **2**:233.

Andersson, J., F. Melchers, C. Galanos and O. Lüderitz. 1973. The mitogenic effect of lipopolysaccharide on bone marrow-derived mouse lymphocytes. Lipid A as the mitogenic part of the molecule. *J. Exp. Med.* **137**:943.

Baserga, R. 1969. *Biochemistry of cell division* (ed. R. Baserga) p. 3. C. C. Thomas, Springfield, Ill.

dePetris, S., G. Karlsbad and B. Pernis. 1963. Localization of antibodies in plasma cells by electron microscopy. *J. Exp. Med.* **117**:849.

Gitler, C. 1972. Plasticity of biological membranes. *Ann. Rev. Biophys. Bioeng.* **1**:51.

Greaves, M. F. and N. M. Hogg. 1971. Immunoglobulin determinants on the surface of antigen binding T and B lymphocytes in mice. *Progr. Immunol.* **1**:111.

Jerne, N. K. 1967. Waiting for the end. *Cold Spring Harbor Symp. Quant. Biol.* **32**:591.

Jerne, N. K., A. H. Nordin and C. Henry. 1963. The agar plaque technique for recognizing antibody-producing cells. *Cell-bound antibodies* (ed. B. Amos and H. Koprowski) p. 109. Wistar Institute Press, Philadelphia, Pa.

Kepes, A. and S. Beguin. 1966. Peptide chain initiation and growth in the induced synthesis of β-galactosidase. *Biochim. Biophys. Acta* **123**:546.

Krakoff, I. H., N. C. Brown and P. Reichard. 1968. Inhibition of ribonucleoside diphosphate reductase by hydroxyurea. *Cancer Res.* **28**:1559.

Lerner, R. A., P. J. McConahey, I. Jansen and F. Dixon. 1972. Synthesis of plasma membrane-associated and secretory immunoglobulin in diploid lymphocytes. *J. Exp. Med.* **135**:136.

Loor, F., L. Forni and B. Pernis. 1972. The dynamic state of the lymphocyte membrane. Factors affecting the distribution and turnover of surface immunoglobulins. *Eur. J. Immunol.* **2**:203.

Lüderitz, O., C. Galanos, V. Lehmann, M. Nurminen, E. T. Rietschel, G. Rosenfelder, M. Simon and O. Westphal. 1972. Lipid A, chemical structure and biological activity. *J. Infect. Dis.,* Endotoxin Suppl., **128**:S17.

Maaløe, O. and N. O. Kjeldgaard. 1966. *Control of macromolecular synthesis,* p. 91. W. A. Benjamin, New York.

Marchalonis, J. J., R. E. Cone and J. L. Atwell. 1972. Isolation and partial characterization of lymphocyte surface immunoglobulins. *J. Exp. Med.* **135**:956.

Melchers, F. 1970. Biosynthesis of carbohydrate portion of immunoglobulins. Kinetics of synthesis and secretion of [³H]leucine-, [³H]galactose- and [³H]mannose-labelled myeloma protein by two plasma cell tumors. *Biochem. J.* **119**:765.

———. 1971a. Biosynthesis of the carbohydrate portion of immunoglobulins. Incorporation of radioactive fucose into immunoglobulin G_1 synthesized and secreted by mouse plasma cell tumor MOPC 21. *Biochem. J.* **125**:241.

———. 1971b. The secretion of a Bence-Jones type light chain from a mouse plasmacytoma. *Eur. J. Immunol.* **1**:330.

———. 1972. Difference in carbohydrate composition and a possible conformational difference between intracellular and extracellular immunoglobulin M. *Biochemistry* **11**:2204.

Melchers, F. and J. Andersson. 1973. Synthesis, surface deposition and secretion of immunoglobulin M in bone-marrow-derived lymphocytes before and after mitogenic stimulation. *Transpl. Rev.* **14**:76.

Miller, J. F. A. P. and G. F. Mitchell. 1969. Thymus and antigen-reactive cells. *Transpl. Rev.* **1**:3.

Miller, R. G. and R. A. Phillips. 1969. Separation of cells by velocity sedimentation. *J. Cell. Physiol.* **73**:191.

Pantelouris, E. M. 1968. Absence of thymus in a mouse mutant. *Nature* **217**:370.

Parkhouse, R. M. E. and B. A. Askonas. 1969. Immunoglobulin M biosynthesis. Intracellular accumulation of 7S subunits. *Biochem. J.* **115**:163.

Parkhouse, R. M. E. and F. Melchers. 1971. Biosynthesis of the carbohydrate portions of immunoglobulin M. *Biochem. J.* **125**:235.

Peacock, A. W. and C. W. Dingman. 1968. Molecular weight estimation and separation of ribonucleic acid by electrophoresis in agarose-acrylamide composite gels. *Biochemistry* **7:**668.

Rabellino, E., S. Colon, H. M. Grey and E. R. Unanue. 1971. Immunoglobulins on the surface of lymphocytes. I. Distribution and quantitation. *J. Exp. Med.* **133:**156.

Rifkind, R. A., E. F. Osserman, K. C. Hsu and C. Morgan. 1962. The intracellular distribution of gamma globulin in a mouse plasma cell tumor (X5563) as revealed by fluorescence and electron microscopy. *J. Exp. Med.* **116:**423.

Shapiro, A. L., M. D. Scharff, J. V. Maizel, Jr. and J. W. Uhr. 1966. Polyribosomal synthesis and assembly of H and L chains of gamma globulins. *Proc. Nat. Acad. Sci.* **56:**216.

Shimizu, A., F. W. Putnam, C. Paul, J. R. Clamp and I. Johnson. 1971. Structure and role of five glycopeptides of human IgM immunoglobulins. *Nature New Biol.* **231:**73.

Singer, S. J. and G. L. Nicholson. 1972. The fluid mosaic model of the structure of cell membranes. *Science* **175:**720.

Taylor, R. B., P. H. Duffus, M. C. Raff and S. dePetris. 1971. Redistribution and pinocytosis of lymphocyte surface immunoglobulin molecules induced by anti-immunoglobulin antibody. *Nature New Biol.* **233:**225.

Tsanev, R. 1973. Cellular reprogramming and cellular differentiation. *Biochemistry of cell differentiation* (ed. A. Monroy and R. Tsanev) p. 177. Academic Press, New York.

Vitetta, E. S. and J. W. Uhr. 1972. Cell surface immunoglobulin. V. Release from murine splenic lymphocytes. *J. Exp. Med.* **136:**676.

Williamson, A. R. and B. A. Askonas. 1966. Biosynthesis of immunoglobulins: The separate classes of polyribosomes synthesizing heavy and light chains. *J. Mol. Biol.* **23:**201.

Wilson, J. D., G. J. V. Nossal and H. Lewis. 1972. Metabolic characteristics of lymphocyte surface immunoglobulins. *Eur. J. Immunol.* **2:**223.

Regulation of the Immune Response at the Single Cell Level

E. Diener, K-C. Lee,* R. E. Langman, N. Kraft, and V. H. Paetkau*
MRC Transplantation Group and Departments of Immunology and Biochemistry*
University of Alberta, Edmonton, Alberta, Canada

B. Pernis
Basel Institute of Immunology, Basel, Switzerland

Traditionally immune triggering has been examined from three different viewpoints:

Antigen Receptors as Mediators of Antigen Recognition

The humoral immune response to most, but not all, antigens depends on the co-operation of two cell classes, the thymus-derived T cells and the bone marrow-derived B cells. Both of these cell types have been shown to specifically recognize antigen. There is strong evidence that the receptors on B cells are IgM immuno-globulins (Warner, Byrt and Ada 1970), whereas there is conflicting evidence regarding the nature of T cell receptors (Marchalonis, Atwell and Cone 1972; Vitetta, Uhr and Boyse 1973).

It is commonly thought that immunocompetent cells express receptors with only one specificity.

The Cell Membrane as a Dynamic Structure

The plasma membrane of the cell surface provides an anchorage for the receptors and participates in the regulation of their dynamic state. Besides immunoglobulin receptors, the lymphocyte surface also carries receptors for other ligands such as phytomitogens.

The Cytoplasm

The cytoplasm receives signals of receptor-antigen interaction and translates them into biochemical events.

While each of these aspects has been studied morphologically and biochemically in great detail, the mechanisms which govern their interaction remain unclear.

The first step in triggering at the level of the single immunocompetent cell is the interaction between antigen molecules and immunoglobulin receptors. This paper first discusses the immunological significance of antigen binding cells and then focuses attention on mechanisms of antigen-receptor interaction with a view to examining their relevance in immune triggering.

411

MATERIALS AND METHODS

Experimental Animals

CBA/H mice of both sexes, 80–120 days of age, were used. Rabbits maintained at the Basel Institute of Immunology have been described elsewhere (Kelus and Pernis 1971).

T Cell-depleted Mice (ATXBM)

CBA/H mice were thymectomized at 4 weeks of age. One week later they were given 950 rads of gamma radiation from a ^{137}Cs source and injected intravenously with 2×10^6 viable syngeneic bone marrow cells each. Mice were used as spleen donors 4–10 weeks after injection of bone marrow. Spleens from mice with visible thymic remnants at autopsy were rejected.

Separation of Rabbit Peripheral Lymphocytes

Peripheral blood from rabbits was mixed at a ratio of 2:1 with a 3% purified pig skin gelatin (Eastman, organic chemicals) in Hanks balanced salt solution and kept at 37°C. After this the mixture was incubated for 30 minutes, then white cells which remained in the supernatant were pipetted off and washed three times in medium 199 at 37°C.

Staining of Rabbit Lymphocytes with Antiallotype Antibodies

Washed lymphocytes were incubated for different time periods and at different temperatures at a concentration of 2×10^7 cells/ml in tissue culture medium 199 (Difco Labs. Detroit, Mich.) containing 10% (v/v) heat-inactivated fetal calf serum (FCS) and 100 U/ml penicillin and streptomycin. Antiallotype immunoglobulins were added to cell cultures at a final concentration of 500 μg of protein/ml.

Antigen Binding Test

Suspensions of mouse spleen cells (2×10^7 cells/ml) were incubated for various times and at different temperatures in the presence of tritiated polymerized flagellin of *S. derby*. The cells were then washed, smeared on gelatin-coated slides and processed for radioautography as described previously (Diener and Paetkau 1972). Slides were screened until at least 30 labeled cells were registered or, for slides with few labeled cells, until 2×10^6 cells had been screened. A cell was scored as labeled if the number of silver grains in the emulsion overlying it was ten times higher than the background.

Tissue Culture

Cell suspensions of mouse spleen cells were prepared and washed once in Leibovitz medium (Grand Island Biological Co. New York) containing 10% (v/v) FCS and cultured in Eagle's minimal essential medium (Grand Island Biological Co. New York) containing 10% (v/v) FCS (Grand Island Biological Co. New York) as described by Diener and Armstrong (1969). Purified polymerized flagellin (POL) from *S. adelaide* (250 ng/ml of cell suspension) was used for immunization and *S. derby* was used as the indicator strain for enumerating antibody-forming cells (AFC) (Diener 1968). Antibody-forming cells to sheep erythrocytes (SRC) were detected by the plaque technique of Cunningham and Szenberg (1968).

Cell Separation

Cell separation by velocity sedimentation was carried out as described by Diener, Kraft and Armstrong (1973).

Antigen

Purified polymerized flagellin (POL) was prepared from *Salmonella adelaide* (strain SW 1338, H antigen fg; 0 antigen 35) according to the method of Ada et al. (1964). Sheep erythrocytes were collected in Alsevers solution and stored at 4°C. The cells were washed in sterile Leibovitz medium containing 10% (v/v) FCS and suspended in Eagle's minimal essential medium for addition to tissue cultures.

Preparation of Radiolabeled Polymerized Flagellin

Salmonella derby (H-antigen fg; 0 antigen 1, 4, 12), which shares the H but not the 0 antigen with *S. adelaide,* was used to prepare biosynthetically ^{3}H-labeled POL according to the method of Diener and Paetkau (1972). [^{125}I]POL was prepared according to the method of Ada et al. (1964).

Phytomitogens

Concanavalin A (ConA) was obtained from Miles Yeda Ltd., Rehovot, Israel. Phytohemagglutinin (PHA) (reagent grade MRC10, #K4415) was purchased from Wellcome Reagents Ltd., Beckenham England).

Antisera

Allotype-specific antisera were kindly provided by Dr. A. S. Kelus. They were prepared according to Kelus and Pernis (1971).

Microscopy

The cell preparations were examined under a Leitz Orthoplan microscope equipped with the Opah-Fluor vertical illuminator (E. Leitz, GmbH. Wetzlar, Germany). The cells were viewed alternatively in phase contrast and under specific illumination for Rhodamine. The data obtained in this manner were confirmed on a cell smear fixed with methanol.

RESULTS

The Significance of Antigen-binding Cells

Extensive studies have quantitated the number of antigen-binding cells in various phases of the immune response and in the tolerant state (Ada and Cooper 1971; Möller and Sjöberg 1972). Furthermore Ada and Byrt (1969) have demonstrated that immunocompetence could be specifically deleted from a cell population by the use of highly radiolabeled antigen ("irradiation suicide"). Thus there is little doubt that some antigen-binding cells can be equated with immunocompetent lymphocytes. A question remains, however, as to the exact relationship between the binding of an antigen to a lymphocyte and its subsequent capacity to participate in an immune response. The velocity sedimentation technique has allowed us to compare populations of immunocompetent and of antigen-binding cells present in the normal and immunized mouse spleen. The velocity sedimentation method is known to separate cells mainly on the basis of size differences (Armstrong and Kraft 1973). Previous work with this technique has permitted size changes of immuno-competent cells to be followed during the initial phase of the immune response. Spleen cells from CBA/H mice were separated by sedimentation at unit gravity in a fetal calf serum and phosphate-buffered saline gradient and the different cell fractions tested for their comparative ability to bind antigen and to initiate an immune response in vitro. Polymerized flagellin of *Salmonella adelaide* labeled with ^{125}I was used as the antigen. This antigen is known to trigger humoral immune responses in B cells in the absence of both thymus-derived and accessory cells (Diener, O'Callaghan and Kraft 1971; Diener, Shortman and Russell 1970). In Fig. 1 the total numbers of antibody-forming and of antigen-binding cells in each velocity fraction are expressed as a percentage of the maximum in any one fraction. In an unprimed spleen cell population the distribution profile of antigen-binding cells corresponded entirely with that of in vitro immunological activity (Fig. 1a). However this was not the case with cells from animals primed with the antigen 24 hours previously (Fig. 1b). Here some antigen-binding cells differed from immunocompetent cells in that a proportion (30–65%) of the former had failed to transform into larger cells upon antigenic stimulation. The possibility exists that these antigen-binding lymphocytes of small size are not identical with immunocompetent cells or, if they are identical, that they have been rendered unresponsive because of high avidity receptors for the antigen.

The Antigen-binding Cell and the Phenomenon of Capping

The circumstantial evidences discussed in the previous section for the identity of antigen-binding cells with immunocompetent cells has directed our attention to the immune triggering process at the single cell level.

The use of ^{125}I as a radioactive tracer to accurately study the topographic distribution of surface bound antigen on a lymphocyte proved inadequate since the high

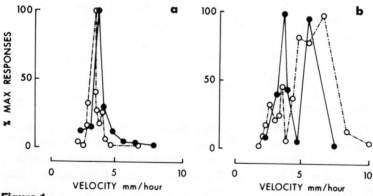

Figure 1
Comparison of sedimentation velocity values of antigen-binding cells with immunocompetent cells from **(a)** spleen of normal mice, **(b)** mice immunized 24 hours previously. From Diener et al. (1973) with permission of *Cellular Immunology*.

(•———•) Antigen-binding cells;
(o– – –o) in vitro essay for immunocompetence.

energy of the isotope produces long tracks on a photographic film. For this reason we have biosynthesized a tritium-labeled protein antigen from *Salmonella* flagella (Diener and Paetkau 1972). This purified antigen used in polymeric form ([³H]POL) was highly suitable in labeling antigen-binding cells. Lymphocytes from either the spleen or lymph nodes of unimmunized CBA/H mice were incubated at 4°C in the presence of different concentrations of [³H]POL for a period of 30 minutes. Radioautographs of such cells revealed a uniform distribution across the cell membrane of distinct silver grains, which upon prolonged incubation at 4°C tended to fuse and form grain clusters (Fig. 2a).

Work by others with fluorescein-conjugated anti-immunoglobulin on receptor-bearing B cells had led to the discovery of the "capping" phenomenon (Loor, Forni and Pernis 1972; Taylor et al. 1971). The immunoglobulin receptors coalesce to form a single macro-aggregate at one pole of the cell when incubated with anti-immunoglobulin at temperatures greater than 15°C. The use of tritiated antigen thus permitted us to investigate whether receptor capping would also occur upon interaction with antigen. Our data (Fig. 2b) confirmed that polymerized flagellin, like anti-immunoglobulin, causes lymphocyte receptors to cap. In contrast to anti-immunoglobulin, however, the clearance of receptors by the capping mechanism was followed within hours by the formation of new receptors. This is best illustrated in an experiment in which cells were stimulated with unlabeled POL and newly formed receptors were traced with [³H]POL (Table 1). Moreover upon prolonged incubation of the cells at 37°C in the presence of [³H]POL, there was a marked increase in receptor density (Fig. 3). This difference between antigen-induced and anti-immunoglobulin-induced capping most likely reflects the need for a ligand to bind to the receptor's antigen combining site in order to trigger immunoglobulin production, a condition that is met by antigen but probably not by anti-immuno-globulin.

Antigen Receptors as Mediators of the Triggering Signal

Our attention was now focused on the possibility that the single receptor, upon specific interaction with antigen, might convey the triggering signal. The species of choice to test this hypothesis was the rabbit since rabbit IgM immunoglobulin, unlike that of the mouse, carries clearly identified allotypic markers on different parts of the molecule: the variable (Aa locus allotypes *a*, 1,2,3) and the constant (Ms 1,2,4,5,6) region of the μ chain, on the constant region of the light chain (Ab locus allotypes *b*, 4,5,6), and on the κ light chain in association with the hinge region (Ms 3) (Fig. 4). Two rabbits were selected as alternate donors of peripheral lymphocytes to match the specificities of the available antiallotype immunoglobulins. Lymphocytes were incubated for various periods of time and under different temperature conditions in suspensions containing 2×10^7 cells/ml in the presence of 500 μg/ml of either unconjugated or fluorochrome-(Rhodamine) conjugated anti-allotype antibody of known specificity. First, capping of receptors was induced with unconjugated antiallotype antibody. After the cells were washed, they were further incubated in the presence of Rhodamine-conjugated antibody of the same specificity. At different times of incubation cell samples were screened by fluorescent micros-copy, according to criteria outlined elsewhere (Loor et al. 1972), for the appear-ance of newly formed surface immunoglobulins. Results from a representative experiment (Table 2) indicate that the reformation of surface immunoglobulin

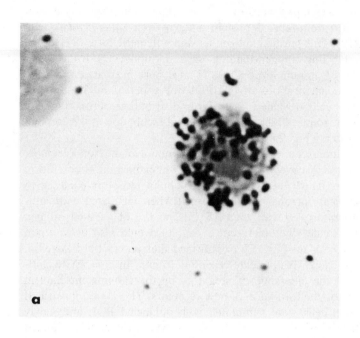

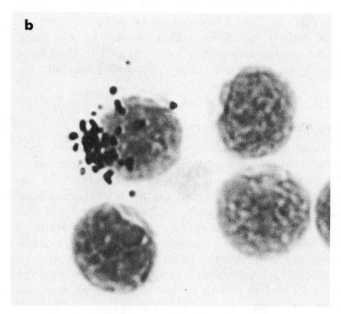

Figure 2
Radioautographs of lymph node cells from unimmunized mice with bound,
tritiated polymerized flagellin. 2400 ×. From Diener and Paetkau (1972) with
permission of *Proc. Nat. Acad. Sci.*

(a) Uniform distribution of antigen after incubation at 0° for 30 min; (b)
aggregation of antigen on one cell pole after incubation at 37°C for 15 min.

Table 1

Antigen Binding Following Antigen-induced Capping

Pre-treatment	Treatment	Treatment time (hr)	Labeled cells/10^6	% Caps
[³H]POL			18	94
POL	[³H]POL	2	1	
		4	8	0
		6	20	25

Cells were pretreated (30 min at 0°C, then 15 min at 37°C) with antigen (500 ng/ml) to induce capping. The cells were then washed and treated with [³H]POL at 37°C for various times.

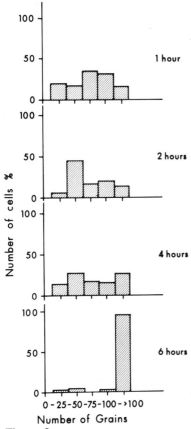

Figure 3

Grain count distribution profile of antigen-binding cells from unimmunized mice. Lymph node cells were incubated in vitro with 250 ng/ml of tritiated polymerized flagellin for 1, 2, 4 and 6 hr. The proportions of lymphocytes that bound the antigen are placed into categories according to the number of grains overlying the cells in the radioautograph. From Diener and Paetkau (1972) with permission of *Proc. Nat. Acad. Sci.*

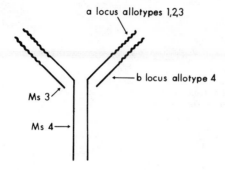

a locus allotypes 1,2,3

b locus allotype 4

Ms 3

Ms 4

Figure 4

Schematic diagram indicating the approximate location of allotypic markers on the rabbit IgM molecule. Only those alleles which were involved in the experiments described are listed.

following immunoglobulin clearance by "capping" was indeed stimulated by antibody directed either against allotypic markers on the variable region of the μ heavy chain or the constant region of the light chain. However no such stimulation could be observed when capping was induced with an antibody directed against either hinge region or μ chain constant region allotypic markers. It may be argued that the failure to trace newly formed surface immunoglobulins could be due to the fact that the latter failed to emerge sufficiently above the cell surface membrane in order to have the allotypic marker on the constant region of the μ chain accessible for labeling by the appropriate anti-immunoglobulin. This possibility could be ruled out, however, in experiments when cells were stimulated first with unconjugated antibody against a or b locus allotypic markers, followed by tracing with anti-μ chain constant region antibody of newly formed surface immunoglobulin. In conclusion the data indicate that, in the rabbit, stimulation of surface immunoglobulin production depends on anti-immunoglobulin binding to the F(ab')$_2$ portion of a surface IgM molecule. We have preliminary data indicating that binding of a ligand to certain parts of the constant region of the μ chain (possibly Fc portion) not only fails to stimulate but inhibits the cell from further immunoglobulin production.

Table 2

Stimulatory Effect of Antiallotype Antibodies on the Formation of Surface Immunoglobulins by Rabbit Lymphocytes

Incubation time (hr)	Temp. (C°)	Antibody	Labeled cells/10³ screened	Antibody	Labeled cells/10³ screened	Antibody	Labeled cells/10³ screened
0.5	4	anti-$a1$*	74	anti-$b4$*	44	anti-Ms3*	71
2.0	37		100		179	Ms4	33
3.0	37		129		200		72
4.5	37		290		276		41
10.5	37		132		196		18

A representative experiment out of 4 with similar data but somewhat different timing of incubation at 37°C. Considering all 4 experiments the differences between groups (anti-$a1$; anti-$b4$) and group (anti-Ms3; anti-Ms4) are statistically significant ($p < 0.01$).

* Cells incubated with Rhodamine-conjugated antibody after capping had been induced with unlabeled antibody. The percentage of caps in all groups was between 80 and 95%.

Table 3
Rate of Antigen Binding in the Presence of Immunogenic and
Tolerance-inducing Concentrations of Antigen [³H]POL

Antigen conc. ($\mu g/ml$)	Incubation, temperature conditions	No. antigen-binding cells/10^6 screened	% Caps
0.250[a]	0°(30 min) + 37°(1 hr)	26	94
0.250	37°(6 hr)	14	48
20[b]	0°(30 min) + 37°(5 min)	15	35
20	37°(6 hr)	26	4

Data from Diener and Paetkau (1972).

[a] Immunogenic concentration of antigen.

[b] Tolerance-inducing concentration of antigen.

The Significance of Antigen-Receptor Capping

Besides our evidence that the single antigen receptor is involved in the transmission of a distinct triggering signal, we had reason to assume the phenomenon of receptor capping plays an additional but essential part in immune induction. For example, concentrations of POL known to induce immunological tolerance in vitro (Diener and Armstrong 1969) inhibited cap formation (Table 3). Furthermore it was found that phytomitogens, such as phytohemagglutinin (PHA) and concanavalin A (ConA), not only prevented capping induced by POL but also caused a marked inhibition of the immune response to this and to another antigen such as sheep erythrocytes (SRC) in vitro (Lee et al. 1973) (Table 4). However it was found that ConA acted differently on the immune response depending on the presence or absence of thymus-derived T cells. When T cells were present as a normal constituent of the spleen, ConA suppressed the immune response to POL and to SRC.

Table 4
Effect of ConA and PHA on [³H]POL-induced Cap
Formation and on the Immune Response In Vitro

	Conc. ($\mu g/ml$)	% Caps	Labeled cells/10^6 scanned	Immune response[a] AFC/culture ± SEM
Control		80	24	8555 ± 540
ConA	0.1	40	27	11,110 ± 460
	1.0	30	19	434 ± 32
	10.0	19	18	51 ± 18
PHA	10^{-2} [b]	21	29	29 ± 6

Mouse spleen cells were incubated with various concentrations of ConA or PHA at 0°C for 30 min. ³H-labeled POL (250 ng/ml) was added and the incubation continued at 37°C for 1 hr. Data from Lee et al. (1973).

[a] Mouse spleen cells were cultured in vitro with various concentrations of ConA or PHA in the presence of an immunogenic concentration of POL (250 ng/ml). Numbers of antibody-forming cells (AFC) were measured after 4 days of culture.

[b] Expressed as the final dilution of the PHA solution purchased.

Table 5

Effect of ConA on the Immune Response of Normal and
ATXBM Mouse Spleen Cells to POL and SRC In Vitro

	Anti-SRC response (AFC/culture ± SEM)		Anti-POL response (AFC/culture ± SEM)	
	Control	+ ConA	Control	+ ConA
Normal	3941 ± 620	256 ± 21	4158 ± 150	100 ± 18
ATXBM	100 ± 30	3075 ± 342	1105 ± 154	2108 ± 164
ATXBM (no antigen)		56 ± 9		100 ± 40

Spleen cells (2×10^7 cells/ml) from normal and from ATXBM mice were incubated with
ConA ($10 \mu g/ml$) for 30 min at 0°C before the addition of POL and SRC (final concentration
250 ng/ml and 0.1% v/v respectively). Each value represents the mean number of AFC per
culture ± standard error of the mean. Data from Lee et al. (1973).

In contrast when cells were tested from adult thymectomized, irradiated and bone
marrow reconstituted mice (ATXBM mice are deprived of T cells), ConA was no
longer inhibitory to the immune response in vitro (Table 5); yet the phytomitogen
inhibited antigen-induced capping in a manner that was probably independent of
T cells. These findings are difficult to reconcile with the notion that receptor capping
by an immunocompetent cell is a prerequisite for immune induction. Perhaps capping exists as an immune regulatory mechanism that keeps the cell surface cleared
of antigen that might otherwise accumulate and reach tolerance-inducing levels.

DISCUSSION

The prime target for an antigen molecule is the immunocompetent cell. This cell is
defined by its capacity to specifically react to an antigenic determinant by means
of cell proliferation and synthesis of secretory products such as antibodies. The
central question pertaining to the identification of an immunocompetent cell concerns its possible identity with the antigen-binding cell. The "irradiation suicide"
experiments by Ada and Byrt (1969) have been interpreted to indicate the selective binding of the radiolabeled antigen to relevant immunocompetent cells and
their subsequent destruction by irradiation. Our use of cell separation techniques
together with radioautographic identification of antigen binding cells adds additional
quantitative evidence in support of this interpretation.

An adequate functional marker for immunocompetence as far as the B cell is
concerned is the antigen-induced production of antibody molecules, the specificity
of which is identical with that of the antigen recognition site. We can now detect
this specific marker by the use of [³H]POL. This T cell independent antigen permits
immunocompetent B cells to be identified by their increased antigen binding capacity after prior contact with antigen. In confirmation of studies with anti-immunoglobulin, [³H]POL also causes cell surface receptors to aggregate into a polar
"cap," a phenomenon consistent with a model which describes the cell surface
membrane as a fluid lipid-protein mosaic (Singer and Nicolson 1972). The discovery of ligand-induced cell surface dynamics has encouraged the suggestion that
this phenomenon may represent the mechanism whereby triggering of immunocompetent cells occurs. This interpretation gained further support by our finding that

inhibition of receptor distribution by either tolerance-inducing concentrations of antigen or by phytomitogens would coincide with inhibition of the immune response. However conditions could be obtained such that phytomitogens would inhibit capping without affecting the immune response. Furthermore some preparations of anti-immunoglobulin fail to stimulate receptor formation on mouse lymphocytes in spite of their capacity to cause receptor capping (Taylor et al. 1971). Whatever the true significance of receptor redistribution, it is unlikely to play a significant role in immune triggering. This has prompted us to investigate the possibility that interaction of surface immunoglobulin receptors with a ligand has to involve certain distinct portions of the receptor molecule. The use of rabbit antisera that would specifically bind to allotypic markers on different parts of the immunoglobulin receptor on rabbit lymphocytes has indeed provided evidence in favor of this assumption. Triggering of receptor reappearance was induced by the binding of anti-immunoglobulin to parts of the $F(ab')_2$ portion of the IgM receptor molecule. No stimulation occurred, however, when such binding involved the constant region of the μ chain and an allotypic marker associated with the hinge region μ the receptor. The fact that all of our antiallotype sera were stimulatory in vitro as far as blastogenesis is concerned suggests the necessity for two separate triggering signals or set of signals for cell proliferation and for antibody production.

Our findings favor those molecular theories which postulate that antigen receptors mediate triggering signals independently from each other, perhaps by antigen-induced allosteric changes (Bretscher and Cohn 1968). They are less compatible with the idea that cross-linking of antigen receptors by multivalent ligands provides the necessary conditions for immune triggering, conditions which are thought to be facilitated by a T cell-derived helper factor (Feldmann and Basten 1971). If receptor cross-linking provided the only prerequisite for immune induction, triggering should occur regardless of the manner by which a ligand binds to the receptor molecule, an expectation which is incompatible with our data.

Adopting the former hypothesis, we may thus speculate that the triggering of an immunocompetent cell depends on the sum total of individual interactions between surface receptors and antigenic determinants. This could imply that each such interaction triggers the activation of certain specific enzymes. Interactions of antigen with a large number of receptors could lead to an additive rise of these enzymes to levels required for immune induction. In view of available data on in vitro induced tolerance (Diener and Armstrong 1969), one may further postulate that excessive levels of such enzymes could have an inhibitory influence on the cell concerned.

REFERENCES

Ada, G. L. and P. Byrt. 1969. Specific inactivation of antigen-reactive cells with ^{125}I-labeled antigen. *Nature* **222**:1291.

Ada, G. L. and M. G. Cooper. 1971. The *in vivo* localization patterns and the *in vitro* binding to lymphocytes of normal and tolerant rats by *Salmonella* flagellin and its derivatives. *Ann. N.Y. Acad. Sci.* **181**:96.

Ada, G. L., G. J. V. Nossal, J. Pye and A. Abbot. 1964. Antigens in immunity. I. Preparation and properties of flagellar antigens from *Salmonella adelaide*. *Aust. J. Exp. Biol. Med. Sci.* **42**:267.

Armstrong, W. D. and N. E. Kraft. 1973. The early response of immunocompetent cells as analysed by velocity sedimentation separation. *J. Immunol.* **110**:157.

Bretscher, P. and M. Cohn. 1968. Minimal model for the mechanism of antibody induction and paralysis by antigen. *Nature* **220**:444.

Cunningham, A. J. and A. Szenberg. 1968. Further improvements in the plaque technique for selecting single antibody forming cells. *J. Immunol.* **14**:599.

Diener, E. 1968. A new method for the enumeration of single antibody producing cells. *J. Immunol.* **100**:1062.

Diener, E. and W. D. Armstrong. 1969. Immunological tolerance *in vitro*. Kinetic studies at the cellular level. *J. Exp. Med.* **129**:591.

Diener, E. and V. H. Paetkau. 1972. Antigen recognition: Early surface receptor phenomena induced by binding of a tritium labeled antigen. *Proc. Nat. Acad. Sci.* **69**:2364.

Diener, E., N. Kraft and W. D. Armstrong. 1973. Antigen recognition: I. Immunological significance of antigen binding cells. *Cell. Immunol.* **6**:80.

Diener, E., F. O'Callaghan and N. Kraft. 1971. Immune response *in vitro* to *S. adelaide* H-antigens not affected by anti-theta serum. *J. Immunol.* **107**:1775.

Diener, E., K. Shortman and P. Russell. 1970. Induction of immunity and tolerance *in vitro* in the absence of phagocytic cells. *Nature* **225**:731.

Feldmann, M. and A. Basten. 1971. The relationship between antigenic structure and the requirement for thymus-derived cells in the immune response. *J. Exp. Med.* **134**:103.

Kelus, A. S. and B. Pernis. 1971. Allotypic markers of rabbit IgM. *Europ. J. Immunol.* **1**:123.

Lee, K-C., R. E. Langman, E. Diener and V. H. Paetkau. 1973. Antigen recognition: III. The effect of phytomitogens on antigen-receptor capping and the immune response *in vitro*. *Europ. J. Immunol.* (in press).

Loor, F., L. Forni and B. Pernis. 1972. The dynamic state of the lymphocyte membrane. Factors affecting the distribution and turnover of surface immunoglobulins. *Europ. J. Immunol.* **2**:203.

Marchalonis, J. J., J. L. Atwell and R. E. Cone. 1972. Isolation of surface immunoglobulin from lymphocytes from human and murine thymus. *Nature New Biol.* **235**:240.

Möller, E. and O. Sjöberg. 1972. Antigen binding cells in immune and tolerant animals. *Transpl. Rev.* **8**:26.

Singer, S. J. and G. L. Nicolson. 1972. The fluid mosaic model of the structure of cell membranes. *Science* **175**:720.

Taylor, R. B., W. Philip, H. Duffus, M. C. Raff and S. dePetris. 1971. Redistribution and pinocytosis of lymphocyte surface immunoglobulin molecules induced by anti-immunoglobulin antibody. *Nature New Biol.* **233**:225.

Vitetta, E. S., J. W. Uhr and E. A. Boyse. 1973. Immunoglobulin synthesis and secretion by cells in the mouse thymus that do not bear theta antigen. *Proc. Nat. Acad. Sci.* **70**:834.

Warner, N. L., P. Byrt and G. L. Ada. 1970. Blocking of the lymphocyte antigen receptor site with anti-immunoglobulin sera *in vitro*. *Nature* **226**:942.

Hormone Receptors and Membrane Glycoproteins during In Vitro Transformation of Lymphocytes

Morley D. Hollenberg and Pedro Cuatrecasas

Department of Pharmacology and Experimental Therapeutics
and the Department of Medicine
The Johns Hopkins University School of Medicine, Baltimore, Maryland 21205

Emergence of Insulin Receptors with Mitogenic Stimuli

The stimulation by plant lectins such as concanavalin A (ConA) or phytohemagglutinin (PHA) of lymphocyte blast transformation provides a striking example of how specific cell surface stimuli can affect mitotic activity. Recently it has been observed (Fig. 1, Krug, Krug and Cuatrecasas 1972) that during ConA-mediated blast transformation in human lymphocytes, there is a dramatic appearance of cell surface receptors for insulin.

It has not been possible to detect significant specific binding of insulin to untransformed, column-purified human circulating lymphocytes, whether this is measured in whole lymphocytes, broken cell preparations, or detergent-solubilized membrane preparations. These results contrast with those obtained (Krug et al. 1972) using human lymphocytes (RPMI 6237) maintained in long term culture, or in cells obtained from patients with acute lymphocytic leukemia, where specific insulin binding is readily observed in both whole and broken cell preparations. The insulin binding that appears in transformed lymphocytes is specific in that it is of high affinity $K_D \simeq 10^{-9}$ M, it is saturable with respect to insulin (Fig. 1), and it is not affected by unrelated peptides such as glucagon and ACTH. In addition as with the fat cell (Cuatrecasas 1971a), the number of available insulin binding sites is augmented about threefold by digesting transformed lymphocytes with phospholipase C (EC 3.1.4.3) (Krug et al. 1972). This effect on insulin binding is not observed when untransformed lymphocytes are treated with phospholipase C. It is likely that the insulin binding sites which appear in association with transformation are either synthesized de novo or are initially present in a form incapable of binding insulin.

What might be the significance of the appearance of these receptors? They are clearly not present early (0–24 hours) in the process of transformation when the activation of cellular RNA and protein synthesis has already occurred. Furthermore their emergence is not a consequence of cell division since stimulated, multinucleated lymphocytes, treated with cytochalasin B to block cytoplasmic cleavage

423

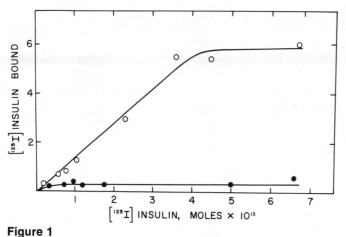

Figure 1

Specific binding of insulin to transformed (○) and untransformed (●) lymphocytes as a function of insulin concentration. Fresh peripheral human lymphocytes, isolated by gravity sedimentation after filtration through nylon fiber columns, were transformed with ConA (40 μg/ml) (Krug et al. 1972) and insulin binding was measured (Cuatrecasas 1971b; Krug et al. 1972) in 0.2-ml aliquots of Hanks buffer–0.1% albumin at a cell concentration of 15–20 × 10^6 per ml. Insulin bound is in 10^{16} × moles per 10^6 cells. Data from Krug et al. (1972).

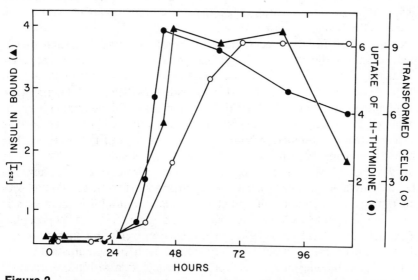

Figure 2

Binding of insulin to lymphocytes during transformation. Column-purified human peripheral lymphocytes were transformed with ConA and insulin binding (10^{16} × moles/10^6 cells) was measured at timed intervals after the initial ConA stimulus. Incorporation of [³H]thymidine (6.7 mCi/μmole, expressed as 10^4 × cpm, was measured as described previously (Krug et al. 1972). The number of transformed cells present (× 10^{-5}) per 10^6 cells was measured microscopically. Data from Krug et al. (1972).

424

after nuclear division, possess about ten times more binding sites for insulin (in proportion to their increased surface area) than do the control transformed lymphocytes (Krug et al. 1972). The time course of appearance of receptors also demonstrates that this change precedes the morphologic changes of transformation (Fig. 2). It is possible that these receptors are related to the continuation of some process occurring after the initial transformation stimulus but before cell division.

The appearance of insulin receptors does not seem to depend specifically on ConA as the mitogenic stimulus. The plant lectin, phytohemagglutinin, and the chemical reagent, periodate (Novogrodsky and Katchalski 1972), can cause lymphocyte transformation presumably by interacting with different cell surface components. Nonetheless insulin receptors do appear when human lymphocytes are transformed by either of these latter two compounds (Fig. 3). Thus the appearance of receptors may be independent of the nature of the initial stimulus and may reflect a more general requirement of the lymphocyte transformation process. It will be interesting in future work to determine if such receptors also appear when lymphocytes are subjected to a more physiologic stimulus, namely a specific antigen.

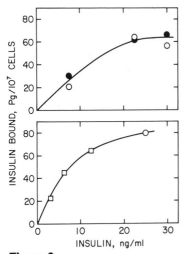

Figure 3

Insulin binding by transformed human peripheral lymphocytes. *Above:* Cells transformed by ConA (o) and by PHA (•). Column-purified lymphocytes were stimulated either by ConA or by PHA and insulin binding (Cuatrecasas 1971b) was measured 72 hr after the addition of lectin when thymidine incorporation was maximal. *Below:* Cells transformed by ConA (o) and by periodate (□). Unfractionated peripheral human lymphocytes, collected by gravity sedimentation, were stimulated by ConA or by periodate. Periodate stimulation was initiated by exposing lymphocytes (10^7/ml in phosphate-buffered saline) to periodate (7.5×10^{-4}M) for 10 min at 24°C and washing with an equal volume of Medium 199. Cells (10^6/ml) were then incubated at 37°C for 72 hr in Medium 199 containing 10% v/v fresh human serum and 2 mM glutamine, supplemented with 100 U/ml each of penicillin and streptomycin and equilibrated with 5% CO_2 in air. Insulin binding was measured at 72 hr.

Changes in Other Hormone Receptors

The appearance of insulin receptors during transformation raised the possibility that receptors for other hormones might also change during the process of mitogenesis. There is now evidence that receptors for growth hormone are pesent in both stimulated lymphocytes (Bockman and Sonenberg pers. commun.) and in cultured human lymphocytes (Lesniak et al. 1973). We have recently observed that receptors for another growth factor, nerve growth factor, also appear in transformed lymphocytes (Fig. 4). On the other hand it has not been possible to detect receptors for other biologically active polypeptides, such as glucagon and epidermal growth factor, in either transformed or untransformed lymphocytes. The emergence of receptors thus appears to be a relatively specific process, which may be essential to the subsequent division and differentiation of the lymphocyte.

Changes in ConA and WGA Binding Sites

The appearance of the hormone receptors indicated that other changes in membrane glycoproteins would undoubtedly accompany the process of lymphocyte transformation. The plant lectins ConA (which binds to α-D-gluco- and mannopyranoside residues) and wheat germ agglutinin (WGA) (which binds to N-acetyl-D-glucosamine residues [Sharon and Lis 1972]) were used as probes to detect alterations in the cell surface during transformation (Krug, Hollenberg and Cuatrecasas 1973). Both unstimulated and stimulated lymphocytes (from which the ConA originally added has been dissociated) bind ConA in a complex manner, suggesting the presence of multiple binding sites (Fig. 5). The binding of ConA is shown in relation to its mitogenic activity. Maximal stimulation of DNA synthesis

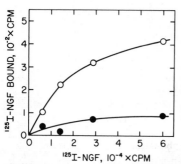

Figure 4

Binding of nerve growth factor (β subunit, β-NGF) to transformed (o) and untransformed (•) lymphocytes as a function of β-NGF concentration. Column-purified peripheral human lymphocytes were cultured for 72 hr in the presence or absence of ConA (40 μg/ml) as described by Krug et al. (1972). Binding of [^{125}I]NGF (100 μCi/μg prepared by the method of Cuatrecasas (1971b) but omitting the use of sodium metabisulfite) was then measured in 0.2 ml aliquots (2 $\times$ 10^7 cells/ml) by the filter filtration technique (Cuatrecasas 1971b; Banerjee et al. 1973), where maximal displacement of [^{125}I]NGF is achieved with 10 μg/ml native NGF. Binding is expressed as total cpm bound per sample, corrected for the radioactivity not displacd by 10 μg/ml native NGF.

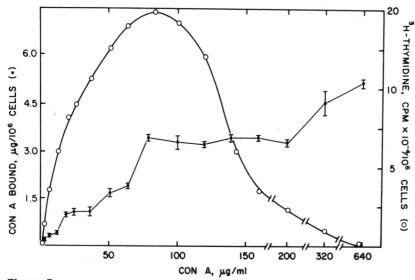

Figure 5
Mitogenic activity and binding of ConA to untransformed lymphocytes as a
function of lectin concentration. Data from Krug et al. (1973).

is obtained at the point where one major class of binding sites appears to be
saturated (about 75 μg/ml). Increasing the ConA concentration further leads to
a diminution of mitogenic activity as another class of binding sites becomes apparent
(150–640 μg/ml). In contrast only one class of binding sites is discernable for
WGA (Fig. 6). No mitogenic activity is observed for WGA either at concentrations
shown in Fig. 6 or for concentrations well above those required to saturate the
WGA binding sites.

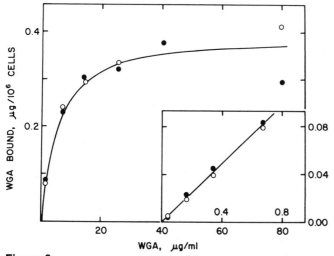

Figure 6
Binding of WGA to untransformed (o) and transformed (•) lymphocytes. Data
from Krug et al. (1973).

During transformation the amount of ConA bound per cell increases at all concentrations of lectin; the general shape of the binding curve is unchanged from that depicted in Fig. 5 (Krug et al. 1973). This relative increase in ConA binding contrasts with the lack of change in WGA binding (Fig. 7). The increase in ConA binding during mitogenesis closely parallels the increase in cell surface area. It is thus evident that new ConA binding sites appear to keep the average density of such sites unchanged, whereas the density of WGA binding sites must fall since the number of such sites remains unchanged.

Insulin-like Activity of Mitogenic Stimuli

It is pertinent, in view of the appearance of insulin receptors in transformed lymphocytes and the affinity with which ConA and WGA bind to these cells, that both of these lectins have potent insulin-like activity and that both can interact directly with insulin receptors (Cuatrecasas and Tell 1973).

In the isolated fat cell these lectins are as effective as insulin in stimulating glucose oxidation (Fig. 8) (Cuatrecasas and Tell 1973; Czech and Lynn 1973); they are also as effective as insulin in suppressing epinephrine-mediated lipolysis (Cuatrecasas and Tell 1973). The effects of the lectins are completely reversed by those specific sugars (N-acetylglucosamine, mannopyranoside) known to inhibit the binding of ConA and WGA to cell surfaces. It is also pertinent that two other potent mitogenic stimuli, PHA and pokeweed mitogen, also mimic the action of insulin in isolated fat cells in the processes of glucose transport and lipolysis (unpublished).

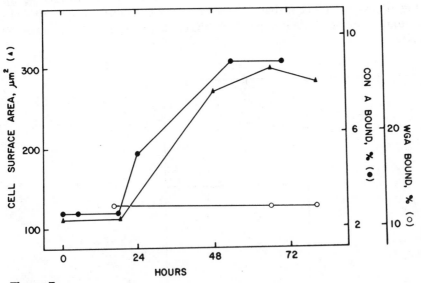

Figure 7
Binding of ConA and WGA during in vitro transformation by ConA. The proportion of lectin added to the medium (0–1 μg/ml) which is bound specifically is described. The mean surface area was calculated from the observed mean cell diameters. Data from Krug et al. (1973).

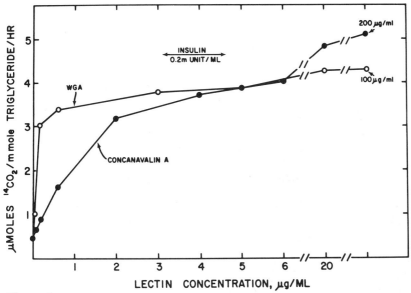

Figure 8
Effect of ConA and WGA on the conversion of [^{14}C]glucose to ^{14}CO$_2$ by isolated fat cells. Data from Cuatrecasas and Tell (1973).

Perturbation of Insulin Binding

ConA and WGA also affect the specific binding of insulin to fat cell and liver
membranes (Table 1) (Cuatrecasas and Tell 1973; Cuatrecasas 1973). At low
concentrations (1–20 μg/ml) WGA enhances insulin binding, whereas at higher
concentrations (80–300 μg/ml) insulin binding is blocked. ConA reduces the binding of insulin at both low (2.5 μg/ml) and high (600 μg/ml) concentrations of
lectin (Table 1). Agarose adsorbents containing ConA or WGA have been used to
purify by affinity chromatography the insulin receptor glycoprotein from detergent-
solubilized preparations of cell membranes (Cuatrecasas 1972).

Inhibition of Adenylate Cyclase

Physiological concentrations of insulin can inhibit adenylate cyclase activity in
isolated liver and fat cell membranes (Illiano and Cuatrecasas 1972) and in fat
cell ghosts (Hepp and Renner 1972). This effect of insulin is also mimicked by
low concentrations of the lectins (Fig. 9). ConA (5–50 μg/ml) effectively inhibits
adenylate cyclase activity in the presence or absence of epinephrine. Concentrations
of ConA greater than 50 μg/ml markedly reverse the inhibition of basal enzyme
activity and cause stimulation of this enzyme. WGA (2–50 μg/ml) also decreases
both the basal and the epinephrine-stimulated activities of adenylate cyclase; no
stimulation of enzyme activity is observed.

It is possible that the mitogenicity of ConA might be related to its insulin-like
actions and more specifically to its effects on adenylate cyclase activity or a closely
related biochemical process in the cell membrane. Those concentrations of ConA
which lower adenylate cyclase activity in fat cell membranes (Fig. 9) correspond

Table 1

Effect of Agglutinins on Specific
Binding of Insulin to Cell Membranes

	Specific [^{125}I]insulin bound*	
	fat cells	*liver cells*
No addition	830 ± 60	18,200 ± 810
Wheat germ agglutinin		
0.3 μg per ml	1500 ± 110	18,300 ± 220
1.2 μg per ml	2010 ± 160	21,900 ± 180
5 μg per ml	820 ± 40	32,800 ± 290
20 μg per ml	590 ± 30	29,500 ± 510
80 μg per ml	510 ± 40	16,000 ± 320
300 μg per ml	400 ± 20	10,100 ± 600
Concanavalin A		
2.5 μg per ml	880 ± 40	16,600 ± 310
40 μg per ml	760 ± 50	15,300 ± 440
160 μg per ml		14,300 ± 390
600 μg per ml	440 ± 20	4900 ± 90

Fat and liver cell membranes were incubated for 50 min at 24°C in 0.2
ml of Krebs-Ringer-bicarbonate buffer containing 0.1% (w/v) albumin
and the plant lectin at the indicated concentration. [^{125}I]insulin (1.1 ×
10^5 cpm) was added and specific binding was determined after incubating
for 50 min at 24°C. The liver and fat cell membrane concentrations were
680 and 41 μg protein per ml, respectively. Data from Cuatrecasas
(1973).

* cpm; average ± standard error of the mean of three replications.

to those concentrations which stimulate lymphocytes (Fig. 5). On the other hand
those concentrations of ConA which stimulate adenylate cyclase activity correlate
well with those concentrations of lectin which inhibit lymphocyte transformation
and lead to cell death. It is noteworthy that compounds which elevate lymphocyte
cyclic AMP concentrations inhibit lymphocyte transformation and that lymphocytes
exposed for 24 hours to PHA have decreased levels of cyclic AMP (Smith et al.
1971a,b). These considerations are important in view of recent demonstrations of
the relations between cyclic AMP and cell growth (Burger et al. 1972; Otten,
Johnson and Pastan 1971; Perry, Johnson and Pastan 1971).

DISCUSSION

The significance of the striking cell surface changes we observe during transformation is as yet not known. The appearance of new hormone receptors may in part
reflect the development of new cellular capabilities or the requirement for further
hormonal action to complete the process of transformation once initiated by a
variety of stimuli. The changes in lectin binding during transformation, which in
part may be due to the emergence of glycoprotein components of hormone recep-

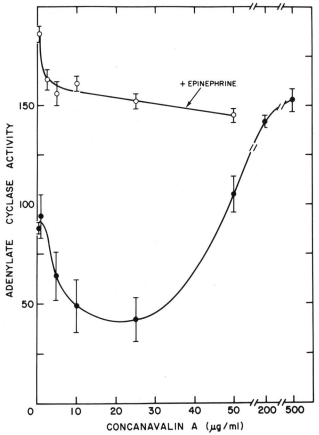

Figure 9

Effect of ConA on the basal and the epinephrine- (1 μM) stimulated adenylate cyclase activities of fat cell membranes. Activity is expressed as pmoles of cyclic AMP produced per min per mg protein (average of triplicates $\pm$ SEM). Data from Cuatrecasas and Tell (1973).

tors, also imply that dramatic changes in the composition of other membrane glycoproteins must occur during in vitro lymphocyte transformation.

The inability to detect a significant number of insulin receptors on nylon column-purified, resting, peripheral lymphocytes (Krug et al. 1972) contrasts with other studies reported recently (Gavin et al. 1972, 1973). In our previous studies it was demonstrated that insulin binding can be detected in unfractionated peripheral lymphocytes, but not in lymphocytes purified by passage through nylon columns. These columns remove circulating macrophages, which bind very large amounts of insulin (unpublished observations), and probably a proportion of those lymphocytes bearing surface immunoglobulins (B cells). The amount of insulin binding detected by Gavin et al. (1972, 1973) in unstimulated cell populations is 1/300 the amount bound by cells in permanent culture (Fig. 1, Gavin et al. 1973). This small amount of insulin binding (an average of about three receptors per cell) could well be attributed to the presence of small numbers (0.3% or less) of macrophages or spontaneously transformed lymphocytes in the unstimulated peripheral lymphocyte population.

Whether both T and B cells bind insulin when stimulated remains to be determined since the nylon column-purified preparation, while enriched in T cells, is not entirely free from B cells. It may be significant in this respect that acute lymphocytic leukemia lymphoblasts possess insulin receptors equivalent in number to those found in normal cells transformed in vitro with ConA (Krug et al. 1972). In contrast we fail to detect insulin receptors in chronic lymphocytic leukemia cells. These cells are also refractory to stimulation by plant lectins (PHA and ConA, Robbins 1964) and they bind considerably less lectin than do normal cells (unpublished and Novogrodsky and Biniaminov 1972).

It will be important to determine whether the insulin-like activity, which is a property of various stimuli that transform lymphocytes, is a fundamental, initial biochemical event in mitogenesis and whether the insulin receptors that appear during transformation are involved in regulating the terminal events of mitogenic transformation, morphologic alterations and cell division.

It is important that mitogenic stimuli have recently been demonstrated to cause immediate, very large elevations of cyclic GMP in lymphocytes (Hadden et al. 1972). Insulin has similarly been shown (Illiano et al. 1973) to cause rapid and significant elevations of this cyclic nucleotide in sensitive tissues (fat cells, liver) but not in resting lymphocytes, which lack insulin receptors and are apparently insensitive to this hormone. Thus in this activity, at least, certain lectins can also be considered to be insulin-like. Recent evidence suggests that there may be important reciprocal relationships between cyclic AMP and cyclic GMP (George et al. 1970; Lee, Kuo and Greengard 1972; Hadden et al. 1972; Illiano et al. 1973). Although it is not yet known by what mechanisms these cyclic nucleotides are regulated in a concerted fashion, it is quite possible that such changes may be intimately related to the processes of mitogenesis. The possibility must be considered that mitogenic stimuli are artificially eliciting fundamentally insulin-like biochemical responses (e.g., inhibition of adenylate cyclase, elevation of cyclic GMP, suppression of cyclic AMP), which are capable of activating a cell that lacks receptors for insulin or for other normally occurring stimuli having insulin-like activity. In this hypothesis it is anticipated that antigenic activation of the lymphocyte would similarly occur by selectively triggering an insulin-like response via highly specific receptors for that antigen on the surface of the lymphocyte.

The possibility must be considered that the emergent lymphocyte surface structures detected by insulin binding may not be true receptors for insulin, but for another, yet unrecognized but closely related, hormone or growth factor in serum that may be important in regulating the terminal process of mitogenesis. In this respect it may be of considerable importance that the dissociation constant for insulin binding to emergent lymphocyte receptors is about 10^{-9} M, whereas in the fat cell and in liver membranes the K_D is about 10^{-10} M. In addition it has recently been demonstrated that the small polypeptide, somatomedin (sulfation factor, thymidine factor), has potent insulin-like properties in fat cells and can effectively compete with insulin for binding to receptors in liver, fat and chondrocyte membranes (Hintz et al. 1972). Somatomedin has also been shown to inhibit adenylate cyclase activity in broken cell preparations (Tell et al. 1973). Another peptide with insulin-like and growth-promoting activities, which has properties in common with somatomedin, has recently been isolated from calf serum and from buffalo rat hepatoma cells in culture (Pierson and Temin 1972; Dulak and Temin 1973a, b). It is unlikely that the emergent structures are receptors for the growth-promoting or trophic peptide, epidermal growth factor (Hollenberg and Cuatrecasas 1973), since this peptide does not demonstrate cross-reactivity with insulin for binding to receptors. The relation of the emergent receptors for nerve growth factor and growth hormone to other serum factors must also be considered.

Acknowledgments

This work was supported by grants from the American Cancer Society (BC-63), National Science Foundation (GB34300), National Institute of Arthritis and Metabolic Diseases (AM14956), and The Kroc Foundation. M.D.H. is a recipient of a Postdoctoral Fellowship from the Medical Research Council, Canada. P.C. is a recipient of USPHS Research Career Development Award AM 31464.

REFERENCES

Banerjee, S. P., S. H. Snyder, P. Cuatrecasas and L. A. Greene. 1973. Nerve growth factor receptor binding in sympathetic ganglia. *Proc. Nat. Acad. Sci.* (in press)

Burger, M. M., B. M. Bombik, B. McL. Breckenridge and J. R. Sheppard. 1972. Growth control and cyclic alterations of cyclic AMP in the cell cycle. *Nature New Biol.* **239:**161.

Cuatrecasas, P. 1971a. Unmasking of insulin receptors in fat cells and fat cell membranes. Perturbation of membrane lipids. *J. Biol. Chem.* **246:**6532.

———. 1971b. Insulin-receptor interactions in adipose tissue cells: Direct measurement and properties. *Proc. Nat. Acad. Sci.* **68:**1264.

———. 1972. Affinity chromatography and purification of the insulin receptor of liver cell membranes. *Proc. Nat. Acad. Sci.* **69:**1277.

———. 1973. Interaction of concanavalin A and wheat germ agglutinin with the insulin receptor of fat cells and liver. *J. Biol. Chem.* **248:**3528.

Cuatrecasas, P. and G. P. E. Tell. 1973. Insulin-like activity of concanavalin A and wheat germ agglutinin. Direct interactions with insulin receptors. *Proc. Nat. Acad. Sci.* **70:**485.

Czech, M. P. and W. S. Lynn. 1973. Stimulation of glucose metabolism by lectins in isolated white fat cells. *Biochim. Biophys. Acta* **297:**368.

Dulak, N. C. and H. M. Temin. 1973a. A partially purified polypeptide fraction from rat liver cell conditioned medium with multiplication-stimulating activity for embryo fibroblasts. *J. Cell. Physiol.* **81:**153.

———. 1973b. Multiplication-stimulating activity for chicken embryo fibroblasts from rat liver cell conditioned medium: A family of small polypeptides. *J. Cell. Physiol.* **81:**161.

Gavin, J. R., III, J. Roth, P. Jen and P. Freychet. 1972. Insulin receptors in human circulating cells and fibroblasts. *Proc. Nat. Acad. Sci.* **69:**747.

Gavin, J. R., III, P. Gorden, J. Roth, J. Archer and D. N. Buell. 1973. Characteristics of the human lymphocyte insulin receptor. *J. Biol. Chem.* **248:**2202.

George, W. J., J. B. Polson, A. G. O'Toole and N. D. Goldberg. 1970. Elevation of guanosine 3':5'-cyclic phosphate in rat heart after perfusion with acetylcholine. *Proc. Nat. Acad. Sci.* **66:**398.

Hadden, J. W., E. M. Hadden, M. K. Haddox and N. D. Goldberg. 1972. Guanosine 3':5'-cyclic monophosphate: A possible intracellular mediator of mitogenic influences in lymphocytes. *Proc. Nat. Acad. Sci.* **69:**3024.

Hepp, K. D. and R. Renner. 1972. Insulin action on the adenyl cyclase system: Antagonism to activation by lipolytic hormones. *FEBS Letters* **20:**191.

Hintz, R. L., D. R. Clemmons, L. E. Underwood and J. J. Van Wyk. 1972. Competitive binding of somatomedin to the insulin receptors of adipocytes, chondrocytes, and liver membranes. *Proc. Nat. Acad. Sci.* **69**:2351.

Hollenberg, M. D. and P. Cuatrecasas. 1973. Epidermal growth factor: Receptors in human fibroblasts and modulation of action by cholera toxin. *Proc. Nat. Acad. Sci.* (in press)

Illiano, G. and P. Cuatrecasas. 1972. Modulation of adenylate cyclase activity in liver and fat cell membranes by insulin. *Science* **175**:906.

Illiano, G., G. P. E. Tell, M. I. Siegel and P. Cuatrecasas. 1973. Guanosine 3':5'-cyclic monophosphate and the action of insulin and acetylcholine. *Proc. Nat. Acad. Sci.* **70**:2443.

Krug, U., M. D. Hollenberg and P. Cuatrecasas. 1973. Changes in the binding of concanavalin A and wheat germ agglutinin to human lymphocytes during *in vitro* transformation. *Biochem. Biophys. Res. Commun.* **52**:305.

Krug, U., F. Krug and P. Cuatrecasas. 1972. Emergence of insulin receptors on human lymphocytes during *in vitro* transformation. *Proc. Nat. Acad. Sci.* **69**:2604.

Lee, T. P., J. F. Kuo and P. Greengard. 1972. Role of muscarinic cholinergic receptors in regulation of guanosine 3':5'-cyclic monophosphate content in mammalian brain, heart muscle, and intestinal smooth muscle. *Proc. Nat. Acad. Sci.* **69**:3287.

Lesniak, M. A., J. Roth, P. Gorden and J. R. Gavin, III. 1973. Human growth hormone. Radioreceptor assay using cultured human lymphocytes. *Nature New Biol.* **241**:20.

Novogrodsky, A. and M. Biniaminov. 1972. Binding of concanavalin A to rat, normal human, and chronic lymphatic leukemia lymphocytes. *Blood* **40**:311.

Novogrodsky, A. and E. Katchalski. 1972. Membrane site modified on induction of the transformation of lymphocytes by periodate. *Proc. Nat. Acad. Sci.* **69**:3207.

Otten, J., G. S. Johnson and I. Pastan. 1971. Cyclic AMP levels in fibroblasts: Relationship to growth rate and contact inhibition of growth. *Biochem. Biophys. Res. Commun.* **44**:1192.

Perry, C. V., G. S. Johnson and I. Pastan. 1971. Adenyl cyclase in normal and transformed fibroblasts in tissue culture. *J. Biol. Chem.* **246**:5785.

Pierson, R. W. and H. M. Temin. 1972. The partial purification from calf serum of a fraction with multiplication-stimulating activity for chicken fibroblasts in cell culture and with non-suppressible insulin-like activity. *J. Cell. Physiol.* **79**:319.

Robbins, J. H. 1964. Tissue culture studies of the human lymphocyte. *Science* **146**:1648.

Sharon, N. and H. Lis. 1972. Lectins: Cell-agglutinating and sugar-specific proteins. *Science* **177**:949.

Smith, J. W., A. L. Steiner and C. W. Parker. 1971b. Human lymphocyte metabolism. Effects of cyclic and noncyclic nucleotides on stimulation by phytohemagglutinin. *J. Clin. Invest.* **50**:442.

Smith, J. W., A. L. Steiner, W. M. Newberry, Jr. and C. W. Parker. 1971a. Cyclic adenosine 3',5'-monophosphate in human lymphocytes. Alterations after phytohemagglutinin stimulation. *J. Clin. Invest.* **50**:432.

Tell, G., P. Cuatrecasas, J. J. Van Wyk and R. L. Hintz. 1973. Somatomedin: Inhibition of adenylate cyclase activity in subcellular membranes of various tissues. *Science* **180**:312.

Inhibition of Protein and DNA Synthesis in Mammalian Cells by the *Ricinus communis* Agglutinins

Stuart Kornfeld, Wendy Eider, and Walter Gregory

Departments of Medicine and Biochemistry
Washington University School of Medicine, St. Louis, Missouri 63110

The carbohydrate-binding proteins of plants, termed lectins or agglutinins, possess a variety of biologic activities. These include stimulation of mitogenesis in resting lymphocytes (Nowell 1960), insulin-like activity toward fat cells (Cuatrecasas and Tell 1973), inhibition of phagocytosis (Berlin 1972), and the release of insulin from pancreatic islet cells (Lockhart Ewart et al. 1973). In addition it has been known for many years that the lectins of *Ricinus communis* (castor bean) and *Abrus precatorius* are highly toxic. Recently several groups have shown that these lectins inhibit the synthesis of protein and DNA by a variety of cell types (Lin et al. 1971; Olsnes and Pihl 1972a; Onozaki et al. 1972). Onozaki et al. (1972) have demonstrated that this toxicity can be blocked by the haptene galactose. This finding suggests that the initial step in lectin-induced toxicity is the binding of the lectin to a carbohydrate receptor on the plasma membrane of the target cell. As part of a study on the role of cell surface receptors in plant lectin function, we have investigated the relationship of the cell membrane binding of *Ricinus communis* lectins to the subsequent inhibition of macromolecular synthesis in the target cell.

MATERIALS AND METHODS

Purification of Ricinus *Agglutinins*

Ricinus communis beans were purchased from a local seed house. Four hundred g beans in one liter PBS (0.15 M NaCl–0.005 M sodium phosphate buffer pH 7.5) were homogenized in a Waring blendor, using four 30-second pulses at top speed. The homogenate was stirred at 4°C for one hour and then filtered through cheese cloth. The filtrate was centrifuged for 10 min at 20,000 g and the supernatant fluid brought to 75% saturation with solid ammonium sulfate. Following centrifugation the precipitate was dissolved in a minimal amount of PBS and dialyzed overnight at 4°C vs. 6 liters PBS. The dialyzed material was centrifuged at 20,000 × g for 10 min to remove the precipitate which had formed, and the supernatant fluid was then applied to an ovomucoid Sepharose column (a column 1.6 × 10.5 cm con-

435

taining 10 mg/ml ovomucoid could absorb approximately 150 mg of agglutinin). The affinity column was washed extensively with PBS and the agglutinins then eluted with 0.1 M galactose in PBS. The eluted material was dialyzed against 0.02 M sodium phosphate buffer pH 5.05 and applied to a column (2.0 × 7.4 cm) of phosphocellulose that had been equilibrated with the same buffer. The column was washed with the starting buffer and eluted with a linear gradient (400 ml) from 0.02 M phosphate pH 5.05 to 0.2 M phosphate pH 8.0. Under these conditions the applied material eluted as two major peaks, each of which had a shoulder (Fig. 1). The fractions were pooled as noted, and each of the four pools was rerun on the phosphocellulose column under similar conditions. The fractions were called *Ricinus* PHA I–IV. Fractions II–IV gave single bands on disc gel electrophoresis. On the basis of their elution position from a calibrated Sephadex G-150 column, *Ricinus* II had an apparent molecular weight of 60,000 and *Ricinus* IV a molecular weight of 118,000. Other investigators have found similar agglutinins in castor beans (Gurtler and Horstmann 1973; Nicolson and Blaustein 1972; Tomita et al. 1972).

Crystalline ricin and anti-ricin antibodies were kindly provided by Dr. Elvin Kabat.

Iodination of Ricinus *Agglutinins*

Ricinus PHAs II and IV were iodinated with ^{125}I by the chloramine-T method using a 10-second exposure to the chloramine-T (Hunter 1967).

Preparation of Cells

Human lymphocytes were isolated from peripheral blood by the Ficoll-Hypaque method as previously described (Mendelsohn et al. 1971). Mouse L1210 leukemic lymphoblasts were prepared from the spleens of mice which had been previously injected with L1210 cells.

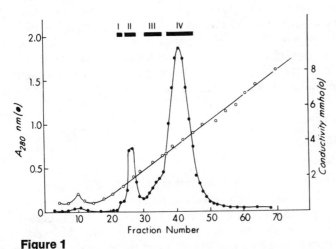

Figure 1
Phosphocellulose chromatography of *Ricinus* PHAs. 300 mg of *Ricinus* PHA was applied to a phosphocellulose column (2.0 × 7.4 cm) that had been equilibrated with 0.02 M sodium phosphate buffer pH 5.05. The details of the procedure are given in the text. The heavy lines indicate the fractions which were pooled and rerun on the phosphocellulose column.

(●——●) Absorption at 280 nm; (o——o) conductivity.

RESULTS AND DISCUSSION

Binding of *Ricinus* PHAs to Various Cell Types

The ability of [125]I-*Ricinus* PHA II and IV to bind to L1210 cells is shown in Fig. 2, with the data being plotted in a double reciprocal fashion according to the method of Steck and Wallack (1965). In both instances a straight line was obtained, indicating that there was only one type of receptor molecule. In other experiments the binding of extremely low concentrations of [125]I-*Ricinus* PHA II was measured (down to 0.001 μg/ml) and similar data were obtained. From these and similar experiments we have calculated the number of cell surface receptors for *Ricinus* PHAs II and IV on three different cell types (Table 1). While there are similar numbers of receptors for both lectins on human lymphocytes and L1210 cells, there are significantly more receptors for *Ricinus* II than for *Ricinus* IV on human erythrocytes. In addition *Ricinus* PHA IV bound to these various cell types with a greater affinity than did *Ricinus* PHA II.

As noted by previous investigators the binding of *Ricinus* PHA II and IV to cell membranes can be effectively blocked in the presence of the sugars galactose and

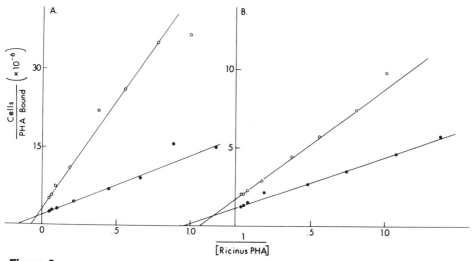

Figure 2

Binding of [125]I-*Ricinus* PHAs to L1210 cells and the effect of each agglutinin on the others binding. The binding reactions were carried out in PBS which contained in 0.2 ml: from 1–25 μg [125]I-*Ricinus* PHA II (Exp. A) or PHA IV (Exp. B), 1 mg bovine serum albumin, 4×10^6 L1210 cells and, as noted, 12.5 μg unlabeled *Ricinus* PHA IV in Exp. A and 29 μg unlabeled *Ricinus* PHA II in Exp. B. After 45 min incubation at room temperature, the cells were washed twice with PBS and the amount of bound [125]I-*Ricinus* PHA determined. The data are plotted by the method of Steck and Wallack (1965) according to the equation:

$$c/[\text{PHA bound}] = (1/Kn)(1/[\text{PHA}] + 1/n)$$

where [PHA] = concentration free PHA (μg), n = number of PHA binding sites per cell expressed as micrograms of PHA bound per cell, and c = number of cells. K is the affinity constant of PHA.

A, (●——●) [125]I-*Ricinus* PHA II alone; (○——○) + unlabeled *Ricinus* PHA IV.
B, (●——●) [125]I-*Ricinus* PHA IV alone; (○——○) + unlabeled *Ricinus* PHA II.

Table 1
Binding of *Ricinus* PHA II and IV to Various Cell Types

	Cell type	Binding sites per cell*	Association constant ($\times 10^6$ M^{-1})
Ricinus PHA II	Human erythrocytes	2.24×10^6 (3)	1.6
	Human lymphocytes	4.06×10^6 (2)	3.0
	L1210 leukemia		
	(in DBA-2 mice)	2.42×10^6 (2)	3.7
Ricinus PHA IV	Human erythrocytes	1.12×10^6 (6)	12.5
	Human lymphocytes	3.35×10^6 (4)	9.7
	L1210 leukemia	3.53×10^6 (1)	9.1

Cells were incubated with ^{125}I-labeled PHAs as described in legend to Fig. 2.
* Numbers in parentheses indicate the number of separate binding studies performed.

lactose (Drysdale et al. 1968; Nicolson and Blaustein 1972; Pardoe and Uhlenbruck 1970; Takahashi et al. 1962; Tomita et al. 1972; Waldschmidt-Leitz and Keller 1969). However when the *Ricinus* PHAs were incubated with lymphocytes or L1210 cells prior to the addition of lactose, it was found that a significant percentage of these lectins could no longer be released from the cells. Thus after 30 minutes of incubation with human lymphocytes, 92% could be released with 0.01 M lactose, whereas only 65% could be released after a 2.5-hour incubation. Repeated washing of the cells with 0.4 M lactose did not remove the residual lectin, nor did treatment of the cells with β-mercaptoethanol, trypsin, or changes in the pH of the washing buffers. This irreversible binding occurred at all temperatures (4, 24, 37°C) and was not inhibited by the presence of 0.03 M sodium azide. The fact that the total amount of lectin associated with the cells did not increase during this time suggested either that the lectin was remaining bound to the membrane receptor while a conformation change occurred resulting in higher affinity binding, or that if the lectin entered the cell, the membrane receptor was not available to bind another molecule of *Ricinus* PHA. To resolve this point we are utilizing thin section electron microscopy to determine the fate of ferritin-conjugated *Ricinus* PHA after it binds to L1210 cells. With this technique it should be possible to establish whether or not the lectin enters the cell. This irreversible binding did not occur with erythrocytes since 0.01 M lactose could release greater than 98% of the bound *Ricinus* PHA from these cells after a 3-hour incubation at 24°C.

To determine whether the *Ricinus* PHAs bound to unique receptor sites or shared a common receptor with other lectins, competitive inhibition studies were performed. In these experiments the concentration of each competing lectin added was sufficient to saturate its binding sites. As shown in Table 2, [125]I-*Ricinus* PHA II binding to L1210 cells was inhibited by *Ricinus* PHAs II and IV and by *Abrus* PHA, but not by the other lectins tested.[1] In fact mushroom PHA, soybean PHA,

[1] In this experiment a 23-fold excess of unlabeled *Ricinus* PHA II inhibited the binding of [125]I-*Ricinus* PHA II to erythrocytes only 49%. The explanation of this apparent low level of inhibition can be found by examining the binding curves for *Ricinus* PHA II (see Fig. 2). In these experiments the amount of *Ricinus* PHA II bound increased approximately tenfold when the concentration of the lectin was increased from 0.75 μg/0.2 ml to 18 μg/0.2 ml. Thus while the [125]I-*Ricinus* PHA was diluted 23-fold with unlabeled *Ricinus* PHA, there was a tenfold increase in the amount of binding, resulting in a net inhibition of approximately 50%.

Table 2

Ability of Various Agglutinins to Inhibit
Ricinus PHA II Binding to Cells

	L1210	*Erythrocytes*
	(% of control)	
No addition	100	100
17.5 μg *Ricinus* PHA II	31	49
19.8 μg *Ricinus* PHA IV	58	24
19.2 μg *Abrus* PHA	22	25
22.2 μg mushroom PHA	155	75
20.6 μg wheat germ agglutinin	85	31
19.1 μg soybean PHA	127	134
17.1 μg lentil PHA	141	122
16.2 μg *Ph. vulgaris* E-PHA	96	62

[125]I-labeled *Ricinus* PHA II (0.75 μg) was incubated in 0.9% NaCl–0.01 M NaHCO$_3$ containing 5 mg/ml of bovine serum albumin, either L1210 cells or human erythrocytes, and unlabeled agglutinins at the indicated amounts in a total volume of 0.2 ml. The reaction was begun with the addition of the cells and the incubations continued for 30 min at room temperature. The cells were then harvested, washed with buffer, and the inhibition of ^{125}I-PHA binding determined.

and lentil PHA actually enhanced *Ricinus* PHA II binding. Since mushroom PHA and *Ph. vulgaris* E-PHA bind to galactose-containing cell membrane oligosaccharide receptors (Kornfeld and Kornfeld 1970; Presant and Kornfeld 1972), these data indicate that the *Ricinus* PHA II binds to a distinct group of galactose-containing oligosaccharides. Similar results were obtained using [125]I-*Ricinus* PHA IV. However the data derived from standard competitive inhibition curves (Fig. 2) indicate that the two *Ricinus* PHAs probably bind to different receptors since they are not competitive inhibitors of each other's binding.

Inhibition of Protein and DNA Synthesis by *Ricinus* PHAs

When the *Ricinus* PHAs (and *Abrus* PHA) were incubated with mouse L1210 cells, a marked inhibition of protein and DNA synthesis resulted. *Ricinus* PHA II was greater than 100-fold more potent in this regard than *Ricinus* PHA IV (Fig. 3).[2] In this experiment the concentration of *Ricinus* PHA II that caused a 50% inhibition of DNA synthesis was 0.0037 μg/ml. Assuming that at most 50% of the added lectin was bound by 2.5×10^6 target cells, it can be calculated that there were approximately 7500 molecules of lectin bound per cell. Thus 50% inhibition of DNA synthesis occurred when only 0.3% ($7500/2.42 \times 10^6$) of the available receptors were occupied. By the same type of calculation, 50% inhibition of DNA synthesis resulted when there were 8.3×10^5 *Ricinus* PHA IV molecules per cell, which is equivalent to 23.6% saturation of the available receptor sites. Inhibition of DNA synthesis in normal human lymphocytes stimulated with mitogens occurred at similar concentrations of *Ricinus* PHA II and IV.

[2] *Ricinus* PHA II was also 100 times more toxic toward mice than *Ricinus* PHA IV. Thus the minimum quantity of *Ricinus* PHA II required to kill a 20-g mouse in 72 hours was 0.15 μg compared to 15 μg of *Ricinus* PHA IV.

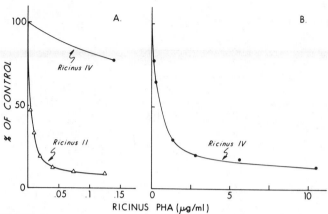

Figure 3
Inhibition of DNA synthesis in L1210 cells by *Ricinus* PHA II and IV. L1210
cells (5×10^6) were incubated in TC 199 medium containing various concentrations of *Ricinus* PHA II and IV as shown in a total volume of 2.0 ml. The
cells were incubated with the lectins for 3 hr at 37°C in 5% CO_2–95% air and
then 3 μCi of [³H]thymidine was added for 20 min. The amount of [³H]thy-
midine incorporated into trichloroacetic acid-precipitable material was then
determined. The results are expressed as the percentage of the control cultures.

Figure 4 shows the time sequence of the toxic effects of *Ricinus* PHA II on
L1210 cells. After rapid binding of the lectin, first protein synthesis is inhibited,
then DNA synthesis, and much later and to a lesser extent RNA synthesis is inhibited. A similar sequence of inhibition occurred with the use of *Ricinus* PHA IV.
To prove that the inhibition of protein synthesis was not the result of impaired transport of labeled precursors into the cells, we determined the effect of *Ricinus* PHA IV
on the uptake of α-aminoisobutyric acid (AIB) by human lymphocytes. In these
experiments 1.25 μg/ml *Ricinus* PHA IV actually enhanced AIB uptake by 50%
over control levels, proving that the inhibition of protein synthesis could not be explained by a decreased membrane amino acid transport system.

While binding of the lectins to the cells occurs quite rapidly, there is a significant
lag between lectin binding and the inhibition of protein synthesis (Fig. 4). At the
concentration of *Ricinus* PHA II used in this experiment (0.055 μg/ml) half-
maximal binding of the lectin occurred in about 7 minutes, whereas inhibition of
protein synthesis was not detected for the first 30 minutes. The duration of the lag
period is dependent on the concentration of lectin used, but even at high lectin concentrations the lag period is not abolished. Thus with 44.5 μg/ml *Ricinus* PHA II
the lag period was 20 minutes. The explanation for this lag period is unclear. It
could reflect the time required for the lectin to induce a rearrangement of the plasma
membrane or the time required for the lectin to be transported into the cell.
Alternatively the lag period could represent the time required for the lectin to be
converted from an inactive molecule to one that turns off protein and DNA synthesis. Finally the lag period may result from the fact that the lectin has to inactivate an intracellular component of protein synthesis that is not normally rate
limiting. In this case most of the component would have to be inactivated before
protein synthesis decreased.

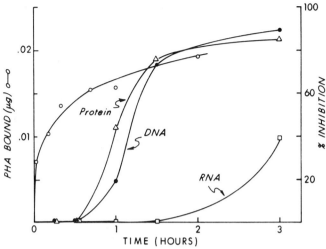

Figure 4

Time course of *Ricinus* PHA II binding to L1210 cells and the inhibition of protein, DNA and RNA synthesis. L_{1210} cells (5×10^6) were incubated alone or with 0.11 μg *Ricinus* PHA II in 2.0 ml cultures. At the times indicated duplicate cultures were incubated for 15 min with either 3 μCi [³H]leucine, 3 μCi [³H]uridine, or 3 μCi [³H]thymidine to label protein, RNA or DNA respectively. The incorporation of labeled material into TCA-precipitable material was then determined. Similar cultures were incubated with 0.11 μg of ¹²⁵I-*Ricinus* PHA II for the indicated times. The cells were then harvested, washed with buffer, and the amount of bound lectin determined.

    (o———o) ¹²⁵I-*Ricinus* PHA II bound;
    (△———△) [³H]leucine incorporation;
    (●———●) [³H]thymidine incorporation;
    (□———□) [³H]uridine incorporation.

The cell toxicity induced by the *Ricinus* PHAs can be prevented by the addition of lactose to the cultures (Table 3, Expt. 1), indicating that the lectins must bind to the cell membrane receptors in order to exert their toxic effects. However if the *Ricinus* PHA is allowed to incubate with the target cell for one hour or more prior to the addition of the lactose, then the haptene does not fully protect the cells from subsequent damage. This is not surprising since after an hour's incubation a significant amount of the bound *Ricinus* will no longer be eluted from the cell by lactose.

Relation of *Ricinus* PHAs to Ricin

Ricin is a highly toxic protein which has been isolated in crystalline form from castor beans (Gurtler and Horstmann 1973; Ishiguro et al. 1964a,b; Kabat et al. 1947; Olsnes and Pihl 1972b). Its relation to the *Ricinus* lectins has been a subject of considerable debate. Kabat et al. (1947) reported that crystalline ricin was a poor agglutinin in saline, but that it became a much more potent agglutinin when tested in dilute rabbit sera. Based on these data Kabat et al. concluded that ricin is itself a potent agglutinin. On the other hand Ishiguro et al. (1964b) reported that

Table 3

Prevention by Lactose of *Ricinus* PHA
and Ricin-induced Cell Toxicity

	[^3H]*thymidine incorporated (cpm/2 hr/culture)*
Experiment 1	
No addition	450
L-PHA	59,400
L-PHA + *Ricinus* PHA II (0.025 µg/ml)	1100
L-PHA + *Ricinus* PHA II + lactose	55,000
L-PHA + *Ricinus* PHA IV (0.5 µg/ml)	6800
L-PHA + *Ricinus* PHA IV + lactose	48,800
Experiment 2	
No addition	31,000
Ricin (0.0025 µg/ml)	16,500
Ricin + lactose	32,400
Ricin (0.0075 µg/ml)	11,450
Ricin + lactose	28,600

In Experiment 1 human peripheral blood lymphocytes were isolated, cultured and stimulated with the mitogen L-PHA as previously described (Mendelsohn et al. 1971). The *Ricinus* PHAs and lactose (0.01 M) were added to the cultures (2.0 ml) at 0 time. On day 3 the cultures were pulse-labeled with 3 µCi [^3H]thymidine for 2 hr and the acid-precipitable counts determined. In Exp. 2 the cultures (2.0 ml) contained 5×10^6 L1210 cells and crystalline ricin and lactose (0.01 M) as noted. After a 3-hr incubation the cultures were pulse-labeled with 3 µCi [^3H]thymidine for 30 min and the acid-precipitable counts determined.

their preparation of crystalline ricin was completely devoid of hemagglutinating activity. These investigators did not state whether they tested their material in dilute rabbit sera. To resolve this question a crude extract of *Ricinus communis* was passed over an ovomucoid Sepharose affinity column to adsorb out all the agglutinin activity. The agglutinin was then eluted from the column with 0.1 M galactose and the various fractions were tested for toxicity in the mouse system (Table 4). In this experiment greater than 98% of the toxic principal present in the extract adsorbed to the column and could be eluted with the haptene galactose. When the extract was passed over the column in the presence of 0.1 M galactose, neither the agglutinin nor the toxin adsorbed to the affinity column. In other experiments we found that the toxicity of crystalline ricin (kindly provided by Dr. Elvin Kabat) toward L1210 cells could be completely blocked by the haptene lactose (Table 3, Expt. 2). These data strongly suggest that crystalline ricin is a galactose-binding protein. In addition antibodies prepared against crystalline ricin cross-react with a line of identity with *Ricinus* PHA II and IV. Based on immunochemical studies Kabat et al. (1947) proposed that the castor bean contains two proteins, which react identically with antisera and agglutinate erythrocytes, but only one of which (the "ricin") is toxic. Our studies support this view. Thus *Ricinus* PHA II seems to be "ricin" and *Ricinus* PHA IV, with one-hundredth the toxicity, may be equivalent to the nontoxic hemagglutinin.

Table 4

Absorption of the Toxin and Agglutinin of *Ricinus communis* by Ovomucoid Sepharose

	Protein (*mg*)	Hemagglutinating units	Mouse toxicity ($LD_{100}/72\ hr$)
Experiment 1			
Extract applied	5.9	20,800	2.8 μg
Nonabsorbing material	4.2	12	160.0 μg
Galactose eluate (0.1 M)	1.7	19,000	0.6 μg
Experiment 2			
Extract applied			
in 0.1 M galactose	5.6	20,000	2.8 μg
Nonabsorbing material	5.0	19,200	5.0 μg

Crude extracts of *Ricinus communis* beans were prepared as described in the text. The 20,000 g supernatant was then centrifuged for 1 hr at 80,000 $\times$ g. This supernatant was applied to a column (1 $\times$ 6.5 cm) of ovomucoid Sepharose. After extensive washing with 0.9% NaCl–0.005 M sodium phosphate pH 8.0, the column was eluted with 0.1 M galactose. In Exp. 2 the crude extract was brought to 0.1 M galactose before it was applied to the column. All fractions were titered for hemagglutinating activity and mouse toxicity (the μg of protein required to kill a mouse in 72 hr). The various fractions were dialyzed against PBS to remove the haptene galactose prior to assay.

Mechanism of Action of *Ricinus* Lectins

These data demonstrate that when a small number of *Ricinus* PHA II molecules bind to galactose-containing receptors on the target cell membrane, an irreversible inhibition of protein and DNA synthesis occurs after a fixed lag period. The basic question of whether or not the lectin must enter the cell to exert its toxic effect remains unresolved. Onozaki et al. (1972) have reported that *Ricinus* PHA covalently bound to large Sepharose beads is still able to inhibit DNA synthesis in rat ascites tumor cells, suggesting that the interaction of this lectin with the plasma membrane may be sufficient to cause the subsequent inhibition of macromolecular synthesis. However such experiments have to be interpreted with caution. To achieve comparable inhibition six times greater amounts of Sepharose-bound lectin than free lectin were used, and since the lectin-Sepharose beads settle to the bottom of the culture tubes along with the cells, the local concentration is even higher. Under these conditions it is difficult to exclude the possibility that small amounts of lectin detached from the beads and bound directly to the cells. Furthermore such experiments do not rule out the possibility that one of the subunits of the bound *Ricinus* PHA entered the cell while the other subunits remained attached to the Sepharose beads.

An alternative mechanism for lectin-induced toxicity is that the lectin (or one of its subunits) enters the cell after binding to the cell membrane receptor and either interferes directly with protein synthesis or activates a substance which can, in turn, inhibit protein synthesis. The findings of Olsnes and Pihl (1972a) support this mechanism. These investigators have found that ricin and abrin are potent inhibitors of peptide chain elongation in a cell-free system from rabbit reticulocytes. In their studies protein synthesis was markedly inhibited by ricin at a concentration of 0.2 $\mu g/ml$. Treatment of the toxin with β-mercaptoethanol increased the in vitro inhibitory effect by greater than tenfold while markedly reducing the in vivo toxicity (Olsnes and Pihl 1972b). In the in vitro system there is a lag of several minutes before the inhibition of protein synthesis becomes manifest, similar to the lag period we have observed using intact cells. This supports the concept that the lag represents the time required to inactivate a component of protein synthesis that is not normally rate limiting. The concentration of β-mercaptoethanol-treated ricin that inhibits protein synthesis in this system is very comparable to the concentration of *Ricinus* PHA II (0.0037 $\mu g/ml$) that causes a 50% inhibition of protein synthesis in intact L1210 cells.

In other experiments Olsnes and Pihl (1972a) showed that ricin inhibits protein synthesis without causing polysome disaggregation, indicating that the toxin prevented the completion or release of nascent peptide chains. In contrast to these results Lin et al. (1972) have reported that ricin and abrin cause the degradation of polysomes of rat liver and Ehrlich ascites tumor cells. These workers also found that ricin had no direct effect on polysomes but acted indirectly by increasing the ribonuclease activity in the postmicrosomal supernatant fractions of the treated cells. The reason for the discrepancy in these various reports is not clear.

SUMMARY

Two agglutinins, which bind to distinct galactose-containing cell membrane receptors, have been isolated from *Ricinus communis* beans by affinity chromatography. The number of cell surface receptors for these two agglutinins has been determined on three cell types. Both agglutinins cause an irreversible inhibition of protein and DNA synthesis in L1210 cells and human lymphocytes, although *Ricinus* PHA II is 100 times more toxic than *Ricinus* PHA IV. A 50% inhibition of protein synthesis in L1210 cells occurs when *Ricinus* PHA II occupies less than 0.3% of the cell's 2.4×10^6 membrane receptors. *Ricinus* PHA IV must occupy about 24% of its receptor sites (3.5×10^6 per L1210 cell) to inhibit protein synthesis by 50%. Even at high concentrations of *Ricinus* PHA II, there is a fixed lag period between the time of lectin binding to the cell membrane and the onset of inhibition of protein synthesis. Lectin binding to membrane receptors is essential for the inhibition of protein synthesis to occur since the toxic effects can be blocked with the haptene lactose.

Evidence is presented which suggests that ricin, the toxic protein of the castor bean, is itself a carbohydrate-binding protein.

Acknowledgments

This research was supported in part by grant R01 CA 08759 from the U.S. Public Health Service. S.K. is a recipient of the USPHS Research Career Development Award AM 50298.

REFERENCES

Berlin, R. D. 1972. Effect of concanavalin A on phagocytosis. *Nature New Biol.* **235**:44.

Cuatrecasas, P. and G. Tell. 1973. Insulin-like activity of concanavalin A and wheat germ agglutinin—direct interaction with insulin receptors. *Proc. Nat. Acad. Sci.* **70:** 485.

Drysdale, R. G., P. R. Herrick and D. Franks. 1968. The specificity of the haemagglutinin of the castor bean, *Ricinus communis. Vox. Sanguinis* **15:**194.

Gurtler, L. G. and H. J. Horstmann. 1973. Subunits of toxin and agglutinin of *Ricinus communis. Biochim. Biophys. Acta* **295:**582.

Hunter, W. M. 1967. *Handbook of Experimental Immunology* (ed. D. M. Weir) pp. 608–654. Blackwell Publ. Co., London.

Ishiguro, M., T. Takahashi, K. Hayashi and M. Funatsu. 1964a. Biochemical studies on ricin. II. Molecular weight and some physico-chemical properties. *J. Biochem.* (Tokyo) **56:**325.

Ishiguro, M., T. Takahashi, G. Funatsu, K. Hayashi and M. Funatsu. 1964b. Biochemical studies on ricin. I. Purification. *J. Biochem.* (Tokyo) **55:**587.

Kabat, E. A., M. Heidelberger and A. E. Bezer. 1947. A study of the purification and properties of ricin. *J. Biol. Chem.* **168:**629.

Kornfeld, R. and S. Kornfeld. 1970. The structure of a phytohemagglutinin receptor site from human erythrocytes. *J. Biol. Chem.* **245:**2536.

Lin, J. Y., K. Liu, C. C. Chen and T. C. Tung. 1971. Effect of crystalline ricin on the biosynthesis of protein, RNA and DNA in experimental tumors. *Cancer Res.* **31:**921.

Lin, J. Y., C. C. Pao, S. T. Ju and T. C. Tung. 1972. Polyribosome disaggregation in rat liver following administration of the phytotoxic proteins, abrus and ricin. *Cancer Res.* **32:**943.

Lockhart Ewart, R. B., S. Kornfeld and D. M. Kipnis. 1973. Stimulation of insulin release from isolated rat islets by mushroom phytohemagglutinin in vitro. *Clin. Res.* **21:**622.

Mendelsohn, J., A. Skinner, Sr. and S. Kornfeld. 1971. The rapid induction by phytohemagglutinin of increased α-aminoisobutyric acid uptake by lymphocytes. *J. Clin. Invest.* **50:**818.

Nicolson, G. and J. Blaustein. 1972. The interaction of *Ricinus communis* agglutinin with normal and tumor cell surfaces. *Biochim. Biophys. Acta* **266:**543.

Nowell, P. C. 1960. Phytohemagglutinin: An initiator of mitosis in cultures of normal human leukocytes. *Cancer Res.* **20:**462.

Olsnes, S. and A. Pihl. 1972a. Ricin—A potent inhibitor of protein synthesis. *FEBS Letters* **20:**327.

———. 1972b. Treatment of abrin and ricin with β-mercaptoethanol. Opposite effects on their toxicity in mice and their ability to inhibit protein synthesis in a cell-free system. *FEBS Letters* **28:**48.

Onozaki, K., M. Tomita, Y. Sakurai and T. Ukita. 1972. The mechanism of the cytotoxicity of *Ricinus communis* phytoagglutinin toward rat ascites tumor cells. *Biochem. Biophys. Res. Commun.* **48:**783.

Pardoe, G. I. and G. Uhlenbruck. 1970. Characteristics of antigenic determinants of intact cell surfaces. *J. Med. Lab. Technol.* **27:**249.

Presant, C. and S. Kornfeld. 1972. Characterization of the cell surface receptor for the *Agaricus bisporus* hemagglutinin. *J. Biol. Chem.* **247:**6937.

Steck, T. L. and D. F. H. Wallack. 1965. The binding of kidney bean phytohemagglutinin by Ehrlich ascites carcinoma. *Biochim. Biophys. Acta* **97:**510.

Takashashi, T., G. Funatsu and M. Funatsu. 1962. Biochemical studies on castor bean hemagglutinin. I. Separation and purification. *J. Biochem.* (Tokyo) **51:**288.

Tomita, M., T. Kurokawa, K. Onozaki, N. Ichiki, T. Osawa and T. Ukita. 1972. Purification of galactose-binding phytoagglutinins and phytotoxin by affinity chromatography using Sepharose. *Experientia* **28:**84.

Waldschmidt-Leitz, V. E. and L. Keller. 1969. Über Ricin: Reinigung und differenzierung der Wirkungen. *Z. Physiol. Chem.* **350:**503.

Plasma Membrane Interactions and the Control of Cell Division

A. C. Allison

Clinical Research Centre, Harrow, Middlesex, England

In multicellular organisms cell division and differentiation are controlled not only by mechanisms operating within individual cells but also by influences spreading from cell to cell. Apart from the long-range effects exerted by hormones and nerve impulses, it is now clear that electrical stimuli and small molecules, but not macromolecules, can travel freely among adjacent cells. In addition many plant and invertebrate cells and some mammalian cells are linked by true cytoplasmic bridges. Although the physiological roles of such junctional systems are still poorly understood, it seems likely that they are involved, among other processes, in the short-range control of cell division. Recent information on the mobility of proteins within membranes, on the properties of ionophores and on the disposition of particles visualized in membranes by freeze-fracturing electron microscopy provide clues to the structural basis of junctional communication between adjacent cells.

An important mechanism limiting the multiplication of malignant cells is that exerted by the immune system. This is illustrated by the raised incidence of tumors, especially those that are virus-induced, in immunosuppressed animals and human subjects. Immunological surveillance is linked to the capacity of immunocompetent cells to kill tumor cells or other target cells with antigens different from those in the host animal. Our recent observations show that the mechanism of target cell killing by sensitized lymphocytes is similar to that allowing functional coupling between many different cell types.

Intercellular Bridges

Apart from fertilization itself, the well-known formation of syncytia in developing muscle and bone cells and in the syncytiotrophoblast (see Poste and Allison 1973) illustrates the role of cell fusion in ontogeny. The observations of Fawcett (1971) on the syncytial nature of male germ cells can be cited as an example of the role of cytoplasmic bridging in maintaining synchrony of cell division. The male germ cells develop as clones in which all the cellular units of common lineage remain joined by intercellular bridges. This has been shown in a range of animal forms,

447

extending from fruit flies to man. In mammals these bridges develop as a consequence of incomplete cytokinesis at each of several successive germ cell divisions. A thin layer of dense material is deposited on the inner aspect of the limiting membrane of the bridge, stabilizing its dimensions. These connections, which are large enough to allow passage of organelles and diffusible materials, are usually occupied by cytoplasm indistinguishable from that of the conjoined cells. Recent studies have shown that the number of germ cells in each syncytium is large and that the bridges do not remain unchanged throughout spermiogenesis. At clearly defined times in the spermatogenic cycle, they develop transverse septa that result in transient isolation of the cellular units within the syncytium.

Type A spermatogonia, which divide to renew the stem cell population, appear to undergo complete cytokinesis. However as the spermatogonium becomes committed to progressive differentiation, it is thought that the progeny of all subsequent divisions remain connected by intercellular bridges. From the number of mitotic and meiotic divisions occurring in rat spermatogenesis, there should be at least 256 units in spermatid syncytia. As Fawcett (1971) has suggested, the intercellular bridges may serve to maintain synchrony of division and differentiation, which is a striking feature of groups of adjacent cells in seminiferous tubules.

In insects each germ cell clone is separately encapsulated, whereas in human seminiferous tubules, although there is no encapsulation, small patches of germ cells, each developing synchronously, are observed. Each patch probably represents a germ cell clone. The synchrony observed in the clones may well result from the passage of controlling materials from one cell to another. This would be analogous to the tendency for nuclei from cells in different stages of the mitotic cycle to become synchronized when they are fused to form heterokaryons (Rao and Johnson 1970 and this volume; Sei-Ichi 1972). However when groups of spermatogonia or of spermatocytes prepare for division, the bridges connecting them develop transverse membranous septa that appear to isolate cell territories within the syncytium. When these septa are present in a bridge, the synchrony of mitotic events in conjoined cells is imperfect. The septa disappear soon after reconstitution of daughter cell nuclei and are very rarely observed between cells with interphase nuclei. The maintenance of synchrony through several cycles of cell division, despite the development of transient asynchrony during mitosis itself, underlines the importance of cytoplasmic bridges in the control of cell division. This in turn supports observations, reviewed by Harris (1968), that cytoplasmic factors control DNA synthesis and cell division as well as differentiation.

Functional Coupling of Cells

Much commoner in mammalian cells than the presence of complete cytoplasmic bridges is the form of communication known as functional coupling, in which cells are joined by specialized areas of contiguous membranes and yet preserve some of their individuality. Two approaches have been used by Loewenstein (1973) and others to examine the permeability of such membrane junctions. One is to inject molecules with fluorescent labels into cells with the aid of a micropipette and follow cell-to-cell diffusion by microscopical examination. The second is to introduce a current of small inorganic ions into a cell and determine, by electrical measurement with intracellular microelectrodes, whether part of the current flows into adjacent cells. Fluorescein (molecular weight 376) readily passes from cell to cell through communicating junctions, and molecules of molecular weight up to about 10,000 appear sometimes to be able to traverse these junctions.

Studies of the analogous electrotonically coupled segmental axons of the crayfish, *Procambarus,* by Reese, Bennett and Feder (1971) have shown that microperoxidase (1800 daltons) can pass from cell to cell, whereas horseradish peroxidase (40,000 daltons) cannot.

Evidence is accumulating that metabolic coupling in cultured mammalian cells is also mediated through such junctional regions. Subak-Sharpe, Bürk and Pitts (1966) found that cells of a hamster fibroblast line lacking inosinic pyrophosphorylase activity (IPP⁻ cells) when grown by themselves do not incorporate exogenous tritiated hypoxanthine into their nucleic acid and are therefore unlabeled in radioautographs. These cells can be grown in mixed cultures that have inosinic pyrophosphorylase activity (IPP⁺ cells), and where cell contact is established IPP⁻ cells show some incorporation of tritiated hypoxanthine. Such exchange of material through intercellular channels does not take place when combinations of cells are used which are unable to establish transfer of fluorescein or electrical current through points of contact and do not show plaque-like gap junctions when examined in the electron microscope (Gilula, Reeves and Steinbach 1972; Azarnia, Michalke and Loewenstein 1973). These and other results suggest that the presence of specific areas of cell membrane differentiation, gap junctions, are required for ionic and metabolic coupling between cells.

Gap junctions are characterized by the presence in apposed cell membranes of well-defined plaques containing particles visible by freeze-fracturing electron microscopy. The particles are close-packed, often in hexagonal array, with a 90–100 Å center-to-center spacing. The plaques on either side of the gap are in strict register. Since molecules can pass from cell-to-cell in gap junctions, which are common in mammalian organs, their possible role in the control of cell division is worth considering. The metabolic coupling experiments show that nucleotide derivatives are among those which can pass from cell to cell through gap junctions, so there is no reason to believe that controlling molecules such as cyclic AMP or cyclic GMP would be excluded, although passage of most proteins and nucleic acids cannot occur. Azarnia and Loewenstein (1971) have reported that certain Morris hepatoma cells in culture (from a cloned line) were unable to establish junctional communication, and the same is true of a malignant epithelial strain derived from irradiated embryonic tissues (Loewenstein 1973). However most of the malignant cells examined have shown no discernible abnormality in junctional communication. It seems likely that defective junctional communication is one of several mechanisms that can lead to uncontrolled growth.

Membrane Proteins in Relation to Increased Permeability

The plasma membranes of most mammalian cells have high specific resistance (about $10^4 \Omega cm^2$), of the same order as a lipid bilayer, whereas from the propagation of electrical signals from cell to cell it is clear that the resistance of gap junctions is not much greater than that of the general cytoplasm (about $100 \Omega cm^2$; see Satir and Gilula 1973). In view of the clustering of particles visualized by freeze-fracturing electron microscopy in the gap junctional membrane plaques, it is worth considering whether the presence of these particles might facilitate permeability of ions through this region of the plasma membrane.

This concept receives support from recent studies with cyclic peptide ionophores (ion carriers) in model membrane systems. Most investigations have been carried out with the black membrane system, in which a lipid bilayer separates two aqueous compartments. Passage of ions from one compartment to the second through the bilayer is recorded as an electric current (see Läuger 1972). When the bilayer consists of phospholipid with or without cholesterol, the resistance is very high and conductance very low. The passage of alkali metal cations across the membrane is greatly increased by introducing into the system cyclic peptide ionophores. These are of two kinds. One, of which valinomycin is the prototype, has a ring structure with a central space through which potassium ions can pass. Each ionophore molecule functions independently, as shown by the linear kinetics of ion transport in relation to the number of ionophore molecules (Fig. 1). The second type of ionophore, of which alamethicin is an example, has no central space, and several molecules of the ionophore (under most conditions, six) must come together to form a cluster through which ions can pass. Ion conductance through the bilayer increases in direct proportion to the sixth power of the ionophore concentration.

Since it is known that some membrane glycoproteins extend through the lipid bilayer from the outer to the inner surface, and since they have translational mobility in the plane of the membrane, it is possible that they can act in a manner analogous to that of the alamethicin type of ionophore. The suggestion can be made that agents which cross-link and so form clusters of membrane glycoproteins may generate between the proteins hydrophilic "pores" through which ions can pass (Fig. 2). Such increased passive flux of ions could bring about membrane depolarization, which represents an important trigger mechanism for various cellular reactions, and when excessive brings about cell death through osmotic lysis. To test the hypothesis that clustering of membrane glycoproteins can be related to increased passive ion flux, we have analyzed the mechanism of hemolysis of erythrocytes by Sendai virus.

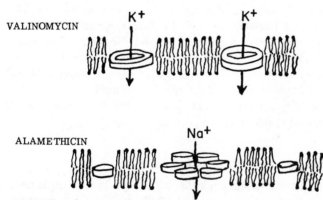

Figure 1

Diagram illustrating the ways in which the two types of ionophore increase passive ion flux across a lipid bilayer. Each molecule of valinomycin allows passage of potassium ions through a central space. Single molecules of alamethicin do not allow increased ion flux, but where six molecules of alamethicin come together to form a cluster an ion-conducting channel is generated.

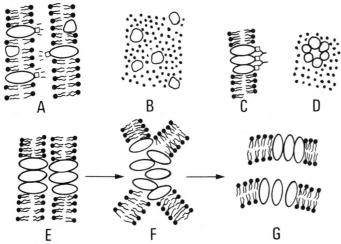

Figure 2

A. Diagram showing the normal arrangement of proteins (ovals with square carbohydrate moieties, kept apart by net negative charges) penetrating through lipid bilayers in apposed membranes. **B.** Cross section showing random arrangement of proteins (visible as particles in freeze-fracturing) among lipid molecules (black dots). **C.** Clustered membrane glycoproteins shown in section across the membrane. **D.** Clustered membrane glycoproteins in tangential section, showing between them an ion-conducting channel. **E–G.** Clustered proteins on apposed membranes interdigitating to produce membrane fusion, which spreads to the adjacent lipid bilayer regions.

Clustering of Proteins, Increased Passive Ion Flux and Virus Hemolysis

Although it has long been known that certain viruses can lyse erythrocytes and other cells, the underlying mechanism has not been established. Recently my colleagues and I have presented evidence that lysis of human erythrocytes by Sendai virus is an osmotic effect associated with greatly increased passive ion flux through the membrane and with clustering of membrane glycoproteins visualized by freeze-fracturing electron microscopy. The results will shortly be published elsewhere (Allison, Farrant and Hammond 1973; Allison, Bächi and Hammond 1973) and only the main conclusions will be given here.

Cell death can be classified into two main types, according to whether the changes are initiated in the plasma membrane or within the cell (Allison 1973b). There are three varieties of cytolysis initiated at the plasma membrane:

1. Some agents such as detergents, including lysolecithin generated by the activity of the enzyme phospholipase A, interact with plasma membrane lipids and rapidly disorganize the bilayer structure. Lysis is not osmotic and protection cannot be achieved by introducing nonpenetrating solutes into the extracellular medium.

2. Antibody and complement produce relatively large holes in plasma membranes, through which ions and relatively small molecules, but not proteins or nucleic acids, can leak. The passive ion flux exceeds the capacity of the active cation transport system to reverse it. Because of the osmotic pressure exerted by the entrapped protein, the cell swells and eventually bursts. As shown by Green et al. (1959), such osmotic lysis can be prevented by introducing into the extracellular

medium sufficient albumin to balance the osmotic pressure exerted by the intracellular protein, even though this manipulation does not prevent ion leakage into and out of the cells. Protection against complement lysis can also be achieved by dextran of high molecular weight (40,000 daltons) but not by dextran of lower molecular weight (10,000 daltons).

3. A third group of agents lyses cells by producing in the plasma membrane small holes, which allow greatly increased passive ion flux but not passage of molecules such as choline or sucrose. Lysis is again osmotic, but in this case it can be prevented by the presence in the extracellular medium of nonpenetrating solutes such as sucrose or low molecular weight dextran (10,000 daltons), which are ineffective in counteracting complement lysis.

Hemolysis by Sendai virus falls into the last-named category since it can be inhibited by the presence of sucrose in the extracellular medium. Attachment of the virus to erythrocytes is followed by fusion of the virus envelope with the plasma membrane. Virus envelope neuraminidase removes the sialic acid residues from membrane glycoproteins, thus reducing their net negative charge and facilitating their close approach to one another by translational mobility within the plane of the membrane. The virus envelope hemagglutinin then cross-links and so forms clusters of membrane glycoproteins by a process within the membrane analogous to agglutination of cells when the virus forms a bridge between the apposed plasma membranes. The clustering of membrane glycoproteins following penetration of the Sendai virus envelope is strikingly demonstrated by freeze-fracturing electron microscopy (Bächi, Aguet and Howe 1973; Allison et al. 1973b).

Within a few minutes of attachment of Sendai virus particles to human erythrocytes, the passive permeability of alkali metal cations through the plasma membrane is greatly increased, and hemolysis follows shortly afterwards unless osmotic protection is provided (Allison et al. 1973b). We have suggested that, by analogy with the effects of ionophores of the alamethicin type, the clustering of membrane glycoproteins and increased passive ion flux may be causally related, although there is at present no direct evidence that this interpretation is correct.

Immunological Control of Cancer Cell Growth

During the past decade it has been shown that most cancer cells have antigens different from those in host cells, so that they elicit an immune response. In the case of tumors induced by oncogenic viruses the antigens are relatively strong (eliciting responses comparable in magnitude to those following exposure to foreign histocompatibility antigens). The antigens are virus-specific, so that all tumors induced by an oncogenic virus, even in different organs and in different mammalian hosts, share common antigens. Probably the tumor cell antigens are coded by the virus genome, although they are distinct from virion antigens. In contrast, tumors induced by carcinogenic chemicals have individually specific antigens, although evidence is accumulating that tumors arising from the same organ (e.g., urinary bladder or gastrointestinal tract) often share antigens.

It has also been found that immunosuppression is associated with an increased incidence of certain types of cancer in man as well as in experimental animals (Allison 1973a). These observations are of theoretical interest and practical importance in medicine. From the theoretical point of view, Burnet (1970) and others have postulated that in normal animals an immunological surveillance mechanism prevents many neoplastic cells from growing into tumors, and observations on immunosuppressed patients and experimental animals provide a test of this hypothesis. From the clinical point of view, it is becoming apparent that the development of neoplasia is a major risk associated with long-term immunosuppressive therapy in kidney transplant recipients.

An example of immunological surveillance is provided by polyoma virus in mice. Polyoma virus is a small DNA virus about 45 nm in diameter, which when injected into newborn mice induces tumors in a high proportion of recipients (salivary and mammary adenocarcinomas, osteosarcomas and others). Normal adult mice are resistant to polyoma oncogenesis even though the virus multiplies extensively in host organs. Many colonies of mice, in laboratories and in the wild state, are infected with polyoma virus, as shown by serum antibodies and virus isolation. Unlike leukemogenic and mammary tumor viruses, polyoma virus is not vertically transmitted from mother to offspring. Indeed, an infected mother transmits antibodies to her offspring, which are then protected for some time against infection by polyoma virus. Only when passive protection has waned do the young animals become horizontally infected by polyoma virus from other mice in the colony. The lateness of the infection explains why tumors hardly ever arise under natural conditions.

A basic question is, therefore, why newborn animals are susceptible to oncogenesis by polyoma and other viruses whereas adult animals are resistant. Two main explanations can be considered. There might be some feature of differentiation by which adult cells are relatively insusceptible to polyoma virus transformation. Alternatively there might be an efficient immune response in the adult against the virus-specific transplantation antigens on the surface of the transformed cells, thereby preventing them from growing into a tumor, whereas the immune response is absent, weak or delayed in newborn animals.

These alternatives can be tested by the use of immunosuppression of adult mice infected by contact or inoculated with polyoma virus. The most efficient immunosuppressive procedure has been thymectomy followed by repeated injections of anti-lymphocytic globulin (ALG). As shown in Table 1, all mice so treated develop polyoma tumors after infection as adults. The importance of cell-mediated immunity in the prevention of tumor growth is emphasized by the results of restoration experiments. Immune serum from infected mice gives some protection when given 24 hours after the virus—apparently because it limits the dissemination of the infection—but when given 7 days after virus it has no protective effect. Likewise normal syngeneic lymphoid cells given after the cessation of the ALG treatment provide no protection. In contrast, transfer of lymphoid cells from syngeneic donors immunized with a polyoma tumor protects the animals very efficiently. Transfer of cells from animals immunized with a tumor induced by another virus is not effective. Treatment of the specifically sensitized lymphoid cells by anti-θ serum and complement, to destroy thymus-dependent (T) lymphocytes, abolishes the protective effect (Allison 1972).

Thus the main reason why adult mice infected with polyoma virus do not develop tumors is because they mount an effective cell-mediated immune response against the tumor cells. This response is mediated by T lymphocytes, so that the mechanism by which these lymphocytes kill tumor cells is relevant to this problem and is discussed further below.

Comparable observations on a variety of other viruses are reviewed by Allison

Table 1

Development of Tumors in CBA Mice Infected as Adults with Polyoma Virus

Preliminary treatment	Restoration at 7 weeks	No. animals	% with tumors
Normal rabbit globulin	None	24	0
Thymectomy + anti-lymphocyte globulin	None	14	100
Thymectomy + anti-lymphocyte globulin	Normal lymphoid cells	10	90
Thymectomy + anti-lymphocyte globulin	Sensitized lymphoid cells	11	0
Thymectomy + anti-lymphocyte globulin	Sensitized lymphoid cells	10	0
Thymectomy + anti-lymphocyte globulin	Sensitized lymphoid cells, treated with anti-θ serum and complement	6	83
Thymectomy + anti-lymphocyte globulin	Antibody (24 hours)	12	17
Thymectomy + anti-lymphocyte globulin	Antibody (7 days)	10	90

(1973a). Even the vertically transmitted leukemogenic viruses, which were long thought to induce tolerance, are now known to elicit immune responses, and the incidence of leukemias and lymphomas can be increased by immunosuppressive treatments.

The increasing use during the past decade of powerful immunosuppressive drugs in man has had obvious clinical benefits, of which the most notable has been kidney transplantation. However the use of long-term immunosuppression in transplantation has had complications, including severe virus infections and lymphoreticular malignancy. Penn and Starzl (1972) have reported that in a group of 7581 and 179 cardiac homograft recipients on immunosuppressive therapy until the end of 1971, 28 had lymphoreticular malignancies, 20 being reticulum cell sarcomas. The incidence of reticulum cell sarcomas in the immunosuppressed patients is at least 100 times higher than in the general population matched for age and sex, and the excess is highly significant statistically. Whether other types of malignancy are also increased in immunosuppressed subjects is currently under investigation.

The Mechanism of Lymphocyte-mediated Cytotoxicity

Various model systems have been used to demonstrate immunologically specific killing by lymphocytes of target cells in culture. One of the most popular systems has been the cytotoxicity of mouse mastocytoma cells tagged with ^{51}Cr by sensitized mouse lymphocytes. The effector cells are T lymphocytes, as shown by their susceptibility to lysis by anti-θ serum and complement, by the effectiveness of educated thymus cells in the reaction and other observations (see Cerottini and Brunner 1973). In this system killing requires contact of lymphocytes and target cells; there is no demonstrable liberation of a soluble cytotoxic factor (lymphotoxin) into the medium. Serum complement is not required for lymphocyte-mediated cytotoxicity.

The mechanism by which tumor cells are killed by lymphocytes has long been debated. Two main hypotheses have been put forward. According to the first, phospholipases from the lysosomes or some other source in the lymphocyte attack the target cell membrane, generate lysolecithin and so lyse the cell. According to the second hypothesis, the lymphocyte itself provides terminal components of the complement sequence, so that complement-like holes are produced in the target cell membrane. Our recent investigations (Allison and Ferluga 1973) provide evidence against both these hypotheses. Target cells were tagged with ^{51}Cr, which is bound to protein, and ^{86}Rb, which remains unbound and behaves like K^+. In the presence of immune lymphocytes, but not of control nonimmune lymphocytes, there was an early and marked increase in ^{86}Rb efflux from target cells and a delayed release of ^{51}Cr. When dextran of molecular weight 10,000 daltons was added to the extracellular medium, there was no significant inhibition of ^{86}Rb efflux, but a marked inhibition of ^{51}Cr release. These results show that contact of a sensitized T lymphocyte with a target cell results in selective leakiness of the plasma membrane of the target cell so that there is greatly increased passive ion flux but not of protein. Secondarily there is osmotic lysis with release of protein (Fig. 3).

The initial contact of the sensitized lymphocyte with the target cell appears to be analogous to the functional coupling described above. Sellin, Wallach and Fischer (1971) observed the passage of fluorescein from target cells to lymphocytes and vice versa. We have confirmed this finding and also observed passage of free [^{3}H] leucine from target cells to sensitized lymphocytes by radioautography; when [^{3}H]leucine was incorporated into protein it was not transferred to lymphocytes (Allison and Ferluga 1973). Thus there is interchange of small molecules, but not macromolecules, between sensitized lymphocytes and target cells; this is not seen with unsensitized lymphocytes.

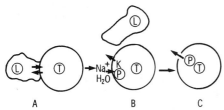

Figure 3
Diagrammatic representation of the killing by a sensitized lymphocyte (*L*) of a tumor cell (*P*). **A.** The uropod of the lymphocyte establishes functional coupling with the target cell (*T*), so that ions and small molecules, but not macromolecules, are exchanged. **B.** The lymphocyte moves away, leaving a gap junctional plaque on the tumor cell membrane. There is increased efflux of K^+ and influx of Na^+ and H_2O, but no loss of protein (*P*) at this stage. **C.** Because of osmotic pressure exerted by entrapped protein, the tumor cell swells and ruptures, releasing protein.

After a period of contact the lymphocytes become detached from target cells and appear to leave the gap junctional region on the membrane of the target cell, which accounts for the observed leakage of ions. Further evidence in support of the importance of functional coupling in target cell killing comes from observations of Asherson, Ferluga and Janossy (1973) that T lymphocytes stimulated by phytohemagglutinin kill target cells, whereas lymphocytes stimulated by concanavalin A do not. Only the former readily establish electrical coupling (Hülser and Peters 1972).

The formation of coupling contacts between tumor cells and sensitized lymphocytes, and perhaps also macrophages, may, in addition, provide a means of inhibiting the growth of the tumor cells even if they are not destroyed. As in the examples discussed above, cyclic nucleotides or other regulatory small molecules might pass to the tumor cells, so bringing about cytostatic effects.

The Possible Role of Intrinsic Proteins in Membrane Fusion

The interpretation that membranes fuse in regions consisting predominantly of lipids poses certain problems. According to the model of Lucy (1969), transition from a bilayer to a micellar configuration may be required for fusion. However there is no evidence that such a transition occurs in biological membranes under physiological conditions or that it would be compatible with membrane integrity. On physico-chemical grounds it is quite possible that the formation of even small micellar areas in the plasma membrane would lead to cytolysis. If a lesser degree of "disorder" is all that is required for fusion, the nature of the disorder has not yet been defined precisely in physico-chemical terms.

It is therefore worth reconsidering, in the light of recent advances in our understanding of the behavior of intrinsic membrane proteins, the possibility that these proteins participate in membrane fusion. Frequently intrinsic membrane proteins are glycoproteins, with carbohydrate groups containing sialic acid. The sialic acid is mainly responsible for the considerable net negative charge of the cell surface. Intrinsic proteins have lateral or (translational) mobility in the plane of the membrane, and can be brought together in clusters by antibodies or lectins which cross-link their carbohydrate moieties, or in other ways.

An instructive example of the role of proteins in deforming membranes and facilitating membrane fusion occurs during assembly and budding of enveloped viruses (see Allison 1971). The membrane becomes infiltrated with viral envelope proteins, which can combine with one another (e.g., the hemagglutinin and neuraminidase of influenza). As a result viral envelope proteins become concentrated or clustered in parts of the plasma membrane, which is distorted outwards until a spherical evagination is formed. Eventually the stalk holding the virus to the cell disappears, the margins of the viral envelope and subjacent plasma membrane fusing to reestablish continuity. In this case membrane distortion and fusion are related to the presence of a high concentration of protein in the membrane.

The question arises whether similar changes could occur with intrinsic membrane proteins. Allison et al. (1973b) have suggested that membrane glycoproteins, which above the transition temperature of the membrane lipids frequently collide as a result of lateral diffusion, are normally kept apart by the negative charges of their sialic acid moieties. If the sialic acid groups are removed, clustering of membrane proteins, e.g., by raising the ionic strength of or lowering the pH of the medium, is facilitated. A convenient way to bring about clustering of membrane proteins under physiological conditions, as shown by freeze-fracturing and etching (Bächi and Howe 1972; Bächi, Aguet and Howe 1973), is incubation of cells with Sendai virus. The hemagglutinin–neuraminidase complex splits off the sialic acid from the membrane glycoproteins and agglutinates the proteins within the membrane, so forming clusters. The clustering has two results: the first is greatly increased passive permeability of cations through the membrane, which results in hemolysis unless the osmotic effects of intracellular protein are counterbalanced by the presence in the extracellular medium of a nonpenetrating solute, as discussed above. The second result of exposure to Sendai virus is cell fusion. It is tempting to speculate that areas of clustered proteins formed in adjacent membranes are able to interdigitate, so initiating fusion which could spread laterally to include lipid bilayer regions of membrane. The general points previously discussed (Poste and Allison 1973), namely, low radius of curvature of membranes at the point of contact of two cells (perhaps the virus itself), and temporary displacement of calcium ions, may facilitate such protein interdigitation, which would not require a transition to a micellar configuration and would be compatible with the maintenance of membrane integrity through the process of fusion.

In other situations evidence is accumulating that clustering of membrane proteins is related to fusion. An example is mucocyst secretion in *Tetrahymena* (Satir, Schooley and Satir 1973). The undischarged mucocyst is a sac-like secretory vesicle limited by a unit membrane. The tip of the organelle lies beneath a special site on the plasma membrane which is shown by freeze-fracturing and etching to be a rosette of 150 Å diameter particles within the plasma membrane. During fracture the particles adhere predominantly to the inner leaflet. Matching this site, the mucocyst membrane develops adherent to its outer leaflet an annulus of 110 Å diameter particles, above the inner edge of which the rosette particles lie. During fusion a depression forms in the rosette, the particles spread outwards and fusion with the annulus follows. The fusion of the fracture faces of the mucocyst and plasma membrane is shown by the appearance of the annulus particles at the lip. Fusion of the plasma membrane and mucocyst membrane is followed by central rupture and discharge of the mucocyst. Satir et al. (1973) quote findings in their laboratory of particle arrays corresponding to the rosette above secretory organelles of several protozoa. These include rosettes above the center of the trichocyst of *Paramoecium* and the haptocyst in the heliozoon *Heliophrys* and a plaque of particles in cubical array above a haptocyst-like secretory vesicle in the helioflagellate *Celiophrys*. The presence of rosette-like aggregates of particles has also been observed in freeze-etch studies of the plasma membranes of rat pulmonary endothelial cells during pinocytosis (Smith, Ryan and Smith 1973). At the base of the developing caveolae particles come together to form rosettes and plaques which adhere predominantly to the cytoplasmic aspect of the outer leaflet. The particles move outwards to form a ring at the lips of the caveolae, which fuse to separate the endocytic vesicle from the plasma membrane.

Although further work is required before any generalizations are made, these results focus attention on the possible role of proteins in membrane fusion. Proteins might also provide specific recognition systems in apposed membranes, and the observation by freeze-etching and fracturing of clusters of proteins in gap junctions (Kreutziger 1968; Gilula, Branton and Satir 1970) and in synapses (Akert et al. 1969) raises the possibility that membrane protein interactions can have two consequences. The first is the establishment of "functional coupling" between cells, as discussed above, and the second is complete membrane fusion. The factors which determine which of these processes occur, and their biological consequences, remain to be investigated. Among the consequences may well be numbered the control of cell division.

REFERENCES

Akert, K., H. Moor, K. Pfenniger and C. Sandri. 1969. *Progr. Brain Res.* **31**:223.

Allison, A. C. 1971. The role of membranes in the replication of animal viruses. *Int. Rev. Exp. Path.* **10**:18.

———. 1972. Immunity and immunopathology of virus infections. *Ann. Inst. Pasteur* **123**:585.

———. 1973a. Immunosuppression and cancer. *Biohazards in biological research* (ed. A. Hellman et al.) p. 274. Cold Spring Harbor Laboratory.

———. 1973b. Lysosomes in cancer cells. *J. Clin. Path.* Suppl. (in press).

Allison, A. C. and J. Ferluga. 1973. Observations on the mechanism of killing of tumour cells by immune lymphocytes. *Nature* (in press).

Allison, A. C., J. Farrant and V. Hammond. 1973a. Increased plasma membrane permeability, osmotic lysis and its prevention. *Nature* (in press).

Allison, A. C., T. Bachi and V. Hammond. 1973b. Observations on the mechanism of haemolysis by Sendai virus. *Nature* (in press).

Asherson, G., J. Ferluga and G. Janossy. 1973. Nonspecific cytotoxic killing by T blasts in the presence of plant mitogens: The requirements for plant agents to reactivate T cells and to bind them to target cells. *Clin. Exp. Immunol.* (in press).

Azarnia, R. and W. R. Loewenstein. 1971. Intercellular communication and tissue growth. V. A cancer cell strain that fails to make permeable membrane junctions with normal cells. *J. Membrane Biol.* **6**:368.

Azarnia, R., W. Michalke and W. R. Loewenstein. 1973. Intercellular communication and tissue growth. VI. Failure of exchange of endogenous molecules between cancer cells with defective junctions. *J. Membrane Biol.* (in press).

Bächi, T. and C. Howe. 1972. Fusion of erythrocytes by Sendai virus studied by electron microscopy. *Proc. Soc. Exp. Biol. Med.* **141**:141.

Bächi, T., M. Aguet and C. Howe. 1973. Fusion of erythrocytes by Sendai virus studied by immunofreeze-etching. *J. Gen. Virol.* (in press).

Burnet, F. M. 1970. The concept of immunological surveillance. *Progr. Exp. Tumor Res.* **13**:1.

Cerottini, J. C. and T. Brunner. 1973. Mechanisms of cell mediated cytotoxicity. *Advances in immunology* (in press).

Fawcett, D. W. 1971. Observations on cell differentiation and organelle continuity in spermatogenesis. Proc. Int. Symp., *The genetics of the spermatozoon* (ed. R. A. Beatty and S. Gluecksohn-Waelsch) p. 37. University Press, Edinburgh.

Gilula, N. B., D. Branton and P. Satir. 1970. The septate junction, a structural basis for intercellular coupling. *Proc. Nat. Acad. Sci.* **67**:213.

Gilula, N. B., O. R. Reeves and A. Steinbach. 1972. Intercellular communication: Metabolic coupling, ionic coupling and cell contacts. *Nature* **235**:262.

Green, H., R. A. Fleisher, P. Barrow and B. Goldberg. 1959. The cytotoxic action of immune gamma globulin and complement on Krebs ascites tumor cells. II. Chemical studies. *J. Exp. Med.* **109**:511.

Harris, H. 1968. *Nucleus and cytoplasm.* University Press, Oxford.

Hülser, D. F. and J. H. Peters. 1972. Contact co-operation in stimulated lymphocytes.

II. Electrophysiological investigations on intercellular communication. *Exp. Cell Res.* **74:**319.

Kreutzger, G. O. 1968. Freeze-etching of intercellular junctions of mouse liver. Proc. 25th Meeting of the Electron Microscope Society of America, San Francisco, p. 234.

Läuger, P. 1972. Carrier mediated ion transport. *Science* **178:**24.

Loewenstein, W. R. 1973. Membrane junctions in growth and differentiation. *Fed. Proc.* **32:**60.

Lucy, J. A. 1969. Lysosomal membranes. *Lysosomes in biology and pathology,* (ed. J. T. Dingle and H. B. Fell) vol. 2, p. 313. North-Holland, Amsterdam.

Penn, I. and T. Starzl. 1972. Malignant tumors arising *de novo* in immunosuppressed organ transplant recipients. *Transplantation* **14:**407.

Poste, G. and A. C. Allison. 1973. The membrane fusion reaction. *Biochim. Biophys. Acta* (in press).

Rao, U. N. and G. T. Johnson. 1970. Mammalian cell fusion studies on the regulation of DNA synthesis and mitosis. *Nature* **255:**159.

Reese, T. S., M. V. L. Bennett and N. Feder. 1971. Cell to cell movement of peroxidases injected into the septate axon of the crayfish. *Anat. Rec.* **169:**409.

Satir, P. and N. B. Gilula. 1973. The fine structure of membranes and intercellular communication in insects. *Ann. Rev. Entomol.* **18:**143.

Satir, B., C. Schooley and P. Satir. 1973. Membrane fusion in a model system. Mucocyst secretion in *Tetrahymena. J. Cell Biol.* **56:**153.

Sei-Ichi, M. 1972. Induction of prophase in interphase nuclei by fusion with metaphase cells. *J. Cell Biol.* **54:**120.

Sellin, D., D. F. H. Wallach and H. Fischer. 1971. Intercellular communication in cell-mediated cytotoxicity. Fluorescein transfer between H–2[a] target cells and H–2[b] lymphocytes *in vitro. Eur. J. Immunol.* **1:**453.

Smith, U., J. W. Ryan and D. S. Smith. 1973. Freeze-etch studies of the plasma membrane of pulmonary endothelial cells. *J. Cell Biol.* **56:**492.

Subak-Sharpe, J. H., R. R. Bürk and J. D. Pitts. 1966. Metabolic co-operation between biochemically marked mammalian cells in tissue culture. *J. Cell Sci.* **4:**353.

Growth Behavior, Contact, and Surface Structure of Cells

S. Hakomori, C. G. Gahmberg, R. Laine, and S. Kijimoto*

Departments of Pathobiology and Microbiology
School of Public Health and School of Medicine
University of Washington, Seattle, Washington 98195

Various growth behaviors of animal cells in vitro are regarded as reflections of the physiological and pathological status of cells in vivo. A possible growth control of animal cells through the surface membrane has been suggested by contact inhibition of cell movement (Abercrombie and Heaysmann 1954) and of cell replication (Dulbecco 1970), susceptibility of cell growth to various serum factors (Todaro, Lazar and Green 1965; Holley and Kiernan 1968; Holley 1972), and stimulation of cell growth by insulin, sialidase (Vaheri, Ruoslahti and Nordling 1972), trypsin (Burger 1970), and by various mitogenic lectins (Nowell 1960; Krüpe et al. 1968).

It has been our primary interest to determine whether any behavioral changes of cells are accompanied by change of surface structure with regard to glycolipid or glycoproteins. To answer this question, several approaches have been taken. The first is a simple comparison of glycolipid composition between genetically related cells with different physiological functions, such as normal versus transformed cells or cells at growing phase versus nongrowing phase. This approach is based on the partial assessment that glycolipid is a component of cell surface membranes (Renkonen et al. 1970; Klenk and Choppin 1970). Although this approach is time-consuming and requires large amounts of cells, it has become increasingly popular since a significant difference in glycolipid composition between normal and transformed cells was noticed (Hakomori and Murakami 1968). Several works from independent laboratories agree that the change is directed towards incomplete synthesis or simplification (Mora et al. 1969; Cheema et al. 1970; Sakiyama, Gross and Robbins 1972; Critchley and Macpherson 1973; Diringer, Strobel and Koch 1972), although in some transformed clones the change is in a different direction (Diringer et al. 1972; Yogeeswaran et al. 1972). The glycolipid changes seem to be indispensable during the process of malignant transformation, as reversible changes have been demonstrated in various cells transformed with various temperature-sensitive mutants of Rous sarcoma virus (unpublished cooperative

---

* Present address: Cancer Research Institute, Hokkaido University, Sapporo, Japan.

work with Drs. J. Wyke and P. Vogt) and with temperature-sensitive polyoma virus (cooperative work with D. Kiehn). Of particular interest, along with this first approach, are the cell contact-dependent changes of glycolipid synthesis (Hakomori 1970; Robbins and Macpherson 1971). The results of our most recent work are described briefly in this paper.

The second approach is the determination of reactivity of cell surface glycolipids to their specific antibodies. Comparison has been made between cells at different growth stages and their physiological functions, as in the first approach. The results of this second approach are also briefly described. The third approach is the comparison of surface labeling patterns with galactose oxidase and tritiated sodium borohydride, followed by gel electrophoresis and thin-layer chromatography.

Recently the change in growth behavior of tumor cells in vitro to that of normal cells in vitro upon addition of dextran sulfate (Goto and Sato 1972) or cyclic 3'5'-adenosine monophosphate (cyclic AMP) (Hsie and Puck 1971; Johnson, Friedman and Pastan 1971) and restored contact orientation with galactose (Kalckar et al. 1973) have been reported. These cells showing reverted growth behavior are also of interest for comparison of surface structures.

Cell Contact-dependent Changes of Glycolipid Synthesis

The chemical quantities of some glycolipids increase when the growth rate of cells decreases on cell-to-cell contact. The cell density-dependent increase of the concentration of ceramide trihexoside in some BHK cells and of disialohematoside in human 8166 diploid cells was reported first (Hakomori 1970). Subsequently a similar increase of hematoside and of various neutral glycolipids, such as globoside and Forssman glycolipids, in various clones of NIL cells has been reported (Robbins and Macpherson 1971; Sakiyama et al. 1972; Critchley and Macpherson 1973). The net synthesis of glycolipids as determined by the amount of galactose incorporated into various glycolipids of NIL cells at different population densities was compared. The enhanced synthesis of hematoside and Forssman glycolipids in NIL2E cells and in NIL2K cells was observed at a relatively early stage of confluency ($1–1.4 \times 10^5/cm^2$ for NIL2E and $0.4–0.6 \times 10^5/cm^2$ for NIL2K cells), and the synthesis rate decreased at complete confluency ($2 \times 10^5/cm^2$ for NIL2E and $0.7–0.9 \times 10^5/cm^2$ for NIL2K cells) (Hakomori and Kijimoto 1972). The enhanced synthesis of Forssman glycolipid antigen on cell contact was demonstrated by increased reactivity of the glycolipid fraction obtained from confluent cells against anti-Forssman antiserum (Kijimoto and Hakomori, 1972). The transformed cells (PyNIL2E) had a very low concentration of higher glycolipids, such as globoside and Forssman, and the immunological reactivity of Forssman glycolipids was not demonstrated. Synthesis of glycolipids of polyoma-transformed cells as determined by galactose incorporation into higher neutral glycolipids was low, and there was no enhancement on increase of cell population density (Fig. 1).

A similar observation with NIL cells was described by Robbins and Macpherson (1971) and by Sakiyama, Gross and Robbins (1972). Critchley and Macpherson (1973) described that glycolipids of sparse quiescent NIL cells were different from confluent and sparse growing cells, suggesting that cell-cell contact may be necessary for change of glycolipid components.

Enhanced glycolipid synthesis on cell contact was studied with cell-free systems by Kijimoto and Hakomori (1971, 1972). Activity of the enzyme for synthesis

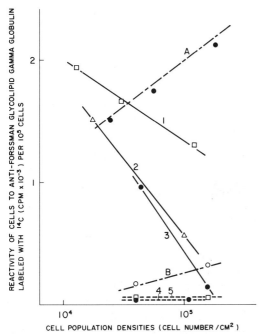

Figure 1
Reactivity of cell surface Forssman antigen in various states of NIL cells and their dependency on cell population densities. **A,** NIL2K cells, separated from monolayer culture by EDTA treatment and then reacted with anti-Forssman antibody. **B,** Polyoma-transformed NIL2E cells, separated from the culture by EDTA treatment. Lines 1, 2 and 3, reactivities of NIL2E, NIL2K and PyNIL2E cells, respectively, grown on glass surfaces and measured as such.

of ceramide trihexoside (UDP-Gal:lactosylceramide-α-galactosyltransferase) enhanced two- to threefold in hamster BHK and NIL cells when the growth of these cells was inhibited at higher cell population densities, as compared to the same enzyme activity of the same cells at lower cell population density. The activity of this enzyme in polyoma-transformed NIL2E and BHK cells was only 10–50% of the activity of growing NIL and BHK cells. The activity of the enzyme for synthesis of lactosylceramide (UDP-Gal:glucosylceramide-β-galactosyltransferase) did not increase as cell population density increased and was not affected by transformation. The activity of α-galactoside was higher in polyoma-transformed cells; thus cell density-dependent synthesis was demonstrated on an enzyme basis, and regulation of the activity of such an enzyme on cell surfaces was assumed to play a key role in defining cell growth behavior. A similar study on the relation of enzymatic activity for synthesis of GM2 ganglioside (UDP-GalNAc:hematoside-N-acetyl-galactosyltransferase) and contact inhibition in mouse fibroblastic cells was not successful (Fischmann et al. 1972). Cell population density-dependent enhancement of the concentration of ganglioside was demonstrated previously in human 8166 cells (Hakomori 1970). However it was not demonstrated in mouse 3T3 cells (Yogeeswaran et al. 1972). Further work on density-dependent chemical quantity of gangliosides is being carried out by us, as we compare actively growing sparse 3T3 cells, sparse but quiescent cells, and confluent contact-inhibited cells. Sparse quies-

Table 1

Ganglioside Pattern of 3T3 Cells at Different Population Densities

	Sparse growing	Sparse quiescent	Confluent
GM3	+++	+++	+++
GM2	+	+	+
GM1	++	++	++
GD1a	±	±	++
GD1b	±	±	±
GT	±	±	±

cent cells were obtained according to Seifert and Paul (1972) using 1% calf serum. The ganglioside pattern, especially the presence or absence of disialoganglioside (GD1a), was noticed to depend on cell density (see Table 1). It is interesting that sparse quiescent cells showed an identical pattern to sparse growing cells but were quite distinctive from confluent cells. This finding suggests that change of ganglioside or glycolipids in general requires cell-to-cell contact but is not simply induced by lower serum concentration. Cell contact which induces enhanced synthesis of some carbohydrate components on cell surfaces may indeed be necessary to change the surface structure, which in turn changes membrane conformation. The changed membrane conformation may initiate a series of membrane-bound enzyme changes, such as enzymes regulating the level of cyclic AMP and protein kinase to inhibit S phase initiation.

Contact-dependent Changes of Glycolipid Reactivity in Membrane

Whereas the net synthesis of Forssman hapten glycolipids is enhanced in contact-inhibited NIL cells, particularly at the early stage of contact inhibition, the surface reactivity of cells to anti-Forssman glycolipid antiserum is significantly decreased when they are contact-inhibited.

Forssman reactivity of NIL cells greatly decreased at confluency (Fig. 1), although net synthesis and the amount of Forssman hapten increased, as described in the previous section. Cells from confluent culture were, however, not reactive for growing cell cultures when previously treated with 1% EDTA (Fig. 1, lines A and B). In other words, the density-dependent decrease of cell reactivity to anti-Forssman antibody was not observed when the cells were treated with EDTA. The reactivity of polyoma-transformed cells was very low and not dependent on cell density (Fig. 1, lines 4 and 5). This accords with the small amount of Forssman glycolipids, as mentioned in the previous section. It is interesting, however, that the reactivity of polyoma-transformed cells separated from the culture by 1% EDTA showed low activity, which increased at higher cell density (Fig. 1, line B) (Hakomori and Kijimoto, 1972).

Because the amount of net synthesis of Forssman glycolipid increased at the confluent phase, when Forssman reactivity of cells decreased, it seems that most of the Forssman glycolipid on cell surfaces is masked at confluency. EDTA may either remove or alter the material that masks Forssman antigen, or further, the membrane may rearrange when the cells are detached. Another possibility is that the antigen is blocked by a complementary receptor site of counterpart cells (either a cell surface enzyme as hypothesized by Roth and White 1972 and Roseman 1970, or an as yet unidentified protein component). In either case the antigen is expressed by EDTA and becomes cryptic in the confluent state. A similar phenomenon could be found for cell surface hematoside, as the chemical quantity, immunogenic activity, and metabolic activity of this glycolipid are valuable cell-to-cell contact and for malignant transformation.

Change of Surface Labeling Pattern and Altered Growth Behavior of Malignant Cells

If growth control is indeed related to and governed by cell surface structure, could the change of surface structure be detected when growth behavior of the cells is changed, even temporarily? A number of studies in the past on transformation-dependent changes of glycolipids and glycoproteins suggest this possibility, but it is not definite. Recently surface labeling of galactosyl and galactosaminyl residues in surface glycoproteins and glycolipids has been developed (Gahmberg and Hakomori 1973). Also evidence shows that contact inhibitability of malignant cells can be restored by growing them under suitable conditions. Goto and Sato (1972) observed that transformed cells became contact-inhibitable when cultured in the presence of dextran sulfate. Remarkable changes in morphology (Hsie and Puck 1971; Johnson et al. 1971) and growth (Sheppard 1971) have also been caused by cyclic 3'5'-adenosine monophosphate (cyclic AMP). In either dextran sulfate or cyclic AMP, cells originally overlapping and disoriented became properly oriented and contact sensitive. In addition Kalckar et al. (1973) described that contact inhibitability increased and that contact orientation of PyNIL cells appeared when NIL and PyNIL cells were cultured in medium in which glucose was replaced with galactose. A difference in surface labeling patterns between normal growth and transformed cells was clearly demonstrated when cells were treated with neuraminidase. PyNIL cells contained the enhanced peak "d." The peak "f," which is also enhanced in transformed cells, is the nonspecific label occurring also in the extract of cells without galactose oxidase (Fig. 2). (The nature of this "f" peak is as yet unidentified.) If confluent NIL cells were treated with trypsin and then labeled, there appeared a greatly enhanced peak "d," which was not markedly labeled without trypsinization.

Confluent 3T3 cells had a remarkably different label pattern from growing 3T3 cells, which had a protein pattern similar to 3T3 cells transformed with SV40 virus (Fig. 3). The protein label of transformed cells changed towards that of normal cells when the cells were grown in the presence of dibutyryl cyclic AMP.

The relative activity of peak "b" was greatly enhanced when contact inhibitability and orientation normalized in the presence of dextran sulfate (Fig. 2E) and dibutyryl cyclic AMP (Fig. 2H, cf. 2C, 2D). The glycoprotein label profile significantly changed in PyNIL cells grown in galactose-replaced medium (2F), which showed obviously different cell contact orientation as described previously (Kalckar et al. 1973).

The relation of carbohydrate structure to their label and the effect of cell contact on the degree of labeling were clearly seen in the labeling pattern of glycolipids. As expected neutral glycolipids of NIL cells were strongly labeled. The label in higher glycolipids, such as globoside or Forssman glycolipids, decreased, and the proportion of label in simpler glycolipids greatly increased in PyNIL cells in agreement

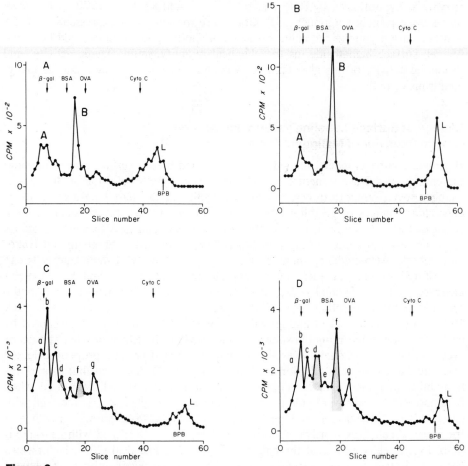

Figure 2
Sodium dodecyl sulfate-polyacrylamide gel electrophoresis of labeled NIL cells. Cells were harvested with EDTA and then processed in different ways: **A,** Confluent NIL cells labeled without neuraminidase treatment; **B,** PyNIL cells labeled without neuraminidase treatment; **C,** confluent NIL cells treated with neuraminidase, then labeled; **D,** PyNIL cells treated with neuraminidase, then labeled; **E,** PyNIL cells grown in the presence of 4 μg/ml dextran sulfate, treated with neuraminidase, then labeled; **F,** PyNIL cells grown

with the change of chemical composition as described in the first section. A significant increase of globoside label and decrease of the label in simpler glycolipids were observed in PyNIL cells grown in galactose medium when contact orientation was restored. Similarly the label in Forssman glycolipid significantly increased in PyNIL cells when contact inhibitability was restored and when contact orientation became obvious by either dextran sulfate or dibutyryl cyclic AMP.

COMMENTS

Knowledge of biochemical events occurring on cell-to-cell contact is still fragmentary, but three facts are indicated by these studies: (1) change of glycolipid

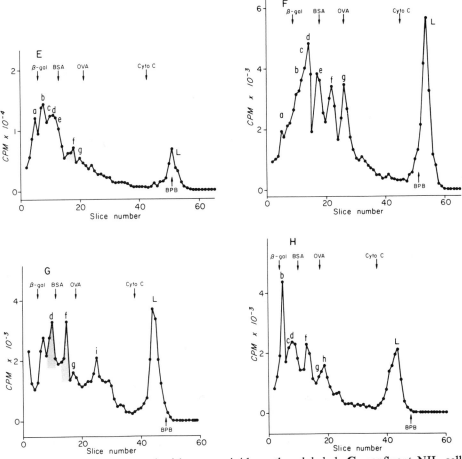

in galactose medium, treated with neuraminidase, then labeled; **G,** confluent NIL cells treated with trypsin, then with neuraminidase, then labeled; **H,** PyNIL cells grown in the presence of dibutyryl cyclic AMP, treated with neuraminidase, then labeled.

β-gal = β-galactosidase; BSA = bovine serum albumin; OVA = ovalbumin; CytoC = cytochrome C; BPB = bromophenol blue tracking dye. Shading under peaks *d* and *f* indicates a change on transformation and on reverted cell behavior. Peaks *B* and *f* contained a nonspecific label.

composition and elongation of the carbohydrate chain ("contact extension") (Hakomori 1970); (2) change of the reactivity of the carbohydrate hapten on cell surfaces, especially the disappearance of reactivity on cell contact and reappearance on EDTA treatment (hereby called "contact masking") (Hakomori and Kijimoto 1972); and (3) structural change of the surface glycoproteins as revealed by surface labeling with galactose oxidase (Gahmberg and Hakomori 1973).

Both "contact extension" and "contact masking" and any other structural changes detectable by surface labeling may induce the changes of membrane conformation and then significantly alter the membrane-bound enzyme system, which affects the level of intracellular activity of enzymes regulating cell growth. Disorganization of these surface contact functions (disfunctional contact) is clearly seen in a variety of transformed cells.

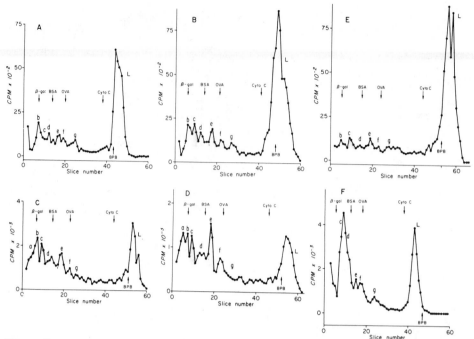

Figure 3

Sodium dodecyl sulfate-polyacrylamide gel electrophoresis of labeled 3T3 cells and their transformants. Cells were harvested with EDTA. **A,** Confluent 3T3 cells treated with neuraminidase, then labeled; **B,** growing 3T3 cells treated with neuraminidase, then labeled; **C,** SV3T3 cells treated with neuraminidase, then labeled; **D,** SV/Py3T3 cells treated with neuraminidase, then labeled; **E,** confluent 3T3 cells treated with trypsin, then with neuraminidase, then labeled; **F,** SV/Py3T3 cells grown in the presence of dibutyryl cyclic AMP, treated with neuraminidase, then labeled. Markers are the same as for Fig. 2. Shading under peaks *c, e,* and *a* indicates a change on transformation and on reverted cell behavior. The peak *e* contained a nonspecific label.

It is indeed noteworthy that the surface labeling profile of transformed cells shifted towards normal when the contact response of the transformed cells was altered by modified culture conditions, suggesting that contact response of cells is due to surface structural changes through which cells can recognize each other. It is now clearly demonstrated that malignant transformed cells have a unique surface structure, which must be the basis of their abnormal behavior, as the unique glycoprotein and glycolipid profile was no longer detectable when cell growth behavior of malignant cells changed towards normal.

The change in surface structure associated with contact inhibitability and orientation, previously described in the foregoing section, is now clearly demonstrated by the surface labeling technique, and all evidence points to the possibility that growth behavior of cells is governed by surface structure and that altered growth behavior is due to the altered surface structure, which leads to disfunctional cell contact and failure of cell-cell recognition. The cell-cell recognition could be between the carbohydrate structure and a certain complementary protein structure on the counterpart cell. These two structures should therefore be present on the same cell surface.

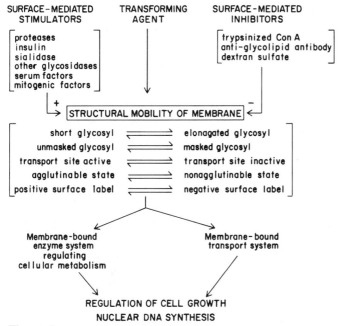

Figure 4

Structural mobility of membrane as regulatory machinery for cell proliferation. The transitional states of cell membrane structures as detected by various methods depend on cell cycles or cell contact. Such changes may be instrumental in regulating a membrane-bound enzyme that in turn regulates cellular metabolism, such as adenylcyclase, phosphodiesterase, protein kinase, etc. On the other hand, the change of surface membrane can be coupled to the change of the transport system. By either mechanism, the structural mobility of membrane directs cell growth or regulates nuclear DNA synthesis.

Such a complementary structure may be instrumental in coding a signal to inhibit the initiation of S phase.

A number of surface membrane changes associated with malignant transformation, such as agglutinability by lectin, incomplete synthesis of glycolipid, surface labeling pattern, etc., are in fact transitional changes of membrane, depending on cell cycle and cell contact. Such structural motility of the surface membrane as shown in Fig. 4 must be instrumental in the regulation of a membrane-bound enzyme system, which in turn regulates cellular metabolic pattern or directs the activity of nutrient transport. Tumor cells are characterized by frozen membrane motility, showing no obvious changes in membrane structure during the cell cycle. This frozen motility of the membrane must be the cause of unregulated tumor cell growth.

Acknowledgments

This work was supported by National Cancer Institute research grants CA12710 and CA10909 and by American Cancer Society grant BC-9C. C.G.G. is supported by International Fogarty Center research fellowship #1F05TW01885-01.

Note added in proof: Increase of GD1a ganglioside in confluent 3T3 cells (see Table 1) is more remarkable at the early stage of confluency as compared to later stages (Yogeeswaran and Hakomori unpubl.).

REFERENCES

Abercrombie, M. and J. E. M. Heaysmann. 1954. Observations on the social behavior of cells in tissue culture. *Exp. Cell Res.* **6:**293.

Burger, M. M. 1970. Proteolytic enzymes initiating cell division and escape from contact inhibition and growth. *Nature* **227:**170.

Cheema, P., G. Yogeeswaran, P. Morris and R. K. Murray. 1970. Ganglioside patterns of three minimal deviation hepatomas. *Fed. Eur. Biochem. Soc. Lett.* **11:**180.

Critchley, D. R. and I. Macpherson. 1973. Cell density dependent glycolipids in NIL2 hamster cells derived from malignant and transformed cell lines. *Biochim. Biophys. Acta* **296:**145.

Diringer, H., G. Strobel and M. S. Koch. 1972. Glycolipids of mouse fibroblasts and virus transformed mouse cell lines. *Hoppe-Seyler's Z. Physiol. Chem.* **353:**1769.

Dulbecco, R. 1970. Topoinhibition and serum requirement of transformed and untransformed cells. *Nature* **227:**802.

Fischmann, P. H., V. MacFarland, P. Mora and R.O. Brady. 1972. Ganglioside biosynthesis in mouse cells: Glycosyltransferase activities in normal and virally transformed lines. *Biochem. Biophys. Res. Commun.* **48:**48.

Gahmberg, C. G. and S. Hakomori. 1973. External labelling of cell surface galactose and galactosamine in glycolipid and glycoproteins of human erythrocytes. *J. Biol. Chem.* **248:**4311.

Goto, K. and H. Sato. 1972. Decrease of saturation density in cultured tumor cells by dextran sulfate. *Gann* **63:**371.

Hakomori, S. 1970. Cell density dependent changes of glycolipid concentrations in fibroblasts and loss of this response in virus transformed cells. *Proc. Nat. Acad. Sci.* **67:**1741.

Hakomori, S. and S. Kijimoto. 1972. Forssman reactivity and cell contacts in cultured hamster cells. *Nature New Biol.* **239:**87.

Hakomori, S. and W. T. Murakami. 1968. Glycolipids of hamster fibroblasts and derived malignant cell lines. *Proc. Nat. Acad. Sci.* **59:**254.

Holley, R. W. 1972. A unifying hypothesis concerning the nature of malignant growth. *Proc. Nat. Acad. Sci.* **69:**2840.

Holley, R. W. and J. A. Kiernan. 1968. "Contact inhibition" of cell division in 3T3 cells. *Proc. Nat. Acad. Sci.* **60:**300.

Hsie, A. W. and T. T. Puck. 1971. Further changes in differentiation state accompanying the conversion of Chinese hamster cells to fibroblastic form by dibutyryl adenosine cyclic 3'5'-monophosphate and hormones. *Proc. Nat. Acad. Sci.* **68:**1648.

Johnson, G. S., R. M. Friedman and I. Pastan. 1971. Restoration of several morphological characteristics of normal fibroblasts in sarcoma cells treated with adenosine 3'5'-monophosphate and its derivatives. *Proc. Nat. Acad. Sci.* **68:**425.

Kalckar, H. M., D. Ullrey, S. Kijimoto and S. Hakomori. 1973. Carbohydrate catabolism and the enhancement of uptake of galactose in hamster cells transformed by polyoma virus. *Proc. Nat. Acad. Sci.* **70:**839.

Kijimoto, S. and S. Hakomori. 1971. Enhanced glycolipid: α-galactosyltransferase activity in contact-inhibited hamster cells and loss of this response in polyoma transformants. *Biochem. Biophys. Res. Commun.* **44:**557.

——— 1972. Contact-dependent enhancement of net synthesis of Forssman glycolipid and hematoside in NIL cells at the early stage of cell-to-cell contact. *FEBS Letters* **25**:38.

Klenk, H. D. and P. W. Choppin. 1970. Glycosphingolipids of plasma membranes of cultured cells and an enveloped virus grown in these cells. *Proc. Nat. Acad. Sci.* **66**:57.

Krüpe, M., W. Wirth, D. Nies and A. Ensgraben. 1968. Untersuchungen über die mitogene Wirkung agglutininhaltiger Extrakte aus verschiedenen Pflanzen auf die kleinen Lymphozyten des peripheren Blutes. *Z. Immunitätsforsch.* **135**:19.

Mora, P., R. O. Brady, R. M. Bradley and V. W. McFarland. 1969. Gangliosides in DNA virus transformed and spontaneously transformed tumorigenic mouse cell lines. *Proc. Nat. Acad. Sci.* **63**:1290.

Nowell, P. C. 1960. Phytohemagglutinins: An initiator of mitosis in cultures of normal human leukocytes. *Cancer Res.* **20**:462.

Renkonen, O., C. G. Gahmberg, K. Simons and L. Kääriäinen. 1970. Enrichment of gangliosides in plasma membranes of hamster kidney fibroblasts. *Acta Chem. Scand.* **24**:733.

Robbins, P. W. and I. Macpherson. 1971. Glycolipid synthesis in normal and transformed animal cells. *Proc. Roy. Soc. London* B **177**:49.

Roseman, S. 1970. The synthesis of complex carbohydrates by multiglycosyltransferase systems and their potential function in intercellular adhesion. *Chem. Phys. Lipids* **5**:270.

Roth, S. and D. White. 1972. Intercellular contact and cell surface galactosyltransferase activity. *Proc. Nat. Acad. Sci.* **69**:485.

Sakiyama, H., S. K. Gross and P. W. Robbins. 1972. Glycolipid synthesis in normal and virus transformed hamster cell lines. *Proc. Nat. Acad. Sci.* **69**:872.

Seifert, W. and D. Paul. 1972. Level of cyclic AMP in sparse and dense cultures of growing and quiescent 3T3 cells. *Nature* **240**:281.

Sheppard, J. R. 1971. Restoration of contact-inhibited growth to transformed cells by dibutyryl adenosine 3′5′-monophosphate. *Proc. Nat. Acad. Sci.* **68**:1316.

Todaro, G., G. K. Lazar and H. J. Green. 1965. Initiation of cell division in a contact-inhibited mammalian cell line. *J. Cell. Comp. Phsyiol.* **66**:325.

Vaheri, A., E. Ruoslahti and S. Nordling. 1972. Sugar uptake and cell division: Stimulation by neuraminidase, insulin, trypsin, and serum. *Nature New Biol.* **238**:211.

Yogeeswaran, G., R. Sheinin, J. Wherrett and R. K. Murray. 1972. Studies on the glycosphingolipids of normal and virally transformed 3T3 mouse cells. *J. Biol. Chem.* **244**:5146.

Cell Cycle Synthesis of Glycolipids Including the Forssman Antigen

Barbara Anne Wolf and P. W. Robbins

Department of Biology, Massachusetts Institute of Technology
Cambridge, Massachusetts 02139

Several laboratories have studied the metabolism of glycolipids in normal and transformed cells. A model has been presented that transformed cells are blocked in the synthesis of the more complex glycolipids. However in some cell lines the glycolipid pattern of transformed cells resembles the pattern observed in rapidly growing normal cells. We were interested in complementing the study of the metabolism of these glycolipids during the growth cycle with a study of their synthesis during the cell cycle.

Unfortunately previous studies on the synthesis of membranes during the cell cycle are conflicting. Some of the conflicts may be due to a loss of synchrony, to perturbations caused by the method of synchronization, to variations among cells in the composition of the membranes, or to variations in the composition of the membranes within a cell. Most experiments employ radioactive labeling rates as a probe for synthesis without any attempt to measure possible turnover, conversion of label, pool sizes, or transport rates. This is especially important in light of recent experiments showing that the transport of uridine, thymidine and an amino acid analog varies with cell cycle (Sander and Pardee 1972). Although membrane fractions were isolated in some studies, no attempt was made to purify individual components. In summary, all types of control have been observed, including absence of regulation (Warmsley, Phillips and Pasternak 1970), synthesis of all components during a limited time in the cell cycle (Glick, Gerner and Warren 1971), and synthesis of different components during different phases (Bosmann and Winston 1970).

On the other hand, studies of the cellular surface architecture during the cell cycle have been consistent. The most active period of the cell cycle seems to be mitosis. Antigenic "sites," which are unreactive during S, become reactive during mitosis (Fox, Sheppard and Burger 1971; Cikes, Friberg and Klein 1972; Pasternak, Warmsley and Thomas 1971). Often increased reactivity is also observed in G_1 and G_2 but it is not clear whether this reflects either a unique state of surface architecture present in G_1 or G_2, or a contaminating population of mitotic cells. At mitosis cells are more susceptible to virus-induced fusion (Stadler and Adelberg

1972), release half the surface-bound heparin sulfate (Kraemer and Tobey 1972), have increased electrophoretic mobility (Kraemer 1967), and permanently expose newly synthesized glycoproteins (Onodera and Sheinin 1970; Glick and Buck 1973).

Recently the structure of the Forssman antigen was clarified (see Hakomori et al. this volume). It is one of the neutral glycolipids present in the NIL hamster line, and the reactivity, synthesis and chemical amounts of this antigen have been found to vary with cell density. In many ways the Forssman antigen is similar to the lectin agglutination sites, which have been extensively studied. Like lectin agglutination sites, this antigen is more readily detected in virally transformed cells (Fogel and Sachs 1964; O'Neill 1968; Robertson and Black 1969) or upon trypsin treatment of normal cells (Burger 1972). Since the wheat germ lectin binding site is one of the determinants exposed during the cell cycle (Fox et al. 1971) and because the Forssman antigen resembles the lectin agglutination sites, we were interested in the behavior of this antigen during the cell cycle.

METHODS AND RESULTS

Synchronization

Clone 2Cl of the NIL hamster line was isolated and characterized by others (Saki-yama et al. 1972) and used 6–10 passages after clonal isolation. Cells were synchronized with a double thymidine block (Tobey et al. 1966) and mitotic collection from monolayer with colcemide (Peterson, Anderson and Tobey 1968). Details and characterization of synchrony will be published elsewhere.

Analysis of Lipid

Cells were labeled for 4 hours with 10 μCi [1-^{14}C]palmitate (59 mCi/mmole) adsorbed to serum (Sakiyama et al. 1972) or with 25 μCi [1-^{14}C]glucosamine-HCl (58 mCi/mmole). Dry cell pellets were extracted with chloroform:methanol 2:1 and analyzed by two-dimensional thin layer chromatography (Gray 1967). Palmitate data is expressed as counts in glycolipid divided by counts in phospholipids plus glycolipids.

The structure and nomenclature of the glycolipids studied are given in Table 1. Unidentified spots were preliminarily classified by [^{32}P]phosphoric acid and [^{14}C]glucosamine labeling but were not further characterized.

Table 1
Glycolipid Nomenclature

Ceramide-Glu	
-Gal	GL-1
Ceramide-Glu-Gal	GL-2
Ceramide-Glu-Gal-Gal	GL-3
Ceramide-Glu-Gal-Gal-GalNAc	GL-4
Ceramide-Glu-Gal-Gal-GalNAc-GalNAc	GL-5
Ceramide-Glu-Gal-NANA	
-NGNA	GM$_3$

Table 2

Incorporation of [^{14}C]Palmitate into Glycolipids

	M	G_1	S			G_2 M	S	G_2 M
			Labeling time in hours					
	0–4	*4–8*	*8–12*	*12–16*	*16–19*	*0–4*	*4–8*
GL-1	1.70	1.50	1.20	1.30	.94	1.70	1.90
GL-2	.03	.04	.03	.03	.03	.04	.02
GL-3*	.08	.06	ND†	ND	ND	<.01	<.01
GL-4	.37	.45	.31	.07	.16	.06	.06
GL-5	.47	.61	.58	.67	.41	.28	.22
GM$_3$	1.05	1.20	1.18	1.15	1.00	1.06	1.20

Numbers are expressed as incorporation into glycolipid divided by incorporation into phospholipids plus glycolipids.

* Only 2 experiments

† Not done

The uptake of [^{14}C]palmitate into glycolipids during 4-hour pulses in shown in Table 2. GL-1, GL-2, and GM$_3$ showed no preferential labeling during any phase of the cell cycle, either in individual experiments or in the average of six experiments. GL-3 represented so few counts compared with a phospholipid that chromatographed near GL-3 that it was difficult to find in most radioautograms. However two experiments exposed for longer periods of time showed labeling of GL-3 only in G$_1$. These observations are suggestive of synthesis of GL-3 only during G$_1$ and experiments are in progress to confirm these observations using a galactose label. Similarly GL-4 showed preferential labeling during G$_1$ and early S. In contrast to the other lipids, the labeling of GL-5 was quite variable and no systematic pattern was observed.

In order to clarify the labeling pattern of GL-5 and to confirm the results with the other glycolipids, synchronized cells were labeled with [^{14}C]glucosamine. Incorporation into the glycolipid represented between 1–2% of the total at all times. Approximately equal incorporation of glucosamine into GM$_3$ occurred during all labeling periods (see Table 3). Incorporation into GL-4 in G$_1$ and early S was

Table 3

Incorporation of [^{14}C]Glucosamine into Glycolipids

	M	G_1	S	G_2 M
		Labeling time in hours		
	0–4	*4–8*	*8–12*	*12–16*
GL-4	.52	.62	.22	.17
GL-5	4.7	5.3	2.8	2.1
GM$_3$	.48	.35	.16	.44

Numbers are expressed as incorporation into glycolipids per μg protein.

higher than in late S and G_2. GL-5 incorporation followed the pattern of total glycolipid incorporation. GL-5 was labeled at all phases but decreased in incorporation in late S and G_2.

Inhibition of Hemolysis by Cell Extracts

The assay used was a modification of the complement fixation assay. Rabbit anti-sheep blood cells were preincubated with antigen before addition of sheep red blood cells plus guinea pig complement. Hemoglobin release was measured spectrophotometrically. Antiserum and crude antigen standards were always included. No lysis occurred without antibody and no anti-complement activity was observed. This assay is highly specific and reacts only with GL-5 (data not shown). It is also highly sensitive, detecting picomoles of antigen. In Table 4 the total amount of antigen per cell in lipid extracts averaged over four determinations is shown. In agreement with radiolabeling data, the amount of antigen gradually increased 2- to 3-fold during the cell cycle.

Binding of Radioactive Antibody to Whole Cells

Rabbit antiserum prepared against purified Forssman glycolipid, kindly supplied by Dr. Hakomori (Hakomori and Kijimoto 1972), was labeled with [^{14}C]formaldehyde (Rice and Means 1971). Binding studies were done at 0°C with intact monolayers preincubated with 0.3 mg/ml bovine serum albumin, 0.3 mg/ml rabbit gamma-globulin (kindly supplied by Dr. Steiner) and then radioactive antiserum (12.5 μg, 1000 cpm/μg) was added. Binding studies with trypsinized cells were carried out similarly except for the following modifications. Monolayers were washed with solution A and treated with 25 μg/ml trypsin for 10 minutes at 37°C. Then 50 μg/ml soybean trypsin inhibitor was added. Cells were incubated on ice and washed by centrifugation. Controls using no cells or clones without antigenic reactivity in the hemolysis assay were at background level in the binding assay.

The binding of radioactive anti-Forssman antiserum to cells before and after trypsin treatment is shown in Fig. 1. The amount of antiserum bound to trypsinized cells increased gradually until 8 hours after mitosis. No additional binding was observed when mitotic cells were trypsinized. As the cell progressed through the

Table 4

Total Forssman Antigen in Extracts of Cells during the Cell Cycle

Time after mitosis	pmoles/cell*
0	.086
4	.11
8	.13
11	.13
14	.21

* Average of 4 determinations

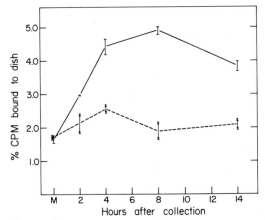

Figure 1
Radiolabeled anti-Forssman antiserum was added to untreated or trypsin-treated monolayers of mitotically harvested cells as described in Methods. Bars or linked X's indicate variations between duplicates.
(——) Trypsin treated; (– – –) untreated.

cell cycle, the proportion of antigen exposed by trypsin treatment increased until 8 hours after mitosis. Similar measurements have shown that 60–70% of the antigen detected in trypsinized growing cells and 70–95% of antigen in trypsinized confluent cells was unavailable to antiserum before trypsin treatment (Hakomori and Kijimoto 1972 and B.A.W. unpublished results).

DISCUSSION

In agreement with Pasternak and colleagues (Warmsley et al. 1970), we found incorporation of radioactivity into the majority of glycolipids during all phases of the cell cycle. This was in contrast to the synthesis of glycolipids during G_2 as observed by others (Bosmann and Winston 1970). However the lymphoma line studied by Bosmann has a short doubling time with almost no G_1 and G_2. On the other hand, using glucosamine and choline labels, Warren and collaborators observed incorporation during G_1. Although plasma membranes were isolated, no attempt was made to fractionate the labeled components. We observed a peak of incorporation of glucosamine into total glycoproteins and into one glycolipid during G_1.

The observation that some complex glycolipids are made during G_1 and early S agrees with studies performed during the growth cycle (Sakiyama and Robbins 1973). GL-3 was not significantly labeled until cells became confluent. GL-4 was labeled somewhat at low densities but incorporation increased at confluence. GL-5 was labeled at all cell densities but the greatest incorporation was in confluent cells. Similarly GL-3 and GL-4 were labeled preferentially during G_1 and early S, while GL-5 was labeled during all phases of the cycle. Thus the incorporation of label into GL-3 and GL-4 is restricted during the cell cycle and in growing cultures whereas the incorporation of label into GL-5 is not.

We have used the term rate of labeling instead of synthetic rate because we cannot exclude the possibility of turnover. Our intent in using fatty acid and hexosamine labels was to reduce the possibility of inaccurate synthetic rates due to turn-over of parts of the molecule (for example, acylation of the sphingosine moiety). However turnover by degradation of the whole molecule cannot be ruled out. In addition the equal rate of labeling of some of the glycolipids throughout the cycle makes the possibility of artifacts due to pool sizes unlikely. We found no differences in the total lipid precursor pools during the cell cycle but cannot exclude compensating alterations in separate pools.

The state of the cellular surface during the cell cycle is another intriguing phenomenon. As stated above, many different antigenic sites are "exposed" during mitosis. Although this may be a consequence of release of substances attached to the surface, like heparin sulfate, or other mucopolysaccharides in the cell coat, some rearrangement of the existing architecture may also be involved. In this sense the mitotic cell seems to resemble a trypsinized cell, for example, in the increased reactivity of certain sites, in the spherical rounding of a flattened cell and in the more tenuous attachment to the substratum. Increasing reactivity due merely to a rounding of a flattened cell is unlikely because G_1 cells expose a higher percentage of their antigen than S cells, although both are completely flattened. Alternately trypsin treatment or mitosis may result in greater mobility of membrane components. A third hypothesis is that mitosis and G_1 are the times in the cell cycle when these antigenic sites are inserted into the membrane. Subsequently they either become diluted in the bilayer as more phospholipid backbone is assembled or covered up by the insertion of membrane proteins or extracellular material. In this context it is interesting that Sander and Pardee (1972) found decreased transport rates in mitosis and early G_1 which quickly returned to interphase specific activity (rate per mg protein) by late G_1 and S. We are currently investigating the release of extracellular material and the insertion of new material which is exposed to the surface.

Whether the Forssman antigen is an important factor in cellular regulation of growth is another interesting question. Like agglutination with lectins, the Forssman antigen is more readily detected in transformed cells than in "normal" contact-inhibited parents. The exposure of the Forssman antigen during mitosis as well as in the transformed cell is consistent with the proposal that the transformed cell surface is locked in the "mitotic phase" of the cell cycle (Burger 1972). However the observations of Shoham and Sachs, reported in this volume, suggest that the mitotic normal cell is similar to the transformed interphase cell in the binding of fluorescent lectin but that the mitotic normal and the mitotic transformed cell are different.

Acknowledgments

We thank D. Holleman, S. Weinzierl and H. Samuelsdottir for their excellent technical assistance. This investigation was supported by National Institutes of Health grant no. 5-RO1-AM-6803-10. B.A.W. is supported by the NIH Program Grant CA12174.

REFERENCES

Bosmann, H. B. and R. A. Winston. 1970. Synthesis of glycoprotein, glycolipid, protein, and lipid in synchronized L51781 cells. *J. Cell Biol.* **45**:23.

Burger, M. M. 1972. Mitotic cell surface changes. *Membrane research* (ed. C. F. Fox), p. 241. Academic Press, New York.

Cikes, M., S. Friberg, Jr. and G. Klein. 1972. Quantitative studies of antigen expression in cultured murine lymphoma cells. *J. Nat. Cancer Inst.* **49:**1667.

Fogel, M. and L. Sachs. 1964. The induction of Forssman-antigen synthesis in hamster and mouse cells in tissue culture as detected by the fluorescent antibody technique. *Exp. Cell Res.* **34:**448.

Fox, T. O., J. R. Sheppard and M. M. Burger. 1971. Cyclic membrane changes in animal cells. *Proc. Nat. Acad. Sci* **68:**244

Glick, M. C. and C. A. Buck. 1973. Glycoproteins from the surface of metaphase cells. *Biochemistry* **12:**85.

Glick, M. C., E. W. Gerner and L. Warren. 1971. Changes in the carbohydrate content of the KB cell during the growth cycle. *J. Cell. Physiol.* **77:**1.

Gray, G. M. 1967. Chromatography of lipids. *Biochim. Biophys. Acta* **144:**511.

Hakomori, S. and S. Kijimoto. 1972. Dependency of Forssman reactivity on cell-to-cell contact in cultured hamster cells. *Nature New Biol.* **239:**87.

Kraemer, P. M. 1967. Configuration change of surface sialic acid during mitosis. *J. Cell Biol.* **33:**197.

Kraemer, P. M. and R. A. Tobey. 1972. Cell cycle dependent desquamation of heparin sulfate from the cell surface. *J. Cell Biol.* **55:**713.

O'Neill, C. H. 1968. An association between viral transformation and Forssman antigen detected by immune adherence in cultured BHK 21 cells. *J. Cell Sci.* **3:**405.

Onodera, K. and R. Sheinin. 1970. Macromolecular glucosamine-containing component of the surface of cultivated mouse cells. *J. Cell Sci.* **7:**337.

Pasternak, C. A., A. Warmsley and D. B. Thomas. 1971. Structural alterations in the surface membrane during the cell cycle. *J. Cell Biol.* **50:**562.

Peterson, D. F., E. C. Anderson and R. A. Tobey. 1968. Mitotic cells as a source of synchronized cultures. *Methods in Cell Physiol.* **3:**347.

Rice, R. H. and B. E. Means. 1971. Radioactive labeling of proteins *in vitro. J. Biol. Chem.* **246:**831.

Robertson, H. T. and P. H. Black. 1969. Changes in surface antigens of SV40-virus transformed cells. *Proc. Soc. Exp. Biol. Med.* **130:**363.

Sakiyama, H. and P. W. Robbins. 1973. The effect of dibutyryl adenosine 3:5′-cyclic monophosphate on the synthesis of glycolipids by normal and transformed NIL cells. *Arch. Biochem. Biophys.* **154:**407.

Sakiyama, H., S. K. Gross and P. W. Robbins. 1972. Glycolipid synthesis in normal and virus-transformed hamster cell lines. *Proc. Nat. Acad. Sci.* **69:**872.

Sander, G. and A. B. Pardee. 1972. Transport changes in synchronously growing CHO and L cells. *J. Cell. Physiol.* **80:**267.

Stadler, J. K. and E. A. Adelberg. 1972. Cell cycle changes and the ability of cells to undergo virus-induced fusion. *Proc. Nat. Acad. Sci.* **69:**1929.

Tobey, R. A., D. F. Peterson, E. C. Anderson and T. T. Puck. 1966. Life cycle analysis of mammalian cells. *Biophys. J.* **6:**567.

Warmsley, A. M., B. Phillips and C. A. Pasternak. 1970. The use of zonal centrifugation to study membrane formation during the life cycle of mammalian cells. *Biochem. J.* **120:**683.

Glycolipids of NIL Hamster Cells as a Function of Cell Density and Cell Cycle

D. R. Critchley, K. A. Chandrabose, J. M. Graham, and I. Macpherson

Department of Tumour Virology, Imperial Cancer Research Fund
London WC2A 3PX

Current thinking has emphasized the possible importance of the cell surface and surface complex carbohydrates in control of properties such as contact inhibition and malignancy (Burger 1971, Kraemer 1971). Interest in glycolipids in this context stemmed from the observation of Hakomori and Murakami (1968) that polyoma virus-transformed BHK21 cells had a different glycolipid composition than the parent cell line. Similar studies on other cell lines (Hakomori, Teather and Andrews 1968; Mora et al. 1969; Brady and Mora 1970; Hakomori, Saito and Vogt 1971) have led to the concept that transformed cells fail to complete the carbohydrate chain of glycolipids. That this is an important characteristic of malignant cells has received support from similar findings in vivo (Rapport et al. 1958; Hakomori et al. 1967; Siddiqui and Hakomori 1970). Of additional interest is the observation that in some cell lines the levels of certain glycolipids increase at high cell density and perhaps significantly are low or absent in virus-transformed cells (Hakomori 1970; Robbins and Macpherson 1971). It has been suggested that the lower concentrations of these density-dependent compounds in transformed cells may be closely related to loss of contact inhibition and the appearance of tumorigenic properties (Hakomori 1971).

We have been studying density-dependent glycolipids of NIL hamster cells in an attempt to elucidate their possible role in growth regulation (Critchley and Macpherson 1973; Critchley 1973).

METHODS

Cell Culture

The cells studied were clones isolated from the NIL2 Syrian hamster cell line (Diamond 1967). NIL8 cells have an epithelial type morphology in dense culture, whereas NIL1 are fibroblastic. Saturation densities ranged from 4 to 15 $\times$ 10^6 cells per 9-cm dish when cells were grown in Dulbecco's modification of Eagle's medium plus 10% heat-inactivated calf serum.

Glycolipid Analysis

Glycolipids labeled after pulsing cells with [1-^{14}C]palmitate (1 μCi/ml) were extracted from the dried cell pellet in 2:1 chloroform:methanol and separated by two-dimensional thin layer chromatography. Lipids were detected by radioautography and quantitated by scintillation counting (Critchley and Macpherson 1973). Results are expressed as cpm in an individual glycolipid $\times$ 10^3/cpm in phospholipid plus glycolipid.

UDP Galactose:Glycolipid Galactosyltransferase Assay

Both ceramide di- and trihexoside syntheses were assayed according to the method of Kijimoto and Hakomori (1971) with the following modifications. After removal of unreacted[^{14}C]UDP galactose the organic phase was saponified to remove label which was apparently in phospholipid and which chromatographed with ceramide trihexoside. Several substrate concentrations were employed to ascertain true maximum velocities because of inhibition at high substrate/enzyme ratios. Ceramide mono- and dihexoside substrates were obtained from Miles Yeda.

UDP Galactose:Glycoprotein Galactosyltransferase Assay

Fetuin active as a galactose acceptor was prepared by the method of Spiro (1964). Assay conditions were similar to those described by Bosmann (1971) except the final concentration of UDP galactose was 30 μM (specific activity 10 μCi/μmole, 0.05 μCi/assay) and 1 mg of acceptor was used per assay. The activity was linear with respect to time up to 30 minutes and cell protein up to 100 μg and was saturating with respect to UDP galactose and fetuin. It was entirely dependent on Mn^{++} and only fetuin with both sialic acid and galactose removed acted as an efficient acceptor.

RESULTS

NIL8 hamster cells contain five neutral glycolipids from a ceramide mono- to pentahexoside and only one ganglioside, sialyldihexosyl ceramide (hematoside). A possible synthetic pathway is shown in Fig. 1.

Incorporation of [1-^{14}C]palmitate shows that levels of the ceramide tri-, tetra-, and pentahexoside increase at high cell density in normal cells but are virtually absent from virus-transformed derivatives. In contrast, incorporation into hematoside is not grossly affected either by cell density or transformation (Critchley and Macpherson 1973). A study of the kinetics of the density dependence indicates that it is not just a difference of dividing compared to nondividing cells (Fig. 2). Thus incorporation into ceramide trihexoside increases while the cells are still dividing logarithmically, although the major increase in labeling of the tetra- and pentahexoside occurs somewhat later when cell growth is reaching a plateau. An attractive conclusion is that the levels of these compounds increase because of increased cell contact.

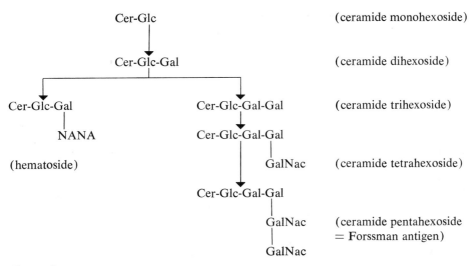

Figure 1

Possible pathway of glycolipid synthesis in NIL cells. Glc, glucose; Gal, galactose; NANA, *N*-acetylneuraminic acid; GalNac, *N*-acetylgalactosamine; Cer, ceramide.

Kinetic Experiments

To study the speed with which glycolipid synthesis responds to cell density, sparse cells were trypsinized and reseeded at high cell density. Cells were pulsed with [1-^{14}C]palmitate, and the label in hematoside, the major glycolipid in sparse cells, compared to that in the density-dependent glycolipid pathway (Table 1). Approximately 9 hours after seeding the ratio of incorporation in hematoside:density-dependent glycolipids had decreased in dense cells, but a similar change was found in cells reseeded sparse. This latter change was apparently transient, for after 18 hours the sparse pattern was reestablished although the hematoside:density-dependent glycolipid ratio continued to decrease in cells seeded dense. This change in

Table 1

Incorporation of [1-^{14}C] Palmitate in NIL8
Cells Seeded from Sparse to Dense Culture

Hours after reseeding	*Cell number* $\times 10^{-6}$	*Hem/Gl*	*Cell number* $\times 10^{-6}$	*Hem/Gl*
0	1.0	4.9		
9	1.0	2.6	5.0	2.4
18	2.1	4.6	9.1	1.8
27	2.5	3.6	11.6	0.9

Sparse cultures of NIL8 cells (1×10^6 cells per 9-cm dish) were harvested (0.25% trypsin in Tris-saline) and either seeded at 1 or 5×10^6 cells per dish. Cells were pulsed with [1-^{14}C]palmitate (1 μCi/ml, 9 hr) over the next 27 hr, and incorporation into density-dependent glycolipids (Gl) compared to hematoside (Hem).

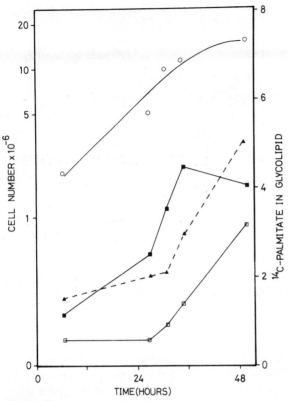

Figure 2
Glycolipids of NIL8 cells growing from sparse to dense culture. Cells were pulsed with [1-14C]palmitate (1 μCi/ml, 6 hr) as cell density increased. Incorporation into ceramide trihexoside (■———■), tetrahexoside (□———□) and pentahexoside (▲–––▲) at various times after the start of the experiment is related to cell number × 10⁻⁶ (o) at the end of each pulse. (Data replotted from Critchley and Macpherson 1973.)

metabolism was inhibited by 2.5 mM hydroxyurea, suggesting that DNA synthesis is necessary for a sparse cell to adopt the dense glycolipid pattern.

Increasing cell contacts by adding transformed NIL8 cells to monolayer cultures of normal cells prelabeled with [1-14C]palmitate did not lead to increased incorporation into density-dependent glycolipids.

In the converse of these experiments, nondividing dense cells were trypsinized to low cell density (Fig. 3). Incorporation into ceramide trihexoside remained high for the first 12 hours after seeding, and the hematoside:ceramide trihexoside ratio was still similar to that of dense cells. After 19 hours DNA synthesis had started and incorporation into ceramide trihexoside was considerably reduced. By 26 hours the cells had approximately doubled in number and the ratio of incorporation into hematoside:ceramide trihexoside was strongly in favor of hematoside. Dense cells reseeded dense also showed some reduction in incorporation into ceramide trihexoside around the onset of DNA synthesis, but because incorporation into hematoside did not drastically increase, the ratio stayed near that of dense cells. Cells seeded to sparse culture in the presence of 3 mM thymidine or 2.5 mM hydroxyurea showed

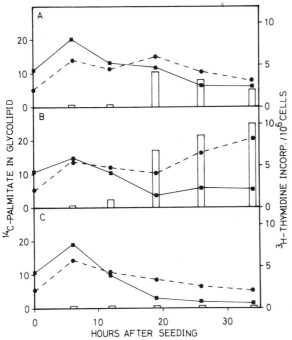

Figure 3

Glycolipids of NIL8 cells seeded from dense to sparse culture. Dense quiescent cultures of NIL8 cells (7×10^6 cells/9-cm dish) were trypsinized and dishes seeded with **(A)** 7×10^6 cells, **(B)** 7×10^5 cells in normal medium, **(C)** 7×10^5 cells in medium containing 2.5 mM hydroxyurea. Cells were pulsed with [1-^{14}C]palmitate (1 μCi/ml, 6 hr) 1–7, 6–12, 13–19, 20–26 and 28–34 hr after seeding. Cells were pulsed with [^{3}H]thymidine (0.1 μCi/ml, 1 hr) between the 5th–6th hr of the [1-^{14}C]palmitate pulse. Cell counts were determined at the end of each pulse period. Incorporation of [1-^{14}C]palmitate into ceramide trihexoside (■——■) and hematoside (●– – –●). Open bars represent [^{3}H]thymidine incorporation $\times 10^{-4}$/10^6 cells.

similar reduction in incorporation into ceramide trihexoside but did not exhibit the stimulated incorporation into hematoside. The time taken for the incorporation into ceramide trihexoside to decrease may be a reflection of the half-life of the enzyme involved in its synthesis. However it is interesting that incorporation into ceramide trihexoside in dense cells also decreased after initiation of DNA synthesis, although labeling of the tetra- and pentahexoside increased 2- to 3-fold. This suggests that ceramide trihexoside may be normally synthesized during early G_1 phase of the cell cycle.

Cell Synchrony

The ratio of incorporation of [1-^{14}C]palmitate into hematoside:ceramide trihexoside in cells blocked in sparse culture by serum deprivation (Bürk 1966) is similar to that of dense cells. The ratio is dramatically increased in cells moved from low serum block to the G_1/S phase boundary by addition of serum plus 2.5 mM hydroxyurea (Fig. 4). Incorporation into ceramide trihexoside remained low during S phase following release from hydroxyurea block although labeling of hematoside

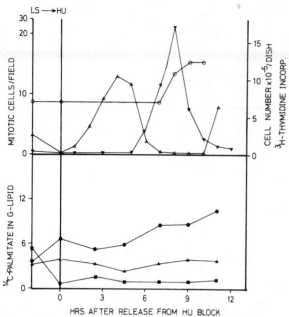

Figure 4

Glycolipids in synchronized sparse NIL8 cells. Sparse NIL8 cells were seeded at
6×10^5 cells/9-cm dish in low serum medium (LS). After 48 hr cells were
moved from LS block by addition of complete medium plus 2.5 mM hydroxyurea (HU). After 17 hr hydroxyurea medium was replaced with complete
medium. [³H]thymidine incorporation (0.1 μCi/ml, 1 hr) ($\triangle$———$\triangle$) cpm $\times$
10^{-4}, cell number ($\circ$———$\circ$), and mitotic cells per microscope field ($\blacktriangledown$———$\blacktriangledown$)
were determined at hourly intervals. LS- and HU-blocked cells were pulsed with
[1-¹⁴C]palmitate (1 μCi/ml, 3 hr). Released cells were pulsed at 0–3, 2–5, 4–7,
6–9, 8–11 hr. Incorporation of [1-¹⁴C]palmitate in hematoside ($\bullet$———$\bullet$),
ceramide trihexoside ($\blacksquare$———$\blacksquare$), and pentahexoside ($\blacktriangle$———$\blacktriangle$).

increased. Throughout these maneuvers incorporation into the tetra- and pentahexoside was little affected. A closer look at the early events on release from low
serum block show that decreased incorporation into ceramide trihexoside begins
well before the onset of DNA synthesis (Fig. 5). Cells blocked at apparently the
same point in G_1 (see Pardee et al. this volume) by isoleucine or glutamine deprivation (Ley and Tobey 1970) incorporated little label into ceramide trihexoside.
We have subsequently found that depletion of glutamine from semiconfluent cells
drastically reduces their ability to synthesize the density-dependent components.
Dense cells blocked by isoleucine deprivation were, however, capable of synthesizing all the glycolipids. It is therefore unclear why sparse, low-serum-blocked
cells synthesize ceramide trihexoside while isoleucine-blocked sparse cells are unable to do so.

Quiescent dense cells produced by serum deprivation entered S phase some 20
hours after addition of fresh medium containing 10% serum, i.e., a somewhat
longer interval than the period required if the cells are sparse (cf. Fig. 5 and 6.).
During the G_1 period a peak of incorporation into ceramide trihexoside occurred 10

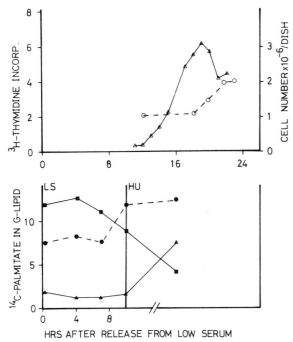

Figure 5

Glycolipids of sparse NIL8 cells in early G_1 phase of the cell cycle. Low serum-blocked cells (LS) were released by addition of complete medium and synchrony assessed by [^{3}H]thymidine incorporation ($\triangle$———$\triangle$) cpm $\times$ 10^{-5} and increase in cell number (o– – –o). Incorporation of [1-^{14}C]palmitate (1 μCi/ml, 4 hr) into glycolipid was studied 0–4, 3–7, and 6–10 hr after release and in cells moved to G_1/S boundary in complete medium + 2.5 mM hydroxyurea (HU). Hematoside (●– – –●), ceramide trihexoside (■———■), and pentahexoside (▲———▲).

hours after release from the serum block (Fig. 6A). Incorporation into trihexoside decreased sometime prior to the onset of DNA synthesis at 21–22 hours, and labeling of hematoside increased. These latter events were confirmed in cells released from hydroxyurea block (Fig. 6B). Similar results were obtained from dense quiescent cells stimulated by addition of fresh medium. Incorporation into tetra- and pentahexoside increased around the same time as enhanced trihexoside synthesis but an additional increase in incorporation occurred during S phase.

UDP Galactose: Glycolipid Galactosyltransferase in NIL1 Cells

To gain a fuller understanding of the control of glycolipid metabolism in NIL cells, we have studied the activity of two of the key enzymes involved in their synthesis. The activity of the UDP galactose:lactosyl ceramide α-galactosyltransferase is low or absent in transformed cells although the synthesis of dihexoside is slightly enhanced (Table 2A). Experiments where NIL1 and NIL1/HSV homogenates were mixed did not reveal the presence of an inhibitor of trihexoside synthesis in transformed cells. In addition the activity of the enzymes is strongly influenced by cell

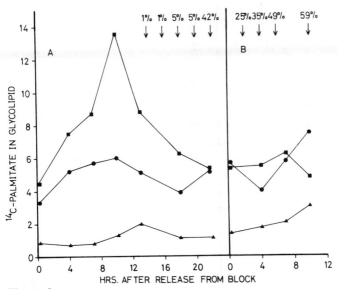

Figure 6
Glycolipids in synchronized dense NIL8 cells. Low serum-blocked dense cells were (A) stimulated by addition of complete medium, (B) moved to the G_1/S boundary by addition of complete medium + 2.5 mM hydroxyurea, then 17 hr later were released. Degree of synchrony was estimated from percent labeled nuclei after a 2-hr pulse with [^{3}H]thymidine (1 μCi/ml) (arrows). Incorporation of [1-^{14}C]palmitate into ceramide trihexoside (■——■), pentahexoside (▲——▲), and hematoside (●——●) was determined as in Fig. 4.

density (Table 2B). There is a twentyfold increase in the synthesis of trihexoside in dense as compared to sparse cells, whereas dihexoside synthesis activity increases only twofold. Subsequent assay of another NIL cell clone NIL8 showed no detectable trihexoside synthetic activity although the cells incorporate [^{14}C]galactose and [^{14}C]palmitate into trihexoside. In addition we have been unable to detect the ceramide tetra- and pentahexoside synthetic activities in either cell line. We have yet to find an explanation for these observations although the presence of an inhibitor released by homogenizing the cell has not been discounted (Duffard and Caputto 1972).

We have looked at the subcellular distribution of the UDP galactose:lactosyl ceramide galactosyltransferase in comparison to the distribution of a glycoprotein galactosyltransferase (Table 3). We have found the glycolipid synthetic activity in mitochondrial, plasma membrane and endoplasmic reticulum fractions but absent from nuclear fractions. The overall low enrichment in these fractions is in agreement with the widespread distribution of the enzyme in the cell. The glycoprotein galactosyltransferase activity was low in nuclei and mitochondria but was enriched 4.5-fold in endoplasmic reticulum and 8-fold in plasma membrane fractions. The specific activity of both enzymes was greatest in the plasma membrane fractions. The possibility that this fraction was contaminated with golgi has not, however, been excluded.

Table 2

Specific Activities of UDP Gal: Glycolipid Galactosyltransferases: Variation with (A) Cell Type and (B) Cell Density

	Glycolipid synthesized (pmoles/mg protein/hr)	
	Dihexoside	Trihexoside
A		
NIL1	650	512
NIL1–HSV	780	22
NIL1 + NIL–HSV	ND	256
B		
NIL1		
0.2×10^7	224	29
1.16×10^7	366	159
2.2×10^7	592	174
3.5×10^7	408	496
5.0×10^7	265	511

A A 50:50 mixture of cell homogenate protein of NIL1 and NIL–HSV was made in an attempt to detect an inhibitor of the enzyme in transformed cells. HSV, hamster sarcoma virus.

B Cells were seeded at 0.2×10^7 cells per roller bottle and harvested by scraping at the densities indicated.

Table 3

UDP Gal: Glycolipid and Glycoprotein Galactosyltransferase Activities in NIL1 Cells

Cell fraction	Ceramide trihexoside synthesis			Glycoprotein galactosyl-transferase Δ
	endogenous	exogenous	Δ	
homogenate	565	850	285	3.5×10^4
nuclei	<20	<20	<20	0.46×10^4
mitochondria	429	821	392	3.5×10^4
endoplasmic reticulum	841	1306	465	16.0×10^4
plasma membrane	1615	2520	905	27.2×10^4

Δ represents the difference between endogenous and exogenous activity. Activities in pmoles/mg protein/hr under the assay conditions specified. Isolation and characterization of the subcellular fractions as described previously by Critchley et al. 1973.

DISCUSSION

Our results on the kinetics of increase in incorporation of [1-^{14}C]palmitate into the density-dependent glycolipids of NIL cells are similar to those reported by Saki-yama, Gross and Robbins (1972), Sakiyama and Robbins (1973a), and Kijimoto and Hakomori (1972). It is clear that the levels of all three glycolipids do not in-crease simultaneously, suggesting that their formation is subject to individual con-trol. This is further borne out from the results on cell synchrony, where for example incorporation of [1-^{14}C]palmitate into ceramide trihexoside, but not in tetra- and pentahexoside, is stimulated in the G_1 phase of the cycle.

Sakiyama and Robbins (1973a) have reported that in one of their clones of NIL cells the tetrahexoside is synthesized mainly in G_1, whereas synthesis of the penta-hexoside did not vary as a function of cell cycle. This correlated with the observa-tion that the levels of tetrahexoside increased sharply when the cells were nearly quiescent, i.e., were in G_0 or G_1, whereas the pentahexoside increased gradually with increasing cell contacts. This rationale does not explain our results as, although the trihexoside is made in increased amounts in G_1, the major increase in [1-^{14}C] palmitate incorporation in trihexoside occurs before the cells have become quies-cent. We have confirmed the data of Kijimoto and Hakomori (1971) relating activity of UDP galactose:lactosyl ceramide galactosyltransferase activity to cell density. We are, however, uncertain whether the activity expressed in vitro neces-sarily relates to that in vivo. The activity apparently explains the increased levels of trihexoside in dense culture in NIL1 cells. Contact may lead to induction of the enzyme although there is a considerable increase in activity before there is ap-preciable cell contact. An additional possibility is that some of the trihexoside formed in dense culture is synthesized by the process of transglycosylation (Roth, McGuire and Roseman 1971; Roth and White 1972). The surface transferase may interact with ceramide dihexoside on neighboring cells, leading to synthesis of trihexoside. The fact that sparse low-serum-blocked cells incorporate high levels of [1-^{14}C]palmitate into trihexoside suggests that such a mechanism is not the major mode of synthesis.

Cells appear to take some considerable time to modify the synthesis of these glycolipids in response to changing cell density. The cell may well have to go through a round of DNA synthesis before it can respond. Our previous data sug-gests that the preformed glycolipids in dense cells are not rapidly degraded before the first cell division (Critchley and Macpherson 1973).

A number of studies have shown that glycolipids are concentrated in the plasma membrane (Klenk and Choppin 1970; Renkonen et al. 1972). We have found that the density-dependent glycolipids of NIL hamster cells are present at, but not con-fined to, the cell surface (Critchley, Graham and Macpherson 1973). The mole-cules are thus in a position to influence cell surface properties. However we have found, similar to the results of Sakiyama et al. (1972), that the presence of the density-dependent glycolipids does not correlate with saturation density. All clones of polyoma and hamster sarcoma virus-transformed cells we have studied have only trace amounts of the tri- and tetrahexoside although some pentahexoside remains. Of eleven tumor cell lines derived from NIL8 cells, ten incorporated only trace amounts of [1-^{14}C]palmitate into tri- and tetrahexoside but all had pentahexoside. One tumor line retained all three density-dependent components (Critchley and

Macpherson 1973). Similar results have been reported by Sakiyama and Robbins (1973b) in a study of the tumors induced by their NIL clones. Brady and Mora (1970) found that a spontaneously tumorigenic mouse cell line had a normal glycolipid pattern. In addition they found no response of mouse cell glycolipids to cell density (Fishman et al. 1972) so the phenomenon may be of limited occurrence or may be exemplified by changes in other cell surface components, e.g. glycoproteins.

If the glycolipid pattern is an important determinant of cell behavior, perhaps these molecules act as recognition sites on the cell surface. Binding of these sites to receptors on neighboring cells may lead to the signal for cessation of cell division. A block at any point in this chain of reactions might result in breakdown of the control mechanism. Thus the growth properties of transformed cells may in part result from a block in synthesis of glycolipids, the receptor(s), and the "mediator(s)" say between cell surface and nucleus. The similar properties of the tumor line with normal glycolipids may be due to lack of receptor(s) and/or "mediator(s)." The failure of cyclic AMP to restore normal glycolipids to transformed cells (Sakiyama and Robbins 1973a) is perhaps because cyclic AMP is one such "mediator" and bypasses the need for the glycolipid trigger. It has recently been reported that addition of a bacterial glycolipid to transformed rat embryo fibroblasts restored some properties of growth control and also augmented the pool size of cyclic AMP (Brailovsky et al. 1973).

The use of purified glycolipids, glycolipid antibodies, and specific glycosidases may help to elucidate the role of glycolipids in control of cell behavior.

REFERENCES

Bosmann, H. B. 1971. Membrane glycoprotein transferases: Time courses of activity changes in a synchronized cell population and enzyme activity half-lives in an asynchronous population. *Arch. Biochem. Biophys.* **145**:310.

Brady, R. O. and P. T. Mora. 1970. Alteration in ganglioside pattern and synthesis in SV40 and polyoma virus-transformed mouse cell lines. *Biochem. Biophys. Acta* **218**:308.

Brailovsky, C., M. Trudel, R. Lallier, and V. N. Nigam. 1973. Growth of normal and transformed rat embryo fibroblasts. Effects of glycolipids from Salmonella minnesota R. mutants. *J. Cell Biol.* **57**:124.

Burger, M. M. 1971. Cell surfaces in neoplastic transformation. *Current topics in cellular regulation* (ed. B. L. Horecker and E. R. Stadtman) vol. 3, p. 135. Academic press, New York.

Bürk, R. R. 1966. Growth inhibitor of hamster fibroblast cells. *Nature* **212**:1261.

Critchley, D. R. 1973. Glycolipids and cancer. *Membrane mediated information* (ed. P. W. Kent). Medical and Technical Publishers, Lancaster. (In press)

Critchley, D. R., and I. Macpherson. 1973. Cell density dependent glycolipids in NIL2 hamster cells, derived malignant and transformed cell lines. *Biochim. Biophys. Acta* **296**:145.

Critchley, D. R., J. M. Graham and I. Macpherson. 1973. Subcellular distribution of glycolipids in a hamster cell line. *FEBS Letters* **32**:37.

Diamond, L. 1967. Two spontaneously transformed cell lines derived from the same hamster embryo culture. *Int. J. Cancer.* **2**:143.

Duffard, R. O. and R. Caputto. 1972. A natural inhibitor of sialyl transferase and its possible influence on this enzyme activity during brain development. *Biochemistry* **11:**1396.

Fishman, P. H., V. W. McFarland, P. T. Mora and R. O. Brady. 1972. Ganglioside biosynthesis in mouse cells: Glycosyl transferase activities in normal and virally-transformed lines. *Biochem. Biophys. Res. Commun.* **48:**48.

Hakomori, S. 1970. Cell density-dependent changes of glycolipid concentrations in fibroblasts, and loss of this response in virus transformed cells. *Proc. Nat. Acad. Sci.* **67:**1741.

―――― 1971. Glycolipid changes associated with malignant transformation. *Dynamic structure of cell membranes* (ed. D. F. Wallach and H. Fischer) p. 65. Springer-Verlag, New York.

Hakomori, S. and W. T. Murakami. 1968. Glycolipids of hamster fibroblasts and derived malignant-transformed cell lines. *Proc. Nat. Acad. Sci.* **59:**254.

Hakomori, S., T. Saito and P. K. Vogt. 1971. Transformation by Rous sarcoma virus: Effects on cellular glycolipids. *Virology* **44:**609.

Hakomori, S., C. Teather and H. Andrews. 1968. Organizational difference of cell surface "hematoside" in normal and virally transformed cells. *Biochem. Biophys. Res. Commun.* **33:**563.

Hakomori, S. I., J. Koscielak, K. J. Block and R. W. Jeanloz. 1967. Immunologic relationship between blood group substances and a fucose containing glycolipid of human adenocarcinoma. *J. Immunol.* **98:**31.

Kijimoto, S. and S. Hakomori. 1971. Enhanced glycolipid: galactosyltransferase activity in contact-inhibited hamster cells, and loss of this response in polyoma transformants. *Biochem. Biophys. Res. Commun.* **44:**557.

―――― 1972. Contact-dependent enhancement of net synthesis of Forssman glycolipid antigens and hematoside in NIL cells at the early stage of cell to cell contact. *FEBS Letters* **25:**38.

Klenk, H. D. and P. W. Choppin. 1970. Glycosphingolipids of plasma membranes of cultured cells and an enveloped virus (SV5) grown in these cells. *Proc. Nat. Acad. Sci.* **66:**57.

Kraemer, P. M. 1971. Complex carbohydrates of animal cells: Biochemistry and physiology of the cell periphery. *Biomembranes* (ed. L. S. Manson) vol. 1, p. 67. Plenum Press, New York.

Ley, K. D. and R. A. Tobey. 1970. Regulation of initiation of DNA synthesis and cell division by isoleucine and glutamine in Gl arrested cells in suspension culture. *J. Cell Biol.* **47:**453.

Mora, P. T., R. O. Brady, R. M. Bradley and V. W. McFarland. 1969. Gangliosides in DNA virus-transformed and spontaneously transformed tumorigenic mouse cell lines. *Proc. Nat. Acad. Sci.* **63:**1290.

Rapport, M. M., L. Graf, V. Skipski and N. F. Alonzo. 1958. Cytolipin H, a pure lipid heptene isolated from human carcinoma. *Nature* **181:**1803.

Renkonen, O., C. G. Gahmberg, K. Simons and L. Kaarianen. 1972. The lipids of the plasma membranes and endoplasmic reticulum from cultured baby hamster kidney cells (BHK 21). *Biochim. Biophys. Acta* **255:**66.

Robbins, P. W. and I. Macpherson. 1971. Glycolipid synthesis in normal and transformed animal cells. *Proc. Roy. Soc. London* B **177:**49.

Roth, S. and D. White. 1972. Intercellular contact and cell-surface galactosyl transferase activity. *Proc. Nat. Acad. Sci.* **69:**485.

Roth, S., E. J. McGuire and S. Roseman. 1971. Evidence for cell-surface glycosyl-transferases. Their potential role in cellular recognition. *J. Cell Biol.* **51:**536.

Sakiyama, H. and P. W. Robbins. 1973 a. The effect of dibutyryl adenosine 3′:5′-cyclic monophosphate on the synthesis of glycolipids by normal and transformed NIL cells. *Arch. Biochem. Biophys.* **154:**407.

———— 1973 b. Glycolipid synthesis and tumorigenicity of clones isolated from the NIL 2 line of hamster embryo fibroblasts. *Fed. Proc.* **32:**86.

Sakiyama, H., S. K. Gross and P. W. Robbins. 1972. Glycolipid synthesis in normal and virus-transformed hamster cell lines. *Proc. Nat. Acad. Sci.* **69:**872.

Siddiqui, B. and S. Hakomori. 1970. Change of glycolipid pattern in Morris hepatomas 5123 and 7800. *Cancer Res.* **30:**2930.

Spiro, R. G. 1964. Periodate oxidation of glycoprotein fetuin. *J. Biol. Chem.* **239:**567.

Composition and Turnover of Membrane Lipids in Semliki Forest Virus and in Host Cells

Ossi Renkonen, Arja Luukkonen, Jaakko Brotherus, and Leevi Kääriäinen

Departments of Biochemistry and Virology, University of Helsinki
Haartmaninkatu 3, 00290 Helsinki 29, Finland

The envelope of Semliki Forest virus (SFV) has proven to be a good model of biological membranes (Simons et al. 1973). This virus consists of a nucleocapsid surrounded by a lipoprotein envelope. The envelope has a bilayer structure as shown by low angle X-ray diffraction (Harrison et al. 1971). The envelope contains two glycoprotein species, probably about 300 pairs per virion, and 17,000 different phospho- and glycolipid molecules as well as 16,000 molecules of free cholesterol per virion (Laine, Söderlund and Renkonen 1973).

The present report describes experiments which show that many BHK cell organelles have different lipid compositions. SFV is then used as a tool to study the factors which may cause these differences.

Lipids of BHK Cells and Their Organelles

BHK cells contain a rare phospholipid, lysobisphosphatidic acid (LBPA) (Fig. 1) in addition to the usual lipids. LBPA of BHK cells was identified by its chromatographic properties, which are identical with those of the authentic sample isolated by Body and Gray (1967) from pig lungs. Carboxylic ester groups, glycerol and phosphorus were present in molar ratios 2.2:2.0:1.0 in the BHK cell lipid. Its hydrolysis with a mixture of acetic acid and water gave the expected cleavage products, lysophosphatidic acid and monoglycerides in a good yield, and acetolysis gave monoglyceride diacetates from the BHK cell lipid as well as from the authentic sample (J. Brotherus, K. Sandelin and O. Renkonen, submitted).

Phospholipids of BHK cell plasma membranes, endoplasmic reticulum (ER), mitochondria and nuclei have characteristic compositions (Table 1). In comparison to the lipids of the whole cell, the lipids of the plasma membranes have a high content of sphingomyelin and phosphatidylserine; on the other hand they have a low content of lecithin, cardiolipin and lysobisphosphatidic acid. Lecithin in turn is the characteristic phospholipid in ER and in nuclei, cardiolipin in mitochondria and LBPA in lysosomes.

So far we have not isolated satisfactory BHK cell lysosomes, but we have induced

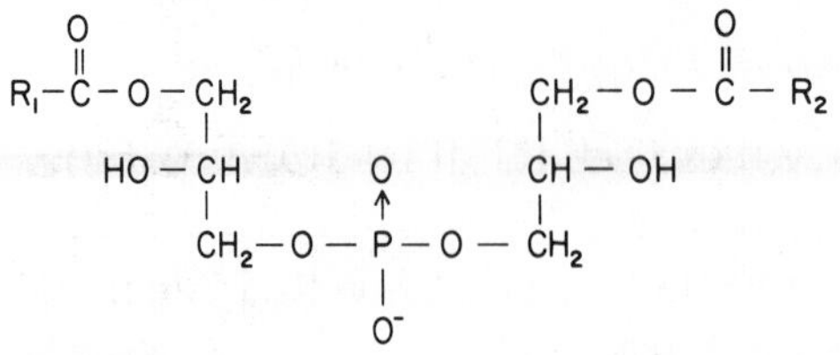

Figure 1
Structure of lysobisphosphatidic acid. The location of acyl groups and the stereochemistry of the glycerol residues is not known. ·

an increase of lysosome-like vacuoles in the cells by incubating them for 2–4 days in Eagle's minimum essential medium containing 0.2% bovine serum albumin. In the incubated cells we found a 4- to 5-fold increase in the activity of acid phosphatase and protease. This was accompanied by a 3- to 6-fold increase in LBPA and a 2-fold increase in sphingomyelin, both of which are lysosomal lipids in rat liver (Henning, Kaulen and Stoffel 1970; Wherrett and Huterer 1972).

Lipids of Semliki Forest Virus

Phospholipids of SFV grown in BHK cells (Table 1) are similar to those of the host plasma membrane, but different from those of the endoplasmic reticulum, mitochondria and nuclei. The similarity between the plasma membrane and the

Table 1
Phospholipid Composition (%) of Organelle Membranes Derived from BHK21 Cells

	Homogenate[a]	Plasma membrane	ER	Nuclei[b]	Mitochondria[b]	SFV[a]
Total phospholipids	100	100	100	100	100	100
Lysobisphosphatidic acid[c]	1.2	0.8	0.9	0.4	0.3	0.4
Cardiolipin	3.3	0.4	2.0	1.7	15.6	0.2
Phosphatidylethanolamine[d]	23	18	14	20	27	23
Phosphatidylcholine	50	44	62	68	48	43
Phosphatidylserine	6.6	8.9	3.6	2.4	1.9	13
Phosphatidylinositol	5.7	2.3	4.3	3.9	3.0	1.6
Sphingomyelin	6.9	18	10	2.5	4.2	16

[a] Taken from Renkonen et al. 1972b.

[b] J. Brotherus et al. unpublished observations.

[c] J. Brotherus, K. Sandelin and O. Renkonen, submitted for publication.

[d] Includes the corresponding plasmalogen.

virus membrane can be quantitated: They have only 7% of dissimilar phospholipids, whereas the ER and the virus have 24%, the nuclei and the virus 28%, and mitochondria and the virus 24% of dissimilar phospholipids. (The fraction of dissimilar lipids is obtained by counting the molecules which are not common to the two sets of lipids, both containing one hundred phospholipid molecules, and dividing their sum by the total number of molecules in the two sets.) Absence of LBPA from the virus suggests that the viral phospholipids differ also from those of the lysosomes.

BHK cells contain a large amount of plasmalogens, or alkenyl-acyl derivatives, among their ethanolamine phosphatides. The ratio of the plasmalogenic and the other types of glyceryl-phosphoryl-ethanolamine lipids is not constant in all BHK cell organelles; the plasma membrane shows a higher ratio (1.43) than the rest of the cell (0.92). SFV lipids are almost identical with those of the host plasma membrane even in their plasmalogen content (Laine, Söderlund and Renkonen 1973).

The viral membranes resemble the plasma membranes more than other BHK cell organelles even in their glycolipid composition (Renkonen et al. 1971, 1972a). The plasma membrane and the viral membrane have also very high cholesterol contents and differ thus from the other host membranes (Renkonen et al. 1971, 1972a).

These data show that SFV lipids and host plasma membrane lipids have nearly identical polar groups. It was obviously important to see whether the identity between the virus and the plasma membrane extends also to the fatty acids and other hydrocarbon chains of their lipids. Our data show that this is the case; each viral phospholipid class has nearly the same composition of hydrocarbon chains as the corresponding lipid class of the plasma membrane (Laine et al. 1972).

Can SFV Lipids Be Varied by Varying the Host?

SFV is an arbo A virus, and it can be propagated in insect cells as well as in mammalian cells. Many insects have membrane lipid compositions rather similar to the mammalian systems, but Diptera are strikingly different (Fast 1966). They have the same phospholipid molecules in their membranes as the mammals, but the relative amounts of the lipids are quite different; in addition there are some lipids in the Diptera which are not found in the mammalian systems.

We analyzed the phospholipids of cultured cells of a mosquito, *Aedes albopictus*, and obtained data which agreed quite well with Fast's results on whole Diptera (A. Luukkonen, M. Brummer-Korvenkontio and O. Renkonen, submitted). *A. albopictus* cells differed from the BHK cells in their phospholipids quite remarkably (Table 2). Phosphatidylethanolamine was the largest phospholipid component in the insect cells, whereas phosphatidylcholine dominated in the BHK cells. Another major difference was that sphingomyelin was present only in very small amounts in the insect cells, though it was the third largest component in BHK cells.

SFV was grown in cultured *A. albopictus* cells (Luukkonen, Kääriäinen and Renkonen unpubl.). The isolated virus had the same density in isopycnic runs on sucrose gradients as the reference SFV from BHK cells. It revealed also the same nucleocapsid protein in polyacrylamide gel electrophoresis (PAGE). Its band of envelope proteins was nearly, but not quite, identical with that of the reference virus in PAGE carried out according to Weber and Osborn (1969). The small

Table 2
Phospholipid Composition (%) of
BHK Cells and *A. albopictus* Cells

	A. albopictus[a]	*BHK*[b]
Total phospholipids	100	100
Lysobisphosphatidic acid	0	1.2
Cardiolipin	3.3	3.3
Phosphatidic acid	0.2	0.6
Phosphatidylethanolamine[c]	53	23
Phosphatidylcholine	24	50
Phosphatidylserine	7.3	6.6
Phosphatidylinositol	6.5	5.7
Ceramide phosphorylethanolamine	4.5	0
Sphingomyelin	0.8	6.9

[a] Unpublished data of A. Luukkonen, M. Brummer-Korvenkontio and O. Renkonen.

[b] Taken from Renkonen et al. 1972b.

[c] Includes the corresponding plasmalogen.

difference is probably caused by different carbohydrate structures of the envelope glycoproteins. The envelope proteins of the insect SFV were further split into two bands, quite as those of the reference virus, on PAGE carried out according to Neville (1971). The purity of the isolated insect SFV was similar to that of the reference virus as judged from electron microscopy and PAGE.

The phospholipids of SFV grown in *A. albopictus* cells proved to be rather different from those of SFV grown in BHK cells (Fig. 2). Almost two-thirds of the phospholipids of the insect cell virus consisted of phosphatidylethanolamine. In contrast, the reference virus derived from BHK cells contained only 23% of this lipid. Other large differences were the high content of phosphatidylcholine and sphingomyelin in the mammalian virus, and the low content of these lipids in the insect virus. The low sphingomyelin content in the latter was interestingly compensated by another sphingolipid, ceramide phosphorylethanolamine.

The insect cells did not contain any trace of the plasmalogenic or the alkyl-acyl forms of phosphatidylethanolamine. It seems safe to assume that even the corresponding SFV was devoid of these ether lipids. This makes the differences between the two viruses even bigger than shown in Fig. 2; in SFV derived from BHK cells only 9% of the total phospholipids are of the usual diacyl type of phosphatidylethanolamine, the rest of the ethanolamine lipids being of the plasmalogenic or alkyl-acyl type (Laine et al. 1972).

Summing up, the differences between the two viruses were so great that only 36% of their phospholipids were common to both of them.

Pulse-Chase Labeling of the Lipids of SFV and Its Host Cells

When BHK cells are grown in the presence of [^{32}P]phosphate, a state of isotope equilibrium is eventually reached where all phospholipids have the same specific

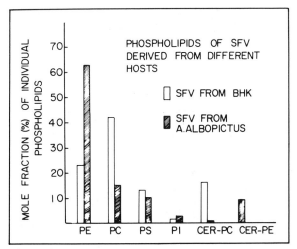

Figure 2

Comparison of phospholipid profiles in SFV derived from cultured BHK cells and from *A. albopictus* cells. PE = phosphatidylethanolamine; PC = phosphatidylcholine; PS = phosphatidylserine; PI = phosphatidylinositol; CER-PC = sphingomyelin; CER-PE = ceramide phosphorylethanolamine.

radioactivity. Under these conditions analysis of the ^{32}P radioactivity provides a measure of the phospholipid phosphorus. We have used this method for analysis of the lipids of SFV and Uukuniemi virus (Renkonen et al. 1972b). If these "equilibrium labeled" cells are pulsed with lipid precursors labeled with ^{3}H or ^{14}C, the specific radioactivities of the phopholipids can be measured from their isotope ratios, ^{3}H:^{32}P or ^{14}C:^{32}P.

Table 3 summarizes our preliminary data, obtained by using [2-^{3}H]glycerol as lecithin precursor in the infected "equilibrium labeled" BHK cells. The cells incorporated [^{3}H]glycerol into their lecithin, and the specific radioactivity decreased then during the chase period. The viral lecithin became radioactive, too. When the pulse was given during the release of the virus and the cells, as well as the virus accumulated in the medium, were collected after three hours, the viral lecithin had only half of the specific activity of the mixed cellular lecithins. This is the result which would be expected with linear release of virus and with linear incorporation of glycerol in cellular lecithin vs. time if rapid mixing of cellular lecithin-glycerol took place.

When the glycerol pulse was followed by a chase period, the viral lecithins had specific activities which were higher than that of the actual host cells. (These are called "cells at the end of virus release" in Table 3.) In comparison to infected control cells at the mid-point of virus release, the virus accumulated in the medium showed slightly higher specific activity in one case, and slightly lower in another. Control cells at the beginning of virus release in turn had slightly higher or similar specific activities than the accumulated virus. These results, too, are in keeping with rapid mixing of the cellular lecithin-glycerol pool.

A similar pulse-chase experiment was carried out by labeling the sphingomyelin-sphingosine for two hours with L-[3-^{14}C]serine. Infected BHK cells revealed quite active turnover of their sphingomyelin (Table 4). The virus accumulated in the

Table 3

Specific Radioactivity of Lecithin in Semliki Forest Virus and in Host Cells

| Timing of pulse | Specific activity of lecithin (cpm ^3H/cpm ^{32}P) | | | | Radio-activity of viral lecithin (cpm ^3H) |
	Cells at beginning of virus release	Cells in middle of virus release	Cells at end of virus release	Virus	
During virus release[a]	0		0.72	0.37	650
0–2 hr before virus release[b]	0.22	0.20	0.14	0.17	1200
4–6 hr before virus release[c]	0.57 ± 0.03	0.54 ± 0.03	0.43 ± 0.02	0.60	1500
24–26 hr before virus release	0.50		0.45	0.49	1100

The cells used were "equilibrium labeled" with [^{32}P]phosphate. Pulses of 2 hr were given and chase was carried out with 10 mM cold glycerol. SFV release begins 4 hr after the infection and continues in a relatively linear manner for 6 hr. The analyzed virus represents the population of particles released during 6 hr.

[a] Virus released during 3 hr was analyzed; pulse duration also 3 hr.

[b] Experiment carried out with [^{14}C]glycerol.

[c] Two TLC separations of the phospholipids were carried out with one cell batch. The data give the mean and the range of the observations.

medium had a higher specific activity than the actual host cells at the end of the experiment. Infected control cells in the middle of virus release had also significantly lower specific radioactivity than the virus, and even at the beginning of the virus release the control cells had slightly lower specific activities in their sphingomyelins than the virus. These results suggest that the sphingomyelin-sphingosine is not as rapidly equilibrated between the different pools of infected cells as the lecithin-glycerol.

Table 4 shows also the behavior of phosphatidylserine and phosphatidylethanolamine in the pulse-chase experiment with [3-^{14}C]serine. Phosphatidylserine of the virus particles accumulated in the medium was of higher specific radioactivity than the cells at any time of the experiment. In contrast, phosphatidylethanolamine of the virus had very nearly the same specific activity as the host cells. The specific activity of the plasmalogenic phosphatidylethanolamine was also similar in the virus and in the cells (Table 4). Interestingly the labeling of the plasmalogenic molecules from serine was a slower process than that of the diacyl phosphatidylethanolamine.

DISCUSSION

The present data suggest that lysobisphosphatidic acid (LBPA) is a lysosomal component in BHK cells. Pure lysosomes were not isolated, but several other

Table 4

Specific Radioactivity of Phospholipids in Semliki Forest Virus and Host Cells

	Specific radioactivity of sphingomyelin (cpm ^{14}C/cpm ^{32}P)				Radio-activity in viral phospholipid (cpm ^{14}C)
	Cells at beginning of virus release	Cells in middle of virus release	Cells at end of virus release	Virus	
Sphingomyelin[a]	5.5 ± 0.5	4.3 ± 0.1	3.7 ± 0.1	5.8 ± 0.1	9400
Phosphatidylserine[a]	5.5 ± 0.4	4.6 ± 0.1	4.6 ± 0.4	5.8 ± 0.4	8000
Diacyl phosphatidyl-ethanolamine	0.67	0.66	0.68	0.68	420
Plasmalogenic phosphatidyl-ethanolamine[b]	0.37	0.46	0.46	0.43	1600

The cells were "equilibrium labeled" with [^{32}P]phosphate and pulsed with L-[3-^{14}C]serine during the infection but before virus release. The pulse duration was 2 hrs; chase was carried out with 10mM cold serine. Other details as in Table 3.

[a] Two TLC separations of the phospholipids were carried out with one batch of cells and virus; the data give the mean and the range.

[b] Calculated as difference between total and plasmalogen-less phosphatidylethanolamine.

organelles were isolated and none of them contained LBPA in as high concentrations as the whole cell homogenate. The evidence of lysosomal location of LBPA in BHK cells rests on artificial production of increased number of lysosome-like vacuoles in BHK cells and the simultaneous increase of two lysosomal enzymes, acid phosphatase and protease, as well as two lipids, sphingomyelin and LBPA, both of which are known to be enriched in the lysosomes of rat liver.

The analysis of phospholipid profiles in mitochondria, nuclei, plasma membranes and endoplasmic reticulum of BHK cells allows the conclusion that Semliki Forest virus (SFV) phospholipids resemble those of plasma membranes but are quite different from those of the other organelles. Particularly significant is the fact that plasma membrane and SFV contain only very small amounts of LBPA, the lysosomal lipid. In other respects the lysosomal and the plasma membrane lipids may be quite similar (Henning, Kaulen and Stoffel 1970). Morphologic evidence suggests also that SFV acquires its lipoprotein envelope as it buds through the host plasma membrane (Acheson and Tamm 1967).

Our data show that SFV lipids can be varied by varying the host cell. In the present experiments two widely different host cells, a mammalian cell and an insect cell, were used to "offer" membrane lipids as different as possible to the envelope proteins of SFV. SFV derived from these two hosts had indeed very different phopholipids. The two sets of viral phospholipids had only one-third of the molecules in common. Previously Klenk and Choppin (1969, 1970) have shown a similar, though less dramatic, difference in phospholipids of parainfluenza virus SV5 derived from BHK cells and from monkey kidney cells. Quigley, Rifkin and Reich (1971) in turn have shown that several different viruses have similar phospholipid profiles when they are grown in identical host cells. All these data suggest that the bulk of the viral lipids is not specified by the membrane proteins, although the possibility still remains that some of the viral lipids are protein specified. SFV envelope proteins are quite typical membrane proteins; they bind detergents (Helenius and Simons 1972), and they are attached to the membrane by lipophilic polypeptide segments (Gahmberg, Utermann and Simons 1972). Therefore it seems unlikely that specific lipid-protein interactions are responsible for the *bulk* differences in the lipids of different organelle membranes found in BHK cells and in other cells.

The pulse-chase experiments conducted with radioactive lipid precursors in infected BHK cells represent an attempt to use SFV for sampling the plasma membrane lipids in pure form. This seems justified as the membrane lipids are believed to be in a fluid state (Singer and Nicolson 1972). The results support the view that the cellular lecithin-glycerol and that of the plasma membrane achieve an equilibrium fairly rapidly. In rat liver cells the same is true for lecithin-choline (Stein and Stein 1969). Intermembrane movement of intact lecithin molecules is well known in in vitro systems (Ehnholm and Zilversmith 1973), and lecithin-glycerol may also move around in form of lysolecithin (Sagami, Minari and Orii 1965; Peterson and Rubin 1969). On the other hand, equilibration based on intermembrane movement of the other glycerol-containing building blocks of lecithin, namely phosphatidic acid, diglycerides or monoglycerides, appears not likely; their reconversion into lecithin would require the machinery of the de novo synthesis, but plasma membrane appears to be devoid of the key enzyme (Victoria et al. 1971). Equilibration based on movement of free glycerol out of the question under the chase conditions used in our experiments.

Our results suggest also that sphingosine, the central building block of sphingomyelin, appears to be more slowly equilibrated between the plasma membrane and the other organelles than the glycerol of lecithin. Accordingly sphingomyelin may move more slowly than lecithin between the different organelles. This explanation is supported also by two recent findings from other laboratories: Ehnholm and Zilversmith (1973) found that the phospholipid exchange protein from beef heart stimulates a much faster exchange of lecithin than of sphingomyelin between liposomes. The observations of Lee and Snyder (1973) suggest that acetate-labeled sphingomyelin of rat liver plasma membranes turns over more slowly than its microsomal counterpart; no such difference was evident between the lecithins of these two pools.

Our data suggest that even the distal building blocks of some membrane lipids, e.g., the serine of phosphatidylserine, may equilibrate rather slowly within the total cellular pool. Their slow equilibration suggests even slower movements of the intact phospholipid molecules between the organelles.

Lecithin is found in all organelles of animal cells. Endoplasmic reticulum contains enzymes catalyzing de novo synthesis of this lipid (Wilgram and Kennedy 1963), whereas other organelles may be devoid of them (Victoria et al. 1971; McMurray and Dawson 1969; Eibl, Hill and Lands 1969). Therefore some form of transportation of lecithin-glycerol is necessary in the cells. The predominance and easy transportation of lecithin makes it useful as a general two-dimensional solvent for specific membrane lipids and proteins. Its rapid availability makes it also a potentially important precursor for other membrane lipids. For instance, Diringer et al. (1972) have suggested that lecithin donates its phosphorylcholine to ceramide, whereby sphingomyelin is formed.

In clear contrast to lecithin, most other phospholipids are not present in all organelles; phosphatidylserine and sphingomyelin are concentrated in the plasma membranes, phosphatidylinositol in the endoplasmic reticulum, cardiolipin in the mitochondria and lysobisphosphatidic acid in the lysosomes. These skewed distributions of the lipids, the lack of specific interactions between the bulk of membrane lipids and the membrane proteins, and the indications of slow equilibration of sphingomyelin and phosphatidylserine but rapid equilibration of lecithin within the cell lead us to the following hypotheses: There is a specific organelle distribution of enzymes of phospholipid metabolism. Lecithin is the transportable raw material of the lipid bilayer in the cellular membranes. Other phospholipids move slowly between the organelles in comparison to the rate of their metabolism.

Acknowledgments

Excellent technical assistance was provided by Mrs. Satu Cankar and Mrs. Anneli Asikainen. The work was conducted under tenure of research grants from Sigrid Jusélius Foundation, Helsinki, and from the Association of Finnish Life Assurance Companies, Helsinki.

REFERENCES

Acheson, N. H. and L. Tamm. 1967. Replication of Semliki Forest virus: An electron microscopic study. *Virology* **32**:128.

Body, D. R. and G. M. Gray. 1967. The isolation and characterization of phosphatidyl glycerol and a structural isomer from pig lung. *Chem. Phys. Lipids* **1**:254.

Diringer, H., W. D. Marggraf, M. A. Koch and F. A. Anderer. 1972. Evidence for a new biosynthetic pathway of sphingomyelin in SV40-transformed mouse cells. *Biochem. Biophys. Res. Commun.* **47**:1345.

Ehnholm, C. and D. B. Zilversmith. 1973. Exchange of various phospholipids and cholesterol between liposomes in the presence of highly purified phospholipid exchange protein. *J. Biol. Chem.* **248**:1719.

Eibl, H., E. E. Hill and W. E. M. Lands. 1969. The subcellular distribution of acyltransferases which catalyze the synthesis of phosphoglycerides. *Eur. J. Biochem.* **9**:250.

Fast, P. G. 1966. A comparative study of the phospholipids and fatty acids of some insects. *Lipids* **1**:209.

Gahmberg, C. G., G. Utermann and K. Simons. 1972. The membrane proteins of Semliki Forest virus have a hydrophobic part attached to the viral membrane. *FEBS Letters* **28**:179.

Harrison, S. C., J. David, J. Jumblatt and J. E. Darnell. 1971. Lipid and protein organization in Sindbis virus. *J. Mol. Biol.* **60**:523.

Helenius, A. and K. Simons. 1972. The binding of detergents to lipophilic and hydrophilic proteins. *J. Biol. Chem.* **247**:3656.

Henning, R., H. D. Kaulen and W. Stoffel. 1970. Isolation and chemical composition of the lysosomal and the plasma membrane of the rat liver cell. *Hoppe-Seyler's Z. Physiol. Chem.* **351**:1191.

Klenk, H.-D. and P. W. Choppin. 1969. Lipids of plasma membranes of monkey and hamster kidney cells and of parainfluenza virions grown in these cells. *Virology* **38**:255.

———. 1970. Plasma membrane lipids and parainfluenza virus assembly. *Virology* **40**:939.

Laine, R. A., H. Söderlund and O. Renkonen. 1973. The chemical composition of Semliki Forest virus. *Intervirology* **1**:110.

Laine, R. A., M.-L. Kettunen, C. G. Gahmberg, L. Kääriäinen and O. Renkonen. 1972. Fatty chains of different lipid classes of Semliki Forest virus and host cell membranes. *J. Virol.* **10**:433.

Lee, T.-C. and F. Snyder. 1973. Phospholipid metabolism in rat liver endoplasmic reticulum. Structural analyses, turnover studies and enzymic activities. *Biochim. Biophys. Acta* **291**:71.

McMurray, W. C. and R. M. C. Dawson. 1969. Phospholipid exchange reactions within the liver cell. *Biochem. J.* **112**:91.

Neville, D. M. 1971. Molecular weight determination of protein-dodecyl sulfate complexes by gel electrophoresis in a discontinuous buffer system. *J. Biol. Chem.* **246**:6328.

Peterson, J. A. and H. Rubin. 1969. The exchange of phospholipids between cultured chick embryo fibroblasts and their growth medium. *Exp. Cell Res.* **58**:365.

Quigley, J. P., D. B. Rifkin and E. Reich. 1971. Phospholipid composition of Rous sarcoma virus, host cell membranes and other enveloped RNA viruses. *Virology* **46**:106.

Renkonen, O., L. Kääriäinen, K. Simons and C. G. Gahmberg. 1971. The lipid class composition of Semliki Forest virus and of plasma membranes of the host cells. *Virology* **46**:318.

Renkonen, O., C. G. Gahmberg, K. Simons and L. Kääriäinen. 1972a. The lipids of the plasma membranes and endoplasmic reticulum from cultured baby hamster kidney cells (BHK21). *Biochim. Biophys. Acta* **255**:66.

Renkonen, O., L. Kääriäinen, R. Petterson and N. Oker-Blom. 1972b. The phospholipid composition of Uukuniemi virus, a non-cubical tick-borne arbovirus. *Virology* **50**:726.

Sagami, T., O. Minari and T. Orii. 1965. Behavior of plasma lipoproteins during exchange of phospholipids between plasma and erythrocytes. *Biochim. Biophys. Acta* **98**:111.

Simons, K., L. Kääriäinen, O. Renkonen, C. G. Gahmberg, H. Garoff, A. Helenius, S. Keränen, R. Laine, M. Ranki, H. Söderlund and G. Utermann. 1973. Semliki Forest virus as a simple membrane model. *Membrane-mediated information* (ed. P. W. Kent) Oxford (in press).

Singer, S. J. and G. L. Nicolson. 1972. The fluid mosaic model of the structure of cell membrane. *Science* **175**:720.

Stein, O. and Y. Stein. 1969. Lecithin synthesis, intracellular transport and secretion in rat liver. *J. Cell Biol.* **40**:461.

Victoria, E. J., L. M. G. van Golde, K. Y. Hostetler, G. L. Sherphof and L. L. M. van Deenen. 1971. Some studies on the metabolism of phospholipids in plasma membranes from rat liver. *Biochim. Biophys. Acta* **239**:443.

Weber, K. and H. Osborn. 1969. The reliability of molecular weight determinations by dodecyl sulfate polyacrylamide gel electrophoresis. *J. Biol. Chem.* **244**:4406.

Wherret, J. R. and S. Huterer. 1972. Enrichment of bis-(monoacylglyceryl) phosphate in lysosomes from rat liver. *J. Biol. Chem.* **247**:4114.

Wilgram, G. F. and E. P. Kennedy. 1963. Intracellular distribution of some enzymes catalyzing reactions in the biosynthesis of complex lipids. *J. Biol. Chem.* **238**:2615.

Alterations in the Pattern and Synthesis of Gangliosides in Tumorigenic Virus-Transformed Cells

Roscoe O. Brady and Peter H. Fishman

Developmental and Metabolic Neurology Branch
National Institute of Neurological Diseases and Stroke
National Institutes of Health, Bethesda, Maryland 20014

It is virtually certain that some aspects of the control of proliferation of cells are mediated by biochemical or physicochemical events on cell surfaces. The primary and oft-cited support for this speculation is the well-known phenomenon of contact inhibition of movement and growth of "normal" cells in tissue culture. After these cells are seeded, they multiply until they form a monolayer covering all of the available subtratum. At this point cell division stops and cellular motion almost completely ceases. On the other hand, cells which have become transformed "spontaneously" or by exposure to tumorigenic viruses or a carcinogenic chemical lose this regulation of growth, and they continue to divide after coming into contact with each other so that multiple layers of cells are formed. The most naive assumption for this unchecked proliferation of cells is that a depletion of some component(s) on the surface of the transformed cells has occurred with a consequent disappearance of a signal from the plasma membrane, which normally mediates the cessation of cell division. Because acidic glycolipids called gangliosides are known to be highly concentrated in membranous elements of cells, we undertook an investigation of the composition and metabolism of these compounds in control and various transformed cell lines. We summarize here our findings in this area with tumorigenic DNA and RNA virus-transformed cells and discuss some preliminary experiments with chemically induced tumorigenic cell lines.

METHODS AND PROCEDURES

Most of the work described in this report was carried out with well-characterized mouse cell lines. "Normal" and transformed lines were derived from the AL/N mouse strain (Takemoto et al. 1968), the Balb/c mouse strain (Aaronson and Todaro 1968) and the Swiss mouse strain (Pollack and Burger 1969). Lipids were isolated and analyzed by standard extraction and thin-layer chromatographic procedures (Folch, Lees and Stanley 1957; Mora et al. 1969; Brady, Borek and Bradley 1969). These determinations were verified by gas-liquid chromatography of the glycolipids (Dijong, Mora and Brady 1971). Various labeled aminosugar precursors were added to the tissue culture medium in certain experiments (Brady, Borek and Bradley 1969; Brady and Mora 1970). Procedures for the determination of glycosyltransferase activities are described in detail elsewhere (Cumar et al. 1970; Fishman et al. 1972).

RESULTS

Contact-inhibited mouse cell lines have a homologous series of gangliosides ranging from monosialosyllactosylceramide (G_{M3}) to disialosyltetrahexosylceramide (G_{D1a}) (Fig. 1). These substances often appear as doublets on thin-layer chromatograms due to the presence of the respective N-acetyl- or N-glycolylneuraminic acid derivatives. The latter lipids migrate slightly slower than the N-acetylneuraminic acid-containing gangliosides. It is also possible that differences in the mobility of the individual gangliosides may be due to variation in the length of the fatty acid component of the ceramide moiety.

The ganglioside pattern is greatly simplified in mouse cells that have been transformed by the tumorigenic DNA viruses, Simian virus 40 and polyoma virus. In the virus-transformed cell lines only monosialosyllactosylceramide is present (Fig.

Ceramide = N-acylsphingosine

$$CH_3-(CH_2)_{12}-CH{=}CH-CH(OH)-CH-C^*H_2OH$$

$$\begin{array}{c} | \\ N-H \\ | \\ C{=}O \\ | \\ (CH_2)_n \\ | \\ CH_3 \end{array}$$

G_{M3} (hematoside):   Cer—Glc—Gal—NANA

G_{M2}:   Cer—Glc—Gal—GalNAc
                                                |
                                            NANA

G_{M1}:   Cer—Glc—Gal—GalNAc—Gal
                                                |
                                            NANA

G_{D1a}:   Cer—Glc—Gal—GalNAc—Gal
                                                | |
                                            NANA   NANA

Figure 1
Structures of gangliosides in contact-inhibited mouse cell lines. * Indicates point of attachment of various substituents. Cer = ceramide, Glc = glucose, Gal = galactose, GalNAc = N-acetylgalactosamine, NANA = N-acetylneuraminic acid.

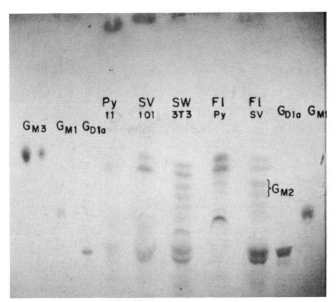

Figure 2
Thin-layer chromatogram of gangliosides in Swiss mouse cell lines. Py11 = polyoma virus-transformed line; SV101 = Simian virus 40-transformed line; SW3T3 = contact-inhibited parent cell line; Fl Py and Fl SV are flat revertant lines from Py11 and SV101, respectively. G_{M3}, G_{M2}, G_{M1}, and G_{D1a} are ganglioside standards. The plate was sprayed with resorcinol to identify the sialoglycolipids. The dark areas in the region of G_{D1a} in the Py11 and SV101 lanes contain material which does not react with resorcinol. From Mora, Cumar and Brady (1971) with permission of *Virology*.

2). The pattern of labeling of gangliosides obtained when [³H]N-acetylmannosamine or [³H]glucosamine is added to the incubation medium is consistent with the alteration in ganglioside content observed in the transformed cells (Fig. 3). The quantity of gangliosides was determined and data from a representative experiment are presented in Table 1. This alteration in the ganglioside content of the tumorigenic DNA virus-transformed cells is summarized in schematic form in Fig. 4. We observed this change in ganglioside composition in 20 out of 21 individually transformed cell lines, which were obtained from a variety of independent sources. The single exception was an SV40 transformant of a Balb/c line selected for bromodeoxyuridine resistance (Brady, Fishman and Mora 1973).

We undertook studies to determine the metabolic alteration that caused the change in the ganglioside composition of the virus-transformed cells. There were two primary possibilities which required investigation. The first was an excessively rapid rate of catabolism of gangliosides larger than G_{M3} in the transformed cells. This possibility was excluded by investigating the catabolism of gangliosides which had been specifically labeled in the sialic acid portion of the molecule (Cumar et al. 1970). On the basis of this information, we directed our attention to an examination of the biosynthetic reactions involved in the formation of gangliosides larger than G_{M3}. The synthesis of these glycolipids occurs in a stepwise fashion and the

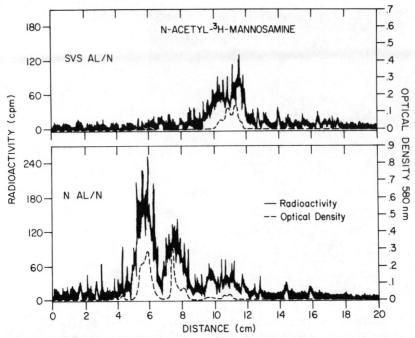

Figure 3

Thin-layer chromatogram of gangliosides from normal (N AL/N) and Simian virus 40-transformed cells (SVS AL/N) grown in the presence of [³H]N-acetyl-mannosamine. After the distribution of labeled gangliosides was determined by scanning the plates for radioactivity, the gangliosides were detected by spraying with resorcinol. From Brady and Mora (1970) with permission of *Biochim. Biophys. Acta*.

conversion of G_{M3} to G_{M2}, the next higher homolog, is catalyzed by the aminosugar transferase indicated in Reaction 1:

Cer-Glc-Gal-NeuNAc (G_{M3}) + UDP-GalNAc

$$\xrightarrow[\text{minyltransferase}]{\text{hematoside:N-acetylgalactosa-}}$$

Cer-Glc-Gal-(NeuNAc)-GalNAc (G_{M2}) + UDP

The activity of this enzyme is drastically reduced in cells which have been transformed by a tumorigenic DNA virus (Table 2). This alteration in enzymatic activity was seen in all 20 of the virus-transformed cell lines that showed the changes in the ganglioside pattern. Other enzymes involved in ganglioside synthesis indicated in Fig. 5 were either normal, increased, or on occasion, slightly decreased in activity (Fishman et al. 1972). However, the common consistent change was the diminished activity of the hematoside:*N*-acetylgalactosaminyltransferase.

It was necessary to learn whether the alteration in ganglioside pattern and synthesis would occur in cells which were not transformed but were productively infected with the tumorigenic viruses. Our experiments showed that there was no change in glycolipid content and metabolism under these conditions. The change in ganglioside composition in these cells occurred only when the transforming virus genome was stably incorporated into the genetic apparatus of the host cell (Mora, Cumar and Brady 1971). We then began a series of experiments to examine the

Table 1

Distribution of Gangliosides in Normal
and Transformed Mouse Cell Lines

	G_{D1a}	G_{M1}	G_{M2}	G_{M3}	Total
			(nmoles/mg of protein)		
N AL/N	1.8	1.5	0.8	0.6	4.7
SVS AL/N	0.16	0.22	0.1	1.9	2.4
Py AL/N	0.1	0.15	0.2	1.8	2.3
Swiss 3T3	2.4	2.6	1.8	4.0	10.8
Py 11*	0.1	0.2	0.05	3.2	3.5
SV101†	0.05	0.1	0.05	3.5	3.7

* Polyoma virus-transformed Swiss mouse cell line.

† Simian virus 40-transformed Swiss mouse cell line. From Mora et al. (1969) and Brady and Mora (1970).

effect of altering various physiological parameters on ganglioside metabolism in control and virus-transformed cells. There was no discernible difference in these components when the cells were harvested at different stages of cell density, varying from sparse to confluent (Fishman et al. 1972). There was no evidence for production of a diffusible inhibitor of the aminosugar transferase by the transformed cells; neither did we find any intracellular accumulation of an inhibitor of these enzymes in the transformed cells (Mora, Cumar and Brady 1971). On the other hand, there was no indication of a factor in the normal cells which would correct this enzymatic deficiency in the transformed cells.

A potentially very important finding was made in an investigation of flat revertant cell lines obtained from SV40- and polyoma virus-transformed cells. These revertant cell lines were obtained by the addition of fluorodeoxyuridine to the culture medium. Some of these cells showed all of the characteristics of the contact-inhibited state, and they were not as tumorigenic as the transformed lines from which they were derived. The pattern of gangliosides and hematoside:*N*-acetylgal-

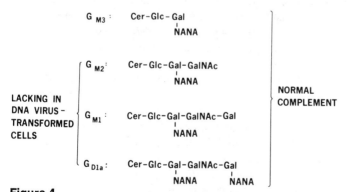

Figure 4

Gangliosides of normal or tumorigenic virus-transformed mouse cell lines. Cer = ceramide; Glc = glucose; Gal = galactose; GalNAc = *N*-acetylgalactosamine; NANA = *N*-acetylneuraminic acid.

Table 2

Uridine Diphosphate N-Acetylgalactosamine:
Hematoside N-Acetylgalactosaminyltransferase
Activity in Mouse Cell Lines

	Glycolipid acceptor	
	$G_{M3}NAc$*	$G_{M3}NGlyc$†
	(Activity as percent of control cells)	
N AL/N	100	100
T AL/N	189	100
SVS AL/N	0	10
Py AL/N	0	0
Swiss 3T3	100	100
SV101	33	30
Py 11	26	2

Data calculated from Cumar et al. (1970).

* N-acetylneuraminylgalactosylglucosylceramide (hematoside).

† N-glycolylneuraminylgalactosylglucosylceramide (hematoside).

actosaminyltransferase activity returned to normal in a revertant line of cells derived from an SV40-transformed cell line (Mora, Cumar and Brady 1971). Two aspects of these studies should be emphasized. The first is the fact that the tumorigenic virus could be rescued from this flat revertant cell line; therefore at least the viral genome is present in the chromosomes of the revertant cell in an unexpressed state. The second is that the number of chromosomes in the revertant cells had increased to the subtetraploid level (Pollack, Wolman and Vogel 1970). These factors seem to be significant and appear to be likely to provide insight into the mechanism of tumorigenic virus transformation, once appropriate experiments exploiting these observations can be devised.

Our attention has also been directed to an investigation of the content and metabolism of gangliosides in tumorigenic RNA virus-transformed cell lines. This study provided substantial support for the concept that changes in gangliosides are really a consequence of viral transformation. Studies of cells transformed with tumorigenic RNA viruses have some advantages over investigations with tumorigenic DNA viruses. The first is that, at the present time, many more animal tumors have been shown to be caused by RNA viruses than DNA viruses. The second is that transformation of cells with RNA viruses can approach 100% if appropriate in vitro conditions are used, whereas transformation with tumorigenic DNA viruses is a much less frequent event. Thus the possibility of selection of a metabolically disadvantaged cell, which is more susceptible to transformation by tumorigenic DNA viruses, can be eliminated in the case of RNA viruses. We therefore undertook a series of investigations with the Moloney strain of murine sarcoma virus using Swiss mouse 3T3 cells. Two days after exposure to the virus, nearly all of the cells appeared transformed in their morphological characteristics. At this time there was no change in the activity of ganglioside synthesizing enzymes. However by four days after infection, the activity of the hematoside:N-acetylgalactosaminyltransferase was only 50% of that in control, mock-infected cells. Sub-

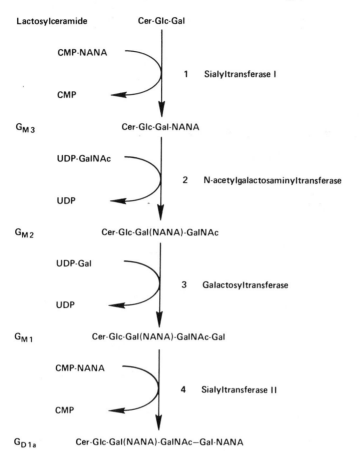

Figure 5

Schematic representation of the reactions involved in the biosynthesis of gangliosides.

sequently there was a further decline in the activity of this enzyme in the transformed cells compared with that in the control cells. Other glycosyltransferases involved in ganglioside formation (Fig. 5) were unchanged or slightly increased in activity (Mora et al. 1973). The content of gangliosides larger than G_{M3} decreased to 20% of that in the control cells by the 6th day after exposure to the murine sarcoma virus. Infection with murine leukemia virus did not cause a change in the glycosyltransferase activities or in the ganglioside pattern.

We have recently observed a slightly different alteration of the pattern of gangliosides in another RNA virus-transformed cell system. When contact-inhibited Balb/c cells, which have a full complement of gangliosides, were transformed with the Kirsten strain of murine sarcoma virus, there was a decrease of gangliosides larger than G_{M2} rather than G_{M3} as observed in the previous studies with DNA and RNA virus-transformed cells. This change is due to a specific block of G_{M2}:galactosyltransferase activity (Fishman, Brady, Todaro and Aaronson, manuscript in preparation). There was no decrease in the activity of hematoside:N-acetylgalactosaminyltransferase activity in these cells.

A final aspect of our studies concerns glycolipid metabolism in chemically trans-

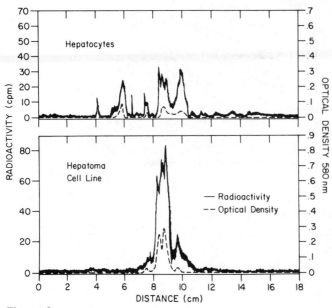

Figure 6

Scans of optical density at 580 nm and radioactivity of thin-layer chromatograms of gangliosides from rat hepatocyte and hepatoma cells grown in the presence of [³H]N-acetylmannosamine. From Brady, Borek and Bradley (1969) with permission of *J. Biol. Chem.*

formed cells. There was a marked difference in the labeling pattern of gangliosides in cultures of a minimally deviated hepatoma cell line compared with that of hepatocytes when [³H]N-acetyl-mannosamine was added to the culture media (Fig. 6). It should be clearly indicated that the large peak of radioactivity in thin-layer chromatogram of the gangliosides isolated from the hepatoma cell line is in G_{M2}, rather than G_{M3} as originally believed. We have carried out preliminary experiments with cell lines chemically transformed in vitro. In these transformed cells, there was a 3- to 4-fold increase in the quantity of G_{M2} compared with that in the non-transformed control cells, which could be attributed to decreased galactosyltransferase activity. However in contrast with the virus-transformed cells, some disialoganglioside (G_{D1a}) appeared to be synthesized by these chemically transformed cells (Fishman, Oshiro, Dipaolo and Brady, unpublished observations).

DISCUSSION

The first question that requires careful consideration concerns the significance of the alteration of ganglioside composition and synthesis in the tumorigenic cells. Although gangliosides are highly enriched in the plasma membrane of cells, the role of these substances in the social behavior of cells is not understood at the present time. The decrease in the sialic acid-containing glycolipids in the virus-transformed cells may lessen the net negative charge on the surface of cells, but the importance of this effect is far from clear. In particular, spontaneously transformed tumorigenic cells also show loss of contact inhibition of growth and movement in culture, yet the ganglioside pattern in these cells is usually similar to that in the contact-inhibited parent cell lines.

On the other hand, we cannot dismiss the dramatic alteration in ganglioside pattern and synthesis since it is clearly a consequence of stable transformation of cells with tumorigenic DNA and RNA viruses. We have seen these effects in 20 out of 21 DNA virus-transformed lines, and other investigators have reported a related change in virus-transformed hamster kidney cells (Hakomori and Murakami 1968). However there are two recent reports of a lack of change in gangliosides after viral transformation of a mouse cell line (Yogeeswaran et al. 1972; Diringer, Strobel and Koch 1972). Thus we must be cautious in our attempts to interpret the importance of this manifestation of the insertion of viral genetic material. The fact that the formation of gangliosides is affected by tumorigenic DNA and RNA virus transformation of most cells, coupled with the fact that this alteration does not occur in lytically infected cells or cells exposed to nontransforming viruses, makes this phenomenon preeminently worthy of exploration. The principal question that must be answered is how the insertion of virus genetic material causes the decrease in the activity of hematoside:N-acetylgalactosaminyltransferase, or in the special case of the RNA virus-transformed cell line, G_{M2}:galactosyltransferase activity. Among the more obvious possibilities is the production of an inhibitor of these enzymes; however we have not been able to obtain conclusive evidence for the formation of such a substance (Cumar et al. 1970; Mora, Cumar and Brady 1971).

Alternative hypotheses would be the presence of a virus-specific protein that represses the formation of the glycosyltransferase messenger RNA or its subsequent translation into transferase protein or integration of the viral genome into the host chromosome at the site which codes for the affected glycosyltransferase. A recent observation in our laboratory of a specific effect on ganglioside synthesis in a cell line containing a temperature-sensitive transforming mutant virus may help us to differentiate between these alternatives. The ganglioside composition is simplified when the effect of the virus is expressed at a low, permissive temperature in culture, whereas the pattern and synthesis of gangliosides returns to that in the nontransformed parent cell line at higher temperatures and the cells no longer manifest other properties of tumorigenic transformation (Fishman, Aaronson and Brady, unpublished observations). It is conceivable that transcription of the cistron for the aminosugar transferase is impaired when a virus-specific repressor protein, or possibly the intercalated viral genome, assumes a particular configuration at the lower permissive temperature. At the higher temperature a change in production or conformation of the repressor or uncoiling of the viral DNA now allows transcription to occur in an apparently normal fashion. Experiments with this temperature-modulated system should provide an exceptional opportunity to obtain insight into the nature of the regulatory event or factor(s) involved in expression of the integrated virus genome, and it is certain that this system will be utilized extensively.

Another aspect of the changes in ganglioside metabolism in virus-transformed cells which requires comment is the return of hematoside:N-acetylgalactosaminyltransferase activity in the flat revertant cells produced by Pollack and his coworkers (1970). The activity of this aminosugar transferase, which was drastically reduced in the SV40 virus-transformed cells, was slightly greater in the revertant cell line than in the parent Swiss 3T3 mouse cell line from which these cells were derived. The ganglioside pattern was virtually the same in the revertant cells as in the original cell line. However the polyoma-revertant cells were phenotypically and karyotypically heterogeneous, and the activity of the depleted aminosugar transferase and the pattern of gangliosides were only partially restored to that in the parent cell line. The block and subsequent restoration of ganglioside synthesis in these cells may be due to a gene-dose effect since the greatest activity of the hematoside:N-acetylgalactosaminyltransferase was found in the subtetraploid SV40-revertant cells. Presumably a sufficient quantity of messenger RNA for this enzyme is formed in these cells to overcome the block caused by the SV40 virus genome, which has been demonstrated in these cells. This hypothesis is consistent with very recent experiments with a revertant cell line from a tumorigenic RNA virus transformant in which the ganglioside pattern did not return to that in the parent cell line (Fishman, Todaro and Brady, unpublished observations). The karyotype of the normal, transformed and reverted cells was essentially diploid in each line.

A brief comment seems appropriate regarding certain similarities between the increase in ganglioside G_{M2} seen in the hepatoma cells and cells chemically transformed in vitro and the lack of ganglioside homologs larger than this component in the transformed cell line produced with the Kirsten strain of murine sarcoma virus. These observations suggest that the mechanism of tumorigenesis in these cells may be related and raises the possibility of activation of a latent virus in the chemically transformed cells (Aaronson, Hartley and Todaro 1969; Huebner and Todaro 1969). This possibility merits investigation in the near future.

SUMMARY

The pattern of gangliosides in tumorigenic DNA and RNA virus-transformed cell lines is markedly simplified from that in their respective contact-inhibited parent cell lines. This change in transformed mouse cells is primarily due to a decrease in the activity of hematoside:N-acetylgalactosaminyltransferase, which catalyzes an early step in the synthesis of the oligosaccharide chain of gangliosides. Recently a block of galactosyltransferase activity has been observed in a conventional cell line transformed by a particular strain of murine sarcoma virus. These cell systems are very useful for investigating the metabolic and compositional alterations which are the consequences of viral carcinogenesis.

REFERENCES

Aaronson, S. A. and G. J. Todaro. 1968. Basis for the acquisition of malignant potential by mouse cells cultivated in vitro. *Science* **162**:1024.

Aaronson, S. A., J. W. Hartley and G. J. Todaro. 1969. Mouse leukemia virus: "Spontaneous" release by mouse embryo cells after long term in vitro cultivation. *Proc. Nat. Acad. Sci.* **64**:87.

Brady, R. O. and P. T. Mora. 1970. Alteration in ganglioside pattern and synthesis in SV40 and polyoma virus transformed mouse cell lines. *Biochim. Biophys. Acta* **218**:308.

Brady, R. O., C. Borek and R. M. Bradley. 1969. Composition and synthesis of gangliosides in rat hepatocyte and hepatoma cell lines. *J. Biol. Chem.* **244**:6552.

Brady, R. O., P. H. Fishman and P. T. Mora. 1973. Membrane components and enzymes in virally transformed cells. *Fed. Proc.* **32**:102.

Cumar, F. A., R. O. Brady, E. H. Kolodny, V. W. McFarland and P. T. Mora. 1970.

Enzymatic block in the synthesis of gangliosides in DNA virus-transformed tumorigenic mouse cell lines. *Proc. Nat. Acad. Sci.* **67:**757.

Dijong, I., P. T. Mora and R. O. Brady. 1971. Gas chromatographic determination of gangliosides in mouse cell lines and in virally transformed derivative lines. *Biochemistry* **10:**4039.

Diringer, H., G. Strobel and M. A. Koch. 1972. Glycolipids of mouse fibroblasts and virus transformed mouse cell lines. *Hoppe-Seyler's Z. Physiol. Chem.* **353:**1759.

Fishman, P. H., V. W. McFarland, P. T. Mora and R. O. Brady. 1972. Ganglioside biosynthesis in mouse cells: Glycosyltransferase activities in normal and virally-transformed lines. *Biochem. Biophys. Res. Commun.* **48:**48.

Folch, J., M. Lees and G. H. S. Stanley. 1957. A simple method for the isolation and purification of total lipids from animal tissues. *J. Biol. Chem.* **226:**497.

Hakomori, S. I. and W. T. Murakami. 1968. Glycolipids of hamster fibroblasts and derived malignant-transformed cells. *Proc. Nat. Acad. Sci.* **59:**254.

Huebner, R. and G. J. Todaro. 1969. Oncogenes of RNA tumor viruses as determinants of cancer. *Proc. Nat. Acad. Sci.* **64:**1087.

Mora, P. T., F. A. Cumar and R. O. Brady. 1971. A common biochemical change in SV40 and polyoma virus transformed mouse cells coupled to control of cell growth in culture. *Virology* **46:**60.

Mora, P. T., R. O. Brady, R. M. Bradley and V. W. McFarland. 1969. Gangliosides in DNA virus-transformed and spontaneously transformed tumorigenic mouse cell lines. *Proc. Nat. Acad. Sci.* **63:**1290.

Mora, P. T., P. H. Fishman, R. H. Bassin, R. O. Brady and V. W. McFarland. 1973. Transformation of Swiss 3T3 cells by murine sarcoma virus is followed by decrease in a glycolipid glycosyltransferase. *Nature* (in press).

Pollack, R. E. and M. M. Burger. 1969. Surface specific characteristics of a contact inhibited cell line containing the SV40 viral genome. *Proc. Nat. Acad. Sci.* **62:**1074.

Pollack, R. E., S. Wolman and A. Vogel. 1970. Reversion of virus-transformed cell lines: Hyperploidy accompanies retention of virus genes. *Nature* **228:**938.

Takemoto, K. K., R. C. Y. Ting, H. L. Ozer and P. Fabish. 1968. Establishment of a cell line from an inbred mouse strain for viral transformation studies: Simian virus 40 transformation and tumor production. *J. Nat. Cancer Inst.* **41:**1401.

Yogeeswaran, G., R. Scheinin, J. R. Wherett and R. K. Murray. 1972. Studies on the glycosphingolipids of normal and virally transformed 3T3 mouse fibroblasts. *J. Biol. Chem.* **247:**5146.

Biological and Biochemical Characterization of Surface Changes in Normal, MSV- and SV40-Transformed, and Spontaneously Transformed Clones of Balb/c Cells

William J. Grimes

Department of Biochemistry, College of Medicine
University of Arizona, Tucson, Arizona 85724

Many cell surface alterations have now been detected when normal cells are transformed by viruses. Low concentrations of plant lectins preferentially agglutinate transformed cells (Burger and Goldberg 1967; Burger 1969; Inbar and Sachs 1969). Transformed cells have an increased ability to transport amino acids, carbohydrates, and nucleosides (Foster and Pardee 1969; Cunningham and Pardee 1969; Hatanaka and Hanafusa 1970; Hatanaka, Augl and Gilden 1970; Isselbacher 1972). Recently observations that SV40-3T3 cells have an increased ability to catalyze carbohydrate transport have been questioned (Romano and Colby 1973). Whatever the mechanism, measurements of 2-deoxyglucose uptake have been shown to be capable of differentiating between normal and transformed cells (Martin et al. 1971; Grimes and Schroeder 1973).

Sialic acid levels in cells transformed by viruses are generally reduced (Ohta et al. 1968; Wu et al. 1969; Grimes 1970; Perdue et al. 1972). And leukemic lymphocytes have reduced sialic acid levels as compared with their normal counterparts (Kornfeld 1969; McClelland and Bridges 1973). While these changes have been observed for many cell systems, exceptions have been noted (Grimes 1973; Hartman et al. 1972). Other workers have reported that many different carbohydrate moieties are reduced when 3T3 cells are transformed by SV40 (Wu et al. 1969).

Alterations in glycoproteins and glycolipids accompany transformation (Meezan, Black and Robbins 1969; Onodera and Sheinin 1970; Buck, Glick and Warren 1970; Cumar et al. 1970; Robbins and Macpherson 1971; Hakomori 1970; Sakiyama and Robbins 1973; Yogeeswaran et al. 1972). And glycosyl transferases which participate in complex polysaccharide syntheses are altered by transformation (Grimes and Robbins 1972; Cumar et al. 1970; Kijimoto and Hakomori 1972; Warren, Fuhrer and Buck 1972).

From the rapidly expanding literature describing transformed cells, it is clear that most of the observed changes are secondary to transformation. Furthermore, the early belief that the simple DNA tumor viruses, SV40 and polyoma, would cause few secondary changes appears not to be tenable. In order to generate a system that may provide answers to biological questions of surface change in transformation, we have initiated studies of primary and secondary normal cells and their spontaneous and viral transformants. We are interested in three aspects of cell surface involvement in cancer: loss of contact inhibition, increased invasiveness, and altered immunogenicity.

MATERIALS AND METHODS

[^{14}C]CMP sialic acid (229 mCi/mmole) and [^{3}H]2-deoxyglucose (7.2 Ci/mmole) were purchased from New England Nuclear, Boston. Ability to transport [^{3}H]2-deoxyglucose was determined as described (Grimes and Schroeder 1973). Concanavalin A was purchased from Miles Laboratories (Kankakee, Ill.). Ability of cells to bind human type O red blood cells was determined by published procedures (Furmanski et al. 1972).

Cell Lines

A31 is a cloned line of Balb/c 3T3 cells and was a gift from Dr. G. Todaro of N.I.H. SVT2 and 3T12 are SV40-transformed and spontaneously transformed Balb/c 3T3 cells, respectively. MSV are mouse sarcoma virus-transformed Balb/c 3T3 cells and were a gift from Dr. L. Culp of Case Western University. MSV cells used in these studies do not produce virus or viral particles and have been shown to lack murine leukemia virus antigens (Aaronson and Rowe 1970). Primary Balb/c mouse fibroblasts were produced by trypsinizing mouse embryos. Clones of A31 cells were prepared by plating cells at densities of 50–100 cells per 35-mm plate. From one to three clones grew per plate and several were selected for further study. Isolation of clones was performed as described (Puck, Marcus and Creceuia 1956). Routine tests for mycoplasma contamination using [^{3}H]uridine were negative (Levine 1972).

Cell Carbohydrate Analysis

Monolayers of cells growing in Bellco roller bottles (1400 cm^2 growing area) were rinsed with 50 ml of solution A (0.8% NaCl, 0.05% KCl, 0.001 M potassium phosphate, pH 7.4) containing 0.01 M EDTA. Cells were removed from the monolayer in 50 ml of solution A-EDTA and collected by centrifugation at $1000 \times g$ for 5 minutes. Pelleted cells were suspended in 10 ml of TBS (0.15 M NaCl–0.02 M Tris HCl, pH 7.4) and recentrifuged. Washed cell pellets were stored frozen in 2 ml of TBS.

Sialic acid levels of cells were determined by described procedures. We have routinely purified hydrolyzed sialic acid on Dowex 1 (formate) columns prior to analysis by the thiobarbituric acid procedure (Grimes 1973; Warren 1959).

Frozen cell pellets (5–10 mg of protein) were suspended in 10 ml of 0.01 M Tris pH 8.0. After 10 minutes cells were homogenized by 10–20 strokes using a tight fitting Dounce homogenizer. One ml containing 0.03 M MgCl$_2$–0.1 M NaCl was immediately added and the nuclei separated by sedimentation at $1000 \times g$ for 5 minutes. Glycoprotein and glycolipid materials were then pelleted by centrifugation at 37.5 k rpm for one hour. The microsomal pellets were suspended in 2 ml of water and an aliquot removed for protein analysis. The preparation was lyophilized and the residue extracted three times with 2 ml of chloroform:methanol (2:1). Chloroform-methanol soluble (glycolipids) and insoluble (glycoproteins) materials were analyzed for carbohydrates by gas–liquid chromatography. These fractions are referred to as Cx and Px, respectively.

Analysis of alditol acetates was performed essentially as described by Albersheim et al. (1967; Jones and Albersheim 1972). Fractions (containing from 1–10 mg of microsomal protein or C:M extractions) were hydrolyzed in 2 N trifluoroacetic acid at 120°C for 90 minutes. 2-Deoxyglucose (50 μg) was then added as an internal standard and the hydrolysate dried using filtered air. The residue was suspended in 0.5 ml of water and extracted five times with one ml aliquots of hexane. NaBH$_4$ in NH$_4$OH was then added to final concentrations of 0.3 N and 1 N, respectively. Following reduction at room temperature for one hour, glacial acetic acid was added and the reaction dried at 40°C. The alditols were extracted into 70% alcohol, transferred to new tubes, and dried. One ml of MeOH was added and the sample again dried. This was repeated a total of five times. After drying overnight in a vacuum desiccator over P$_2$O$_5$, one ml of acetic anhydride was added, the tube sealed, and the reaction placed in an autoclave for acetylation at 120°C for one hour. The sample was carefully dried and desiccated overnight. The derivatized sugars were extracted in several portions of methylene chloride and the pooled extractions dried in 0.3-ml reaction vials. For analysis 50 μl of acetone was added and 5 μl injected into a Hewlett-Packard 5700 chromatograph. We used glass columns (6 ft × 1/8 in.) containing 3% OV225 on Supelcon (Supelco, Bellefonte, Pa.). The columns were programed for 8 minutes at 180°C followed by a 4°C/min increase to a final temperature of 230°C. Responses and retention times were determined on a Hewlett-Packard 3370b integrator.

Tumorigenic Studies

Cells used for tumorigenic studies were harvested in log phase using 0.05% trypsin to effect cell suspension. Cells were washed once and suspended at a concentration of 2.5 × 10^7 cells/ml in 4× MEM, 0.5% fetal bovine serum. Cell samples were diluted in the same medium prior to injection. The backs of 3-week-old male Balb/c mice were injected subcutaneously with 0.2 ml of the cell suspension. Mice were checked for palpable tumors every two days. Cell lines were produced by dissecting the tumor from surrounding tissue and placing excised pieces (20 × 20 mm) into 4× MEM containing 10% fetal bovine serum. Within several days large numbers of cells migrated from the tumor mass forming a monolayer.

RESULTS

Growth characteristics of normal cells and viral or spontaneous transformants are shown in Fig. 1. A31 and PBC cells exhibit contact-regulated control of cell growth and a limited saturation density. 3T12 and MSV cells grow past densities where 3T3 cells are contact inhibited but fail to reach cell densities achieved by SVT2 and clones 3,5,7 and 8. Clones 3 and 8 had cells with a flat morphology. At low densities both clones exhibited growth properties resembling contact inhibition but became very crowded with continued culture. Clones 3 and 8 were readily distinguished from clones 5 and 7 and from SVT2 cells. The latter were composed of

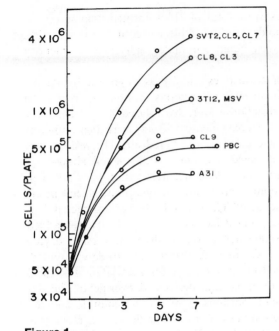

Figure 1

Growth curves of normal and transformed cell lines. Cells were grown on 35-mm petri plates containing 2 ml of 4× MEM and 10% fetal calf serum. On day 0, 5 × 10⁴ cells were added to each plate. Fresh medium was added every other day. A31 is the established line of Balb/c 3T3 cells. Cl 3,5,7,8 and 9 are clones derived from A31 cells. 3T12, MSV and SVT2 cells are spontaneous, murine sarcoma virus and SV40-transformed Balb/c 3T3 cells, respectively. PBC is a secondary line of Balb/c fibroblasts which has been passaged less than five times in culture (Methods).

small spindly cells, with no evidence of contact inhibition. (Photographs of SVT2, cl 5, and cl 8 cells are shown in Fig. 7.)

Cell saturation densities and the ability to transport [³H]2-deoxyglucose are shown in Table 1. The transformed cells SVT2, MSV, cl 5 and cl 7 have elevated abilities to transport the sugar analogue relative to normal A31 and PBC cells. Surprisingly, the non-contact-inhibited cell lines, cl 3, cl 8, and 3T12 cells transport [³H]2-deoxyglucose similarly to normal cells. Increased transport does not appear to be a necessary prerequisite for failure of contact inhibition.

Agglutination has routinely been determined by carefully suspending cells in varying concentrations of lectins and then determining aggregation through a phase microscope (Burger and Goldberg 1967; Inbar and Sachs 1969). Recently a new procedure has been described whereby agglutination is determined by adding human type O red blood cells to a monolayer of cells previously incubated with lectin (Furmanski, Phillips and Lubin 1972). This procedure has been shown to differentiate normal and transformed cells and exhibits protease stimulation of normal cell agglutination, as do the cell suspensions. The procedure using red blood cells provides the more reliable estimate of lectin-stimulated agglutination (Fig. 7). SVT2 and the spontaneous transformants cl 5, cl 7, and 3T12 exhibited the expected high levels of agglutination. The normal cells (A31 and PBC) and especially

Table 1

Balb/c Clones, Saturation Densities
and 2-Deoxyglucose Transport

Cell line	Saturation density (cells/plate)	2-Deoxyglucose (cpm/μg protein)
A31	3.6×10^5	74 ± 22
SVT2	3×10^6	246 ± 48
MSV	1.2×10^6	252 ± 2
Cl 3	3×10^6	131 ± 9
Cl 5	2×10^6	273 ± 17
Cl 7	3×10^6	237 ± 3
Cl 8	3×10^6	97 ± 23
Cl 9	5.5×10^5	21 ± 3
PBC	5×10^5	98 ± 9
3T12	1.2×10^6	80 ± 14

Saturation densities were determined from the curves of Fig. 1.
With several of the transformed lines, a true limiting density
could not be obtained because cells detached upon reaching
high densities. [^{3}H]2-deoxyglucose uptake was measured in
triplicate using cells in log phase. The values shown represent
the mean ± the average deviation.

cl 3 and cl 8 agglutinated red blood cells poorly, even in the presence of high levels
of lectin. MSV cells were less agglutinable than many of the other transformed cell
lines.

Early observations of agglutinability of RNA tumor virus-transformed cells
were negative (Moore and Temin 1971). Burger and Martin (1972) have attributed this result to the effect of increased hyaluronic acid production by Rous
sarcoma virus-transformed chick cells. Lehman and Sheppard (1972) observed
that RNA tumor virus-transformed cells showed increased agglutinability when
determinations were made from 12–24 hours after subculture. Agglutinations
determined in the results of Fig. 2 were from experiments where cells had either
been subcultured for less than 24 hours or cultured for several days. The time
spent in culture had little effect on agglutination. The observations that the non-
producer lines of MSV-3T3 cells are less agglutinable supports similar conclusions
reached by Salzberg and Green (1972).

All of these cell lines were produced from the inbred strain of Balb/c mice and
are suitable for tumorigenic studies (Aaronson and Todaro 1968). Because of
our interest in the immunological reactivity of the cells, these studies used im-
munocompetant mice (Fig. 3). Normal cells did not produce tumors. Large doses
of SVT2 cells produced tumors, which were maintained for several weeks, followed
by tumor regression. 3T12 cells also produced tumors that regressed rapidly. Cl 5,
but not cl 7, was tumorigenic, and again the tumors rapidly regressed. Cl 8 produced
tumors only after a significant delay. MSV cells were highly malignant. None of
the animals that developed tumors after injections of MSV cells showed evidence
of tumor regression. The tumors grew very large (one was $3 \times 3 \times 4$cm), and
metastases to the intestine, spleen, and lungs eventually led to death of the animal.
These tumor formation–regression responses are reproducible and apparently reflect
an inherent property of each cell line.

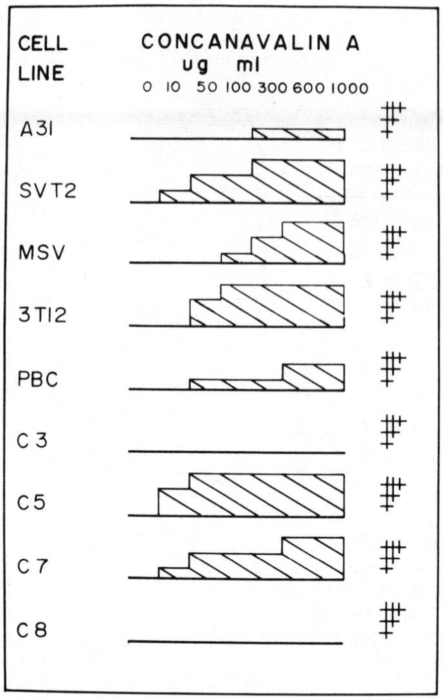

Figure 2
The ability of concanavalin A to cause normal and transformed cells to agglutinate human type O red blood cells. Cells were layered on 35-mm plates
at concentrations of from 1–2 × 10⁵ cells. After 24 hours, agglutinability was
determined by described procedures (Furmanski et al. 1972). A +, ++, and
+++ indicate that 10%, 50%, and 90% of the cells showed hemagglutination.

Cl 8 showed a delayed tumorigenic response relative to other cell lines, suggesting that further cell alterations had occurred in the animal. We isolated tumors from
animals injected with cl 5 and cl 8 (cl 5T and 8T) and remeasured agglutinability.
Cl 8T in contrast to cl 8 is agglutinable (Fig. 4 and 7). We attempted to select for
an agglutinable cell line in vitro by growing cl 8 cells to high densities in culture. All
of these efforts were negative. These spontaneous transformants seem to require
increased agglutinability before they can form tumors, and this conversion only
occurs within the animal. The morphology of cl 8T cells relative to cl 8 cells was
altered. Cl 8T cells resembled cl 5 and cl 7 cells (Fig. 7).

A31 cell agglutination was stimulated by trypsin (Fig. 5). 2°KMSVT cells, a
secondary tumor produced from metastases by MSV cells, were less agglutinable
than transformed cells such as 3T12 and showed increased agglutinability after
trypsin treatment. Thus this cell line (the only line which caused metastases) does
not fully express this cell surface alteration. 3T12 and 3T12T cells were both fully
agglutinable in the absence of trypsin.

One of our primary goals is to determine the biology of complex polysaccharides
in these cell systems. Intracellular organelles have been reported to possess carbohydrate structures (Bosmann et al. 1972; Meyers and Bosmann 1972; Henning
and Uhlenbruck 1973; Keshgegian and Glick 1973). As a control experiment we
determined what effect treating whole cells with neuraminidase would have on sialic

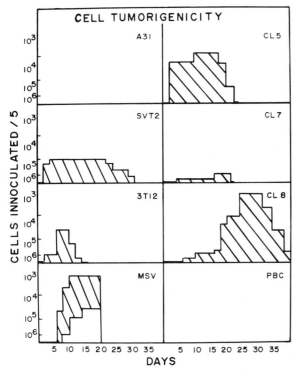

Figure 3

Tumorigenicity studies of normal and transformed cells. Cells at each of the indicated concentrations were injected into three 3-week-old male Balb/c mice. Palpable tumors were scored as positive. The open area under the tumor growth curve in MSV cells represents animal death. All of the MSV-bearing animals that developed tumors have died.

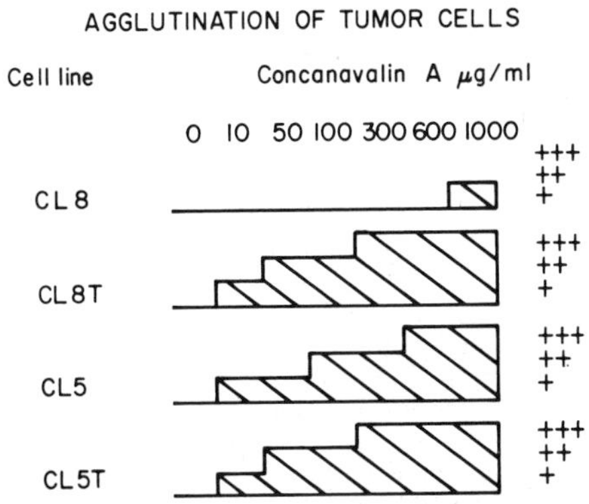

AGGLUTINATION OF TUMOR CELLS

Cell line   Concanavalin A μg/ml

Figure 4

Agglutinability of cl 5 and 8 and of cell lines (5T and 8T) derived from mouse tumors. Agglutination was determined as described in Methods and in the legend of Fig. 2.

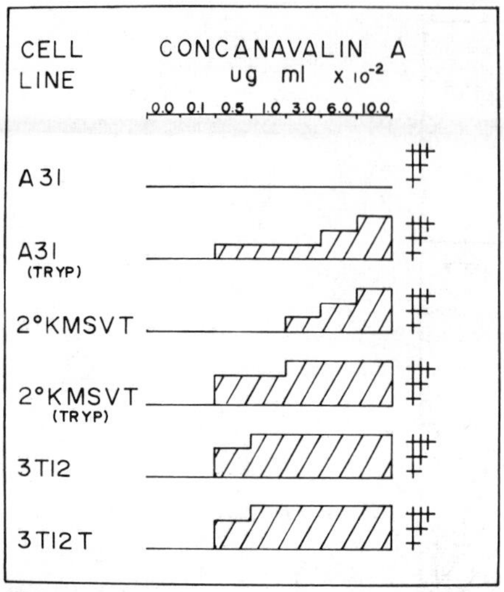

Figure 5

Agglutination of normal and transformed cells. Agglutination was determined
as described in Methods. 2°KMSVT cells are from a metastasis of MSV cells in
one of the mice. A31(tryp) and 2°KMSVT(tryp) agglutinations were determined after treating the monolayer of cells with 0.001% trypsin for 2 min at
room temperature. The monolayer was washed with 0.02% soybean trypsin
inhibitor and agglutination determined as described. 3T12T cells were produced
from a tumor caused by injecting 3T12 cells into mice.

Table 2

Sialic Acid Composition of Cellular Organelles

Cell fraction	Sialic acid level (μg/mg protein)		SA Neur./Control
	Control	Neur.	
Whole cells	2.62	1.28	0.49
Nuclei	0.64	0.35	0.54
Mitochondria	8.74	4.54	0.52

SVT2 cells (2–3×10^8) were suspended in 20 ml of TBS. Aliquots were
removed for determining total cell sialic acid levels. Ten ml of the suspended cells was treated with 250 units of neuraminidase (*V. cholerae*,
Calbiochem, San Diego) at 37°C for 30 min. The remaining 10 ml of cells
served as a control. The cells were collected by centrifugation and an
aliquot of the supernatant solution was removed for determining released
sialic acid. Control and neuraminidase-treated cells were rinsed once in
0.15 M sucrose and homogenized in the same. Mitochondria and nuclei
were prepared as described (Bosmann et al. 1972; Meyers and Bosmann
1972). The mitochondrial preparations from neuraminidase and control
cells were washed five times prior to determining sialic acid levels. Sialic
acid was determined on samples which had been hydrolyzed in 0.1 N
H_2SO_4 for one hr at 80°C as described in Methods.

acid levels of intracellular organelles (Table 2). Removing 49% of the cellular sialic acid by treating suspended cells with neuraminidase removed a like percentage of the sialic acid of washed mitochondria and nuclei produced from these cells. Thus cell surface contamination is a major consideration for studies of intracellular organelle carbohydrates. In a repeat experiment removing 40% of the total cell sialic acid caused loss of 41% of the sugar associated with mitochondrial fractions.

We have studied carbohydrates in the cell clones. Sialic acid and sialic acid transferase levels are reduced in SVT2 cells and in clones 5,7, and 8 (Table 3). In the case of cl 8, reduced sugar and transferase levels are observed prior to the further changes leading to agglutination and tumorigenesis. Apparently this particular parameter is easily altered but is not by itself a sufficient change for tumor formation.

The carbohydrate compositions of glycolipids and glycoproteins of all the cells were determined using alditol acetate derivatives of the sugars (Fig. 6). Preliminary results are described in Table 4. The transformed cell lines show definite reductions in all carbohydrate levels relative to A31 cells. This result confirms the prediction by Wu et al. (1969). The carbohydrates of other cell lines are of obvious interest. It is likely that all transformed lines do not share these changes. We have already reported that Py3T3 cells do not have reduced sialic acid levels (Grimes 1973).

DISCUSSION

Phenomena such as ability to transport 2-deoxyglucose, agglutination, carbohydrate levels and glycolipid compositions do not predict the behavior of cells in the

Table 3

Balb/c Clones, Sialic Acid and
Sialic Acid Transferase Levels

Cell line	Sialic acid ($\mu g/mg$ protein)	Sialic acid transferase (cpm/mg enzyme)
A31	5.3 ± 0.7	$27,579 \pm 968$
SVT2	1.6 ± 0.3	$15,262 \pm 2268$
Cl 5	1.5 ± 0.3	$16,321 \pm 224$
Cl 7	2.1 ± 0.5	$11,274 \pm 778$
Cl 8	2.6 ± 0.5	5915 ± 1381

Sialic acid levels were determined on duplicate or triplicate cultures of cells as described in Methods. Sialic acid transferring ability was determined using desialized bovine submaxillary mucin as acceptor (Grimes 1973). Reactions contained 0.1–0.2 mg enzyme protein; 25,000 cpm [^{14}C]CMP-sialic acid (229 mCi/mmole); 0.1% triton X100; and 10 mM $MgCl_2$ and $MnCl_2$ in 0.05 M sodium cacodylate pH 6.5. After one hr incubation, reaction products were precipitated by adding 1% phosphotungstic acid in 0.5 N HCl. Particulate material was collected by filtration through a 0.3-μ filter and washed several times with cold 5% TCA. The filter was placed in scintillation vials along with 1 ml water and 10 ml Patterson-Green solution. After storage overnight in the cold, radioactivity was determined using a Beckman LS250 scintillation counter.

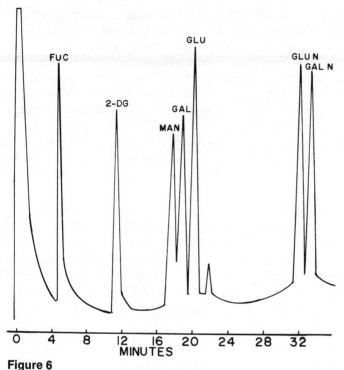

Figure 6

One μg of each of the standard sugars fucose, 2-deoxyglucose, mannose, galactose, glucose, glucosamine, and galactosamine were derivatized and injected into the gas chromatograph as described in Methods.

animal. Many of these changes must therefore be secondary to transformation. The clones we have studied arise spontaneously and with fairly high frequency. In the several clones that we studied, ability to be agglutinated seems to be a necessary part of tumorigenesis. It may be that transformation by SV40 selects for, rather than causes, these changes when transforming 3T3 cells. If such a selection occurred, it would explain why there are so many changes in cells transformed by the simple DNA virus.

The necessity of studying the behavior of normal and transformed cells in the animal must be emphasized. The goal is to be able to predict consequences of the presence of a cell in the animal based on an understanding of the molecular biology of the cell. For cancer, ability to produce tumors is no more important than the fate of the cells in the animal, i.e., mestastases or regression. The differing patterns of tumor growth and regression is intriguing. In preliminary studies of the immunogenicity of the tumor cells, it was noted that once animals had rejected tumors formed by injecting 3T12 cells (an early and rapid process) the animals were not protected from tumors caused by cl 5 cells. It may be that the immunological reactivities of the various non-virally transformed cells are different. Since all of these cells are derived from inbred Balb/c mice, we can compare parameters of cell–animal relationships in cancer. And we believe that studies of these cells using cytotoxic lymphocytes will be of crucial importance for determining cell surface changes, which are important both for tumor growth and tumor regression.

Table 4
Carbohydrate Compositions of Glycoproteins and Glycolipids of Normal and Transformed Cells

Cell line	Fraction	Fucose	Mannose	Galactose	Glucose	Glucosamine	Galactosamine
SVT2	Cx	—	—	3.99 ± 0.40	4.97 ± 0.55	—	1.17 ± 0.35
	Px	6.50 ± 1.23	5.84 ± 1.52	6.27 ± 1.08	—	6.60 ± 1.80	0.81 ± 0.06
Cl 5	Cx	—	—	5.92 ± 0.43	7.17 ± 0.69	—	1.84 ± 0.26
	Px	9.56 ± 1.98	8.93 ± 2.30	12.85 ± 1.84	—	11.68 ± 0.53	1.26 ± 0.70
Cl 7	Cx	—	—	5.74 ± 0.25	7.11 ± 1.68	—	1.69 ± 0.26
	Px	8.37 ± 0.86	7.46 ± 0.83	9.78 ± 0.63	—	10.09 ± 1.37	0.97 ± 0.10
Cl 8	Cx	—	—	7.52 ± 0.12	17.72 ± 0.51	—	2.37 ± 0.45
	Px	11.88 ± 1.63	11.61 ± 2.5	14.29 ± 2.4	—	17.73 ± 4.0	0.43 ± 0.5
A31	Cx	—	—	22.28 ± 5.7	40.56 ± 5.1	—	12.82 ± 1.2
	Px	—	27.41 ± 0.86	28.97 ± 2.3	—	27.15 ± 12.1	4.64 ± 0.21

Carbohydrate compositions of glycolipid (Cx) and glycoprotein (Px) fractions from microsomes of Balb/c cells. Carbohydrate analysis was determined on duplicate cell preparations as described in Methods. Values given are µg sugar/mg protein.

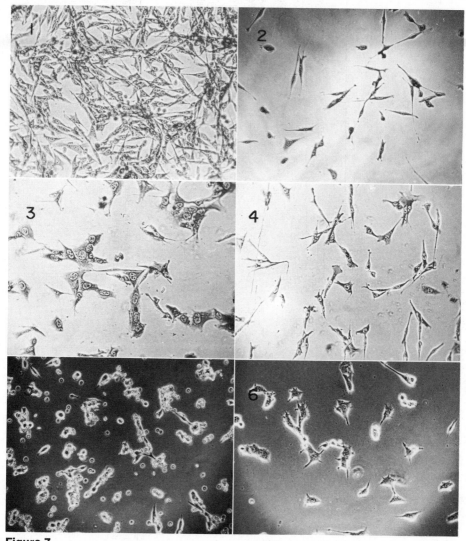

Figure 7
Morphologies of cell clones. Photographs 1–4 represent SVT2, cl 5, cl 8, and cl 8T cells, respectively. Cells have been stained with Wright-Giemsa (Harlco-Philadelphia, Penn.). Photographs 5 and 6 represent cl 8T and cl 8 after incubation first with 300 μg/ml of concanavalin A and then human type O red blood cells as described in Methods.

SUMMARY

A variety of clones of normal Balb/c cells exhibiting properties associated with spontaneous transformation have been studied, along with normal cells and their viral transformants. Carbohydrate compositions, sialic acid transferase levels, and ability to transport 2-deoxyglucose have been compared with biological parameters of cell behavior, including agglutination, contact inhibition, and tumorigenicity. Spontaneously transformed cells and virally transformed cells differ widely in many of these parameters, indicating that several pathways may lead to cell transformation.

Acknowledgment

This work was supported by grants CA12753 from the National Cancer Institute and DRG 1214 from the Damon Runyon Memorial Foundation.

REFERENCES

Aaronson, S. and W. Rowe. 1970. Non-producer clones of murine sarcoma virus transformed Balb/c 3T3 cells. *Virology* **42**:9.

Aaronson, S. and G. Todaro. 1968. Basis for the acquisition of malignant potential by mouse cells cultivated *in vitro*. *Science* **162**:1024.

Albersheim, P., D. Nevins, P. English and A. Karr. 1967. A method for the analysis of sugars in plant cell-wall polysaccharides by gas-liquid chromatography. *Carbohyd. Res.* **5**:340.

Bosmann, H., M. Myers, D. Dehond, R. Ball and K. Case. 1972. Sialic acid residues on the surface of isolated rat cerebral cortex and liver mitochondria. *J. Cell Biol.* **55**:147.

Buck, C., M. Glick and L. Warren. 1970. A comparative study of glycoproteins from the surface of control and Rous sarcoma virus transformed hamster cells. *Biochemistry* **9**:4567.

Burger, M. 1969. A difference in the architecture of the surface membrane of normal and virally transformed cells. *Proc. Nat. Acad. Sci.* **62**:994.

Burger, M. and A. Goldberg. 1967. Identification of a tumor specific determinant on neoplastic cell surfaces. *Proc. Nat. Acad. Sci.* **57**:359.

Burger, M. and G. Martin. 1972. Agglutination of cells transformed by Rous sarcoma virus by wheat germ agglutinin and concanavalin A. *Nature New Biol.* **237**:9.

Cumar, F., R. Brady, E. Kolodny, V. McFarland and P. Mora. 1970. Enzymatic block in the synthesis of gangliosides in DNA virus-transformed tumorigenic mouse cell lines. *Proc. Nat. Acad. Sci.* **67**:757.

Cunningham, D. and A. Pardee. 1969. Transport changes rapidly initiated by serum addition to contact inhibited 3T3 cells. *Proc. Nat. Acad. Sci.* **64**:1049.

Foster, D. and A. Pardee. 1969. Transport of amino acids by confluent and nonconfluent 3T3 and polyoma virus-transformed 3T3 cells growing on glass cover slips. *J. Biol. Chem.* **244**:2675.

Furmanski, P., P. Phillips and M. Lubin. 1972. Cell surface interactions with concanavalin A: Determination by microhemadsorption. *Proc. Soc. Exp. Biol. Med.* **140**:216.

Grimes, W. 1970. Sialic acid transferases and sialic acid levels in normal and transformed cells. *Biochemistry* **9**:5083.

———— 1973. Glycosyltransferase and sialic acid levels of normal and transformed cells. *Biochemistry* **12**:990.

Grimes, W. and P. Robbins. 1972. Virus control of the synthesis of glycosidic linkages: Glycoprotein and glycolipid sialic acid transferases from normal and SV40 transformed Balb/c cells. *Biochemistry of the glycosidic linkage* (ed. R. Piras and H. Pontis) p. 113. Academic Press, New York.

Grimes, W. and J. Schroeder. 1973. Dibutyryl cyclic adenosine 3', 5'-monophosphate, sugar transport, and regulatory control of cell division in normal and transformed cells. *J. Cell Biol.* **56**:487.

Hakomori, S. 1970. Cell density-dependent changes of glycolipid concentrations in fibroblasts, and loss of this response in virus-transformed cells. *Proc. Nat. Acad. Sci.* **67**:1741.

Hartman, J., C. Buck, V. Defendi, M. Glick and L. Warren. 1972. The carbohydrate content of control and virus transformed cells. *J. Cell. Physiol.* **80**:159.

Hatanaka, M. and H. Hanafusa. 1970. Analysis of a functional change in membrane in the process of cell transformation by Rous sarcoma virus: Alteration in the characteristics of sugar transport. *Virology* **41**:647.

Hatanaka, M., C. Augl and R. Gilden. 1970. Evidence for a functional change in the plasma membrane of murine sarcoma virus-infected mouse embryo cells. *J. Biol. Chem.* **245**:714.

Henning, R. and G. Uhlenbruck. 1973. Detection of carbohydrate structures on isolated subcellular organelles of rat liver by heterophile agglutinins. *Nature New Biol.* **242**:120.

Inbar, M. and L. Sachs. 1969. Interaction of the carbohydrate-binding protein concanavalin A with normal and transformed cells. *Proc. Nat. Acad. Sci.* **63**:1418.

Isselbacher, K. 1972. Increased uptake of amino acids and 2-deoxy-D-glucose by virus-transformed cells in culture. *Proc. Nat. Acad. Sci.* **69**:585.

Jones, T. and P. Albersheim. 1972. A gas chromatographic method for the determination of aldose and uronic acid constituents of plant cell wall polysaccharides. *Plant Physiol.* **49**:926.

Keshgegian, K. and M. Glick. 1973. Glycoproteins associated with nuclei of cells before and after transformation by a ribonucleic acid virus. *Biochemistry* **12**:1221.

Kijimoto, S. and S. Hakomori. 1972. Contact-dependent enhancement of net synthesis of Forssman glycolipid antigen and hematoside in NIL cells at the early stage of cell-to-cell contact. *FEBS Letters* **25**:38.

Kornfeld, S. 1969. Decreased phytohemagglutinin receptor sites in chronic lymphocytic leukemia. *Biochim. Biophys. Acta* **192**:542.

Lehman, J. and J. Sheppard. 1972. Agglutinability by plant lectins increases after RNA virus transformation. *Virology* **49**:339.

Levine, E. 1972. Mycoplasma contamination of animal cell cultures: A simple rapid detection method. *Exp. Cell Res.* **74**:99.

Lowry, O., N. Rosebrough, A. Farr and R. Randall. 1951. Protein measurement with the folin phenol reagent. *J. Biol. Chem.* **193**:265.

Martin, S., S. Venuta, M. Weber and H. Rubin. 1971. Temperature dependent alterations in sugar transport in cells infected by a temperature-sensitive mutant of Rous sarcoma virus. *Proc. Nat. Acad. Sci.* **68**:2739.

McClelland, D. and J. Bridges. 1973. The total N-acetyl neuraminic acid content of human normal and lymphatic leukaemic lymphocytes. *Br. J. Cancer* **27**:114.

Meezan, E., H. Wu, P. Black and P. Robbins. 1969. Comparative studies on the carbohydrate-containing membrane components of normal and virus transformed mouse fibroblasts. II. Separation of glycoproteins and glycopeptides by Sephadex chromatography. *Biochemistry* **8**:2518.

Meyers, M. and H. Bosmann. 1972. Mitochondrial autonomy: Depressed protein and glycoprotein synthesis in mitochondria of SV-3T3 cells. *FEBS Letters* **26**:294.

Moore, E. and H. Temin. 1971. Lack of correlation between conversion by RNA tumor viruses and increased agglutinability of cells by concanavalin A and wheat germ agglutinin. *Nature* **231**:117.

Ohta, N., A. Pardee, B. McAuslan and M. Burger. 1968. Sialic acid contents and controls of normal and malignant cells. *Biochim. Biophys. Acta* **158**:98.

Onodera, K. and R. Sheinin. 1970. Macromolecular glucosamine-containing component of the surface of cultivated mouse cells. *J. Cell Sci.* **7**:337.

Patterson, M. and R. Greene. 1965. Measurement of low energy beta emitters in aqueous solution by liquid scintillation counting of emulsions. *Ann. Chem.* **37**:854.

Perdue, J., R. Kletzien and V. Wray. 1972. The isolation and characterization of plasma membrane from cultured cells. *Biochim. Biophys. Acta* **266:**505.

Puck, T., P. Marcus and S. Creceuia. 1956. Clonal growth of mammalian cells *in vitro*. *J. Exp. Med.* **103:**273.

Robbins, P. and I. Macpherson. 1971. Glycolipid synthesis in normal and transformed animal cells. *Proc. Roy. Soc. London* B **177:**49.

Romano, A. and C. Colby. 1973. SV40 virus transformation does not specifically enhance sugar transport. *Science* **179:**1238.

Sakiyama H. and P. Robbins. 1973. Glycolipid synthesis and tumorigenicity of clones isolated from the NIL 2 line of hamster embryo fibroblasts. *Fed. Proc.* **32:**86.

Salzberg, S. and M. Green. 1972. Surface alterations of cells carrying RNA tumor virus genetic information. *Nature New Biol.* **240:**116.

Todaro, G. and H. Green. 1963. Quantitative studies of the growth of mouse embryo cells in culture and their development into established lines. *J. Cell Biol.* **17:**299.

Warren, L. 1959. The thiobarbituric acid assay of sialic acids. *J. Biol. Chem.* **234:**1971.

Warren, L., J. Fuhrer and C. Buck. 1972. Surface glycoproteins of normal and transformed cells: A difference determined by sialic acid and a growth dependent sialyl transferase. *Proc. Nat. Acad. Sci.* **69:**1838.

Wu, H., E. Meezan, P. Black and P. Robbins. 1969. Comparative studies on the carbohydrate-containing membrane components of normal and virus transformed mouse fibroblasts. I. Glucosamine-labeling patterns in 3T3, spontaneously transformed 3T3, and SV40 transformed 3T3 cells. *Biochemistry* **8:**2509.

Yogeeswaran, G., R. Sheinin, J. Wherrett and R. Murray. 1972. Studies on the glycosphingolipids of normal and virally transformed 3T3 mouse fibroblasts. *J. Biol. Chem.* **247:**5146.

The Effect of Polyprenols on Cell Surface Galactosyltransferase Activity

J. Kevin Dorsey and Stephen Roth

Department of Biology, The Johns Hopkins University
Baltimore, Maryland 21218

Glycosyltransferases have been found on the surface of neural retina cells (Roth, McGuire and Roseman 1971), blood platelets (Jamieson, Urban and Barber 1971), mouse fibroblasts (Bosmann 1972; Roth and White 1972), and intestinal epithelia (Weiser 1973). These enzymes usually catalyze the transfer of a sugar moiety from a sugar nucleotide to a specific glycoprotein, glycolipid, or saccharide acceptor. Although it is likely that cell surface glycosyltransferases originate in the contiguous Golgi apparatus, their function is apparently more than a biosynthetic one. For example, aspirin and chlorpromazine, inhibitors of platelet aggregation, also inhibit platelet glucosyltransferase activity (Jamieson et al. 1971). This suggests that an early event in hemostasis might be the adherence of platelets to collagen via cell surface enzymes and supports the hypothesis that enzyme–substrate complexes between glycosyltransferases and cell surface carbohydrates are important in intercellular recognition (Roseman 1970). Additional support for this hypothesis was obtained from studies on embryonic chick neural retina cells. Addition of exogenous galactose acceptors or enzymatic modification of surface carbohydrates alters the self-recognition of these cells (Roth et al. 1971).

We have recently been investigating the possibility that the glycosyltransferases detected on the surface of Balb/c mouse fibroblasts might play a role in growth control in vitro. Transfer of sugars to endogenous acceptors varied according to the degree of cell contact when intact 3T3 cells were assayed (Roth and White 1972). If cell contact was diminished during the assay period by stirring the cell suspension, the rate of galactose transfer was reduced compared to an unstirred aliquot of the same 3T3 cell suspension. This result was not observed when transformed cells were assayed. The data were interpreted as follows: transfer of galactose by 3T3 cells can only occur between a transferase on one cell and an acceptor on an adjacent cell (trans-glycosylation), whereas transformed cells are capable of transferring galactose to acceptors on the same cell (cis-glycosylation).

An outgrowth of these preliminary experiments is the speculation that, upon contact, mutual glycosylation of cell surface carbohydrates might control growth in vitro. Conversely, loss of growth control can then be thought of as the failure of cells to glycosylate on contact. Consistent with this interpretation is the observation that normal cells from confluent cultures have less available acceptors than do normal cells from sparse cultures or malignant cells at any culture density.

A necessary condition for glycosylation to occur on the cell exterior is an available supply of the sugar donor. This presents a problem since sugar nucleotides do not readily penetrate cell membranes. This is a problem not restricted to surface glycosylation; normal glycoprotein biosynthesis occurs via transferases located in the Golgi (Schachter et al. 1970), an organelle whose biosynthetically active surface is topologically external to the cytoplasm. This difficulty has been obviated somewhat by the discovery of lipid intermediates in the glycosyltransferase reaction (for a recent review see Lennarz and Scher 1972). Carbohydrate derivatives of the polyisoprenols, dolichol and retinol (vitamin A) are capable of transferring their sugar moiety to various glycoprotein acceptors (Baynes 1973; Behrens and Leloir 1970; DeLuca et al. 1972, 1973; Helting and Peterson 1972; Parodi et al. 1972).

The present report describes the effect of polyprenols on the transfer of galactose to endogenous cell surface acceptors and exogenous acceptors by intact mouse fibroblasts. Ficaprenol stimulated galactose transfer to endogenous acceptors twofold when intact, normal (3T3) Balb/c mouse fibroblasts were assayed. Retinol and dihydrophytol were half as effective as ficaprenol in stimulating endogenous galactose transfer. Intact, malignant Balb/c 3T12 cells did not show increased galactose transfer to endogenous acceptors upon the addition of ficaprenol, retinol or dihydrophytol. When N-acetylglucosamine was added as an exogenous acceptor, the results obtained with intact normal and malignant cells were the same. All three alcohols slightly stimulated the synthesis of N-acetyllactosamine.

METHODS

Cells

Balb/c 3T3 (3T3) and Balb/c 3T12 (3T12) cells were cultured in Dulbecco's modified Eagle's medium with 10% calf serum and without antibiotics. Cells were harvested from petri dishes with 0.1% crude trypsin (Difco 1:250). Some characteristics of these cells are summarized in Table 1.

In several experiments cells were grown in medium containing serum which had been depleted of retinol by irradiation. Uncovered petri dishes containing calf serum

Table 1
Characteristics of Mouse Fibroblasts

	3T3	3T12
Saturation density (cells/cm²)	5.1×10^4	3.6×10^5
Malignancy	no	yes
Endogenous galactosyltransferase activity	requires* contact	does not* require contact

* Roth and White 1972

were exposed to an ultraviolet lamp for one hour at a distance of 5 inches. These conditions completely destroyed known samples of retinol as measured by the decrease in absorption at 325 mμ.

Transferase Assays

Methods for the assay of galactosyltransferase activity have been reported in detail in other publications (Roth et al. 1971; Roth and White 1972). Briefly, cells were harvested and washed in a salts solution (medium J) containing 10 mM MnCl$_2$ and 10 mM NaN$_3$. A suspension of 10^7 intact cells per ml was added to a tube containing uridine diphosphate-[^{14}C]galactose and, when specified, exogenous acceptor. After incubating at 37°C for the appropriate time, the reaction was stopped by the addition of 5 μl EDTA (0.2 mM) to 25 μl aliquots of the reaction mixture. Incorporation of galactose was determined by liquid scintillation counting of electrophoretically immobile radioactivity, as previously described (Roth et al. 1971; Roth and White 1972).

Retinol (Eastman Kodak) was prepared fresh for each experiment and was added to incubation mixtures in dimethylsulfoxide. Control incubations received dimethylsulfoxide only. Preparation of retinol solutions was carried out in dim light.

Ficaprenol and dihydrophytol were the gift of W. J. Lennarz; dolichol was generously supplied by F. W. Hemming. The polyprenols were added to the assay as a solution in dimethylsulfoxide. Concentrations were determined by assuming that ficaprenol was predominantly C$_{55}$ and that dolichol was C$_{95}$.

RESULTS

When solutions of polyprenols in dimethylsulfoxide were added to a galactosyltransferase assay mixture of either 3T3 or 3T12 cells, the results shown in Fig. 1 were obtained. Transfer of galactose to endogenous acceptors on 3T12 cells was unaffected by the addition of ficaprenol, dihydrophytol, or retinol (Fig. 1B). 3T3 endogenous galactosyltransferase activity was stimulated twofold by ficaprenol and, to a lesser extent, by retinol and dihydrophytol (Fig. 1A). Transfer of galactose to the exogenous acceptor, N-acetylglucosamine, was slightly stimulated by addition of polyprenols to assay mixtures of either intact 3T3 or intact 3T12 cells (Fig. 1C,D).

Addition of retinyl acetate, retinoic acid, or dolichol had no effect on endogenous galactosyltransferase activity in whole 3T3 cells (Table 2). The same results were obtained when 3T12 cells were assayed.

Several experiments were performed on 3T3 and 3T12 cells grown for at least six days in medium prepared with irradiated serum. Addition of retinol to assays of these cells gave the same results as cells grown in normal medium.

DISCUSSION

The results presented in this report demonstrate that polyprenols stimulate cell surface galactosyltransferase activity of mouse fibroblasts. A twofold increase in

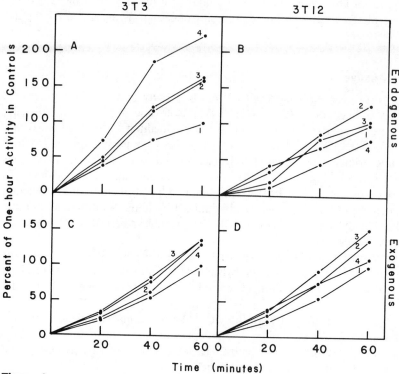

Figure 1

Transfer of [^{14}C]galactose from UDP-[^{14}C]galactose to endogenous and exogenous acceptors. Cells (1 × 10^6) in 0.1 ml medium J, containing 10 mM MnCl$_2$ and 10 mM NaN$_3$, were added to the reaction mixture which had been previously evaporated to dryness. The reaction mixture consisted of UDP-[^{14}C]galactose (10.8 mμmoles, 0.2 μCi) and, where appropriate **(C,D)**, N-acetylglucosamine (0.25 μmole). Immediately after addition of the cells 5 μl of a 1 mM solution of the appropriate polyprenol in dimethylsulfoxide was added and the tubes were incubated at 37°C. Control incubations received 5 μl of dimethylsulfoxide. Aliquots of 25 μl were withdrawn at various times and added to tubes containing 5 μl EDTA (0.2 mM) to stop the reaction. A portion (20 μl) of this mixture was electrophoresed and assayed as previously described (Roth et al. 1971; Roth and White 1972). Samples containing the exogenous acceptor N-acetylglucosamine **(C,D)** were first centrifuged, and 20 μl of the supernatant only was electrophoresed. Additions are indicated on the graphs as follows: 1, dimethylsulfoxide; 2, dihydrophytol; 3, retinol; 4, ficaprenol. The data are expressed in each graph as percents relative to the (100%) value obtained for the control incubation after one hour at 37°C. The actual data obtained for 25 μl aliquots of each control at one hour were: **A,** 10.7 pmoles; **B,** 3.6 pmoles; **C,** 102 pmoles; **D,** 35.2 pmoles.

galactose incorporation to endogenous acceptors on 3T3 cells was obtained if ficaprenol was added to the assay mixture. When malignant 3T12 cells were assayed under identical conditions, no increase was observed and, in fact, galactose incorporation decreased slightly. However when galactose transfer to the exogenous acceptor, N-acetylglucosamine, was measured, N-acetyllactosamine formation

Table 2

Effect of Retinol Analogs and Dolichol on 3T3
Endogenous Galactosyltransferase Activity

	% control activity*
Retinol	137
Retinoic acid	84
Retinyl acetate	96
Dolichol	104

The compounds tested were added as solutions in
dimethylsulfoxide to a final concentration of 5×10^{-5} M. The procedure described for endogenous
activity in the legend to Fig. 1 was used.

* Activity is expressed as percent of a control incubation receiving dimethylsulfoxide only. Control
value was 50 pmoles per 10^6 cells per hour.

increased to the same extent in both 3T3 and 3T12 cells upon the addition of
retinol, ficaprenol, or dihydrophytol.

The effect of polyprenols is apparently specific. Retinyl acetate or retinoic acid
did not stimulate endogenous activity in either 3T3 or 3T12 cells. Dolichol also
had no effect. It is interesting to note that Behrens et al. (1971) have shown that
uridine diphosphate galactose was not a sugar donor for dolichol phosphate in a
liver microsome system.

In order for surface or intracellular glycosylation to occur, a sugar donor must
be readily accessible. Since sugar nucleotides are apparently unable to permeate
cell membranes, membrane intermediates in glycoprotein biosynthesis have been
sought. Likely candidates are the sugar phosphate and pyrophosphate derivatives
of polyprenols. Accordingly in preliminary experiments we have found that bacitracin at a final concentration of 125 μg per ml inhibits endogenous activity by
50%. This result suggests (Stone and Strominger 1971) that some of the cell
surface reactions involving a galactose transfer occur through a terpenylpyrophos-
phate intermediate.

If, in fact, polyisoprenoid glycolipids are intermediates in glycoprotein biosynthesis and cell surface glycosylation, then the question of donor specificity must
be considered. Glycosyltransferase reactions are commonly thought to use a sugar
nucleotide donor. For any given sugar, the nucleotide is specified (e.g., all known
galactosyltransferases use uridine diphosphate as the nucleotide linked to galactose). Retinol glycolipids containing glucose (DeLuca et al. 1972), mannose (De-
Luca et al. 1973), galactose (Helting and Peterson 1972), or glucuronic acid
(DeLuca et al. 1973) have been identified. Dolichol glycolipids of glucose
(Behrens et al. 1970, 1971; Parodi et al. 1972), mannose (Alam et al. 1971;
Baynes 1973; Behrens et al. 1971; Richards et al. 1971) and N-acetylglucosamine
(Behrens et al. 1971) have also been found. The possibility that different lipid inter-
mediates of a given sugar are used for different reactions, or that different sugars
require different lipids, remains open.

The nature of the actual lipid intermediate in the reaction studied here is difficult to ascertain. The significant, but small, stimulation of galactose incorporation
with added retinol, coupled with a lack of assayable difference in cells grown in

irradiated or normal serum, could indicate that retinol is not the major intermediate in galactose transfer. However it is unlikely that the cells are totally retinol deficient when grown for a few generations in medium prepared with irradiated serum. Furthermore a lipid used in glycoprotein synthesis would presumably be regenerated, and only minute quantities would be required to compete with any exogenously added lipid. The marked stimulation of 3T3 endogenous galactosyltransferase activity by ficaprenol could be especially interesting in view of the finding that 3T12 cells do not respond similarly. However only a narrow concentration range was tested in these two cell lines.

Because these experiments were carried out with intact cells, the data indicate that some cell surface transferase activity may proceed through a lipid intermediate, as previously suggested for some intracellular glycosylations. We are investigating the possibility that some of the diverse effects of retinol on mucopolysaccharide synthesis in cell (Kochar et al. 1968) and organ culture (Fell and Mellanby 1953) and on developmental systems are mediated by retinol glycolipid intermediates in the transferase reaction.

Acknowledgments

This study was supported by a research grant from the American Cancer Society. JKD was a postdoctoral fellow of the National Institutes of Health. This is contribution No. 734 from the McCollum-Pratt Institute.

REFERENCES

Alam, S. S., R. M. Barr, J. B. Richards and F. W. Hemming. 1971. Prenol phosphates and mannosyltransferases. *Biochem. J.* **121**:19P.

Baynes, J. W. 1973. Ph.D. thesis, The Johns Hopkins University.

Behrens, N. H. and L. F. Leloir. 1970. Dolichol monophosphate glucose: An intermediate in glucose transfer in liver. *Proc. Nat. Acad. Sci.* **66**:153.

Behrens, N. H. and L. F. Leloir. 1970. Dolichol monophosphate glucose: An intermediate monophosphate in sugar transfer. *Arch. Biochem. Biophys.* **143**:375.

Bosmann, H. B. 1972. Cell surface glycosyltransferases and acceptors in normal and RNA- and DNA-virus transformed fibroblasts. *Biochem. Biophys. Res. Commun.* **48**:523.

DeLuca, L., A. J. Barber and G. A. Jamieson. 1972. Possible role of vitamin A in platelet-collagen adhesion. *Fed. Proc.* **31**:242.

DeLuca, L., N. Maestri, G. Rosso and G. Wolf. 1973. Retinol glycolipids. *J. Biol. Chem.* **248**:641.

Fell, H. B. and E. Mellanby. 1953. Metaplasia produced in cultures of chick ectoderm by high vitamin A. *J. Physiol.* **119**:470.

Helting, T. and P. A. Peterson. 1972. Galactosyl transfer in mouse mastocytoma: Synthesis of a galactose-containing, polar metabolite of retinol. *Biochem. Biophys. Res. Commun.* **46**:429.

Jamieson, G. A., C. L. Urban and A. J. Barber. 1971. Enzymatic basis for platelet-collagen adhesion as the primary step in haemostasis. *Nature New Biol.* **234**:5.

Kochar, D. M., J. T. Dingle and J. A. Lucy. 1968. The effects of vitamin A (retinol) on cell growth and incorporation of labelled glucosamine and proline by mouse fibroblasts in culture. *Exp. Cell Res.* **52**:591.

Lennarz, W. J. and M. G. Scher. 1972. Metabolism and function of polyisoprenol sugar intermediates in membrane-associated reactions. *Biochim. Biophys. Acta* **265:**417.

Parodi, A. J., N. H. Behrens, L. F. Leloir and H. Carminatti. 1972. The role of polyprenol-bound saccharides as intermediates in glycoprotein synthesis in liver. *Proc. Nat. Acad. Sci.* **69:**3268.

Richards, J. B., P. J. Evans and F. W. Hemming. 1971. Dolichol phosphates as acceptors of mannose from guanosine diphosphate mannose in liver systems. *Biochem J.* **124:** 957.

Roseman, S. 1970. The synthesis of complex carbohydrates by multiglycosyltransferase systems and their potential function in intercellular adhesion. *Chem. Phys. Lipids* **5:**270.

Roth, S. 1973. A molecular model for cell interactions. *Quart. Rev. Biol.* (in press).

Roth, S. and D. White. 1972. Intercellular contact and cell-surface galactosyl transferase activity. *Proc. Nat. Acad. Sci.* **69:**485.

Roth, S., E. J. McGuire and S. Roseman. 1971. Evidence for cell-surface glycosyltransferases: Their potential role in cellular recognition. *J. Cell Biol.* **51:**536.

Schachter, H., I. Jabbal, R. L. Hudgin, L. Pinteric, E. J. McGuire and S. Roseman. 1970. Intracellular localization of liver sugar nucleotide glycoprotein glycosyltransferases in a Golgi-rich fraction. *J. Biol. Chem.* **245:**1090.

Stone, K. J. and J. L. Strominger. 1971. Mechanism of action of bacitracin: Complexation with metal ion and C_{55}-isoprenyl pyrophosphate. *Proc. Nat. Acad. Sci.* **68:**3223.

Weiser, M. M. 1973. Glycosyltransferases and endogenous acceptors of the undifferentiated cell surface membrane. *J. Biol. Chem.* **248:**2542.

Rous Sarcoma Virus Transformation of the Chick Cell Surface

Gary G. Wickus, Philip E. Branton, and P. W. Robbins

Department of Biology, Massachusetts Institute of Technology
Cambridge, Massachusetts 02139

In considering the control of cell proliferation it is obviously interesting and important to be aware of biochemical differences between the cell surfaces of "normal" and "transformed" cells. It seems likely that growth control in normal cells is mediated through reactions that originate from contacts between cell surfaces and that chemical and physical changes in the cell surface following transformation play a key role in the loss of normal growth control.

Until recently most biochemical comparisons of normal and transformed plasma membranes have utilized cloned established lines transformed with DNA viruses (e.g., BHK/PyBHK, 3T3/SV40, 3T3). Work with these systems has revealed transformation-associated differences in agglutinability by plant lectins and has generated numerous biochemical studies with isolated plasma membranes. However work with these established lines presents the biochemist with a severe problem. Transformations by DNA viruses tend to be relatively rare events, and it is usually necessary to clone a transformed cell and grow it for at least 20 generations before biochemical comparisons can be carried out with its untransformed "parent." This extensive growth, coupled with the drift in karyotype which often occurs in established lines, has introduced serious questions concerning the validity of many of the biochemical comparisons of normal and transformed established lines.

With these facts in mind we have turned our attention to the transformation of chick fibroblasts by Rous sarcoma virus. This system offers the advantages of complete, rapid transformation and the availability of excellent temperature-sensitive mutants. We have used TS-68, a mutant of the Schmidt-Ruppin strain of Rous sarcoma virus isolated by Kawai and Hanafusa (1971). Chick cells infected with this mutant virus have the normal cell phenotype at 41°C and the Rous transformed phenotype at 36°C. Virus production at the two temperatures is normal.

Methionine Pulse-Labeling of TS-68 Transformed Cells

Figure 1 presents photomicrographs of secondary chick embryo fibroblasts infected with TS-68. At 41°C (A) the typical parallel arrangement of normal fibroblasts is

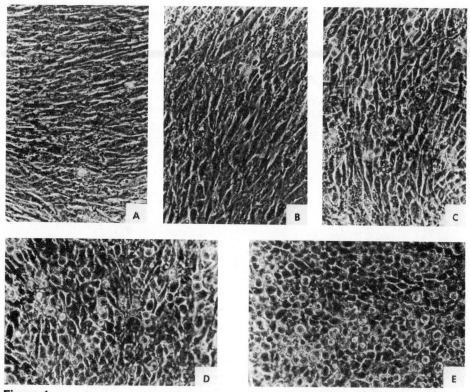

Figure 1
Temperature shift experiment (41→36°C) for TS-68-infected chick embryo fibroblasts. Cells were grown and infected as previously described in medium 199 supplemented with 4% calf serum, 1% heat-inactivated chick serum, and 10% tryptose phosphate broth (see Wickus and Robbins 1973). **(A)** Cells at 41°C; **(B)** 1.5 hours, **(C)** 3 hours, **(D)** 11 hours, **(E)** 24 hours after shift to 36°C.

seen. After the shift to 36°C the cells become somewhat more refractile within 1½ hours (B). This tendency is clearly evident after 3 hours (C). Fully transformed morphology with loose attachment to the plastic substrate and many rounded cells is present at 11 hours (D) and 24 hours (E) after the temperature shift.

TS-68-infected cells were pulse-labeled with high specific activity [^{35}S]methionine immediately before and for 3-hour periods after the shift in temperature from 41 to 36°C. Following the labeling periods monolayers were washed with phosphate-buffered saline and disrupted in SDS. The SDS extracts were processed and fractionated on 8% polyacrylamide slab gels as previously described (Wickus and Robbins 1973). In the experiment illustrated in Fig. 2 duplicate petri plates of TS-68-infected cells were labeled for 3 hours before the temperature shift (A and B) and 0–3 hours (C and D), 3–6 hours (E and F), 6–9 hours (G and H), and 9–12 hours (I and J) after the shift from 41 to 36°C. Two major changes occur in the labeling pattern. Appropriate control experiments with untransformed cells and wild-type SR-Rous cells demonstrate that both of these changes are associated with transformation and are not produced by the temperature shift per se; i.e., both changes are present in wild-type Rous cells at 41 and 36°C and are not found in untransformed cells at either temperature.

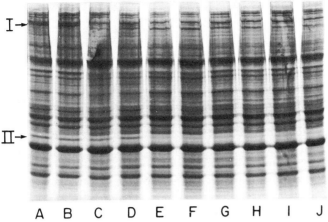

Figure 2
Radioautogram of an 8% polyacrylamide slab gel electropherogram of CEF fibroblasts infected with TS-68. Cell monolayers were pulse-labeled with [^{35}S]methionine at 3-hour intervals following a shift from the nonpermissive to the permissive temperature. The pulse-labeled cell monolayers were disrupted in SDS. **(A,B)** 3-hour pulse at the nonpermissive temperature (41°C); **(C,D)** 3-hour pulse immediately after shift from 41 to 36°C; **(E,F)** 3-hour pulse from 3–6 hours after shift to 36°C; **(G,H)** 3-hour pulse from 6–9 hours after shift to 36°C; **(I,J)** 3-hour pulse from 9–12 hours after shift to 36°C.

The 45,000 daltons protein indicated at II in Fig. 2 is a prominent component of plasma membranes prepared from normal chick cells. Although a plasma membrane component, it is not available to the lactoperoxidase-catalyzed iodination reaction and is thus not an "external" plasma membrane component. As shown previously by Wickus and Robbins (1973) this 45,000 molecular weight protein is present in lower concentrations in plasma membranes from Schmidt-Ruppin Rous-transformed cells. Figure 2 shows clearly that the reduction in the synthesis of component II occurs 3–6 hours after the induction of transformation by the temperature shift. Little or no change is evident 0–3 hours after the shift. We therefore conclude that decreased synthesis of this protein may occur secondarily to the very earliest transformation events.

Even more interesting than the decrease in the 45,000 molecular weight protein is the rapid decrease in the component indicated as I in Fig. 2. Untransformed cells have two dark bands in this high molecular weight region. The concentration of the lower of these two bands is substantially reduced following the temperature shift and concomitantly a band of slightly lower molecular weight appears. This change is complete even in the 0–3 hour pulse period. This shift in protein pattern is obviously one of the earliest transformation-specific changes to take place in the cell.

At present the intracellular location of component I is unknown, but our hypothesis is that it may be a cell surface protein bound loosely to the external face of the cell. It does not appear in plasma membrane vesicles prepared by the method of Perdue, Kletzien and Miller (1971) but does remain with cell ghosts prepared by a modification of the method of Brunette and Till (1971). It may be that the zinc ion treatment of the cell surface and the generally gentler isolation procedure of Brunette and Till may lead to retention of components loosely bound to the cell surface.

Iodination of the Cell Surface

A further interesting and suggestive fact is that the major protein of the chick cell surface which can be iodinated by the lactoperoxidase-^{125}I system is a component with electrophoretic properties similar to those of component I. The results of a typical ^{125}I experiment are presented in Figure 3. Confluent monolayers were washed with phosphate-buffered saline and treated with lactoperoxidase and [^{125}I]iodide by the method in use currently in our laboratory (see Wickus and Robbins 1973). Following iodination the plates were treated with SDS and electrophoresis was carried out as described above. In Fig. 3 lane A was derived from a plate that contained complete medium but no cells. The radioactive iodine bands represent serum proteins that adsorb strongly to plastic culture dishes. Lane B was derived from an iodinated monolayer of normal chick fibroblasts. It can clearly be seen that most of the radioactive iodine is associated with protein(s) of similar electrophoretic mobility to component I of Fig. 2. Lane C demonstrates that most of this radioactivity is removed by treating the labeled cells for 15 minutes with 5 μg/ml trypsin. This finding confirms the assumption that the high molecular weight iodinatible protein is located on the external face of the plasma membrane. Lane D shows that iodination immediately after trypsin treatment gives a very similar pattern to that obtained by iodination followed by trypsin treatment (lane C). It is of interest that the trypsin treatment did not lead to the "exposure" of many new iodinatible sites on the cell surface. In contrast to our results, Phillips and Morrison (1973) have found that treatment of human erythrocytes with 25 μg/ml

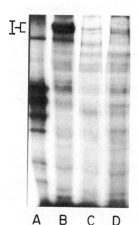

A   B   C   D

Figure 3

Radioautogram of an 8% polyacrylamide slab gel electropherogram of adsorbed serum proteins and surface proteins from uninfected chick embryo fibroblast monolayers labeled with [125]I. **(A)** Serum proteins (calf serum) adsorbed to plastic petri dish (Falcon) containing no cells; **(B)** surface and serum proteins from normal CEF monolayer; **(C)** labeled CEF monolayer treated with 5 μg/ml trypsin in PBS (+ Ca^{++} + Mg^{++}) for 15 minutes at 39° C; **(D)** trypsiniezd CEF monolayer labeled with ^{125}I immediately after trypsin treatment as in C.

trypsin for 15 minutes leads to approximately a 10-fold increase in cell surface proteins available to the lactoperoxidase-catalyzed iodination reaction.

Since transformed chick cells, like trypsinized cells, have little "component I-like" iodinatible surface protein (data not shown), a major effort is being made to establish the relationships among the high molecular weight methionine-labeled proteins and the iodinatible proteins of the normal chick cell surface. Cells labeled with glucosamine also show bands in this high molecular weight region. For this reason we are interested in the question of whether or not the surface protein(s) which disappear following transformation are glycoproteins.

DISCUSSION

It is impressive that within three hours of induction of transformation by temperature shift the transformed phenotype is already completely expressed with respect to incorporation of methionine into high molecular weight, possibly extracellular proteins. What is the mechanism for this change in labeling patterns? One intriguing possibility is that the normal component I protein is still made after transformation but that it is hydrolyzed rapidly by the transformation-specific protease described by Unkeless et al. (1973) and Ossowski et al. (1973). These authors have demonstrated that a specific fibrinolytic activity is produced by transformed cells. If this fibrinolysin is able to act on specific cell surface components, it may produce the transformed cell surface phenotype in much the same way that trypsin treatment will "transform" the cell surface.

One finding not in harmony with the assumption that component I is being cleaved by the transformation-specific fibrinolysin is the finding by Unkeless et al. that, when TS-5-infected chick cells are shifted from 40 to 34°C, significant fibrinolytic activity begins to be seen only 8 hours or more after the temperature shift. On the other hand, this result could simply be a reflection of the difference between TS-5 and TS-68. Kawai and Hanafusa (1971) have demonstrated that following a temperature shift TS-68-infected cells can acquire the completely transformed phenotype within 8 hours. Cells infected with the Martin mutant TS-5 may not respond as quickly to a temperature shift. A second point to be considered is that Unkeless et al. were measuring extracellular fibrinolytic activity. The enzyme might act more rapidly and completely on endogenous components of the cell surface.

At present we have no experimental evidence either for or against the postulate that component I is cleaved by proteolytic enzymes. It is possible that transformation simply produces a rapid shut-off of the synthesis of component I and an equally rapid turning on of the synthesis of the somewhat lower molecular weight protein that appears soon after the temperature shift. There are a number of experiments which we are currently carrying out to clarify these points.

Other changes in the cell surface occur more slowly following the induction of transformation. A prominent example is the decrease in the synthesis of component II, a protein found in plasma membrane vesicles (see Wickus and Robbins 1973). It seems likely that the change in glycolipid pattern reported by Hakomori, Saito and Vogt (1971) occurs even more slowly following the initial transforming event. The kinetics for the shift in glycoprotein pattern reported by Warren, Critchley and Macpherson (1972) obviously should be investigated with great care since this transformation-dependent change clearly involves protease-sensitive cell surface proteins.

Acknowledgments

We thank S. Weinzierl and H. Samuelsdottir for their excellent technical assistance. This investigation was supported by National Institutes of Health grant no. 5-R01-AM06803-10. G.G.W. is a fellow of the Damon Runyon Memorial Fund for Cancer Research. P.E.B. is a fellow of the National Cancer Institute of Canada.

REFERENCES

Brunette, D. M. and J. E. Till. 1971. A rapid method for the isolation of L-cell surface membranes using an aqueous two-phase polymer system. *J. Membrane Biol.* **5:**215.

Hakomori, S., T. Saito and P. K. Vogt. 1971. Transformation by Rous sarcoma virus: Effects on cellular glycolipids. *Virology* **44:**609.

Kawai, S. and H. Hanafusa. 1971. The effects of reciprocal changes in temperature on the transformed state of cells infected with a Rous sarcoma virus mutant. *Virology* **46:**470.

Ossowski, L., J. C. Unkeless, A. Tobia, J. P. Quigley, D. B. Rifkin and E. Reich. 1973. An enzymatic function associated with transformation of fibroblasts by oncogenic viruses. *J. Exp. Med.* **137:**112.

Perdue, J. F., R. Kletzien and K. Miller. 1971. The isolation and characterization of plasma membrane from cultured cells. I. The chemical composition of membrane isolated from uninfected and oncogenic RNA virus-converted chick embryo fibroblasts. *Biochim. Biophys. Acta* **249:**419.

Phillips, D. R. and M. Morrison. 1973. Changes in accessibility of plasma membrane protein as the result of tryptic hydrolysis. *Nature New Biol.* **242:**213.

Unkeless, J. C., A. Tobia, L. Ossowski, J. P. Quigley, D. B. Rifkin and E. Reich. 1973. An enzymatic function associated with transformation of fibroblasts by oncogenic viruses. *J. Exp. Med.* **137:**85.

Warren, L., P. Critchley and I. Macpherson. 1972. Surface glycoproteins and glycolipids of chicken embryo cells transformed by a temperature-sensitive mutant of Rous sarcoma virus. *Nature* **235:**275.

Wickus, G. G. and P. W. Robbins. 1973. Plasma membrane proteins of normal and Rous sarcoma virus-transformed chick-embryo fibroblasts. *Nature New Biol.* **245:**65.

Functional Membrane Changes and Cell Growth: Significance and Mechanism

Arthur B. Pardee, Luis Jiménez de Asúa, and Enrique Rozengurt

Department of Biochemical Sciences, Princeton University
Princeton, New Jersey 08540 and Imperial Cancer Research Fund Laboratories
Lincoln's Inn Fields, London WC2A 3PX

The unrestricted growth of malignant cells in vivo and of transformed cells in culture is strikingly different from the regulated growth of normal and untransformed cells. This difference poses fundamental questions of growth control and at the same time provides material for experimental attack. Our objective here is to summarize several sorts of experiments performed with fibroblasts and designed to examine the biochemical basis of cellular growth control.

Our point of view is based on early ideas (Pardee 1964) that the cell membrane exerts a selective control upon the transport of regulatory molecules, which might be either positive or negative effectors of growth. As originally expressed, this concept stressed the potential importance of membrane transport in the complex process of cellular growth control. Pursuing this functional approach, we have examined the transport ability of a variety of normal and transformed cell lines under different growth conditions, as well as the influences of serum factors and insulin on several transport systems. More recently we have been examining these problems in relation to cAMP, which acts as a "second messenger" between molecules at the cell surface and the processes inside the cell that are affected by them.

Our results are sufficiently diverse that for clarity we express them in the framework shown in Fig. 1. This scheme also serves to organize, according to our view, information obtained by a large number of investigators. This literature has recently been reviewed (Pardee and Rozengurt 1974). In broad outline we show five levels of cellular growth control: (1) exogenous factors known to affect the cell growth pattern; (2) the membrane state which is changed by these factors; (3) cAMP, whose concentration is altered by the membrane state; (4) transport regulated by cAMP concentration; and (5) effects of transported compounds upon the cell cycle at what we call the restriction (R) point.

A particularly striking example of changes at all these levels is seen when cells infected with temperature-sensitive virus mutants are shifted between permissive and nonpermissive temperatures. Not only are growth and morphological changes observed, but also changes of the cell surface (Bader 1972), cAMP level (Otten et al. 1972) and transport activity (Martin et al. 1971).

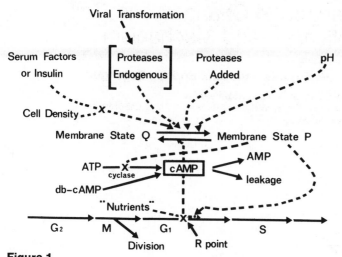

Figure 1
A scheme for the roles in animal cell division of external factors, membrane
states, and cAMP transport. Regulatory connections are shown by dashed lines.
These terminate in an X when inhibition is indicated or with an arrow when
stimulation is indicated. See text for details.

Agents Affecting Cell Growth

Virus-transformed cells have permanently changed growth patterns; they grow to
high densities and their growth is no longer as dependent on serum as are untransformed cells. Transformation dramatically alters the cell surface in several ways
(see Burger 1971), which include functional changes. For instance, polyoma-
transformed 3T3 cells transported certain amino acids approximately threefold
faster than did untransformed 3T3 cells (Foster and Pardee 1969).

Alteration of a great variety of surface properties upon introduction of only a
few viral genes leads one to suspect a common underlying event. (This event
could be different for different viruses.) One such event is proposed to be production of a protease. Protease inhibitors have been reported to restore the growth of
transformed cells to something like the normal pattern (Schnebli and Burger
1972). Several transformed cell lines were found to produce a protease that depends on serum for its activation (Ossowski et al. 1973). The activation of division
and creation of surface changes in untransformed cells by added proteases, as mentioned above, is consistent with the idea that an endogenous protease could modify
the surface of transformed cells and thereby activate their division.

A different kind of growth control is seen when untransformed cells cease to
divide as they become crowded. Their saturation density is dependent on the serum
concentration in the medium. Furthermore factors in serum can transitorily release
cells from this arrested growth (Paul, Lipton and Klinger 1971). The basis for
"density-dependent" inhibition is unknown. According to one hypothesis, dense
cell populations do not receive an adequate supply of nutrients; serum factors raise
the internal nutrient concentration by making transport systems more active

(Holley 1972). An earlier hypothesis suggests that the actual contact of a cell with its neighbors is inhibitory (see Stoker 1971; Dulbecco 1971). The stimulation by serum factors would then be attributed to their ability to diminish these inhibitory contacts. In either hypothesis, serum factors would act on the cell surface. There is no direct evidence for serum, but insulin, which stimulates growth of several cell lines (Clarke et al. 1970; Jiménez de Asúa et al. 1973), probably acts at the cell surface; and receptor sites for it have been detected on human fibroblasts (Gavin et al. 1972).

Other compounds that act on the cell surface also stimulate growth. These include proteases, either soluble (Sefton and Rubin 1970) or bound to Sepharose beads (Burger 1971) and neuraminidase (Vaheri et al. 1972). Their effects, like those of serum and insulin, are transitory and generally permit only one round of cell division after the agent is removed.

Growth is much affected by pH (Ceccarini and Eagle 1971). These effects of pH might also be at the cell surface (Rubin 1971).

The growth of untransformed cells depends on the positive influence of a proper substratum (Clarke et al. 1970). Conceivably this "anchorage dependence," a surface effect, is mediated by increased transport activity; attached CHO cells had a three-fold higher rate of aminoisobutyrate transport than did these cells growing in suspension (Foster and Pardee 1969). Malignant cells could have sufficiently high transport to grow in suspension without a substratum (Glinos and Werrlein 1972).

Transformation, serum, and other agents that relax growth control have many similar effects on the cell (see below, and Pardee and Rozengurt 1974). A complex set of effects has been called the pleiotypic response by Hershko et al. (1971). These authors have proposed, as we do, that the primary effects are on the cell surface and act through a "pleiotypic mediator" (which we consider to be cAMP) so as to modify many properties of the cell.

Membrane States

The numerous differences in membranes of transformed and untransformed cells (see Burger 1971; Pardee and Rozengurt 1974 for reviews) suggest that the membrane might exist in alternative states that regulate the ability of the cells to grow. A classical case of two alternative membrane states is seen in the nerve cell, which can have either high or low permeability to sodium and potassium ions. Dulbecco (1971) has suggested different membrane states for growing and quiescent cells.

For simplicity we show in Fig. 1 two alternative states for fibroblast membranes, although other states can be envisaged as well. The P state is defined by functional (enzyme and transport), architectural, and chemical properties that are present in transformed cell membranes. We propose that this membrane is in a configuration productive for cells. It is associated with a lower intracellular concentration of cAMP which, in turn, influences the membrane state, as will be seen later. The Q state is defined by surface properties present in quiescent cells. Growing untransformed cells transiently have their membrane in the P state during M and early G_1, since several membrane changes take place during this interval (see Pardee and Rozengurt 1974). This change in membrane state is proposed to be fundamental for the cell to proceed to the next cycle. Serum, insulin, and proteases added to cells with membranes in the Q state bring about changes towards the P state. Not all the properties of the P state are proposed to be required simultaneously in order to permit one cycle of growth, but they are probably necessary to stabilize the membrane in a permanent P state.

Transport activities and phospholipid turnover rates change earlier than other cell properties after the cells are stimulated by addition of serum (Cunningham and Pardee 1969) or insulin (Rozengurt and Jiménez de Asúa 1973). Transforming viruses also produce transport changes before most other changes are seen (Hatanaka and Hanafusa 1970; Stoker 1971). Thus transport activity provides the investigator with a rapid, convenient test for detecting a change in membrane state.

The property of the membrane state of BHK cells that appears at present most closely related to growth control is the activity of adenyl cyclase. Bürk (1968) early demonstrated a lower activity of this enzyme in transformed cell membranes. From this observation and others on the effect of phosphodiesterase inhibitors on growth, he proposed the now well-established role of cAMP in growth control. Strong support for the role of adenyl cyclase in growth control is shown in Fig. 2 (Jiménez de Asúa et al. 1973). Here insulin at very low concentrations is seen to inhibit the adenyl cyclase activity of isolated BHK cell membranes. This effect provides a possible mechanism for the changes in cAMP levels produced by insulin. These observations do not exclude effects of phosphodiesterase and leakage on the cAMP levels.

Cyclic AMP Effects on Growth and Transport

The role of cAMP as a "second messenger" for hormone action is now well known. Cyclic AMP and probably cGMP are also strongly implicated in the regulation of

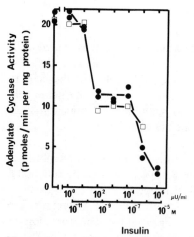

Figure 2
Adenylate cyclase activity of BHK membranes as a function of insulin concentration. (□———□) Amorphous porcine insulin or (●———●) recrystallized porcine (glucagen-free) insulin added to the membranes at the time of assay for adenylate cyclase. From Jiménez de Asúa et al. (1973) with permission of *Proc. Nat. Acad. Sci.*

cell growth and morphology. Cells that have been transformed or stimulated to grow with various agents generally contain lower internal cAMP (see Pardee and Rozengurt 1974 for summary). Cyclic AMP changes can be observed within minutes (Burger et al. 1972; Otten, Johnson and Pastan 1972; Rozengurt and Jiménez de Asúa 1973) after treatment.

The growth stimulation by serum seems in large part to depend on decreased cAMP concentration. This was shown by opposite actions of serum and dibutyryl cAMP (db-cAMP) on growth (Rozengurt and Pardee 1972; Frank 1972; Froehlich and Rachmeler 1972). Dibutyryl cAMP appears to be a moderately effective analog of cAMP; its activity is enhanced by other compounds such as testosterone propionate or theophylline. It affects intact cells, unlike cAMP which apparently does not penetrate the membrane and is more sensitive to phophodiesterases. Figure 3 illus-

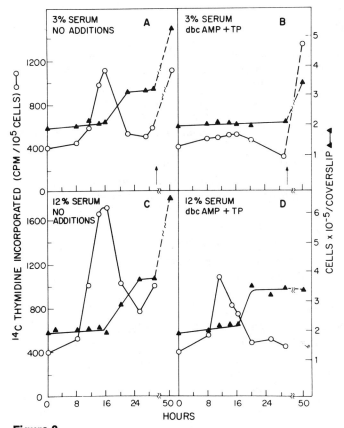

Figure 3
Opposite effects of serum and db-cAMP on the rate of [^{14}C]thymidine incorporation and cell number. Twenty-four hours after seeding CHO cells they were transferred to F-10 supplemented with 4×10^{-5} M L-isoleucine, 10^{-3} M L-glutamine, and 0.3 mM db-cAMP and 30 μM testosterone propionate. Sixty hours later they were put into fresh medium reinforced with isoleucine and glutamine and with or without db-cAMP and testosterone propionate. After 30 hours the cultures exposed to 3% serum were switched to normal medium. All were assayed at intervals for thymidine incorporation and cell number. From Rozengurt and Pardee (1972) with permission of *J. Cell. Physiol.*

trates the preferential inhibition in G_1 of growth of CHO cells. The db-cAMP–serum antagonism is also shown in this figure. After 60 hours in medium containing db-cAMP and testosterone, the cells were transferred to medium containing high or low serum and with and without the inhibitory compounds. On the one hand, the inhibitors were effective in blocking resumption of DNA synthesis and growth at low serum. On the other hand, high serum overcame this inhibition for one round of replication. Apparently a factor initially adequate for growth initiation in the high serum medium is depleted during the first cycle. The concentration of this factor in the low serum medium might be insufficient to counteract the effect of db-cAMP.

Antagonistic effects have also been found when insulin largely replaces serum for stimulating growth of BHK cells. Insulin added to medium containing very low serum gave a large stimulation of growth, caused marked morphological changes in the culture, and allowed the cells to grow in suspension. These effects were all counteracted by monobutyryl cAMP and theophylline (Jiménez de Asúa et al. 1973).

Serum factors lower the concentration of cAMP; when they are depleted either in sparse or confluent 3T3 cultures, the cAMP level rises (Seifert and Paul 1972), and when serum is added it drops (Otten, Johnson and Pastan 1972). These results are consistent with action of serum factors on the cell membrane because synthesis of cAMP by adenyl cyclase, degradation by phosphodiesterase (at least in part), and escape by leakage from the cell are all phenomena that are affected by the state of the membrane.

The state of the membrane could directly affect transport of metabolites, since transport systems are located within the membrane. But also the membrane state could indirectly alter transport rates by changing the cAMP concentration, which in turn affects the rates of some transport processes. This possibility was suggested by the observation that treatment of CHO cells with db-cAMP and testosterone reduced transport rates for some amino acids (Rozengurt and Pardee 1972).

Measurements of nucleoside transport provide a suitable system for studying the effect of cAMP on transport (Rozengurt and Jiménez de Asúa 1973). Within 30 minutes of serum addition, uridine transport into mouse fibroblasts is increased and this is accompanied by a decrease in cAMP (Fig. 4). When the cAMP concentration was raised over 50-fold after prostaglandin E1 (a stimulator of adenyl cyclase [Otten, Johnson and Pastan 1972]) and theophylline (an inhibitor of phosphodiesterase) were added, there was little decrease in uridine transport below the basal level. Kinetics of the transport changes are shown in Fig. 5. The rate of transport became maximal in 30 minutes. Its rise was blocked if prostaglandin E1 and theophylline were added 10 minutes before addition of serum. These compounds caused the transport rate to return to the basal level within 5 minutes, and the rate rose rapidly after their removal. Similar effects were obtained when insulin was used instead of serum.

Thus at least one transport process appears to be activated at low cAMP concentrations. But phosphate transport was not altered by prostaglandin and theophylline when tested in parallel experiments (Rozengurt and Jiménez de Asúa 1973). Since phosphate transport is stimulated by factors in serum (Cunningham and Pardee 1969), there must be another, cAMP-independent mechanism. Whether this second mechanism depends on alteration of the membrane state directly or by

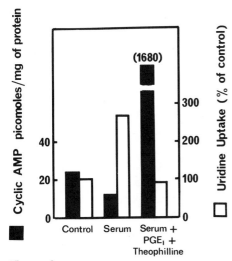

Figure 4

Effect of serum, PGE_1 and theophylline on cyclic AMP concentration and uridine transport. Mouse embryo fibroblasts were maintained in 0.2% serum medium during 4 days and transferred to fresh 10% serum medium for 30 min, without and with PGE_1 and theophylline, at 25 μg/ml and 1 mM, respectively. Cyclic AMP was measured by the method of Gilman; the samples were purified by Dowex-50 chromatography. For uridine uptake the cultures were pulse-labeled with [^{3}H]uridine (2.5 μCi/ml, 29 mCi/mmole) for 5 min, washed thoroughly and the acid-soluble radioactivity was measured by liquid scintillation counting.

influence of another intracellular mediator such as cGMP is unknown. The result suggests early cAMP-independent processes of growth control.

The membrane state is proposed to affect the concentration of cAMP, but cAMP can also affect the membrane state. The effect of cAMP on the membrane is shown by transport and other changes; and furthermore the membrane contains protein kinases sensitive to cAMP activation (Johnson et al. 1972; Rubin and Rosen 1973). These enzymes provide a receptor system potentially capable of modifying the membrane by phosphorylation of specific proteins. These reciprocal effects of membrane state and cAMP taken together suggest an autocatalytic mechanism for altering the total membrane in an all-or-none way.

As the simplest example, suppose the cell membrane is in the Q state. When a few molecules of serum factor become attached, nearby parts of the membrane would be converted to the P state with its lower cyclase activity. As a result, cAMP concentration inside the cell would decrease. Since this lower cAMP favors the P state, more of the membrane would shift to the P state, and then yet more cyclase activity would be lost. By this autocatalytic process the entire membrane would shift to the P state. The membrane would remain in the P state until events that determine the cAMP level or membrane state have become sufficiently different to start an autocatalytic shift in the other direction.

Either internal processes that occur at stages of the cell cycle or external influences could shift the membrane state. A change in state might occur only when both a cyclical event and an external one act in conjunction. For instance, trans-

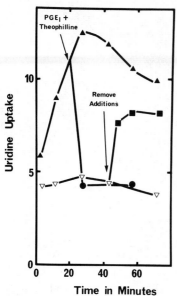

Figure 5

Effect of prostaglandin E1 and theophylline on the time course of the uridine transport increase induced by serum in mouse embryo fibroblasts. The cells were made quiescent by incubating 5 days in a medium containing 0.2% serum. The conditions were as follows: (▲) 10% serum medium; (▽) 25 μg/ml PGE_1 + 1 mM theophylline added 10 min before serum; (■) PGE_1 + theophylline were removed after 40 min exposure in 10% serum medium; (●) PGE_1 + theophylline were added 20 min after the cells were switched to 10% serum medium. The cultures were pulse-labeled with [³H]uridine and acid-soluble radioactivity was measured as described for Fig. 4.

formation might lead to full conversion to the P state only after mitosis (Burger 1971) because only then would the combined effects of virus and mitosis achieve a sufficiently low cAMP level. Conversely, after serum is removed cells might have to pass through M where cAMP is low (Burger et al. 1972) to G_1 where cAMP appears to be high (Sheppard and Prescott 1972) before the cAMP concentration reaches the critical growth inhibitory level.

Transport and Growth

Uncontrolled growth of cells in culture and of cancer cells in the whole organism has been proposed to result from changed membrane properties that alter transport of growth regulatory substances (Pardee 1964; Holley 1972). This hypothesis provides a testable basis for growth control. The two predictions that (1) transformed cells and (2) untransformed cells that have stopped growing should have different membrane transport properties from growing untransformed cells have both been confirmed (Foster and Pardee 1969; Cunningham and Pardee 1969). Demonstrations of changed transport rates following transformation for various compounds have now been obtained in a number of laboratories (see Pardee and Rozengurt 1974 for review).

The hypothesis raises a number of questions. Are growth regulatory substances present as such in the medium? Can transport systems modify the intracellular concentrations of these substances sufficiently to control growth? What substances limit growth under particular sets of conditions? Does the membrane change cause specific transport changes or is there a general decrease of all transport activities? What is the mode of growth control by the transported compound(s), nutritional or regulatory?

Substances from the medium can greatly affect cell growth, as demonstrated by the exacting balance of nutrients required in culture media. When growth stops at high cell density, added substances can reinitiate cell division. Serum was early shown to permit further growth of several cell lines. Whether serum itself contains a growth limiting substance, or whether it provides stimulatory factors that increase the uptake of nutrients (or both) is not clear. However, the ability of very small amounts of highly purified serum factors to stimulate growth (Paul, Lipton and Klinger 1971) favors the idea that serum is primarily stimulatory. Fresh nutrients stimulate some cell lines to grow further, but not others (Rovera and Baserga 1973). One needs to account for this difference in response.

Both nutrients and the factors that stimulate transport could be depleted in old cultures, and this combination of deficiencies could limit transport activity and growth. These depletions could be especially severe in the medium that lies extremely close to a confluent layer of cells (Stoker pers. commun.). Thus it appears quite likely that growth of dense cultures ceases because the cells are unable to take up an adequate concentration of components from the medium. The increases of transport activity brought about by virus transformation or added serum factors correlate well with the ability to resume growth.

The transport changes could be either general for all compounds, resulting from a general change in the properties of the membrane, or they could be specific changes in individual transport systems. The general change hypothesis in its simple form appears unlikely (Foster and Pardee 1969). Polyoma-transformed 3T3 cells more actively transported only some amino acids, not all, as compared to growing or nongrowing untransformed 3T3 cells. Furthermore confluent cells after serum addition did not have higher transport for adenosine, as contrasted to uridine and phosphate (Cunningham and Pardee 1969). Measurements of internal pools of the transported compounds were made in conjunction with growth studies of human lung cells (Griffiths 1972). Growth stopped when amino acids in the medium were far from exhausted and the internal pools were quite high. Thus, the available evidence does not support a general transport decrease in confluent cells.

A specific decrease in the transport of one essential component could limit growth if the decrease were sufficiently great. The transport system for glucose has been extensively studied; it is considerably higher in transformed cells, and its activity drops considerably when cells stop growing (Hatanaka and Hanafusa 1970; Sefton and Rubin 1971; Vaheri et al. 1972; Grimes and Schroeder 1973). However, the internal concentration of glucose only drops moderately after human lung cells become confluent, and higher extracellular glucose does not restore growth (Griffiths 1972). No known compound has been shown to drop to a very low level in a quiescent cell culture. Thus, proof of the role of specific inhibition of transport in growth control is lacking, but seems well worth further exploration.

A specific limitation of growth by very low concentration of a component may occur with media that contain limiting amounts of serum (Bürk 1970; Temin 1971) or isoleucine (Ley and Tobey 1970). Isoleucine is not totally absent, being supplied either by turnover of proteins or as a contaminant of leucine (Everhart

1972). Growth does not abruptly stop, as when most nutrients are omitted, but the cells proceed slowly until they reach G_1. This result suggests that growth might also be blocked at a specific part of the cell cycle when components other than isoleucine or serum are at a very low concentration inside the cell.

Holley (1972) has proposed that the diminished transport of any essential nutrient into cells at confluence halts growth. On this hypothesis transformed cells would have high transport activities at confluence and so would continue to receive an adequate supply of the limiting component. Holley (1972) has suggested that the substances that limit growth are nutrients. Since neither their identities nor their functions are known, we prefer to call them components of the medium. They could be supplied as known compounds or in serum.

According to the original hypothesis of Pardee (1964), a component of the medium exerts a regulatory effect within the cell to an extent depending on its transport into the cell. This could be true either for known nutrients or for components of serum. The demonstration that cells stop in a definite part of their growth cycle when some extracellular components are provided at very low concentrations suggests that cells have developed a regulatory mechanism responsive to moderate decreases in the (internal) concentrations of certain compounds. These compounds could, for instance, act as inducers to allow synthesis of proteins needed for initiation of DNA synthesis. The control could be stronger if factors in serum were important both within the cell and also for entry of the limiting compounds into the cell; this dual control would provide a sharper cutoff of cell growth when the serum factors became limiting.

Finally it is likely that the effectiveness of controls depends upon the environment in contact with the cells. Media of different compositions could limit in different ways. In particular controls operating on cells in culture might be less complex and even dependent on different substances than they are for cells in the whole animal. The latter are bathed in fluid that contains a variety of substances carried in blood, derived from different organs, and of varying labilities. These substances may be effective at different concentrations in vivo than in cell culture. In vivo there could be components with negative growth control effects that are absent in the sera generally in use (Houck and Hennings 1973). These differences in control might account for the lack of correlation sometimes found between transformed cells' growth patterns in culture and in the intact animal.

We conclude that the hypothesis of growth limitation by transport deficiencies is a promising one. It needs much further study for its verification; specific components that limit transport as well as their modes of action need to be identified.

Site of Growth Restriction

Untransformed cells that have stopped growing owing to a variety of causes including high density, serum deprivation, and nutrient insufficiencies have been found to be in G_1 rather than at random places in the cell cycle. A demonstration that the cells are blocked at the same point in G_1 would be important for several reasons. One is that this result suggests existence of a special growth regulating event. Second, it indicates the point in the cycle (e.g. mid-G_1 rather than S for instance) at which one should look for biochemical phenomena of growth control. Third, if nutritional shortages act at the same point as serum and density effects the relation between the three are strengthened, in accord with Holley's (1972) suggestion.

We have preliminary evidence that there is indeed the same control by serum deprivation, isoleucine or glutamine starvation. We have used three methods to study the timing of these events relative to the initiation of DNA synthesis. One is to stop growth by depriving cells of one of these factors and then, after putting them into complete medium, measuring how much time is needed before DNA synthesis starts. The minimal time required was about 8 hours in each of the three cases. A second test is to starve the cells for one factor and then to switch them to starvation for the second factor. One would expect them to synthesize DNA only if the second starvation creates a block earlier in G_1 than does the first starvation, because the block point due to the second factor's absence would have been passed during the first starvation. In none of the nine possible cases was there a major escape of the cells into the S phase (Table 1). This result is consistent with the blocks induced by serum, isoleucine or glutamine deprivation all being at the same point. A third test is to starve cells for one factor, then interpose a recovery period in complete medium (Bürk 1970; Temin 1971), and finally place them under starvation conditions and see how much DNA they make. Preliminary experiments indicated that the interval in complete medium that restored ability to make DNA during the final starvation was similar under the three conditions shown in Table 1.

From this work we consider that there may be a special event in G_1 at a point we call R for restriction of growth; normal cells cannot accomplish this R event unless both serum and nutrients are adequate.

In contrast to these findings with untransformed cells, 3T3 cells transformed with SV40 were not stopped in G_1 by db-cAMP (Smets 1972). Similarly, the aminonucleoside of puromycin stopped lung fibroblasts in G_1, but not HeLa cells (Studzinski and Gierthy 1973). These results suggest that the G_1 control process has become ineffective for these cells. Either their R-point control is not applied because transport is so high that intercell nutrient concentrations never become much reduced, or their control mechanism itself might be altered by transformation, and possibly by subsequent cell selection.

Table 1

Ability of BHK Cells to Synthesize DNA
after a Shift of Starvation Conditions

First starvation	Thymidine uptake (% of control) during second starvation for		
	Serum	Isoleucine	Glutamine
Serum	9.7	2.4	2.2
Isoleucine	20.4	4.0	4.0
Glutamine	13.1	2.9	1.3

BHK cells growing in 4 × Eagle's medium plus 10% calf serum were transferred into (a) medium with 0.25% serum plus 0.1 mM ornithine and 0.1 mM hypoxanthine (Bürk 1970) or (b) medium lacking isoleucine, or (c) medium lacking glutamine. After incubation for 52 hr, the cells were rinsed; then fresh media, either complete (control) or each of the above, all plus 0.025 mM adenosine, was added to various cultures. [³H]thymidine (0.5 μM, 3 × 10⁵ cpm/ml) was added and incubation was continued for 22.5 hr. The cells were rinsed, extracted with 5% TCA, and the insoluble counts were determined.

Properties of the R Point

Since growth appears to stop at a special point in G_1, one wishes to know why this point is extrasensitive to environmental factors and what biochemical events are critical at this stage of the cycle. What general sort of process might be required? Smith and Martin (1973) have recently presented a hypothesis regarding the duration of the cell cycle, including as an extreme case cessation of growth, that accounts for the very different lengths of G_1 and of the cycle for different cells in a culture. They divide the cycle into two parts, A and B, that have different time dependences. Each cell proceeds through part B (which includes the last part of G_1, S, G_2, M, and the first part of G_1) in a rather constant length of time. Then it enters part A, from which it can escape with a constant probability per unit time into the next part B. Thus the greatest number of cells move out of A during the earliest time interval, and successively smaller fractions later escape from A. The escape is according to the semi-log frequency rule that also applies to radioactive decay; it permits quantitative prediction of the distribution of G_1 times. The average cell would spend much of its G_1 period in state A and not progressing toward S; the events leading to DNA synthesis would only occur subsequently near the end of G_1.

What differences have already been described between cell populations in various parts of G_1? A higher cAMP concentration in mid-G_1 has already been mentioned. Lower transport activity for several compounds was found in early G_1 (Sander and Pardee 1972). The transport activities fall twofold at mitosis and then after about one hour rise to their initial specific activities. In cells blocked by isoleucine starvation (Sander and Pardee 1972) or excess db-cAMP (Rozengurt and Pardee 1972), the transport activity is also at about half normal. It recovers about 3 hours after addition of isoleucine, which is hours before DNA synthesis commences.

Protein synthesis is required during G_1 for progress into S (see Doida and Okada 1972). Furthermore when protein synthesis is inhibited in early G_1 but not late G_1, a deterioration of the cell's ability to proceed afterwards has been observed (Highfield and Dewey 1972). This delay could result from lability of substances that must gradually be synthesized during G_1. Further relations between the state of the cell and the conditions that allow escape from the A state are important studies for the future.

The nature of the molecular event that switches each normal cell to growth after mitosis is unknown. We suggest that the biochemical basis for this event could be a transitory, probabilistic (Smith and Martin 1973) decrease of cAMP to below a critical concentration. This could occur in G_1 at any moment, with a probability that depends on the concentration of factors in the medium. Fluctuations of the number of serum factor molecules bound to a cell could create variations localized in the membrane. These membrane changes would correspondingly inactivate adenyl cyclase molecules and thereby briefly decrease cAMP. Once a critical low cAMP concentration was reached, even for quite a short time, our scheme (Fig. 1) predicts that the cells should be permanently switched to growth. This is because of the autocatalytic sequence, discussed above, in which when low cAMP converts some of the membrane to the low adenylate cyclase (P) state, this low enzyme activity in turn further lowers cAMP, which further lowers cyclase; cAMP would thus soon reach equilibrium at a decreased concentration. Then the cells would be committed to go on to divide.

SUMMARY

In our view uncontrolled growth leading to malignancy is brought about by permanent transformation of the cell surface (as is caused by some viruses). Transient release of growth control due to actions of extracellular agents such as serum is suggested to result from temporary surface changes. These agents rapidly shift the membrane from an extreme state Q, typical of nongrowing cells, toward a state P found in transformed cells. These surface changes in turn inhibit adenyl cyclase and lower the cAMP concentration inside the cell. Low cAMP increases transport of critical compounds through the membrane. Although transport can limit the availability of compounds from the medium, we think it unlikely that nutrients limit growth by inadequately making substrates available for metabolism. Growth control probably is exerted at a specific restriction point R in G_1. This specificity would not be expected from simple nutritional deprivation. Rather, we propose that when a crucial component from medium drops below a critical internal concentration, it functions in a regulatory way to halt a specific cell cycle process at R. The control of cell growth is proposed to have evolved so that an allosteric control comes into play when extracellular conditions are appropriate in terms of the entire organism.

Our scheme shows cell division control to be based on a sequence of positive and negative controls. To summarize them: cell density inhibits serum effects $(-)$; serum factors (or transformation) inhibit cyclase $(-)$; cyclase stimulates cAMP formation $(+)$; cAMP (or db-cAMP) inhibits transport of some factors from the medium $(-)$; transport stimulates growth $(+)$. High cell density would thus give $(- - + - +)$ or a net negative effect. This scheme suggests that control of growth is much more complex, and alterable in more ways, than initially imagined.

Acknowledgments

This work was supported from Grant AI-CA 04409 from the U.S. Public Health Service. A.B.P. was a Scholar of the American Cancer Society. L.J. de A. was a fellow of the John Simon Guggenheim Memorial Foundation.

REFERENCES

Bader, J. P. 1972. Temperature-dependent transformation of cells infected with a mutant of Bryan Rous sarcoma virus. *J. Virol.* **10:**267.

Burger, M. M. 1971. Cell surfaces in neoplastic transformation. *Current topics in cellular regulation,* vol. 3, p. 135. Academic Press, New York.

Burger, M. M., B. M. Bombik, B. McL. Breckenridge and J. R. Sheppard. 1972. Growth control and cyclic alterations of cyclic AMP in the cell cycle. *Nature New Biol.* **239:**161.

Bürk, R. R. 1968. Reduced adenyl cyclase activity in a polyoma virus transformed cell line. *Nature* **219:**1272.

———— 1970. One-step growth cycle for BHK21/13 hamster fibroblasts. *Exp. Cell Res.* **63:**309.

Ceccarini, C. and H. Eagle. 1971. pH as a determinant of cellular growth and contact inhibition. *Proc. Nat. Acad. Sci.* **68:**229.

Clarke, G. D., M. G. P. Stoker, A. Ludlow and M. Thornton. 1970. Requirement of serum for DNA synthesis in BHK21 cells: Effects of density, suspension and virus transformation. *Nature* **227:**798.

Cunningham, D. D. and A. B. Pardee. 1969. Transport changes rapidly initiated by serum addition to "contact inhibited" 3T3 cells. *Proc. Nat. Acad. Sci.* **64**:1049.

Doida, Y. and S. Okada. 1972. Effects of actinomycin D and puromycin on the cell progress from M to G$_1$ and S stages in cultured mouse leukemia L5178Y cells. *Cell Tissue Kinetics* **5**:15.

Dulbecco, R. 1971. Growth control in cell cultures. *Ciba Fndn. Symp.* (ed. G. E. W. Wolstenholme and J. Knight) p. 71. Churchill Livingstone, London.

Everhart, L. P. 1972. Effects of deprivation of two essential amino acids on DNA synthesis in Chinese hamster cells. *Exp. Cell Res.* **74**:311.

Foster, D. O. and A. B. Pardee. 1969. Transport of amino acids by confluent and nonconfluent 3T3 and polyoma virus-transformed 3T3 cells growing on glass cover slips. *J. Biol. Chem.* **244**:2675.

Frank, W. 1972. Cyclic 3′:5′-AMP and cell proliferation in cultures of embryonic rat cells. *Exp. Cell Res.* **71**:238.

Froehlich, J. E. and M. Rachmeler. 1972. Effect of adenosine 3′, 5′-cyclic monophosphate on cell proliferation. *J. Cell Biol.* **55**:19.

Gavin, J. R., III, J. Roth, P. Jen and P. Freychet. 1972. Insulin receptors in human circulating cells and fibroblasts. *Proc. Nat. Acad. Sci.* **69**:747.

Glinos, A. D. and R. J. Werrlein. 1972. Density dependent regulation of growth in suspension cultures of L-929 cells. *J. Cell. Physiol.* **79**:79.

Griffiths, J. B. 1972. The effect of cell population density on nutrient uptake and cell metabolism: A comparative study of human diploid and heteroploid cell lines. *J. Cell Sci.* **10**:515.

Grimes, W. J. and J. L. Schroeder. 1973. Dibutyryl cyclic adenosine 3′, 5′ monophosphate, sugar transport, and regulatory control of cell division in normal and transformed cells. *J. Cell Biol.* **56**:487.

Hatanaka, M. and H. Hanafusa. 1970. Analysis of a functional change in membrane in the process of cell transformation by Rous sarcoma virus; alteration in the characteristics of sugar transport. *Virology* **41**:647.

Hershko, A., P. Mamont, R. Shields and G. M. Tomkins. 1971. Pleiotypic response. *Nature New Biol.* **232**:206.

Highfield, D. P. and W. C. Dewey. 1972. Inhibition of DNA synthesis in synchronized Chinese hamster cells treated in G$_1$ or early S phase with cycloheximide or puromycin. *Exp. Cell Res.* **75**:314.

Holley, R. W. 1972. A unifying hypothesis concerning the nature of malignant growth. *Proc. Nat. Acad. Sci.* **69**:2840.

Houck, J. C. and H. Hennings. 1973. Chalones-specific endogenous mitotic inhibitors. *FEBS Letters* **32**:1.

Jiménez de Asúa, L., E. S. Surian, M. M. Flawia and H. N. Torres, 1973. Effect of insulin on the growth pattern and adenylate cyclase activity of BHK fibroblasts. *Proc. Nat. Acad. Sci.* **70**:1388.

Johnson, E. M., T. Ueda, H. Maeno and P. Greengard. 1972. Adenosine 3′, 5′-monophosphate-dependent phosphorylation of a specific protein in synaptic membrane fractions from rat cerebrum. *J. Biol. Chem.* **247**:5650.

Ley, K. D. and R. A. Tobey. 1970. Regulation of initiation of DNA synthesis in Chinese hamster cells. II. Induction of DNA synthesis and cell division by isoleucine and glutamine in G$_1$-arrested cells in suspension culture. *J. Cell Biol.* **47**:453.

Martin, G. S., S. Venuta, M. Weber and H. Rubin. 1971. Temperature-dependent alterations in sugar transport in cells infected by a temperature-sensitive mutant of Rous sarcoma virus. *Proc. Nat. Acad. Sci.* **68**:2739.

Ossowski, L., J. C. Unkeless, A. Tobia, J. P. Quigley, D. B. Rifkin and E. Reich. 1973. An enzymatic function associated with transformation of fibroblasts by oncogenic viruses. II. Mammalian fibroblast cultures transformed by DNA and RNA tumor viruses. *J. Exp. Med.* **137:**112.

Otten, J., G. S. Johnson and I. Pastan. 1972. Regulation of cell growth by cyclic adenosine 3′, 5′-monophosphate. *J. Biol. Chem.* **247:**7082.

Otten, J., J. Bader, G. S. Johnson and I. Pastan. 1972. A mutation in a Rous sarcoma virus gene that controls adenosine 3′, 5′-monophosphate levels and transformation. *J. Biol. Chem.* **247:**1632.

Pardee, A. B. 1964. Cell division and a hypothesis of cancer. *Nat. Cancer Inst. Monogr.* **14:**7.

Pardee, A. B. and E. Rozengurt. 1974. Role of the surface in production of new cells. *Biochemistry of cell walls* (ed. C. F. Fox). Medical and Technical Publ. Co., London. (in press).

Paul, D., A. Lipton and I. Klinger. 1971. Serum factor requirements of normal and simian virus 40-transformed 3T3 mouse fibroblasts. *Proc. Nat. Acad. Sci.* **68:**645.

Rovera, G. and R. Baserga. 1973. Effect of nutritional changes on chromatin template activity and non-histone chromosomal protein synthesis in WI-38 and 3T6 cells. *Exp. Cell Res.* **78:**118.

Rozengurt, E. and L. Jiménez de Asúa. 1973. Role of cyclic AMP in the early transport changes induced by serum in quiescent mouse embryo fibroblasts. *Proc. Nat. Acad. Sci.* **70:** (In press).

Rozengurt, E. and A. B. Pardee. 1972. Opposite effects of dibutyryl adenosine 3′:5′-cyclic monophosphate and serum on growth of Chinese hamster cells. *J. Cell. Physiol.* **80:**273.

Rubin, C. S. and O. M. Rosen. 1973. The role of cyclic AMP in the phosphorylation of proteins in human erythrocyte membranes. *Biochem. Biophys. Res. Commun.* **50:**421.

Rubin, H. 1971. pH and population density in the regulation of animal cell multiplication. *J. Cell Biol.* **51:**686.

Sander, G. and A. B. Pardee. 1972. Transport changes in synchronously growing CHO and L cells. *J. Cell. Physiol.* **80:**267.

Schnebli, H. P. and M. M. Burger. 1972. Selective inhibition of growth of transformed cells by protease inhibitors. *Proc. Nat. Acad. Sci.* **69:**3825

Sefton, B. M. and H. Rubin. 1970. Release from density dependent growth inhibition by proteolytic enzymes. *Nature* **227:**843.

———. 1971. Stimulation of glucose transport in cultures of density-inhibited chick embryo cells. *Proc. Nat. Acad. Sci.* **68:**3154.

Seifert, W. and D. Paul. 1972. Levels of cyclic AMP in sparse and dense cultures of growing and quiescent 3T3 cells. *Nature New Biol.* **240:**281.

Sheppard, J. R. and D. M. Prescott. 1972. Cyclic AMP levels in synchronized mammalian cells. *Exp. Cell Res.* **75:**293.

Smets, L. A. 1972. Contact inhibition of transformed cells incompletely restored by dibutyryl cyclic AMP. *Nature New Biol.* **239:**123.

Smith, J. A. and L. Martin. 1973. Do cells cycle? *Proc. Nat. Acad. Sci.* **70:**1263.

Stoker, M. G. P. 1971. Tumour viruses and the sociology of fibroblasts. *Proc. Roy. Soc. London B* **181:**1.

Studzinski, G. P. and J. F. Gierthy. 1973. Selective inhibition of the cell cycle of cultured human diploid fibroblasts by aminonucleoside of puromycin. *J. Cell. Physiol.* **81:**71.

Temin, H. M. 1971. Stimulation by serum of multiplication of stationary chicken cells. *J. Cell. Physiol.* **78:**161.

Vaheri, A., E. Ruoslahti and S. Nordling. 1972. Neurominidase stimulates division and sugar uptake in density-inhibited cell cultures. *Nature New Biol.* **238:**211.

Cyclic AMP and Malignant Transformation

I. Pastan, W. B. Anderson, R. A. Carchman,
M. C. Willingham, T. R. Russell, and G. S. Johnson
Laboratory of Molecular Biology, National Cancer Institute
National Institutes of Health, Bethesda, Maryland 20014

Cell culture is now widely accepted as a method to study how normal cells become transformed into cancer cells. When grown in culture normal embryo cells or fibroblasts are characterized by their flattened, elongated shape, firm adherence to substratum, moderate growth rate, and inability to grow in soft agar. Most normal cells also show contact inhibition of movement and density-dependent inhibition of growth. Following transformation these properties are changed. The "transformed cells" tend to change their shape, become less adherent to substratum, grow rapidly, lose density-dependent inhibition of growth and contact inhibition of movement, and are able to grow in soft agar. Our laboratory has focused on the possible role of cyclic AMP in controlling some of these processes.

The first approach we pursued was to treat transformed cells with an active cyclic AMP derivative, dibutyryl cyclic AMP, or agents which elevate cyclic AMP levels (phosphodiesterase inhibitors such as theophylline, or adenylate cyclase activators such as prostaglandin E_1) and to examine what happens to the transformed cells. In general we find that treatment with dibutyryl cAMP quickly produces an increase in adhesiveness to substratum (Johnson, Morgan and Pastan 1972) and a decrease in motility (Johnson and Pastan 1972a). A few hours later the cells begin to flatten and become elongated (Johnson and Pastan 1972b). With some cell lines, for example the Kirsten strain of murine sarcoma virus-transformed normal rat kidney (NRK) cells, many of the cells look almost normal after dibutyryl cAMP treatment (Fig. 1). With other transformed lines the morphologic response is less striking (Johnson and Pastan 1972b). The effect of dibutyryl cAMP on adhesiveness or cellular morphology does not require new protein synthesis, for these responses are not prevented by treatment of the cells with cycloheximide (Johnson et al. 1972; Johnson and Pastan 1972b). Dibutyryl cAMP (Johnson and Pastan 1972b) or prostaglandin E_1 (Johnson and Pastan 1971) treatment slows the growth rate of both normal and transformed cells. In addition such treatment decreases the saturation density of contact-inhibited 3T3 cells, but does not restore density-dependent inhibition of growth to transformed cells (Johnson and Pastan 1972b). Treatment of contact-inhibited 3T3 cells (Willingham, Johnson and Pastan 1972)

563

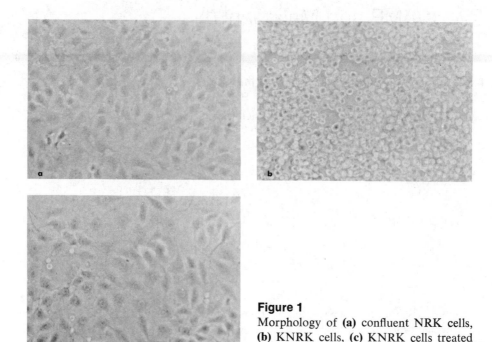

Figure 1
Morphology of **(a)** confluent NRK cells,
(b) KNRK cells, **(c)** KNRK cells treated
for 4 days with 1 mM dibutyryl cAMP.

or density-inhibited primary chicken embryo fibroblasts (Russell and Pastan, pers. commun.) with dibutyryl cAMP at the time of replating (early G_1 phase) prevents the synchronous wave of DNA synthesis that is seen in untreated cells.

Another approach to the study of the role of cyclic AMP in transformation is to measure cyclic AMP levels in normal and transformed cells. Two basic observations have emerged from such studies. The first is that there is a good correlation between the doubling time and the level of cyclic AMP in cells. Rapidly growing cells have low cyclic AMP levels and slowly growing cells have high levels (Otten, Johnson and Pastan 1971). We find the levels of cyclic AMP become particularly high as cells become growth inhibited due to density-dependent inhibition of growth. We have observed this rise of cyclic AMP in 3T3 cells, NRK cells and human fibroblasts (Fig. 2 and Otten, Johnson and Pastan 1971, 1972; Carchman et al. pers. commun.). On the other hand cyclic AMP levels fail to rise in a variety of transformed cells even when they grow to very high densities (Otten, Johnson and Pastan 1971).

The second observation about the relationship between cyclic AMP and growth comes from experiments in which the growth rate has been altered and cyclic AMP levels measured. Those agents which stimulate contact-inhibited cells to grow all lower cyclic AMP levels. These include fresh serum, insulin, and trypsin (Otten, Johnson and Pastan 1972). Some conditions which inhibit cell growth, such as serum depletion (Seifert and Paul 1972) or treatment with prostaglandin E_1 (Johnson and Pastan 1971), raise cyclic AMP levels.

How do cyclic AMP levels become elevated as normal cells grow together? To answer this we have measured the activity of the enzymes of cyclic AMP metabolism in untransformed, contact-inhibited NRK cells at various stages of growth.

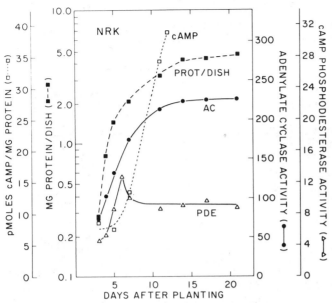

Figure 2

Influence of increasing NRK cell population density on cyclic AMP concentration, adenylate cyclase activity, and cyclic AMP phosphodiesterase activity. Cells were plated in 100-mm petri dishes at 1×10^5 cells per dish and grown at 37°C. Growth media were changed every 48 hr. The increase in cell population density is indicated by the change in total mg protein per dish (■– – –■). Cyclic AMP levels (□....□) are given as pmoles cAMP per mg protein. Adenylate cyclase activity (●——●) was measured with 0.2 mM ATP and 5 mM Mg++ and is expressed as pmoles cyclic AMP formed per 10 min per mg protein. Cyclic AMP phosphodiesterase activity (△——△) was determined with 0.12 μM cyclic AMP as substrate and is expressed as pmoles cyclic AMP hydrolyzed per min per mg protein.

The results of these studies are illustrated in Fig. 2. During the early phases of growth both adenylate cyclase and cyclic AMP phosphodiesterase activities are increasing, but the levels of cyclic AMP remain low. As the cells become heavier the phosphodiesterase stops rising, but the adenylate cyclase continues to rise and concomitantly the levels of cAMP begin to rise and continue to do so as the cells become crowded together.

What happens to the enzymes of cyclic AMP metabolism after transformation? Such studies are best done in cells very recently transformed to avoid possible effects of prolonged culture. For such studies cells transformed by the RNA tumor viruses are employed, for these viruses rapidly transform just about all the cells in a culture. We first studied chick embryo cells transformed by either the Bryan high titer strain (RSV-BH) or the Schmidt-Ruppin strain (RSV-SR) of Rous sarcoma virus. Transformation with either of these viruses leads to a fall in cyclic AMP levels (Otten et al. 1972; Carchman and Pastan pers. commun.) and a decrease in adenylate cyclase activity (Anderson, Johnson and Pastan 1973; Anderson, Lovelace and Pastan 1973). However the two viruses inactivate adenylate cyclase by different mechanisms. In cells transformed with RSV-BH the K_m [ATP] and

Table 1

Cyclic AMP Level and Adenylate Cyclase Activity in Chick
Embryo Fibroblasts Transformed by Rous Sarcoma Viruses

		Adenylate cyclase	
	cAMP (pmoles/mg NA)	K_m [ATP] (mM)	V_{max} (pmoles/10 min/ mg protein)
Untransformed	50	0.2	405
RSV-BH-transformed	25	1.0	375
RSV-SR-transformed	25	0.2	208

the Mg^{++} concentration dependence of the adenylate cyclase is altered, whereas with RSV-SR-transformed cells, the V_{max} of the enzyme is lowered but there is no change in K_m [ATP] (Table 1).

Temperature-sensitive transformation mutants have been isolated from Bryan high titer and Schmidt-Ruppin strains of Rous sarcoma virus and the Kirsten strain of murine sarcoma virus. Cells infected with these viruses appear normal when grown at 40–41°C, but appear transformed when shifted to a lower temperature. We have employed cells transformed by these mutant viruses to examine the relationship between the fall in cyclic AMP levels, the inactivation of adenylate cyclase, and the expression of the transformed phenotype. In Table 2 we summarize the results of these studies. With RSV-BH-Ta the fall in cyclic AMP, inactivation of adenylate cyclase, and expression of the transformed phenotype is very rapid. Some changes can be seen as early as 10 minutes after the temperature shift. Cells transformed by RSV-SR-T5 require some hours to develop changes in all three parameters. Finally, cells transformed by the mutant of Kirsten sarcoma virus take 24–

Table 2

Time Required to Produce Changes in Cells Transformed
by Temperature-sensitive Mutant Viruses

	Decreased cAMP level	Decreased adenylate cyclase activity	Change in morphology	Increased hexose uptake
CEF-RSV-BH-Ta	10–30 min	10 min	10–30 min[a]	3–4 hours[a]
CEF-RSV-SR-T5		6–8 hours	12–18 hours[b]	> 5 hours[b]
NRK-KSV-T6	24–48 hours		24–48 hours	

CEF-RSV-BH-Ta: Chick embryo fibroblasts transformed by mutant of Bryan high-titer strain Rous sarcoma virus.

CEF-RSV-SR-T5: Chick embryo fibroblasts transformed by mutant of Schmidt-Ruppin strain Rous sarcoma virus.

NRK-KSV-T6: Rat kidney cells transformed by mutant of Kirsten strain murine sarcoma virus.

[a] Bader 1972.

[b] Martin et al. 1971.

48 hours to transform morphologically. It also takes about the same time to detect a fall in cyclic AMP in these cells.

The results presented so far indicate that transformed fibroblasts have low levels of cyclic AMP and that some of the properties of these cells are due to their low cyclic AMP levels. However other properties of transformed cells are probably not due to low cyclic AMP levels but related to other actions of the "transforming virus." How might one establish which effects of transformation are solely due to low cyclic AMP levels? The approach we chose was to isolate host cell mutants defective in cyclic AMP metabolism. Since adhesiveness to substratum is a property of cells that we know is strikingly affected by treatment with dibutyryl cAMP, we set up a selection for mutants of 3T3 cells that were normally adhesive at 39°C but had decreased adhesiveness when the temperature was lowered. A number of phenotypically similar isolates have been obtained (Willingham, Carchman and Pastan pers. commun.). The characteristic appearance of cells from one of these isolates before and after a temperature shift is shown in Fig. 3. When such cells are removed from a 39°C incubator, they have normal cAMP levels (40 pmoles per mg nucleic acid). Within 30 seconds after being placed at 23°C the cAMP levels have begun to fall and by 2 minutes have reached a low of around 20 pmoles/mg nucleic acid. The fall in cAMP is principally due to its release into the medium. After 2–5 minutes the cells become less adherent and can readily be removed from the dish by spraying the dish with medium from a pipette. At this time the cells are still mostly flattened, although a few are beginning to retract their processes. However by 15 minutes after the temperature shift at least one-half of the cells have retracted their processes and rounded up. Pretreatment of the cells with dibutyryl cAMP prevents the morphologic response to temperature shift. These studies allow us to conclude that the flattened shape and high adherence of normal fibroblasts require the presence of adequate levels of cyclic AMP. Another interesting aspect of these mutants is that the low cyclic AMP levels only persist for about one hour. Then

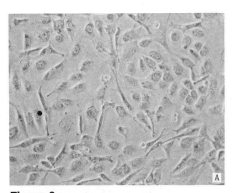 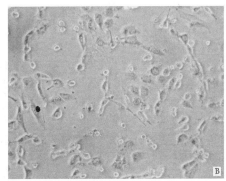

Figure 3
Effect of temperature shift on a cyclic AMP mutant of 3T3 cells. This figure shows two Polaroid pictures of the same field of subconfluent cells grown at 39.5°C in 10% calf serum Dulbecco-Vogt's medium. These were photographed at **(A)** 30 seconds and **(B)** 20 minutes after removal from the 39.5°C incubator with an inverted phase-contrast microscope (100× magnification) at room temperature (reference dot marks the same cell in each picture). Note that in (*B*) many of these cells have retracted their processes and some have detached from the plastic. Of the 130 cells in this field, at least 80 (61%) show definite morphologic change in 20 minutes.

the levels begin to rise, overshoot and finally after a few hours, return to normal. Concomitant with the rise in cyclic AMP the cells reattach firmly to the substratum, extend their processes, and become flattened.

Based on studies on cyclic AMP metabolism and viral transformation, the following working model has emerged. We suggest that some transforming viruses bring into the cell a single "transforming gene"; different viruses may have different "transforming genes." The products of these "transforming genes" each modify adenylate cyclase activity, either directly or indirectly, but in different ways. For example, transformation by RSV-BH alters the K_m [ATP] and the Mg^{++} response of the membrane-bound adenylate cyclase, whereas RSV-SR transformation results in a decreased V_{max}. Each of these modifications leads to decreased cyclic AMP levels. As a consequence of low cyclic AMP levels growth rate increases, the cell shape changes, and adhesiveness is decreased.

We propose the schematic model depicted in Fig. 4 to represent the mechanism for RSV-BH transformation of chick embryo fibroblasts. In this model the product of the "transforming gene" (temperature-sensitive in the Ta mutant) alters some component of the plasma membrane. The catalytic portion of adenylate cyclase is intimately associated with the plasma membrane and can "sense" and respond to alterations in its environment. The alteration in membrane environment induced by RSV-BH results in decreased enzyme activity. Since inhibitors of RNA and protein synthesis do not affect the rapid decrease in adenylate cyclase activity and cyclic AMP levels or the morphological changes of transformation resulting from the temperature shift of chick embryo fibroblasts infected with RSV-BH-Ta, this product appears to be relatively stable. The "transforming gene" product may modify other cellular processes as a consequence of membrane changes. These processes would then be independent of cyclic AMP levels. One such process

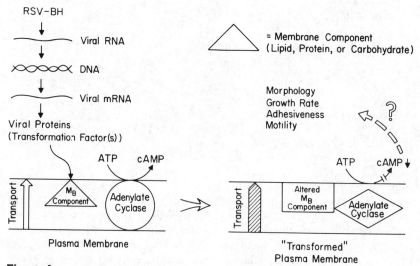

Figure 4
Schematic model for RSV-BH transformation of chick embryo fibroblasts.

altered in transformation, which does not appear to be under cyclic AMP control, is enhanced hexose uptake (Gazdar et al. 1972).

Other workers have found that glycolipids and glycoproteins are altered after transformation (Buck, Glick and Warren 1970; Hakomori 1970; Hakomori, Saito and Vogt 1971; Mora et al. 1969; Wu et al. 1969). Therefore it seems possible that alterations in the structure of a glycolipid or glycoprotein may be responsible for decreased adenylate cyclase activity. This idea is consistent with the finding that the modified ganglioside pattern of transformed cells is not affected by treatment with dibutyryl cAMP (Sakiyama and Robbins 1973). Elucidation of the nature of the intracellular factors that normally control adenylate cyclase activity will be required to clarify the mechanisms by which these viral products change adenylate cyclase activity.

Acknowledgment

The studies on Kirsten sarcoma virus-transformed NRK cells were performed in collaboration with Dr. E. Scolnick of the National Cancer Institute.

REFERENCES

Anderson, W. B., G. S. Johnson and I. Pastan. 1973. Transformation of chick embryo fibroblasts by wild-type and temperature-sensitive Rous sarcoma virus alters adenylate cyclase activity. *Proc. Nat. Acad. Sci.* **70:**1055.

Anderson, W. B., E. Lovelace and I. Pastan. 1973. Adenylate cyclase activity is decreased in chick embryo fibroblasts transformed by wild type and temperature sensitive Schmidt-Ruppin Rous sarcoma virus. *Biochem. Biophys. Res. Commun.* **52:**1293.

Bader, J. P. 1972. Temperature-dependent transformation of cells infected with a mutant of Bryan Rous sarcoma virus. *J. Virol.* **10:**267.

Buck, C. A., M. C. Glick and L. Warren. 1970. A comparative study of glycoproteins from the surface of control and Rous sarcoma virus transformed hamster cells. *Biochemistry* **9:**4567.

Gazdar, A., M. Hatanaka, R. Herberman, E. Russell and Y. Ikawa. 1972. Effects of dibutyryl cyclic adenosine phosphate plus theophylline on murine sarcoma virus transformed non-producer cells. *Proc. Soc. Exp. Biol. Med.* **141:**1044.

Hakomori, S. 1970. Cell density-dependent changes of glycolipid concentrations in fibroblasts, and loss of this response in virus-transformed cells. *Proc. Nat. Acad. Sci.* **67:** 1741.

Hakomori, S., T. Saito and P. K. Vogt. 1971. Transformation by Rous sarcoma virus: Effects on cellular glycolipids. *Virology* **44:**609.

Johnson, G. S. and I. Pastan. 1971. Change in growth and morphology of fibroblasts by prostaglandins. *J. Nat. Cancer Inst.* **47:**1357.

———. 1972a. Cyclic AMP increases the adhesion of fibroblasts to substratum. *Nature New Biol.* **236:**247.

———. 1972b. Role of 3′,5′-adenosine monophosphate in regulation of morphology and growth of transformed and normal fibroblasts. *J. Nat. Cancer Inst.* **48:**1377.

Johnson, G. S., W. D. Morgan and I. Pastan. 1972. Regulation of cell motility by cyclic AMP. *Nature* **235:**54.

Martin, G. S., S. Venuta, M. Weber and H. Rubin. 1971. Temperature-dependent alterations in sugar transport in cells infected by a temperature-sensitive mutant of Rous sarcoma virus. *Proc. Nat. Acad. Sci.* **68:**2739.

Mora, P. T., R. O. Brady, R. M. Bradley and V. W. McFarland. 1969. Gangliosides in DNA virus-transformed and spontaneously transformed tumorigenic mouse cell lines. *Proc. Nat. Acad. Sci.* **63**:1290.

Otten, J., G. S. Johnson and I. Pastan. 1971. Cyclic AMP levels in fibroblasts: Relationship to growth rate and contact inhibition of growth. *Biochem. Biophys. Res. Commun.* **44**:1192.

————. 1972. Regulation of cell growth by cyclic adenosine 3′,5′-monophosphate. Effect of cell density and agents which alter cell growth on cyclic adenosine 3′,5′-monophosphate levels in fibroblasts. *J. Biol. Chem.* **247**:7082.

Otten, J., J. P. Bader, G. S. Johnson and I. Pastan. 1972. A mutation in a Rous sarcoma virus gene that controls adenosine 3′,5′-monophosphate levels and transformation. *J. Biol. Chem.* **247**:1632.

Sakiyama, H. and P. W. Robbins. 1973. The effect of dibutyryl adenosine 3′,5′-cyclic monophosphate on the synthesis of glycolipids by normal and transformed NIL cells. *Arch. Biochem. Biophys.* **154**:407.

Seifert, W. and D. Paul. 1972. Levels of cyclic AMP in sparse and dense cultures of growing and quiescent 3T3 cells. *Nature New Biol.* **240**:281.

Willingham, M., G. S. Johnson and I. Pastan. 1972. Control of DNA synthesis and mitosis in 3T3 cells by cyclic AMP. *Biochem. Biophys. Res. Commun.* **48**:743.

Wu, H. C., E. Meezan, P. H. Black and P. W. Robbins. 1969. Comparative studies on the carbohydrate-containing membrane components of normal and virus-transformed mouse fibroblasts. I. Glucosamine-labelling patterns in 3T3, spontaneously transformed 3T3, and SV40 transformed 3T3 cells. *Biochemistry* **8**:2509.

Cyclic AMP
and Cell Proliferation

J. R. Sheppard and S. Bannai

Dight Institute for Human Genetics
University of Minnesota, Minneapolis, Minnesota 55455

Investigations concerning the pharmacology and biochemistry of cyclic adenosine 3'5'-monophosphate (cyclic AMP) have spanned many years (Robison, Butcher and Sutherland 1971); however only recently has its broad biological significance been recognized. A report of its possible inhibitory effect on cell proliferation was first published in 1968 (Bürk 1968); simultaneous independent studies (Ryan and Heidrich 1968; Heidrich and Ryan 1970) directly measured the inhibitory effect of cyclic AMP on the growth of cultured HeLa and L cells.

While recent evidence indicates a relationship does exist between cyclic AMP and cell proliferation (Hsie and Puck 1971; Johnson et al. 1971; Sheppard 1971, 1972a; Powell et al. 1971; Froelich and Rachmeler 1972; Frank 1972; Yoshikawa-Fukada and Nojima 1972; Rozengurt and Pardee 1972; Reddi and Constantinides 1972), the precise character of that relationship is far from clear. Recent investigations, however, suggest that the role of cyclic AMP may be far more basic than had previously been supposed.

Cyclic AMP and the Pleiotypic Response

In order to account for the apparently coordinated response of a number of essential biochemical functions to changes in the cell's environment, Tomkins (Hershko et al. 1971; Kram, Mamont and Tomkins 1973) has proposed the existence of a regulatory system—the pleiotypic program—which adjusts the growth rate to variable environmental conditions. The concerted alteration of such processes as transport and the synthesis and degradation of protein in response to, for example, the availability of nutrients, prompts the hypothesis that some common mediator may govern a wide range of cellular functions.

The similarity of this phenomenon to the stringent response in bacteria, shown by Cashell (1969) to be mediated by guanosine 5'-diphosphate, 2'- or 3'-diphosphate (ppGpp) in *E. coli,* led Mamont et al. (1972) to investigate whether that nucleotide might be the eukaryotic pleiotypic mediator as well. In the absence of any evidence of significant amounts of ppGpp in animal cells, speculation has

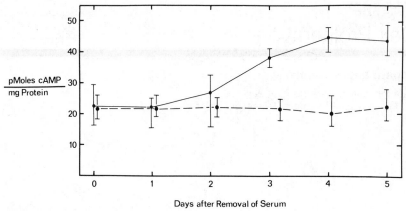

Figure 1

Effect of serum-free medium on cyclic AMP levels in neuroblastoma. Neuro-blastoma cells were grown under the usual conditions and at the beginning of the experiment the normal medium was replaced with serum-free medium. The cAMP levels were measured on consecutive days following exposure to serum-free medium. From Sheppard and Prasad (1973) with permission of *Life Science*.

focused on cyclic AMP as a leading candidate for the inhibitory pleiotypic mediator. (Another nucleotide, cyclic GMP, is discussed by Goldberg et al. in this volume as functioning in the capacity of a positive metabolic regulator.)

An accretion of cellular cyclic AMP occurs under conditions of serum deprivation in neuroblastoma cells (Fig. 1 from Sheppard and Prasad 1973). We have also observed that serum or glucose deprivation increases cyclic AMP levels in baby hamster kidney cells (Table 1). As more evidence of a direct correlation between generalized growth inhibition due to nutrient depletion and elevated cyclic AMP levels accumulates, it does not illuminate the nature of the relationship which obtains between the two phenomena, i.e., whether increased concentrations of cyclic AMP are the condition or the consequence of restrained growth. Thus we

Table 1

Effect of Deficient Medium on
Baby Hamster Kidney Cells

	cAMP *(pmoles/mg protein)*
Normal medium	23 ± 3
Serum-free medium	54 ± 5
Glucose-free medium	67 ± 5

Baby hamster kidney (BHK) cells were plated at 10^5 cells per 60-mm dish and allowed to settle and resume growth in complete medium for 24 hr. At this time experimental plates were exposed to deficient medium for the next 48 hr. Cyclic AMP levels and protein were then determined as described in Frank (1972).

measured cyclic AMP levels in mouse neuroblastoma cells treated with several agents known to inhibit growth by disrupting macromolecular synthesis (Sheppard and Prasad 1973) (Table 2). None of these inhibitors (vinblastine sulfate for microtubule assembly, cytochalasin B for microfilament assembly, fluorodeoxyuridine for DNA synthesis and actinomycin D for RNA synthesis) had a significant effect on cyclic AMP levels. We conclude that these nonspecific inhibitors of growth are not associated with increased cyclic AMP levels. Moreover the administration of cyclohexamide, which inhibits protein synthesis, was followed by a decrease in cellular cyclic AMP levels, suggesting that protein synthesis is a requirement for the maintenance of a normal, basal cyclic AMP level (Sheppard and Prasad 1973; Sheppard and Prescott 1972). (Interestingly chloramphenicol, a bacterial protein synthesis inhibitor, interferes with the stringent response under the control of the mediator ppGpp.)

It may be, then, that increased cyclic AMP concentrations are a condition of inhibited growth due to specific environmental conditions. The elevation of cellular cyclic AMP levels in consequence of serum and glucose starvation may, in fact, key a generalized stress response in the cell. In *E. coli,* for example, elevated cyclic AMP levels induced by glucose deprivation reverse catabolite repression and stimulate the synthesis of enzymes involved in metabolizing secondary energy sources (e.g., lactose) (Perlman and Pastan 1968). Replenishment of the glucose supply suppresses cyclic AMP levels and leads to repressed synthesis of secondary catabolic enzymes. Also the exogenous addition of dibutyryl cyclic AMP impairs cellular transport systems in fibroblasts (Grimes and Schroeder; Hauschka, Everhart and Rubin 1972), while growth stimulation of confluent normal cells by serum and trypsin both depresses cyclic AMP levels (Sheppard 1972a) and facilitates glucose transport (Sefton and Rubin 1971). These studies suggest that cyclic AMP, functioning as a generalized regulator of growth activity, may respond to the limited availability of an essential nutrient (e.g., glucose) by inhibiting the transport of other key nutrients, thus preventing unbalanced growth.

Table 2

Effect of Metabolic Inhibitors on Axon Formation
and Cyclic AMP Levels in Neuroblastoma

Condition	% Differentiated	pmoles cAMP/mg protein
Control	11 ± 3.0*	23.4 ± 3.7
Vinblastine sulfate (0.01 μg/ml)	0	25.1 ± 4.3
Cytochalasin B (0.1 μg/ml)	5 ± 1.0	23.2 ± 3.8
Cycloheximide (5 μg/ml)	7 ± 1.0	15.5 ± 5.2
Actinomycin D (5 μg/ml)	5 ± 1.0	25.0 ± 2.9
Fluorodeoxyuridine (2 μmoles)	5 ± 1.0	22.1 ± 3.5

The mouse neuroblastoma cells were exposed to the metabolic inhibitor for 24 hr after which the number of differentiated cells were counted and the cellular cAMP levels measured.

* Standard deviation

In transformed cells unrestrained growth may result from a membrane change which permits accelerated transport of nutrients, thus keeping cyclic AMP levels low and the rate of growth high. The considerable evidence that transport (e.g., of glucose, amino acids and phosphate) is heightened in transformed cells, while cyclic AMP is depressed, lends weight to this conjecture.

Cyclic AMP and Differentiation

In addition to the implication that cyclic AMP is that phase of the pleiotypic program associated with the inhibition of growth, there is also evidence attributing to cyclic AMP some responsibility for the control of differentiation. In experiments with mouse neuroblastoma cells (selected for use because of the detectability of a "differentiated" trait—in this case, axonal elongation), we measured cyclic AMP levels in the cells after treatment with agents that Prasad has shown to induce "morphological differentiation." Exposure of the neuroblastoma cells to serum-free media, prostaglandin E1 and phosphodiesterase inhibitors (see Prasad and Kumar, this volume) resulted in increased cyclic AMP levels, axon development and eventual growth inhibition. Nevertheless, because conditions exist (e.g., in the presence of vinblastine sulfate) under which the differentiating stimuli—while increasing cyclic AMP levels—do not induce axon outgrowth, it seems that cyclic AMP may be a condition of differentiation, but incapable, in the absence of other propitious conditions, of activating it.

High cyclic AMP concentrations are also associated with differentiation in less complex systems. Such responses involve cell specialization (e.g., bacterial flagella formation) (Yokota and Gors 1970) or stimulation of biochemical processes involved in progress through the cell cycle (e.g., formation of the fruiting body in slime molds) (Bonner 1970).

Indeed cyclic AMP seems to be involved in certain of the biochemical events that take the cell through its mitotic cycle. Willingham, Johnson and Pastan (1973) have observed the exertion of inhibitory growth effects at two points in the cell cycle of 3T3 fibroblasts. Our results (Sheppard and Prescott 1972; Burger et al. 1972) have revealed a reduction in the cellular cyclic AMP level during the mitotic phase (Table 3). The reasons for the depletion of cyclic AMP levels during mitosis are unknown. It is possible, however, that it functions to induce some differential activity (e.g., microtubule assembly) required for division.

Cyclic AMP Metabolism in Fibroblasts

In attempting to isolate the mechanisms responsible for regulating cyclic AMP levels, attention has centered on two enzymes: cyclic nucleotide phosphodiesterase, which degrades cyclic AMP, and adenylate cyclase, which catalyzes its synthesis from ATP.

The increased levels of cyclic nucleotide phosphodiesterase-specific activity recorded for transformed cells have been regarded as a possible explanation of the depressed cyclic AMP levels also found in such cells (Sheppard 1973). It is presently unclear, however, why transformed cells exhibit a high activity of the enzyme. No apparent difference has been observed in substrate binding constants, variations in pH optima or inhibition kinetics between normal and transformed cells with respect to kinetic parameters recorded for cyclic nucleotide phosphodiesterase

Table 3

Variation of Cyclic AMP
Levels Over the Cell Cycle

	pmoles cAMP/mg protein	
	3T3	3T6
G_1	18	10
S	23	12
G_2	22	10
M	13	6

The cells were synchronized by inhibition of DNA synthesis (e.g., 1
mM thymidine or 0.5 mM hydroxyurea). 3T3 cells could also be synchronized by simply replating at a
lower density. Cyclic AMP levels were
measured by the procedures described
in Sheppard (1972).

(Table 4). It is possible that the increased activity is attributable to some activator
substance, such as has been proposed by Cheung (1970).

Adenylate cyclase is also found in a higher basal specific activity in transformed
cell lines (Table 5). This data, derived from a broken cell assay, is paradoxical in
view of the fact that transformed cells have lower levels of cyclic AMP. The
observed decreased sensitivity of the transformed cell's adenylate cyclase to hormones reported by Peery (Peery, Johnson and Pastan 1971) has suggested an
alteration in the cell's hormone receptor site or surface structure of the transformed
cell.

The Plasma Membrane and Cell Division

Considerable indirect evidence can be adduced in support of the hypothesis that the
uncontrolled growth of transformed cells is the consequence of an alteration in the
membrane that facilitates nutrient transport. Pardee (1964) and Wallach (1968)
have proposed that the development of neoplasms may be traceable to aberrations

Table 4

Kinetic Studies of cAMP Phosphodiesterase in
Normal and Transformed Cells

	V_{max} (10^{-11} moles cAMP degraded) (mg prot/min)	K_m (μmoles)
3T3	3.0 ± 0.5	1.0
Py3T3	6.8 ± 0.6	1.2
3T6	5.5 ± 1.4	1.0

Phosphodiesterase was measured by the method of
O'Dea, Haddox and Goldberg (1971) on cells grown
as described in Sheppard (1972).

Table 5

Adenylate Cyclase Specific Activity from Homogenates of Normal and Transformed Cells

	pmoles cAMP formed (min/mg prot)		
	Basal	F^- (10mM)	PGE_1 (10 μg/ml)
3T3	6 ± 2	20 ± 4	10 ± 3
Py3T3	12 ± 4	38 ± 7	29 ± 6
3T6	32 ± 7	94 ± 11	49 ± 7

Adenylate cyclase was measured by the method of Krishna, Weiss and Brodie (1968) on cells grown as described in Sheppard (1972).

in the cell membrane transport mechanism, a suggestion which is itself rendered indirectly more plausible by reported changes in the surface architecture, agglutinability, chemical composition, transmembrane potential and antigenicity of the transformed cell membrane (Burger 1971).

We are currently investigating the possibility that the activity of membrane-bound enzymes is coordinated in some manner. Stimulation or inhibition of one membrane-bound enzyme might precipitate an effect upon adjacent enzymes. Cyclic AMP, for example, inhibits membrane ATPase (Chandhary and Frenkel 1970; Mozsik 1970; Luly, Barnabei and Tria 1972), whereas insulin, which stimulates Na^+/K^+ ATPase (Hadden et al. 1972), inhibits adenylate cyclase (Illiano and Cuatrecasas 1972) and depresses cellular cyclic AMP levels (Sheppard 1972b).

Transport of cellular nutrients is thought to be at least partially associated with membrane ATPase activity and may also be related to cellular cyclic AMP levels. For example, glucose transport, itself stimulated by insulin, seems also to be related to cyclic AMP. It is accelerated in transformed cells (Hatanaka, Huebner and Gilden 1969; Isselbacher 1971) and confluent normal cells stimulated to further growth by serum and trypsin (Sefton and Rubin 1971), whereas cyclic AMP levels are reduced in both cases (Sheppard 1972a). In *E. coli* glucose has been identified as suppressing cyclic AMP formation (Perlman and Pastan 1968). It is thought that cyclic AMP may interact with a protein in the membrane responsible for phosphorylating glucose and facilitating its transport. Mutant strains of the bacteria lacking this membrane protein have a cyclic AMP level that is not sensitive to glucose.

Only peripheral evidence for the effect of glucose on cyclic AMP levels in animal cells is available (e.g., the increased activity of glucose transport in transformed cells, and increased glucose transport after administration of serum and trypsin). Nevertheless these facts, when considered in the light of the independent evidence that cyclic AMP is stress-responsive in cultured cells, seem to be further supportive of the increasingly compelling hypothesis that cyclic AMP is the pleiotypic mediator. The precise mechanism according to which glucose exerts its putative effect on cyclic AMP is unknown. It is possible that, as Robison et al. (1971) have suggested, it may function as a primitive hormone, similar in action to insulin, certain of the prostaglandins and the α-adrenergic catecholamines.

Postulating the existence of a pleiotypic mediator and hypothesizing that cyclic AMP might serve that function assigns it a much more basic regulatory position in the biochemistry of the cell than that accorded it initially. An initial working hypothesis held that cell contact or interaction led to an elevated cyclic AMP level, which inhibited further cell division (Johnson et al. 1971). This contact inhibition–membrane interaction theory was devised to explain the inhibitory effects of dibutyryl cyclic AMP on the growth of transformed cells and to account for the differential basal concentrations of cyclic AMP in normal and transformed cells.

The membrane interaction hypothesis lends itself to the corollary proposition that the lower cyclic AMP levels and unrestrained growth, which are characteristics of transformed cells, are the consequence of a breakdown in some contact-induced, membrane-mediated process through which cyclic AMP concentrations increase in normal cells. Thus if contact with adjacent cells inhibited proliferation by inducing growth-restraining cyclic AMP increases (possibly through membrane stimulation of adenylate cyclase activity), then cyclic AMP levels should rise as the normal contact-inhibited cells reach confluence. Cyclic AMP levels should remain constant in confluent transformed cells because of the postulated membrane alteration which makes proliferation of these cells unresponsive to cell contact. Our study, however, has revealed no noticeable change in the cyclic AMP levels of either normal or transformed cells after reaching confluence (Sheppard 1972a). Accordingly, while an elevated level of cyclic AMP may be—and quite likely is—a condition or prerequisite of contact inhibition or density-dependent inhibition of growth, it seems apparent it is *not* the cause. A much more likely situation is that cyclic AMP, as the negative pleiotypic mediator, responds to cell contact (not confluency), through a process mediated by the membrane, to regulate growth and differentiation of the cell.

REFERENCES

Bonner, J. T. 1970. Induction of stalk cell differentiation by cyclic AMP in the cellular slime mold *Dictyostelium discoideum. Proc. Nat. Acad. Sci.* **65**:110.

Burger, M. M. 1971. *Current topics in cellular regulation* (ed. B. L. Horecker) vol. 3, p. 135–193. Academic Press, N.Y.

Burger, M. M., B. M. Bombik, B. M. Breckenridge and J. R. Sheppard. 1972. Control growth and cyclic alterations of cyclic AMP in the cell cycle. *Nature New Biol.* **239**:161.

Bürk, R. R. 1968. Reduced adenylate cyclase activity in a polyoma virus transformed cell line. *Nature* **219**:1272.

Cashel, M. 1969. Control of ribonucleic acid synthesis in *Escherichia coli. J. Biol. Chem.* **244**:3133.

Chandhary, A. H. and A. W. Frenkel. 1970. Effect of 3′,5′-cyclic monophosphoric acid on certain light-induced reactions and on ATPase activity of isolated chromatophores from *Rhodospirillum rubrum. Biochem. Biophys. Res. Commun.* **39**:238.

Cheung, W. U. 1970. Cyclic 3′,5′-nucleotide phosphodiesterase. *Biochem. Biophys. Res. Commun.* **38**:533.

Frank, W. 1972. Cyclic 3′:5′ AMP and cell proliferation in cultures of embryonic rat cells. *Exp. Cell Res.* **71**:238.

Froelich, J. E. and M. Rachmeler. 1972. Effect of adenosine 3′,5′-cyclic monophosphate on cell proliferation. *J. Cell Biol.* **55**:19.

Grimes, W. J. and J. L. Schroeder. 1973. Dibutyryl cyclic adenosine 3'5' monophosphate, sugar transport, and regulatory control of cell division in normal and transformed cells. *J. Cell Biol.* **56:**487.

Hadden, J. W., E. M. Hadden, E. E. Wilson, R. A. Good and R. G. Coffey. 1972. Direct action of insulin on plasma membrane ATPase activity in human lymphocytes. *Nature New Biol.* **235:**174.

Hatananka, M., R. J. Huebner and R. V. Gilden. 1969. Alterations in the characteristics of sugar uptake in mouse cells transformed by murine sarcoma viruses. *J. Nat. Cancer Inst.* **43:**1091.

Hauschka, P. V., L. P. Everhart and R. W. Rubin. 1972. Alteration of nucleoside transport of Chinese hamster cells by dibutyryl adenosine 3':5'-cyclic monophosphate. *Proc. Nat. Acad. Sci.* **69:**3542.

Heidrich, M. L. and W. L. Ryan. 1970. Nucleotides on cell growth *in vitro. Cancer Res.* **30:**376.

Hershko, A., P. Mamont, R. Schields and G. M. Tomkins. 1971. Pleiotypic response. *Nature New Biol.* **232:**206.

Hsie, A. W. and T. T. Puck. 1971. Morphological transformation of Chinese hamster cells by dibutyryl adenosine cyclic 3':5'-monophosphate and testosterone. *Proc. Nat. Acad. Sci.* **68:**358.

Illiano, G. and P. Cuatrecasas. 1972. Modulation of adenylate cyclase activity in liver fat cell membranes by insulin. *Science* **175:**906.

Isselbacher, K. J. 1971. Increased uptake of amino acids and 2-deoxy-D-glucose by virus-transformed cells in culture. *Proc. Nat. Acad. Sci.* **69:**585.

Johnson, G. S., R. M. Friedman and I. Pastan. 1971. Restoration of several morphological characteristics of normal fibroblasts in sarcoma cells treated with adenosine-3':5'-cyclic monophosphate and its derivatives. *Proc. Nat. Acad. Sci.* **68:**425.

Kram, T., R. Mamont and G. M. Tomkins. 1973. Pleiotypic control by adenosine 3':5'-cyclic monophosphate: A model for growth control in animal cells. *Proc. Nat. Acad. Sci.* **70:**1432.

Krishna, G., B. Weiss and B. B. Brodie. 1968. A simple sensitive method for the assay of adenyl cyclase. *J. Pharmacol. Exp. Therap.* **163:**379.

Luly, P., D. Barnabei and E. Tria. 1972. Hormonal control *in vitro* of plasma membrane-bound (Na$^+$-K$^+$)-ATPase in rat liver. *Biochem. Biophys. Acta* **282:**447.

Mamont, P., A. Hershko, R. Kram, L. Schacter, J. Lust and E. M. Tomkins. 1972. Pleiotypic response in mammalian cyclic AMP levels and the morphological differentiation of mouse neuroblastoma cells. *Biochem. Biophys. Res. Commun.* **48:** 1378.

Mozsik, G. 1970. Direct inhibitory effect of adenosine monophosphates on Na$^+$-K$^+$-dependent ATPase prepared from human gastric mucosa. *Eur. J. Pharmacol.* **9:**207.

O'Dea, R. F., M. K. Haddox and N. D. Goldberg. 1971. Interaction with phosphodiesterase of free and kinase-complexed cyclic adenosine 3':5'-monophosphate. *J. Biol. Chem.* **246:**6183.

Pardee, A. B. 1964. Cell division and a hypothesis of cancer. *Nat. Cancer Inst. Monogr.* **14:**7.

Peery, C. V., G. S. Johnson and J. Pastan. 1971. Adenyl cyclase in normal and transformed fibroblasts in tissue culture. *J. Biol. Chem.* **246:**5785.

Perlman, R. L. and I. Pastan. 1968. Regulation of β-galactosidase synthesis in *Escherichia coli* by cyclic adenosine 3',5'-monophosphate. *J. Biol. Chem.* **243:**5420.

Powell, J. A., E. A. Duell and J. J. Voorhees. 1971. Beta-adrenergic stimulation of endogenous epidermal cyclic AMP formation. *Arch. Dermatol.* **104:**359.

Reddi, P. K. and S. M. Constantinides. 1972. Partial suppression of dibutyryl cyclic AMP and theophylline. *Nature* **238:**286.

Robison, G. A., R. W. Butcher and E. W. Sutherland, ed. 1971. *Cyclic AMP*. Academic Press, N.Y.

Rozengurt, E. and A. B. Pardee. 1972. Opposite effects of dibutyryl adenosine 3′:5′-cyclic monophosphate and serum on growth of Chinese hamster cells. *J. Cell. Physiol.* **80:**273.

Ryan, W. L. and M. L. Heidrich. 1968. Inhibition of cell growth *in vitro* by adenosine 3′,5′-monophosphate. *Science* **162:**1484.

Sefton, B. M. and H. Rubin. 1971. Stimulation of glucose transport in cultures of density-inhibited chick embryo cells. *Proc. Nat. Acad. Sci.* **68:**3154.

Sheppard, J. R. 1971. Restoration of contact-inhibited growth to transformed cells by dibutyryl adenosine 3′:5′-cyclic monophosphate. *Proc. Nat. Acad. Sci.* **68:**1316.

————. 1972a. Difference in the cyclic adenosine 3′,5′-monophosphate levels in normal and transformed cells. *Nature New Biol.* **236:**14.

————. 1972b. *Membranes and viruses in immunopathology* (ed. S. Day and R. A. Good) Academic Press, N.Y.

————. 1973. Cyclic AMP metabolism in normal and transformed fibroblasts. *Biochem. Biophys. Res. Commun.* (in press).

Sheppard, J. R. and K. N. Prasad. 1973. Cyclic AMP levels and the morphological differentiation of mouse neuroblastoma cells. *Life Science II,* **12:**631.

Sheppard, J. R. and D. M. Prescott. 1972. Cyclic AMP levels in synchronized mammalian cells. *Exp. Cell Res.* **75:**293.

Wallach, D. F. H. 1968. Cellular membranes and tumor behavior. A new hypothesis. *Proc. Nat. Acad. Sci.* **61:**868.

Willingham, M. E., G. S. Johnson and I. Pastan. 1973. Control of DNA synthesis and mitosis in 3T3 cells by cyclic AMP. *Biochem. Biophys. Res. Commun.* **48:**743.

Yokota, T. and J. S. Gors. 1970. Requirement of adenosine 3′,5′-cyclic phosphate for flagella formation in *Escherichia coli* and *Salmonella typhimurium. J. Bacteriol.* **103:**513.

Yoshikawa-Fukada, M. and T. Nojima. 1972. Biochemical characteristics of normal and virally transformed mouse cell lines. *J. Cell. Physiol.* **80:**421.

Cyclic AMP and the Differentiation of Neuroblastoma Cells in Culture

K. N. Prasad and S. Kumar

Department of Radiology, University of Colorado Medical Center
Denver, Colorado 80220

The cellular factor(s) which trigger neural differentiation and the mechanisms which regulate the expression of differentiated functions remain elusive. During the past few years some new insight into those factors has been obtained using mouse neuroblastoma cell culture as an experimental model. Although the neuroblastoma cell culture may not be a perfect model to study the regulation of neural differentiation, it has proven extremely useful in elucidating the regulation of differentiated functions. Serum-free medium (Seeds et al. 1970) and 5-bromo-deoxyuridine (Schubert and Jacob 1970) cause neurite formation in mouse neuroblastoma cell culture. We have shown (Prasad and Hsie 1971; Prasad 1972a; Prasad and Sheppard 1972a; Prasad and Vernadakis 1972; Waymire, Weiner and Prasad 1972; Prasad and Mandal 1973) that adenosine 3',5'-cyclic monophosphate (cAMP) induces many differentiated functions in mouse neuroblastoma cell culture. Preliminary data indicate that cAMP may play a similar role in human neuroblastoma cell culture.

Features of Neuroblastoma Cells

The procedures for culturing and maintaining mouse neuroblastoma cells were previously described (Prasad 1971). These cells have a relatively high rate of glycolysis (Sakamoto and Prasad 1972). They contain tyrosine hydroxylase (TH) (Schubert et al. 1969), choline acetyltransferase (ChA), acetylcholinesterase (AChE) (Augusti-Tocco and Sato 1969), and catechol-o-methyltransferase (COMT) (Blume et al. 1970; Prasad and Mandal 1972), but lack tryptophan hydroxylase (Prasad et al. 1973b; Amano, Richelson and Nirenberg 1972). Mouse neuroblastoma tumors have a 5th band of muscle lactate dehydrogenase, which is absent from the brain (Prasad, Prasad and Prasad 1973). Four types of clone (Prasad et al. 1973b) have been isolated from mouse neuroblastoma tumors. These include (1) clones with TH but no ChA, (2) clones with ChA but no TH, (3) clones with neither TH nor ChA, and (4) clones with both TH and ChA. The first three types of clone have also been isolated by other investigators (Amano et al. 1972). Cells from all of these clones have a doubling time of about 18–24 hours, show spontaneous morphological differentiation varying from 1 to 15 percent, and produce malignant tumors when injected subcutaneously into male A/J mice.

A clone of human neuroblastoma cells (IMR-32) obtained from the American Tissue Culture Association initially showed no tyrosine hydroxylase activity (Prasad, Mandal and Kumar 1973a). But recently an extremely low level of TH (5 $\pm$ 2 pmoles/mg protein) has become detectable. Also a relatively high amount of choline acetyltransferase is demonstrable in this culture (Prasad, Mandal and Kumar 1973b).

Drug Treatment

Dibutyryl- and monobutyryl-derivatives of cAMP and many 8-substituted analogs of cAMP were used. Prostaglandin (PG)E_1 increases the intracellular level of cAMP by stimulating adenylate cyclase. Theophylline, 4-(3-butoxy-4-methoxybenzyl)-2-imidazolidinone (R020-1724), 4-(3-dimethoxybenzyl)-2-imidazolidinone (R07-2956) (Sheppard and Wiggan 1971), and papaverine (Triner et al. 1970) increase the cAMP level by inhibiting the cAMP phosphodiesterase (PDE) activity. Therefore in addition to analogs of cAMP and PGE$_1$, the above PDE inhibitors were used to investigate further the role of cAMP in growth and differentiation. The procedures for making solutions and treating the culture with an individual drug were previously described (Prasad and Hsie 1971; Prasad 1972a; Prasad et al. 1972).

Morphological Differentiation and Cyclic AMP

Analogs of cAMP Prasad et al. 1971; Prasad 1972b; Furmanski, Silverman and Lubin 1971), PGE$_1$ (Prasad 1972a), and PDE inhibitors (Prasad et al. 1972) cause morphological differentiation of neuroblastoma cells as shown by the formation of long neurites and by an increase in size of the nucleus and soma (Fig. 1b, c). Some cells that did not form long neurites did show an increase in size of soma and nucleus. Table 1 shows the relative potency of various agents in causing morphological differentiation. PGE$_1$, PDE inhibitors and 8-benzylthio cAMP are more affective than $N^6,O^{2'}$-dibutyryl cyclic adenosine 3',5'-monophosphate (dibutyryl cAMP). For a period of 4 days, no significant cell death occurred after the above treatments. The viability of attached cells as determined by the uptake of supravital stain (trypan blue in 1% saline) was similar to that of control cells (90–95%).

The number of morphologically differentiated cells after PGE$_1$ treatment is time and concentration dependent (Prasad 1972a). A significant increase in the number of differentiated cells was noted 24 hours after treatment and cell division continued up to the third day. This indicates that the inhibition of cell division temporally follows the induction of morphological differentiation and thus may be secondary to differentiation. The kinetics of morphological differentiation and growth after treatment with dibutyryl cAMP or PDE inhibitors (Prasad and Sheppard 1972; Prasad and Hsie 1971) were similar to those after PGE$_1$ treatment.

Cyclic AMP, 5'-AMP, theophylline, some 8-substituted analogs of cAMP (8-hydroxyethylthio-, 8-amino-, 8-hydroxyethylamino and 8-methylamino cAMP), adenosine triphosphate, adenosine diphosphate and butyric acid inhibited cell growth without causing morphological differentiation. These data indicate that the

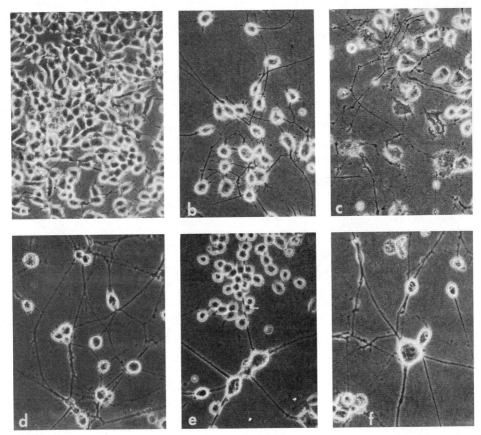

Figure 1
Phase contrast micrographs of mouse neuroblastoma cells of clone $NBA_{2(1)}$ in culture. Cells (50,000) were plated in Falcon plastic dishes (60-mm) and prostaglandin (PG)E_1 (10 μg/ml) and 4-(3-butoxy-4-methoxybenzyl)-2-imidazolidinone (R020-1724, 200 μg/ml) were added separately 24 hr later. The medium and drug were changed at 3, 5, 8 and 11 days after treatment. **(a)** Control culture, cells grow in clumps and some of them have short cytoplasmic processes. **(b)** R020-1724-treated culture 4 days after treatment; cells have long neurites and are larger than the controls. **(c)** PGE_1-treated culture 4 days after treatment also shows morphological differentiation. **(d)** PGE_1-treated culture 14 days after treatment; the remaining cells maintain their differentiated phenotype. **(e)** R020-1724-treated culture 14 days after treatment; a few cells are dividing in the presence of drug. **(f)** R020-1724-treated culture, in which the drug was removed 3 days after treatment and the micrograph was taken 8 days later; many cells maintain their differentiated phenotype. 78.6 $\times$

inhibition of cell division need not necessarily produce the expression of morphological differentiation.

Like cAMP, guanosine 3′,5′-cyclic monophosphate (cGMP) is also present in mammalian cells (Hardman, Robison and Sutherland 1971). Therefore the effects of cGMP on neuroblastoma cell cultures were examined. Neither cGMP, nor $N^2,0$-dibutyryl cGMP caused morphological differentiation, although both agents inhibited cell division.

Table 1

Effect of Various Cyclic AMP Agents on
Mouse Neuroblastoma Cells in Culture

	"Differentiated" cells (% total cells)
Control	9.0 ± 2.0*
Monobutyryl cyclic AMP (0.5 mM)	47 ± 5.0
Dibutyryl cyclic AMP (0.5 mM)	51.0 ± 4.8
8-Benzylthio cyclic AMP (400 μg/ml)	66.0 ± 4.5
Prostaglandin E_1 (10 μg/ml)	72.0 ± 5.0
R020-1724 (200 μg/ml)	71.0 ± 5.4
Papaverine (25 μg/ml)	79.0 ± 2.5

Quantitative estimation of differentiated cells after treatment with
various cAMP agents. Cells (50,000) were plated in Falcon plastic
dishes (60-mm) and treated with drug 24 hr after plating. The
morphologically differentiated cells (cytoplasmic processes were
greater than 50 μm long) were scored 3 days after treatment. At
least 300 cells were counted and the number of differentiated cells
were expressed as percent of total cells. Each value represents an
average of 6–8 samples. The data are summarized from Prasad
(1973).

* Standard deviation

Irreversibility of Growth Inhibition and Morphological Differentiation

The morphological differentiation and inhibition of growth induced by dibutyryl
cAMP, PGE$_1$ or R020-1724 for the most part are irreversible, provided the drug is
present in the medium for at least 3–4 days (Prasad and Hsie 1971; Prasad 1972a;
Prasad, Waymire and Weiner 1972). When the differentiated cells, which are
treated for 3 days with R020-1724 or PGE$_1$, were removed from dishes with
Viokase solution and replated in separate dishes, cells attached and formed long
neurites within 24 hours even though no drug was present during this period. The
number of morphologically differentiated cells in the newly plated dishes was
similar to those in which drug was not removed. This indicates that the cellular
factors that control the expression of the differentiated phenotype remain functional
after subculturing. We have evidence to indicate that an elevated level of cAMP is
one of the important cellular factors controlling the expression of the differentiated
phenotype.

The cultures maintained in the presence of R020-1724 or PGE$_1$ for 15 days
showed many floating, dead cells after 5 days. This may be due to the fact that the
differentiated cells die in culture as a function of time. Most of the attached cells
maintained their differentiated phenotype (Fig. 1d) even 15 days after treatment.
At this time the treated cultures had two or three clones which appeared to be
dividing in the presence of drug (Fig. 1e). The neurite formation for the most
part was irreversible. When PGE$_1$ or R020-1724 was removed 4 days after treatment, many differentiated cells were seen 8 days later (Fig. 1f). Dibutyryl cAMP
also induces irreversible neurite formation in human neuroblasts (Macintyre et al.
1971). The irreversibility of morphological differentiation and growth in mouse
neuroblastoma cells is consistent with the fact that the mature neurons are incapable

of division. The above findings are in contrast to the observations made on non-nerve cells in which cAMP effects are reversible at all times soon after the removal of the drug (Hsie and Puck 1971; Johnson et al. 1971; Sheppard 1971).

Requirements for Expression of the Differentiated Phenotype

Vinblastine sulfate and cytochalasin B (which interfere with the assembly of microtubules and microfilaments, respectively) and cycloheximide (which inhibits protein synthesis) completely block the axon formation induced by cyclic AMP, whereas actinomycin D (an inhibitor of RNA synthesis) does not. Thus the expression of the differentiated phenotype requires at least the assembly of microtubules and microfilaments and the synthesis of new protein (Prasad 1972d), but does not require new RNA synthesis. Since the inhibitors used in this study are known to affect several other cellular parameters in addition to those mentioned here, a different interpretation of these data cannot be excluded at the present time.

Sensitivity of Neuroblastoma Clones to Cyclic AMP

Cells of most clones are sensitive to cAMP in causing morphological differentiation. Also some clones, irrespective of their neuronal cell type, are sensitive to PGE_1 but not to R020-1724 and vice versa (Prasad et al. 1973b; Prasad 1972b). The clone which is insensitive to R020-1724 is also unresponsive to dibutyryl cAMP.

Tumorigenicity of Differentiated Cells

Control cells when injected subcutaneously produce tumors in all A/J mice, whereas the tumorigenicity of differentiated cells (4 days after treatment) is either partially or completely abolished (Prasad 1972c). Uncloned cells are used since they are more nearly duplicate to the in vivo condition. Since some cells are responsive to PGE_1 but not to PDE inhibitor and vice versa, PGE_1 was combined with R020-1724 to maximize the effect on differentiation. Indeed cells treated as above lost completely a prime feature of malignancy, their tumorigenicity (Table 2).

Regulation of Enzymes

Tyrosine hydroxylase (TH), a rate-limiting enzyme in the biosynthesis of catecholamines, is markedly increased (Table 3) by some analogs of cAMP and papaverine (Waymire et al. 1972; Prasad et al. 1972). The morphologically differentiated neuroblastoma cells induced by X-ray (Prasad 1971), serum-free medium (Seeds et al. 1970), and cytosine arabinoside (Kates et al. 1971) show no change in TH levels (Prasad et al. 1972; Kates et al. 1971). Sodium butyrate, which inhibits cell division without causing morphological differentiation, increases the enzyme activity (Prasad et al. 1972; Waymire et al. 1972) and cAMP level, the latter by about twofold (Prasad, Gilmer and Kumar 1973). These data suggest that morphological differentiation and TH activity are independently regulated, and our hypothesis is that cAMP may be involved in the regulation of TH activity; this has been confirmed by recent studies on the TH level of mouse adrenal gland (Guidotti and Costa 1973) and mouse neuroblastoma cells (Richelson 1973).

Table 2

Incidence of Tumors after Subcutaneous Injection of
Control and Differentiated Neuroblastoma Cells

Treatment	Number of animals	Incidence of tumors (% total)
Control cell treated with or without solvent	30	100
Dibutyryl cAMP	15	50
R020-1724	15	40
8-Benzylthio cAMP	15	60
PGE_1	15	25
PGE_1 + dibutyryl cAMP	15	0
PGE_1 + 8-benzylthio cAMP	15	0
PGE_1 + R020-1724	16	0

Cells (10^5) were plated in Falcon plastic dishes (60-mm) and treated with drugs 24 hr later. Dibutyryl cyclic AMP (0.5 mM), 4-(3-butoxy-4-methoxybenzyl)-2-imidazolidinone (R020-1724, 200 µg/ml), 8-benzylthio cyclic AMP (400 µg/ml) or prostaglandin (PG)E_1 (10 µg/ml) were added individually or in combination with PGE_1 (10 µg/ml). After 4 days of incubation the control and differentiated cells (0.25 × 10^6) were injected subcutaneously into male A/J mice (6–8 weeks old). Cell viability in the control and drug-treated cultures was 90–95%.

Table 3

Tyrosine Hydroxylase Activity and Differentiation
of Mouse Neuroblastoma Cells in Culture

Treatment	Activity (pmoles product/ 30 min/10^6 cells)
Control, log phase	15.1 ± 1.9*
Control, confluent phase	11.2 ± 0.7
Serum-free medium	17.3 ± 0.4
Dibutyryl cyclic AMP (0.25 mM)	473 ± 17
8-Methylthio cyclic AMP (0.3 mM)	587 ± 9
Papaverine (0.13 mM)	977 ± 46
Sodium butyrate (0.5 mM)	300 ± 12

Neuroblastoma clone P2 was used in this study. This clone has both tyrosine hydroxylase and choline acetyltransferase. Neuroblastoma (0.5 × 10^6) were treated with X-rays 24 hr after drug treatment. Each value represents an average of at least 4 samples. The data are taken from Waymire, Weiner and Prasad (1972).

* Standard deviation

The ChA that synthesizes acetylcholine is markedly increased in differentiated cells induced by cAMP and is also increased after X-irradiation, 5'-AMP and sodium butyrate treatment (Table 4). The maximal increase in the ChA level coincides with cessation of cell division (Prasad and Mandal 1973). These data indicate that the level of ChA and morphological differentiation are independently regulated, and that cAMP is not necessarily involved in the regulation of ChA. Data on AChE (Prasad and Vernadakis 1972) suggest a similar mode of regulation. On the other hand cAMP is not involved in the regulation of COMT, because the enzyme activity in cyclic AMP-induced differentiated cells does not change (Prasad and Mandal 1972).

Changes in Nucleic Acid and Protein Contents

Since a marked increase in the size of soma and nucleus is seen during cAMP-induced differentiation of neuroblastoma cells, changes in the total content of nucleic acid and protein were investigated 3 days after treatment. Table 5 shows that total DNA of differentiated cells is markedly decreased (Prasad et al. 1973a), but total RNA and protein are increased by about two- to threefold. The pronounced reduction in DNA per cell is interpreted as evidence that most of the differentiated cells accumulate in the G_1 phase of the cycle. The increase in RNA and protein is consistent with observations made during differentiation and maturation of mammalian neurons.

It is generally presumed that blocking of cells in the G_1 phase allows the expression of the differentiated phenotype. This does not appear to be the case in mouse neuroblastoma cells because sodium butyrate-treated cells are blocked in the G_1 phase of the cell cycle, but no expression of morphological differentiation occurs.

Table 4

Effect of Various Agents on Choline Acetyltransferase
Level of Neuroblastoma Cells

Treatment	Activity (pmoles/15 min/10^6 cells)
Control (exponential)	260 ± 35*
Control (confluent)	300 ± 34
Dibutyryl cyclic AMP (0.5 mM)	1300 ± 72
Prostaglandin E_1 (10 μg/ml)	880 ± 100
R020-1724 (200 μg/ml)	1280 ± 160
5'-AMP (0.25 mM)	1320 ± 80
Butyric acid (0.5 mM)	760 ± 100
600 rads	1640 ± 144

Neuroblastoma cells (0.5 × 10^6) were plated in Falcon plastic flasks (75-cm^2) and each drug was added 24 hours later. Fresh growth medium and drug were added 2 days after drug treatment and the choline acetyltransferase was analyzed 3 days later. Each value represents an average of 5–6 samples. The data are taken from Prasad and Mandal (1973).

* Standard deviation

Table 5

Nucleic Acid and Protein Contents in Cyclic AMP-induced
Differentiated Mouse Neuroblastoma Cells in Culture

Treatment	DNA (pg/cell)	RNA (pg/cell)	Protein (pg/cell)
Control	13.3 ± 1.5*	15.3 ± 1.0	500 ± 29
Dibutyryl cAMP	6.6 ± 0.6	33.6 ± 2.5	1580 ± 122
PGE$_1$	6.0 ± 1.6	24.4 ± 1.9	870 ± 47
R020-1724	6.7 ± 1.2	33 ± 1.8	1016 ± 54
Na butyrate	5.3 ± 1.0	31.2 ± 3.9	1479 ± 111

Cells (0.5×10^6) were plated in large Falcon plastic flasks (75-cm^2) and dibutyryl cyclic AMP (0.5 mM), prostaglandin (PG)E$_1$ (10μg/ml), 4-(-3-butoxy-4-methoxy-benzyl)-2-imidazolidione (R020-1724) (200 μg/ml), and Na butyrate (0.5 mM) were added separately 24 hr later. The total nucleic acid and protein contents were assayed 3 days after treatment. Each value represents an average of 4–6 samples. The data are presented from Prasad, Gilmer and Kumar (1973).

* Standard deviation

Membrane Changes during Differentiation

Mouse neuroblastoma cells agglutinate in the presence of the glycoproteins of plant origin concanavalin A (ConA) and wheat germ agglutinin (WGA), whereas morphologically differentiated cells induced by R020-1724 do not (Prasad and Sheppard 1972a). Agglutinability by WGA is lost earlier than that by ConA during morphological differentiation of neuroblastoma cells. Trypsin treatment unmasked only a portion of ConA agglutinin receptor sites but completely failed to unmask WGA receptor sites. The differentiated cells for the most part did not renew cell division after being treated with trypsin. Even though R020-1724-induced differentiated cells do not agglutinate in the presence of ConA and WGA, dibutyryl cAMP- or PGE$_1$-induced differentiated cells do. Thus it was concluded that changes in the agglutinin sites are not necessarily linked with the differentiation or growth of neuroblastoma cells in culture. Burger (1969) has suggested that in fibroblasts the changes in agglutinin sites are linked with the growth rate and malignancy, but our data indicate that this may not be true in neuroblastoma cell culture.

Levels of Adenylate Cyclase, cAMP and PDE

The differentiated cells induced by PGE$_1$ and R020-1724 have a higher level of adenylate cyclase (Table 6), cAMP (Fig. 2) and PDE (Fig. 3). Therefore a working hypothesis has been proposed (Prasad and Kumar 1973) that the levels of adenylate cyclase, cAMP and PDE increase during differentiation of mouse neuroblastoma cells in culture: the reverse may be true during malignant transformation of nerve cells. The fact that dibutyryl cAMP causes neurite formation in the chick embryo dorsal root ganglion and mouse sensory ganglion (Roisen et al. 1972; Hass et al. 1972) indicates cAMP may be involved in the differentiation of normal nerve cell culture.

Norepinephrine and epinephrine do not affect the level of adenylate cyclase in the homogenate of control cells, but cause a marked increase in differentiated cells.

Table 6

Effect of Catecholamines and Prostaglandin E_1 on Adenylate
Cyclase Activity in Homogenates of Neuroblastoma Cells

	Adenylate cyclase activity (pmoles/mg protein)					
Treatment	Basal	DA	NE	EP	NaF	PGE_1
Control	15 ± 1.4*	32 ± 1	18 ± 2.4	19 ± 1.9	26 ± 1.9	36 ± 3.4
R020-1724	21 ± 1	36 ± 4	42 ± 5	46 ± 5.1	38 ± 4.2	46 ± 4.3
PGE_1	22 ± 1.5	46 ± 3.4	51 ± 5	41 ± 2.1	38 ± 2.5	60 ± 8

Cells (0.5×10^6) of clone $NBA_{2(1)}$ were plated in large Falcon plastic flasks (75-cm²) and prostaglandin (PG)E_1 (10 μg/ml) and 4-(3-butoxy-4-methoxybenzyl)-2-imidazolidinone (R020-1724, 200 μg/ml) were added separately 24 hr later. The drug and medium were changed 2 days after treatment and the adenylate cyclase activity was assayed 3 days after treatment. The control cells were treated with an equivalent volume of solvent. For the assay of enzyme the cells were washed twice with PDS buffer, 5 ml Tris-HCl pH 7.5 was added to a flask, and the cells were removed from the flask surface using a Pasteur pipette. The contents of 4–6 flasks were collected in 5 ml Tris buffer and homogenized. The adenylate cyclase activity in the homogenate was determined by a modification (Perkins and Moore 1971) of the procedure described earlier (Krishna, Weiss and Brodie 1968). The dopamine (DA), norepinephrine (NE) and epinephrine (EP) were added separately in the incubating mixture at a concentration of 10^{-4} M. PGE_1 (0.1 μg/ml) and sodium fluoride (NaF, 10.0 mM) were also added separately in the incubating mixture. Each value represents an average of 6–8 samples.

* Standard deviation.

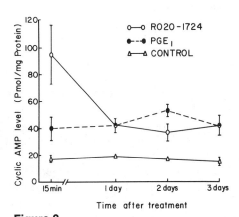

Figure 2

Changes in the cyclic AMP level during morphological differentiation. Cells (0.5×10^6) of clone $NBA_{2(1)}$ were plated in large Falcon plastic flasks (75-cm²) and prostaglandin (PG)E_1 (10 μg/ml) and 4(-3-butoxy-4-methoxy-benzyl)-2-imidazolidinone (R020-1724, 200 μg/ml) were added separately 24 hr later. The control cultures received an equivalent volume of alcohol. The drug and medium were changed 2 days after treatment and the level of cAMP was determined 3 days after treatment. Cells were washed twice with PDS buffer and 2 ml 5% TCA was added to each flask. The content was removed by a rubber policeman and homogenized. The cAMP level was measured according to Gilman's method (1971). Each value represents an average of 6–8 samples. The bar at each point is standard deviation.

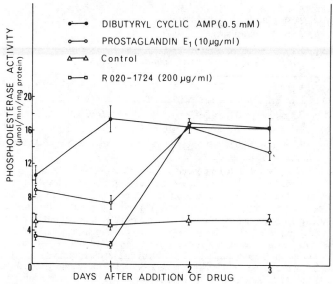

Figure 3

Changes in the level of cyclic AMP phosphodiesterase (PDE) during morphological differentiation. Neuroblastoma cells (0.5×10^6) of clone $NBA_{2(1)}$ were plated in large Falcon plastic flasks (75-cm^2). Dibutyryl cyclic AMP, prostaglandin E_1 and 4-(-3-butoxy-4-methoxybenzyl)-2-imidazolidinone (R020-1724) were added individually for a period of 2 hr, 1, 2, and 3 days. Fresh growth medium and drug solution were added 2 days after treatment with drug. The PDE activity was analyzed according to the method of O'Dea, Haddox and Goldberg (1971). Each value represents an average of 8–10 samples. The vertical bars are standard deviations. From Prasad and Kumar (1973) with permission of *Proc. Soc. Exp. Biol. Med.*

The fact that the brain adenylate cyclase undergoes an age-dependent increase in its sensitivity to norepinephrine (Schmidt and Robison 1970) suggests the following: the unresponsiveness of adenylate cyclase to norepinephrine is an embryonic feature which is reexpressed in neuroblastoma cells; and this defect of malignancy is corrected in cAMP-induced differentiated cells.

Dopamine stimulates the adenylate cyclase activity both in control and differentiated cell homogenates. In differentiated cells dopamine increases the enzyme activity at a much lower concentration (10^{-6} M) than does norepinephrine (10^{-4} M). Isoproterenol (10^{-4} and 10^{-5} M), a stimulator of β-adrenergic receptor, decreases the adenylate cyclase activity in differentiated cells. As opposed to neuroblastoma cells the cAMP levels in glial cultures markedly increase after treatment with norepinephrine and isoproterenol, but are unaffected after dopamine treatment (Clark and Perkins 1971; Gilman and Nirenberg 1971). This stimulatory effect of catecholamines on the cAMP level is apparently due to an increase in the adenylate cyclase activity. Although response to norepinephrine in glial and differentiated neuroblastoma cells appears similar, the response to dopamine markedly differs between these cell types. These data indicate that the regulation of adenylate cyclase in glial cells may in part be different from that in neuroblastoma cells. The dopamine-sensitive adenylate cyclase present in mouse neuroblastoma cells has also been demonstrated in mammalian basal and peripheral ganglia (Kebabian and Greengard 1971; Kebabian, Petzold and Greengard 1972) and in mammalian retina (Brown and Makman 1972).

Differentiation of Human Neuroblastoma Cell Cultures

Both mouse and human neuroblastoma tumors contain more than one neuronal cell type (Prasad et al. 1973b). Human neuroblastoma cell cultures also show morphological differentiation after treatment with dibutyryl cAMP (2.0 mM), papaverine (1–10 μg/ml)(Fig. 4b) and PGE$_1$ (10 μg/ml). The time of expres-

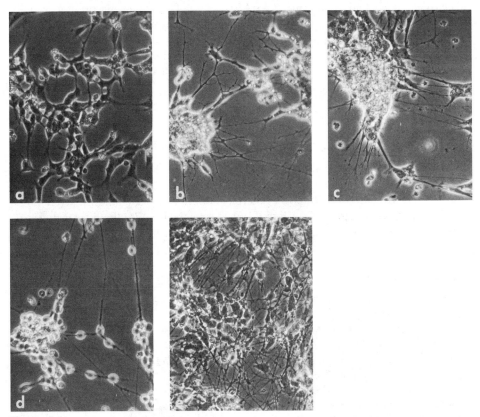

Figure 4
Phase contrast micrographs of human neuroblastoma cells in culture (IMR-32 clone). Cells were plated in Falcon plastic dishes (60-mm^2) and papaverine (2.5 μg/ml), sodium butyrate (0.5 mM), serum free-medium (SFM) and 5-bromodeoxyuridine (2.5 μM) were added individually 4 days after plating. The drug and medium were changed every 2–3 days and the cultures were maintained for 10–13 days. **(a)** Control culture shows that cells grow in clumps and exhibit no spontaneous morphological differentiation (cytoplasmic processes greater than 50 μm in length). **(b)** Papaverine-treated culture 10 days after treatment shows the formation of extensive neurites. Much cell death occurred during this period. **(c)** Sodium butyrate-treated culture 10 days after treatment also shows the formation of extensive neurites. Some cell death occurred during this period. **(d)** SFM-treated culture 3 days after treatment and **(e)** 5-BrdU-treated cultures 10 days after treatment show an extensive neurite formation. 75$\times$

sion and extent of differentiation are concentration dependent. The combination of PGE_1 (10 μg/ml) and papaverine (2.5 μg/ml) allows the expression of the differentiated phenotype much earlier and to a much greater extent than that produced by each individual agent. Among all cyclic AMP agents, papaverine was most potent in causing morphological differentiation. To our great surprise, sodium butyrate induces neurite formation in a dose-dependent fashion (Fig. 4c). This is in contrast to mouse cells in which sodium butyrate reversibly inhibits cell division. Serum-free medium (Fig. 4d) and 5-bromodeoxyuridine (Fig. 4e) cause morphological differentiation similar to that observed in mouse cells.

The control culture had an extremely low level of tyrosine hydroxylase (5 $\pm$ 2.0 pmoles/mg protein), but dibutyryl cAMP-(0.5 mM) and sodium butyrate-(0.5 mM) treated cells had about 550 $\pm$ 102 and 72 $\pm$ 19 pmoles/mg protein, respectively. Thus the response of human and mouse neuroblastoma cell cultures to various cAMP agents are qualitatively similar.

In summary, cAMP appears to be involved in the differentiation of mouse neuroblastoma cell cultures. Preliminary data indicate cAMP may play a similar role in human neuroblastoma cell cultures.

Acknowledgments

This work was supported by USPHS NS-09230, CA-12247 and DRG-1182 from Damon Runyon Memorial Fund for Cancer Research. I thank Drs. J. E. Pike of UpJohn Company, H. Sheppard of Hoffman-La Roche and R. K. Robins of ICN Nucleic Acid Research Institute for generous supply of prostaglandins, R020-1724 and 8-substituted analogs of cyclic AMP, respectively. We thank April Montgomery, Marianne Gaschler and Katrina Gilmer for their technical help.

REFERENCES

Amano, T., E. Richelson and M. Nirenberg. 1972. Neurotransmitter synthesis by neuroblastoma clones. *Proc. Nat. Acad. Sci.* **69:**258.

Augusti-Tocco, G. and G. Sato. 1969. Establishment of functional clonal lines of neurons from mouse neuroblastoma. *Proc. Nat. Acad. Sci.* **64:**311.

Blume, A., F. Gilbert, S. Wilson, J. Farber, R. Rosenberg and M. Nirenberg. 1970. Regulation of acetylcholinesterase in neuroblastoma cells. *Proc. Nat. Acad. Sci.* **67:**786.

Brown, J. H. and M. H. Makman. 1972. Stimulation by dopamine of adenylate cyclase in retinal homogenates and of adenosine 3′,5′-cyclic monophosphate formation in intact retina. *Proc. Nat. Acad. Sci.* **69:**539.

Burger, M. M. 1969. A difference in the architecture of the surface membrane of normal and virally transformed cells. *Proc. Nat. Acad. Sci.* **62:**994.

Clark, R. B. and J. P. Perkins. 1971. Regulation of adenosine 3′:5′-cyclic monophosphate concentration in cultured human astrocytoma cells by catecholamines and histamine. *Proc. Nat. Acad. Sci.* **68:**2757.

Furmanski, P., D. J. Silverman and M. Lubin. 1971. Expression of differentiated functions in mouse neuroblastoma mediated by dibutyryl cyclic adenosine monophosphate. *Nature* **233:**413.

Gilman, A. 1971. A protein binding assay for adenosine 3′:5′-cyclic monophosphate. *Proc. Nat. Acad. Sci.* **67:**305.

Gilman, A. G. and M. Nirenberg. 1971. Effect of catecholamines on the adenosine 3′,5′-cyclic monophosphate concentrations of clonal satellite cells of neurons. *Proc. Nat. Acad. Sci.* **68:**2165.

Guidotti, A. and E. Costa. 1973. Involvement of adenosine 3':5'-monophosphate in the activation of tyrosine hydroxylase elicited by drugs. *Science* **179**:902.

Hardman, J. G., G. A. Robison and E. W. Sutherland. 1971. Cyclic nucleotides. *Ann. Rev. Physiol.* **33**:311.

Hass, D. C., D. B. Hier, B. G. W. Aranson and M. Young. 1972. On a possible relationship of cyclic AMP to the mechanism of action of nerve growth factor. *Proc. Soc. Exp. Biol. Med.* **140**:45.

Hsie, A. W. and T. W. Puck. 1971. Morphological transformation of Chinese hamster cells by dibutyryl adenosine cyclic 3':5'-monophosphate and testosterone. *Proc. Nat. Acad. Sci.* **68**:358.

Johnson, G. S., R. M. Friedman and I. Pastan. 1971. Restoration of several morphological characteristics of normal fibroblasts in sarcoma cells treated with adenosine 3':5'-cyclic monophosphate and its derivatives. *Proc. Nat. Acad. Sci.* **68**:425.

Kates, J. R., R. Winterton and K. Schlessinger. 1971. Induction of acetylcholinesterase activity in mouse neuroblastoma tissue culture cells. *Nature* **224**:345.

Kebabian, J. W. and P. Greengard. 1971. Dopamine-sensitive adenyl cyclase: Possible role in synaptic transmission. *Science* **174**:1346.

Kebabian, J. W., G. L. Petzold and P. Greengard. 1972. Dopamine-sensitive adenylate cyclase in caudate nucleus of rat brain and its similarity to the "dopamine receptor." *Proc. Nat. Acad. Sci.* **69**:2145.

Krishna, G., B. Weiss and B. B. Brodie. 1968. A simple sensitive method for the assay of adenyl cyclase. *J. Pharm. Exp. Therap.* **163**:379.

Macintyre, E. H., J. P. Perkins, C. J. Wintersgill and A. E. Vatter. 1971. The responses in culture of human tumor astrocytes and neuroblasts to $N^6,O^{2'}$-dibutyryl adenosine 3',5'-cyclic monophosphoric acid. *J. Cell Sci.* **11**:639.

O'Dea, R. F., M. K. Haddox and N. D. Goldberg. 1971. Interaction with phosphodiesterase of free and kinase-complexed cyclic adenosine 3':5'-monophosphate. *J. Biol. Chem.* **246**:6183.

Perkins, J. P. and M. M. Moore. 1971. Adenylate cyclase of rat cerebral cortex. *J. Biol. Chem.* **246**:62.

Prasad, K. N. 1971. X-ray-induced morphologic differentiation of mouse neuroblastoma cells *in vitro*. *Nature* **234**:471.

―――. 1972a. Morphological differentiation induced by prostaglandin in mouse neuroblastoma cells in culture. *Nature New Biol.* **236**:49.

―――. 1972b. Neuroblastoma clones: Prostaglandin vs. dibutyryl cyclic AMP, benzylthio cyclic AMP, phosphodiesterase inhibitors and X-ray. *Proc. Soc. Exp. Biol. Med.* **140**:126.

―――. 1972c. Cyclic AMP-induced differentiated mouse neuroblastoma cells lose tumourgenic characteristics. *Cytobios* **6**:163.

―――. 1972d. Effect of cytochalasin B and vinblastine on X-ray, dibutyryl cyclic AMP and prostaglandin-induced differentiation of mouse neuroblastoma cell culture. *Cytobios* **5**:265.

―――. 1973. Role of cyclic AMP in the differentiation of mouse neuroblastoma cell culture. *Role of cyclic nucleotides in carcinogenesis* (ed. H. Gratzner and J. Schultz) pp. 207–237. Academic Press, New York.

Prasad, K. N. and A. W. Hsie. 1971. Morphologic differentiation of mouse neuroblastoma cells induced *in vitro* by dibutyryl adenosine 3':5'-cyclic monophosphate. *Nature New Biol.* **233**:141.

Prasad, K. N. and S. Kumar. 1973. Cyclic AMP 3',5'-AMP phosphodiesterase activity during cyclic AMP-induced differentiation of neuroblastoma cells in culture. *Proc. Soc. Exp. Biol. Med.* **142**:406.

Prasad, K. N. and B. Mandal. 1972. Catechol-*o*-methyltransferase activity in dibutyryl cyclic AMP, prostaglandin and X-ray-induced differentiated neuroblastoma cell culture. *Exp. Cell Res.* **74**:532.

————. 1973. Choline acetyltransferase level in cyclic AMP and X-ray-induced morphologically differentiated neuroblastoma cells in culture. *Cytobios* (in press)

Prasad, K. N. and J. R. Sheppard. 1972a. Neuroblastoma cell culture: Membrane changes during cyclic AMP-induced morphological differentiation. *Proc. Soc. Exp. Biol. Med.* **141:**240.

————. 1972b. Inhibitors of cyclic nucleotide phosphodiesterase-induced morphological differentiation of mouse neuroblastoma cell culture. *Exp. Cell Res.* **73:**436.

Prasad, K. N. and A. Vernadakis. 1972. Morphological and biochemical study in X-ray and dibutyryl cyclic AMP-induced differentiated neuroblastoma cells. *Exp. Cell Res.* **70:**27.

Prasad, K. N., K. Gilmer and S. Kumar. 1973. Morphologically "differentiated" mouse neuroblastoma cells induced by non-cyclic AMP agents: Levels of cyclic AMP, nucleic acid and protein. *Proc. Soc. Exp. Biol. Med.* **143:**1168.

Prasad, K. N., B. Mandal and S. Kumar. 1973a. Human neuroblastoma cell culture: Effect of 5-bromodeoxyuridine on morphological differentiation and levels of neural enzymes. *Proc. Soc. Exp. Biol. Med.* (in press)

————. 1973b. Demonstration of cholinergic cells in human neuroblastoma and ganglioneuroma. *J. Pediat.* **82:**677.

Prasad, R., N. Prasad and K. N. Prasad. 1973. Esterase, malate and lactate dehydrogenase activity in murine neuroblastoma. *Science* **181:**450.

Prasad, K. N., J. C. Waymire and N. Weiner. 1972. A further study on the morphology and biochemistry of X-ray and dibutyryl cyclic AMP-induced differentiated neuroblastoma cells in culture. *Exp. Cell Res.* **74:**110.

Prasad, K. N., S. Kumar, K. Gilmer and A. Vernadakis. 1973a. Cyclic AMP-induced differentiated neuroblastoma cells: Changes in total nucleic acid and protein contents. *Biochem. Biophys. Res. Commun.* **50:**973.

Prasad, K. N., B. Mandal, J. C. Waymire, G. J. Lees, A. Vernadakis and N. Weiner. 1973b. Basal level of neurotransmitter synthesizing enzymes and effect of cyclic AMP agents on the morphological differentiation of isolated neuroblastoma clones. *Nature New Biol.* **241:**117.

Richelson, E. 1973. Stimulation of tyrosine hydroxylase activity in an adrenergic clone of mouse neuroblastoma by dibutyryl cyclic AMP. *Nature New Biol.* **242:**175.

Roisen, F. J., R. A. Murphy, M. E. Pichichero and W. G. Braden. 1972. Cyclic adenosine monophosphate stimulation of axonal elongation. *Science* **175:**73.

Sakamoto, A. and K. N. Prasad. 1972. Effect of DL-glyceraldehyde on mouse neuroblastoma cell culture. *Cancer Res.* **32:**532.

Schmidt, M. J. and G. A. Robison. 1970. Cyclic AMP, adenyl cyclase, and the effect of norepinephrine in the developing rat brain. *Fed. Proc.* **29:**479.

Schubert, D. and F. Jacob. 1970. 5-Bromodeoxyuridine-induced differentiation of a neuroblastoma. *Proc. Nat. Acad. Sci.* **67:**247.

Schubert, D., S. Humphreys, C. Baroni and M. Cohen. 1969. *In vitro* differentiation of a mouse neuroblastoma. *Proc. Nat. Acad. Sci.* **64:**316.

Seeds, N. W., A. G. Gilman, T. Amano and M. W. Nirenberg. 1970. Regulation of axon formation by clonal lines of a neural tumor. *Proc. Nat. Acad. Sci.* **66:**160.

Sheppard, J. R. 1971. Restoration of contact-inhibited growth to transformed cells by dibutyryl adenosine 3′:5′-cyclic monophosphate. *Proc. Nat. Acad. Sci.* **68:**1316.

Sheppard, H. and G. Wiggan. 1971. Analogues of 4-(3, 4-dimethoxybenzyl)-2-imidazolidinone as potent inhibitors of rat erythrocyte adenosine cyclic 3′,5′-phosphate phosphodiesterase. *Mol. Pharmacol.* **7:**111.

Triner, L., Y. Vulliemoz, I. Schwartz and G. G. Nahas. 1970. Cyclic phosphodiesterase activity and the action of papaverine. *Biochem. Biophys. Res. Commun.* **40:**64.

Waymire, J. C., N. Weiner and K. N. Prasad. 1972. Regulation of tyrosine hydroxylase activity in cultured mouse neuroblastoma cells. Elevation induced by analogs of adenosine 3′,5′-cyclic monophosphate. *Proc. Nat. Acad. Sci.* **69:**2241.

Lack of Correlation between Catecholamine Analog Effects on Cyclic AMP Levels and Adenylate Cyclase Activity and the Stimulation of DNA Synthesis in Mouse Parotid Gland

John P. Durham

Department of Biochemistry, University of Glasgow
Glasgow G12 8QQ, Scotland

Renato Baserga

Department of Pathology and Fels Research Institute
Temple University, Philadelphia, Pennsylvania 19140

Fred R. Butcher

Division of Biological Sciences, Brown University
Providence, Rhode Island 02912

A single intraperitoneal injection of isoproterenol (IPR) produces, after a period of about 24 hours, a marked increase in DNA synthesis in the salivary glands of both rats (Barka 1965) and mice (Baserga 1966). The ability to stimulate DNA synthesis is altered by modification of the IPR molecule, but no analog is able to induce DNA synthesis without causing a prior secretion of α-amylase (Kirby, Swern and Baserga 1969). Among the earliest responses of the mouse parotid gland to IPR administration are increases in the level of cyclic AMP (Guidotti, Weiss and Costa 1972; Malamud 1972) and the activity of adenylate cyclase (Malamud 1969). These results have led to the suggestion that in the parotid gland cyclic AMP may play a role in the stimulation of DNA synthesis elicited by IPR. In contrast, cyclic GMP and not cyclic AMP has been implicated as the intracellular mediator of the mitogenic influences of phytohemagglutinin and concanavalin A in lymphocytes (Hadden et al. 1972).

We have extended the number of catecholamine analogs investigated and used them to study further the relationships between the stimulation of DNA synthesis and changes in cyclic AMP levels and adenylate cyclase activity in mouse parotid.

RESULTS AND DISCUSSION

Effect of IPR on Cyclic AMP Concentration

The time curve of cyclic AMP levels following IPR injection (Fig. 1) shows several interesting features. First, the response is very rapid, a large increase in cyclic AMP being observed at 2.5 minutes after injecting IPR. Second, the response is biphasic, with peaks of cyclic AMP concentration at both 2.5 minutes (or earlier) and 15 minutes. Third, the effect of IPR is short-lived and cyclic AMP levels return to control values within one hour and do not rise above control levels again in the period up to the onset of DNA synthesis (not shown). This rapid fall occurs in spite of the presence of considerable amounts of unmetabolized IPR in the mouse salivary glands for at least an hour after its injection (Baserga, Sasaki and Whitlock

595

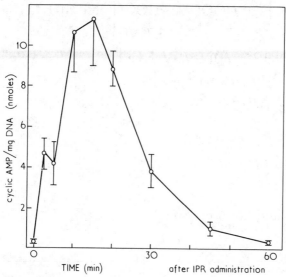

Figure 1

Effect of isoproterenol on the concentration of cyclic AMP in mouse parotid. IPR (3 mg) was injected into 27–30 g male Fels A or Porton mice, aged 3–4 months and starved 2 hr prior to beginning the experiment. At the times indicated mice were killed by cervical dislocation, the parotid gland rapidly dissected and frozen in liquid nitrogen. Glands from three animals were combined for the determination of cyclic AMP both by the protein kinase activation (Butcher 1971) and protein kinase binding (Johnson et al. 1972) assays. For the measurement of cyclic AMP by the protein kinase activation assay, glands were homogenized in 5% TCA, centrifuged, the TCA removed with ether, the aqueous layer lyophilized and chromatographed on Whatman 3MM with isopropanol:NH_4OH:H_2O (70:15:15 v/v). For other determinations glands were homogenized in 0.3 N perchloric acid, centrifuged, neutralized with KOH, applied to an 0.7 × 3 cm column of Dowex AGl-X8 (formate form), which was then washed with H_2O and cyclic AMP eluted with 1.5 N formic acid. Recoveries were determined by the addition of labeled cyclic nucleotide to the homogenization medium. Duplicate assays were always performed. Bars indicate the standard error of the mean value from 3 or 4 groups of animals.

1969). Schramm and Naim (1970) have shown that by 2.5 hours after IPR administration rat parotid adenylate cyclase activity was depressed to approximately one-half of control values. In mouse parotid we observed no corresponding decrease in cyclic AMP levels below the control.

Effect of Isoproterenol on DNA Synthesis

The effect of varying concentrations of IPR on the stimulation of DNA synthesis and α-amylase secretion and the levels of cyclic AMP is shown in Fig. 2. Even very low doses of IPR (0.05 mg) are sufficient to induce almost maximal secretion of α-amylase and to cause significant increases in both cyclic AMP levels and DNA synthesis. As the dose of IPR is increased, the rises in both cyclic AMP levels and

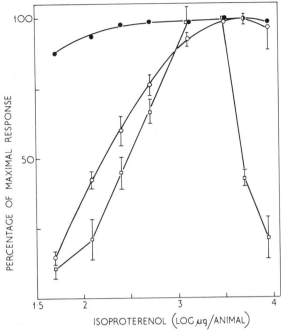

Figure 2
The extent of stimulation of cyclic AMP levels, DNA synthesis and α-amylase secretion with varying doses of isoproterenol. Cyclic AMP was measured at 15 min, α-amylase activity at 2 hr and DNA synthesis at 26 hr after IPR administration. The bars represent the standard errors. Six animals were used for the determination of both DNA synthesis and cyclic AMP levels and three animals for α-amylase. DNA synthesis and α-amylase activity were determined as described by Kirby, Swern and Baserga (1969). Other conditions were as described in Fig. 1.
    ($\bullet$———$\bullet$) α-Amylase; ($\circ$———$\circ$) DNA specific activity; ($\square$———$\square$) cyclic AMP.

DNA synthesis parallel one another, reaching a maximum response at about the same concentration of IPR. However as the dose of IPR is increased further, the stimulation of DNA synthesis remains maximal but the increase in cyclic AMP concentration is greatly diminished.

Effect of Inhibitors of Adenylate Cyclase

Propranolol and dichloroisoproterenol (DCI) are β-adrenergic antagonists which block the activation of isolated adenylate cyclase by IPR (Rosen, Erlichman and Rosen 1970), and chlorpromazine prevents the stimulation of adenylate cyclase by a variety of hormones (Wolff and Jones 1970). These three compounds all inhibited the IPR stimulation of cyclic AMP levels to a greater extent than its stimulation of DNA synthesis (Table 1). This was especially clear with propranolol and DCI (in H_2O), which almost completely inhibited the increase in cyclic AMP levels at concentrations that had almost no effect upon DNA synthesis.

Table 1

Effect of β-Adrenergic Antagonists on Isoproterenol-induced Increase in Cyclic AMP Level and DNA Synthesis

	Cyclic AMP				DNA synthesis	
	pmoles/mg DNA		% Increase		Specific activity (cpm/mg DNA)	% Increase
	2.5*	15*	2.5*	15*		
IPR	4164	14,393	100	100	44,897	100
250 μg propranolol + IPR	483	4171	0	26	43,828	98
250 μg DCI in HCl + IPR	2419	7390	52	50	31,488	69
2 mg DCI in H_2O + IPR	522	853	0	2	37,610	83
100 μg chlorpromazine + IPR	4931	4895	120	32	38,839	86
H_2O	501	513			1629	

Propranolol, DCI and chlorpromazine were injected 10 min prior to the administration of IPR (1.5 mg). DNA synthesis was determined on 4 animals by the method of Kirby, Swern and Baserga (1969). Cyclic AMP was measured as described in Fig. 1.

* Time after administration of IPR (min).

Effect of IPR Analogs on Cyclic AMP Levels and DNA Synthesis

Thirty-one analogs have been investigated and 17 were found to produce a marked stimulation of DNA synthesis in mouse parotid. All of these 17 also caused α-amylase secretion. The effects of a selection of these analogs on stimulation of DNA synthesis and on cyclic AMP levels are shown in Tables 2 and 3, respectively. The D(+) and and L(−) isomers of IPR (compounds II and III) were both as effective as racemic IPR in inducing DNA synthesis. However D(+)-IPR is far less effective in activating rat erythrocyte adenylate cyclase than is L(−)-IPR (Sheppard and Burghardt 1971) and this is reflected in the effect of these two compounds on cyclic AMP levels. L(−)-IPR produces a large increase in these levels but D(+)-IPR induces only a small increase and then only at 15 minutes. Indeed as the D(+)-IPR was only 99% pure, the 1% L(−)-IPR would almost be sufficient to cause the observed increase.

		R_1	R_2	R_3	R_4	R_5	R_6
I, II, III	Isoproterenol (IPR)	OH	OH	H	HOH	H	$CH(CH_3)_2$
IV	Orciprenaline	H	OH	OH	HOH	H	$CH(CH_3)_2$
V	Protokylol	OH	OH	H	HOH	H	1
VI	WG253	OH	OH	H	HOH	H	2
VII	R-007-XL	OH	OH	H	HOH	2	2
VIII	Isoetharine	OH	OH	H	HOH	CH_2CH_3	CH_2CH_3
IX	Soterenol	OH	$NHSO_2CH_3$	H	HOH	H	$CH(CH_3)_3$
X	MI39	H	OH	H	HOH	H	$CH(CH_3)_2$
XI	PI39	OH	H	H	HOH	H	$CH(CH_3)_2$
XII	1-Phenyl-2-iso-propylaminoethanol	H	H	H	HOH	H	$CH(CH_3)_2$
XIII	3, 4-Dihydroxy-α-iso-propylaminoacetophenone	OH	OH	H	O	H	$CH(CH_3)_2$
XIV	Isopropyldopamine	OH	OH	H	H	H	$CH(CH_3)_2$
XV	Dichloroisoproterenol (DCI)	Cl	Cl	H	HOH	H	$CH(CH_3)_2$

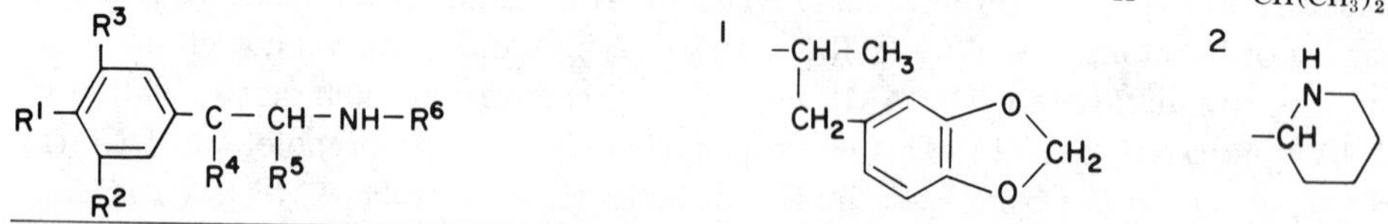

Figure 3

Structural formulas of isoproterenol and its analogs.

Table 2

Effect of Isoproterenol and Its Analogs
on DNA Synthesis in Mouse Parotid

Compound	Amount administered (mg)	Injection medium	DNA specific activity (cpm/mg DNA)
I	3	H_2O	43,339 ± 5151
II	3	H_2O	39,878 ± 6215
III	3	H_2O	46,275 ± 2307
IV	3	H_2O	40,845 ± 8230
V	3	H_2O	31,277 ± 5943
VI	1.5	H_2O	12,882 ± 3901
VII	1.5	H_2O	367 ± 165
VIII	3	H_2O	2862 ± 421
IX	2	H_2O	11,251 ± 3273
X	3	H_2O	43,201 ± 4558
XI	3	H_2O	35,393 ± 9538
XII	2	0.1 N HCl	45,271 ± 7253
XIII	3	H_2O	33,213 ± 2841
XIV	3	H_2O	25,732 ± 5103
XV	2	H_2O	31,290 ± 3317
XVI	2	0.1 N HCl	251 ± 89
Control		H_2O	1589 ± 321
Control		O.1 N HCl	524 ± 276

DNA synthesis was determined as described in Fig. 2.

Increasing the size of the substituent on the amino group did not greatly affect the ability either to stimulate DNA synthesis or to increase cyclic AMP levels. Therefore only protokylol (V), the compound with the largest substituent group tested, is included from these compounds. However binding of the substituent on the amino group to the α-carbon atom, as in WG 253 (VI), significantly reduced the effectiveness of the compound. Decreasing the size of the substituent, as in 1-(3,4-dihydroxyphenyl)-2-ethylaminoethanol (VII) rendered the analog unable to affect either DNA synthesis or cyclic AMP levels.

Altering the position of the hydroxyl groups on the phenyl ring did not affect the ability to stimulate DNA synthesis, but the increase in cyclic AMP produced was now very much smaller (orciprenaline, IV). Greater modification of the substituents on the phenyl ring (IX to XII and XV) or alterations at the β-carbon atom (XIII, XIV) results in a series of analogs able to stimulate DNA synthesis almost as effectively as IPR. None of these compounds produced any increase in cyclic AMP levels, except perhaps isopropyldopamine (XIV). That these compounds produce a true stimulation of DNA synthesis was confirmed by showing that they did not produce any significant change in the acid-soluble pool of thymidine nucleotides, but did lead to an increase in the number of mitotic figures at 28–36 hours after administration.

The results of Tables 1, 2 and 3 clearly show that there is no necessary correlation between the ability of a compound to raise the level of cyclic AMP in parotid gland and its ability to stimulate DNA synthesis in this tissue. A number of ob-

Table 3

Effect of Isoproterenol and Its Analogs
on Cyclic AMP Level in Mouse Parotid

	Cyclic AMP	
	(pmoles/mg parotid DNA)	
Compound	2.5*	15*
I	4608	11403
II	391	1031
III	1824	9281
IV	1468	2134
V	6487	4858
VI	1224	1880
VII	401	320
VIII	319	220
IX	356	464
X	326	287
XI	232	230
XII	457	299
XIII	369	323
XIV	623	607
XV	185	142
Control (H$_2$0)	351	329
(0.1 N HCl)	438	412

Cyclic AMP was determined as described in Fig. 1 and compounds were administered as in Table 2.

* Time after administration of compound (min).

jections might be made to this statement. First, measuring cyclic AMP levels at 2.5 and 15 minutes after agonist injection leaves a very long period between these measurements and the onset of DNA synthesis. However it is known that the initiation of the sequence of events leading to DNA synthesis occurs soon after IPR injection (Sasaki, Litwack and Baserga, 1969). So if cyclic AMP is involved in "triggering" the sequence of events leading to DNA synthesis, it must rise soon after agonist injection. More time points have been investigated for compounds XII and XIII and there is no increase in cyclic AMP induced by these compounds at any of the times studied.

Second, large increases in cyclic AMP may not be needed to induce DNA synthesis. In this case one would expect IPR to fully induce DNA synthesis at very much lower doses than is in fact observed. Also DCI (XV, in H$_2$O) and PI39 (XI) stimulate DNA synthesis yet decrease the levels of cyclic AMP, and a higher dose of compound XII has been found to have the same effect. Third, an increase in cyclic AMP level in the whole cell may not be necessary, but only a redistribution so that its concentration at some specific site is increased. No complete answer can be provided to the problem of compartmentalization; but cyclic AMP levels are reduced by as much as 60% by DCI (XV, in H$_2$O) and yet a strong stimulation of DNA synthesis is observed. A localized increase in cyclic AMP level is still most likely to occur through an activation of adenylate cyclase activity. The effect of IPR and its analogs on this enzyme and the distribution of the enzyme in parotid was therefore studied.

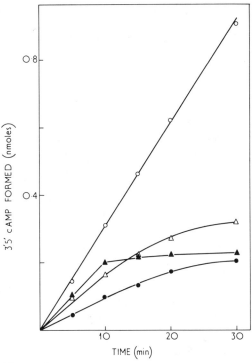

Figure 4

Adenylate cyclase activity in parotid gland homogenates. Adenylate cyclase activity was measured by the method of Bar and Hechter (1969). The assay mixture contained: 40 mM Tris-HCl pH 7.6 at 34°C; 5 mM MgCl$_2$; 10 mM phosphoenolpyruvate; 2.5 mM cyclic AMP; 5 mM theophylline; 5 mM ATP; α-^{32}P-ATP, 1–2 μCi; pyruvate kinase, 2 units; myokinase, 1 unit; BSA, 0.25 mg/ml. The reaction was started by adding the enzyme preparation (10–200 μg protein), followed by incubation at 34°C. The total assay volume was 40 μl. To terminate the reaction 5 μl 0.25 μCi/ml [^{3}H]cyclic AMP was added and the tube placed in a boiling water bath for 3 min. The tubes were centrifuged and 20-μl aliquots applied to polyethyleneimine-impregnated, cellulose thin layer plates, which were developed with 0.3 M LiCl. The cyclic AMP spots were cut out, eluted in liquid scintillation vials with 1 ml 0.1 N HCl/0.2 N NaCl and counted in Triton/toluene liquid scintillation fluid. Blank values varied from 0.01–0.03% of the added radioactivity. All estimations were performed in triplicate.

Parotid glands were homogenized in 0.32 M sucrose containing 10 mM Tris-HCl pH 7.6 at 34°C–5 mM MgCl$_2$–0.1 mM EDTA. Protein: IPR (in vivo), 131 μg; others, 105 μg.

(o– – –o) plus 10 mM NaF; (●– – –●) control; (△– – –△) plus 1 mM IPR; (▲– – –▲) 3 mg IPR injected 2.5 min prior to killing animal.

Adenylate Cyclase Activity

Malamud (1969) has previously shown that IPR causes a rapid activation of parotid adenylate cyclase, with a peak stimulation at 2.5 minutes (or earlier) and a return to baseline levels by 25 minutes. However no data was presented on the activity of the enzyme between 5 and 25 minutes after IPR administration. As the maximum increase in cyclic AMP concentration was observed 15 minutes after

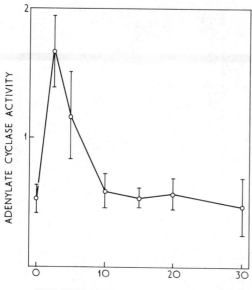

TIME AFTER ISOPROTERENOL INJECTION (min)

Figure 5

Stimulation of adenylate cyclase activity after a single injection of isoproterenol. Animals were killed at the times indicated after injecting 3 mg IPR and adenylate cyclase activity determined as described in Fig. 4, assay time 7.5 min. Adenylate cyclase activity is expressed as nmoles/mg protein/7.5 min. The vertical bars indicate the standard error of the mean value determined on three animals for each time point.

IPR injection (Fig. 1), the adenylate cyclase activity after administration of this compound was reinvestigated. Figure 4 shows the time course of the adenylate cyclase reaction with homogenates of parotid gland. In the presence of fluoride the reaction is linear for at least 30 minutes. But when IPR is administered in vivo or added to the incubation mixture, or in the absence of a stimulating agent, the reaction is linear for no more than 10 minutes. If the adenylate cyclase activity is measured at various times after the administration of IPR to the whole animal, a single peak of increased activity is observed corresponding to the first peak in cyclic AMP concentration (Fig. 5).

Effect of IPR Analogs upon Adenylate Cyclase Activity

The modification of 1-phenyl-2-isopropylaminoethanol (XII) resulting in its ring hydroxylation in the liver has been reported (Labows, Swern and Baserga 1971). Since such modification might be essential to convert certain of the IPR analogs into active forms, the effect of administering the analogs in vivo upon the adenylate cyclase activity measured in vitro was determined. Only compounds that produce a clear rise in cyclic AMP levels in vivo, such as IPR (I) and protokylol (V), had any stimulatory effect upon the adenylate cyclase activity, and DCI decreased the activity (Table 4). The effect of even IPR upon the adenylate cyclase activity was quite small, so that a minor activation of the enzyme would not be picked up.

Catecholamines have been shown to have a greater effect on parotid adenylate cyclase activity in vitro (Schramm and Naim 1970). To study the effect of IPR

Table 4

In Vivo Effect of Isoproterenol and Its
Analogs on Adenylate Cyclase Activity

Compound	Activity (pmoles cAMP/ min/mg protein)
I	134 ± 14
V	113 ± 16
IX	93 ± 12
X	90 ± 8
XI	82 ± 4
XII	81 ± 10
XIII	92 ± 6
XIV	87 ± 2
XV	73 ± 6
Control	84 ± 8

Compounds were injected as described in
Table 2 2.5 min prior to killing the animal.
Adenylate cyclase activity was measured as
described in Fig. 4, incubation time 7.5 min.
Estimations were performed in triplicate on
duplicate animals.

and its analogs on partially purified adenylate cyclase and to determine the cellular
localization of this enzyme, the parotid gland was fractionated in aqueous sucrose
media.

Subcellular Fractionation of Parotid Gland

One side of the compartmentalization problem discussed earlier is, is adenylate
cyclase confined to the plasma membrane or does it have other, intracellular locali-
zations as well? If present within the cell then the observations that, after a single
injection of labeled IPR considerable amounts of unmodified IPR are present in
salivary glands for a long period of time and that 26% of the radioactivity is
localized within the nucleus (versus only 2–3% in other tissues) (Malamud and
Baserga 1967), could be of great significance. Adenylate cyclase activity has been
found in nuclei purified from rat liver (Soifer and Hechter 1969) and rat ventral
prostate (Liao, Lin and Tymoczko 1971). But on the other hand immobilized
hormones, which are incapable of entering cells, can still exert many of their effects
on target cells, indicating that their primary site of action is the cell membrane
(Schimmer, Uedak and Sato 1968; Cuatrecasas 1969).

The result of the subfractionation of the parotid gland by differential centrifugation is shown in Table 5. The crude nuclear preparation contained appreciable
adenylate cyclase activity, although less than the whole homogenate. The relative
activity of adenylate cyclase in the homogenate and nuclear fraction was similar to
the relative activities of the two plasma membrane marker enzymes in these two
fractions, suggesting that the adenylate cyclase activity of the nuclear preparation
might not actually reside in the nuclei. Pure nuclei were prepared by the method of
Blobel and Potter (1966). These nuclei contained no measurable adenylate cyclase
activity, so that the total activity of this enzyme in the nuclei is less than 0.5% of

Table 5
Cellular Localization of Adenylate Cyclase in Parotid Gland

Fraction	Protein % Total recovery	Relative specific activity 5'-Nucleotidase	Relative specific activity Na+/K+-ATPase	Adenylate cyclase Specific activity (nmoles/mg protein/15 min)	%	Total activity (nmoles/15 min)	%
Homogenate	100	1	1	4.10 ± 0.65	1	410	
Nuclei	26.2 ± 2.9	0.4	0.7	2.52 ± 0.54	0.6	66.1	16
Mitochondria + RER	3.1 ± 0.3	4.1	6.4	26.5 ± 4.8	6.5	82.3	20
Smooth membranes	3.2 ± 0.2	13.2	14.5	55.1 ± 8.6	13.4	174	43
Ribosomes	2.4 ± 0.1	4.1	1.3	7.51 ± 1.20	1.8	17.8	4.4
Supernatant	56.4 ± 5.9	0.6	< 0.1	< 0.1			
Total recovery (%)	91.3	112	97				83

Parotid glands from 10 animals were homogenized gently in an 0.011-in. clearance homogenizer in 0.32 M sucrose–0.04 M Tris-HCl pH 7.6 at 34°C–0.005 M $MgCl_2$. The homogenate was centrifuged 15 min at 650 g (nuclei), followed by centrifugation of the supernatant for 10 min at 15,000 g (mitochondria + RER), 1 hr at 40,000 g (smooth membranes) and 1 hr at 215,000 g (ribosomes). Each sediment was washed once and recentrifuged. Adenylate cyclase was determined as in Fig. 4 with the addition of 10 mM NaF, incubation time 15 min. 5'-Nucleotidase and Na+/K+-ATPase were measured by the methods of Durham and Galanti; their specific activities in the homogenate were 3.2 and 0.4 μmole nucleotide hydrolyzed/hr/mg protein, respectively.

the whole gland activity. However this technique removes much of the protein content from parotid nuclei (Durham and Galanti unpubl.) and could remove adenylate cyclase activity. Therefore the adenylate cyclase activity of the crude nuclear pellet and of tissue slices was localized using the cytochemical technique of Howell and Whitfield (1972). In neither case was adenylate cyclase activity found either in the nuclear membrane or within the nucleus (Revis and Durham unpubl.).

The mild homogenization procedure used results in large rough endoplasmic reticulum (RER) fragments, which sediment almost totally at 15,000 g, whereas smooth membranes sediment upon centrifugation at 40,000 g. The 40,000 g sediment contains 14-fold purified plasma membranes as determined by the marker enzymes 5'-nucleotidase and Na$^+$K$^+$-ATPase and a similar 14-fold increase in adenylate cyclase activity. Both the 15,000 g and 215,000 g sediments contained adenylate cyclase activity. On subfractionation of these pellets using the discontinuous sucrose gradient technique of Neville (1960), the activity concentrated at the interface of the two sucrose layers, as did the plasma membrane marker enzymes, suggesting that the activity in these sediments resides in plasma membrane fragments. Thus it seems probable that adenylate cyclase activity is confined primarily, if not entirely, to the plasma membrane.

This cellular localization and the fact that long after IPR administration parotid cells contain unaltered IPR, yet cyclic AMP levels have returned to control values, suggest that once IPR has entered the cell it may no longer be able to affect the adenylate cyclase system.

Pyrogallol, an inhibitor of catechol-O-methyltransferase, potentiates the action of IPR in the salivary glands. This suggests that the ability of IPR to stimulate DNA synthesis in salivary glands alone may result from its concentration within the cells of these glands in an unmetabolized form (Baserga, Sasaki and Whitlock 1969). These facts together lead to the conclusion that there is a possibility that the site of action of IPR in stimulating DNA synthesis could be intracellular.

Table 6

Effect of Isoproterenol and Its Analogs on
Partially Purified Adenylate Cyclase Activity

Compound	Conc. (mM)	Activity (pmoles cAMP/ 10 min)
I	1	430 ± 23
I	0.1	410 ± 41
V	1	354 ± 18
IX	1	183 ± 4
X	1	175 ± 14
XI	1	173 ± 10
XII	1	189 ± 12
XIII	1	191 ± 10
XIV	1	199 ± 6
XV	1	148 ± 5
Control		180 ± 11

Adenylate cyclase activity was measured as described in Fig. 4, incubation time 10 min. The 40,000 g sediment, prepared as in Table 5, was the source of enzyme; 24 μg protein was used per assay.

Effect of Isoproterenol and IPR Analogs on Plasma
Membrane Adenylate Cyclase Activity

The effect of IPR analogs that do not raise cyclic AMP levels in vivo, but do cause stimulation of DNA synthesis, on the partially purified plasma membrane adenylate cyclase activity is shown in Table 6. Isoproterenol (I) and protokylol (V), two compounds which raise cyclic AMP levels in vivo, produce a clear stimulation of adenylate cyclase. Isopropyldopamine (XIV) stimulates activity very slightly and DCI (XV) clearly decreases activity. None of the other analogs had any measurable effect.

CONCLUSIONS

Evidence has been presented that an increase in the level of cyclic AMP is not the mechanism whereby the effects of catecholamine analogs on stimulation of DNA synthesis in mouse parotid gland are mediated. This evidence is threefold:

1. The IPR induced increases in cyclic AMP can be prevented without affecting the stimulation of DNA synthesis.
2. A number of IPR analogs stimulate DNA synthesis but have no effect on cyclic AMP levels.
3. These analogs have no effect on adenylate cyclase activity.

Acknowledgments

During the course of part of this work F.R.B. was a USPHS postdoctoral fellow (F02-CA-43,880) in the laboratory of Dr. Van R. Potter and J.P.D. a postdoctoral fellow of the Damon Runyon Memorial Fund for Cancer Research. J.P.D. is presently supported by the Scottish Hospital Endowments Research Trust (HERT 422). Financial support was provided for F.R.B. in part by USPHS departmental grant CA-07175, training grant CA-5002 through the National Cancer Institute (University of Wisconsin) and an American Cancer Society Grant IN 45-L (at Brown University) and for R.B. by USPHS Grant DE-02678 from the National Institute of Dental Research.

REFERENCES

Bar, H. P. and O. Hechter. 1969. Adenyl cyclase assay in fat ghost cells. *Anal. Biochem.* **29:**476.

Barka, T. 1965. Stimulation of DNA synthesis by isoproterenol in the salivary gland. *Exp. Cell Res.* **39:**355.

Baserga, R. 1966. Inhibition of stimulation of DNA synthesis by isoproterenol in submandibular glands of mice. *Life Sci.* **5:**2033.

Baserga, R., T. Sasaki and J. P. Whitlock, Jr. 1969. The prereplicative phase of isoproterenol-stimulated DNA synthesis. *The biochemistry of cell division* (ed. R. Baserga) pp. 77–90. C. C. Thomas, Springfield, Illinois.

Blobel, G. and V. R. Potter. 1966. Nuclei from rat liver: Isolation method that combines purity with high yield. *Science* **154:**1662.

Butcher, F. R. 1971. A rapid filter disk assay for picomole amounts of cyclic AMP using a cyclic AMP dependent protein kinase. *Hormone and Met. Res.* **3:**336.

Cuatrecasas. P. 1969. Interaction of insulin with the cell membrane: The primary action of insulin. *Proc. Nat. Acad. Sci.* **63**:450.

Guidotti, A., B. Weiss and E. Costa. 1972. Adenosine 3′,5′-monophosphate concentrations and isoproterenol induced synthesis of deoxyribonucleic acid in mouse parotid gland. *Mol. Pharmacol.* **8**:521.

Hadden, J. W., E. M. Hadden, M. K. Haddox and N. D. Goldberg. 1972. Guanosine 3′:5′-cyclic monophosphate: A possible intracellular mediator of mitogenic influences in lymphocytes. *Proc. Nat. Acad. Sci.* **69**:3024.

Howell, S. L. and M. Whitfield. 1972. Cytochemical localization of adenyl cyclase activity in rat islets of Langerhans. *J. Histochem. Cytochem.* **20**:873.

Johnson, M. E. M., N. M. Das, F. R. Butcher and J. N. Fain. 1972. The regulation of gluconeogenesis in isolated rat liver cells by glucagon, insulin, dibutyryl cyclic adenosine monophosphate and fatty acids. *J. Biol. Chem.* **247**:3229.

Kirby, K. C., D. Swern and R. Baserga. 1969. The effect of structural modifications of the isoproterenol molecule on the stimulation of deoxyribonucleic acid synthesis in mouse salivary glands. *Mol. Pharmacol.* **5**:572.

Labows, J., D. Swern and R. Baserga. 1971. Isoproterenol stimulated DNA synthesis: Requirement for OH groups on the phenyl ring for activity. *Chem. Biol. Interact.* **3**:449.

Liao, S., A. H. Lin and J. L. Tymoczko. 1971. Adenyl cyclase of the cell nuclei isolated from rat ventral prostate. *Biochim. Biophys. Acta* **230**:535.

Malamud, D. 1969. Adenyl cyclase: Relationship to stimulated DNA synthesis in parotid glands. *Biochem. Biophys. Res. Commun.* **35**:754.

———. 1972. Amylase secretion from mouse parotid and pancreas: Role of cyclic AMP and isoproterenol. *Biochim. Biophys. Acta* **279**:373.

Malamud, D. and R. Baserga. 1967. On the mechanism of action of isoproterenol in stimulating DNA synthesis in salivary glands of rats and mice. *Life Sci.* **6**:1765.

Neville, D. M., Jr. 1960. The isolation of a cell membrane fraction from rat liver. *Biochem. Biophys. Cytol.* **8**:113.

Rosen, O. M., J. Erlichman and S. M. Rosen. 1970. The structure activity relationships of adrenergic compounds that act on adenyl cyclase of the frog erythrocyte. *Mol. Pharmacol.* **6**:524.

Sasaki, T., G. Litwack and R. Baserga. 1969. Protein synthesis in the early prereplicative phase of isoproterenol-stimulated synthesis of deoxyribonucleic acid. *J. Biol. Chem.* **244**:4831.

Schimmer, B. P., K. Uedak and G. H. Sato. 1968. Action of ACTH coupled to cellulose on isolated adrenal cells. *Biochem. Biophys. Res. Commun.* **32**:806.

Schramm, M. and E. Naim. 1970. Adenyl cyclase of rat parotid gland. *J. Biol. Chem.* **245**:3225.

Sheppard, H. and C. R. Burghardt. 1971. The effect of alpha, beta and dopamine receptor-blocking agents on the stimulation of rat erythrocyte adenyl cyclase by dihydrophenethylamines and their β-hydroxylated derivatives. *Mol. Pharmacol.* **7**:1.

Soifer, D. and O. Hechter. 1969. Adenyl cyclase in rat liver nuclei. *Biochim. Biophys. Acta* **230**:539.

Wolf, J. and A. P. Jones. 1970. Inhibition of hormone-sensitive adenyl cyclase by phenothiazines. *Proc. Nat. Acad. Sci.* **65**:454.

The Yin Yang Hypothesis of Biological Control: Opposing Influences of Cyclic GMP and Cyclic AMP in the Regulation of Cell Proliferation and Other Biological Processes

**Nelson D. Goldberg, Mari K. Haddox, Earl Dunham,
Carlos Lopez, and John W. Hadden**

Departments of Pharmacology and Pathology
University of Minnesota Medical School, Minneapolis, Minnesota 55455

With the development of our understanding of biological regulatory mechanisms there is an increasing awareness of the role that cyclic 3',5'-adenosine monophosphate (cyclic AMP) may play as a major controlling component. The list of cellular events that appear to be influenced in one way or another by cyclic AMP has grown to such proportions that this compound is now viewed by many as a key regulatory effector of the particular process peculiar to a given type of cell or of the event of importance at a given point in time in the maturation or survival of a cell or organism. It is understandable that a certain amount of skepticism should be voiced against such a sweeping concept and omnipotent role for a single biological component.

An objection, though of a somewhat different nature than the general skepticism just described, was lodged by this laboratory a number of years ago. Our objection was voiced with the understanding that cyclic AMP was present in almost every living animal (and probably plant) cell, and because of its established role as a regulatory molecule that it more than likely would be shown to control a specific cellular process in each of the cells in which it resides. The disagreement was with the concept that cyclic AMP could, single-handedly, through bidirectional changes (i.e., elevation or reduction) in its cellular steady-state levels account for both the positive and negative regulatory influences imposed upon biological processes. This concept, which implies that a critical steady-state level of cyclic AMP is required for "normal" function or for the basal activity of a system, seemed incompatible with the idea that (in mammalian tissues) cyclic AMP was a "second messenger" of hormone action, since hormones are usually thought of as agents that modulate rather than maintain the activity of a given cellular process.

The challenge to this "unitary" concept of control through cyclic AMP was also made on experimental grounds. It was found that metabolic effects of insulin which are antagonistic to those of agents such as glucagon and epinephrine (i.e., are expressed through an elevation of cellular cyclic AMP concentration) were manifest without (Goldberg et al. 1967) or before (Nichols and Goldberg 1972) any detectable reduction of tissue cyclic AMP levels. These objections motivated a search

609

for a cellular component involved in promoting cellular events that could be considered to be opposite to those that occur when the levels of cyclic AMP in the cell are increased. The pursuit of such a component has led to a series of discoveries linking another cyclic nucleotide, cyclic 3',5'-guanosine monophosphate (cyclic GMP), to the action of a number of different biologically active agents that produce cyclic AMP-antagonistic cellular responses.

One cellular process now believed to be affected dramatically by cyclic AMP is cell proliferation, and evidence for an opposing influence of cyclic GMP, in at least a certain stage of this process, has recently been uncovered (Hadden et al. 1972). The following report contains some of the information we have obtained implicating cyclic GMP as a regulatory factor in cell proliferation. However, because of the very recent emergence of cyclic GMP as a possible regulatory component, the overall biological importance we have assigned to this cyclic nucleotide, along with the evidence we have accumulated in support of our view, will be described first.

The Yin Yang or Dualism Hypothesis of Biological Control through Cyclic GMP and Cyclic AMP

In a number of different biological systems examined thus far, we have found that hormones or other biologically active substances that promote the cellular accumulation of cyclic GMP produce cellular responses antagonistic to those that occur when the concentration of cyclic AMP is increased in the same tissues or cells. From these observations we have concluded that there are biological systems in which cyclic GMP and cyclic AMP may have opposing or antagonistic regulatory influences. We believe this concept of biological regulation through opposing actions of the two cyclic nucleotides is well described by an ancient oriental concept embodied in the term Yin Yang, which symbolizes a dualism between two opposing natural forces (Goldberg et al. 1973a). A feature inherent to the concept that may apply to certain biological systems is that the dual opposing forces can enter into a mutual interaction resulting in synthesis. According to the hypothesis (Fig. 1) the opposing actions of cyclic GMP and cyclic AMP are expressed in systems which are susceptible to both stimulatory and inhibitory controlling influences or those comprised of antagonistic events. The latter are termed "bidirectionally" controlled systems. In contrast to the type of system that may be under bidirectional control, it seems likely that some regulated cellular events may be unopposed by an antagonistic event or susceptible to only one type of regulatory influence, a positive or stimulatory one. We have designated these as "monodirectionally" controlled systems and suggested that the steroidogenic mechanism in the adrenal cortex may represent an example, since there is no known physiological antagonist of the stimulatory effect of ACTH on steroidogenesis. In this system it is possible that (a) the two cyclic nucleotides express the actions of different stimulatory hormones intracellularly, (b) they promote different stages of the steroidogenic process, or (c) they direct the synthesis of different types of steroids (i.e., mineralocorticoids *vs.* glucocorticoids). The existence of systems that may be under monodirectional type of control remains to be established, but there is little doubt that there are a number that are controlled in a bidirectional manner.

From our present understanding of cellular events influenced by cyclic AMP,

BIOLOGICAL REGULATION THROUGH
cyclic GMP and cyclic AMP

I Bidirectionally Regulated Systems

Antagonistic influences on opposing cellular events
(Yin Yang or Dualism Hypothesis)

A−Type

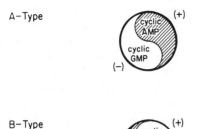

B−Type

II Monodirectionally Regulated Systems

Facilitory influences on unopposed cellular events

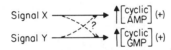

Figure 1

The proposed influences of cyclic GMP and cyclic AMP in the regulation of cell functions. Bidirectionally controlled systems are those comprised of opposing cellular processes. Depending upon the type of bidirectional system an increase in the concentration of cyclic AMP (A-type) or cyclic GMP (B-type) represents the facilitatory influence while the suppressive influence and/or opposing cellular event is promoted by an increase in the cellular level of the other cyclic nucleotide. A monodirectionally controlled system is an unopposed cellular process which only responds to a stimulatory signal. Cyclic GMP or cyclic AMP may represent intracellular mediators of the same or different biological signals but both cyclic nucleotides, in this case, would promote an analagous or the same cellular event or different steps in the same overall process.

there appear to be at least two types of bidirectionally controlled systems (Fig. 1): those *stimulated* by an elevation in the level of cellular cyclic AMP (A-type) and those *suppressed* by an increase in the concentration of cyclic AMP (B-type). According to the Yin Yang hypothesis an elevation of cellular cyclic GMP concentration promotes the opposing cellular event or provides the opposing regulatory influence in each type of bidirectionally controlled system. A reciprocal change in the levels of the two cyclic nucleotides should provide for maximum expression of the incoming signal. The extent to which there may be a lowering of the cellular concentration of the cyclic nucleotide not linked to the predominant incoming signal will depend upon what the prevailing level may be relative to the basal concentration and probably other factors such as the degree of stimulation from opposing signals.

Development of the Evidence in Support of the Dualism Hypothesis

Cyclic GMP was found to occur in mammalian urine by Price et al. (1963) and when sufficiently sensitive analytical procedures for its detection were developed, it was found to be a naturally occurring component of mammalian and other animal tissues (Goldberg, Dietz and O'Toole 1969; Ishikawa et al. 1969) and to be present in bacteria (Goldberg et al. 1973a). The concentration of this cyclic nucleotide (10^{-8} to 10^{-7} moles/kg wet weight) in most tissues and cells is generally 1/10 to 1/50 the concentration of cyclic AMP. The analytical difficulties associated with the quantitation of the extremely low tissue levels of cyclic GMP account to a great extent for the slow rate of progress in the field.

After establishing its ubiquity a major effort was put forth to uncover its biological importance. The approach taken was directed toward identifying a biologically active agent that could promote the cellular accumulation of cyclic GMP at the same time that a definable response could be demonstrated in cell function. Although correlating changes that take place in cellular steady-state levels of a component with alterations in cell function may not constitute definitive proof of cause and effect, it seemed that such information could provide the first step in helping to define a biological role for this cyclic nucleotide. Little progress was made for quite some time in our attempts to find an agent, altered endocrine condition, or nutritional state that would stimulate cellular cyclic GMP accumulation.

Cyclic GMP and Cholinergic Stimulation

The possibility that cholinergic action might be linked to an action of cyclic GMP arose from an observation that imidazole, which is known to produce cholinergic-like effects in a number of biological systems, could inhibit phosphodiesterase-catalyzed hydrolysis of cyclic GMP while having only a slight stimulatory effect on the degradation of cyclic AMP (Goldberg et al. 1970). Upon examining the changes in the concentration of the two cyclic nucleotides in rat myocardium following treatment with acetylcholine, the first definable hormone-induced alteration in cell function related to an elevation of tissue cyclic GMP levels was established (George et al. 1970). In these experiments it was demonstrated that the depression of cardiac contractility induced by acetylcholine occurred coincident with an elevation (ca. 2- to 3-fold) in the concentration of tissue cyclic GMP and that no change or a small delayed decrease occurred in the concentration of cyclic AMP. Also found at that time was that the stimulatory effect of isoproterenol on cardiac contractility, which is known to occur in conjunction with an elevation of myocardial cyclic AMP levels, was also accompanied by a decrease in the concentration of cyclic GMP (ca. 50%). One conclusion that can be drawn from these results is that the response of a tissue to an agent that promotes cyclic AMP generation is opposed by the influence of a neurohumor that stimulates the accumulation of cellular cyclic GMP.

Cholinergic stimulation has now been shown in our laboratory to be associated with the rapid accumulation of cellular cyclic GMP in a number of different tissues and cells, including the rat uterus, rabbit lung, human peripheral blood lymphocytes, and mouse cerebellum (Goldberg et al. 1973a). The same relationship between cholinergic action and cyclic GMP has been found by other investigators in ductus deferens (Schultz et al. 1972a), intestinal smooth muscle (Lee, Kuo and Greengard 1972), submaxillary gland (Schultz et al. 1972b), and thyroid (Yamashita and Field 1972). In the cases where a tissue or cell response to the agent is detectable, the action produced by cholinergic stimulation is opposite to the effect promoted by cyclic AMP or agents known to stimulate its accumulation. The thyroid is an exception to the latter and may represent an example of a monodirectionally controlled system. Furthermore it appears that only the muscarinic type of cholinergic action, and not the nicotinic type, is linked to cyclic GMP generation since atropine can block the functional response as well as the increase in cyclic GMP levels, and cholinergic effects of the nicotinic variety (i.e., neuronal stimulation of skeletal muscle contraction) are not associated with tissue accumulation of cyclic GMP (Goldberg et al. 1973a; Lee et al. 1972). It should also be pointed out that a number of cholinergic effects presumed to be linked to an enhanced generation of cyclic GMP have recently been uncovered in cells never before known to be susceptible to cholinergic modulation. Furthermore in each instance the cholinergic effect has been linked directly or indirectly with cyclic GMP. In the lymphocyte we have found that acetylcholine can promote RNA and protein synthesis (Hadden et al. 1973), and Strom et al. (1972) have noted that this agent can enhance the cytotoxic action of these cells. Colombo, Ignarro and Chart (1973) and Zurier, Tynan and Weissman (1973) have found that lysosomal enzymes are released from leukocytes upon exposure to cholinergic agents or cyclic GMP.

After uncovering the relationship between cyclic GMP and cholinergic action, experiments were conducted to determine if other agents that promote cyclic AMP-antagonistic cellular events also stimulate cellular cyclic GMP accumulation.

Other Hormone-induced Changes in Cyclic GMP Levels

From experiments with the rat uterus (Fig. 2), which undergoes relaxation upon exposure to cyclic AMP or substances (e.g., epinephrine) that increase the levels of this cyclic nucleotide (Triner et al. 1971), it was found that three agents, oxytocin, serotonin and prostaglandin $F_{2\alpha}$ ($PGF_{2\alpha}$), which like cholinergic agents stimulate uterine smooth muscle contractility, also increase (2- to 5-fold) the concentration of cyclic GMP in this tissue within seconds.

Similar effects of $PGF_{2\alpha}$ were observed in bovine vein, which is stimulated to contract in response to this agent (Fig. 3). In the latter case the increases that occurred in tissue cyclic GMP levels were accompanied by a decrease, no change, or a small increase in cyclic AMP concentration. In spite of the qualitative variability in the changes that occurred in the levels of cyclic AMP, there was a significant increase in the ratio of cyclic GMP to cyclic AMP in every case (Fig. 4). An identical situation (i.e., consistent increases in cyclic GMP concentration with no consistent direction of change in those of cyclic AMP after $PGF_{2\alpha}$-induced contraction) was uncovered in experiments conducted with canine veins (not shown). The cyclic GMP to cyclic AMP ratio was, however, increased significantly in the latter case as well (Fig. 4). In contrast to the increase in cyclic GMP/cyclic AMP ratio associated with contraction, the ratio was reduced (i.e., with respect to control) during PGE_2-induced relaxation of canine veins that had first been exposed to norepinephrine to increase vascular tone.

Another bidirectionally controlled (B-type) cellular process that has been shown to be suppressed when the concentration of cyclic AMP is increased intracellularly and thought, therefore, to be activated as a result of a lowering of cyclic AMP levels, is platelet aggregation. Epinephrine- or collagen-induced aggregation can now be shown to be associated with increases in the levels of cellular cyclic GMP at

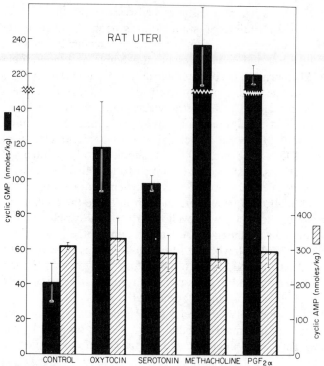

Figure 2

Increases in uterine cyclic GMP concentrations by agents that stimulate its contractility. Uteria from diethylstilbesterol-treated rats were equilibrated (2 min) in Munsick's media, then exposed to oxytocin (0.1 units/ml), serotonin (10^{-6}M), methacholine (4×10^{-5}M) or prostaglandin F_{2a} (PGF_{2a}) (10^{-6}M) for 45 sec (when contraction was marked) before they were quick-frozen in Freon-12 ($-150°$C). Cyclic GMP was determined by the method of Goldberg and O'Toole (1971) and cyclic AMP by the procedure of Gilman (1970).

a time when the very first sign of aggregation is detectable (Goldberg et al. 1973b; White et al. 1973). There is no significant change in the levels of cyclic AMP when aggregation is induced by either epinephrine or collagen. Another platelet aggregating agent, phorbol myristate acetate, has also recently been shown to elevate the cyclic GMP concentration of platelets (White et al. 1973 and cf. Estensen et al. this volume).

The results described above demonstrate that hormone-induced alterations in cell function, definable as responses that oppose those believed to be mediated by cyclic AMP, can be correlated with relatively rapid accumulation of cellular cyclic GMP. Other supportive evidence for the Dualism Hypothesis can also be found in changes brought about in cyclic GMP levels by other agents or conditions in cases where the function altered is not altogether definable. For example, we have noted that adrenalectomized rats exhibit elevated levels of tissue cyclic GMP in lung and kidney, while glucocorticoid administration decreases the levels in these tissues to below those found in nonadrenalectomized animals (Goldberg et al. 1973a). This observation may relate in some way to the antimitotic activity associated with

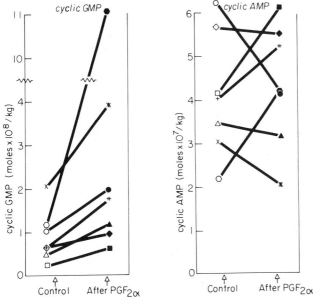

Figure 3

$PGF_{2\alpha}$-induced elevation of cyclic GMP levels in bovine vein. The veins were dissected shortly after the animals were slaughtered and paired helical strips cut from bifurcations of the common dorsal digital vein. The strips were suspended (0.5–1.0 g tension) and superfused (10 ml/min) with Krebs physiological salt solution (37°C). The veins were frozen between stainless steel blocks ($-190°$) 30–180 sec after perfusion with 7×10^{-6}M $PGF_{2\alpha}$ which produced a response averaging 60% of a near maximal contraction elicited by 105 mM KCl. Each of the 7 symbols represents paired helical strips from a single animal. A modified version of the radioimmune method of Steiner et al. (1972) was used for the analysis of cyclic GMP (Goldberg et al. 1973b).

glucocorticoid action in certain tissues and cells since the initiation of cell division, which will be discussed later, appears to be associated with an elevation in cellular cyclic GMP concentration.

Reciprocal Changes in Nucleotide Levels

In mouse cerebellar tissue following electroconvulsive shock treatment, the levels of cyclic AMP increase approximately 4-fold within 30 seconds during the time that the animals exhibit tonic and clonic convulsive behavior (Goldberg et al. 1973a). The cyclic GMP concentration remains relatively unaffected during this period of hyperactivity. At later time periods (60 seconds) when the animals become noticeably depressed (i.e., postical depression), the cyclic AMP levels decline toward control and the cyclic GMP concentration increases about fourfold. These results only point out the reciprocal nature of the changes that can occur in the two cyclic nucleotides under certain circumstances and the possible relationship of these changes to contrasting states of activity in the central nervous system.

An example of the reciprocal manner in which the levels of the two cyclic nucleotides may be affected, which suggests that "Dualism" may be universal in nature and extend to even the simplest of organisms, is a situation regarding the changes

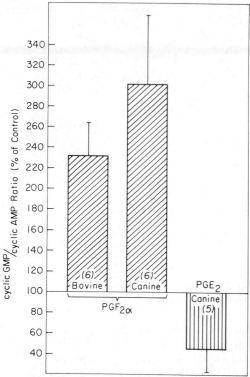

Figure 4
Reciprocal effects on the cyclic GMP/cyclic AMP ratio in bovine and canine veins with $PGF_{2\alpha}$, a venoconstrictor and PGE_2, a venodilator. Conditions were similar to those described in the legend of Fig. 3 except that the canine helical strips were prepared from segments of the saphenous vein and norepinephrine (45–170 nM) was contained in the perfusion media (to provide vascular tone) before and during the infusion of PGE_2 (37–77 nM). The concentration of $PGF_{2\alpha}$ used to contract the canine reins was 2 μM.

found to occur in *E. coli* under certain conditions.[1] It is now well recognized, as a result of the work of Makman and Sutherland (1965) and Pastan and Perlman (1970), that cyclic AMP has a regulatory function in bacteria related to the control of gene expression and in turn the biosynthesis of certain non-constitutive enzymes required for the metabolism of nutrients other than specific carbohydrates including glucose. When glucose serves as the carbon source, the levels of cellular cyclic AMP are generally found to be inversely related to the concentration of carbohydrate in the media. Upon examining the changes that occur in the levels of cyclic GMP and cyclic AMP in *E. coli* grown in a culture containing a concentration of glucose predetermined to become limiting during the progression of the growth curve, it was found (Fig. 5) that the levels of cyclic GMP decline to about 1/5 of the initial level with the cessation of bacterial growth, while the concentration of cyclic AMP, as would be expected, undergoes a steady increase. The levels of cyclic GMP were minimal and those of cyclic AMP maximal when stationary growth was achieved.

[1] This work was conducted in collaboration with Dr. Robert Bernlohr in the Department of Microbiology at the University of Minnesota.

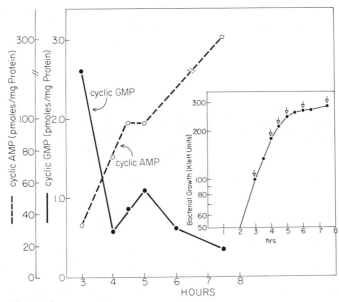

Figure 5
Changes in bacterial cyclic nucleotide concentrations during log and stationary growth on glucose. *E. coli* (K12 lambda sensitive) were grown at 37°C in a media containing (per liter) 14 g K_2HPO_4, 6 g KH_2 PO_4, 2 g $(NH_4)_2SO_4$ and 0.2 g $MgSO_4$. Glucose was present initially at a concentration of 20 mM. Samples were removed from the culture at the times indicated and filtered under vacuum, washed quickly with growth media, and the filter containing the organism emersed in 5% trichloroacetic acid. Cyclic GMP was determined by the method of Goldberg and O'Toole (1971) and cyclic AMP by the procedure of Gilman (1970).

Similar progressive reciprocal changes in the levels of the two cyclic nucleotides during log and stationary growth (Goldberg et al. 1973a) were observed when succinate was used as the carbon source. Succinate, unlike glucose, is a poor repressor of inducible enzyme synthesis and a poor suppressor of cyclic AMP levels in these organisms, the latter reflected by considerably higher steady-state levels of cyclic AMP levels during log phase growth and commensurately smaller increases in the relative levels of cyclic AMP as the stationary phase of growth was approached. The addition of glucose to a culture grown on succinate led to an immediate lowering of the cellular cyclic AMP concentration. The anticipated reciprocal increase in the cyclic GMP concentration also occurred (30 min), but only after a dramatic lowering that was evident for the first 15 minutes after glucose addition (Goldberg et al. 1973a). In general the overall changes that we have observed in the levels of the two cyclic nucleotides in *E. coli* could be described as reciprocal under most of the conditions examined. The results of the latter experiments indicate that a more complex relationship may exist under some conditions. It may also be worthwhile noting that there appears to be as much of a relationship between the changes found to occur in the levels of the two cyclic nucleotides and the growth rate of the organism as there is with regard to the induction of non-constitutive enzymes. Such an interpretation may be especially applicable in the case of the cyclic nucleotide changes found to occur during growth on succinate, where during the transition from log to stationary phase growth reciprocal changes in the cyclic nucleotide levels occurred independent of significant change in inducible enzyme levels.

Cell Proliferation

An important chapter in the story of cell proliferation is presently taking shape with regard to the regulatory role cyclic AMP may play in this process. The advances in this field stem primarily from the investigations being conducted by Dr. Ira Pastan and his colleagues and Dr. Jack Sheppard. Contributions from the laboratories of both of these investigators appear in chapters of this volume and should be referred to for a more detailed description of the subject. From studies they and others have carried out with various cell lines grown in culture, it appears that relatively high levels of intracellular cyclic AMP are associated with inhibiting or preventing the initiation of cell proliferation and promoting differentiation. Transformation of "normal" appearing cells that exhibit "density-dependent inhibition" of growth to cells with uncontrolled growth characteristics appears to be related to a lowering of cellular cyclic AMP concentration.

A simplistic interpretation of the results according to the concept of regulation involving cyclic AMP alone is that elevated levels of cyclic AMP inhibit, and decreased concentrations of cyclic AMP promote or permit, cell proliferation. Within the framework of the Yin Yang Hypothesis the intracellular signal for the initiation of cell proliferation would not be a passive one represented by a lowering of cyclic AMP levels, but an active one, which we believe may be represented by an elevation of the cellular cyclic GMP level.

Effects of Mitogens on Lymphocytes

The observation that first served to link cyclic GMP with the process of cellular proliferation arose from an investigation dealing with the effects of mitogenic agents on the proliferation of human peripheral blood lymphocytes. The lymphocyte is a cell whose normal functions include replication upon exposure to antigen. Certain plant lectins (i.e., phytomitogens) such as phytohemagglutinin and concanavalin A, mimic the effect of antigen; however, unlike antigen they stimulate a larger population of predominantly thymus-dependent lymphocytes to divide.

Previous studies by other investigators (Smith, Steiner and Parker 1971) have demonstrated that cyclic AMP or agents that stimulate its generation, when introduced before or concomitant with phytohemagglutinin, prevent mitogen-induced proliferation. It would appear, therefore, that in lymphocytes, as in other cell lines grown in culture, increases in cellular cyclic AMP concentration can inhibit the induction of cell proliferation.

When we examined the changes that occur in lymphocyte cyclic nucleotide levels following phytohemagglutinin-induced mitogenesis, we found that the concentration of cyclic GMP in these cells increased over tenfold within 20 minutes of stimulation (Fig. 6) (Hadden et al. 1972). The levels of cyclic AMP were increased about 70–80% when a relatively impure preparation of phytohemmagglutinin which exhibited agglutinating activity was used, but were unaltered when a more purified preparation with minimal agglutinating activity was employed. Increases in lymphocyte cyclic GMP concentration of over tenfold with no accompanying alteration in cyclic AMP levels were also found after stimulation with concan-

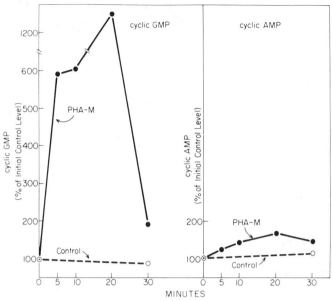

Figure 6
The effect of a partially purified preparation of phytohemagglutinin (PHA-M) on lymphocyte cyclic GMP (*left*) and cyclic AMP (*right*) concentrations. Human peripheral blood lymphocytes (2.5×10^7 cells) were incubated in 1 ml Hank's balanced salt solution with (●——●) or without (○——○) PHA-M (250 μg/ml) for the times indicated before the addition of an equal volume of 20% trichloroacetic acid. Cyclic GMP was measured by the method of Goldberg and O'Toole (1971) and cyclic AMP by the method of Gilman (1970).

avalin A. From these observations it was proposed (Hadden et al. 1972) that, at the time of initiation of proliferation (i.e., during G_0 or early G_1), a marked and temporally discrete increase in cellular cyclic GMP concentration may serve as the active intracellular signal to induce proliferation, while the elevation of cyclic AMP concentration may be viewed as a regulatory influence that limits or inhibits mitogenic action. It was also suggested at that time that the elevation of cellular cyclic GMP levels may be involved in the induction process in systems in which a fall in cellular cyclic AMP has been observed in association with the initiation of proliferation. The hypothesis, as originally stated, did not take into account the possible influences of the two cyclic nucleotides with regard to the division process itself once initiation occurred (i.e., later G_1 through M phases of the cell cycle). Extension of the hypothesis to include all aspects of the process must await information about changes that occur in the levels of both cyclic nucleotides at all stages of the cell cycle.

The results of other investigations conducted with the lymphocyte system may be viewed as supportive of the proposed involvement of cyclic GMP as a positive effector of the proliferative process and also point out the possible modulating role hormonal agents may have. It was pointed out earlier that acetylcholine has been found to stimulate cyclic GMP accumulation in the lymphocyte. The magnitude of the increases induced by the hormone appears to be considerably smaller (ca. 2- to 3-fold) than those associated with mitogenic action (ca. 10-fold). Treatment of lymphocytes with acetylcholine alone has been reported (Hadden et al. 1973) to stimulate the incorporation of labeled precursors into RNA and protein but not into DNA. The neurohormone alone does not, therefore, appear to be mitogenic. On the other hand, phytohemagglutinin-induced incorporation of labeled precursors into RNA, protein and DNA is enhanced in the presence of the cholinergic agent (Hadden et al. 1973). One interpretation of these results is that the small increase induced by acetylcholine in lymphocyte cyclic GMP concentration is not in itself great enough to trigger the initiation of proliferation, but combined with the signal generated by the mitogen the increase in cyclic GMP resulting from the combined action of the two agents, in a certain population of the cells, serves effectively to initiate the process.

Additional evidence favoring the view that cyclic GMP may participate as a regulator of the proliferation process in the manner proposed derives from the observation that cyclic GMP itself can stimulate [3H]uridine incorporation into RNA in isolated lymphocyte nuclei[2] (Fig. 7). Since increased RNA synthesis appears as one of the earliest events following initiation, a stimulatory effect of cyclic GMP on this process would be consistent with the role attributed to it earlier (Hadden et al. 1972) as a "membrane to nuclear signal." Certain features of the cyclic GMP-induced stimulation of [3H]uridine incorporation into RNA are noteworthy. First, the stimulation is detectable with extremely low concentrations of the cyclic nucleotide (10^{-11} M), with a peak effect at 10^{-10} M. Furthermore the stimulation appears to occur only in a discrete concentration range of the cyclic nucleotide (10^{-11} to 10^{-9} M). Cyclic AMP appears to be a poor substitute for cyclic GMP in stimulating the process; the peak affect occurs at a concentration of cyclic AMP (10^{-6} M) that is 10^4-fold greater than found with cyclic GMP. The nuclear RNA induced by cyclic AMP may be of a different type and induced by a mechanism quite different than that induced by cyclic GMP. It should also be pointed out that in the absence of calcium no nuclear RNA synthesis is detectable and cyclic GMP has no influence to promote RNA synthesis. It has recently been suggested (Goldberg, O'Dea and Haddox 1973) that the mechanism underlying the intracellular expression of cyclic GMP action has an obligatory requirement for increased translocation (i.e., from the plasma membrane) or transport of calcium into the cell. This concept is consistent with the fact that each of the cellular processes promoted (smooth muscle contractility, platelet aggregation, cell proliferation, and suppression of cardiac muscle performance) that have been shown to be linked to an accumulation of cellular cyclic GMP all have a recognized dependence on calcium. Also worthwhile noting is that the steroidogenic and secretory system in the adrenal cortex, which we have designated as a "monodirectionally" controlled system and, therefore, influenced in a positive manner by cyclic GMP, also has a requirement for calcium. Furthermore calcium has been shown to be inhibitory to epinephrine- and glucagon-promoted activation of adenylate cyclase (Birnbaumer, Pohl and Rodbell 1969), whereas Schultz et al. (1973) have demonstrated a dependence on calcium for acetylcholine-promoted accumulation of cyclic GMP, and Hardman et al. (1972) have shown that calcium can serve as an activator of guanylate cyclase activity. Of note in this regard is that there appears to be an obligatory requirement for calcium during the induction of proliferation in lymphocytes by phytohemagglutinin and that within 15 minutes of PHA stimulation, lymphocytes show a

[2] These experiments were conducted in collaboration with Dr. Gerald Meetz.

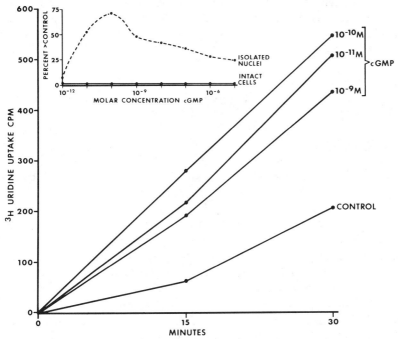

Figure 7

Stimulation by cyclic GMP of RNA synthesis in isolated intact lymphocyte nuclei. Human peripheral blood lymphocytes isolated on ficoll-hypaque gradients were homogenized (10 min) in glass homogenizers with a motor-driven teflon pestle (0°C) in 250 mM sucrose. The nuclei were isolated and incubated according to the procedure of Allfrey et al. (1964). Incubations were carried out with 5×10^6 nuclei per tube (1 ml) with 5 μCi of [^{3}H]uridine. Following incubation for the times indicated (30 min incubation for the concentration curve with cyclic GMP) the nuclei cultures were acidified with 10% trichloroacetic acid. The precipitate obtained upon centrifugation was washed twice with 10% TCA, then 80% ethanol before they were analyzed for [^{3}H]uridine. Under the incubation conditions [^{3}H]uridine incorporation into acid-insoluble material of intact lymphocytes was not stimulated by cyclic GMP. The stimulation of [^{3}H]uridine incorporation into the isolated nuclei was dependent upon their viability as determined by basal uptake ratio and the radioactivity incorporated was digestible by ribonuclease.

progressive increase in ^{45}Ca accumulation (Hadden et al. 1973). These observations indicate that mitogen-induced increases in cyclic GMP may involve a special relationship to calcium influx, and it might be envisioned that calcium influx may perpetuate the influence of the "trigger-type" rise in cyclic GMP seen with the mitogens.

Cyclic GMP Levels during Fibroblast Proliferation

Further support for the proposed involvement of cyclic GMP in proliferation derives from results we have obtained in fibroblast proliferation models (Goldberg et al. 1973b). Insulin has been shown to stimulate cell division (Timen 1967) and is thought to have a common mechanism of action relating to growth factors of the somatomedin type. Insulin action to promote cell division has been shown to be associated with a lowering of the cellular cyclic AMP levels in mouse fibroblasts (Sheppard 1972), and the latter has been viewed (Cuatrecasas and Tell 1973) as the key regulatory event promoted by insulin that leads to the initiation of cell division.

The changes that occur in the concentration of cyclic GMP after insulin treatment of confluent monolayers of mouse 3T3 fibroblasts are shown in Fig. 8. There is a concentration-dependent increase in the levels of this cyclic nucleotide 20 minutes after exposure to the hormone, ranging from about 10- to 40-fold with concentrations of insulin from 0.1 to 1000 munits/ml. With 10 munits/ml an increase of about 10-fold was detected 3 minutes after exposure, and by 20 minutes the concentration of cyclic GMP was 20-fold greater than in the control, confluent monolayers not exposed to insulin.

If the cellular steady-state levels of cyclic GMP represent a major determinant in promoting cell proliferation, it would be expected that the concentration of this cyclic nucleotide would be greater in cells during log phase growth than when they reach confluency and proliferation ceases. To obtain such information studies have been initiated with cultured chick embryo fibroblasts, which under the conditions employed (5% calf serum without change of medium) reach stationary phase of growth at 4 days following inoculation and remain stationary for 2 days before degeneration of the monolayer begins (average viability, 86%). The cellular levels of cyclic GMP the first and second day after inoculation were 0.72 and 0.67 pmoles/μg DNA, respectively. The levels decreased dramatically at day 3, 4 and

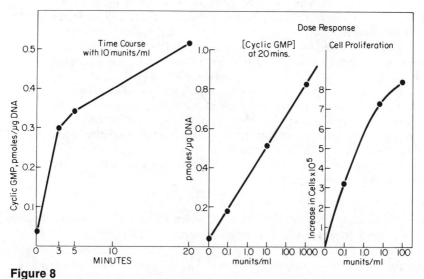

Figure 8
Insulin-induced accumulation of cyclic GMP in mouse 3T3 fibroblasts (Swiss) grown in culture. The cells were grown to confluency (3 days) in 10% calf serum. After addition of insulin cells were washed twice at the times indicated with a balanced salt solution before 10% trichloroacetic acid was added. Cyclic GMP was determined by a modification (Goldberg et al. 1973b) of the radio-immune method of Steiner et al. (1972). Cell number (in triplicate samples) increased 2.1 times in 48 hr when compared to controls.

5 as the cells entered and reached stationary phase to levels of 0.025, 0.09 and 0.035 pmoles/μg DNA, respectively. The decline in cellular cyclic GMP concentration with the cessation of proliferation in these experiments is reminiscent of the decrease observed in *E. coli* upon approaching and reaching stationary growth phase and the reciprocal of the change that occurs in the levels of fibroblast cyclic AMP observed by Otten, Johnson and Pastan (1971) when the cells become confluent.

Other observations reported at this conference by investigators involved in collaborative efforts dealing with the role cyclic GMP may play in the regulation of cell division are consistent with the earlier proposal that it may represent an intracellular effector involved in initiating the process. In the report by Estensen et al. it is demonstrated that phorbol myristate acetate-induced division of mouse 3T3 fibroblasts is associated with an extremely rapid, temporally discrete (within 60 seconds) increase of over tenfold in cyclic GMP concentration, and in the report by Voorhees et al. it is shown that this agent produces a similar effect on the levels of cyclic GMP in mouse epithelium, whereas the increases in cyclic GMP levels found to be induced by histamine in epithelium develop more slowly.

Conclusion

Clearly the information obtained to date represents only the very beginning of the story regarding the possible regulatory role cyclic GMP may play in the control of cell proliferation and other cellular events. The finding of marked increases in cellular cyclic GMP in cells induced to divide and increases that are in general of a lesser magnitude associated with modulating actions of hormones suggests, but in no way proves, that cyclic GMP actually mediates the mitogenic signal or the hormone-induced alterations in cell function. A great deal of further investigation will, of course, be required to establish whether cyclic nucleotides are playing an active regulatory role or are merely reflecting the progress of ongoing events. The ultimate resolution of these questions will rely upon a more complete knowledge of the mechanism of cyclic nucleotide generation, action and interaction, and the relationships that these events may have to others within the cell that are associated with the differentiative and the proliferative processes.

REFERENCES

Allfrey, V. G., V. C. Littan and A. E. Mirsky. 1964. Methods for the purification of thymus nuclei and their application to studies of nuclear protein synthesis. *J. Cell Biol.* **21**:213.

Birnbaumer, L., S. L. Pohl and M. Rodbell. 1969. Adenyl cyclase in fat cells. I. Properties and the effects of adrenocorticotropin and fluoride. *J. Biol. Chem.* **244**:3468.

Colombo, C., L. J. Ignarro and J. J. Chart. 1973. Enzyme release from polymorphonuclear leukocyte lysosomes: Inhibition by catecholamines and cyclic 3′,5′-adenosine monophosphate, acceleration by cholinergic agents and cyclic 3′,5′-guanosine monophosphate. *Fed. Proc.* **32**:291.

Cuatrecasas, P. and G. P. E. Tell. 1973. Insulin-like activity of concanavalin A and wheat germ agglutinin—direct interactions with insulin receptors. *Proc. Nat. Acad. Sci.* **70**:485.

George, W. J., J. B. Polson, A. G. O'Toole and N. D. Goldberg. 1970. Elevation of guanosine 3',5'-cyclic phosphate in rat heart after perfusion with acetylcholine. *Proc. Nat. Acad. Sci.* **66**:398.

Gilman, A. G. 1970. A protein binding assay for adenosine 3',5'-cyclic monophosphate. *Proc. Nat. Acad. Sci.* **67**:305.

Goldberg, N. D. and A. G. O'Toole. 1971. Analysis of cyclic 3',5'-adenosine monophosphate and cyclic 3',5'-guanosine monophosphate. *Methods of Biochemical Analysis,* vol. 10, pp. 1–39. Wiley, New York.

Goldberg, N. D., S. B. Dietz and A. G. O'Toole. 1969. Cyclic guanosine 3',5'-monophosphate in mammalian tissues and urine. *J. Biol. Chem.* **244**:4458.

Goldberg, N. D., R. F. O'Dea and M. K. Haddox. 1973. Cyclic GMP. *Recent advances in cyclic nucleotide research* (ed. P. Greengard and A. G. Robison) vol. 3. Raven Press, New York (in press).

Goldberg, N. D., C. Villar-Palasi, H. Sasko and J. Larner. 1967. Effects of insulin-treatment on muscle 3',5'-cyclic adenylate levels *in vivo* and *in vitro*. *Biochim. Biophys. Acta* **148**:665.

Goldberg, N. D., W. D. Lust, R. F. O'Dea, S. Wei and A. G. O'Toole. 1970. A role of cyclic nucleotides in brain metabolism. *The role of cyclic AMP in neuronal function. Advances in Biochemical Psychopharmacology* (ed. E. Costa and P. Greengard) vol. 3, pp. 67–87. Raven Press, New York.

Goldberg, N. D., M. K. Haddox, D. K. Hartle and J. W. Hadden. 1973a. The biological role of cyclic 3',5'-guanosine monophosphate. *Fifth Int. Congr. Pharmacol.* pp. 146–169. Karger, Basel.

Goldberg, N. D., M. K. Haddox, R. Estensen, J. G. White, C. Lopez and J. W. Hadden. 1973b. Evidence for a dualism between cyclic GMP and cyclic AMP in the regulation of cell proliferation and other cellular processes. *Cyclic AMP in immune response and tumor growth* (ed. L. Lichtenstein and C. Parker) Springer-Verlaag, New York (in press).

Hadden, J. W., E. M. Hadden, M. K. Haddox and N. D. Goldberg. 1972. Guanosine 3',5'-cyclic monophosphate: A possible intracellular mediator of mitogenic influences in lymphocytes. *Proc. Nat. Acad. Sci.* **69**:3024.

Hadden, J. W., E. M. Hadden, G. Meetz, R. A. Good, M. K. Haddox and N. D. Goldberg. 1973. Cyclic GMP in cholinergic and mitogenic modulation of lymphocyte metabolism and proliferation. *Fed. Proc.* **32**:1022.

Hardman, J. G., T. D. Chrisman, J. P. Gray, J. L. Suddath, and E. W. Sutherland. 1972. Guanylate cyclase: Alteration of apparent subcellular distribution and activity by detergents and cations. *Fifth Int. Congr. Pharmacol.* (abstracts) p. 227–228.

Ishikawa, E., S. Ishikawa, J. W. Davis and E. W. Sutherland. 1969. Determination of guanosine 3',5'-monophosphate in tissues and of guanyl cyclase in rat intestine. *J. Biol. Chem.* **244**:6371.

Lee, T. P., J. F. Kuo and P. Greengard. 1972. Role of muscarinic cholinergic receptors in regulation of guanosine 3',5'-cyclic monophosphate content in mammalian brain, heart muscle, and intestinal smooth muscle. *Proc. Nat. Acad. Sci.* **69**:3287.

Makman, R. S. and E. W. Sutherland. 1965. Adenosine 3',5'-phosphate in *Escherichia coli*. *Biol. Chem.* **240**:1309.

Nichols, W. K. and N. D. Goldberg. 1972. The relationship between insulin and apparent glucocorticoid promoted activation of hepatic glycogen synthetase. *Biochim. Biophys. Acta* **279**:245.

Otten, J., G. S. Johnson, and I. Pastan. 1971. Cyclic AMP levels in fibroblasts: Relationship to growth rate and contact inhibition of growth. *Biochem. Biophys. Res. Commun.* **44**:1192.

Pastan, I. and R. Perlman. 1970. Cyclic adenosine monophosphate in bacteria. *Science* **169**:339.

Price, D F.., R. Lipton, M. M. Melicow and T. D. Price. 1963. Isolation of adenosine 3′,5′-monophosphate and guanosine 3′,5′-monophosphate. *Biochem. Biophys. Res. Commun.* **11**:300.

Schultz, G., J. G. Hardman, J. W. Davis, K. Schultz and E. W. Sutherland. 1972a. Determination of cyclic GMP by a new enzymatic method. *Fed. Proc.* **31**:440.

Schultz, G., J. G. Hardman, K. Schultz, C. E. Baird, M. A. Parks, J. W. Davis and E. W. Sutherland. 1972b. Cyclic GMP and cyclic AMP in ductus deferens and sub-maxillary gland of the rat. *Fifth Int. Congr. Pharmacol.* (abstracts) p. 206.

Schultz, G., J. G. Hardman, L. Hurwitz and E. W. Sutherland. 1973. Importance of calcium for the control of cyclic GMP levels. *Fed. Proc.* **32**:773.

Sheppard, J. R. 1972. Difference in the cyclic adenosine 3′,5′-monophosphate levels in normal and transformed cells. *Nature New Biol.* **236**:14.

Smith, J. W., A. L. Steiner and C. W. Parker. 1971. Human lymphocyte metabolism. Effects of cyclic and noncyclic nucleotides on stimulation by phytohemagglutinin. *J. Clin. Invest.* **50**:442.

Steiner, A. L., A. W. Pagliara, L. R. Chase and D. M. Kipnis. 1972. Radioimmunoassay for cyclic nucleotides. II. adenosine 3′,5′-monophosphate and guanosine 3′,5′-mono-phosphate in mammalian tissues and body fluids. *J. Biol. Chem.* **247**:1114.

Strom, T. B., A. Deisseroth, J. Morganroth, C. B. Carpenter and J. P. Merrill. 1972. Alterations of the cytotoxic action of sensitized lymphocytes by cholinergic agents and activators of adenylate cyclase. *Proc. Nat. Acad. Sci.* **69**:2995.

Timen, H. M. 1967. Studies on carcinogenesis by avian sarcoma viruses. IV. Differen-tial multiplication of uninfected and of converted cells in response to insulin. *J. Biol. Chem.* **69**:377.

Triner, L., G. G. Nahas, Y. Vulliemoz, N. I. A. Overweg, M. Verosky, D. V. Habif and S. R. Ngri. 1971. Cyclic AMP and smooth muscle function. *Ann. N. Y. Acad. Sci.* **185**:458.

White, J. G., N. D. Goldberg, R. D. Estensen, M. K. Haddox, and G. H. R. Rao. 1973. Rapid increase in platelet cyclic 3′,5′-guanosine monophosphate (c-GMP) levels in association with irreversible aggregation, degranulation, and secretion. *Amer. Soc. Clin. Invest.* Abstract #329.

Yamashita, K. and J. B. Field. 1972. Elevation of cyclic guanosine 3′,5′-monophosphate levels in dog thyroid slices caused by acetylcholine and sodium fluoride. *J. Biol. Chem.* **247**:7062.

Zurier, R. B., N. Tynan and G. Weissman. 1973. Pharmacologic regulation of lysosomal enzyme release from human leukocytes. *Fed. Proc.* **32**:744.

Phorbol Myristate Acetate: Effects of a Tumor Promoter on Intracellular Cyclic GMP in Mouse Fibroblasts and as a Mitogen on Human Lymphocytes

Richard D. Estensen, John W. Hadden, Elba M. Hadden, Francoise Touraine, Jean-Louis Touraine, Mari K. Haddox, and Nelson D. Goldberg

Departments of Pharmacology and Pathology
University of Minnesota Medical School, Minneapolis, Minnesota 55455

Phorbol myristate acetate (PMA) is one of several biologically active substances that can be isolated from croton oil. Interest in croton oil or in PMA stems from their action as promoters in the two-stage carcinogenesis model first described by Berenblum (1941). The essential features of two-stage carcinogenesis are as follows: (1) One application of an initiator (a carcinogen of some type) alone to mouse skin will produce few or no tumors. (2) Many applications of a promoting substance alone will produce inflammation followed by epithelial hyperplasia but rarely produce tumors. (3) One application of an initiator followed by many applications of a promoter produces the appearance of both benign and malignant epithelial tumors (Van Duuren 1969; Hecker 1968).

Several observations have indicated that some of the effects of PMA may be mediated by interaction with cell membranes. Sivak and Van Duuren (1971) observed that tritium-labeled PMA could be localized in an ATPase-containing fraction of cell homogenates and subsequently demonstrated (Sivak, Mossman and Van Duuren 1972) that the addition of PMA to isolated plasma membrane fractions enhanced the activity of both ATPase and 5'-nucleotidase. PMA added to isolated liver cell membranes at concentrations of 2 ng/ml will cause a 10–20% decrease in intrinsic fluorescence of these membranes (Estensen and Sonnenberg unpubl.). A decrease of a similar magnitude has also been observed when human red blood cell membranes are exposed to human growth hormone and can be interpreted as a change in conformation of some membrane proteins (Sonnenberg 1971).

Recent evidence indicates that membrane interactions may be sufficient to produce a mitogenic response since Greaves and Bauminger (1972) have demonstrated that PHA does not have to enter lymphocytes in order to induce proliferation. Hadden and coworkers (1972) have suggested that a mitogenic signal may be produced at the membrane when lectins interact with lymphocytes to produce both early and striking increases in cellular cyclic 3',5'-guanosine monophosphate (cyclic GMP). They also suggested that similar increases in cellular cyclic GMP may also serve to signal the initiation of cell division in other types of cells. This proposal led to the investigation of PMA effects in two models of proliferation.

RESULTS AND DISCUSSION

PMA Effects on Mouse Fibroblasts

Sivak (1972) has demonstrated that PMA can produce an increase in confluent density of a wide variety of cell types including Balb/c 3T3, Swiss 3T3, SV101 FL, and secondary mouse embryo fibroblasts. The cell chosen for our experiments was a Balb/c 3T3 originally supplied to us by Dr. George Todaro. The usual density of these cells at confluency in 10% serum was 30,000 cells per cm^2. Our experiments were performed on monolayers grown to confluency in 3 days. Plates were exposed to 0.0025% dimethylsulfoxide (DMSO) or to the same volume of DMSO containing sufficient PMA to result in concentrations of 100 ng/ml in the medium. Exposure to PMA uniformly resulted in an increase in cell density, whereas DMSO control plates showed no change. Changes in appearance of cells treated with PMA resembled those observed by Sivak, Ray and Van Duuren (1969) in that cells seemed to lose some contact with neighboring cells and to have refractile borders by phase contrast microscopy.

In our first experiments we measured cyclic GMP after exposure of confluent monolayers to PMA after 30 seconds, 3, 5, 10 and 20 minutes. Only at the earliest time period after exposure to PMA was an change (a tenfold increase) detectable in intracellular cyclic GMP concentration. DMSO-treated plates demonstrated no changes in cyclic GMP levels at any time. This finding led us to investigate the effects of PMA at much earlier time periods: 15, 30, 45 and 60 seconds. PMA causes a dramatic increase of intracellular cyclic GMP concentrations within seconds and a greater than 20-fold increase at 45 seconds (Fig. 1). At 60 seconds cyclic GMP levels have begun to decrease. Cyclic AMP levels did not change in this time period (not shown). Confluent Balb/c 3T3 cells exposed to DMSO or to PMA (100 ng/ml) at the same time that other plates were sampled for cyclic GMP were observed for 48 hours and then counted. DMSO-exposed cells remained at the confluent density (30,000 cells/cm^2) seen at the beginning of the experiment, and those exposed to PMA doubled their density in 48 hours (60,000 cells/cm^2). Two conclusions may be drawn from these data: (1) that PMA can raise intracellular cyclic GMP markedly; (2) that the PMA-induced elevation in cyclic GMP level occurs at a much earlier time than has been observed for other mitogenic agents (Goldberg et al. this volume). The reason for the latter finding is not known but could conceivably represent a much more rapid association rate with PMA than with other mitogens. In addition these observations are consistent with the hypothesis (Hadden et al. 1972) that increases in intracellular levels of cyclic GMP are among the earliest changes associated with initiation of cell division and therefore may serve as an important part of the mitogenic signal.

PMA Effects on Human Lymphocytes

During the progress of this work we became aware that PMA also acted on another type of cell, the platelet, and could affect another definable cellular process, platelet aggregation. Zucker and coworkers (1972) had reported that nanogram quantities

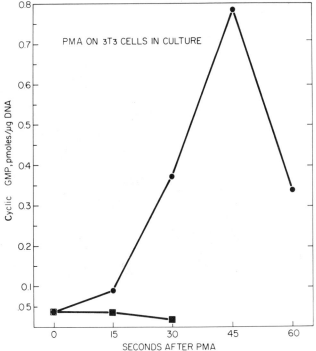

Figure 1
Elevation of cyclic GMP in 3T3 cells after exposure to PMA. All experiments
were carried out on Balb/c 3T3 cells grown in Dulbecco's modified Eagle's
medium with 10% calf serum. The analysis of cyclic GMP was carried out by
radioimmunoassay after partial purification of cyclic GMP (Goldberg et al.
1973). (■——■) Cells exposed to 0.0025% DMSO (volume, final concentration) alone. (●– – –●) Cells exposed to 0.0025% DMSO (volume) and 100
ng/ml PMA (final concentration).

of PMA caused platelet aggregation. We examined the possibility that this PMA-
induced response might be associated with an increase in intracellular cyclic GMP,
and we found (Goldberg et al. 1973) that within 15 seconds of exposure to 100
ng/ml of PMA there was a twofold increase in cyclic GMP levels and by 30
seconds the level was fourfold that of controls. By 60 seconds the platelet concentration of cyclic GMP began to decrease. These observations have led us to believe
that PMA may raise intracellular cyclic GMP in association with promoting a
number of different cellular responses in a wide variety of cell types and therefore
may serve as a useful probe for cyclic GMP-related actions.

If this were true, then PMA might be expected to be a mitogenic stimulus in cell
types other than mouse fibroblasts. The system that we chose to examine was the
human peripheral blood lymphocyte. This cell or, more properly, population of cell
types undergoes proliferation as part of its biological response. In our initial experiments we examined the effects of PMA on lymphocyte "transformation" by examining the morphologic and biochemical responses of human lymphocytes. In
each case we also examined the response of the same cells to PHA, which served as
a positive control. In Fig. 2 human lymphocytes exposed to no mitogen (2a), PHA

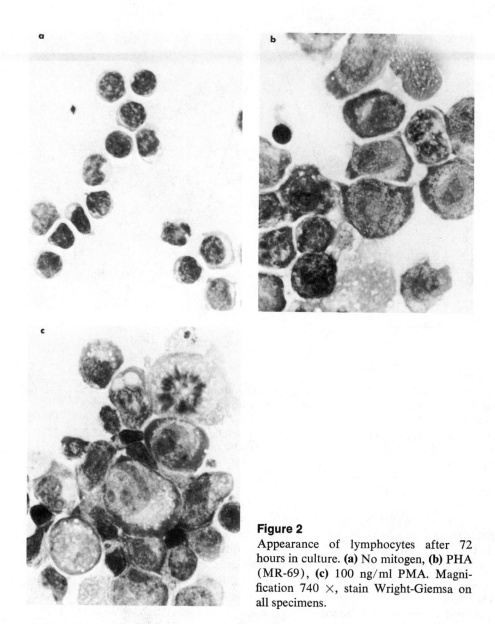

Figure 2
Appearance of lymphocytes after 72 hours in culture. **(a)** No mitogen, **(b)** PHA (MR-69), **(c)** 100 ng/ml PMA. Magnification 740 ×, stain Wright-Giemsa on all specimens.

(2b), and PMA (2c) are seen as they appear after 72 hours in culture. The cells not exposed to mitogen have the typical appearance of small, untransformed lymphocytes. The cells exposed to PHA show the changes that accompany blast transformation. The change induced in PMA-treated cultures resembles in some respects the changes that occur in PHA-treated cultures. Nonetheless there are some differences that can be observed. Mitosis is a less frequent event in PMA-exposed lymphocytes than in PHA-treated lymphocytes at 72 hours (not shown). The cytoplasm in PMA-treated cells is vacuolated and there seems to be more cytoplasm per cell than in cells exposed to PHA. These morphologic changes in response to

PMA were accompanied by increases of the order of about 100-fold over control in the uptake of [³H]thymidine into acid-insoluble material. This increase in [³H]thymidine uptake at 72 hours varies somewhat from donor to donor, but is in general about 20% less than that observed in response to PHA.

When PHA and PMA were added to lymphocytes simultaneously, the resultant [³H]thymidine incorporation was slightly greater than the sum of the uptake in response to each agent separately. The response of lymphocyte [³H]thymidine uptake at 72 hours to varying concentrations of PMA is seen in Fig. 3. The range of PMA concentrations found to be effective is from about 10 ng/ml to 100 ng/ml. This range is 5 to 10 times that needed to stimulate density-inhibited Balb/c 3T3 cells to proliferate (Sivak 1972).

Since Peters and Hausen (1971) had shown that PHA-P caused a marked increase in the rate of facilitated diffusion of glucose in bovine lymphocytes, we examined the possibility that PMA might cause a similar effect on this process. Addition of PMA caused a marked increase in uptake of 2-deoxyglucose in lymphocytes within 5 minutes after its addition to cultured lymphocytes (Fig. 4). When the results with PMA are compared to the effect of PHA-P on the same process, it was found that both agents produced about a twofold increase within 5 minutes in the rate of 2-deoxyglucose uptake.

PMA, then, resembles other lymphocyte mitogens in producing morphologic transformation, increasing the incorporation of [³H]thymidine into acid-insoluble material, and increasing the rate of 2-deoxyglucose uptake. Nonetheless the fact that when added to the same culture at the same time a summation of PHA and PMA effect on the [³H]thymidine uptake was observed indicated that these agents may be acting on a different cell population. In order to test this hypothesis the technique adapted by Zoschke and Bach (1971) of eliminating cells that respond to a mitogenic stimulus was employed. Our experiment was designed as follows:

1. Lymphocytes were stimulated by a mitogen or left unstimulated.
2. All cultures were then exposed to 1 mM bromodeoxyuridine for 22 hours followed by a two-hour exposure to light to eliminate dividing cells.

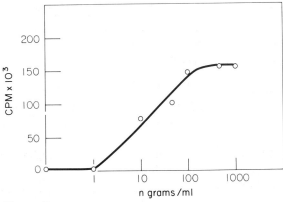

Figure 3
Relationship of [³H]thymidine uptake by human lymphocytes to dose of PMA. Uptake into acid-insoluble fraction determined as described previously (Hadden et al. 1972).

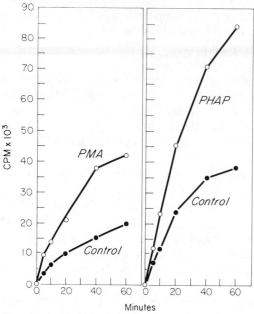

Figure 4

Effect of PMA and PHA-P on uptake of 2-deoxyglucose. Purified human lymphocytes (see Hadden et al. 1972) were washed in glucose-free media and resuspended at 2×10^6 cells/ml in media containing 100 μM [³H]2-deoxyglucose (1 μCi). Lymphocytes were exposed to 100 ng/ml PMA or to ~ 25 μg/ml PHA-P (maximal mitogenic doses of each agent). At the times indicated cells were centrifuged, washed with balanced salt solution, treated with an equal volume of cold 10% TCA and the total (acid soluble and insoluble) radioactivity determined.

3. Subsequently the bromodeoxyuridine was removed and the cells washed twice and resuspended in fresh medium.

4. The remaining cells were left unstimulated or were stimulated by a mitogen and the [³H]thymidine uptake was measured 144 hours after the start of the experiment (72 hours after the second stimulatory period).

The results of these experiments are seen in Table 1. Of the cells that responded to PMA about 95% of the response could be eliminated by bromodeoxyuridine treatment. When these cells were exposed to PHA the response was almost that of the control, which had no mitogen stimulation for the first 72 hours and was then exposed to PHA. Although the PHA-responding cell could not be as effectively suppressed, the results indicate that PMA also acted to enhance the uptake of [³H]thymidine over that seen with PHA alone. It is important to note that addition of PHA following initial stimulation with PHA failed to increase the cells response at 144 hours. The results support the interpretation that PHA and PMA act on different subpopulations of lymphocytes. Measurements of PMA effect on cyclic GMP levels are underway.

Table 1

Effect of PMA and PHA on Lymphocytes

Initial treatment	Second treatment	cpm/culture (in thousands)
PHA	0	65
None	PHA	178
PHA	PHA	55
PMA	0	5
None	PMA	98
PMA	PHA	158
PHA	PMA	130

Human lymphocytes 4×10^5 were exposed to mitogen (or no mitogen), then 1 mM bromodeoxyuridine and light, then to a second application of mitogen (or no mitogen). Lymphocytes were exposed to [³H]thymidine for 24 hr (120–144 hr) washed, and processed as indicated previously (Hadden et al. 1972) for acid-insoluble radioactivity. Maximal mitogenic doses were used for PHA (MR-69) and PMA.

CONCLUSION

The results of our experiments have demonstrated that PMA causes a rapid and striking increase in intracellular cyclic GMP in Balb/c 3T3 cells and under our experimental conditions also produces a twofold increase over control in cell number. Unlike other agents that increase cyclic GMP concentration in fibroblasts (Goldberg et al. 1973), PMA produces an increase in cyclic GMP that is maximal within the first minute and has returned to control levels by 3 minutes. This very rapid rise and decay in cyclic GMP concentrations may be the result of differences in mechanisms of action of these compounds or different rates at which the same process may be affected. This difference also serves to indicate that measurement of cyclic GMP levels at a few points may not be sufficient to demonstrate very rapid changes.

In addition to its mitogenic action on fibroblasts PMA is also a potent mitogen for lymphocytes. PMA apparently acts on a different population of cells than PHA. Vaheri and coworkers (this volume) have conversely demonstrated B lymphocyte mitogens are also effective agents for induction of cell proliferation in fibroblasts. This similarity of action in different cell types suggests also that meaningful comparisons may be made between the mechanism of mitogenic action in both lymphocytes and fibroblasts. In these cell types measurements of cyclic GMP have indicated that an increase in concentration of this molecule is one of the earliest responses to a mitogen.

PMA can also be shown to increase cyclic GMP levels in the platelet and induce platelet aggregation. In this respect it resembles other platelet aggregating agents that also increase cyclic GMP (Goldberg et al. 1973). This finding suggests that many of the effects of PMA may be mediated by the action to increase intracellular cyclic GMP. While increases in cyclic GMP levels in lymphocytes have not yet been measured after PMA treatment, we would speculate that PMA should produce a rapid increase. PMA action, then, seems to be associated with its ability to increase intracellular cyclic GMP levels. In this regard the agent has been shown to enhance polymorphonuclear leukocyte chemotaxis. Other agents that can be shown to increase intracellular cyclic GMP in other systems can also produce enhanced chemotaxis (Estensen, Hill, Goldberg and Quie, unpubl.). PMA effects on a wide variety of cell types and function make this agent useful as an indicator or probe of those cell functions that may be modified by changing cyclic GMP concentrations.

Acknowledgments

This work has been supported in part by USPHS Grant CA12607-02 and in part by funds of the American Heart Association in conjunction with the Minnesota Heart Association. J. W. H. is an established investigator of the American Heart Association. R.D.E. is an Elsa A. Pardee investigator. The authors wish to thank Dr. Andrew Sivak for the generous gift of phorbol myristate acetate.

REFERENCES

Berenblum, I. 1941. The cocarcinogenic action of croton resin. *Cancer Res.* **1**:44.

Goldberg, N. D., M. K. Haddox, R. Estensen, C. Lopez and J. W. Hadden. 1973. Evidence for a dualism between cyclic GMP and cyclic AMP in the regulation of cell proliferation and other cellular processes. *Cyclic AMP in immune response and tumor growth* (ed. L. Lichtenstein and C. Parker). Springer-Verlag, New York. (in press).

Greaves, M. F. and S. Bauminger. 1972. Activation of T and B lymphocytes by insoluble phytomitogens. *Nature New Biol.* **235**:67.

Hadden, J. W., E. M. Hadden, M. K. Haddox and N. D. Goldberg. 1972. Guanosine 3':5'-cyclic monophosphate: A possible intracellular mediator of mitogenic influences in lymphocytes. *Proc. Nat. Acad. Sci.* **69**:3024.

Hecker, E. 1968. Cocarcinogenic principles from the seed oil of *Croton tiglium* and from other Euphorbiaceae. *Cancer Res.* **28**:2338.

Peters, J. H. and P. Hausen. 1971. Effect of phytohemagglutinin on lymphocyte membrane transport. II. Stimulation of facilitated diffusion of 3-O-methylglucose. *Eur. J. Biochem.* **19**:509.

Sivak, A. 1972. Induction of cell division: Role of cell membrane sites. *J. Cell. Physiol.* **80**:167.

Sivak, A. and B. L. Van Duuren. 1971. Cellular interaction of phorbol myristate acetate in tumor promotion. *Chem-Biol. Interaction* **3**:401.

Sivak, A., B. T. Mossman and B. L. Van Duuren. 1972. Activation of cell membrane enzymes in the stimulation of cell division. *Biochem. Biophys. Res. Commun.* **46**:605.

Sivak, A., F. Ray and B. Van Duuren. 1969. Phorbol ester tumor-promoting agents and membrane stability. *Cancer Res.* **29**:624.

Sonnenberg, M. 1971. Interaction of human growth hormone and human erythrocyte membranes: Studies by intrinsic fluorescence. *Proc. Nat. Acad. Sci.* **68**:1051.

Van Duuren, B. L. 1969. Tumor-promoting agents in two-stage carcinogenesis. *Progr. Exp. Tumor Res.* **11**:31. Karger, Basel/New York.

Zoschke, D. C. and F. H. Bach. 1971. Specificity of allogenic cell recognition by human lymphocytes in vitro. *Science* **172**:1350.

Zucker, M. B., S. J. Kim, S. Belman and W. Troll. 1972. Phorbol ester—A potent new platelet aggregating agent. *Third Congr. Int. Soc. Thrombosis Hemostasis*. Washington, D.C.

Imbalanced Cyclic AMP and Cyclic GMP Levels in the Rapidly Dividing, Incompletely Differentiated Epidermis of Psoriasis

John J. Voorhees, Nancy H. Colburn, Marek Stawiski, and Elizabeth A. Duell

Department of Dermatology
University of Michigan Medical School, Ann Arbor, Michigan, 48104

Mari Haddox and Nelson D. Goldberg

Department of Pharmacology
University of Minnesota Medical School, Minneapolis, Minnesota, 55414

Common proliferative skin diseases appear to be the result of abnormally modulated homeostasis in epithelium (epidermis) of skin. We have used the genetic skin disease psoriasis, which is prototypic of these disorders, as an "experiment of nature" in an attempt to understand the control of the normal proliferation-differentiation relationship (keratinization) and the perturbed metabolic regulation in proliferative skin diseases. Psoriasis is characterized by glycogen accumulation, excessive proliferation and altered differentiation (specialization) in involved epidermis (IE) versus uninvolved epidermis (UE) (Halprin and Taylor 1971). A second approach to an understanding of the molecular basis of epidermal proliferation has been in vivo and in vitro use of the potent tumor promoter and epidermal mitogen, tetradecanoyl-phorbol-acetate (TPA) (Hecker 1971). TPA produces cellular hyperplasia and altered specialization of epidermis, but glycogen content has not been examined (Raick 1973).

Our studies are based on earlier observations of others. Bullough and Laurence (1961) showed that epinephrine inhibited the G_2 phase of the epidermal cell cycle in the mouse. Epinephrine is known to raise adenosine 3′,5′-monophosphate (cyclic AMP) levels in many tissues (Robison, Butcher and Sutherland 1968). Ryan and Heidrick (1968) found that cyclic AMP inhibited growth of tumor cells in vitro and Bürk (1968) showed that cells transformed with polyoma virus have decreased adenylate cyclase activity. Bürk suggested that this might lead to a lower cellular cyclic AMP level and decreased specialization of cellular function. Epinephrine is known to promote glycogen breakdown by stimulating cyclic AMP synthesis (Robison, Butcher and Sutherland 1968).

These facts led us to the hypothesis that perhaps a certain "normal" level of cyclic AMP must be maintained in epidermis by epinephrine, or an as yet unidentified hormone, in order to maintain a normal cell cycle rate and glycogen content (Voorhees and Duell 1971). A corollary of this proposition and the above facts is that "below normal" epidermal cyclic AMP levels might permit excessive cellular proliferation and glycogen accumulation and prohibit normal epidermal specialization (keratinization). This notion is not entirely original as Iversen (1969) had suggested a possible role for cyclic AMP in epidermal differentiation, but did not consider glycogen accumulation. In the fall of 1970 experiments were initiated to test our hypothesis that cyclic AMP is one key mediator of epidermal homeostasis (the maintenance of a normal proliferation-specialization relationship and normal glycogen content). The purpose of this communication is to review these experiments and present a revised hypothesis of epidermal growth control, which derives from recent determinations of cellular guanosine 3′,5′-monophosphate (cyclic GMP) levels in psoriasis and TPA- and histamine-treated mouse epidermis.

METHODS

Cell Cycle Assay

The G_2 phase of the mouse epidermal cell cycle is approximately 4.8 hours. A description of in vitro assay of $G_2 \rightarrow M$ flow has been previously published (Voorhees et al. 1973).

Epidermal Procurement

All epidermal samples are obtained by keratoming (Voorhees et al 1972a). The purity of epidermis is monitored by paraffin or frozen section histology.

Incubation of Epidermal Slices

Keratome strips of mouse epidermis were sliced and preincubated for 20 minutes at 37°C in Krebs-Ringer phosphate buffer before drugs were added. Incubations were stopped by dropping the tissue into liquid nitrogen.

Cyclic AMP Assays

Early work utilized the assay of Brooker, Thomas and Appleman (1968). The assay of Gilman (1970) is currently used. The trichloroacetic acid supernatant fractions containing cyclic AMP are partially purified by anion exchange column chromatography. Cyclic AMP phosphodiesterase digestion results in the disappearance of 88% of what is assayed as cyclic AMP in the assay of Brooker, Thomas and Appleman and 95% in the Gilman assay.

Cyclic AMP Phosphodiesterase Assay

This enzyme was assayed using the method of d'Armiento, Johnson and Pastan (1972) except that the buffer pH was 7.5. The enzyme was assayed in the whole homogenate, the 17,000g supernatant or precipitate obtained from epidermis of rat or man.

Cyclic GMP Analysis

Cyclic GMP assays were performed in the laboratory of Professor Nelson Goldberg by the fluorometric enzymic cycling technique of Goldberg and O'Toole (1971) for psoriasis specimens and by a modification (Goldberg et al. 1973) of the radioimmunoassay of Steiner, Parker and Kipnis (1972) for mouse epidermal slices.

Statistical Design

Computations and statistical analyses were performed on the IBM 360/67 computer. Interpretation of statistical significance was done as previously described (Voorhees et al. 1972a).

RESULTS AND DISCUSSION

Criteria of a Cyclic AMP-mediated Event

Our data show that β-adrenergic agonists slow the flow of cells through the $G_2 \to M$ phase of the cell cycle via a rise in cyclic AMP. The four criteria of Robison, Butcher and Sutherland (1971) that permit such a statement have within the limits of our experimental system been fulfilled (Voorhees et al. 1973). Therefore we will summarize our findings as follows:

1. The stimulation of adenylate cyclase in broken cell preparations of epidermis by the β-adrenergic agonist, isoproterenol (IPR), is blocked by the β-antagonist, propranolol (Duell et al. 1971).
2. IPR stimulation of cyclic AMP synthesis in intact epidermis and inhibition of mitosis within 5 hours are both blocked by propranolol (Voorhees et al. 1972b).
3. Theophylline inhibits epidermal cyclic AMP phosphodiesterase and mitosis. IPR inhibition of epidermal mitosis is potentiated by theophylline (Voorhees et al. 1972b).
4. Exogenously added dibutyryl cyclic AMP inhibits epidermal mitosis, thus mimicking an IPR-induced endogenous rise in intraepidermal cyclic AMP. Adenosine 5'-monophosphate and sodium butyrate do not inhibit epidermal mitosis at molar concentrations at which dibutyryl cyclic AMP does inhibit mitoses (Voorhees, Duell and Kelsey 1972).

Marks and Rebien (1972) have confirmed certain of these four criteria.

The physiological significance of a cyclic AMP-induced $G_2 \to M$ delay is unclear. Mueller (1971) has suggested that certain G_2 events may play a role in the recognition of the chromatin segments which are to become active or inactive in the subsequent G_1 period. Therefore the G_2 phase of the cell cycle might be involved in the stabilization of phenotype and the initiation of differentiative changes in a cell lineage. Perhaps the cyclic AMP concentration in G_2 may be one of the controls involved in these events. Nonetheless it seems likely that the control of cell proliferation must lie prior to DNA synthesis in the G_1 phase of the cell cycle (Prescott 1973). TPA apparently stimulates the $G_1 \to S$ flow of the epidermal cell cycle (Baird, Sedgwick and Boutwell 1971). We have performed several experiments designed to detect a $G_1 \to S$ delay in the TPA-stimulated epidermis after mice were given doses of IPR intraperitoneally which raised intraepidermal cyclic AMP levels six- to twelvefold. These experiments have yielded suggestive but no conclusive evidence of a $G_1 \to S$ delay. Since a variety of factors may be influencing these results, similar experiments using epidermal cells in culture and radioautography are underway.

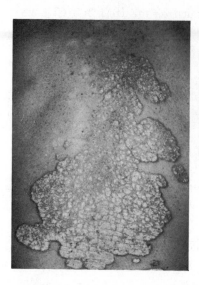

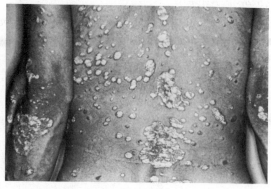

Figure 1
Back and arms of two different patients having psoriasis of moderate severity. Note the sharp margins of the psoriasis lesions. The white and gray areas in the lesions are due to variable loss of dead white scale.

Psoriasis—General Comments

A patient having psoriasis of moderate severity is shown in Fig. 1. The global prevalence ranges from 1 to 3% without preference for either sex (Voorhees et al. 1973). Several clinical and histologic features are especially pertinent to our approach to the molecular basis of the epidermal abnormalities of this disease: (1) epidermal hyperplasia; (2) altered differentiation; (3) genetic (possibly multifactorial); (4) lesions are sporadic and disappear spontaneously only to reappear in the same spot or elsewhere later; (5) lesions can be produced by damage to skin.

Figure 2 shows the light microscopic difference between hyperplastic psoriasis epidermis and normal epidermis. The altered differentiation is also apparent. Whether the defective differentiation is primary or secondary to the rapid cellular

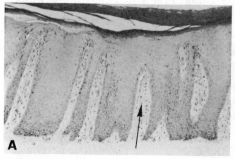

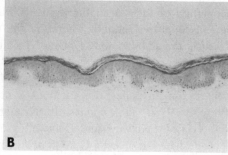

Figure 2
Light microscopy of a psoriasis lesion **(A)** such as shown in Fig. 1. Note the epidermal hyperplasia and altered tissue specialization as compared to uninvolved epidermis **(B)** which is analogous to the normal appearing areas of the patients in Fig. 1. The arrow in *A* indicates the extent of mesenchymatous contamination in a keratome biopsy. However inspection indicates that approximately 75% of the involved tissue biopsy is epidermal in origin. Stained with hematoxylin and eosin; magnification 32.8 ×.

proliferation is unknown and no a priori evidence exists to favor one possibility over the other. The changes in cell specialization in the lesion can be summarized as follows: (1) reduced number of visible tonofilaments; (2) variable amount of visible keratohyalin; (3) possibly reduced histidine-rich protein; (4) markedly decreased periodic acid-silver stainable cell surface glycoproteins; (5) increased Sudan black B stainable lipid content; (6) increased glycogen content; (7) variable nuclear retention in stratum corneum (Halprin and Taylor 1971).

The apparently paradoxical relationship between rapid epidermal proliferation and glycogen accumulation is not unique to psoriasis since it is seen in several epidermal proliferative diseases. Prominent among these diseases is ordinary eczema, in which glycogen storage is seen in the epidermis (Voorhees et al. 1973). Although in general glycogen accumulation is seen not in the proliferative compartment but in the specializing compartment, its presence is still paradoxical because as Halprin and Taylor (1971) point out, the specializing compartment in psoriasis epidermis is metabolically very active. Perhaps glycogen must be utilized for normal keratinization to occur and to provide glucose residues for conversion to the sugar moieties of cell surface glycoproteins, which Mercer and Maibach (1968) found reduced in psoriasis epidermis.

Figure 3A shows the markedly increased labeling index of psoriasis epidermis

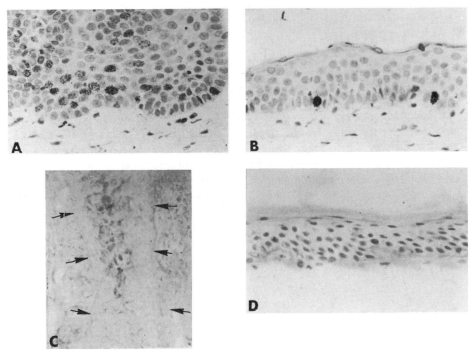

Figure 3
Increased labeling index in involved psoriasis epidermis as determined by incubating human biopsies in vitro with tritiated thymidine **(A)**. Uninvolved psoriasis epidermis has sparse labeling **(B)**. In **C,** PAS-stainable, amylase-hydrolyzable material (glycogen) can be seen in the central portion of a hyperplastic epidermal peg (outlined by arrows). **(D)** PAS stain of uninvolved epidermis. 150 ×

in comparison with uninvolved areas (Fig. 3B). The increased glycogen content can be seen in Fig. 3C, which represents a fivefold increase in comparison with uninvolved areas (Fig. 3D) when measured biochemically (Halprin, Ohkawara and Levine 1973). The increased labeling index (Fig. 3A) is a reflection of what may be approximately a twelvefold shortened cell cycle (Weinstein and Frost 1971). This interpretation has been questioned by Flaxman (1972), who feels that the increased labeling index may be due mainly to differences in the size of the proliferating population, rather than to differences in length of the cell cycle. However the recent work of Hegazy and Fowler (1973), showing that all germinative cells in the unplucked and plucked mouse epidermis are labeled, suggests that the cell cycle in involved psoriasis epidermis is probably shortened.

Psoriasis and Cyclic AMP

Figure 4 presents our original hypothesis for the molecular basis of the epidermal abnormalities of psoriasis. This model was developed from our experiments, which demonstrated that β-adrenergic agonists could inhibit cell division via cyclic AMP and the assumption that cyclic AMP could control epidermal glycogen metabolism as had been shown to occur in liver and muscle (Robison, Butcher and Sutherland 1968). Recent evidence in support of this assumption has been obtained by Halprin and associates (personal communication 1973), who, based on preliminary observations, show that epinephrine produces a dramatic decrease in the glycogen content of both involved and uninvolved psoriasis epidermis. However the decrease is greater in UE than IE.

To test the hypothesis shown in Fig. 4 we measured cyclic AMP levels in IE and UE epidermis of 25 patients by the method of Brooker, Thomas and Appleman (1968) and in another 25 patients by the method of Gilman (1970). Since the IE and UE are dissimilar in many respects, we expressed our data in terms of picomoles of cyclic AMP per mg wet weight, μg DNA and mg protein. Based on DNA, the most reliable denominator, the decrease in cyclic AMP levels in IE was 43% ($p < 0.001$) by the Brooker assay and 36% ($p < 0.001$) by the Gilman assay (Voorhees et al. 1973). We consider DNA to be the most reliable denominator since the DNA concentration per cell should be relatively constant. Also in 50

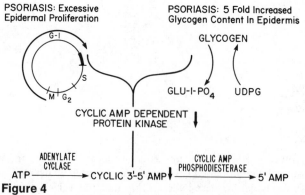

Figure 4

Hypothesis for the role of cyclic AMP in the aberrant regulation of cell cycle control and glycogen metabolism in psoriasis epidermis.

patients the amount of DNA extracted was only 6.7% greater in IE than UE. These data derived by two different assays in rigorous histologically controlled specimens from a total of 50 patients are compatible with the hypothesis depicted in Fig. 4.

Cyclic AMP levels in epidermis from 25 normal subjects, matched for sex but not age, were compared with the levels in uninvolved epidermis of the 25 psoriasis patients by the Brooker assay. An apparent but statistically insignificant ($0.20 > p > .10$ based on DNA) increase was seen in the epidermal cyclic AMP levels of uninvolved areas of psoriasis patients in comparison with normal volunteers. A similar comparison is currently being conducted using the Gilman assay. A significant increase in cyclic AMP concentration of uninvolved epidermis of psoriasis patients in comparison with normal volunteers might provide a valuable clue to the pathogenesis of the rapid epidermal kinetics of psoriasis.

We then attempted to answer the question as to why there is a decrease in cyclic AMP levels in IE versus UE. We considered three possible explanations: (1) cyclic AMP is leaking out of defective IE membranes; (2) cyclic AMP biosynthetic capacity of IE is defective; (3) catalysis of cyclic AMP by cyclic AMP phosphodiesterase in IE is increased. Preliminary experiments to test for a difference in the rate of diffusion of cyclic AMP from the tissue into the medium have been performed on IE and UE from five psoriasis patients. After incubation of the tissue with IPR (1×10^{-6} M) for 8 minutes, there was no significant difference in the cyclic AMP present in the medium or in the tissue when comparing IE with UE.

Psoriasis and Adenylate Cyclase

To test the possibility of defective cyclic AMP biosynthetic capacity, the ideal approach would be to prepare a membrane fraction from IE and compare its ability to convert radioactive ATP to cyclic AMP with that of UE. However the ratio of healthy to dying membranes (plasma membrane of cornifying cells) and the susceptibility to damage by homogenization may vary when IE is compared with UE. Thus meaningful interpretation of the data would be very difficult. The adenine prelabeling technique is useful in many systems but seems inappropriate in this situation since the ATP pools may not be labeled to the same extent in the IE and UE samples. Therefore we elected to examine adenylate cyclase activity indirectly. Cyclic AMP formation in epidermal slices incubated in a theophylline-free medium containing 1×10^{-6} M IPR was very similar in IE and UE from 23 psoriasis patients. These data definitely do not exclude but argue against a gross defect in cyclic AMP synthesis as the cause of decreased cyclic AMP in IE (Voorhees et al. 1973). In contrast Wright et al. (1973), using a broken cell preparation without an apparent ATP regenerating system or the adenine prelabeling technique, have reported a substantial deficiency in adenylate cyclase activity in IE versus UE. The discrepancy between these data and ours may be the result of differences in the methods used and the assumptions inherent in the different assay methods for measuring the ability of epidermis to generate cyclic AMP.

Psoriasis and Cyclic AMP Phosphodiesterase

Initially the high K_m enzyme has been assayed in both IE and UE from 18 psoriasis patients with no detectable difference in enzymatic activity. Recently the low K_m enzyme has been assayed in both particulate and soluble fractions of IE and

UE from six psoriasis patients. The preliminary results shown in Table 1 indicate that the soluble enzyme present in the psoriasis lesion may have an increased V_{max} and therefore might account for the decreased levels of cyclic AMP present in IE versus UE. The increased activity may be due to induction of a new enzyme (Weiss, Shein and Uzunov 1973), loss of an inhibitor, or stimulation of the enzyme by cyclic GMP (Beavo, Hardman and Sutherland 1971). There appears to be no difference in the V_{max} of the particulate enzyme in IE versus UE, but there may be a difference if the percent of viable membranes relative to total protein in the two fractions is not the same.

Psoriasis and Cyclic GMP

Prior to the preliminary low K_m phosphodiesterase studies just mentioned, we were unable to explain the decreased cyclic AMP in IE. In the spring of 1972 we met Professor Goldberg, who told us of the evidence he was accumulating in support of the concept that cyclic GMP may be involved in promoting cellular events that are antagonistic to those promoted when tissue levels of cyclic AMP are elevated. Of particular pertinence were the results of cell transformation experiments he had conducted in collaboration with Dr. John Hadden at the University of Minnesota. Hadden and Goldberg had found striking increases in the cellular cyclic GMP concentrations associated with mitogen-induced lymphocyte transformation without changes in cyclic AMP levels (Hadden et al. 1972). However increased cyclic GMP levels have been associated with decreased cyclic AMP levels in the heart (George et al. 1970). It seemed possible that a chronically proliferating psoriasis epidermis might also have elevated cyclic GMP levels, which could decrease the cyclic AMP levels by stimulation of cyclic AMP phosphodiesterase (Beavo, Hardman and Sutherland 1971). Consequently histologically monitored IE and UE were analyzed for cyclic GMP. An increase of 94% (p = 0.001) was found in IE versus UE (Voorhees et al. 1973). The effect of cyclic GMP on cyclic AMP hydrolysis in epidermis is under current study.

Tetradecanoyl-Phorbol-Acetate, Histamine and Cyclic GMP

Tetradecanoyl-phorbol-acetate is a potent epidermal mitogen and tumor promoter (Hecker 1971) and histamine is associated with rapidly dividing healing wounds and certain tumors (Kahlson and Rosengren 1970). Furthermore we had shown

Table 1
Kinetic Parameters of the Low K_m Cyclic Phosphodiesterase Obtained from Six Psoriasis Patients

	Uninvolved		Involved	
17,000 × g	K_m (10^{-7} M)	V_{max} (pmoles/mg protein/min)	K_m (10^{-7} M)	V_{max} (pmoles/mg protein/min)
Pellet	5.6	23	5.3	32
Supernatant fraction	4.6	28	12.5	111

that high doses of exogenous histamine would stimulate $G_2 \rightarrow M$ cell cycle flow in epidermis (Voorhees et al. 1972c). Elevated cyclic GMP concentration is associated with rapid proliferation in psoriasis epidermis and other systems. Since TPA and histamine are epidermal mitogens, we tested the possibility that treatment of epidermis with these two agents might produce an elevation of cyclic GMP. The effects of TPA and histamine on the cellular cyclic GMP concentrations in epidermal slices at 0.5, 1 and 3 minutes of incubation are shown in Table 2. TPA produced a rapid, greater than twofold increase in cyclic GMP concentration at 0.5 minutes of incubation—the shortest technically feasible assay time. Since TPA has been shown to produce increases of over tenfold in cellular cyclic GMP in other cells (Goldberg et al. 1973) within seconds after exposure of cells, it is possible that a more detailed examination of the steady state changes of this cyclic nucleotide at even earlier time periods may reveal even greater changes than occur at 0.5 minutes. Histamine produced a less rapid but more dramatic increase in cyclic GMP concentration of about 16-fold relative to control at 3 minutes. Assays of cyclic AMP in these TPA- and histamine-treated epidermal slices revealed no significant shifts in cyclic AMP concentrations at 0.5 to 3 minutes of incubation.

Cell Proliferation—Specialization and Cyclic AMP and Cyclic GMP

The decreased concentration of cyclic AMP in rapidly dividing epidermis versus uninvolved areas seems analogous to the low cyclic AMP levels that Otten et al. (1972) and Sheppard (1972) have demonstrated in rapidly dividing "malignant" cells in culture. Whether this apparent analogy has biological significance is a problem for the future, as we have not examined endogenous cyclic AMP levels in malignant epidermis and psoriasis does not undergo spontaneous malignant transformation.

An apparently "benign" phenotype can be conferred on transformed cells in

Table 2

Effects of Phorbol Ester and Histamine on Concentrations
of Cyclic GMP in Hairless Mouse Epidermal Slices

Minutes of incubation	0.5	1.0	3.0	% ↑ At maximum
	femtomoles/ μg DNA			
Control	2.5		2.2	
Phorbol ester[a]	6.3 ± 1.5	6.1 ± 0.6	1.8 ± 0.5	168 (0.5 min)
Histamine[b]	3.4 ± 1.2	4.4 ± 0.7	38.5 ± 4.5	1538 (3 min)
	femtomoles/ mg protein			
Control	80		94	
Phorbol ester[a]	200 ± 12	196 ± 20	30 ± 8	130 (0.5 min)
Histamine[b]	113 ± 24	139 ± 51	1160 ± 367	1233 (3 min)

[a] The phorbol ester (tetradecanoyl-phorbol-acetate) concentration was 75 ng/ml.

[b] The histamine concentration was 10^{-3} M.

culture by treating them with cyclic AMP (Johnson, Friedman and Pastan 1971; Hsie, Jones and Puck 1971; Prasad 1972; Prasad and Sheppard 1972; Prasad et al. 1973). When cyclic AMP is removed from the culture medium, the "malignant" phenotype reappears (Johnson et al. 1971; Hsie et al. 1971; Prasad 1972; Prasad and Sheppard 1972).

We have recently obtained analogous results in a double-blind clinical study in which papaverine, a known cyclic AMP elevating agent, was applied to the lesions of 45 psoriasis patients (Stawiski et al. unpublished 1973). Papaverine treatment produced a statistically significant restoration to a clinically normal epidermal phenotype. When papaverine treatment was discontinued, the psoriasis lesions reappeared. Apparently cyclic AMP elevation decreases growth rate and increases expression of specialized function, whether the rapidly dividing cells being treated are benign or malignant.

The increased level of cyclic GMP in the excessive epidermal proliferation of psoriasis is analogous to the elevated levels of cellular cyclic GMP associated with a number of other rapidly proliferating systems (Goldberg et al. 1973b). Our observation that the epidermal mitogens TPA and histamine produce rapid elevation of cyclic GMP is compatible with the postulate that mitogen-induced elevation of cyclic GMP is an early event that triggers subsequent cellular proliferation. Although the finding of elevated cyclic GMP in psoriasis and mitogen-stimulated epidermis adds support to the accumulating evidence linking cyclic GMP elevation to rapid proliferation, no association of cyclic GMP with the extent of cell specialization or differentiation has been reported for benign or malignant cells.

Dualism and Pleiotypism

Goldberg et al. (1973a) have proposed the "dualism" theory of biological regulation through opposing actions of cyclic AMP and cyclic GMP. In psoriasis and mitogen-treated epidermis, cyclic GMP may be the primary and active signal while a fall in cyclic AMP, should it occur, would be permissive. The data obtained for psoriasis and mitogen-stimulated epidermis can be interpreted in terms of the "dualism" concept as it applies to a bidirectionally regulated system (Goldberg et al. 1973b).

Figure 5 depicts our concept of the role of cyclic AMP and cyclic GMP in the fine control of epithelial proliferation and specialization. This model would also seem to fit the concept of negative and positive pleiotypism proposed by Hershko et al. (1971). Those investigators proposed that cyclic AMP serves as a pleiotypic mediator. Our hypothesis, which is modeled after that of Hadden et al. (1972), would extend the proposal of Hershko in that cyclic AMP would be the negative pleiotypic mediator and cyclic GMP the positive pleiotypic mediator in our systems—psoriasis and the mitogen-stimulated epidermis. Viewed in this way, human epidermis is under dominant control of cyclic AMP, which keeps the normal epidermis under "shift-down" conditions. In psoriasis or mitogen-stimulated epidermis positive pleiotypism is operative with the epidermis existing under "shift-up" conditions, mediated by cyclic GMP.

SUMMARY

The epidermis of psoriasis is characterized by excessive cell proliferation, reduced cell specialization, and glycogen accumulation. Assay of cyclic AMP levels in 50

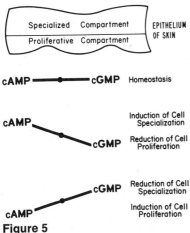

Figure 5
Balance between epidermal proliferation and specialization regulated by cyclic AMP and cyclic GMP levels: a hypothesis.

patients demonstrated that the concentration of cyclic AMP is decreased in the epidermis of the psoriasis lesion. Our results suggest that this decrease might occur as a result of elevated cyclic AMP hydrolysis and not because of any gross defect in cyclic AMP synthesis. Such an association of low cyclic AMP with rapid proliferation and loss of specialized function has been reported for other systems.

Cyclic GMP is substantially increased in the psoriasis lesion. In addition the epidermal mitogens tetradecanoyl-phorbol-acetate and histamine produced rapid elevation of cyclic GMP in epidermal slices. Elevated cyclic GMP has also been found to be associated with rapid cellular proliferation in other systems.

We subscribe to the "dualism" theory of biological regulation and postulate that the excessive proliferation and reduced specialization of psoriasis epidermis is triggered and maintained by an imbalance in the levels of epidermal cyclic AMP and cyclic GMP.

Acknowledgments

This investigation was supported in part by the National Institute of Arthritis and Metabolic Diseases research grant AM 15740-01, the Irene Heinz and John La-Porte Given Foundation, General Research Support grant RR-05383-09, Babcock Dermatological Endowment Fund and American Cancer Society institutional research grant IN-40J and grants NB-05979 and HE-07939. We wish to thank Branka Baic, David Chernin, Janet Dohler, Kathleen Englehard, Emmet Hayes and William Kelsey for expert technical assistance.

REFERENCES

Baird, W. M., J. A. Sedgwick and R. K. Boutwell. 1971. Effects of phorbol and four diesters of phorbol on the incorporation of tritiated precursors into DNA, RNA, and protein in mouse epidermis. *Cancer Res.* **31**:1434.

Beavo, J. A., J. G. Hardman and E. W. Sutherland. 1971. Stimulation of adenosine 3′,5′-monophosphate hydrolysis by guanosine 3′,5′-monophosphate. *J. Biol. Chem.* **246**:3841.

Brooker, G., L. J. Thomas, Jr. and M. M. Appleman. 1968. The assay of adenosine 3′,5′-cyclic monophosphate and guanosine 3′,5′-cyclic monophosphate in biological materials by enzymatic radioisotopic displacement. *Biochemistry* **7**:4177.

Bullough, W. S. and E. B. Laurence. 1961. Stress and adrenaline in relation to the diurnal cycle of epidermal mitotic activity in adult male mice. *Proc. Roy. Soc. London* **154**:540.

Bürk, R. R. 1968. Reduced adenyl cyclase activity in a polyoma virus-transformed cell line. *Nature* **219**:1272.

D'Armiento, M., G. S. Johnson and I. Pastan. 1972. Regulation of adenosine 3′:5′-cyclic monophosphate phosphodiesterase activity in fibroblasts by intracellular concentrations of cyclic adenosine monophosphate. *Proc. Nat. Acad. Sci.* **69**:459.

Duell, E. A., J. J. Voorhees, W. H. Kelsey and E. Hayes. 1971. Isoproterenol-sensitive adenyl cyclase in a particulate fraction of epidermis. *Arch. Derm.* **104**:601.

Flaxman, B. A. 1972. Replication and differentiation in vitro of epidermal cells from normal human skin and from benign (psoriasis) and malignant (basal cell cancer) hyperplasia. *In Vitro* **8**:237.

George, W. J., J. B. Polson, A. G. O'Toole and N. D. Goldberg. 1970. Elevation of guanosine 3′,5′-cyclic phosphate in rat heart after perfusion with acetylcholine. *Proc. Nat. Acad. Sci.* **66**:398.

Gilman, A. G. 1970. A protein binding assay for adenosine 3′:5′-cyclic monophosphate. *Proc. Nat. Acad. Sci.* **67**:305.

Goldberg, N. D. and A. G. O'Toole. 1971. Analysis of cyclic 3′,5′-adenosine monophosphate and cyclic 3′,5′-guanosine monophosphate. *Methods of biochemical analysis* (ed. D. Glick). John Wiley and Sons, New York.

Goldberg, N. D., R. F. O'Dea and M. K. Haddox. 1973. Cyclic GMP. *Advances in cyclic nucleotide research* (ed. P. Greengard and G. A. Robison), vol. 3. Raven Press, New York. (in press).

Goldberg, N. D., M. K. Haddox, D. K. Hartle and J. W. Hadden. 1973a. The biological role of cyclic GMP. *Fifth Int. Congr. Pharmacol.* Karger, Basel, Switzerland. (in press).

Goldberg, N. D., M. K. Haddox, R. Estensen, C. Lopez and J. W. Hadden. 1973b. Evidence for a dualism between cyclic GMP and cyclic AMP in the regulation of cell proliferation and other cellular processes. L. M. Lichtenstein. *Cyclic AMP in immune response and tumor growth* (ed. L. M. Lichtenstein et al.) Springer-Verlag, New York. (in press).

Hadden, J. W., E. M. Hadden, M. K. Haddox and N. D. Goldberg. 1972. Guanosine 3′:5′-cyclic monophosphate: A possible intracellular mediator of mitogenic influences in lymphocytes. *Proc. Nat. Acad. Sci.* **69**:3024.

Halprin, K. M. and J. R. Taylor. 1971. The biochemistry of skin disease: Psoriasis. *Advance Clin. Chem.* **14**:319.

Halprin, K. M., A. Ohkawara and V. Levine. 1973. Synthesis of glycogen in the psoriatic lesion. *Arch. Derm.* **107**:706.

Hecker, E. 1971. Isolation and characterization of the cocarcinogenic principles from croton oil. *Methods in cancer research* (ed. H. Busch) p. 439. Academic Press, New York.

Hegazy, M. A. H. and J. F. Fowler. 1973. Cell population kinetics of plucked and unplucked mouse skin. *Cell and Tissue Kinetics* **6**:17.

Hershko, A., P. Mamont, R. Shields and G. M. Tomkins. 1971. Pleiotypic response. *Nature New Biol.* **232**:206.

Hsie, A. W., C. Jones and T. T. Puck. 1971. Further changes in differentiation state accompanying the conversion of Chinese hamster cells to fibroblastic form by dibutyryl adenosine cyclic 3′:5′-monophosphate and hormones. *Proc. Nat. Acad. Sci.* **68**:1648.

Iversen, O. H. 1969. Chalones of the skin. *Homeostatic regulators* (ed. G. E. W. Wolstenholme and J. Knight) p. 29. J. & A. Churchill, London.

Johnson, G. S., R. M. Friedman and I. Pastan. 1971. Restoration of several morphological characteristics of normal fibroblasts in sarcoma cells treated with adenosine-3′:5′ cyclic monophosphate and its derivatives. *Proc. Nat. Acad. Sci.* **68**:425.

Kahlson, G. and E. Rosengren. 1970. Histamine formation as related to growth and protein. *Biogenic amines as physiological regulators* (ed. J. J. Blum) p. 223. Prentice-Hall, Englewood Cliffs, New Jersey.

Marks, F. and W. Rebien. 1972. The second messenger system of mouse epidermis. I. properties and β-adrenergic activation of adenylate cyclase in vitro. *Biochim. Biophys. Acta* **284**:556.

Mercer, E. H. and H. I. Maibach. 1968. Intercellular adhesion and surface coats of epidermal cells in psoriasis. *J. Invest. Derm.* **51**:215.

Mueller, G. C. 1971. Biochemical perspectives of the G_1 and S intervals in the replication cycle of animal cells: A study in the control cell growth. *The cell cycle and cancer* (ed. R. Baserga) vol. 1, p. 269. Marcel Dekker, New York.

Otten, J., J. Bader, G. S. Johnson and I. Pastan. 1972. A mutation in a Rous sarcoma virus gene that controls adenosine 3′,5′-monophosphate levels and transformation. *J. Biol. Chem.* **247**:1632.

Prasad, K. N. 1972. Morphological differentiation induced by prostaglandin in mouse neuroblastoma cells in culture. *Nature New Biol.* **236**:49.

Prasad, K. N. and J. R. Sheppard. 1972. Inhibitors of cyclic nucleotide phosphodiesterase induce morphological differentiation of mouse neuroblastoma cell culture. *Exp. Cell Res.* **73**:436.

Prasad, K. N., B. Mandal, J. C. Waymire, G. J. Lees, A. Vernadakis and N. Weiner. 1973. Basal level of neurotransmitter synthesizing enzymes and effect of cyclic AMP agents on the morphological differentiation of isolated neuroblastoma clones. *Nature New Biol.* **241**:117.

Prescott, D. M. 1973. The mechanism of regulation of cell reproduction. *Cancer, the misguided cell,* p. 76. Pegasus Publ., *New York.*

Raick, A. 1973. Ultrastructural, histological, and biochemical alterations produced by 12-O-tetradecanoyl-phorbol-13-acetate on mouse epidermis and their relevance to skin tumor promotion. *Cancer Res.* **33**:269.

Robison, G. A., R. W. Butcher and E. W. Sutherland. 1968. Cyclic AMP. *Annu. Rev. Biochem.* **37**:149.

————. 1971. *Cyclic AMP,* p. 36. Academic Press, New York.

Ryan, W. L. and M. L Heidrick 1968. Inhibition of cell growth in vitro by adenosine 3′,5′-monophosphate. *Science* **162**:1484.

Sheppard, J. R. 1972. Difference in the cyclic adenosine 3′,5′-monophosphate levels in normal and transformed cells. *Nature New Biol.* **236**:14.

Steiner, A. L., C. W. Parker and D. M. Kipnis. 1972. Radioimmunoassay for cyclic nucleotides. *J. Biol. Chem.* **247**:1106.

Voorhees, J. J. and E. A. Duell. 1971. Psoriasis as a possible defect of the adenyl cyclase-cyclic AMP cascade. *Arch. Derm.* **104**:352.

Voorhees, J. J., E. A. Duell and W. H. Kelsey. 1972. Dibutyryl cyclic AMP inhibition of epidermal cell division. *Arch. Derm.* **105**:384.

Voorhees, J. J., E. A. Duell, L. J. Bass, J. A. Powell and E. R. Harrell. 1972a. Decreased cyclic AMP in the epidermis of lesions of psoriasis. *Arch. Derm.* **105**:695.

Voorhees, J. J., E. A. Duell, W. H. Kelsey and E. Hayes. 1972b. Effects of alpha and beta adrenergic stimulation on cyclic AMP formation and mitosis in epidermis. *Clinical Res.* **20:**419.

Voorhees, J. J., E. A. Duell, L. J. Bass, J. A. Powell and E. R. Harrell. 1972c. The cyclic AMP system in normal and psoriatic epidermis. *J. Invest. Derm.* **59:**114.

Voorhees, J., W. Kelsey, M. Stawiski, E. Smith, E. Duell, M. Haddox and N. Goldberg. 1973. Increased cyclic GMP and decreased cyclic AMP levels in the rapidly proliferating epithelium of psoriasis. *The role of cyclic nucleotides in carcinogenesis* (ed. J. Schultz and H. G. Gratzner) vol. 6, pp. 325–373. Academic Press, New York.

Weinstein, G. D. and P. Frost. 1971. Methotrexate for psoriasis. *Arch. Derm.* **103:**33.

Weiss, B., H. Shein and P. Uzunov. 1973. Induction by norepinephrine of a specific molecular form of cyclic 3′,5′-AMP phosphodiesterase in cloned rat astrocytoma cells (C-2A). *Fed. Proc.* **32:**679 (abs.)

Wright, R. K., S. H. Mandy, K. M. Halprin and S. L. Hsia. 1973. Defects and deficiency of adenyl cyclase in psoriatic skin. *Arch. Derm.* **107:**47.

Hormonal Regulation of Cyclic AMP in Aging and in Virus-Transformed Human Fibroblasts and Comparison with Other Cultured Cells

Maynard H. Makman, B. Dvorkin, and Elaine Keehn

Departments of Biochemistry and Pharmacology
Albert Einstein College of Medicine, Yeshiva University, Bronx, New York 10461

There is evidence that in a number of cell cultures adenylate cyclase activity and adenosine cyclic 3′,5′-monophosphate (cyclic AMP) levels may be dependent on the growth rate and/or cell density. Hormone-stimulated adenylate cyclase activity may be enhanced by cell-cell interaction or contact, as indicated by studies of cultured fibroblasts, HeLa cells, Chang's liver cells, and rat hepatoma (HTC) cells (Makman 1971c). In these studies transfer of cells from suspension to stationary culture and increase in density of cells in stationary culture resulted in increased adenylate cyclase activity. Also for at least some types of cells basal cyclic AMP levels are higher in slowly growing than in more rapidly growing cells (Otten, Johnson and Pastan 1971; Sheppard, 1972). Other studies have indicated that growth of cultured fibroblasts may be inhibited by dibutyryl cyclic AMP or prostaglandin E_1 (PGE_1) (Johnson, Friedman and Pastan 1971; Otten, Johnson and Pastan 1972). Thus the adenylate cyclase system may be modulated in a complex manner by proliferation or growth of cells and by cell density or contact, and in turn, cyclic AMP may itself regulate cell density or growth.

This paper presents further studies designed to more fully evaluate the influence of epinephrine and PGE_1 on cyclic AMP formation by, and growth of, cultured cells. Marked differences in the hormonal responsiveness of various cell cultures with respect to magnitude and time course of cyclic AMP formation have been found. In particular cyclic AMP formation by early and late passage and SV40 virus-transformed WI-38 human fibroblasts has been studied. Inhibition of growth of lens epithelial cells by epinephrine and cyclic nucleotides is reported and comparison of this response is made to the response of WI-38 cells to these agents.

METHODS

Cell Cultures

Cultures were grown routinely on plastic dishes (Falcon Plastics). Chang's liver, HeLa (AT), astrocytoma (rat, clone C-6) and rabbit lens epithelial cells were grown in Dulbecco's modified Eagle's medium for monolayer culture with 5% calf and 5% fetal calf serum under 10% CO_2 in air. Strain WI-38 and WI-38 VA 13B 2RA cells and also in later experiments some cultures of lens epithelial cells were grown in Eagle's basal medium with Earle's salts for diploid culture (BME) with 10% fetal calf serum in an atmosphere of 5% CO_2 in air. In most instances 25 mM Hepes buffer was also present in the medium. Cells were removed from culture dishes with 0.25% trypsin–0.04% EDTA for serial passage.

Starter cultures of WI-38 cells and of SV40 virus-transformed WI-38 (WI-38 VA 13B 2RA) were obtained from Dr. Hayflick. The SV40-transformed cells had been previously obtained from Dr. Girardi (Girardi, Jensen and Koprowski 1965) and maintained in continuous culture in Dr. Hayflick's laboratory. Rabbit lens explants were prepared as described by Shapiro et al. (1969) and cells were propagated in continuous culture for periods up to 2 years.[1]

Adenylate Cyclase Assay

Cells were lysed in 20 mM glycylglycine buffer (pH 7.8)–5 mM $MgSO_4$ and aliquots of the cell lysate incubated for 20 minutes at 30°C with [α-^{32}P]ATP in the presence of $MgSO_4$, caffeine, unlabeled cyclic AMP, phosphoenolpyruvate, pyruvate kinase and Tris-HCl buffer pH 7.6 (Makman 1971c) followed by chromatographic isolation of the [^{32}P]-cyclic AMP formed (Brown and Makman 1972).

Cyclic AMP Assay

For measurement of cyclic AMP concentration in intact cells, acetic acid extracts of cells were prepared essentially as described by Sherline, Lynch and Glinsmann (1972), with rapid removal of medium followed by addition of aqueous acetic acid solution (glacial acetic acid diluted 1:1 [v:v] with water) to the culture dishes. The acetic acid suspensions containing cellular debris were heated for 3 minutes at 90°C and centrifuged; aliquots of the supernatant fluids were dried at 80°C in small tubes and then assayed for cyclic AMP by a modification (Brown and Makman 1973) of the method of Gilman (1970). For measurement of cyclic AMP in the culture medium, medium was added to an equal volume of glacial acetic acid and then processed as for cells.

In studies where dibutyryl cyclic AMP was added to cell cultures, cyclic AMP content of washed cells represents total of cyclic AMP plus monobutyryl cyclic AMP since both have essentially the same activity in the binding assay used. The wash procedure used for cells grown in cyclic AMP or analogue appeared to remove all extracellular cyclic AMP, since appreciable increases in cellular levels required continued incubation and did not occur in all cells or all conditions studied.

For studies of cyclic AMP formation in a number of instances cells were incubated in a medium designated "salts plus glucose" containing NaCl (6.8 mg/ml), KCl (0.4 mg/ml) $NaHCO_3$ (2.2 mg/ml) and glucose (1 mg/ml) (concentrations identical to those in the complete BME).

Protein was determined by the method of Lowry et al. (1951) and DNA by the diphenylamine reaction as modified by Burton (1968).

---

[1] These studies were done in collaboration with E. Levine and H. L. Kern.

Table 1

Hormone-stimulated Adenylate Cyclase Activity of Cell Lysates

| Addition to assay | *pmoles/mg cyclic AMP formed* | | | |
| | *WI-38 fibroblasts (passage 32)* | *Rabbit lens epithelium* | | |
		*RLE-1**	*RLE-2**	*RLE-3**
None	63	14	3620	230
Epinephrine, 20 μM	357	86	8340	590
Prostaglandin E_1, 1 μg/ml	238	54		240
Glucagon, 2 μg/ml	71	64		1330
Adrenocorticotropin, 0.5U/ml	71			
NaF, 8 mM	405	114	13,130	1430

* Cultures derived from different animals and grown in continuous tissue culture.

RESULTS

We have previously reported the presence of catecholamine-stimulated adenylate cyclase activity in HeLa cells, Chang's liver cells, mouse 3T3 and 3T6 fibroblasts (Makman 1970) and in rabbit lens epithelial cells (Makman and Kern 1971, 1972) and some human diploid fibroblasts (Makman 1971c). In addition we found that the enzyme in HeLa and rat hepatoma (HTC) cells is stimulated by glucagon and in these two cells lines and in certain fibroblast lines by prostaglandin E_1 as well (Makman 1971c).

The human embryonic lung fibroblast strain WI-38 represents an interesting culture system for study of hormone receptors and cyclic AMP since the limited capacity for doubling displayed by these cells has been well documented and has been proposed as a model for the study of aging in vitro (Hayflick and Moorhead 1961; Hayflick 1965). We find that the WI-38 strain fibroblasts contain adenylate cyclase activity, which is stimulated by epinephrine and by prostaglandin E_1 but not by glucagon or ACTH (Table 1).

It should be noted that different human diploid fibroblast cultures may vary markedly in presence and magnitude of hormonal response (Makman 1971c). We have also noted such differences in culture of lens epithelial cells which we have prepared from rabbit lenses under apparently identical conditions (Makman, Kern and Levine unpubl.). Lens epithelium obtained directly from experimental animals contains adenylate cyclase, which is stimulated by glucagon and PGE_1[2] as well as by epinephrine. The adenylate cyclase activities of three lens cultures, each with morphological characteristics similar to normal lens epithelium (Shapiro et al. 1969), nevertheless differ markedly with respect to hormone-stimulated adenylate cyclase activity (Table 1).

The magnitude of hormonal stimulation of cyclic AMP levels in intact cells grown in stationary culture increases, in some instances rather appreciably, with increase in cell density, as shown in Table 2 for Chang's liver cells, WI-38 fibroblasts and

[2] M. H. Makman and H. L. Kern, unpublished studies. Preliminary report presented at the annual meeting of the Association of Research in Vision and Ophthalmology, April, 1972.

Table 2
Influence of Cell Density on Hormone-stimulated Cyclic AMP Levels

Cell culture	Mg protein per 60-mm culture dish	Basal	3 μM Epinephrine			0.3 μg/ml PGE$_1$, 10 min
			10 sec	30 sec	1 min	
Chang's liver	0.81 (sparse)	1.9	9.7	6.3	6.9	
Chang's liver*	1.5 (contact)	2.6	33.7		18.3	
Chang's liver*	2.2 (multilayer)	1.9	39.7	25.0	18.2	121
WI-38, passage 15–28	0.24 (sparse	14			23	
WI-38, passage 15–28	0.80 (contact)	24			351	281
WI-38, passage 15–28	1.4 (dense)	9			71	231
WI-38, passage 15–28*	1.2 (dense)	12			110	
WI-38, passage 40–48	0.12 (sparse)	9				
WI-38, passage 40–48	0.33 (light)	17				
WI-38, passage 40–48	0.91 (contact)	8			115	195
WI-38, passage 40–48*	0.83 (contact)	12				192
SV40-transformed WI-38	0.39 (light)	5.3			194	429
SV40-transformed WI-38	0.82 (contact)	2.2			429	872
SV40-transformed WI-38*	0.90 (contact)	2.9			476	1,178
SV40-transformed WI-38	1.1 (contact)	3.0			194	379

* Theophylline (0.5 mM) added 30 min prior to obtaining basal value or adding hormone. Incubations were carried out in complete medium with serum.

WI-38VA (SV40 virus-transformed) fibroblasts. Similar results are obtained with HeLa and lens epithelial cells where glucagon response is also enhanced with increased cell density (data not shown). It should be noted that after reaching maximal value at high cell density, values eventually decrease in old cultures, a phenomenon most clearly evident with lens cultures, which can be maintained in a viable nonproliferating state without subculture for 4–6 weeks. These findings are in accord with our previous studies of adenylate cyclase and cell density referred to above (see introduction).

A number of interrelated factors appear to contribute to both the magnitude of the maximal response and the overall time course of the response of various cultured cells to epinephrine and PGE_1. The magnitude of the maximal response of cells to catecholamines varies markedly with cell type (Table 3). In contrast to the 0.17–1.0 minute time for maximal response to epinephrine, the response to PGE_1 develops more gradually and the maximal response is sustained for a longer time period.

A major factor contributing to the response of intact cells to catecholamines is a phenomenon that we have termed catecholamine receptor inactivation (Makman 1971a,b). This phenomenon consists of a selective loss of capacity of adenyl cyclase to be stimulated by catecholamines when intact cells are exposed continuously to catecholamines (Makman 1971b, 1972). This process is caused by concentrations of β-adrenergic agonists that initially stimulate the cyclase and it is dependent on time and temperature of incubation. The inactivation is evident upon assay for adenylate cyclase activity of lysates of washed cells previously exposed to catecholamine. Such lysates retain their normal basal activity and stimulation by NaF, glucagon, and PGE_1 even with complete loss of catecholamine stimulation. The inactivation process is reflected in the extremely transient response of intact Chang's liver, HeLa, and lens epithelial cells to catecholamines (Table 3). In these cells, as well as in 3T6 fibroblasts (Makman 1971a) and lymphoid cells (Makman 1971b), inactivation is apparently complete by 20–120 minutes as evidenced by assay of adenylate cyclase activity of cell lysates. Cyclic AMP levels in these cells are about twice basal levels at 180 minutes, are not stimulated by readdition of epinephrine, but are by addition of PGE_1 (data not shown). However as described below, not all cells may exhibit so marked an inactivation process.

At both low and high cell density the initial responses of WI-38VA cells to epinephrine (0.17–1 minute) and PGE_1 (10–30 minutes) are greater than are those of either early or late passage WI-38 fibroblasts, whether incubations are in complete (BME plus serum) medium or in salts plus glucose (Tables 2, 4–7). Even at low cell density (Table 5, exp. II) the response of WI-38VA cells to PGE_1 is greater than that of WI-38 at high cell density (Table 4, Exp. II and III) and the responses to epinephrine are about the same. The greater responsiveness of WI-38VA is also evident after longer incubations with hormones (150 minutes) (Tables 4 and 5). In contrast basal intracellular levels of cyclic AMP in WI-38VA are not density dependent, being about 4 pmoles/mg protein at both low and high cell density, levels which are lower than those in WI-38 cells at high density (Table 2).

The responses of both normal and transformed WI-38 cells to PGE_1 and to epinephrine obtained in these studies are very large, PGE_1 responses being several orders of magnitude greater than that reported for L929 fibroblasts (Otten et al. 1972) or neuroblastoma cells (Gilman and Nirenberg 1971a) and epinephrine

Table 3

Comparison of Time Course of Response of Lens, Chang's Liver, and HeLa Cells to Prostaglandin E_1 and Epinephrine

Cell culture and conditions	Cyclic AMP content of	pmoles cyclic AMP/mg cell protein Minutes of exposure to epinephrine or prostaglandin						
		0	0.17	0.5	1.0	10	20	40
Rabbit lens epithelium								
PGE$_1$, 3 μg/ml	cells	3.8	6.1		8.6	13.2		
Epinephrine, 3 μM	cells	3.8	66.8		450	55.8		
Chang's liver								
PGE$_1$	cells	1.7				3.4		
Epinephrine, 3 μM	cells	2.8	39.7	25.0	17.5	5.4	5.2	4.2
	medium	9.8						
Epinephrine, 3 μM (in salts + glucose)	cells	3.3	43.5	25.6	21.5	10.0		8.5
	medium	11.6	18.5	26.1	32.4	36.7		33.8
HeLa AT								
Epinephrine, 3 μM	cells	1.8	19.3	12.8	11.8	3.5	2.8	3.0

Incubations in 1 ml complete medium with serum except for the experiment where incubation in salts plus glucose solution is indicated. Theophylline (0.5 mM) was added 30 min prior to hormone or prostaglandin.

Table 4

Time Course of Response of WI-38 Fibroblasts to Prostaglandin E_1 and Epinephrine

Agent added	Minutes present	Medium	Increase in cyclic AMP due to PGE_1 or epinephrine (pmoles/mg cell protein)		
			Cells	Medium	Total
I. Passage 40					
Epinephrine	1	SG	324	19	343
	30	SG	899	1424	2323
	150	SG	498	2254	2752
	150	CM	178	1650	1828
II. Passage 40					
Epinephrine	1	SG	605	14	619
	150	SG	475	4173	4648
	150	CM	414	2788	3202
	450	CM	278	2868	3146
PGE_1	30	SG	1770	2083	3853
	150	SG	1262	6013	7275
	150	CM	322	1833	2155
III. Passage 29					
Epinephrine	1	SG	521	47	568
	150	SG	363	4405	4768
	150	CM	277	3165	3442
	450	CM	240	1935	2175
PGE_1	30	SG	2097	1560	3657
	150	SG	1240	6130	7370
IV. Passage 18					
Epinephrine	1	SG	384	72	456
	30	SG	331	62	393
	150	SG	186	325	511

Epinephrine, 3 μM; PGE_1, 300 μg/ml; SG, salts plus glucose; CM, complete medium with serum. 0.5 mM Theophylline and SG or CM present 30 min prior to experiment (1.0 ml medium in all cases). Cell proteins averaged 0.20, 0.46 and 0.39 mg protein/60-mm diameter dish for I, II and III, respectively, and 0.06 mg/35-mm diameter dish for IV. Basal values for cyclic AMP averaged 26 and 158 pmoles/mg cell protein for cells and medium respectively for all incubations in SG.

responses being up to several times the already large response reported for astrocytoma cells (Clark and Perkins 1971; Gilman and Nirenberg 1971b).

The epinephrine response of WI-38 and WI-38VA cells is also distinguished by its prolonged time course, similar to that found to occur in astrocytoma in the studies just cited. This is quite unlike the transient response of many other cells (Table 3) and indicates that the catecholamine receptor inactivation process is much less rapid or evident in WI-38 and WI-38VA than in the other cells.

A considerable fraction of the cyclic AMP produced by WI-38 and WI-38VA cells in response to either epinephrine or PGE_1 is recovered in the medium, and as the time of incubation is increased the medium may contain the major (accumu-

Table 5

Time Course of Response of SV40-transformed WI-38 Fibroblasts to Prostaglandin E_1 and Epinephrine

Agent added and time of addition (or control incubation) prior to termination	Increase in cyclic AMP due to PGE_1 and/or epinephrine (pmoles/mg cell protein)		
	Cells	Medium	Total
I. 300 μg/ml PGE_1 10 min	471	1012	1483
30 min	905	7320	8225
60 min	490	11,760	12,250
II. 3 μM Epinephrine 1 min	621	26	647
150 min	9428	14,566	23,994
150 min complete medium	9226	15,333	24,559
300 μg/ml PGE_1 30 min	13,590	3866	17,456
Control incubation 120 min, then 300 μg/ml PGE_1 30 min	16,100	1680	17,780
Control incubation 120 min, then 0.3 μg/ml PGE_1 30 min	8190	2714	10,908
300 μg/ml PGE_1 150 min	12,320	9202	21,522
0.3 μg/ml PGE_1 150 min	6500	13,296	19,796
III. 300 μg/ml PGE_1 150 min	6940	4981	11,951
300 μg/ml PGE_1 plus 25 μM cycloheximide 150 min	7735	4978	12,353
Control incubation 120 min, then 300 μg/ml PGE_1 30 min	8340	981	9321
25 μM Cycloheximide 150 min + 300 μg/ml PGE_1 last 30 min	12,350	1708	14,058
Control incubation 149 min, then 3 μM epinephrine 1 min	248	15	263
300 μg/ml PGE_1 150 min, then 3 μM epinephrine last 1 min	7410	4997	12,407
3 μM epinephrine 150 min	1507	6855	8362
3 μM epinephrine 150 min + 300 μg/ml PGE_1 last 30 min	9375	9765	19,140

Incubations for 30 min with 0.5 mM theophylline in 0.5 ml (Exp. I) or 1.0 ml (Exp. II, III) of either salts plus glucose solution or where specified complete (BME) medium with 10% fetal calf serum preceeded the further incubations and additions indicated above. Cyclic AMP values represent increases over basal levels of approximately 10 and 42 pmoles/mg protein for cells and medium respectively. Average value for mg cell protein per dish were 0.12/35-mm diameter dish and 0.25 and 0.44/60-mm diameter dish for Exp. I, II and III respectively.

lated) fraction of cyclic AMP (Tables 4, 5). Without taking this fraction into account it is possible to greatly underestimate the response of a cell to hormone or other stimulating agent. Significant increases in cyclic AMP in the medium following hormonal stimulation also occurred in Chang's liver cells (Table 3).

WI-38 cells are sensitive to low concentrations (0.03 μg/ml) of PGE_1, with maximal responses produced by 0.3–6.0 μg/ml (Table 6). Very high concentrations of PGE_1 (300–600 μg/ml) are also maximally effective and appear to increase the ratio of intracellular to extracellular cyclic AMP formed (Tables 6, 7).

The marked response of WI-38VA to PGE_1 was not related to a protein component with rapid turnover since prior incubation of cells for 2 hours with cycloheximide not only did not decrease the response but actually enhanced it (Table 5). The reason for this enhancement is not known. No enhancement of response occurred when cells were exposed to PGE_1 and cyclohexamide for 2.5 hours (Table 5), suggesting further that induction of phosphodiesterase by cyclic AMP did not occur to a significant extent with respect to limiting the PGE_1 response during this time period. Induction of phosphodiesterase by cyclic AMP in transformed mouse fibroblasts (d'Armiento, Johnson and Pastan 1972; Maganiello and Vaughan 1972) and astrocytoma cells (Uzunov, Shein and Weiss 1973) has been reported.

The response of WI-38VA to epinephrine for 150 minutes plus PGE_1 for the last 30 minutes (19,140 pmoles/mg) is essentially the sum of values for separate incubations with epinephrine (8363 pmoles/mg) and PGE_1 (9321 pmoles/mg)

Table 6

Response of Normal and Transformed WI-38 Fibroblasts to Prostaglandin E_1

Cell culture	Incubation for 10 min with PGE_1 $\mu g/ml$	Increase in cyclic AMP due to PGE_1 (pmoles/mg cell protein)		
		Cells	Medium	Total
WI-38 (passage 18)	0.03	94	60	154
(0.64 mg	0.30	130	100	230
protein/dish)	6.0	163	224	387
	150	682	74	756
WI-38 (passage 39)	0.03	341	256	597
(0.31 mg	0.30	797	939	1736
protein/dish)	6.0	741	613	1354
	150	779	693	1472
	300	1187	355	1542
WI-38 VA	0.6	7137	1432	8569
	6.0	8410	1753	10,163
(0.24 mg	30	10,864	2207	13,071
protein/dish)	150	10,341	1571	11,912
	300	9463	1026	10,589
	600	9819	409	10,228

Cells grown in 35-mm diameter plastic dishes, incubated in 0.5 ml glucose-salt solution containing 0.5 mM theophylline for 30 min prior to addition of PGE_1. Basal values for cyclic AMP just prior to PGE_1 addition averaged 15 pmoles/mg in WI-38 cells (plus 31 pmoles/mg in the medium) and 6 pmoles/mg in WI-38VA cells (plus 12 pmoles/mg in the medium).

Table 7

Factors Influencing Cyclic AMP Levels in Cultured Cells

Cell culture and hormone addition	pmoles cyclic AMP/mg cell protein				
	Balanced salts plus (30 min)			salts only (30 min)	complete medium + serum†
	control	+ serum† last 5 min	+ 0.2 mg/ml iodoacetate last 15 min		
Chang's liver					
control	5.3		5.8		
epinephrine, 3 μM, 10 sec	30.6*		10.0	34.2	
Astrocytoma (rat, C-6)					
control	5.0	4.3	6.5	5.2	6.1
epinephrine, 3 μM, 10 sec	291	141	449	422	115
SV40 transformed WI-38					
control	6.2	3.4	5.0	2.8	3.5
epinephrine, 3 μM, 10 sec	159	72	46	119	35
PGE_1, 300 μg/ml, 10 min	4682	2540	1730	6139	1178
WI-38 (passage 20)					
control	13.3	5.5			
epinephrine, 3 μM, 10 sec	96				
PGE_1, 300 μg/ml, 10 min	1925	937		1549	
WI-38 (passage 44)					
control	11.9	9.1			
epinephrine, 3 μM, 10 sec	108				
PGE_1, 3 μg/ml, 10 min	659	192			
PGE_1, 300 μg/ml, 10 min	1487	570		877	

Cells were incubated for 30 min with 0.5 mM theophylline in the medium indicated prior to addition of hormone.

* Values of 34.3, 37.6 and 30.2 were obtained for cells incubated for 15 min with 0.75 mM ouabain, 1 mM 2,4-dinitrophenol and 0.1 μg/ml oligomycin, respectively, prior to exposure to epinephrine for 10 seconds.

† Serum where present was either 5% fetal calf plus 5% calf for astrocytoma or 10% fetal calf for WI-38. Complete culture media used were as indicated in the text.

(Table 5). Also in studies by S. Morris in this laboratory, additions of maximal concentrations of PGE_1 (10 minutes), epinephrine (30 seconds) or both (epinephrine for only the last 30 seconds) resulted in 460, 360 and 900 pmoles intracellular cyclic AMP/mg, respectively, for WI-38 cells, and in 580, 630 and 1130 pmoles intracellular cyclic AMP/mg for WI-38VA cells. Thus the epinephrine and PGE_1 responses in these cells clearly represent independent components of the adenylate cyclase system.

Although both early and late passage WI-38 cells are very responsive to both epinephrine and PGE_1, the early passage WI-38 cells at low density show a considerably smaller response to epinephrine than do late passage cells (e.g., Table 4, Exp. I and IV). Early passage WI-38 also appear to show a density-dependent increase in basal intracellular levels of cyclic AMP (average increase from 14 to 24 pmoles/mg protein). Late passage cells may have somewhat less increase in levels

Table 8

Influence of Serum on Epinephrine- and Prostaglandin-
stimulated Cyclic AMP Levels in Fibroblasts

Preincubation in salts plus glucose 30 min with	*pmoles cellular cyclic AMP/mg cell protein*			
	WI-38 (passage 27)		*WI-38 (passage 28)*	
	Basal	*3 μM epinephrine 10 sec*	*Basal*	*30 μg/ml PGE₁ 10 min*
Control	30	126	22	705
Fetal calf serum (A), 2%		112		
(A), 10%		83		309
(A), 30%		33		
(B), 10%		45		297
(C), 10%		68		167
(D), 10%		65		
(E), 10%		54		
Calf serum, 10%		76		
Gamma-globulin-free calf serum, 10%		46		

with increase in density (e.g., Table 2). Possible difference in levels at low density and in changes with increase in density will require additional experiments at a variety of cell densities and at different times after medium change to fully evaluate their significance.

Studies have been carried out concerning optimal conditions for obtaining hormonal response. Exogenous substrate in the medium does not appear to be necessary for the large responses of cells to hormone since omission of glucose from the simple salts plus glucose medium is generally without influence (Table 7). Although iodoacetate inhibits the hormonal response of WI-38VA and Chang's liver cells, this is probably not related to decrease in cellular ATP since 2,4-dinitrophenol and oligomycin did not alter the response of Chang's liver cells to epinephrine (Table 7).

Responses to epinephrine and PGE₁, as well as basal levels of cyclic AMP in cells and medium, are greater when WI-38 cells are incubated in the simple salts plus glucose solution rather than in complete medium (Table 4). Addition of serum to the salts plus glucose solution results in responses essentially the same as are obtained in complete medium (Table 7). Comparable inhibition of basal and of epinephrine- and PGE₁-stimulated cellular cyclic AMP levels are produced by serum (Table 7), and this inhibition is exerted within 5 minutes of addition of serum (data not shown). Inhibition was produced by all batches of fetal calf serum tested, was manifest over a concentration range of 2–30%, and was also produced by calf serum (Table 8).

Exposure of lens epithelial cells, WI-38 cells (Table 9) and Chang's liver cells (data not shown) to epinephrine for 1–6 days results in intracellular levels of cyclic AMP approximately twice the basal levels. Growth of lens epithelial cells is markedly inhibited by epinephrine and by either cyclic AMP or dibutyryl cyclic AMP in the presence of theophylline, and to a lesser but still appreciable extent in the absence of theophylline (Table 9). Since WI-38 and WI-38VA cells show more marked and prolonged initial stimulation of cyclic AMP levels by epinephrine, it was thought that epinephrine might also inhibit growth of these cells. However neither epinephrine nor dibutyryl cyclic AMP altered the growth rate of WI-38 cells (Table 9). Inhibition of growth of early and late passage WI-38 and WI-38VA cells was obtained when cells were grown with a high concentration of PGE_1 (Table 9). These results indicate that WI-38 cells are inherently less sensitive than are lens epithelial cells to cyclic AMP-mediated inhibition of growth. Although 5-hydroxytryptamine has been reported to stimulate proliferation of certain fibroblast cultures (Boucek and Alvarez 1970), we did not observe this with WI-38 cells (Table 9).

DISCUSSION

The magnitude and duration of the stimulation of adenylate cyclase of intact cells in culture by hormone and by prostaglandin differ greatly in different cells and also are very much dependent upon the conditions of incubation and cell growth. Responses are enhanced by increased cell density or cell-cell contact, in some instances related to decreased rate of cell proliferation. In Chang's liver cells a marked decrease in hormone-stimulated adenylate cyclase occurs during the S phase of the cell cycle (Makman and Klein 1972), a phenomenon which may contribute to the cell density-dependent increase in activity in this cell line. Increased cell density to the point of maximal cell contact, but not necessarily to the point of cessation of growth, appears to enhance the adenylate cyclase system in a number of different cell cultures. Enhancement is evident with respect to hormonal stimulation of adenylate cyclase activity of cell homogenates and of cyclic AMP levels in intact cells, as well as increased basal intracellular levels of cyclic AMP in some instances. It is postulated that this cell culture phenomenon reflects a normal developmental process that occurs in vivo, an increase in or appearance of hormone-sensitive adenylate cyclase at a relatively late stage in cell maturation. Considered in this light, the enhancement of adenylate cyclase with increased cell density is primarily a phenomenon of development or differentiation, which secondarily in some instances may in turn regulate rate of growth.

Full consideration of the magnitude of hormone and PGE_1 stimulation of cyclic AMP production clearly requires consideration of cyclic AMP in the medium as well as in the cells. The importance of this has not generally been realized. Changing the culture medium to a simple salts plus glucose solution (without serum) resulted in marked and rapid release of cyclic AMP into the medium by WI-38 and WI-38VA cells, and in increased cellular levels of cyclic AMP. The significance of release of cyclic AMP into the medium following medium change or hormonal stimulation with respect to cell function is not clear at present.

The inhibition of basal and epinephrine- and PGE_1-stimulated levels of cyclic AMP by serum indicates that the catalytic function of the adenylate cyclase, rather than specific hormone receptors, is involved in this effect. Also removal of serum from the medium increases total (cell plus medium) cyclic AMP. We find similarly that incubation of mouse thymocytes with phytohemagglutinin or concanavalin A decreases basal and also catecholamine- and PGE_1-stimulated adenylate cyclase activity (Makman 1971b,c). It is possible that concanavalin A and serum glycoproteins interact similarly with cell surface constituents, in turn directly or indirectly altering the entire adenylate cyclase system.

The different degree of loss by different cells of catecholamine responsiveness with continuous exposure of cells to catecholamines appears to be a major deter-

Table 9

Cyclic AMP Content and Growth of Cells Exposed Continuously
to Hormones, Prostaglandin E_1, or Nucleotides

Cell culture, duration and type of treatment	Cyclic AMP (pmoles/mg protein)	Increase in cell protein ($\mu g/cm^2$)
SV40 transformed WI-38, 2 days		
1. control	6.2	20
PGE$_1$, 200 μg/ml	36	10
2. control	3.1	21
epinephrine, 3 μM	4.2	23
5-hydroxytryptamine, 3 μM	3.3	21
WI-38 (passage 20), 2 days		
1. control	4.8	23
PGE$_1$, 200 μg/ml	17	16
2. control	5.4	25
epinephrine, 3 μM	9.5	25
5-hydroxytryptamine, 3 μM	7.1	23
WI-38 (passage 44), 2 days		
control	6.2	20
PGE$_1$, 200 μg/ml	36	10
WI-38 (passage 25), 5 days		
control	6.7	18
dibutyryl cyclic AMP, 1 mM	59	18
WI-38 (passage 39), 5 days		
control	11.2	19
dibutyryl cyclic AMP, 1 mM	58	18
Rabbit lens epithelium, 5 days		
control	7.2[9.8]*	14[10.0]*
dibutyryl cyclic AMP, 0.5 mM	281 [258]*	10[6.3]*
cyclic AMP, 0.5 mM	99 [113]*	10[3.8]*
epinephrine, 20 μM	14 [18]*	12[2.5]*
5-AMP, 0.5 mM	6.2	14

Experiments were begun 1–2 days after initial planting of cells on 35-mm or 60-mm
diameter plastic dishes. Medium was changed and agents readded every 24 hr. Values
for cyclic AMP were obtained 24 hr after the last treatment and included monobutyryl
cyclic AMP for cells treated with dibutyryl cyclic AMP.

* Values in brackets are for cells grown in the presence of 1 mM theophylline.

minant of total responsiveness of cells to catecholamines. The much more prolonged response to epinephrine of WI-38 and WI-38VA cells, as compared to that
of Chang's liver and HeLa cells, may also be of interest with respect to cyclic AMP-
dependent regulatory processes. Such processes could be different in cells capable
of achieving higher and more sustained intracellular levels of cyclic AMP in
response to hormones.

It is of interest that the SV40 virus-transformed WI-38 cells have consistently
lower basal cyclic AMP levels than normal WI-38 but nevertheless greater hormonal

responsiveness. This divergence indicates that cell transformation could have a complex effect on cyclic AMP-dependent processes in a cell with variable exposure to hormone, i.e., a transformed cell in vivo. It is also clear that one can make no assumptions concerning the level of adenylate cyclase or hormonal stimulation based upon basal levels of cyclic AMP alone.

Although some transformed or malignant cells have adenylate cyclase systems markedly stimulated by epinephrine (WI-38VA, rat C-6 astrocytoma) and by PGE$_1$ (WI-38VA), normal WI-38 cells at early and late passage show responses to these agents which, although less than those of WI-38VA, are still quite marked. The greater epinephrine responsiveness (with respect to cyclic AMP formation) of low density WI-38 cells at late passage than of those at early passage might indicate a role of adenylate cyclase in the aging process. Against this, however, is the finding that early and late passage WI-38 and WI-38VA cells appear to be equally resistant to growth inhibition by agents that increase intracellular cyclic AMP. In contrast, lens epithelial cells are much more sensitive to growth inhibition by epinephrine and dibutyryl cyclic AMP. This inhibition in lens cells might reflect a physiological response since administration of epinephrine to animals in vivo produces an inhibition of labeled thymidine incorporation into lens epithelium (Leeson and Voaden 1970; Grimes and von Sallman 1972). The basis for the insensitivity of WI-38 cells to growth inhibition by cyclic AMP and the mechanism of the inhibition produced by catecholamines in the lens epithelial cells are two problems of interest that will require further investigation.

Acknowledgments

This work was supported by grants 5R01 CA13176 and 1P01HD 07173 from the National Institutes of Health. We thank J. E. Pike of the Upjohn Company for the generous gift of prostaglandin E$_1$.

REFERENCES

Boucek, R. J. and T. R. Alvarez. 1970. 5-Hydroxytryptamine: A cytospecific growth stimulator of cultured fibroblasts. *Science* **167:**898.

Brown, J. H. and M. H. Makman. 1972. Stimulation by dopamine of adenylate cyclase in retinal homogenates and of adenosine 3':5'-cyclic monophosphate formation in intact retina. *Proc. Nat. Acad. Sci.* **69:**539.

———. 1973. Influence of neuroleptic drugs and apomorphine on dopamine-sensitive adenylate cyclase of retina. *J. Neurochem.* **21:**477.

Burton, K. 1968. Determination of DNA concentration with diphenylamine. *Methods in enzymology* (ed. L. Grossman and K. Moldave) vol. 12B, p. 163–166. Academic Press, New York.

Clark, R. B. and J. P. Perkins. 1971. Regulation of adenosine 3':5'-cyclic monophosphate concentration in cultured human astrocytoma cells by catecholamines and histamine. *Proc. Nat. Acad. Sci.* **68:**2757.

d'Armiento, M., G. S. Johnson and I. Pastan. 1972. Regulation of adenosine 3':5'-cyclic monophosphate phosphodiesterase activity in fibroblasts by intracellular concentrations of cyclic adenosine monophosphate. *Proc. Nat. Acad. Sci.* **69:**459.

Gilman, A. G. 1970. A protein binding assay for adenosine 3':5'-cyclic monophosphate. *Proc. Nat. Acad. Sci.* **67:**305.

Gilman, A. G. and M. Nirenberg. 1971a. Regulation of adenosine 3':5'-cyclic monophosphate metabolism in cultured neuroblastoma cells. *Nature* **234:**356.

————. 1971b. Effect of catecholamines on the adenosine 3′:5′-cyclic monophosphate concentrations of clonal satellite cells of neurons. *Proc. Nat. Acad. Sci.* **68**:2165.

Girardi, A. J., F. C. Jensen and H. Koprowski. 1965. SV40 transformation of human diploid cells: Crisis and recovery. *J. Cell. Comp. Physiol.* **65**:69.

Grimes, P. and L. von Sallmann. 1972. Possible cyclic adenosine monophosphate mediation in isoproterenol-induced suppression of cell division in rat lens epithelium. *Invest. Ophthalmol.* **11**:231.

Hayflick, L. 1965. The limited *in vitro* lifetime of human diploid cell strains. *Exp. Cell Res.* **37**:614.

Hayflick, L. and P. S. Moorhead. 1961. The serial cultivation of human diploid cell strains. *Exp. Cell Res.* **25**:585.

Johnson, G. S., R. M. Friedman and I. Pastan. 1971. Restoration of several morphological characteristics of normal fibroblasts in sarcoma cells treated with adenosine 3′:5′-cyclic monophosphate and its derivatives. *Proc. Nat. Acad. Sci.* **68**:425.

Leeson, S. J. and M. Voaden. 1970. A chalone in the mammalian lens. II. Relative effects of adrenaline and noradrenaline on cell division in the rabbit lens. *Exp. Eye Res.* **9**:67.

Lowry, O. H., N. J. Rosebrough, A. L. Farr and R. J. Randall. 1951. Protein measurement with the Folin phenol reagent. *J. Biol. Chem.* **193**:265.

Makman, M. H. 1970. Adenyl cyclase of cultured mammalian cells. Activation by catecholamines. *Science* **170**:1421.

————. 1971a. Hormone sensitive adenyl cyclase of cultured human and mouse lbroblasts and of cells of malignant origin. *Fed. Proc.* **30**:458 (abstr.)

————. 1971b. Properties of adenylate cyclase of lymphoid cells. *Proc. Nat. Acad. Sci.* **68**:885.

————. 1971c. Conditions leading to enhanced response to glucagon, epinephrine, or prostaglandins by adenylate cyclase of normal and malignant cultured cells. *Proc. Nat. Acad. Sci.* **68**:2127.

————. 1972. Catecholamine receptor inactivation in Chang's liver cells due to continued exposure to catecholamines. *Fed. Proc.* **31**:883 (abstr.)

Makman, M. H. and H. L. Kern. 1971. Adenyl cyclase of lens, cultured lens epithelial cells and retina: Stimulation by catecholamines. *Fed. Proc.* **30**:1205 (abstr.)

————. 1972. Catecholamine-stimulated adenylate cyclase of lens. Distribution, changes during lens maturation, and presence in cultured lens epithelial cells. *Ophthalmol. Res.* **3**:15.

Makman, M. H. and M. I. Klein. 1972. Expression of adenylate cyclase, catecholamine receptor and cyclic AMP-dependent protein kinase in synchronized cultures of Chang's liver cells. *Proc. Nat. Acad. Sci.* **69**:456.

Manganiello, V. and M. Vaughan. 1972. Prostaglandin E_1 effects on adenosine 3′5′-cyclic monophosphate concentration and phosphodiesterase activity in fibroblasts. *Proc. Nat. Acad. Sci.* **69**:269.

Otten, J., G. S. Johnson and I. Pastan. 1971. Cyclic AMP levels in fibroblasts: Relationship to growth rate and contact inhibition of growth. *Biochem. Biophys. Res. Commun.* **14**:1192.

————. 1972. Regulation of cell growth by cyclic adenosine 3′5′-monophosphate: Effect of cell density and agents which alter cell growth on cyclic adenosine 3′5′-monophosphate levels in fibroblasts. *J. Biol. Chem.* **247**:7082.

Shapiro, A. L., I. M. Siegel, M. D. Scharff and E. Robbins. 1969. Characteristics of cultured lens epithelium. *Invest. Ophthalmol.* **8**:393.

Sheppard, J. R. 1972. Differences in the cyclic adenosine 3′5′-monophosphate levels in normal and transformed cells. *Nature New Biol.* **236**:14.

Sherline, P., A. Lynch and W. H. Glinsmann. 1972. Cyclic AMP and adrenergic receptor control of rat liver glycogen metabolism. *Endocrinology* **91**:680.

Uzunov, P., H. M. Shein and B. Weiss. 1973. Cyclic AMP phosphodiesterase in cloned astrocytoma cells: Norepinephrine induces a specific enzyme form. *Science* **180**:304.

Sequential Biochemical Events in Preparation for DNA Replication and Mitosis

Robert A. Tobey, Lawrence R. Gurley, C. E. Hildebrand, Robert L. Ratliff, and Ronald A. Walters

Cellular and Molecular Radiobiology Group, Los Alamos Scientific Laboratory
University of California, Los Alamos, New Mexico 87544

Regulation of mammalian cell proliferation in vivo is an extraordinarily complex process mediated by a multiplicity of interacting operations of ill-defined nature. To facilitate detection of major regulatory events, it would be desirable to obtain a simpler in vitro model system in which large quantities of proliferating cells could be reversibly induced to enter a nonproliferating, biochemically stable state similar to the G_1-arrested or G_0 state observed in vivo. We have developed a technique at this Laboratory in which mammalian cells grown in vitro may be reversibly arrested in G_1 by limiting the concentration of isoleucine in the culture medium (Tobey and Ley 1971; Enger and Tobey 1972; Tobey 1973). The properties of these cells arrested in the G_1 state at least superficially mimic properties of G_0 cells in vivo in that (1) cells are arrested in the G_1 phase of the cell cycle; (2) cells remain viable during arrest for prolonged periods; (3) cells maintain high levels of biosynthetic activity for various cellular macromolecules except DNA; (4) cells do not enter a state of gross biochemical imbalance during arrest; and (5) cells can be stimulated to reenter readily the proliferative state.

In the work reported here we have employed cells reversibly arrested in G_1 by the isoleucine deprivation technique to yield information on events which are prerequisite to initiation of DNA synthesis and which represent early evidence for conversion of cells from a nonproliferating to a proliferating state. Additional data are presented describing processes that appear to be intimately involved in chromosome condensation during entry into mitosis.

METHODS

Cell Culture and Synchronization

Chinese hamster cells (line CHO) were maintained free of PPLO in F-10 medium supplemented with 10% calf and 5% fetal calf sera and antibiotics. Cells in suspension culture were synchronized and accumulated in G_1-arrest following maintenance in isoleucine-deficient medium for 36 hours, as described previously (Tobey and Ley 1971; Tobey 1973). Resumption of cell-cycle traverse was accomplished by addition of twice the normal F-10 concentration of isoleucine to the arrested culture or by resuspension of arrested cells in fresh medium containing twice the normal concentration of isoleucine. In certain experiments cells prepared by mitotic selection or released from isoleucine-mediated G_1-arrest were treated for 9 hours with 10^{-3} M hydroxyurea to resynchronize cells in very late G_1 near the G_1/S boundary (Tobey and Crissman 1972). Reversal of hydroxyurea effects was accomplished by placing washed cells in fresh, hydroxyurea-free medium. In certain experiments cells were synchronized by mitotic selection without use of drugs or trypsin, as described previously (Tobey, Anderson and Petersen 1967).

Histone Isolation and Purification

Histone fractions were prepared from labeled cells by the first method of Johns (1964), as previously described for cultured cells by Gurley and Hardin (1968), except that 0.14 M 2-mercaptoethanol was present in all solvents and solutions used for extraction and recovery of arginine-rich histones in order to prevent dimerization of histone f3 (Smith, Delange and Bonner 1970). Lysine-rich (f1 and f2b) and arginine-rich (f2a1, f2a2, and f3) histone fractions were subjected separately to preparative gel electrophoresis, as described previously by Gurley and Walters (1971), using the method of Panyim and Chalkley (1969) adapted for use with a Canalco Prep-Disc apparatus. Purified histone fractions were removed from the bottom of the gel by a cross-flow of buffer and collected in 2-ml fractions for liquid scintillation counting, as described previously (Gurley and Walters 1971; Gurley et al. 1973a,b,c).

Isolation of DNA-Lipoprotein Complexes

A modification of the sarkosyl crystal technique (Earhart et al. 1968; Hanoaka and Yamada 1971) was employed for isolation of DNA-lipoprotein complexes from CHO cells, as described in detail elsewhere (Hildebrand and Tobey 1973). Briefly washed cells were treated with a solution of sodium lauroyl sarcosinate (sarkosyl) 0.1% final concentration at 4°C for 3 min to bring about cell lysis. The lysates were then sheared by vortex mixing for 20 sec, followed by addition of $MgCl_2$ to produce sarkosyl crystals. The crystal suspension was then layered onto a sucrose step-gradient comprised of 20% and 47% sucrose solutions. The discontinuous gradients were centrifuged at 15,000 rpm for 20 min at 4°C, during which time sarkosyl crystals containing the DNA-lipoproteins sedimented to the 20/47% sucrose interphase, forming the characteristic layer designated the "M band" (Earhart et al. 1968). Gradient fractions were collected into 10% trichloroacetic acid (TCA) in 0.01 M pyrophosphate containing carrier bovine serum albumin. After several hours in an ice bath, samples were filtered through Whatman GF-82 filter discs and washed with TCA/pyrophosphate solution. The air-dried filters were counted in a liquid scintillation spectrometer.

Determination of Deoxyribonucleoside Triphosphate Pool Levels

The technique employed for analysis of deoxyribonucleotide pools has been described in detail elsewhere (Walters, Tobey and Ratliff 1973). Washed cell suspensions were treated with 0.5 M $HClO_4$ and following centrifugation the supernatant was neutralized with KOH. Samples were then centrifuged, the pelleted $KClO_4$ was discarded, and the supernatant was utilized in the pool assay. The assay employed for determination of dATP and dTTP was essentially that of Lindberg and Skoog (1970), and the measurement of dCTP and dGTP was performed essentially as described by Skoog (1970) except that (1) *Micrococcus luteus* DNA polymerase was used instead of *Escherichia coli* DNA polymerase, and (2) α-^{32}P-labeled deoxyribonucleoside triphosphates were employed instead of tritium-labeled deoxyribonucleotides. Neutralized cell extracts were added to solutions containing Tris-HCl, $MgCl_2$, 2-mercaptoethanol, appropriate template—poly[d(A-T)] for determination of dATP or dTTP pools and poly[d(I-C)] for determination of dCTP or dGTP pools—and *M. luteus* DNA polymerase. After incubation for 1 hr at 37°C, the samples were collected on a Whatman GF/B glass filter disc, precipitated with a solution containing 5% TCA in pyrophosphate, washed, dried, and counted in a liquid scintillation spectrometer.

The rationale for this method is as follows. Utilizing poly[d(A-T)] as a primer, DNA polymerase incorporates dATP and dTTP into an acid-insoluble product in a strictly alternating sequence so that equimolar amounts of dATP and dTTP are polymerized. When one deoxyribonucleotide is present in limiting amount (as in our cell extracts), the extent of polymerization is dependent on amount of this nucleotide. When the other nucleotide is labeled and is present in excess, the extent of polymer formation can be measured by incorporation of radioactivity into an acid-precipitable form, and the amount of radioactivity in the polymer is directly proportional to amount of the limiting nucleotide in the pool from the cell extract. Similar arguments apply to incorporation of dGTP or dCTP utilizing poly[d(I-C)] as template.

RESULTS

Cell Synchronization Techniques

Cells transferred to medium containing suboptimal quantities of isoleucine are arrested in G_1 but remain capable of synthesizing various cellular macromolecules without entering a state of gross biochemical imbalance (Tobey and Ley 1971; Enger and Tobey 1972; Tobey 1973). Reversal of arrest is accomplished by restoration of isoleucine. In Fig. 1A are results obtained when cells maintained for 36 hours in isoleucine-deficient medium were resuspended in fresh isoleucine-containing medium supplemented with [^{3}H]thymidine. At intervals thereafter aliquots were removed for determination of the fraction of labeled cells scored radioautographically (triangles) and fraction of cells dividing (circles). Since the divided fraction represents $N/N_0 - 1$, a true population doubling would be indicated by an increase from 0 to 1 on the scale provided in Fig. 1A. For comparison cells prepared by mitotic selection and resuspended in medium containing [^{3}H]thymidine were examined for labeled fraction (diamonds) and divided fraction (squares). It is readily apparent that, in both cultures, cells initially in early G_1 began entering S phase at 4 hours and began dividing at 12 hours, with essentially the entire population traversing the cell cycle. Progressive loss of synchronization arising from variations in rate of cycle progression by individual cells in the population prevents detailed study of late interphase events. It was necessary to synchronize cells at a later stage in the cell cycle for investigation of processes in late interphase. This was accomplished by resynchronizing cells prepared by mitotic selection or released from isoleucine-mediated G_1-arrest at the G_1/S boundary through treatment with

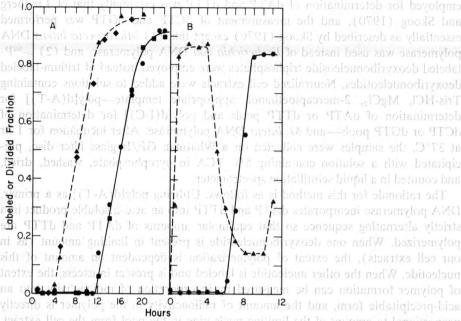

Figure 1

Techniques employed to synchronize CHO cells. **(A)** Cells synchronized by the isoleucine deficiency technique were resuspended in fresh complete medium containing twice the normal concentration of isoleucine at $t = 0$. [³H]Thymidine was added to a final concentration of 0.05 μCi/ml. At intervals aliquots were removed for determination of the labeled fraction (▲) and scored radioautographically, and the divided fraction (•) was determined with an electronic cell counter. Cells prepared by mitotic selection were placed in suspension culture along with [³H]thymidine (0.05 μCi/ml), and aliquots were removed at intervals for radioautographic determination of the labeled fraction (♦) and determination of the divided fraction (■).

(B) Cells were synchronized by the isoleucine deficiency/hydroxyurea technique, then washed and resuspended in fresh complete medium at $t = 0$. At intervals 5-ml aliquots were pulse-labeled for 15-min periods in a 37°C shaker water bath with 10 μCi [³H]thymidine, after which radioautographs were prepared and the fraction of labeled cells (▲) was determined. Cell number determination was accomplished with an electronic cell counter (•). The divided fraction represents $N/N_0 - 1$. [³H]Thymidine (6 Ci/mmole) was purchased from Schwarz BioResearch.

hydroxyurea (Tobey and Crissman 1972). Following removal of hydroxyurea most cells entered the S phase immediately thereafter and commenced dividing in highly synchronous fashion within 6 hours (Fig. 1B). Synchronization techniques described in Fig. 1 were utilized in the various studies that follow.

Cell Cycle-specific Changes in Histone Phosphorylation

A currently attractive concept suggests that reversible chemical modifications of histones are involved in modulating the physical state of chromatin, thereby exerting control of biological activity (Allfrey 1971; Bradbury and Crane-Robinson

1971). This concept suggests that changes in histone modification patterns might be expected as cells undergo transitions in their metabolic states, both in proliferating and in nonproliferating cultures. Support for this notion was provided by the observations of Balhorn et al. (1972) of a linear correlation between growth rate in a variety of tumors and extent of f1 phosphorylation. That is, f1 phosphorylation capacity is directly related to proliferative capacity.

With these thoughts in mind, we began a study of the kinetics of histone phosphorylation in both nonproliferating cultures and at varying stages of the cell cycle in proliferating cultures. Histone f2a2 phosphorylation was observed to occur in G_1-arrested cultures as well as in all phases of the cell cycle in proliferating cells (Gurley et al. 1973a,b). In contrast phosphorylation of histone f1 was virtually absent in nonproliferating cultures (Gurley et al. 1973a). The f1 phosphorylation rate also was very low in the early stages after reversal of isoleucine-mediated G_1-arrest (Gurley et al. 1973a) but steadily increased prior to entry into S phase (Fig. 2). Extrapolation of the phosphorylation rate curve in Fig. 2 to zero indicated

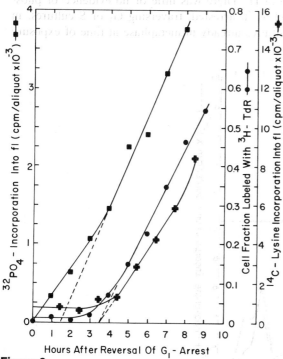

Figure 2

Phosphorylation and synthesis of histone f1 in synchronized CHO cells. After 36 hr in isoleucine-deficient medium, isoleucine (twice the normal F-10 concentration) was added back to the arrested culture. At intervals aliquots of 500 ml at 300,000 cells/ml were pulse-labeled for 1-hr periods with either 10 μCi $^{32}PO_4$ or 125 μCi [^{14}C]lysine and the amount of label associated with electrophoretic f1 fractions was calculated to reveal patterns of phosphorylation of f1 (■) or synthesis of f1 (crosses). Aliquots from a culture synchronized in parallel were exposed to 1 μCi/ml [^{3}H]thymidine for 1-hr periods and the labeled fraction was scored radioautographically (●). [^{14}C]Lysine (310 μCi/mmole) was purchased from Schwarz BioResearch.

that f1 phosphorylation preceded entry into the DNA synthetic phase of the cell cycle by approximately 2 hours. Note in Fig. 2 that newly synthesized f1 appeared simultaneously on chromatin with entry into S phase. These results clearly indicate that *phosphorylation* of f1 precedes *synthesis* of this histone and also initiation of DNA synthesis by approximately 2 hours. Further studies not shown indicated that phosphorylation of f1 continued at an increased rate through late interphase and into mitosis (Gurley et al. 1973b).

Histone phosphorylation was studied also in cells traversing from G_2 into mitosis (Fig. 3). The culture was prepared in the following manner. Exponentially growing monolayer cultures, prelabeled for 52 hr with [^{3}H]lysine, were treated with Colcemid and $H_3{}^{32}PO_4$ for 2 hr, and then the accumulated cells arrested in metaphase were dislodged selectively from the monolayer with a mechanical shaker. All metaphase-arrested cells accumulated in this sample had been exposed to $^{32}PO_4$ as they moved from G_2 into M. Note in Fig. 3 that, in addition to the phosphorylated f1 and f2a2 species normally seen in mid- to late-interphase and mitotic samples (Gurley et al. 1973a,b,c), there were two additional phosphorylated histones: f3 and a slowly migrating subfraction of f1. There was little or no evidence of phosphorylation of the f3 or f1 subfraction in arrested, traversing G_1 or S cultures, in traversing G_2-rich cultures, or in cultures already in metaphase at time of exposure

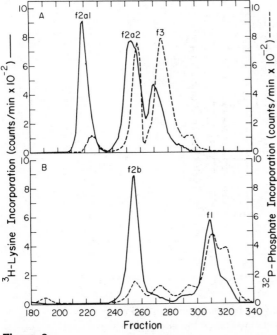

Figure 3

Preparative electrophoresis profile of histone phosphorylation in CHO cells traversing from G_2 into M. Cells prelabeled with [^{3}H]lysine for 52 hr and grown in monolayers were exposed to 0.12 μg/ml Colcemid and 20 μCi/ml $^{32}PO_4$ for 2 hr, after which accumulated metaphase cells were removed by mitotic selection and histones were isolated and analyzed by preparative gel electrophoresis.

(———) [^{3}H]lysine label;

(– – –) label from $^{32}PO_4$ (i.e., phosphorylated histones).

to labeled phosphate (Gurley et al. 1973a,b,c). Furthermore in cultures grown in $^{32}PO_4$ during the G_2 to M transition and then allowed to reenter G_1 in the absence of label, there was a nearly total dephosphorylation of the f3 and f1 subfractions (Gurley et al. 1973c). Taken together the data obtained previously and the results shown in Fig. 3 indicate that histone f3 and f1 subfraction are phosphorylated specifically as cells cross the G_2/M boundary and are dephosphorylated as cells move out of mitosis into G_1.

Cell Cycle-specific Changes in DNA-Lipoprotein Complexes

A growing body of evidence in both bacteria (review by Klein and Bonhoeffer 1972) and in mammalian cells (Hanoaka and Yamada 1971; Mizuno et al. 1971a,b; Hildebrand and Tobey 1973) suggests involvement of DNA-lipoprotein complexes in some aspect of genome replication, although the nature of their role is unclear. In a previous publication we described techniques for isolation and characterization of complexes containing DNA and lipoprotein (Hildebrand and Tobey 1973). (In line with current convention we shall utilize the terms "DNA-lipoprotein" and "DNA-membrane" interchangeably to refer to the complexes, although we have not yet demonstrated unequivocally that the lipoprotein portion is indeed membrane.) These complexes were resistant to the effects of RNase, but treatment with DNase or heat in excess of 37°C released DNA from the complexes while leaving the lipoprotein fragments intact. Both Pronase and sodium dodecyl sulfate released DNA from the complexes and solubilized the lipoprotein component. Less than 10% of the DNA in the complexes could be attributed to nonspecific attachment. Pulse-labeled, newly replicated DNA was located initially in close proximity to the complexes, but it was possible to chase a significant portion of the pulse-labeled DNA away from the complexes. Finally results suggested that, if the complexes serve as initiation sites for DNA replication as well as attachment points of DNA to membrane, only a relatively small fraction of DNA replication sites are membrane-associated (Hildebrand and Tobey 1973).

In view of specific properties of these complexes, we utilized highly synchronized cultures to examine their involvement in cell cycle events (Hildebrand and Tobey 1973). In one experiment cells whose DNA had been prelabeled with [^{14}C]thymidine for two generation times were obtained from monolayers by mitotic selection and allowed to traverse the cell cycle in suspension culture in label-free medium. At intervals thereafter aliquots were removed from the culture for determination of the divided fraction (Fig. 4A, triangles) and for determination of the percent of old [^{14}C]thymidine prelabeled DNA associated with lipoprotein (i.e., old DNA in the M band) by the affinity of this material to sarkosyl crystals (sarkosyl crystal assay, see Fig. 4A, circles). From a nonprelabeled culture synchronized in parallel, aliquots were removed and pulse-labeled with [^{3}H]thymidine to determine radioautographically the labeled fraction (i.e., fraction of cells in S phase) at various times after synchronization (Fig. 4A, squares). A small amount of DNA was associated with membrane (or membrane fragments) even during mitosis and early G_1, with a sudden enhanced association commencing at approximately 2 hr prior to initiation of DNA replication. An approximately twofold increase in amount of DNA attached to membrane was observed in the mid-S phase as compared to values in M or early G_1.

Cells whose DNA had been prelabeled with [^{14}C]thymidine were synchronized by

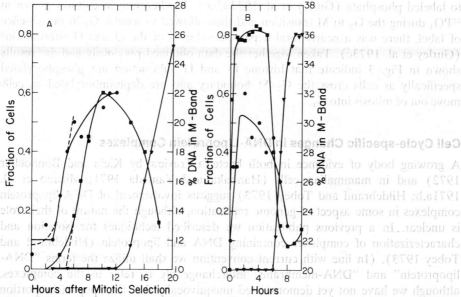

Figure 4

DNA-membrane complexes in synchronized CHO cells. **(A)** Cells prelabeled for 33 hr with 0.006 μCi/ml [^{14}C]thymidine were synchronized by mitotic selection and were placed in spinner flasks at $t = 0$. At intervals the fraction of cells in the M band was calculated by the sarkosyl crystal assay ($\bullet$), while the fraction of labeled cells ($\blacksquare$) was determined from radioautographs prepared from an unlabeled culture synchronized in parallel and pulse-labeled with [^{3}H]thymidine as in Fig. 1B. Cell number determination ($\blacktriangledown$) was made with an electronic counter.

(B) Cells synchronized by mitotic selection were treated with 10^{-3} M hydroxyurea between 1 and 10 hr after selection: then the washed cells were resuspended in fresh complete medium at $t = 0$. The labeled fraction ($\blacksquare$) and divided fraction ($\blacktriangledown$) were calculated as in Fig. 1B. The fraction of DNA in the M band ($\bullet$) was calculated by the sarkosyl crystal technique from a culture prelabeled with [^{14}C]thymidine and synchronized in parallel. The divided fraction represents $N/N_0 - 1$.

combined mitotic selection and hydroxyurea (as in Fig. 1B) to examine DNA-membrane complex formation during late interphase. Following removal of hydroxyurea the divided fraction (Fig. 4B, triangles) and the percent of old [^{14}C]thymidine prelabeled DNA associated with membrane (Fig. 4B, circles) were determined. The fraction of cells in S phase was determined from samples pulse-labeled with [^{3}H]thymidine (Fig. 4B, squares) taken from a nonprelabeled culture synchronized by means of the mitotic selection/hydroxyurea treatment technique. At the time of removal of hydroxyurea (time 0 sample in Fig. 4B when cells were accumulated at the G_1/S boundary), the percent of DNA in the M band was already at a high level, indicative of enhancement of DNA complexed to membrane prior to entry into S phase (shown in Fig. 4A). There was a further enhancement of DNA associated with membrane after hydroxyurea removal, continuing through S and into G_2 and decreasing rapidly at approximately the time that cells began dividing (Fig. 4B). Thus the results demonstrate an enhanced association of

DNA complexed to membrane commencing approximately 2 hr prior to entry into S phase and persisting at a high level through G_2.

Cell Cycle-specific Changes in Levels of Deoxynucleoside Triphosphate Pools

Before a cell can commence synthesizing DNA, adequate levels of deoxyribonucleotides obviously must be present. To determine the kinetics of deoxyribonucleoside triphosphate acid-soluble pools as a function of proliferation stage and cell cycle position, a series of experiments were performed with synchronized CHO cells. A culture prepared by mitotic selection was allowed to traverse the cell cycle in suspension culture, and at intervals thereafter aliquots were removed for determination of the quantities of each of the four deoxyribonucleoside triphosphates in acid-soluble pools. Additional aliquots were removed and pulse-labeled with [³H]thymidine and the labeled cell fraction was determined radioautographically to indicate the fraction of cells synthesizing DNA. The levels of dATP, dGTP, and dCTP were extremely low during G_1, with only a small quantity of dTTP present at this time (Fig. 5). However shortly before cells began entering S phase, the levels of all four deoxyribonucleotides began increasing. Extrapolation of the curves representing deoxyribonucleotide contents back to the baseline G_1 values revealed that the increase began at from 0.3 to 0.6 hr prior to initiation of DNA synthesis. Although not shown, the levels of all four deoxyribonucleoside triphosphates in isoleucine-mediated, G_1-arrested cultures were equivalent to the low values obtained in traversing cells in early G_1 shown in Fig. 5 (Walters et al. 1973). To demonstrate changes in pool levels in late interphase, cells were resynchronized at the G_1/S boundary by combined mitotic selection and hydroxyurea treatment. Shortly

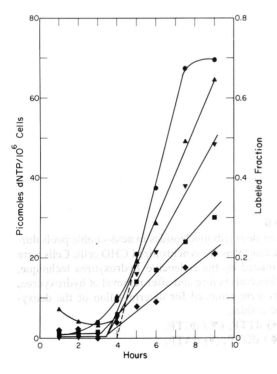

Figure 5

Levels of deoxyribonucleotides in acid-soluble pools during early interphase in synchronized CHO cells. Cells were synchronized by mitotic selection, and at intervals the levels of (▲) dTTP, (▼) dCTP, (■) dATP and (♦) dGTP were determined. The fraction of cells pulse-labeled with [³H]thymidine was determined radioautographically as in Fig. 1B.

before removal of hydroxyurea, the levels of dTTP, dCTP, and dGTP were already elevated even though DNA synthesis had not commenced (Fig. 6). In contrast the level of dATP remained at the very low G_0/early G_1 level. Within 3 minutes after removal of hydroxyurea, the level of dATP increased ninefold and DNA synthesis commenced simultaneously. After a subsequent early drop in levels of dTTP, dGTP, and dATP, the levels of these deoxyribonucleotides continued to increase throughout S and G_2, whereas the dCTP content increased across the S phase, then began decreasing as cells entered G_2.

These trends continued as evidenced from the data presented in Fig. 7A. Mitotic cells possessed high levels of dTTP, dATP, and dGTP, while the level of dCTP continued to decrease and was low in mitotic cells (time 0 sample). As cells progressed rapidly out of mitosis, the levels of all four deoxyribonucleotides also dropped precipitously to their characteristic low G_1 values.

Because the level of dATP remained at the low G_1 level, whereas the other deoxyribonucleotides increased as the result of accumulation at the G_1/S boundary following treatment with hydroxyurea (Fig. 6), the kinetics of change in nucleotide pools was followed in cells selected in mitosis and allowed to progress to G_1/S in the presence of hydroxyurea (Fig. 7B). The results indicate that, as expected, dATP remains in limited supply throughout the period of treatment with hydroxy-

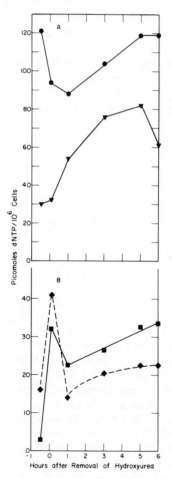

Figure 6
Levels of deoxyribonucleotides in acid-soluble pools during late interphase in synchronized CHO cells. Cells were synchronized by the isoleucine/hydroxyurea technique, and at intervals before and after removal of hydroxyurea, aliquots were removed for determination of the deoxyribonucleotides.
A (●) dTTP; (▼) dCTP.
B (◆) dGTP; (■) dATP.

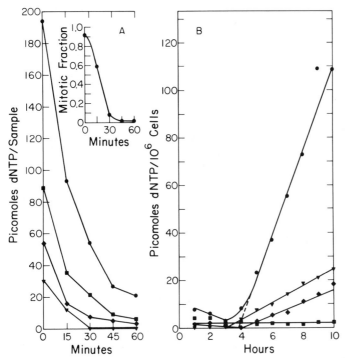

Figure 7
Levels of deoxyribonucleotides in acid-soluble pools in CHO cells escaping from mitosis and traversing early interphase in the presence of hydroxyurea. **(A)** As cells progressed out of mitosis following mitotic selection, the levels were determined at the indicated intervals. The insert shows the rate of progression of cells out of mitosis.

(B) Cells were synchronized by mitotic selection and at intervals aliquots were removed for determination.

(●) dTTP; (■) dATP; (♦) dGTP; (▼) dCTP.

urea. The other three nucleotides began increasing in quantity at approximately the same time as in cultures not exposed to hydroxyurea (Fig. 5). These results suggest a specific effect of hydroxyurea on accumulation of dATP in the CHO cell, thereby preventing the cells from initiating DNA synthesis.

DISCUSSION

By combining the results described in this review with previously published data from this Laboratory, it is possible to construct a map illustrating temporal relationships between biochemical events and the CHO cell cycle (Fig. 8). The temporal sequence of events shown in Fig. 8 obviously applies specifically to line CHO Chinese hamster cells grown under our cultivation conditions. Histone f1 phosphorylation was virtually absent in G_1-arrested cells, and f1 was phosphorylated at only a very low rate during the early stages after resumption of proliferation (Gurley et al. 1973a). Extensive phosphorylation of f1 commenced at 2 hours prior to initiation of DNA replication and continued at increased rates throughout late interphase into mitosis. Since *phosphorylation* of f1 preceded *synthesis* of this histone,

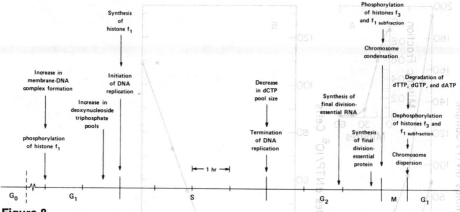

Figure 8

Temporal map of biochemical events in the CHO cell cycle.

it is apparent that the f1 species phosphorylated 2 hours prior to entry into S phase was not newly synthesized, but instead was a preexisting f1 species synthesized during a previous proliferation cycle. It appears that phosphorylation of old f1 prior to initiation of DNA synthesis represents one of the earliest indications that a cell has embarked upon a program of proliferation. Stated differently, phosphorylation of f1 represents the first detectable sign that the cells are in G_1 and are preparing for DNA replication and division.

At precisely the time at which phosphorylation of f1 commences in preparation for DNA replication, there is a sudden enhancement in amount of DNA associated with presumptive membrane lipoprotein. From studies on the physical state of chromatin, Bradbury, Carpenter and Rattle (1973) have suggested that f1 may play an integral role in chromatin organization and condensation into chromosomes. We suggest the possibility that initiation of f1 phosphorylation may alter the structural configuration of chromatin beginning 2 hours prior to entry into S phase, allowing an enhanced association of chromatin with lipoprotein. Thus phosphorylation of f1 and concomitant increased association of DNA with lipoprotein may be coordinated and thereby explain the temporal identity of the two processes. Further support for the notion that the two processes may be linked is provided by the observation that elevated levels of DNA complexed to lipoprotein and f1 phosphorylation continued through late interphase and decreased at approximately the time cells began dividing (Gurley et al. 1973c). Phosphorylation of f1 would then be expected to be involved intimately in proliferation.

Why should phosphorylation of histone f1 commence prior to initiation of DNA synthesis and continue throughout interphase into mitosis? Continued phosphorylation of f1 throughout most of the cycle might suggest that cell cycle progression is dependent upon continuous phosphorylation of histone f1 and might further suggest that continuous subtle modifications in chromatin structure are induced by phosphorylation of f1 throughout most of interphase. In regard to the latter, Pederson (1972) has obtained interesting results with synchronized HeLa cells and BSCb green monkey cells. Pederson found that accessibility of DNA to DNase and actinomycin varied at different stages of the cell cycle, suggesting the existence of a "chromosome cycle" in eukaryotic cells involving continuous alterations in chromatin structure during interphase, culminating in chromosome condensation during mitosis. The chromosome cycle was interrupted in nonproliferating cells (Pederson, 1972) as was phosphorylation of f1 (Gurley et al. 1973a). In view of our findings and those of Bradbury et al. (1973) relating chromatin/chromosome structural changes to f1, we suggest that phosphorylation of f1 may be involved in the cell cycle-specific changes in chromatin accessibility observed by Pederson.

Following phosphorylation of f1 and enhanced association of DNA with membrane 2 hours before S, the levels of all four deoxyribonucleoside triphosphates increased, starting at from 0.3 to 0.6 hours before initiation of DNA replication. Genome replication was prevented by inhibiting the increase in just one of the deoxyribonucleotides, dATP, with hydroxyurea. Within several minutes after removal of hydroxyurea, the level of dATP increased dramatically, and concomitantly cells began synthesizing DNA. Although the levels of dGTP, dATP, and dTTP remained high through S, G_2, and into M, the level of dCTP began to drop as soon as cells completed the S phase and entered G_2. Thus it appears that, in the CHO cell, the pool level of only dCTP is roughly correlated with DNA synthesis. The precipitous drop in level of all four deoxyribonucleotides just prior to or coincident with reentry into G_1 represents a specific process rather than simple diffusion to the cytoplasm during nuclear membrane disaggregation in prophase and subsequent destruction by cytoplasmic phosphatases. Had the latter process occurred, the drop in nucleoside triphosphate activity should have taken place at the beginning of mitosis such that cells prepared by mitotic selection (which are predominantly in metaphase [Tobey et al. 1967]) should have contained minimal rather than maximal levels of nucleotides.

As cells continue progression through the proliferation cycle, preparations are made for mitosis and cell division. At approximately 70 minutes prior to entry into mitosis, the last species of RNA essential for completion of mitosis and cell division is synthesized (Tobey et al. 1966a), while at 12 to 15 minutes prior to entry into M the final synthesis of division-essential protein occurs (Tobey et al. 1966b; Walters and Petersen 1968). Phosphorylation of histone f3 and the f1 subfraction at precisely the time that cells crossed the boundary from G_2 into M and subsequent dephosphorylation of these histones as cells reentered G_1 (Gurley et al. 1973c) strongly suggest that phosphorylation of these histones plays an integral role in chromosome condensation. Based upon nuclear magnetic resonance studies of calf thymus deoxyribonucleoprotein, Bradbury et al. (1973) have speculated that phosphorylation of f1 may trigger chromosome condensation. Taken together these data suggest that phosphorylation of f3 and f1 to yield the f1 subfraction may represent the final interphase operation, terminating in visible chromosome condensation and entry into prophase.

Reentry into G_1 (or G_0) is accompanied by rapid degradation of the deoxyribonucleotides and dephosphorylation of histone f3 and f1 subfraction concomitant with chromosome decondensation. Unlike the total dephosphorylation of f3 and the f1 subfraction, species of f1 and f2a2 phosphorylated in the previous interphase carry over a portion of their phosphate moieties into early G_1 (Gurley et al. 1973c).

The experiments in this report represent an initial attempt to isolate and identify some of the processes indicative of proliferation as a first step toward understanding proliferative control mechanisms. All of the preceding data indicate that regulation of proliferation and maintenance of progression throughout the cell cycle, even in our relatively simple in vitro model system, involve a complex series of coordinated processes. Thus the diagrammatic temporal representation of events in Fig. 8 presents a grossly oversimplified picture.

Our studies indicate that, once a decision is made to reenter the proliferation cycle, a sequence of preprogrammed operations is initiated long before actual replication of DNA commences. Once cells are committed to a program of proliferation, prevention of DNA replication does not necessarily prevent initiation of preparative events. For example, in cells resuming cycle traverse after reversal of isoleucine-mediated G_1-arrest in the presence of hydroxyurea (i.e., cells traverse to the G_1/S boundary but do not synthesize DNA), three processes situated prior to entry into S phase are initiated in preparation for DNA synthesis. They are phosphorylation of histone f1 (unpublished observations), concomitant increase in membrane-associated DNA, and an increase in amount of three of the four deoxyribonucleotides at the normal time. That is, all these processes, presumably prerequisite to initiation of genome replication, are sequentially "turned on" in preparation for initiation of DNA synthesis, indicating that the regulatory mechanisms were still operative. It will be interesting in future experiments to determine whether or not there is a precise, obligatory sequence of preparative events. Must histone f1 phosphorylation always precede the increase in deoxyribonucleotide pool levels, or can the order of occurrence be inverted? If it becomes possible to inhibit specifically the initiation of histone f1 phosphorylation, will the nucleotide levels increase nevertheless? Problems of this nature which relate to specific regulatory mechanisms may soon be amenable to biochemical analysis.

Acknowledgments

The authors would like to acknowledge the excellent technical assistance of Mr. Joseph G. Valdez, Mr. John L. Hanners, Mrs. Evelyn W. Campbell, and Mrs. Phyllis C. Sanders in the performance of these experiments. This work was performed under the auspices of the U.S. Atomic Energy Commission.

REFERENCES

Allfrey, V. G. 1971. Functional and metabolic aspects of DNA-associated proteins. *Histones and nucleohistones* (ed. D. M. P. Philips) pp. 241–294. Plenum Press, New York.

Balhorn, R., M. Balhorn, H. P. Morris and R. Chalkley. 1972. Comparative high resolution electrophoresis of tumor histones: Variation in phosphorylation as a function of cell replication rate. *Cancer Res.* **32**:1775.

Bradbury, E. M. and C. Crane-Robinson. 1971. Physical and conformational studies of histones and nucleohistones. *Histones and nucleohistones* (ed. D. M. P. Phillips) pp. 85–134. Plenum Press, New York.

Bradbury, E. M., B. G. Carpenter and H. W. E. Rattle. 1973. Magnetic resonance studies on deoxyribonucleoprotein. *Nature* **241**:123.

Earhart, C. F., G. Y. Tremblay, M. J. Daniels and M. Schaechter. 1968. DNA replication studied by a new method for the isolation of cell membrane-DNA complexes. *Cold Spring Harbor Symp. Quant. Biol.* **33**:707.

Enger, M. D. and R. A. Tobey. 1972. Effects of isoleucine deficiency on nucleic acid and protein metabolism in cultured Chinese hamster cells. Continued RNA and protein synthesis in the absence of DNA synthesis. *Biochemistry* **11**:269.

Gurley, L. R. and J. M. Hardin. 1968. The metabolism of histone fractions. I. Synthesis of histone fractions during the life cycle of mammalian cells. *Arch. Biochem. Biophys.* **128**:285.

Gurley, L. R. and R. A. Walters. 1971. Response of histone turnover and phosphorylation to X-irradiation. *Biochemistry* **10**:1588.

Gurley, L. R., R. A. Walters and R. A. Tobey. 1973a. The metabolism of histone fractions. VI. Differences in the phosphorylation of histone fractions during the cell cycle. *Arch. Biochem. Biophys.* **154**:212.

————. 1973b. Histone phosphorylation in late interphase and mitosis. *Biochem. Biophys. Res. Commun.* **50**:744.

————. 1973c. Cell cycle-specific changes in histone phosphorylation associated with cell proliferation and chromosome condensation. *J. Cell Biol.* (in press).

Hanoaka, F. and M. Yamada. 1971. Localization of the replication point of mammalian cell DNA at the membrane. *Biochem. Biophys. Res. Commun.* **42**:647.

Hildebrand, C. E. and R. A. Tobey. 1973. DNA-membrane complexes in Chinese hamster cells: Cell-cycle studies. *Fed. Proc.* **32**:640a (abstr.).

Johns, E. W. 1964. Studies on histones. VII. Preparative methods for histone fractions from calf thymus. *Biochem. J.* **92**:55.

Klein, A. and F. Bonhoeffer. 1972. DNA replication. *Ann. Rev. Biochem.* **41**:301.

Lindberg, U. and L. Skoog. 1970. A method for the determination of dATP and dTTP in picomole amounts. *Anal. Biochem.* **34**:152.

Mizuno, N. S., C. E. Stoops and R. L. Pfeiffer, Jr. 1971b. Nature of the DNA associated with the nuclear envelope of regenerating liver. *J. Mol. Biol.* **59**:517.

Mizuno, N. S., C. E. Stoops and A. A. Sinha. 1971a. DNA synthesis associated with the inner membrane of the nuclear envelope. *Nature New Biol.* **229**:22.

Panyim, S. and R. Chalkley. 1969. High resolution acrylamide gel electrophoresis of histones. *Arch. Biochem. Biophys.* **130**:337.

Pederson, T. 1972. Chromatin structure and the cell cycle. *Proc. Nat. Acad. Sci.* **69**:2224.

Skoog, L. 1970. An enzymatic method for the determination of dCTP and dGTP in picomole amounts. *Eur. J. Biochem.* **17**:202.

Smith, E. L. and R. J. Delange and J. Bonner. 1970. Chemistry and biology of the histones. *Physiol. Rev.* **50**:159.

Tobey, R. A. 1973. Production and characterization of mammalian cells reversibly arrested in G_1 by growth in isoleucine-deficient medium. *Methods in cell biology* (ed. D. M. Prescott) vol. 6, p. 67. Academic Press, New York.

Tobey, R. A. and H. A. Crissman. 1972. Preparation of large quantities of synchronized mammalian cells in late G_1 in the pre-DNA replicative phase of the cell cycle. *Exp. Cell Res.* **75**:460.

Tobey, R. A. and K. D. Ley. 1971. Isoleucine-mediated regulation of genome replication in various mammalian cell lines. *Cancer Res.* **31**:46.

Tobey, R. A., E. C. Anderson and D. F. Petersen. 1966b. RNA stability and protein synthesis in relation to the division of mammalian cells. *Proc. Nat. Acad. Sci.* **56**:1520.

————. 1967. Properties of mitotic cells prepared by mechanically shaking monolayer cultures of Chinese hamster cells. *J. Cell. Physiol.* **70**:63.

Tobey, R. A., D. F. Petersen, E. C. Anderson and T. T. Puck. 1966a. Life cycle analysis of mammalian cells. III. The inhibition of division in Chinese hamster cells by puromycin and actinomycin. *Biophys. J.* **6**:567.

Walters, R. A. and D. F. Petersen. 1968. Radiosensitivity of mammalian cells. II. Radiation effects on macromolecular synthesis. *Biophys. J.* **8**:1487.

Walters, R. A., R. A. Tobey and R. L. Ratliff. 1973. Cell cycle dependent variations of deoxyribonucleoside triphosphate pools in Chinese hamster cells. *Biochim. Biophys. Acta* (in press)

Relationships between Nuclear Protein Phosphorylation and Gene Activation in the Cell Cycle of Synchronized HeLa S-3 Cells

V. G. Allfrey, J. Karn, E. M. Johnson, and G. Vidali

The Rockefeller University, New York, New York 10021

In higher organisms, as in prokaryotes, the utilization of DNA as a template for ribonucleic acid synthesis involves both a restriction and a selective activation of particular genetic loci. The mechanisms through which such control is achieved depend, at least in part, on associations between DNA and proteins that influence the structure of the genetic material and its interactions with RNA-polymerizing enzymes.

Examples of DNA-associated proteins that regulate RNA synthesis in bacteria include the repressor protein, which inhibits transcription of the *lac* operon (Gilbert and Müller-Hill 1967; Riggs et al. 1968), and the cyclic AMP receptor protein, which promotes transcription of *lac* and *gal* messenger RNAs (Zubay, Schwartz and Beckwith 1970; Anderson et al. 1971). The first is a relatively acidic protein with a high specificity for particular nucleotide sequences in DNA, whereas the latter is a highly basic protein with less discrimination in its DNA binding properties.

In higher organisms the corresponding elements of genetic control—i.e., suppression of template activity of most of the DNA and activation of RNA synthesis at particular genetic loci—also require the intervention of proteins with different properties, DNA binding affinities, and contrasting effects on transcription. The proteins concerned include the histones, the basic, suppressive structural components of chromatin, and other more acidic proteins that show strong indications of involvement in the positive control of RNA synthesis at specific loci of the genome (Kleinsmith et al. 1966a; Gilmour and Paul 1969; Benjamin and Goodman 1969; Kamiyama and Wang 1970; Allfrey et al. 1971; Teng et al. 1971; Spelsberg et al. 1971; Baserga and Stein 1971; Kamiyama et al. 1972; Kostraba and Wang 1972; Stein et al. 1972; Rickwood et al. 1972; Shea and Kleinsmith 1973).

The present report deals with the latter class of acidic nuclear proteins and emphasizes changes in their metabolism during the replicative cycle of synchronously growing mammalian cells. Particular attention will be paid to the post-synthetic modification of such proteins by enzymatic phosphorylation and dephosphorylation of seryl and threonyl residues in their polypeptide chains.

The question arises as to whether the nuclear phosphoproteins are altered at times of gene activation and repression during the cell cycle. It is known that the synthesis of RNA in synchronously dividing cells is suppressed during mitosis (Pfeiffer and Tolmach 1968; Johnson and Holland 1965; Farber, Stein and Baserga 1972), whereas the synthesis of particular messenger RNAs (e.g., histone mRNAs) is restricted to the late G_1 and early S phases of the cell cycle (Robbins and Borun 1967; Gallwitz and Mueller 1969; Breindl and Gallwitz 1973).

The synthesis of the nuclear acidic proteins, unlike that of the histones, proceeds throughout the cell cycle (Stein and Baserga 1970; Stein and Borun 1972; Borun and Stein 1972). In synchronously dividing HeLa S-3 cells there is an increased rate of synthesis and accumulation of these proteins in the nucleus preceding the onset of DNA synthesis (Stein and Borun 1972). The amount of nuclear protein synthesized, transported, and retained in the acidic chromosomal fraction is greater immediately after mitosis and later in G_1 than in the S or G_2 phases of the cell cycle (Borun and Stein, 1972).

The modification of the nuclear acidic proteins by phosphorylation adds an additional complexity to this analysis. Our preliminary studies of ^{32}P-phosphate uptake into HeLa nuclear proteins during the cell cycle indicate that the rate of phosphorylation is maximal in the early S phase and decreases in the late S and G_2 phases when RNA synthesis is also reduced (Allfrey at al. 1973). The suppression of nuclear acidic protein phosphorylation during M has recently been reported by Platz, Stein and Kleinsmith (1973).

The present study is a detailed analysis of the uptake and turnover of ^{32}P-phosphate in the phenol-soluble, nuclear acidic proteins of synchronously growing HeLa S-3 cells.

MATERIALS AND METHODS

Cell Culture and Synchronization

HeLa S-3 cells were maintained in suspension culture at 2–6×10^5 cells per ml by daily dilution with fresh Joklik-modified Minimal Essential Medium containing 10% fetal calf serum and supplemented with 2.5 units/ml penicillin G, 2.5 μg/ml streptomycin, and 20 units/ml mycostatin.

Synchronization was achieved by the double thymidine block method (Bootsma, Budke and Vos 1964; Puck 1964), exposing the cells to 2 mM thymidine for 14 hours, to normal medium for the following 9 hours, and to 2 mM thymidine for an additional 14 hours. Cells were harvested by centrifugation at $1500 \times g$ for 4 minutes and resuspended in one-fifth of the original volume of thymidine-free medium. The time of release from the thymidine block is taken as starting at this resuspension. The cells were centrifuged and resuspended in the original volume of thymidine-free medium.

The cell cycle was monitored by measurements of cell concentration, mitotic index, and [^{3}H]thymidine incorporation rate. Rates of thymidine incorporation into DNA were determined by incubating 1-ml aliquots of the cell suspension in the presence of 5 μCi of [^{3}H]thymidine of specific activity 20 Ci/mmole for 30 minutes at 37°C. The cells were transferred to Millipore filters, washed, and counted.

Isotopic Labeling of the Nuclear Proteins

At different times after removal of the thymidine block, $4–8 \times 10^7$ cells were harvested by centrifugation and gently resuspended in 5 ml culture medium containing 2 mCi of carrier-free [^{32}P]orthophosphate. After 15 minutes incubation at 37°C, the cells were centrifuged and the cell pellet stored at $-80°C$ prior to isolation of the nuclei.

Estimates of turnover in the nuclear phosphoprotein fraction were based on measurements of isotope retention in the nuclear proteins of cells which had been prelabeled with [^{32}P]phosphate and [^{14}C]leucine. One liter of cells containing $4–6 \times 10^5$ cells/ml was taken after the first thymidine block and exposed to 25 mCi of [^{32}P]orthophosphate and 0.5 mCi of L-[U-^{14}C]leucine (specific activity 316 mCi/mmole) for 9 hours in the thymidine-free medium and for 14 hours in the second thymidine block. The cells were harvested by centrifugation, washed in nonradioactive medium, and resuspended in 1 liter of thymidine-free medium for the cold-chase experiments. At the indicated times after removal of the second thymidine block, 100 ml aliquots of the suspension were withdrawn for isolation of the nuclei and preparation of the nuclear proteins.

Isolation of Nonhistone Nuclear Proteins

Nuclei and chromatin (from mitotic cells) were isolated by a modification of the method of Hancock (1969). Cells ($4–8 \times 10^7$) were homogenized in 5 ml 80 mM NaCl–20 mM EDTA containing 1% Triton X100, pH 7.2. The nuclei were purified by differential centrifugation and analyzed. They were free of cytoplasmic or whole cell contamination, as judged by phase-contrast microscopy. The protein to DNA ratio of the final nuclear pellet was about 2.8 to 1.

The nuclear phosphoprotein fraction was prepared by the method of Shelton and Allfrey (1970), as detailed by Teng et al. (1971). The nuclei were extracted twice with 0.14 M NaCl and twice with 0.25 N HCl. The residue was washed with 1:1 chloroform-methanol containing 0.2 N HCl and with 2:1 chloroform-methanol-HCl. The residue was then suspended in 0.1 M Tris-HCl pH 8.4, containing 0.01 M EDTA and 0.14 M 2-mercaptoethanol. The suspension was gently mixed with an equal volume of phenol (saturated with the buffer) and allowed to stand for 14 hours at 2°C. The combined phenol extracts containing the nuclear phosphoproteins were dialyzed against a series of urea-containing buffers (Teng et al. 1971) to restore the proteins to the aqueous phase.

Electrophoretic Analysis of Nuclear Protein Fractions

Histones were characterized by electrophoresis in 15% polyacrylamide gels containing 6.25 M urea, following the method of Panyim and Chalkley (1969). The nonhistone protein fractions were dialyzed overnight against 0.01 M sodium phosphate buffer pH 7.4, containing 0.1% SDS and 0.14 M 2-mercaptoethanol. The proteins were separated by electrophoresis in 10% polyacrylamide gels containing 0.1% SDS as described by Teng et al. (1971).

The protein bands were stained with 1% Fast Green in 7% acetic acid–35% methanol. Densitometric analysis of the stained gels was carried out in a Gilford spectrophotometer equipped with a linear transport device. Estimates of the molecular weights of individual protein bands were made using plots of mobility vs.

molecular weight for proteins of known molecular weight over the range 12,000–160,000, all measured under identical electrophoretic conditions.

For measurement of isotope distribution in different protein bands, the gels were sliced transversely in 1-mm slices, each of which was dissolved in 200 μl of 50% H_2O_2 and mixed with scintillation fluid for counting. ^{32}P and ^{14}C activities were measured by scintillation spectrometry.

Pulse-labeling of ATP Pools with [^{32}P]phosphate

At different times after removal of the thymidine block, cells were harvested by centrifugation and gently resuspended in 5 ml culture medium containing 2 mCi of carrier-free [^{32}P]orthophosphate. After 15 minutes incubation at 37°C, the cells were centrifuged and extracted with 4 ml cold 0.6 N HClO$_4$. After reextraction with 0.2 N HClO$_4$, the extracts were combined, neutralized, and subjected to chromatography on Dowex-1 (formate) as described by Hurlbert (1957). The ATP fraction was eluted and further purified by chromatography on Whatman No. 3 paper in ethanol : 1 M ammonium acetate : water, 66.5 : 30: 3.5 (v/v/v).

Spots corresponding in R$_f$ to authentic ATP standards were cut out and the ATP eluted and analyzed by measuring absorbancy at 259 nm and ^{32}P activity.

Measurement of Cyclic AMP-dependent Protein Kinase Activity

The levels of cyclic AMP-dependent protein kinase activity were measured in high-speed supernatant fractions obtained after homogenizing HeLa cells in 0.32 M sucrose containing 0.1% Triton X100. The homogenate was centrifuged for 60 minutes at $100,000 \times g$ and aliquots of the supernatant containing 20 μg protein were incubated at 31°C with 100 μg of histone F1 as substrate and [γ-^{32}P] ATP in the presence or absence of 10^{-6} M cyclic AMP as described previously (Maeno, Johnson and Greengard 1971; Johnson and Allfrey 1972). The reaction was stopped by the addition of 5% TCA containing 0.25% sodium tungstate and the precipitate was washed and assayed for ^{32}P activity.

Chemical Analyses

Protein was determined by the method of Lowry et al. (1951). DNA was determined by the diphenylamine reaction as modified by Burton (1956), using highly polymerized calf thymus DNA as a standard. The amino acid composition of the phosphoprotein fractions was determined by ion exchange chromatography after the method of Spackman, Stein and Moore (1958). For determination of alkali-labile phosphate, the proteins were first dialyzed exhaustively against distilled water and hydrolyzed in 1.0 N NaOH at 100°C for 5 minutes.

Inorganic phosphate released into the supernatant was analyzed as the phosphomolybdate complex after precipitation of the protein and extraction into 1 : 1 isobutanol-benzene, as described previously (Kleinsmith et al. 1966b; Teng et al. 1971).

RESULTS AND DISCUSSION
Composition of HeLa Cell Nuclei (or Chromatin) at Different Stages in the Cell Cycle

Nuclear protein and DNA contents have been examined in synchronously growing HeLa S-3 cells at different times after release from a double thymidine block.

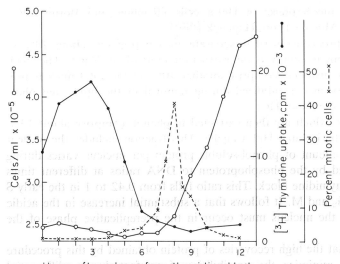

Figure 1
Changes in cell number, mitotic index and [³H]thymidine uptake into DNA at
different stages in the cell cycle. The cells were synchronized by the double
thymidine block method and aliquots of the suspension were analyzed as described in Materials and Methods. The growth parameters are plotted against
the time of release from the thymidine block.

The degree of synchrony achieved is illustrated in Fig. 1, which plots three parameters of growth: cell number, mitotic index, and rate of thymidine uptake into
DNA. The S phase, as measured by the rate of DNA labeling, begins immediately
upon release of the cells from the thymidine block and lasts for 5 to 6 hours.
Maximal rates of DNA synthesis occur at 3 hours. By 8 hours the majority of the
cells have entered mitosis. New cells are first evident at about 7.5 hours; by 12
hours the population has almost doubled.

The timing of these events is highly reproducible and was determined for each
of the isotope labeling experiments to be described. On the basis of this data we
consider the G_2 period of the cell cycle to extend from 5.5–7.5 hours, and the G_1
phase of the following cycle is taken as the period from 8.5–12 hours.

In order to monitor changes in nuclear proteins during the cell cycle, we have
counted the number of cells in the population and analyzed for DNA content and
protein distribution in the recovered nuclei and in each of the nuclear extracts. The
results show that the DNA recovery in the isolated nuclei (or chromatin fractions)
is very high at all stages in the cell cycle, with an average recovery of 86.1 ± 1.7%.
This minimizes the risks of artefact due to the selection of a small or variable fraction of the nuclei for analysis.

These are significant variations in the protein to DNA ratio of the nuclei during
synchronous cell growth, ranging from 2.2 to 1 to as high as 3.3 to 1. This ratio
drops in the late S phase and increases during the G_1 period of the following cycle.
The change in protein to DNA ratio is largely the result of varying proportions of
the nonhistone nuclear proteins, because the histone to DNA ratios are not appreciably altered during the cell cycle but remain constant at about 1.1 to 1. This
is consistent with observations that histone and DNA synthesis proceed con-

comitantly throughout the S phase of HeLa cells (Robbins and Borun 1967; Spalding, Kajiwara and Mueller 1966; Hancock 1969).

In the fractionation procedure used to separate nuclear proteins, about 40% of the total nuclear protein is removed on extraction in 0.14 M NaCl. The acid-soluble proteins, mainly histones, comprise another 40% of the total nuclear proteins. Little, if any, protein is solubilized during removal of the lipids and phospholipids in chloroform-methanol-HCl.

The residual proteins, which are then extracted in phenol, comprise about 13% of the total protein in the isolated HeLa nuclei. This fraction includes the nuclear phosphoproteins. The amount of phenol-soluble protein per nucleus varies during the cell cycle, as judged by the phosphoprotein to DNA ratios at different times after release from the thymidine block. This ratio falls from 0.42 to 1 in the early S phase to 0.28 to 1 in G_2 and M. It follows that a substantial increase in the acidic protein complement of the nucleus must occur in the prereplicative phase of the cycle.

It should be noted that the high recoveries of protein obtained by this procedure (approximately 96%) minimize the possibilities of artefact due to differential extraction of proteins depending upon stages in the cell cycle. As a further check we have compared the recovery of phenol-soluble proteins from metaphase chromosomes (prepared from cells blocked in mitosis by exposure to vinblastine sulfate for 16 hours) and from nuclei obtained from an unsynchronized cell population. No indications of differential extractability were obtained.

Characterization of the Nuclear Phosphoprotein Fraction

The nuclear phosphoprotein fraction comprises a heterogeneous mixture of proteins differing in molecular weight, amino acid composition, and degree of phosphorylation.

The molecular size heterogeneity is indicated by differences in electrophoretic mobility in SDS-polyacrylamide gels. A complex banding pattern reveals the presence of multiple polypeptide chains ranging in molecular weight from 18,000 to 170,000 daltons (Fig. 2). At least 21 major bands are detectable; each of these, in turn, may include different protein species of similar or identical molecular weights.

An alignment of the banding patterns of the nuclear phosphoproteins obtained at different stages of the cell cycle is presented in Fig. 2. The results show a remarkable uniformity in the relative concentrations of the different protein bands. Some minor differences are detectable by densitometry, in agreement with the findings of Bhorjee and Pederson (1972), but on the whole it is the stability of the protein pattern, rather than its variability, which attracts attention. This observation on a dividing, but nondifferentiating, cell population contrasts with findings that the nuclear acidic protein complement changes appreciably in cells during the course of differentiation (LeStourgeon and Rusch 1971, 1973; Hill, Poccia and Doty 1971; Conner and Patel 1972; Vidali et al. 1973).

It has been noted previously that the synthesis of nuclear acidic proteins proceeds throughout the HeLa cell cycle (Stein and Baserga 1970; Stein and Borun 1972; Borun and Stein 1972), with increasing rates of synthesis in late G_1 (Stein and Borun 1972). This is in accord with the increase in phosphoprotein to DNA ratio in the prereplicative phase that we have observed. The reproducibility of the

Hours after release
from thymidine block

0 1.5 3 4.5 6 7.5 9 10.5 12

Figure 2
Electrophoretic patterns in 0.1% SDS–10% polyacrylamide gels of HeLa nuclear acidic proteins prepared at different stages of the cell cycle. The proteins were extracted by the phenol procedure from nuclei isolated at the indicated times after release from the thymidine block. Note the uniformity in protein banding patterns throughout the cycle.

gel patterns at different times suggests that the synthesis and accumulation of many of the major nuclear acidic proteins are under close coordinate control.

The amino acid analyses of the HeLa phosphoprotein fractions (Table 1) confirm the impression of constant proportionality of the various components throughout the cell cycle, although one would expect to detect only gross differences by this technique. The average amino acid composition of the nuclear phosphoprotein fraction shows a clear predominance of the acidic amino acids, aspartic and glutamic acids (21 mole percent) over the basic amino acids, lysine, arginine, and histidine (15 mole percent). This is in accord with amino acid analyses of the corresponding protein fractions from other cell types (Teng et al. 1971).

Nuclear Protein Phosphorylation during the Cell Cycle

The rate of incorporation of [^{32}P]phosphate into the nuclear proteins varies at different stages in the cell cycle. Comparisons were made by selecting aliquots of the cell suspension at different times after release from the thymidine block and pulse-labeling for 15 minutes in the presence of [^{32}P]orthophosphate. The cells were frozen immediately and the nuclear phosphoprotein fraction was isolated. The specific activity of the phosphoprotein fraction was determined and plotted against time (Fig. 3). It should be pointed out that the major advantage of the phenol isolation procedure is the fact that contaminating ^{32}P-labeled nucleic acids are left in the aqueous phase as the phosphoproteins move into the phenol phase. All tests for nucleic acid contamination have proven negative (Teng et al. 1971).

Two peaks of phosphate incorporation in the cell cycle are observed. The first occurs early in the S phase (between 1.5 and 3 hours) and the second peak occurs in early G_1 (at about 10 hours). The rate of phosphate uptake appears to be somewhat greater in S than in G_1; an uptake ratio of about 1.2 to 1.0 was consistently observed. The phosphorylation of the nuclear acidic proteins is markedly reduced in the late S and G_2 phases of the cell cycle and remains low into the M

Table 1

Amino Acid Composition of the Nuclear Phosphoprotein
Fraction at Different Stages of the HeLa S-3 Cell Cycle

	Hours after release of cells from thymidine block				
	0	2	5	9	12
	*(moles/100 moles total amino acids)**				
Lysine	6.3	6.8	6.4	6.6	6.3
Histidine	2.2	2.4	2.5	2.4	2.3
Arginine	6.0	5.7	5.8	6.4	5.9
Aspartic acid	9.7	9.5	9.7	9.8	9.6
Threonine	5.4	5.6	5.1	4.7	5.5
Serine	7.3	7.3	7.3	7.2	6.7
Glutamic acid	12.5	12.5	12.7	11.9	11.5
Proline	4.6	4.6	5.1	4.6	4.9
Glycine	9.0	9.5	9.3	9.2	9.0
Alanine	7.2	7.0	7.2	7.3	7.0
Valine	6.2	6.3	6.4	6.2	5.9
Methionine	1.1	1.5	1.7	2.0	1.5
Isoleucine	5.8	4.8	4.7	4.9	5.8
Leucine	9.7	9.4	8.9	9.3	10.4
Tyrosine	2.9	3.3	3.1	3.4	3.6
Phenylalanine	3.8	3.9	3.7	4.0	3.9

* Values not corrected for hydrolytic losses.

period. Thus the rate of [^{32}P]phosphate incorporation into the nuclear phosphoproteins is high at periods of intense RNA synthesis and low when transcription is
suppressed.

The possibility that estimates of the rate of nuclear protein phosphorylation
might be in error due to injurious effects of the thymidine double block was tested
by comparing ^{32}P uptakes in control (unsynchronized) cultures and in cultures exposed to 2 mM thymidine for 2- or 4-hour periods before pulse-labeling with ^{32}P.
The phosphate uptake into the nuclear proteins of control and thymidine-treated
cells is virtually identical (Table 2). Moreover no differences in gel electrophoretic
patterns could be discerned. Other control experiments have established that
cytoplasmic protein fractions do not contribute significantly to the radioactivity of
the nuclear fractions isolated in phenol. This possibility was tested by preparing
the nuclear phosphoproteins from unlabeled cells homogenized in the presence of
a ^{32}P-labeled post-nuclear supernatant from cells that had been incubated in the
usual way with 2 mCi of [^{32}P]orthophosphate. Less than 4.8% contamination was
observed (Table 2).

It is clear from Fig. 2 that the nuclear phosphoprotein fraction includes many
components of different molecular weights. The distribution of [^{32}P]phosphate in
different size classes of nuclear phosphoproteins after pulse-labeling for 15 minutes
was determined by radioassay of the multiple bands separated by SDS-polyacrylamide gel electrophoresis. The results are shown in Fig. 4. The staining pattern
shown at the bottom of the figure is aligned with the corresponding densitometer

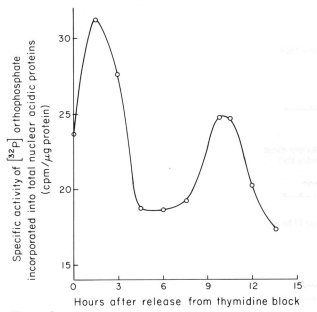

Figure 3
Changes in the rate of nuclear protein phosphorylation at different times in the cell cycle. Aliquots of HeLa cell suspension were taken at the indicated times and pulse-labeled for 15 min with [32P]orthophosphate. The nuclear phosphoprotein fraction was isolated and its 32P activity measured. The specific activity is plotted against time after release from the thymidine block. This is maximal in S and G_1 and minimal in the G_2–M period.

Table 2
Tests for Contamination of Nuclear Phosphoproteins by Cytoplasmic Proteins and for Possible Effects of Thymidine on Protein Phosphorylation

Conditions of experiment	Spec. act. phenol-soluble proteins (cpm/mg protein)
Nonradioactive nuclei + 32P-labeled cytoplasm[a]	992
Cells incubated without thymidine[b]	20,590
Cells exposed to 2 mM thymidine for 2 hr	19,930
Cells exposed to 2 mM thymidine for 4 hr	20,050

[a] Nuclei from 4×10^7 nonradioactive cells were mixed with the post-nuclear supernatant fraction obtained from 4×10^7 cells, which had been labeled for 15 min in the presence of 2 mCi of [32P]orthophosphate. The nuclei were then reisolated and the phenol-soluble proteins were prepared and counted.

[b] Cells (4×10^7) were pulse-labeled for 15 min in the presence of 2 mCi of [32P]orthophosphate in the presence or absence of 2 mM thymidine as described in Materials and Methods.

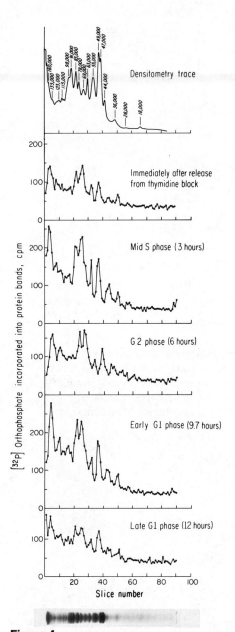

Figure 4

Distribution of [^{32}P]phosphate in the nuclear acidic proteins of HeLa S-3 cells at different stages of the cell cycle. Aliquots of the cell suspension were pulse-labeled at the indicated times by exposure to [^{32}P]orthophosphate for 15 min. The nuclear phosphoproteins were isolated and separated by electrophoresis in 0.1% SDS–10% polyacrylamide gels. The protein banding pattern is shown at the bottom of the figure and the corresponding densitometer tracing is shown in the top panel. Molecular weights are indicated above the major protein peaks. Intermediate panels show the patterns of [^{32}P]phosphate distribution in the nuclear proteins at different times after release from the thymidine block. Note the heterogeneity of the labeling pattern, the increase in rate of ^{32}P uptake in mid-S and early G$_1$, and the relatively low activity in the G$_2$ period.

tracing in the top panel. The other panels compare the distribution and specific activities of the nuclear phosphoproteins at the indicated stages in the cell cycle.

The marked heterogeneity in [^{32}P]phosphate incorporation into the proteins at different regions of the gel is evident. In some cases the peaks of ^{32}P activity do not coincide exactly with the positions of the major protein bands. This is good evidence that minor bands, not visible because of their low concentrations, will contribute disproportionately to the ^{32}P uptake measurements.

The ^{32}P activity of the individual protein bands follows quite closely the cell cycle-dependent changes described for the total phosphoprotein fraction in Fig. 3. Labeling is greatest in the S and G$_1$ phases and depressed in the G$_2$ to M phase. Some minor differences in the rate of labeling of different bands can be observed, but on the whole the pulse-labeling experiments indicate a parallel metabolic response of many diverse nuclear proteins to events occurring at different stages of the cell cycle.

The variations observed in the rates of ^{32}P incorporation into the nuclear phosphoproteins at different stages in synchronous cell growth do not appear to result from fluctuations in the specific activity of the cellular ATP pools. The incorporation of [^{32}P]phosphate into ATP pools was measured in cells at corresponding stages of the cycle, and the specific activity of the cellular ATP pool remained relatively uniform in the course of the experiment (Table 3). The minor variations observed are not likely to account for the large differences in specific activity of the phosphoprotein fraction at different stages. The latter variations are more likely to represent alterations in protein kinase activities in nuclei at different phases of the cell cycle.

Histone Phosphorylation in the Cell Cycle

The rate of histone phosphorylation was determined by pulse-labeling with [^{32}P]phosphate at different times after release from the thymidine block. The histones were extracted from the isolated nuclei (or chromatin) fractions and further purified by electrophoresis in polyacrylamide gels. The peak of histone phosphoryl-

Table 3
[^{32}P]Phosphate Incorporation
into ATP Pools of HeLa Cells

Time after release from thymidine block (hr)	Spec. act. ATP* (cpm/pmole)
0	9.7
2	10.8
5	10.5
9	11.3

* Cells (4×10^7) were incubated for 15 min in the presence of 1 mCi of [^{32}P]orthophosphate and the ATP was isolated and analyzed as described in Materials and Methods.

ation was observed in mid S (3 hours), which coincides with the peak of DNA synthesis (Fig. 1). Correlations between histone phosphorylation and DNA synthesis have been noted before (Ord and Stocken 1968; Balhorn et al. 1972; Oliver et al. 1972; Gurley, Walters and Tobey 1973). The phosphorylation of histone F1 is known to be dependent upon cell cycle position and to be active in the S phase (Balhorn et al. 1972; Oliver et al. 1972; Gurley et al. 1973).

Some differences exist between the timing of histone phosphorylation and phosphate uptake into the nonhistone proteins of the nucleus. The peak of histone phosphorylation (Fig. 5) occurs somewhat later in the cell cycle than does the corresponding peak for nuclear phosphoproteins (Fig. 3). Moreover no major peak for histone phosphorylation occurs in G_1, as it does for the nuclear acidic protein fraction. These differences argue strongly against the view that the variable rates of phosphorylation of the nuclear proteins simply reflect differences in the specific activities of *nuclear* ATP pools at different phases of the cell cycle.

Some preliminary experiments have been carried out to determine whether corresponding changes in kinase activity toward histone F1 could be detected at different stages in synchronous growth. The soluble protein kinase activity of HeLa cell homogenates was measured at different times after release from the thymidine

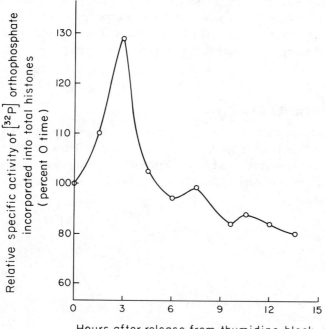

Figure 5
Altered rates of histone phosphorylation at different stages of the cell cycle. Aliquots of the HeLa cell suspension were withdrawn at the indicated times after release from the thymidine block and pulse-labeled for 15 min with [^{32}P]orthophosphate. The specific activities of the histone bands purified in polyacrylamide gels by the method of Panyim and Chalkley (1969) were determined and plotted against time. Note the peak of histone phosphorylation rate in the mid-S phase. Unlike the phosphorylation of the nonhistone nuclear proteins (see Fig. 3), no peak in ^{32}P uptake occurs in G_1.

Table 4

Specific Enzyme Activity of Soluble Protein
Kinase during the HeLa Cell Cycle

Time after release from thymidine block (hr)	Specific activity[a] — cAMP (units)[b]	Specific activity[a] + cAMP (units)[b]	Activation by cAMP (fold)
0	182	795	4.4
2	125	727	5.8
5	139	563	4.1
9	166	545	3.3
12	94	450	4.8

[a] Assay performed as described in Materials and Methods with histone fraction F1 as substrate in the presence and absence of 10^{-6} M cyclic AMP.

[b] One unit of enzyme activity = 1 pmole of phosphate incorporated per mg of enzyme protein in 5 min at 31°C.

block, both in the presence and absence of cyclic AMP. The activation of histone F1 phosphorylation by cyclic AMP was observed to vary from 3-fold to 6-fold during the cycle, being highest at early S and lowest in early M (Table 4).

Turnover of Phosphoprotein-phosphate at Different Stages of the Cell Cycle

The retention by nuclear phosphoproteins of previously incorporated phosphate groups was compared at different times after release of the cells from the thymidine block. In these tests the acidic nuclear proteins were prelabeled in a 23-hour incubation with [^{32}P]orthophosphate and [^{14}C]leucine, as described in Materials and Methods. After washing to remove the radioactive precursors (and thymidine), the cells were incubated in a nonradioactive medium. Aliquots of the suspension were withdrawn at different times during the cold chase for preparation and analysis of the nuclear phosphoprotein fraction.

The results summarized in Fig. 6 compare the retention of [^{32}P]phosphate and [^{14}C]leucine in the total nuclear phosphoprotein fraction. After a brief initial period in which specific activities increase slightly, radioactivity is lost following exponential decay curves. The rate of ^{32}P loss greatly exceeds that of the ^{14}C label; protein phosphate activity declines with an average half-life of 6.7 hours, while the ^{14}C activity of the proteins takes 25 hours to reach 50% of the original activity. These divergent results indicate that phosphoryl groups in the proteins are subject to removal without a corresponding degradation of the polypeptide chain. The loss of [^{32}P]phosphate from the proteins has also been confirmed by analysis of the alkali-labile phosphate activity in the protein samples.

The rate of ^{32}P turnover varies in different nuclear phosphoproteins. The distribution of [^{32}P]phosphate in different size classes of proteins fractionated by electrophoresis in SDS-polyacrylamide gels is shown in Fig. 7. The staining pattern (shown at the bottom of the figure) and the densitometer tracing (in the top panel) show the positions and molecular weights of the major protein bands. (These are fully comparable with the results presented in Fig. 4 for the ^{32}P pulse-labeling ex-

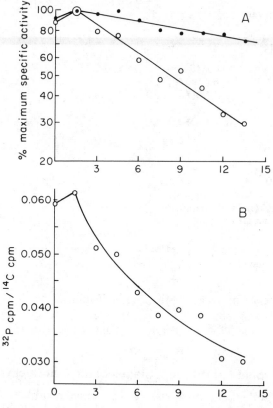

Figure 6

Relative rates of turnover of [^{32}P]phosphate and [^{14}C]leucine in nuclear phosphoproteins of synchronously growing HeLa cells. The acidic nuclear proteins were
prelabeled for 23 hr in the presence of [^{14}C]leucine and [^{32}P]orthophosphate
as described in Materials and Methods. After washing the cells were incubated
under cold-chase conditions and aliquots of the suspension were withdrawn at
the indicated times for preparation and analysis of the nuclear phosphoprotein
fraction.

A, The percent of the maximal specific activity (cpm/mg) is plotted against
time for [^{41}C]leucine (●——●) and for [^{32}P]phosphate (o——o).

B, The ratio of ^{32}P activity to ^{14}C activity is plotted against time. The decreasing ratio indicates that phosphate groups in the protein are subject to
replacement without a corresponding degradation of the polypeptide chain.

periments.) The other panels in Fig. 7 compare the distribution and specific activities of the nuclear phosphoproteins at different stages of the cell cycle.

Each band appears to have a distinctive half-life for its ^{32}P-labeled phosphate
groups. For example, a band appearing at molecular weight 28,000 decays at the
slowest rate, with a half-life of about 12 hours. Another band of MW 55,000
decays most rapidly, with a phosphate half-life of about 5.5 hours. A band at MW
47,000 loses its [^{32}P]phosphate at an intermediate rate, with a half-life of about
9 hours. Many of the bands in the higher molecular weight region of the gel have

half-lives of the order of 6 hours. Such differences in [^{32}P]phosphate metabolism are readily detectable in the long-term, cold-chase experiments but are not at all obvious in the short-term, pulse-labeling experiments. It follows that the retention of [^{32}P]phosphate is a more sensitive index of differential phosphoryl group turnover in different nuclear acidic proteins than is the uptake of [^{32}P]phosphate in short-term incubations.

Phosphate Content of Isolated Nuclear Phosphoproteins

The average phosphorus content of the total nuclear phenol-soluble protein fraction was determined at different times after release from the thymidine block. There are no major fluctuations in the alkali-labile phosphorus content of the proteins prepared at different stages of the cell cycle (Table 5). The steady-state level determined by direct analysis represents a balance between the turnover and replacement of phosphate groups on "old" protein molecules, as well as an incorporation of phosphate into newly synthesized acidic proteins of the nucleus. It is clear from the ^{32}P pulse-labeling experiments that rates of phosphorylation undergo striking changes at different periods of the cell cycle. The studies of isotope retention show that individual nuclear proteins differ in the rates at which they turn over their phosphate groups. The complexity and diversity of these structural modifications are not evident in the overall phosphate level of the acidic protein fraction.

Based on an average phosphorus content of 0.1% (by weight), it can be estimated that a protein of molecular weight 120,000 containing about 73 seryl residues (Table 1) would contain only about four of these in the phosphorylated form.

SUMMARY AND CONCLUSIONS

The phosphoprotein fraction of HeLa S-3 cells is complex and contains proteins ranging in molecular weight from 18,000 to 170,000 daltons. The relative concentrations of the major protein bands do not change appreciably during the cell cycle, but the ratio of acidic protein to DNA increases during G_1 and drops in the late S phase. There are clear differences in rates of phosphorylation of the nuclear proteins at different stages in the cell cycle. Protein phosphorylation is most active at periods when RNA synthesis is high (G_1 and S) and it is minimal in G_2–M when RNA synthesis is suppressed. These changes are not due to changes in the specific activities of the cellular ATP pools, which remain relatively constant. They probably reflect alterations in nuclear protein kinase activities at different phases of the cycle.

The phosphorylation of histones also varies during the cell cycle. Maximum rates of ^{32}P incorporation are observed in mid-S, at the peak of DNA synthesis; this is somewhat later than the peak of phosphorylation of the nonhistone proteins. Moreover no peak of histone phosphorylation occurs in G_1, as it does for the nuclear acidic protein fraction. Tests of kinase activity during the cell cycle using histone F1 as substrate, in the presence or absence of cyclic AMP, show that the activation of histone F1 phosphorylation by cyclic AMP was highest in early S and lowest in early M.

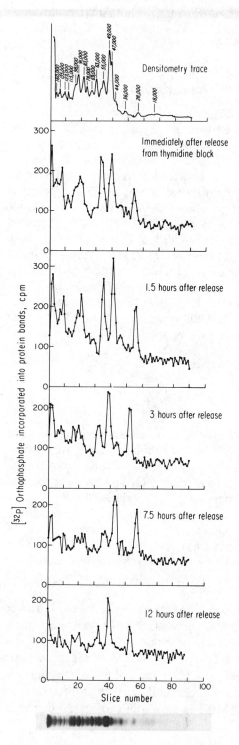

Table 5

Phosphorus Content of Nuclear Phosphoprotein
Fraction at Different Stages of the Cell Cycle
of Synchronized HeLa Cells

Time after release from thymidine block (hr)	Phosphorus content of protein* (percent)
0	0.099 ± 0.003
1.5	0.144 ± 0.024
3.0	0.115 ± 0.018
4.5	0.091 ± 0.014
6.0	0.100 ± 0.008
7.5	0.083 ± 0.001
9.0	0.104 ± 0.007
10.5	0.079 ± 0.004
12.0	0.100 ± 0.009

* Phosphorus determined as alkali-labile phosphate
as described in Materials and Methods. Data presented as average of three values ± SEM.

Phosphate incorporated into the nuclear acidic proteins is not stable, but turns over at rates that differ from one protein species to another. The phosphate groups are subject to exchange without degradation of the polypeptide chain.

If the nuclear phosphoproteins play a role in the control of transcription during the cell cycle, changes in their phosphorylation would offer a mechanism for modifying their structure and influencing their interactions with other components of the chromosome. If their primary role is structural, changes in phosphorylation would affect their binding to histones. In a nondifferentiating cell population, in which the proportions of the major acidic proteins remain relatively stable throughout the cell cycle, changes in phosphate uptake and turnover may be more important than changes in the relative concentrations of the individual protein species.

Figure 7

Differential rates of [^{32}P]phosphate turnover in HeLa nuclear phosphoproteins during synchronous growth. The cell suspension was prelabeled with [^{32}P]phosphate as described in Materials and Methods. After washing the cells were incubated under cold-chase conditions and aliquots were withdrawn at the indicated times for preparation and electrophoretic analysis of the nuclear phosphoprotein fraction. The proteins were separated by electrophoresis in 0.1% SDS–10% polyacrylamide gels. The protein banding pattern is shown at the bottom of the figure and the corresponding densitometer tracing is shown in the top panel. Molecular weights are indicated above the major protein peaks. The intermediate panels show the patterns of ^{32}P distribution in the nuclear proteins at different times after release from the thymidine block. Note the rapid loss of ^{32}P activity from the band at MW 55,000 and the relatively slow decreases in ^{32}P activity of bands at MW 28,000 and 47,000.

REFERENCES

Allfrey, V. G., C. S. Teng and C. T. Teng. 1971. Changes in chromosomal proteins associated with gene activation. Nuclear protein phosphorylation as a possible mechanism for promoting RNA initiation. *Nucleic acid–protein interaction: Nucleic acid synthesis in viral infection* (ed. D. W. Ribbons, J. F. Woessner and J. Schultz) pp. 144–167. North-Holland, Amsterdam.

Allfrey, V. G., E. M. Johnson, J. Karn and G. Vidali. 1973. Phosphorylation of nuclear proteins at times of gene activation. Abstr. 5th Ann. Miami Winter Symp. *Protein phosphorylation in control mechanisms—The role of cyclic nucleotides in carcinogenesis,* pp. 34–36. Univ. Miami Press/Academic Press, New York.

Anderson, W. B., A. B. Schneider, M. Emmer, R. L. Perlman and I. Pastan. 1971. Purification and properties of the cyclic adenosine 3′,5′-monophosphate receptor protein which mediates cyclic adenosine 3′,5′-monophosphate-dependent transcription in *Escherichia coli. J. Biol. Chem.* **246:**5929.

Balhorn, R., J. Bordwell, L. Sellers, D. Granner and R. Chalkley. 1972. Histone phosphorylation and DNA synthesis are linked in synchronous cultures of HTC cells. *Biochem. Biophys. Res. Commun.* **46:**1326.

Baserga, R. and G. Stein. 1971. Nuclear acidic proteins and cell proliferation. *Fed. Proc.* **30:**1752.

Benjamin, W. B. and R. M. Goodman. 1969. Phosphorylation of Dipteran chromosomes and rat liver nuclei. *Science* **166:**629.

Bhorjee, J. S. and T. Pederson. 1972. Nonhistone chromosomal proteins in synchronized HeLa cells. Appearance of histone mRNA in the cytoplasm and its translation in a

Bootsma, D., L. Budke and O. Vos. 1964. Studies on synchronous divisions of tissue culture cells initiated by excess thymidine. *Exp. Cell Res.* **33:**301.

Borun, T. W. and G. S. Stein. 1972. The synthesis of acidic chromosomal proteins during the cell cycle of HeLa S-3 cells. II. The kinetics of residual protein synthesis and transport. *J. Cell Biol.* **52:**308.

Breindl, M. and D. Gallwitz. 1973. Identification of histone messenger RNA from HeLa cells. Appearance of histone mRNA in the cytoplasm and its translation in a rabbit reticulocyte cell-free system. *Eur. J. Biochem.* **32:**381.

Burton, K. 1956. A study of the conditions and mechanism of the diphenylamine reaction for the colorimetric estimation of deoxyribonucleic acid. *Biochem. J.* **62:**315.

Conner, B. J. and G. L. Patel. 1972. Isolation and metabolism of nuclear proteins during early development in the sea urchin. *J. Cell Biol.* **55:**49a.

Farber, J., G. S. Stein and R. Baserga. 1972. The regulation of RNA synthesis during mitosis. *Biochem. Biophys. Res. Commun.* **47:**790.

Gallwitz, D. and G. C. Mueller. 1969. Histone synthesis in vitro on HeLa cell microsomes. Nature of the coupling to deoxyribonucleic acid synthesis. *J. Biol. Chem.* **244:**5947.

Gilbert, W. and B. Müller-Hill. 1967. The *lac* operator is DNA. *Proc. Nat. Acad. Sci.* **58:**2415.

Gilmour, R. S. and J. Paul. 1969. RNA transcribed from reconstituted nucleoprotein is similar to natural RNA. *J. Mol. Biol.* **40:**137.

Gurley, L. R., R. A. Walters and R. A. Tobey. 1973. The metabolism of histone fractions. VI. Differences in the phosphorylation of histone fractions during the cell cycle. *Arch. Biochem. Biophys.* **154:**212.

Hancock, R. 1969. Conservation of histones in chromatin during growth and mitosis in vitro. *J. Mol. Biol.* **40:**457.

Hill, R. J., D. L. Poccia and P. Doty. 1971. Towards a total macromolecular analysis of sea urchin embryo chromatin. *J. Mol. Biol.* **61:**445.

Hurlbert, R. B. 1957. Preparation of nucleoside diphosphates and triphosphates. *Methods in enzymology* (ed. S. P. Colowick and N. O. Kaplan) vol. 3, pp. 785–805. Academic Press, New York.

Johnson, E. M. and V. G. Allfrey. 1972. Differential effects of cyclic adenosine 3′,5′-monophosphate on phosphorylation of rat liver nuclear acidic proteins. *Arch. Biochem. Biophys.* **152:**786.

Johnson, T. C. and J. J. Holland. 1965. Ribonucleic acid and protein synthesis in mitotic HeLa cells. *J. Cell Biol.* **27:**565.

Kamiyama, M. and T. Y. Wang. 1970. Activated transcription from rat liver chromatin by non-histone proteins. *Biochim. Biophys. Acta* **228:**563.

Kamiyama, M., B. Dastugue, N. Defer and J. Kruh. 1972. Liver chromatin nonhistone proteins: Partial fractionation and mechanism of action on RNA synthesis. *Biochem. Biophys. Acta* **277:**576.

Kleinsmith, L. J., V. G. Allfrey and A. E. Mirsky. 1966a. Phosphorylation of nuclear protein early in the course of gene activation in lymphocytes. *Science* **154:**780.

———. 1966b. Phosphoprotein metabolism in isolated lymphocyte nuclei. *Proc. Nat. Acad. Sci.* **55:**1182.

Kostraba, N. C. and T. Y. Wang. 1972. Differential activation of transcription of chromatin by non-histone fractions. *Biochim. Biophys. Acta* **262:**169.

LeStourgeon, W. M. and H. P. Rusch. 1971. Nuclear acidic protein changes during differentiation in *Physarum polycephalum*. *Science* **174:**1233.

———. 1973. Localization of nucleolar and chromatin residual acidic protein changes during differentiation in *Physarum polycephalum*. *Arch. Biochem. Biophys.* **155:**144.

Lowry, O. H., N. J. Rosebrough, A. L. Farr and R. J. Randall. 1951. Protein measurement with the Folin phenol reagent. *J. Biol. Chem.* **193:**265.

Maeno, H., E. M. Johnson and P. Greengard. 1971. Subcellular distribution of adenosine 3′,5′-monophosphate-dependent protein kinase in rat brain. *J. Biol. Chem.* **246:**134.

Oliver, D., R. Balhorn, D. Granner and R. Chalkley. 1972. Molecular nature of F1 histone phosphorylation in cultured hepatoma cells. *Biochemistry* **11:**3921.

Ord, M. G. and L. A. Stocken. 1968. Variations in the phosphate content and thiol/disulfide ratio of histones during the cell cycle. Studies with regenerating liver and sea urchins. *Biochem. J.* **107:**403.

Panyim, S. and R. Chalkley. 1969. High resolution polyacrylamide gel electrophoresis of histones. *Arch. Biochem. Biophys.* **130:**337.

Pfeiffer, S. E. and L. J. Tolmach. 1968. RNA synthesis in synchronously growing populations of HeLa S-3 cells. I. Rate of total RNA synthesis and its relationship to DNA synthesis. *J. Cell. Physiol.* **71:**77.

Platz, R. D., G. S. Stein and L. J. Kleinsmith. 1973. Changes in the phosphorylation of non-histone chromatin proteins during the cell cycle of HeLa S-3 cells. *Biochem. Biophys. Res. Commun.* **51:**735.

Puck, T. T. 1964. Phasing, mitotic delay, and chromosomal aberrations in mammalian cells. *Science* **144:**565.

Rickwood, D., G. Threlfall, A. L. MacGillivray, J. Paul and P. Riches. 1972. Studies on the phosphorylation of chromatin non-histone proteins and their effect on deoxyribonucleic acid transcription. *Biochem. J.* **129:**50p.

Riggs, A. D., R. F. Newby, J. Bourgeois and M. Cohn. 1968. DNA binding of the *lac* repressor. *J. Mol. Biol.* **34:**365.

Robbins, E. and T. W. Borun. 1967. The cytoplasmic synthesis of histones in HeLa cells and its temporal relationship to DNA replication. *Proc. Nat. Acad. Sci.* **57:**409.

Shea, M. and L. J. Kleinsmith. 1973. Template-specific stimulation of RNA synthesis by phosphorylated non-histone chromatin proteins. *Biochem. Biophys. Res. Commun.* **50:**473.

Shelton, K. R. and V. G. Allfrey. 1970. Selective synthesis of a nuclear acidic protein in liver cells stimulated by cortisol. *Nature* **228:**132.

Spackman, D. H., W. H. Stein and S. Moore. 1958. Automatic recording apparatus for use in the chromatography of amino acids. *Anal. Chem.* **30:**1190.

Spalding, J., K. Kajiwara and G. C. Mueller. 1966. The metabolism of basic proteins in HeLa cell nuclei. *Proc. Nat. Acad. Sci.* **56:**1535.

Spelsberg, T. C., L. S. Hnilica and A. T. Ansevin. 1971. Proteins of chromatin in template restriction. III. The macromolecules in specific restriction of the chromatin DNA. *Biochim. Biophys. Acta* **228:**550.

Stein, G. S. and R. Baserga. 1970. Continued synthesis of non-histone chromosomal proteins during mitosis. *Biochem. Biophys. Res. Commun.* **41:**715.

Stein, G. S. and T. W. Borun. 1972. The synthesis of acidic chromosomal proteins during the cell cycle of HeLa S-3 cells. I. The accelerated accumulation of acidic residual nuclear protein before the initiation of DNA replication. *J. Cell Biol.* **52:**292.

Stein, G. S., S. Chaudhuri and R. Baserga. 1972. Gene activations in WI-38 fibroblasts stimulated to proliferate. Role of non-histone chromosomal proteins. *J. Biol. Chem.* **247:**3918.

Teng, C. S., C. T. Teng and V. G. Allfrey. 1971. Studies of nuclear acidic proteins. Evidence for their phosphorylation, tissue-specificity, selective binding to DNA and stimulatory effects on transcription. *J. Biol. Chem.* **246:**3597.

Vidali, G., L. C. Boffa, V. C. Littau, K. M. Allfrey and V. G. Allfrey. 1973. Changes in nuclear acidic protein complement of red cells during embryonic development. *J. Biol. Chem.* **248:**4065.

Zubay, G., D. Schwartz and J. Beckwith. 1970. Mechanism of activation of catabolite sensitive genes: A positive control system. *Proc. Nat. Acad. Sci.* **66:**104.

Histone Methylation and Phosphorylation during the HeLa S-3 Cell Cycle

T. W. Borun, W. K. Paik, H. W. Lee, D. Pearson, and D. Marks

Fels Research Institute and Department of Biochemistry
Temple University School of Medicine, Philadelphia, Pennsylvania 19140

The role that histones play in the economy of the eukaryotic cell has been the subject of investigation since Kossel's initial studies of this class of basic nuclear proteins began nearly a century ago (Kossel 1928). While the precise function of the histones remains a mystery, a number of facts that probably reflect the nature of that function seem to have been convincingly established:

1. There are normally five principal kinds of histone polypeptides in almost all eukaryotes (Johns and Butler 1962; Panyim and Chalkley 1969; Phillips and Johns 1965). Histones f_3 and f_{2a1} have practically the same primary structure throughout the plant and animal kingdom (DeLange and Smith 1971; Panyim et al. 1970, 1971). Histones f_{2b} and f_{2a2} are somewhat more variable in primary structure and histone f_1 is subject to considerable variation in amino acid sequence and polypeptide chain length in different species (Bustin and Cole 1968).

2. In a given cell type constant relative amounts of the five histone polypeptides are synthesized while DNA chains are being polymerized and organized into new chromosomes during S phase of the cell cycle (Borun, Scharff and Robbins 1967; Dick and Johns 1969; Gurley and Hardin 1968; Laurence and Butler, 1965). Since there is normally very little or no net synthesis of these five proteins at earlier or later periods of the cell cycle, histone polypeptide synthesis must be subject to extensive translational and transcriptional control processes. In addition there is convincing evidence that DNA replication and histone synthesis are tightly coupled in mammalian cells by some mechanism of as yet unknown nature, which rapidly terminates further net synthesis of histone polypeptides if DNA replication is artificially inhibited (Borun et al. 1967; Gallwitz and Mueller 1969).

3. After their synthesis histone polypeptides are subject to at least three kinds of modification. These modifications include lysine acetylation (DeLange et al. 1968; Gershey, Vidali and Allfrey 1968; Phillips 1963), methylation of lysine, arginine, histidine and dicarboxylic amino acid residues (Paik and Kim 1971), and phosphorylation of serine and threonine residues (Langan 1968).

701

Just as histone function is obscure, the function of histone modification processes has not been definitively determined. Speculations have credited the modification of histone polypeptide chains with functions ranging from selective regulation of the transcription of genes adjacent to the modified histone molecules (Langan 1968) to facilitation of the transport of newly synthesized histones from their cytoplasmic site of synthesis to the nucleus (Oliver et al. 1972). We have investigated all three types of histone modification during the HeLa S-3 cell cycle, hoping to gain some insight into the function of these modifications in continuously dividing malignant cells. In this paper we shall consider only the processes of histone methylation and phosphorylation since those modifications vary considerably at different stages of the cell cycle, in contrast to histone acetylation levels which do not appear to change appreciably in the HeLa S-3 cell from one mitosis to the next (Zweider and Borun unpubl.). We shall note similarities and differences in the nature of histone methylation and phosphorylation and shall conclude by speculating briefly on the function of histones and their modifications during the proliferation of normal and malignant cells.

Histone Methylation and Methylating Enzyme Activities during the HeLa S-3 Cell Cycle

Estimation of Specificity, Rate, Timing and Stability of Protein Methylations

To estimate the rate, specificity and timing of various protein methylation reactions occurring throughout the HeLa cell cycle in the experiments described below, cell samples of appropriate size were pulse-labeled for 30 minutes with a mixture of L-[methyl-^{3}H]methionine and L-[1-^{14}C]methionine at various points in the cell cycle as described in detail in a previous publication (Borun, Pearson and Paik 1972). The rationale of this sort of labeling procedure is the premise that cellular proteins that have been methylated will incorporate a higher ratio of ^{3}H/^{14}C methionine-derived radioactivity than proteins that have not been methylated. In relatively short pulses, radioactivity derived from L-[1-^{14}C]methionine will be incorporated into proteins almost solely as polypeptidyl methionine residues, whereas the radioactivity derived from L-[methyl-^{3}H]methionine can be incorporated both as polypeptidyl methionine residues and as tritiated methyl groups, which have been transferred to methylated lysines, arginines and histidines, by means of S-adenosyl-methionine-mediated reactions (Comb, Sarkar and Pinzino 1966; Kim and Paik 1965; Lehninger 1970). This index of protein methylations is not subject to complications that could be caused by variations in methionine pool size during the cell cycle. When cells were labeled at an ^{3}H/^{14}C methionine input ratio of 2.4, it was found that the total cellular and various nonhistone nuclear protein fractions consistently incorporated the two kinds of methionine radioactivity at a ratio of about 4–5 at all points in the cell cycle. In contrast, the 0.25 N H_2SO_4-soluble nuclear or "crude histone" fraction exhibited ratios of ^{3}H/^{14}C methionine incorporation of about 8–9 in late mitosis and early G_1. The ratio fell to about 4 in early S, then rose once again in late S and G_2 to 8–9 (Borun et al. 1972). Resolution of the component histone polypeptides within the "crude histone" fraction by acrylamide gel electrophoresis showed that the species-constant histones f_3 and f_{2a1} were methylated at the highest rates, histones f_{2b} and f_{2a2} were methylated at much slower rates, and histone f_1 was not methylated at all. Thus of all the cellular protein fractions we examined by this technique, the histones were the most vigor-

ously methylated class of proteins, and within that class there exists a positive correlation of unknown significance between the rate of these methylations and the
evolutionary stability of the histone polypeptide chains (Borun et al. 1972).

To estimate the metabolic stability or turnover rate of histone methyl groups a
similar sort of experimental labeling procedure was used. Log phase cells were incubated for 12 hours with a mixture of L-[methyl-^{3}H]methionine and L-[1-^{14}C]me-
thionine or L-[U-^{14}C]lysine to label the histone polypeptide chains and their methyl
groups. The cells were then synchronized by selective detachment in mitosis and
cultured at 37°C for the rest of a complete cell cycle. At regular intervals cell
samples were removed, the crude histone fraction was isolated and the histone
polypeptide components of the crude fraction were resolved by acrylamide gel
electrophoresis. The rate at which the ^{3}H/^{14}C ratio fell during the cell cycle following the labeling period indicated that histones f_3 and f_{2a1} lose methyl groups at a
uniform rate of about 1.8% per hour and histones f_{2b} and f_{2a2} lose their methyl
groups at about half that rate throughout the cell cycle (Borun et al. 1972).

The results of these preliminary experiments, suggesting that different histones
were methylated at different rates at specific times in the cell cycle and that the
association of histones with their methyl groups was quite stable, encouraged us to
examine the rate, specificity and extent of methylation of histone lysine, arginine
and histidine residues in a more definitive fashion.

In these experiments we pulse-labeled samples of synchronized HeLa S-3 cells
with L-[methyl-^{14}C]methionine for 60 minutes at regular intervals through the cell
cycle. Histone fractions f_1, f_{2a} (a mixture of about equal parts of histones f_{2a1} and
f_{2a2}), f_{2b} and f_3 were isolated by a modification of the Johns bulk histone extraction
procedure, hydrolyzed and subjected to basic amino acid chromatography as previously described (Borun et al. 1972). These experiments, which are summarized
in Fig. 1, Table 1 and Table 2, allowed a determination of the amount of each kind
of methylated amino acid residue in each histone fraction and the rate of methylation of these residues throughout the HeLa cell cycle. On the average through the
cell cycle, from 36–42% of the total L-[methyl-^{14}C]methionine-derived radioactivity
incorporated into histone fractions f_{2a}, f_{2b} and f_3 was found in methylated lysine,
arginine, and histidine residues (Table 1). The average rate of methylation reactions was about twice as high in fraction f_{2a} as in fraction f_3, which in turn was

Table 1

Average incorporation of L-[methyl-^{14}C]Methionine-derived
Radioactivity into Histone Fractions of HeLa S-3 Cells

Fraction	Total dis- integrations in fraction	Disintegrations in methylated amino acids[a]	% Total disintegrations in methylated amino acids
	(dpm/3 × 10^8 cells)		
f_{2a}	175,000	75,000	42
f_3	118,000	42,000	36
f_{2b}	53,000	19,000	37
f_1	36,000	None	

[a] ε-N-Monomethyllysine, ε-N-dimethyllysine, methylhistidine, and methylated
arginine.

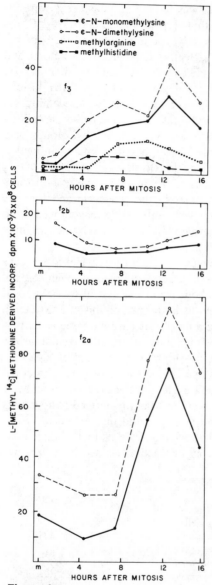

Figure 1

The rates of incorporation of L-[methyl-^{14}C]methionine-derived radioactivity into the methylated amino acids of histone fractions f_{2a}, f_{2b} and f_3 through the cell cycle. Samples of 3×10^8 HeLa S-3 cells were pulse-labeled for 60 min with 30 μCi of L-[methyl-^{14}C]methionine beginning 30 min before the indicated times after mitosis. Then the histone fractions were extracted, hydrolyzed, and chromatographed as indicated in Borun et al. (1972). Thus the points represent the midpoints of the pulse-labeling periods at different times in mitosis, G_1, S, and G_2 phases of the cell cycle.

Table 2
Amount of Lysine, ε-N-Monomethyllysine (MML), and ε-N-Dimethyllysine (DML)
Recovered from Histone Fractions

Fractions after mitosis and time	Lys	MML	DML	DML/MML	% Methylated lysines[a]	% Methylated molecules
	(nmoles recovered per 3 × 10⁸ cells)					
f_{2a} (10.2% Lys)[b]						
Mitosis	1100	31	61	2.0	7.7	77
1.5 hr	980					
4.5 hr	1000	36	59	1.6	8.7	87
7.5 hr	1330	23	59	2.5	5.6	56
10.5 hr	2170	69	103	1.5	7.3	73
16 hr	2430	59	97	1.7	5.2	52
Average	1500	44	76	1.7	6.9	69
f_3 (9% Lys)[b]						75
Mitosis	290	22	16	.72	11.5	64
4.5 hr	530	24	25	1.04	8.3	75
7.5 hr	750	27	29	1.08	7	91
10.5 hr	740	38	31	.83	8.4	82
15.5 hr	640	33	42	1.26	10.3	
Average	630	30	31	1.00	9	100
F_{2b} (14% Lys)[b]						
Average	621	21.5	17	0.80	5.9	84

[a] % Methylated lysines = MML + DML/Lys + MML + DML.

[b] E. W. Johns (see Table 3 of Borun et al. 1972).

methylated at twice as high a rate as histone f_{2b}. Histone f_1 contained no methylated amino acids observable by these methods. The average amounts of methylated lysine residues in histone fractions f_{2a}, f_3 and f_2b also have approximately this same 4:2:1 ratio (Table 2). The rate of amino acid methylation at specific cell cycle times is shown in Fig. 1. It is apparent that the lysine residues of histone fractions f_{2a} and f_3 are mono- and di-methylated at significant rates throughout the cell cycle and these rates reach distinct maxima in late S and G_2. In histone fraction f_3 the methylation rate undergoes a gradual increase while approaching the G_2 maximum. In histone fraction f_{2a} the rate remains more or less constant until about 8 hours after mitosis, at which point it begins to increase rapidly about 3- to 4-fold as the cells approach the next mitosis. Lysine methylation in histone f_{2b} reaches its maximum rate just prior to, during, and just after mitosis. Only histone fraction f_3 contains significant amounts of methylated arginine and histidine residues, each of which comprises about 12% of the average total methylated amino acid radioactivity found to be incorporated into histone f_3. Arginine and histidine are methylated at maximal rates while histones are being synthesized at maximal rates in S phase of the cell cycle, suggesting that these amino acids are modified very soon after they have been polymerized into new histones.

All of the data presented so far are consistent with the idea that histone methylation is a process that affects a very significant percentage of the f_{2a1} and the f_3 polypeptides. This impression was confirmed and extended to histones f_{2b} and possibly to f_{2a2} by calculations of the percent of the total histone molecules which were methylated, summarized in the last column of Table 2. These computations assumed that there was only one site of lysine methylation found in each kind of histone molecule. That assumption is correct for histone f_{2a1}, the principal methylated species of the histone f_{2a} fraction. Recent sequencing data has indicated, however, that there are two sites of lysine methylation in the histone f_3 molecule. Nevertheless it is apparent that, on the average, 69% of the f_{2a} fraction, at least 41% of the f_3 fraction, and 84% of the f_{2b} histones are methylated. From the preceding considerations and the data in Table 2 it is apparent that while methylation affects only certain histones, occurs at high rates in G_2 of the cell cycle, and affects different amino acid residues in different histone polypeptides, it is apparently not a very selective process. That is to say, since it affects very large percentages of the histone polypeptides that are methylated (indeed almost all of histone f_{2a1} molecules may be methylated by the time mitosis is reached), it does not seem likely that methylation has as its function the activation or inactivation of small numbers of genes near the methylated histone molecules. Furthermore while premitotic cells (which have condensed chromosomes) do contain large amounts of methylated histones, these molecules lose their methyl groups so slowly that the G_1 cells (whose chromosomes are no longer condensed) formed by mitosis contain histones that are essentially as methylated as those of the parent cell. Thus histone methylation cannot be the sole cause of premitotic chromosome condensation in any simple fashion, especially since methylation reactions go on throughout the cell cycle. It will be seen in subsequent sections that this point concerning the state of G_1 histones relative to the histones of the parent cell just prior to mitosis is a very important difference between the net result of histone methylation and phosphorylation in the HeLa S-3 cell.

Variations in Histone Methylating Enzyme Activities

Since histones are methylated throughout the HeLa cell cycle with maximum rates of methylation occurring in G_2, well after the histones have been synthesized and incorporated into chromosomes, it is evident that methylation reactions must be affecting histones that are in place on the chromosome. We wondered therefore whether the increased rates of histone methylations that were occuring in G_2 were due to increases in the activity of the enzymes which methylated the histones or whether the increases were due to some premitotic change in the orientation of the regions of the histone molecules which could be methylated. We therefore measured the activity of protein methylase III, a nuclear enzyme that methylates a variety of protein lysyl residues in vitro and a good candidate for the enzyme responsible for most of the methylation of histones in vivo throughout the HeLa cell cycle as described in detail elsewhere (Lee, Paik and Borun 1973). The activity of this enzyme (Fig. 2) increases about fourfold beginning at 6–8 hours after mitosis, the same time that histones f_{2a1} and f_3 were found to begin being methylated at increasing rates (compare Fig. 1 and 2). Since pretreatment of the cells with cycloheximide or actinomycin D at effective concentrations abolishes this scheduled increase in protein methylase III activity, it is very probable that the increase is due to the synthesis of new enzyme molecules. The close temporal correlation between increased histone methylation and increases in protein methylase

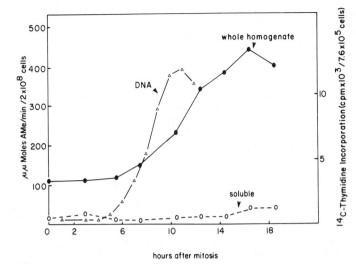

Figure 2
Protein methylase III activity during HeLa S-3 cell cycle. Samples of 2×10^8 cells were harvested by centrifuging at 600 g for 3 min at 3°C, then washed three times with Eagle's Spinner Salts, and finally were resuspended in 6 ml cold deionized water. The suspension was homogenized by an electrically driven teflon glass homogenizer. Three ml whole homogenate was centrifuged at 105,000 $\times$ g for 60 min to obtain the soluble fraction.

The incubation mixture contained 0.1 ml whole homogenate or the soluble fraction, 0.2 ml histone type II-A (2 mg), 0.1 ml 0.5 M Tris-HCl buffer at pH 9.0, and 0.1 ml of S-adenosyl-L-[methyl-^{14}C]methionine. The incubation was carried out at 37°C for 5 min as described in detail in Lee et al. (1973). [2-^{14}C]Thymidine incorporation into DNA was determined as described in Borun et al. (1972).

III (PM III) activity in the HeLa cell and many other proliferating cell systems (Lee et al. 1973), together with the fact that histones were the most methylated class of cellular proteins that we found in the HeLa cell, strongly suggest that one of the principal or *the* principal biological function of HeLa PM III is histone methylation. We therefore investigated the relationship between the appearance of new histones in the HeLa cell and the scheduled increase in PM III activity. Somewhat surprisingly, prevention of the initiation of DNA and histone synthesis or their inhibition after they had begun by treatment of synchronized cells with cytosine arabinoside in G_1 or S phase had no effect on the appearance of increased PM III activity at 6–8 hours after mitosis. The enzyme appeared to increase at the usual time in the inhibited cells and "old" histone molecules were over-methylated from the mono- and di-methyllysine state to the di- and tri-methyllysine state (Table 3; Lee et al. 1973). These results indicate that the synthesis of new substrate is *not* the cause of increased PM III synthesis in the HeLa cell. Rather it appears that this enzyme's induction is cued to some other as yet unknown cell cycle program. Since already methylated histones can be over-methylated, it is apparent that previously methylated histones have not changed their orientation and become inaccessible to the methylating enzymes.

Table 3

Methylation of the ε-Amino Group of Lysine Residues of Histone
in the Presence of Cytosine Arabinoside

	Control				Cytosine arabinoside-treated			
	L*	MML	DML	TML	L	MML	DML	TML
Amount (μμmoles)	1711	6.32	10.04	none	1756	7.12	18.80	1.80
Radioactivity (dpm)		27,220	45,420			15,150	21,620	5570
Specific activity (dpm/μμmole)		4308	4524			2129	1150	3095
% Methylated lysine		37.5	62.5			25.7	67.8	6.5
% Radioactivity		38.6	61.4			35.8	51.1	13.2

The cells (3×10^8) were treated with cytosine arabinoside at the concentration of 40 μg/ml at 1 hr
after synchronization. The control cells without the antimetabolite were also run simultaneously. Both
cells were incubated further for another 11 hr. The cells were then pelleted at 600 g for 3 min and
the pellets were resuspended in 40 ml each of Spinner Salts medium minus methionine plus 2% fetal
calf serum and 50 μCi of L-[methyl-^{14}C]methionine. After 1 hr further incubation the cells were
centrifuged at 600 g for 3 min and were washed three times by resuspending them in 50 ml of
Spinner Salts medium. The cells were partially lysed at 3°C by three washes in 10 ml 80 mM NaCl,
20 mM EDTA, 1% Triton X-100 pH 7.2, and the resulting nuclei were washed twice in 10 ml
0.15 M NaCl, then once in 10 ml 90% ethanol. Unfractionated histone was prepared from these
nuclei by the method published previously (Borun et al. 1972). This histone was hydrolyzed and the
hydrolyzate was analyzed according to the procedure of Borun et al. (1972).

* L, MML, DML and TML represent L-lysine, ε-N-monomethyl-L-lysine, ε-N-dimethyl-L-lysine and
ε-N-trimethyl-L-lysine, respectively.

Studies of Histone Phosphorylation during the HeLa S-3 Cell Cycle

Preliminary Control Experiments

The principal criticism of the results of many earlier investigations of histone phosphorylation using ^{32}P-labeling has been the observation that crude histone extracts
can be contaminated with a variety of materials that incorporate ^{32}P at rapid rates
and obscure authentic histone phosphate uptake and loss. We therefore decided to
determine whether electrophoresis of histone polypeptides into acrylamide gels containing acetic acid and urea, according to the method of Panyim and Chalkley
(1969), was able to remove such contamination. The distribution of ^{32}P radioactivity associated with the histones extracted from log phase HeLa cells labeled for
30 minutes with carrier-free ^{32}P orthophosphate and then separated by electrophoresis on 25-cm long 15% acrylamide gels is shown in Fig. 3. Three regions of
the gel are labeled with ^{32}P. Histone f_1 and f_{2a1} polypeptides at the trailing and leading ends of the histone group in the gel are each unambiguously associated with
radiophosphorus, but the middle region of the gel contains a peak of ^{32}P incorporation which cannot be unequivocally ascribed to either histone f_{2b} or histone f_{2a2} at
this time. We therefore have called this the f_{2b} (f_{2a2}) phosphorylated histone. Pre-
treatments of a crude ^{32}P-labeled 0.25 N H_2SO_4-soluble nuclear extract prior to
electrophoresis with solvents known to extract phospholipids have no effect on the
amount of ^{32}P radioactivity found associated with the three different phosphorylated histones in the gels after electrophoresis (Table 4). When the histone f_1 region
was cut out of the gels, hydrolyzed and then analyzed by amino acid chromatography, over 88% of the ^{32}P in the samples was found in the phosphoserine region
of the amino acid chromatogram. While the much lower levels of radioactivity associated with the other two histone regions of the gels could not be similarly

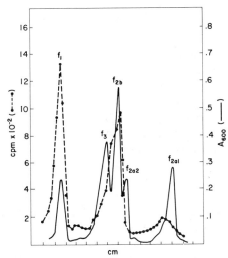

Figure 3

^{32}P incorporation into histone fractions separated by polyacrylamide gel electrophoresis. A 25-cm gel containing 2 M urea was electrophoresed at 190 V for 26 hr. The histones were extracted from 3×10^6 cells, which had been labeled for 30 min with 144 μCi ^{32}P in early S phase. Electrophoresis is from left to right (+ to −). Only the section of the gel containing the histone bands is shown. The lack of exact coincidence between the absorbance and the radioactivity may be due to a decrease in mobility experienced by the small population of molecules that are phosphorylated.

(●———●) ^{32}P cpm. (———) Absorbance at 600 nm of the stained histone bands.

Table 4

Treatment of Histone Preparation with Organic
Solvents to Remove Contaminating Phospholipids

	Radioactivity present (cpm)		
	f_1	f_{2b}	f_{a1}
No treatment	443	201	97
Precipitated with ethanol at 4°C	496	208	99
Washed successively with ethanol and mixture of CHCl$_3$: ethanol:ether	482	250	88

One hundred μg of each histone sample was analyzed on polyacrylamide gels containing 6 M urea, which were electrophoresed for 12 hr at room temperature (1 mA/gel). Each of the histone samples was obtained from 7.7×10^6 cells (from an unsynchronized culture) labeled with 60 μCi ^{32}p for 1 hr.

analyzed, these experiments demonstrate that the radioactivity found associated with histones in this kind of acrylamide gel is almost completely in histone phosphoserines and not in phospholipid or oligonucleotide contamination (Marks, Paik and Borun 1973).

Estimation of Changes in Rate of Phosphorylation, Dephosphorylation and Phosphate Exchange Reactions

To estimate the rate of turnover of phosphate in different phosphorylated histones, we pulse-labeled samples of synchronized HeLa S-3 cells with ^{32}P for one hour in early G_1 and early S phase of the cell cycle. We then "chased" each of the samples in nonradioactive medium for up to 6 hours, removing aliquots for histone extraction and electrophoresis at regular intervals. In both G_1 and S phase the ^{32}P associated with histones f_{2a1} and f_{2b} (f_{2a2}) had half-lives of about one hour, in contrast to the radioactivity associated with the f_1 histones, which was lost at a rate of only about 10–15% per hour throughout the periods tested. Because of these rapid apparent turnovers (which could be due either to dephosphorylation reactions, which lead to a net loss of phosphate, or to exchange reactions, which merely replace the labeled phosphate groups with unlabeled ones with no net change in the amount of a phosphorylated histone), we decided to use a combination of ^{32}P accumulation and pulse-labeling experiments to help estimate what was happening to the net amount of histone phosphates through the cell cycle (Marks et al. 1973).

One kind of estimate of the net result of histone kinase and phosphatase reactions was made by adding ^{32}P to a culture of cells an hour after synchronization in mitosis and allowing the cells to accumulate label for the remainder of a complete cell cycle. We removed samples at appropriate intervals, extracted the histones and then subjected them to analysis by acrylamide gel electrophoresis. The specific activities of histones f_{2a1} and f_{2b} (f_{2a2}) increase slowly in G_1, then began doubling through early S phase until a plateau of specific activity was reached at levels about twice as high as those seen in G_1 (Fig. 4). Histone f_1 polypeptides began accumulating ^{32}P in G_1 prior to the initiation of DNA and histone synthesis, then eventually increased about sixfold in specific activity through S and G_2 of the cell cycle, while approaching mitosis. To estimate how rates of histone phosphorylation reactions varied in relation to DNA and RNA synthesis, we pulse-labeled cells with ^{32}P for 30 minutes during the cell cycle at the times indicated in Fig. 5, extracted the labeled DNA, RNA and histones and then analyzed the histones on acrylamide gels. The phosphorylation rates of histones f_{2a1} and f_{2b} (f_{2a2}) increase about twofold as the cells proceed from G_1 through S into G_2, remaining at the higher level until mitosis is reached (Fig. 5A). These rate increases closely follow and probably explain the increase in histone f_{2a1} and f_{2b} (f_{2a2}) specific activity found in the accumulation experiments described above. The rates of DNA replication and histone f_1 phosphorylation increase in parallel as the cells pass from G_1 into early S of the cell cycle, that is, incorporation of ^{32}P into DNA and histone f_1 polypeptides occurs at very low rates during G_1, then as S phase begins accelerates up to maximum rates along the same curve (Fig. 5B). The maximum rate of histone f_1 phosphorylation is over ten times higher than the G_1 minimum rate. After mid-S has been reached, the rate of DNA replication declines until it is at 30% of its maximal value by 16 hours after mitosis in G_2 of the cell cycle. At this time the histone f_1 phosphorylation rate is still at 60% of its maximum value. Data to be presented in the next section will confirm the impression that the decline in the f_1 phosphorylation rate is more closely correlated with an increase in the number of cells entering and then passing through mitosis than with the G_2 decrease in the DNA replication rate. Neither the histone f_1 phosphorylation rate nor the rates of phosphorylation of histones f_{2a1} and f_{2b} (f_{2a2}) increase in parallel with the increase in RNA synthesis, which occurs in mid-G_1 in the HeLa S-3 cell (Fig. 5B). Thus neither kind of in-

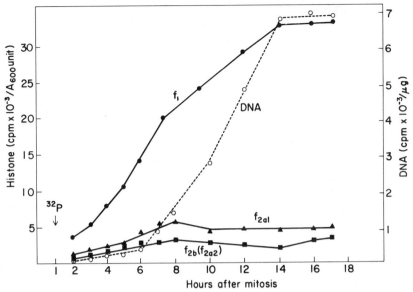

Figure 4

Accumulation of ^{32}P label in histones of HeLa cells that were continuously exposed to ^{32}P throughout the cell cycle. [32]P was added to the cultures 1 hr after mitosis. Aliquots were taken from the cultures at the times indicated, and histones were isolated by acid extraction and separated by electrophoresis on 6 M urea gels (1 ml/gel, 16 hr). The absorbance of the stained histone bands was determined and the specific activity was calculated (cpm per A_{600} unit) after the amount of radioactivity in gel slices was measured. Data from two experiments were normalized and plotted together. In the first experiment cells were labeled (9 μCi ^{32}P/ml) in complete Medium A. In the second experiment cells were labeled (14 μCi ^{32}P/ml) in Medium A containing one-tenth the normal amount of phosphate. Specific activity of DNA determined as described in Marks et al. (1973).

creased histone phosphorylation is obviously related to increased RNA synthesis or "gene activation" in this continuously dividing malignant cell. At the present time it is not clear why histone f_1 accumulates [32]P *prior* to the initiation of DNA replication in G_1 in spite of the fact that the rate of phosphorylation of this protein is very low at that time. Shepherd and his coworkers (1971) have reported a similar phenomenon in G_1 in continuously labeled synchronized CHO (Chinese hamster ovary) cells. It is possible that, since ^{32}P turns over in the histone f_1 fraction at relatively low rates (10–15% per hour), exchange reactions resulting in no net increase or even a decrease in the amount of phosphorylated histone could eventually replace unlabeled histone phosphate groups with labeled ones during long-term accumulation experiments.

To determine whether there was an obligatory couple between histone phosphorylations and DNA replication, we pulse-labeled samples of synchronized cells with ^{32}P in G_1 and S phase, with and without pretreatment with cytosine arabinoside at a level which inhibits DNA and histone synthesis. Histone f_{2a1} and f_{2b} (f_{2a2}) phosphorylations were completely insensitive to cytosine arabinoside treatment in either phase of the cell cycle. The low level of histone f_1 phosphorylation occurring in G_1 was also insensitive to this drug treatment, but inhibition of DNA and histone synthesis in S phase reduced the rate of f_1 phosphorylation to 50% of its normal level.

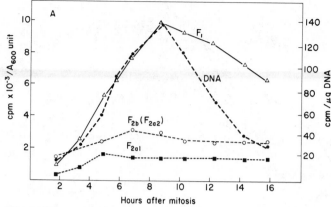

Figure 5A

Specific activity of HeLa histone fractions after ^{32}P pulses at various times during the cell cycle. Aliquots (50 ml) were taken at the indicated times from a culture containing 6.2×10^5 synchronized cells/ml. The aliquot was sedimented, resuspended for one-half hr in 10 ml Medium A containing one-tenth the normal amount of phosphate and 730 μCi ^{32}P, and the histones were isolated. The histones (corresponding to 3.4×10^6 cells/gel) were electrophoresed (6 M urea, 16 hr, 1 mA/gel). The absorbance at 600 nm of the stained histone bands was determined and the specific activity was calculated (cpm/A_{600} unit) after the amount of radioactivity in gel slices was determined. The specific activity of DNA was determined as described in Marks et al. (1973).

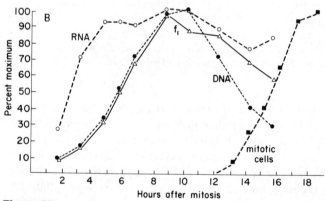

Figure 5B

The temporal relationship of f_1 histone phosphorylation, nucleic acid synthesis and percent cells accumulating in metaphase after colcemide treatment during the HeLa cell cycle. Aliquots of a synchronized culture were pulsed with ^{32}P as described in Fig. 5A. The histones were isolated, electrophoresed and the amount of radioactivity in histone f_1 was determined by counting of gel slices. DNA and RNA were isolated from nuclear pellets remaining after histone extraction and separated as described in Marks et al. (1973). Aliquots were counted to determine the extent of ^{32}P incorporation, and the results are presented as percent of maximum incorporation. Colcemide was added to a concentration of 0.05 μg/ml to a part of the culture at 12 hr after mitosis and the percent of cells accumulating in metaphase was determined by phase contrast microscopy.

712

High Resolution Acrylamide Gel Analysis of Histone f₁ Phosphorylation

Recently Chalkley and his coworkers (Balhorn, Rieke and Chalkley 1971) have resolved the different phosphorylated forms of histone f_1 molecules using a modification of the standard acetic acid-urea acrylamide gel electrophoresis procedure perfected in Dr. Chalkley's laboratory. This new procedure allows the direct determination of the actual *amount* of each kind of phosphorylated f_1 histone molecule. Previously the amounts of phosphorylated histones could only be inferred from the data obtained from the sort of labeling experiments just described. We therefore isolated the histones from samples of control and cytosine arabinoside-treated cells that were removed at points throughout the HeLa S-3 cell cycle and subjected the histones to the improved electrophoresis analysis as described by Balhorn et al. (1971).

Patterns and amounts of electrophoretically resolved HeLa histone f_1 polypeptides at different times in the cell cycle are shown in Fig. 6 and 7. It is evident from these data that there are three principal forms of HeLa f_1 histones, which we have called form I, II and III. Form I appears to be unphosphorylated because it does not incorporate ^{32}P. Forms II and III do incorporate ^{32}P, and form III appears to be converted to form II after alkaline phosphatase digestion. These forms are retarded in migration through the gels to extents which are consistent with form II being a monophosphorylated polypeptide and form III a diphosphorylated molecule (Balhorn et al. 1971). At mitosis over 90% of the total amount of histone f_1 polypeptides are found either in form II or form III. As the cells pass from mitosis into G_1, form III begins to be degraded to forms I and II until by the time S phase begins it is almost all gone. During the next S phase the phosphorylation cycle begins once more and again coverts over 90% of both old and newly synthesized histone f_1 polypeptides into phosphorylated forms II and III. Treatment of G_1 or S phase cells with cytosine arabinoside does not prevent the initiation of the histone f_1 phosphorylation cycle or prevent the continued S phase phosphorylation of previously synthesized histone f_1 molecules that had been dephosphorylated in G_1 (Fig. 7). It is also clear (bottom panel Fig. 7) that while form III or (histone f_1 diphosphate) is produced at the highest rates during DNA and chromosomal replication in S phase, it is retained through G_2 of the HeLa cell cycle. Its presence in these cells is therefore not temporally correlated solely with replication processes but also with the chromosome sorting and condensation processes which precede mitosis.

DISCUSSION

From these data it is clear that histone methylation and phosphorylation are quite complex and different processes in the HeLa S-3 cell. Histone methylation goes on throughout the cell cycle but accelerates in late S and G_2, probably to methylate newly synthesized histones up to some necessary or desirable level while they are in place on premitotic chromosomes. It is also probable that increased histone lysine methylation at these times is due to increased synthesis of HeLa protein methylase III molecules. Histone f_1, which varies considerably in primary structure in various eukaryotes and which is the most easily dissociated histone in nucleoprotein complexes, is not methylated at all. Histones f_{2a1} and f_3, which have almost constant amino acid sequences in various species of eukaryotes, are methylated at the highest rates and to the highest levels during the HeLa cell cycle and both turn-

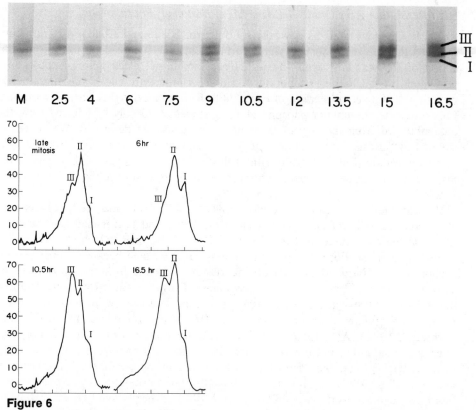

Figure 6
The three forms of histone f_1 resolved on high resolution polyacrylamide gels. The stained gels (25 cm in length, containing 2.5 M urea, run at 200 V for 68.5 hr at 4°C) were aligned, and the f_1 regions were cut out, photographed, and scanned. Forms I, II and III are indicated as well as the time (in hours after mitosis, *M*) at which the histones were isolated.

Top: Photographs of the f_1 region. Electrophoresis is from top to bottom. *Below:* Patterns of the absorbance at 600 nm of the f_1 region of four of the gels shown in A. Electrophoresis is from left to right.

over their methyl groups at a rate of about 1.8% per hour. Histones f_{2b} and f_{2a2}, which vary somewhat in primary structure in different eukaryotes, are methylated at lower rates in the HeLa cell but also have a lower methyl group turnover rate. The net result of these reactions is that very high percentages of histone f_{2a1}, f_3, f_{2b} and possibly f_{2a2} molecules are methylated by the time mitosis is reached. Because histone methyl groups have such slow turnover rates, a HeLa cell that has gone through mitosis into G_1 of the next cell cycle has histones which are about as completely methylated as the histones of its premitotic parent cell. These considerations make it improbable that histone methylation is responsible for *highly selective* gene control or chromosome condensation prior to mitosis. The high stability of the results of histone methylation reactions suggests that these methylations are a form of macromolecular finishing process of unknown function, which confers some useful three-dimensional property on specific regions of histone molecules that remain accessible to the methylating enzymes after modification.

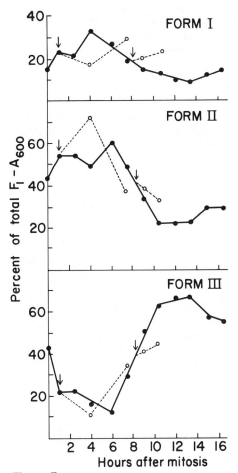

Figure 7
The percent of total histone f_1 found during the cell cycle under the areas designated form I, II, or III described in Fig. 6. (●———●) control; (○– – –○) cytosine arabinoside-treated. The arrows indicate the times at which cytosine arabinoside was added to a portion of the culture. The inhibitor was present from this time until the histones were isolated from the treated samples. The first arrow indicates that an aliquot of cells was treated with cytosine arabinoside in G_1 (from 1 hr after mitosis to the time indicated). In these cells DNA synthesis never began. The second arrow indicates that an aliquot of cells was treated with cytosine arabinoside in S (from 8 hr after mitosis to the time indicated). Although DNA synthesis had begun in these cells, further synthesis was completely inhibited by the cytosine arabinoside.

It is evident from the ^{32}P-labeling experiments described here that there are at least two distinct kinds of histone phosphorylation in the HeLa S-3 cell. Histones f_{2a1} and f_{2b} (f_{2a2}) incorporate ^{32}P throughout the cell cycle at rates that are approximately proportional to the amount of these polypeptides, and once incorporated this label rapidly turns over. These phosphorylations are neither coupled with DNA replication nor obviously related to increased RNA synthesis (or "gene activation") during the cell cycle. Although it is possible that the twofold increase in the rate of phosphorylation of these histones, which occurs as the cells pass from G_1 into S phase, may be related to some aspect of chromosome replication, the function of these phosphorylations remains unknown. The second kind of phosphorylation occurs in histone f_1 polypeptides and is partially coupled with DNA replication. Thus the f_1 phosphorylation rate increases about tenfold as the cell passes from G_1 into S and is reduced to 50% of its normal value if DNA replication is inhibited in S phase. These experiments also indicate that histone f_1 phosphorylation (a) does not appear to be correlated with RNA synthesis increases in G_1 and (b) continues at high rates through G_2 into mitosis of the HeLa cell cycle. The resolution of three different forms of histone f_1 molecules by acrylamide gel electrophoresis and an analysis of the changes in the relative proportions of these forms through the HeLa S-3 cell cycle confirm and considerably expand the ^{32}P-labeling data. Thus it is probable that the increased histone f_1 phosphorylation rate in S and G_2 reflects the net accumulation of the form III (or the putative histone f_1 diphosphate) at those times. It is also probable that the low net rate of f_1 phosphorylation after mitosis, prior to DNA replication in S phase, reflects the nearly complete degradation of form III to the monophosphorylated form II and the nonphosphorylated form I during G_1. This degradation in turn may involve phosphate exchange reactions, which would explain the G_1 accumulation of ^{32}P in histone f_1.

The precise relationship of the histone f_1 phosphorylation cycle to the factors which control proliferation in normal and malignant cells is at the present time not clear. It is evident, however, that the phosphorylation of histone f_1 molecules could be intimately related to (a) the mechanism which transduces surface-mediated proliferative stimuli to the chromosome, opening up that structure to the enzymes responsible for replication, and/or (b) the mechanism by which chromosomes are moved about and separated during replication, segregation and condensation prior to mitosis.

Form III dephosphorylation may be related to the mechanism by which chromosomes are uncoiled in G_1 after mitosis. We are now actively engaged in investigating these possibilities.

REFERENCES

Balhorn, R., W. O. Rieke and R. Chalkley. 1971. Rapid electrophoretic analysis for histone phosphorylation. A reinvestigation of phosphorylation of lysine-rich histone during rat liver regeneration. *Biochemistry* **10**:3952.

Borun, T. W., D. B. Pearson and W. K. Paik. 1972. Studies of histone methylation during the HeLa S-3 cell cycle. *J. Biol. Chem.* **247**:4288.

Borun, T. W., M. D. Scharff and E. Robbins. 1967. Rapidly labeled polyribosome-associated RNA having the properties of histone messenger. *Proc. Nat. Acad. Sci.* **58**:1977.

Bustin, M. and R. D. Cole. 1968. Species and organ specificity in very lysine-rich histones. *J. Biol. Chem.* **243**:4500.

Comb, D. G., N. Sarkar and C. J. Pinzino. 1966. The methylation of lysine residues in protein. *J. Biol. Chem.* **241**:1857.

DeLange, R. J. and E. L. Smith. Histones: Structure and function. *Ann. Rev. Biochem.* **40**:279.

DeLange, R. J., E. L. Smith, D. M. Fambrough and J. Bonner. 1968. Amino acid sequence of histone IV: Presence of ε-N-acetyllysine. *Proc. Nat. Acad. Sci.* **61**:1145.

Dick, C. and E. W. Johns. The biosynthesis of the five main histone fractions of rat thymus. *Biochim. Biophys. Acta* **174**:380.

Gallwitz, D. and G. C. Mueller. 1969. Histone synthesis *in vitro* on HeLa cell microsomes. The nature of coupling to deoxyribonucleic acid synthesis. *J. Biol. Chem.* **244:**5947.

Gershey, E. L., Y. Vidali and V. G. Allfrey. 1968. Chemical studies on histone acetylation. The occurrence of ε-*N*-acetyllysine in the f_{2a1} histone. *J. Biol. Chem.* **243:**5018.

Gurley, J. R. and J. M. Hardin. 1968. The metabolism of histone fractions. The synthesis of histone fractions during the life cycle of mammalian cells. *Arch. Biochem. Biophys.* **128:**285.

Johns, E. W. and J. A. V. Butler. 1962. Further fractionations of histones from calf thymus. *Biochem. J.* **82:**15.

Kim, S. and W. K. Paik. 1965. Studies on the origin of ε-*N*-methyl-L-lysine in protein. *J. Biol. Chem.* **240:**4629.

Kossel, A. 1928. *The protamines and histones.* Longmans Green & Co., London.

Langan, T. A. 1968. Histone phosphorylation: Stimulation by adenosine 3′,5′-monophosphate. *Science* **162:**579.

Laurence, D. J. R. and J. A. V. Butler. 1965. Metabolism of histones in malignant tissues and liver of the rat and mouse. *Biochem. J.* **96:**53.

Lee, H. W., W. K. Paik and T. W. Borun. 1973. The periodic synthesis of *S*-adenosylmethionine:protein methyltransferases during the HeLa S-3 cell cycle. *J. Biol. Chem.* **248:**4194.

Lehninger, A. L. 1970. *Biochemistry,* p. 524. Worth Publishers Inc., New York.

Marks, D., W. K. Paik and T. W. Borun. 1973. The relationship of histone phosphorylation to DNA replication and mitosis. *J. Biol. Chem.* **248:**5660.

Oliver, D., R. Balhorn, D. Granner and R. Chalkley. 1972. Molecular nature of F_1 histone phosphorylation in cultured hepatoma cells. *Biochemistry* **11:**3921.

Paik, W. K. and S. Kim. 1971. Protein methylation. *Science* **174:**114.

Panyim, S. and R. Chalkley. 1969. High resolution acrylamide gel electrophoresis of histones. *Arch. Biochem. Biophys.* **130:**337.

Panyim, S., D. Bilek and R. Chalkley. 1971. An electrophoretic comparison of vertebrate histones. *J. Biol. Chem.* **246:**4206.

Panyim, S., R. Chalkley, S. Spiker and D. Oliver. 1970. Constant electrophoretic mobility of the cysteine containing histone in plants and animals. *Biochim. Biophys. Acta* **214:**216.

Phillips, D. M. P. 1963. The presence of acetyl groups in histones. *Biochem. J.* **87:**258.

Phillips, D. M. P. and E. W. Johns. 1965. A fractionation of the histones of group F_{2a} from calf thymus. *Biochem. J.* **94:**127.

Shepherd, G., B. Noland and J. Hardin. 1971. Histone phosphorylation in synchronized mammalian cell cultures. *Arch. Biochem. Biophys.* **142:**299.

Further Studies on the Involvement of Histone Phosphorylation in Cell Replication

Roger Chalkley, Rod Balhorn, Daryl Granner,*
Nongnuj Tanphaichitr, and Vaughn Jackson

Department of Biochemistry and *Internal Medicine Department
School of Medicine, University of Iowa, Iowa City, Iowa 52242

Phosphorylation of the lysine-rich histone (F_1) has proved to be a complex and at times confusing field of study. Two quite distinct and perhaps unrelated aspects of this phenomenon have been reported. A single residue, serine 38, is phosphorylated to a very small degree in vivo under conditions which are thought to be gene activating (Langan 1969). This residue is substantially phosphorylated upon incubation of isolated F_1 histone and liver cytoplasmic extracts (Langan 1968). On the other hand a massive phosphorylation occurs at several sites located towards the middle of the molecule, only in cells that are involved in cell division (Ingles and Dixon 1967; Sherod, Johnson and Chalkley 1970; Balhorn et al. 1972b). The correlation between extent of phosphorylation and the number of cells in a given population involved in cell division was first suggested by Ord and Stocken (1966) and later confirmed in an exhaustive analysis covering regenerating liver, fetal liver at various stages in development, and in a comparison of numerous tumors with their tissue of origin (Balhorn et al. 1971, 1972a). At this time we know of no exception to the correlation between cell division and histone phosphorylation in mammals. Recently division-associated phosphorylation has been observed in several species of invertebrates, including Tetrahymena and Physarum.

Recent evidence accumulated from studies of a hepatoma tissue culture line, HTC cells, has argued that phosphorylation (at several levels ranging from 1–4 phosphate groups per F_1 molecule) occurs to most of the lysine-rich histone molecules each HTC cell cycle. Both newly synthesized and preexisting histones are phosphorylated (Balhorn et al. 1972d), and these phosphate groups are removed with a half-life of 4–5 hours in HTC cells (Oliver et al. 1972).

Identification of the stage in the cell cycle at which phosphorylation occurs might be expected to throw some light upon the function of such a modification. Unfortunately there is little agreement as to the time of phosphorylation. Lake and Salzman (1972) have argued for substantial phosphorylation during mitosis; Bradbury (1973) favors a premitotic, G_2, event in Physarum; Marks et al. (1973) observe phosphorylation throughout G_1 and S using HeLa cells; Gurley and Walters (1971) have reported on phosphorylation in late G_1, during S and into G_2 (in CHO cells); and finally in this laboratory we find that extensive phosphorylation occurs in HTC cells about 30–60 minutes after synthesis of the histone molecule (Balhorn et al. 1972d). This clearly indicates that phosphorylation occurs during the histone synthetic period (roughly correlated with the S phase), and we have also observed that additional phosphorylation continues at a low level during the rest of the cell cycle so that the overall level of phosphorylated species is maintained in the presence of continuing phosphate turnover. It seems probable that the reduced level of phosphate incorporation in the G_1 phase in HTC cells (Balhorn et al. 1972c) is due to the vastly decreased number of phosphorylatable sites (limited by phosphatase action) relative to those made available by new histone synthesis in the S phase. To what extent differences in cell lines and methods of determining phosphorylation have given rise to the discrepancies is quite unclear at this time.

The uncertainty as to the timing of phosphorylation within the cell cycle and yet the clear dependence of phosphorylation upon the overall processes of cell division led us to pose questions concerning the role of active DNA synthesis and chromosome replication in histone phosphorylation. Is it possible to uncouple chromosome replication and histone phosphorylation, and if so, to what extent? Ideally we would like to specifically inhibit one of these functions and to assay for an effect on the other. At this time we do not know of a method that specifically inhibits histone phosphorylation per se; however several different means of inhibiting DNA synthesis are available and we have been able to study the effect of such inhibitors on histone phosphorylation.

Histone Phosphorylation in the Presence of Inhibitors of DNA Synthesis

We have inhibited DNA synthesis with cycloheximide, which of course acts by inhibiting protein (including histone) synthesis, and with hydroxyurea, which has a much lesser effect on histone synthesis. The cells used in these studies were HTC cells maintained in exponential growth (at cell densities up to ~ 500,000/ml) in modified Swim's 77 medium (Balhorn et al. 1972d). Inhibitor studies were conducted only for those time periods that do not irreversibly affect the cell's ability to survive after removal of the drug. The maximum length of such treatment is 8 hours for cycloheximide and 24 hours for hydroxyurea.

The ability of 10 μg/ml of cycloheximide to inhibit both histone and DNA synthesis is shown in Fig. 1. All five histones are similarly inhibited and in general the level of DNA synthesis is reduced to less than 20% of its original value. The drug is clearly acting very rapidly and its effect is exerted within 10 minutes after administration.

Hydroxyurea likewise rapidly inhibits DNA synthesis; however in HTC cells histone synthesis is not as tightly coupled to DNA synthesis as in HeLa cells and in fact histones are made for several hours. The overall effect on DNA and histone synthesis is dose dependent (Fig. 2). However in no instance is histone synthesis completely inhibited even when DNA synthesis is proceeding at < 1% of the control value. Clearly then additional histone is being added to the chromosome in the absence of DNA replication. This must either be compensated for by a removal of excess histone or else the chromosome must gain an additional complement of histone with an ensuing modification in its net charge. We have tested for histone turnover by labeling cells with [³H]lysine either in advance of, or after, adding hydroxyurea. The subsequent chase period was examined in the presence of the

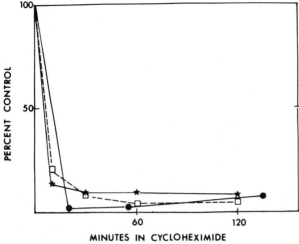

Figure 1

Inhibition of histone and DNA synthesis by cycloheximide. HTC cells in exponential growth were treated with 10 μg/ml cyclyoheximide and aliquots pulsed with [³H]thymidine or [³H]lysine at appropriate times after introduction of the inhibitor. Histone synthesis was estimated by isolation and electrophoresis. (–x–x–) Overall protein synthesis; (●———●) histone synthesis; (–□–□–) DNA synthesis.

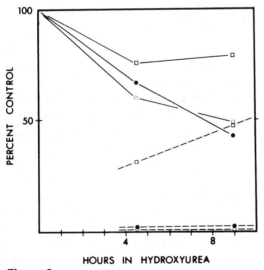

Figure 2

Effect of hydroxyurea on histone and DNA synthesis. Cells were incubated in 1 mM (□), 5 mM (●) or 10 mM (○) hydroxyurea and assayed for DNA synthesis (– – –) or histone synthesis (———) as described in the legend to Fig. 1.

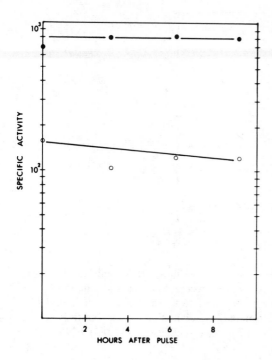

Figure 3
Lack of histone turnover in hydroxy-urea. Cells were labeled for 3 hr with [³H]lysine either before (•) or after (o) adding hydroxyurea (5 mM). After incubation the cells were washed and chased in the presence of 5 mM hydroxyurea. The specific activity of histone was determined by electrophoresis, scanning and counting dissected bands.

inhibitor. The data of Fig. 3 indicate that there is no degree of histone turnover significant in the presence of hydroxyurea. Nor do any of the five histone subfractions turnover selectively in this time period. Clearly then the chromosome must be accruing extra histone. This rather surprising conclusion is supported by the fact that the electrostatic charge on normal nucleohistone at pH 5 is negative (-1.16 μ/sec/V/cm), whereas nucleohistone isolated from cells treated with hydroxyurea (5 mM, 9 hr) has an overall positive charge ($+0.94$) at this pH, presumably due to the excess histone it contains.

We have asked where the extra histone is deposited during the period of histone synthesis in the presence of hydroxyurea. Those parts of the chromosome that had already replicated at the time of the addition of hydroxyurea were labeled by means of a 4-hour preincubation with the density label bromodeoxyuridine. Shortly after the addition of hydroxyurea, [³H]lysine was added to label those histones synthesized in the presence of the drug. Nucleohistone was isolated from such cells and fixed in formaldehyde so that histone would not dissociate upon equilibrium centrifugation in CsCl. The results of such an analysis are shown in Fig. 4. In a control experiment [³H]thymidine is associated with the density-labeled nucleohistone on the denser side of the peak. On the other hand [³H]lysine incorporated into histone (80% of the [³H]lysine is in histone in this experiment) is associated with the less dense side of the nucleohistone peak, as would be expected if the ³H-labeled histone were associated with DNA which had a greater than normal complement of histone. Evidently the newly synthesized histone is deposited on the chromosome in such a way that it is not associated with the post-replicative chromosomal material.

The effect of extended treatment with either cycloheximide or hydroxyurea upon rate of phosphorylation was studied by pulsing the cells with ³²P for short time periods at various intervals after adding the appropriate inhibitor. The results of

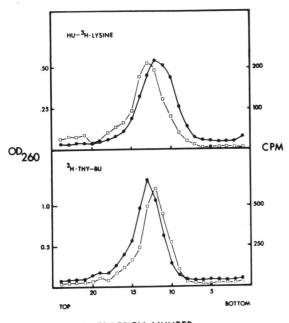

Figure 4

Histone deposition on the chromosome in the presence of hydroxyurea. Cells were grown in the presence of 2×10^{-5} M bromodeoxyuridine for 4 hr. Hydroxyurea was added to a final concentration of 5 mM. After 1 hr [^{3}H]lysine was added for 45 min before the cells were harvested and nucleohistone isolated. Histones were fixed onto DNA using formaldehyde and centrifuged to equilibrium in a CsCl-GuCl gradient at 4°C. (●———●) A_{260}; (□———□) [^{3}H]radiolabel: in the top panel the label is [^{3}H]lysine in histone and in the lower panel it is [^{3}H]thymidine in newly synthesized DNA.

such an experiment are shown in Fig. 5. The rate of phosphorylation falls moderately rapidly in the presence of cycloheximide until a value of about 40% of control is reached after about 80 minutes. This is, however, a much slower response than the inhibition of either DNA or histone synthesis. In contrast, in the presence of hydroxyurea relatively little inhibition is observed during the first 2 hours and maximal inhibition (40–50% of control) is attained only 5–8 hours after the addition of the inhibitor.

In view of these observations and since the half-life for histone dephosphorylation in control cells is 4–5 hours, we had anticipated that in both cases the extent of phosphorylation (as viewed directly by high resolution gel electrophoresis) would decline slightly during the course of extended incubation in the presence of the inhibitors, with the decrease in phosphorylated species being somewhat more pronounced after 8 hours in cycloheximide than during a comparable period in hydroxyurea.

The levels of F_1 phosphorylation actually observed are shown in Fig. 6. To our surprise the level of phosphorylation remained constant in cycloheximide-treated cells for 50 minutes, but during the next 60 minutes the level of the phosphorylated forms of F_1 histone fell sharply. Three hours after the addition of the inhibitor the

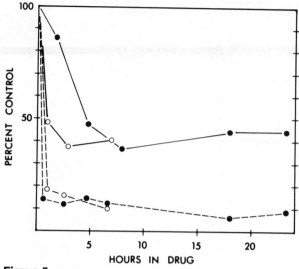

Figure 5

Phosphorylation of lysine-rich histone in the presence of inhibitors of DNA synthesis. Phosphorylation of the F_1 histone was studied in a series of short (20 min) pulses of ^{32}P-phosphate at appropriate times after adding the inhibitor. (•– – –•) DNA synthesis in 5 mM hydroxyurea; (•——•) histone phosphorylation in hydroxyurea; (o– – –o) DNA synthesis in cycloheximide; (o——o) histone phosphorylation in cycloheximide.

microheterogeneity in the F_1 region resembles that found in nonphosphorylated, stationary phase cells. In sharp contrast the level of phosphorylation during extended incubation in the presence of hydroxyurea shows almost no significant change and resembles the exponentially growing HTC cell phosphorylation pattern.

These observations are clearly at variance with predictions based upon normal turnover of the histone phosphate groups, and we looked for a resolution of this conflict in dramatically changed rates of hydrolysis of the phosphate groups in the presence of the two different inhibitors. That our supposition was indeed correct is shown in Fig. 7. The rate of phosphate hydrolysis is reduced in the presence of hydroxyurea ($T_{1/2} = 6.5$ hr) so that the effect of the reduced rate of phosphorylation is nullified and the overall level of phosphorylation remains unchanged. The hydrolysis of phosphate groups proceeds slowly for about 50 minutes in the presence of cycloheximide, after which time the rate increases dramatically to $T_{1/2} = 1.5$ hr, thus accounting for the rapid conversion to the parental, nonphosphorylated form of the F_1 histone after 3–4 hours in the presence of the inhibitor.

It seems unlikely that the unusually rapid turnover of phosphate groups in the presence of cycloheximide is directly related to inhibition of DNA synthesis, since hydroxyurea, which likewise inhibits DNA synthetic activity, actually decreases the rate of phosphate removal. If other agents that block both histone and DNA synthesis yield a similar result, we will be able to exclude a trivial effect of cycloheximide on the activity of the phosphatase. An attractive proposal is that the phosphatase is negatively controlled by a factor which depends upon protein synthesis and which turns over rapidly. Inhibition of the synthesis of this factor by cyclohexi-

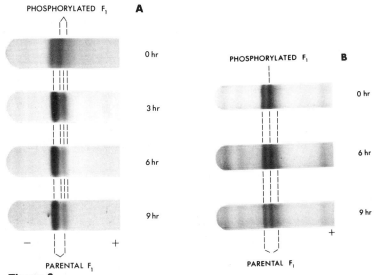

Figure 6

Bulk levels of phosphorylated lysine-rich histone. These were measured as a function of time of incubation in **(A)** cycloheximide (10 μg/ml) and **(B)** hydroxyurea (5 mM).

mide would decrease the amount present and thereby activate the phosphatase. These matters are currently being actively pursued.

Repetitive Phosphorylation during the Cell Cycle

In uninhibited cells there appears to be a peak of phosphorylation in S phase due to extensive modification of histone approximately 30 minutes after deposition of the latter on the chromosome (Balhorn and Tanphaichitr unpubl.). After 3 hours the level of phosphorylation is reduced to about 65%, a level that is maintained throughout the cell cycle until it is once more increased in S phase again. This demands that a measure of phosphorylation proceed in phases of the cell cycle other than S in order to maintain phosphorylation levels in the face of continuing phosphatase activity. Clearly then the possibility exists that a given histone molecule may be repetitively phosphorylated, a point of view which represents a departure from our previous suggestions that were based on shorter-term analyses of phosphorylation changes in S phase cells. Proof of this idea will require bulk isolation of histone from synchronized cells and a direct analysis of the amounts of phosphorylated species in the various phases of the cell cycle, a study which is currently underway.

We wondered if histone could be repetitively phosphorylated in the presence of hydroxyurea. Accordingly cells were labeled with [³H]lysine for 3 hours. After washing out radiolabel the cells were resuspended in fresh medium containing 5 mM hydroxyurea. Samples were withdrawn at appropriate times and histones isolated and assayed for ³H radioactivity in phosphorylated subbands (Fig. 8). The lysine-rich histone was extensively phosphorylated during the initial pulse as is seen in the

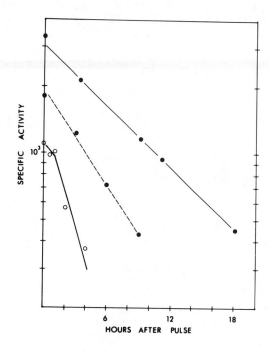

Figure 7
Turnover of F_1-associated phosphate. After < 3 hr pulse of ^{32}P-phosphate turnover was measured ($\bullet$–––$\bullet$) in control cells, ($\bullet$———$\bullet$) in the presence of hydroxyurea, and ($\circ$———$\circ$) in the presence of cycloheximide.

label associated with the phosphorylated electrophoretic bands. If the phosphorylation were nonrepititious, then one would expect the [^3H]lysine-labeled histone to migrate to the parental bands with a half-life of 6.5 hours. However if the molecules are subsequently rephosphorylated, then the distribution of ^3H label will not change so dramatically. It is clear from the analysis of the chase period that the distribution of label does not change greatly during the course of this experiment, and we conclude that histone associated with the chromosome before addition of hydroxyurea continues to be rephosphorylated and to maintain its previous level of phosphorylation even though DNA synthesis has been inhibited for many hours.

Finally we have asked whether histone synthesized in the presence of hydroxyurea is phosphorylated, or if in some way the cell can distinguish a different population of histone molecules. Cells were treated with hydroxyurea and [^3H]lysine added after 9 hours. The newly synthesized histone is distributed among phosphorylated and nonphosphorylated forms in an apparently normal fashion (Fig. 9), indicating that the cell deals with histones synthesized in the presence of hydroxyurea in the same way as in its absence.

SUMMARY

The results described in this paper indicate that inhibition of DNA synthesis per se appears to have relatively little effect on the phosphorylation of the lysine-rich histone. This is most dramatically seen in the studies of phosphorylation in the presence of hydroxyurea where little inhibition of phosphorylation is seen during the first few hours of the block. During this time histone synthesis continues at a reasonable rate, and the later drop in phosphorylation rate probably reflects decreased histone synthetic activity and a concomitant reduction in substrate for phosphorylation. There is a much more rapid and substantial reduction in rate of

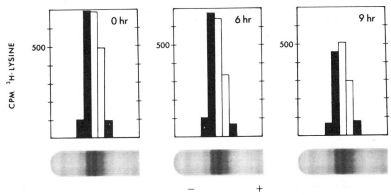

Figure 8
Repetitive phosphorylation in hydroxyurea-treated cells. [³H]Lysine was incorporated into HTC cells for a 3-hr pulse period. The chase was conducted in the presence of hydroxyurea (5 mM). Samples were withdrawn at appropriate times and assayed for [³H]label in phosphorylated histone bands.

phosphorylation during inhibition of DNA synthesis by cycloheximide treatment, but this is probably due in large amount to the unavailability of newly synthesized histone as substrate for the continuing phosphorylation activity. The final rate of phosphorylation after several hours of cycloheximide treatment is probably higher than might have been expected since the phosphatase becomes activated during the course of the experiments and this increases significantly the availability of phosphorylation sites. Salzman and Lake have shown that a nuclear protein kinase, which actively phosphorylates histone, decreases over a several-hour incubation with cycloheximide and this may also contribute to lower phosphorylation rates.

In the presence of hydroxyurea, as in control cells, the system seems to be trying to maintain a level of phosphorylation (approximately 70% in HTC cells) in the face of continued hydrolysis. Why the chromosomal material should need to have

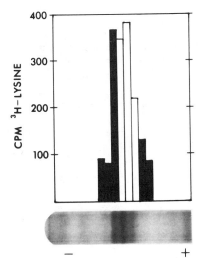

Figure 9
Phosphorylation of histones synthesized in the presence of hydroxyurea.

its lysine-rich histone phosphorylated to this extent on a statistical basis, apparently throughout the cell cycle, is not immediately apparent; but it seems likely that any function for histone phosphorylation will lie in a more general interaction with cell division processes rather than in a specific involvement with the act of DNA replication itself.

Acknowledgments

This work was supported by PHS grants CA-10871 and CA-12191. R.C. is a recipient of a Research Career Development Award GM-46410. D.G. was a Veterans Administration Clinical Investigator.

REFERENCES

Balhorn, R., W. O. Rieke and R. Chalkley. 1971. Rapid electrophoretic analysis for histone phosphorylation. A reinvestigation of phosphorylation of lysine-rich histone during rat liver regeneration. *Biochemistry* **10**:3952.

Balhorn, R., M. Balhorn and R. Chalkley. 1972a. Lysine-rich histone phosphorylation and hyperplasia in the developing rat. *Develop. Biol.* **29**:199.

Balhorn, R., M. Balhorn, H. P. Morris and R. Chalkley. 1972b. Comparative high-resolution electrophoresis of tumor histones: Variation in phosphorylation as a function of cell replication rate. *Cancer Res.* **32**:1775.

Balhorn, R., J. Bordwell, L. Sellers, D. Granner and R. Chalkley. 1972c. Histone phosphorylation and DNA synthesis are linked in synchronous cultures of HTC cells. *Biochem. Biophys. Res. Commun.* **46**:1326.

Balhorn, R., D. Oliver, P. Hohmann, R. Chalkley and D. Granner. 1972d. Turnover of DNA, histones, and lysine-rich histone phosphate in hepatoma tissue culture cells. *Biochemistry* **11**:3915.

Bradbury, M. 1973. *Nature* (in press)

Gurley, G. L. and R. A. Walters. 1971. Response of histone turnover and phosphorylation to X-irradiation. *Biochemistry* **10**:1588.

Ingles, C. J. and G. H. Dixon. 1967. Phosphorylation of protamine during spermatogenesis in trout testis. *Proc. Nat. Acad. Sci.* **58**:1011.

Lake, R. S. and N P. Salzman. 1972. Occurrence and properties of a chromatin-associated F_1 histone phosphokinase in mitotic Chinese hamster cells. *Biochemistry* **11**:4817.

Langan, T. A. 1968. Histone phosphorylation: Stimulation by adenosine 3',5'-monophosphate. *Science* **162**:579.

————. 1969. Phosphorylation of liver histone following the administration of glucagon and insulin. *Proc. Nat. Acad. Sci.* **64**:1276.

Marks, D. B., W. K. Paik and T. W. Borun. 1973. The relationship of histone phosphorylation to deoxyribonucleic acid replication and mitosis during the HeLa S-3 cell cycle. *J. Biol. Chem.* **248**:5660.

Oliver, D., R. Balhorn, D. Granner and R. Chalkley. 1972. Molecular nature of F_1 histone phosphorylation in cultured hepatoma cells. *Biochemistry* **11**:3921.

Ord, M. G. and L. A. Stocken. 1966. Metabolic properties of histones from rat livers and thymus glands. *Biochem. J.* **98**:888.

Sherod, D., G. Johnson and R. Chalkley. 1970. Phosphorylation of mouse ascites of tumor cell lysine-rich hormone. *Biochemistry* **9**:4611.

Specificity of Antibodies to Nonhistone Chromosomal Proteins of Cultured Fibroblasts

Luciano Zardi, Jung-Chung Lin, Robert O. Petersen, and Renato Baserga

Department of Pathology and Fels Research Institute
Temple University School of Medicine, Philadelphia, Pennsylvania 19140

Nonhistone chromosomal proteins are thought to play a major role in the regulation of gene expression (Wang 1968, 1970, 1971; Paul and Gilmour 1968; Teng and Hamilton 1969; Kamiyama and Wang 1971; Spelsberg and Hnilica 1970; Spelsberg et al. 1971; O'Malley et al. 1972; Stellwagen and Cole 1968), including the regulation of cellular proliferation in mammalian cells (Rovera and Baserga 1971; Stein and Baserga 1970; Rosenberg and Levy 1972; Baserga and Stein 1971; Chung and Coffey 1971; Levy et al. 1973). Several authors have reported that the specificity of transcription depends on the source of nonhistone chromosomal proteins (Gilmour and Paul 1969; Wang 1970; Stein et al. 1971) and that nonhistone chromosomal proteins from different sources and physiological conditions have different gel electrophoretic profiles (Elgin and Bonner 1970; Levy et al. 1972; Becker and Stanners 1972; Levy et al. 1973; Tsuboi and Baserga 1972; Yeoman et al. 1973). Three immunological studies suggest organ-specific differences among the nonhistone chromosomal proteins of chicken (Chytil and Spelsberg 1971; Spelsberg et al. 1972) or rat (Wakabayashi and Hnilica 1973), and other studies have demonstrated the low antigenicity of histones (Sandberg and Stollar 1958; Sandberg et al. 1967; Stollar and Ward 1970; Rumke and Sluyser 1966; Sluyser et al. 1969), with the exception of the lysine-rich histone F1 (Bustin and Cole 1968; Kinkade 1969; Bustin and Stollar 1972). Histones exhibit no species specificity as might be expected on the basis of the small differences in their amino acid sequences reported to occur during evolution (De Lange et al. 1968, 1969).

Antibodies to nonhistone chromosomal proteins are one of the most powerful tools available for probing the diversity of nonhistone chromosomal proteins in different organs and species. In the present investigation we report the production of antibodies to chromatin of WI-38 human diploid fibroblasts and of 3T6 mouse fibroblasts. The following points are analyzed in detail: (1) the possibility of obtaining antibodies to chromosomal proteins from different species of cultured fibroblasts, (2) the species specificity of antisera to chromatin, (3) the extent to which antibodies are elicited against chromatin components (histones, nonhistone

729

chromosomal proteins, DNA), and (4) the possibility of isolating antibodies specific for a given chromosomal protein fraction.

METHODS AND MATERIALS

Cell Cultures

3T6 mouse fibroblasts (originally obtained from Dr. Howard Green, Massachusetts Institute of Technology) were grown in Dulbecco's modified Eagle medium (Grand Island Biological Laboratories, Grand Island New York) supplemented with 5% fetal calf serum, as previously described (Tsuboi and Baserga 1972). WI-38 human diploid fibroblasts purchased from Flow Laboratories (Rockville Maryland) were grown in Basal Medium Eagle supplemented with 10% fetal calf serum and antibiotics, as previously described (Rovera and Baserga 1971). 2RA (SV40-transformed WI-38 fibroblasts) were obtained from Dr. Vincent Cristofalo (Wistar Institute) and were grown in MEM supplemented with 5% fetal calf serum. The cells were used for these experiments when they had formed confluent monolayers, either in plastic Falcon flasks (surface of 75 cm²) or in Blake bottles (surface 210 cm²).

Analytical Procedures

Proteins were determined by the method of Lowry et al. (1951) and DNA by the method of Burton (1956) using respectively bovine serum albumin and calf thymus DNA as standards.

Preparation of Chromatin

Chromatin was prepared by a modification of the method of Marushige and Bonner (1966). Approximately 5×10^7 cells, harvested with a rubber policeman, were washed three times in 5 ml 80 mM NaCl, 20 mM EDTA with 1% Triton N-101 at pH 7.4 and three times in 0.15 M NaCl in 10 mM Tris pH 8.00. The clean nuclei were suspended in ice-cold distilled water for 15 minutes and homogenized with a few strikes of a Teflon homogenizer. The chromatin was centrifuged at 27,000 × g for 15 minutes. The pellet was resuspended in 1 ml 10 mM Tris pH.8 and sedimented through 4 ml 1.7 M sucrose for 80 minutes at 37,500 rpm in a Spinco centrifuge with an SW39 rotor. The chromatin pellet was resuspended in 2 ml 10 mM Tris pH.8. The nuclei were free of cytoplasmic contamination by electron microscopy, and the final chromatin preparation contained at the most 0.4% of the total cytoplasmic proteins (Augenlicht and Baserga unpublished).

Immunization Schedule

Chromatin prepared as described above was used to immunize New Zealand white rabbits and chickens. The chromatin was suspended in complete Freund's adjuvant (DIFCO). Four injections, each containing 400 μg DNA as chromatin, were given intradermally to rabbits on the 1st, 8th, 15th and 22nd day. A subcutaneous booster injection was given a week later and serum was obtained one week after that.

Chickens were similarly immunized except that the four injections were given intradermally on the 1st, 4th, 7th and 10th day. A subcutaneous booster injection was given four days later and the antiserum was obtained seven days after the booster. Two groups of rabbits were immunized, one against WI-38 chromatin, the other against 3T6 chromatin. Chickens were immunized only against WI-38 chromatin.

Analysis of Sera

Complement-fixing antibodies in the immune sera were determined by the complement fixation test of Levine and Van Vunakis (1967). Precipitin reactions were carried out according to the method of Benedict, Hersh and Larson (1963). Chicken antiserum to WI-38 chromatin was added to radiolabeled nonhistone chromosomal proteins in 1.5 M NaCl and 0.01 M Tris-HCl buffer at pH 8.4. The reaction of nonhistone chromosomal proteins from WI-38 cells with serum from a nonimmunized chicken served as a blank. The precipitate was collected by centrifugation at $1000 \times g/30$ minutes, rinsed four times with 1.5 M NaCl in 0.01 M Tris-HCl pH 8.4, after which the precipitate was dissolved in 0.1 M NaOH and assayed for radioactivity as described below.

Dissociation of Chromatin

The method used for the dissociation of proteins from DNA and the separation of histones from nonhistones was the one described by Levy, Simpson and Sober (1972).

Preparation of Cytoplasmic Proteins

Total cytoplasmic proteins and cytoplasmic acidic proteins were prepared by suspending the cells in a solution containing 0.02 M NaCl, 0.02 M Tris pH 8.3, 0.005 M $MgCl_2$ at 4°C. The cell suspension was homogenized in a Potter-Elvehjem homogenizer and the homogenate was centrifuged at $12,000 \times g$ for 20 minutes in a Sorvall centrifuge. The proteins in the supernatant were diluted to a final concentration of 1 mg/ml and treated with 10 μg/ml of RNase A, protease free, for 60 minutes at 37°C. Solid urea and guanidine HCl were added to a final concentration respectively of 6 M and 0.4 M and stored at 4°C until the proteins were dissolved. Acidic cytoplasmic proteins were separated from nonacidic cytoplasmic protein by the method of Levy et al. (1972) mentioned above.

Preparation of Nonhistone Chromosomal Proteins Coupled to CNBr-activated Sepharose 4B

The techniques for the preparation of the sepharose coupled to nonhistone chromosomal proteins and for the separation of specific antibodies followed essentially the methodology of Morrison and Koshland (1972). About 15 mg of nonhistone chromosomal proteins from 3T6 cells (labeled with [3H]leucine and prepared as described above) was incubated in 0.1 M carbonate buffer pH 9 and 1.5 M NaCl with 2 g of CNBr-activated sepharose 4B for 30 to 40 hours at 4°C. By this method more than 80% of the proteins in the original solution became bound to the CNBr-sepharose 4B.

Materials

The complete Freund's adjuvants were purchased from DIFCO. The complement, the sheep red blood cells and all other reagents for complement fixation were purchased from GIBCO; [3H]leucine 5 Ci/0.01 mg was purchased from New England Nuclear Corporation.

The CNBr-activated sepharose 4B was purchased from Pharmacia Fine Chemicals. Triton X-100 was purchased from Rohm and Haas (Philadelphia). All other chemicals were of reagent grade.

RESULTS

Production of Antibodies against Chromatin from 3T6 or WI-38 Fibroblasts

When 3T6 chromatin was injected into rabbits, complement-fixing antibodies were produced (Fig. 1). Figure 1 also shows that when rabbit antiserum to 3T6 chromatin was reacted with chromatin from WI-38 cells, complement fixation was much reduced at each of the antiserum dilutions used and for each antigen concentration tested. Despite the well-known reservations about the quantitative aspects of the complement fixation reaction, the results in Fig. 1 do show that rabbit antibodies to 3T6 chromatin can distinguish between the antigens of 3T6 chromatin and those of WI-38 chromatin.

Similarly when chromatin from WI-38 cells was injected into rabbits, the antibodies formed fixed more complement when reacted with WI-38 chromatin than with 3T6 chromatin (not shown). However rabbits immunized with WI-38 chromatin produced antisera with a titer about 10 times higher than the titer of rabbits immunized with 3T6 chromatin.

Similar results were obtained with chicken antiserum. Figure 2 shows the presence of complement-fixing antibodies in the unheated serum of chickens (Kabat and Mayer 1961) injected with WI-38 chromatin. Again species specificity is apparent; i.e., the antiserum reacts much more with WI-38 chromatin than with 3T6 chromatin. Note also that complement-fixing antibodies can be detected in serum from immunized chickens with much smaller concentrations of antigens than required for rabbit antiserum.

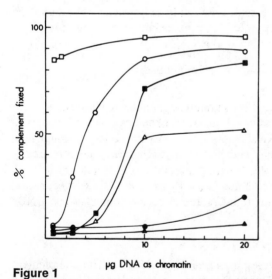

Figure 1
Complement fixation reactions of 3T6 cell chromatin (open symbols) and WI-38 cell chromatin (closed symbols), with rabbit antiserum to 3T6 chromatin. Antiserum dilutions: (□, ■) 1:50; (o, •) 1:400; (△, ▲) 1:600.

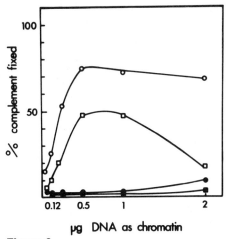

Figure 2
Complement fixation reactions of WI-38 cell chromatin (open symbols) and
3T6 cell chromatin (closed symbols) with unheated chicken antiserum to WI-38
chromatin. Antiserum dilutions (o, •) 1:800; (□, ■) 1:2000.

Complement-fixing Antibodies against Chromatin Components

Complement-fixing antibodies could be directed against only one of the chromatin
components (for instance nonhistone proteins, DNA) or against all of these components, but to a different extent. To determine the extent of immunogenicity of
each chromatin component, complement fixation tests were carried out with antisera to chromatin and individual chromatin components, such as histones, nonhistone protein and DNA.

Figure 3 shows the presence of complement-fixing antibodies against 3T6 non-

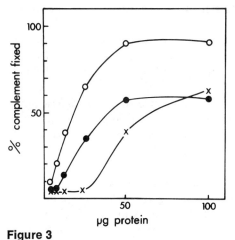

Figure 3
Complement fixation reactions of (o) nonhistone chromosomal proteins from
3T6 cells, (×) acidic cytoplasmic proteins from 3T6 cells, and (•) nonhistone
chromosomal proteins from WI-38 cells with rabbit antiserum to 3T6 chromatin. These curves were obtained with an antiserum dilution of 1:300.

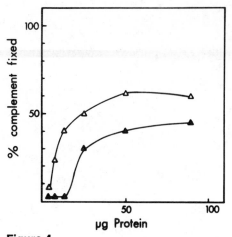

Figure 4

Complement fixation reactions of ($\triangle$) 3T6 histones or ($\blacktriangle$) WI-38 histones with rabbit antiserum to 3T6 chromatin. The antiserum dilution used in these experiments was 1:200.

histone chromosomal proteins in rabbit antiserum to 3T6 chromatin. Figure 3 also shows the complement-fixing activity of this rabbit antiserum when tested against nonhistone chromosomal proteins of WI-38 or against acidic cytoplasmic proteins of 3T6 cells. In both instances complement fixation is reduced to a level approximately half of that obtained with nonhistone proteins from 3T6 cells.

The presence of complement-fixing antibodies against histones in rabbit antiserum to 3T6 chromatin is shown in Fig. 4. Both histones from 3T6 and from WI-38 cells are shown. Complement fixation is approximately the same whether 3T6 histones or WI-38 histones are incubated with rabbit antiserum to 3T6 chromatin. Notice also the lower dilution of the serum in respect to Fig. 3. When DNA was used as an antigen, the complement-fixing activity was much lower than shown in either Fig. 3 or 4 (not shown).

Figure 5 is an attempt to quantitate the data obtained thus far. In this figure we have taken rabbit antiserum to WI-38 chromatin, and from a given concentration of antigen we have plotted the dilution of the antiserum for which the complement fixed is 50% of maximum.

The rabbit antiserum to WI-38 chromatin reacts as expected, at its highest dilution with WI-38 chromatin, and at lower dilutions with nonhistone chromosomal proteins from WI-38, histones from WI-38, 3T6 chromatin, nonhistone proteins of 3T6, histones from 3T6 cells and, finally, DNA. With the reservations about the quantitation of the complement fixation test, Fig. 5 gives evidence for the species specificity of the chromatin components and of the extent of reaction of the individual chromatin components. From this one can conclude that rabbit antiserum to WI-38 chromatin gives a greater complement fixation with WI-38 chromatin than with 3T6 chromatin. Furthermore nonhistone chromosomal proteins from WI-38 give a 50% maximum complement fixation at a greater dilution than histones from the same source. Also as already mentioned above for 3T6 chromatin, although there is a marked difference in the extent of complement fixation with nonhistone chromosomal proteins depending on whether they originate from WI-38

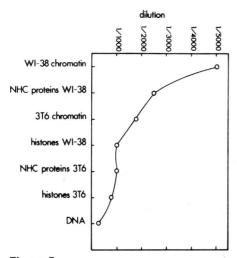

Figure 5
Complement fixation reactions of various chromatin components with rabbit antiserum to WI-38 chromatin. The different chromatins or chromatin components (from 3T6 or WI-38) are listed on the ordinate. The abscissa gives the dilution of the antiserum at which the complement fixed is 50% of maximum, using fixed amounts of antigens.

or from 3T6 cells, the complement-fixing antibodies against histones are present in approximately the same amount, regardless of the source of the histones.

Absorption of Antibodies Reactive with 3T6 and WI-38 Chromatins

We have attempted to separate the antibodies that react with antigens common to both WI-38 and 3T6 chromatin, so that it may be possible to separate both unique antibodies and antigens from either source. Two techniques were tried in order to achieve the separation, the first based on the absorption of antibodies to antigens coupled to CNBr-sepharose, the second on the use of precipitins from chicken antiserum.

The effectiveness of absorption of antibodies by a column of CNBr-sepharose 4B coupled with the respective antigens is shown in Fig. 6A. Nonhistone chromosomal proteins from 3T6 cells were coupled with the immunoabsorbent, as described in Methods and Materials. The absorbed nonhistone chromosomal proteins were then incubated at 4°C with rabbit antiserum to 3T6 chromatin. The immunoabsorbent was removed by centrifugation, leaving a supernatant presumably devoided of antibodies to nonhistone chromosomal proteins from 3T6. The antibodies specifically bound to the immunoabsorbent were then eluted at pH 2.8, as described by Morrison and Koshland (1972). Figure 6A shows the complement fixation test, using as antigen nonhistone chromosomal proteins from 3T6 and as a source of antibodies either the rabbit antiserum to 3T6 chromatin after absorption to the sepharose column or the antibodies eluted from the sepharose column. Nonhistone chromosomal proteins from 3T6 coupled with sepharose immunoabsorbent effectively remove antibodies to nonhistone chromosomal proteins from the rabbit antiserum to 3T6 chromatin. These antibodies can then be eluted from

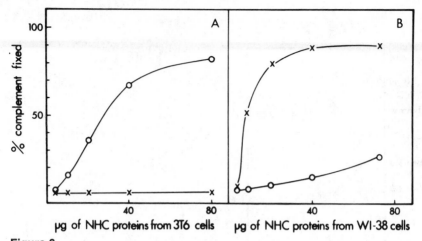

Figure 6

Complement fixation curves of antisera to 3T6 and WI-38 chromatin after specific absorption.

A, Rabbit antiserum to 3T6 chromatin after overnight incubation at 4°C with nonhistone chromosomal proteins from 3T6 cells coupled to CNBr-sepharose 4B. Antigen, nonhistone chromosomal proteins from 3T6 cells. (×) Antiserum after absorption to the CNBr-sepharose; (○) antibodies eluted from the immunoabsorbent. The antiserum used in this reaction was diluted 1–300.

B, Rabbit antiserum to WI-38 chromatin reacted with CNBr-sepharose 4B coupled with nonhistone chromosomal proteins from 3T6 cells. Antigen, nonhistone chromosomal proteins from WI-38. (×) Antiserum to WI-38 chromatin after absorption to sepharose-bound histone chromosomal proteins from 3T6 cells; (○) antibodies eluted from the immunoabsorbent. The serum dilution used in these experiments was 1:4000. The details of the procedures are described in the text and Methods and Materials.

the immunoabsorbent at pH 2.8 and give a reasonably high complement fixation test.

The immunoabsorbent of nonhistone chromosomal proteins from 3T6 cells was then incubated with rabbit antiserum to WI-38 chromatin (Fig. 6B). Complement fixation tests were then carried out using the nonabsorbed antiserum and that fraction of antibodies that had been bound to the sepharose column and had been eluted at pH 2.8. As antigens in these complement fixation tests, nonhistone chromosomal proteins from WI-38 cells were used. The results in Fig. 6B show that most of the antibodies to nonhistone chromosomal proteins of WI-38 cells are not bound to the sepharose column to which nonhistone chromosomal proteins from 3T6 cells are absorbed. However a small fraction of antibodies to WI-38 chromatin capable of giving a positive complement fixation test were bound to the 3T6 sepharose column and were eluted at pH 2.8, as shown in the lower line of Fig. 6B. These results then show that specific antibodies can be effectively removed using a sepharose column to which the appropriate antigens have been coupled.

Another method that was successful in our hands was based on the precipitation of antibodies by chicken precipitins. This was possible because chicken precipitins are still effective in 1.5 M NaCl (Benedict et al. 1963). This high concentration of salt is necessary in testing precipitins against nonhistone chromosomal proteins because of the low solubility of nonhistone chromosomal proteins in low salt concentrations. At a concentration of 1.5 M NaCl, 90% of the nonhistone chromosomal proteins are soluble and yet the chicken precipitins are still active. Antibodies against WI-38 chromatin were obtained by immunization of chickens with WI-38 chromatin. Chicken antiserum to WI-38 chromatin was then incubated with nonhistone chromosomal proteins of WI-38 cells. Figure 7 shows that nonhistone chromosomal proteins from WI-38 cells are effectively precipitated by the chicken antiserum to WI-38 chromatin, whereas nonhistone chromosomal proteins from 3T6 are only slightly precipitated. To demonstrate the specificity of this precipitin reaction, chicken antiserum to WI-38 chromatin was incubated with 3T6 chromatin and the antibody precipitated was then eliminated by centrifugation. Figure 8A shows the complement fixation test using nonhistone chromosomal proteins of WI-38 as antigen, and as antiserum the chicken antiserum to WI-38 chromatin before and after absorption with 3T6 chromatin. Very few antibodies to nonhistone chromosomal proteins of WI-38 cells are absorbed by 3T6 chromatin. On the other hand, most of the antibodies to WI-38 histones are absorbed by 3T6 chromatin, as shown in Fig. 8B.

Difference between Chromatin of WI-38 and Chromatin of 2RA Cells

Figure 9 shows that chicken antiserum to WI-38 chromatin fixes more complement when reacted with WI-38 chromatin than with chromatin from 2RA cells, a cell line originating from WI-38 fibroblasts by transformation with SV40 virus.

DISCUSSION

Chytil and Spelsberg (1971) and Spelsberg et al. (1971) have reported the production of complement-fixing antibodies in rabbits injected with dehistonized chro-

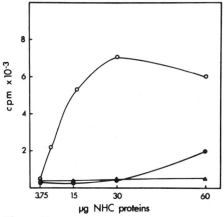

Figure 7

Precipitin reaction of (○) nonhistone chromosomal proteins from WI-38 and (●) nonhistone chromosomal proteins of 3T6 with chicken antiserum to WI-38 chromatin. Triangles represent precipitin reaction of nonhistone chromosomal proteins of WI-38 cells with serum from a nonimmunized chicken. Nonhistone chromosomal proteins of WI-38 and 3T6 cells had the same specific activity.

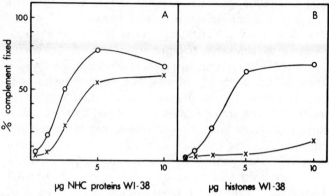

Figure 8

A, Complement fixation reaction of nonhistone chromosomal proteins from WI-38 cells with chicken antiserum to WI-38 chromatin. (o) Original antiserum; (×) antiserum after absorption with 3T6 chromatin and subsequent elimination of the antigen-precipitated antibodies by centrifugation. The antiserum dilution used in this experiment was 1:600.

B, Same experiment as in (A) except that histones from WI-38 were used instead of nonhistone chromosomal proteins. (o) Antiserum before absorption; (×) antiserum after absorption with 3T6 chromatin and subsequent elimination of the antigen-precipitated antibodies by centrifugation. Dilution of the antiserum used in this experiment 1:300.

matin from various tissues of chick. These authors reported that the complement-fixing antibodies in the rabbit serum could distinguish among the dehistonized chromatins of several chick tissues. Their work confirmed on an immunological basis previous reports from a number of laboratories indicating that nonhistone chromosomal proteins of different tissues in the same species, or from different species, could be distinguished from each other by their biochemical characteristics (Elgin and

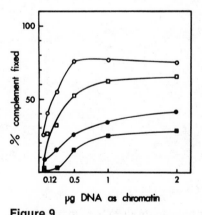

Figure 9

Complement fixation reactions of (o, □) WI-38 chromatin and (•, ■) 2RA chromatin with unheated chicken antiserum to WI-38 chromatin. Antiserum dilutions: (o, •) 1:800, (□, ■) 1:1200.

Bonner 1970; Levy et al. 1972, 1973; Becker and Stanners 1972; Tsuboi and Baserga 1972; Barrett and Gould 1973). Our own results confirm the studies of Chytil and Spelsberg (1971), Spelsberg et al. (1971) and Wakabayashi and Hnilica (1973), indicating that antibodies to the chromatin of fibroblasts in culture can be obtained both in the rabbit and in the chicken. We elected to use preparations of chromatin as antigens for two reasons, namely (a) to avoid loss of protein components which might occur during the dissociation of chromatin and the separation of the two major protein fractions, and (b) because the immunogenicity of chromosomal proteins is enhanced when they are injected together with DNA (Chytil and Spelsberg 1971; Sandberg et al. 1967; Stollar and Ward 1970). In rabbit, human chromatin elicits higher levels of antibodies than mouse chromatin, as it would be expected from the phylogenetic relationship of these three species; and in the chicken both precipitating and complement-fixing antibodies are detectable, whereas in the rabbit serum only complement-fixing antibodies can be obtained. Our results also show that both rabbit and chicken antisera to fibroblast chromatin contain antibodies for various components of chromatin. However the antibody titer is higher for nonhistone chromosomal proteins than for histones and is exceedingly low for DNA. In our chromatin preparation the amount of RNA is negligible (Stein et al. 1971) so that we did not deem it necessary to determine the presence of antibodies to chromosomal RNA.

Complement-fixing antibodies to nonhistone chromosomal proteins (obtained by injecting chromatin in either chicken or rabbit) are largely species specific, although some complement-fixing antibodies reacting with nonhistone chromosomal proteins common to WI-38 and 3T6 cells can be detected. On the contrary, complement-fixing antibodies or precipitating antibodies to histones are largely aspecific, and there is a considerable amount of cross reaction between the histones of 3T6 and the histones of WI-38 in respect to their ability to react with antisera to chromatins from these two cell lines. These results with histones confirm the results of previous investigators indicating that histones are only modestly immunogenic and that they lack species specificity except for the F1 histone (Sandberg and Stollar 1958; Sandberg et al. 1967; Stollar and Ward 1970; Rumke and Sluyser 1966; Sluyser et al. 1969; Bustin and Cole 1968; Kinkade 1969; Bustin and Stollar 1972).

Our results also indicate that it is possible to isolate antibodies specific for a given chromatin by absorption with an immunoabsorbent, especially sepharose, to which the appropriate antigen has been bound. It is also possible to remove antibodies reactive with common antigens by precipitation if chicken antiserum, which is rich in precipitins, is used. The methodology described here clearly indicates that antibodies to chromatin can be used for the immunological identification of nonhistone chromosomal proteins specific for similar tissues from different species. The chromatins from SV40-transformed fibroblasts and their normal counterparts can also be distinguished by these immunological methods.

Acknowledgments

This work was supported by USPHS Research Grants CA-08373 and P01 CA 12923 from the National Cancer Institute, by a Damon Runyon Fund Grant 1019C and by American Cancer Society Research Grant IC-88.

REFERENCES

Barrett, T. and H. J. Gould. 1973. Tissue and species specificity on non-histone chromatin proteins. *Biochim. Biophys. Acta* **294**:165.

Baserga, R. and G. Stein. 1971. Nuclear acidic proteins and cell proliferation. *Fed. Proc.* **30**:1752.

Becker, H. and C. P. Stanners. 1972. Control of macromolecular synthesis in proliferating and resting Syrian hamster cells in monolayer culture. III. Electrophoretic patterns of newly synthesized proteins in synchronized proliferating cells and resting cells. *J. Cell. Physiol.* **80**:51.

Benedict, A. A., R. T. Hersh and C. Larson. 1963. The temporal synthesis of chicken antibodies. The effect of salt on the precipitin reaction. *J. Immunol.* **91**:795.

Burton, K. 1956. A study of the conditions and mechanism of the diphenylamine reaction for the colorimetric estimation of DNA. *Biochem. J.* **62**:315.

Bustin, M. and R. D. Cole. 1968. Species and organ specificity in very lysine-rich histones. *J. Biol. Chem.* **243**:4500.

Bustin, M. and B. D. Stollar. 1972. Immunochemical specificity in lysine-rich histone subfraction. *J. Biol. Chem.* **247**:5716.

Chung, L. K. W. and D. S. Coffey. 1971. Biochemical characterization of prostatic nuclei. I. Androgen-induced changes in nuclear proteins. *Biochim. Biophys. Acta* **247**:570.

Chytil, F. and T. C. Spelsberg. 1971. Tissue differences in antigenic properties of non-histone protein-DNA complexes. *Nature New Biol.* **233**:215.

De Lange, R. J., D. M. Fambrough, E. L. Smith and J. Bonner. 1968. Calf and pea histone IV. I. Amino acid compositions and the identical COOH-terminal 19-residue sequence. *J. Biol. Chem.* **243**:5906.

————. 1969. Calf and pea histone IV. II. The complete amino acid sequence of calf thymus histone IV. Presence of E-N-acetyllysine. *J. Biol. Chem.* **244**:319.

Elgin, S. C. R. and J. Bonner. 1970. Limited heterogeneity of the major non-histone chromosomal proteins. *Biochemistry* **9**:4440.

Gilmour, R. S. and J. Paul. 1969. RNA transcribed from reconstituted nucleoprotein is similar to natural RNA. *J. Mol. Biol.* **40**:137.

Kabat, E. A. and M. M. Mayer. 1961. *Experimental immunochemistry,* p. 223. Charles C. Thomas Pub., Springfield, Illinois.

Kamiyama, M. and T. Y. Wang. 1971. Activated transcription from rat liver chromatin by non-histone proteins. *Biochim. Biophys. Acta* **228**:563.

Kinkade, J. M. 1969. Qualitative species differences and quantitative tissue differences in the distribution of lysine-rich histones. *J. Biol. Chem.* **244**:3375.

Levine, L. and H. Van Vunakis. 1967. Micro complement fixation. *Methods in enzymology* (ed. C. H. W. Hirs) vol. 11, p. 929. Academic Press, New York.

Levy, S., R. T. Simpson and H. A. Sober. 1972. Fractionation of chromatin components. *Biochemistry* **11**:1547.

Levy, R., S. Levy, S. A. Rosenberg and R. T. Simpson. 1973. Selective stimulation of non-histone chromatin protein synthesis in lymphoid cells by phytohemagglutinin. *Biochemistry* **12**:224.

Lowry, O. H., N. J. Rosebrough, A. L. Farr and R. J. Randal. 1951. Protein measurement with the folin phenol reagent. *J. Biol. Chem.* **193**:265.

Marushige, K. and J. Bonner. 1966. Template properties of liver chromatin. *J. Mol. Biol.* **15**:160.

Morrison, S. L. and M. E. Koshland. 1972. Characterization of the J-chain from polymeric immunoglobulin. *Proc. Nat. Acad. Sci.* **69**:124.

O'Malley, B. W., T. C. Spelsberg, W. T. Schrader, F. Chytil and A. W. Steggles. 1972.

Mechanisms of interaction of a hormone receptor complex with the genome of a eukaryotic target cell. *Nature* **235**:141.

Paul, J. and R. S. Gilmour. 1968. Organ-specific restriction of transcription in mammalian chromatin. *J. Mol. Biol.* **34**:305.

Rosenberg, S. A. and R. Levy. 1972. Synthesis of nuclear-associated proteins by lymphocytes within minutes after contact with phytohemagglutinin. *J. Immunol.* **108**:1105.

Rovera, G. and R. Baserga. 1971. Early changes in the synthesis of acid nuclear proteins in human diploid fibroblasts stimulated to synthesize DNA by changing the medium. *J. Cell. Physiol.* **77**:201.

Rumke, P. and M. Sluyser. 1966. Antigenicity of histones. *Biochem. J.* **101**:1c.

Sandberg, A. L. and B. D. Stollar. 1958. Cross-reactions of calf thymus histones with polylysine antisera. *J. Immunol.* **100**:286.

Sandberg, A. L., M. Liss and B. D. Stollar. 1967. Rabbit antibodies induced by calf thymus histone-serum albumin complexes. *J. Immunol.* **98**:1182.

Sluyser, M., P. Rumke and A. Hekman. 1969. Antigenicity of histones: Comparative studies on histones with very high lysine content from various sources. *Immunochemistry* **6**:494.

Spelsberg, T. C. and C. A. Hnilica. 1970. Deoxyribonucleoproteins and the tissue-specific restrictions of the deoxyribonucleic acid in chromatin. *Biochem. J.* **120**:435.

Spelsberg, T. C., L. S. Hnilica and A. T. Ansevin. 1971. Proteins of chromatin in template restriction. III. The macromolecules in specific restriction of the chromatin DNA. *Biochim. Biophys. Acta* **228**:550.

Spelsberg, T. C., A. W. Steggles, F. Chytil and B. W. O'Malley. 1972. Progesterone-binding components of chick oviduct. V. Exchange of progesterone-binding capacity from target to nontarget tissue chromatins. *J. Biol. Chem.* **247**:1368.

Stein, G. and R. Baserga. 1970. The synthesis of acidic nuclear proteins in the prereplicative phase of the isoproterenol-stimulated salivary gland. *J. Biol. Chem.* **245**:6097.

Stein, G., S. Chaudhuri and R. Baserga. 1971. Gene activation in WI-38 fibroblasts stimulated to proliferate. Role of non-histone chromosomal proteins. *J. Biol. Chem.* **247**:3918.

Stellwagen, R. and R. Cole. 1968. Chromosomal proteins. *Ann. Rev. Biochem.* **38**:951.

Stollar, B. D. and M. Ward. 1970. Rabbit antibodies to histone fraction as specific reagents for preparation and comparative studies. *J. Biol. Chem.* **245**:1261.

Teng, C-S. and T. H. Hamilton. 1969. Role of chromatin in estrogen action in the uterus. II. Hormone-induced synthesis of non-histone acidic proteins which restore histone-inhibited DNA-dependent RNA synthesis. *Proc. Nat. Acad. Sci.* **63**:465.

Tsuboi, A. and R. Baserga. 1972. Synthesis of nuclear acidic proteins in density-inhibited fibroblasts stimulated to proliferate. *J. Cell. Physiol.* **80**:107.

Wakabayashi, K. and L. S. Hnilica. 1973. The immunospecificity of non-histone protein complexes with DNA. *Nature New Biol.* **242**:153.

Wang, T. Y. 1968. Restoration of histone-inhibited DNA-dependent RNA synthesis by acidic chromatin protein. *Exp. Cell Res.* **53**:288.

———. 1970. Activation of transcription *in vitro* from chromatin by non-histone proteins. *Exp. Cell Res.* **61**:455.

———. 1971. Tissue specificity of non-histone chromosomal proteins. *Exp. Cell Res.* **69**:217.

Yeoman, L. C., C. W. Taylor and H. Busch. 1973. Two-dimensional polyacrylamide gel electrophoresis of acid extractable nuclear proteins of normal rat liver and Novikoff hepatoma ascites cells. *Biochem. Biophys. Res. Commun.* **51**:956.

Ribosome Synthesis during Preparation for Division in the Fibroblast

Howard Green

Department of Biology, Massachusetts Institute of Technology
Cambridge, Massachusetts 02139

Whether a mammalian cell rests or enters a division cycle is determined by external regulating factors, which are different for every cell type. The onset of DNA synthesis is determined more directly by internal regulating factors, which are probably the same or similar for most cell types. For this reason, whatever the cell type and the nature of the stimulus (hormones, wounding, serum factors, etc.), study of the transition from resting to growing state is likely to reveal a common mechanism that prepares for DNA synthesis.

Most resting mammalian cells possess the diploid quantity of DNA and when stimulated to enter a division cycle must replicate their DNA before they divide. Exceptional cell types which rest in the G_2 period of the cycle (Storey and Leblond 1951; Gelfant 1963; Swann 1958) must possess a different regulation system and will not be considered here. The state of a cell resting with a diploid quantity of DNA is not identical with that of a cell in exponential growth in the G_1 period, for the resting cell is in a steady state, whereas the G_1 cell in cycle is not. There are other reasons making it necessary to distinguish the two conditions (Stanners and Becker 1971; Becker, Stanners and Kudlow 1971); one of the most important is that the resting cell in G_0 contains fewer ribosomes than a growing cell even at the earliest time in its G_1 period.

The onset of DNA synthesis in a cell previously resting in G_0 indicates that a division cycle is underway, but we know very little of the earlier events that prepare the cell for DNA synthesis. We should like to know the order in which all changes take place during the transition from resting to growing state and at what point the onset of DNA synthesis is determined. For example, it seems that changes in the centrioles precede DNA synthesis and are an earlier indication that a division cycle is underway (Robbins, Jentzsch and Micali 1968). I will describe here the changes that take place in RNA synthesis during the transition from resting to growing state and suggest that one of the earliest of all changes is an increase in the rate of ribosome synthesis.

743

Induction of a Division Cycle in Resting Fibroblasts by Serum Growth Factors

The importance of serum protein factors for the multiplication of fibroblasts is well known. At least one of these factors is required by normal fibroblasts, or those sensitive to contact inhibition, and may be used to induce division in such cultures (Todaro, Lazar and Green 1965). By raising the concentration of serum in a culture of resting fibroblasts, all the cells can be induced to divide (Bürk 1966; Todaro et al. 1967; Fried and Pitts 1968; Holley and Kiernan 1968; Nordenskjöld et al. 1970). Extensive work on the purification of serum factors has been carried out by Holley and colleagues (Holley and Kiernan 1971; Lipton et al. 1971; Paul, Lipton and Klinger 1971), and by Temin (this volume). Since neoplastic transformants of fibroblasts require little or none of one of these factors, we may call it a normal fibroblast growth factor in order to distinguish it from other factors required by transformants as well as by normal cells (Holley and Kiernan 1971; Clarke and Stoker 1971).

In a confluent monolayer of 3T6 cells resting in 0.5% serum, the number of cells engaged in DNA replication, as determined by nuclear radioautography of thymidine-labeled cells, is usually less than 0.5%. Within 12 hours after adding serum to a concentration of 10%, cells begin to synthesize DNA, and 60–80% of the cells may later be simultaneously engaged in DNA replication. However it usually requires about 35 hours for the total DNA content to double (Fig. 1), so the degree of cell synchrony during the transition is not very high. An additional division cycle may follow before the rate of DNA accumulation declines, as the cells approach a new saturation density.

RNA Content of Resting and Growing Cells

A relation between the growth rate of a cell and its RNA content was first suggested from studies of eukaryotic cells by Caspersson and by Brachet (see Brachet 1950) and was demonstrated quantitatively in studies on bacterial growth (Neidhardt and Magasanik 1960; Kjeldgaard and Kurland 1963; Rosset, Julien and Monier 1966). Though the possession of a resting but metabolically active state, stable over a long-term, is a property more of cells of multicellular animals than of bacteria, the most studied example of a resting state in which there is control of RNA synthesis is that resulting from stringent regulation in *E. coli*. When uncharged tRNA becomes abundant, either through amino acid starvation or through expression of a mutation in an amino-acyl tRNA synthetase, accumulation of RNA stops almost immediately. Estimations of rates of RNA synthesis from nucleoside incorporation rates are complicated by the relatively slow equilibration of nucleoside triphosphate pools with the isotopic precursor, but according to Nierlich (1972) the rates of synthesis of stable RNA (rRNA and tRNA) are lower by about 2.6-fold when amino acids are unavailable. Relief of the stringent condition leads to rapid increase in the rate of synthesis and accumulation of RNA, and growth is resumed.

In mammalian cells the transition from resting to growing state is marked by an increase in total RNA content. This has been shown for liver cells stimulated to divide by partial hepatectomy (Lieberman and Kane 1965), uterine muscle cells stimulated by estradiol (Mueller, Herranen and Jervell 1958; Hamilton 1968), and lymphocytes stimulated by phytohemagglutinin (Forsdyke 1967). The same is true of fibroblasts. When 3T6 cells resting in 0.5% serum are induced to enter a division cycle by the addition of 10% serum, DNA synthesis begins only about 12 hours later, but RNA probably begins to accumulate almost immediately (Fig. 1). The increase in RNA content is over 40% by the time DNA begins to accumulate and twofold by the time the earliest divisions take place (ca. 25 hours). In the other cell types mentioned, it has also been found that RNA accumulation precedes DNA synthesis. Since 80% of the RNA of the cell is ribosomal, an increase in total RNA content is a reflection of an increased number of ribosomes.

Measurement of Rate of RNA Synthesis in Intact Cells

There have been many attempts to estimate the rate of RNA synthesis in whole cells from their incorporation of labeled ribonucleosides. Cells in transition from resting to growing state have been found to increase their rate of incorporation, and the increase usually follows very quickly after the application of the stimulus.

An example of such a response is shown in Fig. 1. A resting culture of 3T3 cells was stimulated at time zero by the addition of fresh serum containing the normal fibroblast growth factor. Though cells began to synthesize DNA only 12 hours later, the rate of incorporation of labeled uridine began to increase in less than 15 minutes and was elevated by about tenfold in 30 minutes. This rate of incorporation was not sustained and over the succeeding hours fluctuated at lower levels, though remaining higher than the initial rate, at least up to the time DNA synthesis began. Consistent with the increase in RNA content demonstrated or expected in other resting cell types stimulated to divide, increased rates of incorporation of labeled nucleosides into RNA have been reported for liver cells (Fujioka, Koga and Lieberman 1963), lymphocytes (Cooper and Rubin 1965; Pogo, Allfrey and Mirsky 1966), and uterine muscle cells (Hamilton 1968).

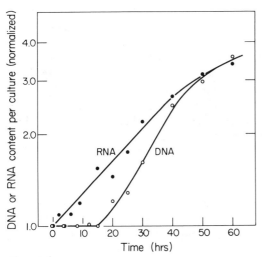

Figure 1
Accumulation of RNA and DNA in cultures of 3T6 following serum stimulation. To confluent cultures of 3T6 resting in 0.5% serum, serum was added to 10% at zero time. Medium containing 10% serum was renewed about every 12 hr thereafter. Ordinate scale is exponential. Adapted from Mauck and Green (1973).

Unfortunately, these studies did not take into account the large and slowly equilibrating cellular nucleotide pool and the possibility that changes in transport rate at the cell membrane could affect the rate of labeling of the nucleotides. This problem was first encountered in experiments on bacteria (Edlin and Neuhard 1967). In studies of animal cells Cunningham and Pardee (1969) showed that within 15 minutes of the addition of fresh serum to a resting culture of 3T3, the rate of uptake of labeled uridine and of ^{32}P-labeled orthophosphate increased three- to fourfold. Since that time estimations of the rate of RNA synthesis from incorporation rates of labeled ribonucleoside have generally taken account of the changing isotope concentration of the nucleotide pool. Most studies have emphasized the effect of changes in transport rate on incorporation rate, though the opposite was first suggested by Edlin and Neuhard (1967) to explain the arrest of uracil transport in bacteria whose RNA synthesis was reduced by amino acid starvation.

The effect of serum on uridine transport by resting fibroblasts, first reported by Cunningham and Pardee (1969), could have resulted, at least in part, from an increase in the rate of RNA synthesis, increased utilization of nucleotides being compensated by an increased transport rate in order to maintain the size of the nucleotide pool.* In order to decide whether changes in uridine transport are independent of changes in the rate of RNA synthesis, it would be valuable to carry out the transport measurements in the presence of sufficient actinomycin D to prevent all RNA synthesis. Since, in contrast to its effect on uridine transport, serum had no effect on adenosine transport (Cunningham and Pardee 1969), the increased uridine transport seemed independent of RNA synthesis and due to a direct effect at the level of the cell membrane. Accordingly it was concluded that the increased incorporation of uridine could be accounted for by the increased rate of transport and that no increase in the rate of transcription had taken place.

This conclusion offered no explanation for the fact that the total RNA content of the cells must have been increasing. From subsequent studies on the rates of RNA synthesis in resting and exponentially growing populations, based on incorporation rates corrected for the specific activity of the nucleotide pools, it was concluded either that there was very little difference in the rate of synthesis of total RNA (Weber and Rubin 1971; Weber and Edlin 1971), or that there was an appreciable difference in the rate of synthesis of ribosomal RNA (Emerson 1971; Weber 1972). Evidently correction factors for the changing radioactivity of the nucleotide pool are not uniformly reliable. The same problem has been encountered in studies of other cell types (Hausen and Stein 1968; Plagemann et al. 1969; Cooper 1972; Kramer, Klapproth and Hilz 1972).

Rate of RNA Synthesis in Ghost Monolayers

A means of avoiding the problem created by the nucleotide pool was found recently through the use of cell monolayers treated with the neutral detergent NP-40. When applied very gently to resting monolayers of 3T6, the detergent does not remove the cells from the vessel surface. It does affect cell membranes drastically, destroying

---

* T-s. Chan and I have studied the size of the ribonucleoside diphosphate and triphosphate pools in serum-stimulated cells in transition from resting to growing state by high-pressure liquid chromatography of cell extracts. The determinations have shown no significant change in the amounts of cellular ribonucleotides even 15 hours after stimulation, when an appreciable proportion of the cells was already engaged in DNA synthesis.

all permeability barriers and emptying the cytoplasmic compartment nearly completely. The ghost monolayer so produced may be washed and its ability to incorporate nucleoside triphosphates measured (Tsai and Green 1973).

Like isolated nuclei, these ghost monolayers continue for a while to synthesize RNA. The incorporation rate is nearly linear for 10 minutes and continues for at least 40 minutes. As the RNA polymerases retain function even following fixation of cells in ethanol-acetone mixtures (Moore 1971; Moore and Ringertz 1973), it is not surprising that they can function after detergent treatment and that, at least for a time, the rate of transcription reflects the activity of the cells at the time detergent was added. It is not clear that the ghost monolayers possess properties different from those of isolated nuclei (Zylber and Penman 1971; Price and Penman 1972), but they may be prepared rapidly from monolayer cultures in transition from resting to growing state and are therefore very convenient for the study of changing rates of RNA synthesis. There is essentially no pool of ribonucleoside triphosphates in the ghost monolayers, since omission of GTP, ATP and CTP reduces incorporation of labeled UTP by 90% (Tsai and Green 1973). Unlabeled nucleosides do not compete with labeled nucleotides for incorporation.

Comparison of the effect of serum stimulation on UTP incorporation by ghost monolayers (Fig. 3) with labeled uridine incorporation by whole cells (Fig. 2) shows significant differences. UTP incorporation by the ghost monolayers rose more steadily with time following serum stimulation. The increase was about twofold by the time DNA synthesis began and the rate was still increasing at 25 hours, when nearly half the cell population had begun DNA synthesis. It seems possible that in intact cells an increase in RNA synthesis resulting from serum stimulation may bring about compensatory changes in rate of transport of uridine, and that an overshoot in the regulating system for transport can produce cyclical changes in incorporation rate.

The nature of the RNA whose rate of synthesis responds to serum stimulation has also been examined in ghost monolayers. (Pre)-ribosomal RNA was determined as RNA whose synthesis was prevented by a low concentration of actinomycin D

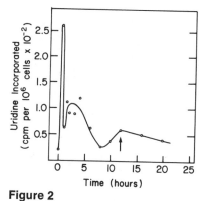

Figure 2

Incorporation of labeled uridine into RNA by intact 3T3 cells following serum stimulation. Resting monolayers were exposed to fresh serum at zero time; at intervals ^{14}C-labeled uridine was added to a culture and after 20 min incorporation into RNA was determined. Additional data on the fluctuation in incorporation rate are given in Todaro, Lazar and Green (1965). Arrow indicates earliest DNA synthesis by radioautography of [^{3}H]thymidine-labeled nuclei.

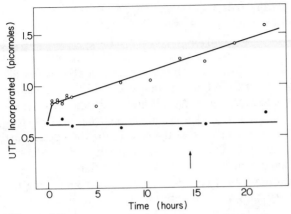

Figure 3
Incorporation of labeled UTP into RNA by ghost monolayers prepared at intervals after serum stimulation of resting 3T6 cells. Ordinate gives incorporation during a 10-minute interval beginning shortly after preparation of the ghost monolayer. (○) Serum stimulated; (●) unstimulated controls. Arrow indicates earliest DNA synthesis in stimulated cultures. From Tsai and Green (1973) with permission of *Nature*.

(0.04 μg/ml) or as RNA synthesized in the presence of α-amanitin. Additional data supporting the validity of these methods will be found in Mauck and Green (1973).

The results obtained by the two methods were very similar. The rate of ribosomal RNA synthesis was increased in ghost monolayers prepared as early as 10 minutes after the addition of serum to resting 3T6 cultures. The rate continued to increase steadily (Fig. 4) though less steeply with time, becoming 2.5–3.5 times higher than the resting rate by the time DNA synthesis began, and 3–4.5 times higher by the time the first cell divisions took place (about 25 hours). The rate of rRNA synthesis is a more sensitive measure of the action of the serum growth factor than is the rate of total RNA synthesis. The reason for this is of particular interest—the overall rate of transcription of heterogeneous nuclear RNA does not respond directly to stimulation by serum. This rate seems to be determined only by the DNA content of the cells, suggesting that it is template limited. The rate of transcription of rRNA, being regulated, is not linked to DNA content (Mauck and Green, 1973).

Relative Contributions of the Two Causes of Ribosome Accumulation

An increase in the number of ribosomes can come about not only through an increase in the rate of ribosome synthesis, but also through a decrease in their rate of destruction. As in bacteria (Meselson et al. 1964), ribosome turnover is arrested in growing fibroblasts (Weber 1972).

The half-life of the ribosomes in resting liver cells is about 5 days (Loeb, Howell and Tomkins 1965; Hirsch and Hiatt 1966). In cultured chick fibroblasts values as short as 35–45 hours have been obtained (Weber 1972). In resting 3T6 cells the half-life is about 2.5 days (Johnson et al., in prep.). Since the RNA content of the resting cell remains constant, half a complement of ribosomes must be resynthesized during the same period, or about 16% of a complement every 15 hours.

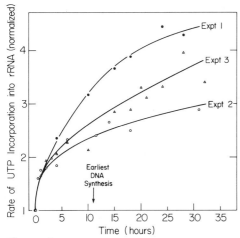

Figure 4

Incorporation of labeled UTP into ribosomal RNA in ghost monolayers prepared at intervals following serum stimulation of resting 3T6 cells. Expt. 1 and 2, actinomycin-sensitive RNA synthesis; Expt. 3, α-amanitin-resistant RNA synthesis. Except for the inhibitors, conditions were the same as for Fig. 3. The results of each experiment are normalized to the level of the resting cultures. From Mauck and Green (1973) with permission of *Proc. Nat. Acad. Sci.*

Following the action of the serum growth factor on resting 3T6 cells, the RNA content rises by a little over 40% in the 15-hour period before DNA accumulation is detected (Fig. 1). If during this period breakdown of the ribosomes were arrested while their rate of synthesis remained constant, the increase in ribosome content could amount to only 16%. The observed increase of 40% is therefore mainly due to an increase in their rate of synthesis. Studies of intact exponentially growing chick fibroblasts have led to the same explanation for their accumulation of rRNA (Emerson 1971; Weber 1972).

Nuclear Size during Transition to the Growing State

Increases are known to occur in the size of nuclei activated by introduction into different cytoplasmic environments (Graham, Armes and Gurdon 1966; Harris 1970; Barry and Merriam 1972). When inactive nuclei are transplanted into eggs or oocytes, their enlargement is spectacular; when they are exposed through cell fusion to cytoplasm of more active somatic cells, the increase in size is more modest but still very appreciable. The increase in size of an inactive nucleus is accompanied by an increase in its mass (Bolund, Ringertz and Harris 1969) and in its rate of transcription (Harris 1970). Since the rate of transcription of ribosomal RNA increases so markedly during the transition of 3T6 cells from resting to growing state, it was of interest to measure the size of the nuclei during the transition.

The length and width of the oval or kidney-shaped nuclei were measured with an ocular micrometer under 2500 × magnification, after fixation and staining of intact monolayers. Those nuclei labeled with tritium, as a result of a short exposure to tritiated thymidine before fixation, were excluded since they were likely to be larger due to accumulated DNA and nuclear proteins deposited during DNA syn-

Table 1
Nuclear Size of 3T6 Cells during Transition from Resting to Growing State

		Time after addition of serum (hr)							
		0	3	5	11	16	19	25
Expt. 1	% Nuclei labeled	1.9	3.0	3.0	14	61	48	25
	Dimensions of non-labeled nuclei	42 × 32 41 × 33	42 × 33	43 × 33	44 × 33	45 × 34	45 × 33	42 × 31
Expt. 2	% Nuclei labeled	2.0	3.0	4.0	15	58	55	22
	Dimensions of non-labeled nuclei	45 × 33 43 × 32	41 × 31	44 × 34	44 × 34	42 × 30	44 × 32	41 × 31

thesis. The values listed in Table 1 are the means of measurements on 25 nuclei. It can be seen that there was no significant change in nuclear size prior to DNA synthesis.* Ribosomal RNA synthesis is a nucleolar function, and it seems that no change in either size of the rest of the nucleus or in the overall rate of nucleoplasmic transcription is necessary to bring about a considerable increase in the rate of rRNA synthesis.

The resting state in the fibroblast appears to be different from that of resting cell types that possess a condensed nucleus. For example, the resting state of the avian erythrocyte nucleus is not reversible unless the nucleus is exposed to foreign cytoplasm containing activating factors (Harris 1970). The resting state of the lymphocyte is also linked to a condensation of the nucleus, but in this case the resting state is reversible. When the lymphocyte enters a division cycle following phytohemagglutinin treatment, the nucleus enlarges and the chromatin becomes less condensed during the 24-hour period preceding DNA synthesis (Tokuyasu, Madden and Zeldis 1968). The resting state of the fibroblast may be designed for easy and rapid reversibility, as DNA synthesis in the stimulated fibroblast begins in about half the time required by the stimulated lymphocyte. This is true not only for established fibroblast lines such as 3T3 and 3T6, but also for human diploid fibroblasts (Rovera and Baserga 1971).

Possibilities for Control of Ribosomal RNA Synthesis during the Transition

The significance of ribosome accumulation in cells in transition from a less active to a more active state has received considerable attention (Tata 1967). I would like to emphasize here the question of how the rate of transcription of rRNA might be regulated in the fibroblast.

The rapidity with which the rate of rRNA synthesis in the fibroblast responds to the serum growth factor may be compared with that of the bacterial response to the alleviation of stringent conditions. Following the relief of amino acid deprivation in *E. coli,* there is an increase in the rate of synthesis of rRNA within 40 seconds (Stamato and Pettijohn 1971), or about 2% of a generation time. Following serum stimulation of resting 3T6 cells, there is an increase in the rate of rRNA synthesis (measured in ghost monolayers) within 10 minutes, or about 1% of a generation time (Mauck and Green 1973).

In studies of different cell types in transition from resting to growing state, other changes in macromolecular synthesis, such as histone acetylation in lymphocytes (Pogo, Allfrey and Mirsky 1966) and acidic nucleoprotein synthesis in human diploid fibroblasts (Rovera and Baserga 1971), have been thought to precede changes in the rate of RNA synthesis. In view of the uncertainty of estimates of RNA synthesis in intact cells, especially when the rate is changing rapidly, the time relation between these changes should be reexamined.

Most likely the fibroblast growth factor of serum initiating division under our conditions is the protein with a molecular weight of 20,000–30,000 described by Paul, Lipton and Klinger (1971). Since, in spite of its large size, the rate of

---

* Preparation of ghost monolayers with NP-40 reduced the length and width of the nuclei by about 10%. Depending on what assumptions are made about the change in the unmeasured third dimension, the volume of the nucleus might decrease as much as 25%. The shrinkage is probably due to loss of material from the nucleus. For example, there is a nucleotide pool in nuclei isolated in a nonaqueous system (Guerney et al. 1972) and this pool must be lost in the ghost monolayers. It is likely that some proteins are lost as well (Fisher and Harris 1962).

transcription of ribosomal genes responds so quickly, the factor probably acts at the cell surface. Even if this action were to take place with no perceptible delay, some restrictions can be placed on the possible intracellular events leading to an increase in rRNA synthesis. For example, a mechanism depending on an increase in the synthesis of a specific protein, itself depending on an increase in the synthesis of its mRNA, is probably excluded, since synthesis of mRNA and its transport to the cytoplasm requires at least 15–20 minutes (Penman, Vesco and Penman 1968). If a change in the rate of synthesis of a specific protein were involved, this change would have to be effected at the level of translation in order for the protein to act in time.

It seems most likely that a small molecule is involved in regulating the transcription of ribosomal genes. Guanosine tetraphosphate and guanosine pentaphosphate are thought to have such a role in bacteria (Cashel 1970; Gallant et al. 1970) but have not been demonstrated in mammalian cells. There is considerable evidence relating cAMP levels to growth rate (Otten, Johnson and Pastan 1971; but see Sheppard, this symposium). During the transition from resting to growing state there is a sharp drop in the level of cyclic AMP. This is true for the transition induced by the normal fibroblast growth factor and that induced by the action of proteolytic enzymes (Otten, Johnson and Pastan 1972; Burger et al. 1972), a very substantial drop in cAMP level taking place within 10 minutes after stimulation. This is sufficiently rapid that consideration could be given to cAMP as a regulator of rRNA synthesis. However the subsequent behavior of the cAMP levels is hardly consistent with such a role. The drop in cAMP content following serum stimulation of 3T3 cells was not sustained for very long and within an hour returned nearly to the initial level (Otten et al. 1972). The same was found for the cAMP level in the course of a growth cycle initiated by trypsin (Burger et al. 1972).

Assuming that the conditions of these experiments were rather similar to our own (the onset of DNA synthesis took place at very nearly the same time), the rate of rRNA synthesis should still have been rising long after the cyclic AMP level had returned to nearly the resting level. Because of the transient nature of the change in cAMP level, Burger et al. (1972) postulated that the initial drop acted like a trigger, setting off other changes more directly linked to the division cycle. Such a role for cAMP would not involve a direct effect on the rate of rRNA synthesis. It remains to be seen whether reciprocal changes in cGMP levels are of the same nature. If mammalian cells do not use guanosine tetraphosphate or pentaphosphate for this purpose either, it would be worthwhile to search for other nucleotides as factors regulating their rRNA synthesis.

REFERENCES

Barry, J. M. and R. W. Merriam. 1972. Swelling of hen erythrocyte nuclei in cytoplasm from *Xenopus* eggs. *Exp. Cell Res.* **71**:90.

Becker, H., C. P. Stanners and J. E. Kudlow. 1971. Control of macromolecular synthesis in proliferating and resting syrian hamster cells in monolayer culture. II. Ribosome complement in resting and early G_1 cells. *J. Cell. Physiol.* **77**:43.

Bolund, L., N. R. Ringertz and H. Harris. 1969. Changes in the cytochemical properties of erythrocyte nuclei reactivated by cell fusion. *J. Cell Sci.* **4**:71.

Brachet, J. 1950. *Chemical embryology*. Interscience Publishers, N.Y.

Burger, M. M., B. M. Bombik, B. M. Breckenridge and J. R. Sheppard. 1972. Growth control and cyclic alterations of cyclic AMP in the cell cycle. *Nature New Biol.* **239**:161.

Bürk, R. R. 1966. Growth inhibitor of hamster fibroblast cells. *Nature* **212**:1261.

Cashel, M. 1970. Inhibition of RNA polymerase by ppGpp, a nucleotide accumulated during the stringent response to amino acid starvation in *E. coli. Cold Spring Harbor Symp. Quant. Biol.* **35**:407.

Clarke, G. D. and M. G. P. Stoker. 1971. Conditions affecting the response of cultured cells to serum. *Ciba Fndn. Symp., Growth control in cell cultures* (ed. G. E. W. Wolstenholme and J. Knight) p. 17. Churchill Livingstone, London.

Cooper, H. L. 1972. Studies on RNA metabolism during lymphocyte activation. *Transplt. Rev.* **11**:3.

Cooper, H. L. and A. D. Rubin. 1965. RNA metabolism in lymphocytes stimulated by phytohemagglutinin: Initial responses to phytohemagglutinin. *Blood* **25**:1014.

Cunningham, D. D. and A. B. Pardee. 1969. Transport changes rapidly initiated by serum addition to "contact inhibited" 3T3 cells. *Proc. Nat. Acad. Sci.* **64**:1049.

Edlin, G. and J. Neuhard. 1967. Regulation of nucleoside triphosphate pools in *Escherichia coli. J. Mol. Biol.* **24**:225.

Emerson, C. P. 1971. Regulation of the synthesis and the stability of ribosomal RNA during contact inhibition of growth. *Nature New Biol.* **232**:101.

Fisher, H. W. and H. Harris. 1962. The isolation of nuclei from animal cells in culture. *Proc. Roy. Soc.* (London) B **156**:521.

Forsdyke, D. R. 1967. Quantitative nucleic acid changes during phytohaemagglutinin-induced lymphocyte transformation in vitro. *Biochem. J.* **105**:679.

Fried, M. and J. D. Pitts. 1968. A resting cell system for biochemical studies on polyoma virus. *Virology* **34**:761.

Fujioka, M., M. Koga and I. Lieberman. 1963. Metabolism of ribonucleic acid after partial hepatectomy. *J. Biol. Chem.* **238**.3401.

Gallant, J., H. Erlich, B. Hall and T. Laffler. 1970. Analysis of the RC function. *Cold Spring Harbor Symp. Quant. Biol.* **35**:397.

Gelfant, S. 1963. A new theory on the mechanism of cell division. *Symp. Int. Soc. Cell Biol., Cell growth and cell division* (ed. R. J. C. Harris) vol. 2, p. 229. Academic Press, New York.

Graham, C. F., K. Arms and J. B. Gurdon. 1966. The induction of DNA synthesis by frog egg cytoplasm. *Develop. Biol.* **14**:349.

Gurney, T., Jr., B. R. Gordon, W. Mangel, E. G. Gurney and E. F. Hughes. 1972. Buoyant densities and RNA polymerase activity of nuclei isolated by a non-aqueous method. *J. Cell Biol.* **55**:100a.

Hamilton, T. H. 1968. Control by estrogen of genetic transcription and translation. *Science* **161**:649.

Hausen, P. and H. Stein. 1968. On the synthesis of RNA in lymphocytes stimulated by phytohemagglutinin. *Eur. J. Biochem.* **4**:401.

Harris, H. 1970. *Cell fusion.* Harvard University Press, Cambridge.

Hirsch, C. A. and H. H. Hiatt. 1966. Turnover of liver ribosomes in fed and in fasted rats. *J. Biol. Chem.* **241**:5936.

Holley, R. W. and J. A. Kiernan. 1968. "Contact inhibition" of cell division in 3T3 cells. *Proc. Nat. Acad. Sci.* **60**:300.

————. 1971. Studies of serum factors required by 3T3 and SV3T3 cells. *Ciba Fndn. Symp., Growth control in cell cultures* (ed. G. E. W. Wolstenholme and J. Knight) p. 3. Churchill Livingstone, London.

Kjeldgaard, N. O. and C. G. Kurland. 1963. The distribution of soluble and ribosomal RNA as a function of growth rate. *J. Mol. Biol.* **6**:341.

Kramer, G., K. Klapproth and H. Hilz. 1972. Determination of specific radioactivity of UTP in HeLa cells by an enzymatic method and its alteration under various growth conditions. *Biochim. Biophys. Acta* **262**:410.

Lieberman, I. and P. Kane. 1965. Synthesis of ribosomes in the liver after partial hepatectomy. *J. Biol. Chem.* **240**:1737.

Lipton, A., I. Klinger, D. Paul and R. W. Holley. 1971. Migration of mouse 3T3 fibroblasts in response to a serum factor. *Proc. Nat. Acad. Sci.* **68**:2799.

Loeb, J. N., R. R. Howell and G. M. Tomkins. 1965. Turnover of ribosomal RNA in rat liver. *Science* **149**:1093.

Mauck, J. and H. Green. 1973. The regulation of RNA synthesis during transition from resting to growing state. *Proc. Nat. Acad. Sci.* (in press).

Meselson, M., M. Nomura, S. Brenner, C. Davern and D. Schlessinger. 1964. Conservation of ribosomes during bacterial growth. *J. Mol. Biol.* **9**:696.

Moore, G. P. M. 1971. DNA-dependent RNA synthesis in fixed cells during spermatogenesis in mouse. *Exp. Cell Res.* **68**:462.

Moore, G. P. M. and N. R. Ringertz. 1973. Localization of DNA-dependent RNA polymerase activities in fixed human fibroblasts by autoradiography. *Exp. Cell Res.* **76**:223.

Mueller, G. C., A. M. Herranen and K. F. Jervell. 1958. Studies on the mechanism of action of estrogens. *Rec. Progr. Hormone Res.* **14**:95.

Neidhardt, F. C. and B. Magasanik. 1960. Studies on the role of ribonucleic acid in the growth of bacteria. *Biochim. Biophys. Acta* **42**:99.

Nierlich, D. P. 1972. Regulation of ribonucleic acid synthesis in growing bacterial cells. I. Control over the total rate of RNA synthesis. *J. Mol. Biol.* **72**:751.

Nordenskjold, B. A., L. Skoog, N. C. Brown and P. Reichard. 1970. Deoxyribonucleotide pool and DNA synthesis in cultured mouse embryo cells. *J. Biol. Chem.* **245**:5360.

Otten, J., G. S. Johnson and I. Pastan. 1971. Cyclic AMP levels in fibroblasts: Relationship to growth rate and contact inhibition of growth. *Biochem. Biophys. Res. Commun.* **44**:1192.

————. 1972. Regulation of cell growth by cyclic adenosine 3′,5′-monophosphate. *J. Biol. Chem.* **247**:7082.

Paul, D., A. Lipton and I. Klinger. 1971. Serum factor requirements of normal and simian virus 40-transformed 3T3 mouse fibroblasts. *Proc. Nat. Acad. Sci.* **68**:645.

Penman, S., C. Vesco and M. Penman. 1968. Localization and kinetics of formation of nuclear heterodisperse RNA, cytoplasmic heterodisperse RNA and polyribosome-associated messenger RNA in HeLa cells. *J. Mol. Biol.* **34**:49.

Plagemann, P. G. W., G. A. Ward, B. W. J. Mahy and M. Korbecki. 1969. Relationship between uridine kinase activity and rate of incorporation of uridine into acid-soluble pool and into RNA during growth cycle of rat hepatoma cells. *J. Cell. Physiol.* **73**:233.

Pogo, B. G. T., V. G. Allfrey and A. E. Mirsky. 1966. RNA synthesis and histone acetylation during the course of gene activation in lymphocytes. *Proc. Nat. Acad. Sci.* **55**:805.

Price, R. and S. Penman. 1972. A distinct RNA polymerase activity, synthesizing 5.5S, 5S and 4S RNA in nuclei from adenovirus 2-infected HeLa cells. *J. Mol. Biol.* **70**:435.

Robbins, E., G. Jentzsch and A. Micali. 1968. The centriole cycle in synchronized HeLa cells. *J. Cell Biol.* **36**:329.

Rosset, R., J. Julien and R. Monier. 1966. Ribonucleic acid composition of bacteria as a function of growth rate. *J. Mol. Biol.* **18**:308.

Rovera, G. and R. Baserga. 1971. Early changes in the synthesis of acidic nuclear proteins in human diploid fibroblasts stimulated to synthesize DNA by changing the medium. *J. Cell. Physiol.* **77**:201.

Stamato, T. D. and D. E. Pettijohn. 1971. Regulation of ribosomal RNA synthesis in stringent bacteria. *Nature New Biol.* **234**:99.

Stanners, C. P. and H. Becker. 1971. Control of macromolecular synthesis in proliferating and resting syrian hamster cells in monolayer culture. I. Ribosome function. *J. Cell. Physiol.* **77:**31.

Storey, W. F. and C. P. Leblond. 1951. Measurement of the rate of proliferation of epidermis and associated structures. *Ann. N.Y. Acad. Sci.* **53:**537.

Swann, M. M. 1958. The control of cell division: A review. II. Special mechanisms. *Cancer Res.* **18:**1118.

Tata, J. R. 1967. The formation and distribution of ribosomes during hormone-induced growth and development. *Biochem. J.* **104:**1.

Todaro, G. J., G. K. Lazar and H. Green. 1965. The initiation of cell division in a contact-inhibited mammalian cell line. *J. Cell. Comp. Physiol.* **66:**325.

Todaro, G., Y. Matsuya, S. Bloom, A. Robbins and H. Green. 1967. Stimulation of RNA synthesis and cell division in resting cells by a factor present in serum. *Wistar Inst. Symp. Monogr.* **7:**87.

Tokuyasu, K., S. C. Madden and L. J. Zeldis. 1968. Fine structural alterations of interphase nuclei of lymphocytes stimulated to growth activity in vitro. *J. Cell Biol.* **39:**630.

Tsai, R. L. and H. Green. 1973. Rate of RNA synthesis in ghost-monolayers obtained from fibroblasts preparing for division. *Nature New Biol.* **243:**168.

Weber, M. J. 1972. Ribosomal RNA turnover in contact inhibited cells. *Nature New Biol.* **235:**58.

Weber, M. J. and G. Edlin. 1971. Phosphate transport, nucleotide pools, and ribonucleic acid synthesis in growing and in density-inhibited 3T3 cells. *J. Biol. Chem.* **246:**1828.

Weber, M. J. and H. Rubin. 1971. Uridine transport and RNA synthesis in growing and in density-inhibited animal cells. *J. Cell. Physiol.* **77:**157.

Zylber, E. A. and S. Penman. 1971. Products of RNA polymerases in HeLa cell nuclei. *Proc. Nat. Acad. Sci.* **68:**2861.

Distribution of Nuclear RNAs in Growing and Nongrowing Cells

G. D. Birnie, J. Delcour,* D. Angus, G. Threlfall, and John Paul

The Beatson Institute for Cancer Research
Royal Beatson Memorial Hospital, Glasgow G3 6UD, Scotland

When mammalian cells in culture proceed from a stage of active growth to one of mitotic inactivity as a result of either depletion of their culture medium or contact inhibition, the drastic decrease in the rate of synthesis of DNA is accompanied by many changes in cellular metabolism, including decreases in the rates at which exogenous precursors are incorporated into proteins and RNA (Hershko et al. 1971; Levine et al. 1965). So far as protein synthesis is concerned the reduction can be accounted for, in part at least, by an approximately 50% reduction in the population of ribosomes in the nongrowing cells (Stanners and Becker 1971; Emerson 1971). Although changes in transport mechanisms (Weber and Rubin 1971) and nucleoside kinase activity (Hausen and Stein 1968) are implicated in the reduction in the rate of incorporation of exogenous nucleosides into RNA, the actual rate at which ribosomal RNA is synthesized is markedly reduced in contact-inhibited cells (Emerson 1971). In contrast, the rate at which nuclear proteins are phosphorylated is inversely proportional to the rate of growth in Yoshida ascites tumor cells (Riches et al. 1973).

Although cessation of growth is accompanied by a marked decrease in the rate of synthesis of ribosomal RNA, synthesis of heterogeneous nuclear RNA (HnRNA) is affected to a relatively small extent (Weber and Rubin 1971; Emerson 1971). Much of the HnRNA consists of sequences which not only turn over rapidly but are also confined to the cell nucleus (Shearer and McCarthy 1967, 1970; Soeiro et al. 1968). Nevertheless it appears that some, at least, of the HnRNA molecules are precursors of cytoplasmic messenger RNAs (Lindberg and Darnell 1970; Melli and Pemberton 1972). In addition a significant proportion of each HnRNA molecule consists of sequences which hybridize to the reiterated fraction of DNA (Davidson and Hough 1969; Melli and Bishop 1969; Darnell, Wall and Tushinski 1971; Georgiev et al. 1972). Britten and Davidson (1969) have suggested that the reiterated sequences in DNA have a role in the regulation of gene expression, and theories of the mechanism of gene activation (Georgiev 1969; Paul 1972) have

---

* Charge de Recherches au Fonds National Belge de la Recherche Scientifique. Present address: Université de Louvain, Unite de Génétique Moleculaire des Eucaryotes, Naamsestraat 61, B-3000 Louvain (Belgium).

suggested that those sequences in HnRNA which are homologous to the reiterated sequences in DNA are reflections of such regulation. It seemed, therefore, that a comparison of the nuclear RNAs in growing and nongrowing cells might yield some information bearing on the role played by gene transcription in the control of proliferation.

METHODS

LS cells (a substrain of mouse NCTC strain L929 which grows in suspension) were seeded at a density of 10^5 cells/ml in Roux flasks containing 50 ml of Eagle's Minimal Essential Medium (Flow Laboratories) enriched with 2% (v/v) calf serum. No antibiotics were added. Primary mouse embryo fibroblasts were grown from trypsin-disrupted 13-day-old whole mouse embryos in Eagle's Minimal Essential Medium enriched with 10% tryptose phosphate broth and 10% (v/v) calf serum. The cells were passaged twice before being used in an experiment. 3T3 cells (received from Dr. H. Green) were cultured in Eagle's Minimal Essential Medium supplemented with 10% fetal calf serum.

For the gel electrophoresis experiments cytoplasmic and nuclear fractions were prepared as described by Penman (1966) except that the supernatant fraction from the detergent wash of the nuclei was discarded. RNA was extracted from the nuclei by a method based on that of Penman (1966), the essential differences being (1) all extractions were done with a mixture of phenol and chloroform–1% amyl alcohol (1:1 v/v) and chloroform–1% amyl alcohol (Perry et al. 1972), and (2) digestion with DNase I (Worthington RNase-free) was done in 100 mM Hepes-HCl, 10 mM Mg^{++} acetate, 25 mM NaCl, 2 mM $CaCl_2$, pH 7.0. Fragments of degraded DNA were removed after both DNase digestions by gel filtration through Sephadex G50. Cytoplasmic RNA was prepared similarly, except that only one DNase digestion was required. For the hybridization experiments nuclei were prepared either as above or by repeated homogenizing with 25 mM citric acid, then washed free of citric acid. RNA was extracted as above. Purified RNA from liver nuclei was labeled in vitro with [^{3}H]dimethylsulfate (Smith, Armstrong and McCarthy 1967).

Chromatin was prepared and transcribed with RNA polymerase from *E. coli* as described previously (Paul et al. 1972) except that the incubation mixture contained 0.1 M Tris-HCl pH 8, 2.5 mM $MnCl_2$, a mixture of ^{3}H-labeled nucleoside triphosphates (each 0.8 mM), 1000 units of RNA polymerase and 500 μg DNA as chromatin. After incubation for 2 hours at 37°C, [^{3}H]RNA was extracted and purified (Paul et al. 1972).

Hybridizations were done with RNA dissolved in 150 μl 50% formamide–4 × SSC (SSC is 0.15 M NaCl, 0.015 M Na^+ citrate pH 7.0) in sealed vials, each of which contained two nitrocellulose filters (Sartorius, 13 mm) to which was bound 2–5 μg alkali-denatured mouse embryo DNA and one blank filter. After 40 hours at 37°C the filters were washed extensively with 4 × SSC, incubated with RNase (20 μg/ml) in 2 × SSC for 1 hour at 20°C and washed again with 4 × SSC. Radioactivity bound to filters was measured in a liquid scintillation spectrometer.

RESULTS

Growth of LS Cells

A double-label technique was adopted to compare the RNA profile from nuclei and cytoplasm of growing and nongrowing cells. A culture of LS cells in log phase ($4–5 \times 10^5$ cells/ml) was labeled for 20 hours with [2-^{14}C]uridine (0.1 μCi/ml; 25 mCi/mmole). Simultaneously another culture of LS cells which had already reached stationary phase ($12–14 \times 10^5$ cells/ml) was labeled with [5-^{3}H]uridine (1 μCi/ml; 24 Ci/mmole). The cells were mixed immediately after they had been harvested, and nuclear and cytoplasmic RNAs were extracted from the mixed cells. When these RNAs were fractionated, calculation of the ratio of ^{3}H to ^{14}C in each fraction gave a direct measure of the relative contributions from nongrowing and growing cells to the total radioactive RNA in that fraction. In this way any differential effect of uncontrolled variations (in particular, degradation by nucleases) on the RNAs from growing and nongrowing cells during fractionation and extraction procedures is minimized, if not eliminated.

Before the cultures were mixed samples were taken to determine the concentrations, viabilities and mitotic indices of the cells. Cultures were considered to be satisfactory if the change in cell numbers over the labeling period, and the viability and mitotic index of the cells at the end of the labeling period, all lay within the limits shown in Table 1.

Purification of RNA

The double-label technique used is very sensitive to small changes in the relative amounts of the two isotopes. However the long period of labeling used meant that some label, particularly in the case of the growing cells labeled with [^{14}C]uridine, leaked into DNA, so it was essential to ensure that any contamination of RNA preparations with DNA was reduced to insignificant levels. The methods for isolating both cytoplasmic and nuclear RNAs were checked by comparing RNA from cytoplasmic and nuclear fractions of cells that had been labeled for 24 hours with [^{3}H]thymidine. In the case of cytoplasmic RNA, contamination by DNA was very low and was easily eliminated by a brief treatment with DNase I. Complete elimination of DNA from nuclear RNA proved much more difficult. Even extended incubation with DNase under commonly used conditions (Tris-MgCl$_2$–KCl buffer pH 7.5) failed to reduce the contamination with DNA below 1%. Bollum (1965)

Table 1
Criteria for Growing and Nongrowing Cells

	Growing	Nongrowing
Percent increase in numbers over labeling period (20 hr)	50–70	$\leqslant 5$
Percent viable cells[a]	85–95	80–90
Percent mitoses[b]	6–8	$\leqslant 0.2$

[a] As determined by exclusion of naphthalene blue.

[b] Minimum of 500 cells counted.

Table 2
Elimination of DNA from Nuclear RNA Preparations

	Percent DNA in nuclear RNA after treatment with DNAase	
Buffer	*Once*	*Twice*
Tris-Mg++-K+*	15.2	1.3
Hepes-Mg++-Na+-Ca++†	0.7	0.02

* 50 mM Tris-HC1, 5 mM MgCl$_2$, 25 mM KC1 pH 7.5.

† 100 mM Hepes-HC1, 10 mM Mg++ acetate, 25 mM NaCl, 2 mM CaCl$_2$ pH 7.0.

showed that some sequences in DNA are extremely resistant to DNase in the absence of Mn++ or Ca++; Table 2 shows that when 2 mM Ca++ is included in the buffer, the proportion of DNA present in a preparation of nuclear RNA can readily be reduced to under 0.05%.

Gel Electrophoretic Patterns of RNA

The discrimination afforded by this double-label method is illustrated by Fig. 1, which depicts the distribution of ^{3}H and ^{14}C in a gel on which the RNA from the cytoplasmic fraction of a mixture of growing (^{14}C-labeled) and nongrowing (^{3}H-labeled) cells has been electrophoresed. The profiles of [^{3}H]RNA and

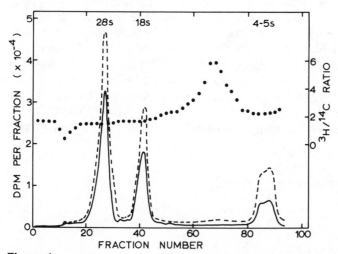

Figure 1
Gel electrophoresis pattern of cytoplasmic RNA from a mixture of growing (^{14}C-labeled) and stationary phase (^{3}H-labeled) LS cells. Electrophoresis was in a 2.6% polyacrylamide gel at 75 V for 2 hr; ^{3}H and ^{14}C were discriminated by liquid scintillation counting using the external standard channels-ratio method to correct for quench.
    (– – –) ^{3}H dpm; (———) ^{14}C dpm; (● ● ●) ratio of ^{3}H to ^{14}C.

[¹⁴C]RNA follow that of the optical density trace but, although there is a large difference in the absolute amounts of [³H]RNA and [¹⁴C]RNA, these profiles appear to be very similar. However when the ratio of ³H to ¹⁴C in each gel slice is calculated, marked differences emerge. Such gels of mixed cytoplasmic RNAs are characterized by a generally low ratio of ³H to ¹⁴C, varying from about 1.5 over 28S and 18S RNA to about 2 over 4–5S RNA. However a markedly higher ratio is seen between the 18S and 4–5S RNA peaks, reaching a maximum of about 6 in the region of the gel corresponding to RNAs of around 8–10S (Fig. 1). Other experiments (not shown) showed that this region of the gel contained RNAs with high turnover rates.

Figure 2 illustrates the pattern exhibited by the RNA from the nuclear fraction. Again the profiles of both [³H]RNA and [¹⁴C]RNA show the same patterns. However the ratio between the labels varies widely from one region of the gel to another. As in gels of cytoplasmic RNA, the ratio of ³H to ¹⁴C is low over 5S RNA (1.7), 18S RNA (2.2) and 28S RNA (3.5), but is notably higher between the 4–5S and the 18S peaks, reaching a maximum of between 6 and 7 in the 8–12S region of the gel. The ³H/¹⁴C ratio is also high at the top of the gel. The RNA in this region is not merely aggregated ribosomal RNA, as is shown by Fig. 3 which indicates HnRNA (greater than 40S) is characterized by a ³H/¹⁴C ratio some four times that in the 18S and 28S RNA regions. Since some, at least, of the HnRNA comprises precursors of cytoplasmic messenger RNA (Melli and Pemberton 1972), it is significant that the only region in the cytoplasmic gel (Fig. 1) to show a ratio of ³H to ¹⁴C greater than that in the ribosomal RNAs is in the region most likely to contain a high proportion of monocistronic messenger RNAs.

Polyadenylate Content of Nuclear RNA

The double-label technique was also used to determine whether there were any differences in the proportion of RNA molecules from growing and nongrowing cells

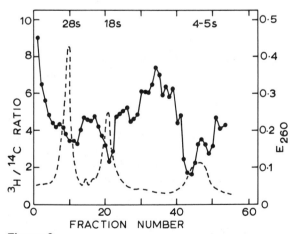

Figure 2

Gel electrophoresis pattern of nuclear RNA from a mixture of growing (¹⁴C-labeled) and stationary phase (³H-labeled) LS cells. The optical density trace (– – –) is that of whole mouse embryo cytoplasmic RNA added as carrier; (●——●) ratio of ³H to ¹⁴C; other details as Fig. 1.

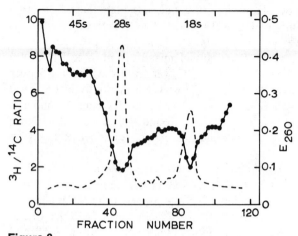

Figure 3

Gel electrophoresis pattern of nuclear RNA from a mixture of growing
(^{14}C-labeled) and stationary phase (^{3}H-labeled) LS cells. The optical density
trace (– – –) is that of whole mouse embryo cytoplasmic RNA added as carrier;
(●———●) ratio of ^{3}H to ^{14}C; other details as Fig. 1 except electrophoresis was
for 4 hr.

which contained poly(A) regions. About 20% of the nuclear RNA from growing
cells bound to oligo(dT)-cellulose, whereas 36% of that from stationary cells did
so. Although any nuclease activity would result in an underestimate of the proportion of nuclear RNA which was polyadenylated, there is no reason to suppose that
the RNA from growing cells is intrinsically more labile during the isolation procedure than the RNA from stationary cells. This result therefore indicates that there
is a significant increase in the proportion of polyadenylated RNA molecules in
nuclei of stationary cells.

Hybridization of Nuclear RNA

Only those species of RNA which are complementary to the repetitious regions of
DNA will hybridize to filter-bound DNA in 50% formamide–4 × SSC at 37°C.
Nevertheless a small but significant proportion of the DNA does form RNase-
resistant hybrids with nuclear RNA under these conditions. Therefore the relative
complexities of the populations of RNA in the nuclei of growing and nongrowing
cells were measured in this system. For these experiments cell cultures were grown
in the presence of [5-^{3}H]uridine (1 μCi/ml) for 20 hours, and the RNAs were
extracted separately from the isolated nuclei. Figure 4 shows the results with RNA
from growing and nongrowing LS cells. It is quite clear that there is a significant
difference in the proportion of the DNA hybridized to the RNAs from the two cultures and, surprisingly, that the RNA from the nongrowing cells appears to have a
considerably higher complexity than that from the growing cells. Similar results
were obtained from primary mouse embryo fibroblast cultures (Fig. 5). In this case
a rapidly growing culture, which had just been fed with fresh medium, is compared
with one which, while not yet stationary, was growing more slowly because of partial
exhaustion of the medium. These observations have been extended to cultures of

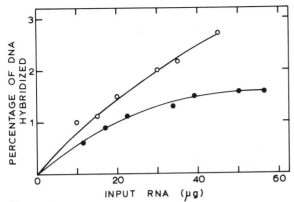

Figure 4
Hybridization saturation curves of nuclear RNA from growing (●———●) and
stationary phase (○———○) LS cells.

3T3 cells in which growth has been limited by contact inhibition, and also to a
whole tissue situation in which nuclear RNA from livers of partially hepatectomized
mice have been compared with normal liver RNA (Table 3).

Table 3 summarizes the results of a number of experiments with LS cells, primary
mouse embryo fibroblasts, 3T3 cells and regenerating mouse liver. Despite some
variation in the figures, the data suggest that the proportion of DNA hybridized to
nuclear RNA under these conditions of low C_0t is inversely proportional to the rate
of growth of the cells from which the nuclear RNA was obtained.

Hybridization of Chromatin-primed RNA

Chromatins isolated from growing and nongrowing LS cells were transcribed with
E. coli polymerase, and the [³H]RNAs synthesized were hybridized to filter-bound
DNA as for the nuclear RNAs. No difference in the hybridization characteristics
of these RNAs could be detected (Fig. 6).

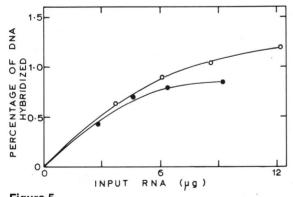

Figure 5
Hybridization saturation curves of nuclear RNA from fed and starved primary
mouse embryo fibroblasts. (●———●) Nuclear RNA from cells labeled immediately after a change of medium; (○———○) nuclear RNA from cells labeled 2
days after a change of medium.

Table 3

Hybridization of Mouse Embryo DNA to Nuclear
RNA from Growing and Nongrowing Cells

	Percent DNA* hybridized to nuclear RNA from	
	Growing	Nongrowing
LS cells	1.3	2.9
	1.5	5.0
	1.6	6.0
	1.7	
Mouse embryo fibroblasts		
Fed	0.9	
	1.2	
Starved (2 days)	1.6	
	1.9	
Starved (6 days)	3.7	
3T3 cells		
Starved (3 days)	2.0	5.9
Mouse liver		
Normal		2.4
		2.8
Regenerating (24 hr)	1.5	
(43 hr)	1.3	

* Saturation estimated from titration experiments as in Fig. 4; if
plateau levels not achieved, saturation estimated by extrapolation
using the double-reciprocal plot method (Bishop 1969).

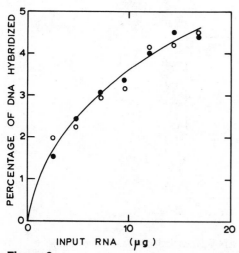

Figure 6

Hybridization saturation curves of RNA synthesized in vitro from chromatins
of growing (•——•) and stationary phase (o——o) LS cells.

DISCUSSION

Stationary phase in suspension cultures of LS cells results mainly from exhaustion of the medium. The data from both the double-label gel electrophoresis experiments and from the experiments in which nuclear RNA from growing and nongrowing cells was hybridized to DNA suggest that such starvation-induced cessation of growth causes significant changes in the distribution of RNA species in the nucleus. These changes could be accounted for by changes in transcription, either in absolute terms (that is, more regions of the genome are transcribed in the stationary-phase cell) or in terms of changes in the relative rates of transcription of different regions of the genome. Alternatively the differences between growing and nongrowing cells could be the result of changes in the relative rates of processing of different species of nuclear RNAs.

In the double-label gels the ratio of ^{3}H to ^{14}C at each point in a gel is a function both of the relative amounts of RNAs labeled with each isotope and of their specific activities. However the cells were labeled continuously for 20 hours and the time required for processing of nuclear RNA is much less than this (see for example Penman 1966; Penman et al. 1968; Soeiro et al. 1968; Emerson 1971). Thus the nuclear RNAs may be assumed to be labeled uniformly; that is, the specific activity of each species is the same so that variations in the ratio of ^{3}H to ^{14}C in the nuclear gels reflect changes in the relative amounts of the various species of RNA. This being so, the high ratio of ^{3}H to ^{14}C in the HnRNA and ribosomal precursor (45S and 32S) RNAs, compared to that in the mature 18S and 28S ribosomal RNAs, indicates that there is less mature ribosomal RNA relative to the amounts of ribosomal precursor RNAs and HnRNA in nuclei of stationary phase cells than in nuclei of growing cells. A reduction in the amounts of mature ribosomal RNAs in the nuclei of stationary phase cells without a corresponding reduction in the amounts of ribosomal precursor RNAs could be the result of a decrease in the rate of processing of the ribosomal precursors to 18S and 28S RNA, or of more rapid transport of the mature RNAs to the cytoplasm, or of a combination of both. Emerson (1971) concluded that the drastic reduction in the rate of accumulation of ribosomal RNA in contact-inhibited chick skin fibroblasts, compared to that in growing cells, could be accounted for by a decreased rate of synthesis of precursor ribosomal RNA, together with a decrease in the stability of mature ribosomal RNA. However in methionine-starved cells, an experimental condition more akin to the present one than contact-inhibited cells, Vaughan et al. (1967) found that the processing of ribosomal precursor RNA was disrupted.

The differences between the nuclear RNAs from growing and stationary phase cells shown by the hybridization experiments cannot be explained by the increase in the proportion of nuclear RNA which is polyadenylated in the stationary cell. The proportion of mammalian DNA shown to consist of stretches of poly(dA) is considerably lower (0.5%) than the proportion of DNA which hybridizes to nuclear RNA (Shenkin and Burdon 1972). These differences can, however, be explained in terms of a change in the distribution of RNA species such as could result from a reduction in the rate of processing of HnRNA in the nuclei of stationary phase cells or from an overall increase in transcription in these cells. The failure to demonstrate any difference between the RNAs transcribed in vitro from chromatins of growing and nongrowing cells suggests that the former explanation is the more likely. The increase in the proportion of nuclear RNAs with poly(A) sequences in stationary phase cells could also be associated with a different pattern of processing of HnRNA. Certainly it has been concluded from competition-hybridization experiments that, when cells are stimulated to proliferate, there are significant changes in the processing of HnRNA as well as an increase in the number of active sites in the genome (Church and McCarthy 1967, 1970; Shearer and Smuckler, 1972).

No definite conclusions relating to differences in transcriptional patterns can be drawn from these experiments because the hybridization technique used allows detection of repetitious sequences only. Evidence is accumulating which suggests that many structural genes in eukaryotes, except notably those coding for histones (Kedes and Birnstiel 1971), are not highly reiterated (Harrison et al. 1972; Bishop et al. 1972; Packman 1972; Suzuki, Gage and Brown 1972; Bishop and Rosbash 1973; Delovitch and Baglioni 1973). As yet there is a lack of evidence bearing on the question of whether control of expression of specific nonreiterated genes is solely by a repression-derepression mechanism at the level of the gene.

Acknowledgments

This work was supported by grants from Medical Research Council and Cancer Research Campaign. J.D. was supported by a fellowship from the Royal Society under the European Exchange Programme. The skilled technical assistance of Mrs. Jenny McGeoch and Miss Elizabeth MacPhail is gratefully acknowledged.

REFERENCES

Bishop, J. O. 1969. The effect of genetic complexity on the time-course of ribonucleic acid—deoxyribonucleic acid hybridization. *Biochem. J.* **113:**805.

Bishop, J. O. and M. Rosbash. 1973. Reiteration frequency of duck globin genes. *Nature New Biol.* **241:**204.

Bishop, J. O., R. Pemberton and C. Baglioni. 1972. Reiteration frequency of haemoglobin genes in the duck. *Nature New Biol.* **235:**231.

Bollum, F. J. 1965. Degradation of the homopolymer complexes polydeoxyadenylate-polydeoxythymidylate, polydeoxyinosinate-polydeoxycytidylate, and polydeoxyguanylate-polydeoxycytidylate by deoxyribonuclease I. *J. Biol. Chem.* **240:**2599.

Britten, R. J. and E. H. Davidson. 1969. Gene regulation for higher cells: A theory. *Science* **165:**349.

Church, R. B. and B. J. McCarthy. 1967. Changes in nuclear and cytoplasmic RNA in regenerating liver. *Proc. Nat. Acad. Sci.* **58:**1548.

———. 1970. Unstable nuclear RNA synthesis following estrogen stimulation. *Biochim. Biophys. Acta* **199:**103.

Darnell, J. E., R. Wall and R. J. Tushinski. 1971. An adenylic acid-rich sequence in messenger RNA of HeLa cells and its possible relationship to the reiterated sites in DNA. *Proc. Nat. Acad. Sci.* **68:**1321.

Davidson, E. H. and B. R. Hough. 1969. High sequence diversity in the RNA synthesized at the lampbrush stage of oogenesis. *Proc. Nat. Acad. Sci.* **63:**342.

Delovitch, T. L. and C. Baglioni. 1973. Estimation of light-chain gene reiteration of mouse immunoglobin by DNA-RNA hybridization. *Proc. Nat. Acad. Sci.* **70:**173.

Emerson, C. P., Jr. 1971. Regulation of the synthesis and the stability of ribosomal RNA during contact inhibition of growth. *Nature New Biol.* **232:**101.

Georgiev, G. P. 1969. On the structural organization of operon and the regulation of RNA synthesis in animal cells. *J. Theoret. Biol.* **25:**473.

Georgiev, G. P., A. P. Ryskov, C. Coutelle, V. L. Mantieva and E. R. Avakyan. 1972. On the structure of the transcriptional unit in mammalian cells. *Biochim. Biophys. Acta* **259**:259.

Harrison, P. R., A. Hell, G. D. Birnie and J. Paul. 1972. Evidence for single copies of globin genes in the mouse genome. *Nature* **239**:219.

Hausen, P. and H. Stein. 1968. On the synthesis of RNA in lymphocytes stimulated by phytohemagglutinin. I. Induction of uridine-kinase and the conversion of uridine to UTP. *Eur. J. Biochem.* **4**:401.

Hershko, A., P. Mamont, R. Shields and G. M. Tomkins. 1971. Pleiotypic responses. *Nature New Biol.* **232**:206.

Kedes, L. H. and M. L. Birnstiel. 1971. Reiteration and clustering of DNA sequences complementary to histone messenger RNA. *Nature New Biol.* **230**:165.

Levine, E. M., Y. Becker, C. W. Boone and H. Eagle. 1965. Contact inhibition, macromolecular synthesis, and polyribosomes in cultured human diploid fibroblasts. *Proc. Nat. Acad. Sci.* **53**:350.

Lindberg, U. and J. E. Darnell. 1970. SV40-specific RNA in the nucleus and polyribosomes of transformed cells. *Proc. Nat. Acad. Sci.* **65**:1089.

Melli, M., and J. O. Bishop. 1969. Hybridization between rat liver DNA and complementary RNA. *J. Mol. Biol.* **40**:117.

Melli, M. and R. E. Pemberton. 1972. New method of studying the precursor-product relationship between high molecular weight RNA and messenger RNA. *Nature New Biol.* **236**:172.

Packman, S., H. Aviv, J. Ross and P. Leder. 1972. A comparison of globin genes in duck reticulocytes and liver cells. *Biochem. Biophys. Res. Commun.* **49**:813.

Paul, J. 1972. General theory of chromosome structure and gene activation in eukaryotes. *Nature* **238**:444.

Paul, J., D. Carroll, R. S. Gilmour, I. A. R. More, G. Threlfall, M. Wilkie and S. Wilson. 1972. Functional studies on chromatin. *Fifth Karolinska Symp. on Research methods in reproductive Endocrinology* (ed. E. Diczfaluscy) p. 277.

Penman, S. 1966. RNA metabolism in the HeLa cell nucleus. *J. Mol. Biol.* **17**:117.

Penman, S., C. Vesco and M. Penman. 1968. Localization and kinetics of formation of nuclear heterodisperse RNA, cytoplasmic heterodisperse RNA and polyribosome-associated messenger RNA in HeLa cells. *J. Mol. Biol.* **34**:49.

Perry, R. P., J. La Torre, D. E. Kelley and J. R. Greenberg. 1972. On the lability of poly(A) sequences during extraction of messenger RNA from polyribosomes. *Biochim. Biophys. Acta* **262**:220.

Riches, P. G., K. R. Harrap, S. M. Sellwood, D. Rickwood and A. J. MacGillivray. 1973. Phosphorylation of nuclear proteins in rodent tissues. *Biochem. Soc. Trans.* **1**:684.

Shearer, R. W. and B. J. McCarthy. 1967. Evidence for RNA molecules restricted to the cell nucleus. *Biochemistry* **6**:283.

———. 1970. Characterization of RNA molecules restricted to the nucleus in mouse L-cells. *J. Cell. Physiol.* **75**:97.

Shearer, R. W. and E. A. Smuckler. 1972. Altered regulation of the transport of RNA from nucleus to cytoplasm in rat hepatoma cells. *Cancer Res.* **32**:339.

Shenkin, A. and R. H. Burdon. 1972. Deoxyadenylate-rich sequences in mammalian DNA. *FEBS Letters* **22**:157.

Smith, K. D., J. L. Armstrong and B. J. McCarthy. 1967. The introduction of radioisotopes into RNA by methylation *in vitro. Biochim. Biophys. Acta* **142**:323.

Soeiro, R., M. H. Vaughan, J. R. Warner and J. E. Darnell, Jr. 1968. The turnover of nuclear DNA-like RNA in HeLa cells. *J. Cell Biol.* **39**:112.

Stanners, C. P. and H. Becker. 1971. Control of macromolecular synthesis in proliferating and resting syrian hamster cells in monolayer culture. I. Ribosome function. *J. Cell. Physiol.* **77**:31.

Suzuki, Y., L. P. Gage and D. D. Brown. 1972. The genes for silk fibroin in *Bombyx mori. J. Mol. Biol.* **70:**637.

Vaughan, M., R. Soeiro, J. Warner and J. Darnell. 1967. The effects of methionine deprivation on ribosome synthesis in HeLa cells. *Proc. Nat. Acad. Sci.* **58:**1527.

Weber, M. J. and H. Rubin. 1971. Uridine transport and RNA synthesis in growing and in density-inhibited animal cells. *J. Cell. Physiol.* **77:**157.

Studies of Poly(A)-bearing RNA in Resting and Growing Human Lymphocytes

Herbert L. Cooper

Cell Biology Section, Laboratory of Biochemistry
National Institute of Dental Research, National Institutes of Health
Bethesda, Maryland 20014

Lymphocytes isolated from the peripheral blood of humans and other animals can be induced to shift in vitro from a nongrowing state to one of active proliferation by the addition of one of a number of mitogenic substances (Oppenheim 1968). Very rapidly after addition of a mitogen such as phytohemagglutinin to a lymphocyte culture, a great many biochemical changes occur (reviewed by Cooper 1973a). Some of these, such as increased synthesis and conservation of ribosomal RNA (Cooper 1970; Cooper and Gibson 1971), the associated increase in total protein (Hirschhorn et al. 1963; Kay, Levinthal and Cooper 1969) and ribosomal protein synthesis (Cooper, unpublished data), and the synthesis of DNA polymerase (Loeb, Agarwal and Woodside 1968) are evidently related to lymphocyte proliferation. In addition many "factors" are elaborated into the culture medium after growth stimulation which are thought to be mediators of the lymphocyte's differentiated function as an immunologically active cell (Lawrence and Landy 1969). All of these changes in activity distinguish the proliferating from the resting lymphocyte. These observations lead us to ask what changes occur in the mode of utilization of the genetic information of the cell so as to provide this great variety of new cellular activities and products. In particular do these functions require the transcription of previously inactive regions of the genome into mRNA? Alternatively are the relevant mRNAs continually transcribed in the resting lymphocyte but not translated because of rapid degradation in the nucleus, failure to be transported to the cytoplasm, or stabilization in the cytoplasm in some quiescent form? Progress in the study of mRNA in lymphocytes has been minimal because of the lack of appropriate methods and limitations in availability of cellular material. The recent discovery of the existence of an extensive polyadenylate sequence on nearly all mRNAs of animals cells, and on a proportion of heterogeneous nuclear RNA molecules (Edmonds, Vaughan and Nakazato 1971; Lee, Mendecki and Brawerman 1971; Darnell, Wall and Tushinski 1971), has given rise to techniques which permit the isolation of putative mRNA from mixtures of RNAs, based on recognition of the poly(A) sequence (Sheldon, Jurale and Kates 1972; Nakazato and Edmonds 1972; Greenberg and Perry 1972a). To date nearly all published studies of mRNA using

769

this technique have involved long-term cultivated cell lines, such as HeLa, L-cell, and plasmacytoma lines. Scant attention has been given to more physiologically representative cell types or to questions relating to the control of cell proliferation under conditions approaching normality. Rosenfeld et al. (1972) have reported the existence of such poly(A)-bearing RNA [poly(A)-RNA] in lymphocytes and have presented some data relating to its metabolism.

In this study a survey of some of the characteristics of poly(A)-RNA in lymphocytes in the resting state and shortly after growth stimulation will be presented. No direct demonstration of the identity of mRNA with poly(A)-RNA will be given, but data from other laboratories make this assumption highly likely, considering the methods used. This type of survey is a necessary first step in answering the questions outlined previously. Several features will emerge that point toward specific early changes in the metabolism of poly(A)-RNA which may be related to the onset of lymphocyte proliferation.

Poly(A)-RNA from Whole Cells

Initial studies were performed on whole cell RNA extracted from intact cells. Resting or growing lymphocytes were labeled with [^{3}H]adenine for 3 hours and RNA was extracted with hot phenol-SDS at low ionic strength in the absence of Mg^{++}. Analysis of the size distribution of the labeled molecules by electrophoresis in acrylamide-agarose gels reveals many of the basic differences in overall RNA metabolism between resting and growing cells (Fig. 1). In resting lymphocytes, even after 3 hours of labeling, the bulk of the radioactivity is found in heterogeneous RNA (predominantly hnRNA) (Fig. 1A, Table 1). Ribosomal RNA and its precursors are labeled to a lesser extent and 32S RNA predominates over 28S RNA, suggesting relatively slow rates for the intermediate steps in rRNA processing. In addition the amount of label appearing in 18S rRNA is only about half that expected on the basis of 32S + 28S labeling (Table 1). This phenomenon has been shown to be the result of degradation of about half of all newly synthesized rRNA molecules before their appearance in the cytoplasm, which we have termed rRNA "wastage" to distinguish it from turnover of cytoplasmic ribosomes (Cooper 1972, 1973b). After 20 hours of PHA stimulation production of rRNA predominates over that of hnRNA under these labeling conditions, and 28S labeling is much more prominent than either 45S or 32S labeling, suggesting both a differential stimulation of rRNA production and an increase in processing rate (Fig. 1B, Table 1). Also the 28:18S labeling ratio shows no deviation from that expected on the basis of equimolar production, which we have shown to be related to abolition of rRNA wastage following growth stimulation (Table 1).

On passage of this material through glass fiber filters on which poly(U) has been immobilized (Sheldon et al. 1972), a small proportion of the total label applied is retained (Table 2). Passage of the resulting filtrate through a second poly(U) filter results in retention of a much smaller proportion of the applied label. This indicates adherence of a specific class of molecules to the poly(U) filters, about 90% of which are retained by each passage through the filter under our conditions. Routine use of two successive filters for each sample then allows a 99% removal of this class of molecules. The work of Sheldon et al. (1972) indicates that adherence of RNA to poly(U) filters is specifically due to possession of a poly(A) sequence by the adhering molecules. We presume, on the basis of that work, that the filter

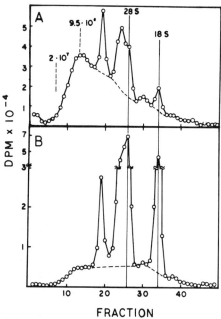

Figure 1

Polyacrylamide gel electrophoresis of whole cell RNA from resting and growing lymphocytes. Lymphocytes were purified from heparinized human venous
blood (Cooper 1968, 1973d) and cultured with or without 5 μg/ml PHA (Wellcome) in Eagle's MEM (suspension modification) with 10% autologous plasma
at a density of 2×10^6 cells/ml in 5% CO_2 atmosphere. After 20 hr cells were
concentrated to 10^7/ml and [2-^{3}H]adenine (20 μCi/ml; 28 Ci/mmole; Schwarz-
Mann) or [U-^{14}C]adenine (2 μCi/ml; 250 mCi/mmole; Amersham) was added
for a 3-hr labeling period. Cultures were then chilled, centrifuged and cells resuspended in low salt buffer (0.05 M NaCOOCH$_3$, pH 5.1; 0.01 M Na$_2$ EDTA,
0.5% sodium dodecyl sulfate; saturated with diethylpyrocarbonate before use).
Two extractions with equal volumes of phenol at 55°C and one with phenol:
CCl$_4$:isoamyl alcohol (48:48:4) followed. The final aqueous phase was made
0.1 M in NaCl and RNA was precipitated with 2 volumes of ethanol at −20°C
for 20 hr. The RNA precipitate was taken up in buffer (0.02 M NaCl, 0.005 M
Tris, pH 7.2, 0.0025 M Na$_2$ EDTA, 20% sucrose). A portion representing the
yield from 5×10^7 cells in 50 μl was applied to 13-cm long cylindrical acrylamide-agarose gels (1.75–0.5%) and subjected to electrophoresis at 2.5 mA/gel
for 2 hr (Dingman and Peacock 1968). Gels were scanned at 260 nm to locate
rRNA (*vertical bars*), then divided into 2-mm segments. Radioactivity in each
segment was determined by scintillation counting in Aquasol.

A, Resting cells [^{3}H]; **B,** 20-hr PHA [^{14}C].

adherent material is a class of poly(A)-bearing RNA present in lymphocytes, as is
the case in other cell types studied. This confirms the previous report of Rosenfeld
et al. (1972).

Whole cell RNA was extracted from resting lymphocytes labeled for 3 hours
with [^{3}H]adenine and poly(A)-RNA retained on poly(U) filters was eluted and
analyzed on polyacrylamide-agarose gels (Fig. 2). The material applied to the

Table 1

[³H]- or [¹⁴C]Adenine Labeling of Lymphocyte RNA

	% Total label		28:18S ratio[†]
	hn-RNA	rRNA*	
Resting	73	27	4.9
20-hr PHA	32	68	2.7

Analysis of data from Fig. 1; 3-hr label.

* Including 45S RNA.

† Theoretical ratio = 2.5–3.0.

filters showed the typical size distribution of whole cell RNA from such cells (Fig. 2A).

The RNA eluted from poly(U) filters showed no peaks of rRNA or its precursors and no 4S RNA, demonstrating the specificity of poly(U) filter retention (Fig. 2B). The poly(A)-RNA had a heterogeneous size distribution, with peak accumulation around a molecular weight of 3–3.5 × 10⁶ daltons, although some molecules as large as 10 × 10⁶ daltons were evidently present. The distribution of poly(A)-RNA differed from that of the hnRNA of the original sample, in that hnRNA had a peak accumulation around a molecular weight of 9.5 × 10⁶ daltons,

Table 2

Specific Binding of Poly(A)-RNA from Lymphocytes to Poly(U) Filters

	pU Adherent DPM	Total DPM Applied	% Retained
Whole cell RNA (resting)		11.4 × 10⁶	
Filter 1	145450		1.20
Filter 2	16030		0.14
Total	161475		
Whole cell RNA (20-hr PHA)		3.5 × 10⁶	
Filter 1	23880		0.68
Filter 2	3086		0.09
Total	26966		
Cytoplasmic RNA (20-hr PHA)		306750	
Filter 1	11800		3.80
Filter 2	903		0.30
Total	12703		

RNA extracts prepared as in Fig. 1 were passed through poly(U) filters as described by Sheldon, Jurale and Kates (1972). RNA adhering to filters was determined by scintillation counting.

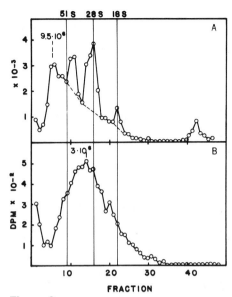

Figure 2
Gel electrophoresis of poly(A)-RNA in whole cell extracts, resting lymphocytes. Resting lymphocytes were labeled with [³H]adenine for 3 hr and whole
cell RNA extracted as described in Fig. 1. Poly(A)-bearing RNA was retained
on poly(U) filters (Sheldon, Jurale and Kates 1972) and eluted in scintillation
vials at 58°C for 10 min in 0.01 M Tris pH 7.5, 0.5% SDS, 50% formamide.
Elution fluid was clarified by centrifugation and made 0.1 M in NaCl. *E. coli*
tRNA (200 μg) was added as carrier and RNA was precipitated with ethanol.
Electrophoresis was performed for 1.5 hr; otherwise as in Fig. 1.

A, Whole cell RNA extract; **B,** poly(A)-bearing RNA contained in whole
cell extract, retained and eluted from poly(U) filters.

with some molecules as large as 2×10^7 daltons (Fig. 1A, 2A). It appears that the
largest hnRNA molecules are relatively under-represented among poly(A)-bearing
RNA molecules. Possible explanations of this finding are:

1. Poly(A)-hnRNA is a distinct subclass of hnRNA, characterized by the size
   distribution observed.
2. The largest hnRNA molecules are those most recently synthesized and are devoid
   of poly(A) segments. They acquire such segments during subsequent processing steps, which include cleavage or partial degradation to smaller molecules.
3. The poly(A)-RNA found in whole cell extracts is predominantly cytoplasmic
   material, which has a smaller average size than hnRNA and contains a much
   higher proportion of poly(A)-bearing molecules than does hnRNA. This explanation is unlikely, since subsequent experiments will show that most of the
   labeled poly(A)-RNA in whole cell extracts labeled for 3 hours is noncyto-
   plasmic.
4. There is preferential failure to retain the largest poly(A)-hnRNA molecules by
   poly(U) filters.

The size distribution of whole cell poly(A)-RNA in resting and growing lymphocytes was compared after a 3-hour labeling period (Fig. 3A). The distributions

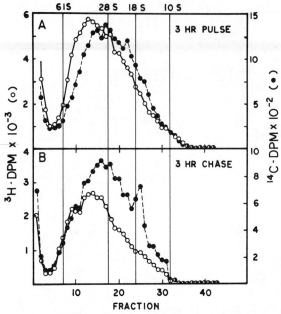

Figure 3

Gel electrophoresis of whole cell poly(A)-RNA from resting and growing lymphocytes. Resting and growing lymphocytes (20-hr PHA) were labeled for 3 hr with [³H]adenine and [¹⁴C]adenine, respectively (see Fig. 1 for details). Half the cells were then chased for 3 hr without radioactive precursor in the presence of 25 μg/ml adenine. Resting and growing cells were combined and poly(A)-bearing RNA was prepared as in Fig. 1 & 2. Gels were assayed by double-label counting technique.

   A, Three-hour pulse; **B,** three-hour pulse followed by 3-hr chase.

   (○) Resting cells, [³H]. (●) 20-hr PHA cells, [¹⁴C]. Approximately 1.5 × 10⁸ cells were used for each original culture.

differed in that following growth stimulation a reproducible shift toward a smaller size range was discernible. The inverse logarithmic relationship between molecular weight and electrophoretic mobility of RNA in acrylamide gels minimizes the apparent difference between the distributions. Calculation of the indicated molecular weights reveals that the peak accumulation has shifted from 3–3.5 × 10⁶ daltons in resting cells to 2–2.6 × 10⁶ daltons after growth stimulation. Thus an average reduction in size of poly(A)-RNA molecules of over 30%, involving a reduction in molecular weight of 10⁶ daltons for the peak fractions, occurs following the shift from the resting to the growing state. The basis of this change is not known. It may reflect the synthesis of a different array of RNAs in the growing cell, or it may be a consequence of an alteration in processing of poly(A)-hnRNA. One possibility is that the normal processing of poly(A)-hnRNA involves progressive size reduction. If the processing of large poly(A)-hnRNA to smaller-sized molecules were more rapid in growing cells, a smaller, average-size distribution would remain after a given length of labeling. This implies that extending the time allowed for processing in resting cells would also result in a smaller molecular size range. However when the 3-hour pulse was followed by a 3-hour chase, there was no significant change in the size distribution of the remaining poly(A)-RNA in either resting or growing cells (Fig. 3B). Thus there is no evidence that the poly(A)-RNA molecules derive from larger poly(A)-containing precursors or that they are subjected to successive shortening after obtaining their poly(A) segment. However the possibility that poly(A)-bearing molecules are processed from larger, non-poly(A)-containing precursors remains.

Cytoplasmic Poly(A)-RNA

RNA was extracted from total cytoplasm prepared from growing lymphocytes after 3 hours of labeling with [³H]adenine. Poly(A)-RNA was then captured on poly(U) filters, eluted, and characterized by gel electrophoresis.

The radioactive material applied to poly(U) filters was almost entirely rRNA (Fig. 4A), although some labeled polydisperse RNA was detectable in the region <18S. In particular a small peak at about 10S was frequently seen (Fig. 4A, inset). Poly(A)-RNA extracted from this preparation was not contaminated by rRNA, despite the overwhelming preponderance of the latter material (Fig. 4B). Its size distribution showed molecules ranging from 2×10^5 to nearly 10^7 daltons. Compared to whole cell poly(A)-RNA (Fig. 2, 3), a greater proportion of the cytoplasmic material was distributed in the region between 10S and 28S (2×10^5 to 2×10^6 daltons). A similar preparation from resting lymphocytes showed the same distribution (Fig. 4C). In both cases a distinct peak was present at 10–12S. On repeated studies this peak was most commonly at $\sim$ 11.5S (MW 2.5×10^5 daltons). The absence of this peak in whole cell poly(A)-RNA preparations (Fig. 2, 3) suggests that only a small proportion of radioactive whole cell poly(A)-RNA derives from the cytoplasm after labeling periods of the sort used here. The appearance of a small shoulder at this point in material from growing cells labeled for 3 hours and then chased for 3 hours (Fig. 3B) suggests that, even with prolonged processing, the bulk of remaining whole cell labeled poly(A)-RNA is poly(A)-hnRNA. From this it is apparent that even a small degree of contamination of cytoplasm preparations by nuclear material may seriously weight the size distribution in favor of larger poly(A)-hnRNA molecules.

The difference in size distribution of newly labeled poly(A)-RNA between resting and growing lymphocytes, seen in whole cell extracts, was absent from the cytoplasmic preparations. This was best seen in double-label studies of poly(A)-RNA prepared from the polysome region of cytoplasm after sucrose density gradient centrifugation (Fig. 5). The 11.5S peak is clearly seen in both resting and growing cell material and the size distributions are the same. Thus whatever the cause of the different size distribution in poly(A)-hnRNA from resting and growing cells, that difference is not retained in the cytoplasmic poly(A)-RNA (presumably mRNA). This observation strengthens the possibility that the size difference in poly(A)-hnRNA is due to some growth-related alteration in processing, rather than to a difference in the genetic regions transcribed.

The effect of growth stimulation on the labeling of whole cell and cytoplasmic poly(A)-RNA was compared (Table 3). PHA treatment caused a 3.8-fold increase in whole cell poly(A)-RNA labeling, and a 9.4-fold increase occurred in cytoplasmic poly(A)-RNA labeling. It should be noted that any contamination of cytoplasmic preparations by nuclear material would tend to reduce the latter value toward that of the whole cell preparation. Thus after a given period of labeling, a

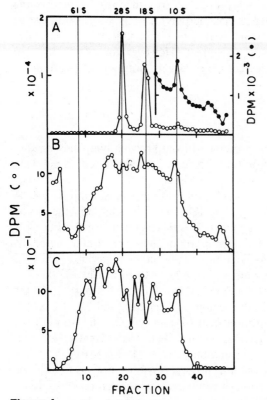

Figure 4

Gel electrophoresis of poly(A)-RNA from cytoplasm of resting and growing lymphocytes. RNA was labeled for 3 hr in resting and growing (20-hr PHA) lymphocytes as in Fig. 1. Cytoplasm was prepared by suspending cells in 0.02 M KCOOCH₃, 0.01 M Tris, pH 7.4, 0.005 M Mg₂COOCH₃, 0.3% Triton-N-101, 0.2 M sucrose, saturated with diethyl pyrocarbonate. After 10 min at 0°C. disruption was completed by 15 strokes with a loose-fitting Dounce homogenizer. Nuclei and intact cells were removed by centrifugation at $800 \times g$ for 5 min. RNA and poly(A)-RNA were prepared and analyzed as described in Fig. 1 & 2.

A, Cytoplasmic RNA from 20-hr PHA lymphocytes; inset = $10 \times$ expanded scale. **B,** Poly(A)-RNA from cytoplasm of 20-hr PHA lymphocytes; yield from $\sim 1.5 \times 10^8$ cells. **C,** Poly(A)-RNA from cytoplasm of resting lymphocytes; yield from $\sim 1.5 \times 10^8$ cells.

greater proportion of new poly(A)-RNA has reached the cytoplasm in growing cells, suggesting that growth stimulation causes an increase in the rate of transport of poly(A)-RNA to the cytoplasm.

Before concluding that the increased labeling of whole cell material indicates increased RNA synthesis, the possibility of growth-related changes in precursor pool specific activity must be considered. It has been shown that nucleosides such as uridine and adenosine are taken up slowly by resting lymphocytes and more rapidly by PHA-stimulated cells, with serious effects on nucleoside triphosphate specific activity (Cooper 1973c). However in unpublished experiments we found

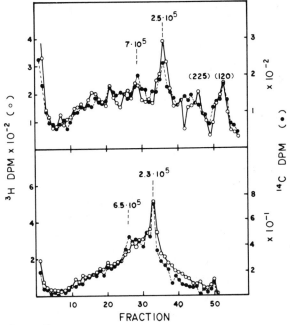

Figure 5
Gel electrophoresis of poly(A)-RNA associated with polyribosomes of resting
and growing lymphocytes. Lymphocytes were labeled for 4 hr with [³H]- or
[¹⁴C]adenine as in Fig. 1. Cytoplasm was prepared from combined resting [³H]
and growing [¹⁴C] cells as in Fig. 4 and layered on a 34-ml linear sucrose
gradient (10–40%) in RSB and centrifuged for 3 hr at 27,000 rpm in the
SW·27 (Spinco) rotor. After fractionation with direct ultraviolet monitoring
(ISCO), the polysome region was pooled and made 0.1 M in NaCl and 0.5%
in SDS. After 5 min at 0°C, μg/ *E. coli* tRNA was added and the mixture precipitated with 2 volumes of ethanol. RNA was extracted from the precipitate
and poly(A)-RNA prepared as in Fig. 1 & 2. Results from two different experiments are shown.
     (○) Resting cells, [³H]adenine. (●) 20-hr PHA cells, [¹⁴C]adenine.

this effect to be considerably reduced with adenine labeling. From our present
information it is probably safe to conclude that synthesis of poly(A)-RNA increases by a minimum of twofold after 20 hours of PHA stimulation.
     The appearance of cytoplasmic poly(A)-RNA after growth stimulation is in-
creased, then, about fivefold over resting cell levels when pool effects are con-

Table 3
Poly(A)-RNA Labeling

	Resting	20-hr PHA	% Change
Whole cell	5559*	20971	380
Cytoplasm	1621	15183	940

* cpm $\times$ μg total RNA^{-1}; 3-hr pulse.

sidered. The difference between whole cell and cytoplasmic stimulation may have two origins:

1. The rate of nuclear processing and transport of poly(A)-RNA is increased. This would amplify the effect of increased synthesis on cytoplasmic labeling.
2. A greater proportion of new poly(A)-hnRNA is conserved and transported as mRNA after growth stimulation.

Evidence supporting the first possibility was obtained from the following experiment. Resting and growing cells were labeled for 3 hours with [³H]adenine, then chased for 20 hours with unlabeled adenine (Fig. 6). The amount of labeled poly(A)-RNA remaining in whole cell and cytoplasmic preparations was determined at various times. The chase was rapidly effective in whole cell preparations in both resting and growing cells (Fig. 6A) and in cytoplasmic preparations of growing cells (Fig. 6B). However in resting cell cytoplasmic preparations, little change occurred for the first 3 hours of chase incubation (Fig. 6B). This may be explained by continued movement of previously labeled poly(A)-RNA from nucleus to cytoplasm during these 3 hours in resting cells due to relatively slow processing and transport. In growing cells more rapid transport would reduce the number of labeled molecules awaiting transport after the pulse. These would be exhausted more rapidly than in resting cells, making any transport delay undetectable in this experiment.

It is also apparent that degradation of cytoplasmic poly(A)-RNA is complex. Although a fraction of the cytoplasmic poly(A)-RNA decayed moderately rapidly,

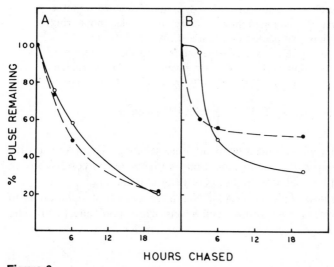

Figure 6
Survival of radioactive poly(A)-RNA during prolonged chase incubation. Lymphocytes were labeled with [³H]adenine for 3 hr and then chased in medium without radioactivity for 20 hr. At various times cells were removed and poly(A)-RNA was prepared from whole cell RNA and from cytoplasmic RNA and measured as in Table 2. Details of preparation as in Fig. 1–4.
    A, Whole cell poly(A)-RNA; **B,** cytoplasmic poly(A)-RNA; (o) resting lymphocytes; (•) 20-hr PHA treatment before pulse label.

a plateau was reached in both resting and growing cells, indicating the presence of a stable cytoplasmic species.

Distribution of Stable Cytoplasmic Poly(A)-RNA among Gradient Fractions

The observation that a relatively stable cytoplasmic poly(A)-RNA is synthesized in resting lymphocytes (Fig. 5) led us to examine the distribution of this material among different fractions of gradient-sedimented cytoplasm and to determine the effect of PHA treatment on that distribution. Resting lymphocytes were labeled for 4 hours with [³H]adenine. They were then chased for 18 hours with unlabeled adenine. After dividing the cells into two portions, one received PHA for three hours, the other received nothing. Cytoplasm was prepared and sedimented on linear, 10–40% sucrose gradients (Fig. 7). The distribution of labeled RNA (predominantly rRNA), when expressed as percent of total label, shows an increase in the proportion of polysomal label after PHA stimulation for only 3 hours and corresponding reduction in labeled 80S ribosomes. It appears that preexisting labeled 80S ribosomes are rapidly mobilized into polysomes upon growth stimulation. Gradient regions were pooled (Fig. 7, horizontal bars), and the proportion of total poly(A)-RNA in each pool was determined (Table 4). It is evident that a shift in the distribution of long-lived poly(A)-RNA occurred on PHA stimulation, in which some of the stable material originally present in the 40S + ribosome region reappeared in the polysome region.

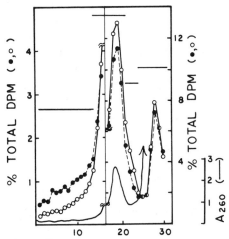

Figure 7

Sucrose density gradient sedimentation of lymphocyte cytoplasm. Resting lymphocytes (7.5 × 10⁸) were divided into 2 equal cultures and labeled with either [³H]adenine or [¹⁴C]adenine for 6 hr. They were then chased for 18 hr without radioactivity in the presence of unlabeled adenine. The ¹⁴C-labeled culture only then received 5 µg/ml PHA for 3 hr. Cultures were combined, cytoplasm prepared as in Fig. 4, and sedimented on a 10–40% linear sucrose gradient (3 hr, 27,000 rpm, Spinco SW·27 rotor). Fractions (1.2 ml) were collected and 0.1 ml of each assayed for acid-insoluble radioactivity. Fractions were then pooled as indicated by horizontal bars. Poly(A)-RNA was prepared from pooled fractions as in Fig. 5.

(o) Control [³H]; (•) PHA-treated [¹⁴C].

Table 4

Effect of PHA on Distribution of Prelabeled
Poly(A)-RNA among Sucrose Gradient Fractions

| Gradient region | Fraction of total label | | % Change |
	− PHA	+ PHA	
Polysomes	0.40	0.50	+ 25
Monosomes	0.35	0.28	− 20
40S	0.12	0.09	− 25
Supernatant	0.13	0.12	

6-hr label, 18-hr chase, 3-hr ± PHA. Poly(A)-RNA prepared
from pooled gradient regions shown in Fig. 7.

In further experiments of this type, using isokinetic sucrose gradients to obtain
good separation of ribosomes and ribosomal subunits (Noll 1967; McCarthy,
Stafford and Brown 1968), preliminary data suggest that stable poly(A)-RNA
moves first from the 40S region into the 80S ribosome region and subsequently into
the polysome region. Further studies are currently in progress.

DISCUSSION

These studies have shown that both the hnRNA and cytoplasmic RNA of lymphocytes include molecules containing poly(A) sequences. The largest hnRNA molecules, however, appear to be deficient in such sequences. This observation differs
from that of Greenberg and Perry (1972b), whose data show that in continuously
growing L-cells, the size distribution of poly(A)-bearing nuclear RNA is larger
than that of hnRNA. Whether technical or systematic differences cause this disparity remains to be resolved. In our hands poly(A)-hnRNA from HeLa cells
showed the same distribution as that from lymphocytes. One possible source of
difference may be the use of low doses of actinomycin D by other workers to
abolish rRNA labeling. The possibility exists that these low doses of the drug may
interfere with the normal processing of hnRNA.

Prolongation of the processing time by a chase incubation did not cause a change
in the size distribution of poly(A)-hnRNA, although loss of label, presumably due
to turnover, occurred. Apparently no progressive shortening or cleavage of poly(A)-
hnRNA molecules occurred with extended processing time. This suggests that, for
the bulk of hnRNA, once the poly(A) segment is attached, the molecule is thereafter either degraded or conserved (possibly transported to the cytoplasm), but not
appreciably altered in size.

The data suggest that only a small proportion of the whole cell poly(A)-RNA
labeled after 3 hours has reached the cytoplasm. These cytoplasmic molecules have
a different size distribution from poly(A)-hnRNA, with smaller molecules more
heavily represented. It is not known whether this results from a selection of smaller
molecules for transport to the cytoplasm or is due to remodeling of a small fraction
of poly(A)-hnRNA.

Growth stimulation altered the size distribution of whole cell poly(A)-RNA—
presumably poly(A)-hnRNA—but had little effect on the size distribution of cyto-

plasmic poly(A)-RNA. Thus if growth stimulation involves the appearance of new kinds of mRNAs in the cytoplasm, these do not have a distinctive size distribution. The difference in size distribution between poly(A)-hnRNA from resting and growing cells most probably reflects a difference in processing after growth stimulation, although the nature of this difference is unclear.

Both the rates of synthesis and of processing and transport of newly synthesized poly(A)-RNA to the cytoplasm were clearly increased after growth stimulation. The increase in synthesis is not unexpected; that of transport was less predictable and may represent an important control point in the onset of cell growth. Future studies must be directed toward determining how soon after growth stimulation this occurs and how the process is regulated.

A discrete peak at ~11.5S (MW 250,000) was always seen in cytoplasmic poly(A)-RNA and in polysomal poly(A)-RNA. The remainder of the distribution was quite heterogeneous and was similar in resting and growing cells. Whereas the overall size distribution of polysome-associated poly(A)-RNA in lymphocytes resembled that reported for HeLa cells and L-cells, the discrete peak at 11.5S has not been described in poly(A)-RNA in those systems (Nakazato and Edmonds 1972; Singer and Penman 1972). No difference in either the size distribution or the relative labeling of different species was detected in any fraction after growth stimulation. Thus there is no evidence, by this admittedly insensitive approach, of the appearance of new mRNAs or of the differential stimulation of synthesis of particular old ones after PHA stimulation.

In considering the earliest events associated with the onset of growth in resting cells, a frequently mentioned but unproven possibility is the activation of preexisting but unused cytoplasmic mRNA. Such "masked" messenger RNA is known to exist and to be activated upon fertilization in the sea urchin ovum (Piatigorsky 1968; Raff et al. 1972), and activation of stable mRNA is suggested in certain differentiating systems, such as the lens epithelium (Craig and Piatigorsky 1973).

In the present study radioactivity in cytoplasmic poly(A)-RNA labeled during a pulse period declined rapidly during the first several hours of chase incubation (Fig. 6). Thereafter a plateau was reached in both resting and growing cells, suggesting that a stable form remained. Addition of PHA to resting lymphocytes treated in this way caused a shift within 3 hours in the distribution of surviving, labeled poly(A)-RNA among sucrose gradient regions. Poly(A)-RNA molecules apparently moved from the 40S and ribosome regions to the polysome region. This was accompanied by an increased number of polysomes formed from preexisting ribosomes.

Since mRNA sedimenting in the 40S region is not actively associated with any protein synthesizing machinery, it is possible that this material is, in fact, "masked" mRNA which is rapidly activated upon growth stimulation. The proteins encoded by this mRNA may serve a variety of control and structural functions essential to the activation of G_0 cells. Alternatively these may be mRNAs specifying one or more of the variety of mediators of cellular immunity elaborated by activated lymphocytes (Lawrence and Landy 1969).

It should be observed that the resting, human, peripheral blood lymphocyte population is composed primarily of T cells, with a smaller fraction of B lymphocytes (Heller et al. 1971; Abdou 1971). Since PHA is known to be primarily a T cell activator under the conditions used (Greaves and Bauminger 1972), it follows that the behavior of masked mRNA postulated here relates to T cell activities.

Acknowledgment

I thank Ms. E. M. Gibson and Mrs. Helen Lee for excellent technical assistance.

REFERENCES

Abdou, N. I. 1971. Immunoglobulin (Ig) receptors on human peripheral leukocytes. II. Class restriction of Ig receptors. *J. Immunol.* **107**:1637.

Cooper, H. L. 1968. Ribonucleic acid metabolism in lymphocytes stimulated by phytohemagglutinin. II. Rapidly synthesized ribonucleic acid and the production of ribosomal ribonucleic acid. *J. Biol. Chem.* **243**:34.

————. 1970. Control of synthesis and wastage of ribosomal RNA in lymphocytes. *Nature* **227**:1105.

————. 1972. Studies on RNA metabolism during lymphocyte activation. *Transpl. Rev.* **11**:3.

————. 1973a. Effect of mitogens on the mitotic cycle: A biochemical evaluation of lymphocyte activation. *Effects of drugs on the cell cycle* (ed. A. Zimmerman, G. Padilla and I. Cameron) p. 137. Academic Press, N. Y.

————. 1973b. Degradation of 28S RNA late in maturation in non-growing lymphocytes and its reversal after growth stimulation. *J. Cell Biol.* (in press).

————. 1973c. Lymphocyte RNA-labelling with ^{3}H-uridine: Correction of data by analysis of UTP pool saturation kinetics. *Proc. 7th Leukocyte Culture Conf.* (ed. F. Daguillard) p. 119. Academic Press, N. Y.

————. 1973d. Purification of lymphocytes from peripheral blood. *Methods in enzymology*, vol. 32 (ed. S. Fleischer, L. Packer, R. Estabrook) Academic Press, N.Y. (in press).

Cooper, H. L. and E. M. Gibson. 1971. Control of synthesis and wastage of ribosomal ribonucleic acid in lymphocytes. II. The role of protein synthesis. *J. Biol. Chem.* **246**:5059.

Craig, S. P. and J. Piatigorsky. 1973. Cell elongation and δ-crystallin synthesis without RNA synthesis in cultured early embryonic chick lens epithelia. *Biochim. Biophys. Acta* **299**:642.

Darnell, J. E., R. Wall and R. Tushinski. 1971. An adenylic-rich sequence in messenger RNA of HeLa cells and its possible relationship to reiterated sites in DNA. *Proc. Nat. Acad. Sci.* **68**:1321.

Dingman, L. and A. Peacock. 1968. Analytical studies on nuclear ribonucleic acid using polyacrylamide gel electrophoresis. *Biochemistry* **7**:659.

Edmonds, M., M. Vaughan and H. Nakazato. 1971. Polyadenylic acid sequences in heterogeneous nuclear RNA and rapidly-labeled polyribosomal RNA of HeLa cells; possible evidence of a precursor relationship. *Proc. Nat. Acad. Sci.* **68**:1336.

Greaves, F. and S. Bauminger. 1972. Activation of T and B lymphocytes by insoluble phytomitogens. *Nature New Biol.* **235**:67.

Greenberg, J. R. and R. P. Perry. 1972a. The isolation and characterization of steady-state labeled messenger RNA from L-cells. *Biochim. Biophys. Acta* **287**:361.

————. 1972b. Relative occurrence of polyadenylic acid sequences in messenger and heterogeneous nuclear RNA of L cells as determined by poly(U)-hydroxyapatite chromatography. *J. Mol. Biol.* **72**:91.

Heller, P., N. Bhoopalam, V. J. Yakulis and N. Costea. 1971. Kappa and lambda receptor sites on single lymphocytes. *Clin. Exp. Immunol.* **9**:637.

Hirschhorn, K., F. Bach, R. L. Kolodny, J. L. Firschein and N. Hashem. 1963. Immune response and mitosis of human lymphocytes in vitro. *Science* **142**:1185.

Kay, J. E., B. G. Leventhal and H. L. Cooper. 1969. Effects of inhibition of ribosomal

RNA synthesis on the stimulation of lymphocytes by phytohemagglutinin. *Exp. Cell Res.* **54:**94.

Lawrence, H. S. and M. Landy. 1969. Mediators of cellular immunity. Academic Press, N.Y.

Lee, S. Y., J. Mendecki and G. Brawerman. 1971. A polynucleotide segment rich in adenylic acid in the rapidly labeled polyribosomal RNA component of mouse sarcoma 180 ascites cells. *Proc. Nat. Acad. Sci.* **68:**1331.

Loeb, L. A., S. S. Agarwal and A. M. Woodside. 1968. Induction of DNA polymerase in human lymphocytes by phytohemagglutinin. *Proc. Nat. Acad. Sci.* **61:**827.

McCarty, K. S., D. Stafford and O. Brown. 1968. Resolution and fractionation of macromolecules by isokinetic sucrose density gradient sedimentation. *Anal. Biochem.* **24:**314.

Nakazato, H. and M. Edmonds. 1972. The isolation and purification of rapidly labeled polysome-bound ribonucleic acid on polythymidylate cellulose. *J. Biol. Chem.* **247:**3365.

Noll, H. 1967. Characterization of macromolecules by constant velocity sedimentation. *Nature* **215:**360.

Oppenheim, J. J. 1968. Relationship of in vitro lymphocyte transformation to delayed hypersensitivity in guinea pigs and man. *Fed. Proc.* **27:**21.

Piatigorsky, J. 1968. Ribonuclease and trypsin treatment of ribosomes and polyribosomes from sea urchins. *Biochim. Biophys. Acta* **166:**142.

Raff, R. A., H. V. Colot, S. E. Selvig and P. R. Gross. 1972. Oogenetic origin of messenger RNA for embryonic synthesis of microtubule proteins. *Nature* **235:**211.

Rosenfeld, M. G., I. B. Abrass, J. Mendelsohn, B. A. Roos, R. F. Boone and L. D. Garren. 1972. Control of transcription of RNA rich in polyadenylic acid in human lymphocytes. *Proc. Nat. Acad. Sci.* **69:**2306.

Sheldon, R., C. Jurale and J. Kates. 1972. Detection of polyadenylic acid sequences in viral and eukaryotic RNA. *Proc. Nat. Acad. Sci.* **69:**417.

Singer, R. H. and S. Penman. 1972. Stability of HeLa cell mRNA in actinomycin. *Nature* **240:**100.

Regulation of Cell Cycle in Hybrid Cells

Potu N. Rao

Department of Developmental Therapeutics, The University of Texas at Houston
M. D. Anderson Hospital and Tumor Institute, Houston, Texas 77025

Robert T. Johnson

Department of Zoology, University of Cambridge, Cambridge, England

The chances of obtaining a viable hybrid between two cells of the same or different species largely depend upon the functional adaptation of the two nuclei to the common cytoplasm (Ephrussi and Weiss 1967). A somatic cell hybrid may be defined as a cell carrying the genomes of both parents in the same nucleus. However such hybrids originate usually as binucleate cells. They remain binucleate until the nuclear fusion takes place, possibly during the first mitosis. Fusion between interphase nuclei is a very rare phenomenon. Most commonly nuclei fuse when they enter mitosis synchronously. In this article we will examine how such nuclear synchrony is achieved when two cells from different points in the cell cycle are fused together. The data presented here indicate that nuclear synchrony in multinucleate cells is mediated through the common cytoplasm and it is achieved either at the time of DNA synthesis or during the initiation of mitosis. Failure to achieve mitotic synchrony may lead to premature chromosome condensation of the lagging nucleus. The cytoplasmic factors responsible for the induction of DNA synthesis and chromosome condensation will be examined.

Cell Fusion

The technique of cell fusion induced by inactivated Sendai virus has proved to be a valuable tool for the study of somatic cell genetics. We applied this technique to study the regulation of DNA synthesis and mitosis by fusing HeLa cells synchronized in one phase of the cell cycle with those in another phase (Rao and Johnson 1970). For example, cells in G_1 phase were fused with S or G_2 cells and G_2 cells were fused with S phase cells. In all these fusions one of the parental population was lightly prelabeled with [³H]thymidine to permit the identification of the hybrids. This made it possible to identify the various classes of multinucleate cells and trace them through DNA synthesis and mitosis (Table 1). Immediately after fusion the proportion of each class of cells, particularly the various classes of mono- and binucleate cells, was determined. DNA synthesis was monitored while cell division was blocked by adding [³H]thymidine and colcemide to the fused cells

Table 1

Classes of Cells Formed by a Random Cell Fusion
between Labeled and Unlabeled Populations

	Fusion classes	
Subpopulation	*Homophasic*	*Heterophasic*
Mononucleate cells (parental types only)	L; U	
Binucleate cells	LL; UU	LU
Trinucleate cells	3L; 3U	2L/U; L/2U
Tetranucleate cells	4L; 4U	3L/U; 2L/2U; L/3U
Pentanucleate*		

L, Prelabeled nucleus; U, unlabeled nucleus.

* The cells with five or more nuclei were too few to be included in this
study.

and recording the changes in the proportion of various classes as a function of time.
For example, among the binucleate cells a decrease in the frequency of LU
(hybrid) class results in a proportionate increase in LL class, indicating the initiation of DNA synthesis in the unlabeled nucleus (see Table 1). The pattern of
mitotic accumulation for each class was obtained by adding only colcemide to
another set of samples from the same fusions.

Regulation of DNA Synthesis

Figure 1 illustrates a rapid induction of DNA synthesis in G_1 nuclei following
fusions with S phase cells, thus virtually eliminating the pre-DNA synthetic period
in the former. The factors responsible for the initiation of DNA synthesis, which are
present in the S phase cytoplasm, migrate to the G_1 nucleus to induce the replication process. However fusion between S and G_2 cells did not result either in the induction of DNA synthesis in the G_2 nucleus or its inhibition in the S nucleus. The
presence of inducing factors alone is not enough to initiate DNA replication. It also
depends on the state of the interphase chromatin. In mammalian cells a second
round of DNA synthesis in one cycle does not seem possible unless the two
chromatids of the G_2 chromosomes are separated either during mitosis or by endo-
reduplication. This also indicates that G_2 cells do not have factors that inhibit DNA
synthesis.

Duration of G_1 Period in Heterophasic Binucleate Cells

From Fig. 1 it is clear that the G_1 period in HeLa cells could be dispensed with if
the factors necessary for the initiation of DNA synthesis are provided by some
means or other. These studies have been extended by Graves (1972) in heterokaryons obtained by fusion of synchronized populations of HeLa with Chinese
hamster (DON) or mouse (NCTC 2472) cells. She observed that when a cell line

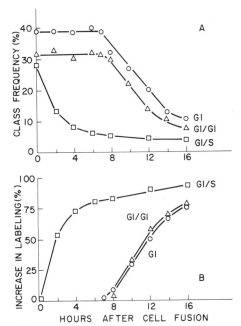

Figure 1

Induction of DNA synthesis in the G_1 nucleus of the G_1/S fusion. **(A)** Class frequencies of G_1, G_1/G_1 and G_1/S as a function of time after fusion and continous incubation with [³H]thymidine. The S nuclei were prelabeled. **(B)** Rate of induction of DNA synthesis in the G_1 nucleus of the G_1/S fused cells. Redrawn from Rao and Johnson (1970) with permission of *Nature*.

with short G_1 period (NCTC) was fused with another having a long G_1 period (HeLa), the nuclei in the heterokaryons exhibited short G_1 period like the mouse cells (Fig. 2). That is, the factors signaling initiation of DNA synthesis in the mouse nucleus also stimulated this process in the HeLa nucleus. This observation once again confirms the previous reports in literature (Rao and Johnson 1970; Johnson and Harris 1969) that DNA replication is due to the presence of some inducing factors, and these factors are not species-specific (Johnson and Harris 1969).

In a series of interesting papers Gordon and Cohn (1970, 1971a,b) have extended our knowledge about the nature of the factors that may be involved in the initiation of DNA synthesis in nuclei of nondividing cells. In their system, which involved fusion between actively proliferating mouse melanocytes and mouse macrophages, they found that the signals for DNA synthesis to begin came from the melanocyte component of the heterokaryon. Both protein and high molecular weight RNA synthesis of the melanocyte were essential for the macrophage nucleus to be rendered inducible into S phase.

The nuclear and/or cytoplasmic factors supplied by the proliferating cell component of the heterokaryon are not known. There is, however, good evidence from a variety of sources (reviewed by Johnson and Rao 1971) indicating that the activation process of quiescent nuclei, either in nuclear transplant or cell fusion situations, is temporally associated with the influx of proteins from the active cytoplasm into

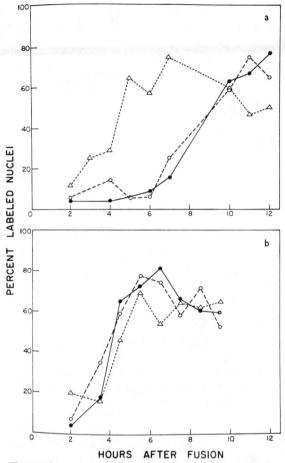

Figure 2

Initiation of DNA synthesis in the nuclei of NCTC-HeLa heterokaryons. **(a)** DNA synthesis in HeLa nuclei included in mononucleate HeLa cells (•); HeLa-HeLa homokaryons (○); and NCTC-HeLa heterokaryons (△). **(b)** DNA synthesis in NCTC nuclei included in mononucleate NCTC cells (•); NCTC-NCTC homokaryons (○); and NCTC-HeLa heterokaryons (△). From Graves (1972) with permission of *Exp. Cell Res.*

the dormant nucleus (Gordon and Cohn 1970; Bolund et al. 1969a,b). In a recent study we (Johnson and Mullinger unpubl.) have examined the influx of protein from the HeLa component of HeLa–avian erythrocyte heterokaryons as a function of the stage of the HeLa cell in the cell cycle at the time of fusion. We find that within 6 hours HeLa protein has entered virtually every erythrocyte nucleus and that this occurs whether the HeLa cells were in early G_1 or early S. DNA synthesis is not induced in erythrocyte nuclei during the 6 hours that they spend in a G_1 environment in heterokaryons, but it is induced in 80% of those that spend a similar length of time in S phase cytoplasm. Thus the major problem still remains of deciphering the role of the inflowing proteins and, in particular, whether they are in part responsible for the induction of DNA synthesis.

Duration of DNA Synthetic (S) Period

The time schedule for the initiation of DNA replication during the cell cycle can be advanced by reducing the G_1 period, but the time required to complete the replication of the genome cannot be shortened. This fact has been clearly demonstrated by Graves' (1972) data on the HeLa–Chinese hamster heterokaryons (Fig. 3). The duration of S period was 6.9 hours in HeLa and 5.1 hours in hamster cells. When two nuclei, one with short and the other with long S period, are brought together

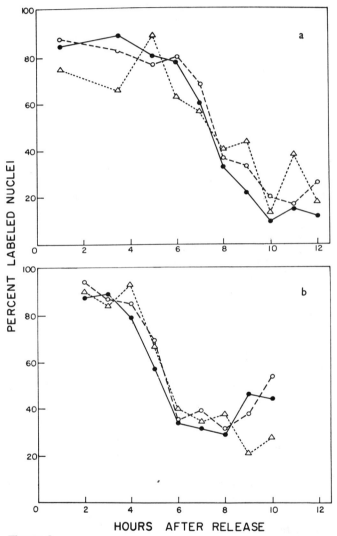

Figure 3

Duration of DNA synthesis in the nuclei of Chinese hamster (DON)-HeLa heterokaryons. **(a)** DNA synthesis in HeLa nuclei included in mononucleate HeLa cells (•); HeLa-HeLa homokaryons (○); and DON-HeLa heterokaryons (△). **(b)** DNA synthesis in DON nuclei included in mononucleate DON cells (•); DON-DON homokaryons (○); and DON-HeLa heterokaryons (△). From Graves (1972) with permission of *Exp. Cell Res.*

by cell fusion, they complete their DNA replication each at its own pace, true to their parental type irrespective of the common cytoplasmic milieu around them.

The data from radioautographic studies suggest that the DNA in a mammalian chromosome is divided into a large number of replicating units or replicons (Taylor 1960; Lima-de-Faria 1959; Painter 1961; Stubblefield and Mueller 1962; Moorhead and Defendi 1963; German 1962; Hsu 1964). The estimated average number of replicons in the nucleus of a mammalian cell may range between 1.8×10^4 to 3.6×10^4, assuming the size of a replication unit to be 50 to 100 μ (Watson 1971). Using density labeling procedures Painter, Jermany and Rasmussen (1966) calculated that there are about 1000 replicating sites undergoing DNA synthesis in a HeLa nucleus at any given time during S phase. Hence all the replication units are not active at the same time, but they are activated according to a heritable sequence (Watson 1971). Heterochromatic regions of mammalian chromosomes are usually late replicating as shown by radioautographic studies (Hsu 1964). The late replicating nature of the heterochromatic X chromosomes in the cells of *Microtus agrestis* can be visualized by fusing them with mitotic HeLa cells, thus inducing premature chromosome condensation in the *Microtus* nucleus. When the autosomes are in S phase, the X chromosomes are in G_1 (Fig. 4) and when the former are in G_2 period, the latter are in S (Fig. 5). The conclusion that can be drawn at this point is that the cytoplasmic inducers of DNA synthesis have no effect on the programed sequence of chromosome replication (Graves 1972). Watson (1971) came to the same conclusion, that "the inducer(s) signaling the start of S phase cannot control the initiation of DNA synthesis in each replication unit." His conclusion is based on the gene amplification studies of Wallace and Birnstiel (1966), delayed replication of a small fraction (0.3%) of the genome during meiosis in lilly (Stern and Hotta 1969), and the correlation between heterochromatinization of chromosomes and their late replication (Lima-de-Faria and Jaworska 1968).

Cytoplasmic Factors for the Initiation of DNA Synthesis

Nuclear transplantation experiments (Graham 1966; Graham, Arms and Gurdon 1966; De Terra 1967; Gurdon 1967) and cell fusion studies (Harris et al. 1966; Johnson and Harris 1969; Bolund et al. 1969a,b; Rao and Johnson 1970) point out the positive role of cytoplasm as a reservoir for the inducers of DNA synthesis during S phase. Attempts have been made to isolate and characterize these cytoplasmic factors. The addition of cytoplasmic extracts from tumor cells actively synthesizing DNA to an in vitro DNA synthesizing system caused a tenfold increase in DNA synthesis, whereas the addition of extracts from normal cells not synthesizing DNA produced little or no increase in DNA synthetic activity (Thompson and McCarthy 1971). Partial purification of cytoplasmic extracts of TLT mouse hepatoma cells indicate that the stimulating factor is a heat-stable cation of low molecular weight. The cytoplasmic factor seems to influence the interaction between DNA polymerase and the DNA template. For the stimulation of DNA synthesis to occur the polymerase, the template, and the cytoplasmic factor must be present (Thompson and McCarthy 1971). However the inducing factors present in the cytoplasm of S phase HeLa cells were found to be heat-labile, but stable to lyophilization or freezing (Kumar and Friedman 1972). They also report that the presence of pancreatic RNase in the incubation mixture did not effect the stimulation of DNA synthesis.

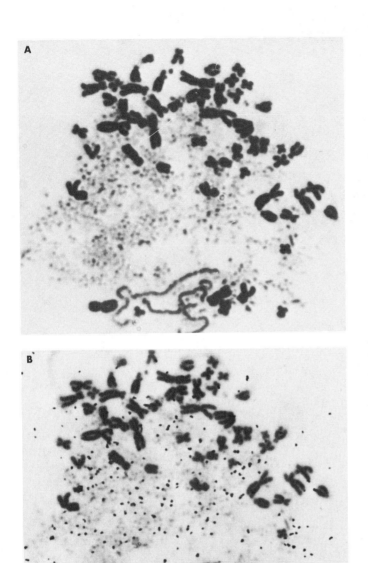

Figure 4
Induction of premature condensation in a HeLa-*Microtus agrestis* heterokaryon.
A random population of female *Microtus agrestis* cells were pulse-labeled with
[³H]thymidine for 15 min prior to fusion with mitotic HeLa cells. **(A)** The
darkly stained chromosomes are of the mitotic HeLa cell. The autosomes of the
Microtus cells present a pulverized appearance. The two intact X chromosomes
of *Microtus* (shown by arrows) are in G_1 phase. **(B)** A radioautograph of the
same cell. Note that the silver grains are localized on the "pulverized" autosomes
of *Microtus* indicating DNA synthesis in these chromosomes, whereas no label
is associated with X chromosomes which are late replicating.

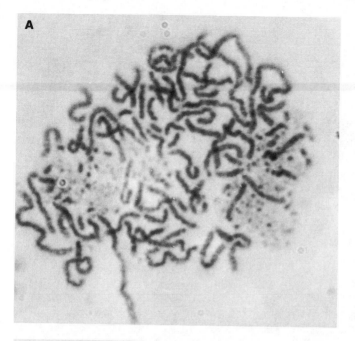

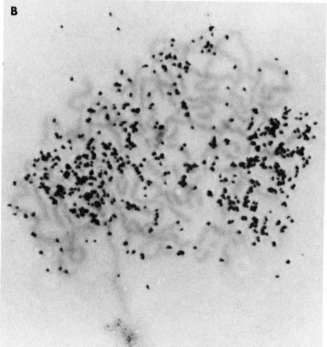

Figure 5

HeLa-*Microtus* heterokaryon. The cells were prepared in the same manner as in Fig. 4. **(A)** The mitotic HeLa chromosomes were widely scattered during preparation and hence they could not be seen in this field. The PCC of the auto-somes of *Microtus* are in G_2 phase as evidenced by their double chromatid structure. The two regions of "pulverized" chromatin represent the two X chromosomes undergoing DNA synthesis. **(B)** A radioautograph of the same cell. The label is predominantly localized on the "pulverized" X chromosomes.

Duration of G$_2$ Period

In the G$_1$/G$_2$ heterophasic binucleate cells the G$_1$ nuclei entered and completed DNA synthesis at about the same time as nuclei in a (G$_1$/G$_1$) homophasic binucleate cell (Rao and Johnson 1970). This suggests that factors present in G$_2$ cells have neither stimulatory nor inhibitory effects on the metabolic processes occurring during G$_1$ or S periods. However the reverse is not true. If G$_1$ is considered as a preparatory period for DNA synthesis, so is G$_2$ for the initiation of mitosis. DNA synthesis is under positive control. On the other hand, initiation of mitosis appears to be regulated by negative, as well as positive, controls. In S/G$_2$ and G$_1$/G$_2$ fusions the G$_2$ nuclei did not enter mitosis until the lagging nuclei in heterophasic cells completed DNA synthesis, whereupon all the nuclei entered mitosis synchronously (Fig. 6). It seems that G$_1$ or S components have an inhibitory effect on the progression of G$_2$ nuclei into mitosis. The completion of DNA synthesis in the lagging nucleus relieves the G$_2$ nucleus from this repression. The explanation for this repression of G$_2$ nucleus by G$_1$ or S cytoplasm may lie in their respective positions in the chromatin condensation cycle. There is an increasing body of evidence suggesting a cyclical pattern of chromatin condensation and decondensation during the cell cycle (Pederson 1972; Alvarez and Valladares 1972). The chromosomes reach their height of condensation during metaphase. Following separation of chromatids during anaphase, the chromosomes go through progressive uncoiling and become most diffuse during DNA synthesis. Subsequent to the replication of DNA, chromosomes go through progressive coiling. These changes in the pattern of chromatin condensation and decondensation during the cell cycle have been confirmed by the technique of differential staining with safranin (Alvarez and Valladares 1972). In a heterophasic cell containing G$_1$/G$_2$ nuclei, the G$_2$ component is preparing for chromatin condensation while its counterpart (G$_1$ or S component)

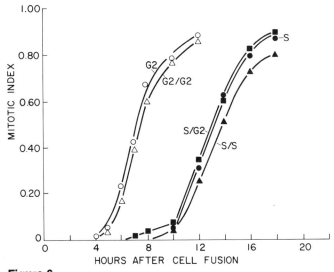

Figure 6

Mitotic accumulation functions for the mono- and binucleate S and G$_2$ classes are compared with the heterophasic S/G$_2$ binucleate class. Note that the pattern of mitotic accumulation in S/G$_2$ is similar to that of the S parent. Redrawn from Rao and Johnson (1970) with permission of *Nature*.

is working against it by preparing for decondensation of chromatin. It is possible that the decondensation factors of G_1 or S component neutralize or even inhibit the action of the condensing factors, thus holding the G_2 nucleus from entering into mitosis.

In a previous study (Rao and Johnson 1970) we pointed out that in a hetero-phasic multinucleate cell increasing the ratio of G_2 to S nuclei resulted in the entry of S nuclei into mitosis earlier than a corresponding homophasic (S/S) multi-nucleate cell. However the advanced entry of the S nucleus into mitosis appears to be due to shortening of its G_2 rather than S period. Graves (1972) has shown that S period cannot be shortened. In our recent studies (Rao and Hittelman unpubl.) we fused two G_2 populations of HeLa cells, one of which was delayed by treating either with X rays (350 rads), cycloheximide (25 μg/ml for 2 hours), or strep-tovitacin (100 μg/ml for 2 hours). Streptovitacin, like cycloheximide, is an in-hibitor of protein synthesis whose effects are not quickly reversible. The kinetics of mitotic accumulation of the hybrids was intermediate between those of normal and retarded populations (Fig. 7). This indicates that G_2 delay induced by X radiation or the inhibition of protein synthesis was partially overcome because of the presence of a normal G_2 component in the fused cell.

The X-ray-induced G_2 delay in cells had been reversed by treating the cells with agmatine or calcium salts following irradiation (Whitfield and Rixon 1962; Whit-field et al. 1962; Whitfield and Youdale 1966). Hence the unirradiated G_2 com-ponent must contain some factors that are homologous to agmatine or calcium in order to reverse the delay in the irradiated G_2 component. The fact that inhibition of protein synthesis blocks most of the G_2 cells from entering into mitosis indicates that at least some of the proteins synthesized during the G_2 period play a role in the

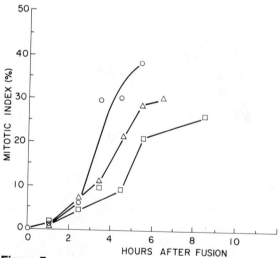

Figure 7

Reversal of X-ray-induced mitotic delay by fusion with G_2 cells. A synchronized G_2 population of HeLa cells, prelabeled with [^{3}H]thymidine, were fused with another unlabeled G_2 population which was exposed to 350 rads of X rays. The mitotic accumulation functions for the binucleate cells containing two labeled nuclei (○), two unlabeled (□); and one labeled and one unlabeled nuclei (△) were plotted as a function of time.

G_2-mitotic transition. The duration of G_2 period can be reduced if these mitotic factors are made available to a cell soon after the completion of DNA replication.

Premature Chromosome Condensation (PCC)

The presence of such factors in mitotic cells can be demonstrated by fusing mitotic cells with interphase cells. Within 30 minutes after fusion the interphase nucleus undergoes rapid transformation resembling that of G_2-prophase transition, leading to the formation of discrete chromosomes. This phenomenon has been termed as premature chromosome condensation or PCC (Johnson and Rao 1970). Because of the striking similarity between PCC induction and prophase, Matsui, Weinfeld and Sandberg (1972) gave yet another name, "prophasing," to this phenomenon. The morphology of the prematurely condensed chromosomes varies according to the position of the interphase cell in the cell cycle at the time of fusion. G_1 PCC are very long, containing one chromatid, whereas G_2 PCC are relatively more condensed and have two chromatids. In comparison to metaphase chromosomes the G_2 PCC are quite extended in length. PCC induction in S phase nuclei results in uneven condensation of chromatin, giving a pulverized appearance. This is best illustrated by the experiment in which mitotic HeLa were fused with cells of *Microtus agrestis* prelabeled with [³H]thymidine for 15 minutes (Sperling and Rao unpubl.). Figure 4A shows that the late replicating, heterochromatic X chromosomes are in G_1 phase as evidenced by its single chromatid structure, while rest of the genome was undergoing DNA replication, which could be identified by the pulverized appearance of the chromatin. On Fig. 5A the autosomes are in G_2 phase, while the X chromosomes present a pulverized appearance. Radioautographs of the same cells reveal silver grains localized in the "gaps" of the pulverized chromosomes, thus providing positive proof that these uncondensed regions are the active sites of DNA replication (Fig. 4B and 5B).

Induction of PCC resembles only the initial stage of mitosis, that is, the prophase. The formation of prematurely condensed chromosomes is not followed by the organization of a spindle or the alignment of chromosomes at the equatorial plate. When the mitotic component is going through anaphase, the PCC are distributed at random between the daughter cells, although it was impossible to detect any precise arrangement on the spindle (Rao and Johnson 1972). The PCC, when incorporated into the daughter nuclei of the dividing cell, undergo decondensation, replication and reenter mitosis in synchrony with the rest of the genome (Rao and Johnson 1972). The PCC is an example where the chromosome condensation is uncoupled from the development of a functional spindle. A spindle can be organized only when the centriole pairs separate and move apart to establish poles. When condensation of chromosomes is prematurely induced in an interphase nucleus rather suddenly, there may not be sufficient time or the proper signals for the maturation and separation of centrioles in order to organize the spindle. Hence the factors responsible for chromosome condensation are different from those for spindle organization.

Factors for Chromosome Condensation

The major difference between an interphase and a mitotic cell lies in the extent of chromatin condensation. Now we can induce chromosome formation in an interphase cell by fusing it with a mitotic cell. The mitotic cell contains factors for chromosome condensation, the nature of which remains to be elucidated. Using mitotic-interphase fused cell system Rao and Johnson (1971) demonstrated that addition of positively charged compounds, particularly Mg^{++} and spermine, promoted PCC induction and the negatively charged compounds tested inhibited this process. This indicates the importance of cations in chromosome condensation. However the addition of inactivated Sendai virus and the positively charged compounds to a pellet of cells from a random population did not lead to chromosome formation. It may be possible that there are factors other than cations. We explored the possibility that some proteins may be involved in bringing about chromosome condensation.

Migration of Mitotic Protein to PCC

We labeled synchronized populations of G_1, S or G_2 HeLa cells with a mixture of ^{3}H-amino acids for 3 hours, after which the radioactive medium was replaced with regular medium containing colcemide to block the cells in mitosis. These prelabeled mitotic cells (G_1^*M, S^*M, G_2^*M) were separately fused with unlabeled interphase cells, chromosome preparations were made and the slides were processed for radioautography. In every fusion involving prelabeled mitotic with unlabeled interphase cells, there was label associated with PCC. Blocking protein synthesis by the addition of cycloheximide (25 μg/ml) to the fusion mixture had no effect either on the induction of PCC or on the migration protein to PCC. The amount of label associated with PCC is proportional to the number of grains on the mitotic chromosomes (Table 2). The label that migrates to PCC is coming from the cytoplasm, but not from the chromosomal component of the mitotic cell. This is substantiated by the absence of any difference in the number of grains on mitotic chromosomes of mononucleate cells versus those associated with PCC. The fusion of prelabeled interphase cells with unlabeled mitotic cells resulted in the migration of a relatively low amount of label to the mitotic chromosomes, suggesting some small degree of nonspecificity. By enzyme digestion it has been established that the label is localized in the protein fraction. Most of these chromosome-specific proteins are synthesized during S and G_2 phases (Rao and Johnson unpubl.).

Table 2

Migration of Labeled Mitotic Protein to
Prematurely Condensed Chromosomes

| | No. grains/karyotype | | |
| | Mitotic chromosomes | | |
Type of fusion	Cells not fusing	Cells fusing	PCC
G_1^*M/I	25.0	24.5	21.2
S^*M/I	67.4	44.1	34.3
G_2^*M/I	69.8	70.0	61.0

Nature of Migratory Proteins

Synchronized S phase HeLa cells were labeled for 3 hours with 10 μCi/ml of tritiated arginine (0.57 Ci/mM), lysine (2.94 Ci/mM) or tryptophan (4.05 Ci/mM), the label was washed off and the cells collected in mitosis following colcemide treatment. These prelabeled mitotic cells were fused with unlabeled interphase cells and the number of grains associated with mitotic chromosomes and PCC were counted (Table 3). The ratio of grains on mitotic chromosomes to PCC was 1:1 when the mitotic cells were prelabeled with tryptophan, whereas it was 2:1 if either arginine or lysine were used for labeling. This indicates that the tryptophan-rich proteins, which are likely to be nonhistone proteins, become associated with PCC more extensively than the arginine- or lysine-rich proteins.

The data presented here indicate that there is a rapid migration of proteins from the mitotic component to the interphase chromatin during the induction of PCC. Chemical analysis of chromosomes reveals that the amount of nonhistone protein bound to metaphase chromosomes is twice that associated with interphase chromatin (Hancock 1969). The migration of cytoplasmic protein to PCC observed here may explain how the metaphase chromosomes become richer in nonhistone proteins. Some workers feel that a high nonhistone protein to DNA ratio of metaphase chromosomes is probably an artifact of isolation procedures (Zirkin and Wolfe 1970). In the present study the migration of proteins to chromosomes during their condensation has been shown without using any isolation techniques. This observation suggests that there is a migration of cytoplasmic proteins to the nucleus and their association with chromatin during G_2 may bring about the condensation of chromosomes, which marks the initiation of mitosis.

SUMMARY

The nature of the regulation for initiation of DNA synthesis and mitosis is examined by fusing synchronized populations of HeLa cells in various phases of the cell cycle. In addition to our studies available literature on this topic is reviewed. On the basis of these various studies the following conclusions can be drawn:

1. Initiation of DNA synthesis is under a positive control and it can be induced in a G_1 nucleus following fusion with an S phase cell.

Table 3
Migration of Protein Labeled with Specific Amino Acids during S*M/I Fusion

	[^3H]arginine	[^3H]lysine	[^3H]tryptophan
No. grains on mitotic chromosomes	19.6	47.1	25.2
No. grains on PCC	10.0	23.0	21.5
Ratio grain count (PCC/M)	0.5	0.5	1.0

2. The cytoplasm of the S phase cell contains factors for the initiation of DNA synthesis, and this stimulating factor appears to be a heat-stable cation of low molecular weight.

3. By appropriate fusions the duration of G_1 period can be reduced but not that of S. In other words the initiation of DNA synthesis can be triggered experimentally, but the sequential pattern of replication of the various replicating units cannot be changed.

4. The initiation of mitosis is influenced by negative as well as positive modes of control. In a binucleate cell, incomplete or lack of DNA replication in one of the nuclei would prevent the other, more advanced, G_2 nucleus from entering into mitosis. This is the negative control. However the fusion between late and early G_2 cells would help the lagging nucleus to enter mitosis earlier than the control. This represents the positive control or the influence of the mitotic factors.

5. The existence of mitotic factors has become even more evident by the observation of premature chromosome condensation (PCC) when mitotic cells are fused with interphase cells.

6. The factors present in mitotic cells that can induce chromosome condensation in interphase nuclei appear to be positively charged compounds such as cations (Mg^{++}), polyamines, and a class of nonhistone proteins. Further studies are necessary to isolate and identify the various types of mitotic factors.

Acknowledgments

This investigation was aided by research grants from the Damon Runyon Memorial Fund for Cancer Research (DRG-1110) and the Medical Research Council of Great Britain.

REFERENCES

Alvarez, Y. and Y. Valladares. 1972. Differential staining of the cell cycle. *Nature New Biol.* **238**:279.

Bolund, L., Z, Darzynkiewicz and N. R. Ringertz. 1969a. Growth of hen erythrocyte nuclei undergoing reactivation in heterokaryons. *Exp. Cell Res.* **56**:406.

Bolund, L., N. R. Ringertz and H. Harris. 1969b. Changes in the cytochemical properties of erythrocyte nuclei reactivated by cell fusion. *J. Cell Sci.* **4**:71.

De Terra, N. 1967. Macronuclear DNA synthesis in *Stentor:* Regulation by a cytoplasmic initiator. *Proc. Nat. Acad. Sci.* **57**:607.

Ephrussi, B. and M. C. Weiss. 1967. Regulation of the cell cycle in mammalian cells: Inferences and speculations based on observations of interspecific somatic hybrids. *Control mechanisms in developmental processes* (ed. M. Locke) suppl. 1, p. 136. Academic Press, New York.

German, J. L. 1962. DNA synthesis in human chromosomes. *Trans. N.Y. Acad. Sci.* **24**:395.

Gordon, S. and Z. Cohn. 1970. Macrophage-melanocyte heterokaryons. I. Preparation and properties. *J. Exp. Med.* **130**:981.

———. 1971a. Macrophage-melanocyte heterokaryons. II. The activation of macrophage DNA synthesis. Studies with inhibitors of RNA synthesis. *J. Exp. Med.* **133**:321.

———. 1971b. Macrophage-melanoma cell heterokaryons. III. The activation of macrophage DNA synthesis. Studies with inhibitors of protein synthesis and with synchronized melanoma cells. *J. Exp. Med.* **134**:935.

Graham, C. F. 1966. The regulation of DNA synthesis and mitosis in multinucleate frog eggs. *J. Cell Sci.* **1**:363.

Graham, C. F., K. Arms and J. B. Gurdon. 1966. The induction of DNA synthesis by frog egg cytoplasm. *Develop. Biol.* **14**:349.

Graves, J. A. M. 1972. DNA synthesis in heterokaryons formed by fusion of mammalian cells from different species. *Exp. Cell Res.* **72**:393.

Gurdon, J. B. 1967. On the origin and persistence of a cytoplasmic state inducing nuclear DNA synthesis in frogs' eggs. *Proc. Nat. Acad. Sci.* **58**:545.

———. 1970. Nuclear transplantation and the control of gene activity in animal development. *Proc. Roy. Soc. London* B **176**:303.

Hancock, R. 1969. Conservation of histones in chromatin during growth and mitosis *in vitro*. *J. Mol. Biol.* **40**:457.

Harris, H., J. F. Watkins, C. F. Ford and G. I. Schoefl. 1966. Artificial heterokaryons of animal cells from different species. *J. Cell Sci.* **1**:1.

Hsu, T. C. 1964. Mammalian chromosomes *in vitro*. VIII. DNA replication sequence in Chinese hamster. *J. Cell Biol.* **23**:53.

Johnson, R. T. and H. Harris. 1969. DNA synthesis and mitosis in fused cells. II. HeLa-chick erythrocyte heterokaryons. *J. Cell Sci.* **5**:625.

Johnson, R. T. and P. N. Rao. 1970. Mammalian cell fusion: Induction of premature chromosome condensation in interphase nuclei. *Nature* **226**:717.

———. 1971. Nucleo-cytoplasmic interactions in the achievement of nuclear synchrony in DNA synthesis and mitosis in multinucleate cells. *Biol. Rev.* **46**:97.

Kumar, K. V. and D. L. Friedman. 1972. Initiation of DNA synthesis in HeLa cell-free system. *Nature New Biol.* **239**:74.

Lima-de-Faria, A. 1959. Differential uptake of tritiated thymidine into hetero- and euchromatin in *Melanoplus* and *Seaale*. *J. Biophys. Biochem. Cytol.* **6**:457.

Lima-de-Faria, A. and H. Jaworska. 1968. Late DNA synthesis in heterochromatin. *Nature* **217**:138.

Matsui, S., H. Weinfeld and A. A. Sandberg. 1972. Fate of chromatin of interphase nuclei subjected to "prophasing" in virus-fused cells. *J. Nat. Cancer Inst.* **49**:1621.

Moorhead, P. S. and V. Defendi. 1963. Asynchrony of DNA synthesis in chromosomes of human diploid cells: *J. Cell Biol.* **16**:202.

Painter, R. B. 1961. Asynchronous replication of HeLa S3 chromosomal deoxyribonucleic acid. *J. Biophys. Biochem. Cytol.* **11**:485.

Painter, R. B., D. A. Jermany and R. E. Rasmussen. 1966. A method to determine the number of DNA replicating units in cultured mammalian cells. *J. Mol. Biol.* **17**:47.

Pederson, T. 1972. Chromatin structure and the cell cycle. *Proc. Nat. Acad. Sci.* **69**:2224.

Rao, P. N. and R. T. Johnson. 1970. Mammalian cell fusion: Studies on the regulation of DNA synthesis and mitosis. *Nature* **225**:159.

———. 1971. Mammalian cell Fusion. IV. Regulation of chromosome formation from interphase nuclei by various chemical compounds. *J. Cell. Physiol.* **78**:217.

———. 1972. Premature chromosome condensation: A mechanism for the elimination of chromosomes in virus-fused cells. *J. Cell Sci.* **10**:495.

Stern, H. and Y. Hotta. 1969. DNA synthesis in relation to chromosome pairing and chiasma formation. *Genetics* **61** (suppl.): 27.

Stubblefield, E. and G. C. Mueller. 1962. Molecular events in the reproduction of animal cells. II. The localized synthesis of DNA in the chromosomes of HeLa cells. *Cancer Res.* **22**:1091.

Taylor, J. H. 1960. Asynchronous duplication of chromosomes in cultured cells of Chinese hamster. *J. Biophys. Biochem. Cytol.* **7**:455.

Thompson, L. R. and B. J. McCarthy. 1971. The effects of cytoplasmic extracts on DNA synthesis *in vitro*. *Fed. Proc.* **30** (part 1): 1177 (abstr.).

Wallace, H. and M. L. Birnstiel. 1966. Ribosomal cistrons and the nucleolar organizer. *Biochim. Biophys. Acta* **114:**296.

Watson, J. D. 1971. The regulation of DNA synthesis in eukaryotes. *Advances in cell biology* (ed. D. M. Prescott, L. Goldstein and E. McConkey) vol. 2, p. 1. Academic Press, New York.

Whitfield, J. F. and R. H. Rixon. 1962. Prevention of post-irradiation mitotic delay in cultures of L mouse cells by calcium salts. *Exp. Cell Res.* **27:**154.

Whitfield, J. F. and T. Youdale. 1966. The effects of calcium, agmatine and phosphate on mitosis in normal and irradiated populations of rat thymocytes. *Exp. Cell Res.* **43:**602.

Whitfield, J. F., R. H. Rixon and T. Youdale. 1962. Prevention of mitotic delay in irradiated suspension cultures of L mouse cells by agmatine. *Exp. Cell Res.* **27:**143.

Zirkin, B. R. and S. L. Wolfe. 1970. The chemical composition of nuclei and chromosomes isolated by the Langmuir trough technique. *Chromosoma* **32:**162.

Interrelationships of Glycolysis, Sugar Transport and the Initiation of DNA Synthesis in Chick Embryo Cells

H. Rubin and D. Fodge

Department of Molecular Biology and Virus Laboratory
University of California, Berkeley, California 94720

The rate of DNA synthesis in cultures of chick embryo cells is decreased by lowering the pH of the medium or reducing serum concentration and is increased by restoring the original conditions (Rubin 1971a; Rubin and Koide 1973a). Both the decrease and increase occur only after a delay of a few hours and reflect changes in the fraction of cells synthesizing DNA at any given time, rather than in the rate of ongoing synthesis in committed cells. In this sense the effects of pH and serum concentration simulate those associated with population density. The rate of sugar transport responds similarly but more quickly to changes in pH, serum concentration or population density (Rubin 1971a; Sefton and Rubin 1971). Unlike DNA synthesis, sugar transport is not an all or none phenomenon, and the response to external conditions probably occurs in all the cells contemporaneously. Although the rate of initiation of DNA synthesis can be altered without changing the rate of sugar transport—by zinc deprivation for example (Rubin and Koide 1973a; Rubin 1972)—we have so far found no treatment which alters sugar transport in healthy cells without ultimately influencing DNA synthesis. In general the changes in DNA synthesis are proportional to the prior changes in sugar transport. It has already been pointed out that the supply of sugar from the medium is unlikely to be the controlling factor in the initiation of DNA synthesis, since large variations in external concentrations of glucose are without effect on the process (Rubin 1971b). It appears, therefore, that sugar transport is itself responsive to an underlying process which might well be the ultimate controlling element in the initiation of DNA synthesis.

The aim of the present investigation was to identify the primary cellular process which responds to external changes and produces effects on both sugar transport and DNA synthesis. The approach was to determine first which components of the medium are required for the initiation as opposed to the continuation of DNA synthesis. Then various metabolic inhibitors were used to identify the biochemical pathways in the chain leading to DNA synthesis. A similar approach was taken for sugar transport. The feature common to both proved to be glycolysis, and the most prominent responding element in glycolysis was the regulatory enzyme

801

phosphofructokinase. Based on these findings, a model is elaborated for the regulation of cell multiplication through control of the glucose catabolism.

MATERIALS AND METHODS

Cells and Culture Techniques

The methods for culturing cells from the carcasses of 10-day-old chick embryos are those used routinely in this laboratory (Rein and Rubin 1968). The growth medium consisted of mixture 199 plus 2% tryptose phosphate broth and 1% chicken serum, designated 199(2–0–1). The rate of DNA synthesis in the cultures was reduced ("turned off") by substituting for 16 hours a medium of pH 6.8 containing 199 and tryptose phosphate broth but no serum and designated 199(2–0–0). The pH was adjusted by varying the concentration of NaHCO$_3$ in the medium (Rubin 1971a). To "turn on" the cells NaHCO$_3$ and/or serum was added to the used medium, or fresh medium at higher pH with or without serum was substituted for the used medium. In one experiment screwcap plastic bottles were used instead of petri dishes. In the other experiments plastic petri dishes were used, usually 60 mm in diameter, except where glycolytic intermediates were to be measured (see below).

For some nutritional experiments mixture 199 was prepared with glucose and amino acids omitted. In other experiments a solution containing only the major inorganic constituents of the medium was prepared and is referred to as Earle's saline. The appropriate components of mixture 199 were restored as indicated. Cells were infected with the Schmidt-Ruppin strain of Rous sarcoma virus according to established procedure (Martin et al. 1971).

Radioisotopes and Chemicals

The following radioisotopes obtained from the New England Nuclear Corp. were used in this study: [^{3}H]thymidine (16.7 Ci/mM), [^{3}H]uridine (5.0 Ci/mM), [^{3}H]leucine (38.8 Ci/mM), and D-[^{3}H]2-deoxyglucose (6.8 Ci/mM). The techniques used for labeling cells and extracting labeled material soluble and insoluble in 5% Cl$_3$CCOOH have been previously described (Rubin and Koide 1973a).

The following chemicals were obtained from Sigma: 2-deoxy-D-glucose (2-dGlc), 3-O-methyl glucose, iodoacetic acid, 2,4-dinitrophenol (DNP) and cycloheximide. Sodium fluoride was obtained from Mallinckrodt.

Measurement of Glycolytic Intermediates

Since the spectrophotometric determination of glycolytic intermediates required large numbers of cells, 3×10^6 cells obtained directly from the embryo or from primary cell cultures were seeded in 100-mm plastic petri dishes in growth medium. The cells were maintained at 38°C with daily medium change. On the fourth day when there were 14 to 26×10^6 cells per dish, the growth medium was replaced with medium 199(2–0–0) at pH 6.8. The cultures were incubated overnight to "turn off" the cells, and NaHCO$_3$ and/or serum was added the next day to turn them on. After 30 minutes of the turn-on period, the cells were washed once with 0.9% NaCl plus 5 mM triethanol amine, pH 7.4, which was immediately replaced with 0.5 ml of ice cold 0.6 N HClO$_4$ for 15 to 30 minutes. The extracts from about 40 mg protein were then made 0.1 M with triethanolamine buffer and brought to pH 7.4 with K_2CO_3. After removing the precipitate of $KClO_4$ the supernatant was analyzed spectrophotometrically for glycolytic intermediates by the coupled enzymatic methods of Minakami and Yoshikawa (1965). The enzymes used in the assay were obtained from Calbiochem as were NADH and $NADP^+$. Phosphoenolpyruvate was obtained from Sigma.

RESULTS

Kinetics of Increase in the Rate of DNA Synthesis

The rate of DNA synthesis in the cultures was lowered by incubating the cells for 16 hours in medium at pH 6.8 without serum. They were stimulated by substituting a serum-containing medium at pH 7.6. The rate of incorporation of [³H]thymidine into DNA showed no increase at 3 hours after stimulation but had increased 7-fold by 5 hours (Fig. 1). It continued to increase up to 19 hours and then declined. A similar pattern was seen when the fraction of cells synthesizing DNA was measured radioautographically by counting nuclei labeled with [³H]thymidine (Table

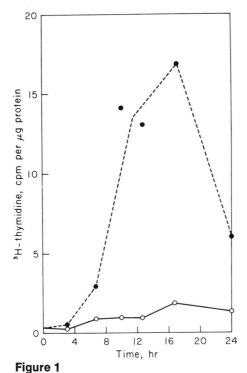

Figure 1
Kinetics of DNA synthesis following "turn-on" of chick embryo cells. Three-day-old cultures of chick embryo cells were "turned off" overnight in serum-free medium at pH 6.8. The medium was then changed to 199 (2–0–1) at pH 7.6 on half the dishes and kept in 199 (2–0–0) at pH 6.8 in the remaining half. The rate of DNA synthesis and the protein content of the cultures were determined at the indicated times.

 o———o medium 199 (2–0–0) pH 6.8
 •– – –• medium 199 (2–0–1) pH 7.6

Table 1

Autoradiography vs. Cell Extraction in the Measurement
of Rate of DNA Synthesis in Stimulated Cells

Hours after turn-on	Autoradiography		Cell Extraction	
	Labeled nuclei field	Factor of increase	[³H]thymidine (cpm/μg protein)	Factor of increase
0	0.93	1.0	0.4	1.0
11	30.7	33.0	16.0	40.0
24	47.0	50.2	17.3	43.0
48	14.5	15.6	5.5	13.6

Four-day-old cultures were "turned off" overnight in serum-free
medium at pH 6.8. The medium was then changed to one at pH 7.5
containing 2% serum and at intervals the cells were labeled with
[³H]thymidine for radioautography or scintillation counting.

1). This showed that incorporation of [³H]thymidine into extractable DNA is a
reliable index of the fraction of cells synthesizing DNA. An interval of 7 hours
from the time of application of the stimulus was adopted as a standard to observe
the effects of various treatments on the initiation of DNA synthesis. This had the
virtues of speed and convenience and allowed observation of the initiation period
of DNA synthesis.

Medium Components Required for Initiation of DNA Synthesis

It has been claimed that only monovalent salts, phosphate and bicarbonate are
required for the initiation of DNA synthesis when serum is restored to serum-
deprived cultures (Temin 1971). We found that removal of vitamins, purines,
pyrimidines and nonessential amino acids was without significant effect on the
initiation of DNA synthesis. Removal of the essential amino acids, however, completely blocked the process (Table 2). Glutamine was required in high concentration unless the medium was supplemented with glutamic acid.

Although the requirement for amino acids was unequivocal, the requirement for
glucose was not so clear. In some experiments it could be entirely removed from
the medium without effect on the initiation of DNA synthesis, and in others even
a 10-fold reduction in external glucose would interfere. A representative experiment is shown in Fig. 2, where in addition to the initiation of DNA synthesis, the
rate of sugar transport is measured using 2-dGlc. Reduction in the glucose content
of the medium had the effect of inducing a slight increase in the rate of uptake
of 2-dGlc. When the relative rate of uptake of 2-dGlc was multiplied by the concentration of glucose present in the medium, the total intake of glucose could be
calculated. A 10-fold reduction in the rate of glucose uptake in the "turned-on"
cultures had little effect on the initiation of DNA synthesis. By contrast the
"turned-off" cultures took up glucose at a much faster rate than the glucose-poor
"turned-on" cultures but synthesized DNA at a much slower rate. This reinforced
the earlier conclusions (Rubin 1971b) that the external supply of glucose is not
the factor restricting the initiation of DNA synthesis in "turned-off" cultures.

Table 2
Amino Acid Requirement for Initiation of DNA Synthesis

Medium	2% Serum	pH	[³H]thymidine (cpm/dish)	Fraction of "turned-on" control
199	+	7.6	8157	1.0
199	−	6.8	1905	0.23
199 minus all amino acids	+	7.6	1540	0.19
199 minus nonessential amino acids	+	7.6	6367	0.79
Earle's saline	−	6.8	1312	0.16
Earle's saline plus essential amino acids	+	7.6	7752	0.95

Three-day-old cultures were "turned off" overnight in serum-free medium at pH 6.8. The medium was replaced with media of the composition indicated above at the appropriate pH. After 7 hr the cells were labeled with [³H]thymidine for one hr and processed for scintillation counting of acid insoluble material.

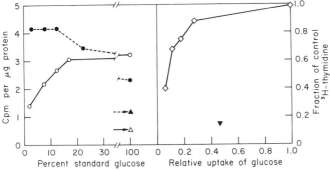

Figure 2
Glucose concentration vs. the initiation of DNA synthesis and uptake of 2-dGlc. Two-day-old cultures were "turned off" overnight as in Fig. 1. The medium was then changed in most of the cultures to 199 pH 7.6 containing no glucose, plus 2% tryptose phosphate broth containing 5.5 mM glucose and 2% dialyzed chicken serum. This basic "turn-on" medium therefore contained glucose 0.1 mM or 2% of the usual glucose concentration. To this medium were added varying concentrations of glucose up to 5.5 mM glucose = 100%. The "turned-off" controls were maintained in 199 (2–0–0) pH 6.8 with 5.5 mM glucose. The rate of 2-dGlc transport into the cells was determined at 3 hr. The rate of DNA synthesis was determined at 7 hr.

	"Turned-on"	"Turned-off"
Left panel		
[³H]2-dGlc	●– – –●	– → ▲
[³H]thymidine	○———○	——→ △
Right panel		
Uptake of glucose vs. rate of DNA synthesis	◇———◇	▼

Metabolic Inhibitors and Initiation of DNA Synthesis

In keeping with the need for essential amino acids in the initiation of DNA synthesis, it was found that inhibition of protein synthesis by cycloheximide blocked the initiation of DNA synthesis (Fig. 3).

Inhibitors of glycolysis regularly and reproducibly inhibited the initiation of DNA synthesis (Table 3), in contrast to the erratic effects of removing glucose from the medium. 2-Deoxyglucose is a glycolytic inhibitor when used in much larger than tracer amounts. The inhibition by 2-dGlc was reversible but not that by NaF and iodoacetic acid. 3-O-Methyl glucose, which is a nonmetabolizable analog of glucose and does not inhibit glycolysis, had no effect on DNA synthesis. 2,4-Dinitrophenol, which uncouples oxidative phosphorylation, also prevented the initiation of DNA synthesis. The effects of all these agents on ongoing DNA synthesis were minimal (Table 4). The inhibitory effects evident at 7 hours after stimulation were still manifest at 16 and 24 hours and did not represent a mere delay in the onset of DNA synthesis.

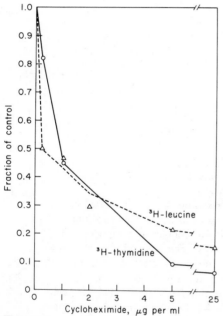

Figure 3
Requirement of protein synthesis for the initiation of DNA synthesis. Two-day-old cultures were turned off overnight as in Fig. 1. Cycloheximide was then added in the indicated concentrations for 30 min. The rate of protein synthesis was determined in half the cultures by labeling from 30–60 min with [³H]leucine in Earle's saline, pH 7.6, plus 2% dialyzed chicken serum. The remaining cultures were changed to 199 (2–0–2) with the appropriate cycloheximide concentrations and the rate of DNA synthesis determined at 17 hr.
o———o [³H]thymidine
△– – –△ [³H]leucine

Table 3
Metabolic Inhibitors vs. Initiation of DNA Synthesis

	Inhibitors (mM)	[³H]thymidine (cpm/µg protein)	Fraction of "turned-on" control
"Turned-off" cells	none	1.1	0.20
"Turned-on" cells	none	5.5	1.0
	2-dGlc 10	0.57	0.10
	NaF 2	6.4	1.2
	5	0.60	0.11
	Iodoacetic acid 0.01	5.0	0.90
	0.02	0.69	0.13
	0.05	0.17*	0.03*
	2,4-DNP 0.1	6.5	1.2
	1.0	1.1	0.2

Four-day-old cultures were "turned off" overnight in serum-free medium at pH 6.8. On all but 2, the medium was changed to 199 (2–0–2) at pH 7.6 and metabolic inhibitors added at the indicated concentration. At 7 hr the cells were labeled with [³H]thymidine and processed for scintillation counting and protein determination.

* Cell sheet severely damaged.

Table 4
Effects of Metabolic Inhibitors on the Initiation
and Continuation of DNA Synthesis

	Inhibitor (mM)	0–7 hours		11.5–18.5 hours	
		[³H]thymidine (cpm/µg protein)	Fraction of "turned-on" control	[³H]thymidine (cpm/µg protein)	Fraction of "turned-on" control
"Turned-off"	none	0.37	0.092	0.80	0.06
"Turned-on"	none	4.0	1.0	13.4	1.0
	2-dGlc 10	0.42	0.10	9.6	0.72
	NaF 5	0.2	0.05	9.5	0.71
	3-MeGlc 10	4.1	1.0	13.5	1.0
	2,4-DNP 1	0.35	0.087	9.0	0.67

Three-day-old cultures were "turned off" in serum-free medium at pH 6.8. They were then "turned on" in 199(2–0–2). Inhibitors were added to some from 0–7 hr and to others from 11.5–18.5 hr after "turn-on." At the end of each period the cells were labeled with [³H]thymidine and processed for scintillation counting and protein determination.

Increase of 2-dGlc Uptake in Turned-on Cells and Response to Metabolic Inhibitors

The rate of 2-dGlc uptake into chick embryo cells increases shortly after stimulating turned-off cultures (Sefton and Rubin 1971; Rubin and Koide 1973b). A strong stimulus utilizing 10% serum was used to accentuate the early rise. Cultures were made in screwcap bottles for better pH control during manipulation. Details of the early rise are shown in Fig. 4. During the first 10 minutes after stimulation there was a 3-fold rise in uptake. Since no attempt was made to label for periods less than 10 minutes, the initial rise could have occurred in a much shorter time. After 10 minutes the increase continued, but at a slower rate.

Cycloheximide has been shown to inhibit the long-term rise in 2-dGlc transport, but there were indications that the early rise was less subject to inhibition (see Fig. 4, Sefton and Rubin 1971). When this matter was reinvestigated under the present conditions of "turn-on," it was found that the early rise occurred independently of protein synthesis (Fig. 5). The failure to interfere with the early rise was a constant feature of all experiments. The form of the rise after the first 30 minutes varied from experiment to experiment. In that of Fig. 5, there was no increase beyond 30 minutes, and the addition of cycloheximide at that time only hastened the decline of 2-dGlc uptake. In other experiments it interfered with increase beyond the first hour.

These results indicated that the increase of sugar transport after "turn-on" was biphasic and suggested the existence of a rapid first stage independent of protein synthesis, followed by a more gradual second stage requiring protein synthesis. In an effort to identify processes other than protein synthesis involved in the rise of 2-dGlc transport, inhibitors of glycolysis were applied and the uptake of [3H]2-dGlc measured at 2 hours. The inhibitors of glycolysis, including high concentrations of

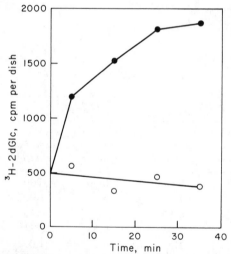

Figure 4
Early kinetics of increase in 2-dGlc transport following "turn-on" of cells. For this experiment only, cultures were made in screwcap plastic bottles rather than petri dishes so the pH could be more carefully controlled in the short time intervals required. After 3 days the cultures were "turned-off" overnight as in Fig. 1. They were "turned-on" by substituting pre-equilibrated 199 (2–0–10) at pH 7.6, allowing minimal exposure of treated and controls to air. [3H]2-dGlc was added in the indicated concentrations for 30 min. The rate of protein synthesis acid soluble material extracted for scintillation counting. This was repeated at 3 more successive 10-min intervals. The points are plotted midway through each interval.
●———● "Turned-on"
○———○ "Turned-off"

2-dGlc itself, not only blocked an increase in 2-dGlc uptake, but forced a decrease in both the "turned-off" and "turned-on" cultures (Fig. 6). Cycloheximide caused only partial inhibition of the increase. Addition of 10 mM glucose to the 5mM glucose already in the medium also partly inhibited the increase in "turned-on" cultures and reduced the level in "turned-off" cultures.

A comparison was made of the concentrations of metabolic inhibitors required to inhibit 2-dGlc uptake and the initiation of DNA synthesis. NaF became an effective inhibitor of both processes at the same concentration (Fig. 7). 2-dGlc and iodoacetic acid at very low concentrations inhibited only the initiation of DNA synthesis, but at higher concentrations inhibited both processes. 2,4-Dinitrophenol inhibited only the initiation of DNA synthesis.

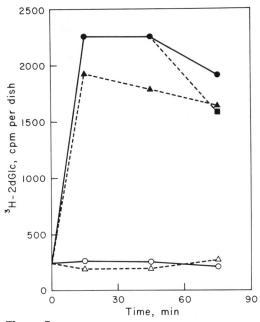

Figure 5

Independence from protein synthesis of the early stage of increase in 2-dGlc transport. Three-day-old cultures were "turned-off" overnight as in Fig. 1. The next morning half the cultures received cycloheximide 20 μg/ml for 30 min. A group of the cycloheximide-treated cultures and a group of the untreated cultures were "turned on" by the addition of chicken serum to a final concentration of 10% and of NaHCO₃ to raise the pH to 7.6. "Turned-on" and "turned-off" cultures, with or without cycloheximide, were labeled with [³H]2-dGlc for 30 min at 30-min intervals. In addition cycloheximide was added to cultures previously without the drug, and the uptake of [³H]2-dGlc examined 30 min later.

	"Turned-on"	"Turned-off"
no cycloheximide	●——●	○——○
cycloheximide at 0 min	▲---▲	△---△
cycloheximide at 60 min	●---■	

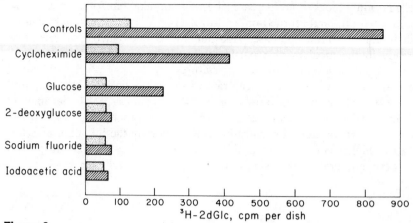

Figure 6

The relation of 2-dGlc transport to glycolysis. Four-day-old cultures were "turned off" overnight as in Fig. 1. The next morning the following materials were added to groups of cultures: cycloheximide 20 μg/ml, glucose 10 mM, 2-dGlc 10 mM, NaF 5 mM, or iodoacetic acid 0.02 mM. After 30 min half the cultures in each group were "turned on" by the addition of 10% serum. Two hr later they were labeled with [³H]2-dGlc.

Shaded bars, "turned-off"

Cross-hatched bars, "turned-on."

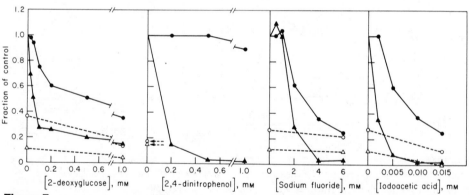

Figure 7

Comparative effects of metabolic inhibitors on 2-dGlc transport and DNA synthesis. Three- and four-day-old cultures were "turned off" overnight as in Fig. 1. The next morning the indicated concentrations of metabolic inhibitors were added to 2 cultures each. After 15 min most of the cultures were "turned on" by the addition of serum to 2% and of NaHCO₃ to pH 7.6. "Turned-off" and "turned-on" cultures with and without metabolic inhibitors were labeled with [³H]2-dGlc at 2 hr or with [³H]thymidine at 7 hr.

	"Turned-on"	"Turned-off"
[³H]2-dGlc	●———●	o-- --o
[³H]thymidine	▲———▲	△-- --△

810

Glycolytic Intermediates after "Turn-on"

The foregoing results raised the possibility that glycolysis is the primary responding process to alterations in serum, pH, and by analogy population density, and that it in turn determines whether DNA synthesis will occur within the experimental period. To determine whether glycolysis does respond to the manipulations used here and, if so, which steps of the pathway are involved, comparison was made of some glycolytic intermediates in "turned-off" and "turned-on" cells at the relatively short interval of 30 minutes after stimulating the former by either raising pH or adding serum.

The data of Tables 5 and 6 show that a crossover point (Chance et al. 1958) occurs at the level of phosphofructokinase, since the concentrations of the substrate, fructose-6-phosphate, decrease in the stimulated cultures and the concentrations of the product, fructose-1,6-diphosphate, increase. Calculations of the mass action ratios for phosphofructokinase indicate the occurrence of a 5- to 8-fold increase in activity in cells stimulated by raising pH or adding serum. Experiments with chick embryo cells transformed by the Schmidt-Ruppin strain of Rous sarcoma virus reveal a 5.5-fold activation of phosphofructokinase at 22 hours after infection and a 25-fold activation at 36 hours (Fig. 8). There was no significant increase in DNA synthesis until 36 hours, so the activation of phosphofructokinase clearly preceded that of DNA synthesis. By 44 hours the glucose was depleted from the medium of the infected cultures and both phosphofructokinase activity and DNA synthesis had decreased to control levels.

Table 5

Effect of pH on Glycolytic Intermediates

	$m\mu moles/mg\ protein$		
	pH 6.7	*pH 7.5*	*pH 7.8*
Glucose-6-phosphate	1.88	1.18	0.98
Fructose-6-phosphate	0.38	0.27	0.22
Fructose-1,6-diphosphate	0.50	1.07	0.905
Dihydroxyacetone phosphate	1.89	1.37	2.16
Glyceraldehyde-3-phosphate	2.03	1.47	1.22
ATP	18.71	18.88	17.52
ADP	1.53	1.95	2.22
AMP	0.88	1.19	0.50
Phosphofructokinase activity (mass action ratio) $\dfrac{(fructose\text{-}1,6\text{-}diphosphate)(ADP)}{(fructose\text{-}6\text{-}phosphate)(ATP)}$	0.107	0.40	0.50

Four-day-old cultures in 100-mm petri dishes were "turned off" overnight in serum-free medium at pH 6.7. Varying amounts of $NaHCO_3$ were then added to achieve the indicated pH. After 30 min the cells were extracted and processed to determine the concentrations of glycolytic intermediates.

Table 6
Effect of Serum Concentration on Glycolytic Intermediates

	Serum			
	0%	*0.5%*	*2.0%*	*10.0%*
		(mμ moles/mg protein)		
Glucose-6-phosphate	0.76	0.88	0.93	0.785
Fructose-6-phosphate	0.66	0.635	0.40	0.35
Fructose-1,6-diphosphate	0.105	0.18	0.32	0.26
Dihydroxyacetone phosphate	0.26	0.24	0.125	0.105
Glyceraldehyde-3-phosphate	0.105	0.18	0.105	0.18
ATP	13.275	16.67	19.60	15.42
ADP	1.74	2.25	2.55	3.24
AMP	0.87	0.90	0.86	0.79
Phosphofructokinase activity (mass action ratio)	0.021	0.038	0.10	0.16

Four-day-old cultures in 100-mm petri dishes were "turned off" overnight in serum-free medium at pH 6.8. The indicated serum concentrations were added and glycolytic intermediates measured at 30 min.

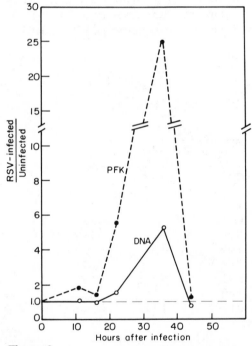

Figure 8
The kinetics of activation of the phosphofructokinase reaction and of DNA synthesis after infection with Rous sarcoma virus. Cultures were seeded in 100-mm plastic petri dishes and half of them infected with the Schmidt-Ruppin strain of Rous sarcoma virus, 8×10^5 focus-forming units. At the indicated times cultures were extracted for analysis of glycolytic intermediates and others exposed to [³H]thymidine to estimate the rate of DNA synthesis.

   ●– – –● Phosphofructokinase activity (mass action ratio)

   o———o [³H]thymidine.

DISCUSSION

The only major low molecular weight constituents of the medium regularly found to be required for the initiation of DNA synthesis are the essential amino acids. Zinc is also required, but it is present in very low concentrations, is not deliberately added to the medium, and must be complexed by chelating agents to demonstrate the consequences of its absence (Rubin 1972). The requirement for amino acids appears to contradict an earlier claim that organic medium constituents were not required to initiate DNA synthesis (Temin 1971). The cells used in the earlier experiments had been preincubated in a medium reinforced with high amino acid concentrations and might therefore have accumulated large reserve pools. Amino acids, though essential to DNA synthesis, do not appear to be involved in regulating its initiation. The rate of uptake of amino acids in density-inhibited cultures is at least half of that in rapidly growing cultures. Yet a deliberate fivefold reduction in concentrations of all amino acids has no effect on DNA synthesis in rapidly growing chick embryo cells, and the addition of a fivefold excess to density-inhibited cultures likewise has no effect.

Taken together, these observations indicate that the supply of amino acids is not the limiting factor in determining density-dependent inhibition of growth. The role of the essential amino acids in the initiation of DNA synthesis is probably to support protein synthesis. The requirement for protein synthesis is demonstrated by the effectiveness of relatively low concentrations of cycloheximide in blocking initiation of DNA synthesis.

Protein synthesis is not required for the early stages of stimulation of 2-dGlc uptake but is required for the later stages of increase. This finding modifies the earlier conclusion that protein synthesis is required for all stages of the increase (Sefton and Rubin 1971). The stimulus in the previous work consisted of adding serum to density-inhibited cultures. The initial rise in 2-dGlc uptake was not as sharp as that observed under the conditions of the present experiments. Even so, it is retrospectively apparent that some early rise did occur despite the presence of cycloheximide (see Fig. 4, Sefton and Rubin 1971). The protein synthesis required for the later stages of increase was thought to involve an increase in the transport machinery itself. The finding that sugar transport responds to changes in rate of glycolysis, and that the latter is controlled by phosphofructokinase, raises the possibility that the requisite protein synthesis involves the key enzymes of the glycolytic pathway. These have been shown in liver to turn over rapidly and to reflect in their activity the growth rate of the cells from which they arise (Weber and Lea 1965).

No short-term requirement for external glucose could be reproducibly demonstrated. Frequently external glucose could be entirely omitted from the medium without affecting the initiation of DNA synthesis. However it was shown by the use of glycolytic inhibitors that glucose catabolism is essential to the initiation of DNA synthesis. The explanation for this paradox probably lies in the fluctuating stores of glycogen present in cultured cells (Wu 1959). Glycogen would, of course, provide an alternate source of substrate for glycolysis, independent of the external supply over a short period of time. At the present time we are ignorant of the factors that regulate the intracellular energy reserves in cultured chick embryo cells.

Glycolysis was shown to be intimately related to the initiation of DNA synthesis and to the transport of 2-dGlc. All three inhibitors of glycolysis—2-dGlc, NaF and iodoacetic acid—completely blocked the initiation of DNA synthesis but had little or no effect on continuing DNA synthesis. 2,4-Dinitrophenol, an agent which uncouples oxidative phosphorylation, had similar effects on DNA synthesis. Although much of the glucose taken up by cultured chick embryo cells is utilized in the production of lactic acid (Bissell, Hatié and Rubin 1972), as much as half the ATP made by the cells still derives from oxidative phosphorylation. It is possible, therefore, that the common denominator for the effects of both types of inhibitors is a reduction in the availability of ATP to carry on the biosynthetic activities of the cell.

Glycolytic inhibitors were the only agents tested which were capable of blocking the early increase in 2-dGlc transport following stimulation of the cells. Each of the inhibitors was effective, although each acts on a different step in the glycolytic pathway. These results, combined with the demonstration that glycolysis is quickly activated by the applied growth stimulus, suggest that the transport of 2-dGlc merely reflects the activity of the glycolytic pathway. Thus the uptake of glucose would be dependent on the rate of its catabolism rather than the other way round.

The minimum dose of the glycolytic inhibitors required to inhibit the initiation of DNA synthesis was of the same order (in the case of NaF) or slightly less (in the cases of 2-dGlc and iodoacetic acid) than that required to inhibit the increased transport of 2-dGlc. If the 2-dGlc transport is an accurate, if somewhat insensitive, indicator of the rate of glucose catabolism, then the initiation of DNA synthesis would appear to be sensitive to small changes in that catabolism. This raises the possibility that glucose catabolism is the ultimate regulatory process for the initiation of DNA synthesis. If so, our results indicate that the enzyme phosphofructokinase* plays a key role in the regulation of DNA synthesis, as it has long been known to do in glycolysis (Lehninger 1970). This enzyme is responsive to a variety of effectors. Perhaps most significantly, it is extremely sensitive to pH (Trivedi and Danforth 1966). Increases of as little as 0.1 to 0.2 pH units can convert the enzyme in vitro from the inactive state to a fully active state. Changes in external pH have similar effects on the activity of the intracellular enzyme of intact red blood cells (Minakami et al. 1964), white blood cells (Halperin et al. 1969) and ascites tumor cells (Wilhelm, Schulz and Hofmann 1971), as well as in chick embryo fibroblasts. The fact that DNA synthesis is similarly correlated with external pH suggests that internal pH may be the signal that regulates the activity of glycolysis and DNA synthesis even in those cases where external pH is unaltered. For example the addition of serum to density-inhibited cells causes a partial depolarization of the cells (Hülser and Frank 1971). Such depolarization would tend to raise the internal pH of the cell (Carter et al. 1967). A small rise in pH might be adequate to trigger the sequence of events which results in the initiation of DNA synthesis.

The rise in activity of phosphofructokinase was particularly dramatic after infection with Rous sarcoma virus. As in the case of the serum and pH-stimulated normal cells, the activation of phosphofructokinase took place hours before an increase in DNA synthesis became detectable. The activation of phosphofructokin-

---

* Others using uniformly labeled glucose have found that the ratio of radioactivity $\frac{\text{fructose-1,6-diphosphate}}{\text{fructose-6-phosphate}}$ increases about 2-fold in chick embryo cells 16 hours after addition of serum and after viral transformation and have suggested that changes in glucose metabolism might be due to activation of phosphofructokinase (M. Bissell, C. Hatié, J. Sudman, and J. Bassham, personal communication).

ase could provide the mechanism for the increased glycolysis observed in tumors (Warburg 1956) and correlated with the growth rates of those tumors (Burk, Woods and Hunter 1967). Of course it leaves unanswered what is the cause of this activation in tumors, but since a number of effectors of the enzyme are known (Lehninger 1970; Trivedi and Danforth 1966), some direction is provided for the search.

Acknowledgments

We gratefully acknowledge the technical work of Toshiko Koide and Cynthia McCreary. This investigation was supported by Public Health Service research grant CA 05619 from the National Cancer Institute.

REFERENCES

Bissell, M., C. Hatié and H. Rubin. 1972. Patterns of glucose metabolism in normal and virus-transformed chick cells in tissue culture. *J. Nat. Cancer Inst.* **49:**555.

Burk, D., M. Woods and J. Hunter. 1967. On the significance of glucolysis for cancer growth with special reference to Morris rat hepatoma. *J. Nat. Cancer Inst.* **38:**839.

Carter, N., F. Rector, D. Campion and D. Seldin. 1967. Measurement of intracellular pH of skeletal muscle with pH-sensitive glass microelectrodes. *J. Clin. Invest.* **46:**920.

Chance, B., W. Holmes, J. Higgins and C. Connelly. 1958. Localization of interaction sites in multi-component systems. Theorems derived from analogues. *Nature* **182:**1190.

Halperin, M., H. Connors, A. Relman and M. Karnovsky. 1969. Factors that control the effect of pH on glycolysis in leukocytes. *J. Biol. Chem.* **244:**384.

Hülser, D. and W. Frank. 1971. Stimulation of embryonic rat cells by a protein fraction isolated from fetal calf serum. I. Electrophysiological measurements at the cell surface membrane. *Z. Naturforschung.* **26b:**1045.

Lehninger, A. 1970. *Biochemistry.* Worth Publishers, New York.

Martin, G., S. Venuta, M. Weber and H. Rubin. 1971. Temperature-dependent alterations in sugar transport in cells infected by a temperature sensitive mutant of Rous sarcoma virus. *Proc. Nat. Acad. Sci.* **68:**2739.

Minakami, S. and H. Yoshikawa. 1965. Studies on erythrocyte glycolysis. I. Determination of glycolytic intermediates in human erythrocytes. *J. Biochem.* **58:**543.

Minakami, S., T. Saito, C. Suzuki and H. Yoshikawa. 1964. The hydrogen ion concentrations and erythrocyte glycolysis. *Biochem. Biophys. Res. Commun.* **17:**748.

Rein, A. and H. Rubin. 1968. Effects of local cell concentrations upon the growth of chick embryo cells in tissue culture. *Exp. Cell Res.* **49:**666.

Rubin, H. 1971a. pH and population density in the regulation of animal cell multiplication. *J. Cell Biol.* **51:**686.

———. 1971b. Growth regulation in cultures of chick embryo fibroblasts. *Ciba Fndn. Symp.* on growth control in cell cultures (ed. G. E. W. Wolstenholme and J. Knight) pp. 127–149.

———. 1972. Inhibition of DNA synthesis in animal cells by ethylene diamine tetra-acetate and its reversal by zinc. *Proc. Nat. Acad. Sci.* **69:**712.

Rubin, H. and T. Koide. 1973a. Inhibition of DNA synthesis in chick embryo cultures by deprivation of either serum or zinc. *J. Cell Biol.* **56:**777.

———. 1973b. Stimulation of DNA synthesis and 2-deoxy-D-glucose transport in chick embryo cultures by excessive metal concentrations and by a carcinogenic hydrocarbon. *J. Cell. Physiol.* **81:**387.

Sefton, B. and H. Rubin. 1971. Stimulation of glucose transport in cultures of density-inhibited chick embryo cells. *Proc. Nat. Acad. Sci.* **68:**3154.

Temin, H. 1971. Stimulation by serum of multiplication of stationary chicken cells. *J. Cell. Physiol.* **78:**161.

Trivedi, B. and W. Danforth. 1966. Effect of pH on the kinetics of frog muscle phosphofructokinase. *J. Biol. Chem.* **241:**4110.

Warburg, O. 1956. On the origin of cancer cells. *Science* **123:**309.

Weber, G. and M. Lea. 1965. The molecular correlation concept of neoplasia. *Advanc. Enzyme Regulation* **4:**115.

Wilhelm, G., J. Schulz and E. Hofmann. 1971. pH-dependence of aerobic glycolysis in Ehrlich ascites tumor cells. *FEBS Letters* **17:**158.

Wu, R. 1959. Regulatory mechanisms in carbohydrate metabolism. V. Limiting factors of glycolysis in HeLa cells. *J. Biol. Chem.* **234:**2806.

Intermittent Amplification and Catabolism of DNA and Its Correlation with Gene Expression

Robert R. Klevecz, Leon N. Kapp, and John A. Remington*

Department of Cell Biology, City of Hope National Medical Center
Duarte, California 91010

Differential replication (amplification) of part of the genetic complement has been observed in certain cells of higher organisms. One well-known instance is the selective amplification of euchromatic sequences of *Drosophila* chromosomes during polytenization (Gall, Cohen and Polan 1971; Dickson, Boyd and Laird 1971). The most well-documented example is amplification of ribosomal DNA (rDNA) sequences occurring in amphibian (Brown and Dawid 1968; Gall 1968) and insect (Gall 1969) oocytes. In amphibians the amplification occurs during meiotic prophase (Gall 1968) by a chromosome copy mechanism (Brown and Blackner 1972), which may involve a ribosomal RNA-DNA complex (Crippa and Tocchini-Valentini 1971; Brown and Tocchini-Valentini 1972; Mahdavi and Crippa 1972; Bird, Rogers and Birnstiel 1973). About 450 reiterated copies of rDNA in a haploid set of chromosomes are amplified approximately 4000 times in the mature oocyte (Brown and Dawid 1968) and are localized extrachromosomally as large circles in multiple nucleoli (Miller 1966; Miller and Beatty 1969). On the other hand, there is little evidence that suggests the amplification of specific gene sequences for specialized proteins in cells of higher organisms. The data of both Bishop, Pemberton and Baglioni (1972) and Harrison et al. (1972) indicate little, if any, reiteration or amplification of hemoglobin DNA in immature red blood cells, and that of Suzuki, Gage and Brown (1972) indicate no persistent amplification of silk fibroin genes in *Bombyx mori*.

A number of other reports also exist which show that replication of rDNA can be divorced from the normal controls of nuclear DNA synthesis. These include findings of (1) preferential rDNA replication during nutritional shift-up in *Tetrahymena* (Engberg et al. 1972), (2) disproportionate rDNA replication in the bobbed phenotype of *Drosophila* (Ritossa 1968), (3) nucleolar DNA replication during the G_2 phase in *Physarum* (Holt and Gurney 1969; Braun and Evans 1969; Guttes and Guttes 1969), (4) differential GC-rich DNA replication in lactating mammary glands (Banerjee and Wagner 1972), and (5) transient

* Present address: Department of Biology, Rensselaer Polytechnic Institute, Troy, New York 12181.

amplification of rDNA during the G_1 phase in *Chlamydomonas* (Howell 1972). It is also interesting that the replication of "DNA puffs" of *Sciaridae,* which are not of nucleolar origin (Meneghini et al. 1971), appears to be under a separable control.

It is significant that in *Chlamydomonas* Howell (1972) has found rapid rDNA degradation occurring in conjunction with differential amplification during the same vegetative cell cycle. In this actively dividing cell the amount of rDNA is determined both by differential synthesis and degradation. Recently Ammermann (1969) and Prescott, Murti and Bostock (1973) have observed differential DNA degradation following polyploidization of non-nucleolar DNA during a single cell cycle in the macronucleus of ciliated protozoa. In addition Pelc (1970) has presented evidence that in a variety of cell types some DNA may be labile and catabolized.

We report here the existence of differential DNA replication and degradation occurring at least once during the cell cycle of aneuploid Chinese hamster fibroblasts. The amplification and rapid degradation of DNA can be readily observed spectrofluorometrically in the cell cycle prior to the replication of bulk DNA. Inhibitor studies indicate that the control of the transient amplification is separable from normal replication for cell division. Coincident with this transient amplification are significant increases in total cellular RNA and protein and increases in the activity of a number of enzymes. These phenomena are discussed in relation to a possible transient amplification of ribosomal and repetitious DNA sequences and to cellular timekeeping mechanisms in differentiating cells.

MATERIALS AND METHODS

Cell Lines

The maintenance and growth characteristics of the cell lines V79 (Klevecz 1972), CHO (Remington and Klevecz 1973), heteroploid Don (Forrest and Klevecz 1972) and WI-38 (Klevecz and Kapp 1973) that were used in this study have been described previously.

Materials

McCoy's 5a medium (Gibco #233–0) was supplemented with 20% fetal calf serum and 100 mg/liter neomycin sulfate. Tritiated thymidine ([³H]dT; New England Nuclear, 6.7 Ci/mmole, 50.4 Ci/mmole and 50.8 Ci/mmole) was prepared in McCoy's 5a medium immediately prior to use. 3,5-Diaminobenzoic acid dihydrochloride (DABA; Aldrich Chemical Co. #11,383–2) was made up in distilled water at 0.4 g/ml and decolorized with charcoal prior to use. Hydroxyurea (HU; Sigma #H–8627) was prepared at 50 mM in McCoy's 5a medium. Cytosine-β-D–arabinofuranoside (ara-C; Calbiochem #251010) was prepared at 250 mg/ml in Hank's balanced salt solution. Actinomycin D (AMD; Merck, Sharp and Dohme) was prepared from powder immediately prior to use at 5 μg/ml in McCoy's 5a medium. Ribonuclease A 2.7.7.16 (RNase A; Worthington Biochemical Corporation #RAF-21A) was prepared at 1 mg/ml in 0.1 M acetate buffer pH 5.1, boiled for 10 minutes prior to use and stored at 4°C. Deoxyribonuclease I 3.1.4.5 (DNase I; Worthington Biochemical Corporation #D) was prepared at 100 μg/ml in 0.1 M acetate buffer with 5 mM $MgSO_4$ pH 5.0 and stored at 4°C.

Synchrony

In the work reported here synchronous cell cultures were established by repeated cycles of mitotic cell detachment from exponential monolayer cultures growing in roller bottles, using the automated cell cycle analyzer described previously (Klevecz 1972; Klevecz and Kapp 1973). Each collection of mitotic cells, detached at hourly or half-hourly intervals, was dispensed into 5 to 15 sterile scintillation vials and allowed to reattach to the surface of the vials. Except where noted, synchronous cultures arrayed through the cell cycle were collected for analysis at a single point in time.

Preparation of Monolayers for Spectrofluorometric DNA Determination

After exposure to radioisotopes or in preparation for chemical analysis, individual vials were washed with 5 ml Hank's balanced salt solution, extracted with two 5-ml washes of 10% trichloroacetic acid (TCA) and one 5-ml wash of 10% potassium acetate in 80% ethanol, all at 4°C. This procedure was accomplished in less than 5 minutes. The monolayers were then extracted once with 10 ml 2:1 ethanol:ether and once with 5 ml ether at room temperature. In some instances the monolayers were then extracted with hot 95% ethanol for 30 minutes.

Spectrofluorometry

Fluorescence was measured in an Aminco-Bowman spectrofluorometer by excitation at 420 nm and emission at 520 (uncorrected). Standard curves using purified p*lac* DNA gave G_1 DNA values of 6.98 pg/cell for the WI-38 cells used in this study. The spectrofluorometer was set to zero using 1 N HCl. Extrapolated DNA standard curves, DNase I digested cell monolayers and DABA alone all gave similar blank readings. DABA (0.5 ml) at a concentration of 0.4 g/ml was added to each vial and reacted for 45 minutes at 60°C. Samples were immediately diluted with 2 ml 1 N HCl. The method was originally described by Kissane and Robins (1958) and in a modified form by Hinegardner (1971).

RESULTS

Synchronous V79 cells growing as monolayers in scintillation vials were pulse labeled with tritiated thymidine for 30 minutes and changes in the rate of incorporation into acid-precipitable counts were determined. At the top of Fig. 1a the overall rate of incorporation through one cell cycle is shown as an average of three experiments, together with the data from a single synchrony experiment in which 30-minute pulses of tritiated thymidine were given at 30-minute intervals.

We had previously noted in this cell line some hint of changes in rate of DNA synthesis through S phase, but these fluctuations were never as apparent or easily detected as they were in diploid fibroblasts. We had also noted that decreasing the duration of the pulse label did not increase the resolution of these bursts in synthesis. If cells were pulse labeled for 2 minutes and immediately fixed and assayed for incorporation into acid-precipitable material (Fig. 1b), the apparent rate of DNA synthesis did not increase through S phase, in contrast to what was found for 30-minute incorporation intervals. That this paradox was a consequence of changes in pool size or in the rate of thymidine transport into the cells was substantiated by

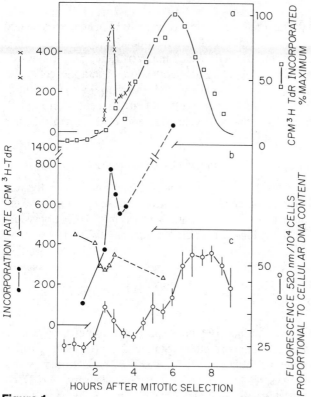

Figure 1

Intermittent DNA synthesis and thymidine incorporation in synchronous V79 Chinese hamster cells.

(a) Mitotic cells were detached at 0.5-hr or 1-hr intervals and dispensed into scintillation vials at concentrations of 2.5 × 10⁴ to 1 × 10⁵ cells/vial. The solid line (——) is the average rate of [³H]dT incorporation from a 30-min pulse with 5 μCi/ml (s.a. 6.7 Ci/mmole) from three independent experiments in which three samples each were taken at hourly intervals. The squares (□) indicate the 30-min pulse rate of incorporation in vials replicate with the experiment shown in c. One set of vials (X) was labeled for 2 min at 2.25 hr after mitotic selection with 100 μCi/ml [³H]dT (s.a. 50.8 Ci/mole). Vials were chased with warm conditioned medium containing 50 μg/ml dT, harvested and quickly fixed in TCA at 0, 1, 5, 10, 15, 20, 40, 60 and 90 min after the chase.

(b) Mitotic cells detached and dispensed as in *a* were pulse labeled for 2 min (△) with 100 μCi/ml [³H]dT (s.a. 50.8 Ci/mmole) at 1, 2, 2.25, 2.5, 2.75, 3 and 5.5 hr after mitotic selection and chased for 30 min each (●) with warm conditioned medium containing 5 μg/ml dT.

(c) Mitotic cells were detached and allowed to progress through the cell cycle. At 9 hr after the first mitotic selection all cultures were harvested and treated as described in the Methods section. The fluorescence/10⁴ cells from four replicate scintillation vials at 0.5-hr intervals through the cell cycle is shown (○) together with the standard error of the mean.

following each 2-minute pulse with a 30-minute chase using 5 μg/ml of unlabeled thymidine (Fig. 1b). Under these circumstances the pattern of incorporation reflected more exactly the changes in DNA synthesis. At the bottom of Fig. 1c the DNA content per cell is shown. It seems clear that the pulse rate of thymidine incorporation in this cell line did not accurately reflect changes in the rate of accumulation of DNA in the cells. DNA content in V79 increases abruptly at the beginning of S phase simultaneously with the increase in the percent of labeled nuclei determined radioautographically. After this brief burst in synthesis there was a decrease in DNA content over the next 2 hours so that the DNA content of cells which are 4 hours into the cycle, and by radioautographic measures 2.5 hours into S phase, was not greatly different from the G_1 DNA value. Thereafter there was a rapid increase in the amount of DNA over the next 3 hours, reaching G_2 values 7 hours after mitosis. The interval of maximum increase in DNA content occurred at the same time as the maximum rate of thymidine incorporation (Fig. 1a).

Further support for the idea that the early replicating DNA was subsequently catabolized was provided by pulse labeling a series of cultures, all of which were 2.5 hours into the cell cycle, for 2 minutes. These cells were chased for intervals of 1 to 90 minutes and then fixed. During this time the amount of thymidine in DNA increased to a maximum, coincident with maximum DNA synthesis, and then fell precipitously back to a low value. In this case the change in tritiated thymidine content in the cells paralleled closely the change in DNA content. When this same protocol was applied to cells 6 hours after mitotic selection or to random exponential cultures, a continuous increase in counts to a plateau was observed during the chase period. When random exponential cultures were pulse labeled and radioautographic grains counted over the ensuing labeled metaphase cells, there was a shift in the maximum grain density distribution from early to late labeling portions of chromosomes in the second cell cycle (Remington and Klevecz 1973). This shift is also evident in published works from other laboratories (Spry 1971) although it is not always commented upon. These results can best be explained as being a consequence of DNA labeling followed by catabolism and reutilization.

The Kissane and Robins technique for measuring DNA content per cell is a modification of the Doebner-Miller quinaldine synthesis and involves initially the formation of apurinic acid and subsequently the reaction of DABA with the aldol condensation product of 2-deoxypentose (Kissane and Robins 1958; Karrer 1950). Synchronous V79 cells (Fig. 2) were fixed and the DNA content per cell assayed after a variety of procedures. Ribonuclease A digestion did not alter the DABA fluorescence. Hydrolysis of the cells with ammonium hydroxide following fixation similarly had no effect on the fluorescence pattern. Extraction in hot ethanol for 1 hour appeared to increase the fluorescent maxima in the cell cycle, suggesting possible quenching by some hot ethanol-soluble material. However the overall pattern was not different from the untreated samples. A similar extraction of random cultures with hot ethanol resulted in a 10% increase in fluorescence. Hydrolysis of the samples in hot 10% TCA for 15 minutes at 90°C reduced the fluorescence to 15–20% of the original value, and hydrolysis for 1 hour reduced fluorescence to 2–5% of the control. Digestion with 1 μg of DNase I for 5 minutes eliminated 50–70% of the fluorescence. Digestion with DNase I at a level of 100 μg for 1 hour removed essentially all the material which reacts with DABA and the maximum remaining fluorescence was less than 2% of the control. If the reaction was run for the standard 45-minute interval and the fluorescent product removed and replaced

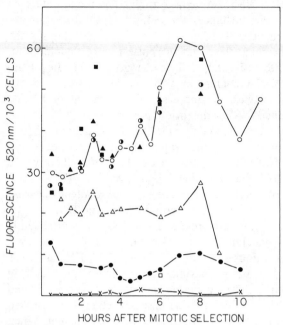

Figure 2

Specificity of DABA for DNA of fixed cell monolayers. Mitotic cells were selected and grown as described in Fig. 1c. Harvested cells were treated as described in the Methods section. Following fixation one set of cultures (o) was reacted with DABA as described in Fig. 1c. Additional sets were treated in the following ways:

(▲) RNase A, 25 μg/vial, 45 min, 37°C

(△) DNase I, 1 μg/vial, 30 min, 37°C

(X) DNase I, 100 μg/vial, 1 hr, 37°C

(•) 10% TCA, 15 min, 90°C

(■) 80% ethanol, 45 min, 60°C.

In each of the above cases the solutions were removed following incubation and the extraction procedure repeated.

(o) 1 N NH_4OH was added to cultures and hydrolysis performed at 60°C for 1 hr. This solution was then evaporated to dryness and the DABA solution added directly.

(□) The fluorescent product from the 6-hr control sample (o) was removed and read, fresh DABA was added to the vial and reincubated at 60°C for 45 min.

with fresh DABA, no more than 5% additional fluorescence was observed. DABA fluorescence is proportional to DNA content and is not reactive with any non-DNA material left in the cells after fixation.

Ara-C is reported to be phosphorylated by cells (Schrecker and Urshel 1968) and the phosphorylated ara-CTP to specifically inhibit the function of polymerase II in *E. coli* (Rama-Reddy, Goulian and Hendler 1971). Synchronous V79 cells were treated with 0.1 mM ara-C immediately after mitotic selection. Another replicate series of cultures was treated with ara-C 2.5 hours after mitotic selection, after the first burst in DNA synthesis. In cultures treated immediately upon selection, the first peak of synthesis of DNA was enhanced over the controls, but overall these cultures failed to synthesize the G_2 amount of DNA. A greater effect was achieved by ara-C when added at 2.5 hours, where there appeared to be several waves of synthesis and degradation but no net increase in DNA content over the G_1 value. Since ara-C is deaminated by cells rather rapidly to the inactive uracil arabinonucleoside (Camiener and Smith 1965), continual resupplementation of the medium with ara-C at 2-hour intervals through the cycle was carried out. Under these circumstances the fluctuations in DNA content were not eliminated, but the net increase in DNA was completely suppressed and the final 9-hour DNA content was not measurably greater than the G_1 value, suggesting that net synthesis or replication for division and the transient increases in DNA content result from two independent processes. Hydroxyurea, which may function to inhibit polymerase by increasing the degree of DNA methylation (Kappler 1970) and has also been reported to inhibit the activity of ribonucleoside reductase (Moore 1969), shows an effect similar to that of ara-C when added at 1 or 5 mM, although in some experiments the intermittent bursts in DNA synthesis were more effectively suppressed by hydroxyurea than by ara-C (Fig. 3). Complete obliteration of intermittent synthesis was accomplished by adding 5 μg/ml of actinomycin D and 5 mM hydroxyurea to the cells immediately following mitotic selection. In this case a constant DNA content per cell, slightly below what we assume to be the G_1 value, was observed throughout the cell cycle. Actinomycin D exerts its primary action by inhibiting RNA polymerase (Reich et al. 1962) and is known to inhibit DNA synthesis as well. However addition of actinomycin D after the initial burst in synthesis did not completely suppress the synthesis of bulk DNA. Several other observations were made incidental to the above. Addition of 3-formyl-rifamycin-sv o-octyl-oxime, at a concentration of 100 μg/ml, had no effect on DNA synthesis.

Bursts in DNA content such as observed here might be expected to have some consequence in gene expression. To observe this the total cellular content of RNA, protein, and the level of several enzymes were assayed (Fig. 4). The first peak in enzyme activity occurs ½ to 1 hour after the initial burst in DNA synthesis as does the increase in total RNA content and protein content per cell. The changes in RNA content must almost certainly be primarily a consequence of changes in the amount of ribosomal RNA per cell. If [³H]deoxyguanosine was used instead of [³H]thymidine to label DNA, it showed relatively 2.5–3 times greater incorporation into the early replicating DNA peak, indicating replication of a GC-rich DNA. Whether or not the increase in RNA content includes other species besides ribosomal RNA is unknown, but it is not unreasonable to think this is so. It has been our speculation that transcription and translation were closely coupled to DNA replication and that peaks in the activity or synthesis of many enzymes were a consequence of an entrainment of enzyme synthesis to DNA replication (Klevecz 1969b). Not all enzymes examined showed maxima at the same time as hexokinase. It is interesting to note that similar enzyme maxima are displayed by a number of glycolytic enzymes. In contrast serine dehydratase, which achieves its maximum level in mid-S phase in CHO (Kapp, Remington and Klevecz 1973) and in V79, can only be induced by treatment with cyclic AMP in mid-S phase. It may be that clusters of genes related by function are replicated and processed intermittently.

Finally the transient amplification and catabolism of DNA is not exhibited exclusively by V79 but also by a number of other cell lines, including CHO and WI-38. In WI-38 (Fig. 5a) fluorescence increases abruptly between 4 and 6 hours

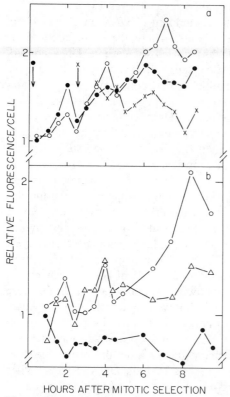

Figure 3

Partial inhibition of DNA synthesis by hydroxyurea or cytosine arabinofurano-side and complete inhibition by hydroxyurea plus actinomycin D.

(a) Following mitotic selection at intervals of 0.5 hr, four replicate vials of V79 cells at a mean concentration of 2.5×10^4 cells/vial were treated at zero time with 0.1 mM of ara-C (●). A second replicate series was treated with 0.1 mM ara-C after 2.5 hr of growth (X). The third set was left untreated (o). Cultures were harvested and analyzed for DNA content as described in the Methods section.

(b) Following mitotic selection at 1 hr and 0.5-hr intervals, four replicate vials of V79 cells at a mean concentration of 2.5×10^4 cells/vial were treated at zero time with hydroxyurea at a concentration of 5 mM (△). A second replicate series was treated with 5 mM hydroxyurea and 5 μg/ml actinomycin D (●) over the same intervals. A third series was left untreated (o).

at the same time as total incorporation from pulse labeling shows a peak. Thereafter there appears to be a plateau period with no further increase in fluorescence until 10 hours into the cell cycle.

Most interesting is the fact that the increase in fluorescence that occurs at 5 hours is almost completely lost between hours 6 and 9. This has been repeatedly observed in experiments with WI-38 and with other cell types, although less clearly. The amount of temporal structure resolvable in S phase varies from cell to cell and is most evident in the human diploid fibroblast WI-38. This is in agreement with our earlier observation in which we found that diploid hamster fibroblasts showed

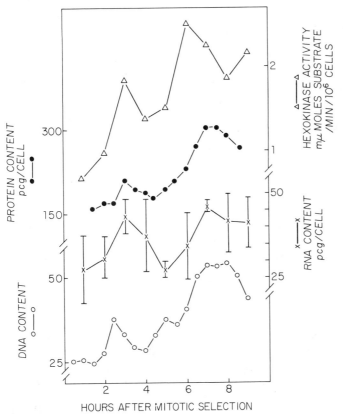

Figure 4

Coordination of RNA, protein and enzyme synthesis with DNA synthesis. The DNA curve of Fig. 1 has been replotted here for comparative purposes.

Following fixation samples (5.8 × 10^5 cells/vial) were hydrolyzed in 0.3 N KOH at 37°C for 6 hr. After neutralization with perchlorate, the amount of RNA was determined by the orcinol reaction (Dische 1955) and assayed in microcuvettes with a Gilford spectrophotometer. Standard errors of the mean from two independent experiments having three samples per time point are shown. Protein content/cell was determined using Folin-phenol reagent according to the method of Lowry et al. (1951). Hexokinase activity was assayed in sonicated cells as described (Klevecz 1972). Except where noted, standard errors do not exceed ± 6%.

multiple maxima in thymidine incorporation but transformed heteroploid cells did not (Klevecz 1969a). Patterns of tritiated thymidine incorporation and DNA content per cell for the heteroploid Don line are shown in Fig. 5b. It is apparent that no significant temporal structure can be detected here. This same uniform increase in DNA content is true of heteroploid V79 cells. This must be a population effect resulting from variations in cell cycle substages among the heterogeneous members of a population, since a uniform aneuploid cell line V79-9 (Fig. 5d), which was cloned from the heteroploid line, displays a pattern of synthesis rather similar to that of the diploid fibroblast WI-38 except that the increases in amount of DNA

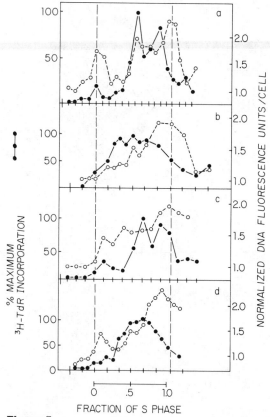

Figure 5

Comparison of DNA replication kinetics in normal, heteroploid and aneuploid cells by [³H]dT incorporation rate and DABA-DNA content. Thymidine incorporation into DNA (●) was determined following a 30-minute pulse at 0.5- or 1.0-hr intervals with 5 μCi/ml (6.7 Ci/mmole) [³H]dT. Values are the average of three replicate samples at each time point and are plotted as a percent of the maximum incorporation rate.

   (a) WI-38 diploid human fibroblasts at passage 27.
   (b) Don heteroploid Chinese hamster fibroblasts.
   (c) CHO established aneuploid Chinese hamster cells.
   (d) V79 cloned aneuploid established Chinese hamster cell line.

   DABA-DNA content per cell (o– – –o) was determined as described in Methods. Fluorescence has been normalized to a value of 1.0 in G_1 for comparative purposes. A value of 1.0 in (a) indicates a DNA content of 6.98 pcg/cell for WI-38. S phase has been normalized to make the change in temporal structure between different cell types more clear. Bars on the abscissa indicate hours after mitotic selection.

following the first burst are not nearly so distinct. Intermittent DNA synthesis is characteristic of normal diploid cells. Moreover there is a change in the overall pattern of replication in established cell lines so that the sharp delineations of early and late replicating DNA are not so apparent. Finally it seems that the Don heteroploid cell line is such a mixed population that any temporal structure in individual cells is obscured.

DISCUSSION

In mammalian cells the rate of DNA synthesis is not constant through S phase, and DNA content does not increase as a simple linear or exponential function of time. Changes in DNA content and rate of synthesis are revealed both by tritiated thymidine incorporation and spectrofluorometric measures, although possible fluctuations in pool size and transport make fluorometry the more reproducible measure. Bursts in DNA synthesis in early S phase are made transient by rapid catabolism. Appreciable increases in DNA content in early S phase can be completely reversed by catabolism.

Transient early amplification appears to be under separate control since inhibitors of *E. coli* polymerase II are ineffective in preventing the bursts in synthesis while effectively stopping the bulk of DNA synthesis. Inhibition of bulk synthesis can also be prevented by the addition of ara-C after the initial burst has occurred. Inhibition of transient synthesis has thus far only been effected using the RNA polymerase inhibitor actinomycin D in combination with hydroxyurea.

All homogeneous cell lines show evidence of several such bursts in each S phase. This phenomenon is displayed most clearly by WI-38, the normal human diploid line, and its observation in this line is favored by a greater overall length of S phase. It is also apparent that the DNA synthetic pattern of established and tumorigenic lines is less discontinuous in time and less precisely ordered than in normal cells.

The functional effects of transiently amplified DNA may be expressed as increases in RNA, protein and enzyme content over a 1-hour interval following the earliest burst in synthesis.

Synchronous initiation of clusters of replicating units with pauses between synthetic bursts can be offered as an explanation for the disparity in length of S phase between early cleavage embryos and somatic cells. For example, *Drosophila* DNA synthesis requires less than 3 minutes in early embryos and 600 minutes in cultured somatic cells (Blumenthal, Kriegstein and Hogness 1974). The rate of travel of the replicative fork is not any slower in somatic cells than it is in early embryos (Painter and Schaefer 1969; Blumenthal, Kriegstein and Hogness 1974), nor do there appear to be sufficiently fewer initiation sites in somatic cells as opposed to early embryos, as judged from radioautographic and electron microscope analysis (Blumenthal, Kriegstein and Hogness 1974), to explain the 200-fold increase in the duration of S phase. In all likelihood the increased length of S phase is a consequence of the fact that in differentiated somatic cells intervals of synthesis are separated periods of very limited replicative activity.

Variations in the Length of S Phase

Since the DNA content of V79 cells 2 hours into S phase or of WI-38 cells 5 hours into S is not measurably different from their G_1 DNA content, it is entirely possible that cycling cells approaching confluency may arrest after completion of the early S synthesis, not before. Reports of unscheduled DNA synthesis in G_1 by unirradiated cells (Djordjevic et al. 1969) may in fact be observations of this early replicating DNA. In this connection the duration of S phase in cultured mammalian cells and in cells in vivo is variable and may in part depend on the method of measurement.

In V79 cells radioautographic analysis of labeled nuclei reveals that S phase begins 1.5 hours after mitosis, while the bulk of the DNA does not begin to accumulate until 6 hours. Often in the literature estimates of S phase duration will vary depending on whether they were made by total radiochemical incorporation, spectrophotometry or radioautography. In the first two cases lower values are reported than if the estimates were made radioautographically. In mouse L cells the duration of S phase has been estimated to be 6.3 hours by spectrophotometry (Mak 1965) and between 12.2 and 13.5 (Cleaver 1967) by radioautography. Bostock and Prescott (1971) observed an 8-hour S phase in rabbit endomitrium by measuring total [³H]dT incorporation, while noting low levels of incorporation over a considerable portion of the cycle. The labeled cells were detectable radioautographically and, if scored for percent of labeled nuclei, gave a value for S phase of 18 hours out of a 27-hour cycle.

In summary, the disparity in length of S can be explained by suggesting that bulk DNA synthesis occurs primarily in the latter part of S phase and that early S phase is a period of intermittent synthesis and catabolism, so that the net DNA synthesis during this period is relatively small. This disparity seems to be especially evident in diploid early passage cells and may reflect the fact that established and transformed cell lines have an altered temporal order of DNA replication such that the intermittent bursts in synthesis and catabolism occur more nearly coincident with bulk DNA synthesis.

A serious objection to the idea of intermittent synthesis and catabolism might be the work of Mueller and Kajiwara (1966) and Amaldi et al. (1972), who found that some of the same DNA sequences that were replicated early in one cell cycle were replicated at the same point in a subsequent cell cycle. Earlier studies used inhibitors of DNA synthesis, however, and this result is not clearly obtained (Comings, personal communication) when simple mitotic selection is used as a method of synchronization. If, as we expect, the order of synthesis of DNA made by polymerase II is fixed and occurs in the same temporal sequence in each cell cycle, then it is this material which is measured by the procedures of Amaldi et al. (1972).

Since increases in ribosomal RNA, total protein and enzyme levels occur in concert with the early amplification of DNA, it is likely that the amplified DNA is in some way functional. The most likely candidate for amplification is ribosomal DNA since total RNA increases rapidly following DNA synthesis. In addition there is precedence in the literature for ribosomal DNA amplification and transient amplification followed by catabolism in the G_1 phase of *Chlamydomonas* (Howell 1972).

The mechanism by which multiple copies of a particular DNA sequence might be made is not yet resolved (Bird, Rogers and Birnstiel 1973), but the requirement for an RNA primer in the initiation of DNA synthesis by polymerase II in phage (Brutlag, Schekman and Kornberg 1971), *E. coli* (Keller, 1972, Sugino, Hirose and Okazaki 1972), as well as higher organisms (Stavianopoulos, Karkas and Chargaff 1971) seems clear.

We are currently pursuing the idea that this periodically synthesized ara-C insensitive DNA represents amplified chromosomal nucleotide sequences which exist extrachromosomally and are not copied by classical, semiconservative mechanisms. Instead it seems more likely that these sequences are copied by way of an RNA intermediate and are then transcribed and subsequently or concomitantly degraded.

Acknowledgments

This work was supported by grants CA-10619 and HD-04699 from the National Institutes of Health. The authors wish to thank Mrs. Karen Messer for her help in preparation of the manuscript.

REFERENCES

Amaldi, F., F. Carnevali, L. Leoni and D. Mariotti. 1972. Replicon origins in Chinese hamster cell DNA. I. Labeling procedure and preliminary observations. *Exp. Cell Res.* **74**:367.

Ammermann, D. 1969. Release of DNA breakdown products into the culture medium of *Stylonychia mytilus* exconjugants (protozoa, ciliata) during the destruction of polytene chromosomes. *J. Cell Biol.* **40**:576.

Banerjee, M. R. and J. E. Wagner. 1972. Gene amplification in mammary gland at differentiation. *Biochem. Biophys. Res. Comm.* **49**:480.

Bird, A., E. Rogers and M. Birnstiel. 1973. Is gene amplification RNA directed? *Nature New Biol.* **242**:226.

Bishop, J. O., R. Pemberton and C. Baglioni. 1972. Reiteration frequency of haemoglobin genes in the duck. *Nature New Biol.* **235**:231.

Blumenthal, A., H. Kriegstein and D. Hogness. 1974. The units of DNA replication in Drosophila chromosomes. *Cold Spring Harbor Symp. Quant. Biol.* **38**: (in press)

Bostock, C. J. and D. M. Prescott. 1971. Shift in buoyant density of DNA during the synthetic period and its relation to euchromatin and heterochromatin in mammalian cells. *J. Mol. Biol.* **60**:151.

Braun, R. and T. E. Evans. 1969. Replication of nuclear satellite and mitochondrial DNA in the mitotic cycle of *Physarum. Biochim. Biophys. Acta* **182**:511.

Brown, D. D. and A. W. Blackner. 1972. Gene amplification proceeds by a chromosome copy mechanism. *J. Mol. Biol.* **63**:75.

Brown, D. D. and I. B. Dawid. 1968. Specific gene amplification in oocytes. *Science* **160**: 272.

Brown, R. D. and G. P. Tocchini-Valentini. 1972. On the role of RNA in gene amplification. *Proc. Nat. Acad. Sci.* **69**:1746.

Brutlag, D., R. Schekman and A. Kornberg. 1971. A possible role for RNA polymerase in the initiation of M13 DNA synthesis. *Proc. Nat. Acad. Sci.* **68**:2826.

Camiener, G. W. and C. G. Smith. 1965. Studies of the enzymatic deamination of cytosine arabinoside. I. Enzyme distribution and species specificity. *Biochem. Pharmacol.* **14**:1405.

Cleaver, J. E. 1967. *Thymidine metabolism and cell kinetics*, vol. 6, p. 126. North-Holland, Amsterdam.

Crippa, M. and P. Tocchini-Valentini. 1971. Synthesis of amplified DNA that codes for ribosomal RNA. *Proc. Nat. Acad. Sci.* **68**:2769.

Dickson, E., J. B. Boyd and C. D. Laird. 1971. Sequence diversity of polytene chromosome DNA from *Drosophila hydei. J. Mol. Biol.* **61**:615.

Dische, Z. 1955. Color reactions of the nucleic acid components. *The nucleic acids* (ed. E. Chargaff and J. Davidson) vol. 1, pp. 285–305. Academic Press, New York.

Djordjevic, B., R. Evans, A. Perez and M. Weill. 1969. Spontaneous unscheduled DNA synthesis in Gl HeLa cells. *Nature* **224**:803.

Engberg, J., D. Mowat and R. Pearlman. 1972. Preferential replication of the ribosomal RNA genes during a nutritional shift-up in *Tetrahymena pyriformis. Biochim. Biophys. Acta* **272**:312.

Forrest, G. L. and R. R. Klevecz. 1972. Synthesis and degradation of microtubule protein in synchronized Chinese hamster cells. *J. Biol. Chem.* **247**:3147.

Gall, J. G. 1968. Differential synthesis of the genes for ribosomal RNA during amphibian oogenesis. *Proc. Nat. Acad. Sci.* **60**:553.

———. 1969. The genes for ribosomal RNA during oogenesis. *Genetics* (supplement) **61**:121.

Gall, J. G., E. H. Cohen and M. L. Polan. 1971. Repetitive DNA sequences in *Drosophila. Chromosoma* **33**:319.

Guttes, E. and S. Guttes. 1969. Replication of nucleolus-associated DNA during the "G2 phase" in *Physarum polycephalum. J. Cell Biol.* **43**:229.

Harrison, P. R., A. Hell, G. D. Birnie and J. Paul. 1972. Evidence for single copies of globin genes in the mouse genome. *Nature* **239**:219.

Hinegardner, R. T. 1971. An improved fluorometric assay for DNA. *Anal. Biochem.* **39**:197.

Holt, C. E. and E. G. Gurney. 1969. Minor components of the DNA of *Physarum polycephalum. J. Cell Biol.* **40**:484.

Howell, S. H. 1972. The differential synthesis and degradation of ribosomal DNA during the vegetative cell cycle in *Chlamydomonas reinhardi. Nature* **240**:264.

Kapp, L. N., J. A. Remington and R. R. Klevecz. 1973. Induction of serine dehydratase activity by cyclic AMP is restricted to S phase in synchronized CHO cells. *Biochem. Biophys. Res. Commun.* **52**:1206.

Kappler, J. W. 1970. The kinetics of DNA methylation in cultures of a mouse adrenal cell line. *J. Cell. Physiol.* **75**:21.

Karrer, P. 1950. *Organic chemistry,* 4th ed. Elsevier Pub. Co., New York.

Keller, W. 1972. RNA primed DNA synthesis *in vitro. Proc. Nat. Acad. Sci.* **69**:1560.

Kissane, J. M. and E. J. Robins. 1958. The fluorometric measurement of deoxyribonucleic acid in animal tissues with special reference to the central nervous system. *J. Biol. Chem.* **233**:184.

Klevecz, R. R. 1969a. Temporal coordination of DNA replication with enzyme synthesis in diploid and heteroploid cells. *Science* **166**:1536.

———. 1969b. Temporal order in mammalian cells. I. The periodic synthesis of lactate dehydrogenase in the cell cycle. *J. Cell Biol.* **43**:207.

———. 1972. An automated system for cell cycle analysis. *Anal. Biochem.* **49**:407.

Klevecz, R. R. and L. N. Kapp. 1973. Intermittent DNA synthesis and periodic expression of enzyme activity in the cell cycle of WI-38. *J. Cell Biol.* **58**:564.

Lowry, O. H., J. Rosenbrough, A. L. Farr and R. J. Randall. 1951. Protein measurement with the folin phenol reagent. *J. Biol. Chem.* **193**:265.

Mahdavi, V. and M. Crippa. 1972. An RNA-DNA complex intermediate in ribosomal gene amplification. *Proc. Nat. Acad. Sci.* **69**:1749.

Mak, S. 1965. Mammalian cell cycle analysis using microspectrophotometry combined with autoradiography. *Exp. Cell Res.* **39**:286.

Meneghini, R., H. A. Armelin, J. Balsamo and F. J. S. Lara. 1971. Indication of gene amplification in Rhynchosciora by RNA-DNA hybridization. *J. Cell Biol.* **49**:913.

Miller, O. L. 1966. Structure and composition of peripheral nucleoli of salamander oocytes. *Nat. Cancer Inst. Monogr.* **23**:53.

Miller, O. L. and B. R. Beatty. 1969. Visualization of nucleolar genes. *Science* **164**:955.

Moore, E. C. 1969. Effects of ferrous ion and dithioerythritol on inhibition by hydroxyurea of ribonucleotide reductase. *Cancer Res.* **29**:291.

Mueller, G. C. and K. Kajiwara. 1966. Early and late replicating deoxyribonucleic acid complexes in HeLa nuclei. *Biochim. Biophys. Acta* **114**:108.

Painter, R. B. and A. W. Schaefer. 1969. Rate of synthesis along replicons of different kinds of mammalian cells. *J. Mol. Biol.* **45**:467.

Pelc, S. R. 1970 Metabolic DNA and the problem of aging. *Exp. Geront.* **5**:217.

Prescott, D. M., K. G. Murti and C. J. Bostock. 1973 Genetic apparatus of *stylonchia sp. Nature* **242:**576.

Rama-Reddy, G. V., M. Goulian and S. S. Hendler. 1971. Inhibition of *E. coli* DNA polymerase II by ara-CTP. *Nature New Biol.* **234:**286.

Reich, E., R. M. Franklin, A. J. Shatkin and E. L. Tatum. 1962. Action of actinomycin D on animal cells and viruses. *Proc. Nat. Acad. Sci.* **48:**1238.

Remington, J. R. and R. R. Klevecz. 1973. Families of replicating units in cultured hamster fibroblasts. *Exp. Cell Res.* **76:**410.

Ritossa, F. M. 1968. Unstable redundancy of genes for ribosomal DNA. *Proc. Nat. Acad. Sci.* **60:**509.

Schrecker, A. W. and M. J. Urshel. 1968. Metabolism of 1-β-D-arabinofuranosylcytosine in leukemia L1210: Studies with intact cells. *Cancer Res.* **28:**793.

Spry, C. J. F. 1971. Mechanism of eosinophilia. V. Kinetics of normal and accelerated eosinopoiesis. *Cell Tissue Kinet.* **4:**351.

Stavianopoulos, J. G., J. D. Karkas and E. Chargaff. 1971. Nucleic acid polymerases of the developing chicken embryo: A DNA polymerase preferring a hybrid template. *Proc. Nat. Acad. Sci.* **68:**2207.

Sugino, A., S. Hirose and R. Okazaki. 1972. RNA-linked nascent DNA fragments in *Escherichia coli. Proc. Nat. Acad. Sci.* **69:**1863.

Suzuki, Y., L. P. Gage and D. D. Brown. 1972. The genes for silk fibroin in *Bombyx mori. J. Mol. Biol.* **70:**637.

Self-Regulation of Growth
in Three Dimensions:
The Role of Surface Area Limitation

Judah Folkman, Mark Hochberg, and David Knighton

Department of Surgery, Children's Hospital Medical Center
and Harvard Medical School, Boston, Massachusetts 02115

For any group of cells to continue growing, the surface area of the aggregate population must be sufficient to permit adequate absorption of nutrients and escape of waste catabolites.

For tumor cells in flat tissue culture, growth continues only in two dimensions. The third dimension becomes constant. It consists of a thin layer of cells that have piled up. Therefore surface area is always directly proportional to the volume of cells (Fig. 1A). Tumor cells in this flat configuration will increase their number indefinitely if they are always provided with open space and sufficient nutrient; in this sense they are analogous to bacterial colonies on agar plates (Hochberg and Folkman 1972). Under these conditions two-dimensional growth is *not* self-limiting.

However tumor cells cultured in suspension (i.e., soft agar) form spheroidal aggregates where growth occurs in three dimensions (Sutherland, McCredie and Inch 1971; Folkman and Hochberg 1973). As the spheroid expands, volume enlarges by the third power, whereas surface area only increases by the second power (Fig. 1B). Eventually a volume is attained at which surface area becomes insufficient to accommodate diffusion of nutrients into the cell population and wastes out of it. We have shown that for a given cell line there is a maximum volume of the spheroid beyond which no further expansion occurs (Folkman and Hochberg 1973) despite frequent change of nutrient and provision of open space. Thus three-dimensional growth *is* self-limiting.

The following in vitro experiments support this concept.

Three-Dimensional Growth In Vitro

B-16 melanoma, V-79 Chinese hamster lung cells, and L-5178Y leukemia cells were suspended in soft agar (0.12%). Single spheroids were removed with a wide-bore pipette and transferred to a new flask containing 10 ml fresh agar. Transfers to a flask containing fresh agar were repeated every 2 or 3 days. This short interval

833

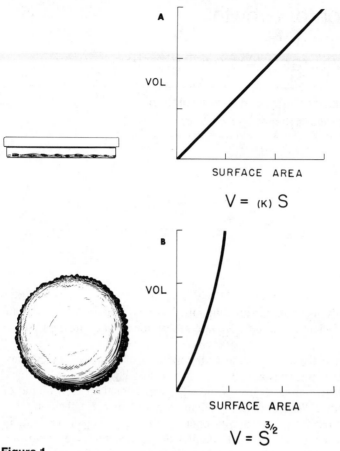

$$V = _{(K)} S$$

$$V = S^{3/2}$$

Figure 1

A, The relationship of surface to volume when growth occurs in two dimensions.

B, The relationship of surface to volume when growth is spheroidal or three-dimensional.

prevented any pH change. The diameter of each spheroid was measured every 2 days by projection of its image. In some experiments media were changed more frequently or continuously.

All spheroids enlarged following an approximately linear growth curve for 5–23 weeks before reaching a critical diameter, beyond which there was no further expansion, i.e., a stage of "population dormancy." For L-5178Y cells the mean diameter of the dormant phase was 3.8 mm ± 0.5 mm at approximately 25 days. For V-79 cells the dormant diameter was 4.0 mm ± 0.8 mm at 175 days (Fig. 2A). For B-16 melanoma cells it was 2.4 mm ± 0.4 mm reached at approximately 100 days. When multiple spheroids were cultured together in one flask, the "dormant" diameter was always smaller than for a single, lone colony; for multiple colonies the final diameter varied inversely with the number of spheroids (Fig. 3).

Spheroids of varying age were incubated with [³H]thymidine, fixed in glutaraldehyde, and prepared in thin Epon sections for radioautography. After a spheroid

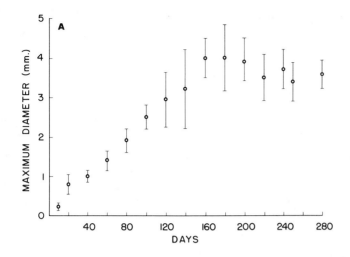

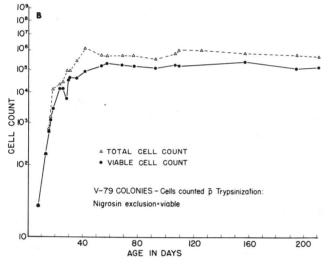

Figure 2

A, Mean diameter and standard deviation of 70 isolated spheroids of V-79 cells. Very old spheroids occasionally shattered and were discarded. Therefore mean diameter after 200 days represents approximately 30 spheroids.

B, The number of total cells and the number of viable cells in V-79 spheroids reach a steady state 130 days before the spheroids stop expanding. From Folkman and Hochberg (1973) with permission of *J. Exp. Med.*

reached approximately 0.3 mm diameter, there was always a central necrotic zone, a middle zone of unlabeled cells, and an outer zone of labeled cells with accompanying mitoses. In spheroids up to 1.0 mm diameter, almost all cells on the surface and five layers deep were labeled with [³H]thymidine (Fig. 4). As the spheroids enlarged further toward the "dormant" state, the ratio of volume of labeled cells to the total volume of the spheroid decreased steadily. In the dormant phase the labeled cells occupied only a single layer on the outermost rim of the spheroid (Fig. 5). Thus from a histological basis, the phase of "population dormancy" appeared to be a

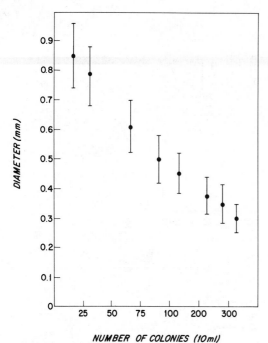

Figure 3
Mean diameter and standard deviation of spheroids of L-5178Y murine leukemic cells in 10 ml soft agar. Diameter is inversely proportional to the number of spheroids.

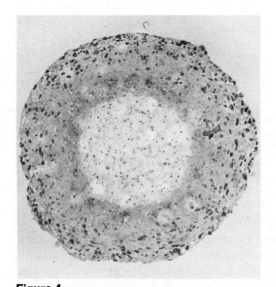

Figure 4
Cross section of V-79 spheroid, 1.0 mm in diameter and 20 days old. Five to seven outer layers of cells are labeled with [³H]thymidine. Volume of labeled cells/total volume = 0.60. From Folkman and Hochberg (1973) with permission of *J. Exp. Med.*

836

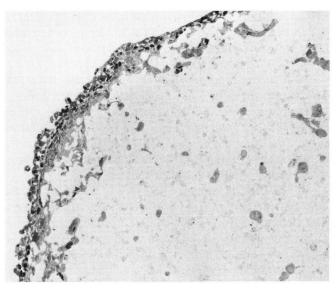

Figure 5
Cross section of V-79 spheroid, 3.2 mm in diameter and 150 days old. Only 1 or 2 outer cell layers are labeled with [³H]thymidine. Volume of labeled cells/ total volume = 0.14. From Folkman and Hochberg (1973) with permission of *J. Exp. Med.*

steady state in which a few proliferating cells on the outer surface balanced a large population of necrotic cells digesting themselves in the center. No amount of nutrient change would make such a spheroid larger. The viability of such a dormant spheroid could always be demonstrated even up to one year: (a) if divided in half, each piece of the spheroid would grow back to the size approximating the previous dormant diameter; (b) the dormant spheroid always reduced pH of fresh medium; (c) the dormant spheroid could be dispersed into free-floating cells, which would form a fresh monolayer.

Furthermore a maximum cell number, similar for all cell lines studied, was contained in each spheroid by the time it became dormant. This was determined by periodic dispersal of the spheroids and counting of cells. For example, for L-5178Y spheroids the maximum viable cell count was 3×10^5 reached at 20 days. For V-79 spheroids it was 2×10^5 reached at 45 days (Fig. 2B). In all spheroids a maximum cell number was reached long before the diameter stopped expanding. This implied that the limit of viable cells that could be supported in such a spheroid was reached early in the trajectory. Further expansion was due to an accumulating necrotic center; i.e., cells were being added to the necrotic center faster than they could be digested or solubilized. The rate of digestion of necrotic debris was determined in separate experiments by prolonged exposure of growing spheroids to a lethal dose of [³H]thymidine (2 μCi/ml). In these experiments the peripheral mitosing cells committed thymidine suicide and the spheroids shrank rapidly over the next 7–10 days.

This "population dormancy" of spheroidal growth is not peculiar to soft agar, but appears also to be a component of the growth curve of solid tumors in vivo.

Growth of Tumors In Vivo

Before looking at experimental evidence of "population dormancy" in vivo, let us introduce the role of tumor vascularization. The growth of solid tumors in vivo may be considered in two stages: the first stage, before vascularization, the second stage after vascularization.

In the first stage tumor cells multiply to form a tiny aggregate in the interstitial fluid. Nutrients and wastes are exchanged by simple diffusion. As this spheroidal population enlarges, its surface area diminishes relative to volume. The surface area becomes inadequate to handle the flux of nutrients and catabolites necessary for further growth. Expansion stops. This may occur at approximately 1 mm^3.

The second stage commences when new capillary sprouts penetrate the perimeter of the tumor nodule. Rapid and prolonged growth is then initiated.

Under usual conditions the induction of neovascularization is so rapid that the separation between the first and the second stage is imperceptible. However the two stages can be dissociated under experimental conditions. When this is accomplished and capillaries are prevented from penetrating the packed tumor cell population, the first stage ends in a state of dormancy. In the absence of capillary penetration, it is the three-dimensional configuration that ultimately restrains further growth in the first stage.

In the following experiment tumor is kept separate from its vascular supply so that the first and second stages of tumor growth are dissociated and can be studied quantitatively. The anterior chamber of the rabbit eye is used. A tiny inoculum of Brown-Pearce tumor is suspended in the anterior chamber of the rabbit eye (Gimbrone et al. 1972). Tumor growth is rapid because of the nutrient properties of the aqueous humour that circulates through the anterior chamber, totally exchanging itself every half-hour. But growth stops at a diameter of approximately 1 mm (Fig. 6). These tiny tumor nodules are not penetrated by new capillaries because of the inability of blood vessel sprouts to make their way through the aqueous fluid. These spheroids remain viable indefinitely and could be transferred from one eye to another. In the dormant state histologic sections were similar to the in vitro agar experiments. There was an outer proliferating layer of cells that incorporated [^{3}H]thymidine, a middle zone of viable cells, and an inner zone of necrotic cells. Whenever such a spheroid was deliberately brought to lie against the blood vessels of the iris, new capillaries penetrated these spheroids within 48 hours. There was immediate, rapid growth and the volume of the spheroid increased 16,000 times over a 2-week period (Fig. 6).

Greene described the phenomenon of avascular tumor spheroids in the anterior chamber in 1941. These floating tumors stayed dormant for up to 2 years, after which they were reimplanted subcutaneously, became vascularized immediately, and grew rapidly. There is an analogous condition in man. Retinoblastoma of the eye often sheds metastatic implants to the anterior chamber. These floating metastases rarely achieve sizes beyond 1–2 mm in diameter.

It could be argued that tumor growth in the anterior chamber is not a good model of the first stage of tumor growth in tissues. A conventional tumor implant would always be surrounded by blood vessels even before it was vascularized by new capillaries (Fig. 7A). Would not blood vessels contiguous to a tumor spheroid be more efficient in supplying fresh nutrient or carrying off wastes than a rapid flow of aqueous humour (or soft agar) around such a spheroid?

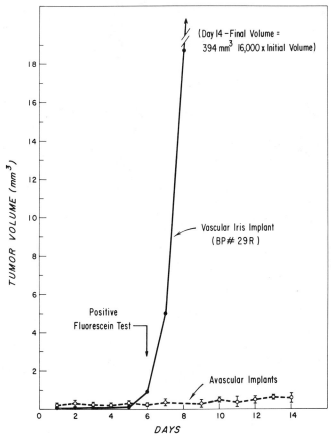

Figure 6

Typical growth curve of Brown-Pearce tumor implanted on the iris. It became vascularized on day 6 as indicated by the positive fluorescein test. For comparison, note the mean daily volumes of 10 tumor implants in the anterior chamber which were unable to become vascularized. From Gimbrone et al. (1973) with permission of *J. Nat. Cancer Inst.*

Until recently it was not possible to counter this objection, because tumors implanted among blood vessels were penetrated by capillaries so quickly (few days) that the dormant phase could not be measured (Fig. 7B). A method of delaying or stopping capillary ingrowth would be needed before tumor dormancy could be demonstrated.

Such a system was discovered in the chick embryo by David Knighton, working in our laboratory. The incubation period for the chick embryo is 21 days. Knighton found that during the first 11 days, no vessels, either in the yolk sac membrane or the chorioallantoic membrane, will send out new capillaries into a tumor, even when the tumor is imbedded amidst these vessels. At about the 11th day capillary proliferation is possible and new capillary sprouts penetrated all tumors over the next 24 hours. During the 11-day refractory period tumors reached a dormant size of approximately 1 mm diameter, regardless of the size of the initial implant (Fig. 8). After the 11-day period there was rapid growth up to 1 cm in diameter or more.

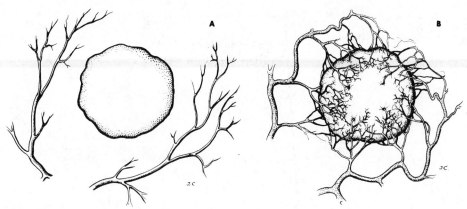

Figure 7

A, Diagram of tumor implant in the chorioallantoic membrane of the chick embryo before it has been penetrated by new capillaries.

B, Tumor implant in the chorioallantoic membrane after penetration by new vessels. In this situation the surface area available for diffusion of nutrients and catabolites has increased markedly.

This was true for all tumors tested in our studies (Walker 256 carcinoma, B-16 mouse melanoma, and mouse adenocarcinoma). The reason for the refractory period is unknown. But it does make clear that even though a tumor spheroid is surrounded by capillaries, the surface area available for diffusion cannot be increased until new capillaries *penetrate* the spheroid (Fig. 7B).

DISCUSSION

Mechanism of Dormancy

Several possible mechanisms may contribute to the dormant state. First, dead cells appear in the center of spheroids less than 400 microns (0.4 mm) diameter. Oxygen cannot diffuse through a layer of packed cells greater than 150–200 microns (Thomlinson and Gray 1955). Therefore cells at a depth of 200 microns or more from the surface of a spheroid would become hypoxic. The greater the diameter, the more cells will find themselves in this hypoxic zone. Second, diffusion of nutrients into the spheroid is proportional to the area of the outer surface of the spheroid, and surface area diminishes in relation to volume as the diameter increases. Similarly the diminishing ratio of surface area to volume may depress the escape of catabolites and necrotic products as the spheroid enlarges. The fact that the number of viable cells becomes constant early and that the proliferating compartment shrinks may be partly the fault of insufficient nutrients and partly the result of accumulation of catabolites. Our data do not decide which is the primary event. But they do make a strong case that spheroidal growth is self-regulatory both in vitro and in vivo.

Implications for Human Cancer

Previously we have shown that a cell-free fraction separable from tumor cells will induce the growth of new capillary sprouts (Folkman 1971; Folkman et al. 1971).

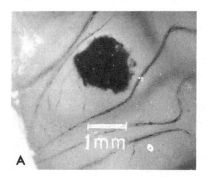

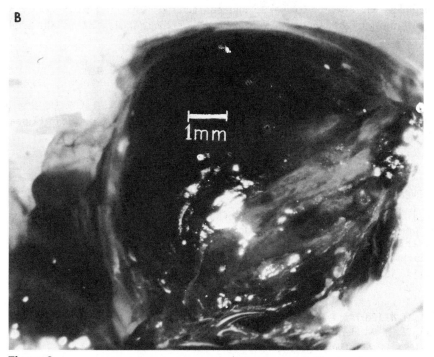

Figure 8
A, Spheroid of melanoma in the chorioallantoic membrane of a 10-day-old
chick embryo. The tumor was implanted on day 7 and has stopped growing at
a diameter of 1 mm. Although vessels surround it, no new vessels have penetrated it.

    B, Same tumor which became vascularized on day 11 and began rapid growth,
reaching a diameter of 10 mm by day 15.

This fraction, called tumor angiogenesis factor (TAF), appears to be a mitogen
specific for endothelial cells of capillaries and small veins (Cavallo et al. 1972).
TAF has been isolated from the cytoplasm and from the nucleus (Tuan et al. 1973)
of tumor cells. It has not been found in normal cells. It has been demonstrated to
diffuse over distances of 3–5 mm into tissue from the edge of a tumor implant or
from an implanted source of the factor (Cavallo et al. 1973; Gimbrone et al. 1973).

    If TAF could be blocked, either by the corresponding antibody or by some other approach, then solid tumors might be held in the avascular or first stage of tumor growth. They should theoretically remain dormant just as they do in soft agar, in the aqueous humour, or in the chick embryo during the refractory state of the chorioallantoic blood vessels. Thus tumor cells in a packed population, subjected to nutrient depletion and catabolite accumulation because of the restrictions of three-dimensional growth in the absence of new capillaries, might be unable to display their malignant potential. Their turnover might be greatly reduced. This in turn might retard the acquisition of more malignant characteristics. And for a number of reasons such a tiny tumor might be more vulnerable to immunotherapy than a vascularized tumor. Whether this blockade of angiogenesis can be accomplished will depend upon further purification and characterization of TAF, still far from complete.

Nevertheless the attempt to isolate and characterize TAF and to demonstrate how solid tumors might behave in the absence of neovascularization should teach us much about the fundamental nature of tumor growth.

Acknowledgments

This work was supported by Public Health Service Grant 5RO1 CA 08185-06 from the National Cancer Institute and by grant IC-28 from the American Cancer Society.

REFERENCES

Cavallo, T., R. Sade, J. Folkman and R. S. Cotran. 1972. Tumor angiogenesis: Rapid induction of endothelial mitoses demonstrated by autoradiography. *J. Cell Biol.* **54:**408.
————. 1973. Ultrastructural autoradiographic studies of the early vasoproliferative response in tumor angiogenesis. *Amer. J. Pathol.* **70:**345.
Folkman, J. 1971. Tumor angiogenesis: Therapeutic implications. *New Eng. J. Med.* **285:**1182.
Folkman, J. and M. Hochberg. 1973. Self-regulation of growth in three dimensions. *J. Exp. Med.* **138:**745.
Folkman, J., E. Merler, C. Abernathy and G. Williams. 1971. Isolation of a tumor factor responsible for angiogenesis. *J. Exp. Med.* **133:**275.
Gimbrone, M. A., Jr., S. B. Leapman, R. S. Cotran and J. Folkman. 1972. Tumor dormancy in vivo by prevention of neovascularization. *J. Exp. Med.* **136:**261.
————. 1973. Tumor angiogenesis: Iris neovascularization at a distance from experimental intraocular tumors. *J. Nat. Cancer Inst.* **50:**219.
Greene, H. S. N. 1941. Heterologous transplantation of mammalian tumors. I. The transfer of rabbit tumors to alien species. *J. Exp. Med.* **73:**461. II. The transfer of human tumors to alien species. *J. Exp. Med.* **73:**475.
Hochberg, M. S. and J. Folkman. 1972. Mechanism of size limitation of bacterial colonies. *J. Infect. Dis.* **126:**629.
Sutherland, R. M., J. A. McCredie and R. Inch. 1971. Growth of multicell spheroids in tissue culture as a model of nodular carcinomas. *J. Nat. Cancer Inst.* **46:**113.
Thomlinson, R. H. and L. H. Gray. 1955. The histological structure of some human lung cancers and the possible implications for radiotherapy. *Brit. J. Cancer* **9:**539.
Tuan, D., S. Smith, J. Folkman and E. Merler. 1973. Isolation of the non-histone proteins of rat Walker carcinoma 256: Their association with tumor angiogenesis. *Biochemistry* **12:**3159.

The Role of Critical Cell Proliferation in Differentiation of Mammary Epithelial Cells

Yale J. Topper and Barbara K. Vonderhaar

National Institute of Arthritis, Metabolism, and Digestive Diseases
National Institutes of Health, Bethesda, Maryland 20014

An intriguing possibility in developmental biology is that cells in a particular state of differentiation cannot, themselves, develop further. Rather they must undergo DNA synthesis and/or mitosis, under certain circumstances, to give rise to daughter cells, qualitatively different from the parent cells, which can then continue development. We shall refer to such proliferation as critical mitosis. This concept was first proposed by Holtzer (Holtzer 1963; Abbott and Holtzer 1965) during his studies on the differentiation of myoblasts and chondroblasts.

Several other systems have subsequently been examined in this context. The synthesis and accumulation of the egg white proteins, lysozyme and ovalbumin, by the oviduct of estrogen-primed immature female chicks has been reported to depend on tubular gland cell proliferation (Oka and Schimke 1969a,b,c; Palmiter and Wrenn 1971). Similarly the estrogen-induced synthesis and secretion of the egg yolk protein, phosvitin, by rooster liver was reported to be dependent on DNA synthesis in vivo (Jailkhani and Talwar 1972). In vitro studies have also led investigators to conclude that critical mitosis is necessary for erythropoietin-induced hemoglobin synthesis in fetal mouse liver (Paul and Hunter 1968) and adult rat bone marrow (Gross and Goldwasser 1970). Likewise the development of the capacity of cultured pancreatic exocrine cells to synthesize characteristic secretory enzymes has been reported to require DNA synthesis and/or mitosis (Wessells 1968; Wessells and Rutter 1969).

In all such cases, resting cells in a protracted G_1 phase of the cell cycle, given the proper stimulus, are triggered to progress through the DNA synthetic phase of the cell cycle (S phase); they then pass through the G_2 phase and mitosis (M phase). The putative competent progeny are thereby formed. These daughter cells may remain in G_1 or they may go through further cycles. As proposed by Holtzer (1963), only one critical mitosis is required for the formation of competent daughter cells. Stimulation of these cells by the appropriate agent will then evoke their unique phenotypic expression.

The model system that we have used for the study of problems of developmental biology, and for inquiry into the question of critical mitosis in particular (Owens,

Vonderhaar and Topper 1973), is the mammary gland of the C3H/HeN mouse. Nonsecretory epithelial cells in explants derived from the glands of late fetal rats (Ceriani 1970) and adolescent as well as mature virgin mice (Stockdale and Topper 1966; Vonderhaar, Owens and Topper 1973; Voytovich and Topper 1967) can be converted into secretory cells in vitro. This can be accomplished in a few days by culturing the tissue fragments in a chemically defined medium in the presence of insulin, prolactin and a glucocorticoid.

Insulin is potentially capable of eliciting a number of responses from these cells in vitro, such as stimulation of DNA synthesis, and enhanced formation of both glucose-6-phosphate dehydrogenase (G-6-PD) and gluconate-6-phosphate dehydrogenase (Gl-6-PD) (Friedberg, Oka and Topper 1970). However freshly isolated cells from nonpregnant mice are insensitive to the hormone. After one day of culture in the presence or absence of insulin, they acquire sensitivity. This makes it possible for them ultimately to manufacture casein and α-lactalbumin, unique secretory products of the mammary gland, in response to insulin, hydrocortisone and prolactin. It was observed (Stockdale and Topper 1966) that, once insulin sensitivity is acquired, the cells always respond to the mitogenic action of insulin before they synthesize the milk proteins in response to the complete hormone complement. The cells in explants from mid-pregnant mice also proliferate in response to insulin prior to forming increased quantities of casein and α-lactalbumin. This does not necessarily mean that cell proliferation is a prerequisite for the formation of these specialized products. It will, in fact, be shown that while such a prerequisite does exist in the case of tissue from virgin mice, it no longer exists once the donor animals attain the mid-pregnant state.

EXPERIMENTAL APPROACH

The use of agents that arrest cells in various phases of the cell cycle is a reasonable way to attack the problem under consideration. M-phase arrest is not desirable, however, since some metaphase cells have a limited general capacity to manufacture macromolecules (Martin, Tomkins and Granner 1969; Fan and Penman 1970). An observed inability of such cells to form a particular product might be attributable, then, to this, rather than to a specific requirement for completion of mitosis. Arrest in G_2 would aid in distinguishing between a requirement for DNA synthesis and a requirement for mitosis. However an adequate G_2 block of mammary gland epithelial cells has not yet been obtained. Agents that inhibit DNA synthesis, i.e., arrest the cells in late G_1, have been most useful. Even so, a number of precautions, noted in the Results section, must be taken.

RESULTS

Table 1 shows that the epithelial DNA content in explants from mature virgin mammary gland increases in the presence of insulin (I), hydrocortisone (F) and prolactin (P). This increment is blocked when either 1-β-D-arabinosylcytosine (ara-C) or fluorodeoxyuridine (FdU) is added at zero time. In the absence of these inhibitors a large increase in casein synthesis occurs after 72 hours, while essentially no such increase takes place in their presence. Note that the Δ casein:DNA ratio is

Table 1

Effect of Ara-C and FdU on Casein Synthesis and the Emergence of α-Lactalbumin in Explants from Mature Virgin Mice

Culture conditions	ng Epithelial DNA/mg wet wt tissue	Casein (cpm/mg wet wt tissue/ 4 hr)	Δ Casein/ ng DNA	Lactose synthetase B-protein activity (pmoles product/mg wet wt/30 min)	B-protein activity/ ng DNA
0 hr	43.2	56 ± 2			
72 hr IFP	99.6	350 ± 44	2.95		
IFP + ara-C added at 0 hr	44.6	79 ± 3	0.52		
0 hr	30.4	109 ± 5			
72 hr IFP	50.8	439 ± 23	6.50		
IFP + FdUR added at 0 hr	31.4	120 ± 6	0.35		
0 hr	43.2			2 ± 1	0.046
120 hr IFP	63.8			33 ± 4	0.517
IFP + ara-C added at 0 hr	41.0			<1	<0.024
IFP + ara-C added at 72 hr	53.6			33 ± 2	0.617
0 hr	33.3			<1	<0.030
120 hr IFP	53.8			41 ± 2	0.762
IFP + FdUR added at 0 hr	26.2			<1	<0.038
IFP + FdUR added at 72 hr	47.5			37 ± 4	0.779

Mammary explants from mature virgin mice were cultured in the presence of insulin, hydrocortisone and prolactin (IFP) for 72 or 120 hr. 1-β-D-Arabinosylcytosine (ara-C; 15 μg per ml) or fluorodeoxyuridine (FdU; 25 μg per ml) was added at the times indicated. Casein synthesis was determined after pulsing with 10 μCi of carrier-free $^{32}P_i$ (Juergens et al. 1965) from 0–4 hr or from 68–72 hr in culture. DNA content of epithelial cells and lactose synthetase B-protein activity was determined as described previously (Vonderhaar, Owens and Topper 1973).

very low when DNA synthesis is prevented. It was shown previously (Vonderhaar, Owens and Topper 1973) that B-protein (α-lactalbumin) activity is virtually undetectable in explants from mature virgin C3H/HeN mice after 72 hours with insulin, hydrocortisone and prolactin, but does appear at 120 hours. It can be seen that the inhibitors, added at zero time, prevent the emergence of B-protein activity. However addition of ara-C or FdU at 72 hours, at a time prior to the emergence of B-protein but after some DNA synthesis occurs, does not depress the B-protein:

DNA ratio. This type of post-mitotic control is critical; in its absence spurious conclusions can be drawn, as will be discussed later in relation to studies with explants from mid-pregnant mice.

The possibility that the effects of ara-C and FdU result from their general toxicity is rendered improbable for several reasons: (1) As stated above, added post-mitotically they do not prevent the emergence of B-protein. (2) They inhibit DNA synthesis by different mechanisms (Heidelberger 1965; Furth and Cohen 1968; Graham and Whitmore 1970), yet have the same effect on the formation of milk proteins. (3) Deoxycytidine prevents the inhibition of both DNA and casein synthesis by ara-C (Table 2). (4) Neither inhibitor blocks the increased ratio, G-6-PD + Gl-6-PD:DNA, which is elicited by IFP (Table 3). These latter results also demonstrate that acquisition of insulin sensitivity, a prerequisite for increased levels of the enzymes (Friedberg, Oka and Topper 1970), is not prevented by the inhibitors.

The experiments reported in Table 4 deal with the question of the hormonal and mitogenic requirements during critical mitosis. For these studies tissue from mature virgin mice that had been primed with estradiol was used; such treatment accelerates the emergence of B-protein activity in vitro, so that it appears after 72 hours with IFP (Vonderhaar, Owens and Topper 1973). Like insulin, serum (free of insulin) can exert a mitogenic effect on virgin mammary gland explants once sensitivity has been acquired (Majumder and Turkington 1971; Oka and Topper 1972). Neither insulin alone, nor serum (fetal calf or porcine) in the presence of hydrocortisone and prolactin, causes B-protein activity to emerge after 48 hours. However those cells made during that initial 48-hour period are fully capable of making B-protein when exposed to IFP and ara-C for an additional 48 hours. Clearly insulin alone can effect critical mitosis. Serum can also promote critical mitosis.

The results in Table 5 demonstrate that ara-C inhibits DNA synthesis and lowers both casein and B-protein production by explants from mice in mid-pregnancy.

Table 2
Reversibility of the Effect of Ara-C by Deoxycytidine

Culture conditions	Epithelial cell number ($\times 10^{-4}$ cells/ mg wet wt)	Casein (cpm/mg wet wt)	Δ Casein/ 10^4 cells
0 hr	2.75		
IF		320	
IFP	3.57	910	165
IFP + ara-C	2.20	400	36.4
IFP + ara-C and dC	3.60	980	183
IFP + dC	3.57	950	176

Mammary explants from mature virgin mice were cultured in the presence of insulin and hydrocortisone (IF) or insulin, hydrocortisone and prolactin (IFP) for 72 hr. 1-β-D-Arabinosylcytosine (ara-C; 15 μg/ml) and/or deoxycytidine (dC; 120 μg/ml) were added at the time of explantation. Casein synthesis and accumulation were determined after a continuous 72-hr exposure of the explants to a [^{14}C]amino acid mixture at 0.25 μCi/ml (Juergens et al. 1965). Epithelial cell number was determined as described previously (Vonderhaar, Owens and Topper 1973).

Table 3

Effect of Ara-C and FdU on Induction of Epithelial Glucose-6-Phosphate Dehydrogenase and Gluconate-6-Phosphate Dehydrogenase in Explants from Virgin Mice

Culture conditions	ng Epithelial DNA/mg wet wt tissue	Combined G-6-PD and Gl-6-PD activities ($\Delta OD_{340} \times 10^3$/ min/mg wet wt)	Enzyme activity/ ng DNA
0 hr	35.6	0.36	.010
72 hr			
IFP	44.3	1.38	.031
IFP + ara-C	31.2	0.92	.029
0 hr	52.6	0.33	0.0063
96 hr			
IFP	98.8	1.20	0.0121
IFP + FdU	52.6	0.55	0.0105

Mammary explants from mature virgin mice were cultured in the presence of insulin, hydrocortisone and prolactin (IFP). 1-β-D-Arabinosylcytosine (ara-C; 15 μg/ml) or fluorodeoxyuridine (FdU; 25 μg/ml) was added at the time of explantation. Enzyme activities and DNA content were determined as described previously (Friedberg, Oka and Topper 1970; Vonderhaar, Owens and Topper 1973).

However the synthesis of these proteins per cell is virtually unaffected by the nucleoside. Pregnancy tissue differs from virgin tissue in this respect. These results point up the importance of the post-mitotic test in studies with inhibitors of DNA synthesis. Previous work with agents such as hydroxyurea (Mayne and Barry 1970) and androgens (Turkington and Topper 1967) led investigators to conclude that critical mitosis is necessary in mid-pregnancy explants. However these agents have since been shown (Owens, Vonderhaar and Topper 1973) to inhibit milk protein production even when added post-mitotically.

Epithelial cells in involuted mouse mammary gland, like those in the gland of the virgin mouse, are virtually devoid of B-protein activity (Table 6). But unlike the cells in the virgin, those in the multiparous animal are themselves competent to make this secretory protein in response to IFP. In this regard they resemble cells in the mid-pregnant mouse.

DISCUSSION

From the results presented it appears securely established that the nonsecretory epithelial cells in the mammary gland of the mature virgin C3H/HeN mouse cannot, themselves, make secretory proteins in response to insulin, hydrocortisone and prolactin. Only their daughter cells, formed either in vitro under appropriate conditions or in vivo during pregnancy, can be converted into producers of milk proteins. Furthermore since the observed total increments in DNA content, which are suf-

Table 4

Hormonal and Mitogenic Requirements for Critical Mitosis

Culture conditions	ng Epithelial DNA/mg wet wt tissue	activity (pmoles Lactose synthetase B-protein product/mg wet wt/30 min)
0 hr	32	Not determined
48 hr		
I	74	<1
SFP	58	<1
72 hr		
IFP	78	32 ± 3
IFP + ara-C		
added at 0 hr	33	<1
96 hr		
I_{48} → IFP + ara-C		
added at 48 hr	60	21 ± 1
SFP_{48} → IFP + ara-C		
added at 48 hr	46	23 ± 1

Mature virgin mice were given a single intramuscular injection of 300 μg of 17-β-estradiol valerate in sesame oil and killed 7 days later. Pooled mammary gland explants were cultured in the presence of insulin (I) or 50% fetal calf serum, hydrocortisone and prolactin (SFP) for 48 hr. The systems were then changed to insulin, hydrocortisone, prolactin (IFP) and 1-β-D-arabinosylcytosine (ara-C; 15 μg/ml) and cultured an additional 48 hr. Lactose synthetase B-protein activity and epithelial DNA content were determined as described previously (Vonderhaar, Owens and Topper 1973).

ficient to permit this transformation in vitro, are often less than 100 percent, probably only one round of cell replication is required for the formation of competent cells. It is not known whether the "critical" component of critical proliferation occurs during S, G_2 or M. Regardless of which phase of the cell cycle is directly implicated, we assume that the ultimate principal sources of milk products are the cells that have completely traversed the cycle and returned to G_1. Nevertheless the eventual elucidation of the chemical nature of the critical event will depend heavily on determination of the particular phase of the cell cycle involved.

Extensive epithelial proliferation occurs in the mammary gland of the adolescent C3H/HeN mouse. As a result the ductal system and primitive alveoli completely penetrate the mammary fat pad. The identity of the mitogenic agent(s) that operates at this time has not been clearly defined. With the attainment of sexual maturity the epithelial proliferation comes to a virtual halt and does not resume until pregnancy begins. The mitogenic agents operative during pregnancy probably are insulin, serum factor(s), or both (Oka and Topper 1972).

We do not know why proliferation stops at maturity. It does not appear to be due to disappearance of the mitogen operative during adolescence since transplantation of ductal fragments from one mature virgin mouse into a cleared fat pad of another

Table 5

Effect of Ara-C on Casein Synthesis and α-Lactalbumin in Explants from Mid-Pregnant Mice

Culture conditions	Epithelial cell number ($\times 10^{-4}$ cells/ mg wet wt tissue)	Casein (cpm/mg wet wt tissue)	Δ Casein/ 10^4 cells	Lactose synthetase B-protein activity (pmoles product/mg wet wt/15 min)	B-protein activity/ 10^4 cells
0 hr	1.21				
72 hr					
IF	Not determined	193			
IFP	2.13	1584	653		
IFP + 45 μg ara-C/ml added at 0 hr	1.10	919	660		
0 hr	4.68			10	2.14
48 hr					
IFP	7.18			73	10.2
IFP + 45 μg ara-C/ml added at 0 hr	4.95			41	8.3
IFP + 90 μg ara-C/ml added at 0 hr	3.70			30	8.1

Mammary explants from mid-pregnant mice were cultured in the presence of insulin and hydrocortisone (IF) or insulin, hydrocortisone and prolactin (IFP). 1-β-D-Arabinosylcytosine (ara-C) was added at the time of explantation. Casein synthesis and accumulation were determined after a continuous 72-hr exposure of the explants to carrier-free $^{32}P_i$ at 1 μCi per ml (Juergens et al. 1965). Lactose synthetase B-protein activity and epithelial cell number were determined as described previously (Vonderhaar, Owens and Topper 1973). Data shown are representative of several experiments.

leads to ductal growth (DeOme et al. 1959). Neither circulating insulin nor serum factor(s) are mitogenic in the mature nonpregnant mouse because the corresponding mammary cells are insensitive to these agents. After isolation these cells acquire the requisite sensitivity (Friedberg, Oka and Topper 1970). Cells in the pregnant animal become sensitive as a result of exposure to increased levels of prolactin (Browning, Larke and White 1962; Bronson, Dagg and Snell 1966; Oka and Topper 1972) or emergence of placental lactogen. These events can account for the proliferation that occurs in explants from virgin mice cultured in the presence of insulin or serum and for the resumption of proliferation that occurs during pregnancy.

It is clear that the mitoses occurring during early development are noncritical, since the cells present in the mature, nonpregnant mouse are themselves incapable of manufacturing milk proteins. In contrast, the mitoses that take place during pregnancy are critical. The competence emanating from critical proliferation during pregnancy is retained by the cells present in the involuted gland. Furthermore mitoses that occur in explants from nonpregnant mice are also critical.

We have considered three interpretations of these observations. (1) Serum in nonpregnant animals may contain an inhibitor that selectively inhibits the "critical" component of mitosis. This agent would not be present in cultures of nonpregnant tissue. The observation that both fetal calf serum and porcine serum can, in fact,

Table 6

Effect of Ara-C on α-Lactalbumin in Explants from Multiparous Mice

Culture conditions	ng Epithelial DNA/mg wet wt tissue	Lactose synthetase B-protein activity (pmoles product/mg wet wt/30 min)	B-protein activity/ ng DNA
Experiment A			
0 hr	30	<1	<0.033
72 hr			
IFP	51	15 ± 2	0.294
IFP + ara-C			
added at 0 hr	32	9 ± 1	0.281
Experiment B			
0 hr	31	<1	<0.032
48 hr			
IFP	48	15 ± 1	0.312
IFP + ara-C			
added at 0 hr	29	10 ± 1	0.344

Following a single pregnancy and a 10–12 day period of lactation, mice were allowed to involute for 4 weeks (experiment A) or 9 weeks (experiment B). Mammary gland explants were cultured in the presence of insulin, hydrocortisone and prolactin (IFP). 1-β-D-Arabinosylcytosine (ara-C; 15 μg/ml) was added at the time of explantation. Lactose synthetase B-protein activity and epithelial DNA content were determined as described previously (Vonderhaar, Owens and Topper 1973).

effect critical mitosis sheds doubt on this possibility. However, definitive disposition of this possibility will require testing of serum from nonpregnant mice. (2) The mitogenic agent itself may determine whether or not mitosis is critical. The mitogen that operates in the immature animal is different from that in the pregnant animal. Also it is clear from considerations of insulin and serum insensitivity and sensitivity (Friedberg, Oka and Topper 1970; Oka and Topper 1972) that the mitogens used with explants from nonpregnant tissue are different from the one that functions in the intact immature animal. Work in progress is aimed at clarification of this possibility. (3) Critical mitosis, in contrast to noncritical mitosis, may only be possible by a mammary cell having characteristics associated with insulin and serum sensitivity. According to this view, mitosis promoted by insulin or serum is not required per se. Rather, the unique properties of a sensitive cell would permit critical mitosis under the influence of any mitogen.

Whatever the detailed nature of the critical component of critical mitosis, it is apparent that it is operative on mammary cells under certain in vitro conditions, and only in particular developmental states within the intact animal. Earlier we discussed other reported examples of the dependence of differentiation upon cell proliferation. The development of still other systems has been observed to be independent of proliferation. Could it be that development of at least some of these systems is, in fact, dependent upon critical mitosis, but that they were examined too late in ontogeny?

Acknowledgment

Some of the data in the Tables was previously published in *J. Biol. Chem. 248*:472 (1973) and *In Vitro 8*:228 (1972).

REFERENCES

Abbott, J. and H. Holtzer. 1965. Critical number of mitoses and the differentiation of chondroblasts and myoblasts. *Anat. Rec.* **151**:439.

Bronson, F. H., C. D. Dagg and G. D. Snell. 1966. Reproduction. *Biology of the laboratory mouse* (ed. E. L. Green) p. 187. McGraw-Hill, New York.

Browning, H. G., G. A. Larke and W. D. White. 1962. Action of purified gonadotropins on corpora lutea in the cyclic mouse. *Proc. Soc. Exp. Biol. Med.* **111**:686.

Ceriani, R. L. 1970. Fetal mammary gland differentiation *in vitro* in response to hormones. II. Biochemical findings. *Develop. Biol.* **21**:530.

DeOme, K. B., L. J. Faulkin, H. A. Bern and P. B. Blair. 1959. Development of mammary tumors from hyperplastic alveolar nodules transplanted into gland-free mammary fat pads of female CeH mice. *Cancer Res.* **19**:515.

Fan, H. and S. Penman. 1970. Mitochondrial RNA synthesis during mitosis. *Science* **168**:135.

Friedberg, S. H., T. Oka and Y. J. Topper. 1970. Development of insulin-sensitivity by mouse mammary gland *in vitro*. *Proc. Nat. Acad. Sci.* **67**:1493.

Furth, J. J. and S. S. Cohen. 1968. Inhibition of mammalian DNA polymerase by the 5'-triphosphate of 1-β-D-arabinofuranosylcytosine and the 5'-triphosphate of 9-β-D-arabinofuranosyladenine. *Cancer Res.* **28**:2061.

Graham, F. L. and G. F. Whitmore. 1970. The effect of 1-β-D-arabinofuranosylcytosine on growth, viability, and DNA synthesis of mouse L-cells. *Cancer Res.* **30**:2627.

Gross, M. and E. Goldwasser. 1970. On the mechanism of erythropoietin-induced differentiation. VII. The relationship between stimulated deoxyribonucleic acid synthesis and ribonucleic acid synthesis. *J. Biol. Chem.* **245**:1632.

Heidelberger, C. 1965. Fluorinated pyrimidines. *Progr. Nucleic Acid Res. Mol. Biol.* **4**:1.

Holtzer, H. 1963. Mitosis and cell transformations. *General physiology of cell specialization* (ed. D. Mazia and A. Tyler) p. 80. McGraw-Hill, New York.

Jailkhani, B. L. and G. P. Talwar. 1972. Induction of phosvitin by oestradiol in rooster liver needs DNA synthesis. *Nature New Biol.* **239**:240.

Juergens, W. G., F. E. Stockdale, Y. J. Topper and J. J. Elias. 1965. Hormonal-dependent differentiation of mammary gland *in vitro*. *Proc. Nat. Acad. Sci.* **54**:629.

Majumder, G. C. and R. W. Turkington. 1971. Stimulation of mammary epithelial cell proliferation *in vitro* by protein factor(s) present in serum. *Endocrinology* **88**:1506.

Martin, D., Jr., G. M. Tomkins and D. Granner. 1969. Synthesis and induction of tyrosine aminotransferase in synchronized hepatoma cells in culture. *Proc. Nat. Acad. Sci.* **62**:248.

Mayne, R. and J. M. Barry. 1970. Biochemical changes during development of mouse mammary tissue in organ culture. *J. Endocrinol.* **46**:61.

Oka, T. and R. T. Schimke. 1969a. Progesterone antagonism of estrogen-induced cytodifferentiation in chick oviduct. *Science* **163**:83.

―――. 1969b. Interaction of estrogen and progesterone in chick oviduct development. I. Antagonistic effect of progesterone on estrogen-induced proliferation and differentiation of tubular gland cells. *J. Cell Biol.* **41**:816.

————. 1969c. Interaction of estrogen and progesterone in chick oviduct development. II. Effects of estrogen and progesterone on tubular gland cell function. *J. Cell Biol.* **43**:123.

Oka, T. and Y. J. Topper. 1972. Is prolactin mitogenic for mammary epithelium? *Proc. Nat. Acad. Sci.* **69**:1693.

Owens, I. S., B. K. Vonderhaar and Y. J. Topper. 1973. Concerning the necessary coupling of development to proliferation of mouse mammary epithelial cells. *J. Biol. Chem.* **248**:472.

Palmiter, R. and J. T. Wrenn. 1971. Interaction of estrogen and progesterone in chick oviduct development. III. Tubular gland cell cytodifferentiation. *J. Cell Biol.* **50**:598.

Paul, J. and J. A. Hunter. 1968. DNA synthesis is essential for increased haemoglobin synthesis in response to erythropoietin. *Nature* **219**:1362.

Stockdale, F. E. and Y. J. Topper. 1966. The role of DNA synthesis and mitosis in hormone-dependent differentiation. *Proc. Nat. Acad. Sci.* **56**:1283.

Turkington, R. W. and Y. J. Topper. 1967. Androgen inhibition of mammary gland differentiation in vitro. *Endocrinology* **80**:329.

Vonderhaar, B. K., I. S. Owens and Y. J. Topper. 1973. An early effect of prolactin on the formation of α-lactalbumin by mouse mammary epithelial cells. *J. Biol. Chem.* **248**:467.

Voytovich, A. E. and Y. J. Topper. 1967. Hormone-dependent differentiation of immature mouse mammary gland in vitro. *Science* **158**:1326.

Wessells, N. K. 1968. Problems in the analysis of determination, mitosis, and differentiation. *Epithelial-mesenchymal interactions* (ed. R. Fleischmajer and R. E. Billingham) p. 132. Williams and Wilkins, Baltimore.

Wessells, N. K. and W. J. Rutter. 1969. Phases in cell differentiation. *Sci. Amer.* **220**:36.

Erythroid Cell Proliferation and Differentiation: Action of Erythropoietin

Paul A. Marks, Linda Cantor, Marvin Cooper, Dinah Singer, George Maniatis, Arthur Bank, and Richard A. Rifkind

Departments of Medicine and of Human Genetics and Development
Columbia University, College of Physicians and Surgeons
New York City, N.Y. 10032

Erythropoietin induction of erythroid cell proliferation and differentiation provides a useful model to investigate several aspects of gene expression during mammalian cell proliferation (Marks and Rifkind 1972). The studies summarized in this paper demonstrate (a) the cellular basis of changes in types of globins formed during fetal development, (b) the target cell of action of erythropoietin, and (c) the effect of erythropoietin on macromolecular synthesis and, specifically, the appearance of biologically active mRNA for globin.

Erythropoiesis in Developing Mouse Fetuses

The systems used in our laboratory for studies of erythroid cell proliferation and differentiation have been fetal mouse yolk sac erythroid cells and liver erythroid cells (Marks and Rifkind 1972) and Friend virus-transformed mouse cells (Friend, Patuleia and DeHarven 1966). To provide background for the fetal mouse erythropoietic systems, we will briefly review the sequence of organ sites of fetal mouse erythropoiesis and the characteristics of globins synthesized by cells differentiating in these different sites.

The mouse has a gestation period of 21 days. The first morphologically identifiable site of erythropoiesis in the developing fetal mouse occurs in the blood islands of the yolk sac at approximately the 8th day of gestation. Precursor cells of circulating nucleated erythroid cells proliferate in these yolk sac blood islands from about the 8th to the 10th day of gestation. Immature nucleated erythroblasts enter the fetal circulation on the 9th day where they continue to proliferate and differentiate through the 13th day of gestation (Marks and Rifkind 1972; Fantoni et al. 1969). The mature circulating erythrocyte of yolk sac blood island origin is nucleated but the cytoplasm contains no mitochondria or ribosomes (de la Chapelle, Fantoni and Marks 1969).

The second site of erythropoiesis in the fetal mouse becomes morphologically detectable in the liver during the 10th day of gestation (Marks and Rifkind 1972). In the fetal liver erythroid cell precursors, referred to as hemocytoblasts, give rise

853

to proerythroblasts, which differentiate through a series of morphologically identifiable stages to non-nucleated reticulocytes. Unlike the terminal differentiation
of yolk sac blood island-derived erythroid cells, the differentiation of liver erythroid
cells is characterized by nuclear expulsion prior to the loss of cytoplasmic ribosomes
and mitochondria. It is of interest to note that the structural changes associated
with the differentiation of the yolk sac erythroid cells are analogous to those
charatceristic of avian and amphibian erythropoiesis. Liver erythroid cell differentiation appears to involve structural changes characteristic of erythropoiesis in
adult mammalian bone marrow.

Thus there are two distinct populations of erythroid cells appearing at different
times during mouse fetal development, namely, the primitive cell line developing in
yolk sac blood islands and the more definitive cell line which appears initially in
the liver.

Direct evidence with respect to the morphological characteristics of yolk sac and
liver erythroid cell lines in the developing fetal mouse is perhaps better than that
available for any other species. These findings of shifting sites of erythropoiesis with
fetal development are analogous to observations in other species (Table 1).

Changes in the types of hemoglobin synthesized during fetal development have
been described not only for the mouse but for several other mammalian and lower
animal species (Marks and Rifkind 1972). Studies in the mouse have provided the
most direct evidence on the relationship between alterations in erythropoietic cell
lines and types of hemoglobin formed. In the erythroid cells of the yolk sac blood
islands of strain C57Bl/6J mice, three hemoglobins are formed (Fantoni, Bank
and Marks 1967). These have been characterized as embryonic hemoglobin E_I,

Table 1

Patterns of Hemoglobin Synthesis Related to Changes
in Erythropoietic Cells during Fetal Development

Species*	Embryonic		Definitive	
	Site	Hemoglobin	Site	Hemoglobin
Mouse	Yolk sac	E_I (xy) E_{II} (αy) E_{III} (αz)	Liver, bone marrow	A $(\alpha\beta)$
Man	? Yolk sac	Gower I (ε) Gower II $(\alpha\varepsilon)$ Portland I $(\zeta\gamma)$	Liver, bone marrow	F $(\alpha\gamma)$ A $(\alpha\beta)$ A_2 (α)
Rabbit	? Yolk sac	E_I $(x\varepsilon)$ E_{II} $(\alpha\varepsilon)$	Liver, bone marrow	A $(\alpha\beta)$
Chicken	Yolk sac	E P $(\alpha^A?)$	Yolk sac, bone marrow	A $(\alpha?)$ D $(\alpha?)$
Frog	Liver	Type I Type II	Liver, bone marrow	Frog I Frog II

* See following references for details: mouse and man (Marks and Rifkind 1972); rabbit
(Steinheider, Medleris and Ostertag 1971): chicken and frog (Ingram 1972).

composed of x and y chains, embryonic hemoglobin E_{II}, composed of α and y chains, and embryonic hemoglobin E_{III}, composed of α and z chains. There is no detectable synthesis of β globin in yolk sac erythroid cells. Structural studies of the globin chains suggest that α, β, x, y and z chains are controlled by separate genes (Steinheider, Medleris and Ostertag 1971).

Mouse fetal liver erythroid cells synthesize a single type of hemoglobin with a globin composition indistinguishable from that of the hemoglobin ($\alpha_2\beta_2$) present in the adult of this species (Fantoni, Bank and Marks 1967). The change in pattern of hemoglobin synthesis from embryonic to adult type of hemoglobin during mouse fetal development is associated with the substitution of liver erythropoiesis for yolk sac erythropoiesis. This cellular basis for changing patterns of hemoglobin synthesis during fetal development is a more general phenomenon in the animal kingdom (Table 1) (Marks and Rifkind 1972). Evidence indicates that an analogous shift from a primitive to a definitive erythroid cell line is associated with the change in the types of hemoglobin synthesis in man, chick, tadpole, rabbits, and probably other species (Table 1). In addition to this pattern of changes in erythropoietic site and cell lines, which appear to determine changes in hemoglobins formed, a further suggestion of pattern may be discerned in the structural alterations in the hemoglobins synthesized in primitive and definitive cell lines. In mouse, man, chicken, goat and sheep, conversion from embryonic to adult type hemoglobins includes the substitution of one globin chain (Table 1).

Target Cell of Action of Erythropoietin

Erythropoietin is the erythropoietic hormone. It has not been prepared in pure, biologically active form. Analysis of partially purified preparations suggest that it has a molecular weight of 46,000 and is a glycoprotein (Goldwasser and Kung 1972).

Central to our understanding of the action of erythropoietin in inducing erythroid cell proliferation and differentiation is identification and isolation of the erythropoietin responsive cell.

Fetal mouse liver has proved to be a suitable tissue to approach the problem of identification of the erythropoietin responsive cell. On day 11 of gestation, approximately 80% of the erythroid cells present in the fetal liver are at a very immature stage and morphologically classified as proerythroblasts (Rifkind, Chui and Epler 1969). These cells are very active in incorporation of uridine into RNA and leucine into protein (Marks and Rifkind 1972). This is indicated by a rate of uptake of [³H]uridine and [³H]leucine by these cells several times higher than that in morphologically more differentiated cells in the same fetal livers and in morphologically comparable cells on subsequent days of gestation. Erythropoiesis in the fetal liver proceeds as a heterogeneous population with regard to cell stage, unlike yolk sac erythroid cells which differentiate in a relatively homogeneous cohort of cells. Fetal liver is a transitional site of erythropoiesis (by birth, the liver is no longer a site of erythropoiesis) involving a self-perpetuating precursor cell, which yields differentiating erythroblasts at least over a short period of time. The disappearance of metabolically active proerythroblasts may reflect the cellular basis for the loss in capacity for sustained erythropoiesis in the fetal liver, as these cells may be the erythropoietin responsive cells.

The identification and purification of the erythropoietin responsive cell from fetal liver was facilitated by understanding certain aspects of erythropoietin action. The hormone-induced increase in hemoglobin formation does not result from a direct effect on the rate of hemoglobin synthesis per cell (Chui et al. 1971). The erythropoietin-stimulated increase in hemoglobin synthesis is due to an increase in the number of cells synthesizing hemoglobin. The hormone acts to maintain the number of immature proerythroblasts and basophilic erythroblasts in the population and to increase the total number of hemoglobin-forming cells. Erythropoietin is required for renewal of the immature population of erythroid cell precursors under these conditions in vitro. In erythroid cells of fetal liver the first detectable effect of erythropoietin is a selective stimulation of RNA synthesis in the most immature cells, the proerythroblasts (Chui et al. 1971). The hormone has no effect on RNA formation in more differentiated erythroid cells or in nonerythroid cells in fetal livers.

The stimulation of RNA synthesis by erythropoietin was used as a criterion for the identification of the erythropoietin responsive cell. These erythropoietin responsive cells are included in a class of precursor cells designated proerythroblasts on the basis of cytological criteria.

There is considerable evidence that the precursor cell responsive to erythropoietin is distinct from the pluripotential hematopoietic stem cell. Stephenson and Axelrad (1971), for example, employing velocity sedimentation of mouse fetal liver cells, obtained a partial separation of erythropoietin responsive cells from hematopoietic spleen colony-forming cells. Exploitation of antigen differences between erythroid cell precursors and more differentiated erythroid cells (Minio et al. 1972) provided an effective basis for isolating a population of immature erythroid precursors. In the presence of complement and antiserum to mature mouse erythroid cells, polychromatophilic erythroblasts and more differentiated cells are hemolyzed. The resulting population of purified erythroid cell precursors can be cultured. These precursor cells require erythropoietin to be maintained in culture and proliferate and differentiate to erythroblasts with the initiation of hemoglobin synthesis (Table 2) (Cantor et al. 1972).

Further evidence that erythropoietin induces proliferation of erythroid cell pre-

Table 2

Effect of Erythropoietin on Erythroid Cell Precursors

	Hr in culture	Number of cells per culture ($\times 10^{-3}$)				
		Pro*	Baso*	Poly*	Ortho*	Total
No addition	0	80	220	0	0	300
No addition	24	5	20	40	10	75
Erythropoietin	24	45	70	370	20	505
No addition	48	0	0	5	10	15
Erythropoietin	48	10	50	300	350	710

See Cantor et al. (1972) for details of methodology.

* Pro, proerythroblast; Baso, basophilic erythroblast; Poly, polychromatophilic erythroblast; Ortho, orthochromic erythroblast.

cursors was obtained by studying the effect of the hormone on the formation of erythroid colonies from precursor cells suspended in semi-solid media (Cooper unpubl.). Colonial growth of precursor cells was achieved in 1.8% methylcellulose-containing CMRL 1066 media, supplemented with vitamins and amino acids. The number of colonies developed under these conditions increased with increasing concentration of erythropoietin up to a concentration of 0.17 units per ml, above which the response plateaus. In higher concentrations of erythropoietin the number of cells per erythroid colony is greater than in cultures with lower concentrations of the hormone. The number of colonies per plate is directly proportional to the number of cells in the inoculum between 1.0×10^5 to 8.0×10^5 precursor cells per ml, under conditions of nonlimiting concentrations of erythropoietin. Thus erythropoietin determines the number of precursor cells that will divide and initiate formation of a colony and effects the size of the colony, i.e., the number of cell divisions of a precursor cell.

Effect of Erythropoietin on Macromolecular Synthesis

Erythropoietin stimulation of RNA synthesis occurs within 1 hour of culture of fetal liver erythroid cells with the hormone (Chui et al. 1971). The erythropoietin stimulation of RNA synthesis is not dependent on any hormone-mediated effect on DNA synthesis (Gross and Goldwasser 1970; Djaldetti et al. 1972). Inhibition of DNA synthesis by hydroxyurea, cytosine arabinoside or 5-fluorodeoxyuridine does not prevent the erythropoietin-stimulated RNA synthesis in proerythroblasts.

While the effects of inhibition of protein synthesis may depend on the nature of the inhibitor, cycloheximide inhibition of protein formation does not prevent the early hormone-stimulated RNA synthesis (Gross and Goldwasser 1972). Erythropoietin-stimulated synthesis of RNA in fetal liver proerythroblasts in culture precedes detectable increase in cell number, the appearance of biologically active mRNA for globin, or hemoglobin synthesis (Djaldetti et al. 1972; Maniatis et al. in prep.). An effect of erythropoietin on DNA formation was observed after 8 to 10 hours of culture (Chui et al. 1971). Inhibition of RNA or of protein synthesis eliminates the erythropoietin effect on DNA formation (Gross and Goldwasser 1969).

With the preparation of purified populations of erythropoietin responsive precursor cells, the effect of erythropoietin on the types of RNA synthesized (Terada et al. 1972) has been examined. Erythropoietin stimulates RNA synthesis in cultured erythroid precursor cells within one hour. The increase in RNA synthesis observed early involves hnRNA, 45S, 32S, and 4S, and subsequently, 28S and 18S RNA. No detectable stimulation in the synthesis of a 10S RNA occurs at this early time. In addition no biologically active mRNA for globin was recoverable from these erythroid precursor cells prior to incubation with erythropoietin (Terada et al. 1972). After 10 hours of culture with the hormone, there is stimulation in the synthesis of 10S RNA as measured by [^{3}H]uridine incorporation. There is also a marked increase in globin mRNA as demonstrated by biological activity of the 10S fraction in directing α and β globin synthesis in a Krebs ascites tumor cell-free system (Maniatis et al. in prep.; Terada et al. 1972). The appearance of globin mRNA activity correlated with the stimulation of globin synthesis in these cells, which occurred between 5 and 10 hours of incubation (Table 3).

Table 3

Activity of mRNA for Globin in Precursor
Cells Culture with Erythropoietin

RNA source	Hr in culture	Total protein cpm	Globin cpm
Total Cell	0	51,540	980
	10	35,360	6940
Cytoplasmic fraction	0	32,040	500
	10	39,920	7420

RNA was extracted from precursor cells before and after
culture with erythroproietin for 10 hr. Two-fifths of the cells
were used for extraction of total cell RNA and the remainder
were used for extraction of RNA from the cytoplasmic fraction. The number of cells used were 7.5×10^7 and 6.2×10^7
for 0 time and 10 hr preparations, respectively. The 6 to 16S
fractions were prepared by sucrose gradient centrifugation and
assayed in the Krebs ascites tumor cell-free system as described by Terada et al. (1972).

Erythroid Cell Proliferation and Differentiation in Friend Virus-infected Cell Cultures

Friend, Patuleia and DeHarven (1966) reported establishing cell culture lines of
murine virus-infected cells in which 1–2% of the cells differentiate to erythroid
cells synthesizing hemoglobin. If dimethylsulfoxide (DMSO) is added to cultures
of these virus-infected cells, erythropoietic differentiation is induced (Friend et al.
1971). Unlike normal fetal mouse erythroid cell precursors, the virus-infected
Friend cells do not appear to respond to erythropoietin in vivo (Mirand 1970).
These cells provide a model for examining transformed hematopoietic precursor
cells and the mechanism of their differentiation, as for example, the mode of action
of DMSO. At least two alternative hypotheses may be developed to account for
the finding that over 50% of the cells become hemoglobinized after treatment with
DMSO. All or almost all of the virus-infected cells may be induced by DMSO to
differentiate to erythroid cells. Alternatively DMSO selects for the proliferation of
a small number of variants of the virus-infected cells capable of differentiating to
erythroid cells and synthesizing hemoglobin. These alternatives were examined by
cloning Friend virus-infected cells in the absence and presence of DMSO in semi-
solid medium containing 1.8% methylcellulose in Dulbecco's modified Eagle
Medium supplemented with 15% fetal calf serum (Singer et al. unpubl.). It was
found that DMSO did not alter the cloning efficiency of Friend cells and that almost
all of the colonies contained large numbers of hemoglobinized cells (Table 4).
These observations suggest that virtually all of the virus-infected cells in these cultures are erythroid precursor cells whose differentiation can be induced by exposure
to DMSO.

Hypothesis for Erythropoietin Action

The erythropoietin responsive precursor cell contains no globin mRNA in biologically active form. Erythropoietin is necessary to induce the proliferation of

Table 4
Effect of DMSO on Friend Virus-infected Cells

DMSO	Plating efficiency (%)	Hemoglobinized colonies* (%)	Hemoglobinized cells/colony (%)
—	42.6	8.3	<10
1%	38.2	94.0	>30

* Evaluated by benzidine staining.

normal precursor cells. The transition from precursor cell to differentiated erythroblast involves the appearance of active globin mRNA, which could reflect either initiation of transcription of globin genes or an increase in the rate of processing of transcribed globin mRNA present in the precursor cells in an inactive form, such as a component of hnRNA. The erythropoietin responsive cell is itself differentiated from a progenitor pluripotential stem cell (Fig. 1).

It may be hypothesized that the hormone responsive cell has a receptor recognition site for erythropoietin, possibly on the cell membrane. The initial effect of the hormone on macromolecular synthesis is to stimulate nuclear RNA synthesis. The increase in RNA synthesis leads to an increase in protein formation—not specifically globin—and, as a consequence, increased DNA synthesis and mitosis. Mitosis affords the opportunity for reprogramming the genome with consequent initiation of globin mRNA synthesis and subsequent globin formation.

This reasoning is consistent with the following known facts: erythropoietin acts selectively on proerythroblasts, and stimulation of a variety of RNAs is the earliest detected effect on macromolecular synthesis. Erythropoietin-stimulated DNA synthesis and mitosis is blocked by inhibitors of RNA formation. Globin mRNA is not present in an active form in the erythropoietin responsive cell and appears only after several hours of incubation with the hormone. Inhibition of RNA or of DNA

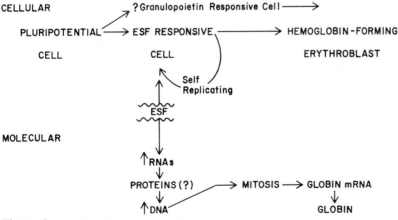

Figure 1
Schematic representation for hypothesis of mechanism of action of erythropoietin induction of erythroid cell differentiation and globin gene transcription. *ESF,* erythropoietin.

synthesis prevents erythropoietin stimulation of globin formation. This hypothesis can be specifically tested with the isolated erythropoietin responsive cell. The critical issues of the site of erythropoietin action and the mechanism of initiation of globin mRNA transcription and its relation to mitosis can be examined in both normal erythroid cell precursors responsive to erythropoietin and virus-infected erythroid cell precursors induced to differentiate by DMSO.

Acknowledgments

Studies presented in this paper are from the laboratories of the authors and were supported in part by grants from the National Institute of General Medical Sciences (GM-14552 and GM-18153), National Cancer Institute (CA 13696) and National Science Foundation (GB-4631, GB-27388X). A. B. is a scholar of the American Cancer Society. G. M. is a Hirschl Trust Scholar.

REFERENCES

Cantor, L. N., A. J. Morris, P. A. Marks and R. A. Rifkind. 1972. Purification of erythropoietin-responsive cells by immune hemolysis. *Proc. Nat. Acad. Sci.* **69**:1337.

Chui, D., M. Djaldetti, P. A. Marks and R. A. Rifkind. 1971. Erythropoietin effects on fetal mouse erythroid cells. I. Cell population and hemoglobin synthesis. *J. Cell Biol.* **51**:585.

de la Chapelle, A., A. Fantoni and P. A. Marks. 1969. Differentiation of mammalian somatic cells: DNA and hemoglobin synthesis in fetal mice. *Proc. Nat. Acad. Sci.* **63**:812.

Djaldetti, M., H. Preisler, P. A. Marks and R. A. Rifkind. 1972. Erythropoietin effects on fetal mouse erythroid cells. II. Nucleic acid synthesis and the erythropoietin-sensitive cells. *J. Biol. Chem.* **247**:731.

Fantoni, A., A. Bank and P. A. Marks. 1967. Globin composition and synthesis of hemoglobins in developing fetal mice erythroid cells. *Science* **157**:1327.

Fantoni, A., A. de la Chapelle, D. Chui, R. A. Rifkind and P. A. Marks. 1969. Control mechanisms of the conversion from synthesis of embryonic to adult hemoglobin. *Ann. N.Y. Acad. Sci.* **165**:194.

Friend, C., M. C. Patuleia and E. DeHarven. 1966. Erythrocytic maturation in vitro of murine (Friend) virus-induced leukemic cells. *Nat. Cancer Inst. Monogr.* **22**:505.

Friend, C., W. Scher, J. G. Holland and T. Sato. 1971. Hg synthesis in murine virus-induced leukemic cells *in vitro*: Stimulation of erythroid differentiation by dimethylsulfoxide. *Proc. Nat. Acad. Sci.* **68**:378.

Goldwasser, E. and C. K.-H. Kung. 1972. The molecular weight of sheep plasma erythropoietin. *J. Biol. Chem.* **247**:5159.

Gross, M. and E. Goldwasser. 1969. On the mechanism of erythropoietin-induced differentiation. V. Characterization of the ribonucleic acid formed as a result of erythropoietin action. *Biochemistry* **8**:1795.

————. 1970. On the mechanism of erythropoietin-induced differentiation. VII. The relationship between stimulated deoxyribonucleic acid synthesis and ribonucleic acid synthesis. *J. Biol. Chem.* **245**:1632.

————. 1972. On the mechanism of erythropoietin-induced differentiation. XI. Stimulated RNA synthesis independent of protein synthesis. *Biochim. Biophys. Acta* **287**:514.

Ingram, V. M. 1972. Embryonic red blood cell formation. *Nature* **235**:338.

Marks, P. A. and R. A. Rifkind. 1972. Protein synthesis: Its control in erythropoiesis. *Science* **175:**955.

Minio, F., C. Howe, K. C. Hsu and R. A. Rifkind. 1972. Antigen density on differentiating erythroid cells. *Nature New Biol.* **237:**187.

Mirand, E. A. 1970. Nonerythropoietin-dependent erythropoiesis. *Regulation of hematopoiesis* (ed. A. S. Gordon), vol. 1, p. 635. Appleton-Century-Crofts, New York.

Rifkind, R. A., D. H. K. Chui and H. Epler. 1969. An ultrastructural study of early morphogenetic events during the establishment of fetal hepatic erythropoiesis. *J. Cell Biol.* **40:**343.

Steinheider, G., H. Medleris and W. Ostertag. 1971. Mammalian embryonic hemoglobins. *Syntheses, Struktur und Funktion des Hamoglobins* (ed. Martin and Nowicki) p. 225–35. J. F. Lehmanns Verlag, Munchen.

Stephenson, J. R. and A. A. Axelrad. 1971. Separation of erythropoietin-sensitive cells from hemopoietic spleen colony-forming stem cells of mouse fetal liver by unit gravity sedimentation. *Blood* **37:**417.

Terada, M., L. Cantor, R. A. Rifkind, A. Bank and P. A. Marks. 1972. Globin mRNA activity in erythroid precursor cells and the effect of erythropoietin. *Proc. Nat. Acad. Sci.* **69:**3575.

Erythroid Differentiation
in Murine Erythroleukemia

Joseph LoBue, A. S. Gordon, A. Weitz-Hamburger,
P. Ferdinand, and J. F. Camiscoli

Laboratory of Experimental Hematology, Biology Department
New York University, New York, N.Y. 10003

T. N. Fredrickson

Department of Pathobiology, University of Connecticut
Storrs, Connecticut 06268

W. D. Hardy, Jr.

The Sloan-Kettering Institute for Cancer Research
New York, N.Y. 10021

Studies of erythropoiesis have often involved perturbations of the process by (a) removal of mature cells from the peripheral circulation to increase the demand for differentiated cells; (b) addition of mature cells to the peripheral circulation, decreasing demand; and (c) administration of erythropoietin, steroids and other hormones to increase numbers of cells undergoing differentiation. It would be convenient in such studies to employ animals that are able to accumulate erythroid precursors so that the response to a given manipulation is exaggerated and hence less equivocal. A model system employing such an accumulation of differentiable precursors has been developed in conjunction with our work (Camiscoli et al. 1972; Fredrickson et al. 1972; LoBue et al. 1972; Weitz-Hamburger et al. unpubl.) on erythropoiesis, using a particular strain of Rauscher murine leukemia virus (RLV), which we refer to as RLV-A. In order to indicate the potential usefulness that we feel this tool affords, we shall describe the pathogenesis of the anemia characteristic of virally infected mice and some experimental results obtained with treatment of these animals with phenylhydrazine (PHZ) or erythropoietin.

MATERIALS AND METHODS

General

Animals and Virus Preparation

Balb/c mice of an isolated colony maintained at the University of Connecticut equally divided as to sex were used in all studies.

Undiluted viremic plasma (0.1 ml) from mice bearing a transplantable leukemia (Fredrickson et al. 1972) was inoculated either intraperitoneally or intravenously. The transplantable leukemia was originally isolated from peripheral blood cells of mice inoculated with a strain of Rauscher Leukemia Virus (RLV) supplied by Dr. Michael Chirigos of the National Cancer Institute.

For pathogenesis studies, to insure attainment of progressive stages of the disease, 90 male and female mice were inoculated either intraperitoneally at 1 and 21 days of age or intravenously at 21 days of age. Eighty-five uninoculated mice (including both sexes) served as controls. All other mice used in this study were inoculated intraperitoneally as weanlings.

Hematological Parameters

Hematocrits, red and nucleated blood cell and differential counts were determined by standard techniques. Reticulocyte percentages were calculated from dried blood films treated with new methylene blue (Brecher 1949).

Preparation of Spleen and Bone Marrow

At sacrifice one femur was rapidly dissected out of each animal, split lengthwise and the contained marrow aspirated into a preweighed double oxalated capillary tube (Aloe 0.8 mm × 32 mm). The tube was reweighted to determine marrow weight and the marrow expelled into fetal calf serum, suspended and counted. The suspension was then concentrated by centrifugation and smears made. A sample of spleen, obtained after recording total splenic weight, was treated similarly. Spleen and bone marrow milligram cellularity was determined as previously described (Fruhman and Gordon 1955; LoBue et al. 1963). Smears of suspended marrow and spleen cells were stained with Wright-Giemsa or LoBue et al.'s (1963) modification of Ralph's benzidine technique (1941). Histological sections of spleen and liver were prepared from specimens fixed in Lillie's neutral formalin and stained with hematoxylin and eosin.

Bioassay of Erythropoietin

Plasma samples were assayed for erythropoietic activity in the post-hypoxic polycythemic mouse assay (Camiscoli and Gordon 1970). Each mouse received a single intraperitoneal injection of 0.5 ml test material diluted in 0.5 ml isotonic saline.

Effect of Erythroid Perturbation on the Course of RLV-induced Erythroleukemia

Experimental Hemolytic Anemia

Thirty-three mice received 0.1 ml (50 mg/kg) of a 1% solution of phenylhydrazine hydrochloride (PHZ) subcutaneously daily for 5 days. Animals received virus 1 day after cessation of PHZ treatment and were sacrificed in groups every 2 weeks for 16–18 weeks. Twenty mice not inoculated with virus, but treated similarly with PHZ, were sacrificed every 2 weeks for 16–18 weeks.

Controls

Thirty-six male and female mice, inoculated with RLV at the same time as PHZ-treated mice, were sacrificed in groups every 2 weeks for 16–18 weeks as a test of viral potency. Thirty male and female mice, sacrificed in groups every 2 weeks, provided normal hematopoietic data over a 4-month period (Table 1).

"Therapy"

Response to PHZ during the Course of RLV Disease

The following studies were designed to evaluate the potential therapeutic benefit of PHZ treatment during various stages of RLV-induced anemia. Three groups of

Table 1

Representative Hematological Data for Serially Sacrificed Untreated Balb/c Mice

Week	HCT (%)	RBC ($\times 10^{-9}/mm^3$)	WBC ($\times 10^{-3}/mm^3$)	Retic. (%)	Sp. Wt. (mg)
0	48.0 ± 0.1[a]	8.07 ± 0.93	5.66 ± 0.26	3.1 ± 0.2	108 ± 10
2	47.0 ± 1.0	7.86 ± 0.40	5.46 ± 0.57	2.8 ± 0.3	128 ± 30
10	45.3 ± 0.3	6.90 ± 0.15	6.51 ± 0.44	3.7 ± 0.6	83 ± 10
12	46.0 ± 1.0	7.16 ± 0.81	5.15 ± 0.22	3.6 ± 0.2	106 ± 5
14	46.0 ± 1.0	7.05 ± 0.26	5.60 ± 0.20	3.5 ± 0.2	124 ± 4
16	45.6 ± 0.6	7.31 ± 0.31	6.55 ± 0.80	4.5 ± 0.7	159 ± 11

Data summarized from Weitz-Hamburger et al. (1973).

HCT = hematocrit; RBC = red cell count; WBC = white cell count; Retic = reticulocyte count; Sp. Wt. = spleen weight

[a] Mean ± SE_m (five mice)

mice were established. Group I consisted of mice with hematocrits of approximately 20%; group II, approximately 30%; and group III, approximately 40%. Half the animals in each group received 0.1 ml of 1% PHZ daily. Group I received two daily doses; group II, three daily doses; and group III, four daily doses. The remaining virus-inoculated mice were untreated controls. Reticulocyte counts, hematocrits and smears of tail vein blood were taken immediately before PHZ treatment and 1 day, 1 week, and 2 weeks after treatment, and every 2 weeks thereafter. Moribund mice were sacrificed every 2 weeks and advanced erythroleukemia established by the usual criteria.

Effect of Splenectomy on the PHZ Response

This study was designed to determine the source of the reticulocytes released in response to PHZ during the midcourse of RLV-induced anemia. Erythroid cells are generally produced in spleen, bone marrow, and liver of RLV-infected animals. However erythroid hepatic involvement is seen only in animals with advanced disease (i.e., hematocrits 20% or less, LoBue et al. 1972). Thus any reticulocytes produced in response to PHZ treatment during the mid-RLV disease were presumed to be of splenic or marrow origin. To test this ten male and female erythroleukemic mice, with hematocrits of 30%, were splenectomized 1 week before receiving three daily doses of PHZ. Hematocrits and reticulocyte counts were made before and 1 week after splenectomy. Mice were sacrificed 1 day after cessation of PHZ treatment and hematocrits and reticulocyte percentages ascertained. Ten male and female sham splenectomized RLV-infected mice given PHZ, and ten RLV-infected splenectomized mice served as controls. In addition ten splenectomized uninfected mice were given PHZ. All blood parameters determined for the experimental group were examined for these groups.

Role of Erythropoietin in Ameliorating the Course of RLV Anemia

Erythropoietin Levels in RLV Anemic Mice Treated with PHZ

Although infected mice fail to produce erythropoietin in response to RLV-induced anemia (Camiscoli et al. 1972; Ebert, Maestri and Chirigos 1972), the possibility nonetheless existed that increased production of erythropoietin might occur after the application of additional anemic stress and that this might be an important factor in the "therapeutic" response invoked by PHZ. Therefore 15 male and female RLV-infected mice, whose hematocrits averaged about 30%, received 0.1 ml of a 1% solution of PHZ subcutaneously daily for 3 days. Two days later the mice were sacrificed, plasma collected for erythropoietin assay, spleen weights recorded, and hematocrits and reticulocyte counts made.

Erythropoietin Levels in Bled, RLV-infected Mice

RLV-infected anemic mice were bled to determine whether hemorrhagic stress would also increase erythropoietin production. Fifteen male and female RLV-infected mice were bled 0.2 ml daily for 3 days from the infraorbital sinus, receiving 0.6 ml of saline to maintain normal blood volume. They were sacrificed 2 days after the last bleeding. Plasma was collected for erythropoietin assay and reticulocyte counts, hematocrits and spleen weights determined. Ten normal mice, similarly bled, were controls.

Effect of Erythropoietin in RLV Anemic Mice

Ten male and female RLV-infected mice, with a mean hematocrit of 33.2% ± 2.0, received 0.2 International Units of erythropoietin in saline suspension subcutaneously for 13 of 17 days (days 1–3, 6–10, and 13–17). The hormone was human urinary erythropoietin (HUE-B-1) obtained from the National Heart and Lung Institute and was derived from patients with anemia due to hookworm infestation. Controls consisted of RLV-inoculated and uninoculated mice not given erythropoietin. Microhematocrits and reticulocyte counts were made 3, 6, 10, 13 and 17 days after initiation of treatment.

RESULTS AND DISCUSSION

Pathogenesis

Although a hemolytic component exists (Morse, Fredrickson and LoBue unpubl.), Balb/c mice infected with RLV-A develop a number of lesions consistent with a maturation block of erythroid precursors. Erythoid cell cycle kinetics are unaltered (Table 2; LoBue et al. 1972) and the spleen accumulates erythroblasts in increasing numbers with a concomitant reduction in lymphoid cells. Moreover the mice become increasingly anemic with no evidence of compensatory reticulocytosis (Fig. 1). These changes can be easily distinguished in histologic sections showing increasing numbers of erythroblasts with progression of the disease in splenic red pulp and, in line with their diminished proportion, lymphoid follicular structures, which become compressed and atrophic (Fig. 2). In the final stages of the disease the splenic pulp becomes a sheet of erythroblasts and immature hemic precursor elements (LoBue et al. 1972; Fig. 3). The number of cells per mg of splenic tissue is not increased[1] (Table 3), presumably because increases in erythroid precursors are accommodated by tremendous increases in total splenic size. This splenic enlargement occurs gradually and serves as a good indicator for staging the disease.

---

[1] Splenic milligram cellularity is actually reduced. We interpret this as resulting from accumulation of immature erythroid precursors and progenitor cells which are larger in size than the more mature elements.

Table 2

Average DNA Synthesis Time and Generation Time for Erythroblasts from Balb/c Mice

	Nonleukemic male		Erythroleukemic male		Nonleukemic female		Erythroleukemic female	
	DNA	Gen.	DNA	Gen.	DNA	Gen.	DNA	Gen.
Marrow								
Proerythroblasts	3.4	8.4	3.1	6.3	2.7	6.3	3.9	9.4
Basophilic erythroblasts	3.3	6.9	3.2	7.2	4.1	7.8	3.4	7.1
Polychromatophilic erythroblasts	4.4	10.2	4.2	10.0	4.8	9.9	5.1	11.0
Spleen								
Proerythroblasts	3.4	8.9	3.5	6.4	3.7	6.7	3.9	7.1
Basophilic erythroblasts	4.1	8.3	4.2	8.0	3.8	6.6	3.7	6.3
Polychromatophilic erythroblasts	4.5	11.4	4.6	12.4	4.5	9.5	4.5	11.2

Data summarized from LoBue et al.(1972). Estimates (in hr) were obtained by [^{14}C]-[^{3}H]thymidine double labeling.

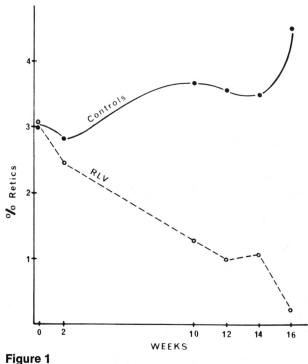

Figure 1

Alterations in reticulocyte counts in representative control and RLV-treated mice. Each point on the curves represents the mean values for five mice. Tabulated data from which the control curve was constructed may be found in Table 1.

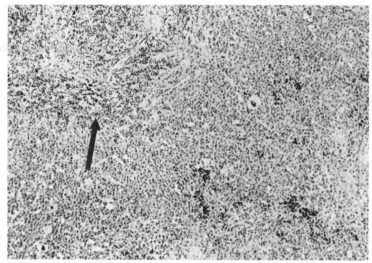

Figure 2
Spleen from moderately advanced RLV-anemic mouse showing atrophic lymphoid follicle (arrow) and accumulation of erythroid precursors. Tissue sections for Fig. 2–5 and 7 were stained with hematoxylin and eosin. 60×

Concomitant with splenic changes the liver becomes filled with erythroblasts around portal triads and within sinusoids (Fig. 4), but these changes occur considerably later than in the spleen. Within femoral and sternal marrow no drastic changes are observed histologically (Fig. 5, Table 3). Large numbers of erythroid precursors are observed within peripheral blood with time (Fig. 6) and a distinct leukocytosis

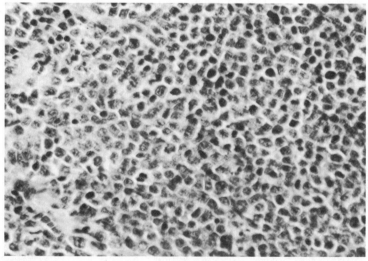

Figure 3
Spleen from terminal RLV-anemic mouse. 600 ×

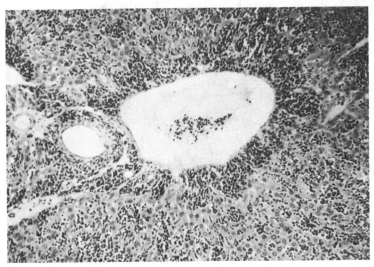

Figure 4
Liver from terminal RLV-anemic mouse showing erythroid precursors in sinusoids and periportal areas. 60 ×

is also observed in some advanced cases. Because of the large number of nucleated cells in the blood, stasis is apparent within vessels of alveolar walls of the lung and occasionally thrombosis of pulmonary arteries occurs (Fig. 7). Infected mice

Table 3
Marrow and Spleen Cellularity and Erythroid Counts for Balb/c Mice

	Nonleukemic male	Erythroleukemic male	Nonleukemic female	Erythroleukemic female
Marrow cells (millions/mg)	2.3 ± 0.2[a]	1.5 ± 0.1	2.8 ± 0.1	1.5 ± 0.2
Erythroid count (thousands/mg)				
Proerythroblast	8.3 ± 3.5	16.4 ± 4.7	7.7 ± 2.1	7.9 ± 2.7
Basophilic erythroblast	40.2 ± 4.8	52.1 ± 12.5	55.3 ± 4.9	30.0 ± 5.1
Polychromatophilic erythroblast	238.1 ± 31.0	275.4 ± 47.6	358.6 ± 52.4	201.8 ± 30.3
Orthochromatic normoblast	271.0 ± 46.9	287.5 ± 74.9	371.5 ± 38.6	227.1 ± 20.7
Spleen weight (g)	0.09 ± 0.004	2.01 ± 0.5	0.10 ± 0.003	0.81 ± 2.9
Spleen cells (millions/mg)	4.3 ± 0.5	2.8 ± 0.1	4.5 ± 0.7	2.5 ± 0.1
Erythroid count (thousands/mg)				
Proerythroblast	24.7 ± 19.5	34.5 ± 7.7	12.7 ± 3.3	52.9 ± 17.9
Basophilic erythroblast	44.2 ± 28.7	156.4 ± 39.7	19.0 ± 4.6	144.0 ± 58.7
Polychromatophilic erythroblast	121.9 ± 36.8	688.9 ± 135.2	84.7 ± 32.6	354.2 ± 123.0
Orthochromatic normoblast	66.9 ± 21.7	771.8 ± 128.2	97.1 ± 17.4	212.5 ± 55.3

Data summarized from LoBue et al. (1972).

[a] Mean ± SE_m

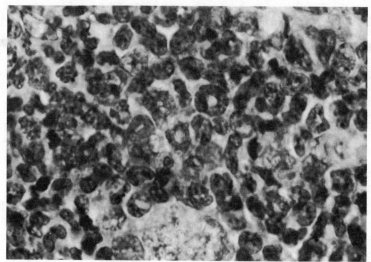

Figure 5
Sternal marrow from terminal RLV-anemic mouse. Active myelopoiesis, erythropoiesis and megakaryocytopoiesis may be readily observed. 600 ×

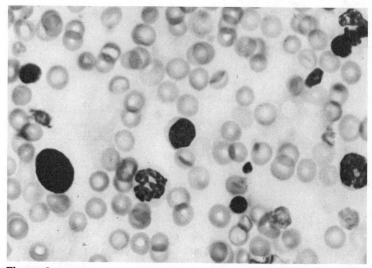

Figure 6
Peripheral blood from terminal RLV-anemic mouse showing peripheralization
of erythroid precursors. Wrights-Giemsa stain. 600 ×

presumably die from affects of severe anemia (the packed cell volume is about
10–15% in terminal cases) since hemorrhage or indications of sepsis, the complications of leukemia in humans, are not evident.

It is of considerable interest that in addition to erythroblasts (Fig. 8) granulocytic precursors, megakaryocytes and monocytoid cells also become infected with
virus (Fig. 9–11).

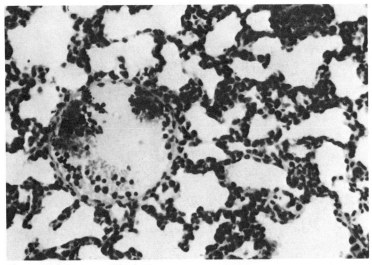

Figure 7
Lung from terminal RLV-anemic mouse indicating thrombosis of pulmonary vessels. 150 ×

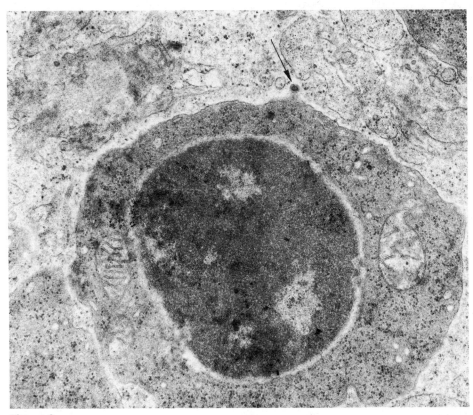

Figure 8
Erythroblast from spleen of week 12 PHZ-pretreated RLV mouse showing viral budding from plasma membrane (at arrow). Glutaraldehyde fixation. 6805 ×

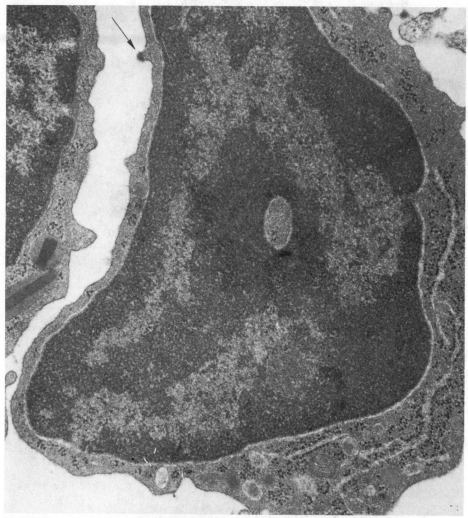

Figure 9
Promyelocyte from the bone marrow of week 2 RLV mouse showing viral budding from plasma membrane (at arrow). Glutaraldehyde fixation. 30,700 ×

Plasma obtained from mice at all stages of the disease had little erythropoietic activity when measured in the exhypoxic polycythemic mouse bioassay. All plasma pools contained less than 0.05 International Units (I.U.) of erythropoietin per 0.5 ml when compared to calibrated secondary standards.

Thus the disease induced with RLV-A, which we describe here, is quite different from that usually attributed to RLV (Siegler 1968), which entails (a) marked necrosis and hemorrhage of splenic tissue with subsequent growth of erythroblastic foci in surviving animals, and (b) a short time to death, usually within 40–50 days after infection. We regard this "typical" Rauscher disease to be relatively unsuitable for investigations of maturation defects in precursors infected with oncornaviruses because of the large degree of tissue damage involved. On the other hand,

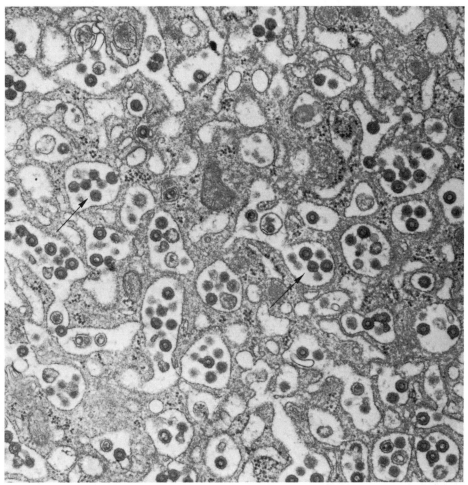

Figure 10
Megakaryocyte from the bone marrow of week 2 PHZ-pretreated RLV mouse showing intravesicular virions (at arrows) and viral budding. Glutaraldehyde fixation. 31,500 ×

the isolate that we use provides considerable opportunity to exploit accumulations of infected precursors in the spleen, from which they can be mobilized under proper experimental conditions (such as PHZ and erythropoietin treatment, described below).

The difference in response of Balb/c to RLV-A is possibly associated with a low rate of viral replication. Serial passage of plasma from infected mice is not accompanied by changes in response in vivo and splenic extracts have been uniformly poor sources of virus. Our viral stocks were tested for presence of spleen focus-forming virus and lymphatic leukemia virus and has been found to contain the latter only. Hence splenic foci are not induced with this isolate.

Origin of the isolate invites some speculation as to its reduced virulence. Balb/c mice were given graded doses of RLV originally obtained from the National Cancer Institute; mice receiving low doses developed myelogenous leukemia and leukemic cells could be transplanted into Balb/c mice (Fredrickson et al. 1972). Plasma from mice with implanted leukemic cells was found to contain the RLV-A isolate and

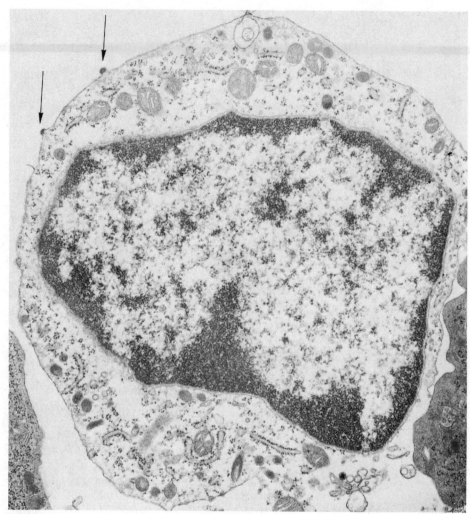

Figure 11
Monocytoid cell from bone marrow of week 2 PHZ-pretreated RLV mouse showing viral budding from plasma membrane (at arrows). In addition to these mononuclear elements, lymphocytes also became infected with virus. Glutaraldehyde fixation. 20,000 ×

this has been our method of propagating virus stocks. Occurrence of myelogenous leukemia has been observed (Pluznik, Sachs and Resnitsky 1966) in RLV infection and would be a reasonable response since myeloid precursors do become infected. Thus RLV-A and possibly all strains of the virus infect a wide variety of hematopoietic precursors. The predominant effect is upon erythroid cell differentiation, but under special circumstances myelogenous responses can also be observed. Thus preliminary studies in our laboratory indicate that continuous stimulation with erythropoietin permits manifestation of a myeloid response that would otherwise not occur or be masked. Further investigations may show that the type of leukemia induced by RLV-A is dependent upon the experimental conditions imposed on the infected mouse.

Effect of Phenylhydrazine-induced Anemia on RLV Disease

Effect of PHZ Alone

Following PHZ treatment mice were anemic and splenic weights increased. Histologically the splenic red pulp was moderately hypertrophied with erythroblasts accounting for 66.7% of the total cell population. These changes were temporary and in mice given PHZ alone, peripheral blood, splenic and marrow hemograms returned to normal by 4 weeks after cessation of treatment. Thus the hematological effect of PHZ in normal Balb/c mice was quite conventional. Plasma (0.5 ml) from mice sacrificed 1 day after cessation of PHZ treatment produced twice the mean iron incorporation into peripheral red cells of exhypoxic mice as 0.05 I.U. of erythropoietin.

Effect of PHZ and RLV

Pretreatment of mice with PHZ before viral infection ameliorated the course of the disease. Spleen weights of virus-infected, PHZ-pretreated mice were slightly greater than those of virus controls until the onset of splenomegaly occurred among the latter about weeks 10–12 (Fig. 12). Splenomegaly indicative of endstage disease

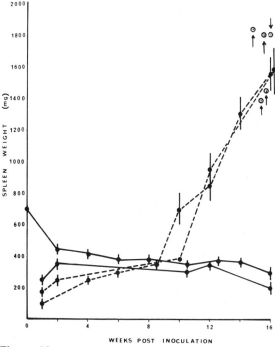

Figure 12

Changes in spleen weight following inoculation of mice with RLV. PHZ refers to animals pretreated with phenylhydrazine. Dotted circles at ends of arrows indicate animals for which PHZ-pretreatment had no protective effect. In Figs. 12 and 13 results of two experiments are superimposed to emphasize reproducibility. In this and subsequent Figures, vertical lines denote ± 1 standard error of the mean. Data summarized from Weitz-Hamburger et al. (1973). (———) PHZ; (– – –) control.

was observed in only five of 33 PHZ-treated animals and only after week 14 (Fig. 12). This decreased incidence of frank disease was similarly reflected in the peripheral blood. Although erythroblasts initially appeared in the peripheral blood in response to the induced hemolytic anemia, none were seen in any PHZ-treated mice after 2 weeks, with the exception of the five that developed late splenomegaly. Hematocrits and red cell counts rose following cessation of PHZ treatment but never achieved normal levels (Fig. 13). Although reticulocyte counts increased in response to PHZ-induced anemia, they fell below normal at week 2 and remained so throughout the study. Plasma (0.5 ml) from RLV-PHZ mice possessed erythropoietic activity equivalent to at least 0.05 I.U. of erythropoietin for 6 weeks after cessation of PHZ treatment (Table 4). Mice treated with RLV alone showed the typical erythroleukemic course indicated earlier.

Hence treatment with PHZ prior to RLV infection was ameliorative in that the erythroid maturation "block" was temporarily overcome and normal erythropoiesis proceeded. However, although treated, infected mice were capable of carrying on normal erythropoiesis, delayed progression of the disease and not permanent "cure" may be the final outcome. This is presently being evaluated by a long-term study now in progress.

An important question, however, remains, namely: how does PHZ exert this "protective" effect? It would be tempting to invoke a PHZ-induced alteration in numbers of viral target cells as the cause of this "protection." However splenic CFU and proerythroblasts are known to be increased in PHZ-induced hemolytic anemia (Hodgson, Bradley and Telfer 1972; Rencricca et al. 1969; Stohlman 1972), and

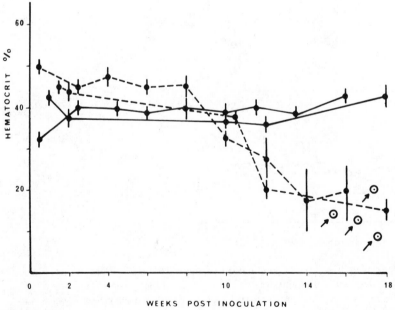

Figure 13

Hematocrit values in mice following inoculation with RLV. Dotted circles as in Fig. 12. Data summarized from Weitz-Hamburger et al. (1973). (————) PHZ; (– – –) control.

Table 4

Erythropoietic Activity of Pooled Plasma from Balb/c Mice Inoculated with RLV and Pretreated with PHZ

Weeks post inoculation	Hematocrit range of donor mice	Mean percent ^{59}Fe peripheral RBC incorporation			
		Plasma	0.9% NaCl	0.05 I.U. Ep	0.20 I.U. Ep
0	20–30	11.25 ± 2.75[a]	0.70 ± 0.15	3.22 ± 0.70	Not done
4–6	30–40	14.80 ± 0	1.17 ± 0.25	4.34 ± 0.74	1 .47 ± 2.42
8	30–40	0.93 ± 0.19	0.70 ± 0.15	3.22 ± 0.70	Not done
10	30–40	0.72 ± 0	0.70 ± 0.15	3.22 ± 0.70	Not done
12	30–40	0.93 ± 0.17	0.78 ± 0.14	4.70 ± 0.86	9.86 ± 2.08
14	30–40	1.56 ± 0.5	1.79 ± 0.34	6.24 ± 1.14	12.51 ± 1.50
16	30–40	0.87 ± 0.19	1.79 ± 0.34	6.24 ± 1.14	12.51 ± 1.50

[a] Mean ± SE_m.   Ep = erythropoietin.

erythropoietin (in this instance, produced by PHZ treatment) is believed to increase ERC numbers (Reissmann and Samorapoompichit 1970). All three of these cell types have been implicated as potential RLV target cells (Plusnick, Sachs and Resnitsky 1966; Seidel 1973; Tambourin and Wendling 1971). Thus any protective effect of PHZ would have to be interpreted in light of an increased number of RLV target cells. This would seem unlikely. Moreover effects on target cell number do not seem relevant since PHZ-pretreated RLV mice show a similar degree of viral infection as RLV-alone mice when this is judged by both electron microscopy (Ferdinand 1973) and direct immunofluorescence using antibodies to MuLV gs antigen (Fig. 14, 15).

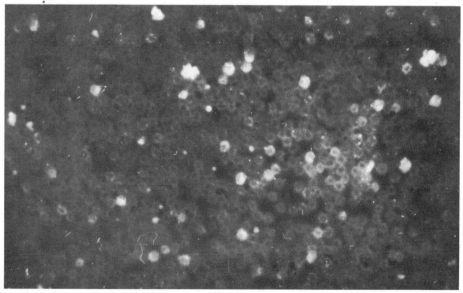

Figure 14

Presence of MuLV gs antigen in mature circulating erythrocytes of RLV-infected mice at 6 weeks post-inoculation. 235 ×. From Weitz-Hamburger et al. 1973 with permission of *Cancer Research*.

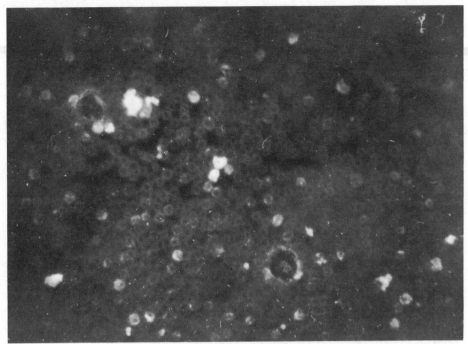

Figure 15
Presence of MuLV gs antigen in mature circulating erythrocytes of PHZ-pretreated, RLV-infected mice at 6 weeks post-inoculation. 237.5 ×

A possible explanation for the PHZ effect is that the reticuloendothelial system (RES) was stimulated by PHZ-induced hemolysis and this altered infectivity. Pertinent evidence that hyperstimulation of the RES affects viral oncogenesis was presented by Heller (1965), who found that pretreatment of mice with an extract of shark oil capable of producing a tenfold increase in phagocytic activity retarded the development of Friend leukemia.

On the other hand PHZ induced increased levels of plasma erythropoietin in pretreated RLV mice for up to 6 weeks after RLV infection. And although the significance of erythropoietin in this disease remains to be determined, these higher levels of erythropoietin might account for this increased erythroid maturation. That delayed progression of the disease may occur in RLV-PHZ mice could be related to the fact that the rise in plasma erythropoietin levels is only temporary.

PHZ "Therapy"

Response to PHZ during the Course of RLV-induced Disease

PHZ treatment was poorly tolerated by RLV-anemic mice and five of seven mice in group I (20% hematocrit), one of eight in group II (30% hematocrit), and four of eight in group III (40% hematocrit) died during treatment compared to 20% mortality in control mice given five daily doses of PHZ. Among all surviving mice, however, the response with regard to accelerated erythropoiesis was dramatic, as illustrated in Fig. 16 and 17 for group II mice.

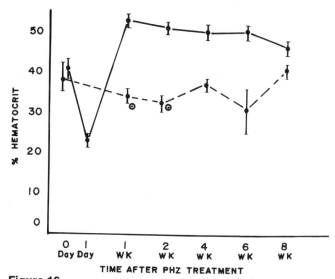

Figure 16

Alterations in hematocrit following PHZ treatment of RLV-anemic mice. Dotted circles: mice not responding to PHZ. Data summarized from Weitz-Hamburger et al. (unpubl.). (——) PHZ-treated; (– – –) virus control.

Thus under conditions of the anemia caused by PHZ, increased erythroid maturation was induced in RLV anemic animals. Preliminary evaluation of anti-gs direct immunofluorescence indicates that a large proportion of the peripheral erythrocytes and reticulocytes observed following PHZ treatment was virally infected.

Our finding of enhanced erythroid differentiation is similar to the observation of Gallien-Lartigue et al. (1969), who demonstrated endogenous hematopoietic recovery after exposure to whole body X-irradiation in Swiss mice made splenomegalic by infection with the anemia-producing strain of Friend virus. Both hematocrits and reticulocytes in severely splenomegalic mice rose above pre-irradiation

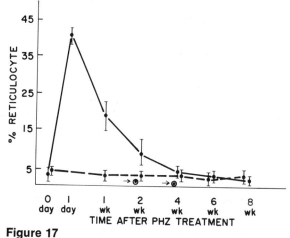

Figure 17

Reticulocyte response to PHZ treatment of RLV-anemic mice. Dotted circles as in Fig. 16. Data summarized from Weitz-Hamburger et al. (unpubl). (——) PHZ; (– – –) control.

Table 5
Changes in Reticulocyte Count and Hematocrit in Splenectomized RLV-infected and Normal Mice

	Pre-Splenectomy		Pre-PHZ treatment		Post-PHZ treatment	
	Hct	Retic	Hct	Retic	Hct	Retic
Normal-splnx[a]	52.0 ± 1.3[b]	3.1 ± 0.3	51.4 ± 0.8	2.4 ± 0.2	23.8 ± 1.3	5.5 ± 0.5
RLV-splnx	33.5 ± 1.5	2.5 ± 0.3	28.2 ± 1.1	2.8 ± 0.5		
Sham-splnx RLV-PHZ	33.5 ± 1.5	2.5 ± 0.3	28.2 ± 1.1	2.8 ± 0.5	19.2 ± 1.0	14.4 ± 1.9
Splnx RLV-PHZ	33.5 ± 1.5	2.5 ± 0.3	28.2 ± 1.1	2.8 ± 0.5	17.5 ± 1.0	4.6 ± 1.0
RLV-PHZ			30.0 ± 1.9	2.0 ± 0.3	19.2 ± 2.1	27.0 ± 2.5

[a] Splnx = splenectomy.   [b] Mean ± SE_m (ten mice).

values, but Friend disease gradually recurred, although the course of the disease was considerably retarded. Their study did not establish, however, whether hematopoietic recovery was the result of the differentiation of erythroid cells infected with FLV or due to proliferation of clones of uninfected erythroid cells. Similarly Friend et al. (1971) have shown that in vitro maturation of Friend leukemic erythroblasts could be effected by addition of dimethylsulfoxide (DMSO) to culture medium without loss of either cytoplasmic C-type particles (Sato, Friend and deHarven 1971) or the ability to produce subcutaneous reticulum cell sarcomas in DBZ/2 mice (Friend et al. 1973). Fischinger et al. (1972) have also reported reversion in vitro of Moloney sarcoma virus (MSV)-infected 3T3 cells to what appeared to be morphologically and functionally normal cells.

Effect of Splenectomy on the Response to PHZ

The data in Table 5 indicates that anemic, RLV-infected, splenectomized mice failed to mount as great a reticulocyte response to PHZ-induced hemolysis as intact RLV-infected mice. These findings indicate, as suspected, that many of the reticulocytes produced by RLV mice in response to the PHZ were of splenic origin.

Role of Erythropoietin in Ameliorating RLV Disease

RLV-Anemic Mice Injected with PHZ

Bioassay of plasma obtained from RLV-anemic mice made more anemic with PHZ indicated that these mice produced considerable erythropoietin (Table 6). The equivalent of 0.20 I.U. of erythropoietin were found in 0.5 ml plasma 2 days after cessation of PHZ treatment.

Bled RLV-infected Mice

Bled RLV-anemic mice responded as PHZ-treated mice, with increased spleen weights, reticulocytosis, and decreased red cell counts and hematocrits 2 days after cessation of bleeding. Erythropoietin levels were elevated to about 0.05 I.U. per 0.5 ml of plasma (Table 7). This was somewhat lower than levels obtained after PHZ-induced hemolysis, but PHZ treatment lowered hematocrits more than phlebotomy did. Normal uninfected mice, similarly bled, also had reduced hemato-

Table 6

Bioassay for Erythropoietin Activity in Plasma of
Mice with RLV Induced Anemia Given PHZ

Days post treatment	Mean percent ^{59}Fe peripheral RBC incorporation			
	Plasma	*0.9% NaCl*	*0.05 I.U. Ep*	*0.20 I.U. Ep*
0	1.63 ± 0.23[a]	1.30 ± 0.14	7.86 ± 1.14	20.04 ± 3.90
1	14.49 ± 1.55	0.86 ± 0.18	8.72 ± 1.09	14.05 ± 1.75
2	15.90 ± 5.46	0.86 ± 0.18	8.72 ± 1.09	14.05 ± 1.75
3	41.70 ± 4.38	1.30 ± 0.14	7.86 ± 1.14	20.04 ± 3.90
5	13.90 ± 1.03	0.91 ± 0.09	5.41 ± 1.44	16.17 ± 0.79
10	1.71 ± 0.36	1.30 ± 0.14	7.86 ± 1.14	20.04 ± 3.90

Data from Weitz-Hamburger et al. (unpubl.)

[a] Mean ± SE_m (five mice)

crits, red cell counts, reticulocytosis and mild splenomegaly 2 days after phlebotomy. These mice had detectable levels of plasma erythropoietin, somewhat lower than those of RLV-anemic bled mice.

Effect of Exogenous Erythropoietin in RLV-Anemic Mice

Increased erythropoiesis was initially induced by exogenous erythropoietin, although to a lesser extent than in mice with a similar degree of anemia treated with PHZ (Fig. 18, 19). However by day 10, after having received seven injections of erythropoietin, mice appeared relatively refractory to the effects of exogenous erythropoietin. From days 10–17 there was no significant difference between hematocrits of erythropoietin-treated and control RLV-infected mice. Reticulocyte levels also declined steadily. Moreover by day 17 spleens of erythropoietin-treated mice began to enlarge rapidly and animals soon succumbed to advanced RLV disease. Twenty-seven days after initiation of treatment, only 1 of 10 erythropoietin-treated mice was alive, as compared to four of five controls. A similar exacerbation of RLV disease by erythropoietin has also been reported by Snodgrass, Yumas and Hanna (1973).

Table 7

Bioassay for Erythropoietin Activity in Mice with
RLV-induced Anemia Two Days post Phlebotomy

Group	Mean percent ^{59}Fe peripheral RBC incorporation			
	Plasma	*0.9% NaCl*	*0.05 I.U. Ep*	*0.20 I.U. Ep*
1	7.68 ± 0.60[a]	1.30 ± 0.23	5.46 ± 1.02	24.47 ± 1.52
2	8.39 ± 1.29	1.30 ± 0.23	5.46 ± 1.02	24.47 ± 1.52
3	5.70 ± 1.35	1.30 ± 0.23	5.46 ± 1.02	24.47 ± 1.52
Normal control	2.07 ± 0.28	1.08 ± 0.17	4.42 ± 1.63	15.20 ± 2.10

Data from Weitz-Hamburger et al. (unpubl.).

[a] Mean ± SE_m

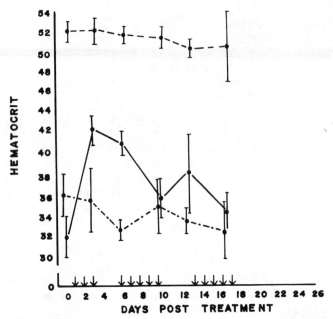

Figure 18

Changes in hematocrit during Ep treatment in RLV-anemic mice. In Fig. 18 and 19 arrows above abscissa indicate time of Ep injection. From Weitz-Hamburger et al. (unpubl.). (———) Ep-treated RLV-anemic mice; (—·—) mice given RLV only; (– – –) uninoculated controls.

Hence we have definitely established that quantities of erythropoietin considerably in excess of those normally produced endogenously in RLV-anemic mice will promote differentiation. Two questions are immediately apparent: (1) Why does not the animal produce sufficient hormone to compensate for the anemia unless there is an additional anemic stress imposed? and (2) Are non-responsive precursors (potential ERC's, Lajtha 1972) also present within spleens of infected mice? A possible explanation for the first question is debilitation, due to viral infection, of areas of the kidney associated with regulation of erythropoietin production. Studies using fluorescent antibody have shown that gs antigen becomes trapped within glomerular tufts in some infected mice (Hardy unpubl.). However, as indicated RLV-A-infected mice can produce erythropoietin in fairly substantial quantities when presented with a sudden, severe reduction in red cell mass. Possibly such erythropoietin production is of extrarenal origin (Gordon and Zanjani 1970). Regarding the second question, work currently in progress in our laboratory suggests that once red cell precursors are mobilized from the spleen by erythropoietin, a considerable lag period occurs before the precursor compartment is able to respond again to erythropoietin. This could be due to a rapid multiplication of potential ERC's or other undifferentiated precursors unresponsive to erythropoietin due directly or indirectly to treatment-induced drainoff of responsive cells. Infected mice given erythropoietin, therefore, develop anemia very rapidly, particularly soon after infection when splenic accumulation of erythroid precursors is not yet very great.

Studies involving DMSO-induced erythroid differentiation in vitro have also indicated a residium of non-responsive precursors (Friend et al. 1971).

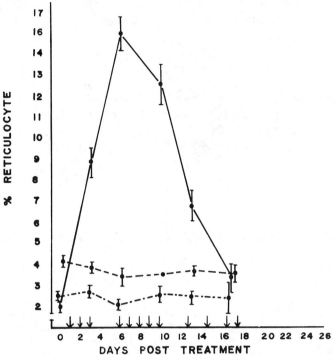

Figure 19

Reticulocyte response in Ep-treated RLV-anemic mice. From Weitz-Hamburger et al. (unpubl.). (——) Ex-treated RLV-anemic mice; (—·—) mice given RLV only; (– – –) unincoluated controls.

SUMMARY

The role of the glycoprotein hormone erythropoietin in the pathophysiology of an RLV-induced murine erythroleukemia is described. The RLV disease is characterized by spleno-hepatic erythroblastosis, peripheralization of erythroid precursors, severe terminal anemia, and faulty erythropoietin production. DNA synthesis and generation time were similar for normal and erythroleukemic red cell precursors. Hence one basic defect in the disease appears to be faulty erythroid maturation and release.

Stimulation of erythropoiesis by phenylhydrazine (PHZ)-induced hemolytic anemia prior to virus inoculation resulted in a greatly reduced incidence of RLV disease. Measurements of gs antigen by direct immunofluorescence established that while most erythroblasts in PHZ-treated mice were infected, normal maturation to erythrocytes occurred. To evaluate the therapeutic potential of PHZ, RLV-anemic mice were treated with this agent. This induced massive reticulocytosis and erythrocytosis.

RLV-infected mice made more anemic by PHZ or bleeding were capable of producing erythropoietin and this was in sharp contrast to untreated RLV-infected mice, which made little erythropoietin in response to the virally induced anemia. This suggested that the erythropoietin-generating system in these animals is impaired. To elucidate the role of erythropoietin in promoting the erythroid differentiation induced by PHZ, RLV-infected mice were given exogenous erythropoietin. This induced erythroid maturation in these animals, indicating that virally infected erythroid precursors were still capable of responding to factors governing normal erythropoiesis.

Acknowledgments

This investigation was supported by USPHS Research Grant 1-R01-CA12815-02 from the National Cancer Institute, by USPHS Research Grant 5-R01-HL03357-16 and Training Grant 5-T01-HL05645-08 from the National Heart and Lung Institute, and by funds supplied by the University of Connecticut Research Foundation.

REFERENCES

Brecher, G. 1949. New methylene blue as a reticulocyte stain. *Amer. J. Clin. Pathol.* **19:** 895.

Camiscoli, J. F. and A. S. Gordon. 1970. Bioassay and standardization of erythropoietin. *Regulation of hematopoiesis* (ed. A. S. Gordon) p. 369–393. Appleton-Century-Crofts, New York.

Camiscoli, J. F., J. LoBue, A. S. Gordon, P. Alexander, E. F. Schultz, A. Weitz-Hamburger and T. N. Fredrickson. 1972. Absence of plasma erythropoietin in mice with anemia induced by Rauscher leukemia virus. *Cancer Res.* **32:**2843.

Ebert, P. S., N. E. Maestri and M. Chirigos. 1972. Erythropoietic response of mice to infection with Rauscher leukemia virus. *Cancer Res.* **34:**41.

Ferdinand, P. 1973. C-type particles and MuLV gs antigen in Rauscher leukemia virus infected erythroleukemic mice. M.S. dissertation, New York University.

Fischinger, P. J., S. Nomura, P. T. Peebles, D. K. Haapala and R. H. Bassin. 1972. Reversion of murine sarcoma virus transformed mouse cell variants without a rescuable sarcoma virus. *Science* **176:**1033.

Fredrickson, T. N., J. LoBue, P. Alexander, Jr., E. F. Schultz and A. S. Gordon. 1972. A transplantable leukemia from mice inoculated with Rauscher leukemia virus. *J. Nat. Cancer Inst.* **48:**1597.

Friend, C., W. Sher, J. G. Holland and T. Sato. 1971. Hemoglobin synthesis in murine virus induced leukemic cells *in vitro*. Stimulation of erythroid differentiation by dimethylsulfoxide. *Proc. Nat. Acad. Sci.* **68:**379.

Friend, C., W. Sher, H. Preisler and J. G. Holland. 1973. Studies on erythroid differentiation of Friend virus induced murine leukemia cells. Proc. 5th Int. Symp. Comp. Leukemia Res. Karger, Basel. (in press).

Fruhman, G. J. and A. S. Gordon. 1955. Quantitative effects of corticosterone on rat bone marrow. *Proc. Soc. Exp. Biol. Med.* **88:**130.

Gallien-Lartigue, O., P. Tambourin, F. Wendling and F. Zajdela. 1969. Spontaneous hematopoietic recovery of Friend virus infected mice after heavy X-irradiation. *J. Nat. Cancer Inst.* **42:**1061.

Gordon, A. S. and E. D. Zanjani. 1970. Some aspects of erythropoietin physiology. *Regulation of hematopoiesis* (ed. A. S. Gordon) p. 413–457. Appleton-Century-Crofts, New York.

Heller, J. 1965. The effect of a new RES stimulant on neoplasia. *RES: Morphology, immunology and regulation*. 4th Int. Symp. on Res, p. 375–390. Nisha, Kyoto.

Hodgson, G., T. R. Bradley and P. Telfer. 1972. Haemopoietic stem cells in experimental haemolytic anemia. *Cell and Tissue Kinetics* **5:**282.

Lajtha, L. G. 1972. Kinetics of proliferation and differentiation in haemopoiesis. *Regulation of erythropoiesis* (ed. A. S. Gordon et al.) p. 47–50. Il Ponte, Milan.

LoBue, J., P. Alexander, Jr., T. N. Fredrickson, E. F. Schultz, A. S. Gordon and L. I. Johnson. 1972. Erythrokinetics in normal and disease states. Virally-induced murine erythroblastosis: A model system. *Regulation of Erythropoiesis* (ed. A. S. Gordon et al.) p. 89–101. Il Ponte, Milan.

LoBue, J., B. S. Dornfest, A. S. Gordon, J. Hurst and H. Quastler. 1963. Marrow distribution in rat femurs as determined by cellular enumeration and FE^{59} labeling. *Proc. Soc. Exp. Biol. Med.* **112**:1058.

Pluznik, D. H., L. Sachs and P. Resnitsky. 1966. The mechanism of leukemogenesis by the Rauscher leukemia virus. *Nat. Cancer Inst. Monogr.* **22**:3.

Ralph, P. H. 1941. The histochemical demonstration of hemoglobin in blood cells and tissue smears. *Stain Technol.* **16**:105.

Reissmann, K. R. and S. Samorapoompichit. 1970. Effect of erythropoietin on proliferation of erythroid stem cells in the absence of transplantable colony-forming units. *Blood* **36**:287.

Rencricca, N. J., V. Rizzoli, D. Howard, P. Duffy and F. Stohlman, Jr. 1969. Proliferative state of stem cells in erythroid hyperplasia. *Blood* **34**:836.

Sato, T., C. Friend and E. deHarven. 1971. Ultrastructural changes in Friend erythroleukemia cells treated with DMSO. *Cancer Res.* **31**:1402.

Seidel, H. J. 1973. Target cell characterization for Rauscher leukemia virus *in vivo*. 5th Int. Symp. Comp. Leukemia Res., Karger, Basel. (in press)

Siegler, R. 1968. Pathology of murine leukemias. *Experimental leukemia* (ed. M. A. Rich) p. 51–95. Appleton-Century-Crofts, New York.

Snodgrass, M. J., J. M. Yumas and M. G. Hanna. 1973. Histoproliferative effects of Rauscher leukemia virus on lymphatic tissue. IV. Lactic dehydrogenase virus potentiation of erythropoietic response. *Int. J. Cancer* (in press)

Stohlman, F., Jr. 1972. Control mechanisms in erythropoiesis. *Regulation of erythropoiesis* (ed. A. S. Gordon et al.) p. 71–87. Il Ponte, Milan.

Tambourin, P. and F. Wendling. 1971. Malignant transformation and erythroid differentiation by polycythemia inducing Friend virus. *Nature* **234**:230.

Weitz-Hamburger, A., T. N. Fredrickson, J. LoBue, W. D. Hardy, Jr., P. Ferdinand and A. S. Gordon. 1973. Inhibition of erythroleukemia in mice by induction of hemolytic anemia prior to infection with Rauscher leukemia virus. *Cancer Res.* **33**:104.

Regulation by Colony-Stimulating Factor of Granulocyte and Macrophage Colony Formation In Vitro by Normal and Leukemic Cells

Donald Metcalf

Cancer Research Unit, Walter and Eliza Hall Institute
Melbourne 3050, Australia

The various hematopoietic populations are generated by the proliferative activity of self-replicating hematopoietic stem cells (see review Metcalf and Moore 1971). Initially these stem cells generate line-specific progenitor cells from which eventually are produced the four major subpopulations of hematopoietic cells (erythroid, granulocytic-monocytic, megakaryocytic and lymphoid). Since blood cell production is continuous throughout life and involves the formation of a variety of highly specialized end cells, hematopoiesis is potentially a superb model in which to study the general biological problems of regulation of cell proliferation and differentiation.

The complexity of the hematopoietic system requires that simple in vitro culture systems be developed that permit the growth of selected types of cells to be analyzed in detail. A semi-solid agar culture has been developed (Bradley and Metcalf 1966; Ichikawa, Pluznick and Sachs 1966) that supports the proliferation of colonies of granulocytes and monocyte-macrophages and allows an analysis of many aspects of the regulation of proliferation and differentiation in these two populations. The two basic features of this system are that (a) the colonies are clones initiated by the specific progenitors of granulocytes and monocytes, the colony-forming cells (CFC), and (b) colony formation is totally dependent on stimulation by a naturally occurring specific regulator, colony-stimulating factor (CSF).

Agar cultures are prepared as 1-ml cultures in 35-mm plastic petri dishes using modified Eagle's medium containing 15% serum in 0.3% agar (Metcalf 1970). This medium will support granulocytic colony formation by cells from all mammalian species, although different sources of CSF must be used for cells from different species. For convenience of counting, the number of cells cultured is usually restricted so that final colony numbers range between 5 and 100 per dish. CSF is supplied to the cultures most simply by mixing a liquid source of CSF with the cell suspension in agar medium before gelling occurs. For some purposes, underlayers containing various feeder layer cells or medium conditioned by such cells can be used as the source of CSF. Cultures are usually incubated for 7 days in 10% CO_2 and colony counts performed on unstained cultures at 25 × magnification.

COLONY-FORMING CELLS

In the adult mouse or human, CFC's are restricted to the bone marrow (1 per 500 cells in the mouse, 1 per 3000 cells in the human), spleen (1–5 per 10^5 cells in mouse and human), and blood (0.1–0.2 per 10^5 cells). CFC's have been partially separated from other marrow cells using adherence columns, buoyant density separation, and velocity sedimentation (Worton, McCulloch and Till 1969; Haskill, McNeill and Moore 1970; Metcalf, Moore and Shortman 1971; Moore, Williams and Metcalf 1972). In monkey marrow most CFC's are medium-sized cells, 9–11 μ in diameter, with a round, excentric nucleus and a slightly basophilic but agranular cytoplasm (Moore et al. 1972), and the cloning efficiency of such CFC's is high (30–100%). In adult mouse marrow the CFC's appear to be more heterogeneous than in the human or monkey and cannot effectively be separated from other marrow cells. Most CFC's are in active cell cycle and 30–50% can be killed in vitro by short exposures to high concentrations of tritiated thymidine (Lajtha et al. 1969; Iscove, Till and McCulloch 1970; Metcalf 1972; Moore et al. 1972; 1973b).

It is still not resolved whether CFC's possess the capacity for self-replication but, if so, this capacity appears to be severely restricted in comparison with that of hematopoietic stem cells (Metcalf and Moore 1971; Dicke et al. 1973; Bradley et al. 1972).

It is a characteristic of agar cultures of marrow or spleen cells that individual colonies vary widely in size. For example in 7-day cultures of mouse bone marrow, colonies in the same dish can vary in size from 50–2000 cells. All such cultures also contain more numerous smaller aggregates ("clusters"), which appear to be formed by the proliferative activity of the immediate progeny of CFC's (Metcalf 1969). Colonies can contain pure populations of granulocytic or macrophage cells or mixtures of both. Progressive differentiation of colony cells occurs during colony growth so that eventually colonies contain populations of mature segmented polymorphs or mature phagocytic macrophages, the latter having a full complement of membrane receptor sites for the Fc fragment of IgG (Metcalf, Bradley and Robinson 1967; Cline, Warner and Metcalf 1972). Differentiation does not proceed synchronously in all colonies and at 7 days of incubation some colonies may be composed wholly of blast cells, some wholly of mature end cells, whilst most will contain cells at varying stages of differentiation.

Table 1 shows data from an experiment in which cultures of 10,000 C57BL mouse bone marrow cells were stimulated by the addition of 0.1 ml of a 1:6 dilution of pooled serum from C57BL mice injected 3 hours previously with 5 μg of endotoxin. This form of CSF is useful in stimulating the development by 7 days of a mixture of granulocytic, macrophage and colonies containing both cell populations (Metcalf 1971a). Colony numbers reached a maximum level by day 5 and did not change significantly up to day 14. However there was a progressive change with time in the proportion of colonies wholly composed of macrophages. With the exception of a few granulocytic colonies that matured to populations of polymorphs and then disintegrated, the vast majority of colonies exhibited a sequential population change in which monocytes and macrophages appeared in granulocytic colonies

Table 1

Transformation of Granulocytic to Macrophage
Colonies on Continued Incubation

Day of incubation	Mean no. colonies*	Percent of colonies†		
		Granulocytic	Mixed granulocytic, macrophage	Macrophage
7	22 ± 5	48	20	32
8	21 ± 6˙	21	37	42
9	20 ± 5	19	15	66
11	21 ± 6	6	12	82
13	22 ± 4	2	0	98
14	22 ± 5	0	2	98

All cultures contained 10,000 C57BL bone marrow cells stimulated by 0.1 ml of a 1:6 dilution of pooled serum from C57BL mice injected 3 hr previously with 5 μg endotoxin.

* Mean colonies ± standard deviations of four replicate cultures.

† 40 sequential colonies analyzed at each time point.

and then increased in number so that eventually colonies were composed entirely of macrophages. This granulocyte to macrophage transformation is a basic feature of the agar culture system, although culture conditions can be modified to alter the proportion of colonies terminating as polymorph or macrophage colonies (see below). There are also significant species differences in the timing of the transformation, the process commencing at 3–5 days of incubation in mouse colonies, but being delayed until after day 14 in human colonies.

Single cell transfer studies using granulocytes from early colonies have demonstrated that single colony granulocytes can transform without division to macrophages or generate small clones of macrophages (Metcalf 1971b). Furthermore cultures of single CFC's obtained by micromanipulation from fractions of monkey bone marrow enriched for CFC's have shown that individual CFC's are able to generate colonies containing both granulocytes and macrophages, proving that monocyte-macrophages and granulocytes are closely related populations with a common progenitor cell, the CFC (Moore et al. 1972).

While most CFC's appear able to generate end populations of either granulocytes or macrophages, depending on the culture conditions used, a minority subpopulation of CFC's which appears able only to generate granulocytic colonies, can be separated by buoyant density separation (Janoshwitz et al. 1971).

Heterogeneity amongst CFC's is also evident from the variable shape, size and composition of colonies generated by these cells at 7 days of incubation. Colonies with the diverse morphology of those shown in Fig. 1 can occur side by side in the same culture dish. Thus a colony like (b) may contain 1000 myeloblasts, while next to it may be a colony like (d), containing an equal number of granulocytic cells, all of which are metamyelocytes or polymorphs. If CFC's were identical and after each division a fixed stochastic probability existed that the two daughter cells would differentiate to more mature cells, it would rarely be possible to generate large colonies with such extremes of differentiation patterns. However it is common-

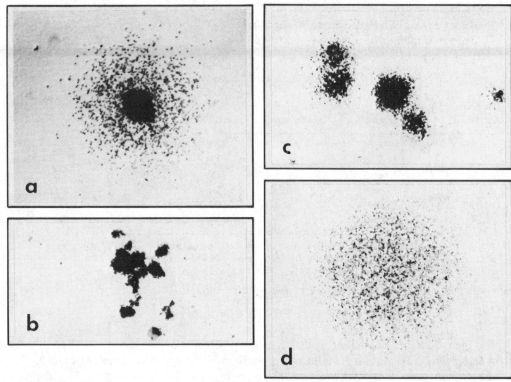

Figure 1
Unstained colonies in 7-day cultures of mouse bone marrow cells showing the wide range of possible morphological forms: **(a)** and **(b)** granulocytic colonies stimulated by endo-toxin serum, **(c)** macrophage colonies stimulated by human urine, **(d)** granulocytic colony stimulated by medium conditioned by murine myelomonocytic leukemic cells.

place in mouse marrow cultures that 10–20% of colonies are of these extreme types, and the conclusion seems inescapable that the CFC's forming these colonies are not identical and must vary significantly in their capacity to generate progeny that will differentiate to more mature forms.

The size attained by individual colonies will be dictated to a large degree by the rapidity with which daughter cells differentiate to nondividing metamyelocytes, polymorphs or macrophages. Furthermore CFC's do not begin to proliferate synchronously in the cultures and some CFC's do not commence proliferation for up to 4 days after initiation of the cultures (Metcalf 1969). These factors account for much of the variation in colony size observed by 7–14 days of incubation. Despite this it is possible to have wide variations in size between colonies which contain cells at the same stage of differentiation and which began proliferation at the same time, suggesting strongly that there may also be significant variability in the number of divisions of which individual CFC's are capable.

The extreme variability between CFC's, which is clearly apparent in any agar culture of mouse marrow cells, makes it impossible to accept the formalized schemes of granulopoiesis which have been proposed from in vivo studies on tritiated thymidine-labeled populations. The generation of granulocytes and monocytes is likely to be far more complex and variable than envisaged in such schemes.

The clonal nature of colonies also makes it possible to conclude that the progeny of at least some CFC's tend to behave uniformly in their responsiveness to differentiation pressures, implying that a particular pattern of responsiveness may be a heritable characteristic within the clone. A detailed analysis of clonal variability in normal or preleukemic populations can be expected, therefore, to give useful insights into the development of myeloid leukemia—a process that appears to involve the sequential emergence of successively more abnormal clones of granulocytic cells—and in fact considerable information has already been gathered on this process (Moore et al. 1973b; Moore 1973; Metcalf 1973).

COLONY-STIMULATING FACTOR

Colony-stimulating factor (CSF) is the operational name given to the specific factor required for granulocyte and macrophage colony growth in vitro. CSF is demonstrable in normal serum and urine and is extractable from all tissues in the body in higher concentrations than in the serum—organs in the mouse having the highest extractable CSF content—being, in order, submaxillary gland (male), submaxillary gland (female), lung, thymus and kidney (Robinson, Metcalf and Bradley 1967; Metcalf and Stanley 1969; Stanley et al. 1972; Sheridan and Stanley 1971). Cells from a wide variety of normal and neoplastic tissues are able to synthesize and/or release CSF in liquid culture (conditioned medium) (Pluznik and Sachs 1966; Ichikawa et al. 1966; Bradley and Sumner 1968; Paran, Ichikawa and Sachs 1968; Metcalf 1971c; Bradley et al. 1971). The actual cells producing CSF have not been identified and two general alternatives are still tenable: (a) that many different cells have the capacity to synthesize CSF, or (b) that certain cells common to all tissues, e.g. macrophages, fibroblasts, or endothelial cells, produce the CSF extractable from these tissues. In this context monocytes and macrophages have been shown to contain CSF (Moore and Williams 1972; Golde and Cline 1972; Chervenick and LoBuglio 1972; Moore et al. 1973a), whereas purified populations of granulocytes do not.

Differences in structure, antigenicity and biological activity exist between CSF's from different species. In general CFC's are most effectively stimulated by CSF from the same species and in some species combinations of heterologous CSF's can be without activity. CFC's from the mouse are exceptional in that they respond to stimulation by CSF's from all species. The situation with regard to CSF is made more complex by the observation that CSF's extracted from different tissues also differ from one another in physical properties and antigenicity (Stanley, Bradley and Sumner 1971; Sheridan and Stanley 1971; Sheridan and Metcalf 1972, 1973, and unpubl.). It is uncertain which CSF's are biologically more important than others, but some evidence suggests that the CSF produced by cells adherent to the inner walls of the marrow cavity may be highly important since the capacity to produce CSF of bone stromal cells fluctuates significantly during the regeneration of CFC populations following whole body irradiation and this is not paralleled by fluctuations in CSF production in other tissues (Chan and Metcalf 1972, 1973).

Small amounts of highly purified CSF have been obtained from two sources, human urine (Stanley and Metcalf 1972) and mouse lung conditioned medium (Sheridan and Metcalf 1973 and unpubl.). Both are neuraminic acid-containing glycoproteins but the molecular weight of human urine CSF is 45,000, whereas the common form of CSF obtained from mouse lung has a molecular weight of 15,000. CSF's produced by L-cells and mouse embryo cells also appear to be glycoproteins of generally similar type (Stanley et al. 1971; Austin, McCulloch and Till 1971; Landau and Sachs 1971).

The concentration of CSF in biological fluids can be bioassayed using cultures of mouse bone marrow cells. Samples (0.1 ml) of varying dilutions of pooled serum from C57BL mice, injected 3 hours previously with 5 μg endotoxin, were added to replicate cultures containing varying numbers of C57BL bone marrow cells. A sigmoid dose-response relationship exists between colony numbers at day 7 of incubation and CSF concentration (Fig. 2), a fact that can be used to determine CSF levels with some precision (Stanley et al. 1972). The assay system is highly sensitive and will detect urine CSF at a molecular concentration of 10^{-11}–10^{-12} M (approximately 100 pg per ml) and lung CSF at even lower concentrations.

CSF is required continuously in the cultures for progressive colony growth (Metcalf and Foster 1967; Paran and Sachs 1968) and is necessary for survival in vitro of CFC's or for retention of their capacity to proliferate (Metcalf 1970). In cultures of CFC's from all species other than the mouse, some colony formation occurs in the absence of added CSF. This is not due to the autonomy of CFC's from these species but to the presence of significant numbers of CSF-producing cells in the marrow population being cultured.

These CSF-producing cells can be separated from CFC's by differential centrifugation, and the CFC's in such preparations are then as dependent on CSF for proliferation as are CFC's in cultures of unfractionated mouse marrow cells (Moore et al. 1972; Moore and Williams 1972; Moore et al. 1973a).

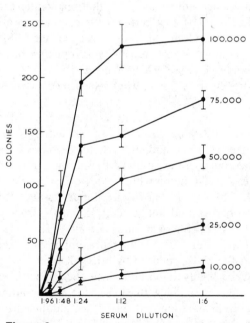

Figure 2

Dose-response relationship between CSF concentration and colony numbers produced by varying numbers of C57BL bone marrow cells. CSF source used was 0.1 ml of serial dilutions of pooled serum from C57BL mice injected 3 hr previously with 5 μg endotoxin. Vertical bars are standard deviations of mean values from six replicate cultures.

The stimulating action of CSF on granulocyte and macrophage colony formation appears to be highly specific and no example has yet been encountered in which semi-purified CSF preparations have stimulated growth in vitro by other cell types (Metcalf and Moore 1973). Although most CFC's are in cell cycle in vivo, they appear to pass out of cycle after being placed in culture and a variable lag period occurs before they recommence division. The average length of the lag period is shortened by increasing the concentration of CSF (Metcalf 1970). CSF also has a concentration-dependent effect on the mean length of the cell cycle of colony cells, cell cycle times being shortened by increasing CSF concentrations (Metcalf and Moore 1973). Although colonies grow at a more rapid rate in the presence of higher concentrations of CSF, it is likely that this is not so much the result of CSF action on cell cycle times as the effect of CSF on differentiation of colony cells. In the presence of high concentrations of CSF, differentiation to nondividing progeny is delayed and daughter cells remain capable of further divisions (Metcalf and Moore 1973).

The most purified preparations of urine or lung CSF form a single band on poly-acrylamide electrophoresis and appear to be free of contaminating protein. When such preparations were used to stimulate colony formation by mouse bone marrow cells, the formation of both granulocytic and macrophage colonies was stimulated. The proportion of granulocytic colonies present in such 7-day cultures was dependent on the concentration of CSF used (Table 2). In an attempt to determine whether a subpopulation of CSF molecules existed with a specific or selective capacity to stimulate colonies of only one morphological type, highly purified CSF from human urine was rerun in a fine resolution polyacrylamide gel and each CSF fraction tested for its capacity to stimulate both granulocyte and macrophage colony formation by C57BL mouse bone marrow cells. Care was taken to analyze cultures containing equivalent concentrations of CSF to avoid concentration-dependent effects on colony morphology. The incidence of granulocytic and macro-phage colonies was similar in cultures stimulated by the various fractions comprising the single peak in the polyacrylamide gel (Fig. 3), suggesting strongly that the same population of CSF molecules was able to stimulate the proliferation of both

Table 2

Effect of CSF Concentration on Morphology of Colonies
Stimulated by Highly Purified Lung CSF

| Lung CSF conc. | Mean no. colonies/ culture | Percent of colonies* | | |
		Granulocytic	Mixed granulocytic, macrophage	Macrophage
1:1	>120	52	27	21
1:10	>120	11	32	57
1:100	>120	5	29	66
1:1000	34 ± 3	0	4	96

Each culture contained 75,000 C57BL bone marrow cells stimulated by 0.1 ml of a 100,000-fold purified preparation of medium conditioned by C57BL lung tissue kindly supplied by Dr. J. W. Sheridan. Colony counts at day 7.

* 40 consecutive colonies classified from each type of culture.

granulocytic and macrophage colonies. CSF preparations have been pretreated with a wide variety of proteolytic enzymes and carbohydrases, including neuramidase. Although pretreatment with several of these enzymes resulted in a reduction of biological activity, and in the case of neuramidase in a marked reduction in the charge of molecule, in no case did pretreatment alter the morphological spectrum of colonies stimulated by the enzyme-treated CSF (Stanley and Metcalf 1971; Sheridan and Metcalf unpubl.).

The CSF-CFC target cell system is therefore intriguing in that individual CFC's can give rise to populations of granulocytes and macrophages, and cellular proliferation in both alternative pathways of differentiation is stimulated by the same regulator, CSF. As noted above the switch mechanism determining whether cells remain in the granulocytic series or enter the monocyte-macrophage pathway is partly determined by CSF concentration. In part, however, this switch is controlled by lipoproteins present in all normal sera, termed CSF inhibitors, which can be bioassayed in marrow cultures because of their capacity to exert a species-specific blocking action on CSF (Chan, Metcalf and Stanley 1971; Chan 1971). In the presence of CSF inhibitors, colony cells transform prematurely to the monocyte-macrophage pathway of differentiation. Preincubation of CFC's with CSF inhibitors, followed by culture of the CFC in an inhibitor-free system, also causes premature macrophage transformation (Chan 1971). It is of interest in this context that marrow cultures from mouse strains having high serum-inhibitor levels, e.g. Balb/c, develop a high proportion of macrophage colonies under cultural conditions in which marrow cultures from low inhibitor strains (e.g. C57BL) develop a high proportion of granulocytic and mixed colonies (Chan 1971).

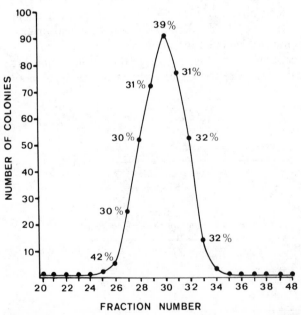

Figure 3

Colony-stimulating activity of polyacrylamide fractions of highly purified human urine CSF, indicating the percentage of granulocytic and mixed colonies in cultures stimulated by equivalent concentrations of CSF from each fraction.

However the true in vivo situation appears to be rather more complex. Marrow and spleen cells were cultured in parallel from high and low inhibitor strains, using varying concentrations of CSF (Table 3). It has been observed that spleen CFC's tend to form granulocytic colonies more often than do the apparently similar CFC's in the marrow (Metcalf and Stevens 1972); this can be seen in the data from the low inhibitor strain C57BL mice, where the effects on colony morphology of reducing CSF concentrations are also evident. Although the bone marrow cells from Balb/c mice gave rise to the expected high proportion of macrophage colonies, somewhat surprisingly Balb/c spleen cells gave rise to as high a proportion of granulocytic colonies as was seen with cells from the low inhibitor C57BL mice. If circulating inhibitors do modify the differentiation capacity of CFC's in vivo, it is evident that this influence must be conditioned by the microenvironment in which CFC's are located. Alternatively, spleen CFC's may not be exactly equivalent to marrow CFC's either in their capacity for differentiation or in their responsiveness to modifying factors such as CSF inhibitors.

Recently a system has been developed for forcing colony populations to remain in the granulocytic pathway. It was observed that CFC's in cultures of whole C57BL blood formed an unusually high proportion of granulocytic colonies, which at 7 days were entirely composed of a remarkably uniform population of well-differentiated metamyelocytes and polymorphs ("polymorph colonies"). In this system,

Table 3

Morphology of Colonies Produced by Bone Marrow and Spleen Cells from C57BL and Balb/c Mice

Strain	Cells cultured	Endotoxin serum dilution	Mean colonies per 10^5 cells	Percent of colonies*		
				Granulocytic	Mixed granulocytic, macrophage	Macrophage
C57BL	Bone marrow	1:6	171	38	24	38
		1:18	140	20	26	54
		1:54	48	14	31	55
	Spleen	1:6	2	54	8	38
		1:18	2	43	36	21
Balb/c	Bone marrow	1:6	144	20	6	74
		1:18	83	12	26	62
		1:54	32	13	4	83
	Spleen	1:6	1	50	33	17
		1:18	1	30	20	50

Cultures contained 75,000 bone marrow cells or 300,000 spleen cells and were stimulated by 0.1 ml of various dilutions of pooled serum from C57BL mice injected 3 hr previously with 5 μg endotoxin. Colonies counted at day 7.

* 40 consecutive colonies classified from each type of culture.

0.2 ml of freshly drawn axillary blood is mixed with 5 ml agar medium before clotting can occur and 1-ml aliquots of the mixture are then plated in four replicate culture dishes, each containing 0.1 ml of a 1:6 dilution of endotoxin serum. When 25,000 C57BL bone marrow cells were added to each ml of the suspension of blood in agar medium, no change was observed in the expected number of colonies generated by the added marrow cells, but a significant increase was observed in the incidence of polymorph colonies in such cultures (Table 4). The active component in whole blood which causes colonies to remain in the granulocytic pathway has not yet been investigated.

It should be pointed out that CSF's from different sources can stimulate recognizably different types of colonies. Thus the CSF produced by mouse myelomonocytic leukemic cells, which stimulates typical granulocytic and macrophage colonies, also stimulates the formation of small numbers of very large dispersed colonies composed of highly differentiated granulocytes (see type B, Fig. 1). Similarly lung CSF in high concentrations stimulates the formation of characteristic granulocytic colonies in which the tight central region is very small in relation to the overall diameter of the colony (Sheridan and Metcalf 1973). In view of the chemical variability of CSF, it is possible that some of these different colonies may be stimulated by unusual molecular forms of CSF.

Table 4

Development of Polymorph Colonies from Marrow
Cells Grown in Presence of Whole Blood

Cells cultured	Mean no. colonies/ culture*	Percent of colonies†			
		Polymorph	Mixed granulo-cytic	Mixed granu-locytic, Macro-phage	Macro-phage
Whole blood	1.0 ± 0.5	43	17	15	25
Bone marrow	61 ± 6	3	17	31	49
	65 ± 3	0	19	30	51
Bone marrow plus whole blood	64 ± 8	19	8	35	38
	61 ± 1	17	10	35	38
	57 ± 7	19	3	30	48
	69 ± 5	10	16	21	53
	70 ± 9	29	12	31	28
	70 ± 15	41	6	22	31

Each culture contained 0.04 ml axillary blood and/or 25,000 bone marrow cells from 3-month-old C57BL mice and was stimulated by 0.1 ml of a 1:6 dilution of a serum pool from C57BL mice injected 3 hr previously with 5 μg endotoxin.

* Colony counts on day 7 ± standard deviations.

† 40 sequential colonies scored from each culture. "Polymorph" colonies composed entirely of metamyelocytes and polymorphs; "mixed granulocytic" colonies contained granulocytic cells at all stages of differentiation; "mixed granulocytic and macrophage" colonies contained granulocytes and macrophages.

Although colonies grown from all species exhibit variability in general morphology, the extremes of morphology and patterns of differentiation are far more evident in cultures of mouse cells than in any other. For this reason and because of the responsiveness of mouse CFC's to most forms of CSF, the mouse seems the most useful source of CFC's for analytical work on mechanisms affecting the differentiation and growth of granulocytic and macrophage colonies.

Influence of CSF on Growth of Human Leukemic Cells

Although human urine CSF is a poor proliferative stimulus for human marrow cells, antisera prepared against urine CSF block the stimulating activity of white cell underlayers or medium conditioned by the human white cells (Moore et al. 1973a). This indicates that CSF is necessary for the growth of human colonies, but (a) urine CSF may be a modified molecule which, while still antigenically cross-reactive, has lost its biological activity for human CFC's, or (b) other factors may be produced by white cells and be necessary, in addition to CSF, for human colony growth.

Bone marrow or blood cell suspensions from patients with chronic myeloid leukemia (CML) contain abnormally large numbers of CFC's (Paran et al. 1970; Brown and Carbone 1971; Shadduck and Nankin 1971; Chervenick et al. 1971; Moore et al. 1973b). The colonies generated are composed of Ph-positive cells but are similar in general appearance to normal colonies. CML colony cells exhibit good differentiation although the granulocytic populations tend to show premature macrophage transformation. Furthermore CML CFC's differ from normal CFC's in that they are of lighter buoyant density and have longer cell cycle times or are more often in a G_0 noncycling state (Moore et al. 1973b).

The leukemic cells from patients with acute myeloid or myelomonocytic leukemia (AML or AMML) have a characteristically different pattern of proliferation in agar. Cells from 20% of patients fail to proliferate in agar, whereas in 60–70% of patients the bone marrow and blood contain no detectable CFC's, but the leukemic cells form large numbers of small clusters (less than 40 cells by day 8 of incubation) (Paran et al. 1970; Robinson, Kurnick and Pike 1971; Greenberg, Nichols and Schrier 1971; Brown and Carbone 1971; Iscove et al. 1971; Moore et al. 1973b). If karyotypic abnormalities are present in the leukemic population, these are also evident in the proliferating cluster cells (Moore and Metcalf 1973). AML cluster-forming cells again differ from normal cluster-forming cells in being of lighter buoyant density and having longer cell cycle times or being more often in a G_0 noncycling state (Moore et al. 1973b). Unlike the situation with colonies grown from CML cells, the cells in clusters produced by AML cells exhibit poor differentiation and many cells are abnormal in morphology.

When a patient with AML enters remission, this is associated with the reappearance of CFC's in the bone marrow. These CFC's are of normal density and cell cycle status and give rise to normal-sized colonies in which the cells differentiate normally and exhibit a normal karyotype (Moore et al. 1973b). The culture data therefore indicate clearly that the AML patient has a double population of normal and leukemic cells and that in the transition from relapse to remission status there is a temporary reappearance of normal cells that had previously been suppressed in the relapse or pretreatment phases of the disease.

Several surveys of CSF levels in patients with AML have shown that serum CSF levels are elevated in approximately one-third of samples and are elevated in all patients at some stage in their disease (Foster et al. 1968; Metcalf et al. 1971; Metcalf and Chan 1973). Twenty-four-hour excretion rates of CSF in the urine were also elevated in 50–60% of samples tested (Robinson and Pike 1970; Metcalf et al. 1971). In a study in progress on patients with chronic myeloid leukemia an even higher proportion of serum samples has been observed to contain elevated CSF levels (40–50%), whilst levels in patients in the acute transformation stage of CML were elevated in 80% of samples. Conversely serum CSF inhibitor levels were abnormally low or undetectable in 60% of sera from patients with AML and in an even higher proportion of sera from the acute transformation stage of CML (Chan et al. 1971; Metcalf et al. 1971; Metcalf and Chan 1973).

It appears, therefore, that both AML and CML are characterized by extended periods in which there is an imbalance in the CSF-CSF inhibitor regulatory system, leading to a net excess of stimulation of granulopoiesis and monocyte formation. It is of considerable importance, therefore, to determine whether the leukemic cells in both diseases remain responsive to these regulatory factors.

An analysis of the behavior in culture of leukemic cells from some 150 patients with AML and CML have shown that, in every case, cultures stimulated by white cell underlayers or conditioned medium exhibited more colonies or clusters and more rapid colony or cluster growth than corresponding unstimulated cultures (Moore et al. 1973b; Moore unpubl.). A typical example of the response of cultured peripheral blood cells from a patient with CML to stimulation by an underlayer of normal human peripheral blood cells is shown in Fig. 4. The histograms represent an analysis of all cell aggregates present at day 8 of incubation in the stimulated and unstimulated cultures of 200,000 cells. No colonies were present in the unstimulated cultures and stimulation caused the development of colonies and larger numbers of clusters, with an overall increase in the size of the aggregates. Leukemic CFC's, separated by density centrifugation from CSF-producing cells in the cultured cell suspension, were as unable to proliferate in unstimulated cultures as were normal CFC's (Moore et al. 1973b).

These observations clearly indicate that the leukemic cells in AML or CML remain responsive to CSF. Recently studies have been undertaken to determine quantitatively the responsiveness of AML and CML cells to varying concentrations of CSF, compared with normal human CFC's. In these studies serial dilutions of CSF-containing medium conditioned by Rhesus monkey lung tissue were placed in 1-ml underlayers of 0.05% agar. The cells for culture were placed in a 1-ml overlayer in 0.3% agar and the number of cells cultured was restricted to 10,000 or 25,000 per dish to minimize the possibility of significant endogenous CSF production. The practical problem in determining colony counts in this type of experiment is that colony size decreases progressively in cultures containing progressively smaller concentrations of CSF. Total plate counts were performed at day 4 of incubation, scoring all aggregates of three or more cells regardless of size, a procedure which avoided any subjective element in deciding lower size limits for colonies. Total aggregate counts at each CSF concentration were expressed as percentages of the maximum aggregate count. The data for normal marrow cells are shown in Fig. 5 and indicate a sigmoid dose-response relationship between CSF concentration and total aggregate numbers. For comparison are shown corresponding data from cultures from six patients with CML and six patients with untreated AML or AML in relapse. In both groups it is clear that the leukemic cells did

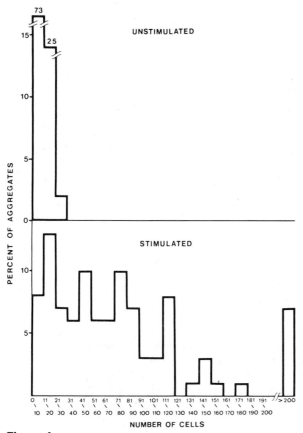

Figure 4

Size distribution analysis of all aggregates present at 8 days in cultures containing 200,000 peripheral blood cells from a patient with chronic myeloid leukemia. Note the larger size of aggregates in the culture stimulated by an underlayer of normal peripheral blood cells. Total aggregate counts: unstimulated 290, stimulated 455.

respond progressively to stimulation by increasing concentrations of CSF, but both dose-response curves were abnormal. CML cells appeared to be slightly less responsive than normal cells, whereas AML cells were more responsive than normal at low CSF concentrations but had a curiously flat overall dose-response curve. Of some interest was the fact that cultures from six patients with AML in remission gave entirely normal dose-response curves.

Monkey lung conditioned medium is likely to differ in some respects from human CSF, and in future experiments the relative responsiveness of AML and CML cells will need checking using human CSF. Despite this reservation it is of some interest that the present results clearly discriminate between the responsiveness patterns of AML and CML cells, and it is intriguing to speculate on the possible connection between these differences and the fact that CSF levels are more consistently elevated in CML than AML, since CML cells appear less responsive to CSF than normal or AML cells.

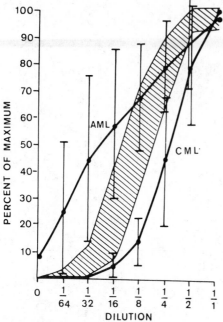

Figure 5

Responsiveness of normal and leukemic human marrow cells to stimulation by underlayers containing varying dilutions of monkey lung conditioned medium. Data based on total aggregate counts at day 4 of incubation. Normal range indicated by hatched area. *AML*, acute myeloid leukemic cells; *CML*, chronic myeloid leukemic cells. Vertical bars are standard deviations of mean values.

Since serum CSF concentrations in AML or CML are commonly 5 to 10 times higher than the maximum concentrations of CSF used in these dose-response studies, it is probable that CSF represents a highly significant proliferative stimulus for the leukemic populations in vivo and must have an important influence on the proliferative behavior of the leukemic populations during the course of these diseases.

CSF as a Humoral Regulator In Vivo

Despite the fact that granulopoiesis and macrophage formation in vitro are wholly dependent on stimulation by CSF, this by itself is not sufficient proof that CSF acts in vivo as a regulator of these cells. It could be argued that CSF is merely necessary for cell survival in vitro and that this is a wholly in vitro artefact because of the use of suboptimal culture media. There has been some confusion created in the literature by descriptions of serum-stimulated colony formation which imply that serum proteins in general are active in stimulating colony formation. In fact, however, if the minute amounts of CSF present in serum (1 ng per ml) are first removed, serum is completely devoid of any capacity to stimulate granulocyte or macrophage colony formation.

While it is quite true that CSF is necessary for survival in vitro of CFC's, the suggestion that CSF is some type of essential nutrient is basically improbable because of the activity of CSF at such low concentrations as 10^{-11}–10^{-12} M and because CSF activity appears to be restricted to cells of the granulocytic and monocyte-macrophage series (Stanley and Metcalf 1972; Metcalf and Moore 1973).

The most cogent single argument in favor of a regulatory role in vivo for CSF is the fact that CSF is demonstrable in all normal sera in concentrations at least ten times those required for detectable activity in vitro. Furthermore serum CSF levels rise as high as fifty times normal levels in situations where increased granulopoiesis and monocyte formation are occurring, such as (a) following the injection of antigens, (b) during the acute phases of viral or bacterial infections, or (c) preceding and during the regeneration of granulocytic and monocytic populations following damage by irradiation, cyclophosphamide or antineutrophil serum (Metcalf 1971; Quesenberry et al. 1972; Morley et al. 1972; Shadduck et al. 1972). Conversely serum CSF levels are abnormally low in germ-free mice in which levels of granulopoiesis are below those in conventional animals (Metcalf, Foster and Pollard 1967; Metcalf and Stevens 1972). Persuasive indirect evidence for an in vivo role for CSF has been the observation that urine CSF levels fluctuate in an out-of-phase relationship with granulocyte levels in both dogs and patients with cyclic neutropenia (Dale et al. 1971; Mangalik and Robinson 1973).

More direct evidence of CSF action in vivo was obtained by a study of the effects of injecting semi-purified CSF preparations into adult mice. This was followed by rises in the levels of marrow CFC's and cluster-forming cells, increased granulocyte formation, and the production of a monocytosis (Metcalf and Stanley 1971). No changes were observed in lymphopoiesis or erythropoiesis. Of some interest was the observation that CFC's from CSF-injected mice exhibited a reduced lag period before initiating proliferation in vitro, an effect similar in general nature to the observed action of CSF on CFC's in vitro (Metcalf 1970).

The evidence for CSF action in vivo is still not as strong as that for erythropoietin and further direct studies on the in vivo effects of purified CSF are certainly indicated. However there is sufficient direct and indirect evidence that CSF acts in vivo in essentially the same manner as in vitro to accept CSF as an important regulator of granulopoiesis and monocyte-macrophage formation.

What is by no means clear is whether CSF is a *humoral* regulator in the conventional sense of classical hormones such as the pituitary hormones. The multiorgan origin of CSF and the evidence for significant local production of CSF within the marrow raise the possibility that circulating CSF levels may not necessarily be the dominant factor in determining local CSF concentrations in the marrow or spleen, and the contribution made by circulating CSF requires further investigation in a variety of normal and experimental situations.

SUMMARY

The clonal proliferation of granulocytes and monocyte-macrophages in agar requires stimulation by the glycoprotein colony-stimulating factor (CSF) and there is direct and indirect evidence that CSF acts in vivo as a regulator of granulopoiesis and monocyte formation. CSF controls entry of colony-forming cells (CFC's) into the cycling state, reduces overall cell cycle times, and influences differentiation to mature granulocytes and macrophages. In humans with acute or chronic myeloid leukemia, CSF levels are commonly elevated and since the leukemic cells retain a near-normal responsiveness to CSF, disturbances in CSF and CSF inhibitor levels must influence the emergence and progressive proliferation of the leukemic populations in these diseases.

Acknowledgments

This work was supported by the Carden Fellowship Fund of the Anti-Cancer Council of Victoria and the Australian Research Grants Committee.

REFERENCES

Austin, P. E., E. A. McCulloch and J. F. Till. 1971. Characterisation of the factor in L-cell conditioned medium capable of stimulating colony formation by mouse marrow cells in culture. *J. Cell. Physiol.* **77:**121.

Bradley, T. R. and D. Metcalf. 1966. The growth of mouse bone marrow cells *in vitro*. *Aust. J. Exp. Biol. Med. Sci.* **44:**287.

Bradley, T. R. and M. A. Sumner. 1968. Stimulation of mouse bone marrow colony growth *in vitro* by conditioned medium. *Aust. J. Exp. Biol. Med. Sci.* **46:**607.

Bradley, T. R., E. R. Stanley and M. A. Sumner. 1971. Factors from mouse tissues stimulating colony growth of mouse bone marrow cells *in vitro*. *Aust. J. Exp. Biol. Med. Sci.* **49:**595.

Bradley, T. R., P. Fry, M. A. Sumner and E. McInerny. 1972. Factors determining colony forming efficiency in agar suspension cultures. *Aust. J. Exp. Biol. Med. Sci.* **50:**813.

Brown, C. H., and P. P. Carbone. 1971. *In vitro* growth of normal and leukemic human bone marrow. J. Natl. Cancer Instit. **46:**989.

Chan, S. H. 1971. Influence of serum inhibitors on colony development *in vitro* by bone marrow cells. *Aust. J. Exp. Biol. Med. Sci.* **49:**553.

Chan, S. H. and D. Metcalf. 1972. Local production of colony stimulating factor within the bone marrow: Role of non-hematopoietic cells. *Blood* **40:**646.

Chan, S. H., and D. Metcalf. 1973. Local and systemic control of granulocytic and macrophage progenitor cell regeneration after irradiation. Cell Tissue Kinet. **6:**187.

Chan, S. H., D. Metcalf and E. R. Stanley. 1971. Stimulation and inhibition by normal human serum of colony formation *in vitro* by bone marrow cells. *Brit. J. Haematol.* **20:**329.

Chervenick, P. A. and A. F. LoBuglio. 1972. Human blood monocytes: Stimulators of granulocyte and mononuclear formation *in vitro*. *Science* **178:**164.

Chervenick, P. A., A. L. Lawson, L. D. Ellis and S. F. Pan. 1971. *In vitro* growth of leukemic cells containing the Philadelphia (Ph) chromosome. *J. Lab. Clin. Med.* **78:**838.

Cline, M. J., N. L. Warner and D. Metcalf. 1972. Identification of the bone marrow colony mononuclear phagocyte as a macrophage. *Blood* **39:**326.

Dale, D. C., C. H. Brown, P. Carbone and S. M. Wolff. 1971. Cyclic urinary leukopoietic activity in grey collie dogs. *Science* **173:**152.

Dicke, K. A., M. J. Van Noord, B. Maat, U. W. Schaefer and D. W. Van Bekkum. 1973. Attempts at morphological identification of the haemopoietic stem cell in primates and rodents *Haemopoietic stem cells*, CIBA Symp. (ed. G. E. W. Wolstenholme). Associated Scientific Publishers, Amsterdam. (in press)

Foster, R., D. Metcalf, W. A. Robinson and T. R. Bradley. 1968. Bone marrow colony stimulating activity in human sera. *Brit. J. Haematol.* **15:**147.

Golde, D. W. and M. J. Cline. 1972. Identification of the colony-stimulating cell in human peripheral blood. *J. Clin. Invest.* **51:**2981.

Greenberg, P. L., W. C. Nichols and S. L. Schrier. 1971. Granulopoiesis in acute myeloid leukemia and preleukemia. *New Eng. J. Med.* **284**:1225.

Haskill, J. S., T. A. McNeill and M. A. S. Moore. 1970. Density distribution analysis of *in vivo* and *in vitro* colony forming cells in bone marrow. *J. Cell. Physiol.* **75**:167.

Ichikawa, Y., D. H. Pluznik and L. Sachs. 1966. *In vitro* control of the development of macrophage and granulocytic colonies. *Proc. Nat. Acad. Sci.* **56**:488.

Iscove, N. N., J. E. Till and E. A. McCulloch. 1970. The proliferative states of mouse granulopoietic progenitor cells. *Proc. Soc. Exp. Biol. Med.* **134**:33.

Iscove, N. N., J. S. Senn, J. E. Till and E. A. McCulloch. 1971. Colony formation by normal and leukemic marrow cells in culture: Effect of conditioned medium from human leukocytes. *Blood* **37**:1.

Janoshwitz, H., M. A. S. Moore and D. Metcalf. 1971. Density gradient segregation of bone marrow cells with the capacity to form granulocytic and macrophage colonies *in vitro*. *Exp. Cell Res.* **68**:220.

Lajtha, L. G., L. V. Pozzi, R. Schofield and M. Fox. 1969. Kinetic properties of haemopoietic stem cells. *Cell Tissue Kinet.* **2**:39.

Landau, T. and L. Sachs. 1971. Characterization of the inducer required for the development of macrophage and granulocyte colonies. *Proc. Nat. Acad. Sci.* **68**:2540.

Mangalik, A. and W. Robinson. 1973. Cyclic neutropenia: The relationship between urine granulocyte colony stimulating activity and neutrophil count. *Blood* **41**:79.

Metcalf, D. 1969. Studies on colony formation *in vitro* by mouse bone marrow cells. I. Continuous cluster formation and relation of clusters to colonies. *J. Cell. Physiol.* **74**:323.

———. 1970. Studies on colony formation *in vitro* by mouse bone marrow cells. II. Action of colony stimulating factor. *J. Cell. Physiol.* **76**:89.

———. 1971a. Acute antigen-induced elevation of serum colony stimulating factor (CSF) levels. *Immunology* **21**:427.

———. 1971b. Transformation of granulocytes to macrophages in bone marrow colonies *in vitro*. *J. Cell. Physiol.* **77**:277.

———. 1971c. Inhibition of bone marrow colony formation *in vitro* by dialysable products of normal and neoplastic haemopoietic cells. *Aust. J. Exp. Biol. Med. Sci.* **49**:351.

———. 1972. Effect of thymidine suiciding on colony formation *in vitro* by mouse hematopoietic cells. *Proc. Soc. Exp. Biol. Med.* **139**:511.

———. 1973. Human leukaemia. Recent tissue culture studies on the nature of myeloid leukaemia. *Brit. J. Cancer* (in press).

Metcalf, D. and S. H. Chan. 1973. Abnormal regulation of granulopoiesis in human acute granulocytic leukemia. *Unifying concepts of leukemia* (ed. R. M. Dutcher and L. Chieco-Bianchi) *Bibl. Haemat.* **39**:878.

Metcalf, D. and R. Foster. 1967. Behavior on transfer of serum stimulated bone marrow colonies. *Proc. Soc. Exp. Biol. Med.* **126**:758.

Metcalf, D. and M. A. S. Moore. 1971. *Haemopoietic cells.* North-Holland, Amsterdam.

———. 1973. Regulation of growth and differentiation in haemopoietic colonies growing in agar. *Haemopoietic stem cells,* CIBA Symp. (ed. G. E. W. Wolstenholme) Associated Scientific Publishers, Amsterdam. (in press)

Metcalf, D. and E. R. Stanley. 1969. Quantitative studies on the stimulation of mouse bone marrow colony growth *in vitro* by normal human urine. *Aust. J. Exp. Biol. Med. Sci.* **47**:453.

———. 1971. Haematological effects in mice of partially purified colony stimulating factor (CSF) prepared from human urine. *Brit. J. Haematol.* **21**:481.

Metcalf, D. and S. Stevens. 1972. Influence of age and antigenic stimulation on granulocyte and macrophage progenitor cells in the mouse spleen. *Cell Tissue Kinet.* **5**:433.

Metcalf, D., T. R. Bradley and W. Robinson. 1967. Analysis of colonies developing *in vitro* from mouse bone marrow cells stimulated by kidney feeder layers or leukemic serum. *J. Cell. Physiol.* **69**:93.

Metcalf, D. R. Foster and M. Pollard. 1967. Colony stimulating activity of serum from germfree normal and leukemic mice. *J. Cell. Physiol.* **70**:131.

Metcalf, D., M. A. S. Moore and K. Shortman. 1971. Adherence column and buoyant density separation of bone marrow stem cells and more differentiated cells. *J. Cell. Physiol.* **78**:441.

Metcalf, D., S. H. Chan, F. W. Gunz, P. Vincent and R. B. M. Ravich. 1971. Colony stimulating factor and inhibitor levels in acute granulocytic leukemia. *Blood* **38**:143.

Moore, M. A. S. 1973. *In vitro* studies in myeloid leukemias. *Excerpta medica reviews in leukemia and lymphoma* (ed. F. J. Cleton, D. Crowther and J. S. Malpas) Excerpta Medica, Amsterdam. (in press)

Moore, M. A. S. and D. Metcalf. 1973. Cytogenetic analysis of human acute and chronic myeloid leukemic cells cloned in agar culture. *Int. J. Cancer* **11**:143.

Moore, M. A. S. and N. Williams. 1972. Physical separation of colony stimulating cells from *in vitro* colony forming cells in monkey hemopoietic tissue. *J. Cell. Physiol.* **80**:185.

Moore, M. A. S., N. Williams and D. Metcalf. 1972. Purification and characterisation of the *in vitro* colony forming cell in monkey hemopoietic tissue. *J. Cell. Physiol.* **79**:283.

————. 1973a. *In vitro* colony formation by normal and leukemic human hemopoietic cells: Interaction between colony-forming and colony-stimulating cells. *J. Nat. Cancer Inst.* **50**:591.

————. 1973b. *In vitro* colony formation by normal and leukemic human hemopoietic cells: Characterization of the colony-forming cells. J. Nat. Cancer Inst. **50**:603.

Morley, A., P. Quesenberry, P. Bealmear, F. Stohlman and R. Wilson. 1972. Serum colony stimulating factor levels in irradiated germfree and conventional GFW mice. *Proc. Soc. Exp. Biol. Med.* **140**:478.

Paran, M. and L. Sachs. 1968. The continued requirement for inducer for the development of macrophage and granulocyte colonies. *J. Cell. Physiol.* **72**:247.

Paran, M., Y. Ichikawa and L. Sachs. 1968. Production of the inducer for macrophage and granulocyte colonies by leukemic cells. *J. Cell. Physiol.* **72**:251.

Paran, M., L. Sachs, Y. Barak and P. Resnitsk... 1970. *In vitro* induction of granulocyte differentiation in hematopoietic cells from ukemic and nonleukemic patients. *Proc. Nat. Acad. Sci.* **67**:1542.

Pluznik, D. H. and L. Sachs. 1966. The induction of clones of normal mast cells by a substance in conditioned medium. *Exp. Cell Res.* **43**:553.

Quesenberry, P. J., A. A. Morley, K. A. Rickard, M. Garrity, D. Howard and F. Stohlman. 1972. Effect of endotoxin on granulopoiesis and colony-stimulating factor. *New Eng. J. Med.* **286**:227.

Robinson, W. A. and B. L. Pike. 1970. Leukopoietic activity in human urine: The granulocytic leukemias. *New Eng. J. Med.* **282**:1291.

Robinson, W. A., J. E. Kurnick and B. L. Pike. 1971. Colony growth of human leukemic peripheral blood cells *in vitro*. *Blood* **38**:500.

Robinson, W., D. Metcalf and T. R. Bradley. 1967. Stimulation by normal and leukemic mouse sera of colony formation *in vitro* by mouse bone marrow cells. *J. Cell. Physiol.* **69**:83.

Shadduck, R. K. and H. R. Nankin. 1971. Cellular origin of granulocyte colonies in chronic myeloid leukaemia. *Lancet* **2**:1097.

Shadduck, R. K., N. G. Nunna, F. Mandarino and F. Yurechko. 1972. Correlative studies of CSF and granulocyte tumour in rats. *In vitro culture of hemopoietic cells* (ed. D. W. Van Bekkum and K. A. Dicke) p. 31. Radiobiological Institute TNO, Rijswijk.

Sheridan, J. W. and D. Metcalf. 1972. Studies on the bone marrow colony stimulating factor (CSF): Relation of tissue CSF to serum CSF. *J. Cell. Physiol.* **80**:129.

————. 1973. A low molecular weight factor in lung-conditioned medium stimulating granulocyte and monocyte colony formation *in vitro. J. Cell. Physiol.* **81**:11.

Sheridan, J. W. and E. R. Stanley. 1971. Tissue sources of bone marrow colony stimulating factor. *J. Cell. Physiol.* **78**:451.

————. 1972. Purification and properties of human urine colony stimulating factor Evidence for a peptide component. *Aust. J. Exp. Biol. Med. Sci.* **49**:281.

————. 1972. Purification and properties of human urine colony stimulating factor (CSF). *Cell differentiation* (ed. R. Harris and D. Viza) p. 149. Munksgaard, Copenhagen.

Stanley, E. R., T. R. Bradley and M. A. Sumner. 1971. Properties of the mouse embryo conditioned medium factor(s) stimulating colony formation by mouse bone marrow cells grown *in vitro. J. Cell. Physiol.* **78**:301.

Stanley, E. R., D. Metcalf, J. S. Maritz and G. F. Yeo. 1972. Standardised bioassay for bone marrow colony stimulating factor in human urine: Levels in normal man. *J. Lab. Clin. Med.* **79**:657.

Worton, R. G., E. A. McCulloch and J. E. Till. 1969. Physical separation of hemopoietic stem cells from cells forming colonies in culture. *J. Cell. Physiol.* **74**:171.

Factors Affecting Normal and Leukemic Hemopoietic Cells in Culture

J. E. Till, H. A. Messner, G. B. Price, M. T. Aye, and E. A. McCulloch

Ontario Cancer Institute and Institute of Medical Science
University of Toronto, Toronto, Ontario, Canada

The regulation of growth and differentiation in organized populations of cells, such as those of the blood-forming system, requires intercellular communication. If this communication involves mediators (such as hormones) that can act at long range, it can be studied in vivo. If, however, the communication is short range, then analysis in vivo may be difficult and a cell culture system is required. The most desirable culture models are those in which three different components of intercellular communication can be demonstrated: the chemical mediators of communication, the producer cells which release them, and the target cells which respond to them. Culture models exhibiting all three of these components are rare; one that possesses them is the model of human granulopoiesis in culture used for much of the work outlined in this paper. This particular model possesses an additional attribute that makes it of interest to the clinician as well as to the cellular or molecular biologist: it utilizes cells freshly explanted from human marrow or peripheral blood. Thus the results obtained may not only contribute to an understanding of biological organization, but also provide new insights into the cellular or molecular basis for human hematological diseases. Studies that illustrate cellular, molecular, and clinical aspects of the system are described below.

Interacting Cell Populations in Cultures of Human Marrow

The key to the development of a new method to study granulopoiesis in culture was provided by Bradley and Metcalf (1966) and by Pluznik and Sachs (1965), who reported the formation of colonies in cultures of cells derived from murine marrow and showed that the colonies contained granulocytes and macrophages (Bradley and Metcalf 1966; Ichikawa, Pluznik and Sachs 1966). Successful colony formation depended on two features of the experimental system: first, the marrow cells had to be suspended in semi-solid medium; and second, the cultures had to contain certain stimulatory factors. The semi-solid medium, agar or methylcellulose, localized the progeny of proliferating progenitor cells so that they could be recognized as members of a single colony. The stimulatory substances could be derived from a number of sources, including medium conditioned by cells in culture (Pluznik and Sachs 1966), mouse sera (Robinson et al. 1967), tissue extracts (Bradley et al. 1969), and human urine (Robinson et al. 1969). The latter source yielded large quantities of material suitable for chemical studies and a factor that stimulates colony formation by mouse marrow cells has been isolated and characterized. This colony-stimulating factor (CSF) appears to be a glycoprotein with a molecular weight of approximately 45,000 (Stanley and Metcalf 1972). It is similar to, but not identical with, factors obtained from other sources such as medium conditioned by excised lungs from endotoxin-treated mice (Sheridan and Metcalf 1973).

The culture system was quickly applied to the study of human marrow (Senn, McCulloch and Till 1967). However significant differences from the findings with mouse marrow were encountered. Mouse marrow cells generally form few, if any, granulocyte or macrophage colonies unless appropriate stimulators are added to the cultures. In contrast cells from human marrow usually form colonies in culture in the absence of added stimulators. Nonetheless media conditioned by cultures of human peripheral leukocytes (Iscove et al. 1971), human spleen cells (Paran et al. 1970), or human embryo kidney cells (Brown and Carbone 1971) have been found to affect colony formation by human marrow; their addition usually results in a small increase in colony-forming efficiency and marked increases in colony size. Unlike human urinary CSF, the active materials in these various conditioned media have not yet been extensively purified or characterized; they will be referred to as colony-stimulating activities (CSA).

It seemed likely that the colony formation observed in cultures of human marrow cells in the absence of exogenous stimulators could be attributed to substances produced endogenously by cells present in the marrow inoculum (Moore and Williams 1972). This hypothesis was tested using a simple cell separation procedure (Messner, Till and McCulloch 1973). Human marrow was allowed to adhere to the surface of glass or plastic culture dishes. The cells decanted after two cycles of adherence were considered to be nonadherent cells (NA cells), and populations remaining on the dishes after the first adherence cycle and extensive washing were considered to be adherent cells. In contrast to the behavior of unseparated marrow cell populations, NA cells usually failed to form colonies in the absence of added CSA. However granulocytic colonies appeared when CSA was added; the number of colonies formed increased linearly with the amount of added CSA, up to a maximum colony count. Adherent cells were also effective in promoting colony formation by populations of NA cells. Thus granulocytic colonies were observed if NA cells were plated over adherent cells, although the latter population was incapable of colony formation even in the presence of exogenous CSA. Further, the number of colonies formed was dependent on the number of adherent cells added to the populations of NA cells.

These studies, and those recently reported by others (Moore, Williams and Metcalf 1973a; Haskill, McKnight and Galbraith 1972), indicate the presence in normal human marrow of at least two cell populations that interact in culture. The first population contains the progenitors of granulocytic colonies, capable of colony formation only when stimulated by CSA, and is referred to as CSA-dependent colony-forming units (CFU-C). The second population contains cells capable of producing CSA, referred to in this paper as adherent CSA-producing cells (AC cells). It is likely that AC cells are morphologically similar to monocytes

(Chervenick and LoBuglio 1972; Golde and Cline 1972; Moore, Williams and Metcalf 1973a).

The separation procedure has permitted evaluation of three components in human marrow suspensions that affect the efficiency of colony formation: CFU-C, AC cells, and CSA. CFU-C may be assayed by determining the colony forming efficiency of either unseparated marrow or populations of NA cells in the presence of appropriate concentrations of exogenous CSA. Both AC cells and CSA may be assayed by mixing increasing quantities of either AC cells or CSA with populations of NA cells and measuring the efficiency of colony formation.

The assays may be applied to studies of hematological disease, particularly acute myeloblastic leukemia (AML). When described in terms of behavior in culture, cell populations from the marrow of untreated patients with AML show considerable variation. A common pattern is a decrease in the efficiency of colony formation (Senn, McCulloch and Till 1967; Harris and Freireich 1970; Brown and Carbone 1971; Iscove et al. 1971; Greenberg et al. 1971; Moore, Williams and Metcalf 1973b), in the capacity of peripheral blood cells to produce CSA (Greenberg et al. 1971; Robinson, Kurnick and Pike 1971), or in AC cell function (Messner et al. 1973). Each of these changes is reversed when the patients are in remission after chemotherapy (Harris and Freireich 1970; Greenberg et al. 1971; Brown and Carbone 1971; Cowan et al. 1972; Moore, Williams and Metcalf 1973b; Messner et al. unpubl.). For other untreated patients colony formation is increased (Paran et al. 1970; Robinson, Kurnick and Pike 1971; Greenberg et al. 1971); also ability to produce CSA need not be correlated with colony-forming ability (Moore, Williams and Metcalf 1973a).

The availability of an AC cell assay has also made it possible to examine the properties of cells of this class. In a preliminary series of experiments (Messner et al. unpubl.) the sedimentation velocity of AC cells has been determined using unit gravity sedimentation (Miller and Phillips 1969). When human marrow was separated by this procedure, two distinct peaks of AC cells were detected, with sedimentation velocities at 4–5 mm per hour and 7–9 mm per hour.

In summary, the culture model of human granulopoiesis outlined above has several attractive features. All three of the components of intercellular communication referred to above can be demonstrated: the target cells (CFU-C), the producer cells (AC cells), and the chemical mediators (CSA). Each component can be measured and tested for heterogeneity or for modifications associated with hematological disease.

Heterogeneity of Colony-Stimulating Activity

An initial characterization of the CSA in medium conditioned by cultures of human peripheral leukocytes, prepared in serum-containing medium and assayed on cultures of mouse marrow cells, revealed a factor of molecular weight 32,000–34,000, with properties similar to those of purified human urinary CSA (Austin 1971). However leukocyte-conditioned medium prepared in the absence of serum was found to contain not only this factor, but another CSA of low molecular weight (less than 1300) (Price, McCulloch and Till 1973). This low molecular weight CSA (LMW-CSA) was found to be extractable by organic solvents such as chloroform or diethyl ether and appeared to be trypsin sensitive. Unlike the CSA of higher molecular weight, the LMW-CSA did not promote colony formation by mouse marrow cells; its activity was most clearly demonstrated using human NA cell populations.

LMW-CSA was purified by passage through a UM-10 ultrafiltration membrane (Amicon) and extraction into chloroform and diethyl ether; it was then treated with dansyl chloride and examined by thin-layer chromatography (solvent: methyl-ethyl ketone, pyridine, water, glacial acetic acid, 70:15:15:2). A single dansylated spot was obtained close to the solvent front, and colony-stimulating activity was associated with material recovered from this spot (Price et al. unpubl.). LMW-CSA prepared from leukemic peripheral blood cells and subjected to the same procedure also yielded a single dansylated spot after thin-layer chromatography. However the biological activities of the LMW-CSA's prepared from the peripheral leukocytes of normal individuals and certain patients with leukemia appeared to be different. Both preparations were tested on NA cells from the marrow of four nonleukemic individuals and on that of two patients with AML. The LMW-CSA derived from normal cells stimulated colony formation by NA cells from nonleukemic individuals. However the latter cells responded poorly, if at all, to LMW-CSA from leukemic peripheral blood cells. When both preparations of LMW-CSA were tested on NA cells from a patient with AML in relapse, whose marrow contained many CFU-C, both preparations were equally effective in stimulating colony formation. This pattern of response (stimulation by LMW-CSA from both normal and leukemic cells) has also been observed to develop in another patient with AML during remission induction with chemotherapy (cyclophosphamide, arabinosylcytosine and vincristine, Cowan et al. 1972). The pattern persisted in complete remission. However the LMW-CSA from leukemic cells has not been entirely specific for leukemic NA cells; CFU-C from the marrow of three patients with sideroblastic anemia have also been found to respond to LMW-CSA from both normal and leukemic peripheral blood cells.

In summary, two different classes of CSA's have been detected, both capable of promoting colony formation by human CFU-C. One class consists of macromolecules, probably glycoproteins; the other consists of small, hydrophobic, trypsin-sensitive molecules. Within both classes there is evidence of further heterogeneity. The significance in vivo of this heterogeneity, and indeed of the CSA's themselves, remains to be determined.

Leukemic Cells in Suspension Culture

Studies of hematopoietic cells in culture are becoming increasingly important to clinical hematologists. The CFU-C/AC cell system described above provides a culture method for investigating human granulopoiesis and may be of importance in the study of specific granulopoietic defects such as neutropenia (Senn et al. 1973). Also since granulopoiesis is disturbed in leukemia, the disease state is reflected in changes in levels of CFU-C and AC cells. However the leukemic transformation probably effects a class of pluripotent stem cells since chromosome markers in chronic (Whang et al. 1963) and acute (Jensen and Killman 1967) leukemia are found in both granulocytes and erythroblasts. Findings with assays for CFU-C and AC cells, therefore, may reflect secondary effects of leukemia on granulopoiesis rather than primary defects in hemopoiesis at the level of pluripotent stem cells.

A more direct approach to leukemia became possible when it was found that populations of leukemic peripheral blood cells, containing a high proportion of blast cells, would proliferate in suspension culture (Aye, Till and McCulloch 1972). In two cases, where marker chromosomes were present in the original suspension, these were the predominant karyotypes identified after 10 days in culture.

Studies of leukemic perpiheral blood cells were facilitated by the finding that they were readily stored at $-70°C$ in 5% dimethylsulfoxide and were suitable for repeated studies of growth kinetics and cultural requirements (Aye et al. unpubl.). We have used this approach to examine peripheral blood cells from nine patients with AML. In all instances the rate of incorporation of [³H]thymidine by cells in suspension culture was improved when medium conditioned by cultures of normal peripheral leukocytes was added, the same conditioned medium that will also stimulate granulocytic colony formation in cultures of normal marrow. Two lines of evidence suggest that the material promoting [³H]thymidine incorporation by leukemic cells in suspension may be similar to CSA. First, 12 different conditioned media were assessed for their capacity to stimulate colony formation by NA cells from normal marrow and to promote [³H]thymidine incorporation by cells from a patient with AML. The two activities were found to be correlated. Second, a crude CSA preparation was filtered through Sephadex G-75 and the fractions assayed on both normal NA cells and cells from leukemic peripheral blood. Both assays yielded a similar elution profile with a single peak of activity.

Studies with highly purified CSA are required before its activity on leukemic cells can be verified; nonetheless our preliminary studies indicate that material obtained from normal leukocytes may influence the growth of leukemic cells in culture. If similar growth requirements exist in vivo, studies of the stimulator molecules and the cells that produce them may yield new avenues of approach to therapy.

CONCLUSION

The studies outlined above provide examples of the use of cell culture methods to investigate possible regulatory mechanisms involved in normal granulopoiesis in man and the changes in regulation that occur when granulopoiesis is defective. The results obtained indicate that cell-to-cell interactions occur in culture, even in populations derived from patients with leukemia, and that at least some of these interactions are mediated by diffusible factors that can be isolated and characterized. The field is still in its infancy, and convincing evidence has yet to be obtained in support of the view that the intercellular communication observed in culture provides meaningful information about physiological regulatory processes operating in vivo. Comparisons of cell populations from normal individuals with those from patients with leukemia provides one way to test this view. It should be possible, through studies of the kind described in this paper, to determine if properties of the cultured cell populations are relevant to mechanisms of disease and, in this way, to obtain new insights into these mechanisms.

Acknowledgments

The work discussed in this paper was supported by grants from the Medical Research Council (MT-1420), the Ontario Cancer Treatment and Research Foundation (236) and the National Cancer Institute of Canada. G.B.P. is a postdoctoral fellow of the Damon Runyon Memorial Fund; M.T.A. is a fellow of the Medical Research Council of Canada. We are grateful to Dr. J. S. Senn for his contribution to part of the work.

REFERENCES

Austin, P. 1971. Studies on conditioning factor activity for marrow cells in culture. Ph.D. thesis, University of Toronto.

Aye, M. T., J. E. Till and E. A. McCulloch. 1972. Growth of leukemic cells in culture. *Blood* **40:**806.

Bradley, T. R. and D. Metcalf. 1966. The growth of mouse bone marrow cells *in vitro*. *Aust. J. Exp. Biol. Med. Sci.* **44:**287.

Bradley, T. R., D. Metcalf, M. Sumner and E. R. Stanley. 1969. Characteristics of *in vitro* colony formation by cells from haemopoietic tissues. *Hemic cells in vitro* (ed. P. Farnes) p. 22–35. Williams and Wilkins, Baltimore.

Brown, C. H. and P. P. Carbone. 1971. *In vitro* growth of normal and leukemic human bone marrow. *J. Nat. Cancer Inst.* **46:**989.

Chervenick, P. A. and A F. LoBuglio. 1972. Human blood monocytes: Stimulators of granulocyte and mononuclear colony formation *in vitro*. *Science* **178:**164.

Cowan, D. H., A. Clarysse, H. Abu-Zahra, J. S. Senn and E. A. McCulloch. 1972. The effect of remission induction in acute myeloblastic leukemia on efficiency of colony formation in culture. *Series Haemat.* **5:**179.

Golde, D. W. and M. J. Cline. 1972. Identification of the colony-stimulating cell in human peripheral blood. *J. Clin. Invest.* **51:**2981.

Greenberg, P. C., W. C. Nichols and S. L. Schrier. 1971. Granulopoiesis in acute myeloid leukemia and pre-leukemia. *New Eng. J. Med.* **284:**1225.

Harris, J. and E. J. Freireich. 1970. *In vitro* growth of myeloid colonies from bone marrow of patients with acute leukemia in remission. *Blood* **35:**61.

Haskill, J. S., R. D. McKnight and P. R. Galbraith. 1972. Cell-cell interaction *in vitro*: Studied by density separation of colony-forming, stimulating, and inhibiting cells from human bone marrow. *Blood* **40:**394.

Ichikawa, Y., D. H. Pluznik and L. Sachs. 1966. *In vitro* control of the development of macrophage and granulocyte colonies. *Proc. Nat. Acad. Sci.* **56:**488.

Iscove, N. N., J. S. Senn, J. E. Till and E. A. McCulloch. 1971. Colony formation by normal and leukemic human marrow cells in culture: Effect of conditioned medium from human leukocytes. *Blood* **37:**1.

Jensen, M. K. and S. Killman. 1967. Chromosome studies in acute leukemia. Evidence for chromosomal abnormalities common to erythroblasts and leukemic white cells. *Acta Med. Scand.* **181:**47.

Messner, H. A., J. E. Till and E. A. McCulloch. 1973. Interacting cell populations affecting granulopoietic colony-formation by normal and leukemic human marrow cells. *Blood* (in press)

Miller, R. G. and R. A. Phillips. 1969. Separation of cells by velocity sedimentation. *J. Cell. Physiol.* **73:**191.

Moore, M. A. S. and N. Williams. 1972. Physical separation of colony stimulating cells from *in vitro* colony forming cells in hemopoietic tissue. *J. Cell. Physiol.* **80:**195.

Moore, M. A. S., N. Williams and D. Metcalf. 1973a. *In vitro* colony formation by normal and leukemic human hematopoietic cells: Interaction between colony-forming and colony-stimulating cells. *J. Nat. Cancer Inst.* **50:**591.

————. 1973b. *In vitro* colony formation by normal and leukemic human hematopoietic cells: Characterization of the colony-forming cells. *J. Nat. Cancer Inst.* **50**:603.

Paran, M., L. Sachs, Y. Barak and P. Resnitzky. 1970. *In vitro* induction of granulocytic differentiation in hemopoietic cells from leukemic and nonleukemic patients. *Proc. Nat. Acad. Sci.* **67**:1542.

Pluznik, D. H. and L. Sachs. 1965. The cloning of normal "mast" cells in tissue culture. *J. Cell. Comp. Physiol.* **66**:319.

————. 1966. The induction of clones of normal mast cells by a substance from conditioned medium. *Exp. Cell Res.* **43**:553.

Price, G. B., E. A. McCulloch and J. E. Till. 1973. A new human low molecular weight granulocyte colony stimulating activity. *Blood* (in press)

Robinson, W. A., J. E. Kurnick and B. L. Pike. 1971. Colony growth of human leukemic peripheral blood cells *in vitro*. *Blood* **38**:500.

Robinson, W. A., D. Metcalf and T. R. Bradley. 1967. Stimulation by normal and leukemic mouse sera of colony formation *in vitro* by mouse bone marrow cells. *J. Cell. Physiol.* **69**:83.

Robinson, W. A., E. R. Stanley and D. Metcalf. 1969. Stimulation of bone marrow colony growth *in vitro* by human urine. *Blood* **33**:396.

Senn, J. S., E. A. McCulloch and J. E. Till. 1967. Comparison of colony-forming ability of normal and leukaemic human marrow in cell culture. *Lancet* **2**:597.

Senn, J. S., H. A. Messner and E. R. Stanley. 1973. Mechanisms in neutropenic granulopoiesis studied in cell culture. *Exp. Hematol.* (in press)

Sheridan, J. W. and D. Metcalf. 1973. A low molecular weight factor in lung-conditioned medium stimulating granulocyte and monocyte colony formation *in vitro*. *J. Cell. Physiol.* **81**:11.

Stanley, E. R. and D. Metcalf. 1972. Purification and properties of human urinary colony stimulating factor (CSF). *Cell differentiation* (ed. R. Harris and D. Viza) p. 272–276. Munksgaard, Copenhagen.

Whang, J., E. Frei, J. H. Tjio, P. P. Carbone and G. Brecher. 1963. The distribution of the Philadelphia chromosome in patients with chronic myelogenous leukemia. *Blood* **22**:664.

Control of Growth and Differentiation in Normal Hematopoietic and Leukemic Cells

Leo Sachs

Department of Genetics, Weizmann Institute of Science, Rehovot, Israel

Studies on the mechanism that controls the growth and differentiation of normal hematopoietic cells, can be of value in elucidating the block in cell differentiation that occurs during leukemogenesis. We have, therefore, developed a tissue culture system in which the growth and differentiation of normal lymphocytes involved in the homograft reaction, mast cells, macrophages and granulocytes, can be studied in mass culture (Ginsburg and Sachs 1963, 1965; Sachs 1964). We also developed a tissue culture cloning assay for two types of these normal hematopoietic cells: macrophages and granulocytes in rodents (Pluznik and Sachs 1965, 1966; Ichikawa et al. 1966) and for granulocytes in humans (Paran et al. 1970). We have shown that various types of cells, including fibroblasts, release a protein inducer, which we now call MGI (Landau and Sachs 1971a,b), that can induce in vitro the formation of colonies with mature differentiated macrophages and granulocytes from normal undifferentiated hematopoietic cells (Pluznik and Sachs 1965, 1966; Ichikawa et al. 1966, 1967; Lagunoff et al. 1966; Paran and Sachs 1968, 1969; Paran et al. 1968, 1969, 1970; Sachs 1970). We have also shown that some, but not all, myeloid leukemic cells from men (Paran et al. 1970) and mice (Fibach et al. 1972, 1973) can be induced to undergo normal cell differentiation by MGI. Erythropoietin cannot substitute for MGI (Paran and Sachs 1968).

The present paper summarizes our studies (Paran et al. 1970; Fibach et al. 1972, 1973; Inbar et al. 1973a,b; Guez and Sachs unpubl.) on (1) the control of normal differentiation of myeloid leukemic cells, and (2) the chemical characterization of MGI.

Control of Normal Differentiation of Myeloid Leukemic Cells

Differentiation to Macrophages and Granulocytes

The experiments with mouse leukemia were carried out with a tissue culture line of myeloblastic leukemic cells (Ichikawa 1969). When this line was cloned in soft agar, about 2% of the cells formed colonies of undifferentiated blast cells. Addition of conditioned medium (CM) from fibroblasts that contains MGI (Pluznik and Sachs 1966) increased the cloning efficiency to 15%. About 60% of the colonies were compact and contained only undifferentiated blast cells (D⁻ colonies) and about 40% (D⁺ colonies) contained a dispersed periphery with cells in various stages of differentiation. Depending on the batch of CM, the differentiation was to mature macrophages or to both mature macrophages and granulocytes. The addition of purified MGI from a batch of CM that induced differentiation only to macrophages also induced differentiation in D⁺ colonies only to macrophages and no differentiation in D⁻ colonies. It will be of interest to determine whether there are different cofactors for the induction of myeloid leukemic cell differentiation to macrophages and granulocytes.

Isolation of 3-times purified clones has shown the hereditability of the D⁺ and D⁻ property in these myeloid leukemic cells. D⁺ clones at 10,12,14 and 16 days after seeding in agar with CM contained about 10, 30, 60 and 90% differentiated cells, respectively. The difference in colony type was not associated with a difference in the growth rate of the cells and both D⁺ and D⁻ colonies were leukemic in mice. With appropriate batches of CM or serum from mice treated with endotoxin, the 3-times purified D⁺ clones at 12 days after seeding gave 45% progeny colonies with both macrophages and granulocytes, in addition to 15% with only granulocytes and 40% with only macrophages. The results with clones have shown that the differentiation to macrophages and granulocytes can occur in the progeny of a single cell. The granulocytes showed all stages of differentiation to mature granulocytes (Fig. 1). As in the case of normal differentiation of normal cells, the mature macrophages and granulocytes induced from the leukemic cells stopped multiplying. Treatment with MGI may thus be of therapeutic value.

Genetic Segregation in D⁺ and D⁻ Clones

Four D⁺ and four D⁻ clones were tested about 40 cell generations after seeding for the third clone isolation to determine whether D⁺ clones can segregate D⁻ progeny and vice-versa. Four thousand cells from each clone were seeded in eight petri dishes with conditioned medium. The results (Table 1) show that three of the D⁺ clones gave 0.4, 0.6 and 0.8% D⁻ progeny and two of the D⁻ clones gave 0.25 and 1.2% D⁺ progeny. Isolation of three revertant colonies of each type has shown that D⁺ gave D⁺ and D⁻ gave D⁻ progeny.

Normal Differentiation in Aneuploid Cells

Counts of chromosome numbers were made in clones grown in mass culture or in agar. The same modal chromosome number was found under both conditions. The modal number in the eight clones (Table 1) were 40, 41 or 42 for D⁺ and 40 or 41 for D⁻. One abnormal chromosome was found in both the D⁺ clones and in one of the D⁻ clones with a modal number of 40. The chromosome counts have thus shown that normal differentiation can occur in D⁺ clones that are no longer diploid. The nondiploid character of these cells has been confirmed by an analysis of their Giemsa banding pattern. It will be of interest to determine whether, as in some other types of tumor cells (Rabinowitz and Sachs 1970; Hitotsumachi et al. 1971, 1972; Yamamoto et al. 1973a,b), there are identifiable chromosome differences between normal, D⁺ and D⁻ cells.

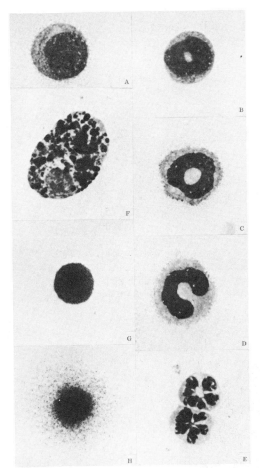

Figure 1
Induction of cell differentiation in myeloid leukemic cells to mature macrophages and granulocytes. **A,** Undifferentiated blast cells; **B–E,** stages in the differentiation to mature granulocytes; **G,** mature macrophage; **G,** D⁻ colony; **H,** D⁺ colony. A to F stained with May-Grunwald Giemsa, G and H unstained.

Differences in the Mobility of ConA Binding Sites in D⁺ and D⁻ Clones

The carbohydrate-binding protein concanavalin A (ConA) has been used as a probe to study the mobility of carbohydrate-containing sites on the surface membrane of D⁺ and D⁻ cells. Changes in distribution of membrane sites can be induced by ConA (Inbar and Sachs 1973; Inbar et al. 1973a). With the appropriate site mobility this induction of a change in distribution resulted in a concentration of ConA membrane site complexes on one pole of the cell to form a cap (Fig. 2). D⁺ and D⁻ clones showed about 50% and 5% cells with caps, respectively, although both types of cells bound a similar number of ConA molecules. Treatment of cells with trypsin increased cap formation from 5% to 40% in D⁻ cells, but did not change the percentage of cells with caps in D⁺ cells (Fig. 3). The results show a

Table 1
Chromosome Numbers and Segregation of D+ and D− Clones

| Clone no. | Parental clones | | | No. and type progeny colonies | |
	Modal no. chromo-somes	Abnormal chromo-somes	Clone type	D+	D−
1	40	1 ST*	D+	835	0
2	40	1 LT*	D+	957	6
3	41	none	D+	625	5
4	42	none	D+	721	3
5	40	1 LT	D−	0	751
6	41	none	D−	0	1030
7	40	none	D−	3	1202
8	41	none	D−	10	851

To determine the type of progeny, 4000 cells per clone were seeded in eight 50-mm petri dishes with 15% conditioned medium. The modal chromosome numbers were found in at least 75% of the cells.

* ST, subtelocentric chromosome; LT, abnormal large telocentric chromosome.

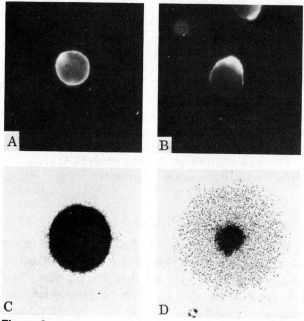

Figure 2

Binding of fluorescent concanavalin A to single cells and D+ and D− colonies grown in soft agar. Cells were incubated with 100 μg fluorescent ConA/ml (saturation conditions) for 15 min at 37°C, the cells washed with PBS and the fluorescence determined on a drop of living cells. **A,** D− cell with a fluorescent ring; **B,** D+ cell with a fluorescent cap; **C,** compact colony of D− cells; **D,** dispersed colony of D+ cells.

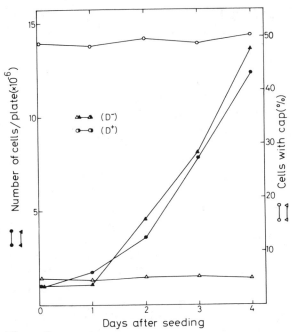

Figure 3
Cap formation in D⁻ and D⁺ cells with 100 μg fluorescent ConA/ml after
treatment of the cells with 1, 10 or 100 μg purified trypsin for 15 min at 37°C.

difference in the mobility of ConA binding sites in these two types of cells and
suggest a difference in the fluid state of these carbohydrate-containing structures on
the surface membrane. This suggests that a gain of the ability of myeloid leukemic
cells to undergo normal differentiation is associated with an increase in the fluidity
of structures on the surface membrane where the ConA sites are located.

It will be of interest to determine whether the difference between D⁺ and D⁻
cells is due to a difference in surface membrane or intracellular receptors for MGI.
It will also be of interest to determine the nature of the block in myeloid leukemic
cells in which differentiation has been induced only to immature granulocytes
(Metcalf et al. 1969) and to further study natural inhibitors of MGI (Ichikawa et
al. 1967; Paran et al. 1969; Mintz and Sachs 1973).

Chemical Characterization of MGI

Requirement of Low Molecular Cofactor That Can Be Substituted by Adenine

MGI can be released by cells in culture into the tissue culture medium (Pluznik
and Sachs 1966) so that conditioned medium from cultured cells can be used as
a source for purification and characterization of MGI. We have shown that MGI
is a protein and that purification resulted in a loss of biological activity (Landau
and Sachs 1971a,b). Activity was regained by addition of a low molecular weight
cofactor that was also present in conditioned medium, and the cofactor can be
substituted by adenine or adenine-containing nucleotides (Landau and Sachs
1971a,b). The following experiments described the purification of the protein MGI

to homogeneity in SDS-polyacrylamide gel electrophoresis, its molecular weight, amino acid composition and labeling by acetylation (Guez and Sachs unpubl.).

Purification of MGI

MGI was purified from the serum-free conditioned medium from a cloned line of mouse cells by the following steps: lyophilization, ultrafiltration with a Diaflo XM-50 membrane, chromatography on hydroxylapatite and DEAE cellulose, and then gel filtration on Sephadex G-150 (Table 2). All bioassays after Diaflo ultrafiltration were carried out with addition of 1.25 ml per petri dish of the low molecular weight cofactor in conditioned medium. As in previous experiments (Landau and Sachs 1971a) the peaks of biological activity from the hydroxylapatite and DEAE cellulose columns were at 0.08 M and 0.01 M phosphate buffer, respectively. There was also a sharp peak of biological activity after the last step of gel filtration on Sephadex G-150. The peak obtained after Sephadex filtration gave a single band on polyacrylamide gel electrophoresis without or with SDS (Fig. 4), was 1800-fold purified, and had an inducing activity of 0.5 ng protein per colony (Table 2). Bioassay of the purified protein with different concentration of the cofactor from conditioned medium or the addition of adenine have shown (Fig. 5) that the concentration of cofactor used, 1.25 ml per petri dish to calculate the colony-inducing activity of purified MGI, was at the optimum. The yield of purified protein was about 0.5 mg from 10 liters conditioned medium.

Induction by Purified MGI of Macrophage and Granulocyte Colonies from Normal Bone Marrow Cells

The purified protein induced the development of both macrophage and granulocyte colonies in bioassays with bone marrow cells from adult mice. Three to 5 μg purified MGI with 1.25 ml cofactor or 2.5 μg adenine per petri dish induced the formation of 10% and 20% granulocyte colonies, respectively.

Table 2

Steps in the Purification of MGI

Solution	mg protein/ ml solution	No. colonies/mg protein	Recovery (%) Activity	Recovery (%) Protein	Degree of purification	ng protein per colony
Serum-free CM*	0.2	1×10^3	100	100	1	1000
Lyophilized CM	0.3	4.9×10^3	100	23	4.9	204
Diaflo (XM-50) ultrafiltered CM	3.9	1.5×10^4	90–100	10	15	67
Hydroxylapatite peak, 0.08 M	1.6	4×10^4	50–70	0.5	40	25
DEAE-cellulose peak, 0.1 M	0.3	0.6×10^6	30–50	0.15	600	1.7
Sephadex G-150 peak	0.01	1.8×10^6	15–20	0.05	1800	0.55

* CM = conditioned medium containing MGI.

Figure 4
Polyacrylamide gel electrophoresis with SDS. *Right,* Purified MGI;
left, bovine serum albumin.

Molecular Weight, Amino Acid Composition and
Labeling of Purified MGI

The mobility of purified MGI in a 10% acrylamide gel with SDS was similar to
that of bovine serum albumin (Fig. 6). This indicates that MGI has a molecular
weight equivalent to 68,000. Similar results were obtained with 7.5% acrylamide
containing SDS. The results of amino acid analysis of purified MGI (Table 3) have
indicated that there was no detectable cystine, cysteine, methionine or arginine.
The thiobarbitoric assay for sialic acid was negative. There was no detectable
hexosamine in the amino acid analyzer and no detectable sugars by gas-liquid
chromatography and thin-layer chromatography. Digestion of MGI with proteolytic enzymes has shown that molecules with a lower molecular weight can also be
biologically active. It will be of interest to compare the purified MGI from the

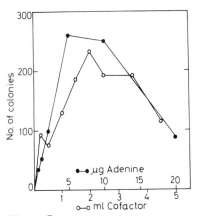

Figure 5
Induction of colony formation by purified MGI in the presence of (o———o) cofactor from conditioned medium (bioassay with horse serum), or (●———●)
adenine (bioassay with calf serum).

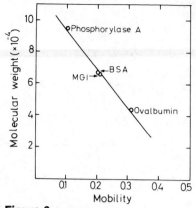

Figure 6

Molecular weight determination of purified MGI by polyacrylamide gel electrophoresis with SDS, expressed as relative mobility against log molecular weight.

mouse with MGI from human sources (Paran et al. 1970; Stanley and Metcalf
1969; Mintz and Sachs 1973) to determine what part of this protein carries the
biological activity and how this differs from erythropoietin (Goldwasser and Kung
1972) and thymosin (Goldstein et al. 1792).

The biological active peak from DEAE cellulose column was labeled with
[³H]acetic anhydride. The labeled protein had the same colony-inducing activity
as the unlabeled protein. As with the unlabeled protein, the labeled MGI gave a
sharp peak of biological activity after gel filtration on Sephadex G-150 and on
polyacrylamide gel electrophoresis (Fig. 7).

Table 3

Amino Acid Composition of Purified MGI

	Amino acid residue (%)
Lysine	4.85
Histidine	1.71
Arginine	
Aspartic acid	10.74
Threonine	7.32
Serine	9.47
Glutamic acid	12.66
Proline	5.10
Glycine	8.39
Alanine	9.74
Cystine (half)	
Valine	8.09
Methionine	
Isoleucine	3.70
Leucine	8.04
Tyrosine	3.72
Phenylalanine	6.41

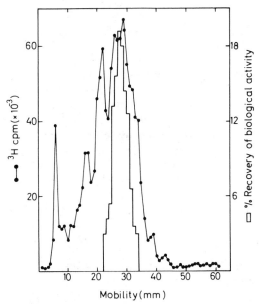

Figure 7

Polyacrylamide gel electrophoresis of labeled MGI from the active peak from DEAE cellulose. (o———o) ^{3}H cpm; histogram, % recovery of activity.

SUMMARY

We have previously developed an in vitro cloning assay for normal hematopoietic cells. We have shown with this system that the development of colonies in which normal undifferentiated blast cells differentiate to mature macrophages or granulocytes requires a protein inducer that can be obtained, among other sources, from conditioned medium from fibroblasts. This protein inducer (MGI) has been purified (molecular weight 68,000) and labeled by acetylation without loss of biological activity. Molecules with a smaller molecular weight can also be biologically active. MGI required for its biological activity a small molecular weight cofactor that is also present in conditioned medium and this cofactor can be substituted by adenine. The availability of purified labeled protein now makes it possible to determine the location of MGI binding sites on the target and other cells and its mode of action.

Our results obtained with myeloid leukemic cells have shown that some myeloid leukemic cells, in mice and men, can be induced to differentiate normally (D$^+$) by MGI, whereas other myeloid leukemic cells (D$^-$) are not inducible. D$^+$ leukemic cells have been induced to differentiate to both macrophages and granulocytes and normal differentiation was induced even in cells that were no longer diploid. D$^-$ cells can segregate some D$^+$ progeny and D$^+$ cells can segregate some D$^-$ progeny. D$^+$ and D$^-$ cells differ in the mobility of ConA binding sites on the surface membrane. This, therefore, provides an experimental system for further elucidation of the genetic and chemical control of normal differentiation in leukemic cells.

Acknowledgments

This work was supported by grants from the Talisman Foundation and the Jerome and Estelle R. Newman Assistance Fund, New York.

REFERENCES

Fibach, E., M. Hayashi and L. Sachs. 1973. Control of normal differentiation of myeloid leukemic cells to macrophages and granulocytes. *Proc. Nat. Acad. Sci.* **70:**343.

Fibach, E., T. Landau and L. Sachs. 1972. Normal differentiation of myeloid leukemic cells induced by a differentiation-inducing protein. *Nature New Biol.* **237:**276.

Ginsburg, H. and L. Sachs. 1963. Formation of pure suspensions of mast cells in tissue culture by differentiation of lymphoid cells from the mouse thymus. *J. Nat. Cancer Inst.* **31:**1.

————. 1965. Destruction of mouse and rat embryo cells in tissue culture by lymph node cells from unsensitized rats. *J. Cell. Comp. Physiol.* **66:**199.

Goldstein, A. L., A. Guha, M. M. Zatz, M. A. Hardy and A. White. 1972. Purification and biological activity of thymosin, a hormone of the thymus gland. *Proc. Nat. Acad. Sci.* **69:**1800.

Goldwasser, E. and C. K. Kung. 1972. The molecular weight of sheep plasma erythropoietin. *J. Biol. Chem.* **25:**5159.

Hitotsumachi, S., Z. Rabinowitz and L. Sachs. 1971. Chromosomal control of reversion in transformed cells. *Nature* **231:**511.

————. Chromosomal control of chemical carcinogenesis. *Int. J. Cancer* **9:**305.

Ichikawa, Y. 1969. Differentiation of a cell line of myeloid leukemia. *J. Cell. Physiol.* **74:**223.

Ichikawa, Y., D. H. Pluznik and L. Sachs. 1966. *In vitro* control of the development of macrophage and granulocyte colonies. *Proc. Nat. Acad. Sci.* **56:**488.

Ichikawa, Y., D. H. Pluznik and L. Sachs. 1967. Feedback inhibition of the development of macrophage and granulocyte colonies. I. Inhibition by macrophages. *Proc. Nat. Acad. Sci.* **58:**1480.

Inbar, M. and L. Sachs. 1973. Mobility of carbohydrate containing sites on the surface membrane in relation to the control of cell growth. *FEBS Letters* **32:**124.

Inbar, M., H. Ben-Bassat and L. Sachs. 1973a. Difference in the mobility of lectin sites on the surface membrane of normal lymphocytes and malignant lymphoma cells. *Int. J. Cancer* **12:**93.

Inbar, M., H. Ben-Bassat, E. Fibach and L. Sachs. 1973b. Mobility of carbohydrate-containing structures on the surface membrane and the normal differentiation of myeloid leukemic cells to macrophages and granulocytes. *Proc. Nat. Acad. Sci.* (in press)

Landau, T. and L. Sachs. 1971a. Characterization of the inducer required for the development of macrophage and granulocyte colonies. *Proc. Nat. Acad. Sci.* **68:**2540.

————. 1971b. Activation of a differentiation-inducing protein by adenine and adenine-containing nucleotides. *FEBS Letters* **17:**339.

Lagunoff, D., D. H. Pluznik and L. Sachs. 1966. The cloning of macrophages in agar: Identification of the cells by electron microscopy. *J. Cell. Physiol.* **68:**385.

Metcalf, D., M. A. S. Moore and N. L. Warner. 1969. Colony formation *in vitro* by myelomonocytic leukemic cells. *J. Nat. Cancer Inst.* **43:**983.

Mintz, U. and L. Sachs. 1973. Differences in inducing activity for human bone marrow colonies in normal serum and serum from patients with leukemia. *Blood* **42:**331.

Paran, M. and L. Sachs. 1968. The continued requirement for inducer for the development of macrophage and granulocyte colonies. *J. Cell. Physiol.* **72:**247.

————. 1969. The single cell origin of normal granulocyte colonies *in vitro*. *J. Cell. Physiol.* **73:**91.

Paran, M., Y. Ichikawa and L. Sachs. 1968. Production of the inducer for macrophage and granulocyte colonies by leukemic cells. *J. Cell. Physiol.* **72:**251.

Paran, M., Y. Ichikawa and L. Sachs. 1969. Feedback inhibition of the development of macrophage and granulocyte colonies. II. Inhibition by granulocytes. *Proc. Nat. Acad. Sci.* **62:**81.

Paran, M., L. Sachs, Y. Barak and P. Resnitzky. 1970. *In vitro* induction of granulocyte differentiation in hematopoietic cells from leukemic and non-leukemic patients. *Proc. Nat. Acad. Sci.* **67:**1542.

Pluznik, D. H. and L. Sachs. 1965. The cloning of normal "mast" cells in tissue culture. *J. Cell. Comp. Physiol.* **66:**319.

––––––. 1966. The induction of clones of normal "mast" cells by a substance from conditioned medium. *Exp. Cell Res.* **43:**553.

Rabinowitz, Z. and L. Sachs. 1970. Control of the reversion of properties in transformed cells. *Nature* **225:**136.

Sachs, L. 1964. The analysis of regulatory mechanisms in cell differentiation. *New perspectives in biology,* p. 246. Elsevier, Amsterdam.

––––––. 1970. *In vitro* control of growth and development of hematopoietic cell clones. *Regulation of hematopoiesis,* vol. 1, p. 217. Appleton-Century-Crofts, New York.

Stanley, E. R. and D. Metcalf. 1969. Partial purification and some properties of the factor in normal and leukemic human urine stimulating mouse bone marrow colony growth *in vitro. Aust. J. Exp. Biol. Med.* **47:**467.

Yamamoto, T., Z. Rabinowitz and L. Sachs. 1973a. Identification of the chromosomes that control malignancy. *Nature New Biol.* **243:**247.

Yamamoto, T., M. Hayashi, Z. Rabinowitz and L. Sachs. 1973b. Chromosomal control of malignancy in tumors from cells transformed by polyoma virus. *Int. J. Cancer* **11:**555.

Regulation of Hemopoietic Stem Cell Differentiation and Proliferation by Hemopoietic Inductive Microenvironments

J. J. Trentin, J. M. Rauchwerger, and M. T. Gallagher

Division of Experimental Biology, Baylor College of Medicine
Houston, Texas 77025

Spleen and bone marrow hemopoietic colony studies in lethally irradiated mice transfused with limited numbers of bone marrow cells have revealed that the direction of differentiation of pluripotent stem cells is controlled by their interaction with hematopoietic inductive microenvironments (HIM) of the spleen and bone marrow reticuloendothelial stroma (Curry, Trentin and Wolf 1964, 1967; Curry and Trentin 1967; Curry, Trentin and Cheng 1967; Wolf and Trentin 1968; Trentin 1970). As indicated by the types of hemopoietic colonies formed from pluripotent stem cells, four types of inducing environments are found within the hemopoietic tissues. The four types of HIM are erythroid, neutrophilic granuloid, eosinophilic granuloid, and megakaryocytic. Each type induces colonies of one line of differentiation. This single line of differentiation is usually maintained for 8 to 10 days. After this period the colonies may outgrow the original HIM and encounter an HIM of a different type. When this occurs the pluripotent stem cells, which multiply within the colonies, are induced to differentiate along the direction coded for by the new HIM, resulting in "mixed" colonies. Karyotypic marker studies have shown that such colonies still consist of only one clone even though they now contain a second line of differentiation (Trentin 1971). If, however, one examines histological sections of the spleen or marrow at or before 8 days, the great majority of the microscopic colonies contain only one line of hemopoietic differentiation.

The four types of HIM occur in different proportions in mouse spleen than in mouse bone marrow stroma. Erythroid HIM predominate in the spleen, giving an erythroid colony to granuloid colony ratio of about 3:1. Granuloid HIM predominate in the bone marrow, giving an erythroid:granuloid (E:G) colony ratio of about 0.5:1 (Curry and Trentin 1967; Curry, Trentin and Wolf 1967; Wolf and Trentin 1968). This fact provided the basis for probably the most graphic early confirmation of the concept of HIM. In a study by Wolf and Trentin a bone marrow plug was implanted by trocar into the spleen of each mouse and a colony assay performed. Many of the colonies were half in the splenic stroma and half in the bone marrow stroma. Colonies of this type usually showed erythroid differentiation in the spleen stroma and granuloid differentiation in the marrow plug (Wolf and Trentin 1968). This gave direct visual evidence of the inductive influence of the organ stroma on the differentiation of pluripotent stem cells.

Much experimental evidence for the existence of HIM has been compiled, but relatively little is known about how they function. The mechanism of the granuloid HIM is still obscure. However some experiments designed to test the role of erythropoietin in spleen colony formation have given some insight into the workings of the erythroid HIM. It has been shown in many laboratories that endogenous erythropoietin production can be shut off in irradiated recipients of bone marrow by hypertransfusion-induced polycythemia. When such irradiated recipients are made polycythemic, no erythroid colonies develop in their spleens although normal numbers of granuloid colonies develop. What happens to the missing erythroid colonies? Do they divert to other lines of differentiation? A detailed microscopic analysis of colonies in such spleens revealed that the stem cells did not divert to other lines of differentiation. Very small and undifferentiated presumptive erythroid colonies took the place of the usually large erythroid colonies (Curry, Trentin and Wolf 1967). These are colonies of erythropoietin-sensitive stem cells, since on administration of exogenous erythropoietin these small undifferentiated colonies rapidly increase in size and transform into well-differentiated erythroid colonies (Curry, Trentin and Wolf 1967). If erythropoietin is injected daily for only the first 4 days of colony development and the polycythemia is reinforced by an additional blood transfusion on the 4th day, by the 8th day the differentiated erythroid cells have departed from the colony—which is again composed of undifferentiated erythropoietin-sensitive cells, but is now much larger in size and cell numbers (Curry, Trentin and Wolf 1967). These experiments indicate that the role of the erythroid HIM is to induce pluripotent stem cells to become erythropoietin-sensitive stem cells. They further indicate that erythropoietin stimulates both rapid proliferation and functional maturation of the erythropoietin-sensitive stem cells. Morphological "differentiation" is presumably secondary to the functional maturation stimulated by erythropoietin. This places erythropoietin in the role of a hormone rather than an inducer. Hormones in general stimulate mitosis and/or secretion on the part of cells that have been induced into a state of hormone responsiveness by a prior local induction phenomenon classically involving specific mesenchyme as the inducer. The results of these early studies (Curry, Trentin and Wolf 1967) concur well with the conclusion presented by Dr. Marks at this conference, arrived at by different approaches, that erythropoietin stimulates mitosis as well as hemoglobulin synthesis on the part of the erythropoietin sensitive stem cell (Marks et al. this volume).

Since these earlier studies, several later experiments have provided confirmation of the HIM concept. The genetically determined macrocytic anemia of the Steel anemic mice has provided some interesting data along these lines. Steel anemic mice were shown to have normal stem cells (McCulloch et al. 1965) and normal or elevated levels of erythropoietin (Bernstein, Russell and Keighley 1968). Furthermore it has been demonstrated by two groups that transplantation of intact HIM (whole spleens from normal littermates) will cure the anemia, whereas transfusions of spleen cell suspensions or extracts will not (Bernstein 1970; Gallagher, McGarry and Trentin 1971; Trentin 1971). The response to whole spleen transplants seemed to be a dose-dependent phenomenon. Whereas a single spleen produced no improvement, transplants of 2, 3, and 6 intact spleens per anemic mouse produced progressively greater and more rapid increases in hematocrit (Gallagher, McGarry and Trentin 1971; Trentin et al. 1971). With these results in mind we performed spleen and bone marrow colony assays in lethally irradiated Steel anemic mice and their normal littermates to determine the number and types of HIM present.

In the marrow no erythroid colonies but normal numbers of granuloid colonies were found (Gallagher, McGarry and Trentin 1971; Trentin 1971; Trentin et al. 1971). The low erythroid to granuloid colony ratios in the Steel anemic mouse spleen and bone marrow, as compared to normal littermates, indicate a defect or absence of the erythroid HIM in the Steel anemic mouse. In the spleen the E:G colony ratio was about 0.5:1 vs. about 3:1 in normal littermates. Whereas there are greatly reduced numbers of erythroid colonies in the Steel anemic mice, they were not replaced by the small undifferentiated colonies of erythropoietin-sensitive stem cells, as they were in irradiated, marrow-transfused polycythemic mice of normal strains. These findings provide convincing evidence of two things: (1) that the Steel anemia is caused by a lack of functional erythroid HIM; (2) that the function of the erythroid HIM is to convert the pluripotent stem cell into the erythropoietin-sensitive stem cell.

More recently we have used still another method to test the HIM concept. This method involves xenogeneic (rat to mouse) bone marrow transplantation. In the several isologous spleen colony studies done in the rat, colonies differentiate solely along the erythroid line or remain partly or wholly undifferentiated (Dunn and Elson 1970; Rauchwerger, Gallagher and Trentin 1973a,b). We thus presumed that the rat spleen possesses only erythroid HIM and visualized an opportunity, by means of xenogeneic grafting, to further test the concept that the HIM determine the line of differentiation of pluripotential stem cells. Thus rat marrow transplanted to irradiated mice might produce other types of colonies besides erythroid.

Control isologous mouse marrow transfused into lethally irradiated mice gave erythroid, granuloid and megakaryocytic types of spleen colonies with an E:G spleen colony ratio of 2.2:1 for 600 colonies in 4 different mouse strains (Table 1). Control isologous rat marrow transfused into lethally irradiated rats gave rise only to erythroid and undifferentiated colonies (E:G colony ratio 490:0) (Table 2). When rat marrow was transfused into lethally irradiated mice, both erythroid and granuloid colonies were formed. The resulting E:G colony ratios were approximately 3:1 for 303 colonies (Table 3). The rat origin of such colonies was established by karyotype analysis. These results further support the concept of the determining influence of the hemopoietic inductive microenvironments with respect to the direction of differentiation of pluripotent stem cells. They indicate that mouse HIM can induce rat pluripotent stem cells to differentiate along lines seen in the mouse spleen, but not encountered in the rat spleen. Colonies of rat origin grew more rapidly in mouse spleens than in rat spleens, and at a rate comparable to isologous mouse colonies (Rauchwerger, Gallagher and Trentin 1973b), suggesting that the mouse HIM, or other host factors, also regulate the rate of growth of spleen colonies. The high proportion of undifferentiated colonies and of undifferentiated cells in erythroid colonies in spleens of irradiated rats given isologous rat marrow suggests that irradiated rats have lower levels of endogenous erythropoietin than do irradiated mice. This in itself could account for a slower growth rate of erythroid spleen colonies in rats than in mice.

Table 1

Isologous Spleen Colonies in Mice of Different H-2 Genotypes after Irradiation

Strain	No. of spleens	Mean colony count per spleen		Total microscopic count	Spleen colony differential*					
		Gross	Microscopic		Ery	Gran	Meg	Un	Mx	E:G
C57 × A (H-2^b/H-2^a)	21	7.2	16.3	343	173	88	43	16	23	~2
C3H (H-2^k)	6	10	18.3	110	66	25	8	3	8	2.6
A (H-2^a)	2	8	27.5	55	28	19	3	0	5	1.5
Balb/c (H-2^d)	4	8	23	92	60	19	9	2	2	3.2

Colonies were counted 8 days after 950–1100 R ^{137}Cs irradiation of whole body and 5×10^4 bone marrow cells. Reproduced from Rauchwerger et al. (1973b) with permission of The Williams & Wilkins Co.

* Ery, erythroid; Gran, granuloid; Meg, megakaryocytic; Un, undifferentiated; Mx, mixed; E:G, erythroid to granuloid colony ratio

Table 2

Isologous Spleen Colonies in Lewis Rats after X-Irradiation

Colony age (days)	No. of spleens	Mean colony count per spleen		Total microscopic count	Spleen colony differential*					
		Gross	Microscopic		Ery	Gran	Meg	Un	Mx	E:G
11	20	11.4	26	518	398	0	0	120	0	398:0
12	5	13.1	25	125	92	0	0	33	0	92:0
11 and 12 combined	25									>490:0

Colonies were counted 11 and 12 days following 1000 R X-irradiation of whole body and 10^6 rat bone marrow cells. From Rauchwerger et al. (1973b) with permission of The Williams & Wilkins Company.

* Ery, erythroid; Gran, granuloid; Meg, megakaryocytic; Un, undifferentiated; Mx, mixed; E:G, erythroid to granuloid colony ratio

Table 3
Xenogeneic (Rat) Spleen Colonies in Mice of Different H-2 Genotypes after Irradiation

Strain	Rat BM dose	No. of spleens	Mean colony count per spleen		Total microscopic count	Spleen colony differential*					
			Gross	Microscopic		Ery	Gran	Meg	Un	Mx	E:G
C57 × A (H-2^b/H-2^a)	10^5–2 × 10^7	41	0	0	0	0	0	0	0	0	—
C3H (H-2^k)	1.25 × 10^5	5	5	17	85	44	16	0	1	24	2.7
	2.5 × 10^5	2	9	15.5	31	15	4	0	1	12	3.8
A (H-2^a)	5 × 10^5	2	11	20	40	19	6	0	0	15	3.2
Balb/c (H-2^d)	2.5 × 10^5	5	7.3	19	95	52	18	0	3	22	2.9
	5 × 10^5	2	14	26	52	32	11	0	0	9	2.9

Colonies were counted 8 days following 950–1100 R ^{137}Cs irradiation of whole body and rat bone marrow cells.

From Rauchwerger et al. (1973b) with permission of The Williams & Wilkins Company.

* Ery, erythroid; Gran, granuloid; Meg, megakaryocytic; Un, undifferentiated; Mx, mixed; E:G, erythroid to granuloid colony ratio

931

Acknowledgments

This work was supported by USPHS Grants CA 12093, CA 03367, CA 05021 and K6 CA 14, 219.

REFERENCES

Bernstein, S. E. 1970. Tissue transplantation as an analytic and therapeutic tool in hereditary anemia. *Amer. J. Surg.* **119**:448.

Bernstein, S. E., E. S. Russell and G. Keighley. 1968. Two hereditary mouse anemias (Sl/Sl^d and W/W^v) deficient in response to erythropoietin. *Ann N.Y. Acad. Sci.* **149**:475.

Curry, J. L. and J. J. Trentin. 1967. Hemopoietic spleen colony studies. I. Growth and differentiation. *Develop. Biol.* **15**:395.

Curry, J. L., J. J. Trentin and V. Cheng. 1967. Hemopoietic spleen colony studies. III. Hemopoietic nature of spleen colonies induced by lymph node or thymus cells, with or without phytohemagglutinin. *J. Immunol.* **99**:907.

Curry, J. L., J. J. Trentin and N. Wolf. 1964. Control of spleen colony histology by erythropoietin, cobalt, and hypertransfusion. *Exp. Hematol.* **7**:80.

———. 1967. Hemopoietic spleen colony studies. II. Erythropoiesis. *J. Exp. Med.* **125**:703.

Dunn, C. D. R. and L. A. Elson. 1970. Quantitative studies of haemopoietic spleen colonies in rats treated with cytotoxic chemicals. *Brit. J. Haematol.* **19**:755.

Gallagher, M. T., M. P. McGarry and J. J. Trentin. 1971. Defect of splenic stroma (hematopoietic inductive microenvironments) in the genetic anemia of Sl/Sl^d mice. *Fed. Proc.* **30**:689.

McCulloch, E. A., L. Siminovitch, J. E. Till, E. S. Russell and S. E. Bernstein. 1965. The cellular basis of the genetically determined hemopoietic defect in anemic mice of genotype Sl/Sl^d. *Blood* **26**:399.

Rauchwerger, J. M., M. T. Gallagher and J. J. Trentin. 1973a. "Xenogeneic resistance" to rat bone marrow transplantation. I. The basic phenomenon. *Proc. Soc. Exp. Biol. Med.* **143**:145.

———. 1973b. Role of the hemopoietic inductive microenvironments (HIM) in xenogeneic bone marrow transplantation. *Transplantation* **15**:610.

Trentin, J. J. 1970. Influence of hematopoietic organ stroma (hematopoietic inductive microenvironments) on stem cell differentiation. *Regulation of hematopoiesis* (ed. A. Gordon) pp. 161–186. Appleton-Century-Crofts, New York.

———. 1971. Determination of bone marrow stem cell differentiation by stromal hemopoietic inductive microenvironments (HIM). *Amer. J. Pathol.* **65**:621.

Trentin, J. J., M. P. McGarry, V. K. Jenkins, M. T. Gallagher, R. S. Speirs and S. N. Wolf. 1971. Role of inductive microenvironments on hemopoietic (and lymphoid?) differentiation, and role of thymic cells in the eosinophilic granulocyte response to antigen. *Morphological and functional aspects of immunity* (ed. Lindahl-Kiessling, Alm and Hanna) p. 289–298. Plenum Publ. Co., New York.

Wolf, N. S. and J. J. Trentin. 1968. Hemopoietic colony studies. V. Effect of hemopoietic organ stroma on differentiation of pluripotent stem cells. *J. Exp. Med.* **127**:205.

Study of Some of the Factors Influencing the Proliferation and Differentiation of the Multipotential Hemopoietic Stem Cells

M. Tubiana, E. Frindel, H. Croizat, and F. Vassort

Institut de Radiobiologie Clinique
Institut Gustave Roussy, 94800 Villejuif, France

Hemopoietic stem cells provide a convenient system for studying quantitatively the regulatory mechanisms of the progenitor cells in a mammalian tissue. They represent the only population of normal cells in which one can follow in vivo after a stress the number and the proliferative status of the viable cells. Furthermore hemopoietic stem cells constitute the critical tissue in most cancer treatments. Their behavior and proliferation kinetics under therapy are therefore of paramount importance. The aim of this paper is to discuss some of the factors that influence the proliferation and differentiation of pluripotential stem cells in the light of recent experimental data.

In the steady state most of the pluripotential stem cells are quiescent (or are cycling with a very long G_1) and the production of differentiated cells is achieved by the proliferation of unipotential committed stem cells and of maturing cells. However under stimulation or during the regeneration of hemopoietic tissues, stem cells are triggered to proliferate and to differentiate.

The understanding of the bone marrow homeostasis requires a knowledge of the time course of the response to stimulation in conditions such that the origin or the nature of the signal can be identified. This has been the purpose of the three sets of experiments that will be reviewed here, in which the proliferative status of the stem cells of mouse bone marrow was studied after (a) selective killing of maturing hemopoietic cells, (b) partial body irradiation, and (c) antigenic stimulation. The first set of experiments provides data on the feedback mechanism between the differentiated cells and the stem cells, the second helps to study long-range stimulating factors, and the third demonstrates the complexity of the feedback mechanisms and the possible role of the thymus. In all these experiments the number of pluripotential stem cells was measured by the spleen colony technique (Till and McCulloch 1961) and we shall refer to these as colony-forming units (CFU). The percentage of CFU's in DNA synthetic phase, or S phase, was evaluated by an in vitro suicide technique, that is, incubation with lethal doses of [³H]thymidine prior to their injection into lethally irradiated recipient mice (Becker et al. 1965). This method is not extensively used, probably because the percentage of CFU's in S is evaluated by the difference between two sets of data and therefore requires, in order to be significant, a large number of animals.

A thorough discussion of bone marrow kinetics would require a study not only of the maturing cell compartment and of the pluripotential stem cell compartment but also of the committed stem cells or unipotential stem cells. In fact it appears that under some conditions, such as after myeleran administration, the CFU and the committed stem cell compartments evolve differently (Blackett, Roylance and Adams 1964; Lamerton and Blackett this volume). Unfortunately only very few authors have as yet studied simultaneously the two types of stem cells after a stress and we shall, in this paper, limit our discussion to the CFU's and the recognizable differentiated bone marrow cells.

Selective Killing of Maturing Cells

We formerly observed (Vassort, Frindel and Tubiana 1971) that injections of hydroxyurea, which selectively kills cells in S, triggers the quiescent stem cells into cycle. These experiments were conducted on C3H mice in which 20% of the CFU's were in cycle. Therefore the stimulation either could have been the consequence of a feedback mechanism within the stem cell compartment itself or could have been dependent on the depletion of the bone marrow maturing cell compartment. Furthermore hydroxyurea has a toxic effect that could have modified the response of the stem cells.

In order to elucidate these points, experiments were performed under conditions such that only the maturing bone marrow cells were damaged in order to avoid any direct effect on the CFU compartment (Vassort et al. 1973). In the C57Bl/6 mouse virtually all stem cells are quiescent. Therefore the injection of 1 mCi [³H]thymidine has no killing effect on the CFU's but selectively kills cells in S in the differentiated compartment. Less than 4 hours after the administration of [³H]thymidine the number of CFU's begins to decrease (Fig. 1). Subsequently the entry of formerly quiescent stem cells into S phase is observed. The rise in the number of CFU's in S is rapid and the peak of the proportion of cells in S is over 40%, which means that at least half of the quiescent stem cells have entered into the cell cycle.

These experiments seem to indicate that it is the damage to the compartment of differentiated cells that causes a differentiation of stem cells. The signal that effects differentiation of CFU's also either triggers simultaneously proliferation or provokes indirectly a recruitment into cycle of quiescent cells due to the depletion of the CFU compartment caused by differentiation.

Similar results using [³H]thymidine were obtained in the C3H strain of mice, in which above 20% of the CFU's were in S (Fig. 2). Immediately after [³H]thymidine administration all CFU's in S phase are killed. Later the percentage of cells in S comes back to 20%, which corresponds to the progression of the surviving cycling cells through the cycle. When this cohort of surviving cells traverses the other phases of the cycle, the S compartment is partially depleted. The second peak occurs after a time interval equal to the duration of the cell cycle. At 12 hours there is an over-shoot, which demonstrates a recruitment of quiescent stem cells into cycle. In order to analyze in these C3H mice the kinetics of the passage of quiescent cells into cycle, a computer model was used (Vassort et al. 1973), which assumes that the rate of progression of surviving cycling CFU's through cycle is unchanged. Three para-

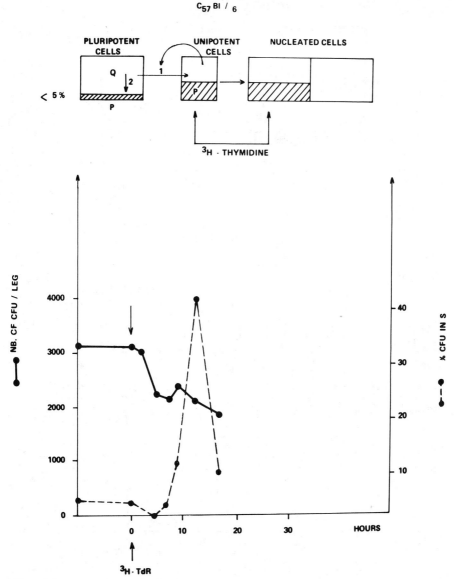

Figure 1

Time course changes of the number of CFU (•———•) and of the proportion of CFU in S phase (•– – –•) after an administration of 1 mCi of radioactive thymidine to C57B1/6 mice. In this strain of mice virtually all CFU are quiescent. The in vivo suicide does not act on CFU but on differentiated cells.

meters were introduced: (1) the percentage of cells entering into the proliferative compartment, (2) the time of their entry into cycle, and (3) the duration of the flux. These parameters were adjusted by trial and error so as to obtain a satisfactory agreement with the experimental data (Vassort et al. 1973).

The assumption that the kinetics of the recruitment of quiescent cells is identical in both the C3H and the C57 mice gives a good fit with the experimental data. This

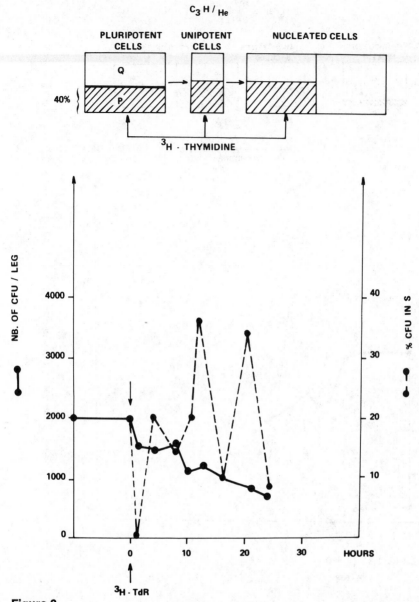

Figure 2

Time course changes of the number of CFU (•———•) and of the proportion of CFU in S phase (•– – –•) after an administration of 1 mCi of radioactive thymidine to C3H mice. In this strain of mice 20% of CFU are normally in S phase. [³H]thymidine acts both on CFU compartment and on differentiated cells.

suggests that the mechanism of the feedback is identical for the two strains and is influenced neither by the proliferative status of the CFU's nor by the killing of CFU's. In both strains about half of the quiescent CFU's enter into the cycle after the same time interval.

Besides this observation the main point which comes out of this experiment is the rapidity with which the feedback mechanism operates. The incorporation of [³H]thymidine into the differentiated bone marrow cells and the expression of the lesion caused by intranuclear irradiation of cell DNA takes some time. Yet less than four hours after intraperitoneal injection of [³H]thymidine the CFU's begin to differentiate, and less than four hours later about half of the CFU's begin to enter into the S phase. In mice the duration of the cell cycle is about 8 hours for the stem cells (Vassort et al. 1973) and for the differentiated bone marrow cells (Frindel, Tubiana and Vassort 1967). The duration of G_1 in the differentiated cells is 2 hours. If one assumes the same duration for stem cells, the quiescent CFU's are triggered into the cell cycle at the time when the number of CFU's begins to decrease.

The signal for triggering the entry into S of stem cells seems to be intense but of relatively short duration (a few hours). In interpreting these data it should be remembered that in vitro a cell stimulator provokes the entry into S of CFU's in less than 2 hours (Byron 1972).

These data provide evidence for an autoregulation of the number of CFU's. It is impossible to know whether it is the same signal which induces the CFU differentiation and proliferation or whether the CFU proliferation is caused by the depletion of the CFU compartment. The sequence of events suggests the latter interpretation, but it may also be that the time interval after which a cell differentiates is shorter than the time interval after which it enters into cycle. The signal may originate from the maturing cell compartment or from the committed stem cell compartment. Data obtained after erythropoietic injection suggest that this substance acts on eryththropoietic committed stem cells and that it is the depletion of this compartment which causes subsequently a stimulation of the CFU's (Morse, Rencricca and Stohlman 1969).

Finally, our data do not give any indication about the nature of the signal and of the feedback mechanism. The stem cell differentiation and proliferation may be controlled either locally by cell-to-cell interaction or abscopally by long-range humoral factors. Subtotal irradiation provides opportunity for studying abscopal mechanisms.

Subtotal Irradiation

Subtotal irradiation has been used during the last decade (Tsuya et al. 1961; Kurnick and Nokay 1964; Carsten and Bond 1968; Croizat, Frindel and Tubiana 1970; Gidali and Lajtha 1972) for studying the indirect effect of irradiation on an undisturbed area of bone marrow. The data obtained (Croizat et al. 1970; Gidali and Lajtha 1972) seems to support the concept of some form of humoral regulation of the stem cell population.

In 1970 we carried out (Croizat, Frindel and Tubiana 1970) experiments in which a dose of 150 rads was delivered to mice with one leg shielded (that is, one-tenth of the hemopoietic tissues). This subtotal irradiation induced in the protected leg a small but significant decrease in the CFU number and, a few hours later, an increase in the proliferative activity of the CFU's (Fig. 3). The stimulus for proliferation appears less intense but more sustained than after [³H]thymidine or hydroxyurea administration and the proportion of CFU's in S reaches its maximum 24 hours after the irradiation. It should be noted that the percentage of CFU's in S is back to normal on the second day (Croizat et al. 1970) at a time when there is

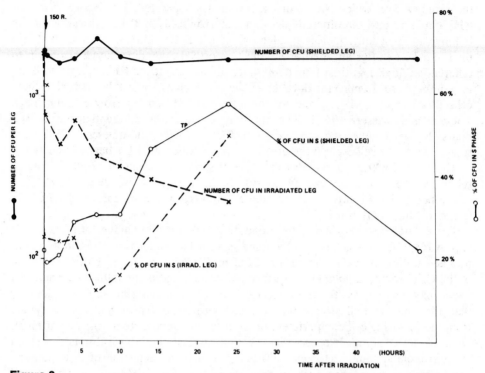

Figure 3

Time course changes of the number of CFU and proportion of CFU in S phase in the irradiated and in the protected bone marrow. (Subtotal irradiation dose: 150 rads, one hind leg shielded).

still an important reduction of the number of CFU's in the organism but when the number of CFU's in the shielded leg is back to normal. Later a similar observation was made by Gidali and Lajtha (1972). This suggests that either the regulation of the CFU's is local or that the stimulus for proliferation is only temporary.

After subtotal iterative irradiations (Fig. 4) the number of CFU's decreases slowly but continuously in the protected leg. In spite of that, the percentage of the CFU's in S comes back to the normal value at a time when there are less than 2% of the initial number of CFU's in the whole body and about 10% of the initial number in the protected leg. This further underlines the fact that the proliferative status of the CFU compartment is not simply regulated by the local or total number of CFU, but it is also influenced by other factors, such as the rate of decrease.

More recently we concentrated on the early events after irradiation. In the protected areas a decrease in the number of CFU's is observed 15 minutes after irradiation. The magnitude of the decrease is proportional to dose. Fifteen minutes after irradiation the number of CFU's in the shielded leg, expressed as the percentage of the pre-irradiation number, is 90% for a dose of 150 rads, 70% for 500 rads, 63% for 1000 rads, and 55% for 1500 rads.

An indirect killing effect being unlikely, this diminution can be the consequence of either an increased rate of migration and/or an increased rate of differentiation.

Migration from unirradiated hemopoietic sites, with consequent seeding in the

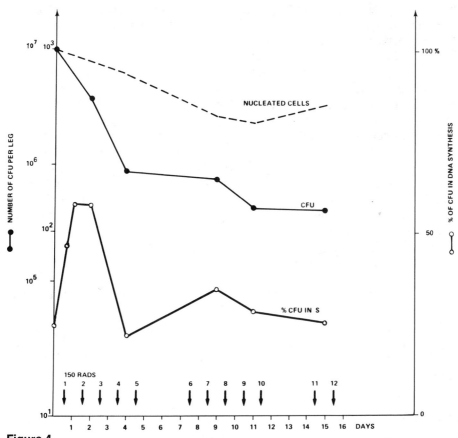

Figure 4
Time course changes of the number of bone marrow nucleated cells and CFU, and the proportion of CFU in S phase in the shielded leg. Subtotal irradiation (on hind leg shielded), dose per session: 150 rads, 5 sessions per week. In spite of the reduction in CFU number the increase of the proportion of CFU in S is only temporary.

irradiated bone marrow or spleen, has been demonstrated for CFU's (Hanks 1964; Fujioka et al. 1967; Hellman and Grate 1968) and for differentiated bone marrow cells (Tubiana et al. 1964). In the present sets of experiments some migration was observed but its magnitude is small and does not increase with the dose. Fifteen minutes after irradiation the number of CFU's in the blood is reduced to one-tenth of its pre-irradiation level, which corresponds to the proportion of protected bone marrow. (Before irradiation the number of CFU's is 80.7 ± 22 per ml of blood, and 15 minutes after 500 rads it is 10.2 ± 4.7 ml.)

However the increase in the number of CFU's in the spleen is very slow; for a dose of 500 rads it rises from 50 CFU's 15 minutes after irradiation to 130 CFU's 6 hours irradiation. After 1000 rads the increase, if any, is even smaller: from 2 CFU's 15 minutes after irradiation, to 4.5 CFU's 6 hours later. If migration is not increased, the decrease is therefore likely to be caused by differentiation. It should be remembered that with the technique used, in which the number of colonies in the recipient mice is measured 10 days later, our results do not mean that the missing cells are already differentiated 15 minutes after irradiation, but they suggest that they have received a signal that has induced the process of differentiation. It seems, furthermore, that the importance of this signal for differentiation is correlated with both the dose received by the irradiated bone marrow (Croizat et al. unpubl.) and the proportion of irradiated bone marrow (Gidali and Lajtha 1972).

Another point comes out of this study. If the early decrease in the number of CFU's is large (such as after 1000 rads or 1500 rads), there is no further decrease during the subsequent hours. On the other hand if the initial decrease is smaller, as after 500 rads, this number continues to decrease. This is in agreement with data from Boggs, Chervenick and Boggs (1972) suggesting that after a small reduction in the number of CFU's the surviving CFU's proliferate and differentiate, whereas when the number of stem cells in the body is reduced below 10% of its initial level, the stem cells proliferate but do not differentiate. In our experiment the number of CFU's in the protected area is always 10%, whatever the dose, but the arrest of in-hibition may be triggered by the further reduction below that level. The mechanism for such an arrest is far from clear. It may be local or general. It seems from our data that the differentiation is not inhibited when the decrease in CFU number is slow, as after a dose of 500 rads, or as after iterative subtotal irradiation with 150 rads.

In summary, partial body irradiation demonstrates the importance of an abscopal effect that induces mainly a differentiation of CFU's and also a temporary increase of CFU proliferation. This feedback mechanism is triggered less than 15 minutes after irradiation. Its nature and the relative importance of local (Maloney and Patt 1969; Rencricca et al. 1970) to general regulation processes are still open to discussion.

Antigenic Stimulation

Previous experiments have shown that pluripotential stem cells may respond when differentiated bone marrow cells or committed stem cells are damaged. The study of antigenic stimulation illustrates the complexity of some of these regulatory mechanisms.

In our animal quarters various strains of mice have different proportions of cells in S phase. For instance in the C57Bl strain the proportion is rather low, about 5–10%. In a Balb/c strain it is about 50%. It is likely that in the strain where most CFU's are quiescent, the hemopoietic homeostasis is maintained by unipotential stem cells, but the question arises as to why in some strains the help of multipotential stem cells is required. In order to investigate this problem we compared the same strain, namely C3H, in three conditions: In the germ-free (axenic) mice practically all CFU's are quiescent and the number of CFU's per leg is 3.5×10^3. In conven-tional animal quarters (holoxenic mice) the percentage of cells in S is 25% and the number of CFU's per leg is 2.7×10^3 (Croizat et al. 1970). When axenic mice are transferred to the conventional animal house the proportion of CFU's in S rises to 25% and their number is 4.1×10^3 per leg (Croizat et al. 1970). This suggested to us the possible role of antigenic stimulation in CFU proliferation. Relatively little work has been published on this subject. It has been demonstrated that antigenic stimulation causes an increase in circulating CFU's and in splenic CFU's (Monette et al. 1972; McNeill 1970; Barnes and Loutit 1967), but these authors did not study the proliferative status of bone marrow CFU's. Their observations and ours lead us to study the effect of a challenge with an antigen such as Salmonella O or peroxydase on the entry into the cell cycle of quiescent CFU's. A rise of the percentage of CFU's in S in axenic mice was observed (Fig. 5) lasting about 2 weeks after the injection of antigen. During this time interval about 50–80% of the CFU's are cycling. The long duration of this proliferative activity contrasts with the burst observed after selective killing and suggests that the mechanism for CFU stimulation is different. Furthermore after antigenic injection the number of CFU's increases (Croizat et al. 1970), which suggests that the rate of proliferation is higher than the rate of differentiation.

In neonatally thymectomized mice the number of CFU's is decreased as the number of lymphocytes in blood and lymphnode is decreased. The challenge with an antigen has no effect on the proliferative status of the CFU's in these mice (Horsch et al. 1971). If a thymus is grafted, a stimulatory effect of an antigenic injection is again observed (Horsch et al. 1971).

In conventional mice thymectomized at 8 weeks of age the proportion of CFU's in S falls from about 20% to practically zero in less than one week and remains very low during the 8 months of observation (Frindel and Croizat 1973). For the mice

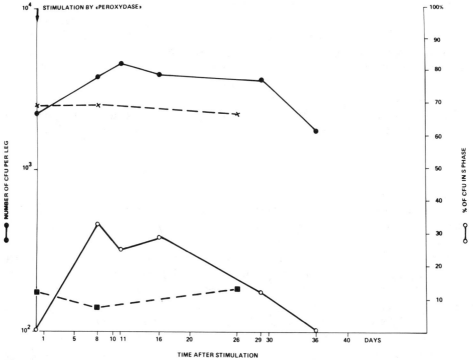

Figure 5

Time course changes of a number of CFU and of the proportion of CFU in S phase after an antigenic stimulation by peroxydase in 2-month-old C3H mice.

(●———●) CFU number in axenic control mice

(x– – –x) CFU number in axenic neonatally thymectomized mice

(o———o) Percentage of CFU in S phase, axenic control mice

(■– – –■) Percentage of CFU in S phase. Neonatally thymectomized mice

In thymectomized mice antigen injection does not provoke an increase in the proportion of CFU in S phase.

thymectomized at 8 weeks the injection of an antigen does not stimulate the CFU's. On the other hand, the CFU's in these animals respond normally to bleeding and the increase of the proportion of CFU's in S phase is neither reduced nor delayed (Frindel and Croizat 1973). Preliminary data suggest that CFU's also respond to CSF, which is known to be a factor that stimulates the granulocytic series. It seems, therefore, that thymectomy does not interfere with the CFU regulating mechanisms related to the erythropoietic or granulocytic series but that thymus is involved in the CFU regulatory mechanism of at least some of the immunocompetent cells and in the differentiation of CFU's towards antigen responsive cells (ARC). Furthermore these data show that most of the CFU proliferation observed in this strain of mice is normally related to proliferation of the lymphocytic series.

These observations fit nicely with some of the previous data in the literature. It is known that immunocompetent cells are derived from the CFU's (Ford et al. 1966) and that removal of the thymus at birth results in lymphopenia and in a reduction of the number of CFU's (Resnitzky, Ziperi and Trainin 1971). Mice thymectomized at birth contain very few ARC's (Muller and Mitchell 1967); however when ARC's from normal mice are injected into neonatally thymectomized mice, they yield a normal number of plaque-forming cells (Weiss et al. 1967). When marrow suspensions are injected into thymectomized, irradiated recipients, there is, after an antigenic challenge, only a small increase in the number of plaque-forming cells; whereas in the same circumstances the number of plaque-forming cells increases exponentially with time in irradiated, nonthymectomized recipient mice or in thymectomized irradiated mice in which a cell-impermeable diffusion chamber containing thymus was implanted (Osoba 1968).

Our data further suggest that the progenitor cells of the ARC's are the CFU's and that in the absence of the thymus the depletion of the ARC compartment does not stimulate the CFU's to proliferate and/or differentiate. It may be that there is no signal from the ARC compartment to the CFU compartment or that the CFU's are not responsive to this particular signal. However it is still difficult to know whether in thymectomized mice the response of CFU to ARC stimulus is completely suppressed or whether it is only lowered or delayed.

One of the modes of action of the thymus may be a humoral factor. Preliminary data suggest that when a cell-impermeable diffusion chamber containing thymus is implanted into a thymectomized mouse, the proportion of CFU's in S phase comes back to the normal value. Similar long-range humoral factors secreted by some other tissue in the response of CFU compartment to the depletion of the erythrocytic or granulocytic series have not yet been observed or postulated; but they might exist and this could contribute to explain why in some circumstances there is no increase in CFU proliferation in spite of a reduction in the number of CFU's.

Although little is known about the homeostatic regulation of bone marrow cells, the data available underline both the complexity of the mechanisms involved and the surprising rapidity with which they act.

REFERENCES

Barnes, D. W. H. and J. F. Loutit. 1967. Effects of irradiation and antigenic stimulation on circulating haemopoietic stem cells of the mouse. *Nature* **213**:1142.

Becker, A. J., E. A. McCulloch, L. Siminovitch and J. E. Till. 1965. The effect of differ-

ing demands for blood cell production on DNA synthesis by hemopoietic colony forming cells of mice. *Blood* **26**:296.

Blackett, N. M., P. J. Roylance and K. Adams. 1964. Studies of the capacity of bone
marrow cells to restore erythropoiesis in heavily irradiated rats. *Brit. J. Haemat.*
10:453.

Boggs, S., P. A. Chervenick and P. D. Boggs. 1972. The effect of post-irradiation bleeding or endotoxin on proliferation and differentiation of hemopoietic stem cells. *Blood*
40:375.

Byron, J. W. 1972. Evidence for a β-adrenergic receptor initiating DNA synthesis in
haemopoietic stem cells. *Exp. Cell Res.* **71**:228.

Carsten, A. L. and V. P. Bond. 1968. CFU content of the X-ray exposed and shielded
femur. *Exp. Haemat.* **15**:95.

Croizat, H., E. Frindel and M. Tubiana. 1970. Proliferative activity of the stem cells in
the bone-marrow of mice after single and multiple irradiations (total or partial-body
exposure). *Int. J. Radiat. Biol.* **18**:347.

Croizat, H., E. Frindel, M. Tubiana and J. C. Salomon. 1970. Antigenic stimulation of
DNA synthesis in haemopoietic stem cells of axenic mice. *Nature* **228**:1187.

Ford, C. E., H. S. Micklen, E. P. Evans, J. F. Gray and D. A. Osden. 1966. The inflow
of bone marrow cells to the thymus: Studies with part-body irradiated mice injected
with chromosome marked bone marrow and subjected to antigen stimulation. *Ann.
N.Y. Acad. Sci.* **129**:283.

Frindel, E. and H. Croizat. 1973. The possible role of the thymus in CFU proliferation
and differentiation. *Biomedicine* (in press).

Frindel, E., M. Tubiana and F. Vassort. 1967. Generation cycle of mouse bone marrow.
Nature **214**:1017.

Fujioka, S., K. Hirashima, T. Kumatori, F. Takaku and K. Nakao. 1967. Mechanism of
hematopoietic recovery in the X-irradiated mouse with spleen or one leg shielded.
Radiat. Res. **31**:826.

Gidali, J. and L. G. Lajtha. 1972. Regulation of haemopoietic stem cell turnover in
partially irradiated mice. *Cell Tissue Kinetics* **5**:147.

Hanks, G. E. 1964. In vivo migration of colony forming units from shielded bone marrow in the irradiated mouse. *Nature* **203**:1393.

Hellman, S. and H. E. Grate. 1968. Kinetics of circulating haemopoietic CFU in the
mouse. *Effects of radiation in cellular proliferation and differentiation*, p. 187. Int.
Atomic Energy Ag., Vienna.

Horsch, A., H. Croizat, J. C. Salomon and E. Frindel. 1971. Role du thymus dans l'ac-
tivité des cellules souches médullaires chez les souris axeniques. *Nouv. Rev. Franc.
Hématol.* **11**:835.

Kurnick, N. B. and N. Nokay. 1964. Abscopal effects and bone-marrow repopulation in
man and mouse. *Ann. N.Y. Acad. Sci.* **114**:528.

Maloney, M. A. and H. M. Patt. 1969. Origin of repopulating cells after localized bone
marrow depletion. *Science* **165**:71.

McNeill, T. A. 1970. Antigenic stimulation of bone marrow colony forming cells. Effect
in vivo. *Immunology* **18**:61.

Monette, F. C., B. S. Morse, D. Howard, E. Niskanen and F. Stohlman. 1972. Hemopoietic stem cell proliferation and migration following Bardetella Pertussis vaccine.
Cell Tissue Kinetics **5**:121.

Morse, B. S., N. J. Rencricca and F. Stohlman, Jr. 1969. The effect of hydroxyurea on
differentiated marrow erythroid precursors. *Proc. Soc. Exp. Biol. Med.* **130**:986.

Muller, J. F. A. A. and G. F. Mitchell. 1967. Cellular basis of the immunological defects
in thymectomized mice. *Nature* **214**:994.

Osoba, D. 1968. Thymic control of cellular differentiation in the immunological system.
Proc. Soc. Exp. Biol. Med. **127**:418.

Rencricca, N. J., V. Rizzoli, D. Howard, P. Duffy and F. Stohlman. 1970. Stem cell migration and proliferation during severe anemia. *Blood* **36:**764.

Resnitzky, P., D. Ziperi and N. Trainin. 1971. Effect of neonatal thymectomy on hemopoietic tissue in mouse. *Blood* **37:**634.

Till, J. E. and E. A. McCulloch. 1961. A direct measurement of the radiation sensitivity of normal mouse bone marrow cells. *Radiat. Res.* **14:**213.

Tsuya, A., V. P. Bond, T. M. Fliedner and L. E. Feinendegen. 1961. Cellularity and deoxyribonucleic acid synthesis in bone marrow after total and partial body irradiation. *Radiat. Res.* **14:**618.

Tubiana, M., E. Frindel, G. L. Kozinets and N. Dalle-Chavaudra. 1964. Migration des cellules hématopoiétiques médullaires chez les souris irradiées et normales. *Rev. Fr. Clin. Biol.* **9:**742.

Vassort, F., E. Frindel and M. Tubiana. 1971. Effects of hydroxyurea on the kinetics of colony forming units of bone marrow in the mouse. *Cell Tissue Kinetics* **4:**423.

Vassort, F., M. Winterholer, E. Frindel and M. Tubiana. 1973. Kinetic parameters of bone marrow stem cells using *in vivo* suicide by tritiated thymidine or by hydroxyurea. *Blood* **41:**789.

Weiss, N. S., G. F. Mitchell and J. F. A. P. Miller. 1967. Cellular basis of the immunological defects in thymectomized mice. III. Proliferation and differentiation of antigen-sensitive cells in neonatally thymectomized mice. *Nature* **214:**995.

The Survival Value of the Dormant State in Neoplastic and Normal Cell Populations

Bayard D. Clarkson, M.D.

Memorial Sloan-Kettering Cancer Center
and Cornell University Medical College, New York, New York 10021

There are several types of human cancer which can now be cured with cytotoxic drugs with fair regularity, whereas other types, which are also very responsive to the available drugs, have proven extremely difficult to cure (National Conference on Cancer Chemotherapy 1972). One of the major reasons for failure in the latter instances is the presence of dormant tumor cells, that is cells in an extended G_1 or G_0 state, in which state they are relatively invulnerable to clinically permissible doses of cytotoxic agents (Clarkson 1973). There are also many solid forms of cancer in which there is usually little appreciable influence of chemotherapy on survival, and in these, too, the dormant state is an important factor in their refractoriness. However the proliferative behavior of solid tumors is more complex than that of disseminated hematologic neoplasms, and other factors such as variability in vascularization and nutrient supply and inherent resistance to available cytotoxic agents are also important in determining their unresponsiveness (Greene 1941; Tannock 1970; Folkman 1971). In this paper attention will therefore be directed towards examining the role of the dormant state in a responsive disseminated neoplasm—acute leukemia—in which it is felt to be a major cause of therapeutic failure.

PATHOGENESIS OF ACUTE LEUKEMIA AND POTENTIAL CURABILITY

Leukemia is characterized by execssive proliferation of the leukemic cells, and it is first necessary to understand the essential difference between their proliferative behavior and that of the normal hematopoietic cells before considering how it might be possible to destroy the leukemic population without killing the host. Normal hematopoiesis, as exemplified by a simplified model of erythropoiesis, is shown in Fig. 1. The system is finely regulated, and cell production equals cell loss due to aging and other factors (Stohlman 1970; Erslev 1972). The stem cells produce daughters both to maintain the stem cell pool and to furnish a constant flow of cells committed to differentiate into mature red cells. The latter process is amplified by a series of maturation divisions so that some 16 to 32 reticulocytes are produced by each

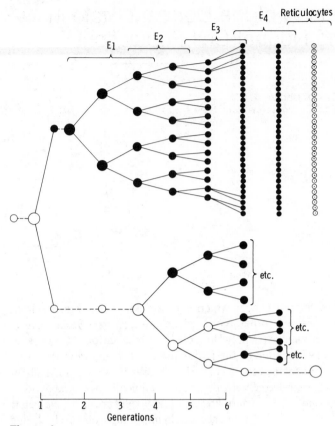

Figure 1
In this scheme of normal steady state erythropoiesis, the following assumptions
have been made: (1) Commitment to differentiation occurs only after division;
(2) the number of uncommitted stem cells remains stable, although temporary
fluctuations may occur; (3) erythropoiesis is entirely effective (i.e., there is no
death until the cells are fully mature); (4) stem cells may remain inactive (i.e.,
dormant) for variable lengths of time; (5) once a daughter becomes committed,
it immediately begins a series of maturation divisions; (6) the exact number of
divisions occurring at each level of maturation is uncertain, but here five divisions are assumed to occur between the pronormoblast (E_1) and polychromatic
normoblast (E_3) maturation compartments. This results in 32 orthochromatic
normoblasts (E_4), which then mature directly to reticulocytes without undergoing further division. (o) Stem cell; (•) committed cell; (– – –) dormant state.

earliest committed precursor. Although fluctuations may occur, on the average half
of the daughters resulting from stem cell division remain as stem cells and half
become committed to differentiation, and thus the total number of cells remains
constant (Clarkson and Fried 1971).

In order to maintain a steady state, only a minority of the stem cells need be in
the division cycle at any time (Lajtha, Oliver and Gurney 1962; Becker et al. 1965;
Haas, Bohne and Fliedner 1971). The rate of production can be increased temporarily by mobilizing more stem cells to divide in response to increased cell loss

(e.g., hemorrhage, hemolysis), but after restoration of homeostasis, the sizes of both the stem cell and maturation compartments return to their previous levels.

In contrast, acute leukemic populations are generally characterized by progressive expansion (Fig. 2). Cytogenetic and cytokinetic evidence favors the concepts that acute leukemia usually arises from (neoplastic) transformation of a single stem cell with unlimited proliferative potential, that the leukemic population is separate and distinct from the normal hematopoietic population, and that the leukemic cells have a hereditable defect(s) rendering them incapable of normal maturation and relatively unresponsive to regulatory mechanisms that tightly control the size of the normal population (Kiossoglou, Mitus and Dameshek 1965; Sandberg 1966; Jensen 1967; Zuelzer and Cox 1969; Whang-peng et al. 1969; Clarkson 1969, 1972; Clarkson et al. 1970; Moore and Metcalf 1973; Trujillo et al. 1973).

The sequence of events resulting in transformation and the critical lesion(s) responsible for the neoplastic state are still poorly understood, but the crucial behavioral difference between leukemic and normal cell proliferation is that maturation is defective in acute leukemia and the ratio of stem cells to cells committed to differentiation is greater than one; otherwise the population would fail to expand (Clarkson and Fried 1971). The fraction of leukemic cells remaining as stem cells, here defined as having the potential to proliferate indefinitely or to reach a lethal number of cells, varies among different populations and may also change during the course of disease or may be altered by therapy. Partial (but usually abnormal) maturation is common in acute leukemia, and, as in normal hematopoiesis, once a cell becomes committed to differentiation, it probably can undergo only a limited number of divisions and is incapable of reproducing the whole population. The rate of population growth may vary greatly in different patients and will depend not only on the ratio of stem cells to committed cells but also on the fraction of resting or dormant cells, the cell cycle duration of actively proliferating cells, the number of maturation divisions occurring among committed cells with finite life spans, and the rate of cell loss.

When the leukemic mass reaches a critical size, which may vary greatly among different populations, production of normal cells is inhibited, probably at the stem cell level, and the rate of growth of the leukemic population may also decrease. The mechanism of inhibition is poorly defined, but it is undoubtedly complex and dependent on the interplay between stimulatory and inhibitory factors and probably interactions between different types of cells in the marrow. When the leukemic mass is reduced sufficiently by therapy, the normal cells are releasd from inhibition, and, because they generally proliferate faster than the leukemic cells, they repopulate the marrow before the surviving leukemic cells can do so (Clarkson et al. 1967a, 1970; Clarkson 1969, 1972).

If one accepts the proposition that in acute leukemia the normal and leukemic cells coexist in the marrow as two separate populations, then if one can destroy all the leukemic cells while sparing enough normal cells to allow the host to survive, it should be possible to cure the disease, providing reinduction of a new leukemia does not occur. To effect a cure it should not be necessary to kill the entire leukemic population, but only all leukemic stem cells since the cells committed to maturation are capable of only a limited number of divisions before dying spontaneously.

The lethal number of leukemic cells in man is generally between one and several trillion (Clarkson and Fried 1971). It is usually only necessary to cause about a three decade reduction to achieve a complete remission according to conventional

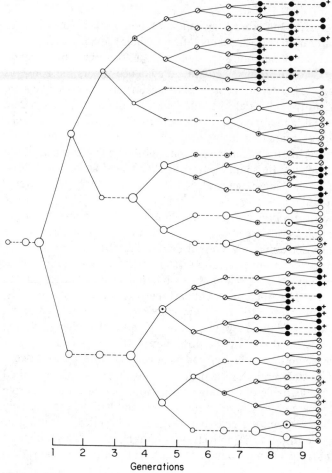

Figure 2

In the scheme of human acute leukemic cell proliferation, the following assumptions have been made: (1) Not all leukemic cells have unlimited proliferative integrity, and the fraction that retains this critical (stem cell) property may change as the disease progresses; (2) once a leukemic cell becomes committed to maturation (albeit imperfectly), it and its progeny are destined to die as in normal hematopoiesis; (3) the number of divisions occurring after commitment to maturation is unknown, but here it is assumed that a committed stem cell divides only once, producing two partially mature (Type II) committed cells; (4) the Type II cells can then divide a maximum of three times before maturing to Type III cells, which are no longer capable of division and have a limited life span (5) cell death may occur at any stage of maturation; (6) both stem cells and committed cells may remain dormant (i.e., in G_1 or G_0) for variable periods of time; (7) although cells of different sizes are capable of dividing, they must achieve a certain minimal threshold mass before they can do so. For simplicity only the stem cells in the diagram are shown to exhibit size variability. The proportions of cell types during generations 7, 8, and 9 approximate those observed in acute myelomonocytic leukemia (Clarkson and Fried 1971). (o) Uncommitted cell; (⊙) stem cell committed to maturation and death; (⊘) Type II committed cell; (●) Type III committed cell; (– – –) dormant state; (+) cell death.

criteria, since several billion leukemic cells scattered throughout the body generally cause no adverse effects to the host and since one often cannot distinguish them from normal primitive cells on morphological grounds.

In the nonlymphoblastic forms of acute leukemia, mainly because of the lack of effective drugs with sufficient selectivity against leukemic myeloblasts and related cell types, the degree of cell kill is ordinarily not great, and in most remissions the leukemic population is barely below the detectable threshold (Clarkson 1972). This is reflected by the generally short remission duration and survival time, and it is only the rare patient who survives more than two years. However in acute lymphoblastic leukemia, there is reason to believe that aggressive therapy with cytotoxic drugs may come close to completely eradicating the leukemic population (Holland and Glidewell 1972; Simone et al. 1972; Clarkson 1973). Some patients may in fact already have been cured, although the occasional occurrence of late relapses emphasizes the necessity to reserve final judgment until many years after discontinuing treatment.

IMPORTANCE OF THE DORMANT STATE IN PREVENTING CURE

Probably the major reason for failure to cure lymphoblastic leukemia is that a small but significant fraction of the leukemic population remains for long periods in a dormant state (Clarkson and Fried 1971). Development of drug resistance, which used to be a significant factor during treatment with single agents, is a less important reason for failure now. This is especially true in lymphoblastic leukemia, for which many lymphocidal drugs with different mechanisms of action are now available and therapists are learning to use them more effectively in combinations. The question of whether the long-term dormant cells are in a distinct G_0 state or merely in an extended G_1 phase is still unclear and must await precise biochemical definition of the distinction and then proper categorization of individual dormant cells. In any case the dormant cells are completely resistant to agents, such as arabinosylcytosine, that kill only proliferating cells when used in permissible therapeutic doses, and they are relatively insensitive to cell cycle phase nonspecific agents such as cyclophosphamide (Bruce, Meeker and Valeriote 1966; Skipper 1968, 1971; Bhuyan, Scheidt and Fraser 1972). The dormant state is important in preventing eradication of leukemic cells not only in the marrow but also in pharmacologic sanctuaries, especially the central nervous system, into which most cytotoxic drugs fail to penetrate in effective lethal concentrations (Kuo et al. 1973). For meningeal leukemia, therapy is presently restricted to intrathecal injections of S-phase lethal drugs and/or cranial irradiation because most of the cell cycle nonspecific agents are neurotoxic.

That human tumor cells commonly remain dormant for long periods has been clearly demonstrated by prolonged continuous infusions of [^{3}H]thymidine and radioautographic analysis of serial samples of the tumors. The experimental conditions must of course be such that all cells passing through the cell cycle have their nuclei labeled above the background threshold; this is easy to verify if no unlabeled cells are found in mitosis. My colleagues and I have performed over 25 such studies in patients with different types of leukemia, lymphoma and other types of cancer (Clarkson, Ota and Karnofsky 1962; Clarkson, Fried and Ogawa 1969; Clarkson et al. 1965, 1967b, 1970, 1971; Ogawa et al. 1970; Ohara et al. 1971; Kuo et al.

1973). In every case a significant fraction of the neoplastic population has been found to be in a prolonged dormant state, although the individual kinetic parameters, such as the cell cycle time of the proliferating cells and the durations of the phases of the cycle, varied considerably among different cases and sometimes with the stage of disease in individuals who were studied repeatedly. One illustrative study will be described.

J. T. was a 26-year-old man with acute lymphoblastic leukemia whose marrow was filled with leukemic cells and who had numerous circulating blasts at the time the kinetic study was done (Fig. 3). A continuous intravenous infusion of [³H]thymidine was given at a constant rate for 20 days (Sp. Ac. 6 Ci/mm; 4 mCi/24 hr) (Clarkson et al. 1971). All mitotic cells in the marrow were labeled after the third

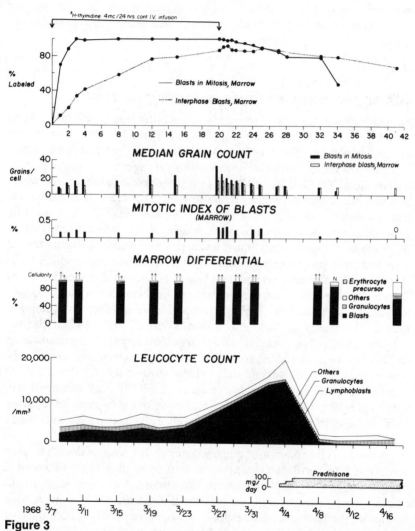

Figure 3
Labeling pattern of mitotic and interphase leukemic cells in marrow during and after continuous intravenous infusion of [³H]thymidine (4 mCi/24 hr) in J. T., a patient with acute lymphoblastic leukemia. Radioautographs exposed 12 weeks.

day, but at the end of the 20 days' infusion, only 89% of the interphase lymphoblasts in the marrow were labeled and 94% of those in the blood. The median grain count halving time of the mitotic leukemic cells after the end of the infusion (before treatment was started) was 147 hours, whereas that of the interphase marrow leukemic cells was 297 hours. This reflects the intermitotic times of the labeled cells, and the fact that the grain count halving time of the interphase cells was twice that of the mitotic cells indicates that many daughter cells entered a dormant state after their parents divided. Unlabeled cells (i.e., cells remaining dormant for more than 20 days) were of course not included in these estimates, and if one considered the whole leukemic population, the mean intermitotic time would have been even longer than 12 days. This is one of the longest intermitotic times we have observed in human leukemic populations, the shortest being about 2 days and most falling between these extremes. The leukemic population in J. T. had apparently nearly reached its maximum expansion at the time the kinetic study was performed, and presumably the cells would have been proliferating faster at an earlier stage of disease.

One can estimate the turnover rate of the dormant state, expressed as the percent of dormant cells resuming active proliferation per day, using a simple compartmental model together with the measured rate of increase of the [^{3}H]thymidine labeling index (LI) of the interphase cells during the latter portion of the infusion after the rapidly proliferating cells have become fully labeled (Fried unpubl.). The labeling index during the last 12 days approaches 100% exponentially, and the rate constant corresponding to this exponential is equal to the turnover rate of the dormant cells. Using a least squares fitting procedure, Dr. Jerrold Fried in our laboratory computed this to be about 11% per day.

The nuclear areas of the marrow lymphoblasts were measured on radioautographic smears as previously described (Clarkson et al. 1970) and their labeling indexes plotted as a function of their nuclear areas (Fig. 4). Twenty-four hours after the start of the infusion, 10.8% of the marrow blasts were labeled, and, as invariably noted in other studies, almost all of these were large cells. Very few intermediate-sized cells were labeled and none with nuclei less than $75\,\mu^2$. No loading dose of [^{3}H]thymidine was given, and since it takes at least several hours for the isotopic concentration to build up to label cells above threshold, and since the duration of the S phase of leukemic cells during advanced disease is usually around 20 hours (Clarkson et al. 1967a; Killmann 1968; Clarkson and Fried 1971), it can be assumed that most of these labeled cells had entered S for the first time since the start of the infusion. (About 30% of mitoses were still unlabeled at 24 hours (Fig. 3) and probably represent either cells with long G_2 periods or cells which were part way through S before the isotopic concentration had become sufficiently high to label them). The labeling intensity of the large cells was almost the same as that of the smaller ones (Fig. 4), thereby providing confirmatory evidence that very few large cells were going through a second S phase during the first 24 hours. The labeling index of 10.8% at 24 hours is only slightly higher than the flash [^{3}H]thymidine labeling index of the marrow lymphoblasts (8%, determined in vitro just prior to starting the infusion), and it therefore seems clear that relatively few of the labeled cells could have divided within the first 24 hours.

At the end of the infusion 99–100% of the lymphoblasts with nuclear areas > $125\,\mu^2$ were labeled, but the labeling index of the smaller cells fell progressively and only 64% of the smallest cells were labeled (Fig. 4). Cells of all sizes were more

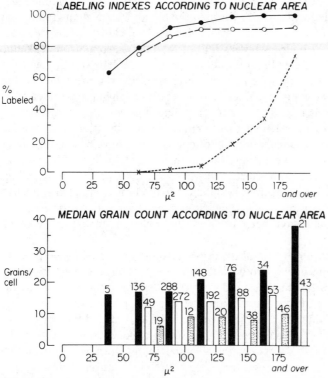

Figure 4

Relation of nuclear size of marrow leukemic cells in J. T. to [³H]thymidine labeling frequency and intensity 24 hr after start (x– – –x and shaded bar); at end of 20 days [³H]thymidine infusion (●———● and black bar); and 7 days after the end of the infusion (o– – –o and open bar). The total labeling index of the whole population of leukemic cells was 10.8% at 24 hr, 89% at the end of the infusion, and 87% 7 days after stopping. The areas of the nuclei of 800 or more leukemic cells were measured for each sample for the labeling indexes according to nuclear area (*top*). The number of labeled cells counted in each nuclear size category for labeling intensity are indicated above each column of grains per cell (*bottom*). Radioautographs exposed 12 weeks.

highly labeled at 20 days than at 24 hours, indicating that most of them had passed through several cell cycles in the interim. The larger cells had significantly higher median grain counts (MGC) than the small ones, probably largely because a significant number of them were going through S and had a higher DNA content at the time the infusion was ended. A possible alternative explanation is that the large cells represent a stable, rapidly dividing subpopulation that underwent more divisions during the 20-day period. However this seems much less likely and the data are more compatible with the interpretation that there is continual interchange between cells of all sizes due to growth and division.

Seven days after ending the infusion, both the labeling indexes and median grain counts of cells in all size categories fell (Fig. 4), thus providing further evidence that there was continuing division of labeled cells of all sizes with dilution of label. About 10% of the large cells (> 125 μ^2) at this time were unlabeled. Since the overall labeling index had only dropped slightly by 7 days (to 87%), only a few of the labeled cells could have diluted their label below threshold by division, and some of the unlabeled cells must have enlarged and started proliferating; in confirmation of this the labeling index of the mitotic cells 7 days after ending the infusion had fallen to 86% (Fig. 3).

The size distribution of the total population of marrow lymphoblasts did not change significantly during the infusion nor in the seven days thereafter (Fig. 5). The unlabeled cells at the end of the infusion, which had been dormant for the entire 20-day period, were significantly smaller, but during the next 7 days some of them enlarged and prepared to divide and their size distribution shifted to more closely approach that of the whole population.

Based on the rate of change in the nuclear size distribution of unlabeled cells

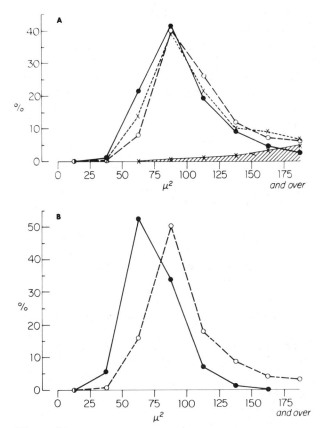

Figure 5

A, Nuclear size distribution of the total population of leukemic cells in the marrow of J. T. 24 hours after starting the [³H]thymidine infusion (x– – –x); at the end of the 20-day infusion (•——•); and 7 days later (o– – –o). 10.8% of the cells were labeled 24 hr after starting the infusion and their size distribution is also shown under the shaded area.

B, Nuclear size distribution of the *unlabeled* cells at the end of the [³H]thymidine infusion (•——•), and 7 days later (o– – –o) before treatment was started. The areas of 800 nuclei were measured in each sample for the total population and 400 in each sample for the unlabeled cells.

after the end of the infusion, Dr. Fried computed lower and upper bounds for the turnover rate of dormant cells. (Because the size distribution of dormant and proliferating cells overlap, a precise estimate could not be obtained using cell size data.) The minimum and maximum possible rates were 7.5% and 18% per day, and the earlier estimate of 11% per day obtained from the continuous labeling data falls within these limits and is in fact close to their average.

Taking the rate of 11% turnover per day to be correct, and assuming cells leave the dormant state randomly, it would require 20 days for 90% of the cells originally present to leave the dormant compartment and 20 days for each successive decade. If one arbitrarily defines the "effective" dormant compartment as comprising those cells remaining dormant for at least 20 days, then the effective dormant compartment in this case constitutes approximately 10% of the whole population. Assuming several trillion leukemic cells are present, then the dormant compartment contains $> 10^{11}$ cells, and it would require at least 220 days for all cells to leave the dormant state if their turnover rate remained constant.

If the leukemic population expands at an exponential rate, this rate of efflux from the dormant compartment might be expected to be maintained indefinitely, since daughters of proliferating cells will enter the dormant compartment at the same rate as previously dormant cells are leaving it. However growth of tumor populations is generally assumed to follow a Gompertzian pattern (i.e., the growth rate slows as the size of the population increases) (Laird 1965; Clarkson 1973), and as the population expands, the turnover rate would probably decrease.

On the other hand, at an earlier stage of disease or after the leukemic population has been greatly reduced by therapy, the average duration of G_1 may decrease. Following treatment of leukemia, myeloma and other human tumors, an increased growth rate has frequently been noted as evidenced by higher flash [^{3}H]thymidine labeling indexes and mitotic indexes as well as changes in other proliferative parameters (Lampkin, Nagao and Mauer 1971; Salmon and Smith 1972; Sheehy et al. 1973; Clarkson 1973). However the extent of population reduction necessary to obtain this effect and the time taken for the surviving previously dormant cells to begin rapid proliferation apparently varies greatly and cannot be predicted in individual cases. In no case so far studied have all the residual dormant cells immediately begun to proliferate once therapy has been interrupted, and it is likely in most tumors that some of them may remain dormant for several months or longer.

In the preceding discussion it has of course been assumed that at least some of the small dormant cells have stem cell capability (i.e., have the potential to reproduce the whole population). While there is no proof of this in human leukemia, small dormant spontaneous AKR thymic leukemic lymphoblasts, upon transplantation to young normal AKR mice, are capable of proliferating and causing death from leukemia at about the same time as do large, actively proliferating leukemic cells (Rosen, Perry and Schabel 1970; Omine and Perry 1972). It therefore seems probable that the long-term dormant subpopulation contains within it many leukemic cells with stem cell capability and that all of them must be destroyed to effect a cure.

It is ordinarily not possible to study the proliferative behavior of small numbers of surviving leukemic cells in patients since one usually cannot distinguish leukemic blasts from normal ones; this is especially true after chemotherapy, which may sublethally damage some cells and alter their morphology. If one assumes that to cure the disease it is necessary to kill all leukemic cells, or at least all those with stem cell capability, then it is of great importance to understand why some cells survive therapy and to devise ways to prevent them from reproducing the disease.

SURVIVAL VALUE OF THE DORMANT STATE IN ESTABLISHED HUMAN HEMATOPOIETIC CELL LINE

In order to obtain quantitative information on small surviving cell fractions, it has been necessary to turn to an in vitro model system. The system we have chosen is an established cell line of hematopoietic blast cells (SK-L7), which was derived from the peripheral blood of a child with acute leukemia and has been growing continuously for over 7 years (Clarkson, Strife and deHarven 1967). Whether the line originated from leukemic cells or normal lymphocytes is uncertain, but this is irrelevant for the purpose of the experiments to be described since the line has the properties of a pure stem cell population (i.e., the cells do not differentiate and the majority have unlimited proliferative potential). The cells produce malignant tumors when injected into newborn or suitably conditioned heterologous hosts (Southam et al. 1969a,b), and they presumably carry EB virus (Glaser and Nonoyama 1973), although this has never been demonstrated in this particular cell line (Clarkson 1967). SK-L7's proliferative kinetics and response to cytotoxic drugs is similar to most of the other established lines of human lymphoblastoid cells derived from both normal and leukemic subjects which we have studied (Todo et al. 1971).

The line was originally diploid but it now has a stem line of 47 chromosomes with an extra one in the D group, and about 20% of the cells are aneuploid (Traganos unpubl.). Eighteen sublines developing from cloned cells were also analyzed in an attempt to obtain a karyotypically stable line for use in the studies to be reported. However within a few weeks after their establishment, all of these sublines also contained 15–20% polyploid cells in addition to the stem line with 47 chromosomes. The polyploid cells presumably have only limited proliferative potential. The spontaneous death rate in the cultures cannot be measured directly, but on the basis of models derived from computer analysis of the kinetic data, Dr. Fried has estimated that the death rate is around 1% per hour and that it remains relatively constant during the different growth phases of the cultures, at least until the cells reach stationary phase.

General Growth Characteristics of SK-L7 Cells

In the experiments to be reported the cells were routinely grown in McCoy 5a medium (modified) supplemented with 30% fetal calf serum (FCS) at 37°C in National Incubators in an atmosphere of 5% CO_2 and 95% air and diluted twice per week with fresh medium unless otherwise noted. The cells do not adhere to glass and grow in small clumps easily dispersed by shaking, and accurate cell counts can be obtained with a hemocytometer. Viability is estimated by trypan blue dye exclusion. Depending on the experimental conditions, the maximum cell density obtainable is between 1 and 5×10^6 cells per ml medium; growth is also limited by the available bottom surface area of the culture vehicle and will generally not exceed 10^7 cells/cm^2.

The cytokinetic parameters vary according to the population density (Todo et al.

1971). At a starting concentration of 2×10^5 cells/ml growth ceases after about 2.5 doublings at around 1.2×10^6 cells/ml. Previous measurements of cell cycle parameters begun with cells at a starting density of 2×10^5/ml showed an average doubling time of about 1 day; median cell cycle time = 18 hours; mean S phase = 12 hours; mean G_2 = 1.8 hours; M = 0.5 hours; and mean G_1 = 3.7 hours, but highly variable. If the measurements are begun when the cells are approaching their saturation density, there is a progressive lengthening of S and even more so of G_1, and an increasing fraction of cells remain in a greatly prolonged G_1.

Continuous [³H]Thymidine Labeling

When log phase cells were continuously exposed to [³H]thymidine (0.01 μCi/ml), starting at 2×10^5 cells/ml, and allowed to grow without feeding or dilution for 4 days to reach a maximum density of 1.2×10^6 cells/ml, 95% of the cells were labeled at 1 day and 99.2% after 4 days (Todo et al. 1971). The nuclear size distributions of the whole population and that of the labeled cells at 1 hour and unlabeled cells at 4 days are shown in Fig. 6. As in the case of leukemic populations in vivo, the majority of unlabeled cells at 4 days were small cells. Progressively fewer cells were labeled on continuous exposure to [³H]thymidine in cultures approaching stationary phase (Todo et al. 1971).

If cultures are deliberately maintained at low cell densities by repeated feeding and dilution (starting at about 10^4 cells/ml and not allowed to exceed a density of

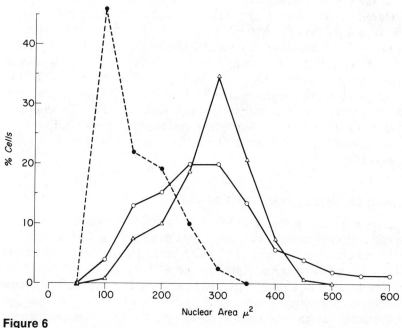

Figure 6

Nuclear size distribution of SK-L7 cells in logarithmic growth phase during continuous exposure to [³H]thymidine (0.01 μCi/ml). (o———o) Whole population; (△———△) labeled nuclei after 1-hr exposure; (•– – –•) unlabeled nuclei after 96-hr exposure. 56% of the cells were labeled at 1 hr and 99.2% after 96 hr.

around 6×10^5 cells/ml of media or per cm² of bottom surface area), the cells achieve their maximum growth rate. Under these conditions the fastest doubling times observed have been 14–16 hours and almost all the cells divide rapidly. However even at these maximum growth rates a small fraction of the population remains dormant, and when the population is sufficiently large the dormant cells can be detected by appropriate labeling and drug sterilization experiments.

The 30-minute [³H]thymidine labeling index (LI) of cells growing at these low densities is usually about 60%, and their mitotic index (MI) about 2% (excluding early prophase). When such rapidly growing cells were continuously exposed to [³H]thymidine (Sp.Ac. 6 Ci/mM; 0.01–0.05 μCi/ml) and maintained in exponential growth phase by appropriate dilutions and biweekly feeding with fresh media containing the same concentration of [³H]thymidine, there was no detectable effect on their growth rate at 0.01 μCi/ml and only slight inhibition at 0.05 μCi/ml, at least during the first 14 days. Samples prepared for radioautographs at daily intervals showed 96–98% were labeled after 24 hours and only rare unlabeled cells could be found after 48 hours. Of 6500 cells counted for each sample, only 9 unlabeled cells were found at 4 days (0.14%) and only 4 or 5 (0.07 to 0.08%) at 5, 6, and 7 days. No unlabeled cells were found among 6500 cells at 8 days and thereafter (<0.02%). Since some of the unlabeled cells were multinucleated or otherwise abnormal and undoubtedly had limited proliferative potential, the fractions of unlabeled cells with stem cell capability were even lower than these values. There were, of course, too few unlabeled cells in these experiments to do size distribution studies.

Cloning Efficiency and Lack of Influence of Dead Cells

Single cell cloning experiments showed a cloning efficiency of better than 50% when carefully done; the average results of three experiments are shown in Table 1. In these experiments single cells were obtained with a micropipette, placed in a tiny drop on a piece of a coverslip, and carefully examined microscopically to be certain only one cell was present. The drop was flushed into a Kimble 4-dram (15-ml) vial containing 10 ml of media and the culture was then fed twice weekly with 1/2 fresh media, taking care not to disturb the lower half. Growth can first be detected around

Table 1

Cloning Efficiency of Single SK-L7 Cells and Doubling Times

	Cultures growing/ cells cloned	Cloning eff'cy (%)	Doubling time (hr)	% Cultures with stated doubling times
Single cell in 30% FCS	23:35	66	14–17	70
			18–20	17
			23–29	13
Single cell in 10% FCS	24:35	69	15–17	38
			18–20	42
			21–28	20

10^4 cells/ml, but for the doubling time estimates the cells were allowed to reach at least 10^5 cells/ml and thereafter to grow continuously in order to obtain accurate cell counts. The doubling times shown in Table 1 are the averages required for one cell to reach $1-2 \times 10^6$ cells per vial (20–21 doublings) and assume there is no lag phase (i.e., the single cell begins to divide immediately). Since the doubling times were slightly faster in 30% than in 10% FCS, all drug sterilization and survival experiments were carried out using 30% FCS.

The presence of many dead cells has no detectable influence on the ability of one or several cells to grow (Table 2). In these experiments SK-L7 cells were irradiated with 3000 rads 1 day previously and single or several living cells were obtained as before and either kept as controls or added to vials containing 5×10^6 killed cells. The doubling times were similar in the comparable control and dead cell cultures and no inhibitory nor stimulatory effect of the dead cells was detected.

Sterilization of Cultures by Cytotoxic Drugs

Unless otherwise noted drugs were added to 500-ml Erlenmeyer flasks containing 5×10^6 exponentially growing cells at a density of between $1-2 \times 10^5$ cells/ml of media supplemented with 30% FCS. In these experiments the untreated controls had doubling times of about 16 hours and hence were growing at or near their maximum rate when exposed to the drugs. At the time of drug removal the cells were centrifuged, washed twice with drug-free media, and resuspended a third time in 10 ml of fresh media in 15-ml vials; they were then fed twice weekly with 1/2 fresh media until continuous growth was detected or held routinely for 8 or 10 weeks before discarding. Growth of surviving cultures almost invariably occurred within the first 6 weeks; only rare cultures grew between 6 and 10 weeks and none thereafter. Once continuous growth was clearly apparent, the cultures were discarded and their long-term growth characteristics were not followed routinely. However in most instances the growth rate of the surviving cultures, at least during their initial period of regrowth, was similar to that of untreated controls.

Table 2
Effect of Dead Cells on Growth of Living SK-L7 Cells

	Cultures growing/ cultures started	% Cultures growing
5×10^6 dead SK-L7 cells	0:9	0
Single living cell + 5×10^6 dead cells	5:12	42
Single living cell	5:11	45
Two living cells + 5×10^6 dead cells	4:6	67
Two living cells	4:7	57
3 or 4 living cells + 5×10^6 dead cells	16:16	100
3 or 4 living cells	12:12	100

Sterilization with Cell Cycle Phase Nonspecific Drugs

Seven cell cycle phase nonspecific cytotoxic drugs were tested for their ability to sterilize cultures during one hour's exposure. These agents are lethal to both proliferating and dormant cells, although they may have greater activity against the former (Bruce, Meeker and Valeriote 1966; Skipper 1968, 1971; Bhuyan, Scheidt and Fraser 1972). All seven agents sterilized the cultures within 1 hour, although relatively high concentrations were required to do so (Table 3).

Experimental Conditions using S-Phase Specific Agents

Arabinosylcytosine (Ara-C) and hydroxyurea (HU) were chosen for study as drugs which are lethal only to cells in S (Skipper 1968, 1971). Since it has been suggested that these drugs may also inhibit passage of cells from G_1 into S, suicidal doses of [³H]thymidine were also tested for comparison. During the periods of drug exposure the cultures were fed daily with 1/2 new media containing the same concentration of fresh drug to be certain that the long survival times were not due to drug degradation, although in fact no significant degradation was found to occur.

Persistence of levels of unchanged drug were determined serially by adding tritiated Ara-C and [³H]thymidine to growing cultures or to media containing 30% FCS without cells (Balis et al. unpubl.). One-half fresh media containing the same isotopic concentration (plus fresh drug in the case of Ara-C) was added either daily or every 3 or 4 days. Samples were taken daily for the first 4 days and then at 7, 11, and 14 days for chromatographic separation of the parent compounds and their

Table 3

Drugs Sterilizing Cultures after Exposure for 1 Hr

	Molar conc.	No. cultures surviving/ No. exposed	% Surviving
Daunorubicin	10^{-7}	15/15	100
	10^{-6}	11/32	34
	10^{-5}	0/32	0
Adriamycin	10^{-6} or less	29/29	100
	10^{-5}	21/30	70
	10^{-4}	0/9	0
Nitrogen mustard (HN_2)	10^{-5} or less	30/30	100
	10^{-4}	0/20	0
Actinomycin D	10^{-7}	1/10	10
	10^{-6} or more	0/14	0
BCNU	10^{-4} or less	38/38	100
	10^{-3}	0/12	0
CCNU	10^{-5} or less	30/30	100
	10^{-4}	16/19	84
	10^{-3}	0/10	0
CH_3-CCNU	10^{-5} or less	20/20	100
	10^{-4}	14/15	93
	10^{-3}	0/8	0

degradation products, and the strips encompassing the appropriate spots were counted in a liquid scintillation system. There was virtually complete recovery of unchanged [³H]thymidine (Sp. Ac. 54 Ci/mM) at concentrations between 5 and 50 μCi/ml of media at all times in all experiments, and there was no difference between the cultures containing cells and the cell-free media. Tritiated Ara-C was added to growing cultures and cell-free media containing nonradioactive Ara-C at concentrations of 10^{-7}, 10^{-6}, 10^{-5}, and 10^{-4} M. Recovery of unchanged Ara-C was consistently better than 80% in all samples, and less than 20% Ara-U was found in any. No differences were found between the cultures containing cells and the cell-free media.

To be certain that there was no significant cell loss at the time of drug removal in the experiments using small numbers of cells, in preliminary experiments untreated cells were washed immediately with fresh media in a manner identical to the treated cultures at the time of drug removal, and growth invariably occurred at the anticipated times.

Inhibition of [³H]Thymidine Incorporation by Ara-C and Hydroxyurea

The concentrations of Ara-C and HU necessary to inhibit [³H]thymidine incorporation were first determined. Aliquots containing 5×10^5 rapidly growing cells were incubated at various drug concentrations for 1 hour and then [³H]thymidine (5 μCi/ml of cell suspension) was added for another hour. A 1000-fold excess of unlabeled thymidine was then added and the tubes immediately placed in ice. The cells were centrifuged at 4°C, the supernatant removed, resuspended in phosphate-buffered saline and filtered onto a 0.45μ Millipore filter, which was then washed sequentially with: saline washes of the centrifuge tube $\times$ 2; 5 ml 5% TCA; 5 ml 2% TCA $\times$ 3; and 95% alcohol. The filter was partially dried by vacuum and then placed in a scintillation vial and thoroughly dried in an oven. A scintillation "cocktail" consisting of toluene and PPD was added and the vials counted in a Beckman LS-200B liquid scintillation system. It was found that 10^{-6} M of Ara-C and 10^{-4} M of HU were necessary to inhibit [³H]thymidine incorporation by more than 50% and 10^{-5} M of Ara-C and 10^{-3} M of HU were necessary to inhibit incorporation by 90% or more.

Sterilization with S-Phase Specific Agents

SK-L7 cells were able to grow when exposed continuously to 10^{-6} M Ara-C, although at a significantly slower rate than untreated cultures, whereas Ara-C 10^{-3} M and 10^{-4} M caused rapid cell death and 10^{-5} M only slightly delayed killing (Fig. 7). Similar results were found with HU, except that the concentration necessary to achieve a comparable degree of cell kill was approximately 1.5 $\log_{10}$ higher than with Ara-C.

Cultures, each containing 5×10^6 exponentially growing cells, were continuously exposed to increasing concentrations of Ara-C, HU, or [³H]thymidine for varying periods to see how long it would take to sterilize them. These experiments are not yet completed, but the results to date are summarized in Table 4. The percent survival values are based on a minimum of 10 cultures treated for each time period shown; in most cases a total of 20 or 30 cultures were treated in 2 or 3 experiments, and sometimes for critical times a total of 50 or 60 cultures in 5 or 6 separate experiments. There was some variability in different experiments, but if the endpoint is taken as the day on which less than 50% of the cultures survived and following

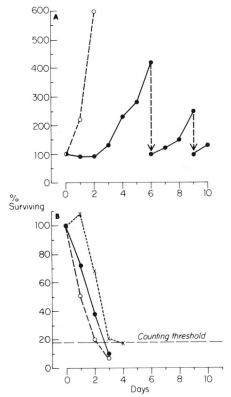

Figure 7

Rate of kill of SK-L7 cells in their logarithmic growth phase during continuous exposure to different concentrations of arabinosylcytosine. A total of 5×10^6 cells were exposed to each drug concentration shown at a starting density of 10^5 cells/ml of media. The counts shown are viable cells counts (as estimated by trypan blue dye exclusion) and are given as the % of the initial cell count. **A**, (●———●) Ara-C 10^6M; (○– – –○) untreated cells; arrow indicates dilution with new media. **B**, (x– – –x) Ara-C 10^{-5}M; (○– – –○) 10^{-4}M; (●———●) 10^{-3}M.

which there were progressively fewer survivors, it can be seen that the majority of cultures survived 9 or 10 days' exposure to 10^{-5} M Ara-C, 10^{-3} M HU, and 5 μCi/ml [³H]thymidine, and at least 6 or 7 days' exposure at the next higher drug concentrations.

The increasing lethality of higher concentrations of Ara-C is more obvious when fewer cells are treated (Table 5). Five $\times 10^4$ cells were sterilized in 1 day with Ara-C 10^{-3} M, whereas it took 4 days to sterilize $>50\%$ of the cultures with 10^{-5} M.

Inhibition of DNA Synthesis with Different Concentrations of Ara-C

As might be expected in view of the above results, increasing concentrations of Ara-C (between 10^{-6} M and 10^{-3} M) cause progressively greater inhibition of DNA synthesis. Dr. Xenophon Yataganas in our laboratory has developed a quantitative fluorescent acriflavine feulgen staining method and with the use of a

Table 4

Cultures Surviving (%) Sterilization with S-phase Specific Drugs (5×10^6 cells per culture)

	Molar conc. (or μCi/ml)	Days of continuous exposure to drug													
		1	2	3	4	5	6	7	8	9	10	11	12	13	14
Arabinosyl-cytosine	10^{-3}	100	100	93	85										
	10^{-4}	100	100	96	67	100	96	57	33	7	10	0			
	10^{-5}	100	100	100	100	100	100	100	90	100	100	25			
Hydroxyurea	10^{-2}			100	93	43	51	45	14	0	22	0	10		
	10^{-3}			100	100	100	79	90	83	55	41	35	10	20	10
[³H]thymidine	50	100	100	100	100										
(sp.ac. 20–54	10	100	100	100	100	75	75	92	100	100	100	88			
Ci/mM)	5	100	100	100	100	100	100	100	100	100	100	70			

Table 5

Cultures Surviving (%) Arabinosylcytosine Treatment (5×10^4 cells per culture)

Molar Conc.	Days of continuous exposure to Ara-C						
	1	2	3	4	5	6	7
10^{-3}	0	0	0	0			
10^{-4}	82	71	24	24			
10^{-5}	100	93	55	45			
10^{-6}	100	100	100	100	80	88	100

flow microfluorimeter* the fluorescence intensities of single cells can be shown to correspond to their DNA contents (Yataganas et al. unpubl.) (Fig. 8). The fluorescence intensity, which is proportional to the DNA content, is shown on the abscissa and the number of cells per channel on the ordinate. The DNA distribution of the whole population of untreated exponentially growing asynchronous SK-L7 cells is shown in the upper left. The first peak represents cells in G_1 or very early S, and the second peak cells in late S, G_2 or mitosis; the area in between represents cells in mid-S.

Figures 8 to 10 show that with increasing concentrations of Ara-C from 10^{-6} to 10^{-4} M, the movement of cells through S was inhibited progressively more completely. Exponentially growing cells were used and 1/2 fresh media containing Ara-C at the initial concentration was added daily in all these experiments. Only viable cells take the fluorescent stain, and since the fluorescence intensity may vary with cell density, approximately equal numbers of *viable* cells were analyzed for their DNA distribution in all samples.

At 10^{-6} M (Fig. 8) the cells initially accumulated in S, and the mitotic index (MI) fell from a control value of 2% to a nadir of 0.2% at 8 hours. However the block was only partial and the cells continued to progress slowly through S. The mitotic index began to rise again after 16 hours, and there was little cell death as

* Cytofluorograf, Bio/Physics Systems, Inc., Baldwin Place Road, Mahopac, N.Y. 10542.

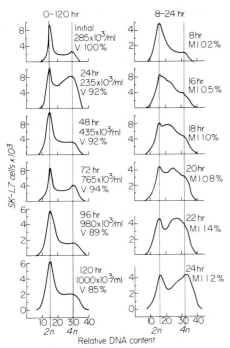

Figure 8

Acriflavine feulgen DNA distribution of SK-L7 cells in logarithmic growth phase during continuous exposure for 120 hr to arabinosylcytosine 10^{-6}M. The channel number in the abscissa expresses fluorescence intensity, which corresponds to the relative DNA content, and the number of cells per channel is shown on the ordinate. The mitotic index (*MI*) remained almost constant after 24 hr around 1%. The total cell counts indicated at 0–120 hr are viable cell counts; and the % viability (*V*) of each sample is also shown.

the viability remained good and the viable cell count only fell slightly at 24 hours. After 48 hours the DNA distribution pattern of the population had returned almost to normal, and the viable cell count thereafter increased progressively, although at a slower rate than in untreated control cultures.

At 10^{-5} M (Fig. 9) the cells were initially blocked earlier in S, but many progressed slowly through S and some were still able to reach mitosis. However the cells reaching mitosis were clearly abnormal and most of the cells accumulated in S where they presumably died. The viable cell count fell progressively and only 24% of the cells still excluded trypan blue at 120 hours, by which time the viable cell count had fallen to about 10% of the pretreatment value.

At 10^{-4} M (Fig. 10) the cells were blocked very early in S. The number of cells in the S region fell significantly by 24 hours due to death of cells already in S and failure of other cells to move through S. No mitoses were found and the viable cell count fell progressively to less than 10% of the pretreatment level by 96 hours. The increase in the G_2 and late S peak at 72 and 96 hours probably represents cells, already in mid or late S when the drug was added, which had managed to complete or almost complete S but were unable to divide. The latter presumably died within a few more days, and it seems virtually certain that the few cells which can survive

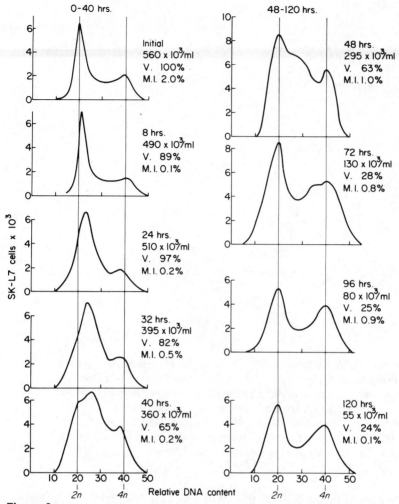

Figure 9

Acriflavine feulgen DNA distribution of SK-L7 cells in logarithmic growth phase during continuous exposure for 120 hr to arabinosylcytosine 10^{-5}M. See legend Fig. 8.

4 days' exposure to this concentration of Ara-C are those within the G_1 or very early S region. The results with 10^{-3} M were similar to those found with 10^{-4} M, although movement of cells through S is blocked even more completely.

Relation of Population Size to Survival

Between the range of 500 to 5 million cells, the smaller the total number of cells treated, the shorter the time required for sterilization with the same concentration (10^{-4} M) of Ara-C (Table 6). We are presently expanding these experiments to try to define more accurately the relation between population size and survival time. However because of volume limitations it is technically difficult to perform these sterilization experiments with very large numbers of cells while at the same time making certain that their density does not exceed that required for the cells to be

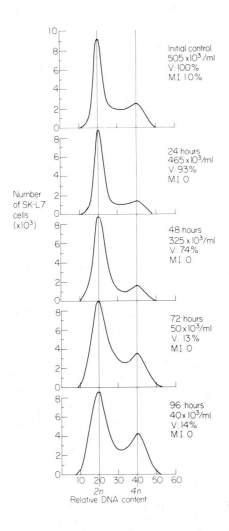

Figure 10

Acriflavine feulgen DNA distribution of SK-L7 cells in logarithmic growth phase during continuous exposure for 120 hr to arabinosylcytosine 10^{-4}M. See legend Fig. 8.

growing at their maximum rate when the drug is added. Previous experiments have clearly shown that as cells approach their saturation density and their growth slows, significantly longer periods of exposure to the same concentrations of S-phase active drugs are required to achieve sterilization (Hryniuk, Fischer and Bertino 1969; Gee et al. 1970; Clarkson and Fried 1971). In these earlier experiments the survival times for logarithmically growing SK-L7 cells were shorter than those found in the present experiments; this is probably mainly because different incubators were used and the atmospheric conditions and pH were less well controlled and we were unable to detect as few surviving cells. If it can be ascertained that all population sizes treated are growing at the same maximum rate when the drug is added, it will be of interest to see whether or not the relation between population size and survival time is linear. Because in the recent experiments with very rapidly growing cultures a few cells were found to survive exposure to cytocidal concentrations of S-phase active drugs for a week or longer, we have not attempted to repeat these sterilization experiments with cultures approaching their saturation densities, but presumably some cells in the latter would survive even longer periods of drug exposure.

Table 6

Cultures Surviving (%) Arabinosylcytosine Treatment

Total no. cells/ culture	Days of continuous exposure to Ara-C (10^{-4}M)			
	1	2	3	4
5×10^6		100	100	100
5×10^5		100	88	94
5×10^4		71	24	24
5×10^3	25	13	0	0
5×10^2	0	0	0	0

Separation and Properties of G_1 Subpopulation

Using a sucrose gradient system Dr. Yasuharu Mitomo and Mrs. Annabel Strife in our laboratory have been able to separate a fairly pure fraction of G_1 cells from the whole population of SK-L7 cells using a modification of a method previously reported by Perry and his coworkers (Rosen, Perry and Schabel 1970; Omine and Perry 1972; Yataganas et al. unpubl.). In these experiments the cells were started at a concentration of about 8×10^4 cells/ml in McCoy media containing only 5% FCS and grown for 3 days prior to the separation, aiming at a density of around 6×10^5 cells/ml on the day of separation in order to avoid having to centrifuge the cells since this causes increased clumping. The cells grow slower with this reduced supplement of FCS; their doubling time is generally about 22 hours and their 30-minute [³H]thymidine labeling index is between 30 and 45%. Just prior to separation, which was carried out at 4°C, [³H]thymidine was added (1 μCi/ml of media), the cells were concentrated by removing some of the overlying media, and 5×10^7 cells were resuspended in 100 ml and loaded into the bottom of the 2000-ml separatory chamber. A continuous sucrose gradient in McCoy media containing only 1% FCS was then carefully introduced at the bottom and 50-ml fractions were taken from the top of the chamber. In some cases G_1 cells obtained from a previous experiment were allowed to regrow and then the procedure was repeated.

The cells are separated on the basis of size and the size distribution of the different fractions can be measured with a Cytograf* (Yataganas et al. unpubl.). The early fractions contain the smallest cells and the late fractions the largest ones (Fig. 11). The gradient and the number of cells in each fraction in a typical experiment are shown in Fig. 12. In the best separations the [³H]thymidine labeling index of the early fractions as determined by radioautography is around 5% or less. The labeling indexes then increase to a maximum of about 70% and then decline in the later fractions, which include more cells in G_2. The DNA distributions of the whole population and of different pooled fractions, measured by fluorescence of acriflavine-feulgen stained cells, are shown at the top. The earliest fractions contain mostly dead cells and are not shown. Fractions 6–18 contain mostly G_1 cells, together with a small percentage of cells in early S; the latter have not yet increased their DNA content sufficiently to be recognized by microfluorimetry, but it is evident that they have already entered S since they are clearly labeled with [³H]thymidine on the radioautographs. Fractions 19–22 contain more cells in early and middle S, frac-

* Cytograf, Bio/Physics Systems, Inc., Baldwin Place Road, Mahopac, N.Y. 10542.

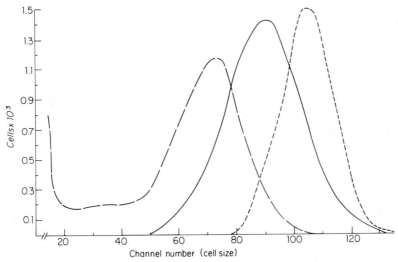

Figure 11

Cytograf determination of size distribution of pooled early (9–17) fractions (— —) and late (36–39) fractions (– – –) of SK-L7 cells recovered from separator as compared with the whole population (———). The initial increase before channel number 20 represents dead cells.

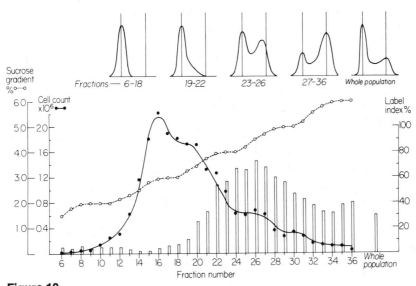

Figure 12

Sucrose gradient separation of total of 5 × 10⁷ SK-L7 cells. (o– – –o) Gradient determined by refractometer; (•———•) cell count × 10⁶ per fraction; [³H]thymidine labeling index of whole population and each fraction is indicated by bars. Mostly dead cells are found in fractions earlier than 6. The microfluorimetric measurements of acriflavine feulgen-stained cells in various pooled fractions and in the whole population are shown at the top.

tions 23–26 mostly cells throughout S, and fractions 27–36 predominantly cells in late S, G_2 and M.

We are presently studying the growth potential and other properties of the separated G_1 cells. It is already clear that their cloning efficiency and growth rate in both liquid media and methylcellulose are the same as that of cells in the later fractions or of the whole population. We plan to see if the growth requirements of the smallest G_1 cells in the earliest fractions differ from those in late G_1 and also to compare early G_1 cells from rapidly growing cultures with those from cultures approaching their saturation density. Eventually we plan to extend these studies to dormant leukemic cells freshly obtained from patients.

DISCUSSION

By means of continuous infusions of [^{3}H]thymidine and radioautographic analysis, it has been shown in all human tumors so far studied in vivo that significant fractions of the tumor populations remain dormant for long periods, probably at least several months. In these studies the critical measurements were not done to try to distinguish true resting (G_0) cells from prereplicative (G_1) cells (Epifanova and Terskikh 1969; Pasternak, Warmsly and Thoms 1971; Stein and Baserga 1972), and we have therefore used the more general and inclusive term "dormant" to refer to cells in both categories. For the purpose of this discussion, it will be assumed that all dormant cells are in G_1, but that the duration of this phase is highly variable (Brown 1968; Prescott 1968; Fried 1970).

The dormant tumor cells are small cells which are invulnerable to S-phase active cytotoxic drugs and relatively so to cell cycle phase nonspecific agents. Since there is strong presumptive evidence based on analogy with animal and cell culture experiments that they can reproduce the whole neoplastic population, they constitute a major obstacle to curative therapy which is especially critical in tumors such as lymphoblastic leukemia, which are moderately responsive to the available drugs.

The dormant state is not a unique feature of neoplastic cells, but rather a property shared with their normal ancestors that greatly increases any population's probability of survival. In some normal populations such as the hematopoietic system, which must stand ready to rapidly increase the production of blood cells in response to increased demand, the majority of stem cells are normally resting (Lajtha, Oliver and Gurney 1962; Becker et al. 1965; Rosse 1970; Haas, Bohne and Fliedner 1971; Yoshida and Osmond 1971). When the stem cell compartment is depleted by injury, it repopulates itself; but unlike neoplastic populations, the level of repopulation is tightly regulated, and except perhaps for a transient rebound above the baseline level during recovery, the compartment does not exceed its original size. One is ordinarily not concerned with quantitation of very small numbers of surviving normal stem cells since their reduction below a certain point is incompatible with the host's survival. However in the case of neoplastic populations, if one assumes that a single tumor (stem) cell can reproduce the whole population, then it is very important to define precisely the behavioral characteristics of the last few surviving cells if we are to devise better ways to destroy them or otherwise prevent them from reproducing the disease.

Since suitable methods are not available to study small numbers of leukemic cells in vivo, we have investigated the survival value of the dormant state in an established line of human hematopoietic cells (SK-L7); the system is sufficiently sensitive to detect 1 or 2 surviving cells among several million killed cells. The system is not a perfect model for studying the proliferative kinetics of human tumors in vivo since the cultured cells grow faster, and, in contrast to most human neoplasms in which there is usually considerable partial maturation and spontaneous cell loss, the SK-L7 cells show little or no evidence of maturation and almost all of them are capable of reproducing the whole population. Nevertheless there are many similarities and we believe much can be learned about the behavior of dormant cells in this simple in vitro system. When SK-L7 cells are well nourished and kept at a low cell density, they grow at their maximum rate with a doubling time of about 16 hours. If they are allowed to approach their saturation density, their cell cycle time increases, especially the G_1 phase, and an increasing number of cells remain in G_1 for longer and longer times. Providing they are fed and agitated frequently to prevent the cell aggregates from becoming too large and inhibiting diffusion of nutrients, they eventually reach a plateau, during which some cells continue to proceed very slowly through the cell cycle and growth is balanced by cell death (Todo et al. 1971). Unlike the commonly used 3T3 line (Todaro, Lazar and Green 1965; Holley and Kiernan 1968), SK-L7 cells are thus not completely arrested in G_1 (or G_0) when they reach their saturation density, and in this respect their behavior more closely resembles that of spontaneous tumors in vivo, especially that of widely dispersed ones such as ascitic tumors and disseminated hematologic neoplasms (Clarkson 1969, 1973; Sheehy et al. 1973).

Just as in spontaneous tumors in vivo, the SK-L7 cells that are in a prolonged G_1 phase are mostly very small cells. Although the size distribution of cells initiating DNA synthesis varies considerably (Fig. 6), it appears that they must attain a certain minimal mass before they can do so. Even when the cells are growing at low cell densities at their maximum rate, a very small fraction of the population, probably less than 0.01%, may remain dormant for a week or longer and then repopulate the culture. At least over a limited range, the duration of survival is related to the total size of the population, but the exact nature of the relationship is not yet clear. When the population density increases and growth slows, the size of the dormant fraction increases as does the duration of the dormant state, and this further enhances the population's probability of survival.

Because dormant tumor cells are a major cause of therapeutic failure in many human tumors, it is very important that we develop fuller understanding of their properties and learn ways to stimulate them to divide in order to destroy them more readily with cytotoxic drugs, induce them to differentiate, or otherwise render them incapable of reproducing the tumor. Because of the enormous complexity of human tumors, it is likely that better understanding of ways to manipulate dormant cells will first emerge from well-defined in vitro systems.

SUMMARY

Significant fractions of leukemic and other neoplastic populations can remain dormant in vivo for long periods, probably many months. The dormant state is probably the major reason for failure to cure such diseases as lymphoblastic leukemia, which are moderately responsive to chemotherapy, since the dormant cells are invulnerable to drugs active against proliferating cells and relatively resistant to cell cycle phase nonspecific agents. The dormant state is not unique to neoplastic cells, but rather is a property shared with their normal ancestors which greatly increases any population's probability of survival. Even in a stem cell population growing at its maximum rate in vitro with a doubling time of 16 hours, a tiny fraction of the population may remain dormant for a week or more, and then can subsequently repopulate the culture. At least over a limited range, the duration of survival is related to the size of the population. When the population density increases and growth slows, the size of the dormant fraction and the duration of the dormant state increases, further enhancing the population's probability of survival.

Acknowledgments

This work was supported in part by USPHS Research Grants CA-05826 and CA-08748 from the National Cancer Institute, American Cancer Society Grant DT-1D, Hearst Foundation Fund, and the United Leukemia Fund.

REFERENCES

Becker, A. J., E. A. McCulloch, L. Siminovitch and J. E. Till. 1965. The effect of differing demands for blood cell production on DNA synthesis by hematopoietic colony-forming cells of mice. *Blood* **26**:296.

Bhuyan, B. K., L. G. Scheidt and T. J. Fraser. 1972. Cell cycle phase specificity of antitumor agents. *Cancer Res.* **32**:398.

Brown, J. 1968. Long G_1 or G_0 state: A method of resolving the dilemma for the cell cycle of an in vivo population. *Exp. Cell Res.* **52**:565.

Bruce, W. R., B. E. Meeker and F. A. Valeriote. 1966. Comparison of the sensitivity of normal hematopoietic and transplanted lymphoma colony-forming cells to chemotherapeutic agents administered in vivo. *J. Nat. Cancer Inst.* **37**:233.

Clarkson, B. 1967. Formal discussion: On the cellular origins and distinctive features of cultured cell lines derived from patients with leukemias and lymphomas. *Cancer Res.* **27**:2483.

———. 1969. A review of recent studies of cellular proliferation. *Human tumor cell kinetics* (ed. S. Perry) *NCI Monog.* No. 30, p. 81.

———. 1972. Acute myelocytic leukemia in adults. *Cancer* **30**:1572.

———. 1973. Cell-cycle kinetics: Clinical applications. *Antineoplastic and immunosuppressive agents for the handbook of experimental pharmacology* (ed. D. Johns and A. C. Sartorelli), Springer-Verlag, New York. (in press)

Clarkson, B. D. and J. Fried. 1971. Changing concepts of treatment in acute leukemia. *Med. Clinics of North America* **55**:561.

Clarkson, B. D., J. Fried and M. Ogawa. 1969. Magnitude of proliferating fraction and rate of proliferation of populations of leukemic cells in man. *Recent results in cancer research: normal and malignant cell growth* (ed. R. J. M. Fry et al.) vol. 17. p. 175–185. Springer-Verlag, New York.

Clarkson, B., K. Ota and D. A. Karnofsky. 1962. Incorporation of tritiated thymidine (TdR-H³) by human cancer cells in vivo. *Proc. Amer. Ass. Cancer Res.* **3**:311.

Clarkson, B. D., A. Strife and E. deHarven. 1967. Continuous culture of 7 new cell lines (SK-L1 to 7) from patient with acute leukemia. *Cancer* **20**:926.

Clarkson, B., K. Ota, T. Ohkita and A. O'Connor. 1965. Kinetics of proliferation of cancer cells in neoplastic effusions in man. *Cancer* **18**:1189.

Clarkson, B. D., T. Ohkita, K. Ota and J. Fried. 1967a. Studies of cellular proliferation in human leukemia. I. Estimation of growth rates of leukemic and normal hematopoietic cells in two adults with acute leukemia given single injections of tritiated thymidine. *J. Clin. Invest.* **46**:506.

Clarkson, B. D., Y. Sakai, T. Kimura, T. Ohkita and J. Fried. 1967b. Studies of cellular proliferation in human leukemia. II. Variability in rates of growth and cellular differentiation in acute myelomonoblastic leukemia and effects of treatment. 21st Ann. Symp. Fundamental Cancer Res., The University of Texas M. D. Anderson Hospital and Tumor Institute at Houston. Reprinted from *The proliferation and spread of neoplastic cells,* p. 295–334.

Clarkson, B. D., J. Fried, A. Strife, Y. Sakai, K. Ota, T. Ohkita and R. Masuda. 1970. Studies of cellular proliferation in human leukemia. IV. Behavior of normal hematopoietic cells in three adults with acute leukemia given continuous infusions of ^{3}H-thymidine. *Cancer* **26:**1.

Clarkson, B., A. Todo, M. Ogawa, T. Gee and J. Fried. 1971. Consideration of the cell cycle in chemotherapy of acute leukemia. *Recent results in cancer research: current concepts in the management of leukemia and lymphoma* (ed. J. E. Ultmann et al.) vol. 36. p. 88–118. Springer-Verlag, New York.

Epifanova, O. and V. Terskikh. 1969. On the resting period in the cell life cycle. *Cell and Tissue Kinetics* **2:**75.

Erslev, A. J. 1972. Production of erythrocytes. *Hematology* (ed. W. J. Williams et al.) p. 162–177. McGraw-Hill, New York.

Folkman, J. 1971. Tumor angiogenesis. Therapeutic Implications. *New Eng. J. Med.* **285:**1182.

Fried, J. 1970. A mathematical model to aid in the interpretation of radioactive tracer data from proliferating cell populations. *Mathem. Biosciences* **8:**379.

Gee, T. S., A. Todo, A. Strife and B. D. Clarkson. 1970. Effect of hydroxyurea (HU) and arabinosylcyosine (Ara-C) on logarithmic (log) and stationary (stat) phase human lymphoid (SK-L7) cells in vitro. *Proc. Amer. Ass. Cancer Res.* **11:**29.

Glaser, R. and M. Nonoyama. 1973. Persistence of Epstein-Barr Virus in EB negative hybrid cells. *Proc. Amer. Ass. Cancer Res.* **14:**73.

Greene, H. S. N. 1941. Heterologous transplantation of mammalian tumors. I. The transfer of rabbit tumors to alien species. *J. Exp. Med.* **73:**461.

Haas, R., F. Bohne and T. M. Fliedner. 1971. Cytokinetic analysis of slowly proliferating bone marrow cells during recovery from radiation injury. *Cell and Tissue Kinetics* **4:**31.

Holland, J. F. and O. Glidewell. 1972. Chemotherapy of acute lymphocytic leukemia of childhood. *Cancer* **30:**1480.

Holley, R. W. and J. A. Kiernan. 1968. "Contact Inhibition" of cell division in 3T3 cells. *Proc. Nat. Acad. Sci.* **60:**300.

Hryniuk, W. M., G. A. Fischer and J. R. Bertino. 1969. S-Phase cells of rapidly growing and resting populations. Differences in response to methotrexate. *Mol. Pharmacol.* **5:**557.

Jensen, M. K. 1967. Chromosome studies in acute leukemia. III. Chromosome constitution of bone marrow cells in 30 cases. *Acta Med. Scand.* **182:**629.

Killmann, S. 1968. Acute leukemia. The kinetics of leukemic blast cells in man. *Ser. Haemat.* **1:**38.

Kiossoglou, K. A., W. J. Mitus and W. Dameshek. 1965. Chromosomal aberrations in acute leukemia. *Blood* **26:**610.

Kuo, A., X. Yataganas, J. Galicich and B. Clarkson. 1973. Proliferation kinetics of central nervous system (CNS) leukemia. *Proc. Amer. Ass. Cancer Res.* **14:**67.

Laird, A. K. 1965. Dynamics of tumour growth: Comparison of growth rates and extrapolation of growth curve to one cell. *Brit. J. Cancer* **19:**278.

Lajtha, L., R. Oliver and C. W. Gurney. 1962. Kinetic model of a bone marrow stem cell population. *Brit. J. Haematol.* **8:**442.

Lampkin, B. C., T. Nagao and A. M. Mauer. 1971. Synchronization and recruitment in acute leukemia. *J. Clin. Invest.* **50:**2204.

Moore, M. A. S. and D. Metcalf. 1973. Cytogenetic analysis of human acute and chronic myeloid leukemic cells cloned in agar culture. *Int. J. Cancer* **11:**143.

National Conference on Cancer Chemotherapy, 1972. *Cancer* **30:**1474.

Ogawa, M., J. Fried, Y. Sakai, A. Strife and B. D. Clarkson. 1970. Studies of cellular proliferation in human leukemia. VI. The proliferative activity, generation time and emergence time of neutrophilic granulocytes in chronic granulocytic leukemia. *Cancer* **25**:1031.

Ohara, K., J. Fried, M. D. Dowling, Jr., E. S. Bittar and B. D. Clarkson. 1971. Studies of cellular proliferation in human leukemia. VII. Cytokinetic behavior of neoplastic cells in a patient with reticulum cell sarcoma in a leukemic phase. *Cancer* **28**:862.

Omine, M. and S. Perry. 1972. Use of cell separation at 1g for cytokinetic studies in spontaneous AKR leukemia. *J. Nat. Cancer Inst.* **48**:697.

Pasternak, C., A. Warmsly and D. Thom. 1971. Structural alterations in the surface membrane during the cell cycle. *J. Cell Biol.* **50**:562.

Prescott, D. M. 1968. Regulation of cell production. *Cancer Res.* **28**:1815.

Rosen, P. J., S. Perry and F. M. Schabel, Jr. 1970. Proliferative capacity of leukemic cells in AKR leukemia. *J. Nat. Cancer Inst.* **45**:1169.

Rosse, C., 1970. Two morphologically and kinetically distinct populations of lymphoid cells in the bone marrow. *Nature* **27**:73.

Salmon, S. E. and B. A. Smith. 1972. Induction of tumor-susceptibility to cycle-active agents in 1gG multiple myeloma. *Clinical Res.* **20**:570.

Sandberg, A. A. 1966. The chromosomes and causation of human cancer and leukemia. *Cancer Res.* **26**:2064.

Sheehy, P. F., J. Fried, R. Winn and B. D. Clarkson. 1973. Cell cycle changes in ovarian cancer after arabinosylcytosine. *Cancer* (in press)

Simone, J., R. J. A. Auer, H. O. Hustu and D. Pinkel. 1972. "Total therapy" studies of acute lymphocytic leukemia in children. Current results and prospects for cure. *Cancer* **30**:1488.

Skipper, H. E. 1968. Cellular kinetics associated with "curability" of experimental leukemias. *Perspectives in leukemia.* Grune and Stratton, New York.

⸻. 1971. The cell cycle and chemotherapy of cancer. *The cell cycle and cancer* (ed. R. Baserga) vol. 1, p. 355–387. Marcel Dekker, New York.

Southam, C. M., J. H. Burchenal, B. Clarkson, A. Tanzi, R. Mackey and V. McComb. 1969a. Heterotransplantation of human cell lines from Burkitt's tumors and acute leukemia into newborn rats. *Cancer* **23**:281.

⸻. 1969b. Heterotransplantability of human cell lines derived from leukemia and lymphoma into immunologically tolerant rats. *Cancer* **24**:211.

Stein, G. and R. Baserga. 1972. Nuclear proteins and the cell cycle. *Advances in cancer research* (ed. G. Klein et al.) vol. 15, p. 287–300. Academic Press, New York.

Stohlman, F., Jr. 1970. Regulation of red cell production. *Formation and destruction of blood cells* (ed. T. J. Greenwalt and G. A. Jamieson) p. 65–84. J. B. Lippincott, Philadelphia and Toronto.

Tannock, I. F. 1970. Population kinetics of carcinoma cells, capillary endothelial cells, and fibroblasts in a transplanted mouse mammary tumor. *Cancer Res.* **30**:2470.

Todaro, G. J., G. K. Lazar and H. Green. 1965. The initiation of cell division in a contact-inhibited mammalian cell line. *J. Cell. Comp. Physiol.* **66**:325.

Todo, A., A. Strife, J. Fried and B. D. Clarkson. 1971. Proliferative kinetics of human hematopoietic cells during different growth phases in vitro. *Cancer Res.* **31**:1330.

Trujillo, J. M., A. Cork, J. S. Hart, S. L. George and E. J. Freireich. 1973. Clinical implications of aneuploid cytogenetic profiles in adult acute leukemia. *Cancer* (in press)

Whang-peng, J., E. J. Freireich, J. J. Oppenheim, E. Frei, III and J. H. Tjio. 1969. Cytogenetic studies in 45 patients with acute lymphocytic leukemia. *J. Nat. Cancer. Inst.* **42**:881.

Yoshida, Y. and D. Osmond. 1971. Identity and proliferation of small lymphocyte precursors in culture of lymphocyte rich fractions of guinea pig bone marrow. *Blood* **37**:73.

Zuelzer, W. W. and D. E. Cox. 1969. Genetic aspects of leukemia. *Seminars Hematol.* **6**:271.

A Comparison of Proliferative Characteristics of Bone Marrow and Tumors and Response to Cytotoxic Agents

L. F. Lamerton and N. M. Blackett

Biophysics Division, Institute of Cancer Research: Surrey
Belmont, Sutton, Surrey, England

The number and proliferative characteristics of the stem cells of normal tissues and of tumors and the changes that occur during treatment are vital factors in the design and application of chemotherapeutic procedures. However in most tumors and in a number of normal tissues, notably bone marrow, stem cells cannot be recognized morphologically and functional methods must be employed for their study.

In this paper we would like to discuss various methods of assay of stem cells, both normal and malignant, and the information that can be inferred about their proliferative state. The normal tissue we shall be considering is the bone marrow, which is an important critical normal tissue in chemotherapy. In fact we know a great deal more about the early progenitors for this tissue than we do for tumors, which can be discussed only in rather general terms.

The Bone Marrow

The proliferative properties and relationship of the various subpopulations in the bone marrow are illustrated in Fig. 1. This diagram applies to both red cell and granulocyte series and shows the mature cells, the recognizable maturing cells in two classes, nondividing and dividing, the stem cells which represent a very small proportion of the total cell population, and a population of cells between the stem cells and the first recognizable precursors, about which a great deal still must be learned. As is the case for most, if not all, normal tissues, the stem cells in the unstimulated state are dividing slowly with intermitotic time in the rat and mouse of the order of days, in comparison to the rapid rate of division of the later, recognizable, precursors of both red cell and granulocyte series, where the intermitotic time is about 10 hours. Hematopoietic tissue provides a number of different processes of compensation for cell loss or damage. In the case of the red cell system at least three processes can be identified: (1) There can be a speeding up of the rate of division of the stem cells (Becker et al. 1965). (2) A mechanism can operate at the level of the dividing maturing cells, leading to some shortening of the intermitotic time, but mainly to an increase in transit time through the maturation compartments

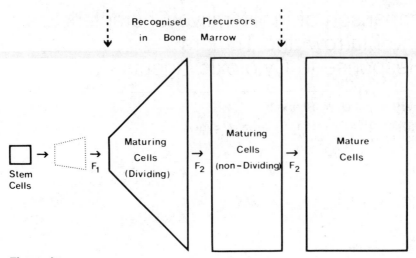

Figure 1

Diagrammatic representation of maturation pathway of hemopoietic cells.

so that more divisions are inserted and the "amplification factor" of this compartment is increased (Tarbutt 1969). (3) A further compensatory mechanism comes into operation in the red cell system when the animal is made anemic. Here the increase in blood level of erythropoietin acts not on the stem cells, but on cells just prior to the first recognizable precursors, increasing the inflow into the maturation compartments.

With the granulocyte system much less is known about the processes of compensation, and the relationship of the earlier progenitors of the red cell and granulocyte series has not yet been clarified. However it is reasonable to assume that a number of processes of compensation operate as for the red cell series.

This capacity of the bone marrow to compensate for injury at different levels has the advantage that a considerable fall in stem cell numbers can be tolerated without a serious reduction in rate of mature cell production, but it also means that a depletion of stem cell numbers to the critical limit may occur in cancer therapy procedures, with little change in rate of production of mature cells.

Assays of the Early Progenitor Cells of the Bone Marrow

Any discussion of the nature and characteristics of the stem cells of the bone marrow is bound up with the means by which they are assayed. It is generally considered that the technique of colony formation in the spleens of marrow-depleted mice, following intravenous administration of a single cell suspension, is the most appropriate for measuring bone marrow stem cells, and the CFU (colony forming unit) is generally identified with the bone marrow stem cell. There is a great deal of argument in favor of this, but there are other assay techniques which may be as appropriate, or more appropriate, for special purposes. For instance, when one is interested in the recovery capacity of the bone marrow during therapeutic procedures by radiation or other cytotoxic agents, it may be more appropriate to measure the capacity for producing functional blood cells. Such an assay will not necessarily give the same result as the spleen colony technique, since variation in ultimate family size and in rate of proliferation will affect the repopulating assay without necessarily affecting the spleen colony assay.

The two repopulating assays that have been used are erythroid repopulating ability (ERA) and granulocyte repopulating ability (GRA), the first being based on a radioactive iron method for assessing production of red cells, and the second based on an endotoxin mobilization technique for assessing production of granulocytes. For details and discussion of these techniques, see references by Blackett (1968) and Hellman et al. (1970).

Another bone marrow assay becoming of increasing importance and interest is the agar colony unit (ACU) yielding only granulocyte colonies (see Metcalf this volume).

It would be very useful if each of these assays could be used with both mice and rats. However neither CFU nor ACU assay techniques can yet be applied reliably in rats and, in our own experience, the GRA assay presents some considerable difficulties in the mouse. We have felt it important to work on both rats and mice to explore the relationship between the different assays and possible species differences.

Comparison of the Various Assays

The spleen colony assay has the best claim for providing a measure of the earliest progenitor cells of the bone marrow, since the CFU's have been shown to be capable of self-maintenance and also of differentiating along the red cell, granulocyte and megakaryocyte pathways. The extent to which the ERA, GRA and ACU assays can be said to measure stem cells, defined in terms of capacity for self-maintenance as well as providing a large number of mature cells, has not yet been established. One approach to the problem of the relationship between the cells measured by these different assays is to compare their response to a range of cytotoxic agents, both proliferation-dependent and independent.

In our hands the CFU and ERA assays in the mouse give similar results for response to most cytotoxic agents. An example is given in the following section, where the results using repeated Ara-C injections indicate no difference between CFU and ERA in normal and speeded-up bone marrow. However the two repopulating ability assays, ERA and GRA, do show some differences when studied in the rat. Figure 2 shows the ERA and GRA response to single doses of dimethylmyleran, which is a proliferation-independent agent, and Fig. 3, the response to cyclophosphamide, which has a degree of proliferation dependence. For dimethylmyleran the responses were very similar, but for cyclophosphamide they were different.

Insofar as the repopulating abilities measure the capacity of the animal to produce mature cells, this is an important result, suggesting that CFU (or ERA) assays do not necessarily give a good indication of the response of the granulocyte system to treatment, which is in general more important to the animal than that of the erythroid system.

A second point of interest from Fig. 2. is that under dimethylmyleran, the ERA, although showing much less initial depletion than with cyclophosphamide, did not recover its control value nearly as quickly. Hellman et al. (1970) showed a difference in the recovery curves of ERA and GRA following radiation. The fact that

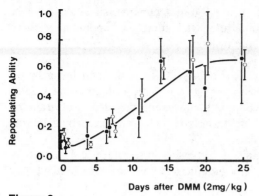

Figure 2

Recovery of (o) erythroid and (•) granulocytic repopulating ability after a single dose of dimethylmyleran (2 mg/kg body weight).

rate of stem cell recovery may depend on the particular drug used as well as on the degree of initial depletion is also a factor that must be taken into consideration in chemotherapy.

A number of other differences between ERA and GRA in the rat have also been found. Repeated doses of methotrexate have a greater effect on ERA than on GRA (Constable and Blackett 1972), and the increase in proliferation rate following depletion by dimethylmyleran is greater for the ERA (unpublished). Also the ratio of ERA to GRA is much less for bone marrow cells than spleen cells. Although the possibility cannot be ruled out that these differences are due to changes in the pathway of differentiation for the two cell types, the more likely explanation would seem to be that there are two separate populations, and the response to agents of known proliferation dependence suggests that the erythroid repopulatiing cells are proliferating the more rapidly.

The ACU assay is of particular importance because it can be used in man. However there is still much uncertainty regarding its relationship to CFU. Dicke,

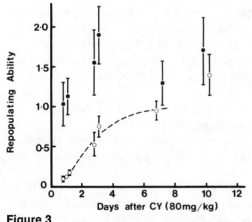

Figure 3

Recovery of (o) erythroid and (•) granulocytic repopulating ability after a single dose of cyclophosphamide (80 mg/kg body weight).

Platenburg and van Bekkum (1971), using a technique somewhat different from that of other workers, claims that their ACU technique measures CFU. Other workers have reported that the ACU have little or no ability for self-maintenance and proliferate more rapidly than CFU. In our own experiments on the effects of cytotoxic agents, we have found the same sensitivity for ACU and CFU to vinblastine in normal C57BL mice, but a large difference when the marrow is speeded up during regeneration (Millard, Blackett and Okell 1973). We have also found a difference between ACU and CFU in the effect of myleran, which is not a proliferation-dependent agent (Blackett and Millard 1973). The evidence points to ACU and CFU being different populations, but little for certain can be said about the proliferative and other characteristics of the ACU.

To avoid constant qualification in this paper, the term "stem cell" will be used in general discussion to include CFU, ERA, GRA and ACU, but the particular assay employed will be stated in the presentation of the data.

The Kinetic State of the Bone Marrow Stem Cells

One method of obtaining information on the proliferative characteristics of the bone marrow stem cells is to observe their fall in numbers, using an appropriate stem cell assay, while a phase-specific cytotoxic drug is given continuously or repeatedly at a dose level sufficient to kill cells as they enter the drug-sensitive phase of the cell cycle (Blackett 1968; Bruce et al. 1969).

Our data for mice are shown in Fig. 4. Normal animals were studied and also animals that had received whole-body radiation at 35 rad/day for 2 weeks previously in order to determine the effect on proliferation rate of a severe depletion of the stem cell compartment (Twentyman and Blackett 1970). The drug used was Ara-C, and the fall in CFU and in ERA was measured during the course of re-

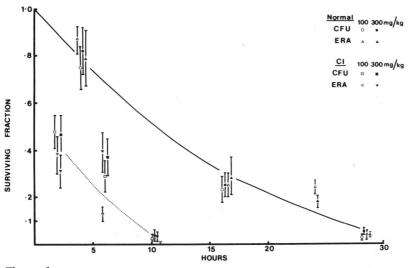

Figure 4

Effect of repeated administration of cytosine arabinoside on spleen colony and erythroid repopulating ability of C57BL mice. Administration every 4 hr at 100 mg/kg and 300 mg/kg to normal mice and continuously irradiated mice.

peated administrations (every 4 hours) at dose levels 100 and 300 mg/kg. The data show no difference between the effects at these two dose levels.

The data for rats are shown in Fig. 5. For biochemical reasons Ara-C is not very effective in rats, and it was necessary to use methotrexate, another phase-specific drug, given every 4 hours at a dose level of 30 mg/kg, which is on the plateau of the dose-response curve. Data are shown for the ERA response of normal animals, of continuously irradiated animals, and also of animals that had previously been made anemic by injections of phenylhydrazine (Blackett 1968).

These results demonstrate that in mice there is no difference in response between CFU and ERA and that in rats a severe, maintained anemia did not change the repopulating ability or the rate of cell proliferation. In both mice and rats continuous irradiation produced a marked increase in proliferation rate since the cells pass through the sensitive phase much more quickly than in normal animals.

The interpretation of these curves in terms of estimating the absolute cell proliferation rate depends on whether the proliferation rate of the stem cells is altered by the treatment. There is some evidence that the proliferation rate may be reduced due to the hold up of cells in the cell cycle. There is evidence for this with methotrexate in mice (Bruce et al. 1969) and also for Ara-C with L cells cultured in vitro (Graham and Whitmore 1970). It has also been suggested that depletion of maturing cells, which is much greater with Ara-C due to their higher rate of proliferation, may induce stem cells to proliferate more rapidly. However there is little direct evidence for this and our attempts to obtain such evidence in rats have not been successful (see later in paper). Even if such an effect can be demonstrated, it

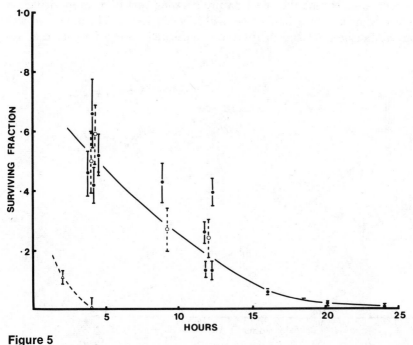

Figure 5

Effect of repeated administration of methotrexate on erythroid repopulating ability of rats. Administration every 4 hr at 30 mg/kg for normal rats (•), continuously irradiated rats (△) and phenylhydrazine-induced anemic rats (○).

will have to occur within a period of 24 hours to significantly alter the slope of the curves shown in Fig. 4 and 5. An early increase in proliferation of stem cells following severe depletion of the stem cells has been reported by Lahiri and van Putten (1972), and Tubiana (this volume) has shown that a single dose of tritiated thymidine in vivo, which causes little reduction in CFU survival, causes a marked increase in the in vitro thymidine suicide within a few hours. Whether or not similar changes occur with phase-specific cytotoxic agents as with radiation has yet to be determined. However the shape of the curves (Fig. 4 and 5) does not suggest an increase in proliferation unless this has developed within a short time of the administration of the first dose.

If Ara-C does not alter the proliferation rate of stem cells, our results indicate a dividing population in which the great majority have an intermitotic time of less than 35–45 hours for mice and 25–35 hours for rats. We have obtained thymidine suicide values of about 20% for both rats and mice which would suggest DNA synthesis times in the range 5–9 hours.

A number of workers have suggested that normal bone marrow stem cells consist of two populations, one relatively rapidly dividing, and the other in a "resting" or "arrested" state from which they can be recruited into division by the appropriate stimulus (Bruce et al. 1969; Lajtha, Pozzi and Schofield 1969; Vassort et al. 1973; Tubiana et al. this volume). This model would be consistent with the data of Fig. 4 and 5 only if the stem cell population had speeded up during the course of the experiments, and then it would be necessary for the recruitment to be effectively complete by 30 hours.

No final conclusion can yet be drawn about the kinetic state of the normal stem cell population. However it can no longer be assumed that a long intermitotic time must necessarily involve periods of rest or arrest and, by implication, a substantial spread in cell cycle time. Recent work by Hegazy and Fowler (1973) on the basal cells of mouse skin has demonstrated that an intermitotic time of as long as 5 days does not necessarily include periods of rest or arrest. In unstimulated skin they obtained a labeled mitoses curve with a very clear second peak about 100 hours after the first, indicating a long but very narrow distribution of intermitotic time and thus a slow, continuous progression around the cycle. By stimulation through hair plucking, the intermitotic time could be reduced to about 50 hours, mainly by reduction in G_1.

What Triggers Stem Cells into Faster Division?

A fundamental question relating to stem cell populations of any tissue is how far the rate of cell proliferation is controlled by direct feedback from later compartments, and how far by population size control mechanisms within the stem cell compartment itself. This is of particular interest in studies of chemotherapy, since proliferation-dependent drugs can produce very different extents of cell kill in the stem cell and later dividing compartments of the bone marrow. The data of Fig. 5. show that a maintained anemia in rats led to no change in the proliferative rate of the ERA.

With regard to possible feedback from depletion of the recognizable bone marrow precursors, some current work on rats by Mr. G. Standen in our laboratory is of interest. He has shown that repeated doses of methotrexate, at a level of 0.3 mg/kg per day, result in a substantial depletion of the recognizable red cell and granulocyte precursors in the bone marrow and a great reduction in output of red cells and granulocytes, but no appreciable effect on stem cell population size, as measured by ERA and GRA. The question was then whether there was an increase in the stem cell proliferation rate. This was tested by measuring stem cell survival after a large dose of hydroxyurea. No difference was found between methotrexate-treated and normal animals, indicating a normal rate of proliferation. However it is possible that a stimulus for increased proliferation was present, but that the repeated methotrexate treatment had affected the capacity of the stem cells to respond. To test this it was necessary to determine whether the stem cells in the methotrexate-treated rats could increase their rate of division when directly depleted by an appropriate agent. Preliminary results show that cyclophosphamide administration does, in fact, cause an increase in proliferation rate. It would appear therefore that, at least in the rat, the ERA and GRA cells are neither depleted nor stimulated to more rapid division by depletion of the maturing cells.

Thus we have found so far no evidence for feedback to the stem cell compartment from depletion of the later precursors or mature cells. On the other hand, there are many demonstrations that radiation and other cytotoxic agents that directly deplete the stem cell compartment will cause substantial increases in rate of proliferation. These results do not rule out the possibility of direct feedback on precursor populations later than the stem cells, but suggest that internal size control of proliferation rate is the predominant mechanism for stem cells in rat bone marrow.

Tumors

In considering the response of tumors to therapeutic procedures, we need to be concerned with the population size and characteristics of cells which, under the conditions existing at the time, have the capacity to repopulate the tumor. These cells may be described as tumor stem cells, repopulating cells or clonogenic cells, but none of these terms is ideal. The term "tumor stem cell" will be used here, though with the recognition that it has to be qualified in various ways.

The assay of tumor stem cells has a number of problems in common with the assay of stem cells of normal tissues, as exemplified by the bone marrow. In the same way as changes in the blood count can be a very poor reflection of changes in the population of bone marrow stem cells, so changes in the total tumor mass or volume are a poor indication of changes in the tumor stem cell population. This was first shown by Hermens and Barendsen (1969) following irradiation with 2000 rad of a rat-transplantable tumor, when the maximum reduction in volume was only about 25%, while the tumor stem cell population, as assayed by an in vitro cloning technique, fell to about 1% of its control value. This was due to the persistence of many of the damaged cells during the period of regeneration.

Second, as with bone marrow there are a number of different techniques of stem cell assay, most of them testing the capacity of a single cell suspension from the tumor to form clones in vitro or to regrow tumor in recipient animals in various sites. The first question is whether they will all give the same result for response to a cytotoxic agent. Our own experiments in this area, by my colleagues Dr. Steel, Dr. Hill and Mrs. Courtenay, have been with the mouse B16 melanoma, for which we have developed assay techniques of in vitro cloning, end-point dilution by subcutaneous and intramuscular transplantation, and lung colony production following intravenous injection of a cell suspension. Dose-response data for a single dose of cyclophosphamide are shown in Fig. 6 and it can be seen that there is a good measure of agreement between the various assays. The B16 melanoma is, however, a very rapidly growing, anaplastic tumor where only one or a few cells are required for a take with the in vitro and transplantation techniques. Before it can be assumed that all assays will give the same result, it will be necessary to work on other tumor types, and particularly on more slowly growing, differentiating tumors, where the stem cells may be only a small fraction of the total cell population.

The second question is whether the assay techniques give a valid measure of the change in regrowth capacity of the tumor during the period of damage and repopulation. The problem is that the environment of the cells in the assay procedures may not sufficiently match the cell environment in situ, where it is possible that only a proportion of the potential stem cells are in a situation where they are able to be effective and start focal repopulation.

The situation is well exemplified by the transplantable mouse tumor used by Tannock (1968) in our laboratory. In this tumor a pattern of viable tumor tissue in cords around the blood vessels, and surrounded by necrotic area, is easily recognizable. The radius of the cords is about 100μ, corresponding to the diffusion distance of oxygen. Measurements showed that the cells moved outwards from around the blood vessel to the necrotic area in about 36 hours. Thus in the tumor, growing in its unperturbed state, the effective stem cells are those in the immediate

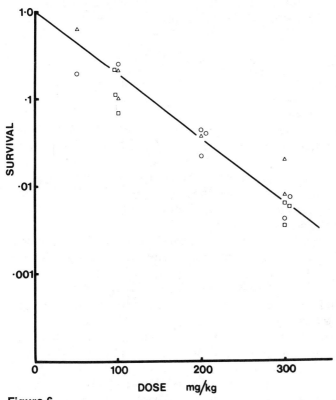

Figure 6

Dose survival curve of B16 melanoma for cyclophosphamide using end point dilution (o), lung colony (△) and in vitro cloning (□) assays.

vicinity of the blood vessels. The progeny of these cells may indeed retain stem cell capacity (though this has not been directly demonstrated) but will have no opportunity to manifest it, since they will soon be lost in the necrotic areas. However, following cell depletion after treatment with cytotoxic agents, the normal processes of cell migration and cell loss will be interfered with. At least some of the potential stem cells will now become effective, but their number will depend on the degree to which the cell loss processes have been modified and how far the capacity of a cell to repopulate the tumor depends on the environment in which it finds itself, particularly its spatial relation to patent blood vessels. These are both subjects on which there is at present little information.

The Kinetic State of Tumor Stem Cells

In tumors such as the B16 melanoma, where every cell may be a stem cell, the intermitotic time of the stem cells is that of the whole tumor and, under these conditions, a relatively short intermitotic time, generally of the order of 12–24 hours with very little spread, is found. With the more slowly growing and differentiating primary tumors of animal or man, the general finding is one of a considerable spread in intermitotic time (Steel 1972), with a substantial fraction of cells having an intermitotic time longer than can be measured by the technique of labeled mitoses. It is possible that the slowly dividing group represents cells that have slowed down because of nutritional deficiency, chromosomal abnormality, or other cause, and have lost reproductive integrity. However it is important to determine how far stem cells are included in this group, and some information about this can be obtained, as in the case of the bone marrow stem cells by using a tumor stem cell assay and studying the change in population during the course of repeated administration of an S-phase specific cytotoxic agent. The information such an experiment will give is the proportion of stem cells that have not entered the S phase within the period of drug administration, assuming that no subpopulation resistant to the particular drug used is present. Such experiments with repeated Ara-C have been done on a few animal tumors for periods of up to 1 or 2 days (Skipper pers. commun.). With the AKR spontaneous mouse leukemia about 3% of stem cells survive a 24-hour period of administration. Since the average duration of $G_1 + G_2$ in this tumor is only about 8 hours, the great majority of proliferating cells would have been killed, and it is evident that the slowly proliferating component of the population contains an appreciable number of stem cells.

These experiments will give no information on the kinetic state of the slowly dividing component of the stem cells, whether they are resting, arrested or slowly progressing around the cycle. Other approaches are needed to obtain such information, such as in the case of acute leukemia in man where there is evidence that the tumor cell can exist in two forms, one of which is a resting cell that can be brought into division (Saunders and Mauer 1969).

With regard to the possibility of change in rate of stem cell proliferation, the available data show little concordance. A lengthening of intermitotic time has been observed with age in ascites tumors but not in solid tumors (see Tubiana 1971). Following irradiation a speeding up of cell proliferation has been found by Hermens and Barendsen (1969) in a transplantable rat tumor. In certain other tumor experiments this has not been shown (Tubiana 1971).

There is a need to try to correlate such effects with tumors of different histological types. In particular we have to study the stem cells of the more slowly growing, differentiating tumors. Here some of the normal controls of cell proliferation may be retained, which could profoundly affect the nature of the proliferative response to treatment by cytotoxic agents.

REFERENCES

Becker, A. J., E. A. McCulloch, L. Siminovich and J. E. Till. 1965. The effect of differing demands for blood cell production on DNA synthesis by hemopoietic colony forming cells of mice. *Blood* **26**:296.

Blackett, N. M. 1968. Investigations of bone marrow stem cell proliferation in normal, anemic and irradiated rats using methotrexate and tritiated thymidine. *J. Nat. Cancer Inst.* **41**:909.

Blackett, N. M. and R. E. Millard. 1973. A differential effect of Myleran on two normal hemopoietic progenitor cell populations. *Nature* **244**:300.

Bruce, W. R., B. E. Meeker, W. E. Powers and F. A. Valeriote. 1969. Comparison of the dose and time survival curves for normal hemopoietic and lymphoma colony-forming cells exposed to vinblastine, vincristine, cytosine arabinoside and amethopterin. *J. Nat. Cancer Inst.* **42**:1015.

Constable, T. B. and N. M. Blackett. 1972. A comparison of the effects of four cytotoxic agents on granulocytic and erythroid repopulating ability of rat bone marrow. *J. Nat. Cancer Inst.* **48**:941.

Dicke, K. A., M. G. C. Platenburg and D. W. van Bekkum. 1971. Colony formation in agar: *in vitro* assay for hemopoietic stem cells. *Cell and Tissue Kinetics* **4**:463.

Graham, F. L. and G. F. Whitmore. 1970. The effect of 1-β-D-arabinofuronasylcytosine on growth and viability and DNA synthesis of mouse L-cells. *Cancer Res.* **30**:2637.

Hegazy, M. A. H. and J. F. Fowler. 1973. Cell population kinetics of plucked and unplucked mouse skin. I. Unirradiated skin. *Cell and Tissue Kinetics* **6**:17.

Hellman, S., H. E. Grate, J. T. Chaffney and R. Carmel. 1970. Hemopoietic stem cell compartment: Patterns of differentiation following radiation or cyclophosphamide. *Hemopoietic cellular proliferation* (ed. F. Stohlman). Grune & Stratton, New York.

Hermens, A. F. and G. W. Barendsen. 1969. Changes of cell proliferation characteristics in a rat rhabdomyosarcoma before and after X-irradiation. *Eur. J. Cancer* **5**:173.

Lahiri, S. K. and L. M. van Putten. 1972. Location of the G_0 phase in the cell cycle of the mouse haemapoietic spleen colony forming cells. *Cell and Tissue Kinetics* **5**:365.

Lajtha, L. G., L. V. Pozzi and R. Schofield. 1969. Kinetic properties of hemopoietic stem cells. *Cell and Tissue Kinetics* **2**:39.

Millard, R. E., N. M. Blackett and S. F. Okell. 1973. A comparison of the effect of cytotoxic agents on agar colony forming cells, spleen colony forming cells and erythroid repopulating ability of mouse bone marrow. *J. Cell. Physiol.* (in press)

Saunders, E. F. and A. M. Mauer. 1969. Re-entry of non-dividing leukaemic cells into a proliferative phase in acute childhood leukaemia. *J. Clin. Invest.* **48**:1299.

Steel, G. G. 1972. The cell cycle in tumours. An examination of data gained by the technique of labelled mitoses. *Cell and Tissue Kinetics* **5**:87.

Tannock, I. F. 1968. The relationship between cell proliferation and the vascular system in a transplanted mouse mammary tumour. *Brit. J. Cancer* **22**:258.

Tarbutt, R. G. 1969. Cell population kinetics of the erythroid system in the rat. The response to protracted anaemia and to continuous γ-irradiation. *Brit. J. Haematol.* **16**:9.

Tubiana, M. 1971. The kinetics of tumour cell proliferation and radiotherapy. *Brit. J. Radiol.* **44**:325.

Twentyman, P. R. and N. M. Blackett. 1970. Red cell production in the continuously irradiated mouse. *Brit. J. Radiol.* **43:**898.

Vassort, F., M. Winterholer, E. Frindel and M. Tubiana. 1973. Kinetic parameters of bone marrow stem cells using *in vivo* suicide by tritiated thymidine or by hydroxyurea. *Blood* **41:**789.

Modification of Cell Cycle Response of Synchronous Mammalian Cells to Ionizing Radiation by Inhibition of Repair

Warren K. Sinclair

Division of Biological and Medical Research Argonne National Laboratory
Argonne, Illinois 60439

Responses to ionizing radiation in mammalian cells vary as a function of the position of the cell during its generation cycle at the time of irradiation (Sinclair 1968a). In addition to lethal damage these responses include at least the following: slowed progression in the cell cycle resulting from a reduction in the rate of DNA synthesis in S cells and consequent slow emergence from S (S cell retention); the G_2 block, which together with S cell retention results in division delay; and cytological and other responses that have important relationships with respect to the subsequent growth of irradiated cells (Sinclair 1967a). Experiments on the association of X-ray induced blocks with biochemical blocks induced in G_2 by inhibitors of protein and RNA synthesis (Tobey, Anderson and Peterson 1966; Doida and Okada 1969; Bacchetti and Sinclair 1970) also bear on the question of radiation interference with the controls for cell division. However I intend to ignore these features in this discussion and concentrate on experiments with synchronous mammalian cells that are concerned with those factors that appear to control the lethal radiation response and its fluctuations during the cell generation cycle. For the sake of clarity I also intend to describe experiments conducted primarily in one mammalian cell system and in doing so I shall inevitably not be able to refer to much original work conducted in other laboratories and upon other cell lines. General references, such as Sinclair 1968a and 1970, should be consulted for further reference to such work.

METHODS

The cells used were sublines of the V79 line of Chinese hamster lung cells which were grown at $37°C$ under defined culture conditions. The cells were synchronized by mitotic selection using a shaking technique which yields a high proportion of cells in mitosis (Sinclair and Morton 1965). The S period was identified by pulse-labeling cultures at intervals with tritiated thymidine [³H]dT. In some cases additional techniques, e.g., hydroxyurea added in G_1 (Sinclair 1965) or [³H]dT killing (Sinclair and Morton 1966), were used to improve the quality of synchrony. Cultures of synchronous cells were irradiated at different stages of the cell cycle with 250 kVp X rays (half value layer 0.9 mm Cu, exposure rate 160 R/min, 0.945 rad/R absorbed dose) at room temperature and subsequently incubated at 37°C for 8 days before scoring visible colonies as a measure of the survival of single cells. Various agents, including cysteamine (MEA), N-ethylmaleimide (NEM), and hydroxyurea (HU), were used to treat the cells, usually when cells were still single or in the small 2-cell microcolonies normally obtained by this method of synchrony (Sinclair and Morton 1965) and at pertinent stages of the cell cycle. Agents were either added to the medium (MEA and HU) or dissolved in phosphate-buffered saline (PBS), as in the case of NEM, and added to the culture. Further details are provided elsewhere (Sinclair 1973).

Assays of cellular sulfhydryl, total protein, and NEM uptake were undertaken by J. Archer (Archer, Long and Sinclair in prep.). Investigations of the effects of NEM on DNA synthesis involved assays of the uptake of [^{3}H]dT and other precursors by radioautographic and liquid scintillation counting (Sinclair, Archer and Blakely in prep.).

RESULTS

The main features of the lethal response to X radiation at different stages of the cell cycle are well known and have been described in detail (e.g. Sinclair 1970, 1972). In cells with a long G_1, e.g. HeLa cells, mitosis and G_2 are the most sensitive periods, but the G_1–S border may be almost as sensitive. In between are two peaks of resistance, one early in G_1 and one late in S. In cells with a short G_1 the first peak is not usually evident, but G_1 is slightly more resistant than mitosis or G_2 and the features in S are the same, survival rising rapidly in S to a maximum in the latter part of S. The maximum differences between the most resistant and the most sensitive cells is about a factor of 3, but both shoulder (n) and slope (D_o) of the survival curve vary (mitotic cells, $n \sim 1$, $D_o \sim 125$ rad; late S cells, $n \sim 8$, $D_o \sim 200$ rad) (Sinclair and Morton 1966). Careful examination of the onset of DNA synthesis (Sinclair 1970) shows that survival rises concomitantly with DNA synthesis, indicating that either DNA synthesis itself or some factor directly related to it is one factor controlling survival (i.e., the lethal response). However maximum survival does not occur at the maximum rate of DNA synthesis (middle of S) or when all of the DNA is synthesized (end of S) but in between, about two-thirds of the way through S, in Chinese hamster cells. In addition when cells are prevented from synthesizing DNA (by adding hydroxyurea to synchronized cells in G_1; Sinclair 1967b), survival first falls sharply and then rises again late in the "cell cycle" (i.e., in the distorted cell cycle resulting from hydroxyurea inhibition in which protein and RNA synthesis continue; Sinclair 1967c) to about the former G_1 value (Fig. 1). These facts suggest that another factor (named Q), which varies as a function of cell cycle position, also controls cell survival.

Thus a model of the cell cycle response *could* include two components as described by Sinclair (1972). It should be noted, however, that experiments performed in Chinese hamster cells relate to that part of the cell cycle starting during the short G_1, so that specific evidence for the factor Q being responsible for changes during G_1 is lacking, except for the suggestive information on nonprotein sulfhydryl (NPSH) variation during the cell cycle of HeLa cells provided by Ohara and Terasima (1969).

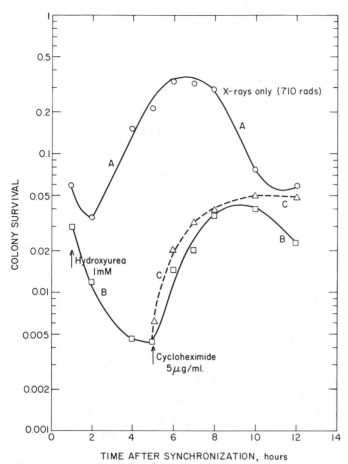

Figure 1

Survival of synchronous Chinese hamster cells **(A)** after 710 rad X irradiation only, **(B)** when hydroxyurea was added in G_1 and 710 rad given at times indicated, **(C)** when hydroxyurea was added to G_1 cells (at 1 hr) and cycloheximide was added at 5 hr (Sinclair 1967b).

My investigations of the identity of Q now *suggest* that it may be some small fraction of the intracellular sulfhydryl. Evidence in support of this was provided by the facts that (1) the aminothiol cysteamine protects cells from lethal damage caused by X irradiation differentially during the cell cycle, the most sensitive cells being protected the most (Sinclair 1968b, 1969); (2) the sulfhydryl binding agent N-ethylmaleimide sensitizes cells differentially during the cell cycle (Sinclair 1972, 1973); and (3) N-ethylmaleimide suppresses the rise in survival late in the "cell cycle" of hydroxyurea-inhibited cells (Sinclair 1972). Put together, these features can be shown, somewhat schematically, in Fig. 2.

The representation in Fig. 2 cannot be considered exact quantitatively. The individual experiments A, B, C, and D were conducted over a period of years under somewhat different synchrony conditions and with sublines differing *slightly* in various properties, including generation times and the parameters of the lethal X-ray response. Curve A, for example, was obtained by carefully derived single cell values

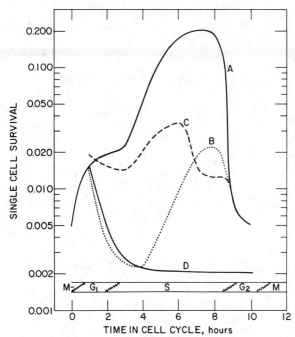

Figure 2

Representation of cell cycle lethal response to X radiation of synchronized Chinese hamster cells. **(A)** X rays only (710 rad); **(B)** 710 rad X rays given to cells inhibited with 1 mM HU in G_1; **(C)** 710 rad X rays given in presence of 0.75 μM NEM for ½ hr; **(D)** 710 rad X rays given to HU-inhibited cells in presence of 0.75 μM NEM for ½ hr. (*Note:* This representation should not be considered exact quantitatively; see text.)

from double synchrony experiments, using both mitotic selection and [³H]dT killing designed for the highest precision (Sinclair and Morton 1966). The other experiments used mitotic selection only, and thus were less well synchronized; and cellular multiplicities in the microcolonies were not specifically determined. An assumed cellular multiplicity of 2 (as in untreated cells) was used in obtaining the "single cell" representation of Fig. 2, but especially in cells treated with hydroxyurea, the cellular multiplicity is probably lower than 2, in which case the lowest survival values on curves B and D should probably be elevated. In fact the lowest values of survival on curves B and D may not differ much from the lowest values found for mitotic and G_2 cells. This view is strengthened by the fact that neither hydroxyurea (Sinclair 1968c) nor N-ethylmaleimide (Sinclair 1973) has any effect on irradiated mitotic cells.

Thus we might generalize, to a first approximation at least, to say that (a) cells completely inhibited from making DNA have a minimum value of survival about equal to, but possibly less than, that of mitotic cells (however at both ends of the cell cycle cells survive about as well as untreated G_1 cells); (b) cells with their intracellular sulfhydryl bound by N-ethylmaleimide survive least well near the end of S, but have about the G_1 value of survival and are not reduced to the minimum value represented by mitotic cells; (c) cells treated with both hydroxyurea and

N-ethylmaleimide have the minimum survival (i.e., of mitotic cells) throughout the whole of the latter part of the "cell cycle."

This generalization is consistent with a model in which a major variant during the cell cycle is the capacity of the cell to repair damage. (Primary radiation damage may also vary during the cell cycle, but I will not attempt to discuss the evidence for this here.) Thus mitotic cells, for example, may be sensitive because they lack the capacity to repair radiation damage. Furthermore the capacity to repair damage appears to be dependent both on DNA synthesis and on the presence of some substances containing sulfhydryl. Thus hydroxyurea and *N*-ethylmaleimide together can strip the cell of all capacity to repair radiation damage, but neither one alone can do it completely. This generalization may only be true to a first approximation, but it may nevertheless account for a large portion of the variation in response expressed as lethal damage during the cell cycle.

Pre- and Post-Irradiation Effect of NEM

Sulfhydryl protective agents such as cysteamine normally must be present during irradiation to exert a protective effect. However SH-binding agents could be effective if added before irradiation, if a stable complex is formed by the agent that prevents intracellular sulfhydryl from functioning in a self-protective way. In bacteria these binding agents are most effective during irradiation (Bridges 1969) but pretreatment effects have also been observed (Lynch and Howard-Flanders 1962). In general post-irradiation effects have not been observed.

In Chinese hamster cells, however, both pre- and post-irradiation effects have been observed (Sinclair 1973). Figure 3 shows that NEM given just before exposure (open square at 8½ hr) and removed is just as effective as NEM during exposure. So also is NEM given immediately after exposure (full square at 8½ hr). With increasing time before irradiation (open squares, X rays at 8½ hr) the pretreatment effect is slowly reduced. With increasing time between irradiation and NEM exposure (full squares) survival increases and in about 2 hours the cells are no longer sensitive to the addition of NEM.

These results suggest that NEM forms a stable complex that renders a repair mechanism in the cell inoperative. Further studies on the post-irradiation response of NEM show that it is temperature dependent, recovery occurring more rapidly at 37°C than at 24°C (Fig. 4). Results at 5°C (not shown) show recovery initially like that at 24°C and then essentially stopping. This observation suggests dependence on metabolism, which is presently being studied in other experiments. An important point to note, already reported (Sinclair 1973) but also evident in Fig. 4, is that recovery is much slower in G_1 cells than in late S cells; i.e., the radiation lesion induced in G_1 remains sensitive to the addition of NEM much longer (6–8 hours) than in late S cells ($\sim$ 2 hours).

NEM and SH Binding

That NEM interferes with repair and thus lowers survival seems clear, but the mechanism by which this interference is accomplished requires investigation. The foregoing would seem to implicate intracellular sulfhydryl, to which NEM binds; however experiments by Harris (1972) using diamide, a specific nonprotein SH-binding agent, show that this agent does not sensitize oxygenated Chinese

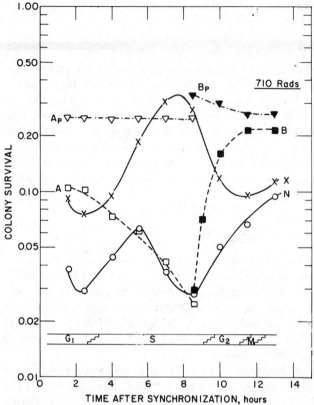

Figure 3

Effect of timing of a 30-minute NEM exposure relative to X irradiation. The survival of synchronous Chinese hamster cells after 710 rad of X radiation alone (curve X) or during a 30-minute exposure to 0.75 μM NEM (curve N). In the curves labeled A and B cells were exposed to 0.75 μM NEM for 30 minutes at the times indicated, fresh medium put on and all plates irradiated at 8.5 hr. Curves Ap and Bp were for cells treated with PBS at the times indicated and irradiated a 8.5 hr. Data from two very similar experiments were averaged. From Sinclair (1973) with permission of *Radiation Research*.

hamster cells, an observation that I have also made in my laboratory (unpublished experiments).

Measurements on the sulfhydryl content by various methods and on NEM binding and protein content at different times in the cell cycle of synchronous Chinese hamster cells have been made in my laboratory (Archer et al. 1973). The results show (Fig. 5) that nonprotein SH (NPSH) is greatest in G_1 (curve C), whereas total SH is greatest in S (curve D). Furthermore NEM uptake or binding is greatest in late S. These assays of gross cellular components are, of course, by no means conclusive, but they do suggest that SH binding is involved with the lethal response to X radiation. Studies on the increase of SH binding with time, now in progress, and its relation to that of increasing sensitization with time (Sinclair 1973) may cast further light on this point. It is evident from the gross sulfhydryl assays represented in Fig. 5, however, that whatever small critical fraction of the total cellular SH is

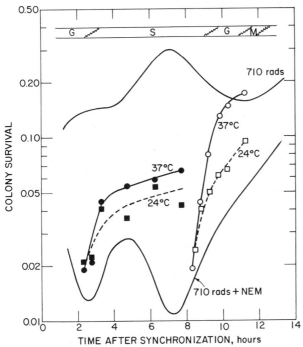

Figure 4

The post-irradiation effect of *N*-ethylmaleimide on G_1 (2½ hr) cells given 710 rad and upon late S (8¼ hr) cells given 710 rad. Cells were maintained at 37°C or 24°C respectively after irradiation.

involved in this response, it is more likely to be in the protein SH fraction rather than in the nonprotein SH. This interpretation would also avoid any conflict with the results of Harris (1972) noted above, which indicate that binding *all* of the cellular NPSH with diamide does not sensitize the cell. However, as noted earlier in Fig. 1 and described by Sinclair (1967c), the arrest of protein synthesis in hydroxyurea-inhibited cells does *not* prevent the rise in survival late in the "cell cycle"; further protein synthesis is therefore not necessary for this fraction of the intracellular SH to increase. There presumably are various other ways in which this fraction can increase in the absence of protein synthesis.

It must also be mentioned that NEM inhibits DNA synthesis (Sinclair, Archer and Blakely in prep.) and recent studies by J. Archer (1973) have shown that NEM binds to DNA, possibly via NH_2 groups, although to a much less extent than to SH (6×10^{-11} μmoles of NEM bound to DNA compared with 10×10^{-10} μmoles NEM bound to SH per cell). Clearly these actions of the agent must be investigated further before the mechanism responsible for control of radiation response is understood.

Other Agents

Another approach to the problem of elucidating the action of NEM is to examine the response of cells to X irradiation in the presence of other agents with similar chemical composition or similar properties to those of NEM.

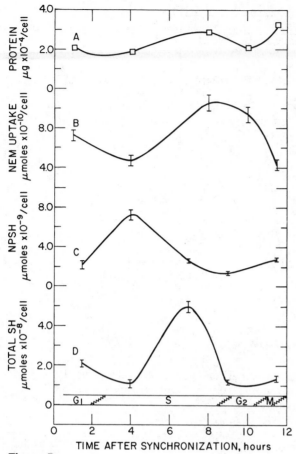

Figure 5

Variation in protein content, NEM uptake, nonprotein sulfhydryl, and total sulfhydryl during the cell cycle of synchronous Chinese hamster cells (Archer, Long and Sinclair 1973).

Figure 6 shows the formulae of four agents of similar structure. N-ethylmaleimide sensitizes, as has been described, but maleimide apparently hydrolyzes too rapidly to bind to SH; nevertheless this agent may be worthy of further study.

I have examined succinimide (Eastman Kodak, Rochester, N.Y.) and N-ethyl-succinimide (NES) prepared for me by General Biochemicals, Chagrin Falls, Ohio. Succinimide sensitized slightly only at very high concentration ($>$ 750 mM). Initial samples of N-ethylsuccinimide (NES) appeared to sensitize at about 2.5 mM but the presence of NEM was detected by UV absorption. Purified samples of NES sensitized only at very high concentrations.* The sensitization apparently caused by these agents is very probably due to an impurity, probably NEM, and neither succinimide or N-ethylsuccinimide sensitizes Chinese hamster cells to X irradiation. These results imply that sensitization is associated with the C=C bond, very probably the result of SH binding.

A reversible SH-binding agent p-chloromercuribenzene sulfonic acid (PCMBS)

---

* I am indebted to N. A. Frigerio for purified samples of NES.

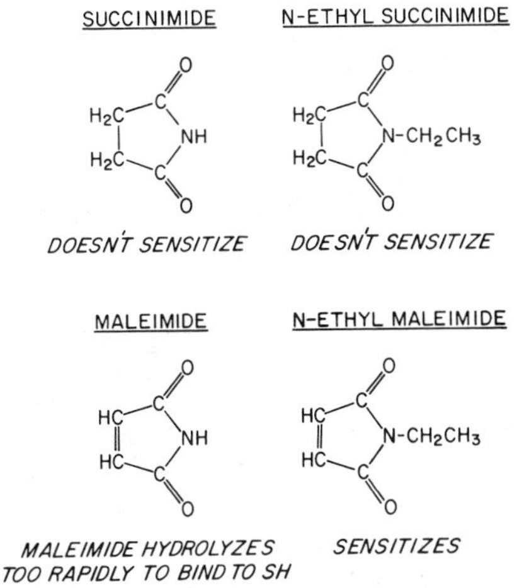

SUCCINIMIDE   N-ETHYL SUCCINIMIDE

DOESN'T SENSITIZE   *DOESN'T SENSITIZE*

MALEIMIDE   N-ETHYL MALEIMIDE

MALEIMIDE HYDROLYZES   *SENSITIZES*
TOO RAPIDLY TO BIND TO SH

Figure 6
Formulae of compounds related to N-ethylmaleimide. Their effect upon the lethal X ray response is noted.

(Sigma Chemical Co., St. Louis, Mo.) is also being explored. Since the binding of this agent is reversible, the effects of residues of medium components are critical and careful rinsing is needed. Preliminary results indicate that there are differences in the effects observed when the agent is used before, during, or after irradiation, but all are effective and the effect is greatest about the end of the S period, as for NEM. Again this points to SH binding being responsible for the observed sensitization. In the case of PCMBS, however, further studies are needed.

DISCUSSION AND CONCLUSION

The two part model for the control of the radiation response in Chinese hamster cells appears to be broadly valid at least to a first approximation. DNA synthesis is one important controlling factor and N-ethylmaleimide interferes with the other, probably, but not certainly, by SH binding. NEM appears to inhibit post-irradiation repair as demonstrated by its profound post-irradiation effects. Experiments need to be conducted in other cell lines with a long G_1 to determine whether the second component Q, affecting survival late in S, is the same factor responsible for variations during G_1. Further studies on the kinetics of post-irradiation repair are likely to be revealing concerning the nature of the radiation lesion induced at different points in the cell cycle.

Acknowledgments

I am indebted to my colleagues Grace Racster and Mel Long for able technical assistance and to Dr. Joy Archer for helpful discussion, comments and the inclusion of some of her data.

This work was supported by the U.S. Atomic Energy Commission.

REFERENCES

Archer, J. 1973. Binding of *N*-ethylmaleimide to DNA. Annual report of the division of biological and medical research, *Argonne Nat. Lab. Report* **7970:**133.

Bacchetti, S. and W. K. Sinclair. 1970. The relation of protein synthesis to radiation induced division delay in Chinese hamster cells. *Rad. Res.* **44:**788.

Bridges, B. A. 1969. Sensitization of organisms to radiation sulfhydryl binding agents. *Advances in radiation biology,* vol. 4 (ed. L. G. Augenstein, R. Mason and M. R. Zelle). Academic Press, New York.

Doida, Y. and S. Okada. 1969. Radiation induced mitotic delay in cultured mammalian cells (L 5178 Y). *Rad. Res.* **38:**513.

Harris, J. W. 1972. On the mechanisms of anoxic resistance. *Rad. Res.* **51:**524 (abstract).

Lynch, J. P. and P. Howard-Flanders. 1962. Effects of pre-treatment with nitric oxide and *N*-ethylmaleimide on the level of sulfhydryl compounds in bacteria and their sensitivity to X irradiation under anoxia. *Nature* **194:**1247.

Ohara, H. and T. Terasima. 1969. Variations of cellular sulfhydryl content during the cell cycle of HeLa cells and its correlation to cyclic X ray sensitivity. *Exp. Cell Res.* **58:**182.

Sinclair, W. K. 1965. Hydroxyurea: Differential lethal effects on cultured mammalian cells during the cell cycle. *Science* **150:**1929.

———. 1967a. Radiation effects on mammalian cell populations *in vitro. Radiation research,* Proc. 3rd Int. Cong. of Rad. Res. (ed. G. Silini) p. 607. North-Holland, Amsterdam.

———. 1967b. X ray survival and DNA synthesis in Chinese hamster cells. I. The effect of inhibitors added before irradiation. *Proc. Nat. Acad. Sci.* **58:**115.

———. 1967c. Hydroxyurea: Effects in Chinese hamster cells grown in culture. *Cancer Res.* **27:** (Part 1) 297.

———. 1968a. Cyclic X-ray responses in mammalian cells *in vitro. Rad. Res.* **33:**620.

———. 1968b. Cysteamine: Differential X ray protective effect on Chinese hamster cells during the cell cycle. *Science* **159:**442.

———. 1968c. The combined effect of hydroxyurea and X rays on Chinese hamster cells *in vitro. Cancer Res.* **28:**198.

———. 1969. Protection by cysteamine against lethal X ray damage during the cell cycle of Chinese hamster cells. *Rad. Res.* **39:**135.

———. 1970. Dependence of radiosensitivity on cell age. *Time and dose relationships in radiation biology as applied to radiotherapy,* NCI-AEC Conf. Carmel 1969, Proc. Brookhaven Nat. Lab. report BNL-50203 (L-57) 97-116.

———. 1972. Cell cycle dependence of the lethal radiation response in mammalian cells. *Current Topics Rad. Res. Quart.* **7:**264.

———. 1973. *N*-ethylmaleimide and the cyclic response to X rays of synchronous Chinese hamster cells. *Rad. Res.* **55:**41.

Sinclair, W. K. and R. A. Morton. 1965. X Ray and UV sensitivity of synchronized Chinese hamster cells at various stages of the cell cycle. *Biophys. J.* **5:**1.

———. 1966. X ray sensitivity during the cell generation cycle of cultured Chinese hamster cells. *Rad. Res.* **29:**450.

Tobey, R. A., E. C. Anderson and D. F. Petersen. 1966. RNA stability and protein synthesis in relation to the division of mammalian cells. *Proc. Nat. Acad. Sci.* **56:**1520.

Biochemistry and Cell Kinetics as Aids in Selecting Combinations of Agents for Cancer Therapy

Glynn P. Wheeler

Biochemistry Department, Southern Research Institute
Birmingham, Alabama 35205

Possible bases for selecting combinations of agents for cancer therapy include the following (Sartorelli 1971):

a. differences in manifestation of clinical toxicity;
b. lack of cross-resistance;
c. differences in pharmacodynamics—period of persistence in blood, tissue distribution, rates of catabolism, etc.;
d. use of a secondary agent to effect activation or prevent deactivation of a primary agent;
e. use of a secondary agent to alter permeability of membranes to primary agents;
f. use of a secondary agent to preferentially overcome toxicity to host tissues;
g. sequential, concurrent, and complementary metabolic blocks;
h. differences in sensitivity and effects in relation to cell and tissue kinetics.

It is likely that several of these factors are operative in any combination of agents, but conscious consideration of each factor should aid in the selection of useful combinations. In the present deliberations emphasis is placed upon the last two factors and their relationships to the cytotoxicities of combinations of agents.

Figure 1 is a greatly simplified chart of the metabolic pathways involved in the synthesis of RNA and DNA, showing the positions along the pathways at which various types of agents interfere and listing exemplary agents[1] (Barranco et al. 1973; Creasey, Bensch and Malawista 1971; Wheeler 1967, 1973; Wheeler and Simpson-Herren 1973; Wheeler et al. 1972). Several examples of combinations of agents that would cause sequential blocks along the pathways can be recognized from this chart, for example, hydroxyurea, which interferes with the reduction of ribonucleotides, plus Ara-C, which interferes with the polymerization of deoxyribonucleotides. A combination of 6-thioguanine, which interferes with the synthesis of

[1] The following abbreviations are used in this report: 6-TG, 6-thioguanine; Meth, methotrexate; 5-FUra, 5-fluorouracil; HU, hydroxyurea; 5-FdU, 5-fluorodeoxyuridine; Ara-C, 1-β-D-arabino-furanosylcytosine; HN2, nitrogen mustard [bis(2-chloroethyl)methylamine]; CCNU, 1-(2-chloroethyl)-3-cyclohexyl-1-nitrosourea.

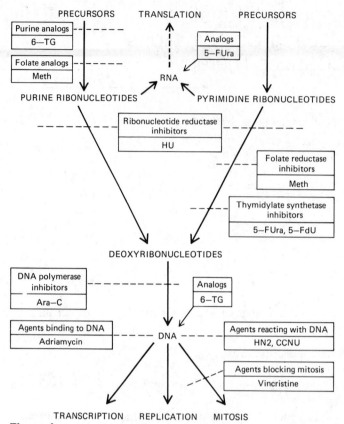

Figure 1

Loci at which selected agents interfere with the biosynthesis and functioning of DNA and RNA.

purine ribonucleotides, plus fluorodeoxyuridine, which inhibits the synthesis of dTMP, would be expected to cause concurrent blocks. A combination of adriamycin, which binds to DNA and interferes with the functioning of DNA, with 5-fluorouracil, which (along with other mechanisms) may be incorporated into RNA and thus perhaps interfere with translation, is an example of complementary blocks. Some agents, such as 6-thioguanine, 5-fluorouracil, and methotrexate, might inhibit at multiple sites along the pathways and might as single agents cause sequential, concurrent, or complementary blocks. Consideration of this chart could lead to many other combinations of two or more of these agents that might cause additive or synergistic cytotoxicities. Needless to say, there are other well-known agents that could be added to almost all of the types of inhibitors that are shown. Therefore many combinations based upon consideration of biochemical knowledge could be suggested.

Figure 2 shows positions in the cell cycle at which cells are most sensitive to the agents listed in Fig. 1 and also positions at which some of the agents block progression of cells through the cycle (Madoc-Jones and Mauro 1973; Wheeler and Simpson-Herren 1973; Wheeler et al. 1970a,b). Using this information as a basis of selecting combinations of agents for cytotoxicity or chemotherapeutic tests, one

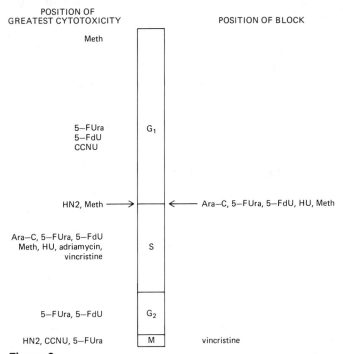

POSITION OF
GREATEST CYTOTOXICITY   POSITION OF BLOCK

Figure 2

Positions in the cell cycle at which selected agents are most cytotoxic and positions where the agents block progression of the cells.

might choose combinations such as: (a) a phase-specific agent and a nonphase-specific agent, (b) two agents having maximum cytotoxic effects when the cells are in different phases, and (c) combinations of agents that have maximum cytotoxic effects when the cells are in the same phase but that cause these effects by different biochemical mechanisms. Although it is not indicated in Fig. 2, it is known that some of the listed agents are cytotoxic to both proliferating and nonproliferating cells, whereas others are cytotoxic only to proliferating cells. Therefore to kill a population of cells consisting of a mixture of proliferating and nonproliferating cells, such as the populations of most solid tumors, one might try a combination of a noncycle-specific agent and a cycle-specific agent (such as CCNU plus Ara-C), or perhaps two noncycle-specific agents that function by different biochemical mechanisms or have different physicochemical properties (such as CCNU plus HN2).

Figure 2 shows that several agents that are highly or maximally toxic to cells in the S-phase also cause a block in the progression of cells at the G_1/S boundary, and therefore these agents might be self-limiting with regard to cytotoxicity, as the cells might not progress to the sensitive phase of the cycle. These agents include Ara-C, methotrexate, hydroxyurea, and possibly 5-FdU. Although vincristine would not prevent cells from progressing from G_1 into S, where they would be most sensitive to this agent, it would prevent cells in G_2 and M from progressing to the most sensitive site. However if cells are retained in a blocked position for a sufficiently long time by an agent, they might die even though they are not at the point of maximum sensitivity. It is logical to use a combination of Ara-C, hydroxyurea, methotrexate, or 5-FdU with HN2 or CCNU, since the former agents would kill cells in S and accumulate other cells at G_1/S, where the cells are highly sensitive to HN2 and CCNU. Similarly, since vincristine would kill cells in S and accumulate the others at M, where the accumulated cells would be highly sensitive to HN2 or CCNU, then vincristine plus HN2 or CCNU would be logical combinations.

Table 1 lists some of the possibly interesting combinations of the agents listed in Fig. 1 and 2 and gives the logic for combining the indicated agents. Many other combinations could be considered, but these particular ones were selected for experiments to determine the cytotoxicities for cultured cells.

Testing Combinations for Cytotoxicity to Cultured Cells

A number of choices must be made in designing an experimental system to determine the cytotoxicities of combinations of agents, particularly if one wishes to obtain results that might be pertinent to the use of the agents for therapy. For some of the choices the selection must be rather arbitrary because of our limited knowledge and the limited techniques available at this time.

Table 1
Some Possibly Interesting Combinations

Combination	Logic
Ara-C, HU, Meth, or 5-FdU + HN2 or CCNU	Agent active in S phase and causing block at G_1/S plus nonphase-specific agent having high cytotoxicity at G_1/S
Ara-C + 6-TG	S phase-specific agent plus agent active in G_1; sequential biochemical blocks
Ara-C, HU, Meth, or 5-FdU + 5-FUra	Agent active in S phase plus nonphase-specific antimetabolite
Ara-C + HU, Meth, or 5-FdU	Two S phase active agents acting by different biochemical mechanisms; sequential and concurrent biochemical blocks
Ara-C + adriamycin or vincristine	S phase-specific agent plus an agent causing high cytotoxicity in S; concurrent and sequential biochemical blocks
HN2 + CCNU	Two nonspecific agents generating different alkylating moieties in addition to a carbamoylating moiety from CCNU
HN2 or CCNU + vincristine	Accumulation of cells at M where they are quite sensitive to HN2 and CCNU; also take advantage of high cytotoxicities in different parts of the cycle
HN2 or CCNU + 5-FUra	Nonspecific alkylating agent plus nonphase-specific antimetabolite; sequential and perhaps complementary biochemical blocks
HN2 or CCNU + 6-TG	Sequential and perhaps complementary biochemical blocks
HN2 or CCNU + adriamycin	Agent reacting with DNA plus agent complexing with DNA

Determination of Cell Survival

Although dye uptake or dye exclusion techniques and changes in cell numbers can be used to measure cell survival, clone formation might be a more satisfactory method. The specificity of the dye techniques is sometimes questioned, and the possible effects of the test agents upon the permeability of the cells for the dye are unknown. The use of cell numbers can be complicated by the lysis or persistence of dead cells and by the clumping of cells. Although the cloning technique has its limitations, it is probably preferable to the other two techniques. Bioassay techniques are quite useful but a suitable bioassay system is not available for the HEp No. 2 cells used in the experiments described below.

Exposure of Cells to the Agent

The cells might be exposed to agents for a specified period of time and then washed prior to plating for cloning, or they might be plated into medium containing the agents. In either case the concentration of the agent during the course of the period of exposure would be dependent upon the chemical and biochemical stability of the agent, and it is possible that one agent might alter the rate of anabolism or catabolism of the other and hence make the predictability of the persistence of that agent uncertain. Since exposing the cells for specific periods of time and then washing them prior to plating defines the maximum period of extracellular exposure to the agent, this method seems preferable to plating the cells in medium containing the agents.

The period of exposure of the cells to the agents is important. If the agent is non-phase-specific and kills the cells by first-order kinetics, one can determine the cell kill after two or three intervals of time and derive an equation that may be used to predict the extent of cell kill after selected times (long or short) of exposure. For phase-specific agents the rate of cell kill is not simply exponential, and the extent of kill will depend upon how many cells progress into the sensitive stage of the cycle during the period of exposure. If the phase-specific agent is also self-limiting by blocking the progression of the cells into the sensitive phase of the cycle, then selection of the proper period of exposure becomes more difficult. If the agent prevents the synthesis of DNA but permits unbalanced growth to occur, then the extent (measured in quantity and/or time) of unbalanced growth that the cells can endure and yet survive is an important factor in selecting a period of exposure. Although different periods of exposure might be logically selected for various agents or combinations of agents, an exposure period equal to one cell-cycle time seems to be a reasonable one for the initial evaluation of combinations of a variety of agents.

One could also present arguments for exposing the cells to some of the agents sequentially rather than simultaneously, and, if sequential exposures were chosen, then it would be necessary to decide upon the order of the agents in the sequence, the interval between the exposures, and the length of time of each exposure. The decisions would be dependent upon the particular agents under consideration. Even though the diversity of the agents is recognized, it is desirable to have a simple, uniform system for initially screening combinations, and for this purpose it is reasonable to expose the cells to the agents simultaneously. For secondary screening of selected combinations, appropriate sequences of exposures might be tested.

Proliferation Status of the Cells

Since nonproliferating cells are not affected by cycle-specific and phase-specific agents, and proliferating cells are affected by nonspecific, cycle-specific, and phase-specific agents, exponentially growing cultures with a growth fraction of 1 should be used for the cytotoxicity tests.

Desired Levels of Cytotoxicity for Individual Agents and for Combinations

In order to characterize the cytotoxicity caused by a combination of agents, it is desirable to use concentrations of the individual agents that cause only moderate inhibition when they are used alone. If one agent alone would kill a high fraction of the cells, then the supplementary effects of a second agent might be difficult to evaluate. Perhaps concentrations of the individual agents that would permit a 70% survival of the cells when the agent is used alone would be satisfactory. It is realized, however, that sometimes multiple biochemical mechanisms of action come into play as the concentration of an agent is increased and that data obtained at low concentrations might not be representative of what would occur at higher concentrations. Therefore data obtained for any concentration of an agent must be interpreted with caution.

One might ask which cells (with regard to the position in the cycle) are killed when only 30% of the population is killed. For HEp No. 2 cells under the experimental conditions indicated below, the cell cycle data are as follows (Wheeler et al. 1970a): $T_C = 26.1$ hr; $T_{G1} = 14.4$ hr; $T_S = 6.9$ hr; $T_{G2} = 3.9$ hr; and $T_M = 0.9$ hr. The corresponding theoretical distribution of cells among the various phases is: G_1, 63.6%; S, 22.8%; G_2, 11.1%; and M, 2.5%. For a 30% kill by an S phase-specific agent, supposedly the cells initially in S at the beginning of the exposure (23%) plus the first additional 7% of the population arriving at G_1/S or entering S would be killed. Agents with the greatest cytotoxic effect to cells in M and at G_1/S would probably kill the cells initially in M (ca. 3%) plus the cells initially in G_2 (ca. 12%) as they progressed into M plus the first 15% of the cells to arrive at G_1/S. Agents with equal cytotoxicity to cells throughout the cycle would kill cells randomly distributed throughout the cycle in proportion to the numbers of cells in the various phases.

Interpretation of Cell Kill Data for Combinations of Agents

If the two agents of a combination are phase-specific and they kill only the cells in precisely the same parts of the cycle, then the combination should kill no more cells than either agent alone. If each agent kills 30% of the cells, the combination would also kill 30%.

If the two agents are phase-specific but they kill cells in exclusively different phases of the cycle, the cell kill of the combination would be the sum of the individual kills. If each agent kills 30% of the cells, the combination would kill 60%.

If the two agents are nonphase-specific, if cells throughout the cycle are equally sensitive to each of the agents, and if each of the two agents kills cells according to first-order kinetics, then several results are possible for the combination. If each agent alone kills 30% of the cells (surviving fraction, 70%) and if the presence of either agent in a cell excludes the entry of the other agent into the cell, the two agents in the combination would kill cells independently of each other, and the combined kill would be 51% (surviving fraction, $0.7 \times 0.7 = 0.49$ or 49%). If the presence of either agent in a cell does not prevent the entry of the other agent but it is still necessary for lethality of the cell for one or the other of the agents to reach the intracellular concentration required for that agent alone to kill the cell, the cell kill of the combination would be between 30% and 51%. If the two agents freely enter the cells so that the intracellular concentrations of the first agent in all of the cells were equal and the intracellular concentrations of the second agent in all of the cells were equal and if the two agents act independently of each other, the expected cell kill would be 51%. If the two agents can freely enter any cell in the population and lethality results when the agents are present intracellularly at less than their individual lethal concentrations, the cell kill may be greater than 51%.

For a combination of a phase-specific agent and a nonphase-specific agent, the cell kill would be between 30% and 51%.

The above predictions are listed in Table 2.

Therefore based upon the above considerations, any combination of two agents that kills more than 60% of the cells when each agent of the combination kills only 30% of the cells when present alone would cause synergistic cell kill. Since in the combinations tested in this study there was no combination consisting of agents specific for two different phases, any combination killing more than 51% of the cells in these experiments would demonstrate synergism or potentiation. If the experimental cell kills for the individual agents are other than 30%, the above values would have to be altered accordingly.

Experimental Methods

The method for determining the cytoxicities of single agents and combinations of agents was the same as that used in a previous study (Wheeler et al. 1972). One hundred thousand HEp No. 2 cells were placed into each of a number of 1-ounce glass prescription bottles with 5 ml of SRI-14 medium. At 24 hours the medium

Table 2

Expected Cell Kill of Combinations of Two Agents When
Each Agent Alone Kills 30% of the Population

Types of agents	*Expected cell-kill*
Two agents specific for same phase	30%
Two agents specific for different phases	60%
Two mutually exclusive,* nonphase-specific agents	51%
Two nonmutually exclusive, nonphase-specific agents causing death when one of the agents reaches its own lethal concentration	30–51%
Two nonmutually exclusive, nonphase-specific agents when all cells contain equal concentrations of the first agent and also equal concentrations of the second agent	51%
Two nonphase-specific agents causing death when each intracellular concentration is less than its own lethal concentration	>51%
A phase-specific agent and a nonphase-specific agent	30–51%

* The presence of one agent in a cell excludes entry of the second agent.

was decanted and replaced with fresh medium. At 48 hours the medium was again decanted and replaced with fresh medium containing the test agent(s), and incubation was continued for 26 hours. The medium and any suspended cells were then decanted into centrifuge tubes, the glass-attached cells were displaced by trypsin and transferred to the centrifuge tubes, and the cells were sedimented centrifugally and washed two times with SRI-14 salts solution. The cells were resuspended in complete SRI-14 medium, the cell count was made with a Coulter Model B electronic particle counter, and after appropriate dilution of the suspension 10 ml of the suspension containing 100 cells was placed into each of three 4-ounce prescription bottles. The bottles were laid on the flat sides, and the cultures were incubated at 37°C for 7 days. The cells were then stained with Giemsa and macrocolonies were counted with the aid of a magnifying glass. Mean values were calculated for each set of three bottles. The plating efficiency for the control cultures was 44%.

Experimental Results and Discussion

Table 3 contains the initial data that have been obtained by the described method. The data for two independent experiments have been combined, and the mean values are given in the table. The observed cell kill values for the individual agents, the calculated cell kills for the combinations (assuming independent action), the observed cell kills for the combinations, and possible interpretations of the results are presented.

The values for the individual agents indicate the difficulty encountered in selecting a concentration of the agent to give a desired extent of cell kill. Based upon previously obtained concentration-cell kill data, concentrations were selected that were expected to kill 30% of the cells. Unfortunately in these initial experiments some of the agents killed a much higher portion of the cells, while others killed essentially none of the cells. With further experimentation it should be possible to obtain more predictable and reproducible responses, but even the limited data presently available indicate some interesting effects of several combinations. Combinations of the various agents yielded all of the possible effects—greater than additive, additive, less than additive, and antagonistic. The combinations yielding these various effects are grouped together in Table 4.

It is of interest to consider together the logic of the combinations given in Table 1 and the observations summarized in Table 4. In tests of combinations of an agent active in S phase and causing a block at G_1/S with a nonphase-specific agent having a high cytotoxicity at G_1/S, a variety of results was obtained: hydroxyurea + HN2 and methotrexate + HN2 gave greater than additive effects; 5-FdU + HN2 was less than additive; and Ara-C + HN2 was antagonistic. When CCNU was used instead of HN2 in combination with each of these four agents, all of the combinations had somewhat less than additive effects.

The combination Ara-C + 6-thioguanine consists of agents that cause sequential biochemical blocks and perhaps act in sequential phases of the cell cycle. This combination gave antagonistic effects, as it killed fewer cells than Ara-C alone. The antagonism might be mutual for these agents, since 6-thioguanine can prevent the progression of cells into S and Ara-C can prevent the incorporation of 6-thioguanine into DNA.

When the two S phase-specific agents hydroxyurea and Ara-C were used in combination with the nonphase-specific agent 5-FUra, hydroxyurea + 5-FUra gave

Table 3

Cell Kill of HEp No. 2 Cells by Combinations of Agents

Agent A	Agent B	Cell kill by A (%)	Cell kill by B (%)	Pre-dicted[a] cell kill by A + B (%)	Ob-served cell kill by A + B (%)	Interpretation
Ara-C	HU	59	(29)	(71)	95	(>Additive)
Ara-C	Meth	59	36	74	77	Additive
Ara-C	5-FdU	59	82	93	86	Additive?
Ara-C	5-FUra	59	50	79	48	Antagonistic
Ara-C	Vincristine	59	2	60	66	Additive
Ara-C	Adriamycin	59	0	59	73	>Additive
Ara-C	6-TG	59	80	92	73	Antagonistic
Ara-C	HN2	59	5	61	45	Antagonistic
Ara-C	CCNU	59	59	83	61	<Additive
HN2	HU	5	(29)	(33)	45	(>Additive)
HN2	Meth	5	36	39	50	>Additive
HN2	5-FdU	5	82	83	77	<Additive
HN2	5-FUra	5	50	52	59	Additive
HN2	Vincristine	5	2	7	23	>Additive
HN2	Adriamycin	5	0	5	(49)[b]	(>>Additive)
HN2	6-TG	5	80	81	70	Antagonistic
HN2	CCNU	5	59	61	30	Antagonistic
CCNU	HU	59	(29)	(71)	(41)	(Antagonistic)
CCNU	Meth	59	36	74	66	<Additive
CCNU	5-FdU	59	82	93	84	<Additive
CCNU	5-FUra	59	50	79	66	<Additive
CCNU	Vincristine	59	2	60	16	Antagonistic
CCNU	Adriamycin	59	0	59	(55)	(Additive)
CCNU	6-TG	59	80	92	64	Antagonistic
5-FUra	HU	50	(29)	(64)	73	(Additive)
5-FUra	Meth	50	36	68	86	>Additive
5-FUra	5-FdU	50	82	91	93	Additive
5-FUra	Vincristine	50	2	51	64	>Additive
5-FUra	Adriamycin	50	0	50	68	>Additive
5-FUra	6-TG	50	80	90	93	Additive

[a] The predictions are based upon the assumption that the kill effects of the agents are independent and additive.

[b] Values in parentheses are means for clone counts falling within a broad range, and therefore the significance and reproducibility of these values are questionable.

additive toxicity, suggesting independent action, and Ara-C + 5-FUra gave evidence of antagonism. Although methotrexate and 5-FdU can kill cells in other phases of the cycle, they are frequently thought of as S-specific agents, and it is interesting to compare their effects with those of hydroxyurea and Ara-C. Considerably greater than additive effects were obtained for methotrexate + 5-FUra, and additive effects were obtained with 5-FdU + 5-FUra (which leaves unanswered the question whether 5-FUra and 5-FdU are killing cells by the same biochemical mechanisms).

Table 4

Summary of Cell Kill Effects

> Additive	Additive
(Ara-C + HU)*	Ara-C + Meth
Ara-C + adriamycin	Ara-C + 5-FdU
(HN2 + HU)	Ara-C + vincristine
HN2 + Meth	HN2 + 5-FUra
HN2 + vincristine	(CCNU + adriamycin)
(HN2 + adriamycin)	(5-FUra + HU)
5-FUra + Meth	5-FUra + 5-FdU
5-FUra + vincristine	5-FUra + 6-TG
5-FUra + adriamycin	

< Additive	Antagonistic
HN2 + 5-FdU	Ara-C + 6-TG
CCNU + Ara-C	Ara-C + 5-FUra
CCNU + Meth	Ara-C + HN2
CCNU + 5-FdU	HN2 + 6-TG
CCNU + 5-FUra	HN2 + CCNU
	(CCNU + HU)
	CCNU + vincristine
	CCNU + 6-TG

* The values for the combinations inclosed in parentheses fall within a broad range, and hence the significance and reproducibility of these results are questionable.

Thus in combinations with 5-FUra, methotrexate gave synergistic toxicity, hydroxyurea and 5-FdU each gave additive toxicity, and Ara-C gave antagonism. When two of these S-phase active agents that act by different biochemical mechanisms were combined, Ara-C + hydroxyurea was quite synergistic, and Ara-C + methotrexate and Ara-C + 5-FdU were additive. These results emphasize the differences among these agents and indicate that they are not simply killing the same complement of the cell population.

Although Ara-C, adriamycin, and vincristine are each most cytotoxic to cells in S phase and although at the concentrations used in the present experiments adriamycin and vincristine killed very few cells, Ara-C + adriamycin was synergistic, and Ara-C + vincristine was additive or perhaps slightly greater than additive. This indicates that the cytotoxicity of an agent that interferes with the synthesis of DNA can be supplemented by the cytotoxicity of an agent (adriamycin) that complexes with DNA or of an agent (vincristine) that interferes with the organization of DNA in mitosis.

Although it might be expected that a combination of HN2 and CCNU would be additive or synergistic, since these agents are progenitors of different alkylating moieties and since CCNU also is a progenitor of a carbamoylating moiety, the combination was found to be antagonistic. The reason for this antagonism is not presently known. In other experiments performed by a slightly different procedure in this laboratory synergism for HN2 + CCNU was observed for certain low concentrations of the agents but not for higher ones. Although HN2 and vincristine individually killed very few cells, the combination HN2 + vincristine killed considerably more cells than would be predicted. On the other hand, CCNU + vincristine was antagonistic. The results for HN2 + vincristine are consistent with the logic set forth in Table 1, but the reason for the antagonistic effects of CCNU + vincristine is not known. When the two nonphase-specific agents HN2 and CCNU were used in combination with the nonphase-specific antimetabolite 5-FUra, HN2 + 5-FUra was additive, and CCNU + 5-FUra was less than additive. Both HN2 + 5-FdU and CCNU + 5-FdU gave less than additive cell kills. These less than additive cell kills indicate there was some "double-kill" by the combined agents. Both HN2 + 6-thioguanine and CCNU + 6-thioguanine were antagonistic, and one might speculate whether or not there was in interaction between the agents. It is notable that 5-FUra + 6-thioguanine gave additive cell kill.

The results of these experiments show that although present knowledge of the biochemical effects and of the cell cycle effects and sensitivities to agents may be used to predict successfully some combinations that give synergistic or additive cytotoxicities, other predictions were unsuccessful. Therefore this methodology serves not only to give evidence of additive or synergistic cytotoxic effects of agents but also to give evidence of our limited knowledge of the mechanisms of action and cell cycle effects of the agents.

Although combinations of agents might have additive or synergistic cytotoxicities for cultured cells, this information does not indicate that they would also have additive or synergistic toxicities for experimental animals or additive or synergistic anticancer activities. Table 5 brings together for several combinataions the observed

Table 5

Comparison of Effects of Combinations of Agents

	Cytotoxicity to HEp No. 2 cells	Toxicity to mice[a]	Therapeutic effect against leukemia L1210[a]
Ara-C + HU	>Additive	<Additive	
Ara-C + 6-TG	Antagonistic	<Additive	Potentiation
Ara-C + CCNU	<Additive	<Additive	Potentiation
Ara-C + cyclophosphamide (HN2)[b]	Antagonistic	<Additive	Possible potentiation
CCNU + cyclophosphamide (HN2)	Antagonistic	<Additive	Possible potentiation
Cyclophosphamide (HN2) + HU	>Additive	<Additive	
Cyclophosphamide (HN2) + 5-FUra	Additive	<Additive	
Cyclophosphamide (HN2) + vincristine	>Additive	<Additive	
5-FUra + Vincristine	>Additive	<Additive	

[a] Results of F. M. Schabel et al.

[b] Cyclophosphamide was used in experiments with mice; nitrogen mustard was used in experiments with cultured cells.

types of effects for cytotoxicity, toxicity to mice (Schmidt et al. 1970; Skipper and Schabel unpubl.), and therapeutic effects against L1210 leukemia (Schabel 1968; Schmidt et al. 1970; Schabel and Skipper unpubl.). Whereas all of the combinations were less than additive for toxicity in mice, the cytotoxicities for cultured HEp No. 2 cells were of all types—greater than additive, additive, less than additive, and antagonistic. It is interesting that the four combinations that give potentiation or possible potentiation against leukemia L1210 give less than additive or antagonistic cytotoxicity against HEp No. 2 cells. This suggests that there is not a positive correlation between potentiated cytotoxicity and potentiated therapeutic effects; conversely it suggests that less than additive cytotoxicity might be a requisite for potentiated therapeutic effects. It is realized, however, that the scant data of the types that are given in Table 5 that are presently available do not permit an evaluation of the utility of cytotoxicity data for predicting therapeutic usefulness. It is also recognized that attempts to correlate cytotoxicities with therapeutic effects are complicated by the multiplicity of dosage levels and administration schedules used for therapy and the presently unpredictable differential effects of the combinations upon host tissues and neoplastic tissues. Perhaps when more data become available, some correlations will become evident; but even though such correlation might not be forthcoming, the initial data for cytotoxicity of combinations presented in this report indicate that such data can be useful for testing current hypotheses relating to the biochemical and cell cycle effects of the individual agents.

Acknowledgments

The author expresses his appreciation to Dr. H. H. Lloyd for discussions relating to mathematical aspects of the text and to Ms. D. J. Adamson for performing the cytotoxicity determinations presented in this paper. These determinations were conducted under Contract PH-43-66-29, Division of Cancer Treatment, NCI, NIH.

REFERENCES

Barranco, S. C., E. W. Gerner, K. H. Burk and R. M. Humphrey. 1973. Survival and cell kinetics effects of adriamycin on mammalian cells. *Cancer Res.* **33**:11.

Creasey, W. A., K. G. Bensch and S. E. Malawista. 1971. Colchicine, vinblastine and griseofulvin—Pharmacological studies with human leukocytes. *Biochem. Pharmacol.* **20**:1579.

Madoc-Jones, H. and F. Mauro. 1973. Cell-cycle kinetics: Site of action of cytotoxic agents in cell life cycle. *Antineoplastic and immunosuppressive agents* (ed. A. C. Sartorelli and D. G. Johns) *Handbook of Pharmacology.* Springer-Verlag, Heidelberg. (in press)

Sartorelli, A. C. 1971. Some biochemical and pharmacologic considerations of agents in the management of acute leukemia. *Current concepts in the management of leukemia and lymphoma* (ed. J. E. Ultmann et al.) *Recent Results in Cancer Res.* **36**:74.

Schabel, F. M., Jr. 1968. In vivo leukemic cell kill kinetics and "curability" in experimental systems. *The proliferation and spread of neoplastic cells,* 21st Ann. Symp. Fundamental Cancer Res. p. 379–408. Williams and Wilkins, Baltimore.

Schmidt, L. H., J. A. Montgomery, W. R. Laster, Jr. and F. M. Schabel, Jr. 1970. Combination therapy with arabinosyl cytosine and thioguanine. *Proc. Amer. Ass. Cancer Res.* **11**:70.

Wheeler, G. P. 1967. Some biochemical effects of alkylating agents. *Fed. Proc.* **26**:885.

———. 1973. Mechanism of action of alkylating agents: Nitrosoureas. *Antineoplastic and immunosuppressive agents* (ed. A. C. Sartorelli and D. G. Johns) *Handbook of Pharmacology.* Springer-Verlag, Heidelberg. (in press)

Wheeler, G. P. and L. Simpson-Herren. 1973. Effects of purines, pyrimidines, nucleosides and chemically related compounds upon the cell cycle. *Drugs and the cell cycle* (ed. A. M. Zimmerman et al.) p. 249–306. Academic Press, New York.

Wheeler, G. P., B. J. Bowdon, D. J. Adamson and M. H. Vail. 1970a. Effects of certain nitrogen mustards upon the progression of cultured H.Ep. No. 2 cells through the cell cycle. *Cancer Res.* **30**:100.

———. 1970b. Effects of 1, 3-Bis(2-chloroethyl)-1-nitrosourea and some chemically related compounds upon progression of H.Ep. No. 2 cells through the cell cycle. *Cancer Res.* **30**:1817.

———. 1972. Comparison of the effects of several inhibitors of the synthesis of nucleic acids upon the viability and progression through the cell cycle of cultured H. Ep. No. 2 cells. *Cancer Res.* **32**:2661.

Signals and Switches:
A Summary

M. G. P. Stoker

Imperial Cancer Research Fund Laboratories
Lincoln's Inn Fields, London WC2A 3PX

Once again this remarkable institution, with wise guidance from Baynard Clarkson and Renato Baserga, has brought together a wide variety of scientists to discuss a common theme, the control of cell proliferation. The result has been a bumper week in which the riches that have been displayed are now only an embarrassment to someone attempting a summary. Fortunately for all, I do not propose to waste your time by inaccurately reviewing 80 odd papers that will be published anyway, but I will simply give a personal overall impression.

Many speakers began by defining the ideal cell system which, not surprisingly, turned out to be the one they had been working with all along. This gives many models, between which comparisons are difficult, and at times I wished we had a Delbrück to persuade us all to concentrate on a narrower front. However most work described has concentrated on three general systems in which growth can be manipulated at will and the cells sampled satisfactorily. These are (1) the hemopoietic system, with which, as Renato Baserga reminded us at the outset, much of the ground work on the cell cycle was done; (2) the lymphoid cells, relative newcomers in growth studies; and (3) the fibroblasts of various types. These last are especially useful because mutants and revertants have been selected and their growth has been manipulated with carcinogens, in particular oncogenic viruses. Fibroblasts have the disadvantage, however, that their behavior cannot readily be referred to a recognizable cell type in vivo.

Though it is clear from the complexity of the findings that each cell type will have to be studied separately as a regulation system, let us see for the sake of simplicity what can be said in general terms about that nonexistent but useful abstraction, the generalized universal animal cell.

We have been concerned with the way the growth of a cell is determined by its environment, and most of the discussion has dealt with three aspects: (1) the incoming signals, (2) the role of the cell surface, and (3) the intracellular switches and responses.

1009

SIGNALS

The environment of the cell consists of other cells and fluid in which stimuli for growth circulate as free molecules of various sizes, such as steroid and protein growth factors. Small metabolites such as amino acids also stimulate growth and are useful in experimental systems, but despite the data from hepatectomized animals it seems hardly likely that starvation is an important and sensitive controlling system in vivo. The steroid hormones are already familiar as regulating agents, but it is now clear that the requirement of cells for serum or for tissue culture medium conditioned by other cells is due to special proteins, usually termed growth factors. Despite our lack of knowledge of their source and specificity, these could now surely be classed as growth hormones. The factors required for the growth of erythroid and granulocytic cells have been particularly well characterized. Their action on cell growth is highly target specific, and in the case of granulocytes is also production specific, since the target cells are the only ones identified which do not produce the factors. Some of the fibroblast growth factors have been characterized from serum or as a product of cultured cells that is closely related to somatomedin. It should soon be possible to label them and study interaction with the target cells. It is reasonable to predict that tissue fluids such as serum from animals will yield very large numbers of proteins or growth hormones specific for different cell types, in addition to the eight described that are active on 3T3 cells.

So far, with very few exceptions, every naturally occurring and active extracellular molecule shows a positive effect on cell growth in vitro, and the absence of a balancing negative regulation system seems rather surprising. Surely it is unlikely that the delicate homeostatic mechanisms operating in vivo depend solely on accelerators and not on brakes as well? As is clear from studies on chalones, however, inhibitors are not at all easy to characterize, at least in complex mixtures that contain stimulating factors.

These growth stimulating factors or hormones discussed at the meeting are apparently fairly stable and circulate at long range in tissue fluids and also tissue culture dishes. What of the short range or micro environmental influences? These must surely play a major role in vivo in morphogenesis, and we have heard particularly interesting evidence that the local environment has a controlling role in erythropoiesis and granulopoiesis.

The idea that local growth depends on direct, tactile, signals from neighboring cells is an attractive one and is reinforced by the acknowledged importance of cell surface changes in growth control. But no hard evidence for such a mechanism has been produced. Indeed density-dependent inhibition of growth can be interpreted in terms of nutrition and limitation of growth-stimulating factors, and I have myself given some evidence to show how diffusion barriers could explain very short-range effects. Thus there is at present no firm evidence that cells sense one another by feel, rather they smell one another.

THE CELL SURFACE

Many days and nights have been spent in contemplation of the cell surface and its mysteries, and no one will now deny that it is closely linked in some way with growth. This is not surprising since the surface must grow too, it must undergo differentiation during cytokinesis, and it is in contact with extracellular factors.

We may consider briefly three aspects of the role of the surface in regulation: as a receiver of incoming signals; as a transmitter or modulator of these signals; and as a controlling organelle in its own right.

Obviously any molecules from the environment which affect cell proliferation first reach the cell surface. We know that some, the steroid hormones, pass straight through to their primary targets in the cytoplasm. The protein growth factors like other protein hormones, attach to their primary targets, which are probably sugar-containing sites on the surface membrane. Though proof is lacking, the activity is probably transmitted by a secondary change in the membrane rather than by penetration of the incoming molecule itself. The evidence is partly based on the changes in membrane chemistry, topography and perhaps fluidity correlated with initiation of growth and with the changed growth pattern of cell variants. More directly, agents that are known to modify the surface, notably phorbol esters, lectins and proteases, rapidly stimulate cell growth. Fibrinolysin T, which as we heard is a natural product of tumor cells, might modify their growth through an effect on the surface.

Despite the specialized nature of the lymphocytes, it looks as if many answers will come from the study of the interaction of mitogens with these cells since the active molecules such as lectins or antiglobulins can be precisely characterized and a great deal is already known of the surface topography of lymphocytes. The rearrangements of components in the two-dimensional surface layer which follow mitogen binding, and the possible consequences in terms of ionophores, exposure of other hormone receptors, or effects on adenyl cyclase have been amongst the most interesting topics discussed at this meeting. Since there is some evidence of restriction of membrane movement imposed by the underlying microtubular skeleton, I have been tempted to wonder whether, despite my previous remarks, cell to cell contact could affect growth by the interaction of receptor sites, imposing a rigidity in lateral movement within the membrane.

It is abundantly clear that the cell membrane can act as a growth control system in its own right through selective transport of nutrients. Transport of certain small molecules is so intimately related to growth patterns that it may indeed be part of the primary control mechanism, perhaps governed by the cyclic nucleotide levels. The problem remaining, however, is to distinguish a primary effect of transport from a response to increased demand.

INTRACELLULAR EVENTS

There have been many important reports during the meeting on the biochemical events which may or may not be primarily and causally concerned with regulation. To sort these out it would be helpful to consider two classes of controls.

One class, which we may call *intracycle* controls, is concerned with the orderly sequence of events in the cell cycle. Thus as shown by the elegant cell fusion experiments, chromosome condensation is actively promoted during the G_2 phase, and mitosis itself is forbidden if DNA synthesis is continuing. We have also heard of the orderly sequence of synthesis and phosphorylation of some nuclear proteins. There must be a large number of interlocked systems, far more complex than those already known in prokaryotes and obviously involving controlled expression of whole sets of genes. We assume that these events are controlled by signals originating from within the cell and are insensitive to extracellular events other than experimental procedures involving drugs as inhibitors. With such a complex, stepwise, biosynthetic process it would be an advantage for events to move inexorably forward, with the cell deaf to the outside world.

But part of the cell cycle, probably early in the G_1 phase, is responsive and constitutes a sort of window where the cell reads and measures its environment, perhaps by some brief alteration in its surface. A special *extracycle* control system therefore exists, in G_1, which is influenced by the environment and which has been named the restriction (or R) point. Lack of a positive growth stimulus at this switch point, or perhaps persistence of a negative one, halts the cycle and leads to the adoption of an alternative stationary state or G_0, involving a series of coordinated metabolic changes named the pleiotypic response. In this state the cell remains receptive to incoming stimuli, which cause a reversion to the growth cycle, beginning in G_1.

For the sake of our generalized animal cell, I shall assume that all the membrane-transmitted stimuli act through one common restriction point in all cell types, of at least cells without grossly abnormal growth regulation, despite the fact that the arguments and lack of correlations heard in recent days leave this in some doubt.

Much of the discussion has centered on likely biochemical events concerned in this important *extracycle* control step or switch. Unfortunately the common experimental arrangement involves a study of the sequence of events that follow a release from the resting state, and some of the changes observed are probably part of the inevitable sequence of intracycle controls, which are simply a consequence of re-initiation of growth. The best that can be done at present is to search for the earliest changes after the stimulus is applied, since these at least are more likely to be causally involved in regulation than are later events. The transition from the growing to the stationary state has apparently been more difficult to study but deserves attention.

Some years ago in our laboratory Bob Bürk first produced evidence that cyclic AMP synthesis and breakdown might be involved in growth regulation, and I have to admit to being guilty of positive discouragement since it seemed far too ubiquitous a mechanism for interpretation. Now, in a variety of systems, we hear of very rapid perturbations of cyclic nucleotide levels, falls in cyclic AMP levels and adenyl cyclase activity, or swift rises in cyclic GMP, sometimes in less than a minute. There are considerable discrepancies between different laboratories and cells, but it seems that (1) mean cyclic nucleotide levels and rates of synthesis are often correlated with growth characteristics of particular cell types; (2) analogs of cyclic AMP affect growth of some cultured cells; (3) agents that affect synthesis and breakdown of cyclic AMP affect growth. There is now a working assumption that the cyclic nucleotides act as internal messengers in growth, as in other regulation systems in the cell.

Cyclic AMP can act through protein phosphorylation, conceivably of nonhistone chromosomal protein which is also shown to be modified very early after growth stimulation, at least in lymphocytes. But we also saw a very rapid effect of dibutyryl cyclic AMP on cell movement and shape, and the cyclic nucleotides might also affect the microtubular or microfilament system and thus affect growth through some form of membrane anchorage.

Finally we should not forget that a mutant was described which apparently lacked the cyclic AMP receptor but was capable of cell proliferation. Perhaps cyclic AMP is not quite so essential to the economy of the cell after all.

Other very early events were also reported, such as initiation of ribosomal RNA synthesis, increased glycolysis, and the increased rates of transport of certain molecules as already mentioned. These too may be involved in control at the restriction point.

To establish a linear and causally related sequence of events from initial action at the cell surface to the final commitment to a new cycle with chromosome replication and mitosis is a problem for the future. It may well need quite new approaches, in addition to the well-tried combination of genetics and biochemistry. Some of the mutants described and especially the variants of virus-transformed cells are already useful, and hopefully others will be isolated soon to assist the task.

In conclusion, I am quite overawed by the complexity of the problems we have discussed, but the pace is so hot that solutions to some of them will no doubt have been found by the time you have returned home. Let us hope that our exchange of views will have helped to speed the process even further.

Index